Enzymatic Reaction Mechanisms

ENZYMATIC REACTION MECHANISMS

Perry A. Frey and
Adrian D. Hegeman

UNIVERSITY PRESS

2007

OXFORD
UNIVERSITY PRESS

Oxford University Press, Inc., publishes works that further
Oxford University's objective of excellence
in research, scholarship, and education

Oxford New York
Auckland Cape Town Dar es Salaam Hong Kong Karachi
Kuala Lumpur Madrid Melbourne Mexico City Nairobi
New Delhi Shanghai Taipei Toronto

With offices in
Argentina Austria Brazil Chile Czech Republic France Greece
Guatemala Hungary Italy Japan Poland Portugal Singapore
South Korea Switzerland Thailand Turkey Ukraine Vietnam

Copyright © 2007 by Oxford University Press

Published by Oxford University Press, Inc.
198 Madison Avenue, New York, New York, 10016

www.oup.com

Oxford is a registered trademark of Oxford University Press

All rights reserved. No part of this publication may be reproduced,
stored in a retrieval system, or transmitted, in any form or by any means,
electronic, mechanical, photocopying, recording, or otherwise, without
the prior permission of Oxford University Press.

Library of Congress Cataloging-in-Publication Data
Frey, Perry A.
Enzymatic reaction mechanisms / Perry A. Frey and Adrian D. Hegeman. p. cm.
ISBN-13 978-0-19-512258-9 ISBN 0-19-512258-5
1. Enzyme kinetics. 2. Enzymes. 3. Chemical reactions. I. Hegeman, Adrian D.
II. Title.
[DNLM: 1. Enzymes–physiology. 2. Coenzymes–physiology. 3. Catalysis.
4. Structure-Activity Relationship. QU 135 F8935e 2006] QP601.F725 2006
572'.744–dc22
                                                         2005029570

9 8 7 6 5 4 3
Printed in the United States of America
on acid-free paper

*To Professor Frank H. Westheimer,
a great scientist, a great scholar, and a great teacher.*

Preface

In planning this book, we gave considerable thought to how we should proceed.

Enzymology is a very large and multifaceted field that interfaces with and stimulates research in peripheral fields such as protein structure, spectroscopy, x-ray crystallography, polymer science, biotechnology, protein folding, cellular metabolism, and biological regulation. All facets cannot be covered in a single volume. Our motivation to write about the chemical mechanisms of enzymatic catalysis guided us in defining the scope of this volume. To bring the material between two covers, it would be necessary to focus on the chemical mechanisms. In making this decision, we did not intend to de-emphasize related areas of enzymology; we simply could not give them as much attention as the reaction mechanisms in the space available for a single volume. Many of these peripheral areas are brought into the discussion of individual enzymes, with less detail, to place the chemical mechanisms in biological or chemical context. However, the emphasis remains on the mechanisms of enzymatic catalysis.

In the first five chapters, we define the scope of the problem of understanding enzymatic catalysis and introduce most of the principles, theory, and concepts. We begin with principles and theories of catalysis and the role of the active site in chapter 1. We continue with an overview of enzyme kinetics in chapter 2. In chapter 3, we discuss classic organic coenzymes and, in chapter 4, metallocoenzymes and cofactors and their contributions to catalyzing enzymatic processes. In chapter 5, we discuss the theory and practice of the development and characterization of specific inhibitors and inactivators of enzymes, an important focus in pharmaceutical and agricultural chemistry. In this part of the book, we include a few brief case studies or sketches of the mechanisms of action of key enzymes, including structures and chemical and spectroscopic results, to exemplify principles.

In the second part of this volume, we focus on types of enzymatic reactions in succeeding chapters. Each chapter includes a discussion of the underlying chemistry and brief case studies exemplifying enzymes in that class, including the most relevant chemical, kinetic, and structural results pertaining to the mechanism. We start with the kinetically simplest reactions and move toward increasing kinetic complexity. We begin with one-substrate, irreversible reactions in chapter 6, the proteases and esterases, and continue with one-substrate, reversible isomerases and mutases in chapter 7. In chapter 8, we discuss decarboxylases and carboxylases, and in chapter 9, we address the lyases in addition and elimination reactions. We continue with the kinetically more complex group transfer enzymes, the phosphotransferases and nucleotidyltransferases, in chapter 10, the ATP-dependent synthetases in chapter 11, and the glycosyltransferases and glycosidases in chapter 12. In chapter 13, we continue with nitrogen and sulfur transferases; in chapter 14, with carbon-carbon ligases and cleavage enzymes; and in chapter 15, with alkyltransferases. The oxidoreductases are the subjects of chapter 16. In chapter 17, we discuss reactions of O_2 and the mechanisms of action of the oxidases and oxygenases. In chapter 18, entitled complex Enzymes, we conclude with discussions of the most complex enzymes, the multienzyme complexes, the modular enzymes and multienzymes proteins, and enzymes that function in energy transduction.

The chapters were written between 2002 and 2005. We found that each chapter became outdated within a few weeks of being written. We chose to proceed with publication because we prefer to have a book in print than one perpetually in preparation. In a work such as this, there are bound to be oversights and mistakes, and we take full responsibility for them. We are receptive to readers' comments and suggestions for improvements, including updating. If the book has a second printing, we will attempt to incorporate corrections and suggestions.

We include brief case studies of the mechanisms of action of more than 100 enzymes in this volume. It has not been possible to provide complete referencing of all the significant research on all of these enzymes. We have included leading references to recent work and selected references to early work that stimulated research on a given enzyme. In selecting enzymes for inclusion, we were guided by the need to incorporate at least one example of an enzyme catalyzing each reaction type. We were also influenced by our own interests, and we seek our readers' indulgence on this matter.

We are grateful to colleagues who contributed in various ways to this book. We are particularly indebted to Professors W. Wallace Cleland, George H. Reed, and Brian G. Fox, who contributed significantly to the various chapters and who read and commented on selected chapters. They bear no responsibility for our mistakes, and we thank them for suggesting improvements and correcting many of our errors.

Internet Enzymology Resources

Today, it would be unnecessarily limiting to work in the field of enzymology without using internet-based resources. Because of their rapid development and their relative impermanence compared with archived literature, these resources are touched on briefly and considered in the context of the main text.

Throughout the text, database accession information is provided in two forms. First, in each figure for which three-dimensional structural information was used, a four-character PDB code is listed in the caption (e.g., IMGO refers to the structure coordinate file for horse liver alcohol dehydrogenase complexed with 2,3-diflurobenzyl alcohol and NAD$^+$). These codes may be used to access archived structure coordinate files and header information from the Brookhaven Protein Data Bank (PDB; Berman et al, 2000), which is hosted by a number of organizations, including the Research Collaboratory for Structural Bioinformatics (RCSB; http://www.rcsb.org/pdb/) and the Enzyme Structures Database (at http://www.ebi.ac.uk/thornton-srv/database/enzymes/). Second, as each enzyme activity in the text is introduced, the enzyme classification (EC) number is provided. The EC number format is EC #.#.#.#, where each refers to a unique aspect of catalysis in four hierarchic categories so that all four numbers refer to a unique activity (e.g., EC 1.1.1.1 refers to the activity of alcohol dehydrogenase). This systematization helps to eliminate confusion resulting from conflicting naming conventions, and it serves as an accession number for enzymes database. BRENDA (http://www.brenda.uni-koeln.de), hosted by the University of Cologne, contains a large amount of information on enzyme substrates, products, and inhibitors and other functional data, which may be accessed using EC numbers or by using cross-listed preferred or alternate nomenclature (Schomburg et al, 2004). Other searchable databases vary over time, but they tend to be maintained and updated (or at least linked to) by several key organizations, including the National Center for Biotechnology Information (NCBI) (http://www.ncbi.nim.nih.gov) and the European Molecular Biology Laboratory–European Bioinformatics Institute (EMBL-EBI) http://www.ebi.ac.uk/). These sites frequently include access to free, often experimental, and downloadable or web-based tools.

The PDB access sites also provide freeware for structure file viewing, although structure viewers are generally widely available. All of the structural figures presented in the text were generated using the free visualization and rendering package MolView (v1.5.0)

(Smith, 1995). An extensive collection of other web-based tools for protein structural and proteomic analysis is available at the ExPASy (Expert Protein Analysis System) proteomics server of the Swiss Institute of Bioinformatics (SIB). Many interesting experimental tools, such as the web-based homology modeling package SWISS-MODEL (http://swissmodel.expasy.org/) exist that may be of interest to anyone exploring the vast reams of genomic data that have been accumulated in the past decade.

Finally, some novel strategies for populating and querying database have been developed that apply specifically to the interface of high-throughput structural biology and enzymology. The *catalytic Site Atlas* (CSA; http://www.ebi.uk/thornton-srv/database/CSA), for example, provides catalytic residue annotation for enzymes in the PDB, with the ultimate goal being the characterization of enzyme structural elements (e.g., active-site residues) that directly pertain to function (Porter et al, 2004).

If successful, this sort of analysis may provide additional insight into the function of uncharacterized gene products that cannot be gained from the analysis of amino acid sequence or domain structure alone.

References

Berman HM, J Westbrook, Z Feng, G Gilliland, TN Bhat, H Weissig, IN Shindyalov, and PE Bourne (2000) *Nucleic Acids Res* **28**, 235.

Gasteiger E, A Gattiker, C Hoogland, I Ivanyi, RD Appel, and A Bairoch (2003) *Nucleic Acids Res* **31**, 3784.

Porter CT, GJ Bartlett, and JM Thornton (2004) *Nucleic Acids Res* **32**, D129.

Schomburg I, A Chang, C Ebeling, M Gremse, C Heldt, G Huhn, and D Schomburg (2004) *Nucleic Acids Res* **32**, D431.

Smith TJ (1995) *J Mol Graphic* **13**, 122.

Contents

Chapter 1. Enzymes and Catalytic Mechanisms, 1
    Catalysis and the Active Site, 1
    Rate Enhancement in Enzymatic Catalysis, 3
    Conformational Mobility in Catalysis, 5
        Substrate-Induced Conformational Changes, 5
        Catalysis of Multistep Reactions, 6
        Structural Mobility in Enzymes, 6
    Acid-Base Catalysis, 9
        Acids and Bases, 9
        Acid- and Base-Catalyzed Reactions, 11
    Nucleophilic Catalysis, 16
    Electrophilic Catalysis, 21
        Catalysis of Enolization, 21
        Imine Formation by Lysine, 23
        Catalysis by Metal Ions, 26
    Hydrogen Bonding, 30
        Strong and Weak Hydrogen Bonds, 30
        Hydrogen Bonding in Catalysis, 32
    Binding Energy in Catalysis, 34
        Binding and Activation Energy, 34
        The Active Site as an Entropy Trap, 36
        Dissecting the Binding Effect in Enzymatic Action, 40
        Stabilization of the Transition State, 41
        Binding the Near Attack Conformation, 46
        Destabilization of Ground States, 48
        Rate Enhancement through Binding of Remote Groups, 48
    Characterization of Active Sites, 53

Competitive Inhibitors: Analogs of Substrates, 53
Group-Selective Chemical Modification, 53
Site-Directed Mutagenesis, 57
Affinity Labeling, 59
Why Are Enzymes Large Molecules?, 62
Sizes of Enzymatic Binding Domains, 62
Catalytic Antibodies, 63

Chapter 2. Kinetics of Enzymatic Reactions, 69

Steady-State Kinetics, 69
One-Substrate Reactions, 70
Two-Substrate Reactions, 74
Three-Substrate Reactions, 89
Isotope Effects, 91
Classes of Isotope Effects, 91
Measurement of Isotope Effects, 95
Transient-Phase Kinetics, 101
Reaction Characteristics, 101
Transient Methods, 102
pH-Rate Profiles, 111
Profile Interpretation, 111
Measurements of pH-Rate Profiles, 111
Allosteric Regulation, 117
Theory, 118
Binding Equations for Cooperative Systems, 120
Aspartate Transcarbamoylase, 123

Chapter 3. Coenzymes I: Organic Coenzymes, 129

Nicotinamide Coenzymes, 129
Structures and Functions of Nicotinamide Coenzymes, 129
Stereospecificity of Hydride Transfer, 132
NAD^+ as a Coenzyme, 134
Thiamine Pyrophosphate, 141
Structure, 141
Reaction Mechanism, 141
α-Lipoamide, 147
Pyridoxal-5'-Phosphate, 148
Enzymatic Reactions Facilitated by Pyridoxal-5'-Phosphate, 149
Pyridoxal-5'-Phosphate–Stabilized Amino Acid Carbanions, 149
Mechanisms of Pyridoxal-5'-Phosphate–Dependent Reactions, 151
Flavin Coenzymes, 158
Structures of Flavin Coenzymes, 158
Mechanisms of Flavin Catalysis, 159
Biotin, 163
Structure and Role as a Carboxyl Carrier, 163

Chemistry of Biotin and N^1-Carboxybiotin, 164
Mechanism of Biotin-Dependent Carboxylation, 164
Phosphopantetheine Coenzymes, 165
Structures of Phosphopantetheine Coenzymes, 165
Mechanism of Phosphopantetheine Action, 165
Folate Compounds, 167
Folate Compounds of One-Carbon Metabolism, 168
Enzymes in Tetrahydrofolate Metabolism, 170
Biological Importance of Folate, 171
Amino Acid–Based Coenzymes, 172
Pyruvoyl Decarboxylases, 172
Methylidene Imidazolinone–Dependent Deaminases, 173
Quinoproteins, 174

Chapter 4. Coenzymes II: Metallic Coenzymes, 189

Vitamin B_{12} Coenzymes, 190
Chemistry of B_{12} Coenzymes, 190
Adenosylcobalamin-Dependent Enzymes, 193
Methylcobalamin-Dependent Enzymes, 199
Heme Coenzymes, 201
Chemistry of Oxygen and Heme, 201
Heme Enzymes, 204
Oxygen Binding and Electron Transfer, 209
Mononuclear Nonheme Iron, 210
Monooxygenases, 210
Dioxygenases, 217
Oxo-Fe$_2$ Complexes, 217
Structures, 218
Reactions of Di-iron Enzymes, 219
Metallopterin Enzymes, 222
Molybdopterin and Tungstopterin, 222
Iron-Sulfur Centers, 227
Structures, 227
Catalytic Functions, 230
S-Adenosylmethionine and Iron-Sulfur Centers, 234
Catalytic Action of S-Adenosylmethionine and [4Fe–4S] Centers, 234
Stoichiometric Reactions of S-Adenosylmethionine and [4Fe–4S] Centers, 236
Divalent Metal Ions, 237
Electrostatic Activation of Coordinated Water, 237
Electrostatic Activation of Enolization, 238
Copper as a Cofactor, 240
Copper Proteins, 240

xiv   Contents

        Other Copper Enzymes, 241
    Nickel Coenzymes, 243
        Nickel in Methanogenesis, 243
        Other Nickel Coenzymes, 245
    Long-Range Electron Transfer, 247
        Biological Electron Transfer, 247
        Marcus Theory, 248

Chapter 5. Enzyme Inhibition, 253
    Two-Substrate Analogs, 254
        Inhibition and Binding, 254
        PALA and Aspartate Transcarbamylase, 254
    Suicide Inactivation, 255
        Thymidylate Synthase, 255
        β-Hydroxydecanoyl Thioester Dehydratase, 260
        γ-Aminobutyrate Aminotransferase, 262
    Kinetics of Slow-Binding and Tight-Binding Inhibition, 268
        Slow Binding, 268
        Tight Binding, 269
    Slow-Binding Inhibition, 270
        Dihydrofolate Reductase, 271
        Prostaglandin H Synthase, 274
    Tight-Binding Inhibition, 280
        HMG-CoA Reductase, 280
        Alanine Racemase, 285
        5-Enolpyruvoylshikimate-3-Phosphate Synthase, 289
        Acetylcholinesterase, 291

Chapter 6. Acyl Group Transfer: Proteases and Esterases, 297
    Chemistry of Acyl Transfer, 297
    Serine Proteases, 300
        Chymotrypsin, 301
        Subtilisin, 311
    Cysteine Proteases, 314
        Papain, 315
        Caspases, 317
    Aspartic Proteases, 317
        Molecular Properties, 318
        Mechanism of Action, 320
    Metalloproteases, 323
        Carboxypeptidase A, 324
        Thermolysin, 327
    Esterases, 328
        Structure and Function, 328
        Phospholipase A_2, 329

Chapter 7. Isomerization, 333
- Aldose and Ketose Isomerases, 333
  - Chemistry, 333
  - Phosphoglucose Isomerase, 334
  - Triosephosphate Isomerase, 335
  - Xylose Isomerase, 341
- Phosphomutases, 341
  - α-Phosphoglucomutase, 341
  - β-Phosphoglucomutase, 343
  - Phosphoglycerate Mutases, 343
- Racemases and Epimerases, 346
  - Proline Racemase, 346
  - Glutamate Racemase, 350
  - Mandelate Racemase, 352
  - UDP-Galactose 4-Epimerase, 355
  - Ribulose-5-P 4-Epimerase, 360
  - UDP-*N*-Acetylglucosamine-2-Epimerase, 361
- Chorismate Mutase, 364
- Δ^5-3-Ketosteroid Isomerase, 366
- Radical Isomerizations, 368
  - Glutamate Mutase, 369
  - Methylmalonyl CoA Mutase, 371
  - Lysine 2,3-Aminomutase, 376
- Newer Isomerases, 379
  - UDP-Galactopyranose Mutase, 379
  - Pseudouridine Synthase, 379

Chapter 8. Decarboxylation and Carboxylation, 387
- Chemistry of Decarboxylation and Carboxylation, 387
- Decarboxylases, 388
  - Pyruvate Decarboxylase, 389
  - Amino Acid Decarboxylases, 394
  - Acetoacetate Decarboxylase, 403
  - Mevalonate Pyrophosphate Decarboxylase, 405
  - Radical-Based Decarboxylases, 407
  - Orotidine Monophosphate Decarboxylase, 414
- Carboxylases, 418
  - Ribulose-1,5-Bisphosphate Carboxylase, 419
  - Phosphoenolpyruvate Carboxylase, 425
  - Vitamin K–Dependent Carboxylase, 426

Chapter 9. Addition and Elimination, 433
- α,β-Elimination/Addition Reactions, 433
  - Cofactor-Independent α,β-Elimination/Addition Reactions, 434
  - Cofactor-Dependent α,β-Elimination/Addition Reactions, 440

xvi   Contents

   β,α-Elimination/Addition Reactions, 456
    Methylidene Imidazolone–Dependent Elimination and Addition, 456
  Carbonic Anhydrase, 462
  Isomerization and Elimination, 465
   Catalytic Process, 465
   Coenzyme B_{12}–Dependent Elimination, 466

Chapter 10. Phosphotransfer and Nucleotidyltransfer, 476
  Chemistry of Phosphoryl Group Transfer, 476
   Phosphomonoesters, 476
   Phosphodiesters, 483
   Phosphotriesters, 483
   Five-Member Ring Phosphoesters, 484
  Enzymatic Phosphoryl Group Transfer, 487
   Single and Double Displacements, 487
   Phosphotransferases, 489
   Protein Phosphorylation: Protein Kinase A, 502
   Phosphomonoesterases, 509
  Enzymatic Nucleotidyl Group Transfer, 521
   Nucleotidyltransferases, 521
   Phosphodiesterases, 539

Chapter 11. ATP-Dependent Synthetases and Ligases, 547
  Ligation and the Energy of ATP, 547
  Activation by Phosphorylation, 548
   Glutamine Synthetase, 548
   Carbamoyl Phosphate Synthetase, 554
  Activation by Adenylylation, 559
   DNA Ligase, 559
   Aminoacyl-tRNA Synthetases, 561
   Ubiquitin, 566

Chapter 12. Glycosyl Group Transferases, 569
  Chemical Mechanisms, 570
   Chemistry of Glycoside Hydrolysis, 570
   Enzymatic Glycosyl Transfer, 573
  Glycosyltransferases, 575
   Sucrose Phosphorylase, 575
   Glycogen Phosphorylase, 577
   Purine Nucleoside Phosphorylase, 584
  Glycosidases, 587
   Families and Structures, 587
   Lysozyme, 589
   T4 Lysozyme, 595

Chapter 13. Nitrogen and Sulfur Transferases, 597
Nitrogen Transfer, 597
Aspartate Aminotransferase, 597
Tyrosine 2,3-Aminomutase, 602
Amidotransfer, 604
Glutamine:PRPP Amidotransferase, 607
Sulfur Transfer, 609
Biotin Synthase, 611
Lipoyl Synthase, 612

Chapter 14. Carbon-Carbon Condensation and Cleavage, 617
Chemistry, 617
Enolization of Acetyl CoA, 619
Acetyl CoA in Ester Condensations, 619
Citrate Synthase, 620
Thiolases, 627
Carbanionic Mechanisms, 630
Transaldolase, 631
Transketolase, 634
Serine Hydroxymethyltransferase, 639
Carbocationic Mechanisms, 645
Farnesyl Pyrophosphate Synthase, 645
Squalene Synthase, 648

Chapter 15. Alkyltransferases, 655
Chemistry of Alkylation, 655
Biological Alkylations, 655
Alkylation Mechanisms, 656
Enzymatic Alkylation, 657
Protein Farnesyltransferase, 657
Catechol *O*-Methyltransferase, 661
S-Adenosylmethionine Synthetase, 665
Methionine Synthases, 670

Chapter 16. Oxidoreductases, 679
Pyridine Nucleotide–Dependent Dehydrogenases, 680
Alcohol Dehydrogenase, 680
Lactate Dehydrogenase, 686
Short-Chain Alcohol Dehydrogenases, 687
Glyceraldehyde-3-P Dehydrogenase, 690
Glutamate Dehydrogenase, 693
Disulfide Oxidoreductases, 694
Dihydrolipoyl Dehydrogenase, 694
Ribonucleotide Reductases, 698
Classes of Ribonucleotide Reductases, 700
Structural Relationships of Ribonucleotide Reductases, 705

Chapter 17. Oxidases and Oxygenases, 710

Oxidases, 710
- D-Amino Acid Oxidase, 710
- Monoamine Oxidases, 716
- Isopenicillin-N Synthase, 718
- Urate Oxidase, 721

Monooxygenases, 722
- Lactate Monooxygenase, 722
- Cytochrome P450 Monooxygenases, 722
- Iron-Methane Monooxygenase, 727
- α-Ketoglutarate–Dependent Oxygenases, 732
- Dopamine β-Monooxygenase, 735
- Copper-Methane Monooxygenase, 737
- Nitric Oxide Synthase, 738

Dioxygenases, 741
- Intradiol Dioxygenases, 741
- Extradiol Dioxygenases, 744

Chapter 18. Complex Enzymes, 749

Multienzyme Complexes, 750
- α-Ketoacid Dehydrogenase Complexes, 750
- Pyruvate Dehydrogenase Complex, 750

Fatty Acid Synthesis, 757
- Acetyl CoA Carboxylase, 757
- Fatty Acid Synthases, 761

Modular Enzymes, 763
- Polyketide Synthases, 763
- Nonribosomal Polypeptide Synthetases, 767

Ribosomal Protein Synthesis, 768
- RNA Polymerase, 768
- The Ribosome, 770

Energy-Coupling Enzymes, 777
- Nitrogenase, 777
- Cytochrome *c* Oxidase, 782
- ATP Synthase, 786
- Myosin and Muscle Contraction, 792

Appendices, 803

Appendix A: Haldane Relationships for Some Kinetic Mechanisms, 803

Appendix B: Inhibition Patterns for Three-Substrate Kinetic Mechanisms, 804

Appendix C: Equations for Number of Occupied Sites in the Binding of a Ligand to a Multisite Macromolecule, 804

Appendix D: Derivation of Steady-State Kinetic Equations by the King-Altman Method, 805

Index, 809

Enzymatic Reaction Mechanisms

1
Enzymes and Catalytic Mechanisms

Catalysis and the Active Site

Enzymes catalyze the biochemical reactions in cells of all organisms. These reactions constitute the chemical basis of life. Most enzymes are proteins—a few are ribonucleic acids or ribonucleoproteins—and the catalytic machinery is located in a relatively small active site, where substrates bind and are chemically processed into products. Illustrations of the molecular structure of chymotrypsin, a typical enzyme, and the location of its active site appear in figs. 1-1A and B. The polypeptide chain is shown as a ribbon diagram (see fig. 1-1A), and the active site is the region in which an inhibitor, the black ball-and-stick model, is bound. The gray ball-and-stick structures are amino acid side chains at the active site that participate in catalysis.

The ribbon diagram shows the individual chains and the α-helices and β-strands as if there were vacant spaces between them; however, very little free space exists in the interior of an enzyme. The packing density in the interior of a protein is typically 0.7 to 0.8, meaning that 70% to 80% of the space is filled and only 20% to 30% is interstitial space (Richards, 1974). That the packing density in hexagonally closest packed spheres is 0.75, similar to a protein, conveys a concept of the interior. The free space inside a protein is so little that in a space-filling model (see fig. 1-1B), the polypeptide chain cannot be discerned, and interactions between active sites and substrate or inhibitors cannot be seen. For this reason, we display structures as ribbon diagrams to facilitate the discussion of ligand binding interactions within an active site.

Chymotrypsin is the most widely studied and one of the best-understood enzymes. It catalyzes the hydrolysis of proteins at the carboxamide groups of hydrophobic amino acid residues, principally phenylalanyl, tyrosyl, and tryptophanyl residues. It also catalyzes

2 Enzymatic Reaction Mechanisms

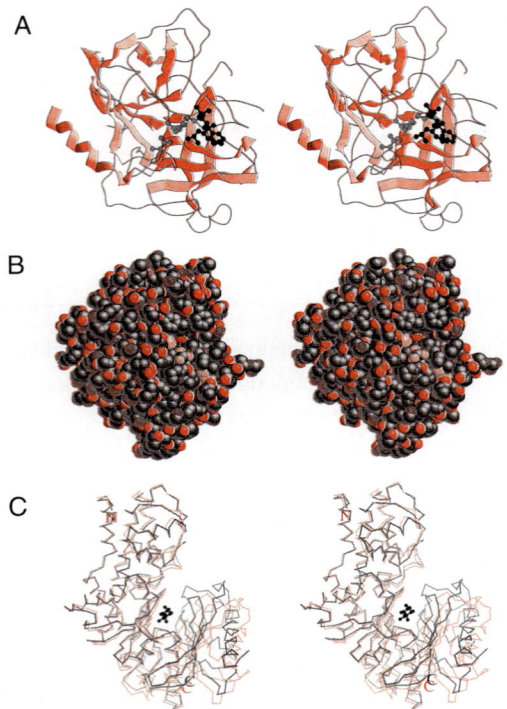

Fig. 1-1. Structures of chymotrypsin and hexokinase. (A) Ribbon diagram of the 1.5-Å resolution structure of bovine chymotrypsin (EC 3.4.21.1) in a complex with the transition-state analog inhibitor N-acetyl-L-leucyl-L-phenylalanine trifluoromethylketone (AcLF-TFK) (PDB 7GCH; Brady et al., 1990). The chain fold includes α-helical segments and β-strands, as well as irregular strands. The black ball-and-stick model is the inhibitor and the gray ball-and-stick segments are amino acid side chains of the catalytic residues Ser195, His57, and Asp102. (B) Space-filling model of the structure showing that the atoms are tightly packed throughout the structure. The space-filling model lacks the hydrogen atoms, and the inclusion of hydrogen would increase the packing density. (C) *Saccharomyces cerevisiae* hexokinase (EC 2.7.1.1) with glucose (black ball-and-stick representation) bound in the active site cleft. The N-terminal domains of the Cα traces of the liganded- (gray, PDB 1HKG, PI isozyme) and apo- (red, PDB 2YHX, PII isozyme) hexokinase structures are aligned to show the apparent motion of the C-terminal half of the enzyme as the whole protein contracts around the substrate (Anderson et al., 1978; Bennett and Steitz, 1980a, 1980b).

the hydrolysis of small substrates, such as acetyltyrosine ethyl ester (ATEE) or acetyltyrosine p-nitroanilide (ATNA). These reactions are practically irreversible, their rates can be measured spectrophotometrically, and they behave kinetically as one-substrate enzymatic reactions.

The overall reaction of ATEE can be written as ATEE → Acetyltyrosine + Ethanol, where the participation of water as a substrate is understood. The chemical steps defining catalysis by chymotrypsin take place at the active site and are reasonably well known. Chymotrypsin is typical in that it brings about the hydrolysis of a peptide bond at a rate about 10^{12} times the spontaneous rate at pH 7. This is an enormous rate enhancement, and the rationalization of rate increases of this and greater magnitude is one of the objectives in mechanistic studies.

The active site of chymotrypsin was first characterized by the identification of Ser195. Chymotrypsin and other serine proteases and esterases are inactivated by diisopropylphosphorofluoridate (DFP) (Jansen and Balls, 1952). Reaction of chymotrypsin with DFP led to the production of HF and diisopropylphosphorylation of the enzyme to form diisopropyl-chymotrypsin (DIP-chymotrypsin). The DIP-moiety in the inactivated enzyme was found to be bound to Ser195 in the amino acid sequence (Hartley and Kauffman, 1966; Hess, 1971). In this way, Ser195 was first identified as a catalytic residue of chymotrypsin. The structure confirmed this assignment (Sigler et al., 1968).

The description of the catalytic action of an enzyme at the atomic level of resolution is the first objective of any mechanistic study. This has been attained in a number of cases. Chymotrypsin was a convenient enzyme for early mechanistic studies because it was commercially available and catalyzed a simple reaction. With the cloning of genes, their expression in systems that produce large amounts of enzymes, and the availability of convenient and fast methods for purifying them, virtually any soluble enzyme can be obtained in a homogeneous state in large enough amounts for detailed studies. In this textbook, we describe studies of about 100 enzymes at various levels of detail.

The most fundamental events of enzymatic catalysis are brought about by the chemistry and dynamics of direct contacts between a substrate and its active site, and they constitute the molecular basis of rate enhancement and catalysis. The contacts constitute binding through hydrophobic, electrostatic, dipolar, and hydrogen bonding. Amino acid side chains that include functional groups provide chemical catalysis by the carboxyl, amino, imidazole, hydroxyl, and thiol groups of glutamate and aspartate, lysine, histidine, arginine, serine, threonine, tyrosine, and cysteine residues, respectively. These groups engage in electrostatic and hydrogen bonding interactions with substrates and also serve as acid-base, nucleophilic, and electrophilic catalysts. Although there are relatively few types of catalytic groups in enzymes, their chemical utility can be extended by microenvironmental effects in enzymatic sites, such as through perturbations of acidity or basicity. In this chapter, we consider the functions of the side chain amino acid groups as well as the overall structure and dynamics of enzymes in catalysis.

Presumably because of the small number of amino acid functional groups available for catalysis, the chemical properties of enzymes have been extended by the evolution of coenzymes, which bind to enzymes and provide physicochemical properties not displayed by the protein structures. Enzymes enhance the catalytic proficiency of coenzymes and cofactors by the same mechanisms they bring to bear on substrates. The coenzymes and cofactors are discussed in chapters 3 and 4.

Rate Enhancement in Enzymatic Catalysis

Enzymes catalyze biological reactions at rates that are often incomprehensibly faster than nonenzymatic counterparts. The actual rate enhancements are known for several classes of enzymatic reactions, but not for all of them. To determine a rate enhancement factor, the rate constant of the nonenzymatic reaction is compared with a comparable rate constant for the enzymatic reaction. This is much more easily said than done. Many problems attend this process, including kinetic problems of comparing rate constants for multisubstrate enzymatic reactions with non-enzymatic counterparts that could require multibody collisions. In many cases, however, rate constants for unimolecular or pseudo-unimolecular enzymatic and nonenzymatic reactions can be compared. A few examples are included in fig. 1-2. The main log scale relates the rate constants and half-times for a collection of first-order and pseudo–first-order nonenzymatic reactions. At the top of this scale the range of k_{cat} values for corresponding enzymatic reactions are coded in gray scale, and this scale is expanded in an off-set. (The parameter k_{cat} is the turnover number for an enzyme, as defined in textbooks of biochemistry. The parameter k_{cat}/K_m is the second-order rate constant for the

4 Enzymatic Reaction Mechanisms

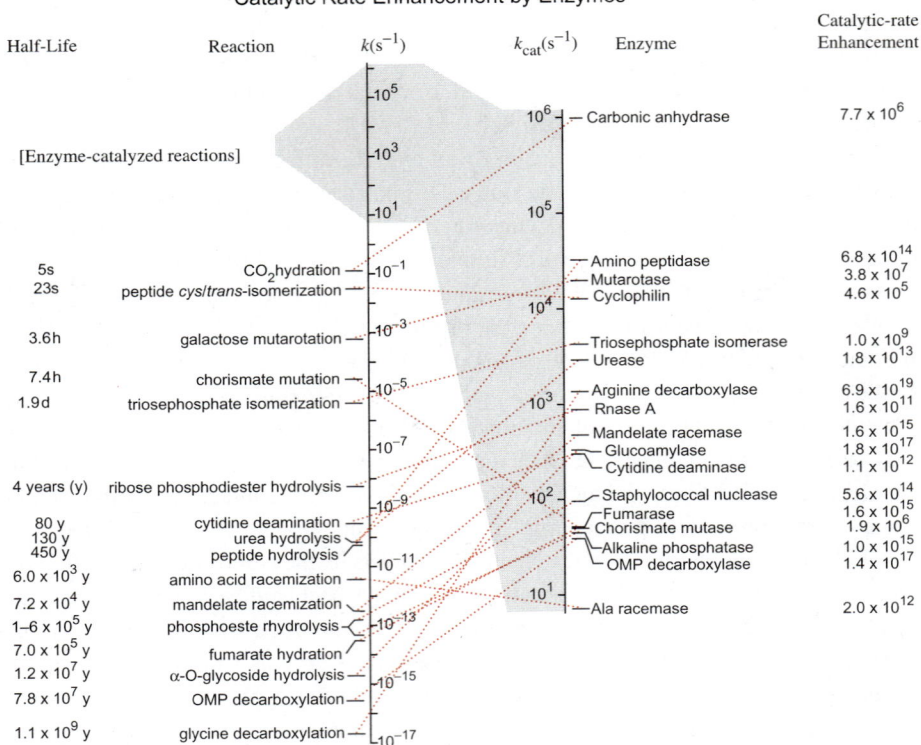

Fig. 1-2. Rate enhancements in enzymatic catalysis. The log scale in the center correlates first-order and pseudo–first-order rate constants for nonenzymatic reactions with turnover numbers (k_{cat}) for enzymes catalyzing similar reactions. The rate constants for the nonenzymatic reactions range from about 10^{-17} to 10^{-1} s^{-1} ($t_{1/2}$ of 1 billion years to 5 seconds). The enzymatic turnover numbers cover a relatively narrow range of 10 to 10^7 s^{-1}, and they are shaded at the top of the scale and expanded in the scale on the right. The enzymes listed on the expanded scale catalyze reactions similar to the nonenzymatic reactions and are connected by the dotted red lines to those reactions on the center scale. The rate enhancement factors, calculated by the ratios of enzymatic to nonenzymatic rate constants, are listed in the column on the right for each enzyme and range from 4.6×10^5 for cyclophilin to 6.9×10^{19} for arginine decarboxylase. The range of enzymatic turnover numbers is limited by the rate of diffusion. The lower values of rate enhancement arise from the high rates of the corresponding nonenzymatic reactions. For example, peptide cis/trans-isomerization does not involve the cleavage of a covalent bond and proceeds with a half-time of 23 s non-enzymatically. Because of the speed of the nonenzymatic reaction, the enzymatic rate constant is only 4.6×10^5 times higher.

reaction of a substrate with an enzyme to produce the product. The parameter K_m is the concentration of substrate that elicits one-half the maximal enzymatic rate. These parameters are derived in chapter 2.)

Rate constants for the nonenzymatic reactions range from 10^{-17} to 10^{-1} s^{-1}, corresponding to half-times of a billion years to 5 seconds. The enzymatic rate constants (k_{cat}) range from about 10 to 10^7 s^{-1}. The range of nonenzymatic rate constants is ten orders of magnitude larger than of enzymatic rate constants. A plot of log k_{cat} against log K_{TS}, where K_{TS} is a measure of efficiency in binding the transition state, displays a slope of only −0.16 (Bruice and Benkovic, 2000). The reason for this apparent compression in enzymatic rates might be regarded as arising simply from the leveling that would be compatible with any correlation of linked cellular processes. However, the upper limit of enzymatic rates is near the diffusional rate constants for enzyme-substrate binding of 10^7 to 10^8 M^{-1} s^{-1} (see chap. 2), and many enzymatic rates are diffusion-limited. In this scenario a leveling of enzymatic rates relative to nonenzymatic rates is inevitable.

The rate enhancement factors for the enzymatic reactions in fig. 1-2 range from 5×10^5 for the *cis*-, *trans*-isomerization of prolyl residues in a protein to 7×10^{19} for the decarboxylation of an amino acid. The mechanism of prolyl isomerization, which does not involve cleaving any covalent bonds, may be similar in the enzymatic and nonenzymatic processes. However, the mechanisms of nonenzymatic and enzymatic decarboxylation of amino acids are not the same. The nonenzymatic decarboxylation is not potentiated by a coenzyme (pyridoxal-5′-phosphate), which contributes most of the catalytic efficiency. The participating coenzyme is an element of the enzymatic process, but it nevertheless changes the mechanism. The rate enhancement factors represent valid differences that may not refer to the same mechanisms.

Any interpretation of the fine points of rate enhancement factors cannot diminish the impact of their magnitudes. Rate enhancements typical of enzymes are stupendous by any standard regardless of reaction mechanism. The rate enhancements in fig. 1-2 represent those that have been measured, not those of all enzymes. The most difficult part of measuring large rate enhancements is obtaining the nonenzymatic rates of reactions that are too slow to measure. Rate enhancement factors for reactions that cannot be observed in the absence of an enzyme cannot be measured.

The objective in mechanistic enzymology is to determine the exact reaction mechanisms and the physicochemical phenomena contributing to the enormous rates. The ultimate goal is to explain the rate enhancements to within an order of magnitude.

Conformational Mobility in Catalysis

For experimental and theoretical reasons, conformational mobility in the action of enzymes is almost inevitably invoked in detailed mechanistic analysis of enzymes. Most enzymes display catalytic properties that require conformational mobility and could not otherwise be explained. In theory, the normal vibrational properties of the atoms within a protein mandate very high frequency motions within the molecule (Karplus, 2002). Key motions may provide the means by which global conformational changes take place.

Substrate-Induced Conformational Changes

Detailed studies reveal properties of enzymes that seem to require conformational changes. The concept of the substrate-induced conformational change was introduced to explain how hexokinase could bind MgATP while not catalyzing its hydrolysis (Koshland and Neet, 1968).

Hexokinase catalyzes (eq. 1-1) the phosphorylation of glucose to glucose-6-P by MgATP.

$$\text{Glucose} + \text{MgATP} \rightarrow \text{Glucose-6-P} + \text{MgADP} \qquad (1\text{-}1)$$

The kinetic mechanism is a random bi bi sequential type, with preferential binding of glucose followed by MgATP to form the ternary complex (Ning et al., 1969). However, the binary complex of hexokinase-MgATP is formed and is kinetically significant. Any realistic explanation for the specificity of hexokinase in catalyzing phosphoryl group transfer from MgATP essentially exclusively to the C6(OH) group of glucose required a structural explanation. It seemed that without such an accounting, there would be no reason for the complex of hexokinase with MgATP to react so poorly with the chemically similar and ubiquitous H(OH) group of water, which was thought to have free access to the active site.

This view was bolstered by the display of a very low ATPase activity by hexokinase. Lyxose and xylose stimulate the ATPase activity but are not themselves phosphorylated. The differences between the rates of phosphotransfer to glucose and water under various conditions required an explanation, and the substrate-induced conformational change provided that rationale. The binding of glucose to hexokinase was postulated to induce a conformational change in hexokinase to a catalytically active structure that would catalyze the phosphotransfer from MgATP exclusively to the C6(OH) of glucose, which would be perfectly placed in the tertiary structure to accept the phosphoryl group. This explanation for the specificity of hexokinase proved to be infectious and quickly spread to many other enzymes. It became a general concept for understanding the actions of enzymes.

The x-ray crystallographic analysis of hexokinase proved that glucose induced a structural change, as predicted based on the kinetic evidence (Bennett and Steitz, 1978, 1982a, 1982b; Shoham and Steitz, 1982). As illustrated in fig. 1-1C, the structural change brought about by the binding of glucose forces the closure of a cleft and essentially creates the active site structure within the cleft. Before this closure, the disparate catalytic machinery is relatively ineffective at catalyzing phosphotransfer. The location of the active site within the cleft of hexokinase has become a recurrent theme in the structures and actions of many other enzymes. A further example appears in chapter 2, that of the substrate-induced conformational change in adenylate kinase (see fig. 2-4).

Catalysis of Multistep Reactions

Very few enzymatic reaction mechanisms involve a single chemical event and a single transition state. Most proceed in multiple steps that involve the transient formation of reactive intermediates. In a multistep reaction, an enzyme must catalyze more than one chemical process. Even as simple a reaction as peptide hydrolysis by chymotrypsin proceeds in two steps, acylation and deacylation, and the transition states for the two processes cannot be identical, although they may be very similar.

In such reactions as the production of tyrosyl tRNA^{Tyr} by tyrosyl tRNA synthetase, two entirely different reactions are catalyzed at the same site. In the first step, the enzyme catalyzes the reaction of tyrosine with ATP to form tyrosyl adenylate and PP$_i$, a phosphotransfer process. Tyrosyl adenylate remains bound to its site, and in the second step the enzyme catalyzes its reaction with tRNA^{Tyr} to form tyrosyl tRNA^{Tyr} and AMP, an acyltransfer process. The two reactions are entirely different but are catalyzed by one enzyme at one site. The transition states must be different, and the enzyme surely undergoes structural changes in its adaption to catalysis of the two reactions.

The two transition states in a reaction that involves acyltransfer in both steps often are similar, but however closely related the steps are, there are always differences. Recognition of these differences by the enzymatic active site must involve at least minor changes in its structure.

Structural Mobility in Enzymes

Quenching of Tryptophan Fluorescence by Oxygen

The high-frequency structural motions in proteins are revealed by the fluorescence properties of enzymes. In studies of the fluorescence of tryptophan in enzymes whose structures were known at the time, it was found that oxygen could quench the fluorescence of all the tryptophan residues, even those most deeply buried (Lakowicz and Weber et al., 1973). This was an important observation that essentially proved the structural mobility of proteins on the nanosecond time scale.

In fluorescence, a molecule absorbs ultraviolet light and is elevated by this energy to an excited state. The excited molecule returns to its ground state by emitting the energy absorbed from the light. Energy emission can occur by various means, including the emission of heat. In fluorescent molecules, one means of returning to the ground state is to emit a small amount of heat in falling to a lower excited state, and then to emit light of lower energy than had been absorbed, that is, light of longer wavelength. The emitted light is fluorescence. The wavelength of absorbed light is the excitation frequency, generally in the ultraviolet range for organic molecules, and the emitted light is at the emission frequency often in the visible range. The lifetime of the excited state is generally, though not always, on the nanosecond time scale.

In the studies of oxygen quenching of protein fluorescence, the lifetime of excited tryptophan in proteins proved to be a few ns. Quenching of tryptophan fluorescence by oxygen proved to require a molecular collision between the oxygen molecule and the photoexcited tryptophan. It was then found that at sufficiently high solution concentrations of oxygen, which required very high oxygen pressures, the fluorescence of all the tryptophan residues in several enzymes were quenched. This meant that oxygen collided with all of the tryptophan residues within the lifetimes of their excited states. Oxygen molecules must have diffused through the molecular structures of these enzymes on the nanosecond time scale. The enzymes displayed normal biological activities at the high pressures of oxygen used in the experiments.

The dimensions of oxygen are significant. Its molecular volume is significantly greater than the spaces between atoms in a protein. A protein must be structurally mobile to permit a molecule the size of oxygen to diffuse through its internal structure.

Packing Densities of Atoms in Enzymes

The significance of the diffusion of oxygen through protein structures is best appreciated by considering the packing of atoms within their structures. Early in the development of concepts of protein structure, it was thought that the interior of proteins might be like oil droplets because of the dominance of hydrophobic residues. After the structures of a few enzymes were known, it became possible to examine this hypothesis. It immediately became clear that the interiors of proteins could not be oil-like (Klapper, 1971; Richards, 1974). From the crystal structures, the molecular volumes of enzymes could be computed, and from the atomic compositions and volumes the atomic packing densities could be computed. The packing densities turned out to be in the range of 0.75; that is, 75% of the total internal volume was occupied by atoms. This packing density corresponds to hexagonally closest packed spheres, and it is incompatible with a liquid structure. Packing densities typical of liquids are less than 0.5, and those typical of solids are greater than 0.6. The interiors of proteins could not be liquid and might be more like wax. There is much more interstitial space in a liquid than in a solid, and tight packing in a solid would not allow fast diffusion of molecular oxygen unless the structure is dynamic.

In a detailed study of the interior of ribonuclease, the packing density proved to vary within the structure (Richards, 1974). Near the surface in the unoccupied active site the packing density was in the range of 0.60 to 0.65. Deep within the interior, the packing density ranged between 0.70 and 0.85. The high frequency motions in protein structures must be understood in the context of high packing densities, which must place restraints on the allowed motions. A conformational change presumably would not lead to an appreciable increase in the overall volume of an enzyme. Such a change must correspond to a highly ordered reorganization of atoms that remain within van der Waals contact throughout the process.

Nature of Motions

No motion within a protein can be random. If random motions were allowed, a protein could not exist in a definite structure. Even the simplest vibrational or rotational motions within a protein structure must be constrained. Evidence indicates that most of the secondary structure within a protein is fairly stable over time and should limit the global conformational changes that may occur on the catalytic time scale of milliseconds.

Most α-helices and β-strands seem to be fairly stable within a protein molecule. The evidence for this is structural, when two conformations are known from x-ray crystallography, and kinetics, supplied from studies of the rates at which peptide-NH groups undergo exchange with deuterium in D_2O. Amide hydrogens can be exchanged with deuterium within a few minutes at 25°C and pH 7, and this is true of the peptide-NH groups that are not engaged in hydrogen-bonded α-helices and β-strands. Hydrogen-bonded protons in secondary structures undergo exchange very slowly because the exchange mechanism disrupts the hydrogen-bonding network. Observations of peptide-NH exchange by nuclear magnetic resonance (NMR) reveal at least two kinetic classes: those that exchange within a few minutes and those that exchange over a period of hours. Most of the latter group are engaged in the hydrogen-bonding networks of α-helices and β-strands. Most of the secondary structural elements in a protein survive the fast substrate-induced conformational changes and the nanosecond motions that allow oxygen to diffuse through the structures.

High-frequency vibrational displacements of α-helices and domain movements such as those illustrated in fig. 1-3 would allow the transient opening of crevices and expose amino acid side chains in the α-helices and β-sheets to external molecules such as molecular

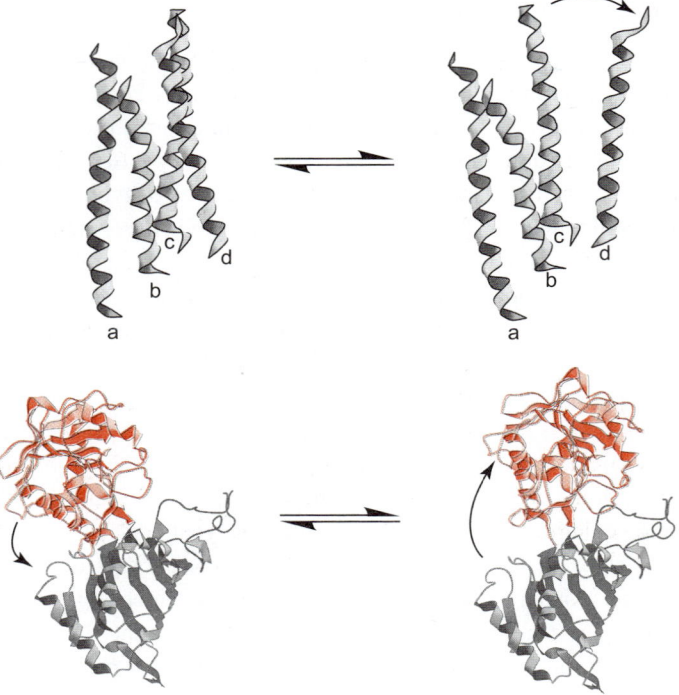

Fig. 1-3. Motions of helices and domains in proteins.

Table 1-1. Time Scales of Motions in Globular Proteins

Motion	Time Scale
Bond vibrational motions	Sub-femtoseconds
Methyl group torsion	Picoseconds
Rotations at β-turns	Nanoseconds to microseconds
Prolyl *Cis-/trans*-isomerization	Seconds

oxygen or solvent. These motions would not disrupt secondary structure. However, a model of protein dynamics should not exclude secondary structural changes of limited scope in protein conformational transitions, which often include modest changes in secondary structure.

Time Regimes of Motions

Among the experimental methods applied, NMR spectroscopy provides the greatest versatility in studying the motions of a protein molecule. Because of the frequency range of NMR experiments and the capacity for NMR transitions to be manipulated by the timing and directionality of irradiation pulses, the time domains for motions in protein molecules can be classified as to type. The extrema are set by methyl groups of alanine, valine, leucine, and isoleucine, which generally spin freely on the picosecond time scale, whereas *cis-, trans*-isomerization of a freely mobile peptide bond can take place on the time scale of seconds or longer. The rates of various classes of motions within a typical globular protein molecule are given in table 1-1.

Structural mobility seems to be important for enzyme specificity and catalysis of multistep reactions. Enzymes are structurally dynamic molecules that can adapt to changing catalytic requirements.

Acid-Base Catalysis

Acids and Bases

Many chemical reactions are catalyzed by acids or bases, or both. Most acids and bases in aqueous solutions are of the Brønsted type; they release or accept protons or hydrogen ions when dissolved in water (structures **1-1** to **1-3**). Brønsted acids release hydrogen ions and Brønsted bases react with and neutralize hydrogen ions. HCl releases H^+ and Cl^- in water, and NaOH reacts with H^+ to form Na^+ and H_2O. Because all buffers release or react with and neutralize hydrogen ions, they are Brønsted acids and bases. Reactions catalyzed by enzymes that are subject to Brønsted acid or base catalysis are illustrated in fig. 1-4 and include the hydrolysis of glycosides, esters, and amides. They also include acyl, phosphoryl, and glycosyl group transfer reactions; dehydration of β-hydroxy carbonyl compounds; and aldimine (Schiff base) and ketimine formation and hydrolysis.

A hydrogen ion, or proton, does not exist as such in aqueous solutions but is coordinated to a water molecule as a hydronium ion (H_3O^+). We refer to them as hydrogen ions or protons, with the understanding that they are hydronium ions in water. Hydroxide ions are also strongly solvated in water. They are encased within three water molecules and react within their solvation spheres, often by mechanisms that involve proton transfers with solvating water molecules. We refer to hydroxide ions as such, with the understanding that they are highly solvated species in water.

10   Enzymatic Reaction Mechanisms

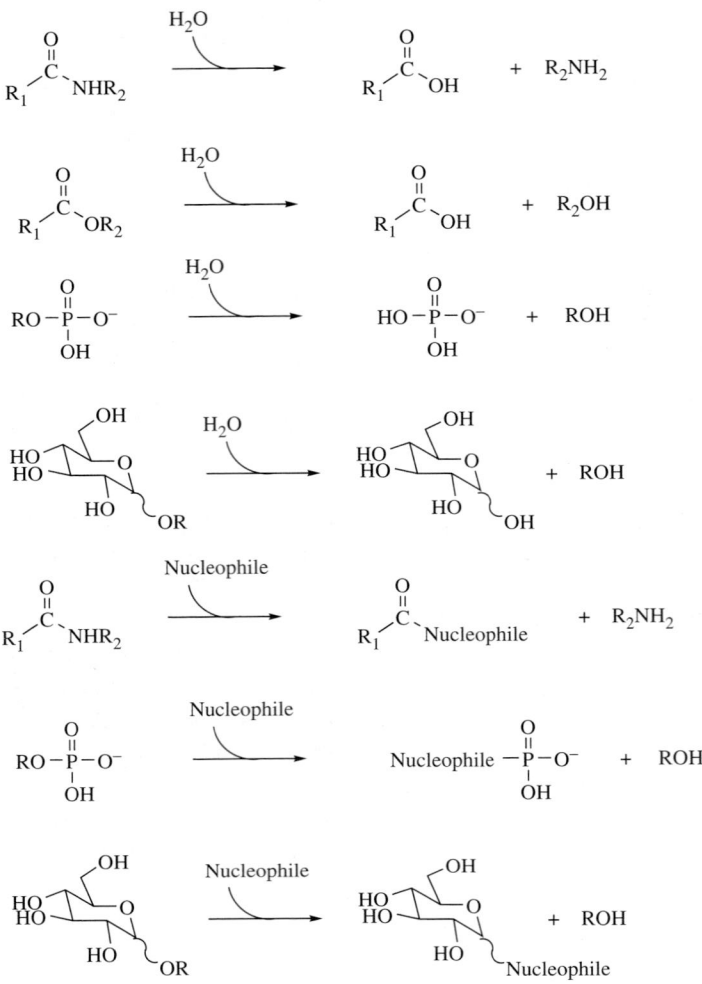

Fig. 1-4. Enzymatic reactions that require only general acid-base catalysis include amide, ester, phosphoester, and glycoside hydrolysis and group transfers. However, general acid-base catalysis is also required in most enzymatic reactions that require coenzymes such as pyridoxal-5′-phosphate, thiamine pyrophosphate, tetrahydrofolate, or biotin. General acid-base catalysis is the most common chemical catalytic process in enzymatic mechanisms, and it is the most important chemical event in the actions of the enzymes that catalyze the reactions shown.

A more general definition of acids and bases is that of Lewis. Lewis acids are defined as molecules or ions that form covalent bonds with nonbonding electrons, and Lewis bases donate nonbonding electrons to form covalent bonds with Lewis acids. This definition includes the hydrogen ion as a Lewis acid and the hydroxide ion as a Lewis base, and they react with each other to form a covalent bond (H_2O). Brønsted acids and bases are special cases of Lewis acids and bases. The Brønsted definition of acids and bases is most widely used for aqueous solutions because hydrogen and hydroxide ions are derived from water and are the strongest acid and base, respectively, in water. The Lewis acid concept is helpful for understanding catalysis by metal ions in aqueous solutions; however, it is especially useful for explaining reactions in nonaqueous media, where hydrogen ions and hydroxide ions do not exist and cannot participate in acid-base catalysis. For example, Lewis acids such as $AlCl_3$ are used in organic chemistry to catalyze electrophilic aromatic substitution reactions. Metal ions act as Lewis acids in biological reactions that take place in

Acid- and Base-Catalyzed Reactions

Specific Acid and Base Catalysis

A specific acid–catalyzed reaction in water is accelerated by the binding of a proton to a functional group of the substrate. A familiar example is the acid-catalyzed hydrolysis of a peptide. The amide functional group is hardly reactive toward water, which makes it well suited to serve as the basic structural unit in proteins. Its stability is due to delocalization of the nonbonding electron pair, as illustrated in structures **1-1** and **1-2**, shown in resonance notation.

The charge separated resonance form contributes about 40% to the true structure, which means that neither the carbonyl (C=O) nor the iminium (C=NH⁺) exists in the composite structure **1-1** to **1-3**. The canonical structures, if they existed, would display significant electrophilic reactivity with water, but the composite does not. Protonation of the peptide carbonyl group dramatically increases its electrophilic reactivity by disrupting the electron delocalization (fig. 1-5) and generating a highly electrophilic species. The protonated species reacts with water to form a series of tetrahedral adducts in rapid, tautomeric equilibrium, one of which eliminates the amine with concomitant formation of the free acid. Reactions catalyzed by hydrogen ions in water are sometimes known as *specific acid catalyzed* in recognition of the fact that they require hydrogen ions, the strongest acid that exists in water. Hydrogen ions react with bases at essentially diffusional rates, so that proton transfer is rarely rate limiting.

Hydroxide ions also cleave peptides. A reasonable mechanism is shown in fig.1-6. In this case, hydroxide ion is a very reactive nucleophile that will undergo addition to the carbonyl group. As a nucleophile, hydroxide reacts through one of the three coordinated water molecules, as illustrated in fig. 1-6, by abstracting a proton from one of the water

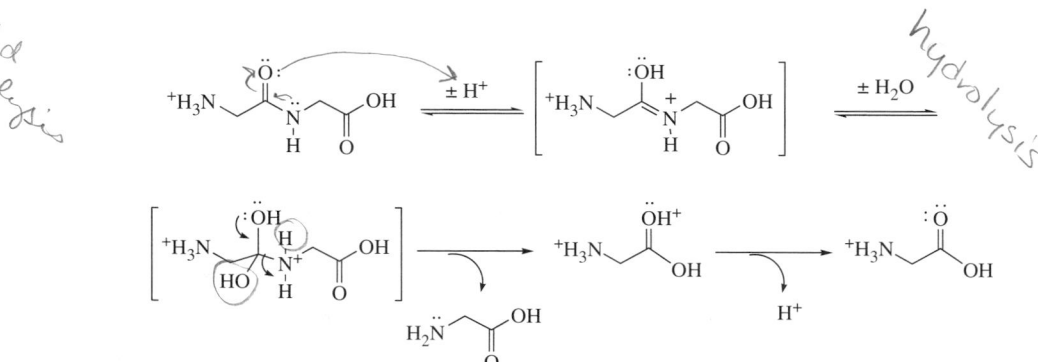

Fig. 1-5. Hydrogen ion catalysis in the hydrolysis of a peptide. Peptide hydrolysis catalyzed by strong acids takes place by specific acid catalysis in which the peptide is first protonated in an equilibrium process. The protonated peptide is highly reactive with water because of the disruption of electron delocalization within the amide linkage and the presence of the positive charge. The addition of water as a nucleophile forms the tetrahedral intermediate, which collapses by cleavage to the two amino acids.

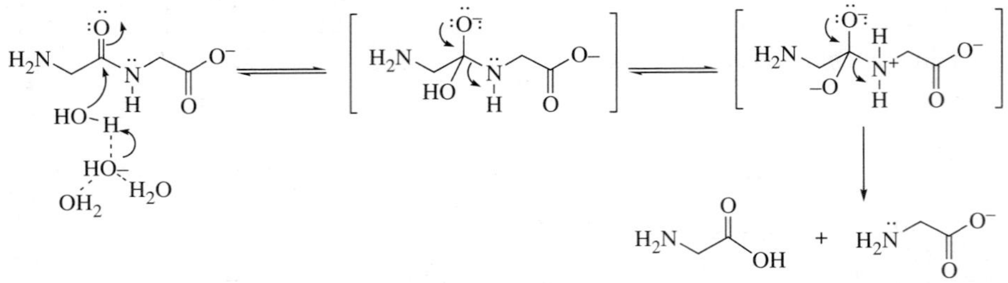

Fig. 1-6. Hydroxide ion catalysis in the hydrolysis of a peptide. The mechanism is written to represent catalysis by hydroxide ion, which initially reacts as a nucleophile adding to the carbonyl group of the peptide. Hydroxide is regenerated in the last step. In a practical reaction, a mole of hydroxide ion is consumed by reaction with the free carboxylic acid produced in the next-to-last step. Hydroxide reacts through intervening molecules of water that hydrate the ion.

molecules that actually attacks the peptide carbonyl group in the transition state. This is known from kinetic isotope effects on the alkaline hydrolysis of acyl compounds (Marlier, 1993; Marlier et al., 1999; Mata-Segreda, 2002). The resulting anionic tetrahedral intermediate can either eliminate hydroxide ion in the reverse direction or the peptide amide group in the forward direction. Proton transfer is concerted with decomposition. Hydroxide ion-catalysis in water is known as *specific base catalysis*.

The pH dependence for the hydrolysis of the peptide bond in *N*-phenylacetyl-glycyl-D-valine follows the rate law of eq. 1-2, in which k_{obs} is the observed first-order rate constant.

$$k_{obs} = k_0 + k_H[H_3O^+] + k_{OH}[OH^-] \tag{1-2}$$

At 37°C in D_2O, the value of k_{H+} is 1.6 10^{-6} M^{-1} s^{-1} and that of k_{OH} is 1.1 10^{-6} M^{-1} s^{-1} (Smith and Hansen, 1998). The specific base–catalyzed hydrolysis in water would be faster because of the likely deuterium kinetic isotope effect, which was not measured in this case. The uncatalyzed hydrolysis by water proceeds with a value of $k_0 = 8.2 \times 10^{-11}$ s^{-1}, corresponding to a half-time of about 240 years.

The hydrolysis of esters is in many respects similar to the hydrolysis of amides and peptides, but it is about 10^4-fold faster because of the greater electrophilic reactivity of the ester group. However, the base-catalyzed hydrolysis of esters does not always involve the formation of a discrete tetrahedral intermediate. When there is a very good leaving group that does not require protonation to leave, the hydroxide ion can displace it in a single step through a tetrahedral transition state (Hess et al., 1998; Shain and Kirsch, 1968).

General Acid and Base Catalysis

Acid-base catalysis by Brønsted acids and bases in addition to hydrogen and hydroxide ions is known as general acid-base catalysis. Such acids and bases can be much weaker than hydrogen and hydroxide ions, respectively, and include intracellular species such as phosphates, carbonates, amino acids, carboxylic acids, amines, and proteins. Nonbiological weak acids and bases include the buffers used in biochemical experiments.

The acidic and basic functional groups in enzymes are among the many weak biological acids and bases. The most common acids in enzymes are the glutamate-γ-COOH and aspartate-β-COOH groups. The most common bases are the imidazole ring of histidine and the ε-amino group of lysine, the guanidine group of arginine. The thiol/thiolate group of cysteine, and the phenol/phenolate group of tyrosine also function as acid-base catalysts. These are shown in table 1-2 together with their pK_a values in aqueous solution.

Because the acidic and basic groups in enzymes are weak, they are less reactive than hydrogen or hydroxide ions. However, their proximity to substrates in active sites of enzyme-substrate complexes allows them to react much faster than 10^{-7} M hydrogen and

Table 1-2. Normal Values of pK_a for Acidic and Basic Amino Acid Groups

Amino Acid	Side Chain	pK_a
Aspartic acid	—CH$_2$COOH	3.9
Glutamic acid	—CH$_2$CH$_2$COOH	4.3
Lysine	—CH$_2$CH$_2$CH$_2$CH$_2$NH$_3^+$	10.5
Histidine	—CH$_2$—(imidazolium, +)	6.0
Tyrosine	—CH$_2$—C$_6$H$_4$—OH	10.2
Cysteine	—CH$_2$SH	8.5

hydroxide ions at pH 7. Moreover, microenvironmental interactions that significantly increase their acid or base strengths can increase their reactivities and effectiveness as catalytic groups. The phenolic group of tyrosine is a very weak acid (pK_a = 10.2), and it does not often act as an acidic or basic group in enzymes. However, in a few enzymes it is an acid-base catalyst. In UDP-galactose 4-epimerase, the pK_a of Tyr149 is decreased to 6.1 by the positive electrostatic field in the active site, and it functions as an acid-base catalyst (see chap. 7).

The pK_a values for the acidic and basic groups in enzymes can be perturbed by microenvironmental effects at the active sites, and the perturbation mechanisms are often essential aspects of enzymatic action. The pK_a of the Lys115 ε-aminium group at the active site of acetoacetate decarboxylase is unusually low, 6.0, and this allows it to react as –NH$_2$ in aldimine (Schiff base) formation in neutral solution (see chap. 8). The low pK_a is thought to be brought about by a positive electrostatic field created by the neighboring Lys116 in the active site (Frey et al., 1971; Highbarger et al., 1996; Kokesh and Westheimer, 1971). If its pK_a were normal (10.5), it would exist as –NH$_3^+$ in neutral solution and could not react rapidly under physiological conditions. The microenvironments at active sites are often less polar than the aqueous medium, and low polarity decreases the acidities of glutamic and aspartic acid side chains.

The pK_a values for side chain carboxylic acid groups in active sites are often higher (pK_a = 6 to 8) than the value of 4.5 observed in aqueous solutions. The ionizations of neutral acids such as —COOH proceed with charge separation (A—H → A$^-$ + H$^+$), and this process is energetically disfavored in a medium of low dielectric relative to that of water. The values of pK_a for such acids are elevated in media that are less polar than water. A list of enzymes that incorporate amino acid side chains with perturbed pK_a values can be found in chapter 2 (see table 2-3).

General acid catalysis and general base catalysis are related to specific acid and base catalysis in that they both involve proton transfer. General acid-base catalysis differs in that proton transfer is an integral part of the transition state, whereas in specific acid-base catalysis, proton transfer is a fast process that precedes the cleavage of other covalent bonds, often in pre-equilibrium steps.

Hydrogen and hydroxide ions are not effective in catalyzing peptide cleavage at pH 7 because their concentrations are so low (10^{-7} M). In contrast, weak Brønsted acids and bases can exist at high concentrations in neutral solutions, so that even though they are less reactive than equal concentrations of hydroxide or hydrogen ions, they can be more reactive at concentrations of = 0.1 M than 10^{-7} M hydroxide or hydrogen ions. In an active site, an acid or base group of the enzyme will exist at a very high effective concentration relative to a substrate that is held in its near vicinity within a Michaelis complex. Moreover, the reaction of such an acid or base with the substrate would not require a bimolecular collision, so that it would be entropically favored.

14   Enzymatic Reaction Mechanisms

Mutarotation of glucose is catalyzed by all Brønsted acids and bases, including hydrogen and hydroxide ions and the acid and base components of buffers. Brønsted and Guggenheim first defined general acid and general base catalysis by using the mutarotation of glucose as the test system (Brønsted and Guggenheim, 1927). Mutarotation essentially entails opening the pyranose ring of either α-D-glucose or β-D-glucose to form the open-chain sugar, which quickly undergoes closure with loss of stereochemistry to form the anomeric mixture. Fig. 1-7 illustrates the mechanisms of specific acid-base and general acid-base catalysis of pyranose ring opening for β-D-glucose. Ring opening limits the rate in each mechanism.

The rate law for mutarotation under particular conditions of pH, buffer concentration, and temperature can be expressed as $v = k_{obs}[\alpha\text{-D-glucose}]$, where the observed first-order rate constant is given by eq. 1-3.

$$k_{obs} = k_0 + k_H[H^+] + k_{OH}[OH^-] + k_{gb}[B] + k_{ga}[AH] \qquad (1\text{-}3)$$

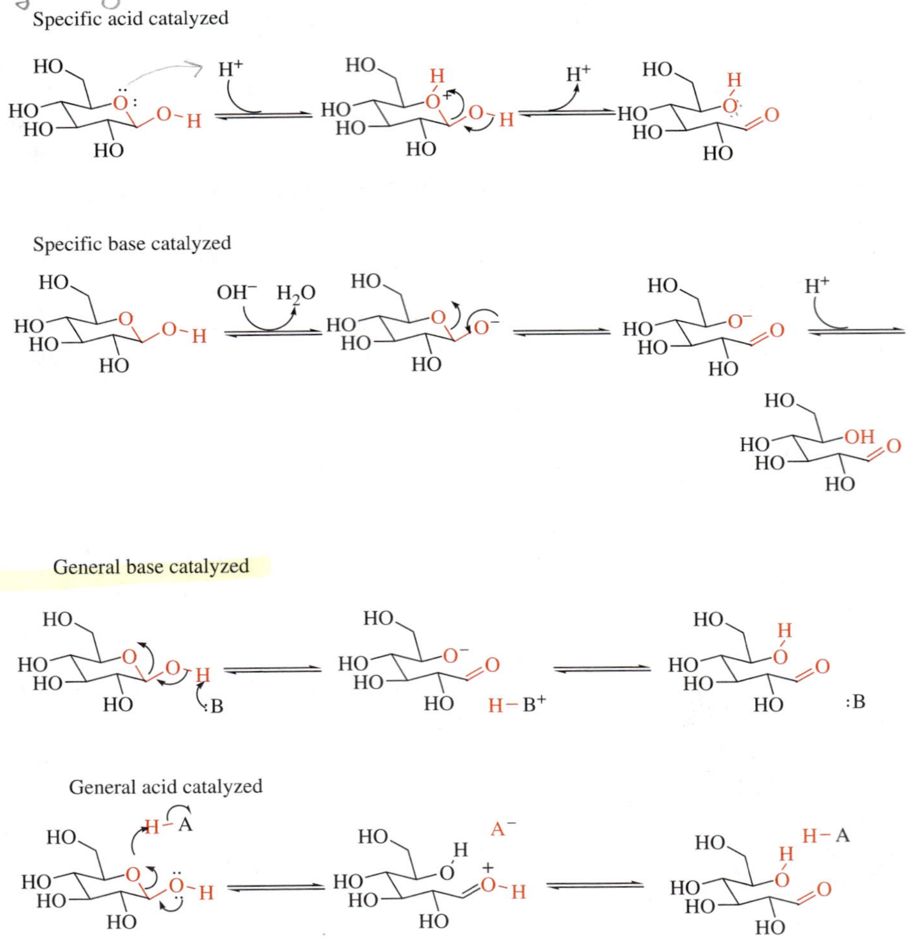

Fig. 1-7. Acid-base catalysis in the mutarotation of glucose. At the top are the mechanisms of specific acid– and specific base–catalyzed mutarotation, in which H+ and OH− ionize the glucose molecule in equilibrium steps preceding the rate-limiting ring opening steps. At the bottom, in general acid and base catalysis, weak acids and bases catalyze the reaction by facilitating the proton transfers in the transition states for ring opening.

The constant k_0 refers to the uncatalyzed process (i.e., the water-catalyzed mutarotation), which contributes significantly to the rate at neutral pH in the absence of buffers. Under various conditions, one or more of the terms in eq. 1-3 can be insignificant. For example, in the absence of buffer components B and AH, the last two terms would be zero, and at extremes of pH, the terms for hydrogen or hydroxide catalysis become insignificant.

The transition states for general acid-base catalysis include proton transfer in the rate-limiting steps. Representations of these transition states are shown in structure **1-4**.

1-4

General base catalyzed   General acid catalyzed

Because bonds to the proton in flight are being broken in the transition states, general acid and general base catalysis generally proceed with solvent deuterium isotope effects in D_2O. The magnitude of the effect depends on the symmetry of the transition state, that is, the extent of proton transfer. The largest deuterium kinetic isotope effect can be expected when the transition state is linear and symmetric, which occurs when the proton in flight is equally bonded to the catalyst and reactant.

Concerted Acid and Base Catalysis

Because an enzyme brings reacting groups together in the Michaelis complex, the possibilities for concerted general acid and general base catalysis are maximized. An early model of concerted catalysis in solution was observed in the mutarotation of tetramethyl glucose (Swain and Brown, (1952). The bifunctional catalyst 1-pyridone, the dominant tautomer of 1-hydroxypyridine, was 10^4 times as effective as an equimolar mixture of phenol and pyridine in catalyzing mutarotation, presumably due to concerted general acid and general base catalysis through a transition state such as that in structure **1-5**:

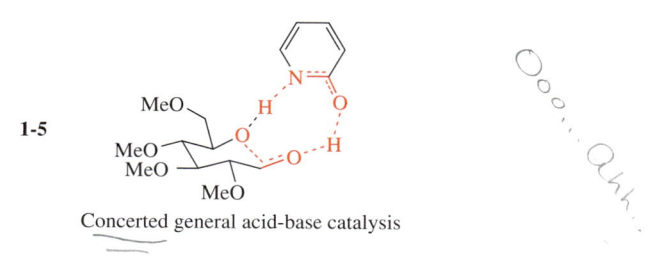

1-5

Concerted general acid-base catalysis

The kinetic advantage of concerted general acid-base catalysis is rarely realized in solution when the acid and base are in separate molecules or ions because it would require a three-body collision. When the acidic and basic groups are in the same molecule, a single collision in proper orientation can lead to concerted catalysis and a markedly enhanced rate. In enzymatic active sites, acid and base groups are often correctly oriented to provide for concerted acid-base catalysis.

Brønsted Catalysis Law

In nonenzymatic reactions, the relative reactivities of acids and bases are correlated by the Brønsted catalysis law, a linear free energy correlation. For general base catalysis, the law is expressed as eq. 1-4:

$$\log k_{gb} = \beta\, pK_a + C \qquad (1\text{-}4)$$

In the equation, k_{gb} is the rate constant for catalysis by a series of bases, pK_a is the negative logarithm of the dissociation constant for the conjugate acids of the bases, C is a constant for the reaction, and β, the coefficient of pK_a, is a measure of the sensitivity of the particular reaction to general base catalysis. A statement of the equation is that for a given reaction facilitated by general base catalysis, the stronger the base the more reactive it is, the quantitative relationship being a proportionality between the logarithms of k_{gb} and K_a.

The sense of the law is easily grasped; if proton abstraction is a fundamentally important part of the transition state, the stronger the base the faster proton transfer takes place. The sensitivity coefficient β is evaluated as the slope of a linear plot of log k_{gb} against pK_a, and it may be large (0.9), small (0.1), or intermediate (0.5) in value, depending on how sensitive the reaction is to general base catalysis. For general acid catalysis, the law is expressed as eq. 1-5:

$$\log k_{ga} = \alpha pK_a + C \tag{1-5}$$

where k_{ga} is the rate constant for catalysis by a series of acids, pK_a is the negative logarithm of the dissociation constant for the acids, C is a constant for the reaction, and α, the coefficient of pK_a, is a measure of the sensitivity of the particular reaction to acid catalysis. Because acid catalysis is increasingly effective with acids of increasing strength, the slope of a linear plot of log k_{ga} against pK_a will be negative, and values of α will be negative. The more negative the value, the more sensitive the reaction is to general acid catalysis. Nonenzymatic reactions may often be models for enzymatic reactions, and their analysis by the Brønsted catalysis law can give important information about the importance of general acid-base catalysis.

The use of the Brønsted catalysis law for enzymatic reactions is complicated by the fact that acid and base catalysts at active sites are parts of the enzyme structure. Alterations of their pK_a values is limited by the amino acid side chains that can be introduced at a given position, and even when they can be changed by site-directed mutagenesis the varied side chain bases can be expected to introduce structural perturbations to the active site that complicate quantitative correlations by the Brønsted catalysis law. In a few cases, these limitations have been overcome by site-directed mutagenesis to delete the catalytic group and chemical rescue of activity by external free acids or bases of varying strength. The first clear success was in the study of base catalysis by Lys208 in aspartate aminotransferase (Toney and Kirsch, 1989, 1992) (see chap. 13). Mutation of Lys208 into alanine led to an inactive enzyme that could not carry out base catalysis. The general base function of the mutated K208A-aminotransferase could be restored (rescued) by primary amines in solution. Presumably the primary amines were binding to the cavity created by deletion of the side chain of Lys208 and reacting in place of the ε-amino group. The effectiveness of the primary amines varied with their pK_a values and steric bulk. After correction for steric requirements, the reactivities of the amines in rescuing catalysis was correlated with their pK_a values, as illustrated in fig. 1-8. The plot is a linear free energy correlation in accord with the Brønsted catalysis law.

Nucleophilic Catalysis

Group transfer reactions in which the transferring group is electrophilic may be subject to nucleophilic catalysis. A nucleophile may react to form a covalent bond with the transferring group in the first step, and the resultant intermediate reacts to transfer the group to another acceptor in the second step. Acyl, phosphoryl, and glycosyl groups among others are subject to enzymatic transfer. There are well-known examples of nucleophilic catalysis in

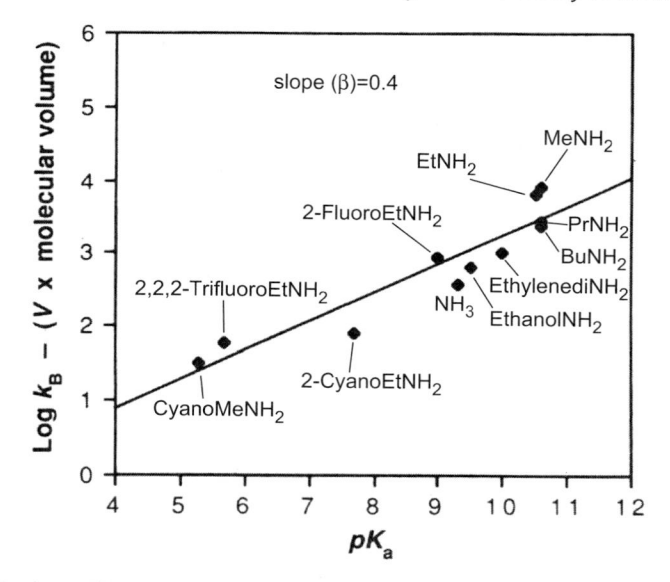

Fig. 1-8. Brønsted catalysis law in the general base catalysis by Lys258 in aspartate aminotransferase. In the reaction of aspartate aminotransferase, a pyridoxal-5′-phosphate–dependent enzyme, the active site Lys208 functions as a base catalyst in part of the mechanism. This function of Lys258 is abolished by site-directed mutagenesis to Ala208. The function of Lys258 can be partially rescued in the variant K208A by addition of free amines. The degree of chemical rescue depends on the basicity of the added amine and its molecular volume. The volume effect can be attributed to the steric limitations of the active site. After correction for the effects of molecular volumes, the observed rate constants can be plotted according to the Brønsted catalysis law as shown . The linear plot has a slope of $\beta = 0.4$, which is consistent with proton transfer being almost half complete in the transition state. (Adapted from Toney and Kirsch, 1989.)

the enzymatic transfer of each of these groups; however, many enzymatic group transfers take place without nucleophilic catalysis. Although nucleophilic catalysis is not a rule in enzymatic group transfer, when it occurs the chemical rules for effective nucleophilic catalysis are followed.

Nucleophilic catalysts possess two essential properties. First, They are very reactive as nucleophiles, more reactive than the ultimate group acceptor. Second, on reaction as nucleophiles they form covalent intermediates in which the group being transferred displays greater electrophilic reactivity than the starting substrate. A thoroughly studied, nonenzymatic example is the imidazole-catalyzed hydrolysis of *p*-nitrophenyl acetate according to eqs. 1-6 and 1-7.

18 Enzymatic Reaction Mechanisms

Imidazole is a much more reactive nucleophile than water and reacts faster with *p*-nitrophenyl acetate to form acetyl imidazole and *p*-nitrophenolate. The intermediate, acetyl imidazole, is much more reactive in its protonated form with water than *p*-nitrophenyl acetate, and it quickly reacts to form acetate and regenerate imidazole. Acetyl imidazole has been observed spectrophotometrically and shown to be kinetically competent as an intermediate.

Imidazole is thought to have special properties, in that it is reactive as a nucleophile, and the intermediate is electrophilically reactive when the imidazole ring is protonated. As shown in eq. 1-8, the ring may be protonated and highly reactive or unprotonated and less reactive in aqueous solutions near neutrality.

$$H_3C-\underset{\underset{O}{\|}}{C}-N\overset{+}{\underset{\text{NH}}{\diagdown}} \overset{pK_a = 6}{\rightleftharpoons} H_3C-\underset{\underset{O}{\|}}{C}-N\underset{\text{N:}}{\diagdown} + H^+ \qquad (1\text{-}8)$$

Despite the fact that imidazole is a nucleophilic catalyst for nonenzymatic acyl group transfer, histidine is not known to participate as a nucleophile in enzymatic acyl group transfer reactions. Histidine is instead an acid-base catalyst in proteases, esterases and acyltransferases. However, histidine is a nucleophilic catalyst in a number of enzymatic phosphotransfer reactions. The most common nucleophilic catalysts for enzymatic acyl group transfer reactions are the thiol group of cysteine and the hydroxyl group of serine. Cysteine transiently accepts acyl groups from thioesters such as acetyl-ACP (acyl carrier protein) and acetyl CoA in the fatty acid synthase and β-ketothiolase, respectively. In these cases, the thioester reactivity is maintained in the intermediate, and the acyl group is covalently bonded to the enzyme, but it is unclear whether reactivity is increased in the covalently bonded ester. The thiolate form of cysteine is the nucleophilic catalyst of peptide hydrolysis catalyzed by cysteine proteases (see chap. 6). Serine is a nucleophilic catalyst in peptide hydrolysis by serine proteases. The intrinsic nucleophilic reactivity of the hydroxyl group in serine is low, but this appears to be overcome by mechanisms that provide a sufficiently strong base to deprotonate the hydroxyl group in the transition state, and the high nucleophilic reactivity is provided by an incipient 3-alkoxide ion in the transition state.

Serine, histidine, lysine, and cysteine are nucleophilic catalysts for enzymatic phosphotransfer. Serine is the catalytic group in alkaline phosphatase and phosphoglucomutase; lysine mediates adenylyl transfer in DNA and RNA ligases and RNA capping enzymes; histidine is the nucleophilic catalyst for nucleoside diphosphate kinase, galactose-1-P uridylyltransferase, succinyl CoA synthetase, and pyruvate phosphate dikinase; and cysteine is the catalyst in phosphotyrosine phosphatase.

The side chain carboxylate groups of glutamate and aspartate are the active site catalysts for glycosyl transferases and glycosidases. Some glycosyltransferases and glycosidases are "retaining"; that is, they catalyze glycosyl transfer with overall retention of configuration at the glycosyl carbon atom. These enzymes use nucleophilic catalysis according to the mechanism in fig. 1-9. The nucleophilic group is a side chain carboxylate group in the active site, which forms a bond to the glycosyl group in step 1, with inversion of configuration at C1. The other carboxylic acid group provides general acid catalysis to facilitate the departure of the leaving group (ROH in fig. 1-9). A molecule of water enters the vacated site, and in the second displacement in step 2, the glycosyl–enzyme intermediate undergoes hydrolysis by essentially the reverse mechanism, in which the carboxylate group resulting from acid catalysis in the first step functions as a Brønsted base in abstracting a proton from water to facilitate the hydrolysis. Step 2 also proceeds with inversion at C1, so that the overall reaction occurs with retention of configuration.

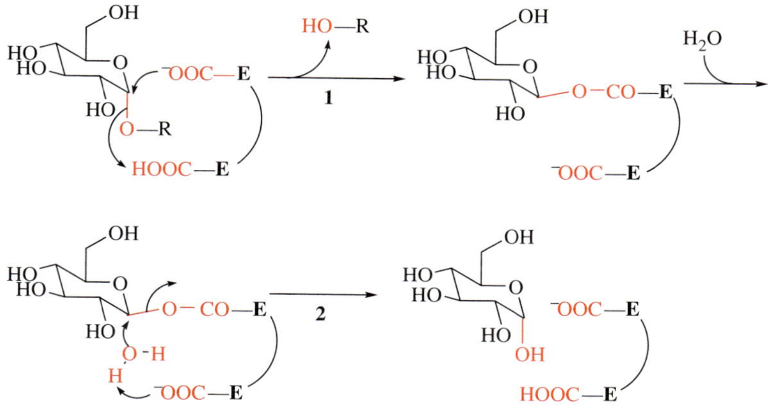

Fig. 1-9. Nucleophilic and general acid-base catalysis in the enzymatic hydrolysis of a glycoside by a "retaining" glycosidase. In a retaining glycosidase, one carboxylate group in the active site reacts as a nucleophile to displace the aglycone and form a covalent glycosyl-enzyme intermediate. The process proceeds with inversion of configuration at C1 of the glycosyl substrate, so that an α-glycoside substrate becomes a β-glycoside intermediate. Hydrolysis of the enzyme-glycoside intermediate also proceeds with inversion of configuration at C1, and the overall hydrolysis proceeds with retention of configuration.

Nucleophiles vary in reactivity and effectiveness as catalysts. An important factor contributing to nucleophilic reactivity is the electron density on the reacting atom. The greater the availability of electrons on that atom, the greater will be its reactivity with an electrophilic center. The reactivities of nucleophiles may be ordered by considering another property that depends on the availability of nonbonding electrons, their propensity to bind protons. As a rule, for a given type of nucleophilic atom, the nucleophilic reactivity increases with increasing basicity. Over a limited range of basicities, the empirical relation of eq. 1-8 holds for a given type of nucleophile, as shown in eq. 1-9:

$$\log k_{nuc} = \beta\, pK_a + C \qquad (1\text{-}9)$$

In the equation, k_{nuc} is the rate constant for the nucleophilic reaction, pK_a refers to the conjugate acid of the nucleophile, β is the sensitivity of the reaction with respect to the reactivity of the nucleophile, and C is a constant. The larger the value of β, the more important is nucleophilic reactivity in reaching the transition state.

The plot in fig. 1-10 exemplifies the dependency of nucleophilic reactivity on the pK_a of the nucleophile, as well as differences among types of nucleophiles. The logarithms of the second-order rate constants for the reactions of the nucleophiles with p-nitrophenylacetate are plotted against the pK_a values of their conjugate acids. At first glance, the points in such a plot can appear unordered. However, in fig. 1-10 the "ordinary" neutral nucleophiles are plotted as solid black squares, the anionic nucleophiles are solid gray triangles, and the α-effect nucleophiles are solid red circles.

This coding unmasks the order within a given type of nucleophile. The result is a set of three nearly parallel lines for three types of nucleophiles. The anionic oxygen nucleophiles are less reactive than the neutral nucleophiles, perhaps because they are more extensively solvated in their ground states and may have to shed some solvation to enter the transition state. The α-effect nucleophiles are more reactive than the ordinary neutral and anionic nucleophiles. When a nucleophilic atom is covalently bonded to another heteroatom with at least one nonbonding electron pair, it displays an enhanced nucleophilic reactivity

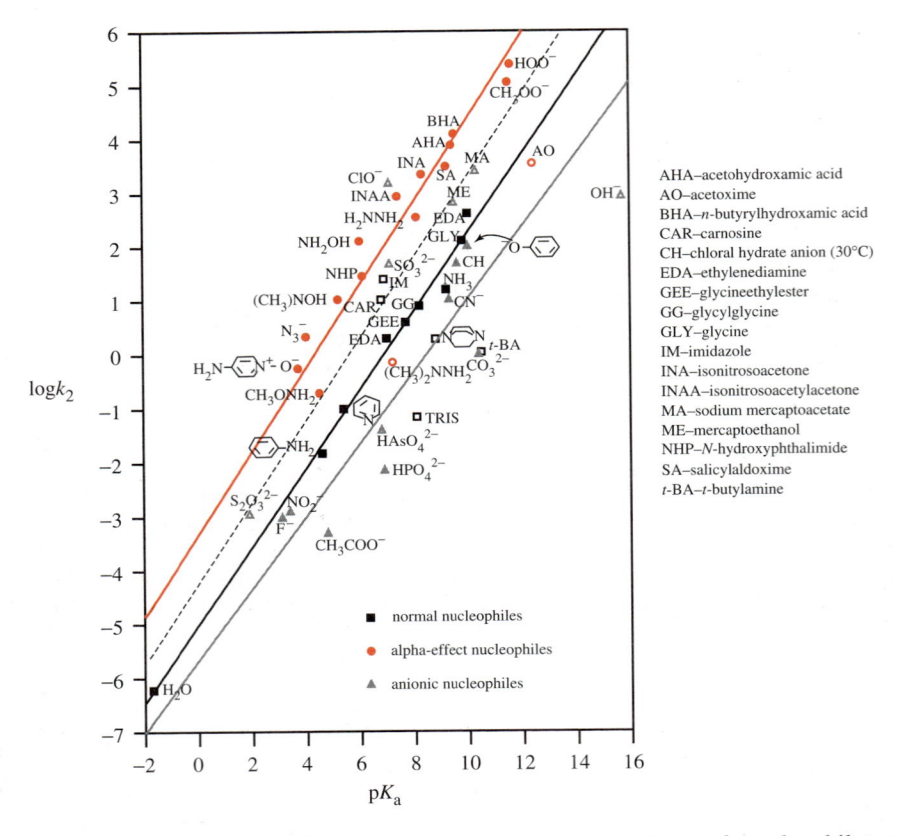

Fig. 1-10. A plot of second-order rate constants for reactions of nucleophiles with *p*-nitrophenyl acetate. The rate at which a nucleophile reacts with *p*-nitrophenylacetate to displace the *p*-nitrophenolate ion depends in part on the basicity of the nucleophile and in part on other effects. The log plot of the second-order rate constants against pK_a for a collection of nucleophiles offers a clear illustration of these effects. The black line and solid black squares represent the reactions of conventional uncharged nucleophiles. The gray line and solid gray triangles represent oxyanionic nucleophiles, which react slower because of the presence of significant solvation spheres. The red line and solid red circles represent α-effect nucleophiles, which react faster because of the effect of an extra electron pair on the second heteroatom. Imidizole and the imidazole ring in carnosine (open black squares) react almost as well as α-effect nucleophiles. The sulfur nucleophiles (open gray triangles) describe a fourth line lying between the conventional neutral nucleophiles and the α-effect nucleophiles. The lines are almost parallel because the plot relates the nucleophiles in each class on the basis of pK_a. Factors such as solvation of charged nucleophiles, α-effect, and intrinsic nucleophilicity of sulfur lead to vertical displacements of the lines. (Adapted from Jencks and Cariuolo, 1960.)

known as the α-*effect* (structure **1-6**). It is as if the neighboring electron pair can increase the available electron density on the reacting atom in the transition state.

α-Effect nucleophiles are not found in the side chains of amino acids; however, imidazole displays enhanced reactivity as a nucleophile and lies between the lines for ordinary neutral and α-effect nucleophiles in fig. 1-10. Its reactivity may be augmented by a resonance enhancement analogous to the α-effect, as illustrated earlier.

Table 1-3. Enzymes that Employ Nucleophilic Catalysis.

Nucleophilic Side Chain	Enzymes
Serine (β-OH)	Serine proteases
	Esterases
	Alkaline phosphatases
Cysteine (β-SH)	Cysteine proteases
	Acyl transferases
	Phosphotyrosine phosphatases
Histidine (β-imidazole)	Nucleoside diphosphate kinase
	Galactose-1-P uridylyltransferase
	Histidine triad (HIT) proteins
	Acid phosphatases
Lysine	DNA ligase
	RNA ligase
	RNA capping enzyme
Glutamate and aspartate (COOH)	Retaining glycosidases
	Coenzyme A transferases

Table 1-3 includes a list of amino acid side chains that are nucleophilic catalysts, together with the enzymes in which they function in this capacity. The list includes most of the reactive functional groups in proteins.

Electrophilic Catalysis

In electrophilic catalysis a substrate reacts chemically with an electrophile to form a compound that displays enhanced reactivity. The side chains of amino acids do not function as electrophilic catalysts. In biocatalysis, the electrophilic catalysts are metal ions, which often form coordination complexes with substrates (MgATP) or with side chains of amino acids. The resulting complexes display enhanced reactivities. The metal ions are regarded as cofactors and are discussed in detail in chapter 4. However, the ε-amino group of lysine can effect electrophilic catalysis by reacting with carbonyl groups of substrates or coenzymes to form imines.

Catalysis of Enolization

Enolization is a very important mechanistic action in enzymology. Enolizations of aldehydes, ketones, esters, and carboxylic acids take place in enzymatic eliminations, aldose-ketose isomerizations, and carbon-carbon bond forming reactions catalyzed by aldolases and carbon-carbon ligases. In each case, a C—H bond adjacent to the carbonyl group of an aldehyde, ketone, ester, or carboxylic acid must be broken by removal of a proton to form an enolate species as an essential step in the mechanism. These compounds are very weak carbon acids and display the high pK_a values listed in table 1-4.

Two major barriers to the enolization of these molecules must be overcome by enzymes as they catalyze reactions involving this process. First is the equilibrium thermodynamic barrier; the acidities of these molecules must be increased at the active site. The need for this can be understood by considering a straightforward calculation based on the ionization of a weak, carbon acid symbolized as HC within an active site, in which the dissociated hydrogen ion is transferred to a base within the active site and symbolized as B: in scheme 1-1.

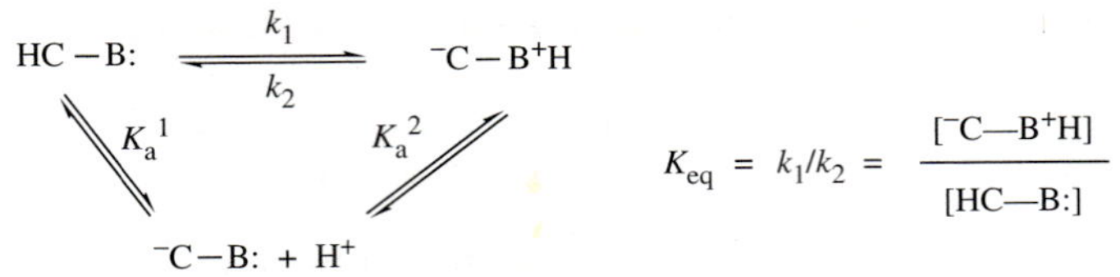

22 Enzymatic Reaction Mechanisms

$$HC-B: \underset{k_2}{\overset{k_1}{\rightleftharpoons}} {}^-C-B^+H$$

$$K_a^1 \diagdown \quad K_a^2 \diagup$$

$$^-C-B: + H^+$$

$$K_{eq} = k_1/k_2 = \frac{[^-C-B^+H]}{[HC-B:]}$$

Scheme 1-1

A typical carbon acid at an active site could be a ketone, with an acid dissociation constant $K_a^1 = 10^{-20}$ ($pK_a = 20$) If the base is a typical enzymatic group, its acid dissociation constant might be $K_a^2 = 10^{-7}$ ($pK_a = 7$). This represents the simplest and most favorable case, in which the ionization state of the carbon acid does not affect the ionization constant for the base and *vice versa*. Within the confines of an active site, this is not likely to be true, but it is a conservative assumption for the present purposes. In the process of scheme 1-1, $K_{eq} = K_a^1/K_a^2 = 10^{-13}$. No more than one molecule in 10 trillion of the enzyme-substrate complex can have the ionized or enolized substrate present at equilibrium.

The consequences of this for the enzymatic rate can be understood as follows. Because the ratio of forward and reverse rate constants also equals the equilibrium constant, it follows that $k_1/k_2 = 10^{-13}$. If the reverse rate constant is very large, say 10^{10} s^{-1}, the forward rate constant would be only 10^{-3} s^{-1}. This means that the value of k_{cat} for this enzyme could not be larger than 10^{-3} s^{-1}, even if the rate constants for subsequent steps were infinitely large.

Table 1-4. Activities of Carbon Acids Expressed as pK_a Values

Compound	pK_a	Ref.	Compound	pK_a	Ref.
H—CH$_2$—alkyl;	≥50		H—CH$_2$—C(=O)—H	17.8	h
H—CH$_2$—COO$^-$	33.5	b			
H—CH(NH$_3^+$)—COO$^-$	28.9	c	H—CH(CH$_2$OPO$_3^{2-}$)—C(=O)—H	16.3	g
H—CH$_2$—CONH$_2$	28.4	b			
H—CH$_2$—COEt	25.6	d	H—CH((CD$_3$)$_2$C-ND$^+$)—COOEt	14	c
H—CH$_2$—COSEt	21.0	e			
H—CH(NH$_3^+$)—COOEt	21.0	f	H—CH$_2$—C(=N$^+$HR)—CH$_3$	~11	j
H—C(H)(OH)—C(=O)—CH$_2$OPO$_3^{2-}$	19.9	g			
			H—CH(C(=O)O$^-$)—C(=O)—	9–10	
H—CH$_2$—C(=O)—CH$_3$	19.3	h			

[a] The carbon pK_a as in this table could not be measured by potentiometric titration because of their high values and the kinectic barriers to the ionization of these acids. The methods by which they were measured and values for additional carbon acids are given in Richard et al, 2002 Bernasconi, 1992; Keefe and Kresge, 1990.
[b] Richard et al., 2002.
[c] Rios et al., 2000.
[d] Amyes and Richard, 1996.
[e] Amyes and Richard, 1992.
[f] Rios and Richard, 1997.
[g] Richard, 1984.
[h] Regenestein and Jencks.
[i] Rios et al., 2001.
[j] Estimated from the relative rates of acetone enolization and formation of an enamine from an iminium ion (Bender and Williams, 1996).

If the reverse rate constant is smaller, the forward rate constant and potential k_{cat} are proportionately smaller. The situation is worsened for weaker carbon acids. In particular, for the enolization of a carboxylate ion to an *aci*-carbanion, a frequent process in enzymology, the pK_a is about 30. The ionization rate would be slower than for a ketone by a factor of 10^{10}. In general, the greater the difference in pK_a between the carbon acid and the enzymatic base, the slower enolization will be. An enzyme can cause a molecule with a low carbon acidity to undergo enolization only by increasing its acid strength by lowering its pK_a. This must be accomplished by provision of an electrophilic environment at the active site to stabilize the enolate form of the substrate.

The second barrier to enolization is kinetic in nature and is often encountered in the hydroxide-catalyzed enolization of carbon acids in aqueous solutions. The kinetic barrier is solvation of the transition state, as illustrated in scheme 1-2.

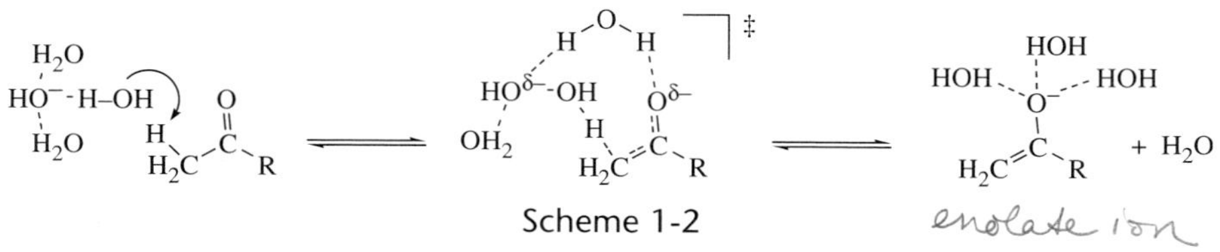

Scheme 1-2

An enolate ion must be solvated just as any other ion must be, including the hydroxide ion. In water, the hydroxide ion is a trihydrate, whereas the carbonyl group of a ketone or ester is much less solvated. In the transition state for hydroxide-catalyzed enolization, the requirements for solvation are shifted from the hydroxide ion, which becomes water, to the carbonyl group, which becomes an enolate ion. In the enolate, the negative charge is substantially localized on the carbonyl oxygen, which brings too little solvation into the transition state. Scheme 1-2 shows that the solvation requirements in the transition state are very different from the ground state, and substantial solvent reorganization is required in the transition state because the developing negative charge is remote from the attacking hydroxide and its solvation sphere. The energy required for solvent reorganization in the transition state slows enolization relative to ionization rates of "normal" acids with the same pK_a values. When the developing negative charge is not remote from the attacking base, the ionization rate is normal, even for carbon acids (Bernasconi, 1987).

This second barrier to enolization is overcome in enzymatic reactions by the design of active sites. The active sites are not aqueous, and the bases in active sites are not hydroxide ions. Evolved structures of active sites include proper solvation of the ground state Michaelis complex, the transition state, and the enolized state. These designs also serve to depress the pK_a values of the enolizing species within the active sites.

In enzymes that catalyze enolization, several means for stabilizing enolates or their equivalents have been elucidated. These include facilitation of imine formation and coordination of divalent metal ions, and strong hydrogen bonds may stabilize enolates in enzymatic sites.

Imine Formation by Lysine

Mechanism of Imine Formation

A number of enzymatic reactions are facilitated by imine formation between the substrate and the ε-amino group of lysine in the active site. In the general case, the chemical mechanism of the reaction would formally require the formation of an enol or enolate ion as a transient intermediate. The difficult steps of these reactions follow the pattern shown in scheme 1-3, where X may be H, CH_2OH, or COO^-.

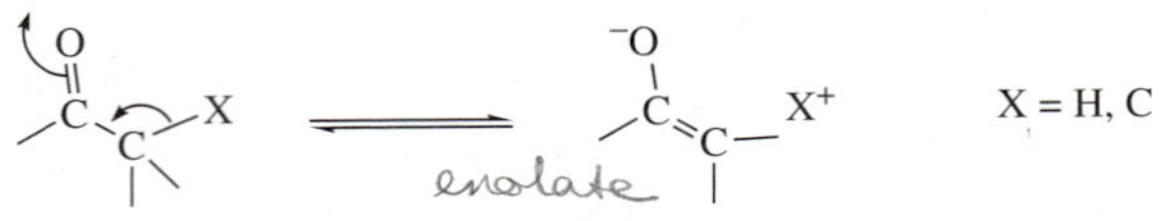

24   Enzymatic Reaction Mechanisms

Scheme 1-3

Enolate ions are very-high-energy intermediates that cannot exist at significant concentrations in the essentially neutral solutions in which enzymes normally function. Imine formation between ketone substrates and the ε-amino group of lysine is one important mechanism for facilitating reactions of this type. The high energy of the enolate has been discussed above in terms of the high values of pK_a for ionization of the corresponding ketone. The pK_a for acetone is 19.3, which corresponds to a standard free energy change of +26 kcal mol^{-1} for ionization at 25°C, a highly unfavorable process. The reaction can be facilitated if the carbonyl group is transformed into the cationic iminium ion resulting from the reaction of the ketone with an amine, as shown in scheme 1-4.

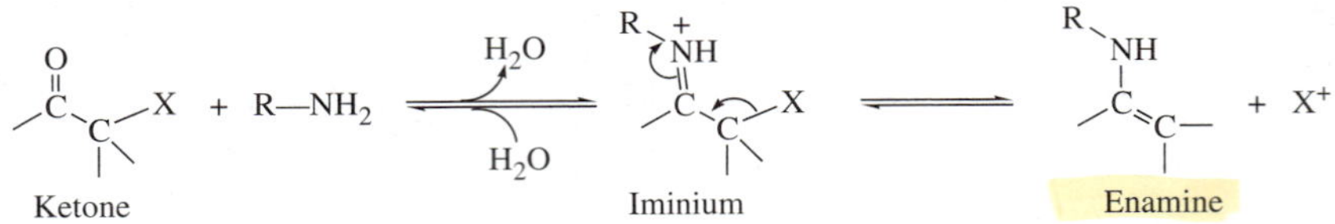

Scheme 1-4

Imine formation is an intrinsically fast reaction that can be catalyzed at enzymatic sites. It begins with nucleophilic addition of an amine to a carbonyl group of an aldehyde or ketone to form a tetrahedral intermediate. The intermediate undergoes acid-catalyzed dehydration to the protonated imine (or iminium ion) according to the mechanism in scheme 1-5 (Jencks, 1969).

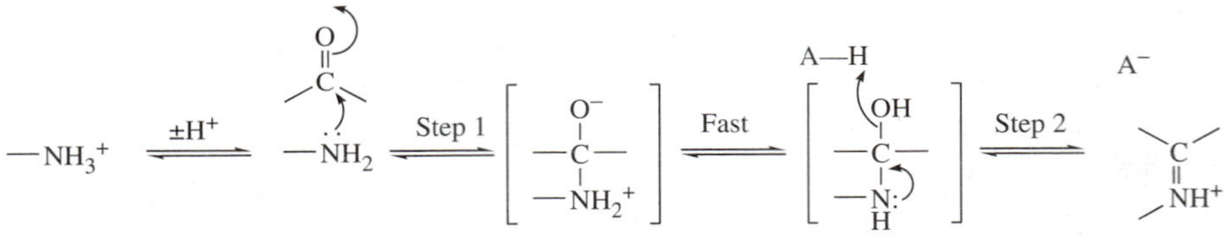

Scheme 1-5

The values of pK_a for protonated imines are in the range of 7 to 8, so that they exist in substantial part in the form of the protonated imine or iminium ion in neutral solutions. The protonated imine is something of a super-ketone, in that the iminium nitrogen carries a positive charge, which serves as an electron sink in promoting the formation of the enamine concomitant with the cleavage of the bond C—X in the sense shown. The barrier to enamine formation from the iminium ion (p$K_a \approx 11$) (see table 1-4) is much smaller than for enolate formation (p$K_a = 19.3$).

Imine formation is subject to general acid catalysis according to the general mechanism in scheme 1-5. After nucleophilic addition of the amino group to the carbonyl group of the ketone to form a carbinolamine as a transient intermediate, dehydration to the protonated imine, or iminium ion, requires acid catalysis. The two-step mechanism leads to the bell-shaped pH-rate profiles characteristic of reactions of this type. At low pH values, step 1 in scheme 1-5 is rate limiting, and the rate decreases with decreasing pH because of the decreasing concentration of the unprotonated amine. This accounts for the acid limb of the bell-shaped profile. At high pH values, step 2 becomes rate limiting because it depends on acid-catalyzed dehydration of the intermediate, and acid concentrations decrease with increasing pH. This accounts for the descending limb at high pH values. The composite of these effects results in an asymmetric but generally bell-shaped pH-rate profile.

The reverse of the mechanism in scheme 1-5 is the hydrolysis of an imine and requires general base catalysis, the converse of the general acid catalysis in the forward direction.

At an active site, both processes should be catalyzed, and there should be an acid-base catalytic group from the side chain of an amino acid suitably positioned to act as an acid in imine formation and as a base in imine hydrolysis.

The electrophilic reactivities of protonated imines are much higher than for ketones and aldehydes because of the positively charged nitrogen, which serves as an electron sink in nucleophilic addition reactions. Imines themselves are not as electrophilic as ketones and aldehydes because nitrogen is less electronegative than oxygen, so that the π bond in the neutral imine is less polarized than that of the carbonyl group. However, on protonation of the imine, the resulting positive charge formally on nitrogen further polarizes the π bond, making it far more reactive than the carbonyl group.

Aldolase

A classic case of a lysyl imine-forming enzyme is type I aldolase, which catalyzes the condensation of dihydroxyacetone phosphate (DHAP) with glyceraldehyde-3-phosphate (GAP) to produce fructose-1,6-bisphosphate (FBP). Other examples are acetoacetate decarboxylase and porphobilinogen synthase. Several structures of FBP aldolase have recently appeared (Choi et al., 2001; Darby et al., 2001; Schurman et al., 2002). Structures of other type I aldolases are available, and that of 1-deoxyribose-5-phospate aldolase is shown in fig. 1-11.

The mechanism of action of muscle aldolase in the direction of FBP formation is illustrated in fig. 1-12. On binding DHAP to the active site, aldolase catalyzes imine formation between the active site lysine and DHAP in step 1 to form the protonated imine (iminium) intermediate. On binding GAP, an enzymatic base abstracts the proton from the pro R position to form the enamine of DHAP in step 2. This base has not been identified in any species of FBP aldolase. However, detailed evidence on the structure and reaction of D-1-deoxyribose aldolase implicated a water molecule bridging Lys201 and Asp102 in mediating the proton transfer (Heine et al., 2002). The enamine reacts as a carbanion equivalent and undergoes addition to the aldehyde group of GAP to form the carbon-carbon bond in step 3. The enzyme catalyzes the hydrolysis of the imine linkage to the product and releases FBP. The reaction is freely reversible but is often written in the direction of flux through the glycolytic pathway.

Several lines of evidence led to the aldolase mechanism. First, aldolase was found to be subject to substrate-dependent inactivation by sodium borohydride (NaBH$_4$). DHAP or FBP could potentiate inactivation by NaBH$_4$, but GAP could not. Second, chemical degradation of aldolase inactivated by NaBH$_4$ in the presence of [1-^{14}C]DHAP led to N^ε-[^{14}C]dihydroxypropyl-lysine. The formation of this product indicated that inactivation by the hydride donor NaBH$_4$ was brought about by reduction of the imine linkage between DHAP and the active site lysine (structure **1-8**). Third, aldolase catalyzed the exchange of the 3-pro-Z hydrogen of DHAP with deuterons in D$_2$O. This suggested that the enzyme-base abstracting the proton in step 2 could undergo exchange with solvent protons at the enamine stage of the overall reaction.

N^ε-2-([2-^{14}C]Dihydroxypropyl-L-lysine

26　Enzymatic Reaction Mechanisms

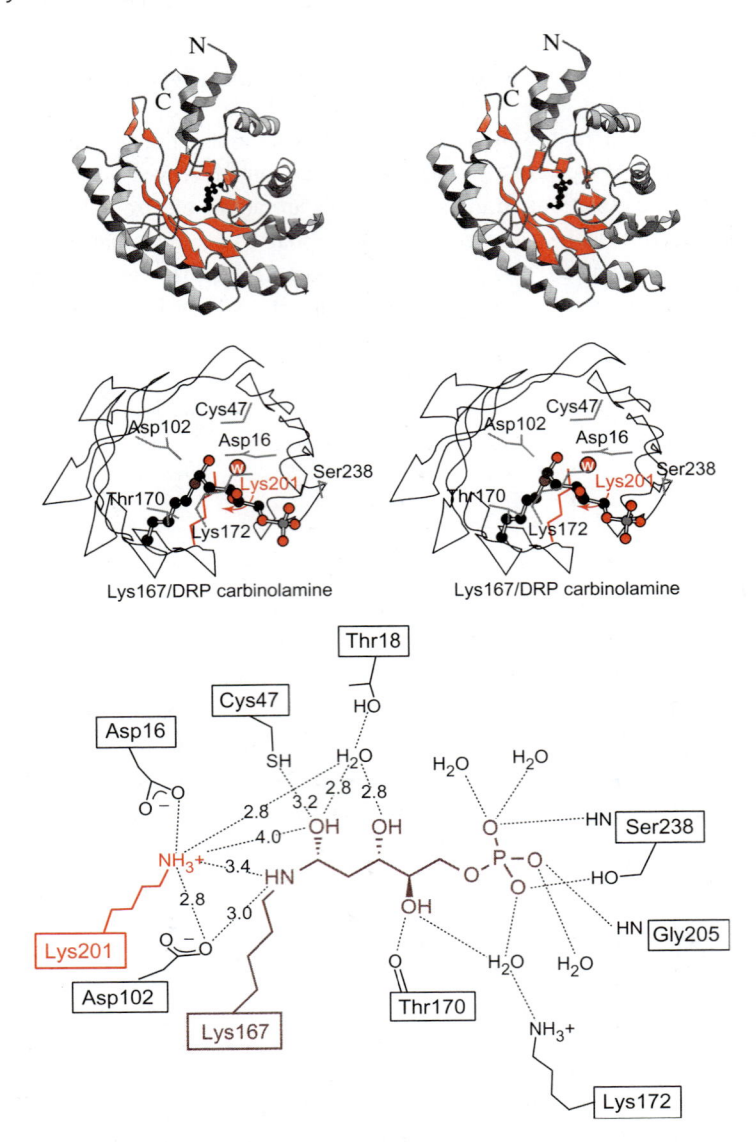

Fig. 1-11. The structure of *Escherichia coli* 1-deoxyribose-5-P aldolase. The cleavage of 1-deoxyribose-5-P by this aldolase (EC 4.1.3.4) follows the same mechanism as in fructose bisphosphate aldolase in fig. 1-12. The structure shown is that of the enzyme in which the substrate is bound as its tetrahedral adduct with Lys167. The chain fold is shown at the top, with the substrate shown in stereo as a black ball-and-stick model. In the middle is a stereographic drawing of the active site showing the adduct of Lys167 and 1-deoxyribose-5-P as a ball-and-stick model. At the bottom is a two-dimensional diagram of the active-site contacts. The images were prepared using PDB coordinate file 1JCL for the model refined to 1.05-Å resolution (Heine et al., 2001).

Catalysis by Metal Ions

Divalent metal ions participate in many enzymatic reactions, generally as electrophilic catalysts. All nucleoside triphosphate (NTP)-dependent phosphotransferases use MgNTP complexes as phosphodonor substrates, and the coordinated Mg^{+2} probably facilitates phosphotransfer by stabilizing the leaving group in the form of MgNDP. Divalent metal ions have long been known to catalyze the nonenzymatic phosphotransfer by ATP (Lowenstein and Schatz, 1961). Many phosphotransfer enzymes contain binding sites for

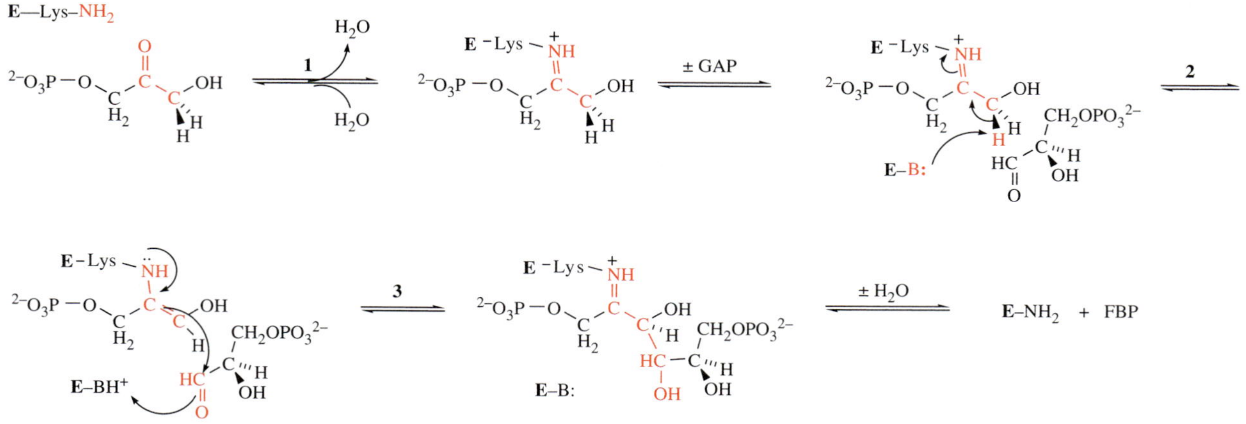

Fig. 1-12. The role of imine intermediates in the chemical mechanism of the reaction of fructose bisphosphate aldolase.

divalent metal ions in addition to those brought into the site by MgATP. For example, two magnesium ions participate in the mechanism of phosphodiester bond formation by DNA polymerase (see chap. 10). Enzymes that facilitate enolate formation in catalyzing aldol reactions, racemizations, and dehydrations sometimes require divalent metal ions that function as electrophilic catalysts. A few nonenzymatic examples of electrophilic catalysis by divalent metal ions can help rationalize catalysis in enzymatic sites.

Magnesium ions catalyze the stereospecific reduction of pyruvate by $NaBH_4$ at the active site of pyruvate kinase, which has oxalacetate decarboxylase activity, as well as the exchange of C3(H) of pyruvate with deuterium in D_2O (Kosicki, 1968; Kosicki and Westheimer, 1968). Magnesium ions can form coordination complexes with pyruvate and polarize the carbonyl group, making it more reactive toward $NaBH_4$. The Mg-pyruvate complex is also more reactive in enolization following the course shown in eq. 1-10, in which Mg^{2+} serves as an electron sink.

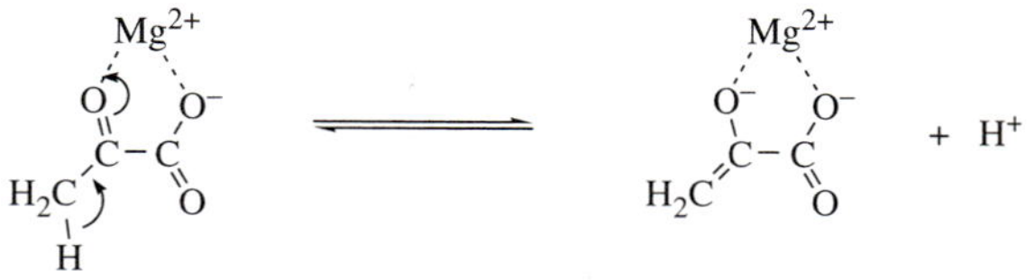

$$\text{(1-10)}$$

Reversal of this process in D_2O leads to the incorporation of deuterium into pyruvate.

Class II Aldolases

Class II aldolases employ electrophilic catalysis by divalent metal ions, typically Zn^{2+}, to stabilize the enolate resulting from the cleavage of FBP and analogous substrates, in place of the lysyl-ketimines of class I aldolases (Horecker et al., 1972; Kobes et al., 1969). A brief mechanism is shown in scheme 1-6.

In this process, coordination of the enediolate to zinc provides the stabilization required to cleave the carbon-carbon bond.

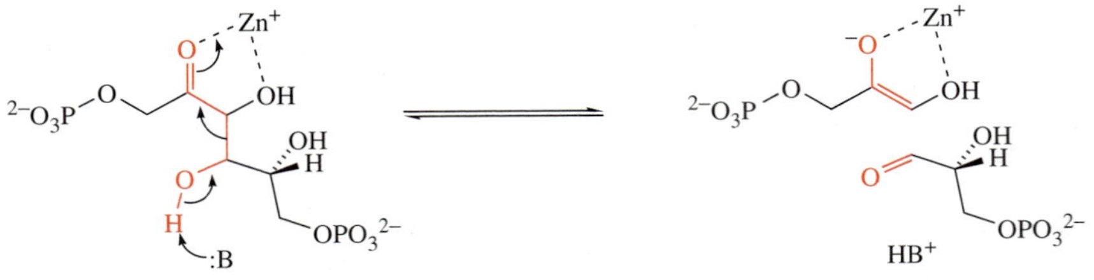

28 Enzymatic Reaction Mechanisms

Scheme 1-6

Electrophilic stabilization of enolates by divalent metal ions potentiates all of the reactions typically observed for enolates in organic chemistry. These include racemization, dehydration of β-hydroxy ketones and β-hydroxy esters, aldose-ketose isomerization in sugars, and aldol cleavage reactions. Examples of enzymes catalyzing these reactions through electrophilic catalysis by divalent metal ions include the racemization of mandelate by mandelate racemase (see chap. 7), the dehydration of 1-phosphoglycerate by enolase (see chap. 9), the isomerization of xylose-5-phospate to xylulose-5-phosphate (see chap. 7), and the cleavage of fructose-1,6-diphosphate into glyceraldehyde-3-P and dihydroxyacetone-P by the zinc-dependent type II aldolase (see chap. 14).

Enolase

Enolase catalyzes the dehydration of 2-phosphoglycerate to phosphoenolpyruvate. This reaction requires base-catalyzed enolization of 2-phosphoglycerate to its *aci*-carbanion, as shown in eq. 1-11a, followed by acid catalyzed dehydration of the *aci*-carbanion to form phosphoenolpyruvate, as shown in eq. 1-11b.

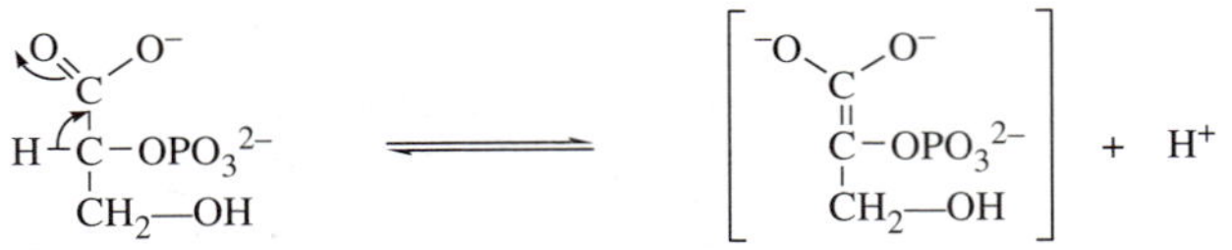

$$\text{(1-11a)}$$

$$\text{(1-11b)}$$

The difficult part of the reaction of enolase is the first chemical step, enolization to the *aci*-carbanion in eq. 1-11a. The pK_a of 2-phosphoglycerate is estimated to be at least 34, which means that the standard free energy change for ionization must be +42 kcal mol^{-1} at 25°C. Fortunately, the situation is not quite so difficult at the active site of enolase because, as illustrated in scheme 1-1 for the general case, the proton from C2 of 2-phosphoglycerate is not released to the solvent, as implied by eq. 1-11a. Instead, it is transferred to a base at the active site, which is Lys345. The energetics of the ionization are determined by the difference in pK_a values for 2-phosphoglycerate and Lys345, which would be more like 28 kcal mol^{-1} if the pK_a of Lys345 is 10. This is still a highly unfavorable reaction.

The structure of the active site in fig. 1-13 shows how many of the barriers to the ionization of 2-phosphoglycerate are overcome by enolase.

The reaction depends on the divalent metal cofactor Mg^{2+}. The enzyme accepts Mn^{2+} as well, and a large body of evidence from EPR experiments implicated two manganese ions in the active site coordinated with a substrate-like inhibitor (Poyner and Reed, 1992; Reed et al., 1996). The structure with the substrate and Mg^{2+} bound at the active site confirmed

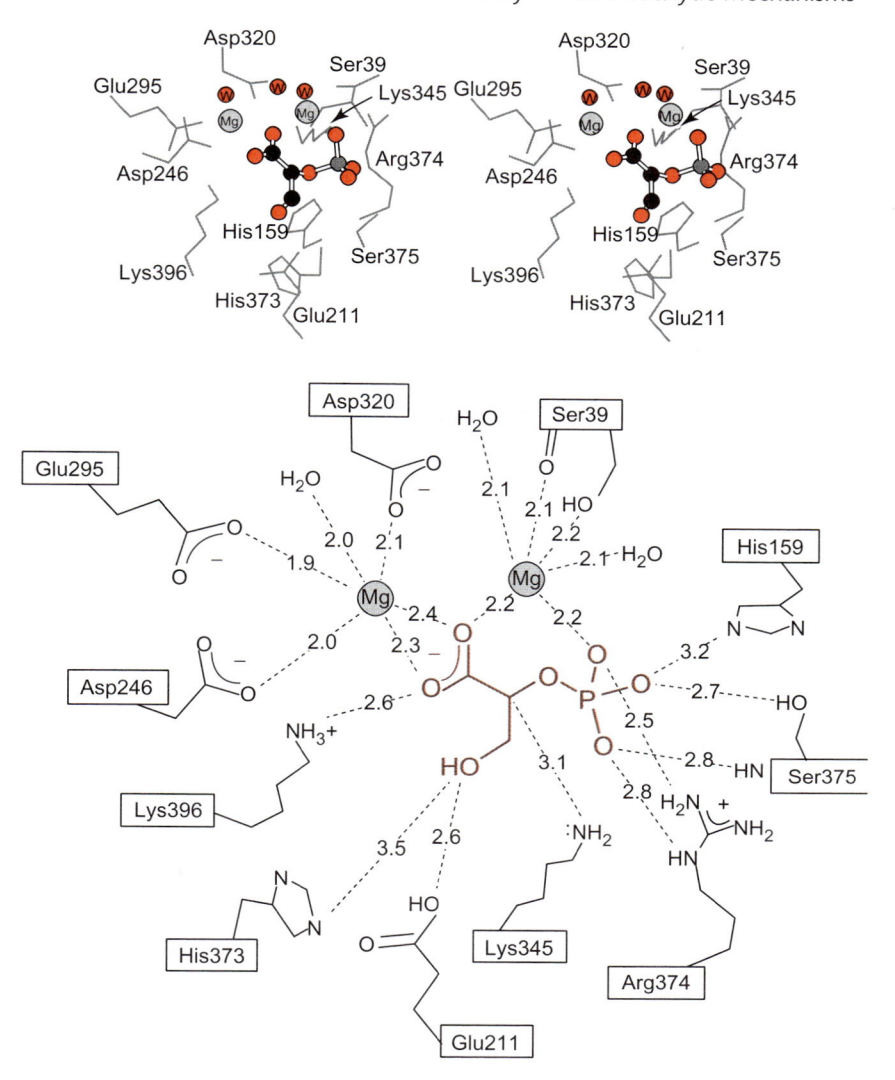

Fig. 1-13. The structure of the active site of yeast enolase with the substrate bound. A stereo-diagram (top) shows of the active site of *Saccharomyces cerevisiae* enolase (EC 4.2.1.11), with a schematic representation of the same provided in the bottom panel. Notice that the carboxylate group of 2-phosphoglycerate is coordinated to two Mg^{2+} ions and two Lys395. These positively charged groups potentiate the ionization of 2-phosphoglycerate to the *aci*-carbanion required to eliminate water in the formation of phosphoenolpyruvate. This figure was prepared using PDB coordinate file 1ONE for the model refined to 1.8-Å resolution (Larsen et al., 1996).

this and proved how magnesium coordinates the substrate (Larsen et al., 1996). The structure also showed the presence of the ε-aminium ion of Lys396 in the site. Two magnesium ions and the Lys396 aminium ion introduces five positive charges into the active site. 2-Phosphoglycerate itself contributes three negative charges. The five positively charged residues interact directly with the carboxylate group of 2-phosphglycerate, both magnesium ions and the ε-aminium ion. These ions focus much of the positive charge on the carboxylate of 2-phosphoglycerate, and it is just what is needed to stabilize the *aci*-carbanion intermediate in eqs. 1-11a and 1-11b. *Aci*-carbanion formation increases the negative charge in a tri-anionic molecule, and most of the increased charge converges on the incipient *aci*-carboxylate group. The two magnesium ions and Lys396 ε-aminium ion are placed in just the right positions to neutralize this increasing charge. This electrostatic stabilization lowers the barrier to *aci*-carbanion formation.

After the *aci*-carbanion is formed, the elimination of water is straightforward, requiring only acid catalysis by the γ-carboxylic acid group (COOH) of Glu211. The driving force for the elimination is the doubly negatively charged carboxylate group of enolized 2-phosphoglycerate, as illustrated in eq. 1-11b.

Hydrogen Bonding

All of the amino acid side chains in table 1-2 can engage in hydrogen bonding as hydrogen-bond donors or acceptors. The serine and threonine hydroxyl groups, the arginine quanidino group, and the main chain peptide carbonyl oxygen and amide nitrogen atoms of an enzyme can engage in hydrogen bonding with substrates at active sites. Three-dimensional structures of many enzyme-substrate and enzyme-inhibitor complexes reveal multiple hydrogen-bonded contacts between these groups and hydrogen-bonding groups of substrates and inhibitors. The cumulative effects of H bonds contribute significantly to the binding process and the immobilization of substrates at active sites.

Most individual H bonds are weak, with electrostatic interactions between the partially positively charged protons of donors and the electronegative and partially negatively charged acceptors, which are generally heteroatoms containing at least one pair of nonbonding electrons. Weak H bonds are illustrated (structure **1-9**) for a main chain carbonyl and an amide nitrogen interacting with an H-bond donor (H—N) and an H-bond acceptor (COO$^-$).

Until recently, all H bonds in enzymes and proteins in general were thought to be of this type. However, a few examples of stronger H bonds have been found in a few enzymes, and these have been postulated to play significant roles in catalysis. Strong H bonds differ qualitatively from weak H bonds, and the formation of strong H bonds takes place under constraints that are described in the following section.

Strong and Weak Hydrogen Bonds

Weak H bonds have been understood to be electrostatic attractive forces between weakly acidic protons and electronegative atoms with at least one nonbonding electron pair. The heteroatoms to which the weakly acidic hydrogen is covalently bonded and hydrogen bonded include oxygen or nitrogen, the principal hydrogen-bonding heteroatoms in proteins. The energies of H bonds classified as weak are generally between 2 and 8 kcal mol^{-1}, referring to the energy difference between a state in which the H bond exists and one in which the participating heteroatoms are at the same distance from each other, but the hydrogen atom is misdirected (structure **1-10**).

1-10 $\overset{\diagdown}{\text{O}}-\text{H}\cdots\text{O}=\overset{\diagup}{\underset{\diagdown}{\text{C}}}$ $\overset{\diagdown}{\underset{\diagup}{\text{O}}}\overset{}{\underset{\text{H}}{}}$ $\text{O}=\overset{\diagup}{\underset{\diagdown}{\text{C}}}$

H-bonded   Unbonded

In biochemistry the strengths of weak H bonds in proteins are sometimes quoted as 1 to 3 kcal mol^{-1}; however, these values generally refer to the difference between two hydrogen-bonded states that have been studied to evaluate the importance of a particular H bond, rather than the difference between an H-bonded and an unbonded state. For example, in the earlier structures, the OH group on the right is not engaged in H bonding, and the structures refer to hydrogen-bonded and unbonded states. If, however, the OH were donating an H bond to a solvent molecule, the structures would represent two hydrogen-bonded states. Because H bonds are stabilizing, the difference would be less when the OH group is hydrogen bonded to solvent than when it is not hydrogen bonded. The former difference would be a measure of the relative strengths of two H bonds. In the following discussion, the strengths of H bonds refer to the difference between hydrogen-bonded and unbonded states.

Shown in fig. 1-14 are qualitative potential energy functions for three types of hydrogen-bonded systems, in which the heteroatoms are symbolized by X and Y and are assumed to have similar proton affinities or, alternatively, similar pK_a values.

Figure 1-14A shows the typical weakly hydrogen-bonded system, in which the heteroatoms are represented as X and Y. The energy wells are interacting but separate, with a high-energy barrier between them. The zero-point vibrational energy of the proton is lower than the barrier, so that the proton is covalently bonded to X and electrostatically attracted to Y. The zero-point vibrational energy of a deuteron in place of the proton is slightly lower because of its higher mass. Weak H bonds are typically found in freely associating species in polar, protic media such as water.

In the hydrogen bond of fig. 1-14B, the heteroatoms are significantly closer together, closer than the sum of their van der Waals radii, as they might be in an internally crowded molecule. Consequently, the energy barrier is decreased and the zero point vibrational energy of the proton is near the level of the barrier. In this case, there is significant covalent bonding of the proton with both X and Y (Gilli et al., 1994, 2000); the proton can

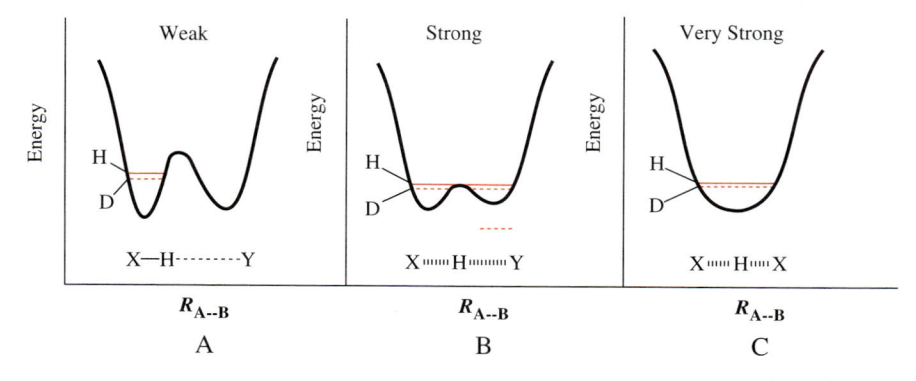

Fig. 1-14. Energy diagrams of weak and strong hydrogen-bonded systems. The three panels describe weak (conventional) hydrogen-hydrogen bonds (A), strong or low-barrier hydrogen bonds (B), and very strong or single-well hydrogen bonds (C). Most hydrogen bonds in proteins and nucleic acids are conventional (A). A few special hydrogen bonds in proteins are low-barrier hydrogen bonds (B). The single-well hydrogen bonds (C) are rare in chemistry and have not been found in proteins.

oscillate between the heteroatoms but may be more strongly attracted to one of them. The X—H and Y—H bond lengths are longer than the typical 0.9 Å for H covalently bonded to O or N, but they may not be equal in length, especially when the heteroatoms are not identical. Such an H bond will be stronger, in the range of 10 to 20 kcal mol^{-1}, than the typical weak H bond (Hibbert and Emsley, 1990; Jeffrey, 1997). Bonds of this type are known as low-barrier hydrogen bonds (LBHBs). Because the zero point energy of a deuteron or triton is lower than that of the proton, a deuteron or triton is more influenced by the barrier, and this leads to deuterium and tritium isotope effects on several properties of the H bond. These isotope effects can be applied to the characterization of LBHBs. They can be found in freely associating species interacting in apolar media or in crystals when the proton affinities (or pK_a values) of the heteroatoms are similar. They can also be found in sterically crowded molecules, in which the heteroatoms are forced together, and in such molecules they can exist even in aqueous/organic solutions (Cassidy et al., 1998; Lin and Frey, 2000; Xhao et al., 1996). LBHBs have been postulated to contribute to enzymatic catalysis (Cleland, 1992; Cleland and Kreevoy, 1994; Gerlt and Gassman, 1993a, 1993b). A few examples of LBHBs have been described in enzymes, in which they appear to stabilize transition states or metastable intermediates, which by the Hammond postulate closely resemble transition states (Choi et al., 2000; Denisov et al., 1994; Frey et al., 1994; Zhao et al., 1996).

In the extreme case of a short H bond, illustrated in fig. 1-14C, the heteroatoms are identical; they are very close together, much closer than allowed by the sum of van der Waals radii; and the proton is equally shared between them. In this case, the X—H bond lengths are much longer than the typical 0.9 Å for a covalently bonded proton, perhaps 1.2 to 1.3 Å, and they are equal in length. This situation corresponds to very strong, symmetric H bonds, sometimes known as single-well H bonds. The energies of such H bonds range from 24 to 40 kcal mol^{-1} (Hibbert and Emsley, 1990). Molecules with symmetric H bonds are rare and have been observed only in crystals of small molecules or in the gas phase. They are unlikely to appear in enzymes and are not discussed further here.

A condition for the formation of a strong or very strong H bond is that the participating heteroatoms should exhibit similar proton affinities; otherwise, the hydrogen will be covalently bonded to one and only weakly attracted to the other. When the proton affinities of the heteroatoms are grossly different, a strong H bond does not exist between them, even if the donor and acceptor are somehow forced to be closer than the sum of their van der Waals radii. In such cases, the H bond is not linear because the proton is out of alignment with the heteroatoms.

Strong and weak hydrogen bonds differ in physical and chemical properties, and these differences can be employed to distinguish them. The relevant physicochemical properties are listed in table 1-5 and include bond lengths, downfield proton NMR chemical shifts, low deuterium fractionation factors for partitioning of deuterium into the H bond in hydrogen-deuterium mixtures, deuterium and tritium isotope effects on the proton chemical shift, and a low ratio of H and D vibrational frequencies.

Hydrogen Bonding in Catalysis

Research indicates that hydrogen bonding can lower activation energies in enzymatic reactions. Most of the documented examples of hydrogen bonding in catalysis involve weak H bonds. The classic case is that of tyrosyl-tRNA synthetase, which catalyzes eqs. 1-12 and 1-13.

$$\mathbf{E} + \text{Tyrosine} + \text{ATP} \rightarrow \mathbf{E} \text{ Tyrosyl-AMP} + \text{PP}_i \qquad (1\text{-}12)$$

$$\mathbf{E} \text{ Tyrosyl-AMP} + \text{tRNA}^{\text{Tyr}} \rightarrow \mathbf{E} + \text{Tyrosyl-tRNA}^{\text{Tyr}} + \text{AMP} \qquad (1\text{-}13)$$

Table 1-5. Physicochemical Properties of Some Low-Barrier Hydrogen Bonds and weak H bonds Involving Nitrogen and Oxygen

Property	Weak H bond O—H- - - -O	LBHB O·· ·H·· ·O	Weak H bond N—H- - - -O	LBHB N·· ·H·· ·O
Lengths	O—H, 0.9 Å	O—H, >1.0 Å	N—H, 0.9 Å	N—H, >1.0 Å
	O - - - O, >2.55 Å	O - - - O, < 2.55 Å	N - - - O, >2.65 Å	N - - - O, < 2.65 Å
δ (ppm)[a]	8–12	17–21	10–14	17–21
Φ[β]	1.0	0.3–0.7	1.0–1.2	0.3–0.7
[δ$_H$ − δ$_{\Delta(D,T)}$][c]	0.0	0.5–1.0	0.0	0.5–1.0

H/D, hydrogen/deuterium, LBHB, Low-barrier hydrogen bond, NMR, nuclear magnetic resource.
[a] Proton NMR chemical shift.
[b] Deuterium fractionation factor. The partitioning constant for the incorporation of deuterium into the hydrogen bond in H/D mixtures.
[c] Isotope effect on the proton chemical shift. The difference between the proton chemical shift and the deuterium or tritium chemical shift.

A thorough study of the effects on k_{cat}/K_m brought about by site-directed mutagenesis to delete single H bonds between enzymatic side chains and either tyrosine or ATP at the active site for tyrosyl-AMP formation indicated that most individual H bonds contributed 1 to 2.5 kcal mol^{-1} of transition state stabilization. These values refer to the energy difference between two hydrogen-bonded states of the substrate in the solvent or the active site, not to the strength of hydrogen bonding in each state. Because of the large number of H bonds between enzymatic side chains and tyrosyl-tRNA^{Tyr}, as illustrated in fig. 2-15 in chapter 2, the summation of individual contributions by weak H bonds provides 15 to 20 kcal mol^{-1} of transition-state stabilization (Fersht, 1987; Wells and Fersht, 1986).

Certain hydrogen bonds in enzymatic catalysis appear to be strong rather than weak. Extensive evidence implicates an LBHB between His57 and Asp102 in the mechanism by which serine 195 reacts as a nucleophilic catalyst in the acylation of chymotrypsin (Cassidy et al., 1997, 2000; Frey et al., 1994; Lin et al., 1998a, 1998b; Neidhart et al., 2001; Westler et al., 2002). The role of the LBHB is shown in fig. 1-15, in which it is postulated to stabilize a metastable tetrahedral intermediate as well as the structurally related transition state.

The LBHB cannot be observed directly as an intermediate in catalysis because of the fleeting existence of tetrahedral intermediates. However, in closely related analogs of tetrahedral intermediates the LBHBs are observed spectroscopically and characterized by the criteria listed in table 1-4. The close contact between His57 and Asp102 in the complex of chymotrypsin with a transition-state analog is consistent with the presence of an LBHB. The structure of the active site in this complex at 1.4-Å resolution is shown in chapter 6.

Chymotrypsin and other serine protease share a property that may provide a glimpse into the transition state even in the absence of an analog of the substrate. A property in common between chymotrypsin at low pH and the transition state is that the imidazole ring of His57 is protonated. At pH values below 7, His57 is in its protonated state, and NMR experiments prove that at pH 4 the proton bridging His57 and Asp102 appears far downfield at 18 ppm (Markley, 1978; Robillard and Shulman, 1972;). This proton is significantly upfield from its position in transition-state analog complexes (18.6 to 18.9 ppm). However, the downfield proton in chymotrypsin displays other spectroscopic and chemical properties of an LBHB that may be not quite as strong as in the transition-state analogs (Westler and Markley, 1996). The structure of the serine protease subtilisin at 0.78 Å resolution allows for the imaging of hydrogen atoms, and the proton bridging His57 and Asp102 (chymotrypsin numbering) at the active site has been imaged and

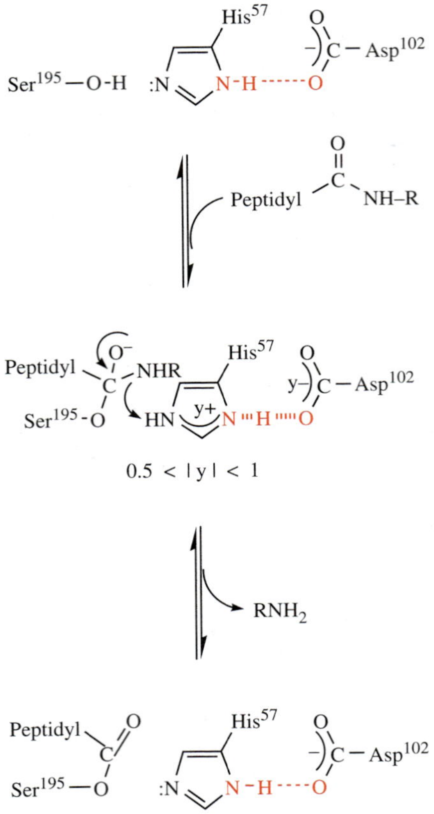

34 Enzymatic Reaction Mechanisms

Fig. 1-15. Acylation of Ser195 of chymotrypsin and the role postulated for an low-barrier hydrogen bond.

modeled as an LBHB (Kuhn et al., 1998). The model of the active site structure of subtilisin at this resolution is shown in chapter 6. Similarly, the structure of pancreatic elastase at 0.95 Å reveals the presence of electron density between histidine and aspartate in the catalytic triad (Katona et al., 2002).

Serine esterases share the active site Ser-His-Asp triad with serine proteases, and considerable evidence implicates strong hydrogen bonding between His and Asp in these enzymes (Massiah et al., 2001; Viragh et al., 2000). Spectroscopic and crystallographic evidence has been presented for the presence of an LBHB in transition-state analog complexes of Δ^5-3-ketosteroid isomerase (Choi et al., 2000; Xhao et al., 1996). This LBHB has been postulated to stabilize an enolate intermediate (see chap. 7). Accumulating evidence suggests that these LBHBs stabilize metastable intermediates and transition states. They may be observed when structural analogs of transition states or metastable intermediates can be studied by NMR spectroscopy and x-ray crystallography.

BINDING ENERGY IN CATALYSIS

Binding and Activation Energy

The binding process itself makes a contribution to catalysis by enzymes. A substrate molecule in an active site experiences desolvation and immobilization, as illustrated in fig. 1-16,

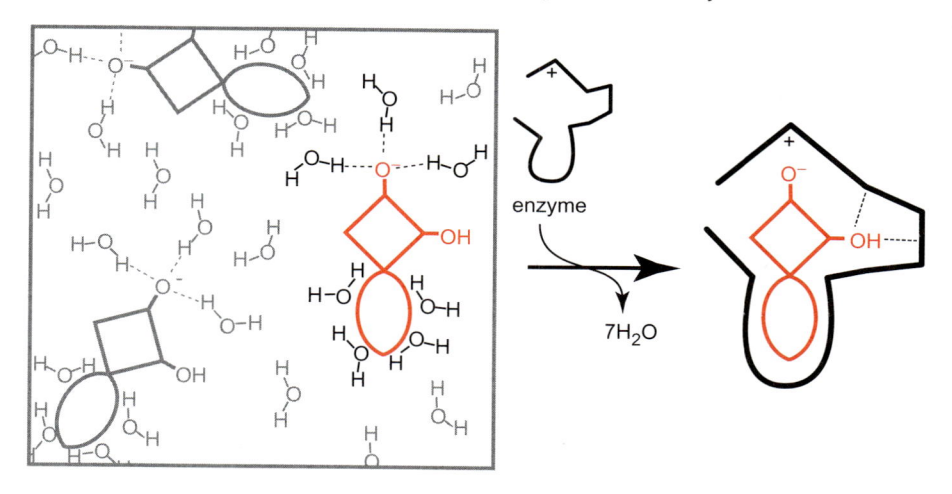

Fig. 1-16. Aspects of substrate binding to an enzymatic site. Substrates in solution tumble without constraints and are generally solvated by water molecules. An enzyme presents a binding site that is sterically and electrostatically complementary, to which the substrate can bind after shedding its solvating water. The active site is preorganized to solvate the substrate in its bound state, and the release of water molecules is one driving force for the binding process, as well as electrostatic, hydrogen-bonded, and van der Waals interactions within the site. The bound substrate cannot tumble independently of the enzymatic site, and both are constrained by one another. Immobilization can facilitate catalysis to a limited degree, whereas the electrostatic and hydrogen bonded interactions can contribute even more.

and the binding interactions that hold it in place impose specific constraints on its behavior. The degrees of freedom a substrate loses on binding to an active site limit its motion, and its contacts with the active site present it with channels of reaction that trace the action of the enzyme.

The means by which binding directs the substrate through steps in the biochemical mechanism include specific van der Waals, hydrophobic, electrostatic, and hydrogen-bonded contacts that lower the activation energies for steps in the reaction. The enzyme-substrate contacts can lead to stabilization of the transition state and destabilization of the ground state. Interactions that stabilize a transition state often also destabilize the corresponding ground state. The two effects are often illustrated in energy level diagrams correlating the free energies of ground states with transition states for enzymatic and nonenzymatic processes. The relationships are never simple enough to be illustrated in a single diagram, because the free energy levels depend on many factors and conditions, including pH, the concentrations of enzyme and substrate, the number of steps in the catalytic mechanism, whether a chemical transformation or a conformational change is rate limiting and whether the rate is diffusion controlled. Each case requires its own diagram of the free energy level.

The three free energy level diagrams in fig. 1-17 illustrate three specific cases for a one-substrate enzymatic reaction proceeding through a single chemical step and one Michaelis complex. In fig. 1-17A, the chemical step is the limiting ($K_m = K_s$) term, and the substrate concentration is lower than K_m; the initial rate $v = (k_{cat}/K_m)[S][E]$. The nonenzymatic reaction proceeds through the same transition state, $S^‡$, as the substrate in $ES^‡$, which is at a lower free energy than $E + S^‡$. The difference is $\Delta G^‡$, the transition-state stabilization energy. The Michaelis complex $(E.S)^*$ is at a higher energy than $E + S$ at all conditions in which $[S] < K_m$. The substrate in $(E.S)^*$ is solvated differently from the free substrate and is immobilized (fig. 1-16), and it may be in a chemically destabilized or activated state.

The free energy difference between E+S and ES* may be regarded as ground-state destabilization, ΔG^*, because of differences in the microenvironment and mobility of the substrate. In this case, the overall difference in activation energy for the enzymatic and nonenzymatic reactions is given as $\Delta G^{\ddagger}_E = \Delta G^{\ddagger}_N - \Delta G^{\ddagger} - \Delta G^*$. That is, the lower activation free energy for the enzymatic reaction arises from both $\Delta G^{\ddagger}$ and ΔG^*, transition-state stabilization and ground-state destabilization.

In fig. 1-17B, the substrate concentration is saturating ($[S] > K_m$) for the same reaction, so that the ES* is at a lower free energy level than E + S. In this case, the energetic effect of ground-state destabilization is overcome—actually paid for—by the high substrate concentration, and the difference in enzymatic and nonenzymatic activation energies is $\Delta G^{\ddagger}_E = \Delta G^{\ddagger}_N - \Delta G^{\ddagger}$. In fig. 1-17C, the transition-state stabilization for the chemical step is sufficient to lower the energy level of ES‡ to that of the Michaelis complex ES*. In this case, diffusion is rate limiting. Further decreases in activation energy for the chemical step would not improve the action of this enzyme. As before, $\Delta G^{\ddagger}_E = \Delta G^{\ddagger}_N - \Delta G^{\ddagger} - \Delta G^*$ when the substrate concentration is subsaturating and $\Delta G^{\ddagger}_E = \Delta G^{\ddagger}_N - \Delta G^{\ddagger}$ at saturating substrate concentrations.

In a later section, we discuss transition-state stabilization by the concept of *transition-state binding*, which occurs through attractive interactions between the enzyme and S‡. Such binding or attraction of the transition state may be regarded as confined to the substrate and enzymatic functional groups that undergo chemical bonding changes in the transition state. In ground-state destabilization, enzyme-substrate contacts in the Michaelis complex lead to distortion of the substrate or a catalytic group, or both, raising the energy toward that of the transition state. Ground-state distortions that raise the free energy of the ES* complex may correspond to some microenvironmental effect such as bond angle strain, compression, desolvation, or altered polarity. The increased energy of the complex due to ground-state destabilization is paid for through weakened binding; that is, formation of the Michaelis complex requires energy in the form of binding free energy. Most of this energy arises from binding interactions between the enzyme and portions of the substrate that are remote from the chemically reacting groups of the substrate and enzyme. Such energy barriers for substrate binding can be overcome—or paid for—by increasing the substrate concentration to near or above the value of K_m. Cellular substrate concentrations are rarely saturating but may approach K_m for many enzymes. Under these conditions, the rate in vivo responds almost proportionally to changes in substrate concentration.

Inasmuch as the structure of any transition state differs from that of the ground state, any structural distortion toward that of the transition state may be regarded as stabilizing the transition state. Should such distortions be allotted to ground-state destabilization? The answer shows how book keeping rules come into play. Because we characterize certain stabilizing interactions as *transition-state binding* in a later section, we assign attractive interactions as *transition-state stabilization*. Other distortions, such as those resulting from charge repulsion, that do not constitute attractive forces but still move the system toward the transition state can be regarded as ground-state destabilization. In succeeding sections, we consider the mechanisms by which substrate binding leads to enzymatic catalysis.

The Active Site as an Entropy Trap

The immobilization brought about by the binding process can promote an enzymatic reaction irrespective of the specific enzyme-substrate contacts that further lower activation energies. This effect could refer to the reaction of two substrate molecules at an active site or to the reaction of a catalytic group at an active site with a single substrate molecule. Conceptualizing this

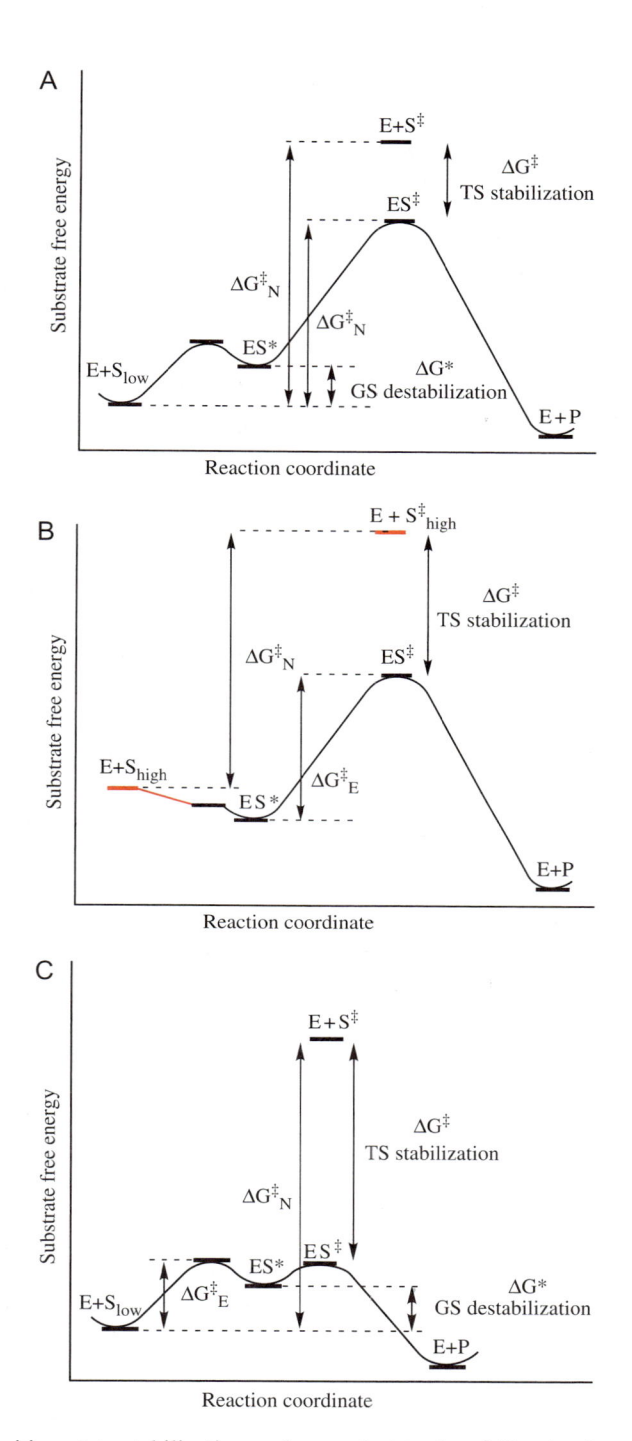

Fig. 1-17. Transition-state stabilization and ground-state destabilization in enzymatic catalysis. The panels illustrate three cases in which activation free energy for a nonenzymatic reaction ($\Delta G^{\ddagger}N$) may be decreased at an active site ($\Delta G^{\ddagger}E$). Two types of interactions between the active site and the substrate can contribute to this effect: transition-state stabilization ($\Delta G^{\ddagger}$) and ground-state destabilization (ΔG^*). Relative energy levels refer to a fixed concentration of enzyme. The energy levels for E + S in A and B refer to a substrate at concentrations below K_m (S_{low}) and above K_m (S_{high}), respectively. When a chemical step is rate limiting (A), the transition state refers to that step and has the highest energy in the profile. Binding to form ES is faster and does not contribute to rate limitation. In case of maximum transition-state stabilization for the chemical step (C), substrate binding becomes rate limiting, and the rate is under diffusion control.

Intramolecular Reactions

One way to study the effect of binding is to consider the consequences of abolishing the diffusion of reacting functional groups. Incorporating them within a single molecule can do this. If both reactant groups are in the same molecule, as they would be in a Michaelis complex, they do not diffuse apart, and their mutual reactivity might be enhanced. Many studies of this effect can be found in the literature, and the best-designed studies make it clear that this is a significant effect. The ingenuity of bioorganic chemists in synthesizing molecules that react internally can lead to complications in the interpretation of kinetic data obtained about reactions.

Consider how to compare the rate of an *intramolecular* reaction, in which both reactant groups are held in the same molecule, with that of an *intermolecular* reaction, which requires the reactants to collide. There is a problem of kinetic order. The rate constant for an intramolecular reaction is first order in units of reciprocal time (e.g., s^{-1}, min^{-1}), whereas that for an intermolecular reaction is higher in order, often second order (e.g., $M^{-1} s^{-1}$, $M^{-1} min^{-1}$). A direct comparison of numerical values of rate constants for a reaction that takes place within a single molecule or between two molecules would be meaningless. When the ratio of the first-order rate constant for the intramolecular reaction to the second-order rate constant for the same reaction between two molecules is computed, the result is a number having the units of concentration. The result may be regarded as an "effective concentration" of the reacting groups in the intramolecular reaction. The situation is further complicated when the two reactions take place by different mechanisms, in which case the rates cannot be meaningfully compared. In any such comparison, it is essential to verify that the same reaction mechanism is followed in the intramolecular and intermolecular reactions.

The numbers generated as effective concentrations have ranged from a few molar to more than 10^{12} M. The confusion for a time prevented the assignment of a definite number for the effect of binding per se in catalysis. A set of data for the intramolecular cleavage of *p*-bromophenyl esters appears in table 1-6 (Bruice and Pandit, 1960a,b).

In the intramolecular reactions, the neighboring carboxylate group reacts as a nucleophile with the ester group to displace *p*-bromophenolate ion, as illustrated in eq. 1-14.

$$\text{(structure with carboxylate and p-bromophenyl ester)} \longrightarrow \text{(phthalic anhydride)} + {}^-\text{O-C}_6\text{H}_4\text{-Br} \quad (1\text{-}14)$$

The resulting anhydride undergoes hydrolysis in the faster step. The first and slowest entry in table 1-6 incorporates the essential advantage of intramolecularity, which can correspond to an effective concentration of upto 10^5 M (see next section). The relative rate constants therefore correspond to effective molarities up to 10^{12}. The effective molarity may be regarded as the concentration of one molecule that would be required to react with a 1 M solution of the other in the intermolecular reaction to give the rate observed in the intramolecular reaction.

Effective concentrations for the entries in table 1-6 cannot correspond to any realistic concept of concentration and must signify something very different. The concentrations of pure liquids and solids range up to only a few tens in molar concentration units, such as

Table 1-6. Rate Enhancement for a Series of Intramolecular Acylation Reactions

Ester		$k/k_{glutartate}$
COOR, COO⁻ (glutarate) → anhydride		1
H₃C, H₃C, COOR, COO⁻ (dimethyl glutarate) → anhydride		20
COOR, COO⁻ (succinate) → anhydride		230
COOR, COO⁻ (maleate) → anhydride		10,000
bicyclic COOR, COO⁻ → bicyclic anhydride		53,000

From WP Jencks, 1969, with permission.

55 M for water. Larger effective molarities must represent rate enhancements that are much larger than the simple effects that would correspond to concentration factors in rate laws such as $v = k[A][B]$.

Loss of Translational and Rotational Entropy

The intramolecular reactions in table 1-6 vary in the degree to which the reacting groups are constrained to interact. Those in which the groups are forced together by the structural framework react fastest. The strain involved in these cases clouds the picture and cannot represent the effect of simply having the functional groups in the same molecule. The largest values of effective molarity cannot be a measure of the rate enhancement arising from simple binding. Because the effect of bringing the reactive groups together without strain should be comparable for all of the reactions in table 1-6, other rate enhancing effects such as internal strain should contribute in addition to the value for simple binding. The 10^{12} M for the maximum effect in table 1-6 should be larger than the value for simply binding the two functional groups in the same molecule.

The entropic consequences of bringing two reactants together can be conceptualized as the loss of net rotational and translational degrees of freedom for the two molecules. This represents a lower entropy of activation ($\Delta S^{\ddagger}$) for the intramolecular reaction relative to its intermolecular counterpart. The energetic consequences of this can be estimated to correspond to a decrease of about 12 kcal mol^{-1} in the free energy at 25°C because of the loss of translational and rotational entropy for the two molecules when they become connected into a single molecule by a bond (Page and Jencks, 1971). If this entropy loss is quantitatively used to lower the activation energy for a reaction between reacting groups, the rate enhancement would be a maximum of 10^8 M (Page and Jencks, 1971). In practice, the maximum is not realized because the potential entropy loss cannot all be captured in the activation energy.

How much of the entropy loss in an intramolecular reaction can lead to rate enhancement? The answer varies with reaction type and conditions. A molecule with two functional groups that can react without straining bonds and that are not forced into van der Waals contact can provide a guide. The reaction of succinic acid to form succinic anhydride in eq. 1-15 served as a guide for comparison with the same reaction between two molecules of acetic acid in eq. 1-16 (Page and Jencks, 1971).

$$\text{succinic acid} \rightleftharpoons \text{succinic anhydride} + H_2O \quad (1\text{-}15)$$

$$2\,CH_3COOH \rightleftharpoons (CH_3CO)_2O + H_2O \quad (1\text{-}16)$$

Careful measurements showed that the ratio of equilibrium constants and forward rates for these reactions gave a value of 10^5 M as the effective concentration of carboxylic acid groups on going from two molecules of acetic acid to one molecule of succinic acid. Only 7 kcal mol^{-1} is realized in this case. The decrement could be attributed to internal degrees of freedom in the transition state and product that were not present in the ground state in the reaction of succinic acid, which minimized the loss of entropy in the transition state. The internal reaction of succinic acid seems to be a reasonable example of one that allows the two carboxylic acid groups to react without forcing them together.

Given these relationships, it would be reasonable to consider the results from the succinic/acetic acid system, 10^5 M, as representative of the rate enhancement that can be expected from the loss of overall translational and rotational entropy brought about by binding a substrate to an active site. The largest values cannot be attributed solely to entropy losses on binding, nor can the very large rate enhancements brought about by enzymes. Nevertheless, the rate enhancement attained in decreasing $\Delta G^{\ddagger}$ by 7 kcal mol^{-1} is a significant part of the overall catalytic process. For most enzymatic reactions, an additional 7 to 17 kcal mol^{-1} or more must be found in decreased enthalpy of activation ($\Delta H^{\ddagger}$).

Dissecting the Binding Effect in Enzymatic Action

Suppose all catalytic functional groups were removed from an active site. Would the enzyme still catalyze its cognate reaction? The answer to this question depends on the reaction mechanism. In the case of trypsin, the answer appears to be yes. Trypsin is a serine protease like chymotrypsin that exhibits maximum activity at lysyl or arginyl groups in a peptide. Mutation of His57 to Ala decreases the activity by 60,000-fold; of Asp102 to Asn by 2500- to 10,000-fold, depending on the substrate; and of Ser195 to Ala by 40,000-fold. However, deletion of *all three groups* (H57A/D102N/S195A) decreases the activity by only 84,000-fold (Corey and Craik, 1989; Craik et al., 1987). In the case of trypsin, the kinetic consequences of mutating the three catalytic residues are not multiplicative or even additive, and the enzyme still displays significant activity in the absence of all of its catalytic residues. Evidently, the binding process accounts for the remaining 10^8 rate enhancement provided by the enzyme. Most likely a significant part of this results from strain induced in the substrate molecule when it is bound at the substrate binding site. If the binding process itself can bring about 10^3 to 10^5 rate enhancement, induced strain may account for the remaining enhancement. Induced strain in this case could represent the consequences of twisting the peptide amide substituent in such a way as to disrupt the orbital overlap between the nonbonding electron pair on nitrogen and the p electrons of the carbonyl group. This would increase the reactivity of the carbonyl group toward nucleophilic attack by Ser195, which is the rate-limiting step in the hydrolysis of a peptide bond by trypsin and chymotrypsin.

In contrast, the removal of essential functional groups from active sites of other enzymes can lead to undetectable activities. In the mechanism of the reaction of galactose-1-P uridylyltransferase, His166 at the active site functions as a nucleophilic catalyst (see chap. 10). Mutation of this histidine to glycine completely abolishes activity (Field et al., 1990), despite the fact that the global structure is largely retained except for the absence of the imidazole ring of His166 (Thoden et al., 1997).

Stabilization of the Transition State

Much evidence supports the theory that enzymes catalyze reactions by binding the transition states more tightly than ground states. This theory has been extant for more than 50 years, and much evidence supporting the concept has appeared during the past 30 years. Before continuing with this theory, we consider an apparent weakness and attempt to clarify the value of the theory despite a qualification in the logic.

The Transition State Theory

In Eyring's theory of absolute rates, a chemical reaction is defined as proceeding through a *transition state*, in which bonds in the substrate that are to be broken and bonds in the product that are to be formed are partially broken and formed, respectively. At the transition state, the system is at its highest energy along the reaction coordinate (Eyring, 1935). The diagram in fig. 1-18 defines the transition state for a simple alkylation reaction as the highest point on the profile of potential energy against the reaction coordinate.

The interval δ at the top of the curve is almost invariant along the reaction coordinate, and species within this interval are regarded as *activated complexes*. The transition state is closely related to the activated complexes, and because the interval is very short and regarded as about 1 Å, or less than the length of most covalent bonds, the transition state and activated complexes are practically identical. The transition state is defined as the structure at the highest point on the reaction coordinate, not as an intermediate or discrete molecule. It is characterized by partially broken and partially formed covalent bonds that do not behave as classic bonds. At the transition state, the vibrational mode corresponding to the reaction coordinate has an imaginary frequency, and there is no resisting force.

The Eyring equation relates the energetics of the transition state to the rate in terms of an activation energy. Because the diagram in fig. 1-18 is drawn in terms of free energies, we write the Eyring equation in terms of thermodynamic parameters.

$$k_r = (kT/h) e^{-\Delta G^{\ddagger}/RT} = (kT/h) e^{-\Delta H^{\ddagger}/RT} \cdot e^{-\Delta S^{\ddagger}/RT} \tag{1-17}$$

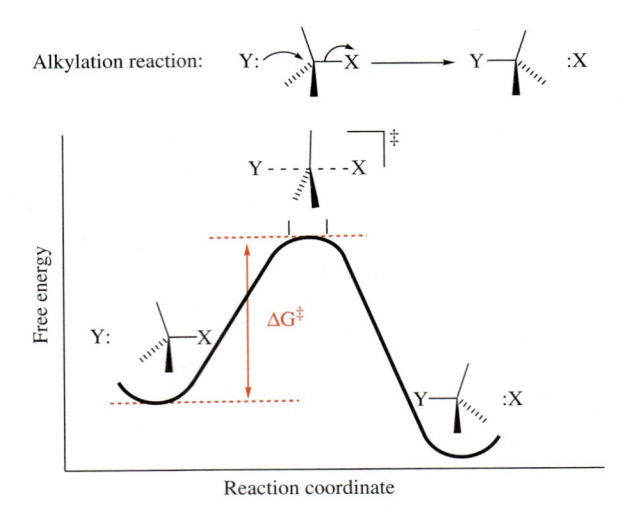

Fig. 1-18. Transition state model for the mechanism of an alkylation reaction. The alkylation shown is a simple, one-step process in which the C—X bond is broken as the Y—C bond is formed. In the transition step, the two bonds are partially broken and formed, and this represents the highest point on the free energy profile. In this case, there is no intermediate in the chemical process.

In eq. 1-17, k_r is the rate constant, $\Delta G^{\ddagger}$ is the free energy of activation, T is the temperature in degrees Kelvin, h is Planck's constant, k is the Botzmann constant, and R is the universal gas constant, The free energy of activation is the difference between the free energies of the reactants and the transition state. In the derivation of the Eyring equation, this energy arises from a kind of equilibrium constant, a quasi-equilibrium constant, between the reactants and transition state. The free energy of activation can be evaluated by plotting the logarithm of k_{obs}/T against $1/T$ (Kelvin) and calculating it from the slope and R. This free energy of activation is analogous to the Arrhenius activation energy E_a, an empirical parameter.

The Eyring equation has been remarkably durable because of its success in accounting for experimental rates in terms of temperature and the kinetic properties of reactions. It can be criticized because the derivation depends on defining a quasi equilibrium between reactants and the transition state. The quasi equilibrium cannot be evaluated in terms of an equilibrium constant independently of eq. 1-17 itself; the transition state is not a molecule that can be measured in terms of a concentration or activity. An equilibrium does not actually exist between the reactants and the transition state because most collisions at that energy pass over the barrier to products, and very few return to the substrates. This raises the question of whether the Eyring equation can correspond to reality. However, the virtually ubiquitous use of the Eyring concept of a transition state, and the equally common use of the Eyring equation in all fields of chemistry attest to its usefulness in understanding reaction rates. In chemistry, the widely accepted test of a theory is its utility, and the transition state theory passes the test.

Enzymatic Rate Enhancement and Transition-State Binding

Accepting transition state theory as a basis for understanding reaction kinetics, it can be shown that an enzyme (or any catalyst) can catalyze a reaction by binding the transition state more tightly than the ground state. This concept made its appearance early in the study of enzymes (Pauling, 1948). It is a straightforward and reasonable concept because of the fact that a binding interaction should be stabilizing; therefore, binding the transition-state should stabilize it, thereby increasing the rate. A quanitative correlation of the rate enhancement in an enzymatic reaction with the binding energy for a transition-state can be derived as illustrated in scheme 1-7 (Wolfenden, 1972).

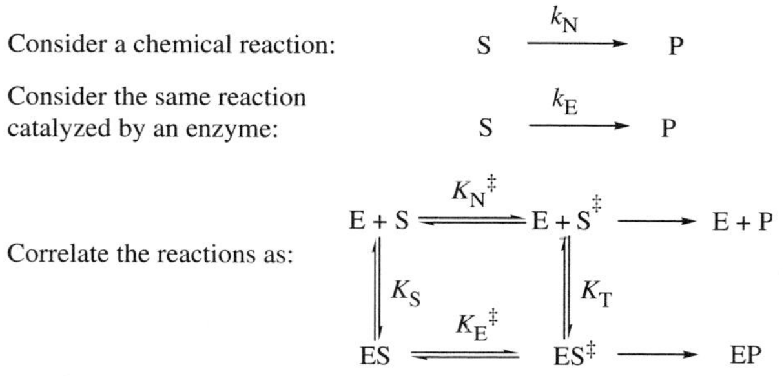

Consider a chemical reaction: $S \xrightarrow{k_N} P$

Consider the same reaction catalyzed by an enzyme: $S \xrightarrow{k_E} P$

Correlate the reactions as:

$$\begin{array}{ccc} E+S & \underset{K_N^\ddagger}{\rightleftharpoons} & E+S^\ddagger \longrightarrow E+P \\ \updownarrow K_S & & \updownarrow K_T \\ ES & \underset{K_E^\ddagger}{\rightleftharpoons} & ES^\ddagger \longrightarrow EP \end{array}$$

Scheme 1-7

In the model of scheme 1-7, the basic Michaelis-Menten kinetic mechanism for a simple reaction is written in parallel with the nonenzymatic mechanism of the same reaction. The enzyme is a spectator in the nonenzymatic reaction on the upper line and does not participate in any way. On the lower line, the ES-complex is the Michaelis complex, the $ES^\ddagger$-complex is of the enzyme-substrate transition-state, and everything takes place at the active site. The substrate transition-state $S^\ddagger$ is the same in the two reactions, with the exception that it is in contact with all the relevant catalytic functional groups of the active site in the $ES^\ddagger$-complex, while being free of all catalytic contacts in the nonenzymatic reaction.

The connections between the two mechanisms allow them to be correlated with equilibrium constants. The result is an equilibrium box of transformations with four equilibrium constants. Two of the constants, $K_N^\ddagger$ and $K_E^\ddagger$, are the nonenzymatic and enzymatic constants for transition-state formation. Another, K_S, is the dissociation constant for binding the substrate at the active site, the constant originally introduced by Michaelis and Menten in their kinetic mechanism for enzyme action. The only conceptually new constant is K_T, the dissociation constant for the substrate transition-state $S^\ddagger$ from the active site $ES^\ddagger$. The thermodynamic cycle in scheme 1-7 leads directly to the highly revealing relationship of eq. 1-18.

$$k_E/k_N = K_E^\ddagger/K_N^\ddagger = K_S/K_T \qquad (1\text{-}18)$$

This relation shows that the ratio of the first-order rate constants for the nonenzymatic and enzymatic reactions equals the ratio of the dissociation constants for the substrate and transition state. According to this, the enzyme simply binds the transition state tightly enough to increase the rate, and the rate enhancement at the active site (k_E/k_N) is determined by the higher affinity of the site for the transition state than for the substrate (K_S/K_T).

Scheme 1-7 refers to an enzymatic reaction in which the rate-limiting step is the chemical transformation of ES to EP, and the rate constant k_E refers to this process. It would not apply to a mechanism in which product dissociation controlled the rate. The rate enhancement refers to a one-step chemical transformation and does not consider how other steps of a multistep mechanism may work. The theory may be extended to these common complications.

A test of the theory would be to measure K_T and K_S independently and compute their ratio, which should be the same as the ratio of rate constants. In principle, this experiment is impossible because the dissociation constant for a transition state cannot be measured.

44   Enzymatic Reaction Mechanisms

However, if an enzymatic reaction can be found, for which a chemically stable analog of the transition state can be synthesized and its affinity for the active site measured, the experiment might be approximated. Results of such studies support the theory and form the basis for a strategy of designing potent inhibitors of enzymes (Mader and Bartlett, 1997; Schramm, 1998).

Cytidine Deaminase

Such a case of designing potent inhibitors of enzymes may be cytidine deaminase (MW 35,000), which catalyzes the reaction of eq. 1-19.

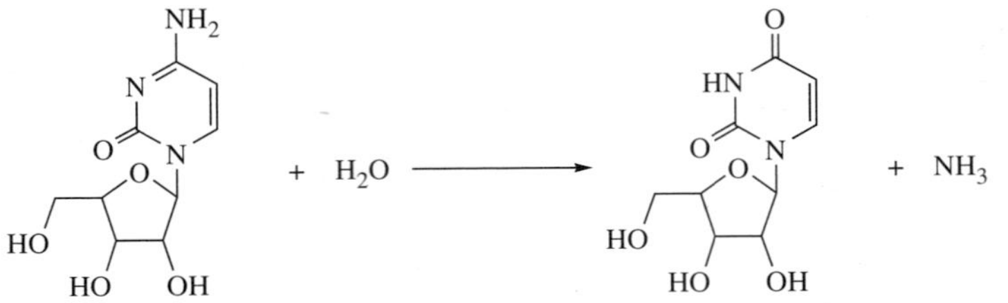

$$\text{cytidine} + H_2O \longrightarrow \text{uridine} + NH_3 \qquad (1\text{-}19)$$

The chemical mechanism of this reaction proceeds by addition of water to C4 of the cytosine ring to form the tetrahedral addition intermediate shown in scheme 1-8.

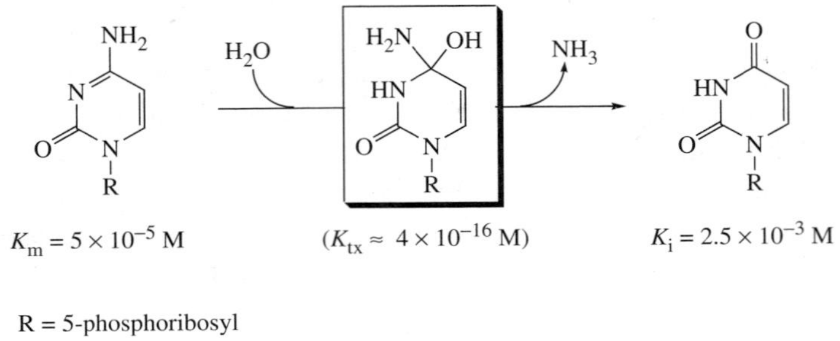

$K_m = 5 \times 10^{-5}$ M $(K_{tx} \approx 4 \times 10^{-16}$ M$)$ $K_i = 2.5 \times 10^{-3}$ M

R = 5-phosphoribosyl

Scheme 1-8

The tetrahedral addition intermediate undergoes elimination of ammonia to form uridine, the product. Scheme 1-8 includes values of K_m for cytidine and K_i for uridine. Kinetic evidence indicates that the K_m can be taken as K_S in eq. 1-18 (Snider et al., 2000). From experimental values for the nonenzymatic and enzymatic rate constants (k_N and k_E), the value of K_T should be 4×10^{-16} M (Frick et al., 1989). This is shown parenthetically as an approximation to K_{tx} beneath the structure of the tetrahedral intermediate in scheme 1-8.

Although the tetrahedral intermediate is not exactly the transition state, there is reason to regard the two as similar to each other. The bonds to the amino and hydroxyl groups in the intermediate are slightly too short for the transition state, but the structures are otherwise essentially the same. The tetrahedral intermediate is metastable and therefore a high-energy intermediate. According to the Hammond postulate, a high-energy intermediate in a reaction is similar to the transition state (Hammond, 1955). The smaller the energy difference between a high-energy intermediate and a transition state, the more similar they are. The situation is illustrated in fig. 1-19, an energy-reaction coordinate diagram for a two-step reaction with a metastable intermediate and two transition states. The dip at the top of the diagram shows the energy level for the metastable intermediate as slightly lower than the energy levels of the flanking transition states. Energy relations such as in fig. 1-19 allow the approximation that metastable intermediates are similar to transition states. The dissociation constant for a high-energy intermediate from an enzyme is an approximation for that of the transition state, that is K_T in eq. 1-18, and vice versa.

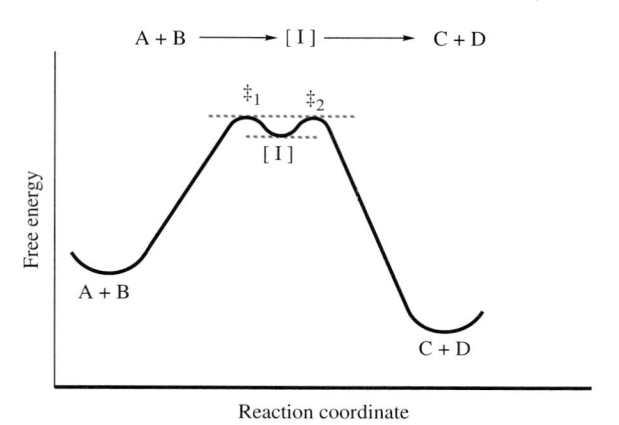

Fig. 1-19. The transition-state model illustrates energy relationships in a two-step reaction that proceeds through a metastable intermediate. The high-energy intermediate is similar in energy to the transition state. The Hammond postulate states that a high-energy intermediate is structurally similar to the transition state. This follows rationally from the fact that it is an intermediate and is similar in energy to the transition state. The diagram also shows that the two transition states for the two steps are also similar.

Rarely is the dissociation constant for an enzymatic intermediate available, especially one for a metastable intermediate. In the case of cytidine deaminase, analogs of the tetrahedral intermediate have been synthesized and studied as inhibitors of the enzyme. In a logical extension of the theory of transition-state binding in scheme 1-7, structural analogs of the transition state can be expected to bind with affinities approaching that of the transition state itself. The more similar the analog, the more similar the dissociation constants. In the case of cytidine deaminase, structural analogs of the tetrahedral intermediate should also be analogs of the transition state. Such analogs do display very low values of K_i, as shown in fig. 1-20. The central compound is similar to the intermediate, with the substitution of H for NH_2, and displays an inhibition constant of 10^{-12} M. The analog with C4 of the pyrimidine ring replaced by P displays an inhibition constant of 10^{-9} M. These are very low values representing high affinity for cytidine deaminase, in accord with theory. The differences from the theoretical value of 4×10^{-16} M for the transition state seem reasonable given the differences in structure. The dramatically lower binding affinity ($K_i = 10^{-3}$ M) for the analog with only two hydrogens on C4 supports this interpretation. It seems that the presence of an amino and hydroxyl group on C4 in the intermediate could account for the difference between 10^{-12} M and 10^{-16} M in affinity.

The structure at the active site of cytidine deaminase shown in fig. 1-21 supports the interpretations of inhibition data in the light of the model in scheme 1-7. The enzyme-inhibitor contacts are extensive and indicate additional interactions in the true intermediate.

The structure shows Zn^{2+} coordinated to the OH group of the transition-state analog (fig. 1-20). Zinc plays an important role in the mechanism by binding the attacking water molecule and lowering its pK_a (see chap. 4), thereby facilitating the formation of a Zn^{2+}-bound hydroxide ion for nucleophilic attack on C4. The effect of divalent metal ions in lowering the pK_a of coordinated water is a widespread phenomenon in enzymatic reaction mechanisms (see chap. 4). The structure of the corresponding complex of inhibitor in fig. 1-20, which lacks the C4(OH), shows the presence of a water molecule intervening between C4 and Zn^{2+} (Xiang et al., 1995). Microcalorimetric studies indicate that most of the binding free energy for the transition-state analog is enthalpic (Snider and Wolfenden, 2001).

The difference in binding free energy between inhibitors with tetrahedral C4 in fig. 1-20 is 10.1 kcal mol^{-1}. The C4(OH) contributes this amount of free energy to stabilizing

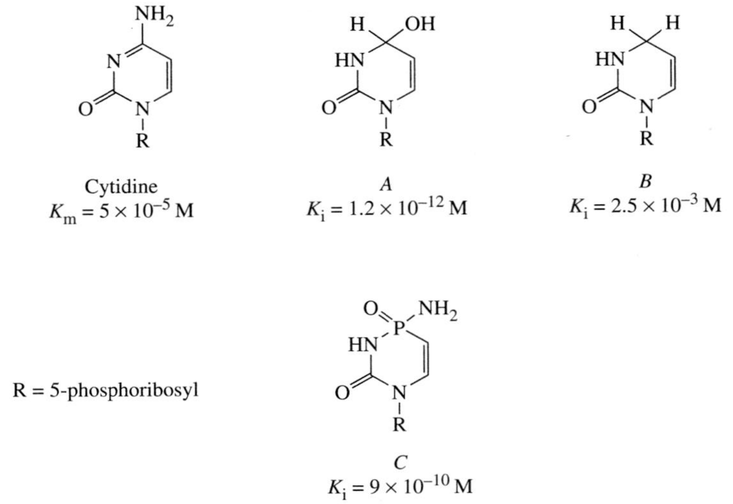

Fig. 1-20. Transition-state analogs as inhibitors of cytidine deaminase. Cytidine and several inhibitors exhibit markedly different binding affinities to cytidine deaminase. The K_m for cytidine is regarded as similar to its dissociation constant K_d, as are the values of K_i for the inhibitors. 3,4-Dihydrouridine in the center is the best inhibitor and is most similar to the tetrahedral intermediate (Frick et al., 1989). The phosphoramidate is also similar to the tetrahedral intermediate (Ashley and Bartlett, 1984). By the Hammond postulate, the metastable tetrahedral intermediate is similar to the transition state.

the complex. The basis for this is of interest. The C4(OH) contributes a hydrogen bond to a glutamate residue, and the structure shows this to be a short hydrogen bond (<2.5 Å). It has been suggested that this might be a low-barrier hydrogen bond (Xiang et al., 1995). This is a reasonable suggestion given the possibility that the Zn^{2+}—OH and the glutamate could have similar pK_a values and are very close in the structure. Because spectroscopic evidence for this contact is not available, a definite assignment cannot be made. Deletion of glutamate by site-directed mutagenesis decreases the activity of cytidine deaminase by a factor of 10^6, verifying the importance of this contact (Carlow et al., 1995).

Many other examples of the strong binding of transition-state analogs can be found in the enzymologic literature. Additional examples in this book include carboxypeptidase in chapter 6 and ribulose bisphosphate carboxylase in chapter 8.

Binding the Near Attack Conformation

The concept of transition-state binding to active sites cannot be regarded as representing a true, reversible binding phenomenon. Transition states cannot dissociate into solution and be measured. The concept should be regarded in the same light as that of the equilibrium between a ground state and activated complex in transition state theory—as a useful theoretical construct.

Another theory holds that enzymes bind substrates so as to promote the formation of near attack conformations (NACs), in which the reacting groups are held closely together in orientations that resemble the transition state (Benkovic and Bruice, 2002). A NAC is defined as a structure in which reacting groups are in close proximity and in orientations that allow the reaction to proceed. A NAC for an acyl transfer reaction in anhydride formation, such as in table 1-6, is defined as one in which the attacking carboxylate group is about 3 Å from a face of the ester group, with one oxygen within 15 degrees of the optimal bonding angle, as illustrated in structure **1-11**.

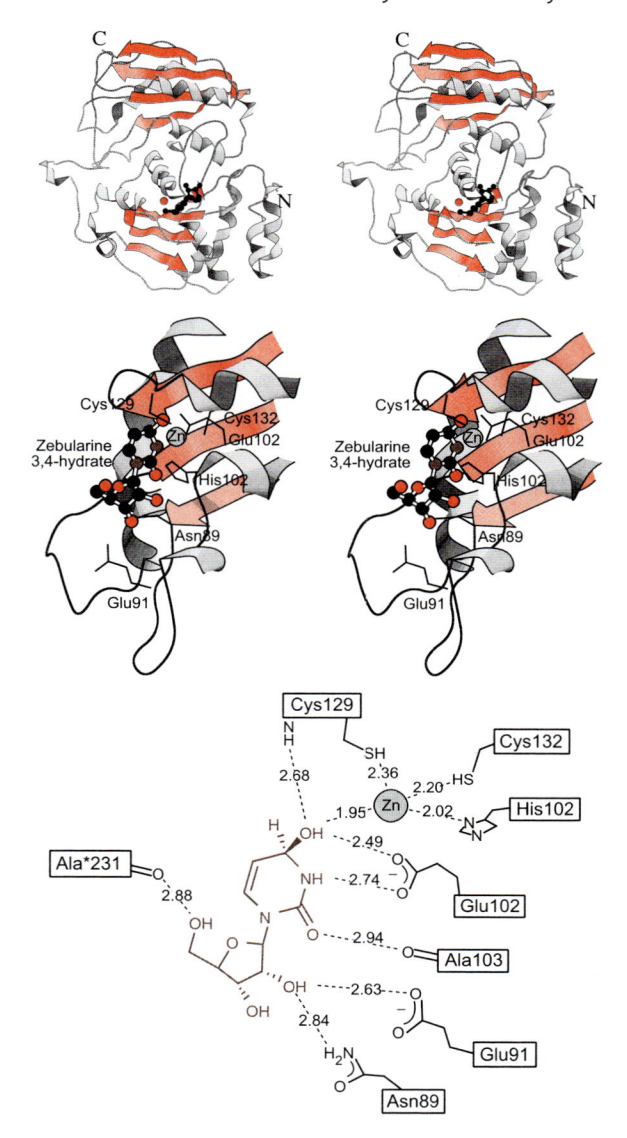

Fig. 1-21. A stereographic ribbon diagram (top) shows a monomer of the homodimeric *E. coli* cytidine deaminase (EC 3.5.4.5), with 3,4-dihydrouridine (zebularine 3,4-hydrate) bound to the active site and shown as a ball-and-stick model. A stereodrawing (middle) shows the inhibitor coordinated with Zn^{2+} at the active site. The close contacts of the inhibitor at the active site are diagrammed at the bottom. Notice the short hydrogen bond between the OH group of the inhibitor and Glu102, an essential amino acid. The structures are generated from PDB 1CTU (Xiang et al., 1997).

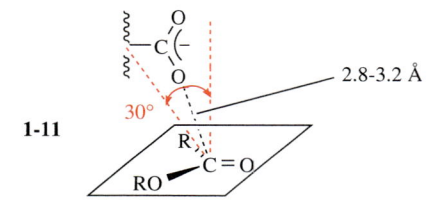

Near Attack Conformation

A NAC is a pretransition-state structure that that faces a small barrier to reaction. In this regard, the energetic relationship with the transition state may be similar to that of a metastable intermediate. The fraction of NACs in a collection of molecules can be calculated, and the results show that the logarithms of the relative rate constants for reaction are proportional to the logarithms of the probability of NAC formation (Bruice and Benkovic, 2000). The higher the fraction of NACs, or transition-state–like structures, the faster the reaction. The major thermodynamic barrier to NAC formation proved to be enthalpy (ΔH), and this was correlated with the enthalpy of activation ($\Delta H^\ddagger = 20$ to 30 kcal mol^{-1}) for a number of nonenzymatic reactions. Entropic effects were much smaller, with values of $T\Delta S^\ddagger$ corresponding to 4.0 to 4.6 kcal mol^{-1}.

The fraction of NACs in enzyme-substrate complexes can be calculated from high-resolution structures. The calculated NACs tend to be structurally and chemically complementary to active sites (Bruice and Benkovic, 2000). The enzymes appear to bind substrates in such as way as to facilitate the formation of transition states.

Destabilization of Ground States

Little information is available about destabilization of ground states in enzymatic catalysis. NACs may be regarded as intrinsically destabilized relative to ground states, and being similar to transition states may be regarded as slightly less unstable than transition states. As pointed out in the discussion of transition state stabilization, certain substrate-destabilizing interactions may be regarded as arising from transition state binding. Clear examples of substrate destabilizing interactions that stabilize a transition state through charge repulsion or induced strain are less documented. An example of charge repulsion that appears in this book includes modulation of the reduction potential of NAD$^+$ bound to UDP-galactose 4-epimerase (see chap. 7). An example of induced steric strain includes the enzymatic cleavage of the cobalt-carbon bond in adenosylcobalamin, the vitamin B$_{12}$ coenzyme (see chap. 4).

Rate Enhancement through Binding of Remote Groups

The Transition State Defined

In structure-function analysis of nonenzymatic, chemical reactions, the activation energy depends on those atoms in a molecule that undergo bonding changes in the transition state. This may involve only a few atoms of a large molecule or all the atoms in a small molecule, depending on the nature of the reaction and the structure of the molecule. However, atoms remote from the reacting center of a typical molecule do not experience bonding changes in the transition state unless they are connected to the reaction center by conjugated multiple bonds, which transmit electronic effects, or unless they sterically interfere with the reacting atoms. In the emergent concept, portions of a molecule that do not impart steric or electronic effects on the reacting atoms do not significantly affect the activation energy.

Enzymatic reactions differ in this respect. Portions of substrate molecules remote from the reacting atoms often make important contributions to catalysis. The most obvious manifestation of this would be the substrate selectivities of enzymes in acting most efficiently on particular molecules. Although the effects of varying structures of substrates on enzymatic rates are often reported in the literature, the activation energies are not often measured. The many reasons for not reporting activation energies are related to the complexities of enzymes and the consequences for temperature studies. The temperature range over which enzymatic rates can be measured without denaturation or structural transitions is very limited. Conformational changes or diffusion also may limit the rates. For these reasons, enzymatic activation energies are not often measured, and those that are reported can be ambiguous. However, enzymatic activation energies can represent the transition state when the chemical step is rate limiting and the enzyme structure is stable over a temperature range of 30°C or more.

Portions of enzyme structures remote from the reacting groups also often influence catalysis. The most obvious biological manifestation of this fact is allosteric regulation and regulation by phosphorylation/dephosphorylation at sites remote from the active site. These interactions alter the structure of the active site. Substrate induced conformational changes also alter the active site structure and may depend on enzyme binding interactions remote from the reacting groups. Although these phenomena are common in enzymology, quantitative aspects are not often reported. The case of acetoacetyl CoA:succinate CoA transferase in the following section is an exception. In that case a thorough kinetic analysis of the influence of remote binding on catalysis is available.

Acetoacetyl CoA:Succinate CoA Transferase

Acetoacetyl CoA:succinate CoA transferase, which we refer to as CoA transferase, catalyzes the reversible transfer of CoA from a β-ketoacid to succinate according to eq. 1-20. Notice the transfer of oxygen (^{18}O) from succinate to acetoacetate.

$$\text{AcAcSCoA} + \text{}^-\text{OOC-CH}_2\text{-CH}_2\text{-COO}^- \rightleftharpoons \text{AcAcCOO}^- + \text{SuccCOSCoA} \quad (1\text{-}20)$$

The reaction proceeds in two distinct steps, with a covalent enzyme-CoA intermediate (Hersh and Jencks, 1967a,b; Moore and Jencks, 1982; Solomon and Jencks, 1969; White and Jencks, 1976). The covalent intermediate has been isolated and characterized. A glutamate residue in the active site is the nucleophilic catalyst that accepts CoA from a substrate to form the intermediate. The two-step mechanism can be represented by eqs. 1-21a and 1-21b.

$$\text{E}-\text{COO}^- + \text{AcAcCOSCoA} \rightleftharpoons \text{E}-\text{C}(=O)\text{SCoA} + \text{AcAcCOO}^- \quad (1\text{-}21a)$$

$$\text{E}-\text{C}(=O)\text{SCoA} + \text{SuccCOO}^- \rightleftharpoons \text{E}-\text{COO}^- + \text{SuccCOSCoA} \quad (1\text{-}21b)$$

In the proof that ^{18}O is transferred from succinate to the enzyme, the reaction was conducted with acetoacetul CoA and [^{18}O]succinate. The enzyme was repurified from the products in the reaction mixture and enzymatically digested into peptides. A specific ^{18}O-labeled peptide was identified by mass spectrometry (Hansen and Boyer, 1967). In the proof that glutamate at the active site binds CoA, reduction of the covalent enzyme-CoA intermediate by NaB^{3}H$_4$ led to an inactive, tritium-labeled protein and the release of CoA. Chemical degradation of the tritium-labeled enzyme allowed the radioactivity to be identified with α-amino-δ-hydroxybutyrate, which arose from borohydride reduction of the CoA ester of glutamate (Solomon and Jencks, 1969). This experiment made use of the fact that borohydride efficiently reduces thioesters, although it hardly reduces oxygen esters.

The CoA transfer mechanism can be formulated as in scheme 1-9, in which anhydride formation between the active site glutamate and the substrate thioester is an essential step.

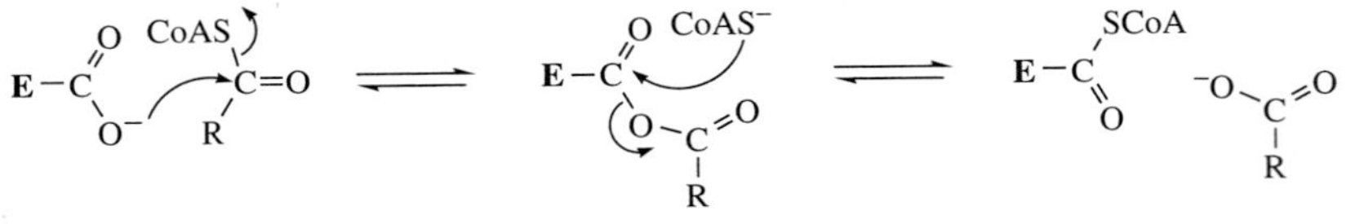

$$\text{E}-\overset{\text{O}}{\underset{\text{O}^-}{\text{C}}} \quad \overset{\text{CoAS}}{\underset{\text{R}}{\text{C}=\text{O}}} \rightleftharpoons \text{E}-\overset{\text{O}}{\text{C}} \quad \overset{\text{CoAS}^-}{\underset{\text{O}-\text{C}=\text{O}}{}} \rightleftharpoons \text{E}-\overset{\text{O}}{\text{C}}\overset{\text{SCoA}}{} \quad ^-\text{O}-\overset{\text{O}}{\underset{\text{R}}{\text{C}=\text{O}}}$$

Scheme 1-9

This explains the ^{18}O-transfer results and the formation of the covalent enzyme-CoA intermediate and accounts for the overall reaction. If the CoA ester on the right side of scheme 1-9 is acetoacetyl CoA, acetoacetate is formed on the left side. Reversal of the process by reaction of the enzyme-CoA intermediate with succinate will produce succinyl CoA and the free enzyme.

From a chemical standpoint, there is no reason for the major part of the CoA molecule to play a role in the mechanism shown in scheme 1-9 or in the overall reaction. Most of the structure is far removed from the reacting groups and has little electronic or steric contact with the reacting thiol group. However, the remote parts of the CoA molecule have an enormous impact on the reaction at every stage. The kinetic evidence is summarized in fig. 1-22 and table 1-7 (Moore and Jencks, 1982). From the value of k_{cat}/K_m for forming the enzyme-substrate anhydride and the second-order rate constant for the nonenzymatic formation of acetic anhydride from acetate and an acetyl thioester, the overall enzymatic rate enhancement is 10^{13}.

Another measure of the role of CoA is the equilibrium constant for reaction of the covalent enzyme-CoA intermediate with an ordinary thiol compound to form a simple enzyme thioester according to eq. 1-22.

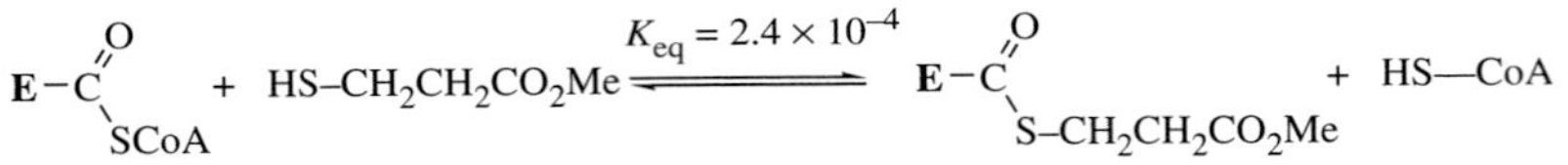

$$\text{E}-\overset{\text{O}}{\underset{\text{SCoA}}{\text{C}}} + \text{HS}-\text{CH}_2\text{CH}_2\text{CO}_2\text{Me} \xrightleftharpoons{K_{eq} = 2.4 \times 10^{-4}} \text{E}-\overset{\text{O}}{\underset{\text{S}-\text{CH}_2\text{CH}_2\text{CO}_2\text{Me}}{\text{C}}} + \text{HS}-\text{CoA} \quad (1\text{-}22)$$

The value of this equilibrium constant shows that the noncovalent contacts between the enzyme and remote parts of CoA stabilize the enzyme-CoA by 5 kcal mol^{-1} at 25°C.

The remote portions of CoA also enhance the kinetic properties of the enzyme-CoA intermediate, which undergoes hydrolysis to the free enzyme and CoA about 1000 times faster than acetyl CoA. Reduction of the thioester group in the enzyme-CoA intermediate by NaBH$_4$ takes place about 1000 times faster than the reduction of acetyl CoA. Acetyl CoA reacts at the same rate in these processes as the enzyme thioester (**E–COS–CH$_2$CH$_2$CO$_2$Me**) on the right side of eq. 1-22.

Figure 1-22 suggests one means in principle by which noncovalent interactions between remote portions of CoA with the CoA transferase could dramatically enhance the reaction rate. This analysis focuses attention on the value of pK_a for the thiol group of CoA. The ratio of the second-order rate constant for the reaction of the free substrate with the enzyme to form the enzyme-CoA and the first-order rate constant for the same process beginning with the noncovalent ES complex is 240 M^{-1}. This number results from the ratio of k_{cat}/K_m to k_{cat}, and this modest number shows that the kinetics of the association of the substrate with the enzyme to form the Michaelis complex does not reveal the basis for the fast rate. This does not mean that binding is unimportant, only that the steady-state kinetic parameters do not explain how binding influences catalysis. The first-order rate constant for transformation of the Michaelis complex into the enzyme-CoA intermediate can be compared directly with the corresponding constant for the formation of succinic anhydride from *p*-nitrothiophenyl succinate, and they differ by only a factor of two, a remarkably

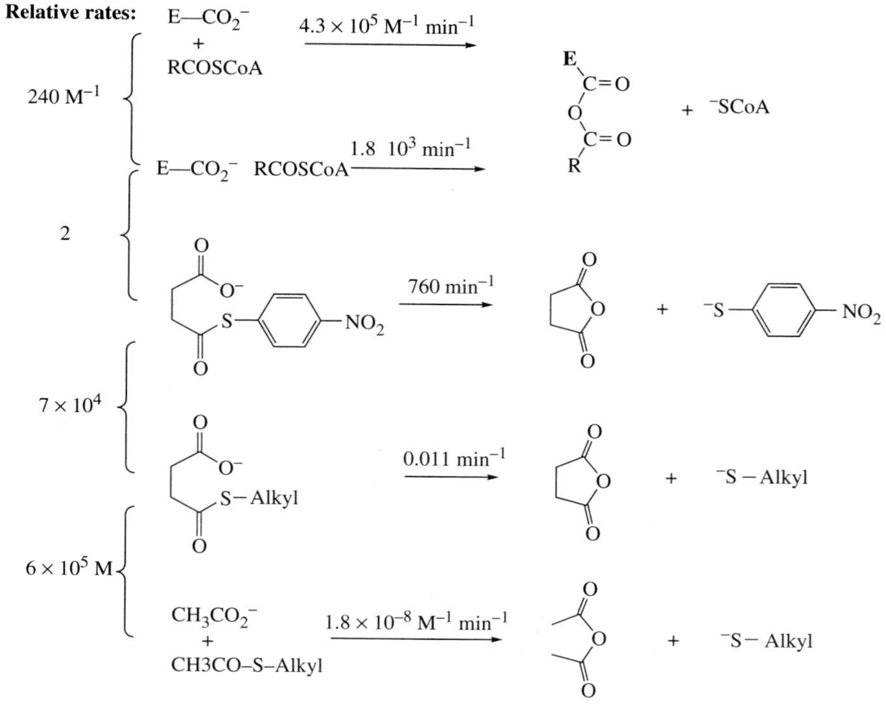

Enzymes and Catalytic Mechanisms   51

Relative rates:

$$\text{E—CO}_2^- + \text{RCOSCoA} \xrightarrow{4.3 \times 10^5 \text{ M}^{-1} \text{ min}^{-1}} \underset{R}{\overset{E}{\underset{O}{\overset{C=O}{|}}}}\overset{C=O}{|} + {}^-\text{SCoA}$$

240 M^{-1}

$$\text{E—CO}_2^- \cdot \text{RCOSCoA} \xrightarrow{1.8 \; 10^3 \text{ min}^{-1}}$$

2

$$\text{succinyl-S-Ar-NO}_2 \xrightarrow{760 \text{ min}^{-1}} \text{succinic anhydride} + {}^-\text{S—}\bigcirc\text{—NO}_2$$

7×10^4

$$\text{succinyl-S-Alkyl} \xrightarrow{0.011 \text{ min}^{-1}} \text{succinic anhydride} + {}^-\text{S—Alkyl}$$

6×10^5 M

$$\text{CH}_3\text{CO}_2^- + \text{CH}_3\text{CO-S-Alkyl} \xrightarrow{1.8 \times 10^{-8} \text{ M}^{-1} \text{ min}^{-1}} \text{acetic anhydride} + {}^-\text{S—Alkyl}$$

$$\frac{4.3 \times 10^5 \text{ M}^{-1} \text{ min}^{-1}}{1.8 \times 10^{-8} \text{ M}^{-1} \text{ min}^{-1}} = 2.4 \times 10^{13} \quad \textbf{Rate enhancement factor}$$

Fig. 1-22. The kinetic role of coenzyme A (CoA) in the reaction of CoA transferase. At the top is the reaction of β-ketoacyl CoA:succinate CoA transferase with a substrate to form the enzyme-anhydride intermediate. The second-order rate constant shown above the arrow is the value of k_{cat}/K_m. The second-order rate constant for the nonenzymatic reaction of acetate with and acetyl thioester is shown at the bottom. The ratio of rate constants is 2.4×10^{13}. The intervening reactions and their associated rate constants suggest that the overall rate enhancement could be accounted for on the basis of two major phenomena: intramolecularity arising from the binding of the substrate to the enzyme, which is an entropic effect, and a hypothetical decrease in the pK_a of CoA when bound to the active site, an enthalpic contribution to catalysis. The pK_a of CoA when bound to the active site is unknown.

Table 1-7. Remote Binding Effects of Coenzyme A in CoA Transferase Reactions

Reaction	k_{cat}/k_m	R.E.	
E–CO$_2^-$ + Succ–CoA	9×10^5	} 3×10^{12}	
E–CO$_2^-$ + Succ–SR	2.9×10^{-7}	} 17	R = —CH$_2$CH$_2$CO$_2$Me
CH$_3$–CO$_2^-$ + Succ–SR	1.7×10^{-8}		
E–COSCoA + Succ	3.8×10^6	} 8×10^8	overall 2.4×10^{14} (binding)
E–COSR + Succ	5×10^{-3}	} 3×10^5	
CH$_3$–CO–SR + Succ	1.7×10^{-8}		7.4 kcal mol^{-1} ↓ in $\Delta G^\ddagger$ through succinate binding

RE, rate enhancement: Succ; succinate.

52   Enzymatic Reaction Mechanisms

small amount. The analogous formation of succinic anhydride from an ordinary thioalkyl succinate is slower by a factor of 7×10^4. This difference further highlights the remarkable fact that *p*-nitrothiophenyl succinate reacts to form succinic anhydride at almost the same rate as the reaction of the Michaelis complex of CoA transferase.

It is as if the *p*-nitrothiophenyl group mimics the enzyme in this process. The difference in rates between the reactions of *p*-nitrothiophenyl succinate and a thioalkyl succinate in fig. 1-22 represents an essential aspect of enzymatic catalysis. This difference can be attributed to the difference in the values of pK_a between a thioalkane ($pK_a = 11$) and *p*-nitrothiophenol ($pK_a = 7$). These pK_a values represent the difference in leaving group reactivities of an alkylthiolate ion and the *p*-nitrophenylthiolate ion. In the microscopic reverse reaction, which is similarly catalyzed by the enzyme, the reacting species would also be the thiolate anions. The overall rate of the enzymatic reaction can be accounted for if the p*K*a of CoA in the complex of covalent acyl CoA transferase and CoA, the central complex in scheme 1-9, is similar to that of *p*-nitrothiophenol. The value of the pK_a for CoA in this complex is not known. It seems unlikely to be determined by studies of the relevant complex because of its kinetic reactivity. However, if a kinetically stable analog of the complex could be prepared that would bind CoA, its pK_a might be measured.

The lower two lines of fig. 1-22 give the ratio of the first-order rate constant of succinic anhydride formation from a thioalkyl succinate to the second-order rate constant for anhydride formation between acetyl CoA and acetic acid. The value of 6×10^5 M represents mainly the entropic advantage in the intramolecular reaction.

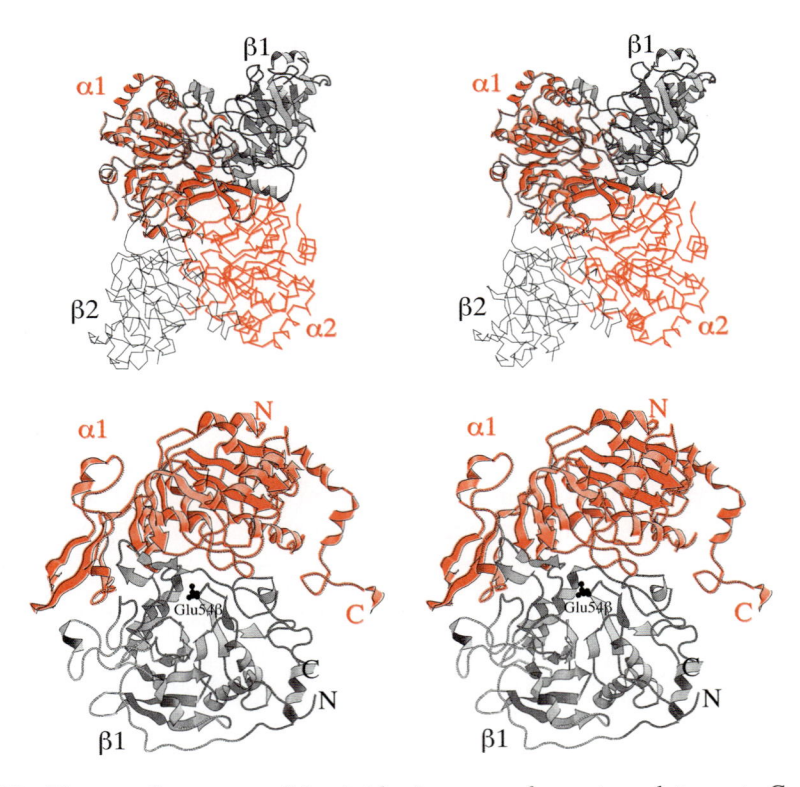

Fig. 1-23. The overall structure of the *Acidaminococcus fermentans* glutaconate CoA transferase (EC 2.8.3.12) $\alpha_2\beta_2$ tetramer is shown in the stereodrawing (top). A stereographic ribbon diagram (bottom) highlights the αβ dimeric portion, showing the active site Glu54β as a ball-and-stick model. The structures are generated from the 2.5-Å model detailed in PDB 1PO1 (Jacob et al., 1997).

A structure for glutaconate CoA transferase in fig. 1-23 shows the active site of the free enzyme and the location of the active site glutamate.

Characterization of Active Sites

Competitive Inhibitors: Analogs of Substrates

A few words about the most common applications of competitive inhibitors can emphasize the widespread role of these compounds in characterizing active sites. In modern enzymology, the most frequent application is co-crystallization with an enzyme to learn about structural contacts at the active site. Co-crystallization with a substrate is not always practical because of its reactivity. In cases of multisubstrate enzymes, co-crystallization with one or the other substrate is often possible and leads to valuable information. In these studies, variation of substrate structure to evaluate specific interactions often leads to competitive inhibitors and their evaluation. Inhibitors constructed as two-substrate analogs can provide valuable information about ternary enzyme-substrate complexes. For all of these reasons, many structures of enzymes in this book show a competitive inhibitor at the active site. The structures focus attention on key interactions, and these contacts can be further investigated by systematically varying the structure of the inhibitor and evaluating the consequences on the inhibition constant and the structure of the enzyme-inhibitor complex. Enzyme-inhibitor contacts are also varied or deleted by site-directed mutagenesis of the enzyme. By evaluating the effects of systematically varying structure in the active site and inhibitor and the effects of variation on the affinity and structure, the nature of enzyme-substrate binding can be deduced.

Group-Selective Chemical Modification

Identification of Essential Amino Acids

In protein chemistry, group selective chemical modification is the reaction of a protein with a chemical reagent that selectively derivatizes a certain type of functional group in amino acid side chains. We use the term *group selective* rather than group specific because no reagent is truly specific for a particular functional group in a protein. Although iodoacetate selectively alkylates sulfhydryl groups of cysteinyl residues in proteins, it also reacts less rapidly with other nucleophilic groups. In the early studies of ribonuclease, iodoacetate alkylated His12 and His119 in the active site to produce carboxymethylhistidinyl residues, and this provided the first evidence for the importance of these amino acids in catalysis. The absence of cysteinyl sulfhydryl groups and the binding affinity of ribonuclease for the anionic iodoacetate simplified this application of group selective chemical modification.

The functional groups in proteins include carboxyl, amino, imidazole, sulfhydryl (thiol), and the hydroxyl groups of serine, threonine, and tyrosine. Group selective modifiers for most of these amino acids are listed in table 1-8. Cysteine, histidine, and lysine can be modified with groups that can later be removed. The ability to remove a modifying group is very useful in certain applications, and this extends the value of chemical modification. Historically, the reaction of a protein with one of the reagents could be monitored by an analytic technique to find out how many of a certain functional group are present in the protein. Today, this type of information comes from translation of cloned genes or from

Table 1-8. Group Selective Chemical Modifying Reagents for Proteins[a]

CYSTEINE

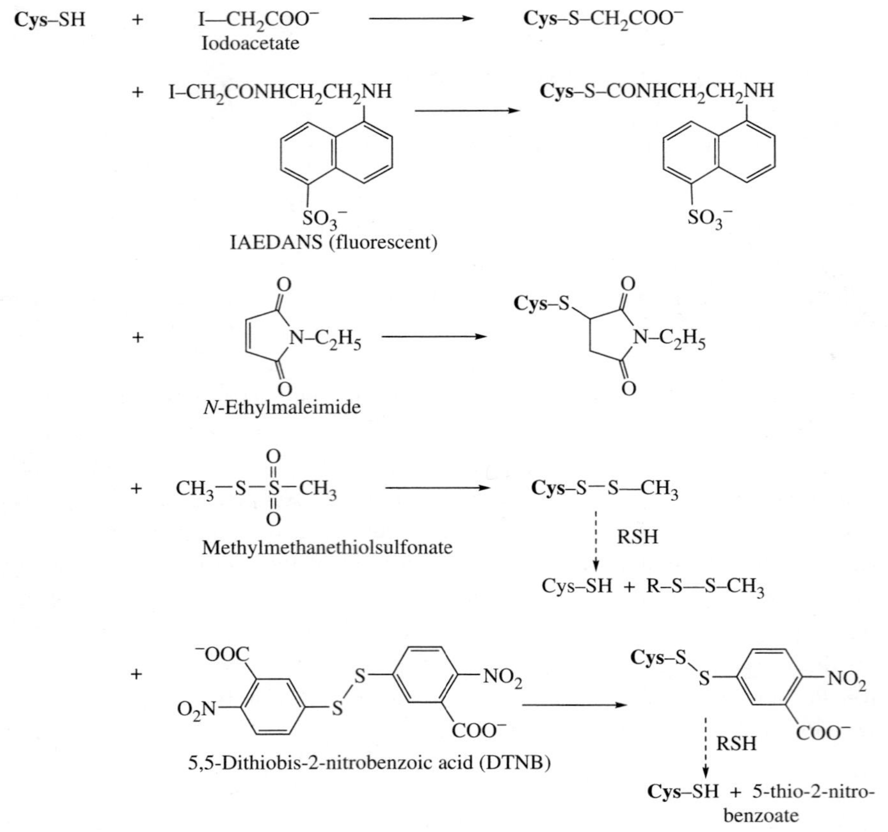

Cys–SH + I—CH$_2$COO$^-$ ⟶ Cys–S–CH$_2$COO$^-$
        Iodoacetate

+ I–CH$_2$CONHCH$_2$CH$_2$NH–(naphthalene-SO$_3^-$) ⟶ Cys–S–CONHCH$_2$CH$_2$NH–(naphthalene-SO$_3^-$)
  IAEDANS (fluorescent)

+ N-Ethylmaleimide (N–C$_2$H$_5$) ⟶ Cys–S–(succinimide)–N–C$_2$H$_5$

+ CH$_3$–S–S(O)$_2$–CH$_3$ ⟶ Cys–S—S—CH$_3$ ⇢ RSH ⟶ Cys–SH + R–S—S–CH$_3$
  Methylmethanethiolsulfonate

+ 5,5-Dithiobis-2-nitrobenzoic acid (DTNB) ⟶ Cys–S–S–(Ar-NO$_2$,COO$^-$) ⇢ RSH ⟶ Cys–SH + 5-thio-2-nitro-benzoate

HISTIDINE

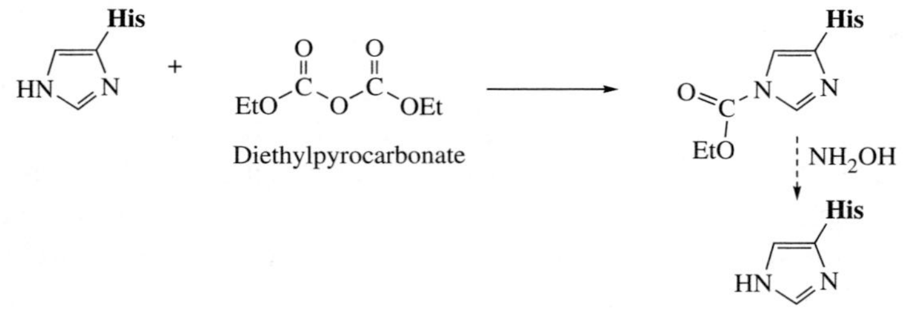

His-imidazole + EtO–C(O)–O–C(O)–OEt ⟶ His-imidazole-N–C(=O)–OEt ⇢ NH$_2$OH ⟶ His
              Diethylpyrocarbonate

LYSINE

Lys–NH$_2$ + R$_1$–C(=O)–R$_2$ —NaBH$_4$→ Lys–NH–CHR$_1$R$_2$

Reductive alkylation; R$_1$ = R$_2$ = H, Reductive methylation

R$_1$ = H, R$_2$ = (pyridoxal: HO, CH$_3$, N, CH$_2$OPO$_3^=$) Pyridoxal phosphate

54

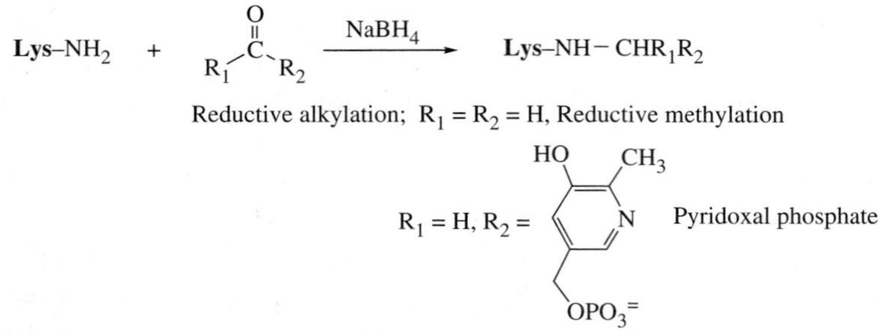

Table 1-8. Group Selective Chemical Modifying Reagents for Proteins[a]—cont'd

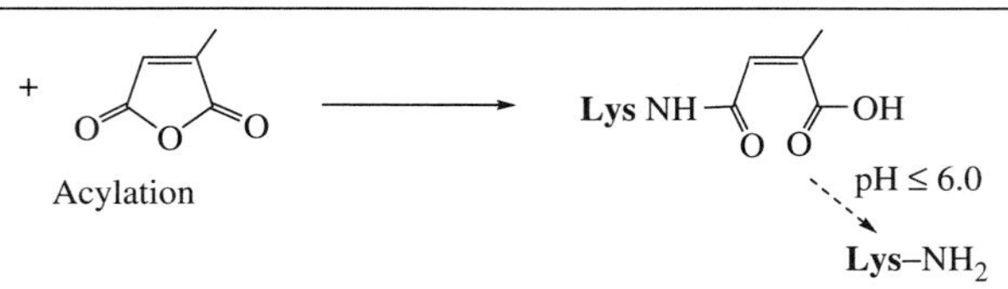

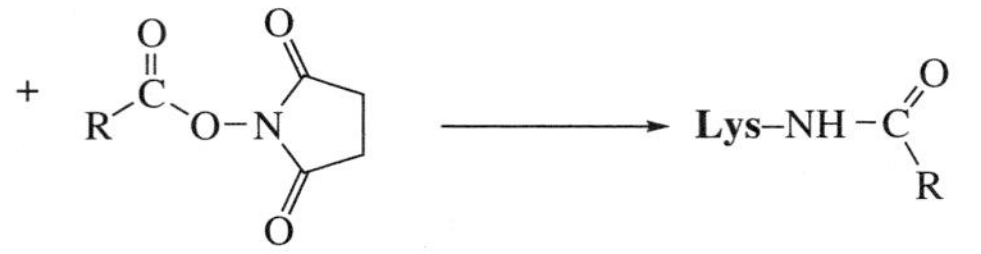

N-Hydroxysuccinimide ester

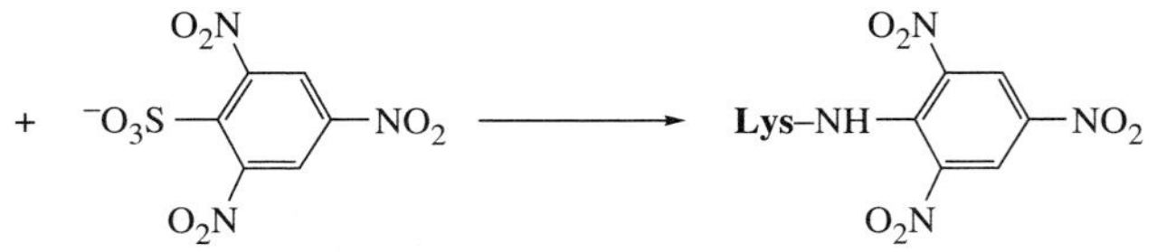

Trinitrobenzenesulfonate

ASPARTATE AND GLUTAMATE

Asp–COOH + R_1-N=C=N R_2 + H_2NCH_2COOMe (or $H_2NCH_2SO_3^-$) ⟶ **Asp**–$CONHCH_2COOMe$

Water soluble carobodiimide

ARGININE

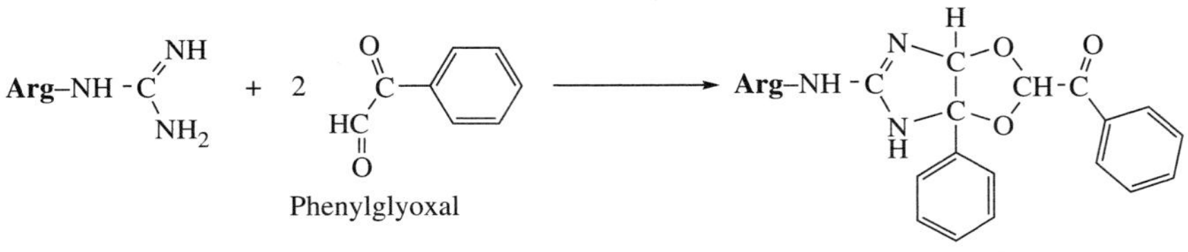

TYROSINE

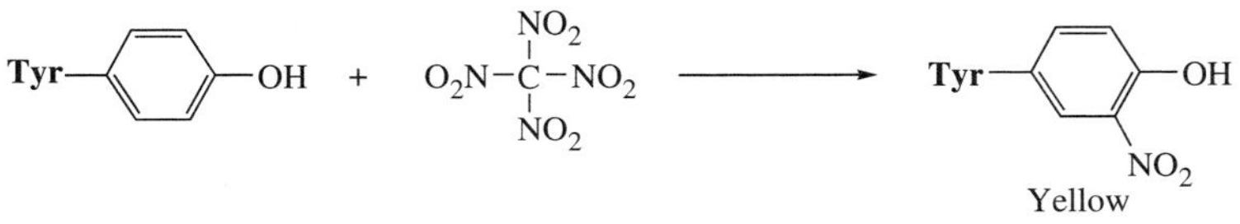

[a] The applications of these and other reagents for the chemical modification of proteins is discussed in the volume by Lunbland (Lundblad, 1995).

quantitative amino acid analysis. However, chemical modification remains important for many modern applications, including quantitative amino acid analysis. The cysteine sulfhydryl groups are routinely protected by alkylation with iodoacetate in preparation for quantitative amino acid analysis.

In enzymology, group selective reagents may be applied to the characterization of active sites. The method has been important historically and is still useful for enzymes that have not been crystallized. It remains a mainstay of studies on membrane bound enzymes and other enzymes that resist crystallization. When used in this way, the number of amino acid side chains of a certain type in an enzyme that are subject to a chemical modification is measured. The experiment can then be repeated in the presence of a competitive inhibitor, which binds to the active site and shields it from the group selective reagent. The difference between the number of modified groups in the absence and presence of the competitive inhibitor indicates the number of such groups in the active site. Because this method often leads to a small difference between large numbers, and errors are significant, very good analytical data must be obtained.

In a variation, there is no need to measure the total number of chemically modified groups. The chemical modification can first be carried out with the inhibitor protecting the active site without measuring the number of groups modified. After removing the competitive inhibitor and chemical reagent by dialysis, ultrafiltration, or gel permeation, the modified enzyme can again be subjected to chemical modification with a radiochemically labeled form of the same chemical reagent. The enzyme can then be subjected to radiochemical analysis to determine the number of radioactive chemical modifiers that have been introduced into the active site. The latter method gives definitive results.

Glu35 and Asp52 in Lysozyme

Chemical modification of lysozyme provides several enduring lessons in enzyme science. Lysozyme catalyzes the hydrolysis of bacterial cell walls by cleavage between N-acetylglucosaminyl residues, and it is the first enzyme structure to be obtained by x-ray crystallography. The structure revealed the presence of Glu35 and Asp52 in the active site cleft. Exhaustive chemical modification of the carboxylic amino acid side chains by the method of water-soluble carbodiimide activation coupled with derivatization by aminomethylsulfonate ($H_2NCH_2SO_3^-$) or glycine methyl ester (H_2NCH_2COOMe) revealed that all were modified except for Glu35 and that lysozyme was inactivated (Hoare and Koshland, 1967; Lin and Koshland, 1969). Scheme 1-10 illustrates the chemical basis for the modification of a protein carboxylic acid group by this method, where glycine methyl ester is used with a water soluble carbodiimide.

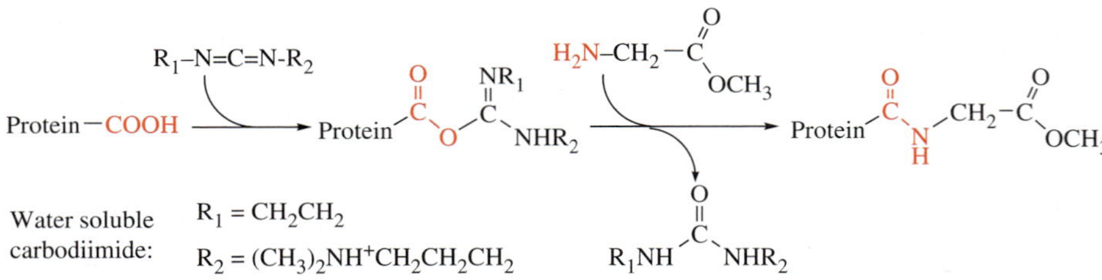

Scheme 1-10

The activity of lysozyme was not protected by a series of nonspecific oligosaccharides or by N-acetylglucosamine, but the α-1,4-trimer of N-acetylglucosamine protected lysozyme from inactivation. The tri-N-acetylaminoglycoside proved to be a slow substrate as well as an inhibitor. A detailed analysis showed that in the presence of the triglycoside all of the carboxyl groups could be chemically modified except for Glu35 and Asp52. The tri-N-acetylaminoglycoside substantially protected enzymatic activity, with 48% retained in the reaction with glycine methyl ester and 56% with aminomethylsulfonate. After removal of the tri-N-acetylaminoglycoside, Asp52 could then be chemically modified, and this destroyed the remaining catalytic activity.

The chemical modification of lysozyme gave important information beyond confirmation of the catalytic importance of Asp52. The structure itself could not prove that Asp52 was essential, and many amino acids found at active sites have not been found to play essential catalytic roles. In a modern study, one might mutate Asp52 by site-directed mutagenesis of the gene and expression of the mutated enzyme and find that activity had been abolished. As it stands, there would be no need to do this. However, the chemical modification experiments revealed more than the essential role of Asp52. Lysozyme with all of its carboxylate groups other than Glu35 and Asp52 chemically derivatized retained about one half of its activity. Modification with aminomethanesulfonic acid led to significantly increased steric bulk at nine carboxylate sites with retention of the overall electrostatics. Modification with glycine methyl ester led to significant increases in bulk at the same sites, but it also completely changed the electrostatics by discharging most of the anionic sites in the enzyme. Still, 48% of the activity remained. The experiments demonstrated the uniquely important role of the active site and the relative unimportance of the overall electrostatic charge of the protein to the structure and function of the active site. This and many other chemical modification experiments showed that most surface functional groups of enzymes are unimportant in catalysis.

Site-Directed Mutagenesis

Site-directed mutagenesis is the only truly specific method for modifying an amino acid residue in a protein. In this technique, the gene specifying the amino acid sequence is altered to change the coding at a specific site. Use of the gene to express the protein leads to a specific change in the amino acid sequence of the protein. Applied to an enzyme, this technique enables the enzymologist to change any amino acid residue in an enzyme to any other amino acid. It is even possible to change a specific amino acid into an unnatural amino acid analog (Ryu and Schultz, 2006). Before the introduction of site-directed mutagenesis, enzymologic research and the characterization of active sites was limited to structure-function analysis of the interactions of enzymes with substrates, substrate analogs, and competitive inhibitors. Site-directed mutagenesis closed the circle on structure-function analysis by making it possible to change the enzyme structure systematically and do structure-function analysis on the enzyme itself.

Complementarity of Site-Directed Mutagenesis and Group-Selective Chemical Modification

In the absence of an enzyme structure, protection from chemical modification by competitive inhibitors complements site-directed mutagenesis in the characterization of active sites. Although site-directed mutagenesis can be regarded as the most specific chemical modification, it cannot by itself identify an active site amino acid residue in the absence of information that connects that residue to the active site. In conjunction with chemical modification, which allows for establishing a connection with the active site through protection by substrates or competitive inhibitors, the two methods can be employed to identify amino acid residues at active sites. This entails the extension of the chemical modification analysis to specific variants of the enzyme, which are generated by site-directed mutagenesis. It also requires the fusion of genetic with chemical information to identify the active site and evaluate the functions of amino acid residues in the mechanism of enzyme action.

A simple case study exemplifies the natural complementation between the two methods. The discovery of galactose mutarotase in the *gal* operon of *E. coli* focused attention on the family of enzymes known as aldose 1-epimerases, which turned out to be related by their amino acid sequences (Bouffard et al., 1994). Two histidine residues (His104 and His175) were found to be conserved. Spectrophotometric and inactivation analysis showed that two histidine residues reacted with diethylpyrocarbonate faster than the others (Beebe and Frey, 1998). Both histidine residues were important for the enzymatic activity, and the substrate protected them from reaction with diethylpyrocarbonate, proving that the two fast-reacting histidine residues were in the active site. Site-directed mutagenesis of His104 to Asn decreased the rapidly modified residues by one, as did site-directed mutagenesis of His175 to Gln, proving that the conserved histidines were the fast-reacting residues in the active site. Specific mutagenesis of the conserved histidine residues coupled with kinetic assays would have proved that they were important for activity, but it would not have linked them to the active site. Their reactivity with diethylpyrocarbonate and protection by galactose made this connection. The enzyme structure subsequently confirmed the location of these histidine residues in the active site, as well as the equally conserved Glu309, and pH rate analysis proved that all three are required for activity and play essential roles in catalysis (Beebe et al., 2003).

Proof of Function

Amino acid residues associated with active sites in x-ray crystallographic structures often play important roles in reaction mechanisms, but not always. Site-directed mutagenesis coupled with functional analysis provides the simplest and most definitive test of the importance and role of an amino acid residue at an active site. Early applications of this technique disproved hypotheses regarding the importance and function of amino acids in well-known enzymes. It often happens that a genetically conserved amino acid residue in an active site really plays no essential role in the reaction mechanism and can be changed to another residue with little or no effect on the activity or mechanism. The essential role of site-directed mutagenesis is to test and evaluate the importance of amino acid residues in active sites. In this application of the method, it is best to mutate candidate residues to several variants to gain the most information. Mutation of an acidic, basic, or nucleophilic amino acid to alanine or glycine, as well as to a sterically conservative residue, is most informative. Histidine can be mutated to alanine or glycine and to asparagine or glutamine; glutamate to alanine, aspartate and glutamine; lysine to alanine, arginine and methionine; cysteine to alanine and serine; arginine to alanine and lysine; and tyrosine to alanine and phenylalanine.

Kinetic Consequences of Mutating a Catalytic Residue

It might be assumed that mutating a catalytic residue in such a way as to delete the functional group would abolish enzymatic activity. However, this is rarely the case, and we have already pointed out that mutation of the active site residues in trypsin lowers the activity by a large factor without totally abolishing it. Nonenzymatic model studies of general acid-base catalysis indicate contributions of 1000- to 10,000-fold enhancements in rate, and these factors are far smaller than typical enzymatic rate enhancements (see fig. 1-1). On that basis, it might be expected a decrease in the values of k_{cat} or k_{cat}/K_m on the same order on mutation of an acid, base, or nucleophilic catalytic group in an enzyme. This is often observed, although even larger factors can result if the mutation has structural consequences or the catalytic mechanism is entirely changed.

Mutation of an important catalytic residue sometimes leads to an unexpectedly small decrease in enzymatic activity. Decreases on the order of 50- or 100-fold seem modest compared to the expected consequences of abolishing a catalytic group. This can happen in a multistep mechanism when the catalytic group in question is not involved in the rate-limiting step. Consider a mechanism consisting of several steps, the third of which is rate limiting and governed by a rate constant that is one-tenth that of the second step. If mutation of an amino acid residue required for the second step decreases the rate constant for that step by a factor of 1000, the second step can become rate limiting. However, the value of k_{cat} for the overall reaction is not decreased to 1/1000 that of the wild-type enzyme; the decrease is more like 1/100 the wild-type value if the rate constant for the third step and all of the other steps of the mechanism are not altered by the mutation. This occurs because of the change in rate-limiting step, not because the catalytic group enhances the rate of the second step only by 100-fold.

Affinity Labeling

Chemical modification can be directed to the active site by incorporating the chemical agent into a molecule that is structurally related to the substrate and binds at the active site. When the structure of the affinity reagent includes most features of the substrate, it will bind very specifically at the active site, and the chemical reactivity of the reagent can be localized to amino acid side chains within that site. This feature of affinity labeling simplifies the chemical modification of groups at the active site, obviating the need for two cycles of modification and the use of a competitive inhibitor to protect the site. However, affinity labels generally require specialized chemical synthesis and are pursued in chemically well-equipped laboratories motivated in this skill.

Kinetics

The kinetics of affinity labeling is related to the steady-state turnover kinetics except for the kinetic order of the rate measurements. In a steady-state turnover study, the initial rate is linear with time because of being zero order with respect to the enzyme; the enzyme is not consumed, and its concentration does not change. In affinity labeling, the enzyme is an ordinary reactant and is consumed in the irreversible chemical modification. The rate of chemical modification by an affinity-labeling agent is first order with respect to the enzyme. The time course of affinity labeling can be followed by monitoring activity or the production of any product, including the modified enzyme itself. All such rates follow first-order kinetics if the system is behaving as it should. An instructive example of affinity labeling is provided by kinetics of the reaction of phenylmethane sulfonyl fluoride (PMSF) with chymotrypsin. The reaction of PMSF with chymotrypsin leads to phenylsulfonylation of Ser195 and produces hydrogen fluoride. The reaction follows pseudo–first-order kinetics and displays saturation behavior with respect to PMSF. A steady-state kinetic analysis shows that the first-order rate constant for affinity labeling follows the same rate law as the Michaelis-Menten rate law for initial rate in a turnover reaction (Fahrney and Gold, 1963).

The inactivation of an enzyme by an affinity label or suicide inactivator (see chap. 5) often proceeds by the mechanism in scheme 1-11, where I is the inactivator, E is the free enzyme, E I is the noncovalent complex between E and I, and E* is the inactivated enzyme.

$$E + I \underset{k_2}{\overset{k_1}{\rightleftharpoons}} E\,I \overset{k_3}{\longrightarrow} E^*$$

$$k_{obs} = \frac{k_3[I]}{[I] + K_m} \qquad K_m = \frac{k_2 + k_3}{k_1}$$

Scheme 1-11

The inactivation follows pseudo–first-order kinetics when $[I] \gg [E]$. The expression for the pseudo–first-order rate constant can be obtained from the rate equation, $d[E^*]/dt = k_3 [E\,I]$; the conservation equation, $[E_0] = [E] + [E\,I] + [E^*]$; and the steady-state approximation, $d[E\,I]/dt = 0$. The expression for k_{obs} shows that inactivation displays saturation kinetics with respect to the concentration of I, and k_{obs} becomes k_3 at saturating I, and K_m is the concentration of I at which k_{obs} is half its maximum value.

This steady-state analysis was first applied to the inactivation of chymotrypsin, trypsin, and acetylcholinesterase by sulfonyl fluorides (Fahrney and Gold, 1963). In many cases, the binding of I to E is an equilibrium process, and K_m is the dissociation constant for I; in scheme 1-11, $k_2 \gg k_3$, and $K_m = k_2/k_1$. This condition corresponds to the kinetic model of Kitz and Wilson (1962) for the inactivation (methylsulfonylation) of acetylcholinesterase by esters of methanesulfonic acid. In variations of the mechanism in scheme 1-11, the rate constant k_3, the maximum value of k_{obs}, is often referred to as k_{inact}.

Chymotrypsin and TPCK

We described the substrate specificity of chymotrypsin in the introduction to this chapter, where we described ATEE as a substrate. A structurally related inactivator of chymotrypsin is *N-p*-toluenesulfonyl-L-phenylalanine chloromethylketone (TPCK). The presence of His57 in the active site was first proved by affinity labeling using TPCK. The structure of TPCK (structure **1-12**) includes many of the characteristics of a substrate for chymotrypsin, and TPCK is an example of an affinity-labeling agent for this enzyme.

The *p*-toluenesulfonyl group of TPCK mimics the leaving peptide group of a substrate, and the chloromethyl group replaces the ethoxy group of ATEE. The chloromethylketone group alkylates any nucleophilic group with which it can interact. TPCK binds to the active site of chymotrypsin in place of a substrate and alkylates His57, inactivating chymotrypsin (Schoellmann and Shaw, 1970; Shaw, 1963). Chemical degradation of the inactive, alkylated chymotrypsin showed that His57 had been alkylated. A competitive inhibitor of chymotrypsin, β-phenylpropionate, protected the enzyme against TPCK. The D-isomer of TPCK did not inactivate chymotrypsin. Performic acid oxidation of TPCK-inactivated chymotrypsin, followed by acid hydrolysis, led to the production of carboxymethylhistidine. The proteolytic degradation of [^{14}C]TPCK-labeled chymotrypsin led to a peptide encompassing His57 and not His40. It was thereby proved that His57 was alkylated by the chloromethylketone group of TPCK. Subsequently, the x-ray crystal structure verified the presence of His57 in the active site (Steitz et al., 1969).

^{13}C-NMR spectroscopic analysis of the complex of chymotrypsin inactivated by TPCK with ^{13}C-enrichment at the carbonyl group of the chloromethylketone suggests that Ser95 has undergone nucleophilic addition to the carbonyl group to form an oxyanionic hemiketal (Finucane and Malthouse, 1992). The x-ray crystal structure of an inactive complex of Gly-Gly-L-phenylalanine chloromethylketone confirms and details the formation of the tetrahedral adduct (MacSweeney et al., 2000). The active site in this complex is shown in fig. 1-24.

The mechanism by which TPCK alkylates His57 was at first thought to follow the course at the top of figure 1-25, in which the chloromethylketone reacted in a conventional alkylation process. However, evidence of hemiketal formation by nucleophilic addition of Ser195

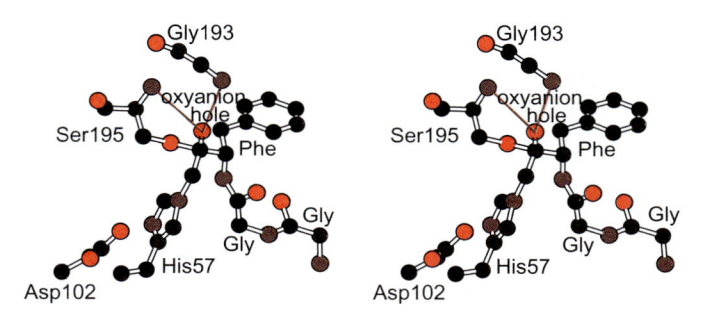

Fig. 1-24. Structure of the active site in chymotrypsin inactivated by a *N*-peptidyl- L-phenylalanine chloromethylketone. A stereodrawing shows a ball-and-stick model of the inactivated complex between GlyGlyPhe-chloromethylketone and the active site of bovine chymotrypsin (3.4.21.1). The structure reveals the cross-linking of the inactivator between His57 and Ser195. The structure was generated from PDB 1DLK from a 2.14-Å resolution model of the complex (MacSweeney et al., 2000).

provided support for alternative mechanisms, in which hemiketal formation with Ser195 preceded alkylation (Kezdy et al., 1964; Powers, 1977). The fact that TPCK did not alkylate anhydrochymotrypsin, in which Ser195 had been chemically dehydrated to dehydroalanine, strongly implied an essential role for Ser195 (Weiner, et al., 1966). Detailed structure-function studies of chloromethylketone inactivation of serine proteases supported the mechanisms involving initial hemiketal formation (Stein and Trainor, 1986). It is still not known whether the hemiketal directly alkylates His57, as in the center of fig. 1-25, or whether the hemiketal oxyanion first displaces chloride to form the epoxyketal, as at the bottom of

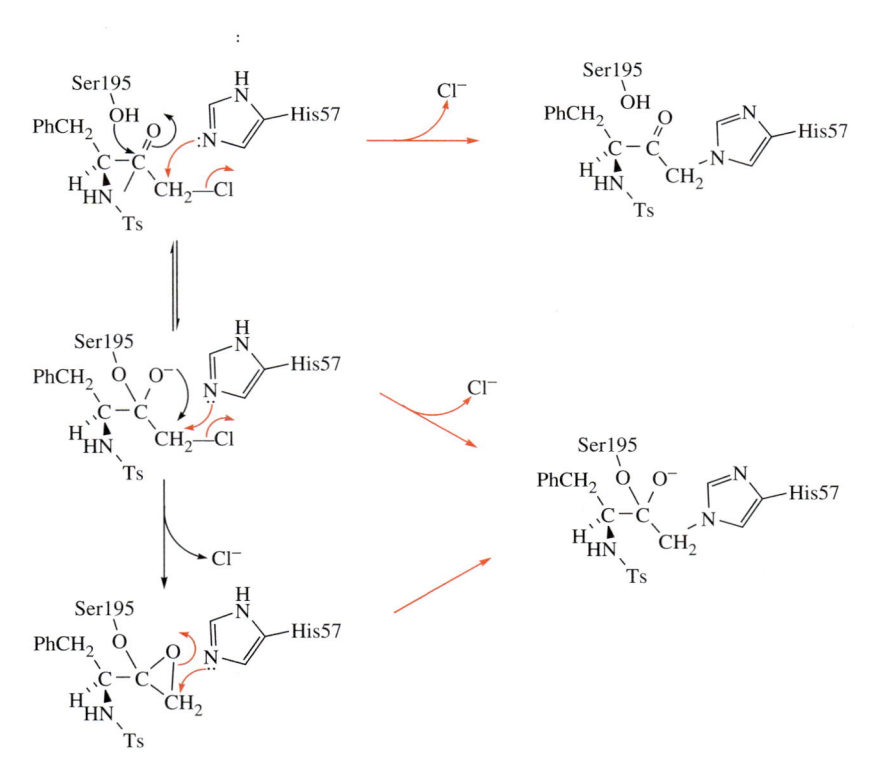

Fig. 1-25. The mechanism of alkylation of His57 by *N-p*-toluenesulfonyl- L-phenylalanine chloromethylketone (TPCK).

fig. 1-25, which alkylates His57. Given that the hemiketal displays a low pK_a (Ficunane and Malthouse, 1992; O'Connell and Malthouse, 1995), the internal displacement of chloride by the hemiketal oxyanion and alkylation of His57 by the epoxyketal seems most attractive. All mechanisms lead to the cross-linked active site hemiketal, with both His57 and Ser195 bonded to the inactivator. Because of the involvement of Ser195 and hemiketal formation in the inactivation mechanism, TPCK can be regarded as a suicide inactivator (see chap. 5), as well as an affinity-labeling agent.

Successes with TPCK in studies of chymotrypsin led to the development of affinity labels for many other enzymes (Baker, 1967; Shaw, 1970; Chowdhry and Westheimer, 1979; Colman, 1983). Inactivators that incorporated the alkylating group —$COCH_2Br$ or the photoalkylating group —$COCH_2N_2$ were favored because they led to carboxymethylamino acids on chemical degradation of the proteins. Authentic carboxymethylamino acids could easily be synthesized for comparison with the degradation products in their identification.

Commercial PMSF, TPCK, and N-p-toluenesulfonyl-L-lysine chloromethylketone (TLCK) are often employed in protein purification to inactivate serine proteases that would otherwise decrease the yield of the purified protein. These inexpensive compounds can improve the yields and quality of enzymes purified from natural sources. Naturally occurring proteolytic inhibitors such as antitrypsin are also often used for this purpose.

Why Are Enzymes Large Molecules?

Sizes of Enzymatic Binding Domains

An active site occupies a fraction of the structure of an enzyme, and one may wonder about the requirement for the balance of the structure. Many reasons for the large sizes of enzymes have been offered, and there may not be a simple answer to the question. In considering the matter, the range of sizes for enzymes comes into consideration. Lysozyme, a very small enzyme composed of a single polypeptide chain with a molecular weight of 14,500, catalyzes the hydrolysis of a glycoside linkage in a single step, with no metabolic regulatory interactions. A good substrate is the β-1,4-hexamer of N-acetylglucosamine, which has a molecular weight of 1320, about one-tenth that of lysozyme. Viewed in this light, lysozyme does not seem to be outsized relative to its substrate.

The largest enzymes include the pyruvate dehydrogenase complex and the ribosome, which consist of several proteins and RNA and have molecular weights of a few million (see chap. 18). These large enzymes catalyze much more complex reactions than lysozyme, and they are also subject to regulation through allosteric effects or covalent modification, or both. If we consider the actions of the *E. coli* pyruvate dehydrogenase (PDH) complex, there are multiple copies of each of three enzymes that catalyze three distinct reactions, each of which follows a multistep mechanism (see chap. 18). The central enzyme must interact with the active sites of the other two enzymes. The activity of the first enzyme to interact with the substrate is regulated, allosterically in the bacterial complexes and by phosphorylation/dephosphorylation in the case of the eukaryotic complexes. If one sums the distinct functions of the *E. coli* complex in terms of elementary reaction steps and regulatory phenomena, it is not difficult to define eleven physicochemical processes that are supported at catalytic rates by the *E. coli* PDH complex. From the molecular weight of 5 million, each function would correspond to 450,000 in molecular weight, about 30 times larger than lysozyme. However, each molecule of this complex consists of 24 sets of active and regulatory sites, so the molecular size for each set would correspond to a molecular weight of 18,750, not so different from the 14,250 for lysozyme. Alternatively, the PDH complex must have 10 binding sites on its three enzymes to carry out its catalytic and regulatory functions. With 24 sets of 10 per molecule, the unit molecular weight would be 20,800. The subunit molecular weight of glycogen phosphorylase is 100,000, and it catalyzes the one-step phosphorolysis of a glycosidic linkage in glycogen. It contains pyridoxal-5′-phosphate (PLP) and is subject to regulation by phosphorylation and dephosphorylation. The sum of binding sites for substrates, PLP, and regulatory enzymes is at least five, so that the molecular weight allotment per binding site is 20,000. It seems that a molecular weight between 15,000 and 20,000 is reasonable for the size of a catalytic or regulatory unit in an enzyme.

Considering that an enzymatic binding site must function in conformational transitions as well as catalysis (often multistep catalysis), the unit size of 15 to 20 kDa seems reasonable. An enzyme typically binds a substrate or a portion of a macromolecular substrate that corresponds to a molecular weight of perhaps 500. The simplest binding site composed from a polypeptide chain is likely to be at least 10 times larger, and a few-fold in additional size would seem reasonable for a site that could undergo controlled conformational changes appropriate to its function.

Catalytic Antibodies

A modern catalytic antibody is the Fab fragment (molecular weight of 50,000) of a monoclonal antibody that has been elicited in response to a hapten molecule designed to resemble the transition state for a specific reaction (Mader and Bartlett, 1997; Schultz et al., 2002). A hapten is a small molecule that elicits the production of antibodies, which bind it specifically with high affinity. To induce antibody production, the hapten must be bonded to a carrier protein and injected into an animal. Monoclonal technology allows the Fab fragment of the antibody to be produced in quantity. According to scheme 1-7, an enzyme can catalyze a reaction by binding the transition state very tightly. To the extent that a hapten structure mimics a transition state, the theory therefore predicts that the corresponding Fab should stabilize the transition state and catalyze the reaction.

Transesterification

In an early effort to produce a monoclonal catalytic antibody, consideration of the tetrahedral intermediate for the transesterification in fig. 1-26 inspired the synthesis of the corresponding chemically stable phosphonate analog for use as the hapten (Wirsching et al., 1991). The tetrahedral phosphonate anion should sterically and electrostatically mimic the metastable tetrahedral intermediate, and an antibody that binds it tightly should, in theory, catalyze the transesterification. The best catalytic antibody catalyzed the reaction at a rate corresponding to an effective molarity of between 10^5 and 10^6 M. Effective molarity refers to the concentration of one substrate required to attain the same rate with the other substrate at 1 M. The rate enhancement for the catalytic antibody is comparable to that observed for intramolecular anhydride formation by succinate relative to intermolecular anhydride formation by acetic acid in eqs. 1-15 and 1-16.

An unexpected complication in the foregoing experiment arose in the mechanistic analysis of the catalytic mechanism. The transesterification turned out to take place by an acylation/deacylation mechanism, in which the acyl donor substrate acylated a histidine residue of the antibody in the first step to form a covalent acyl-antibody intermediate, which then acylated the acceptor substrate to form the transacylation product. Although this was not the expected mechanism, the reaction likely proceeds through two tetrahedral addition intermediates, one in each of the acylation and deacylation processes.

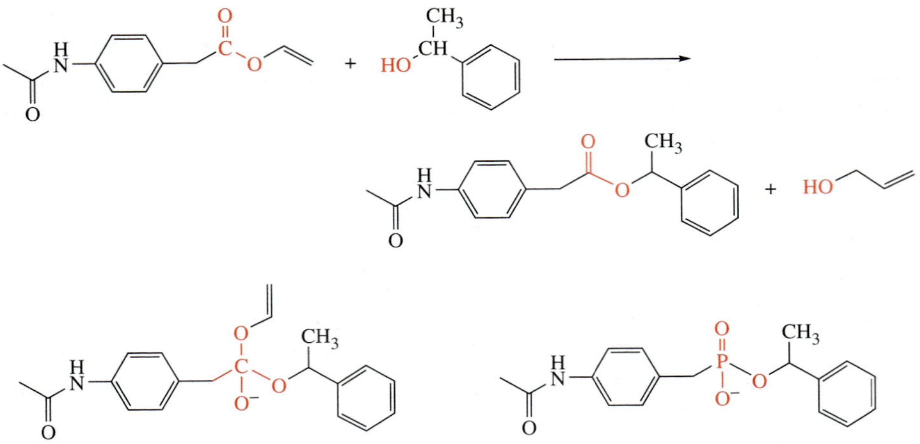

Tetrahedral intermediate   **Phosphonate hapten**

Fig. 1-26. Design for the generation of a catalytic antibody. Reaction of an ester with a carboxylic acid to exchange the acyl group, transesterification, is a slow chemical reaction (top). The structure of the transient tetrahedral addition intermediate is shown at the lower left. A mimic of the metastable, transition-state–like intermediate is the stable phosphonate shown at the lower right. A catalytic antibody generated using this phosphonate as the hapten catalyzes the transesterification reaction by an acylation/deacylation mechanism. A histidine residue in the binding site of the catalytic antibody is the nucleophilic catalyst (Wirsching et al., 1991).

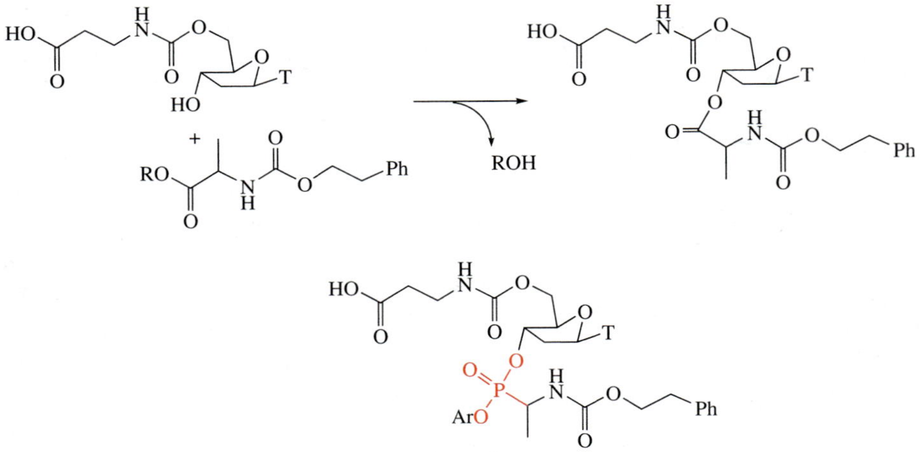

Phosphonate hapten

Fig. 1-27. A transesterifying catalytic antibody was generated by use of the phosphonate hapten (bottom). The design was inspired by the same logic as in fig. 1-26, but the catalytic antibody functioned by a different mechanism, direct transesterification rather than nucleophilic catalysis involving acylation and deacylation of the antibody (Jacobsen et al., 1992).

An analogous study of transesterification avoided the complication of an acylation/deacylation mechanism. The phosphonate hapten in fig. 1-27 was synthesized to elicit a catalytic antibody to catalyze the corresponding reaction. The best catalytic antibody enhanced the reaction as expected at a rate corresponding to an effective molarity of 10^4 and 10^5 M (Jacobsen et al., 1992).

Scope and Limitations

More than a hundred catalytic antibodies have been described that catalyze a variety of reactions. The most striking examples are those that catalyze aldol and Diels-Alder reactions. The aldolase antibodies function by a mechanism similar to that of type I aldolases, with a lysyl residue at the binding site that forms an iminium intermediate analogous to that of aldolase in fig.1-12 (Barbas et al., 1997; Sinha et al., 1998; Wagner et al., 1995). Unlike aldolase, the aldolase antibody displayed broad substrate specificity. Catalytic antibodies that facilitate Diels-Alder reactions are of interest because of the absence of an enzyme that catalyzes such a reaction (Schultz et al., 2002).

Limitations of catalytic antibodies include their limited activities relative to enzymes. In a few cases of chemically fast reactions, in which the enzymatic rate enhancements are modest, the activities of catalytic antibodies approach those of corresponding enzymes. However, in general the activities of catalytic antibodies are many orders of magnitude lower than of enzymes. This may be because of the difficulties associated with truly mimicking a transition state with a stable hapten. The designs of catalytic antibodies do not include a means of programming appropriate conformational changes to accommodate multistep mechanisms and substrate recognition. A major limitation is the absence of practical means for programming the binding and functions of coenzymes and cofactors. Most enzymatic reactions require the actions of the coenzymes and cofactors discussed in chapters 3 and 4.

References

Anderson CM, RE Stenkamp, and TA Steitz (1978) *J Mol Biol* **123**, 15.
Ashley GW and PA Bartlett (1984) *J Biol Chem* **259**, 13621.
Aymes TL and JP Richard (1996) *J Am Chem Soc* **118**, 3129.
Aymes TL and JP Richard (1992) *J Am Chem Soc* **114**, 10297.
Baker BR (1967) *Design of Active-Site Directed Irreversible Enzyme Inhibitors*. John Wiley & Sons: New York.
Barbas CF III, A Heine, G Zhong, T Hoffman, S Gamatikova, R Bjornestedt, B List, J Anderson, EA Stura, IA Wilson, and RA Lerner (1997) *Science* **278**, 2085.
Beebe JA, A Arabshahi, J Clifton, D Ringe, GA Petsko, and PA Frey (2003) *Biochemistry* **42**, 4414.
Beebe JA and PA Frey (1998) *Biochemistry* **37**, 14989.
Bender ML and A Williams (1966) *J Am Chem Soc* **88**, 2502.
Bennett WS Jr and TA Steitz (1978) *Proc Natl Acad Sci U S A* **75**, 4848.
Bennett WS Jr and TA Steitz (1980a) *J Mol Biol* **140**,183.
Bennett WS Jr and TA Steitz.(1980b) *J Mol Biol* **140**, 211.
Bernasconi CF (1987) *Acc Chem Res* **20**, 301.
Bouffard GG, KE Rudd, and SL Adhya (1994) *J Mol Biol* **244**, 269.
Brady K, A Wei, D Ringe, and RH Abeles (1990) *Biochemistry* **29**, 7600.
Brønsted JN and EA Guggenheim (1927) *J Am Chem Soc* **49**, 2554.
Bruice TC and SJ Benkovic (2000) *Biochemistry* **39**, 6267.
Bruice TC and Pandit UK (1960a) *J Am Chem Soc* **82**, 5858.
Bruice TC and Pandit UK (1960b) *Proc Natl Acad Sci U S A* **46**, 402.
Carlow DC, AA Smith, CC Yang, SA Short, and R Wolfenden (1995) *Biochemistry* **34**, 4220.
Cassidy CS, J Lin, and PA Frey (1997) *Biochemistry* **36**, 4576.

Cassidy CS, J Lin, and PA Frey (2000) *Biochem Biophys Res Commun* **273**, 789.
Cassidy CS, J Lin, JB Tobin, and PA Frey (1998) *Bioorganic Chem* **26**, 213.
Choi C, NC Ha, SW Kim, DH Kim, S Park, BH Oh, and KY Choi (2000) *Biochemistry* **39**, 903.
Choi KH, J Shi, CE Hopkins, DR Tolan, and KN Allen (2001) *Biochemistry* **40**, 13868.
Chowdhry V and FH Westheimer (1979) *Annu Rev Biochem* **48**, 293.
Cleland WW (1992) *Biochemistry* **31**, 317.
Cleland WW and M Kreevoy (1994) *Science* **264**, 1887.
Colman RF (1983) *Annu Rev Biochem* **52**, 67.
Corey DR and CS Craik (1992) *J Am Chem Soc* **114**, 1784.
Craik CS, S Roczniak, C Largman, and WJ Rutter (1987) *Science* **237**, 909.
Dalby AR, DR Tolan, and JA Littlechild (2001) *Acta Crystallogr D Biol Crystallogr* **57** (Pt 11), 1526.
Denisov GS, NS Golubev, VA Gindin, HH Limbach, SS Ligay, and SN Smirnov (1994) *J Mol Struct* **322**, 83.
Eyring H (1935) *Chem Rev* **17**, 65.
Fahrney DE and AM Gold (1963) *J Am Chem Soc* **85**, 997.
Fersht AR (1987) *Biochemistry* **26**, 8031.
Field TL, WS Reznikoff, and PA Frey (1989) *Biochemistry* **28**, 2094.
Finucane MD and JP Malthouse (1992) *Biochem J* **286**, 889.
Frey PA, FC Kokesh, and FH Westheimer (1971) *J Am Chem Soc* **93**, 7266.
Frey PA, SA Whitt, and JB Tobin (1994) *Science* **264**, 1927.
Frick L, C Yang, VE Marquez, and R Wolfenden (1989) *Biochemistry* **28**, 9423.
Gerlt JA and PG Gassman (1993a) *J Am Chem Soc* **115**, 11552.
Gerlt JA and PG Gassman (1993b) *Biochemistry* **32**, 11943.
Gilli P, V Bertalasi, V Ferretti, and G Gilli (1994) *J Am Chem Soc* **116**, 909.
Gilli P, V Bertolasi, V Ferretti, and G Gilli (2000) *J Am Chem Soc* **122**, 10405.
Hammond GS (1955) *J Am Chem Soc* **77**, 334.
Hartley BS and DL Kauffman (1966) *Biochem J* **101**, 229,
Heine A, G DeSantis, JG Luz, M Mitchell, CH Wong, and IA Wilson (2002) *Science* **294**, 369.
Hersh LB and WP Jencks (1967a) *J Biol Chem* **242**, 339.
Hersh LB and WP Jencks (1967b) *J Biol Chem* **242**, 3468.
Hess GP (1971) *The Enzymes (3rd Ed)* **3**, 213.
Hess RA, AC Hengge, and WW Cleland (1998) *J Am Chem Soc* **120**, 2703.
Hibbert F and J Emsley (1990) *Adv Phys Org Chem* **26**, 255.
Highbarger LA, JA Gerlt, and GL Kenyon (1996) *Biochemistry* **35**, 41.
Hoare DG and Koshland DE Jr (1967) *J Biol Chem* **242**, 2447.
Horecker BL, O Tsolas, and CY Lai (1972) In Boyer PD (ed): *The Enzymes*, vol 7, 3rd ed. Academic Press: New York, p 213.
Jacob U, M Mack, T Clausen, R Huber, W Buckel, and A Messerschmidt (1997) *Structure (Lond)* **5**, 415.
Jacobsen JR, JR Prudent, L Kochersperger, S Yankovich, and PG Schultz (1992) *Science* **256**, 365.
Jansen EF and AK Balls (1952) *J Biol Chem* **194**, 721.
Jeffrey GA (1997) *An Introduction to Hydrogen Bonding*. Oxford: New York.
Jencks WP (1969) *Catalysis in Chemistry and Enzymology*. McGraw-Hill: New York, p 490.
Jencks WP (1975) *Adv Enzymol Rel Areas Mol Biol* **43**, 219.
Jencks WP (1987) *Catalysis in Chemistry and Enzymology*. Dover Publications: Mineola, NY, pp 615-806.
Jencks WP and J Carriuolo (1960) *J Am Chem Soc* **82**, 1778.
Karplus M (2002) *Acc Chem Res* **35**, 321.
Katona G, RC Wilmouth, PA Wright, GI Berglund, J Hajdu, R Neutz, and CJ Schofield (2002) *J Biol Chem* **277**, 21962.
Kézdy FJ, A Thomson, and ML Bender (1967) *J Am Chem Soc* **89**, 1004.
Kitz R and IB Wilson (1962) *J Biol Chem* **237**, 3245.
Klapper MH (1971) *Biochim Biophys Acta* **229**, 557.
Kobes RD, RT Simpson, RL Vallee, and WJ Rutter (1969) *Biochemistry* **8**, 585.
Kokesh FC and FH Westheimer (1971) *J Am Chem Soc* **93**, 7270.
Koshland DE Jr and KE Neet (1968) *Annu Rev Biochem* **37**, 359.
Kosicki GW (1968) *Biochemistry* **7**, 4310.

Kosicki GW and FH Westheimer (1968) *Biochemistry* **7**, 4303.
Kreevoy MM and T Liang (1980) *J Am Chem Soc* **102**, 3315.
Kuhn P, M Knapp, SM Soltis, G Ganshaw, M Thoene, and R Bott (1998) *Biochemistry* **37**, 13446.
Lakowicz JR and G Weber (1973) *Biochemistry* **21**, 4171.
Larsen TM, JE Wedekind, I Rayment, and GH Reed (1996) *Biochemistry* **35**, 4349.
Lin J, CS Cassidy, and PA Frey (1998a) *Biochemistry* **37**, 11940.
Lin J and PA Frey (2000) *J Am Chem Soc* **122**, 11258.
Lin J, WM Westler, WW Cleland, JL Markley, and PA Frey (1998b) *Proc Natl Acad Sci U S A* **95**, 14664.
Lin TY and DE Koshland Jr (1969) *J Biol Chem* **244**, 505.
Lowenstein JM and MN Schatz (1961) *J Biol Chem* **236**, 305.
Lundblad RL (1995) *Techniques in Protein Modification*, CRC Press: Washington DC.
MacSweeney A, G Birrane, MA Walsh, T O'Connell, JP Malthouse, and TM Higgins (2000) *Acta Crystallogr D Biol Crystallogr* **56**, 280.
Mader M and PA Bartlett (1997) *Chem Rev* **97**, 1281.
Markley JL (1978) *Biochemistry* **17**, 4646.
Markley JL and WM Westler (1996) *Biochemistry* **35**, 11092.
Marlier JF (1993) *J Am Chem Soc* **115**, 5953.
Marlier JF, NC Dopke, KR Johnstone, and TJ Wirdig (1999) *J Am Chem Soc* **121**, 4356.
Massiah MA, C Viragh, PM Reddy, IM Kovach, J Johnson, TL Rosenberry, and AS Mildvan (2001) *Biochemistry* **40**, 5681.
Mata-Segreda JF (2002) *J Am Chem Soc* **124**, 2259.
Moore SA and WP Jencks (1982) *J Biol Chem* **257**, 10893.
Neidhart D, Y Wei, CS Cassidy, J Lin, WW Cleland, and PA Frey (2001) *Biochemistry* **40**, 2439.
Ning J, DL Purich, and HJ Fromm (1969) *J Biol Chem* **244**, 3840.
O'Connell TP and JP Malthouse (1995) *Biochem J* **307**, 353.
Page MI and WP Jencks (1971) *Proc Natl Acad Sci U S A* **68**, 1678.
Pauling L (1948) *Am Sci* **36**, 50.
Powers JC (1977) *Chem Biochem Amino Acids Pept Proteins* **4**, 65.
Poyner RR and GH Reed (1992) *Biochemistry* **31**, 7166.
Reed GH, RR Poyner, RM Larsen, JE Wedekind, and I Rayment (1996) *Curr Opin Struct Biol* **6**, 736.
Regenstein J and WP Jencks (1970) *Handbook of Biochemistry and Molecular Biology*, 2nd ed. Chemical Rubber Co: Cleveland, p J187.
Richard JP, G Williams, AC O'Donoghue, and TL Aymes (2002) *J Am Chem Soc* **124**, 2957.
Richards FM (1974) *J Mol Biol* **82**, 1.
Rios A, TL Aymes, and JP Richard (2000) *J Am Chem Soc* **122**, 9373.
Rios A, J Crugeiras, TL Aymes, and JP Richard (2001) *J Am Chem Soc* **123**, 7949.
Robillard G and RG Shulman (1972) *J Mol Biol* **71**, 507.
Ryu Y and PG Schultz (2006) *Nat Methods* **3**, 263.
Schoellmann G and E Shaw (1963) *Biochemistry* **2**, 252.
Schramm VL (1998) *Annu Rev Biochem* **67**, 693.
Schultz PG, J Yin, and RA Lerner (2002) *Angew Chem Int Ed* **41**, 4427.
Schurmann TS, M Sprenger, and G Schneider (2002) *J Mol Biol* **319**, 161.
Shain SA and JF Kirsch (1968) *J Am Chem Soc* **90**, 5848.
Shaw E (1970) In Boyer PD (ed): *The Enzymes*, vol 1, 3rd ed. Academic Press: New York, p 91.
Shoham M and TA Steitz (1982) *Biochim Biophys Acta* **705**, 380.
Sigler PB, DM Blow, BW Matthews, and R Henderson (1968) *J Mol Biol* **35**, 143.
Sinha SC, CF Barbas III, and RA Lerner (1998) *Proc Natl Acad Sci U S A* **95**, 14603.
Smith RM and DE Hansen (1998) *J Am Chem Soc* **120**, 8910.
Snider MJ, S Gaunitz, C Ridgeway, SA Short, and R Wolfenden (2000) *Biochemistry* **39**, 9746.
Snider MJ and R Wolfenden (2001) *Biochemistry* **40**, 11364.
Solomon F and WP Jencks (1969) *J Biol Chem* **244**, 1079.
Stein R and DA Trainor (1986) *Biochemistry* **25**, 5414.
Steitz TA, R Henderson, and DM Blow (1969) *J Mol Biol* **46**, 337.
Swain CG and JF Brown Jr (1952) *J Am Chem Soc* **74**, 2538.

Thoden JB, FJ Ruzicka, PA Frey, I Rayment, and HM Holden.(1997) *Biochemistry* **36**, 1212.
Toney MD and JF Kirsch (1989) *Science* **243**, 1485.
Toney MD and JF Kirsch (1992) *Protein Sci* **1**,107.
Viragh C, TK Harris, PM Reddy, MA Massiah, AS Mildvan, and IM Kovach (2000) *Biochemistry* **39**, 16200.
Wagner J, RA Lerner, and CF Barbas III (1995) *Science* **270**, 1997.
Wells TN and AR Fersht (1986) *Biochemistry* **25**, 1881.
Westler WM, PA Frey, J Lin, DE Wemmer, H Morimoto, PG Williams, and JL Markley (2002) *J Am Chem Soc* **124**, 4196.
White H and WP Jencks (1976) *J Biol Chem* **251**, 1708.
Wirsching P, JA Ashley, SJ Benkovic, KD Janda, and RA Lerner (1991) *Science* **252**, 680.
Wolfenden R (1972) *Acc Chem Res* **5**, 10.
Xiang S, SA Short, R Wolfenden, and CW Carter (1995) *Biochemistry* **34**, 4516.
Xiang S, SA Short, R Wolfenden, and CW Carter Jr (1997) *Biochemistry* **36**, 4768.
Zhao Q, C Abeygunawardana, P Talalay, and AS Mildvan (1996) *Proc Natl Acad Sci U S A* **93**, 8220.

2
Kinetics of Enzymatic Reactions

At some point in characterizing an enzymatic reaction mechanism, kinetic information is required. This may range from the evaluation of the substrate specificity through comparison of Michaelis-Menten kinetic parameters K_m and V_m for various substrates to the elucidation of the complete kinetic mechanism and evaluation of rate constants for all the steps. In this chapter, we outline the theory and methods of enzyme kinetics and show for a few simple cases the mechanistic information that can be derived.

Steady-State Kinetics

The steady-state kinetic analysis of enzymatic reactions nearly always entails the measurement of initial rates as a function of varying concentrations of a substrate at a fixed enzyme concentration. An initial rate best represents enzyme activity because it is the rate at time zero, before any of the many factors that can decrease enzyme activity come into play. These factors include inhibition by products, changes in pH, denaturation of the enzyme and so forth. In chemical kinetics, a large fraction of the time course for the reaction is usually measured to obtain a large number of data points to determine the kinetic order of the reaction. No problems with denaturation and product inhibition complicate such measurements. In contrast, the progress curve for an enzymatic reaction is generally sensitive to the accumulation of products, which are inhibitory and have to be taken into account. Moreover, the activity of an enzyme sometimes changes because of instability or environmental factors. However, accurate and reproducible initial rates can generally be obtained.

It is possible to follow the full course of an enzymatic reaction by measuring the progress curve, as in conventional chemical kinetics. By fitting the curves to the integrated rate equations the steady-state kinetic parameters for an enzyme can be obtained from a single progress curve (Duggleby, 1995). This method has a number of advantages in principle. In practice, however, the many complications mentioned earlier, especially enzyme stability under reaction condition, have led kineticists to favor the measurement of initial rates at varying substrate concentrations in steady-state kinetic analysis.

The simplifications of initial rate measurements are counterbalanced by the complexities of enzymatic reactions, which may involve one, two, three, or more substrates and comparable numbers of products. It is necessary to simplify the rate measurements by all available means, beginning with the use of initial rates. The steady-state analysis of enzyme kinetics is organized around the number of substrate, cofactor, and product molecules involved, as well as the complete balanced equation for the overall reaction.

One-Substrate Reactions

The enzymes that catalyze reactions of a single substrate are isomerases, lyases, and hydrolases. The isomerases include mutases, racemases, epimerases, and aldose-ketose isomerases and are true one-substrate and one-product reactions. Lyases include decarboxylases, dehydratases, and deaminases; they use a single substrate but produce two or more products. The decarboxylases also incorporate a proton from the medium into the product and may be thought of as two-substrate enzymes, for which the proton is required in eq. 2-1.

$$H^+ + R\text{—}COO^- \rightarrow R\text{—}H + CO_2 \quad (2\text{-}1)$$

However, at a given pH, the concentration of H^+ is constant, it is not regarded as a substrate, and in a kinetic study, the decarboxylation behaves as a one-substrate reaction.

In studies of hydrolases such as proteases, phosphatases, glycosidases, nucleases, and other enzymes in which water is a reactant, its concentration or activity is never varied because such changes are likely to perturb the structure and other properties of the enzyme. The activity of water is taken as unity and is always constant in steady-state kinetic experiments. Hydrolases can be treated kinetically as one-substrate enzymes.

In studies of initial rate as a function of the concentration of a substrate at a given enzyme concentration, a plot of initial rate v against [S] displays saturation kinetics, as illustrated in fig. 2-1. A plot of the reciprocal of initial rate, $1/v$, against the reciprocal of the substrate concentration $1/[S]$ generally gives a straight line (see fig. 2-1, inset) and is the familiar Lineweaver-Burk plot.

The rate law or rate equation corresponding to the data in fig. 2-1 is given by eq. 2-2, where v is initial rate, V_m is the maximum velocity at saturating substrate concentration, and K_m is the concentration of substrate at which v is one half of the value of V_m (fig. 2-1):

$$v = \frac{V_m[s]}{K_m + [s]} \quad (2\text{-}2)$$

The values of v and V_m refer to an experiment with constant enzyme concentration.

Michaelis and Menten first explained saturation kinetics by postulating the reversible formation of an enzyme-substrate complex ES and defining the dissociation constant for this process. A definition of $K_S = \{[E][S]\}/[ES]$, in which E is the free or unliganded enzyme and ES is the enzyme-substrate complex, explained saturation as the case in which all of the enzyme was in the form of ES. The rate at which ES broke down to the product was the maximum velocity V_m, and K_m in eq. 2-2 was the dissociation constant K_S. This defined the simplest mechanism, the reversible equilibrium binding of S to E to form ES, followed by its breakdown to the product.

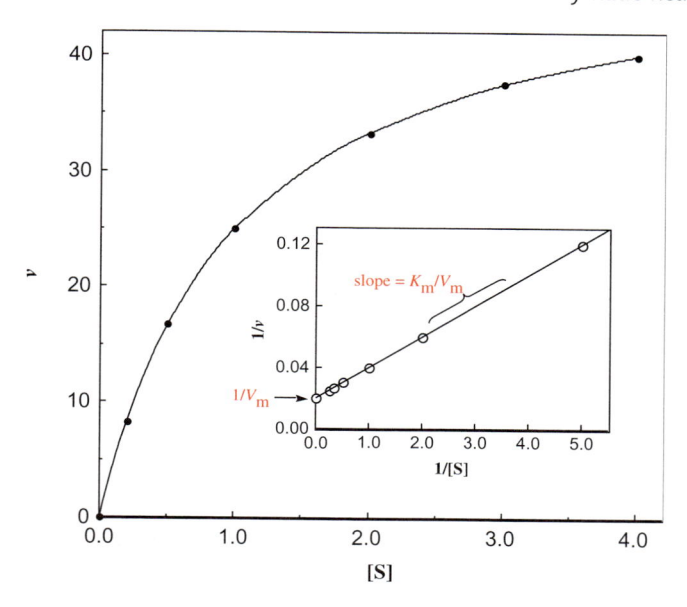

Fig. 2-1. Saturation kinetics in an enzymatic reaction. Typical data for the initial rate v are plotted against substrate concentration [S] for a constant amount of enzyme, demonstrating the saturation effect. In this case, the highest concentration corresponds to about 80% of saturation. The inset shows a double-reciprocal plot, the Lineweaver-Burk plot, of the same data. A value of V_m can be obtained from the intercept and a value of K_m/V_m from the slope. These can be starting values, but the kinetic parameters and the associated standard errors are normally obtained by fitting the data to eq. 2-2.

The most conventional plot of initial rate data is that in the inset of fig. 2-1, the Lineweaver-Burk plot based on eq. 2-3, the reciprocal of eq. 2-2.

$$\frac{1}{v} = \frac{1}{V_m} + \frac{K_m}{V_m} \cdot \frac{1}{[s]} \qquad (2\text{-}3)$$

A plot of $1/v$ against $1/[S]$ allows initial estimates of kinetic parameters to be evaluated as the intercept ($1/V_m$) and the slope (K_m/V_m). In the past, when kinetic parameters were evaluated graphically, many investigators plotted kinetic data as v against $v/[S]$, which is the Eadie-Hofstee plot. This plot is based on eq. 2-4, which can be obtained by algebraic manipulation of eq. 3.

$$v = V_m - \frac{v}{[s]} \cdot K_m \qquad (2\text{-}4)$$

An Eadie-Hofstee plot gave a straight line with a negative slope, in which the ordinate intercept was V_m and the slope was $-K_m$. These terms were more reliable graphical values than those from Lineweaver-Burk plots. Today, parameters are evaluated by fitting data to rate equations such as eq. 2-2 using regression analysis (Cleland, 1979). Because double-reciprocal plots are still useful for deducing kinetic mechanisms in multisubstrate reactions, we employ them in this textbook. However, kinetic parameters should not be determined graphically from such plots, except as preliminary estimates.

Briggs and Haldane introduced a generalization of the kinetic mechanism that accounted for fig. 2-1 and eq. 2-2. It was the steady-state approximation applied to substrate binding and catalysis. The simplest case is shown in scheme 2-1 for a one-substrate and one-product reaction. The Michaelis complex in scheme 2-1 refers to all intermediate complexes, including the enzyme-product complex, and it is sometimes represented as (ES EP).

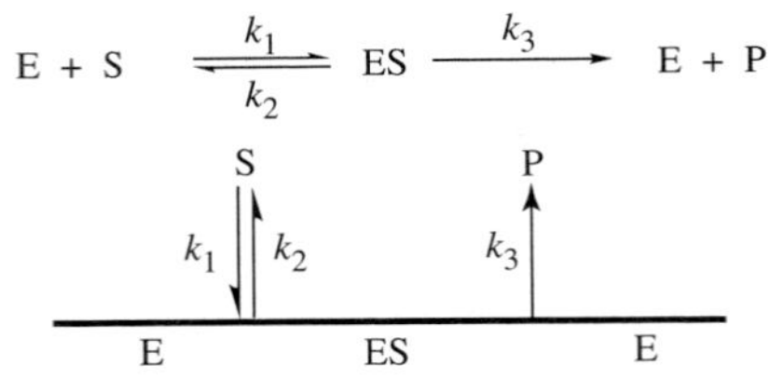

Scheme 2-1

$$E + S \underset{k_2}{\overset{k_1}{\rightleftharpoons}} ES \overset{k_3}{\longrightarrow} E + P$$

A shorthand notation is also shown for illustrative purposes. This notation was introduced by Cleland and is useful for complex mechanisms.

The mechanism in scheme 2-1 is the Briggs-Haldane steady-state elaboration of the basic Michaelis-Menten kinetic mechanism. In the Michaelis-Menten mechanism, the substrate binding and dissociation steps are assumed to be much faster than product formation, so that ES complex formation is at equilibrium. In the Briggs-Haldane formulation, the binding steps are governed by rate constants k_1 and k_2. The constant k_1 has a large value, generally one or two orders of magnitude less than the diffusional rate constant; the value of k_2, however, may be in the range of or lower than that of k_3. This sets up a steady state for the formation and breakdown of the ES complex. The steady-state mechanism is consistent with the rate law of eq. 2-2. Equations 2-5, 2-6, and 2-7 follow directly from the Briggs-Haldane formulation.

Rate equation:

$$v = d[P]/dt = k_3[ES] \qquad (2\text{-}5)$$

Steady-state approximation:

$$d[ES]/dt = k_1[E][S] - (k_2 + k_3)[ES] = 0 \qquad (2\text{-}6)$$

Conservation equation:

$$[E_0] = [E] + [ES] \qquad (2\text{-}7)$$

In these equations, [P] is the concentration of product, and $[E_0]$ is the total enzyme concentration (i.e., sum of the free and liganded forms) at steady state. Another equation, the steady-state approximation of the free enzyme E, can be written, but it is not independent of eqs. 2-6 and 2-7. In general, when there are n forms of an enzyme in a complex kinetic mechanism, there are $n - 1$ independent steady-state equations for the enzyme forms, which are then combined with a conservation equation analogous to eq. 2-7.

The rate eq. 2-2 follows directly from eqs. 2-5 to 2-7, and analogous sets of equations can be written for any more complex mechanism and will always lead to the relevant rate equation pertaining to that mechanism. Matrix algebraic methods are available to solve complex systems. King and Altman (1956) and Cha (1968) have simplified the application of such methods (Segel, 1993). These King-Altman and Cha methods typically are used to deduce rate equations for complex mechanisms. The net rate constant method is a very useful method for unbranched mechanisms (Cleland, 1975).

Equation 2-8 is the rate law for the steady-state mechanism in scheme 2-1 and expresses the kinetic parameters in terms of rate constants.

$$v = \frac{k_3[E_0][S]}{(k_2 + k_3)/k_1 + [S]} \qquad (2\text{-}8)$$

Correlation of eq. 2-8 with eq. 2-2 reveals that $V_m = k_3[E_0]$ and $K_m = (k_2 + k_3)/k_1$. K_m is not the same as the dissociation constant K_S in the Michaelis-Menten mechanism. The more general steady-state mechanism of Briggs and Haldane supersedes that of Michaelis and

Menten. When $k_3 \ll k_2$, the expression for K_m is reduced to k_2/k_1, which is by definition equal to the dissociation constant K_S, and Michaelis-Menten conditions are satisfied. In most cases of enzymatic reactions, this condition is not met. The rate constants for dissociation of the substrate or product, or both, are often smaller than those for chemical steps in the mechanism. In the simple case of scheme 2-1, $K_m = k_3/k_1$ when $k_3 \gg k_2$. In this case, the experimentally measured value of K_m is larger than the dissociation constant K_S. The evaluation of K_S in such cases is possible only by measuring k_1 and k_2 and computing their ratio. The rate constants generally cannot be evaluated from steady-state kinetic parameters. Only when the kinetics satisfies the equilibrium binding condition can this be done; otherwise, transient kinetic data are required together with the steady-state parameters to calculate rate constants.

Many one-substrate enzymes, including lyases and hydrolases, produce two or even three products. In these cases the relationships between the values of K_m and the dissociation constant for the substrate are more complex. The shorthand representation for one example is scheme 2-2, the uni bi case in Cleland's nomenclature.

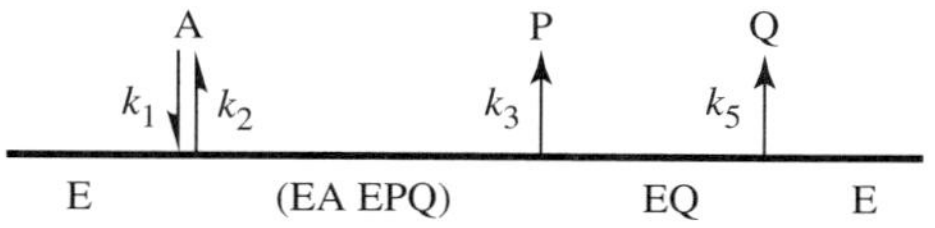

$$v = \frac{k_{cat}[E_0][A]}{[A] + K_{ia}K_a}$$

Scheme 2-2

We designate the Michaelis constant as K_a and the dissociation constant as K_{ia}. In the rate equation in scheme 2-2, the kinetic parameters expressed in terms of rate constants are $k_{cat} = k_3k_5/(k_3 + k_5)$; $K_{ia} = k_2/k_1$; and $K_a = [k_5(k_2 + k_3)]/[k_1(k_3 + k_5)]$. From the relationship between K_a and K_{ia}, we see that when $k_5 < k_2$, $K_a < K_{ia}$; when $k_5 > k_2$, $K_a > K_{ia}$, and when $k_5, k_2 > k_3$, $K_a = K_{ia}$. No simple relationship exists between the values of the Michaelis and dissociation constants for a substrate.

The maximum velocity V_m evaluated in a kinetic experiment is proportional to the enzyme concentration; for scheme 2-1, it is $k_3[E_0]$, and for scheme 2-2, it is $k_3k_5/(k_3 + k_5)[E_0]$. The enzyme concentration independent parameter for the maximum rate is k_{cat}, the turnover number or first-order rate constant for the breakdown of ES to the product. In the case of scheme 2-1, $k_{cat} = k_3$, which can be evaluated experimentally as $V_m/[E_0]$. The units of k_{cat} are reciprocal time, generally s^{-1}. Another useful parameter derived directly from steady-state data is k_{cat}/K_m. The physical significance of this parameter can be understood by considering the mechanism of scheme 2-1 when the substrate concentration is much lower than half-saturation, that is $[S] \ll K_m$. This condition simplifies eq. 2-8 to eq. 2-9.

$$v = \{k_1k_3/(k_2 + k_3)\}[E_0][S] = (k_{cat}/K_m)[E_0][S] \tag{2-9}$$

Under the specified conditions, eq. 2-9 refers to the linear portion of the saturation curve in fig. 2-1 at very low substrate concentrations, well below K_m, and $(k_{cat}/K_m)[E_0]$ is the slope of that portion of the curve.

Physical Significance of Kinetic Parameters

The units of k_{cat}/K_m are those of a second-order rate constant (e.g., $\mu M^{-1} s^{-1}$), and in accord with eq. 2-9, it can be regarded as the second-order rate constant for reaction of the enzyme with the substrate through the first irreversible step in the mechanism. Unlike k_1, which refers strictly to ES complex formation, k_{cat}/K_m refers to the catalytic process and includes the binding of the substrate to form ES as well as the transformation of ES through the first irreversible step, which in 2-1 is product release.

The units of k_{cat} are those of a first-order rate constant, generally s^{-1}. The turnover number is k_{cat}, and this parameter can be thought of as the first-order rate constant for product formation from the ES complex. The parameter k_{cat}/K_m can be applied to interpret kinetic isotope effects and pH effects on enzymatic rates.

Two-Substrate Reactions

Most enzymatic reactions involve two or three substrates and products. The multisubstrate kinetic mechanisms and rate equations are much more complex and diverse. In this section, we describe the more frequently observed kinetic mechanisms for two- and three-substrate reactions. Variations on these mechanisms are possible and sometimes observed. Moreover, the mechanisms in most cases refer to specified reaction conditions and can change when substrate concentrations are high enough ($>>K_m$) to allow secondary binding interactions. Steady-state kinetic analysis is essential for understanding the action of any enzyme, but it should never be undertaken with any particular expectation regarding the outcome.

We adopt the widely used Cleland notation and nomenclature to categorize the reactions and kinetics, and we also make use of Cleland's rules to explain how kinetic mechanisms can be deduced from steady-state data (Cleland, 1970, 1977, 1986, 1990). In the Cleland nomenclature, the prefixes *uni*, *bi*, and *ter* are used to simplify the descriptions of kinetic mechanisms for enzymes with one, two, or three substrates, and the same prefixes are used to describe the number of products in the same reaction. An isomerase is the simplest case of a one-substrate, one-product reaction, and it would be a uni uni reaction. A phosphatase generally has one substrate and two products and would be a uni bi reaction. A simple phosphotransferase involves two substrates and two products and would be a bi bi reaction. The prefixes can be used to encode kinetic mechanisms by breaking them down into product releasing segments. If all the substrates must bind before any products are released, the descriptions are uni uni, uni bi, bi uni, bi bi, ter bi, and ter ter, for one-, two-, and three-substrate mechanisms. When binding steps are ordered, this is specified as ordered bi bi or ordered ter ter, and so forth. It is assumed that substrate and product binding steps are in steady state, after the Briggs-Haldane generalization. However, when kinetic analysis shows them to be at equilibrium, this is specified by describing the mechanism as equilibrium random bi bi or equilibrium random ter ter, and so forth. If any products are released before one or more substrates bind, the mechanism is known as *ping pong* and can be subdivided into product forming segments, as follows. If in a three-substrate mechanism one product is released after the first substrate binds, the mechanism may be encoded as uni uni bi bi ping pong, in which uni uni refers to the binding of the leading substrate and release of the first product. Bi bi refers to the binding of the other two substrates and release of the other products.

The order with which substrates bind and products dissociate are determined by steady-state kinetic analysis; this often gives little information about the chemical mechanisms. However, in the ping pong case, where a product is released from the enzyme before one of the substrates binds, the steady-state kinetics directly implicates a chemical mechanism. In any case, the kinetic mechanism is always important for understanding the action of any enzyme.

Steady-state analysis of a two-substrate reaction proceeds in much the same way as for a single-substrate reaction. Consider a reaction of two substrates A and B to form two products P and Q. Initial rates are measured as $v = dP/dt$ or dQ/dt at varying concentrations of one substrate A, holding everything else constant, including the concentration of B and that of the enzyme. This gives data similar to that obtained in a single-substrate reaction, a saturation curve in a plot of v against [A] and a straight line in a double reciprocal plot of $1/v$ against $1/[A]$. The process is repeated at other fixed concentrations of B, with the same varied concentrations of A, to give additional lines in the double reciprocal plot. Data are obtained to cover the ranges of A and B concentrations that flank estimates of K_m for the two substrates (K_a and K_b). Because the varied concentrations of A are the same in each set of data, they can be plotted as varied [A] at several fixed [B] or as varied [B] at several fixed [A] concentrations. The families of lines generally fall into one of the following three patterns: intersecting and converging at a point to the left of the ordinate; intersecting on the ordinate; or parallel lines. Other patterns are possible—for example, the lines may neither be parallel nor converge to a point—and this suggests complexities in the mechanism, substrate inhibition, or problems with the assays of rate. The most common cases of the three defined patterns are considered in the following sections.

Sequential Binding Mechanisms

The simplest kinetic mechanism for the intersecting pattern is the equilibrium random bi bi mechanism of scheme 2-3.

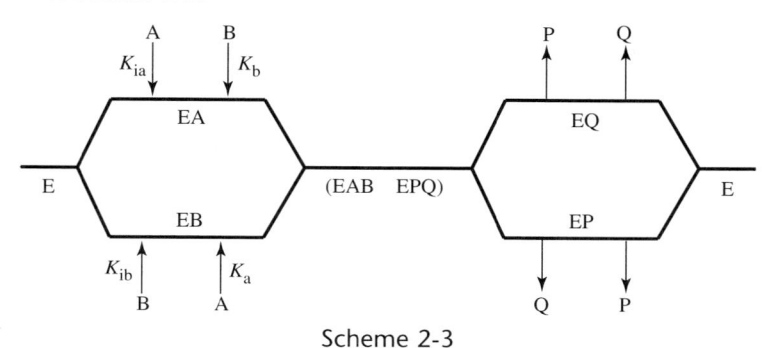

Scheme 2-3

This is a sequential binding mechanism; the substrates A and B bind sequentially and randomly to their respective sites to form a ternary complex, EAB. In the simplest version, conversion of EAB to the ternary product complex EPQ is rate limiting, and the products P and Q dissociate. Each binding step is fast and at equilibrium, the rate-limiting step being the interconversion of the ternary complexes.

Scheme 2-3 describes a mechanism for product formation under conditions in which P and Q are initially absent, and the rate law for this mechanism is eq. 2-10, where K_a and K_b are K_m values for A and B, respectively, and K_{ia} is the dissociation constant for A.

$$v = \frac{k_1[A][B][E_0]}{[A][B] + K_b[B] + K_a[B] + K_{ia}K_b} \tag{2-10}$$

In this mechanism, K_a and K_b are also dissociation constants for A and B, respectively, and K_a may or may not equal K_{ia}. Moreover, $K_{ia}K_b = K_aK_{ib}$, where K_{ib} is the dissociation constant for B.

Equilibrium random binding does not require all binding steps to be equilibrium processes. If the first binding steps are at equilibrium, the second binding steps can be in steady state with the reaction of ternary complexes, and the mechanism will still be equilibrium random. The form of the rate equation will be the same, eq. 2-10, but rate limitation will not reside solely on the interconversion of ternary complexes.

The algebraic form of eq. 2-10 is most commonly observed in rate equations for two-substrate reactions, not because the equilibrium random mechanism is universal, but because it pertains to other kinetic mechanisms as well. In its reciprocal form, eq. 2-10 specifies lines converging at the left of the ordinate in a double reciprocal plot, as shown by eq. 2-11 and in fig. 2-2A.

76   Enzymatic Reaction Mechanisms

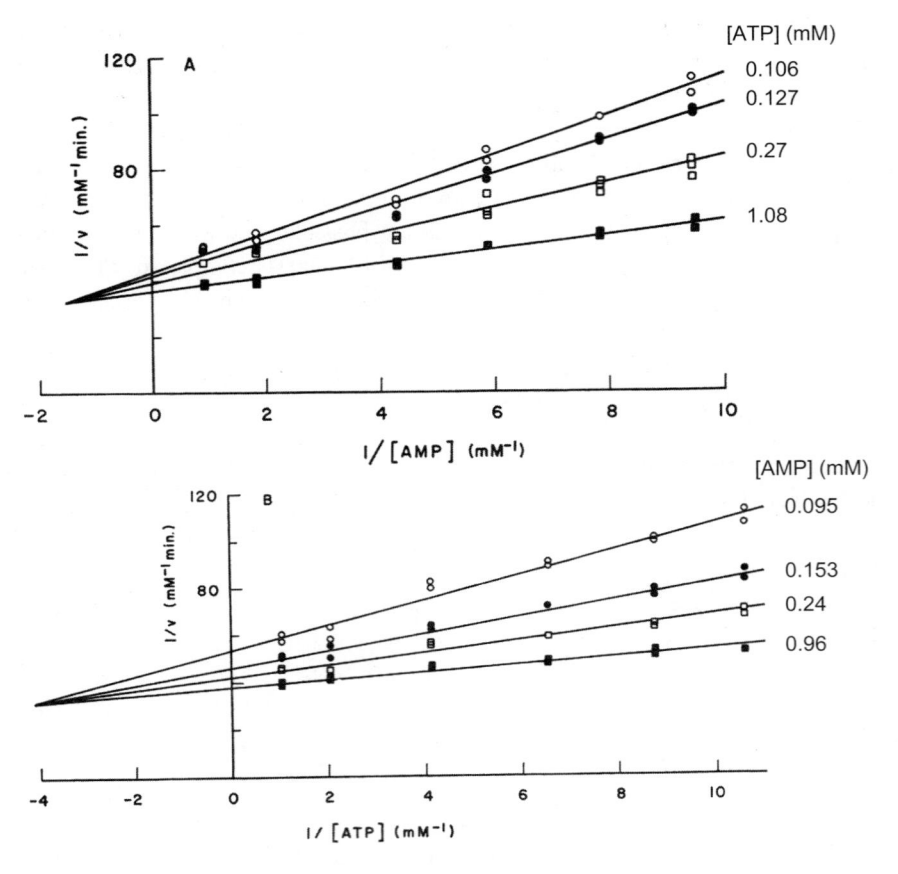

Fig. 2-2. Intersecting pattern in the reaction of adenylate kinase. Adenylate kinase catalyzes phosphoryl group transfer in the reaction of AMP with MgATP to produce ADP and MgADP. The reaction is sequential, and a double-reciprocal plot displays the intersecting pattern. (Adapted from Rhoads and Lowenstein, 1968.)

$$\frac{1}{v} = \frac{1}{k_1[E_0]}\left[1 + \frac{K_b}{[B]}\right] + \frac{1}{k_1[E_0]}\left[K_a + \frac{K_{ia}K_b}{[B]}\right]\frac{1}{[A]} \tag{2-11}$$

Both the intercepts and slopes depend on [B] and vary with different fixed concentrations of B. The reciprocal equation (2-11) can also be factored for plotting data with varying [B] at several fixed [A], with analogous results.

Adenylate Kinase

Adenylate kinase (EC 2.7.4.3) is an important enzyme that catalyzes the reaction of AMP with MgATP to form ADP and MgADP by a random binding mechanism. As in all sequential mechanisms, there are two binding sites, one for AMP (or ADP) and one for MgATP (or MgADP), located adjacent to each other, as shown in fig. 2-3. The phosphoryl group of AMP (red ball and stick) is projected toward the γ-phosphate of AMPPNP (black ball and stick), an unreactive analog of ATP. Site-directed mutagenesis and nuclear magnetic resonance (NMR) experiments decisively implicate the lysine and arginine residues in contact with the substrates in fig. 2-3C as functionally important (Byeon et al.,et al., 1995; Dahnke et al.,et al., 1992; Tsai and Yan, 1991).

The difference in the conformations of the free enzyme and its complex with AMP and AMPPNP exemplifies a substrate-induced conformational change, which is commonly

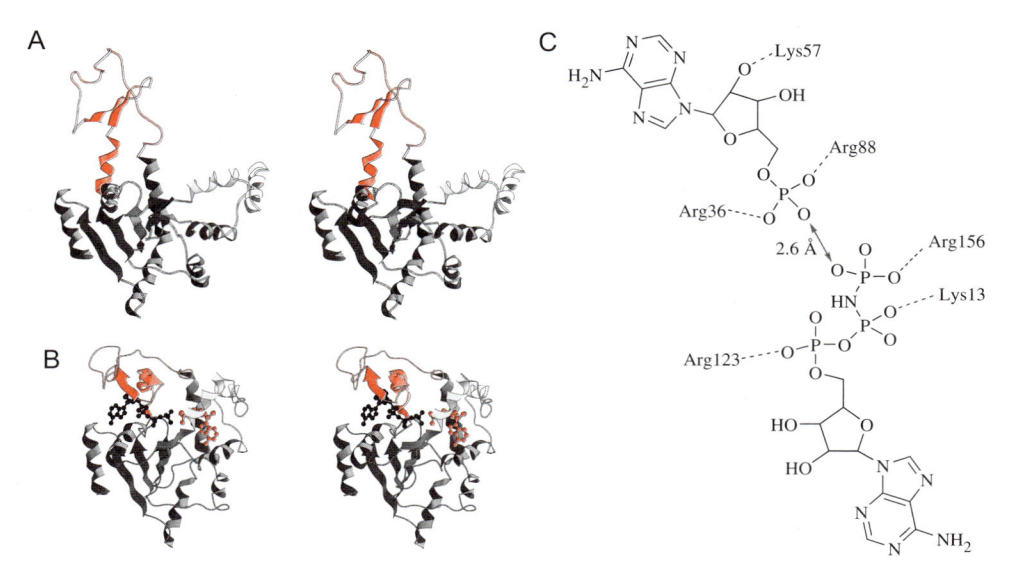

Fig. 2-3. Structures of adenylate kinase and its ternary complex. (A) Structure of *Escherichia coli* adenylate kinase with no substrates or inhibitors bound to the active site (PDB 4AKE; Schulz et al., 1990). (B) Structure with AMP (red ball-and-stick model) and MgAPPNP (black ball-and-stick model), bound at the active site (PDB 1ANK; Althoff et al., 1988). AMPPNP is an analog of ATP in which the oxygen-bridging P_β and P_γ is replaced by NH. AMPPNP is an inhibitor. (C) Results of experiments implicate the lysine and arginine residues in contact with the substrates as functionally important.

observed in enzymatic reactions. As discussed in chapter 1, this phenomenon was originally put forward to explain how phosphotransferases could bind MgATP and not catalyze phosphoryl transfer to hydroxyl groups of the ubiquitously present water. In hexokinase, phosphotransfer occurs only when glucose is in the ternary complex of E.MgATP.Glc (Koshland and Neet, 1968). This has been a durable concept in enzymatic mechanisms.

Neither the sequential kinetics of adenylate kinase nor its structure prove that the phosphoryl group is transferred directly from MgATP to AMP in the ternary complex. This ambiguity remains because the rate equation would have the same form if an additional phosphoryl transfer step intervened between the interconversion of the ternary complexes. For example, the phosphoryl group may be transferred from MgATP to an enzymatic group to form the covalent phosphoenzyme complex E–P*MgADP*AMP, where E–P is a covalent phosphoenzyme. Phosphoryl transfer in a second step to AMP forms the product complex E.MgADP.ADP. Because the hypothetical intervening phosphoenzyme is composed from the same molecules as the other ternary complexes, the rate equation would have the same form as eq. 2-10. The kinetics cannot rule out a phosphoenzyme intermediate in a sequential mechanism. In the case of adenylate kinase the stereochemistry of phosphotransfer rules out a phosphoenzyme intermediate. The enzyme catalyzes the [^{18}O]thiophosphotransfer of eq. 2-12, proving that it proceeds with inversion of configuration at phosphorus (Sheu et al.,et al., 1979).

$$\text{AMP} + {}^{18}\text{O}\overset{S}{\underset{O}{-}}\text{P}-\text{ADP} \longrightarrow \text{AMP}-\overset{S}{\underset{O}{-}}\text{P}\cdots{}^{18}\text{O} + \text{ADP} \qquad (2\text{-}12)$$

Because each phosphotransfer proceeds with inversion of configuration, this is consistent with a single phosphotransfer step. If a thiophosphoenzyme had been formed, two phosphotransfer steps would have been required and would have led to retention.

Ordered Sequential Mechanisms

As in the case of single-substrate reactions, the substrate-binding steps are often not at equilibrium, and substrate and product dissociation are not much faster than the interconversion of ternary complexes. In these cases, substrate binding must be described in terms of rate constants, as in the Briggs-Haldane formulation, not dissociation constants. Nevertheless, the experimental rate equation often has the same form as eq. 2-10 for the equilibrium random mechanism. The ordered bi bi mechanism in scheme 2-4 is an example. The mechanism is shown for the case of initial rate in the forward direction in the absence of products, and eq. 2-10 is the steady-state rate equation, where the kinetic parameters are defined in terms of rate constants as follows: $k_{cat} = k_5 k_7/(k_5 + k_7)$, $K_a = k_5 k_7/[k_1(k_5 + k_7)]$, $K_b = k_7(k_4 + k_5)/[k_3(k_5 + k_7)]$, $K_{ia} = k_2/k_1$.

A   B   P   Q

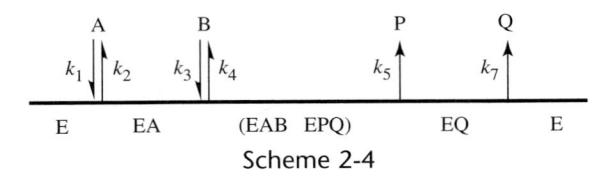

$k_1 \downarrow \uparrow k_2$ $k_3 \downarrow \uparrow k_4$ $k_5 \uparrow$ $k_7 \uparrow$

E   EA   (EAB EPQ)   EQ   E

Scheme 2-4

Although the overall form of the rate equation is the same as that for the equilibrium random mechanism, the kinetic parameters have very different meanings except for K_{ia}, which is the dissociation constant for EA. The two mechanisms can be distinguished by product inhibition patterns and dead-end inhibition experiments described later.

The Theorell-Chance mechanism in scheme 2-5 also displays the intersecting pattern consistent with the rate in eq. 2-10.

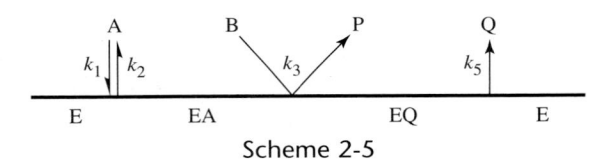

A   B   P   Q

$k_1 \downarrow \uparrow k_2$ k_3 $k_5 \uparrow$

E   EA   EQ   E

Scheme 2-5

In this mechanism, no ternary EAB complex is specified, and the reaction behaves as if EA and B react to form P without B binding to the enzyme. In reality, EAB and EPQ complexes likely exist transiently at very low concentrations that are kinetically not significant. The rate law is eq. 2-10, where $k_{cat} = k_5$, $K_a = k_5/k_1$, $K_b = k_5/k_3$, and $K_{ia} = k_2/k_1$.

The rate equation for the equilibrium ordered bi bi mechanism in scheme 2-6 differs from eq. 2-10 for other sequential mechanisms in that it lacks a term in the denominator. In the mechanism, A and B bind in obligatory order to form the EAB complex, and the first binding step is at equilibrium. Equilibrium binding distinguishes this mechanism from the ordered bi bi version, and they follow different rate equations.

A   B   P + Q

$K_{ia} \downarrow$ $k_3 \downarrow \uparrow k_4$ $k_5 \uparrow$

E   EA   EAB   E

Scheme 2-6

That for the equilibrium ordered mechanism in scheme 2-6 in the absence of products is eq. 2-13, where $k_{cat} = k_5$, $K_b = (k_4 + k_5)/k_3$, and K_{ia} is the dissociation constant for A.

$$v = \frac{k_5 [E_0][A][B]}{[A][B] + K_b[A] + K_{ia} K_b} \qquad (2\text{-}13)$$

The absence of the $K_a[B]$ term results in convergence of the plot of $1/v$ against $1/[B]$ on the ordinate, whereas the plot of $1/[A]$ intersects to the left of the ordinate.

Ping Pong Mechanisms

Parallel lines in double reciprocal plots usually mean that a substrate reacts with the enzyme and produces a product independently of the presence of another substrate. In the two-substrate case, scheme 2-7 is the ping pong bi bi mechanism.

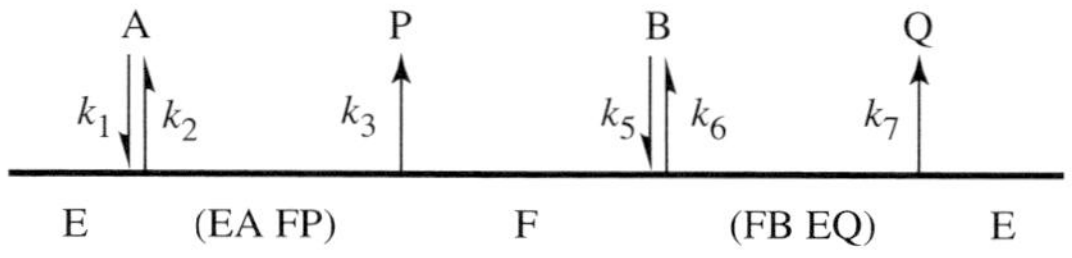

```
       A          P          B          Q
    k₁↓↑k₂      k₃↑       k₅↓↑k₆      k₇↑
  ─────────────────────────────────────────
    E      (EA FP)     F       (FB EQ)    E
```
Scheme 2-7

In this mechanism, the leading substrate A reacts with the enzyme, changing it to a chemically different form designated F in scheme 2-7, and this process leads to the formation and release of the product P. The reaction of E → F may involve an enzymatic group, as in phosphorylation of the enzyme, or a tightly bound coenzymatic group, as in the transformation of pyridoxal-5′-phosphate (PLP) into pyridoxamine-5′-phosphate in a transaminase. All of this occurs in either the presence or absence of the second substrate B and may be regarded as the ping step. In the pong step, the substrate B reacts with the chemically modified enzyme F to form the second product Q and regenerate the original enzyme form E.

Equation 2-14 is the initial rate equation for the ping pong mechanism in scheme 2-7, where $k_{cat} = k_3 k_7/(k_3 + k_7)$, $K_a = k_7(k_2 + k_3)/k_1(k_3 + k_7)$, and $K_b = k_3(k_6 + k_7)/k_5(k_3 + k_7)$.

$$v = \frac{k_{cat}[E_0][A][B]}{[A][B] + K_a[B] + K_b[A]} \qquad (2\text{-}14)$$

The equation differs from eq. 2-10 for most sequential mechanisms by the absence of the constant term $(K_{ia}K_b)$ in the denominator, making the double reciprocal pattern parallel. The absence of any effect of the second substrate on the rate of the first step is the basis for this difference.

The ping pong mechanism often occurs in coenzyme-dependent reactions, in which the coenzyme functions to facilitate the transfer of a group from one substrate to another, as in transaminases. The ping pong mechanism occasionally appears in cases of enzymes that catalyze group transfer reactions such as phosphotransfer and glycosyltransfer. The enzyme forms F differ from E in that the transferred group is covalently bonded to F; it may be bonded either to an amino acid side chain or to a tightly bound coenzyme.

Nucleoside Diphosphate Kinase

Nucleoside diphosphate kinase (EC 2.7.4.6) catalyzes phosphoryl transfer between nucleoside triphosphates and nucleoside diphosphates according to eq. 2-15 by the ping pong bi bi mechanism (Garces and Cleland, 1969; Mourad and Parks, 1966).

$$\text{MgATP} + \text{MgGDP} \rightarrow \text{MgADP} + \text{MgGTP} \qquad (2\text{-}15)$$

The enzyme accepts a variety of nucleotides as substrates, and it is essential for maintaining cellular MgGTP concentrations. The double reciprocal plot in fig. 2-4 illustrates the parallel line pattern for nucleoside diphosphate kinase. MgATP reacts in the first step with the free enzyme E, phosphorylating an active site histidine to form MgADP and the phosphoenzyme intermediate (F in scheme 2-7). In the second step, MgGDP reacts with the phosphoenzyme to form MgGTP and the free enzyme.

80   Enzymatic Reaction Mechanisms

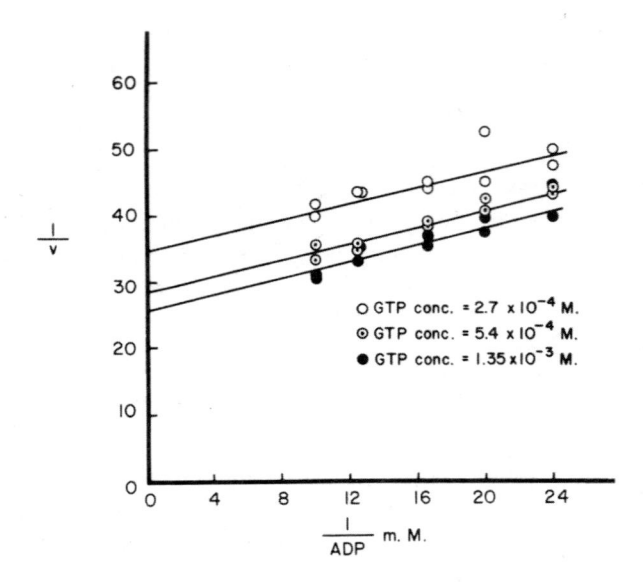

Fig. 2-4. Parallel line pattern for nucleoside diphosphate kinase. Nucleoside diphosphate kinase catalyzes the phosphorylation of nucleoside diphosphates by nucleoside triphosphates. This is a double-reciprocal plot of initial rate in the reaction of MgGDP as the variable substrate at three fixed concentrations of MgATP. (Adapted from Mourad and Parks, 1966.)

Although parallel lines can indicate a ping pong mechanism, it is easy to be misled by an apparently parallel pattern that is imperceptibly converging, and it is important to measure the rates in both directions and to plot the data to verify nonconvergence in both directions (Janson and Cleland, 1974).

Other properties of the ping pong mechanism can validate or exclude it. Because the ping and pong steps are chemically independent, the modified enzyme intermediate (F in scheme 2-7) must be formed by reaction of the leading substrate with the enzyme in the absence of the second substrate. This fact leads to independent tests of the mechanism. We explain them here for the case of nucleoside diphosphate kinase, where MgATP is A, MgGDP is B, and the phosphoenzyme is F in scheme 2-7, and there are analogous tests for all ping pong reactions. First, incubation of Mg[γ-^{32}P]ATP with nucleoside diphosphate kinase leads to the ^{32}P-phosphoenzyme, which can be isolated from excess Mg[γ-^{32}P]ATP by gel filtration and characterized. Chemical problems may complicate the characterization; for example, a [^{32}P]phosphoenzyme may be hydrolytically labile. Methods for overcoming these problems are described in chapter 1.

Second, in a ping pong mechanism, the enzyme catalyzes specific exchange reactions, and if any of these exchanges are not observed or are too slow, the mechanism is excluded. Nucleoside diphosphate kinase catalyzes the exchange reaction MgATP + Mg[^{14}C]ADP $\rightarrow$ Mg[^{14}C]ATP + MgADP. This follows directly from the mechanism, in which the reversal of MgADP formation by the pool of Mg[^{14}C]ADP must produce Mg[^{14}C]ATP according to scheme 2-8, which differs from scheme 2-7 by the exclusion of the second step of the overall reaction.

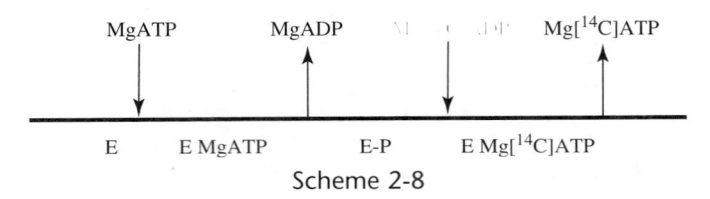

Scheme 2-8

The observation of this exchange does not by itself prove the ping pong mechanism. An analogous exchange reaction of the co-substrate must also occur. Moreover, the maximum rates of these exchange reactions must be compatible with the maximum rates of the forward and reverse reactions. Equation 2-16 shows the relationships among these rates in a ping pong mechanism (Janson and Cleland, 1974).

$$\frac{1}{V_f} + \frac{1}{V_r} = \frac{1}{V_{ex1}} + \frac{1}{V_{ex2}} \tag{2-16}$$

The observation of this relationship implicates a ping pong mechanism, and this mechanism is excluded when this relationship does not hold.

In an isotope exchange experiment, there is no net reaction because it is at equilibrium. Because of this, an apparent K_m value is in reality the dissociation constant for the varied substrate. This is a valid method for measuring the dissociation constants K_{ia}, K_{ib}, and so forth in an enzymatic reaction.

In the case of nucleoside diphosphate kinase, the observation of stereochemical retention of configuration at phosphorus further consolidates the ping pong mechanism. Two nucleophilic displacements on the phosphoryl group being transferred, each proceeding with inversion, leads to overall retention of configuration. Reaction of Mg[^{14}C]ADP with P-chiral Mg[γ-^{18}O]ATPγS with inversion of configuration would lead to epimerization at γ-P of ^{14}C-labeled Mg[γ-^{18}O]ATPγS. However, as illustrated in scheme 2-9, the reaction proceeds with overall retention (Sheu et al., et al., 1979).

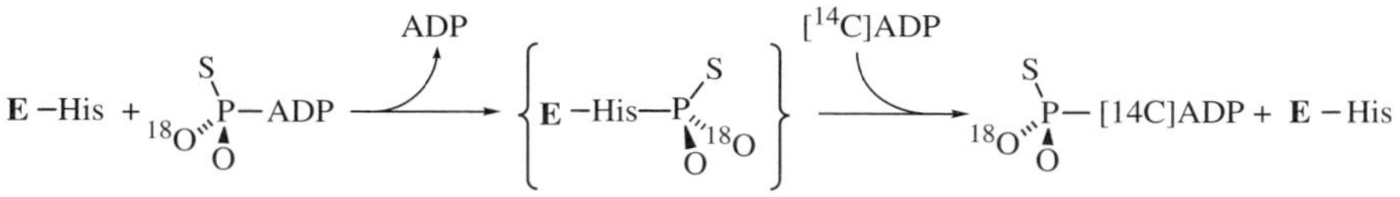

Scheme 2-9

The configuration of the bracketed intermediate in scheme 2-9 was inferred from the overall stereochemistry and was not explicitly determined. Configurational assignment of a chiral P at an enzymatic site is a formidable task. It has been accomplished in the case of galactose-2-phosphate uridylyltransferase (Arabshahi et al., et al., 1988). A detailed discussion of the reaction mechanism can be found in chapter 10.

The homotrimeric structure of nucleoside diphosphate kinase in fig. 2-5 shows a single site for binding a nucleoside triphosphate in each subunit, with His122 in position to accept the γ-phosphoryl group. Phosphorylation of His122 by MgATP produces the phosphoenzyme, and dissociation of MgADP leads to the free phosphoenzyme intermediate. The second substrate, MgGDP in eq. 2-15, binds to the same site and accepts the phosphoryl group to form MgGTP.

Adenylate kinase and nucleoside diphosphate kinase catalyze chemically similar reactions, the transfer of a phosphoryl group from a phosphoanhydride to a phosphate group. However, the mechanisms are very different; one proceeds in two chemical steps through a phosphoenzyme intermediate, and the other proceeds by direct transfer of the phosphoryl group from the donor substrate to the acceptor substrate. The principle of economy in the evolution of binding sites appears to have led to these two mechanisms (Frey, 1982, 1992). According to this principle, *an enzyme evolves with the minimum number of substrate binding sites to allow catalysis of the reaction*. In the case of nucleoside diphosphate kinase, a single nucleotide binding site allows the phosphodonor MgNTP to bind to E and the phosphoacceptor MgGDP to bind to E-P, both at the same site. They can share this site only because the phosphoryl group undergoing transfer is maintained in its reactive form during a changeover between the substrates. This is accomplished by phosphotransfer from MgATP to His122 to form the phosphoenzyme (E-P) and allowing the MgADP

82 Enzymatic Reaction Mechanisms

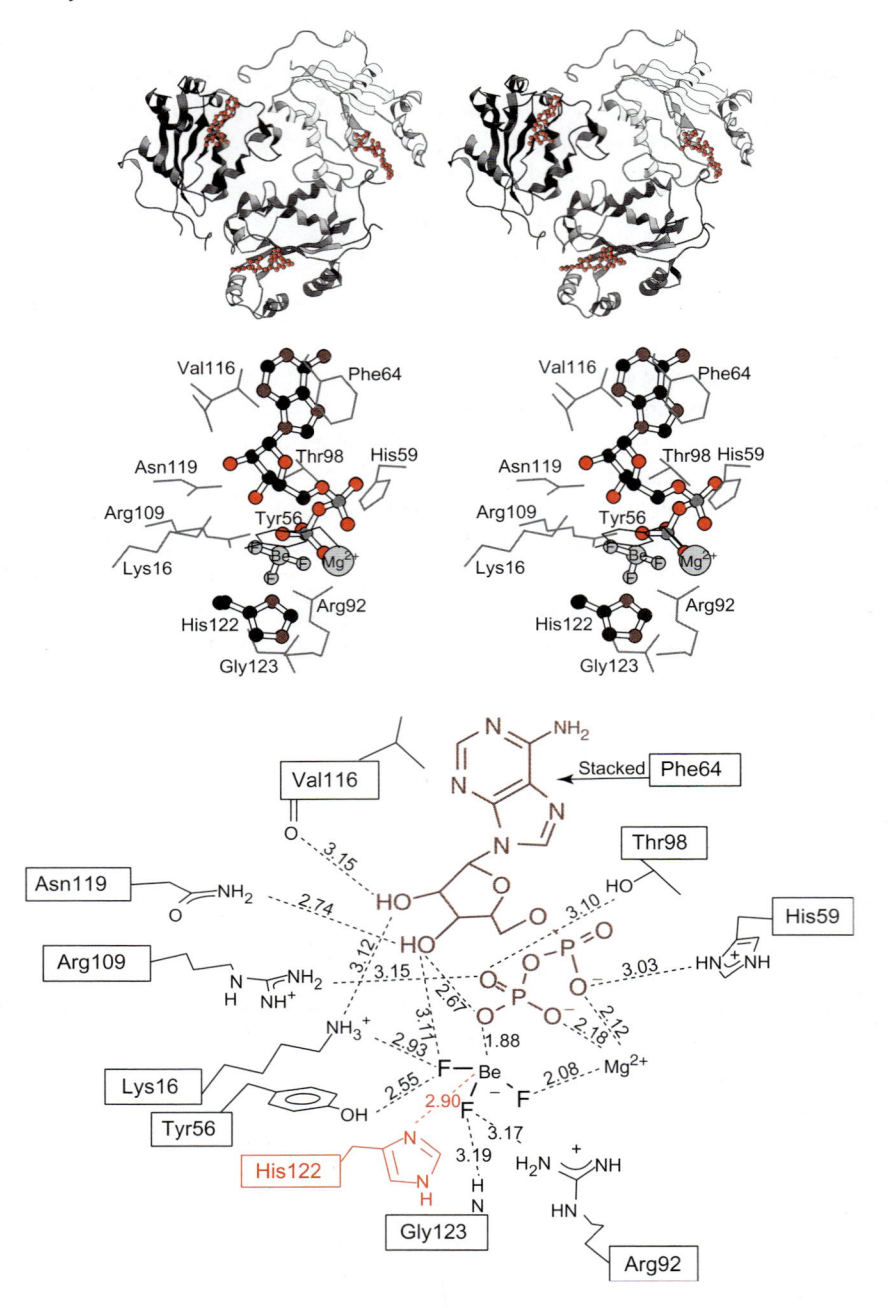

Fig. 2-5. Structure of *Dictyostelium discoideum* nucleoside diphosphate kinase. At the top is a ribbon diagram of the molecular structure of the homotrimeric nucleoside diphosphate kinase, with MgADP and BF$_3$ shown as red ball-and-stick models bound at the active site. BF$_3$ is thought to mimic a phosphoryl group in flight. In the center is a stereodiagram of the active site, showing close contacts of MgADP with amino acid side chains and His122. At the bottom is a two-dimensional diagram of the hydrogen-bonded and electrostatic contacts at the active site. All three images were created using PDB 2BEF (Xu et al., 1997).

to dissociate. MgGDP can then bind to the same site and accept the phosphoryl group. This mechanism is practical when the pair of group donor substrates is sterically and electrostatically similar and when the group acceptor substrates are also similar. Such a mechanism would be impractical for the reaction of adenylate kinase, in which the phosphoryl donor substrates MgATP and ADP are sterically and electrostatically different, and the

Kinetics of Enzymatic Reactions 83

acceptors AMP and MgADP are also different. By the principle of economy, a minimum of two sites is required, one for the donor and another for the acceptor, in adjacent locations as shown in fig. 2-3.

We have shown the magnesium complexes of ATP, ADP, GTP, and GDP as the substrates for adenylate and nucleoside diphosphate kinases because these nucleotides exist as magnesium complexes and they are the true substrates. In later chapters, nucleoside diphosphates and triphosphates will be understood to react as the magnesium complexes.

Two-Site Ping Pong Mechanisms

Two-site ping pong mechanisms are variants of the basic mechanism, in that the group donor and acceptor substrates occupy different sites and bind independently (Northrop, 1969). The group being transferred by the covalent enzyme-substrate intermediate is shuttled from one site to the other. This mechanism first appeared in the reaction of transcarboxylase (EC 2.1.3.1), which catalyzes reaction 2-17.

$$H_3C-\overset{O}{\underset{}{C}}-COO^- + {}^-OOC-\overset{CH_3}{\underset{H}{C}}-COSCoA \rightleftharpoons {}^-OOC-CH_2-\overset{O}{\underset{}{C}}-COO^- + H_2\overset{CH_3}{\underset{}{C}}-COSCoA \quad (2\text{-}17)$$

The carboxyl group transfer is mediated by biotin, which is covalently bonded to transcarboxylase and can move between the two sites (see chap. 3). In effect, the coenzyme is tethered to the enzyme by the pentanoyl group of biotin in an amide linkage to the aminobutyl group of lysine, and the tether allows it to move between the two sites without dissociating from the enzyme, as illustrated in fig. 2-6.

At one site, methylmalonyl CoA transfers its carboxyl group to biotin to form carboxy-biotin-transcarboxylase, and propionyl CoA dissociates. Pyruvate binds at a second site independently of methylmalonyl CoA, and the carboxybiotinyl moiety shuttles to the second site and carboxylates pyruvate to oxaloacetate, which dissociates. The shorthand mechanism appears in scheme 2-10, in which the two sites appear on different lines.

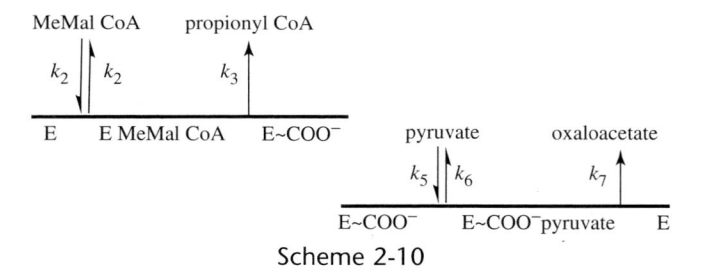

```
         MeMal CoA    propionyl CoA
            k_2 ↕ k_2      k_3 ↑
      ──────────────────────────────
       E      E MeMal CoA    E~COO⁻      pyruvate              oxaloacetate
                                            k_5 ↕ k_6              k_7 ↑
                                        ──────────────────────────────
                                         E~COO⁻     E~COO⁻pyruvate      E
```
Scheme 2-10

E~COO⁻ refers to the carboxybiotinyl-transcarboxylase, in which the carboxybiotin is covalently bonded to a lysyl residue of the enzyme (see fig. 2-6). This arrangement allows the acyl CoAs (substrate and product) to bind at one site and the α-ketoacids (substrate and product) to bind at the other site, both of which can be served by biotin or carboxybiotin. The rate equation for this mechanism is the same as for the conventional ping pong bi bi mechanism (see eq. 2-14), but the product inhibition patterns are reversed for reasons explained in the next section.

Transcarboxylase was the first of a number of multisite ping pong mechanisms to be discovered. Others include all of the biotin-dependent carboxylases and the α-ketoacid dehydrogenase complexes. They all involve the participation of a tethered coenzymatic

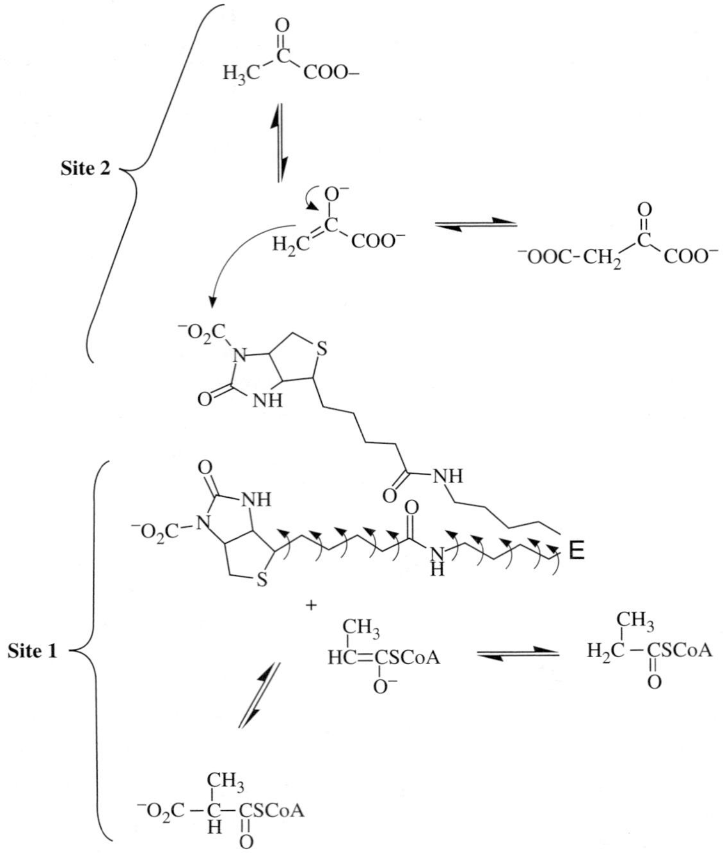

Fig. 2-6. Biotin as a tethered carboxyl carrier in an enzyme. The coenzyme biotin is covalently linked through its carboxyl group to a lysine-e-amino group. The tetramethylene segments of the lysyl and biotinyl side chains allow torsion about every methylene group. In this way, the side chains can function as a mobile tether, which allows the biotinyl group to interact with the two active sites in the two-site ping pong mechanism.

molecule that shuttles groups or reducing equivalents, or both, from one site to another. The pyruvate dehydrogenase multienzyme complex is an example of the α-ketoacid dehydrogenase complexes, in which a lipoic acid moiety is bonded through its carboxyl group to a lysyl residue of the central enzyme. The lipoyl moiety accepts an acetyl group and two electrons from the first enzyme, shuttles the acetyl group to another site to form acetyl CoA, and then shuttles the electron pair to the third enzyme to form NADH in a three-site pong pong mechanism (see chap. 18).

Haldane Relationships

The kinetic parameters for an enzymatic reaction are constrained by the equilibrium constant for that reaction, just as are the rate constants for a simple chemical reaction. For a simple equilibrium between reactant A and product P, the value of K_{eq} is equal to the ratio of the forward and reverse rate constants. Similarly, an enzymatic reaction at equilibrium continues to take place in forward and reverse directions at the same rates. In terms of initial rates v_f and v_r, the net rate is zero whenever $v_r = -v_f$, which is at equilibrium. The rate equations for the forward and reverse directions of a reversible enzymatic reaction can satisfy this condition at equilibrium, and equating the forward and reverse rate equations gives a Haldane relation. Consider the one-substrate case of substrate A in equilibrium with product P, for which the rate equations for the forward and reverse rates are analogous to eq. 2-2, the Michaelis-Menten equation. It is easy to show that $K_{eq} = V_f K_p / V_r K_a$ is the Haldane relationship. One way to check the accuracy of experimental kinetic parameters is to measure the equilibrium constant and compare it with the value calculated from the kinetic parameters using the Haldane relationship. Haldane relationships pertaining to some of the mechanisms discussed earlier appear in Appendix A.

Inhibition Patterns

The primary plots of steady-state kinetic data often do not allow an unambiguous decision regarding the kinetic mechanism in multisubstrate reactions. We have seen that eq. 2-10 is the rate law for three recognized mechanisms. Inhibition studies can distinguish among the mechanisms. Products combine with the enzyme and tend to reverse the reaction. A dead-end inhibitor may resemble a substrate or product but not react. An inhibitor can be a substrate molecule when the substrate concentration is high enough to allow a second molecule to bind in an incorrect site and act as a dead-end inhibitor. An inhibitor can be an alternative substrate that produces a different product and diverts some of the enzyme from catalysis of the primary reaction. A systematic analysis of one or more of these types of inhibition can resolve kinetic ambiguities.

Inhibition analysis proceeds by measurement of initial rates as a function of the concentration of a single substrate at several fixed inhibitor concentrations.. This will generally give a family of straight line in double reciprocal plots ($1/v$ versus $1/[S]$). The patterns of these lines obtained at different inhibitor concentrations identify the inhibition type.

The type of inhibition depends on how the inhibitor concentration appears in the rate equation. The presence of an inhibitor will add at least one factor of the type $(1 + [I]/K_I)$ to the denominator of the rate equation, where $[I]$ is the concentration of inhibitor. This appears as a factor in the double reciprocal form of the equation.

The three types of inhibition kinetics illustrated in fig. 2-7 are commonly observed. In competitive inhibition (see fig. 2-7A), the inhibition factor is a multiplier of the slope in the reciprocal equation and the lines converge on the ordinate. In uncompetitive inhibition the factor is a multiplier of the intercept in the reciprocal equation (see fig. 2-7B) and the lines do not converge. In noncompetitive inhibition the factor multiplies both the slope and intercept (see fig. 2-7C) and the lines converge at the left of the ordinate. Inhibitors display slope or intercept effects or both in double reciprocal plots. Secondary plots of the slopes or intercepts against inhibitor concentrations are often linear. Their slopes give the respective inhibition constants K_{is} and K_{ii} in the equations of fig. 2-7. In practice these parameters are evaluated by statistical fitting of the data to the full rate equation, but the secondary plots are viewed to verify that the inhibition is linear. The plots may on occasion be nonlinear, which indicates that more than one molecule of inhibitor binds and that the correct equation has a different inhibition multiplier, such as $(1 + [I]/K_I + [I]^2/K_i)$. This complex behavior can be sorted out mechanistically (Cleland, 1970, 1977, 1986).

Figure 2-7C shows the intersection left of the ordinate and above the abscissa for noncompetitive inhibition. The intersection may also occur on or below the abscissa. Some kineticists restrict the term *noncompetitive* to the case in which the intersection is on the abscissa, the others being called mixed inhibition. Intersection on the abscissa means that $K_{ii} = K_{is}$. We do not restrict the term *noncompetitive* to this case but use it for all cases in which both slope and intercept effects are observed.

The rate equation is never known in advance in an inhibition experiment. However, as can be seen in fig. 2-7, the inhibition type can give an idea about the rate equation.

86   Enzymatic Reaction Mechanisms

A

$$1/v = \frac{1}{V}\left(1 + \frac{K_b}{[B]}\right) + \frac{1}{V}\left(K_a + \frac{K_{ia}K_b}{[B]}\right)\left(1 + \frac{K_{is}}{[I]}\right)\frac{1}{[A]}$$

B

$$1/v = \frac{1}{V}\left(1 + \frac{K_b}{[B]}\right)\left(1 + \frac{K_{ii}}{[I]}\right) + \frac{1}{V}\left(K_a + \frac{K_{ia}K_b}{[B]}\right)\frac{1}{[A]}$$

C

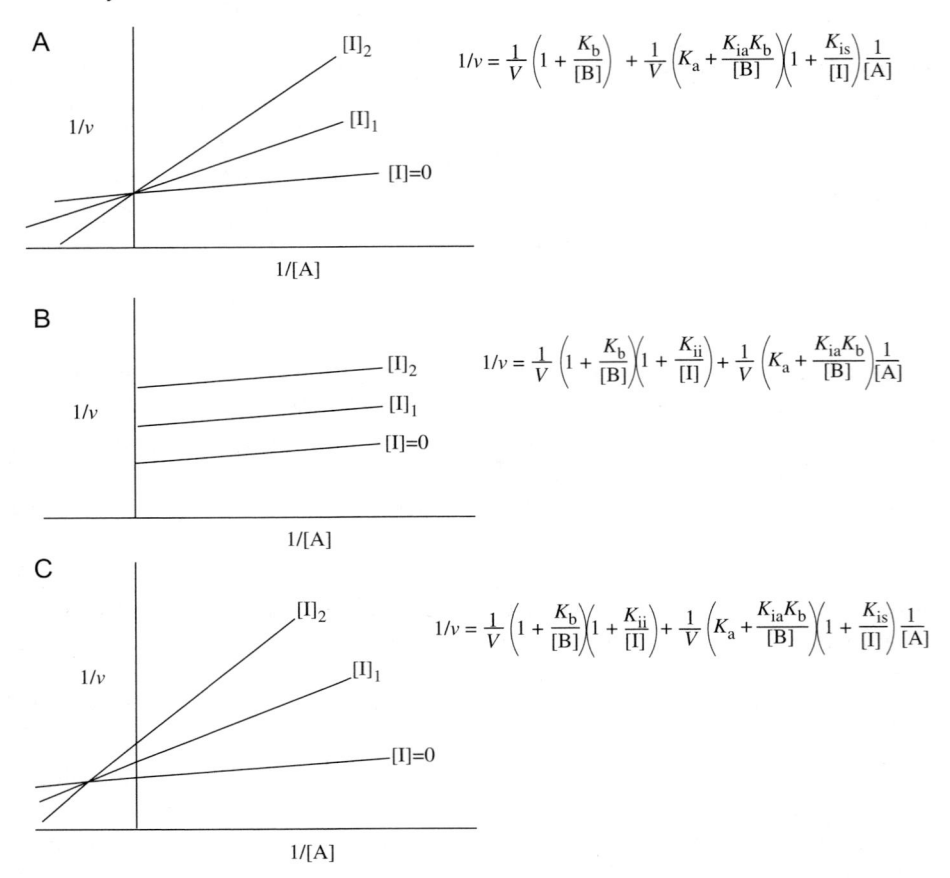

$$1/v = \frac{1}{V}\left(1 + \frac{K_b}{[B]}\right)\left(1 + \frac{K_{ii}}{[I]}\right) + \frac{1}{V}\left(K_a + \frac{K_{ia}K_b}{[B]}\right)\left(1 + \frac{K_{is}}{[I]}\right)\frac{1}{[A]}$$

Fig. 2-7. The plots in parts A, B, and C represent competitive, uncompetitive, and noncompetitive inhibition, respectively, of a two-substrate enzyme. In each case, the variable substrate is A, the fixed substrate is B, and the [I] is fixed for each line and changing in each set of lines. The reciprocal equations are based on eq. 2-10, where $V = k_{cat}[E_0]$, and the inhibition factors $(1 + [I]/K)$ are multipliers for the slope (A), the intercept (B), or both (C). The inhibition constants are K_{is} and K_{ii}, referring to slope and intercept effects, respectively.

Because the rate law is related to the kinetic mechanism, the inhibition pattern contains information about that as well. Inhibition patterns are used in conjunction with Cleland's rules to deduce the kinetic mechanism based on the results of inhibition experiments. The rules are not laws, and there are exceptions, which are in themselves informative when analyzed correctly. The rules may be stated as follows (Cleland, 1970, 1977, 1986):

1. An inhibitor that binds to the same site as the variable substrate gives a slope effect. This is mutually exclusive binding, and in the absence of other interactions, the result is competitive inhibition.
2. An inhibitor that binds to a different site than the variable substrate produces an intercept effect. This results in uncompetitive inhibition if the enzyme form to which the inhibitor binds does not exist when the variable substrate concentration is zero.
3. An inhibitor that binds to a different site than the substrate and can bind at both very high and very low substrate concentrations gives a slope and an intercept effect. This is noncompetitive inhibition.

Special extensions of these rules apply to two-site ping pong reactions, in which the product inhibition patterns are reversed.

Product inhibition patterns are often useful in resolving kinetic ambiguities when the products do not interfere with the assay or alternative assay methods are available.

Table 2-1. Product Inhibition Patterns in Typical Two-Substrate Enzymatic Mechanisms

Kinetic Mechanism	Inhibitor/Varied Substrate			
	Q/A	P/A	Q/B	P/B
Equlibrium random bi bi [b]	C	C (or NC)	NC	C
Ordered bi bi	C	NC (UC sat'n B)	NC	NC
Ping pong bi bi	C	NC	NC	C
Theorell-Chance	C	NC	NC	C

C, competitive, NC, noncompetitive; UC, uncompetitive.
[a]In equilibrium random bi bi mechanisms, two or three competitive patterns are observed, depending on the degree of overlap between the substrate and product containing the transferring piece. If both members of this pair (A,P) cannot bind at the same time, inhibition by one versus the other is competitive. If both can bind simultaneously, inhibition is noncompetitive.

Product inhibition cannot be studied for a product that is the substrate for a coupling enzyme in an assay. An alternative assay is usually possible in these cases. Then the initial rates are measured as in fig. 2-7 at several concentrations of each product and with respect to each substrate as the variable substrate. Cleland's rules are used to interpret the results and derive the mechanism. The typical inhibition patterns for the most common mechanisms of two substrate reactions are given in table 2-1.

Consider the equilibrium random bi bi mechanism in scheme 2-3. Products inhibit competitively with respect to structurally related varied substrates, so that Q is competitive against A as the varied substrate and P is competitive against B. Product Q is noncompetitive against B, and P may inhibit either competitively or noncompetitively with respect to varied A. The products are structurally related to the substrates and can bind to the wrong species of enzyme to form complexes such as EBQ and EAP that do not appear in scheme 2-3. An example is creatine kinase (EC 2.7.3.2), which follows the equilibrium random mechanism at pH 7, but in which the product ADP binds to the complex E.creatine, as well as to E and E.creatine-P. The formation of E.creatine-P.MgADP is inhibitory because it tends to reverse the reaction; however, the formation of complex E.creatine.ADP is not part of the catalytic mechanism, cannot lead to a product, constitutes dead-end complexation, and produces an intercept effect. There are slope and intercept effects, and the inhibition by MgADP is noncompetitive. Because of these types of dead-end complexation, equilibrium random bi bi mechanisms generally display two or three competitive patterns and one or two noncompetitive patterns (see table 2-1).

The situation is different for the ordered bi bi mechanism in scheme 2-4. Here the only competitive pattern is that between the first substrate to bind and the last product to leave, A and Q in scheme 2-4, both of which bind to the free enzyme. The others are all noncompetitive because P binds only to EQ and cannot be competitive with A or B, and Q binds only to E and cannot be competitive with B. This mechanism is characterized by a very useful property that often enables a critical test. In the presence of saturating B, so that the reaction cannot be reversed by P, the product P displays uncompetitive inhibition with respect to A. This occurs because in an ordered mechanism P inhibits by partially reversing the reaction. However, when the concentration of B is very high, it captures every molecule of EA and does not allow any to accumulate. This prevention of reversal eliminates any slope effect by P, and inhibition is uncompetitive because of P binding to EQ at a different site than A and Q.

Dead-end inhibitors bind to enzymes to form complexes that are not part of the catalytic mechanism and do not form products. A dead-end inhibitor is normally an analog of a substrate or product that binds in its place but does not react. Dead-end inhibitors inhibit competitively with respect to their analogous substrates. However, because they cannot form products, they often display uncompetitive inhibition with respect to other substrates.

For example, in the ordered bi bi mechanism (scheme 2-4), an analog of the second substrate B will always display uncompetitive inhibition with respect to the leading substrate A. This occurs because it can bind only to EA and not E, and by rule 2, it will give an intercept effect. Moreover, it cannot bind when A is absent because it binds only to EA, so rule 3 does not apply, and by rule 2, it is uncompetitive. Noncompetitive inhibition applies when the inhibitor can bind at any concentration of A, including very low or zero concentrations.

In the case of two-site ping pong mechanisms, the product inhibition patterns differ from the classic ping pong bi bi case in scheme 2-7. In the classic case, Q is competitive with A because both bind only to the same site in E, and P is competitive with B because both bind only to the same site in F. Then, Q is noncompetitive with B and P with A by rule 3. In the two-site ping pong case, there are two reaction lines for the two sites, as shown in scheme 2-10. Then the leading substrate A and the corresponding product Q can bind separately to their respective sites. We next consider how the product inhibition works out for the inhibition by Q when A is varied. A and Q both bind to E, but at different sites. There is an intercept effect by rule 2. Because Q binds at low and high concentrations of A, rule 3 states that the inhibition is noncompetitive. The competitive component arises from the competition of the occupants of the two sites for the tethered coenzyme, the biotinyl group in the case of transcarboxylase, as shown in fig. 2-7. For similar reasons, P is noncompetitive with B.

Consider inhibition by Q with respect to B, which is found to be competitive (Northrop, 1969). This arises because all substrate-binding steps are at equilibrium. It has been shown that the rate limiting steps for transcarboxylase and all multisite ping pong enzymes to date are the chemical transformations, carboxyl transfer in the case of transcarboxylase. Any molecule that binds to E in the second site, such as product Q, will decrease the amount of enzyme available for reaction with A at site one, which is required for the reaction of B. In the case of transcarboxylase, B (pyruvate) at the first site cannot accept a carboxyl group from Q (oxaloacetate) at the second site, and oxaloacetate specifically prevents B from accepting a carboxyl group from A (methylmalonyl CoA) by tying up the tethered (biotinyl) coenzyme. Similarly, product P is competitive with A.

Inhibition by substrates is occasionally observed as upward curvature in double reciprocal plots at very high concentrations of the varied substrate. This always means that a second molecule of the varied substrate is binding to a form of the enzyme other than its normal partner to form a dead-end complex. These inhibitions can in certain cases be expected and informative regarding the basic kinetic mechanism. For example, in the classic one site ping pong bi bi mechanism substrate inhibition is expected and generally observed in the physiologic direction. Consider the case of nucleoside diphosphate kinase, in which there is a phosphoenzyme (F) intermediate and MgATP is the variable substrate. After the phosphoenzyme is formed and MgADP departs, most of the site is open, and at a low concentration of the acceptor substrate (MgGDP), it would not be surprising if increasing the concentration of MgATP resulted in further binding of MgATP to the vacant portion of the site. This does occur, and substrate inhibition is general and expected for ping pong mechanisms. The inhibition is overcome at very high concentrations of the acceptor (MgGDP), and it is competitive with MgGDP. Inhibition by MgGDP at very high concentrations because of its (weak) binding to the free enzyme is also expected and observed.

Isotope Exchange at Equilibrium

We have discussed isotope exchange between a substrate and product (e.g., A*-P*) in the absence of the second substrate in connection with verifying the ping pong bi bi mechanism.

Such exchanges do not occur in sequential mechanisms because substrate-product interconversion requires the presence of all substrates and products. However, when the substrates and products are present, the kinetics of substrate-product isotope exchange at equilibrium can distinguish between random and ordered sequential binding mechanisms (Cleland, 1972). Several substrate-product exchanges are possible. For example, in the reaction of adenylate kinase, the exchanging pairs are ATP-ADP and AMP-ADP, but not ATP-AMP. The kinetics can also distinguish between ordered sequential and ping pong mechanisms, which are more commonly and easily distinguished by other methods.

The isotope exchange reaction mixture initially contains both substrates and both products, with one of the products (or substrates) containing a radiochemical label such as ^{14}C or ^{32}P. The rate at which the label appears in the corresponding substrate (or product) is then measured as a function of the concentration of a substrate/product pair that is varied in constant ratio. In an ordered mechanism, increasing the concentration of certain substrate/product pairs in constant ratio will lead to complete inhibition of the isotope exchange rate. The rules for isotope exchange at equilibrium are as follows:

1. In ordered mechanisms, if a varied reactant combines with the enzyme between the points at which the isotope exchange partners bind, complete inhibition of isotope exchange will be observed as the concentration of the reactant is increased to infinity.
2. In a random mechanism, isotope exchange at equilibrium cannot be totally inhibited by varying any substrate/product pair.

For example, in the ordered bi bi mechanism with A*-P* exchange being measured, increasing the concentrations of the pairs B/P or B/Q leads to complete inhibition. This occurs because B binds between A and P, and at high concentrations, B ties up the enzyme in complexes that do not allow exchange. Increasing concentrations of pairs A/P or A/Q do not have this effect and do not lead to total inhibition of A/P exchange.

In a random binding mechanism, total inhibition of isotope exchange at equilibrium cannot be induced because more than one route is available to complete the exchange process. The technique of isotope exchange at equilibrium also allows a distinction between fully rapid equilibrium and partial equilibrium binding of substrates. In describing the equilibrium random bi bi mechanism, we mentioned that only the first substrate binding steps have to be at equilibrium, whereas the second steps can be in steady state. In the fully equilibrium mechanism in which all binding steps are at equilibrium, the interconversion of the ternary complexes is fully rate limiting. In that case, all isotope exchange rates are the same because they are all limited by the same step. An observation of different exchange rates for the various possible isotope exchanges means that the second binding steps are in steady state and partially rate limiting.

Three-Substrate Reactions

General Rate Law

A number of important enzymes catalyze the reactions of three substrates to form two or three products. Included in this group are all ATP-dependent synthetases such as glutamine synthetase (EC 2.7.7.42), aminoacyl tRNA synthetases (EC 6.1.1.*), acetyl CoA synthetase (EC 6.2.1.1), succinyl CoA synthetase (EC 6.2.1.5), as well as amino acid dehydrogenases operating in the reductive direction. The α-ketoacid dehydrogenase complexes are also three substrate enzymes, as are monooxygenases, formerly known as mixed function oxidases. The steady-state kinetic analyses of these reactions proceed similarly to the two-substrate enzymes. Initial rates are measured as a function of the concentration of a single varied substrate, with all others being held constant. This is repeated with all three substrates being separately varied. The general rate law for a three substrate reaction in the absence of products is eq. 2-18, where the coefficients (coef. A, B, C) are kinetic constants.

$$1/v = \frac{K_{cat}[E_o][A][B][C]}{[A][B][C] + \{coef.A\}[A] + \{coef.B\}[B] + \{coef.C\}[C] + K_c[A][B] + K_b[A][C] + K_a[B][C] + \text{Const}} \quad (2\text{-}18)$$

Many possibilities for ter-reactant kinetics are possible, including mechanisms with strictly equilibrium binding steps (not yet observed), strictly ordered binding steps, some binding steps in equilibrium and others ordered, and mechanisms with pong pong components. The rate equations differ by the absence of one or more terms from the denominator of eq. 2-18. For example, the rate law for the ping pong mechanisms are missing the constant term, and for the ordered ter ter mechanism, the term in [B] is missing (coef. B = 0).

Examples

Cleland and Viola (1982) describe the possible ter-reactant binding mechanisms, and we do not enumerate them in this chapter. The kinetic mechanisms are elucidated by adaptations of the same methods as for two substrate reactions, determination of the initial rate equation, determination of product and dead-end inhibition patterns, and in cases of ping pong kinetics, characterization of the required exchange reactions and the intermediate. All of the possible kinetic patterns are too numerous to cover in detail here. They include combinations of random and ordered binding of the three substrates in sequential mechanisms, and various sequences of ping pong and sequential components in ping pong mechanisms. The following two cases exemplify the range of mechanisms. Others can be found in the referenced sources (Cleland, 1986, 1991; Segel, 1993).

The ordered ter bi mechanism in scheme 2-11 applies to $NADP^+$-dependent glutamate dehydrogenase (EC 1.4.1.3) (Rife and Cleland, 1980).

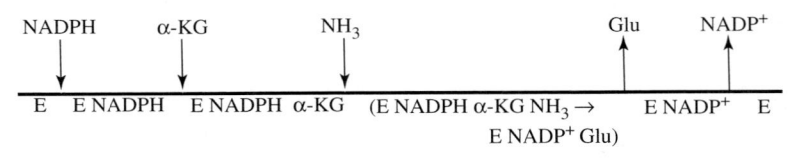

```
   NADPH      α-KG       NH₃                        Glu       NADP⁺
     ↓          ↓          ↓                          ↑          ↑
 ─────────────────────────────────────────────────────────────────────
  E     E NADPH    E NADPH α-KG   (E NADPH α-KG NH₃ →    E NADP⁺    E
                                    E NADP⁺ Glu)
```

α-KG = α-ketoglutarate

Scheme 2-11

The interconversion of the central complexes proceeds by a multistep chemical mechanism, where ammonia reacts with the carbonyl group of α-ketoglutarate in a two-step process leading to formation of the imine. Then NADPH reduces the imine to form the complex [E.NADP⁺.Glu], from which the products dissociate in the order shown. The interconversion of central complexes takes place in three chemical steps, and the steady-state kinetics profile does not give information about them. Detail about this mechanism appears in chapter 16.

A significant difference between ter-substrate and bi-substrate mechanisms is that ping pong kinetics may be seen in three substrate reactions where the intermediate in the ping pong component is derived from two substrates and does not dissociate from the enzyme. In two substrate reactions, a ping pong intermediate is normally a chemically modified form of the enzyme or coenzyme. In three substrate reactions, many examples of ping pong kinetics involving bi-substrate intermediates are those in which substrate-carboxylate groups are activated. These include acyl CoA synthetases and aminoacyl tRNA synthetases.

A specific example is asparagine synthetase (EC 6.3.1.1) from *E. coli*, which catalyzes the reaction of aspartate with ATP and NH$_3$ to form asparagine, PP$_i$, and AMP. The enzyme binds ATP and aspartate randomly, and within the ternary complex the β-carboxylate group of aspartate reacts with the α-P of ATP to displace PP$_I$, which is released. The resultant aspartyl-β-adenylate does not dissociate from the active site. The NH$_3$ then binds and forms asparagine by displacing AMP from the aspartyl-β-adenylate (Cedar and Schwartz, 1969). This is the bi uni uni bi ping pong mechanism shown in scheme 2-12.

As in all reactions to form a derivative of a carboxyl group, the process begins with chemical activation of the β-carboxylate group of aspartate. From a purely chemical standpoint, the formation of AMP and PP$_I$ in the overall reaction may signal activation of aspartate as aspartate-β-pyrophosphate or aspartate-β-adenylate.

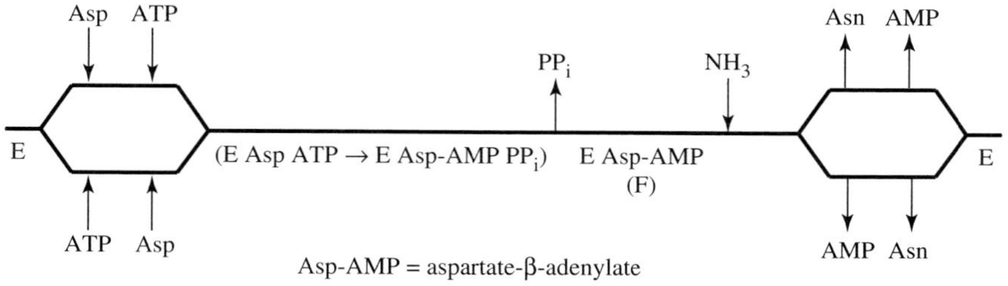

Scheme 2-12

Labeling experiments with [4-^{18}O]aspartate showed that the ^{18}O appeared in AMP and not in PP$_I$, proving that aspartate-β-pyrophosphate could not have been the intermediate and implicating aspartate-β-adenylate (Cedar and Schwartz, 1969). Scheme 2-13 shows the course of ^{18}O transfer to form [^{18}O]AMP. The kinetic mechanism in scheme 2-12 rules out aspartate-β-pyrophosphate because of the departure of PP$_i$ before the binding of NH$_3$.

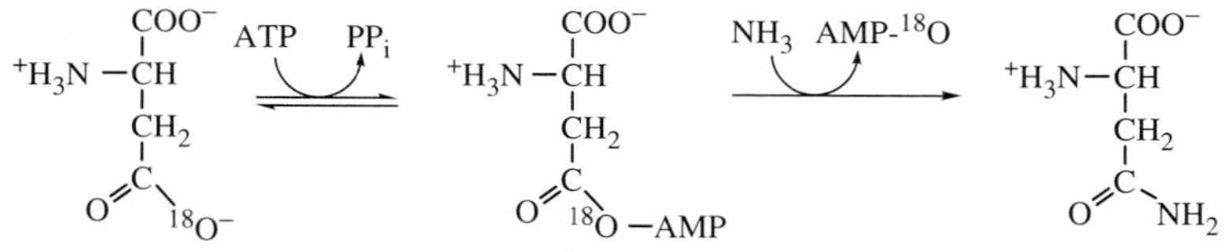

Scheme 2-13

Inhibition Patterns

Inhibitors are used to distinguish kinetic mechanism in three-substrate reactions in the same way as in the two-substrate cases. The same principles apply, but the inhibition patterns are more numerous and complex to interpret, and we do not detail them here. The product-inhibition patterns for a number of common three-substrate kinetic mechanisms are given in Appendix B.

Isotope Effects

Classes of Isotope Effects

Kinetic and Equilibrium Isotope Effects

Isotope effects can contribute important information in the mechanistic analysis of enzymatic catalysis. An isotope effect is any effect on a reaction that is induced by the substitution of a heavy isotope into a reacting molecule. Isotope effects may be induced on either the rate of or the equilibrium constant for a reaction. Effects on the rate are kinetic isotope effects, and effects on the equilibrium constant are equilibrium isotope effects. Most often the kinetic isotope effects provide essential mechanistic information. However, equilibrium isotope effects sometimes also provide direct mechanistic information, and knowledge about them is often essential for interpreting kinetic isotope effects. The most commonly measured kinetic isotope effects in enzymatic studies are those induced by ^{2}H or D for deuterium, ^{3}H or T for tritium, ^{13}C, ^{14}C, ^{15}N, and ^{18}O. We employ the notation of Northrop and Cleland, in which the isotope effect on a kinetic parameter is denoted by the symbol for that parameter, preceded by a superscript designating the isotope. Then, ^DV or ^Dk refers to deuterium kinetic isotope effects on maximum velocity or a rate constant, respectively, and $^Dk = k_H/k_D$. $^TV/K$ and $^{15}V/K$ refer to tritium and ^{15}N kinetic isotope effects on V/K. Because $V = k_{cat}[E_0]$, isotope effects may be reported as $^Dk_{cat}$ or $^Tk_{cat}/K$. Equilibrium isotope effects are similarly denoted by the relevant superscripts preceding the equilibrium constant, so that $^{18}K_{eq}$ means an isotope effect induced by ^{18}O on the equilibrium constant for a reaction.

Spectroscopic Isotope Effects

Heavy isotopes are often also employed in spectroscopic studies such as NMR, electron paramagnetic resonance (EPR), infrared spectroscopy (IR), and Raman spectroscopy. Isotope effects in spectroscopy represent perturbations of chemical bonding and are important in mechanistic studies. The spectroscopic isotope effects may be regarded as structure based, as distinguished from rate-based effects. In IR and Raman spectroscopy, the substitution of a heavy atom decreases the vibrational frequency of the bond, and the change is proportional to the bond order. In NMR spectroscopy, the substitution of a heavy atom can perturb the chemical shift of another atom to which it is bonded. This is known as the *isotope shift*, and its magnitude gives information about the bonding between the two atoms. In EPR, substitution of a heavy isotope differing in nuclear spin alters the nuclear hyperfine splitting patterns of the paramagnetic center and provides structural information about the paramagnetic species. Spectroscopic isotope effects often give essential structural information about the details of interactions at the active sites of enzymes. These include electrostatic and hydrogen bonded interactions that are perturbed in predictable ways by heavy isotopes.

Substitution of heavy atom isotopes is common in various studies of the course of enzymatic reactions. Understanding the kinetic consequences of these alterations is important. The classic studies of the stereochemistry of hydride transfer by NAD$^+$-dependent dehydrogenases required the use of deuterium in place of hydrogen. Phospho groups are tetrahedral about phosphorus but not chiral unless one or two oxygens are replaced with a heavy isotope or sulfur. Chirality induced by such substitutions can enable the stereochemical analysis of enzymatic substitution at phosphorus. Applications of mass spectroscopy, NMR, EPR, and vibrational spectroscopy allow the configuration at chiral phosphorus to be elucidated (Frey, 1982, 1992; Gerlt et al., 1983; Knowles, 1982). Analysis of the stereochemical course of substitution at phosphorus in the reactions of adenylate kinase and nucleoside diphosphate kinase alluded to previously depended on these methods. In this book, we refer to spectroscopic isotope effects on a case by case basis, and in these sections, we deal with the theory and application of kinetic and equilibrium isotope effects.

Magnitudes of Kinetic Isotope Effects

The largest isotope effects involve hydrogen transfer, where a bond to hydrogen is broken in the rate-limiting step. Substitution of deuterium for hydrogen then slows the rate by anywhere from 1.5- to 7-fold, depending on the structure of the transition state. This is a classic primary isotope effect, one in which a bond to the heavy isotope breaks in the transition state. Secondary isotope effects are those in which a bond to the heavy isotope is not broken but is altered in the transition state. The differences between primary and secondary isotope effects can be illustrated for a specific reaction, the oxidation of ethanol to acetaldehyde in eq. 2-19.

$$\underset{\beta\text{-secondary}}{\overset{\alpha\text{-secondary}}{}} \underset{H}{\overset{\text{primary}}{H_3C-C-O}} \underset{\alpha\text{-primary}}{\overset{H}{}} \xrightarrow{[O]} H_3C-\overset{H}{\underset{}{C}}=O \qquad (2\text{-}19)$$

Substitution of D or T for H abstracted from carbon-1 or the hydroxyl group leads to a primary isotope effect, as does the substitution of ^{13}C for carbon-1. Substitution of D for H on carbon-2 or the retained H on carbon-1 leads to a secondary isotope effect. Secondary effects are typically much smaller than primary effects, on the order of one-tenth or less that of a primary effect. Primary kinetic isotope effects are always normal (i.e., $k_H/k_D > 1.0$), whereas secondary isotope effects can be normal or inverse (i.e., an inverse effect is $k_H/k_D < 1.0$).

The physical basis for the primary kinetic isotope effect can be complicated (Huskey, 1991; Schowen, 1991). For most cases, however, the kinetic isotope effect primarily results from a decrease in bond stretching frequency in the transition state. The lost frequency is in effect converted into translational motion for the transfer of the atom or its heavy isotope. Figure 2-8 illustrates the vibrational modes. In the transition state, the bending modes exist but are weakened, the symmetric stretch is weakened or lost, and the restoring force of the asymmetric stretch is abolished.

Most of the isotope effect arises from the difference in zero-point energies for hydrogen and deuterium, called the *ZPE effect*. The plots in fig. 2-9 illustrate this in terms of the differences between the energy at the transition state and the ZPEs for H and D in the ground state. The different ZPEs for the isotopes result in different activation energies to reach the transition state. The primary isotope effect for a heavy atom can vary depending on the structure of the transition state (Westheimer, 1961). In a symmetrical transition state, the transferring atom is equidistant from the donor and acceptor atoms, and one of its vibrational frequencies is essentially converted into translational energy (fig. 2-9A), leading to the maximum classic isotope effect. If the transition state is asymmetrical (fig. 2-9B), with the atom partially or mostly transferred, part of the vibrational frequency is retained or recovered in the transition state. In that case, part of the different zero-point energies will also be retained, diminishing the difference in activation energies and leading to a smaller isotope effect.

Theoretical analyses of isotope effects on chemical equilibria and rates shows that the ZPE effect is dominant for atoms in the first (hydrogen) and second (boron to chloride) periods of the periodic chart (Huskey, 1991). The effect of molecular mass and moment of inertia (MMI) and of the excited-state vibrational component (EXC) become important in

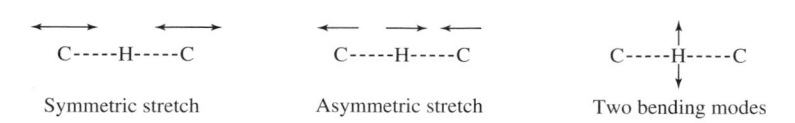

$$\underset{\text{Symmetric stretch}}{\overset{\rightarrow\quad\leftarrow}{C\text{-----}H\text{-----}C}} \qquad \underset{\text{Asymmetric stretch}}{\overset{\leftarrow\quad\rightarrow\quad\leftarrow}{C\text{-----}H\text{-----}C}} \qquad \underset{\text{Two bending modes}}{\overset{\uparrow}{\underset{\downarrow}{C\text{-----}H\text{-----}C}}}$$

Fig. 2-8. Vibrational modes.

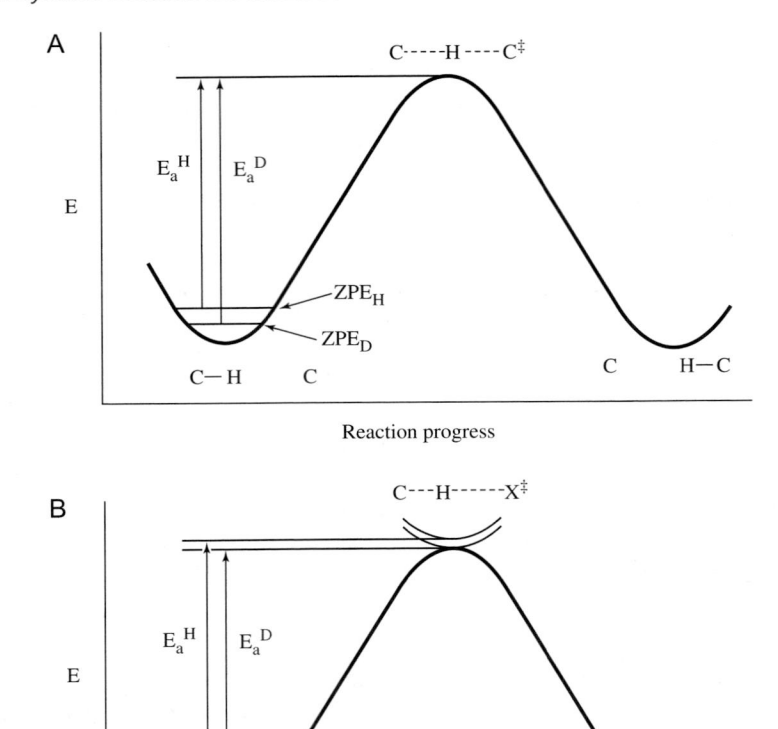

Fig. 2-9. The zero-point energy (ZPE) basis for isotope effects. (A) In a symmetrical transition state, the transferring atom is equidistant from the donor and acceptor atoms, and one of its vibrational frequencies is essentially converted into translational energy. (B) If the transition state is asymmetrical, with the atom partially or mostly transferred, part of the vibrational frequency is retained or recovered in the transition state.

the heavier atoms such as sulfur. A tabulation of Streitweiser semiclassical limits on isotope effects (Huskey, 1991) conveys concepts of the maximum values that can be expected in various situations. Excerpts appear in table 2-2 for atoms that are most relevant to enzymatic reactions. Primary kinetic isotope effects of 2% to 9% can be expected for carbon and nitrogen, depending on the type of bond being broken, with the values for ^{13}C being on the order of one half of those for ^{14}C.

The value of 6.44 for a deuterium kinetic isotope effect (k_H/k_D) is frequently exceeded in practice. In many cases, this may be attributed to nonclassical quantum mechanical tunneling by hydrogen. The hydrogen isotope effects are 10-fold larger than those for carbon and nitrogen because of the small mass of hydrogen and the consequences of this on the ZPE differences. All of the contributing components to the isotope effect—ZPE, MMI, and EXC—can be expressed in terms of vibrational frequencies (Huskey, 1991) and the relative masses of the isotopes determine the frequency differences.

Other aspects of transition-state structure affect the magnitudes of kinetic isotope effects. In hydrogen transfer reactions, the sketches in fig. 2-9 are based on linear transition states, so that a stretching frequency is transformed into translation. When the transition state is not linear, as in the 1,2-hydride shift illustrated in eq. 2-20, the transformed

Table 2-2. Streitwieser Semiclassic Maximum Isotope Effects at 25°C[a]

Reactant Bond	Isotopes[a]	EXC[b]	ZPE[c]	KIE[d]
C—H	H/D	1.00003	6.44101	6.44123
C—H	$^{12}C/^{13}C$	1.00000	1.02095	1.02095
C—H	$^{12}C/^{14}C$	1.00000	1.03931	1.03931
C—C	$^{12}C/^{13}C$	1.00080	1.04797	1.04880
C—C	$^{12}C/^{14}C$	1.00155	1.09174	1.09343
C—N	$^{14}N/^{15}N$	1.00035	1.04396	1.04433
C—N	$^{12}C/^{13}C$	1.00048	1.05979	1.06030
C—N	$^{12}C/^{14}C$	1.00095	1.11500	1.11606
C—O	$^{16}O/^{18}O$	1.00068	1.06605	1.06678
C—O	$^{12}C/^{13}C$	1.00062	1.06076	1.06142
C—O	$^{12}C/^{14}C$	1.00123	1.1699	1.11837
C—S	$^{32}S/^{34}S$	1.00116	1.01271	1.01389
C—S	$^{12}C/^{13}C$	1.00424	1.04551	1.04994
C—S	$^{12}C/^{14}C$	1.00834	1.08731	1.09638

EVC, excited vibrational component; KIE, total maximum semiclassic kinetic isotope effect; ZPE, Zero-point energy.
[a] All isotope effects expressed as K_{light}/K_{heavy} Data from from Huskey 1991.

frequency may be a bending mode. Because bending modes are lower frequency than stretching modes, the maximum kinetic isotope effect will be smaller for a 1,2-hydride shift than for a symmetrical linear hydride transfer (Westheimer, 1961).

$$R-\underset{\underset{O}{\|}}{\overset{H}{\underset{|}{C}}}-\underset{\underset{H}{|}}{\overset{H}{\underset{|}{C}}}-O^- \longrightarrow \left[R-\underset{\underset{O}{\|}}{\overset{H}{\underset{|}{C}}}\cdots\underset{\underset{H}{|}}{\overset{H}{\underset{|}{C}}}=O\right]^{-\ddagger} \longrightarrow R-\underset{\underset{O^-}{|}}{\overset{H}{\underset{|}{C}}}-\underset{\underset{H}{|}}{\overset{H}{\underset{|}{C}}}=O \qquad (2\text{-}20)$$

Secondary kinetic isotope effects involve no bond cleavage to the heavy isotope but a change in bonding occurs in the transition state and usually the product state as well. For example, in eq. 2-20 deuterium in the nontransferring position can display a secondary kinetic isotope effect because the carbon to which it bonds changes from sp^3 to sp^2. The effect is likely to be smaller than the primary effect, which is already diminished by being a 1,2-hydride shift. This difference in bonding is less between the ground and transition states than between the substrate and product, with the latter leading to an equilibrium isotope effect.

In secondary and equilibrium isotope effects, the heavy isotope tends to accumulate in the more strongly bonded position, which is the higher frequency and more stiffly bonded position. An equilibrium isotope effect represents the difference in bonding between the reactant and product. For deuterium substitution in hydrogen transfer, it is expressed as $^DK_{eq} = {^Dk_{for}}/{^Dk_{rev}}$. The secondary kinetic isotope effect represents the analogous difference between the reactant and the transition state.

Equilibrium isotope effects can be calculated from fractionation factors, which are available in tables (Cleland, 1980). Fractionation factors are equilibrium isotope effects relative to a standard compound. The standards are usually water for deuterium and ^{18}O, carbon dioxide for ^{13}C, and aqueous ammonia for ^{15}N. The equilibrium isotope effect for a reaction is the ratio of fractionation factors of the substrate and product.

Measurement of Isotope Effects

Direct Method

When an isotope effect is large, the best measurement is the ratio of rates with and without the heavy isotope. In an enzymatic study, one measures the kinetic isotope effects on

k_{cat} and k_{cat}/K_m by measuring the initial rates as functions of the concentration of the labeled and unlabeled substrates. In general, this method may be used when the isotope effect is ≥ 1.2. Isotope effects between 1.1 and 1.2 can be measured reliably only when the assay method is very accurate. In general, this limits the direct method to measuring primary kinetic isotope effects for deuterium transfer. In some cases, secondary kinetic isotope effects of 1.15 to 1.2 for deuterium transfer can be measured by the direct method. Kinetic isotope effects for ^{13}C, ^{14}C, ^{18}O, or ^{15}N are generally too small to be measured by the direct method.

Internal Competition

The internal competition method depends on changes in the natural abundance of a heavy isotope in a substrate. The natural abundances are 1.1% for ^{13}C, 0.37% for ^{15}N, and 0.20% for ^{18}O. These can be measured accurately only by isotope ratio mass spectrometry, which is applicable only to certain gases, N_2 and CO_2. To measure these isotope effects, the reaction is permitted to proceed to 15% to 30% of completion and then stopped by inactivating the enzyme. The product and unreacted substrate are isolated from the reaction mixture and chemically degraded to CO_2 if the ^{13}C or ^{18}O isotope effect is desired, or to N_2 if the ^{15}N isotope effect is desired. The isotope ratio mass spectrometric analyses then allow the isotope effect to be calculated. Equation 2-21 can be used to calculate a ^{13}C isotope effect when the carbon at the reaction center of the substrate and product can be excised by chemical degradation and quantitatively recovered in the CO_2.

$$^{13}(V/K) = \frac{\log(1-f)}{\left[\log(1-fR_p/R_0)\right]} = \frac{\log(1-f)}{\left\{\log\left[(1-f)(R_s/R_0)\right]\right\}} \quad (2\text{-}21)$$

In eq. 2-21, R_0 is the mass ratio of starting compound, R_p is the mass ratio in product at fraction of reaction f, and R_s is the mass ratio in residual substrate at fraction f. Analogous equations can be used for other isotopes. The results from ratio mass spectrometry can be accurate to 0.01%, so that isotope effects can be measured accurately to five significant figures. Primary kinetic isotope effects of 1 to 9% for ^{13}C, ^{15}N, and ^{18}O can be measured accurately, and many secondary isotope effects also can be measured. Multiple runs and analyses must be done to establish the experimental uncertainty.

The internal competition method can also be used to measure ^3H and ^{14}C isotope effects, but then the analysis is done by radiochemical analyses of the unused substrate and of the unreacted substrate and the product formed at fraction f of the reaction. Because the natural abundance stable and radioactive heavy isotopes are present as trace labels, the internal competition method always gives *V/K* isotope effects and not isotope effects on *V* or k_{cat}.

Remote Label

The remote label method makes use of the isotope ratio mass spectrometer to measure an isotope effect for an atom that cannot be isolated in CO_2 or N_2 by chemical degradation (Cleland, 1990; O'Leary, 1977; O'Leary and Marlier, 1979). This method is most easily understood by considering an example: measuring the ^{18}O isotope effect for an acyl group transfer reaction of *p*-nitrophenyl acetate (PNPA) with ^{18}O in the bridging or nonbridging position of the aceto group. In the bridging position, it would be a primary isotope effect, and in the nonbridging position, it would be a secondary isotope effect. Two samples of PNPA are synthesized, one (the major amount) with natural abundance oxygen (essentially 100% ^{16}O) in the aceto group but with ^{15}N depleted in the *p*-nitrophenyl group ($CH_3CO-O-C_6H_4-^{14}NO_2$) and another with essentially 100% ^{18}O in, for example, the bridging position of the aceto group and 100% ^{15}N in the *p*-nitrophenyl group ($CH_3CO-^{18}O-C_6H_4-^{15}NO_2$). The compounds are mixed to consist of 99.63% of the first and 0.37% of the second. The mixture has natural abundance ^{15}N in the nitro group (i.e., the remote label), and each ^{18}O-labeled molecule contains ^{15}N.

The ^{15}N kinetic isotope effect is measured by the internal competition method, running the reaction to about 15% completion. R_p in eq. 2-21 can be obtained by chemical degradation of *p*-nitrophenol (PNP) to N_2 and by isotope ratio mass spectrometry. To obtain R_0 and R_s, samples of the original substrate and of the re-isolated 85% of unreacted substrate are completely hydrolyzed, the PNP from both is purified, and R_0 and R_p are measured by the same method. The calculated isotope effect (*V/K*) is the product of the primary ^{18}O and secondary ^{15}N isotope effects. The ^{15}N isotope effect is measured by the same method using PNPA containing the natural abundance of ^{15}N. Division of the $^{18}O^{15}N$ isotope effect by the ^{15}N isotope effect gives the ^{18}O isotope effect by the remote label method. Remote ^{15}N labels that have been used include *p*-nitrophenol ($pK_a = 7$), *m*-nitrobenzyl alcohol ($pK_a = 14$, alcohol-like), and the exocyclic amino group of adenine. Remote carbon labels include C1 of glucose and any carboxyl group that can be released as CO_2.

Equilibrium Perturbation

If the substrates and products of an enzyme are mixed in proportions corresponding exactly to equilibrium in the absence of the enzyme, the addition of the enzyme will have no effect on the ratios of reactants to products. However, if in such an experiment one of the original substrates or products contains a heavy isotope that appears in the transition state and elicits an isotope effect, addition of the enzyme will lead to a transient perturbation of the equilibrium mixture. The size of the displacement from equilibrium is a measure of the kinetic isotope effect.

Consider the example of malic enzyme (EC 1.1.1.38) catalyzing the reaction of malate with NAD^+, in which malate-2-*d* reacts with a primary deuterium kinetic isotope effect. In the equilibrium perturbation reaction illustrated in fig. 2-10, the static reaction is written on the left with an equal sign to represent a static reaction formulated at equilibrium concentrations of reactants and products, and A_{340} (NADH) is the equilibrium value. On adding a sample of malic enzyme, the system enters a dynamic state, in which the forward and reverse reactions take place, as represented by the chemical equation on the right side written with reversible arrows. Because malate-2-*d* reacts with a kinetic isotope effect, the forward reaction is initially slower than the reverse, so that the concentration of NADH and A_{340} initially decrease and then return to equilibrium values. The isotope effect can be calculated from the magnitude of the equilibrium perturbation. Because the equations are complex, a computer program has been written to facilitate the calculation (Cleland, 1980).

Intrinsic Isotope Effects

When measuring a kinetic isotope effect for an enzymatic reaction, the value is often not the intrinsic effect for the chemical step in which the heavy isotope participates. To be the intrinsic effect, the isotope-sensitive step must be the sole rate-limiting step, and other steps must be so fast that they are at equilibrium.

When the rate constants for flanking steps are similar to those of the isotope-sensitive step, the observed isotope effect is smaller than the intrinsic effect. The degree to which it is diminished can be quantified by defining forward and reverse commitment factors c_f and c_r,

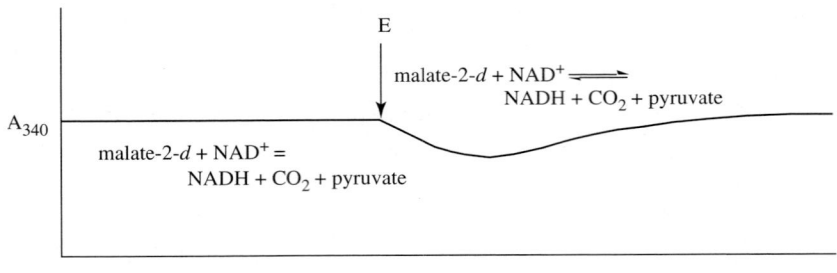

98 Enzymatic Reaction Mechanisms

Fig. 2-10. Equilibrium perturbation in measuring an isotope effect. A solution initially contains the reactants and products of malic enzyme mixed in equilibrium concentrations, with one reactant containing a heavy isotope, and no enzyme present. The A_{340} is monitored unchanged at its equilibrium value. The equality equation on the left represents the system in a static state formulated with equilibrium concentrations of reactants and products. On adding malic enzyme (E), the system enters a dynamic state. The A_{340} transiently decreases and then rises back to the equilibrium value, showing that the isotopically labeled reactant displays a kinetic isotope effect. Absence of a transient displacement of the absorbance at wavelength λ (A_λ) would have indicated the absence of a kinetic isotope effect. The magnitude of the perturbation in A_λ is a measure of the magnitude of the isotope effect.

which are measures of the tendency of the reaction to proceed in the forward and reverse directions from the isotope-sensitive step (Northrop, 1982). These tendencies are determined by the relative values of the rate constants for the isotope-sensitive and flanking steps. The kinetic model in scheme 2-14 exemplifies the situation, where the starred species represent the complexes in the isotope-sensitive step.

$$E + A \underset{k_2}{\overset{k_1}{\rightleftharpoons}} EA \underset{k_4}{\overset{k_3}{\rightleftharpoons}} EA^* \underset{k_6}{\overset{k_5}{\rightleftharpoons}} EPQ^* \underset{k_8}{\overset{k_7}{\rightleftharpoons}} EPQ \underset{k_{10}}{\overset{k_9}{\rightleftharpoons}} EQ \underset{k_{12}}{\overset{k_{11}}{\rightleftharpoons}} E$$

$$c_f = (k_5/k_4)(1 + k_3/k_2) \qquad c_r = (k_6/k_7)(1 + k_8/k_9)$$

Scheme 2-14

We do not normally know all of the rate constants and are often more interested in the relative values represented by the commitment factors, as defined in scheme 2-14. The commitment factors are model-sensitive and differ with variations in the kinetic mechanism (Northrop, 1982). The factors are constructed as the ratio of forward and reverse rate constants for an isotope-sensitive complex (k_5/k_4 for c_f) modified by a factor of one plus the ratio of forward and reverse rate constants for the next complex upstream or downstream ($1 + k_3/k_2$). When k_5 and k_6 are much smaller than the other rate constants, the commitment factors become zero.

The effects of the commitment factors on the observed isotope effects on V/K are given in eq. 2-22 (Northrop, 1982), where $^D k_5$ is the intrinsic isotope effect:

$$^D(V/K) = \frac{^D k_5 + c_f + c_r {}^D K_{eq}}{1 + c_f + c_r} \tag{2-22}$$

Division by $^D K_{eq}$ leads to the equation for the reverse direction. The effects of the commitment factors on $^D V$ are given by eq. 2-23.

$$^D V = \frac{^D k + c_{vf} + c_r {}^D K_{eq}}{1 + c_{vf} + c_r} \tag{2-23}$$

In eq. 2-23, c_{vf} = [k_3 k_5/(k_3 + k_4)][1/k_3+ (1/k_7)(1 + k_8/k_9) + 1/k_9 + 1/k_{11}]. In this case, division by $^D K_{eq}$ does not give the back equation.

To understand the mechanism, we would like the intrinsic isotope effect because it contains information about the geometry of the transition state. In cases of primary deuterium and tritium isotope effects, the intrinsic isotope effect $^D k_5$ can be evaluated by application of Northrop's equation (Northrop, 1982):

$$\frac{^D(V/K)-1}{^T(V/K)-1} = \frac{^D k - 1}{^D k^{1.44} - 1} \qquad (2\text{-}24)$$

This equation depends on the Swain relationship between deuterium and tritium isotope effects, according to which $^T k = (^D k)^{1.44}$ (Swain et al., et al., 1958). One measures both the observed deuterium and tritium isotope effects and computes $^D k$ using eq. 2-24. This method is exact when c_r =1.0 or $^D K_{eq}$ = 1.0. Otherwise, the method can be applied to the reaction in both directions to set limits.

In an alternative approach, multiple isotope effects can be measured to obtain values of, for example, $^D(V/K)$, $^{13}(V/K)_H$, and $^{13}(V/K)_D$. Three equations analogous to eqs. 2-22 and 2-23 with commitment factors can be written, one for each of the isotope effects. Recognizing that the commitment factors are the same in all three equations, one can assume ratios of commitment factors to solve the three equations simultaneously until reasonable and consistent values of the intrinsic isotope effects and commitment factors are obtained.

Three additional methods deserve mention. When by any means the commitment factors are eliminated, intrinsic isotope effects are observed. This can often be accomplished by raising the activation energy for the isotope-sensitive step while holding or lowering the barriers for the other steps. The situation is illustrated in the free energy or reaction progress profile in fig. 2-11. In an enzymatic reaction operating by a mechanism optimized through evolution, the activation barriers for the steps may be similar. Many of the steps are binding and conformational changes, and only one or a few are isotope-sensitive chemical changes. Through evolution, the activation barriers for the chemical steps have been lowered, but evolutionary pressures would not bring them below the barriers for binding.

There are three methods for artificially raising the chemical barriers: mutate an essential amino acid that catalyzes some part of a chemical step but is not very important for substrate binding (Berger et al., 2001); raise or lower the pH to place a catalytic group in the wrong state for catalysis (Cook and Cleland, 1981); or use an alternative substrate that reacts at a much slower rate than the natural substrate (i.e., with a higher activation barrier).

Although all of these methods can in principle unmask intrinsic isotope effects, there can be difficulties and ambiguities. Alteration of the chemical activation barriers by site-directed mutagenesis or by use of an alternative substrate can lead to unintended consequences. These may include the alteration of barriers to conformational changes that are required for catalysis. The ambiguities turned up in studies of UDP-galactose 4-epimerase (EC 5.1.3.2), where the deuterium kinetic isotope effect was not maximized either by an alternative substrate or by mutating the essential acid/base catalyst. The isotope effect had to be unveiled by measurement at low and high pH (Arabshahi et al., et al., 1988) or by mutating a less important amino acid that was not required for an essential conformational change (Berger et al., et al., 2001).

Hydrogen Tunneling

Intrinsic hydrogen isotope effects should be interpreted cautiously with respect to their meaning in assessing the structures of transition states. Because the de Broglie wavelengths

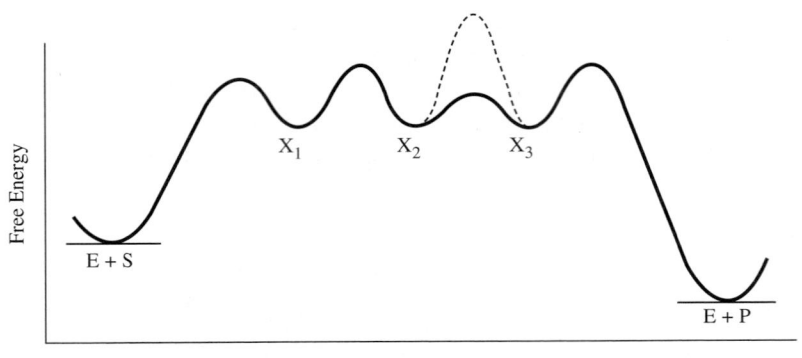

Fig. 2-11. Alteration of the activation barrier for a chemical step in a multistep enzymatic mechanism. In a multistep mechanism, the several enzyme-substrate (ES) complexes designated X_1, X_2, and so on may represent the products of substrate binding, conformational changes, and chemical transformations. In the plot, X_1 may be the Michaelis complex, X_2 a conformationally altered Michaelis complex, and X_3 an enzyme product complex. The chemical reaction is the transformation of X_2 to X_3, a step that is normally not rate limiting because the barrier is lower than those of other steps. This reaction does not display a kinetic isotope effect. If an alternative substrate is used or a specifically mutated form of the enzyme is used, the barrier for the transformation of X_2 to X_3 may be raised (dashed line), making it the rate-limiting step. Then, normal kinetic isotope effects on the chemical step may be observed.

of hydrogen (0.63 Å), deuterium (0.45 Å), and tritium (0.36 Å) are near the distances these atoms traverse in crossing the transition-state barrier (≈ 1 Å), they are subject to tunneling through the barrier (Bahnson and Klinman, 1995; Klinman, 1991). Hydrogen tunneling is possible in any enzymatic reaction in which hydrogen is transferred. This is a quantum mechanical effect, as distinguished from the semiclassical vibrational effect in fig. 2-9. Tunneling imposes consequences on the kinetic isotope effects and can lead to large, nonclassical isotope effects. A narrow barrier as illustrated in fig. 2-12 can favor hydrogen tunneling.

Detailed analyses of hydrogen isotope effects in hydride transfer by alcohol dehydrogenase, hydrogen atom transfer by monoamine oxidase B, and proton transfer by serum amine oxidase uncovered hydrogen tunneling (Cha et al., 1989; Grant et al., 1989; Jonsson et al., 1994). Evidence of tunneling has also appeared in studies of adenosylcobalamin-dependent enzymes, which catalyze hydrogen atom transfer. Reactions with hydrogen tunneling constitute a broad range of reaction types involving hydride, hydrogen atom, and proton transfer.

Violations of the rule of the geometric mean in hydrogen transfer indicate hydrogen tunneling (Bahnson & Klinman, 1995; Klinman, 1991). According to this rule, the observed kinetic isotope effect is the product of primary and secondary effects in the same transition state (Bigeleisen, 1955). This means that the heavy isotopes operate independently on the energy of the transition state. A violation of this rule is evidence of hydrogen tunneling, and for this reason, the measurement of secondary isotope effects is important in studies of tunneling.

In the reaction of yeast alcohol dehydrogenase, hydride transfer displays a primary kinetic isotope effect, and a hydrogen in the same transition state but not transferred displays a secondary kinetic isotope effect. Such a hydrogen would be on in the nontransferring position of the —CH_2OH group or the C4(H) of NAD^+. The first indication of hydrogen tunneling in these reactions appeared in the values of the secondary isotope effects, which were very large (1.22 to 1.35) compared with the equilibrium isotope effects of 0.9 to 1.04 (Cook et al., 1981; Klinman, 1991; Kurz and Frieden, 1980; Welsh et al., 1980).

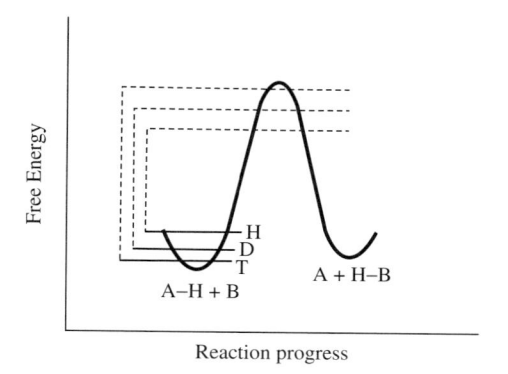

Fig. 2-12. Hydrogen tunneling through a narrow barrier. The dotted lines indicate the large free energy differences associated with protium (H), deuterium (D) and tritium (T) transfer from heteroatom A to heteroatom B.

Coupled motions of the transferred and nontransferred hydrogens in the transition state contributed to this effect but could not fully account for it without invoking hydrogen tunneling (Huskey and Schowen, 1983).

Two additional experimental tests for hydrogen tunneling can be applied (Bahnson & Klinman, 1995; Klinman, 1991). First, a breakdown of an alternative expression of the Swain-Schaad relationship, $^Tk = (^Dk)^{1.44}$, indicates tunneling. The exponent in the alternative relationship $(k_H/k_T) = (k_D/k_T)^{3.26-3.34}$ lies in a narrow range, and a larger experimental value indicates hydrogen tunneling. Accurate values of k_D/k_T can be measured in a single reaction with tritium as a trace label by using deuterium as a trace label together with ^{14}C as a remote label. The ratio of $^3H/^{14}C$ in the initial product is then directly related to k_D/k_T. Second, anomalous temperature effects on the rates of hydrogen and tritium transfer can indicate tunneling. Hydrogen and tritium transfer will display the same Arrhenius prefactor (A_H) when there is no tunneling. The Arrhenius prefactor is obtained in plots of $\ln k$ against $1/T$ based on the equation $\ln k_H = \ln A_H + E_A(H)/RT$. Different Arrhenius prefactors for H and T indicate hydrogen tunneling. The narrow temperature range over which enzymes can be studied hampers the application of this method. In all experiments to investigate hydrogen tunneling, the hydrogen transfer step must be rate limiting.

Hydrogen tunneling is most likely when hydrogen moves a short distance in traversing the transition state. A narrow barrier can bring about this condition (see fig. 2-12). Binding interactions in active sites of enzymes set up conditions for hydrogen tunneling. Because tunneling leads to very large kinetic isotope effects, the structure of a transition state for hydrogen transfer cannot be assigned based solely on the magnitude of the primary kinetic isotope effect. Carbon, with a de Broglie wavelength of 0.18 Å, and other heavier atoms do not tunnel, and the primary and secondary kinetic isotope effects may be indicative of transition-state structure in accord with semiclassical models.

Transient-Phase Kinetics

Reaction Characteristics

The steady-state parameters k_{cat} and k_{cat}/K_m give information about the overall rates but usually do not tell us anything about the rates of individual steps in catalysis. They consist of aggregates of rate constants that cannot be separated into individual constants by steady-state methods. All enzymatic reactions proceed through a series of steps, and even the simplest involve substrate binding and dissociation steps, at least one product-forming step, and a product-dissociation step. In the simple kinetic model shown in scheme 2-15, there are five rate constants in the mechanism.

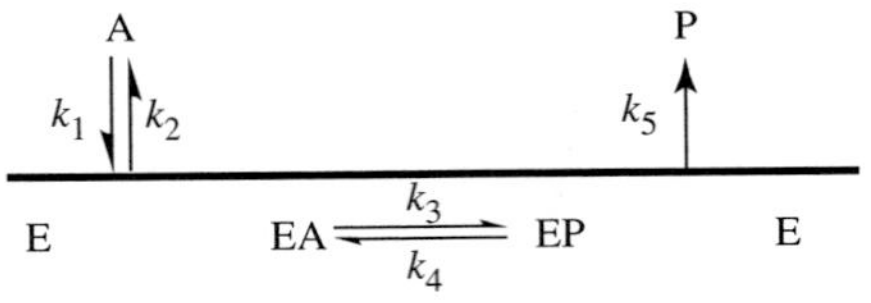

$$E \xrightarrow[k_1]{A} \xrightarrow{k_2} EA \underset{k_4}{\overset{k_3}{\rightleftarrows}} EP \xrightarrow{k_5} E$$

$$v = \frac{k_{cat}[E_0][A]}{[A] + K_{ia}K_a}$$

$$k_{cat} = \frac{k_3 k_5}{k_3 + k_4 + k_5} \qquad K_a = \frac{k_5(k_2 k_4 + k_2 k_5 + k_3 k_5)}{k_2(k_3 + k_4 + k_5)} \qquad K_{ia} = k_2/k_1$$

Scheme 2-15

None of the rate constants can be determined from the steady-state kinetic parameters k_{cat} and k_{cat}/K_m, where $K_m = K_{ia} K_a$. If two of the rate constants contributing to k_{cat} could be measured independently, then the third could be calculated from the expression for k_{cat} in scheme 2-15. If k_1 or k_2 could be measured independently, the other could be calculated from k_{cat}/K_m, and that would complete the evaluation of all of the rate constants in scheme 2-15. In transient kinetic experiments, portions of kinetic mechanisms are studied to evaluate individual rate constants.

Transient kinetic methods differ from steady-state techniques most fundamentally in two ways. First, they generally entail the use of much higher concentrations of enzyme than in steady-state kinetics to allow for the detection of intermediates by physicochemical methods. These concentrations are often in the range of K_m values for substrates, and in a transient kinetic study the substrate concentration may be only slightly higher than or even lower than that of the enzyme. This leads to the second major difference, which is that the steady-state approximation may not be valid. The approximation is always valid when the enzyme concentration is much higher than that of the substrate. It is certainly invalid when the difference is less than 10-fold. This means that the rate equations for even slightly complex mechanisms cannot be solved by the methods used in steady-state kinetics.

Each transient kinetic study must be designed with regard to the chemistry that is taking place. The simple case of scheme 2-15 could be further simplified by the chemistry if the reaction is a hydrolysis or decarboxylation, in which case it could be practically irreversible so that the value of k_4 would be zero. Then fewer rate constants would have to be evaluated. The measurement of any of the rate constants must be tailored to the chemistry and the properties of the enzyme. No generalized approach analogous to steady-state kinetic analysis is available for solving these problems.

Available methods can in principle allow such experiments to succeed when the steps are not too fast for the methods in hand. Most of the methods include rapid mixing flow systems, and the time required for mixing two solutions becomes a limitation on the measurable rates. The mixing devices will allow processes occurring on the time scale of ≥3 ms to be studied. First-order rate constants of up to 200 s^{-1} can be measured by flow methods. Faster processes should be studied in other ways, such as by equilibrium perturbation.

Transient Methods

Burst Kinetics

The chymotrypsin-catalyzed hydrolysis of *p*-nitrophenyl acetate (PNPA) is a simple reaction that can be studied by a transient kinetic method and yields valuable mechanistic information. PNPA is a very poor substrate for chymotrypsin that is easily assayed

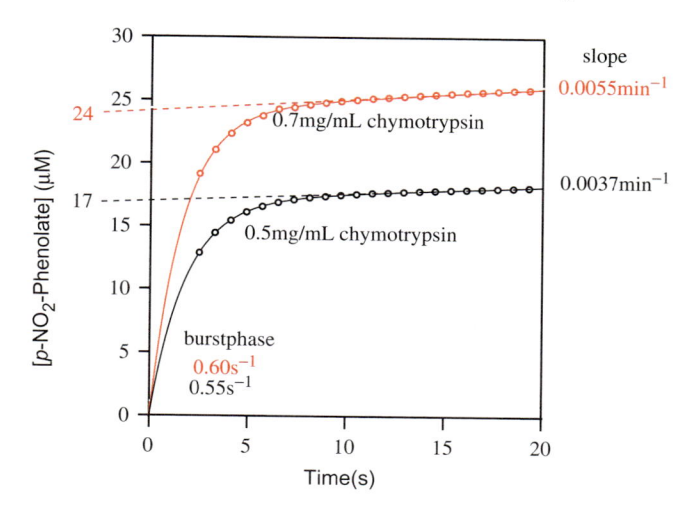

Fig. 2-13. Burst kinetics in the chymotrypsin-catalyzed hydrolysis of *p*-nitrophenyl acetate.

spectrophotometrically because of the yellow color of one of the products, *p*-nitrophenolate ion (PNP⁻). Rapid addition of PNPA to a solution of chymotrypsin (at 0.5 to 1 mg mL^{-1}), followed by observation of the change in absorbance at 400 nm, beginning within a few seconds of mixing, reveals a fairly rapid increase in A_{400} for about 10 seconds, followed by a slower, gradual increase. This biphasic behavior, known as burst kinetics, is illustrated in fig. 2-13. In the experiment, the solutions were mixed by hand within 3 seconds, and the resulting solution monitored for the increase in A_{400}, signaling the formation of PNP⁻. The two kinetic phases represent two steps in the chemical mechanism, acetylation of chymotrypsin to form acetyl-chymotrypsin and PNP⁻, followed by hydrolysis of acetyl-chymotrypsin to acetate and free chymotrypsin.

The quantitative analysis of this simple transient experiment exemplifies some of the complexities that attend transient kinetic studies of enzymes. Given the facts of fig. 2-13 and the Michaelis-Menten behavior of chymotrypsin (Cht), the simplest kinetic mechanism that can be written for the hydrolysis of PNPA is that of eq. 2-25.

$$\text{Cht-OH} + \text{PNPA} \xrightleftharpoons{K_S} \text{Cht-OH/PNPA} \xrightarrow{k_2} \text{Cht-O-acetyl} \xrightarrow{k_3} \text{Cht-OH} + \text{acetate} \quad (2\text{-}25)$$
$$\downarrow$$
$$\text{PNP}^-$$

Because PNPA is a poor substrate, the transient kinetics can be studied on the time scale of seconds and under steady-state kinetics, with [PNPA] >> [Cht]. Even then, the integrated rate equation for the appearance of PNP⁻ with time is complex (Bender et al., 1967) as shown by eq. 2-26, which is based on the further (justified) assumptions that the binding of PNPA is an equilibrium process, acylation and deacylation are irreversible, and $k_2 < k_3$.

$$[\text{PNP}^-] = At + B(1 - e^{-kt})$$

$$A = \frac{\frac{k_2 k_3}{(k_2 + k_3)}[\text{Chto}][\text{PNPAo}]}{[\text{PNPAo}] + \frac{K_S k_3}{(k_2 + k_3)}} \qquad B = [\text{Chto}] \frac{\left(\frac{k_2}{k_2 + k_3}\right)^2}{\left(1 + \frac{K_m^{app}}{[\text{PNPAo}]}\right)^2} \quad (2\text{-}26)$$

At long times, the equation takes the form $[\text{PNP}^-] = At + B$ and corresponds to the second, linear phase in fig. 2-13. The expression for A in eq. 2-26 is the Michaelis-Menten equation for the hydrolysis of PNPA, in which $k_{cat} = k_2k_3/(k_2 + k_3)$, $K_m^{app} = K_S k_3/(k_2 + k_3)$, and B is the extrapolated intercept in fig. 2-13. The intercept is within a few percent of the total concentration of chymotrypsin, and the burst is proportional to the enzyme concentration.

The apparent rate constant k for the formation of acetyl-chymotrypsin can be evaluated from data such as in fig. 2-13 by subtracting the secondary rate from each data point, leaving only the burst data. These data then can be fitted to the first-order rate equation to evaluate k. It was shown that $k \approx (k_2 + k_3)/\{[PNPA_0] + K_S\}$ (Bender et al., 1967). By measuring k over a range of PNPA concentrations, this relationship made it possible to evaluate $(k_2 + k_3)$ and also K_S. The secondary steady-state rates from the same experiments allowed the evaluation of A in eq. 2-26 over the same range of PNPA concentrations. Using the relationship of A with $[PNPA_0]$ in eq. 2-26, $k_2k_3/(k_2 + k_3)$ could be evaluated. Then the rate constants k_2 and k_3 were calculated from the values of $(k_2 + k_3)$ and $k_2k_3/(k_2 + k_3)$ and found to be 0.37 s^{-1} and 1.3×10^{-4} s^{-1}, respectively, at pH 8.2.

The reaction of PNPA with chymotrypsin gave the first kinetic evidence of the intermediate formation of an acyl-chymotrypsin intermediate in catalysis. We now know that all substrates for chymotrypsin react to form acyl-enzyme intermediates (see chap. 6). Acetylation of chymotrypsin by PNPA is at least 1000 times faster than hydrolysis of acetylchymotrypsin, and the second step limits the overall rate in the steady state (second phase). This is true in general for ester substrates of chymotrypsin. However, in the hydrolysis of amide and peptide substrates the acylation step is rate limiting, with hydrolysis of the acylchymotrypsin being much faster (see chap. 6). Consequently, the hydrolyses of amide and peptide substrates by chymotrypsin do not proceed with burst kinetics.

Because PNPA is a poor substrate, chymotrypsin does not catalyze the reaction as fast as the hydrolysis of a specific substrate, and this facilitates the transient kinetic analysis. The technique of using a poor substrate to simplify kinetic analysis was introduced in the section on isotope effects and turns up repeatedly. Similarly, the technique of site-directed mutagenesis to disable some aspect of catalysis to assess its importance or to increase the focus on another aspect of the mechanism is also widely employed.

Stopped-Flow Spectrophotometry

When reactions are too fast to allow observation after hand mixing, similar information may be obtained by the application of flow methods. These methods entail fast, mechanical mixing of two solutions, each of which contains parts of the final reaction mixture. In commercial or home-built mixing apparatuses, the solutions are in syringes connected to a mixing device, and when the two plungers are driven forward together, the solutions are forced through a mixing chamber into a single tube. The reaction proceeds in the emergent solution.

Any of a large number of detection methods can be applied to monitor the rate of reaction in a rapidly mixed solution. In enzymatic studies, stopped-flow absorption or fluorescence spectrophotometry is often used. In this technique, the mixed solution passes through a detection cell for spectrophotometric or fluorescence observation, as shown in fig. 2-14A.

Shortly after mixing, the flow is abruptly stopped, and the absorbance or fluorescence in the detection cell can then be monitored with time. This technique has two advantages for enzyme kinetic studies: the relatively small amounts of enzyme required (compared with other rapid-mix techniques) and the convenience of photometric or fluorescence detection, which allows hundreds or thousands of data points on the progress curve to be obtained in an automated instrument. The only disadvantage is that chromophores or fluorophores must be part of the system being analyzed.

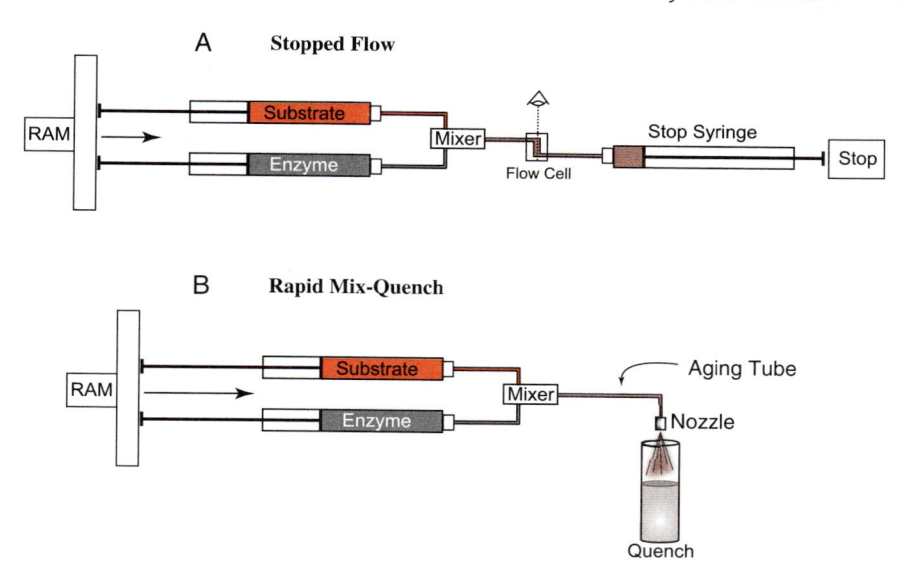

Fig. 2-14. Rapid-flow techniques. (A) In stopped-flow spectrophotometry, reactants are mixed from two syringes in sufficient volumes to flush out a flow cell as the syringes are pushed in a short, rapid burst by a ram. The contents of the flow cell are monitored (e.g., UV-Vis, fluorescence spectrometry) so that the reaction can be characterized as it proceeds (from tens of milliseconds to several minutes). (B) In the rapid mix-quench technique, similarly to stopped flow, reactants are mixed from two (or more) ram-driven syringes. Rather than being monitored directly, the emerging reaction is sent through an aging tube into a quench solution so that a significant amount of material can be collected that reflects the progression of the reaction at a single time point. By varying the speed of the ram or (more typically) the length of the aging tube, a variety of reaction times can be sampled in a series of rapid-mixing experiments and then subsequently analyzed by a wide variety of techniques.

Rapid Mix-Quench

For experiments in which no chromophore or fluorophore is available, rapid mix-quench methods are increasingly applied. In this approach (fig. 2-14B), the two solutions are mixed much as in the stopped-flow apparatus; however, in place of an observation cell, the reaction is quenched by injection into a stopping solution at various times after mixing. The timed, quenched samples are then analyzed for the reaction progress. As a timed-point method, only a limited number of data points are normally measured to define a reaction progress curve. Data from a sufficient number of runs for a statistical analysis can be fitted to first- or second-order progress curves to determine rate constants.

The keys to the success of rapid mix-quench experiments are the development of methods for quenching the reactions and the development of assays for one or more products (analytes) in the quenched solutions. The solutions to these problems are usually tailored to the system. There are a few general approaches to the selection of quenching and assay methods. However, this field is open to the imagination and skill of the investigator, and any method that works is acceptable. Beginning with quenching methods, most enzymatic reactions are stopped by strong acids or bases because of enzyme denaturation. If the analytes are stable in acidic or basic solutions, they can be analyzed in the quenched solutions. When an analyte in the study is labile to acid or base, the reaction can be quenched by injecting the mixed solution into a denaturant such as guanidinium chloride or hot ethanol to inactivate the enzyme without degrading the analyte. For low temperature analytical techniques (e.g., EPR, Mössbauer spectroscopy) the reaction may be stopped by freeze quench (Ballou, 1978). The efficacy of a quenching procedure must be verified before proceeding.

Chemical, enzymatic, and spectroscopic methods have been used to analyze quenched samples. The analyte may often be a substrate for another enzyme for which a convenient assay is available. The quenched sample is prepared for the enzymatic assay by removing or neutralizing the acid, base, or denaturant. New methods for assaying analytes in quenched samples are constantly being developed. Mass spectrometry has been used to analyze pre–steady-state turnover in several enzymatic systems (Gross et al., 2001; Northrop and Simpson, 1998; Paiva et al., 1997; Zechel et al., 1998). Mass spectrometric analysis may prove to be a general method for analysis of rapidly mixed-quenched enzymatic reactions, but each system will ultimately present its own challenges and advantages.

Kinetics of Partial Reactions: Tyrosyl-tRNA Synthetase

Stopped-flow spectrofluorometry and rapid mix-quench kinetics of tyrosyl-tRNA synthetase (EC 6.1.1.1) exemplify the tandem application of transient flow methods. Aminoacyl-tRNA synthetases catalyze the formation of aminoacyl-tRNAs by a bi uni uni bi ping pong kinetic mechanism similar to that in scheme 2-12 for asparagine synthetase. In the overall reaction, the amino acid reacts with ATP and tRNA to produce the aminoacyl tRNA, AMP, and PP_i. In the case of tyrosyl-tRNA synthetase, tyrosine and ATP bind to the enzyme and react to form enzyme-bound tyrosyl AMP and PP_i in the absence of $tRNA^{Tyr}$. The complex E-tyrosyl-adenylate then binds $tRNA^{Tyr}$, the tyrosyl group undergoes transfer to $tRNA^{Tyr}$, and tyrosyl $tRNA^{Tyr}$ and AMP dissociate from the enzyme. The complex of E-tyrosyl-AMP can be isolated and studied structurally and kinetically in isolation from $tRNA^{Tyr}$ and the other components of the overall reaction. Figure 2-15 shows the structure of tyrosyl-tRNA synthetase with tyrosyl-AMP (tyrosyl adenylate) bound in the active site.

The binding of tyrosine elicits a small decrease in the fluorescence of tyrosyl-tRNA synthetase, and the observation of this decrease in stopped-flow fluorescence experiments allowed the binding rate constant (2.4×10^6 M^{-1} s^{-1}) and the dissociation rate constant (24 s^{-1}) to be measured. A further decrease in fluorescence subsequent to tyrosine binding to the enzyme-ATP complex signals the formation of tyrosyl-AMP and PP_I with a rate constant of 18 s^{-1}.

The isolated complex of E-tyrosyl-AMP reacts in reverse with PP_I to regenerate tyrosine and ATP with an increase in the protein fluorescence, the opposite of the fluorescence decrease observed on formation of tyrosyl-adenylate. A stopped-flow fluorescence trace in fig. 2-16A yields the same rate constant as the rate constant for ATP formation obtained under the same conditions in the rapid mix-quench experiment of fig. 2-16B (Fersht et al., 1975). The rapid mix-quench result confirms the interpretation of the stopped-flow fluorescence experiment. This is an excellent example of the care that should be exercised in transient phase studies, especially when employing an indirect method of detection.

Nonsteady-State Analysis

When the steady-state approximation is not valid, the kinetic equations cannot be solved easily or can be solved only for the simplest mechanisms and a few special cases (Hammes and Schimmel, 1970). The simplest approach in such cases is to obtain the relevant data on the appearance and decay of chemical intermediates with time by one of the flow methods. Chemical information and intuition can usually guide the search for intermediates. With the data in hand, one can then consider kinetic mechanisms and choose one or a few

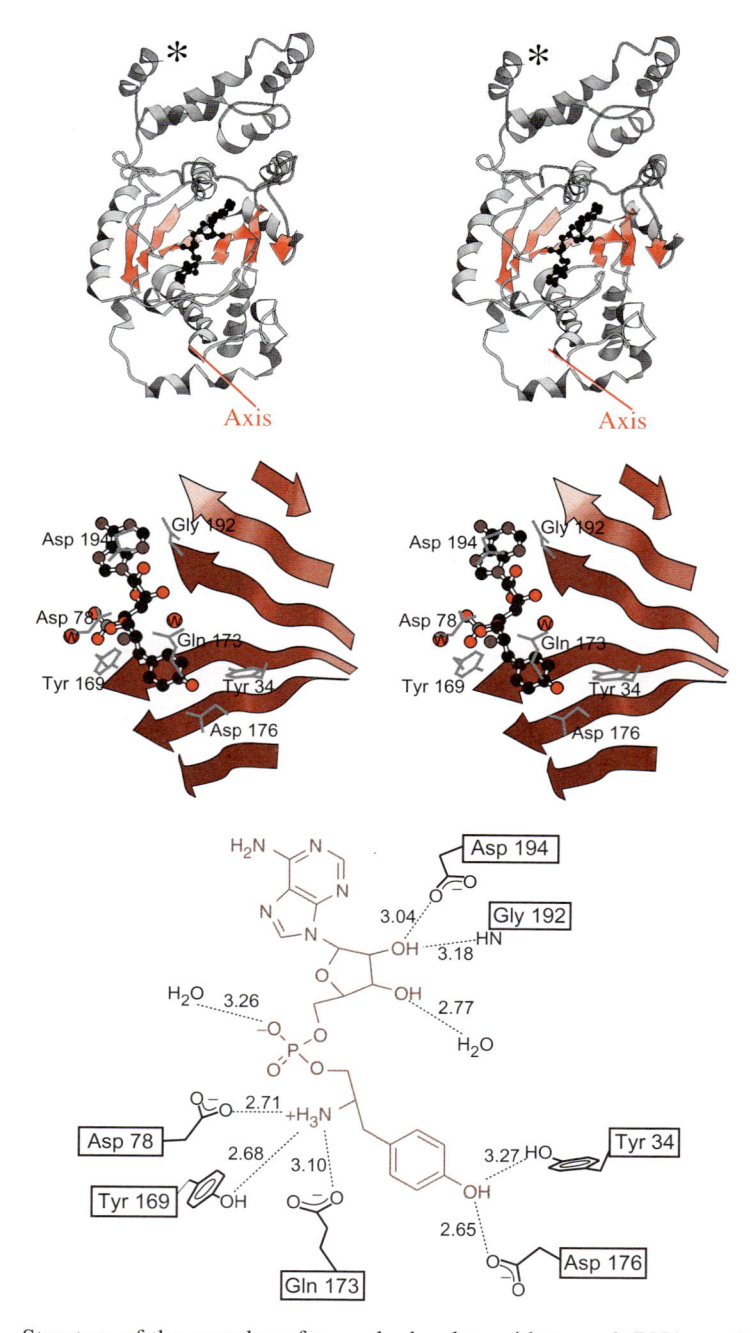

Fig. 2-15. Structure of the complex of tyrosyl-adenylate with tyrosyl tRNA synthetase deletion mutant. Deletion of residues 319 to 418 from the disordered C-terminal tRNA binding domain of the *Bacillus stearothermophilus* tyrosyl tRNA synthetase (EC 6.1.1.1) has allowed refinement of this structural model (PDB 3TS; Brick et al., 1989) to 2.3-Å resolution. The top stereodiagram shows one monomer of the homodimeric synthetase, with the tyrosyl-adenylate shown as a black ball-and-stick model traversing one edge of a central β-sheet (red). The twofold axis of symmetry is also shown in red, and the C-terminal deletion is marked by an asterisk. The second stereogram shows the tyrosyl-adenylate binding site with more detail, including interacting enzyme aminoacid side chains in gray and fixed water molecules as red spheres (W). The bottom figure shows the site schematically, and provides probable contacts (dotted lines) and distances (Å).

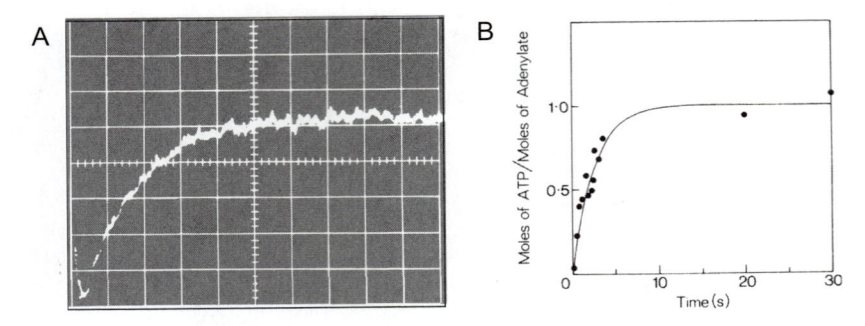

Fig. 2-16. Stopped-flow and rapid mix-quench kinetics of the cleavage of tyrosyl-adenylate by pyrophosphate at the active site of tyrosyl tRNA synthetase. (A) Stopped-flow fluorescence trace after mixing of PP$_i$ with the complex of tyrosyl-tRNA synthetase and tyrosyl-adenylate (k_{obs} = 0.39 s^{-1}). (B) The same reaction was studied in a rapid mix-quench experiment with measurement of the ATP formed. The observed rate constant was 0.37 s^{-1}. (Adapted from Fersht et al., 1975.)

that are compatible with the available information on the transient intermediates. Such a mechanism may involve two intermediates, X$_1$ and X$_2$, as in eq. 2-27, which is governed by the differential eqs. 2-28a to 2-28e.

$$E + A \underset{k_2}{\overset{k_1}{\rightleftharpoons}} X_1 \underset{k_4}{\overset{k_3}{\rightleftharpoons}} X_2 \overset{k_5}{\longrightarrow} E + P \qquad (2\text{-}27)$$

$$-d[A]/dt = k_1[E][A] - k_2[X_1] \qquad (2\text{-}28a)$$

$$-d[E]/dt = k_1[E][A] - k_2[X_1] - k_5[X_2] \qquad (2\text{-}28b)$$

$$-d[X_1]/dt = -k_1[E][A] + (k_2 + k_3)[X_1] - k_4[X_2] \qquad (2\text{-}28c)$$

$$-d[X_2]/dt = -k_3[X_1] + (k_4 + k_5)[X_2] \qquad (2\text{-}28d)$$

$$-d[P]/dt = -k_5[X_2] \qquad (2\text{-}28e)$$

The conservation equations are [A$_0$] = [A] + [X$_1$] + [X$_2$] + [P] and [E$_0$] = [E] + [X$_1$] + [X$_2$]. When the steady-state approximation is invalid, as it is when the [E$_0$] and [A$_0$] are comparable or enzyme is present in modest excess, no simple integrated rate equation can be obtained. However, the differential equations can be solved numerically by computer fitting to experimental data on the concentrations of A, X$_1$, X$_2$, and P with time throughout the course of the reaction. Computational packages are available, several of which are configured for direct application to enzymatic mechanisms such as those previously given, with any number of intermediates. Simulated data for this mechanism appear in fig. 2-17.

In a real experiment, values of [A$_t$], [X$_t$], [X$_2$], and [P$_t$] are measured throughout the course of the reaction, and the mechanism is evaluated for consistency by computer fitting eqs. 2-28 to the data. A satisfactory fit of the experimental data to the computed lines supports the assignment of the mechanism. The method is valid for any number of intermediates that can be quantitatively measured. In general, it is best to obtain analytical measurements for all intermediates in the mechanism; however, if one intermediate is unobservable, but its existence is chemically required and certain, it may be calculated from the concentrations of the other species by use of the mass conservation equations. Data can then be fitted to the required differential equations.

Kinetics of Enzymatic Reactions   109

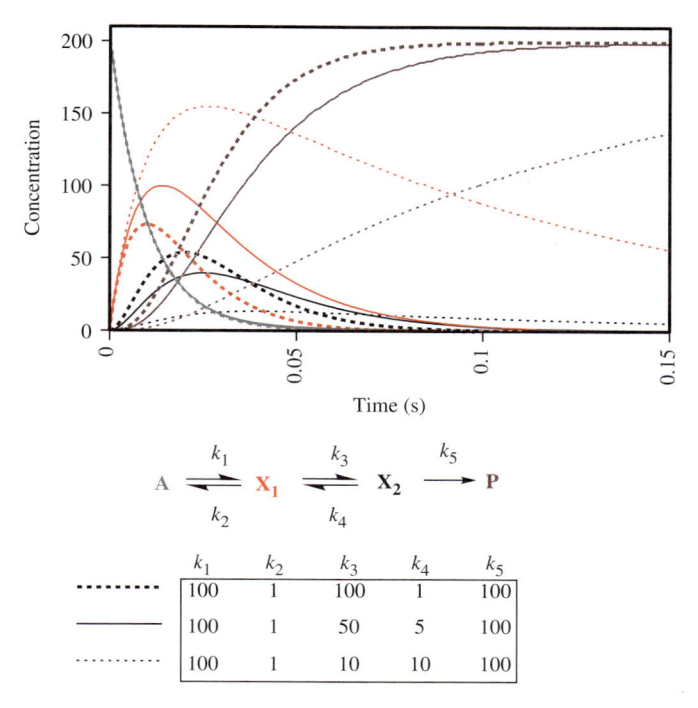

$$A \underset{k_2}{\overset{k_1}{\rightleftharpoons}} X_1 \underset{k_4}{\overset{k_3}{\rightleftharpoons}} X_2 \overset{k_5}{\longrightarrow} P$$

	k_1	k_2	k_3	k_4	k_5
·······	100	1	100	1	100
———	100	1	50	5	100
·······	100	1	10	10	100

Fig. 2-17. Time course of reacting components in a mechanism with two intermediates. This is a simulation of the disappearance of the substrate A, appearance of the product P, and transient appearance of the intermediates X_1 and X_2 in the mechanism of eq. 2-27. The simulation is based on the tabulated assumed values of the rate constants. In a kinetic experiment, [A_t], [X_t], and [P_t] would be measured throughout the time course of the reaction. The data would then be computer fitted to eq. 2-28, governing the mechanism by adjusting the rate constants until a satisfactory fit is obtained. If satisfactory, the rate constants arising from the fit would be accepted as valid.

Relaxation Methods

Many enzymatic rates are too fast to measure by flow methods because of limitations on the mixing time. In favored cases, relaxation methods will allow the measurement of rates up to three orders of magnitude faster than can be observed in flow experiments. Relaxation refers to a return to equilibrium after perturbation. Most equilibria are sensitive to reaction conditions such as temperature or pH. When a reaction at equilibrium is subjected to a sudden change in temperature or pH, the new equilibrium is re-established by relaxation. The time constant τ for the relaxation, a reciprocal rate constant, can contain information about the rates of individual steps (Hammes and Schimmel, 1970). To be valid, the system must be at or near equilibrium. This constraint is less severe than one might think, because at least some steps of many enzymatic processes may be studied near equilibrium. Initial collisional binding of a substrate is sometimes at or near equilibrium before subsequent steps become kinetically significant.

As in any rate measurement, observation of a relaxation requires an analytical technique. Because of the fast rates encountered in enzymatic relaxation experiments, the analytical method must generally be some kind of spectroscopy. Fluorescence is often useful for this purpose. Tryptophan and tyrosine are the most important fluorophores in enzymes, and their fluorescence properties often change when a substrate binds. When a change in the intensity of fluorescence emission can be correlated with the kinetic process under study, fluorescence is then the obvious choice for monitoring relaxations. When the kinetic process involves a chromophore, spectrophotometric detection is the most obvious choice.

The temperature jump relaxation method has allowed many rate constants for substrate binding to be measured. Temperatures of reacting solutions can be suddenly increased by a small electrical discharge. A reaction at equilibrium subjected to a sudden jump in temperature will relax to a new equilibrium position. If this can be monitored, such as by a change in the fluorescence of the protein, the rate constant can be measured. The technique allows rate constants to be measured on the microsecond time scale, so that processes too fast to be observed by flow methods may be observed by the temperature jump technique.

A simple application to substrate binding would make use of a change in tryptophan fluorescence when a substrate binds according to the first step of eq. 2-27. If this step approaches equilibrium $k_2 \geq 10k_3$, the first step can be studied by temperature jump relaxation in a rapidly mixed solution. It can be shown that the time constant for the relaxation is given by eq. 2-29, where E_{eq} and A_{eq} refer to equilibrium concentrations (Hammes and Schimmel, 1970).

$$\tau^{-1} = k_1[E_{eq} + A_{eq}] + k_2 \quad (2\text{-}29)$$

A plot of the observed rate constant for the relaxation against the concentration term at increasing total concentration of A ($[A_0]$) gives a straight line with a positive slope k_1 and the intercept k_2. If k_2 is less than 10-fold larger than k_3 in the mechanism of eq. 2-27, then the plot will level off at high concentrations of the substrate, and the initial slope and intercept can be estimated by curve fitting to a slightly more complex equation (Hammes and Schimmel, 1970).

The association rate constant k_1 for substrate binding is of considerable interest, and a sufficient number of measurements are available to set a range of values commonly observed in enzymatic mechanisms. Tabulated values show that this rate constant is generally in the range of 10^7 to 10^8 M^{-1} s^{-1} (Hammes and Schimmel, 1970). The values are within one to two orders of magnitude smaller than the diffusional controlled limiting rate constant for the collision of small molecules (10^9 M^{-1} s^{-1}). Because of the large sizes of enzymes, a smaller value for the diffusional limit can be expected. In cases in which the value of k_1 is significantly less than 10^8 M^{-1} s^{-1}, a conformational change or other coupled isomerization most likely accompanies binding.

Estimation of Off-Rates by Isotope Partitioning

In a mechanism such as that defined by eq. 2-27, the first step often is not near equilibrium ($k_2 \leq k_3$). In such cases, the on- and off-rate constants cannot be measured directly, but the isotope partitioning method can give an estimate of the value of the off-rate constant k_2 (Rose, 1980). Alternatively, it can prove that $k_2 \ll k_3$; that is, the first step is effectively irreversible under steady-state conditions.

The isotope partition method is a pulse-chase mixing experiment with the enzyme and a radiochemically labeled form of the substrate A*. In the pulse step, the enzyme is mixed with a radioactively labeled form of substrate A* at a concentration near K_m to ensure that part or all the substrate binding sites become occupied as EA*. In the chase step, this solution is mixed with a much larger volume of a solution containing unlabeled substrate A and all other substrates and cofactors of the complete reaction, so that it will proceed forward to products. The chase creates a pool of free substrate in which the radioactive label has been greatly diluted; the specific radioactivity of the free substrate is much lower than that of EA*, and product is formed. After the chase, the partitioning of labeled substrate in EA* into the pool of free, effectively unlabeled A and into the product as P* can be monitored by quenching the reaction with a strong acid. The radioactivity in the isolated substrate and product constitute information about the partitioning of EA*. If $k_2 \ll k_3$, all of the labeled substrate in EA* will appear in the product as P*. If k_2 and k_3 are comparable within an order of magnitude, partitioning into A and P can give information about the value of k_2.

The quantitative analysis of partitioning depends on the kinetic mechanism of the reaction. An analysis for a two-substrate enzyme is available (Rose, 1980). This method applied first to hexokinase using [^{14}C]glucose as the labeled substrate showed that all of the glucose in E.[^{14}C]glucose generated in the pulse could be transformed into [^{14}C]glucose-6-P after the chase with unlabeled glucose and ATP (Rose et al., 1974). In the steady state, about one half of the glucose in ternary complexes is in the form of glucose-6-P, showing that the equilibrium constant for phosphoryl transfer from ATP is about 1 (Wilkinson and Rose, 1979).

pH-Rate Profiles

Profile Interpretation

The effects of pH on enzymatic processes can give information about the ionization states of substrates or amino acid side chains participating in binding or catalysis. The pH dependencies are rarely simple, however, and the enzymatic functional groups often cannot be definitively assigned based solely on the pH profiles. Complementary biochemical, spectroscopic, or structural information can often facilitate interpretations of pH profiles. Here we discuss the proper interpretations of the simplest pH dependencies in enzyme action and summarize some of the complications that can be encountered in pH studies.

Measurements of pH-Rate Profiles

Parameters of Interest

The pH dependencies of V/K, V_m, and K_{ia} for each substrate, and of K_i for inhibitors can give useful information. In this section, V and K refer to the apparent maximum velocity and Michaelis constant, respectively, at a given pH. The pH dependencies of the apparent inhibition constant, $^{app}K_i$, for a competitive inhibitor or $^{app}K_{ia}$ for a substrate give valuable information about the ionization states of the inhibitor or substrate and the enzyme in the binding process.

pH Dependence of Ki

Values of $^{app}K_i$ measured at intervals of at least 0.5 pH unit over the full range of pH at which the enzyme can be assayed are plotted as $-\log\ ^{app}K_i$ (pK_i) versus pH. The profile should be composed of straight-line segments of slope 0, +1, −1, +2, −2, and so on that are connected by curvature at well-defined break points. Data can be obtained at closer pH intervals near the break points. The integral slopes signify integral proton uptake (−1, −2) or release (+1, +2) associated with binding the inhibitor, and data can be fitted to appropriate equations to evaluate pK_a values.

In the general case, described with exceptional clarity by Cleland (1977), a competitive inhibitor may bind to an enzyme more tightly at neutral than at low pH, and the profile

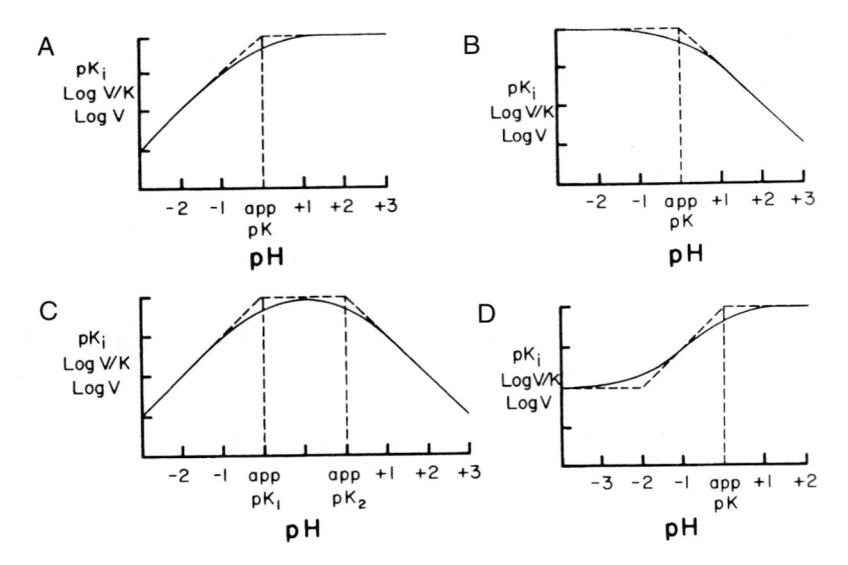

Fig. 2-18. Typical pH profiles for K_i, V/K, and V. The curves are calculated traces for profiles of log V/K and log V for the enzymatic reaction of a substrate and of pK_i for inhibition of an enzyme by a competitive inhibitor. (A) Enzymatic activity is gained or binding of a molecule is increased by the ionization of one group. (B) Enzymatic activity is lost or binding of a molecule is decreased by the ionization of one group. (C) Enzymatic activity or binding of a molecule is optimal when one group is protonated and another is deprotonated. (D) Enzymatic activity or binding of a molecule is decreased but not eliminated by protonation of a group on the enzyme or a binding molecule. (Adapted from Cleland, 1986.)

may be similar to that in fig. 2-18D. In this case, there are four enzyme forms—E, HE, EI, and HEI—that are connected by the interactions defined in scheme 2-16.

$$\text{HE} \underset{\pm I}{\overset{K_j}{\rightleftharpoons}} \text{HEI}$$

$$K_1 \left\| \pm H^+ \right. \qquad \pm H^+ \left\| K_3 \right. \qquad \text{app}K_i = \frac{K_i(1 + [H^+]/K_1)}{(1 + [H^+]/K_3)}$$

$$\text{E} \underset{\pm I}{\overset{K_i}{\rightleftharpoons}} \text{EI}$$

Scheme 2-16

The acid dissociation constants are K_1 and K_3, the inhibitor dissociation constants are K_i and K_j, and they are related as $K_1 K_3 = K_i K_j$. The equation in scheme 2-16 relates the apparent inhibition constant at any pH to the competitive inhibition constant at the higher pH and the two acid dissociation constants. The values of pK_1 and pK_3 are determined from the fit of the curve in fig. 2-18D, and they appear graphically as the intersections of the asymptotes, the higher one being pK_1. When the inhibitor does not bind at low pH, only the right leg of the curve will be observed, as in fig. 2-18A, and only pK_1 and K_i can be evaluated.

Notice that pK_1 refers to the free enzyme unless the ionization results from the inhibitor itself. In that case, E is replaced with I, HE is replaced with HI, and I is replaced with E in scheme 2-16, and pK_1 then pertains to the ionization of the inhibitor. Valuable information about ionizing groups at active sites can be obtained by studying the pH dependence of inhibition constants. The pK_a values of inhibitors generally are known, and if not, acid-base titration, NMR, spectrophotometry, or fluorescence measurements can be used to evaluate them. If a pK_1, as in fig. 2-18D, does not belong to the inhibitor, it can be assigned to the enzyme.

Assignment of pK_1 to a specific amino acid side chain requires additional information. For example, the enzyme structure may indicate a particular amino acid in the active site. This can be mutated to a nonionizing, structurally similar amino acid, such as Glu→Gln, and the pH profile can be repeated. If the ionization is abolished, assignment to that glutamate residue would be indicated. The assignment of pK_1 does not require that the K_i value be the same in the mutated as in the wild-type enzyme. If the ionization of a glutamate residue is important, the $^{app}K_i$ on one side of the break should differ dramatically between the wild-type and mutated enzyme. Analogous plots of log K_{is} or log K_{ib} for the substrates can give similar information about the pH dependence of substrate binding.

pH Dependence of V and V/K

The catalytic parameters in the reaction of a substrate can also be plotted as log V/K or log V against pH. In simple cases the plots describe linear segments that break at one or two pK_a values. A typical plot may display a single break, as in the plots of log V/K in fig. 2-18A or B. The ionizing group may belong to the substrate or the enzyme. As in the case of an inhibitor, the ionization constants of substrates are generally known or can be evaluated. The mechanism in scheme 2-17 is consistent with a plot of log V/K vs pH in fig. 2-18A.

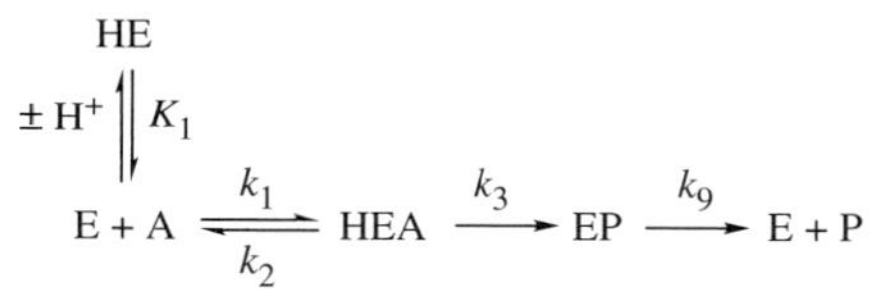

$$v = \frac{V[A]}{[A] + K}$$

$$V = k_3 k_9 [E_0]/(k_3 + k_9)$$

$$V/K = \frac{k_1 k_3 [E_0]}{(k_2 + k_3)(1 + [H^+]/K_1)}$$

Scheme 2-17

The rate equation shows that V/K is pH dependent but not V. V is not pH dependent because the enzyme-substrate complex does not undergo protonation. This may seem improbable, but it occurs more frequently than might be expected because of the tendency for substrate binding to induce closure of the binding site. The rate equation in scheme 2-17 shows that pK_1 refers to the free enzyme and not the enzyme-substrate complex. This pK_a should be the same as one measured by another method, such as in a pH-pK_i profile for a competitive inhibitor.

In a related mechanism, the enzyme-substrate complex also undergoes reversible protonation as shown in scheme 2-18.

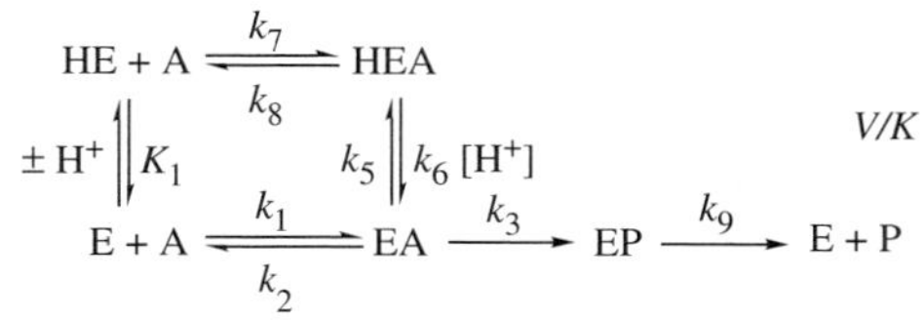

$$V/K = \frac{\frac{k_1 k_3 [E_0]}{(k_2 + k_3)}\left[1 + \frac{k_5 k_7 [H^+]}{k_1(k_5 + k_8)K_1}\right]}{\left[1 + \frac{[H^+]}{K_1}\right]\left[1 + \frac{k_2 k_5 k_7 [H^+]}{k_1(k_2 + k_3)(k_5 + k_8)K_1}\right]}$$

Scheme 2-18

The protonated enzyme binds the substrate, and the EA complex can be protonated to HEA. Because EA is a reactive intermediate, it cannot be assumed that its protonation will occur at equilibrium; k_5 may not be greater than k_3. It is written in scheme 2-18 as a steady-state process, as is the binding of the substrate to HE. These two additional steady-state steps complicate the mechanism and lead to the complex expression for V/K in scheme 2-18. The even more complex expression for V in eq. 2-30, where $K_2 = k_5/k_7$ shows V to be pH dependent, unlike the mechanism in scheme 2-17, in which EA is not subject to protonation.

$$V = \cfrac{\cfrac{k_3 k_9 [E_o]}{(k_3+k_9)}\left[1+\cfrac{k_5 k_8 [H^+]}{k_2(k_5+k_8)K_2}\right]}{1+\cfrac{[k_3 k_5 k_8 + k_9(k_5 k_8+k_3 k_8+k_2 k_8+k_2 k_5)][H^+]/K_2 + k_5 k_8 k_9([H^+]/K_2)_2}{k_2(k_3+k_9)(k_5+k_8)}} \quad (2\text{-}30)$$

Because the denominator of eq. 2-30 is not factored as in the equation for V/K in scheme 2-18, the curve of log V versus pH does not give the true value of pK_2 (Cleland, 1977).

The shape of a plot of log V/K against pH for the mechanism in scheme 2-18 can be much more complex than any of the simple patterns shown in fig. 2-18. Distortions from ideal shapes can be brought about by the relationships among the values of the rate constants, which determine the relative contributions of the bracketed terms in the equation for V/K. Flattening or bumps in the curve may be observed that can be mistaken for the effects of additional ionizing groups. A detailed analysis shows how relationships among the rate constants can account for many of the distortions that may appear (Cleland, 1977).

The most obvious complication in scheme 2-18 is the possibility of "stickiness" in substrate binding. This effect arises when $k_2 \leq k_3$, so that the EA complex goes to product as often as or more often than it dissociates to E + A. The mechanism can lead to a simple plot of log V/K when $k_2 > k_3$ (when the substrate is not sticky), in which case the plot appears as in fig. 2-18A. In this case, both equations for V/K in schemes 2-17 and 2-18 are reduced to $V/K = k_1 [E_0]/(1 + [H^+]/K_1)$. It follows that when an enzyme displays a complex V/K profile, an alternative substrate that is not sticky can be chosen. This is often possible by selecting a "poor" substrate, one that is significantly less reactive than the natural substrate and in which the value of k_3 is low. Such a substrate can allow the true value of pK_1 for the free enzyme to be evaluated from a plot of log V/K versus pH. Another approach is to mutate a residue that is peripheral to the active site and does not play an essential role in catalysis but decreases k_3 or increases k_2. This method alters the active site and can change the value of pK_1. The degree of pK perturbation may be small, and the method may support the assignment to a specific residue.

When a plot of log V/K displays two breaks, as in fig. 2-18C, the mechanism may be similar to that in scheme 2-19. This type of curve is often seen.

$$\begin{array}{c}
\text{HEH} \\
\pm H^+ \updownarrow K_1 \\
\text{HE} + A \underset{k_2}{\overset{k_1}{\rightleftharpoons}} \text{HEA} \xrightarrow{k_3} \text{HEP} \xrightarrow{k_9} \text{HE} + P \\
\pm H^+ \updownarrow K_2 \\
\text{E}
\end{array}
\qquad
\begin{aligned}
v &= \frac{V[A]}{[A]+K} \\
V &= \frac{k_1 k_9 [E_0]}{k_3+k_9} \\
V/K &= \frac{k_1 k_3 [E_0]}{(k_2+k_3)\{1+[H^+]/K_1+K_2/[H^+]\}}
\end{aligned}$$

Scheme 2-19

As in scheme 2-17, the intermediates HEA and HEP do not undergo ionizations. Although V is pH independent, the plot of log V/K versus pH gives the ionization constants for the free enzyme. If the pK_1 and pK_2 are widely separated, the breaks in the log V/K profile will be separated by a plateau, and in this case, the two ionizations can be treated separately, as in figs. 2-18A and B. If, however, the breaks are separated by only 2 pH units, the data must be fitted to the equation in scheme 2-19 to evaluate pK_1 and pK_2.

Indistinguishable pK$_a$ Values

Occasionally, data points in a pH profile such as in fig. 2-18C may be fitted to give identical values for two pK_a values. In principle, it is statistically impossible for two ionizing groups in one molecule to display the same acid dissociation constant. The values of K_a

must differ by a factor of at least 4, meaning that the pK_a values must differ by at least 0.6. The data in a pH profile may not define the pH dependence sufficiently to distinguish the difference. This ambiguity is likely to happen when the difference in pK_a values is less than twofold. The reasons for this are discussed in detail by Cleland (1977). In such cases, the data should be computer fitted with a program that constrains the fit to a difference of 0.6 between pK_1 and pK_2.

Reverse Protonation

When two groups required for enzyme activity display similar values of pK_a, the phenomenon of reverse protonation can complicate the interpretation of function. In general, the mechanism requires one group to be in its acid form and the other in its basic form. In the optimal pH range, the two groups will exist largely in the state, with the group of lower pK_a in its basic form and that with the higher pK_a in its acidic form. However, a small degree of reverse protonation exists because of the contribution of the microequilibria. The group required to be in its basic form in the mechanism is not necessarily that with the lower pK_a. In this case, the reverse protonated form of the enzyme is the active form that binds the substrate and catalyzes the reaction. This may be thought of as inefficient, but it is a means of providing for forward catalysis through the use of binding energy (see chap. 1). A reverse protonated form is at a higher free energy level than the main form, so that if the substrate binds only to the reverse protonated form there will be a strong tendency for the ionizing groups to react if they are engaged in acid-base catalysis. This forward pressure in catalysis must be paid for in the K_m value for the reaction.

Assignment of pK_a Values to Enzymes

The temptation to assign a kinetically determined pK_a to an amino acid side chain based solely on its experimental value and the known pK_a values of amino acids must be resisted. Every amino acid side chain incorporating an acid-base or nucleophilic group has been found to display a grossly perturbed pK_a in one or another class of enzymes. The entries in table 2-3 show the great deviations that have been observed. Additional lines of evidence from biochemical analysis, spectroscopy, and site-directed mutagenesis had to be used to assign the pK_a values. Complementary data of these kinds are always required in making a definitive assignment of pK_a to an amino acid in a protein. Most entries in table 2-3 are for representative examples of enzyme superfamilies. All cysteine proteases, including caspases, display ΔpK_a values similar to those for papain (EC 3.4.22.2). All glycosidases have Glu or Asp residues (or both) with high pK_a values. Mandelate racemase (EC 5.1.2.2) has an active site lysine with a low pK_a similar to that of Lys115 in acetoacetate decarboxylase (EC 4.1.1.4). All members of the short-chain dehydrogenase reductase superfamily have conserved tyrosine residues in their active sites that display depressed pK_a values similar to UDP-galactose 4-epimerase. A list of perturbed pK_a values for proteins has been provided by Harris and Turner (2002).

pH Dependence of Chymotrypsin

Chymotrypsin displays a bell-shaped profile of log k_{cat}/K_M versus pH for the hydrolysis of N-acetyl L-tryptophan ethyl ester in fig. 2-19 (Bender et al., 1964). However, the plot of log k_{cat} against pH (see fig. 2-19) displays only the break on the acid side. This means that the ionization at high pH is blocked in the enzyme-substrate complex. The ionizing groups have been assigned to His57 on the acid side and the N-terminal amino group Ile16 on the

116   Enzymatic Reaction Mechanisms

Table 2-3. Perturbed pK_a Values of Amino Acid Residues in Enzymes

Enzyme	Residue	pK_a	ΔpK_a	Study
Papain	Cys25	3.3	−5.4	Pinitglang, 1997
Papain	His	9.5	+3.0	
Acetoacetate decarboxylase	Lys115	6.0	−4.5	Kokesh, 1971; Frey, 1971
β-Galactosidase	Asp	7.5	+3.5	
UDP-galactose 4-epimerase	Tyr149	6.1	−4.1	Liu, 1997; Berger, 2001

[a] The pK_a of the amino acid side chain in solution minus that of the same residue in the enzyme active site. The normal values of pK_a in solution are given in table 1-2 in chapter 1.

alkaline side. The presence of His57 in the active site was first proved by affinity labeling with TPCK (see chap. 1). Taken together, the chemical, structural, and kinetic data allow the lower pK_a to be assigned to His57.

On the alkaline side, the pK_a of 8.7 has been assigned to Ile16, the N-terminus. When protonated below pH 8.7, Ile16 forms an ion pair with Asp194 that opens the binding site for the substrate. When Ile16 is unprotonated, Asp194 blocks the active site and prevents the substrate from binding. When a substrate or inhibitor is bound to the active site, the ionization of Ile16 is prevented. The kinetic mechanism is represented in scheme 2-20, where His57 can ionize in either the free enzyme or the enzyme-substrate complex, but Ile16 ionizes only in the free enzyme. The rate equation accounts for the plots of k_{cat}/K_M and k_{cat} against pH in fig. 2-19.

$$\begin{array}{c}
HEH + A \underset{k_8}{\overset{k_7}{\rightleftharpoons}} HEHA \\
\pm H^+ \updownarrow K_1 \qquad \pm H^+ \updownarrow K_1' \\
HE + A \underset{k_2}{\overset{k_1}{\rightleftharpoons}} HEA \xrightarrow{k_3} HE + P \\
\pm H^+ \updownarrow K_2 \\
E
\end{array}$$

Scheme 2-20

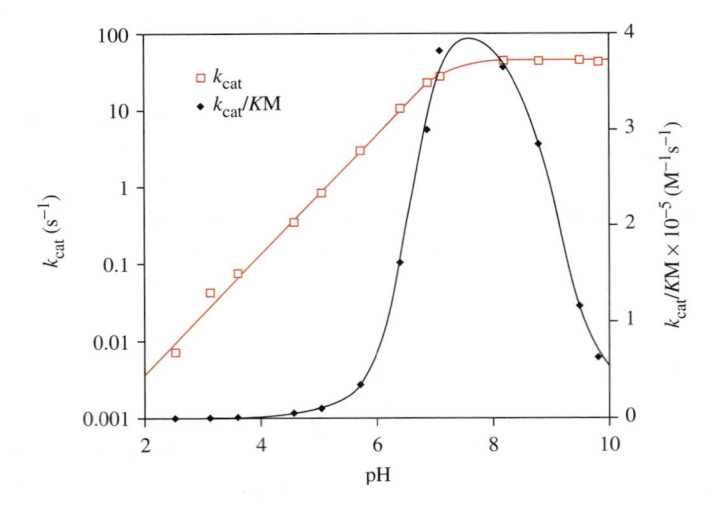

Fig. 2-19. The pH-rate profile for the hydrolysis N-acetyl-L-tryptophan ethyl ester (Bender and Kézdy, 1964).

Allosteric Regulation

We have presented the kinetic behavior of enzymes as if all display classic Michaelis-Menten kinetics and simple ligand binding properties. While this is true of most enzymes, a few do not follow any of the standard rate laws, and the differences are important. These enzymes regulate metabolism, and their special kinetic behavior enables them to control the rates of metabolic processes. The regulatory enzymes appear at key points on metabolic maps, for example at the start of metabolic pathways or at junctions in metabolism, where they can most efficiently control rates or channel metabolites. For example, an enzyme that catalyzes the first step in a biosynthetic pathway may be inhibited by the end product or activated by its own substrate. In these ways, it would function as a control element, stimulating the biosynthesis of the end product when its own substrate is present at high concentrations and decreasing the rate of biosynthesis when the end product is plentiful. In this section, we describe the mechanisms by which these enzymes control their own catalytic activities in the furtherance of their function in regulating metabolism.

The most common and obvious deviation from Michaelis-Menten kinetics is a distortion of the curvature in a saturation plot of activity against substrate concentration. Most regulatory enzymes display the deviations illustrated in red and gray in fig. 2-20. The rectangular hyperbola is typical of nonregulatory enzymes, but the other shapes signal cooperative behaviors, either positive cooperativity or negative cooperativity. The more obvious deviation is positive cooperativity, which results in a sigmoid saturation curve and concavity in the double reciprocal plot shown in red in fig. 2-20. The distortion in the gray saturation curve and convex curvature in the double reciprocal plot is symptomatic of negative cooperativity. Although both types of behavior are observed, positive cooperativity is more frequently found, presumably because it allows for substrate activation in metabolism—a valuable control feature. Cooperative behavior in ligand binding and catalysis arises from intersubunit interactions in regulatory enzymes (Koshland, 1970; Koshland et al., 1966; Monod et al., 1965).

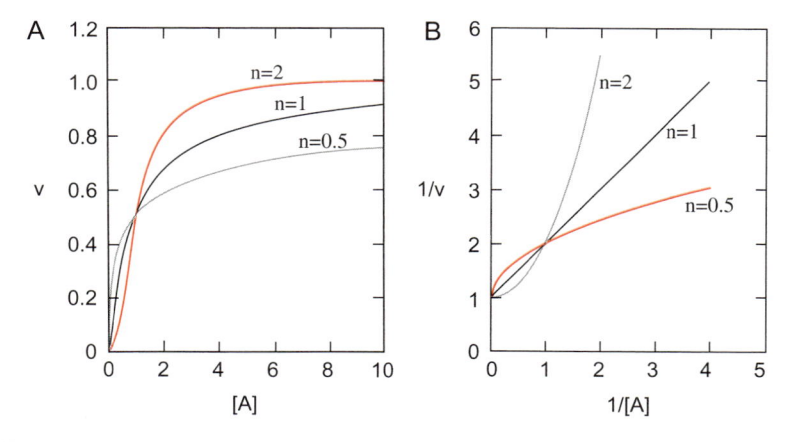

Fig. 2-20. Cooperativity in the kinetics of regulatory enzymes. The curves shown in black represent the conventional Michaelis-Menten kinetic behavior of nonregulatory enzymes. The red curves show the kinetic behavior of enzymes that display positive cooperativity with respect to a substrate A. The saturation curve is sigmoid, and the double-reciprocal plot is concave. The gray curves show the kinetic behavior of enzymes that display negative cooperativity with respect to a substrate A. The saturation curve is nonhyperbolic, and the double-reciprocal plot is convex. (Adapted from Koshland DE Jr [1970] The molecular basis for enzyme regulation. In Boyer PD [ed]: *The Enzymes*, 3rd ed. Academic Press: New York, p 352.)

Theory

Monod, Wyman, and Changeux (MWC) presented the first theory of cooperativity in enzyme kinetics, the concerted transition model in which conformational symmetry was maintained in two conformational states (Monod et al., 1965). This theory explained sigmoid saturation curves quantitatively in terms of a physical model of intersubunit communications. The MWC model also explained the actions of activators and inhibitors, which could exert their effects by binding at sites other than the active site, the allosteric sites. This theory introduced many concepts and explained positive cooperativity. Koshland introduced a ligand-induced transition model that explained both positive and negative cooperativity (Koshland, 1970). The concerted and ligand-induced transition models are encompassed within a general theory, and their relationships with the general theory will become clear.

A few terms and symbols defined here will facilitate the discussion. All proteins that function cooperatively consist of subunits, which may be symbolized in their resting forms as circles (O), in their activated forms as squares (□), and in their inhibited forms as triangles (Δ). Many regulatory enzymes consist of identical subunits. Others consist of catalytic subunits, which bind substrates and catalyze the reactions, and regulatory subunits, which bind metabolic activators and inhibitors and control the activities of the catalytic subunits. *Allosteric regulation* is the modulation of enzyme activity resulting from the interactions of metabolites at sites other than the catalytic sites. *Homotropic effects* are cooperative effects brought about by the substrate, and *heterotropic effects* are brought about allosteric activators or inhibitors interacting at allosteric sites. Metabolic inhibitors are symbolized as I and activators as J and may be referred to as *effectors* of metabolism. *K systems* are allosteric systems in which the effectors modulate activities by changing the values of K_m for substrates, and V systems are those in which the effectors alter the values of V_m.

Concerted and Ligand-Induced Transition Models

The essential features of the concerted transition model in its simplest formulation appear in the upper part of fig. 2-21. The taut (T) states of the enzyme, symbolized by the O_4 species in fig. 2-21, have a low affinity for the substrate and a low activity, whereas the relaxed (R) states, symbolized by $□_4$ species, have a high affinity for the substrate and a high activity. The enzyme exists in essentially these two conformational states, and the only conformational transition is between these two forms. It is intuitively clear that the presence of the substrate will shift the population of enzyme molecules from the T to the R states simply because of their differential affinities for the substrate. Moreover, there are four substrate-binding sites, and the symmetry of the model imposes the same substrate affinity to each subunit in the R state. The binding of substrate to the R state will be positively cooperative. This occurs because the binding of second, third, and fourth substrates to the enzyme shifts the conformational equilibrium further toward the R state and further increases the its apparent affinity for the substrate. The model accounts equally well for cooperative binding of effectors such as activators and inhibitors to allosteric sites. If the R subunits bind an activator J more tightly than the T subunits, an analogous set of conformers and activator-ligated states can be constructed.

In the ligand-induced transition model, the initial conformational change to $O_3□(A)$ can either increase or decrease the energy required to change the conformation of a neighboring subunit. This is because the subunits are in contact, and the interface between them

Kinetics of Enzymatic Reactions 119

Concerted Transition

[Diagram showing square (R state) subunits binding A ligands sequentially: □₄, □₄(A), □₄(A₂), □₄(A₃), □₄(A₄), with equilibria to circle (T state) forms: O₄, O₄(A), O₄(A₂), O₄(A₃), O₄(A₄)]

Ligand-Induced Transition

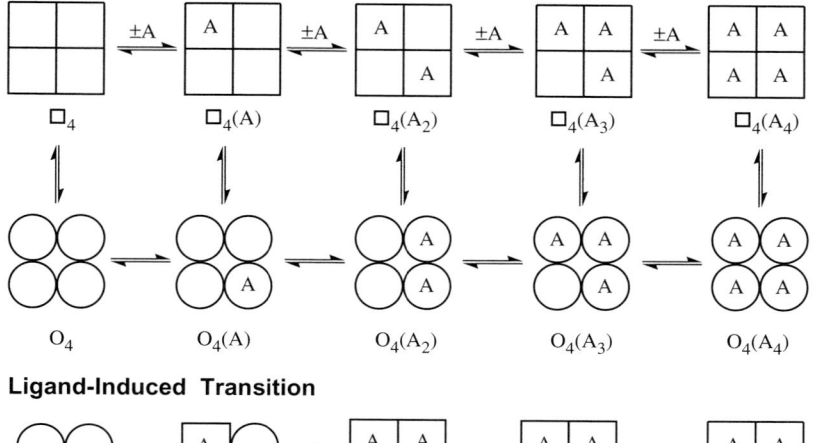

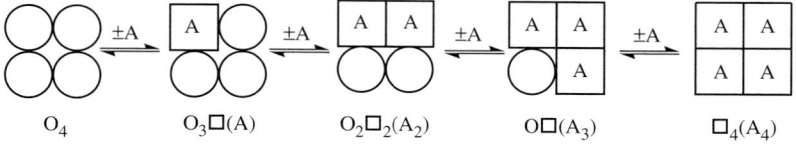

[Diagram showing mixed conformations: O₄, O₃□(A), O₂□₂(A₂), O□(A₃), □₄(A₄)]

Fig. 2-21. The concerted and ligand-induced transition models of cooperativity. In the concerted transition model, the subunits retain symmetry in the two states in that they are all in the T (O) state or all in the R (□) state. All the R states display identical high affinities for A, and all of the T states display identical low affinities for A. In the ligand-induced transition model, the binding of a ligand such as substrate A induces a conformational transition (O→□) in the subunit to which it binds. The affinities for A increase with each substrate that binds, so that the affinity of O₂□₂(A₂) for A is higher than that of O₃□(A).

may be either stabilized or destabilized when the conformation of one of them is changed by the binding of A. When the interface is destabilized, the energy required for the conformational change in a second subunit may be decreased, and the affinity of another subunit for A will be increased. Then the affinity of the enzyme for A increases with each A that is bound until all of the subunits are filled, and this results in positive cooperativity. When the conformational change induced by binding A to one subunit stabilizes the interface with other subunits, the energy required to change the conformation of another subunit increases, the apparent affinity for a second and additional ligands decreases, and the protein displays negative cooperativity. The ligand-induced transition model can explain both positive and negative cooperativity. The concerted transition model can account only for positive cooperativity because the binding of a ligand does not itself induce the conformational change.

General Theory of Cooperativity

The concerted and ligand-induced transition models exclude species that consist of both conformations of subunits that have ligand bound to the O state or that have subunits in the □ state with no ligand bound. These are excluded by the assumptions of the two models. Such assumptions simplify the task of deriving binding equations to which data can be fitted. A more general model that permits these species appears in fig. 2-22. In principle,

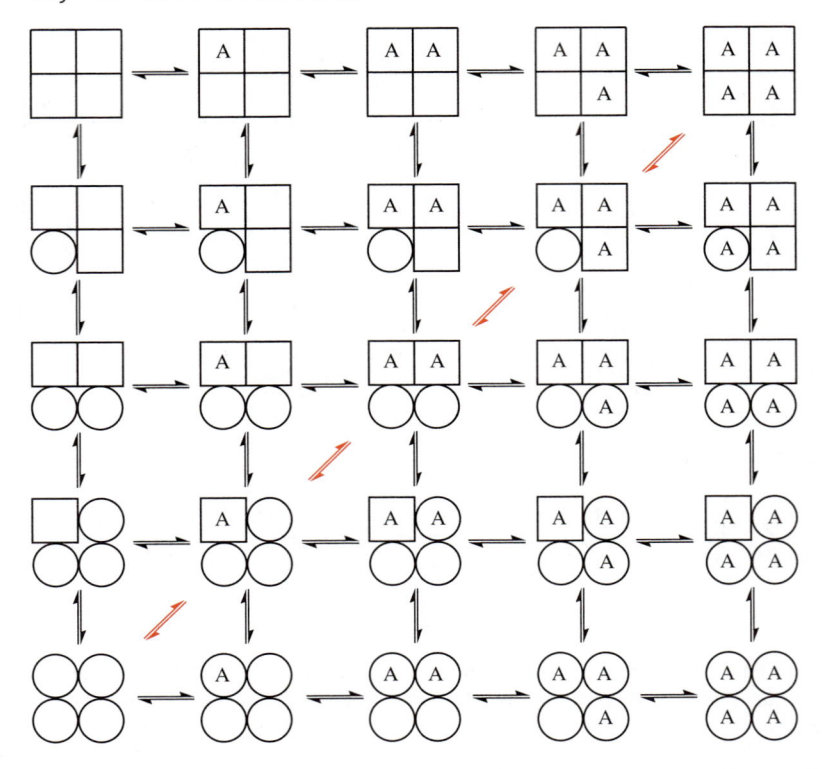

Fig. 2-22. The general model for ligand binding and conformational transitions allows the concerted transition model as a special case in which the entries on the top and bottom lines are interconverted without the participation of the species on the intervening lines. The ligand-induced transition model is represented by the species on the diagonal defined by the red arrows. The general model shows that the ligand-induced and concerted transition models exclude cross-species such as O□$_3$(A$_2$) or O$_3$(A)□(A).

any combination of species can be included in a model for which a binding equation is sought. In practice, it is best to fit data to one of the simple models because sufficient information is not generally available to exclude them or to mandate a more complex model.

The generality of the induced transition model makes it attractive for general use. A method for obtaining binding and kinetic equations to which data may be fitted follows in the next section. However, one must recognize that even the general model in fig. 2-22 does not include all of the possible subunit-ligand interactions. For example, it does not include interactions that lead to association or dissociation of subunits. We take this subject up again in a later section.

Binding Equations for Cooperative Systems

Ligand Binding to a Multisite Macromolecule

Deriving kinetic equations for cooperative systems becomes a complex undertaking for even the simplest kinetic models. This occurs because of the possibility that conformational transitions could partially limit the observed rates, and at less than saturating substrate or other ligand, the number of enzyme species in solutions can be very large.

For these reasons and because the conformational changes in allosteric and cooperative systems usually do not limit rates, investigators generally accept a few simplifying assumptions. It is generally assumed that the conformational changes and ligand binding processes are at equilibrium. A rate equation can be written by incorporating relevant rate constants for turnover into the equation for the average number of occupied sites per enzyme molecule.

A derivation of an equation for the average number of ligands per enzyme molecule with n binding sites as a function of the concentration of the ligand appears in Appendix C. The binding constants $K_1...K_n$ represent free energies ($\Delta G^\circ = -RT \ln K$) and include the energies for any conformational changes and corrections for statistical effects. The form of the final eq. 2-31 gives the average number of occupied sites is of interest.

$$N_x = \frac{\sum_{1}^{n} i L_i^*[A]^i}{1 + \sum_{1}^{n} L_i^*[A]^i} \tag{2-31}$$

The equation can be used to fit binding data. In the simplest case, when all species of enzyme-bound substrate react at the same rate, the binding equation can be related to a rate equation. To see this clearly consider the relationship between the binding equation for $n = 1$ with the Michaelis-Menten equation in the steady-state treatment of a one-substrate reaction, as in eq. 2-8, which is repeated here.

$$v = \frac{k_3[E_0][S]}{(k_2 + k_3)/k_1 + [S]} \tag{2-8}$$

Equation 2-8 can be recast by recalling that when $k_2 > k_3$, the expression for $K_m = (k_2 + k_3)/k_1$ becomes the dissociation constant K_S ($= k_2/k_1$), and from the derivation $v = k_3[EA]$. It is then possible to write eq. 2-8a, from which it follows that $[EA]/[E_0] = [S]/(K_S + [S])$.

$$k_3[EA] = \frac{k_3[E_0][S]}{K_S + [S]} \tag{2-8a}$$

This equation is related to the equation for $N_x = K_1[A]/(K_1[A] + 1) = [A]/(K_1^{-1} + [A])$ for n = 1 derived from eq. 2-31. In accord with expectation, this relationship makes it clear that K_S is the reciprocal of K_1, and N_x may be regarded as the equivalent of $[EA]/[E_0]$. When this is expressed as $k_3[EA]/k_3[E_0]$, the equation for N_x can be written as a Michaelis-Menten eq. 2-8b.

$$k_3[EA] = v = \frac{k_3[E_0][A]}{K_1^{-1} + [A]} \tag{2-8b}$$

It follows that the equation for $n = 1$ from eq. 2-31 can be transformed into the rate eq. 2-32.

$$\frac{v}{[E_0]} = \frac{k_3 K_1[A]}{1 + K_1[A]} \tag{2-32}$$

Generalizing from eq. 2-32, any of the binding equations derived from eq. 2-31 can be recast as kinetic equations for $v/[E_0]$ by multiplying the right side by the rate constant for turnover, generally k_{cat}. When $n = 1$, the equation can be placed in the form of the Michaelis-Menten equation, and a plot of v against $[A]$ is a rectangular hyperbola.

122 Enzymatic Reaction Mechanisms

Let the multisite protein be an enzyme E and the ligand be a substrate A.

Then let all the ligand bind at once in an infinitely cooperative manner.

$$E + nA \rightleftharpoons EA_n \qquad K = \frac{[E][A]^n}{[EA_n]}$$

The fraction of enzyme in the ligand-bound state is given by

$$Y = \frac{[EA_n]}{[E_0]}$$

Combining this with the enzyme conservation equation $[E]_0 = [E] + [EA_n]$,

$$1 - Y = \frac{[E]}{[E_0]}$$

Combining this with the expression for the dissociation constant K,

$$\log \frac{Y}{1-Y} = n\log[A] - \log K \qquad (2\text{-}33)$$

Fig. 2-23. Derivation of the Hill equation.

When $n = 2, \ldots n$, the equation contains higher powers of [A], and plots of v against [A] are nonhyperbolic. When each enzyme form reacts at a different rate, each term in the numerator will require a different rate constant, in which case, the rate equations become even more complex.

The Hill Equation

A very special theoretical case of multisite binding leads to the Hill equation and the Hill coefficient, which is useful for assessing the degree of cooperativity in a real system, in which up to n molecules of a ligand A bind to a protein. Derivation of the Hill equation (eq. 2-33) appears in fig. 2-23. This equation shows that a plot of $\log(Y/1-Y)$ against $\log[A]$, where $Y = [EA_n]/[E_0]$, gives a straight line with a slope of n, the number of binding sites per molecule. The one-step binding scheme in fig. 2-23 is never realized experimentally because of its improbability—in the simplest multisite case of two subunits it requires a three-body collision. The equation serves as a means of assessing the degree of cooperativity in a real system.

When rate measurements are available, they can be plotted according to a variant of the Hill equation, eq. 2-34, in which h is the slope known as the Hill coefficient.

$$\log (v/(V_m - v)) = h \log [A] - \log K \qquad (2\text{-}34)$$

The slope of the plot of $\log v/(V_m - v)$ against $\log [A]$ is often referred to as n, but n is the number of binding sites, which is not the same as the slope. For this and other reasons, designation of the Hill coefficient as h is preferred (Fersht, 1998).

Because the binding scheme in fig. 2-23 represents infinite cooperativity, the Hill coefficient has become useful for assessing the degree of cooperativity in a system. Examples of such plots (fig. 2-24) show the utility of the Hill coefficient. When the Hill coefficient is greater than 1, the system is positively cooperative, and when it is

$$\frac{v}{(V_{max}-v)}$$

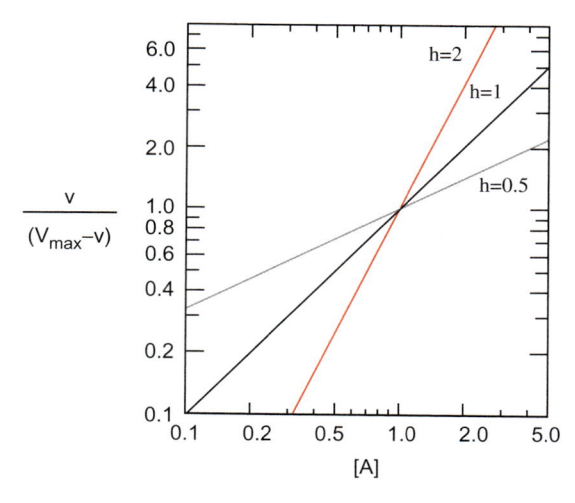

Fig. 2-24. Plots of the Hill equation, showing positive (Hill coefficient [h] > 1) and negative (h < 1) cooperativity. (Adapted from Koshland DE Jr [1970] The molecular basis for enzyme regulation. In Boyer PD [ed]: *The Enzymes*, 3rd ed. Academic Press, New York, p 352.)

less than 1, the system is negatively cooperative. The nearer the value of the Hill coefficient to the number of binding sites per molecule, the more highly cooperative the system. The Hill coefficient for O_2 binding to hemoglobin is typically 2.8, which shows that binding is cooperative. Because there are four binding sites per molecule of hemoglobin, the system is not infinitely cooperative, but it is highly cooperative. If the Hill coefficient for hemoglobin were 1.5, the system would be regarded as slightly cooperative.

Aspartate Transcarbamoylase

The most thoroughly studied, classic example of an allosteric enzyme, aspartate transcarbamoylase (ATCase; EC 2.1.3.2) from *E. coli*, catalyzes the first step in the biosynthesis of the pyrimidine rings of nucleic acids. It catalyzes the reaction of aspartate with carbamoyl phosphate (carbamoyl-P) according to eq. 2-35 to form carbamoylaspartate and P_i.

(2-35)

In subsequent biosynthetic steps, carbamoylaspartate undergoes cyclization between the carbamoyl nitrogen and the β-carboxylic acid group, dehydrogenation between C5 and C6 of the resulting pyrimidine ring, phosphoribosylation, and decarboxylation to form UMP, the precursor of all pyrimidine nucleotides.

Kinetics

The reaction of ATCase is chemically straightforward, an acyl group transfer from a reactive acylating agent to an amino group. However, ATCase is structurally complex and displays cooperative kinetics and allosteric regulation. ATCase is a dodecamer consisting of two types of subunits, two catalytic trimers and three regulatory dimers. The molecular weight of the catalytic subunits is 34,000 and that of the regulatory subunits is 17,000. The catalytic trimers and regulatory dimers can be separated after treating the enzyme with p-hydroxymercuribenzoate, which reacts with cysteine residues (Gerhart and Schachman, 1965). ATCase displays homotropic interactions, that is, positive cooperativity with respect to either aspartate or carbamoyl-P. It also displays heterotropic interactions in the form of inhibition by CTP, an end product of pyrimidine biosynthesis, and activation by ATP, an end product of purine biosynthesis (Gerhart and Pardee, 1962). The separated and purified catalytic trimers are fully active and display conventional saturation kinetics, and they are not inhibited by CTP or activated by ATP, both of which bind to the regulatory subunits.

The ^{13}C-kinetic isotope effects in the reaction of [^{13}C]carbamoyl-P provided valuable information about the kinetic mechanism and the allosteric transition (Parmentier et al., 1992a). The ^{13}C kinetic isotope effect varies with the concentration of aspartate when ATCase is saturated with [^{13}C]carbamoyl-P. Its maximum value is 1.0217 at very low concentrations of aspartate and declines hyperbolically to 1.000 with increasing aspartate concentration. The concentration of aspartate that halves the isotope effect is the same as its K_m value, 4.8 mM. At low aspartate concentrations, transfer seems to be rate limiting, whereas at high concentrations, another step limits the rate. The effect of aspartate concentration on the isotope effect implicates the ordered bi bi kinetic mechanism, with carbamoyl-P as the leading substrate.

Cysteine sulfinate reacts in place of aspartate but displays different kinetics (Parmetier et al., 1992b). Cysteine sulfinate is nearly an isotere of aspartate, in which the carbon of the β-carboxyl group replaced by sulfur. However, cysteine sulfinate does not induce the homotropic interactions in ATCase that are characteristic of aspartate. The ^{13}C-kinetic isotope effect in its reaction or [^{13}C]carbamoyl-P with cysteine sulfinate is both larger (1.039) than with aspartate and independent of cysteine sulfinate concentration. The isotope effect is near the maximum value for a ^{13}C-kinetic isotope effect, indicating that acyl transfer is always rate limiting with this substrate. The isolated catalytic trimers catalyze the reaction with the same isotope effects, 1.024 with aspartate and 1.039 with cysteine sulfinate. These facts suggest that cysteine sulfinate reacts by a random kinetic mechanism with both ATCase and the catalytic trimer, unlike the ordered mechanism in the reaction of aspartate.

The isolated catalytic trimers reside in their active or R conformational state, and the isotope effect for the reaction of cysteine sulfinate is the same for ATCase and the catalytic trimers. The ^{13}C kinetic isotope effect in its reaction with ATCase is independent of its concentration. For these reasons, it seems that cysteine sulfinate is reacting with the preexisting R state of ATCase. This hypothesis is consistent with the concerted transition model. The aspartate concentration dependence of the isotope effect is the same when ATCase is inhibited by CTP or activated by ATP. The isotope effects do not reveal the presence of hybrid forms of ATCase and are consistent with the concerted conformational transition model of Monod, Wyman, and Changeux (1965). The isotope effects failed to detect the participation of hybrid conformational states and supported other work implicating the concerted transition model for ATCase (Howlett et al., 1977).

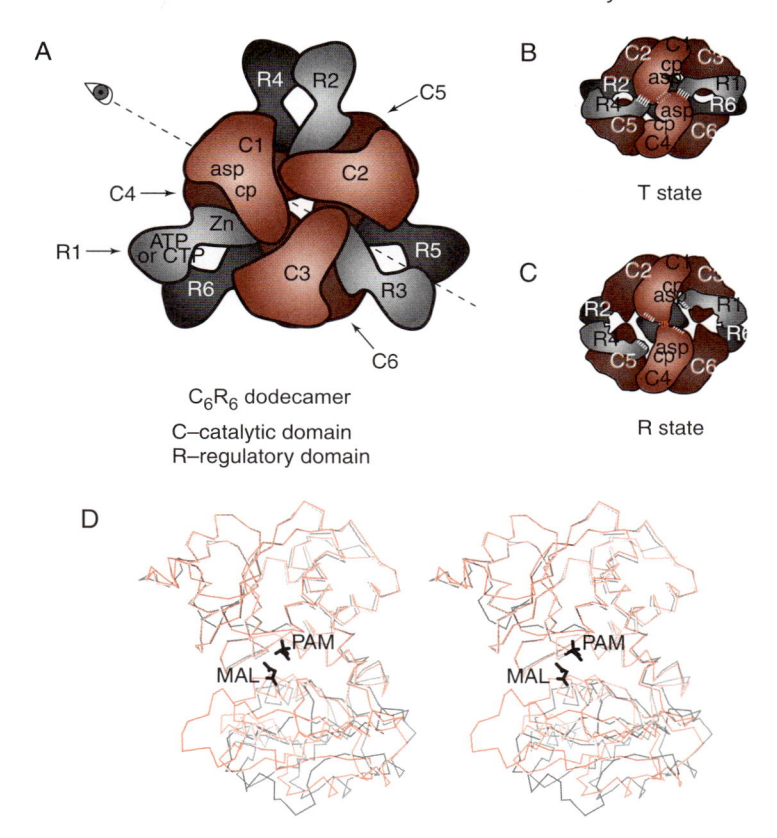

Fig. 2-25. The structure of ATCase from *E. coli*. A schematic representation of the heterododecomeric (C$_6$R$_6$) *E. coli* aspartate transcarbamylase (EC 2.1.3.2) is given in parts A to C. The catalytic domains (C) are red and contain the binding sites for asparagine (asp) and carbamoyl phosphate (cp); the regulatory domains (R) are dark gray and contain the binding sites for Zn^{2+} and ATP or CTP. (A) The complex looking down the threefold symmetric axis. The dotted line is the line of site for B and C, as well as one of three twofold symmetry axes. (B) The T state (CTP bound to R) in which adjacent regulatory and catalytic subunits (e.g., R1 and C4, R4 and C1) have significant contacts (white hatching) that contribute the lower activity of this state by separating the cp and asp domains of the catalytic subunit. (C) In the R state (ATP bound to R), these interactions are disrupted, and the reactant domains are brought closer together. (D) The conformational differences are shown, on which the Cα traces for the catalytic domain structures of representative R state (PDB 7AT1) and T state (5AT1) complexes have been superposed. The traces were aligned using the cp domains of both complexes so that the conformational differences would be observable in the asp domain. The structure of the R state was solved with substrate analogs phosphono acetamide (PAM) and malonate (MAL) bound in the cp and asp binding sites, respectively. (A to D, Adapted from Gouaux and Lipscomb, 1990.)

Structure

The x-ray crystallographic structure of ATCase in fig. 2-25 shows the contacts between the catalytic trimers and the regulatory dimers in the R and T states. The R state is relaxed in that the catalytic trimers are less tightly associated. The structures show directly that CTP and ATP bind to an allosteric site on the regulatory subunits. The conformational changes induced in the catalytic subunit by the binding of allosteric effectors to the regulatory subunit clearly appear in the α-carbon backbone of the catalytic subunit in fig. 2-25.

Reaction Mechanism

Isotope effects are consistent with the expectation that group transfer at the active site of ATCase proceeds by a stepwise mechanism through a tetrahedral intermediate, as illustrated in scheme 2-21 (Parmentier et al., 1992b; Waldrop et al., 1992).

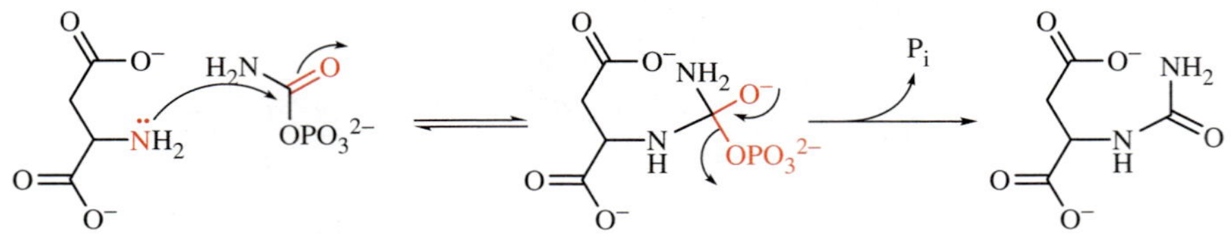

Scheme 2-21

The tetrahedral addition intermediate is likely to be higher in energy than the substrates. According to the Hammond postulate, it is likely to resemble the transition state. Its formation may require acid-base catalysis, and from the crystal structure, a number of amino acid side chains have been identified that might have provided such catalysis. However, despite extensive studies by site-directed mutagenesis and kinetic analysis, a clear picture of acid-base catalysis has not been drawn (Stevens et al., 1991; Waldrop et al., 1992a, 1992b). Evidence from pH studies implicates two functional groups in ATCase that appear to be required for binding the substrates, one with a pK_a of 9.1 that must be protonated and one with a pK_a of 7.0 that must be unprotonated (Turnbull et al., 1992). The first is required for binding carbamoyl-P and the second for binding aspartate. The identities of these groups remain unknown.

References

Althoff S, B Zambrowicz, P Liang, and M Glaser (1988) *J Mol Biol* **199**, 665.
Arabshahi A, GR Flentke, and PA Frey (1988) *J Biol Chem* **263**, 2638.
Bahnson BJ and JP Klinman (1995) *Methods Enzymol* **249**, 373.
Ballou DP (1978) *Methods Enzymol* **54**, 85.
Bender ML and FJ Kézdy (1964) *J Am Chem Soc* **86**, 3704.
Bender ML, FJ Kézdy, and FC Wedler (1967) *Chem Ed* **44**, 84.
Berger E, A Arabshahi, Y Wei, JF Schilling, and PA Frey (2001) *Biochemistry* **40**, 6699.
Bigeleisen J (1955) *J Chem Phys* **35**, 2264.
Brick P, TN Bhat, and DM Blow (1989) *J Mol Biol* **208**, 83.
Byeon L, Z Shi, and MD Tsai (1995) *Biochemistry* **34**, 3172.
Cedar H and JH Schwartz (1969) *J Biol Chem* **244**, 4122.
Cha S (1968) *J Biol Chem* **243**, 820.
Cha Y, CJ Murray, and JP Klinman (1989) *Science* **243**, 1325.
Cleland WW (1970) In Boyer PD (ed): *The Enzymes*, vol 2, 3rd ed. Academic Press: New York, p 1.
Cleland WW (1975) *Biochemistry* **14**, 3220.
Cleland WW (1977) *Adv Enzymol Rel Areas Mol Biol* **45**, 273.
Cleland WW (1979) *Methods Enzymol* **63**, 103.
Cleland WW (1980) *Methods Enzymol* **64**, 104.
Cleland WW (1982) *Methods Enzymol* **87**, 366.
Cleland WW (1986) In Bernasconi CF (ed): *Determination of Rates and Mechanism of Reactions*, vol 6, part 1. Wiley: New York, p 851.
Cleland WW (1990) *FASEB J* **4**, 2899.
Cleland WW and R Viola (1982) *Methods Enzymol* **87**, 353.
Cook PF, NJ Oppenheimer, and WW Cleland (1981) *Biochemistry* **20**, 1817.
Dahnke T, Z Shi, H Yan, RT Jiang, and MD Tsai (1992) *Biochemistry* **31**, 6318.
Duggleby R (1995) *Methods Enzymol* **249**, 61.
Fersht A (1999) *Structure and Mechanism in Protein Science*. Freeman: New York, 1999.

Fersht AR, RS Mulvey, and GLE Koch (1975) *Biochemistry* **14**, 13.
Frey PA (1982) *New Comp Biochem* **3**, 201.
Frey PA (1992) In Boyer PD (ed): *The Enzymes*, vol 20, 3rd ed. Academic Press: New York, 141.
Garces E and WW Cleland (1969) *Biochemistry* **8**, 633.
Gerhart JC and AB Pardee (1962) *J Biol Chem* **237**, 891.
Gerhart JC and HK Schachman (1965) *Biochemistry* **4**, 1054.
Gerlt JA, JA Coderre, and S Mehdi (1983) *Adv Enzymol Rel Areas Mol Biol* **55**, 291.
Gouaux JE and WN Lipscomb (1990) *Biochemistry* **29**, 389.
Grant KL and JP Klinman (1989) *Biochemistry* **28**, 6597.
Hammes GG and PR Schimmel (1970) *The Enzymes*, vol 2, 3rd ed. Academic Press: New York, 67.
Harris TK and GJ Turner (2002) *IUBMB Life* **53**, 85.
Howlett GJ, MN Blackburn, JG Compton, and HK Schachman (1977) *Biochemistry* **16**, 5091.
Huskey P (1991) In Cook PF (ed): *Enzyme Mechanism from Isotope Effects*. CRC Press: Boca Raton, FL, p 37.
Huskey WP and RL Schowen (1983) *J Am Chem Soc* **105**, 5704.
Janson CA and WW Cleland (1974) *J Biol Chem* **249**, 2567.
Jonsson T, DE Edmondson, and JP Klinman (1994) In Yagi K (ed): *Proceedings of the 11th International Symposium on Flavins and Flavoproteins*. De Gruyter: New York, 1994.
King EL and C Altman (1956) *J Phys Chem* **60**, 1375.
Klinman JP (1991) Cook PF (ed): *Enzyme Mechanism from Isotope Effects*. CRC Press: Boca Raton, FL, p 127.
Knowles JR (1982) *Fed Proc* **41**, 2424.
Kokesh FC and FH Westheimer (1971) *J Am Chem Soc* **93**, 7270.
Koshland DE Jr (1970) In Boyer PD (ed): *The Enzymes*, vol 1, 3rd ed. Academic Press: New York, 341.
Koshland DE Jr and KE Neet (1968) *Annu Rev Biochem* **37**, 359.
Koshland DE Jr, G Nemethy, and D Filmer (1966) *Biochemistry* **5**, 365.
Kurz LC and C Frieden (1980) *J Am Chem Soc* **102**, 4198.
Monod J, J Wyman, and J-P Changeux (1965) *J Mol Biol* **12**, 88.
Mourad N and RE Parks Jr (1966) *J Biol Chem* **241**, 271.
Northrop DB (1969) *J Biol Chem* **244**, 5808.
Northrop DB (1982) *Methods Enzymol* **87**, 607.
Northrop DB and FB Simpson (1998) *Arch Biochem Biophys* **352**, 288.
O'Leary MH (1980) *Methods Enzymol* **64**, 83.
Paiva AA, RF Tilton Jr, GP Crooks, LQ Huang, and KS Anderson (1997) *Biochemistry* **36**, 15472.
Parmentier LE, MH O'Leary, HK Schachman, and WW Cleland (1992a) *Biochemistry* **31**, 6570.
Parmentier LE, PM Weiss, MH O'Leary, HK Schachman, and WW Cleland (1992b) *Biochemistry* **31**, 6577.
Pinitglang S, AB Watts, M Patel, JD Reid, MA Noble, S Gul, A Bokth, A Naeem, H Patel, EW Thomas, SK Sreedharan, C Verma, and K Brockelhurst (1997) *Biochemistry* **36**, 9968.
Rhoads D and JM Lowenstein (1968) *J Biol Chem* **243**, 3963.
Rife JE and WW Cleland (1980) *Biochemistry* **19**, 2321.
Rose IA (1980) *Methods Enzymol* **64**, 47.
Rose IA, EL O'Connell, S Litwin, and J Bar Tana (1974) *J Biol Chem* **249**, 5163.
Schulz GE, CW Muller, and K Diederichs (1990) *J Mol Biol* **213**, 627.
Segel IH (1993) *Enzyme Kinetics, Behavior and Analysis of Rapid Equilibrium and Steady-State Enzyme Systems*. Wiley: New York, 1993.
Sheu K-FR, JP Richard, and PA Frey (1979) *Biochemistry* **18**, 5548.
Snider MJ and R Wolfenden (2000) *J Am Chem Soc* **122**, 11507.
Swain CG, EC Stivers, DF Reuwer, and LJ Schaad (1958) *J Am Chem Soc* **80**, 5885.
Tsai MD and HG Yan (1991) *Biochemistry* **30**, 6806.
Turnbull JL, GL Waldrop, and HK Schachman (1992) *Biochemistry*, **31**, 6562.
Waldrop GL, JL Turnbull, LE Parmetier, S Lee, MH O'Leary, WW Cleland, and HK Schachman (1992b) *Biochemistry* **31**, 6592.

Waldrop GL, JL Turnbull, LE Parmetier, MH O'Leary, WW Cleland, and HK Schachman (1992a) *Biochemistry* **31**, 6585.
Welsh KM, DJ Creighton, and JP Klinman (1980) *Biochemistry* **19**, 2005.
Westheimer FH (1961) *Chem Rev* **61**, 265.
Xu Y-W, S Morera, J Janin, and J Cherfils (1997) *Proc Natl Acad Sci U S A* **94**, 3579.
Zechel DL, L Konermann, SG Withers, and DJ Douglas (1998) *Biochemistry* **37**, 7664.

3
Coenzymes I: Organic Coenzymes

Most enzymatic reactions proceed with chemical changes that cannot be brought about by the side chains of amino acid residues. These enzymes function in cooperation with coenzymes and cofactors, which lend physicochemical potentialities not found in amino acids. Many coenzymes are organic molecules incorporating functional groups with chemical properties that enable them to facilitate reactions of certain types. These molecules bind to active sites tailored for them through evolution and equipped to assist in their coenzymatic functions. Many of these coenzymes were derived from vitamins, and in early biochemistry investigations, vitamins and coenzymes were often regarded as closely linked and even synonymous. However, vitamins such as vitamin D are more akin to hormones than to coenzymes, and in modern biochemistry, the newly discovered coenzymes are not related to vitamins and have identities independent of any nutritional origin.

More than 25 biological molecules may be regarded as coenzymatic in nature. In this book, the most common coenzymes and their functions are described in two chapters, the organic coenzymes in this chapter, and the metallo-coenzymes in chapter 4. Each coenzyme and cognate enzyme form a union that allows them to act as a single catalytic entity functioning in concert to bring about a difficult chemical transformation. Each coenzyme provides the chemistry required for a class of enzymatic processes, and the mechanisms of enzymatic catalysis are often revealed through the actions of the participating coenzymes.

Nicotinamide Coenzymes

Structures and Functions of Nicotinamide Coenzymes

Nicotinamide adenine dinucleotide (NAD^+) is the coenzymatic form of the vitamin niacin (vitamin B_1). The structural formula for NAD^+ is shown in fig. 3-1 along with the

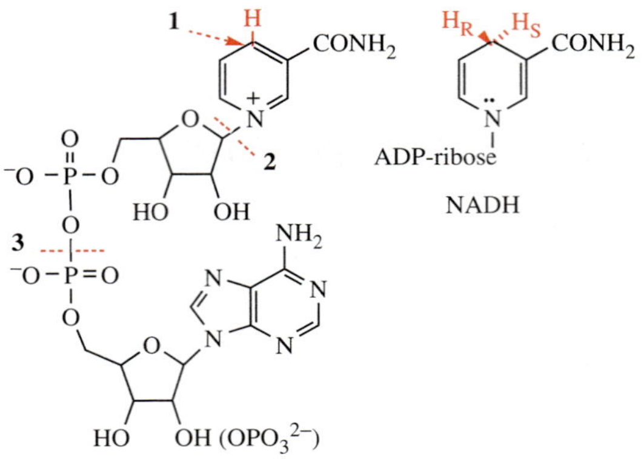

Fig. 3-1. Structures and metabolic cleavage points of nicotinamide coenzymes. The structure of NAD$^+$ and NADP$^+$ are shown with markings to indicate which bonds undergo chemical changes in various enzymatic reactions. Notice the designations H$_R$ and H$_S$ in NADH, which refer to the stereospecificities of various dehydrogenases for transferring hydrogen to and from nicotinamide coenzyme. 4-Pro-R and 4-pro-S specificities of some dehydrogenases are listed in table 3-1.

biochemically reactive bonds and their importance in metabolism. NADH is the reduced form of NAD$^+$ and is produced in the dehydrogenation of substrates. The closely related forms NADP$^+$ and NADPH are phosphorylated at the 2'-hydroxyl group of the adenosyl moiety. NADP$^+$ and NADPH generally participate in biosynthesis (anabolism), whereas NAD$^+$ and NADH generally participate in biodegradative metabolism (catabolism). NAD$^+$ and NADH were formerly known as DPN$^+$ and DPNH, for diphosphopyridine nucleotide, and TPN$^+$ and TPNH, for triphosphopyridine nucleotide.

The most frequent function of NAD$^+$ is as an acceptor of a hydrogen atom and two electrons, a hydride equivalent, in reactions of oxidoreductases, commonly known as *dehydrogenases*. Nicotinamide C4, marked 1 in fig. 3-1, is the locus of hydride transfer in the interconversion of NAD$^+$ and NADH, which participate in redox processes in the cell. In most of these processes, reduction equivalents in the form of NADH from intermediary metabolism are used to produce ATP. Examples include the production of NADH by oxidation of catabolic intermediates in the cytosol, as in the dehydrogenation of glyceraldehyde-3-P by glyceraldehyde-3-P dehydrogenase (EC 1.2.1.12) and NAD$^+$ according to eq. 3-1, and the oxidation of NADH by oxygen through the terminal electron transport system in the cellular membrane to provide the energy for ATP synthesis. Equation 3-2 is a simplified description of the latter, complex process that requires many enzymes and electron carrier molecules.

$$\text{Glyceraldehyde-3-P} + \text{NAD}^+ + P_i \rightarrow \text{3-diphosphoglycerate} + \text{NADH} \quad (3\text{-}1)$$

$$\text{NADH} + 1/2\ O_2 + 3\ \text{ADP} + 3\ P_i \rightarrow 4\ H_2O + 3\ \text{ATP} + \text{NAD}^+ \quad (3\text{-}2)$$

This is one of many ways in which the NAD$^+$/NADH system mediates electron transfer in cells, and it is the most widespread and probably the most important role of the nicotinamide coenzymes in cellular function. Most studies of the mechanism of action of NAD$^+$ have been carried out on enzymes that catalyze the reactions involved in mediating cellular electron transfer. In this role, NAD$^+$ and NADH function as metabolic intermediates, rather than as coenzymes. A number of enzymes, however, use tightly bound NAD⁺ as a coenzyme to facilitate chemically difficult reactions, and we consider this type in subsequent sections.

In the cleavage of the *N*-ribosyl bond marked 2 in fig. 3-1, ADP-ribosyltransferases use NAD⁺ as a substrate and transfer the ADP-ribosyl group to an enzymatic amino acid side chain. This is a significant process in metabolism, for example in the regulation of nitrogenase (EC 1.18.6.1) and the activation of diphtheria toxin. NAD⁺ is also the substrate for poly-ADP-ribosylation. In the phosphoanhydride cleavage shown as 3 in fig. 3-1, DNA ligase (EC 6.5.1.2) from *Escherichia coli* uses NAD⁺ as the substrate for activating the 5′-P of DNA fragments in DNA replication and for sealing nicks in DNA repair. In these capacities, NAD⁺ fills a needed cellular function as a metabolite but not a coenzyme.

In its dehydrogenation functions, NAD⁺ is essentially a hydride acceptor, as illustrated in eq. 3-3, in which the hydride donor is an alcohol. With its pyridinium ring, NAD⁺ is chemically well constituted to serve as a hydride acceptor.

$$ \tag{3-3} $$

The general formulation of eq. 3-3 is extended to related reactions, in which aldehydes or amines are dehydrogenated, through the mechanisms by which enzymes recognize these functional groups. For example, the reaction of glyceraldehyde-3-P dehydrogenase (eq. 3-1) follows the course of eq. 3-4.

$$ \tag{3-4} $$

The sulfhydryl group of an essential cysteine residue undergoes nucleophilic addition to the aldehyde carbonyl group of glyceraldehyde-3-P to form a thiohemiacetal, which then reacts similarly to an alcohol as a hydride donor to NAD⁺. Dehydrogenation leads to the formation of a thioester linkage between the 3-phosphoglyceryl group and the cysteine residue, and phosphate ion reacts as a nucleophile to accept the 3-phosphoglyceroyl group and form 1,3-diphosphoglycerate as the product.

The best known nicotinamide coenzyme–dependent amino acid dehydrogenase is glutamate dehydrogenase (EC 1.4.1.2), which catalyzes the dehydrogenation of glutamate to α-ketoglutarate and ammonia. The basic dehydrogenation mechanism in eq. 3-5 shows that the amino group is initially dehydrogenated to an iminium group.

$$NAD^+ + H-\underset{\underset{COO^-}{|}}{\overset{\overset{CH_2COO^-}{|}}{C}}-NH_2 \underset{NADH}{\rightleftharpoons} H_2N^+=\underset{\underset{COO^-}{|}}{\overset{\overset{CH_2COO^-}{|}}{C}} \underset{H_2O \quad NH_4^+}{\rightleftharpoons} O=\underset{\underset{COO^-}{|}}{\overset{\overset{CH_2COO^-}{|}}{C}} \quad (3\text{-}5)$$

The imine undergoes hydrolysis to the ketone group of α-ketoglutarate, with the concomitant production of ammonia. The processes in eqs. 3-3, 3-4, and 3-5 take place within enzymatic active sites and not in solution.

An important class of alcohol dehydrogenases employs a distinct mechanism for abstracting the hydroxyl proton, which is driven in eq. 3-3 by a base catalyst. These enzymes contain Zn^{2+} in the active site, and the hydroxyl group of the substrate becomes one of the four ligands for the metal ion, others being donated by the side chains of amino acids. Because Zn^{2+} electrostatically stabilizes the alkoxide ion form of the alcoholic ligand, the pK_a of the hydroxyl group is decreased from 15 to approximately 8, so that a significant fraction exists as the alkoxide ion in neutral solution. The alkoxide of a primary or secondary alcohol is a good hydride donor and readily reacts with NAD^+ to produce NADH and a carbonyl group according to eq. 3-6.

$$NAD^+ \quad H-C-O^- -Zn^{2+}- \quad \rightleftharpoons \quad NADH \quad C=O \cdots Zn^{2+}- \quad (3\text{-}6)$$

We discuss the best known example of such an enzyme, alcohol dehydrogenase, in chapter 16.

Stereospecificity of Hydride Transfer

A hallmark of nicotinamide coenzyme enzymology is the stereospecificity of hydrogen transfer. The dehydrogenation of ethanol by alcohol dehydrogenase (ADH, EC 1.1.1.1) proceeds with direct hydrogen transfer to NAD^+ (i.e., without exchange with protons of the solvent). ADH-catalyzed reduction of NAD^+ by 1,1-dideuteroethanol produces NADH that contains 1 deuterium per molecule, and reduction of NAD^+ by ethanol in D_2O produces NADH that contains no deuterium after purification. Hydrogen transferred enzymatically to NAD^+ is incorporated at C4 of the dihydronicotinamide ring. Both C4 of NADH and C1 of ethanol are prochiral centers; they each become chiral when one hydrogen atom is replaced by an alternative atom. Such a substituent can be deuterium, so that the substitution of deuterium for hydrogen stereospecifically at one of these carbon atoms transforms it into a chiral center. The demonstration of this fact for ADH and other nicotinamide coenzyme-dependent dehydrogenases defined the amazing degree of stereospecificity displayed by enzymes (Fisher et al., 1953; Popj·k, 1970; Westheimer et al., 1951). The singular stereochemical projections of 4-H_R and 4-H_S in NADH can be appreciated by viewing the space-filling model of dihydronicotinamide riboside in fig. 3-2.

The stereospecificity of hydrogen and deuterium transfers catalyzed by ADH elegantly exemplify the stereospecificities of NAD^+-dependent dehydrogenases. Those in eqs. 3-7 to 3-9 are typical of a larger body of experiments that proved the concept of prochiral recognition by ADH and other dehydrogenases. Reduction of NAD^+ by 1,2-dideuteroethanol produced $4R$-[nicotinamide-4-$^2H_2$]NADH and 1-deuteroacetaldehyde, showing that deuterium was transferred specifically to the 4-pro-R position of NAD^+ (eq. 3-7).

Coenzymes I: Organic Coenzymes   133

$$\text{ADP-ribose}-\overset{\underset{|}{\text{CONH}_2}}{\text{N}^+}-\text{H} + \text{CH}_3-\overset{\text{OH}}{\underset{\text{D}}{\text{C}}}\text{D} \rightleftharpoons$$

$$\text{ADP-ribose}-\overset{\underset{|}{\text{CONH}_2}}{\text{N:}}\overset{\text{D}}{\underset{\text{H}}{\diagdown}} + \text{CH}_3-\overset{\text{O}}{\underset{\text{D}}{\diagdown\text{C}}} \quad (3\text{-}7)$$

Reduction of 1-deuteroacetaldehyde by NADH, the chemical reverse of eq. 3-7 but using unlabeled NADH, produced NAD$^+$ and S-[1-^{2}H$_1$]ethanol (eq. 3-8). This showed that the hydrogen in the 1-pro-R position of ethanol was subject to being transferred to NAD$^+$ in the reaction of ADH.

$$\text{CH}_3-\overset{\text{O}}{\underset{\text{D}}{\diagdown\text{C}}} + \text{ADP-ribose}-\overset{\underset{|}{\text{CONH}_2}}{\text{N:}}\overset{\text{H}}{\underset{\text{H}}{\diagdown}} \rightleftharpoons$$

$$\text{CH}_3-\overset{\text{OH}}{\underset{\text{D}}{\text{C}}}\text{H} + \text{ADP-ribose}-\overset{\underset{|}{\text{CONH}_2}}{\text{N}^+}-\text{H} \quad (3\text{-}8)$$

Reduction of acetaldehyde by $4R$-[nicotinamide-4-^{2}H$_2$]NADH, produced enzymatically in eq. 3-7, produced R-[1-^{2}H$_1$]ethanol, the enantiomer of the S-isomer (eq. 3-9). The enantiomers of 1-deuteroethanol could be distinguished by the signs of the small specific optical rotations they displayed at the D-line of sodium. This confirmed that ADH specifically transferred hydrogen from the 1-pro-R position of ethanol.

$$\text{CH}_3-\overset{\text{O}}{\underset{\text{H}}{\diagdown\text{C}}} + \text{ADP-ribose}-\overset{\underset{|}{\text{CONH}_2}}{\text{N:}}\overset{\text{D}}{\underset{\text{H}}{\diagdown}} \rightleftharpoons$$

$$\text{CH}_3-\overset{\text{OH}}{\underset{\text{H}}{\text{C}}}\text{D} + \text{ADP-ribose}-\overset{\underset{|}{\text{CONH}_2}}{\text{N}^+}-\text{H} \quad (3\text{-}9)$$

1,4-Dihydronicotinamide riboside

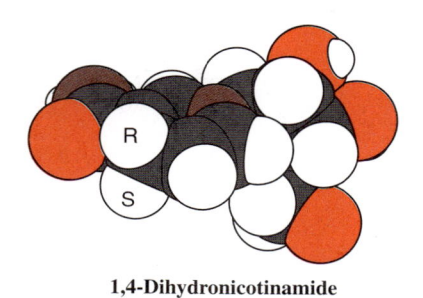

Fig. 3-2. The stereochemical relationship between (R)-4H and (S)-4H in NADH.

134 Enzymatic Reaction Mechanisms

Table 3-1. Pro-*R* and Pro-*S* Stereospecificities for NAD(P)H of Dehydrogenases

Enzyme	NAD(P)H Stereospecificity
Alcohol dehydrogenase	pro-*R*
Malate dehydrogenase	pro-*R*
Lactate dehydrogenase	pro-*R*
Glyceraldehyde-3-P dehydrogenase	pro-*S*
Glycerophosphate dehydrogenase	pro-*S*
Glutamate dehydrogenase	pro-*S*

Each nicotinamide coenzyme-dependent dehydrogenase exhibits a characteristic stereospecificity, both with respect to whether NAD$^+$ accepts hydrogen into the 4-pro-*R* (A-side) or 4-pro-*S* (B-side) position and with respect to the dehydrogenation of its cosubstrate. Several dehydrogenases are listed in table 3-1 together with their stereospecificities with respect to NAD$^+$.

NAD$^+$ as a Coenzyme

In certain enzymes, the NAD$^+$/NADH system plays a catalytic role in facilitating other chemical transformations. The oxidation of an alcoholic group into a ketone has chemical consequences for the surrounding atoms and functional groups. For example, the ketonic group potentiates the cleavage of bonds in its vicinity through its capacity to stabilize a carbanion on an adjacent carbon, thereby decreasing the pK_a value for C—H ionization from more than 40 to less than 20. Carbon-hydrogen ionizations are difficult in general—the value of pK_a for acetone is 19—but many enzymes catalyze such ionizations. However, no enzyme catalyzes the ionization of an unactivated C—H bond. The stabilization of an α-carbanion allows the ketonic group to facilitate a variety of carbon-hydrogen and carbon-carbon bond cleavages, as well as the elimination of good leaving groups. At the conclusion of the transformations, the ketonic group is reduced again to the alcohol or amine by NADH, which has been sequestered at the active site. The types of bond cleavages that are catalyzed by enzymes of this class are illustrated by the bonds highlighted in fig. 3-3.

UDP-Galactose 4-Epimerase

UDP-galactose 4-epimerase (EC 5.1.3.2), the first enzyme to be recognized to use tightly bound NAD$^+$ as a coenzyme, catalyzes the interconversion of UDPGal and UDPGlc (eq. 3-10).

$$\text{UDPGal} \xrightleftharpoons{K_{eq} = 3.5} \text{UDPGlc} \tag{3-10}$$

A substantial body of evidence based on isotopic labeling, spectroscopic, kinetic, inhibition and structural studies support the basic mechanism shown in eq. 3-11, in which the sugar moieties of substrate and product are reversibly dehydrogenated at C4 to produce the tightly bound intermediate UDP-4-ketoglucopyranose. Rotation about the bond linking the glycosyl C1(O) and the β-phosphorus atom of the nucleotide moiety in the intermediate

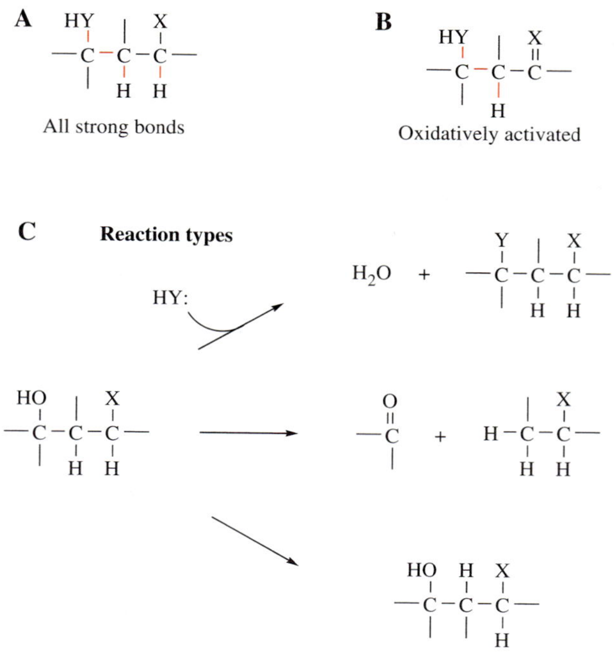

Fig. 3-3. Bond cleavages potentiated by NAD⁺-dependent reversible oxidation of a substrate. (A) The bonds highlighted in red are subject to cleavage after oxidative activation as shown in (B). (C) Several different reaction types may be enhanced by oxidative activation.

allows either face of the C4-carbonyl group to accept a hydride from NADH. UDP-galactose 4-epimerase provides a rare example of nonstereospecific hydride transfer between a substrate and NAD⁺. The enzyme is, however, stereospecific in transferring the hydride to and from the pro-S face of NAD⁺. The mechanism of action of UDPGal 4-epimerase is more extensively discussed in chapter 7.

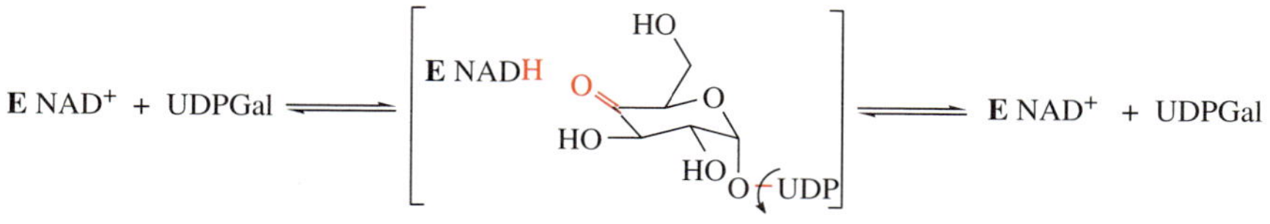

$$\text{E NAD}^+ + \text{UDPGal} \rightleftharpoons [\text{E NADH} \cdots] \rightleftharpoons \text{E NAD}^+ + \text{UDPGal} \quad (3\text{-}11)$$

Nucleotide Sugar 4,6-Dehydratases

The mechanisms in the actions of dTDP-glucose 4,6-dehydratase and CDP-glucose 4,6-dehydratase are more complex than that of the 4-epimerase, although all are members of the short-chain dehydrogenase/reductase family (Jôrnvall et al., 1995). The 4,6-dehydratases (EC 4.2.1.45 and 46) produce the nucleotide derivatives of 4-keto-6-deoxyglucose as intermediates in deoxysugar biosynthesis for the building of cell walls. The tightly bound NAD⁺ mediates hydride-transfer from C4 of the glucosyl moiety of the substrate to C6 of the product by the mechanism in fig. 3-4 (Glaser and Zarkowsky, 1973).

The tightly bound NAD⁺ accepts a hydride from C4 of the glucosyl moiety into the 4-pro-S position of NADH to form the dTDP-4-ketoglucopyranose as the initial intermediate in step 1. Because of electron withdrawal by the 4-keto group, C5(H) is acidic enough to be abstracted by an enzymatic base concomitant with the elimination of the 6-hydroxyl

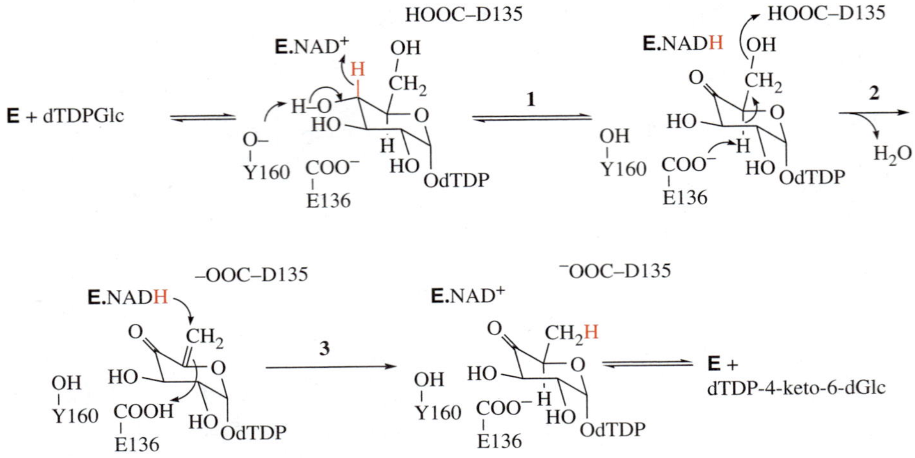

136   Enzymatic Reaction Mechanisms

Fig. 3-4. A mechanism for the action of dTDP-glucose 4,6-dehydratase. The key steps are numbered 1 to 3 and include the dehydrogenation of glucosyl-C4 by NAD⁺ in step 1 and the reduction of the 4-keto-Δ⁵-glucosene at C6 by NADH in step 3. NAD⁺, NADH, and all of the 4-ketosugar intermediates remain tightly bound at the active site throughout the reaction.

group with enzymatic acid catalysis, forming the 4-keto-Δ⁵-glucoseen in step 2. The same hydrogen abstracted from C4 is then transferred as a hydride from NADH to C6 of the 4-keto-Δ⁵-glucosene to form dTDP-4-keto-6-deoxyglucose in step 3. Tyr160, Asp135, and Glu136 serve as the acid-base catalysts for hydride transfer and dehydration (Gerratana et al., 2001; Gross et al., 2001; Hegeman et al., 2001). The structure and mechanism of action are discussed in more detail in chapter 9.

S-Adenosylhomocysteine Hydrolase (EC 3.3.1.1)

S-Adenosylhomocysteine (SAH) undergoes hydrolysis to homocysteine and adenosine according to eq. 3-12 in the presence of this enzyme.

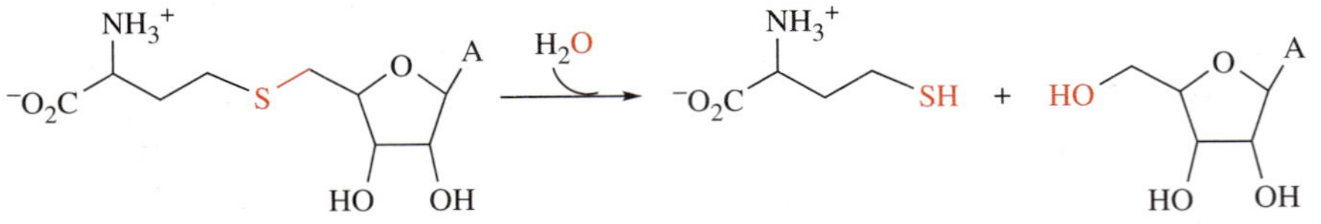

$$\text{(3-12)}$$

From a chemical perspective, this reaction appears to be very difficult. The hydrolysis of sulfides in the laboratory normally requires $ZnCl_2$ and HCl, at high temperatures. However, SAH hydrolase contains tightly bound NAD⁺, which allows the reaction to occur through low-energy intermediates as outlined in fig. 3-5 (Palmer and Abeles, 1976, 1979; Takata et al., 2002).

After binding SAH, the enzyme catalyzes hydride transfer from adenosyl-C3′ to NAD⁺ to form NADH and S-3′-ketoadenosylhomocysteine in step 1 of fig. 3-5. Lysine 185 abstracts the proton from C3(OH) to facilitate hydride transfer. The 3′-keto group increases the acidity of C4′ (H), which is abstracted by Asp130 to drive the elimination of homocysteine from C5′ in step 2. Addition of water in step 3 to C5 of the conjugated

R = CH$_2$CH$_2$CH(NH$_3^+$)COO$^-$

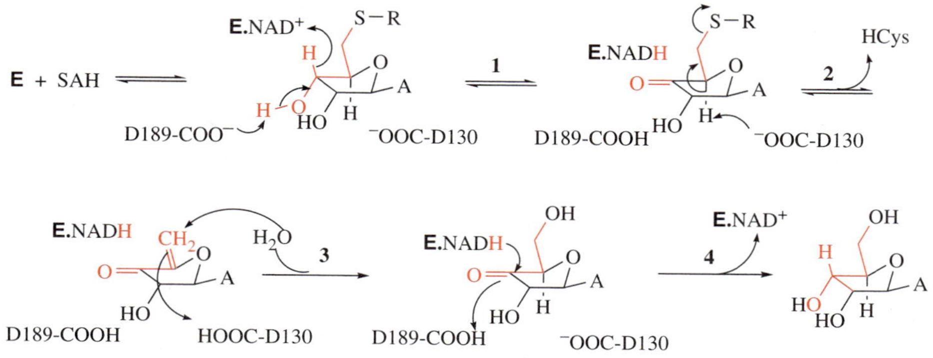

Fig. 3-5. A mechanism for the action of S-adenosylhomocysteine hydrolase. The key steps are numbered 1 to 4 and include the dehydrogenation of SAH at C3′ by NAD$^+$ in step 1 and reduction of the 3′-keto group by NADH in step 4. NAD$^+$, NADH, and all of the 3′-ketoadenosyl intermediates remain tightly bound at the active site throughout the reaction. The x-ray crystallographic structure and results of site-directed mutagenesis implicate Asp189 and Asp130 as the acid-base catalysts (Takata et al., 2002).

α,β-unsaturated ketone from step 2 generates 3′-ketoadenosine at the active site, and reduction of the 3′-keto group by NADH in step 4 produces adenosine.

In addition to the presence of NAD$^+$, more evidence supports the mechanism in fig. 3-5. Chromophores corresponding to NADH and C=C—C=O appear transiently. The enzyme catalyzes the exchange of C4′(H) with solvent hydrogen in SAH, adenosine, or 5′-deoxyadenosine. The enzyme produces 3′-keto-5′-deoxyadenosine and NADH from 5′-deoxyadenosine.

SAH hydrolase is a potential target for antineoplastic agents because of its importance in DNA replication. SAH arises in cells as a by-product of DNA methylation and other methylation reactions by S-adenosylmethionine (SAM). SAH cannot be directly re-methylated to SAM, which is required for DNA methylation. The regeneration of SAM requires hydrolysis of SAH to methionine and adenosine, phosphorylation of adenosine to ATP by kinases, and alkylation of methionine by ATP catalyzed by SAM synthetase (see chap. 15). SAH hydrolase is required for DNA replication, and a good inhibitor of this enzyme could block cell proliferation.

Dehydroquinate Synthase

Dehydroquinate synthase (DHQ; EC 4.2.3.4) is an intermediate in the shikimic acid pathway of aromatic amino acid biosynthesis in plants and microorganisms and is produced by the action of DHQ synthase on 3-deoxy-D-*arabino*-heptulosonate-7-P. In bacteria, DHQ synthase is the product of *aroB*. In microbial eukaryotes such as *Aspergillus nidulans*, DHQ synthase is the N-terminal domain of a multienzyme protein known as AROM, which catalyzes five steps in the biosynthesis of shikimic acid (Moore et al., 1994).

DHQ synthase contains tightly bound NAD$^+$ that undergoes transient reduction, with dehydrogenation at C5 of the substrate (Maitra and Sprinson, 1978). A reasonable mechanism is shown in fig. 3-6A. It appears that the enzyme basically catalyzes the

138 Enzymatic Reaction Mechanisms

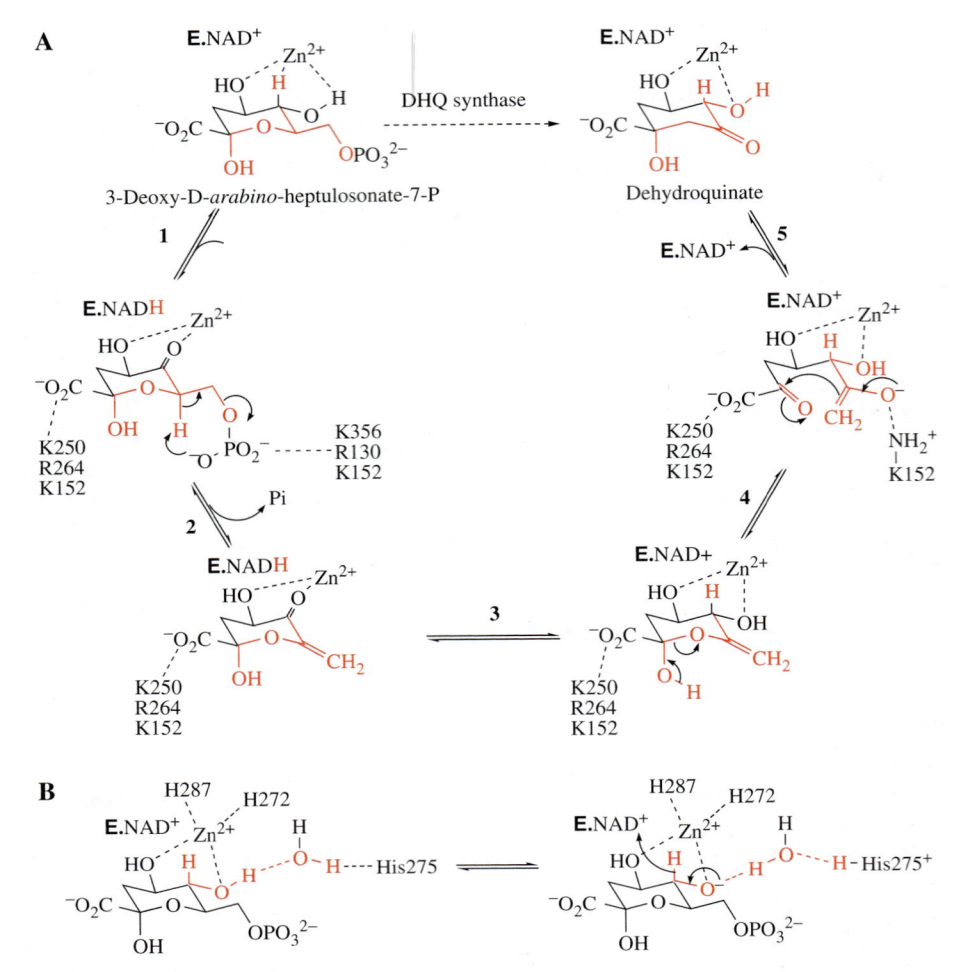

Fig. 3-6. Chemical mechanism for the reaction of dehydroquinate synthase (DHQ). (A) The overall transformation of 3-deoxy-D-*arabino*-heptulosonate-7-P into DHQ is indicated by the dashed arrow. The x-ray crystallographic structures reveal conformational changes attending the binding of the substrate to the open form of the purified complex of DHQ synthase and NAD$^+$ to generate the closed form of the enzyme-substrate complex in step 1 (Carpenter et al., 1998; Nichols et al., 2003). (B) The substrate is coordinated to Zn^{2+}, which is also ligated to His271, His275, and His287. The initial hydride transfer to NAD$^+$ in step 2 is promoted by Zn^{2+}-coordination, with expulsion of the proton to the solvent through an intervening water molecule hydrogen bonded to His275.

dehydrogenation (step 1) and reduction (step 3), and the other steps appear to be either internally catalyzed by the reaction intermediates or to be spontaneous.

DHQ synthase is a zinc metalloenzyme, and Zn^{2+} facilitates hydride transfer in steps 1 and 3 of fig. 3-6A (Nichols et al., 2003). Zinc is coordinated to C4(OH) and C5(OH) of the substrate and to His272 and His287 of the enzyme (fig. 3-6B). Zinc coordination increases the acidity of C5(OH) and facilitates its ionization. The proton released on ionization is relayed through a water molecule to His275. The process is reversed in step 3 of fig. 3-6A.

Step 2 has been postulated to take place through internal catalysis by the phosphate group as in scheme 3-1 (Bender et al., 1989; Widlanski et al., 1989).

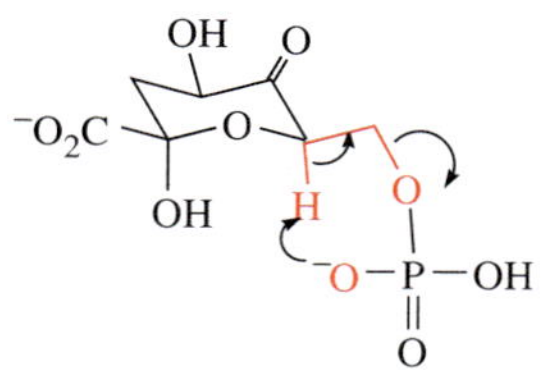

Scheme 3-1

Steps 4 and 5 in fig. 3-6A have been shown to be spontaneous chemical processes (Bartlett and Satake, 1988; Bartlett et al., 1994) although they may be assisted by enzymatic groups at the active site.

Ornithine Cyclodeaminase

The biosynthesis of proline in *Clostridia* terminates with the cyclization of ornithine. The enzyme ornithine cyclodeaminase (EC 4.3.1.12) is NAD$^+$-dependent, but there is no net oxidation in the reaction (Muth and Costilow, 1974). The enzyme has not been studied in detail, and the reaction mechanism is not known. A reasonable mechanism is outlined in fig. 3-7, in which ornithine is initially dehydrogenated at C2 by NAD$^+$ to form NADH and the protonated imine (iminium) at the active site. The intermediate undergoes internal cyclization and transimination with the δ-amino group accompanied by the elimination of the α-amino group.

The resulting imine is then reduced to proline by NADH, which has been sequestered in the active site. Many issues exist regarding this mechanism. Transient reduction of NAD$^+$ has not been demonstrated. In an earlier mechanistic concept, the imine initially formed in the mechanism of fig. 3-7 was postulated to be hydrolyzed to the ketone before cyclization. In any case, oxidative activation by NAD$^+$ provides a remarkably efficient route to proline from ornithine. The x-ray crystallographic structure with ornithine bound at the active site indicates that Arg45, Lys69, and Arg112 bind the substrate-carboxylate group (Goodman et al., 2004). The x-ray crystallographic structure indicates that the reduction of NAD$^+$ proceeds with hydride transfer to the 4-pro-*R* position of NADH.

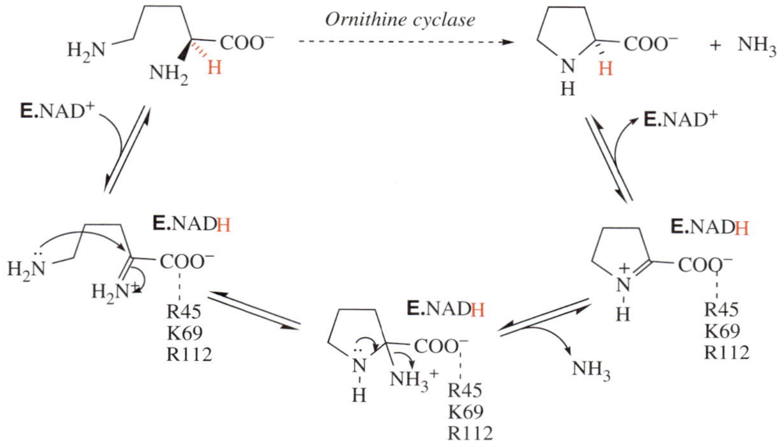

Fig. 3-7. Hypothetical mechanism for NAD$^+$-dependent cyclization of ornithine into proline.

Urocanase

The second step in the metabolic breakdown of histidine is the transformation of *trans*-urocanate into imidazolone propionate by addition of the elements of water. The stereochemistry of deuterium incorporation from D$_2$O is shown in eq. 3-13 (Rétey, 1994).

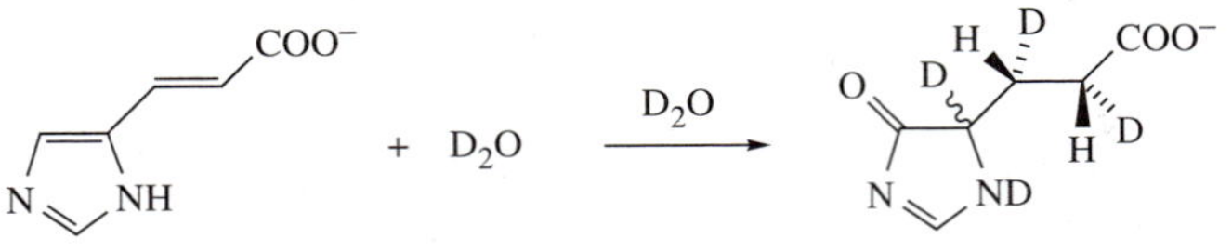

$$\text{(3-13)}$$

Nonstereospecific protonation of the ring adjacent to the carbonyl group suggests that the enol form of the product is released from the active site and spontaneously undergoes ketonization to the predominant tautomer in solution.

Urocanase (EC 4.2.1.49) has been controversial with regard to the role of the coenzyme, which has been identified as NAD$^+$. The mechanistic role of NAD$^+$ is far from obvious. A mechanism put forward that is consistent with available facts is illustrated in fig. 3-8. There is no evidence for reduction of NAD$^+$ to NADH, and such a process would not show promise for potentiating the hydration of urocanate. Instead, NAD$^+$ is proposed to act as an electrophilic catalyst and to form an adduct with imidazole-C4 of urocanate in step 1 of fig. 3-8. Loss of the proton from C4 in step 2 is followed by addition of water to C4, concomitant with protonation at C2 of the side chain in step 3. Protonation at C3 of the side chain in step 4 is followed by dissociation of NAD$^+$ from the adduct and release of *enol*-imidazolone propionate into solution. In support of this mechanism, an adduct formed between the inhibitor imidazole propionate and NAD$^+$ at the active site has been characterized and is consistent with the structures in fig. 3-8. The x-ray crystallographic structure of urocanase with the adduct at the active site is consistent with the mechanism in fig. 3-8 (Kessler et al., 2004).

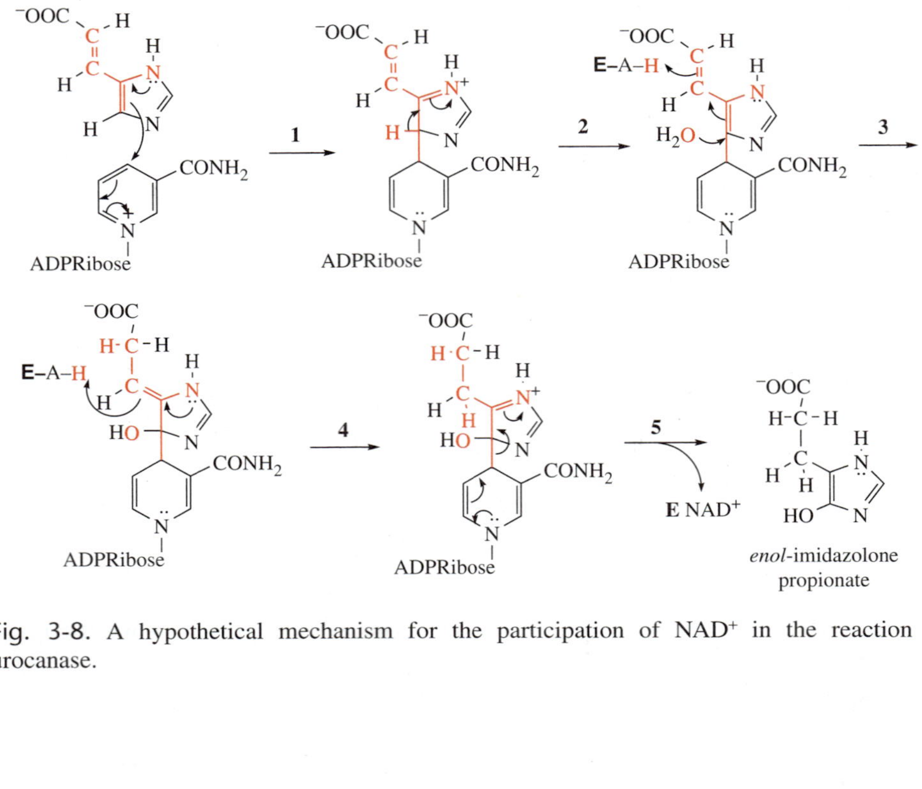

Fig. 3-8. A hypothetical mechanism for the participation of NAD$^+$ in the reaction of urocanase.

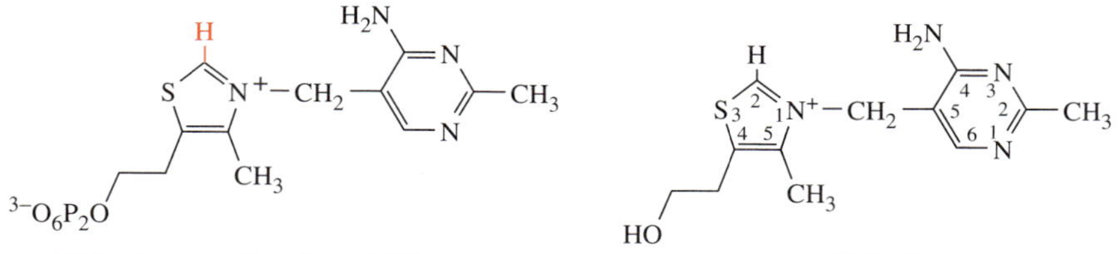

Thiamine pyrophosphate (TPP)
(Thiamine diphosphate [TDP])

Thiamine

Fig. 3-9. The structures of thiamine and thiamine pyrophosphate.

Thiamine Pyrophosphate

Structure

The biologically active form of vitamin B_2 (thiamine) is thiamine pyrophosphate (TPP) (fig. 3-9). TPP serves as the coenzyme in the cleavage of bonds, such as those shown in fig. 3-10. Any enzymatic reaction in which the carbon-carbon bond linking vicinal dicarbonyl groups or the alcoholic and ketonic groups of an α-hydroxyketone is almost certain to be TPP dependent. The ability of TPP to effect these cleavages is not shared by any functional group in a protein, with the exception of the glycyl radical in pyruvate formate lyase. No other coenzyme cleaves bonds of these types. The appearance of thiamine and TPP in biological evolution allowed the chemistry of TPP to become an important part of cellular metabolism.

Reaction Mechanism

Four typical TPP-dependent enzymatic reactions are depicted in fig. 3-11. In later chapters, the mechanisms are described for C—C bond cleavage and decarboxylation. The unifying mechanistic concept for the role of TPP is straightforward and can be set forth here. The function of TPP is to provide a means of avoiding the formation of the acylium ions implied by the mechanisms in fig. 3-12.

The first clue about the mechanism came from an important nuclear magnetic resonance (NMR) experiment in 1957, which showed that C2(H) of 1,5-dimethylthiazolium

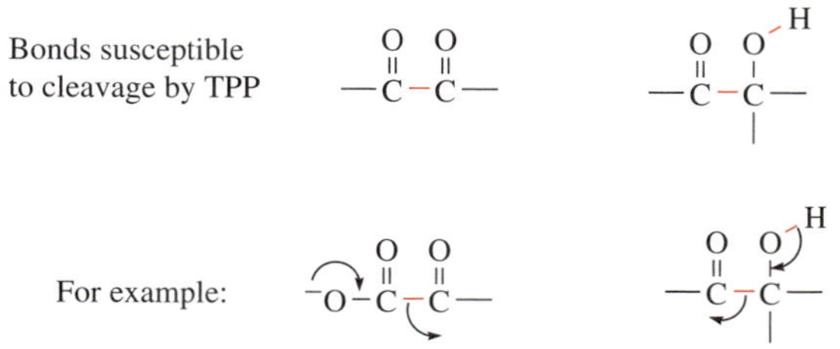

Fig. 3-10. Types of covalent bonds cleaved by action of thiamine pyrophosphate (TPP). Bonds of the type shown in red appear in α-ketoacids, vicinal diketones, and α-hydroxyketones and are cleaved by TPP-dependent enzymes. The nature of the cleavage reactions is illustrated in the lower part of the figure. The chemical properties of the thiazolium ring of TPP in forming adducts with substrates obviate the necessity to produce the unacceptably high-energy acylium ions implied by the electron flow.

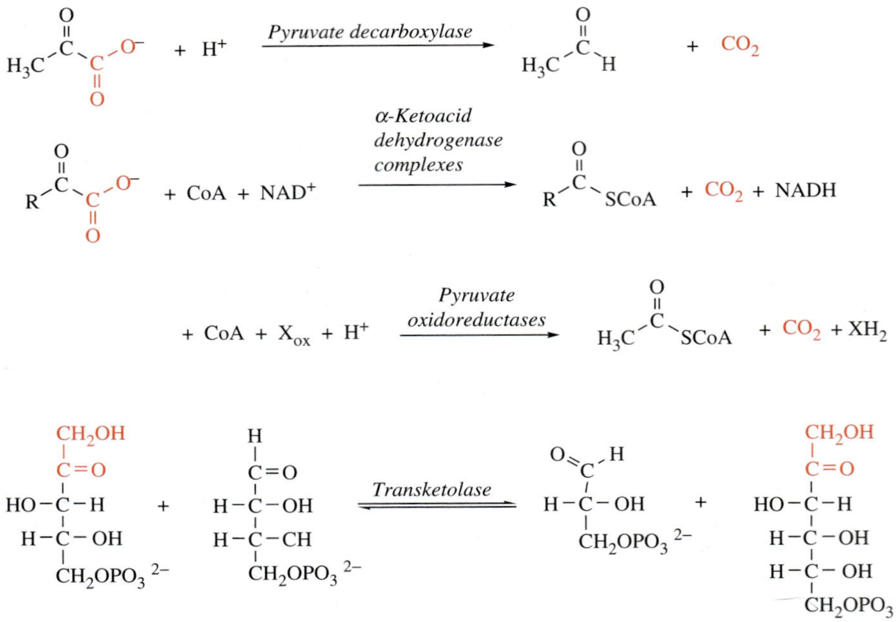

Fig. 3-11. Typical reactions of thiamine pyrophosphate (TPP)–dependent enzymes. Pyruvate decarboxylase and transketolase are TPP-dependent enzymes that do not require other coenzymes or cofactors. Pyruvate oxidoreductases couple the decarboxylation of pyruvate, with its further oxidation to the acetate level; they require other cofactors, including coenzyme A and an electron acceptor such as NADP$^+$, a quinone, or dioxygen. Electron transfer is also mediated by iron-sulfur clusters or flavin coenzymes, or both. The α-ketoacid dehydrogenase complexes consist of at least three proteins and require coenzyme A, NAD$^+$, lipoic acid, and FAD to support the acetyl group transfer and electron transfer, in addition to TPP for decarboxylation.

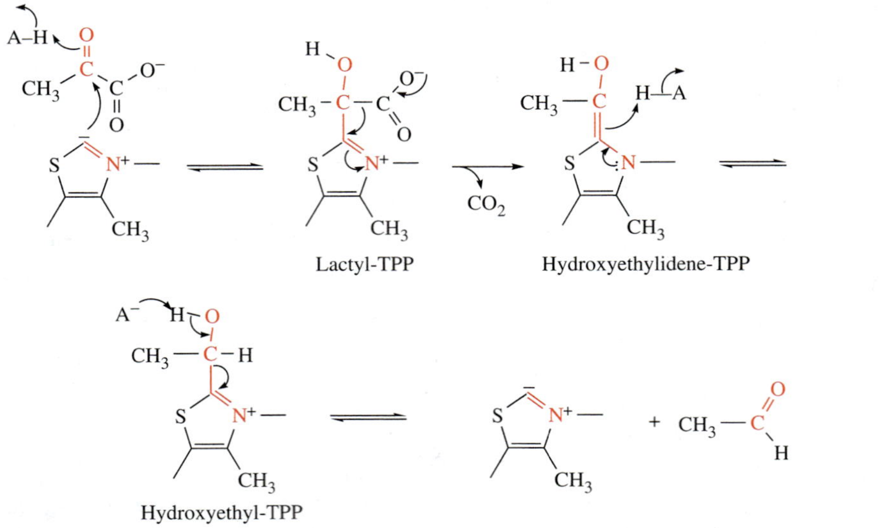

Fig. 3-12. Thiamine pyrophosphate (TPP) catalysis of the decarboxylation of pyruvate. Only the chemically essential thiazolium ring of TPP is explicitly shown in this mechanism, which is intended to focus on the chemical steps.

ion underwent remarkably fast exchange with D$_2$O in neutral solutions (Breslow, 1957, 1958). The exchange was catalyzed by OD$^-$, suggesting the mechanism of eq. 3-14.

$$\underset{CH_3}{\underset{|}{\underset{S\diagdown N^+-CH_3}{\overset{H}{\diagup}}}} \underset{HOD}{\overset{DO^-}{\rightleftharpoons}} \underset{CH_3}{\underset{|}{\underset{S\diagdown N^+-CH_3}{\overset{-}{\diagup}}}} \underset{DO^-}{\overset{D_2O}{\rightleftharpoons}} \underset{CH_3}{\underset{|}{\underset{S\diagdown N^+-CH_3}{\overset{D}{\diagup}}}} \quad (3\text{-}14)$$

The value of pK_a for ionization of C2(H) in the thiazolium ring of TPP in an aqueous solution is about 19 (Washabaugh and Jencks, 1988). The ionization of C2(H) in the active sites of enzymes is orders of magnitude faster than in water, and it is kinetically competent to allow the anionic form to be an intermediate (Kern et al., 1997). We describe the participation of the aminopyrimidine ring of TPP in the ionization of C2(H) at the active sites of enzymes in chapters 8 and 18.

The ready formation of the ylid-like thiazolium-C2 anion suggested a mechanism for the participation of TPP in the reactions of fig. 3-11. Nucleophilic addition of this anion to the carbonyl group of a substrate such as pyruvate would set up the thiazolium-N1 as an electron sink that could facilitate the types of C—C bond cleavages illustrated in fig. 3-10. This proposition leads directly to the mechanism in fig. 3-12 for the decarboxylation of pyruvate. Addition of the thiazolium-C2 anion to the ketonic group of pyruvate produces lactyl-TPP, which is ideally constituted to undergo decarboxylation to hydroxyethylidene-TPP by a mechanism analogous to that for the decarboxylation of a β-ketoacid. Decarboxylation is favored when the coarboxyl group is of the plane of the thiazolium ring. Hydroxyethylidene-TPP undergoes protonation on carbon to hydroxyethyl-TPP, which can eliminate the ylid-like thiazolium-C2 anion and form acetaldehyde.

A key intermediate in fig. 3-12 is hydroxyethylidene-TPP, the stability of which allows the facile decarboxylation of lactyl-TPP. Hydroxyethylidene-TPP is carbanionic in nature, with stabilization provided by its important resonance forms shown in structure **3-1**. The enamine form is likely to be more important to the structure than the charge-separated carbanion, although the polarity of the microenvironment can be expected to influence the relative importance of the two forms.

3-1

Hydroxyethylidene-TPP and closely related species are central intermediates in all TPP-dependent reactions. They react with electrophiles, such as protons and aldehydes, and with oxidizing agents in the manner of enamines and enediols, and these chemical properties explain the mechanisms of TPP-dependent reactions (Jordan, 1999).

The central roles of the enamine-carbanion intermediates are emphasized in fig. 3-13, which shows they are transformed into important biological intermediates. The central carbanion-enamine can react as a carbanion with an aldehyde in transketolase-catalyzed reactions; it can react with the disulfide lipoamide to form *S*-acyldihydrolipoamide and then acyl CoA in the active sites of α-ketoacid dehydrogenase complexes; it can be protonated and ultimately released as acetaldehyde by pyruvate decarboxylase; and it can undergo oxidation, presumably to acyl-TPP, which reacts with CoA to form acyl CoA at the active sites of oxidoreductases.

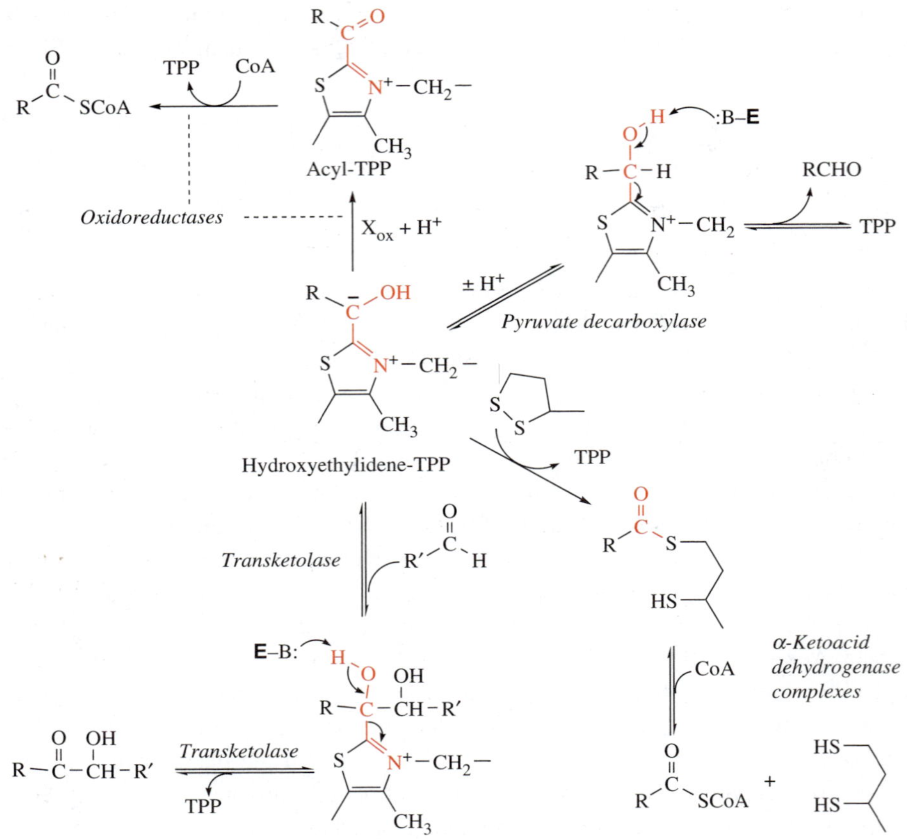

144   Enzymatic Reaction Mechanisms

Fig. 3-13. Enzymatic reactions of hydroxyethylidene–thiamine pyrophosphate (TPP). The major enzymatic fates of hydroxyethylidene-TPP are depicted in its charge-separated carbanionic resonance form.

The evidence supporting the intermediate formation and reactions of hydroxyethyl-TPP and acyl-TPPs is of two types. Representative species such as hydroxyethyl-TPP and acetyl-TPP have been synthesized, characterized, and shown to display the requisite chemical properties of the putative enzymatic species (Gruys et al., 1987, 1989; Holzer et al., 1958; Krampitz et al., 1958). Secondly, hydroxyethyl-TPP, α,β-dihydroxyethyl-TPP, and acetyl-TPP have been trapped in enzymatic reactions as transient species in acid-quenching experiments and chemically and spectroscopically characterized by comparison with authentic samples (Gruys et al., 1989; Tittmann et al 2003, 2005). The carbanion-enamine intermediates are too reactive to purify and study, so that they are presented here as the most reasonable precursors of hydroxyalkyl-TPP and acyl-TPP.

Phosphoketolase

A brief mechanistic study of the action of phosphoketolase (EC 4.1.2.9) has appeared. The reaction is that of xylulose-5-P with phosphate to produce acetyl phosphate and glyceraldehyde-3-P. In [^{3}H]H$_2$O with arsenate in place of phosphate, the enzyme produces [^{3}H]acetate (Goldberger and Racker, 1962). With arsenate, the product would be the hydrolytically labile acetyl arsenate. Arsenate often reacts in place of phosphate in

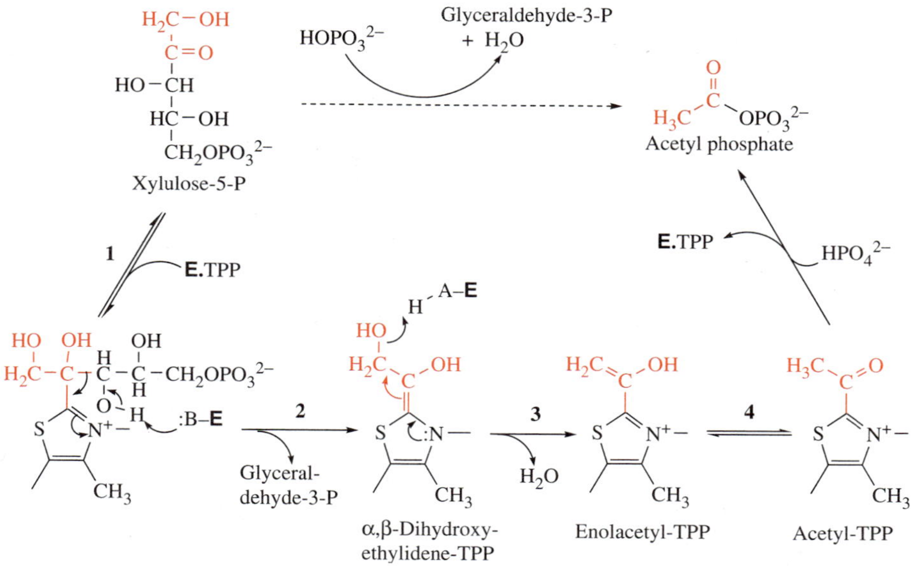

Fig. 3-14. A hypothetical mechanism for the role of TPP in the reaction of phosphoketolase. The overall transformation of xylulose-5-P and phosphate into acetyl phosphate, glyceraldehyde-3-P, and a mole of water is postulated to proceed by the mechanism shown in the lower portion of this figure. The enzyme and reaction mechanism have not been characterized.

phosphorolytic reactions and produces the product of hydrolysis. The enzyme has never been fully characterized in terms of molecular properties. A hypothetical reaction mechanism is pictured in fig. 3-14.

Addition of the TPP-thiazolium-C2 anion to the carbonyl group of xylulose-5-P at the active site produces the TPP-xylulose-5-P adduct in step 1. Abstraction of the C3(OH) proton by an enzymatic base in step 2 produces glyceraldehyde-3-P and α,β-dihydroxyethylidene-TPP at the active site. An enzymatic acid can catalyze the β-elimination of the hydroxyl group in the form of water in step 3 to form enolacetyl-TPP, which can undergo tautomerization to acetyl-TPP in step 4. Acetyl-TPP is an activated form of acetate that reacts with a nucleophilic species, in this case, phosphate to produce acetyl phosphate in the last step.

Pyruvate Oxidases

Pyruvate oxidase (EC 1.2.2.2) is a peripheral membrane of *E. coli*, where it catalyzes the TPP-dependent oxidation of pyruvate to acetate and CO_2, with concomitant reduction of a quinone such as ubiquinone to a hydroquinone (Q_8H_2 in eq. 3-15).

$$CH_3COCOO^- + H_2O + Q_8 \rightarrow CO_2 + CH_3COO^- + Q_8H_2 \qquad (3\text{-}15)$$

The name *pyruvate oxidase* arises from the fact that in the cell, with the quinone and cytochrome b_1 and the *E. coli* electron transport pathway, the terminal electron acceptor is O_2. The enzyme is released by disruption of the membrane and can be purified and crystallized in the absence of detergent (Williams and Hager, 1961). Although detergents are not required to stabilize the protein, the enzyme is subject to activation by lipids (Blake et al., 1978; Hager, 1958). Pyruvate oxidase contains a flavin coenzyme (FAD) that mediates electron transfer to Q_8 from the oxidation of pyruvate. The overall reaction sequence may be described by eqs. 3-15a to 3-15d.

$$\text{E.TPP.FAD} + \text{pyruvate} \rightarrow \text{E.2-lactyl-TPP.FAD} \tag{3-15a}$$

$$\text{E.2-lactyl-TPP.FAD} \rightarrow \text{E.2-hydroxyethylidene} = \text{TPP.FAD} + CO_2 \tag{3-15b}$$

$$\text{E.2-hydroxyethylidene} = \text{TPP.FAD} \rightarrow \text{E.2-acetyl-TPP.FADH}_2 \tag{3-15c}$$

$$\text{E.2-acetyl-TPP.FADH}_2 + H_2O \rightarrow \text{Acetate} + \text{E.TPP.FADH}_2 \tag{3-15d}$$

$$\text{E.TPP.FADH}_2 + Q_8 \rightarrow \text{E.TPP.FAD} + Q_8H_2 \tag{3-15e}$$

In the presence of Q_8, the reduced electron carrier $FADH_2$ transfers two electrons and two protons to form Q_8H_2. $FADH_2$ is an intermediate identified by spectrophotometry; the TPP-intermediates are inferred but not explicitly identified.

The pyruvate oxidase from *Lactobacillus* uses phosphate in place of water in eq. 3-15d and produces acetyl phosphate, and it uses O_2 to oxidize $FADH_2$ in place of a quinone in eq. 3-15e and produces hydrogen peroxide (Hager and Lipmann, 1961; Sedewitz et al., 1984). $FADH_2$ has been identified as a transient species in the action of both enzymes, and a crystallographic structure of the enzyme from *Lactobacillus plantarum* is available (Muller et al., 1994). $FADH_2$ has been identified as an intermediate by spectrophotometry, and both 2-hydroxyethylidene=TPP and 2-acetyl-TPP have been identified as intermediates by NMR spectroscopy (Tittmann et al., 2003).

Pyruvate Oxidoreductases

Pyruvate oxidoreductase (EC 1.2.7.1) from *Klebsiella pneumoniae* catalyzes the decarboxylation and dehydrogenation of pyruvate according to eq. 3-16 (Wahl and Orme-Johnson, 1987).

$$CH_3COCOO^- + \text{CoA} + \text{Flavodoxin}_{ox} \rightarrow CO_2 + CH_3COSCoA + \text{Flavodoxin}_{red} \tag{3-16}$$

The enzyme contains two [4Fe-4S] clusters per molecule, and the iron-sulfur centers presumably couple electron transfer to flavodoxin, the protein-based electron acceptor substrate. The mechanism is not known but is likely to be similar to the putative mechanism for pyruvate oxidase (eqs. 3-15a to 3-15e), with substitution of the iron-sulfur centers for FAD, of flavodoxin for Q_8, and of CoA for water. An analogous enzyme from *Clostridium thermoaceticum* uses ferredoxin as the electron acceptor (Menon and Ragsdale, 1997). Related enzymes from hyperthermophiles also use ferredoxin as the electron acceptor and display specificities for pyruvate, α-ketoglutarate, α-ketoisovalerate, and indole pyruvate (Heider et al., 1996; Mai and Adams, 1994, 1996; Menon et al., 1998). The thermophilic pyruvate oxidoreductase is composed of four different subunits ($\alpha_2\beta_2\gamma_2\delta_2$) of masses approximately 44, 36, 20, and 12 kDa, respectively, with the δ subunit incorporating the two [4Fe-4S] centers (Menon et al., 1998).

Pyruvate oxidoreductase from *Euglena gracilis* contains 2 FAD, 8 Fe, and 8 S^{2-} per dimer (309 kDa) and requires TPP for activity (Inui, 1987). The reaction catalyzed is shown in eq. 3-17.

$$CH_3COCOO^- + NADP^+ + \text{CoA} \rightarrow CO_2 + CH_3COSCoA + NADPH \tag{3-17}$$

The mechanism is not known but is likely to be analogous to the putative mechanism in eqs. 3-15a-e for pyruvate oxidase, with electron transfer from hydroxyethylidene-TPP to the iron-sulfur centers, then to FAD and finally to $NADP^+$.

Acetyl-TPP is a putative intermediate in the reactions of pyruvate oxidase and oxidoreductases. Acetyl-TPP would be produced by oxidation of hydroxyethylidene-TPP in two one-electron steps, most likely through the intermediate formation of a radical in a reaction sequence such as in scheme 3-2. Acetyl group transfer from acetyl-TPP to CoA would follow.

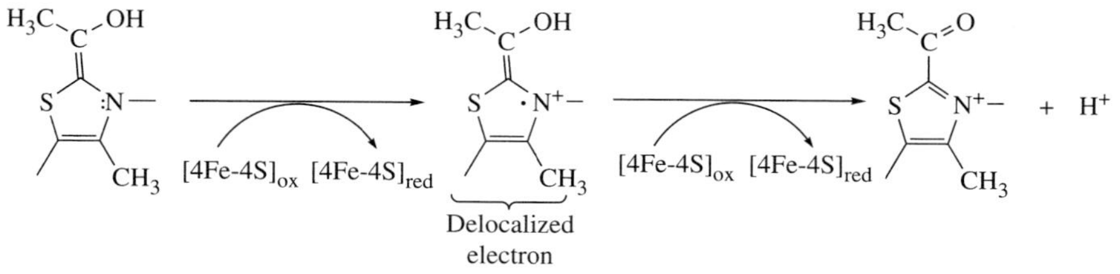

Scheme 3-2

The intermediate radical cation shown is one of the species that could be an intermediate in the actions of pyruvate oxidoreductases, in which one-electron transfer processes are likely, given the involvement of iron-sulfur centers. The unpaired electron on the radical should be widely delocalized over most atoms in the hydroxyethylidene-thiazolium system. The structure can be written as a hybrid of more than seven resonance forms (Frey, 1989, 2001a). Evidence for the transient appearance of this radical in the reaction of pyruvate ferridoxin oxidoreductase (PFOR) has been obtained by EPR spectroscopy (Menon and Ragsdale, 1998; Smith et al., 1994). The radical in the Clostridial oxidoreductase is kinetically competent (Menon and Ragsdale, 1998). The radical in PFOR has been spectroscopically and crystallographically characterized (Chabriere et al., 2001; Mansoorabadi et al., 2006).

That such a radical is readily accessible in the oxidation of hydroxyethylidene-TPP is indicated by the fact that all TPP-dependent enzymes can be assayed by spectrophotometric observation of the reduction of ferricyanide, as exemplified for pyruvate decarboxylase in eq. 3-18.

$$CH_3COCOO^- + H_2O + 2\ Fe(CN)_6^{3-} \rightarrow CO_2 + acetate + 2\ Fe(CN)_6^{4-} + 2\ H^+ \qquad (3\text{-}18)$$

Ferricyanide reduction is a compulsory one-electron process requiring two steps and two moles of ferricyanide to oxidize hydroxyethylidene-TPP. The first step should be analogous to that shown for the oxidation by an iron-sulfur center, and such a one-electron transfer must lead to radical formation. In the second step, ferricyanide should oxidize the radical cation to acetyl-TPP, which is hydrolyzed to acetate with regeneration of TPP.

α-Lipoamide

α-Lipoamide couples electron transfer and acyl group transfer reactions in the α-ketoacid dehydrogenase multienzyme complexes. These complexes catalyze the reactions of α-ketoacids with NAD$^+$ and CoA to produce acyl CoA, NADH, and CO$_2$. The second reaction in fig. 3-11 is that of the pyruvate dehydrogenase complex, which we discuss in chapter 18. The decarboxylation and early steps in dehydrogenation are TPP-dependent, and the overall process requires the participation of five coenzymes: NAD$^+$, TPP, α-lipoamide, CoA, and FAD.

α-Lipoic acid exists in the form of α-lipoamide linkages to the ε-amino groups of lysine residues in lipoyl-bearing domains of the dihydrolipoyl acyltransferase components of these complexes, as illustrated in fig. 3-15. The disulfide group in the dithiolane ring chemically couples the electron transfer and acyl group transfer among three active sites in three enzymes associated with these complexes.

148   Enzymatic Reaction Mechanisms

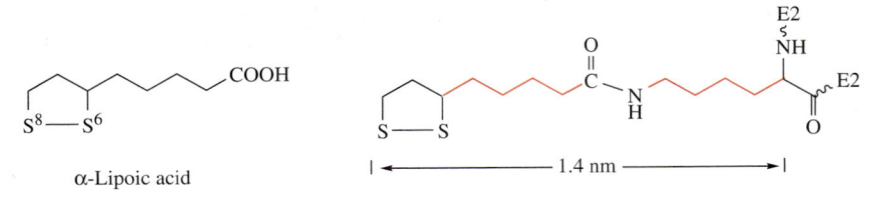

Fig. 3-15. Structures of α-lipoic acid and in conjugation with lysine in an enzymatic site. α-Lipoic acid was originally discovered in research on the pyruvate dehydrogenase (PDH) complex (Reed, 1960), in which it is covalently bonded to lysine residues in lipoyl-bearing domains of the E2 component of this and other α-ketoacid dehydrogenase complexes. The sulfur atoms in the dithiolane ring couple electron transfer and group transfer among the three active sites in these complexes (see chap. 18). Communication among the sites is facilitated by free rotation about the bonds coded red in the lipoamide conjugate, and it is enhanced by the conformational mobility of the lipoyl-bearing domains of the E2 components of these complexes (see chap. 18).

The active sites of the three enzymes are physically separated from one another, and chemical communications among them are facilitated by the structural mobility of the side chains of the lipoyl and lysyl moieties in the conjugate. The bonds highlighted in fig. 3-15 allow rotation, and this accounts in part for the structural mobility of the lipoyl groups. The S^8 of lipoamide is separated from the protein backbone by a maximum of 1.4 nm, so that torsional freedom about bonds in the lysyl and lipoyl side chains can allow the dithionane ring to sweep out a volume of about 2.8 nm in diameter. Protein structural mobility is also important in facilitating chemical communications (see chap.18).

Pyridoxal-5′-Phosphate

Pyridoxal-5′-phosphate (PLP) is the most important coenzyme of amino acid metabolism. The structures of PLP and related forms of vitamin B_6 are given in fig. 3-16A.

Fig. 3-16. Vitamin B_6 coenzymes and the cleavable bonds in pyridoxal-5′-phosphate (PLP) reactions. (A) Structures of vitamin B_6 and its coenzymatic forms. (B) The bonds susceptible to PLP-dependent cleavages in α-amino acids. The δ-amino group in ornithine, γ-amino group in γ-aminobutyric acid, the ε-amino group of lysine, and amino groups in substrates other than amino acids are also cleaved by the actions of PLP-dependent aminotransferases and aminomutases.

Pyridoxal and pyridoxol (or pyridoxine) are regarded as equivalent forms of vitamin B_6. Pyridoxol kinase phosphorylates either compound to the 5′-phosphate, and pyridoxol oxidase catalyzes the oxidation to pyridoxal.

Enzymatic Reactions Facilitated by Pyridoxal-5′-Phosphate

PLP-dependent enzymes conventionally catalyze reactions involving the cleavage of bonds highlighted in fig. 3-16B: A given enzyme cleaves one, two, or three of the highlighted bonds, depending on the overall reaction. Typical PLP-dependent reactions of these types are shown in fig. 3-17, together with the chapters in which the enzymes are discussed. The reaction types include amino acid α-decarboxylation, racemization, transamination, aldol condensation, α,β-elimination, and β,γ-elimination, as well as the β-decarboxylation of aspartate.

Pyridoxal-5′-Phosphate–Stabilized Amino Acid Carbanions

The conventional PLP-dependent reactions depend on the ability of PLP to stabilize one or both of two types of amino acid carbanions shown in fig. 3-18 (Davis and Metzler, 1972;

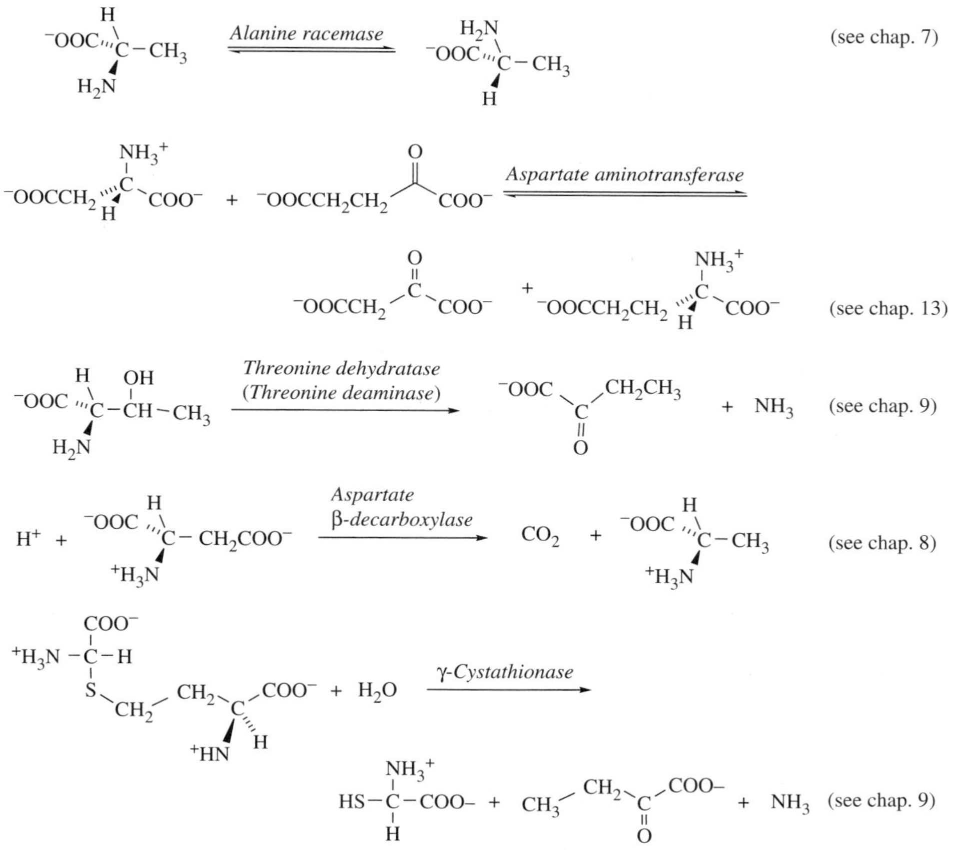

Fig. 3-17. Typical pyridoxal-5′-phosphate–dependent enzymatic reactions are shown with the chapters in which the enzymes are discussed.

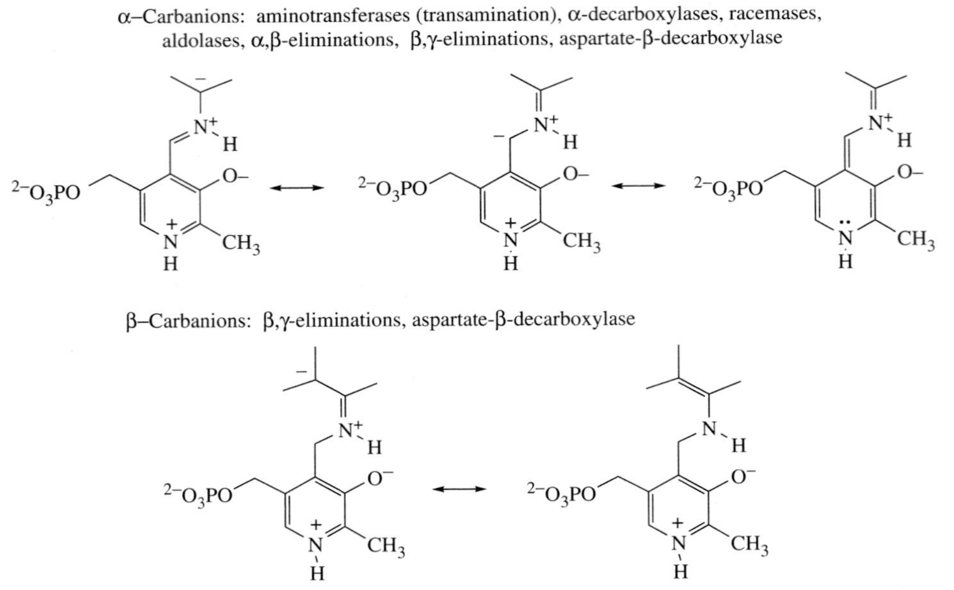

150 Enzymatic Reaction Mechanisms

α–Carbanions: aminotransferases (transamination), α-decarboxylases, racemases, aldolases, α,β-eliminations, β,γ-eliminations, aspartate-β-decarboxylase

β–Carbanions: β,γ-eliminations, aspartate-β-decarboxylase

Fig. 3-18. Structures of pyridoxal-5′-phosphate–stabilized amino acid carbanionic intermediates.

Snell and Di Mari, 1970). The most common species is the α-carbanion at the top of fig. 3-18, which is formed from the amino acid-PLP aldimine (Schiff base) by dissociation of the α-hydrogen of the amino acid or by α-decarboxylation.

The α-carbanion is a resonance-stabilized species in which the electron pair and negative charge are delocalized throughout the conjugated π-bonding system. Three of the many resonance forms appear in fig. 3-18. The other stabilized species are the β-carbanionic species at the bottom of fig. 3-18. These intermediates arise from secondary reactions of the α-carbanionic intermediates and are also delocalized carbanions. They participate in the β,γ-elimination reactions and decarboxylation by aspartate-β-decarboxylase. The α- and β-carbanions of the amino acids themselves are much too high in energy to exist under physiological conditions, but they are stabilized in the PLP derivatives, which allow the types of bond cleavage associated with conventional PLP-dependent reactions.

Aldimine formation of an α-amino acid with PLP facilitates carbanion formation by increasing the acidity of the proton bonded to the α-carbon through two electronic effects. The value of pKa for the α-proton decreases by at least 7 units because of aldimine formation itself, as shown by the relative pK_a values of glycine and its ketimine with acetone in table 1-4 in chapter 1 (Rios et al., 2001).

Additional delocalization of the α-carbanion by delocalization of the electron pair into the pyridine ring may further increase the acidity of the α-proton. This is especially likely when the pyridine ring is protonated, as it is in enzymes such as aminotransferases, PLP-dependent aldolases, and decarboxylases. However, the pyridine ring does not appear to be protonated in alanine racemase (see chap. 7) or in enzymes such as tryptophan synthase that catalyze β-replacement reactions. Inasmuch as protonation of the pyridine ring is not universal in enzyme-PLP complexes, the ionization state of PLP must be established on a case-by-case basis. Irrespective of the ionization state of the pyridine ring, the iminium ion introduced by PLP in an external aldimine increases the acid strength of the α-carbon by at least 7 pK_a units.

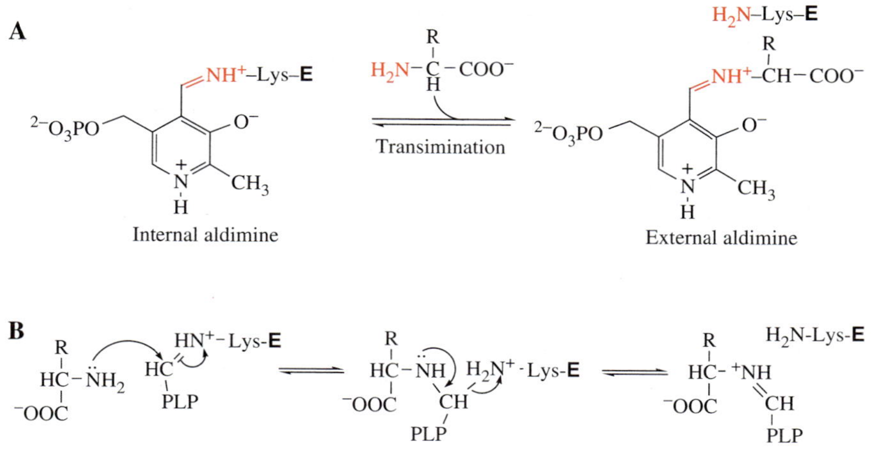

Coenzymes I: Organic Coenzymes 151

Fig. 3-19. Structure of pyridoxal-5′-phosphate (PLP) enzymes. (A) Most PLP enzymes bind PLP covalently through an imine linkage between the aldehyde group of PLP and the ε-amino group of a lysine residue at the active site. (B) The internal aldimines undergo transaldimination with amino acids to form external aldimines much faster than PLP itself would react.

Mechanisms of Pyridoxal-5′-Phosphate–Dependent Reactions

PLP is bound to enzymes through aldimine or Schiff base linkages to ε-amino groups of lysine residues in the active sites. These structures, depicted in fig. 3-19A, are known as *internal aldimines*, and in their protonated states, they readily undergo transimination with the amino groups of substrates to form the *external aldimines* that react in typical of PLP-dependent enzymes.

The internal aldimines maintain PLP in highly reactive states that facilitate the formation of external aldimines. Because of the positively charged nitrogen, a protonated imine is far more electrophilic than the corresponding aldehyde or ketone ($R_2C=NH_2^+$ >> $R_2C=O$) (see chap. 1). The internal aldimines undergo transaldimination with amino acids by the mechanism in fig. 3-19B to form external aldimines much faster than PLP itself would react.

The proton transfer between geminal amino groups in the mechanism of fig. 3-19B is mediated by the 3-aryloxide group of PLP. By facilitating the formation of external aldimines, the internal PLP-aldimines represent the first level of catalysis in PLP-enzymes.

In the first step in the further reaction of an external aldimine, the corresponding α-carbanionic PLP intermediate is formed, either by loss of the α-hydrogen or by α-decarboxylation, illustrated as in scheme 3-3. Loss of CO_2 is shown in scheme 3-3.

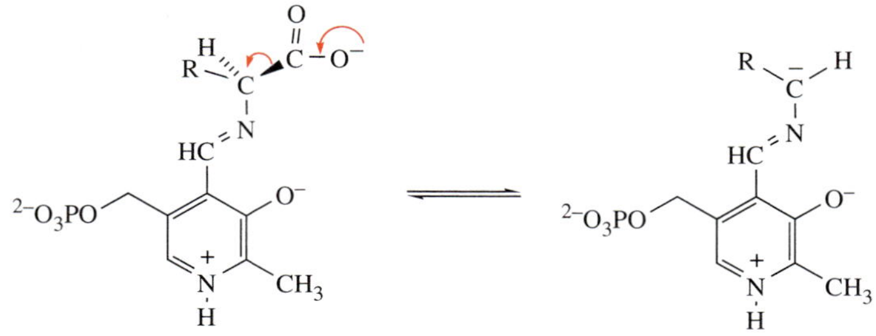

Scheme 3-3

Other steps follow from the stabilization of the α-carbanionic intermediate, which is often transiently observed as a discrete, quinonoid species at 490 to 500 nm. Whether decarboxylation or α-hydrogen abstraction takes place is determined by the binding mode of the amino acid moiety in the active site. The principle is illustrated in fig. 3-20, which depicts aspects of the specificities of PLP-enzymes. Orientation of C_α-substituents and enzymatic catalytic groups in the external aldimine determines whether decarboxylation or removal of the α-hydrogen will take place (Dunathan, 1966; Floss and Vederas, 1982).

Abstraction of the α-hydrogen is the first step in transamination, β-decarboxylation of aspartate, α,β-elimination, β,γ-elimination, and PLP-dependent aldol reactions such as the serine hydroxymethyltransferase reaction (see chap. 14). Decarboxylation is the first step in the reaction of the external aldimine in the action of an amino acid α-decarboxylase.

The reactions of aminotransferases occur in two distinct steps and follow ping pong kinetics. In the first step an amino acid reacts with the internal aldimine (**E=PLP**), transferring the amino group to the coenzyme and forming **E.PMP** and the product α-ketoacid in eq. 3-19a. In the second step, the α-ketoacid cosubstrate reacts with **E.PMP**, the pyridoxamine form of the enzyme, to form another amino acid as a product and regenerate the internal aldimine in eq. 3-19b.

$$R_1CH(NH_2)CO_2^- + \textbf{E=PLP} \rightarrow R_1COCO_2^- + \textbf{E.PMP} \qquad (3\text{-}19a)$$

$$R_2COCO_2^- + \textbf{E.PMP} \rightarrow R_2C(NH_2)CO_2^- + \textbf{E=PLP} \qquad (3\text{-}19b)$$

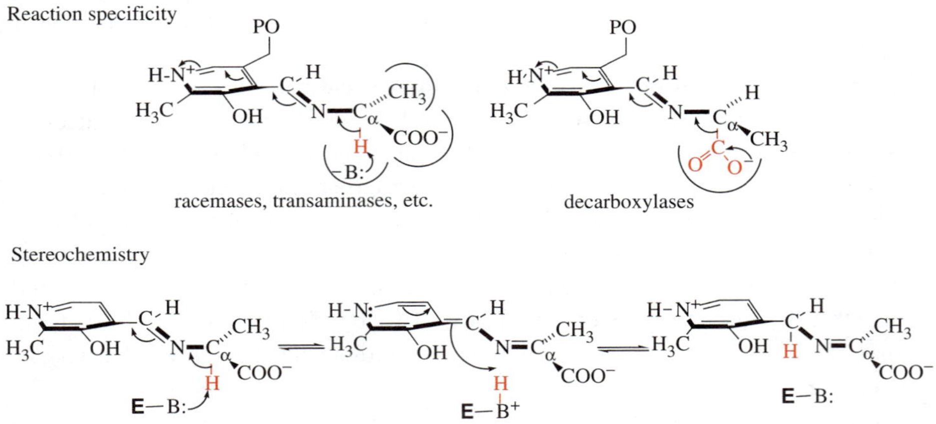

Fig. 3-20. Determinants of reaction specificities and stereospecificities of pyridoxal-5′-phosphate (PLP) enzymes. *Reaction specificity*: Orientation about the N—C_α bond in the external aldimine and the placement of catalytic groups determine whether decarboxylation or removal of the α-hydrogen will take (Floss and Vederas, 1982). Abstraction of the α-hydrogen is facilitated by the orientation at the left, in which the α-carbanion orbital developing from proton abstraction by a well-placed base is aligned for maximum overlap with the π-bonds of the imine and pyridinium ring. Decarboxylation is favored by the placement of the carboxylate group as at the right, in which the α-carbanion orbital developing from decarboxylation attains maximum overlap with the π-bonding system. *Stereospecificity*: In many PLP-dependent enzymes, such as aminotransferases, the α-hydrogen is abstracted and PLP-C4′ is temporarily protonated. In these cases, there is often transfer of the proton from C_α of the amino acid moiety to C4′ of the coenzyme. This is detected in tritium tracer experiments with [2-^{3}H]amino acids, in which tritium can be found stereospecifically incorporated into the C4′ carbon of pyridoxamine-5′-phosphate.

Because the kinetics is ping pong, the two steps can be studied independently. Reaction of [2-³H]amino acid with **E=PLP** often produces tritium labeled **E·PMP**, with tritium stereospecifically incorporated at C4′ of pyridoxamine-5′-phosphate, as shown in fig. 3-20. This arises from suprafacial transfer of tritium by a single enzymatic base, the active site lysine from the internal aldimine, at a rate that exceeds the rate of the exchange of tritium with protons of the medium (Floss and Vederas, 1982). The mechanism of action of aspartate aminotransferase is discussed in detail in chapter 13.

The kinetics of PLP mechanisms have been widely studied because the intermediates can be observed spectrophotometrically. Some of the species are shown in fig. 3-21 together with typical absorption maxima. The spectral signatures of PLP-intermediates in enzymatic reactions are more numerous than those shown in fig. 3-21, and their absorption maxima vary from system to system over ranges of a few nm. The spectrophotometric observation of intermediates makes it possible to characterize them spectrally and to observe their formation and breakdown in transient kinetic studies.

Early transient kinetic studies by the temperature jump technique provided the first and most widely quoted estimates of the rate constants for the formation of enzyme-substrate Michaelis complexes (Fasella and Hammes, 1967). The rate constants turned out to be large (10^7 - 10^8 M^{-1} s^{-1}) as expected, but not as large as the diffusion-limited rate constants for collisions of small molecules (10^9 M^{-1} s^{-1}).

In summation, the widely employed methods for studying the mechanisms of PLP reactions are as follows:

- Steady-state kinetics is most informative for aminotransferases, aldolases, and β-replacement reactions, which are often ping pong systems.
- Deuterium exchange in D$_2$O identifies regiochemistry and stereochemistry of proton transfer.
- Suicide inactivators give clues about the nature of intermediates (see chap. 5).
- Transient phase kinetics can identify intermediates through absorption spectra of transients. X-ray crystallography provides the structures of the enzymes, and complexes with inhibitors and substrates allow identification of catalytic groups and the protonation state of the pyridine ring in PLP.

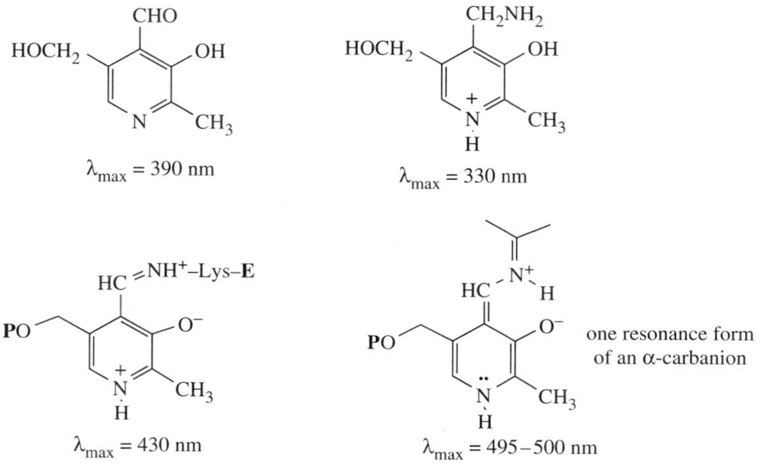

Fig. 3-21. Spectrophotometric properties of some pyridoxal-5′-phosphate compounds.

Tryptophan Synthase

The biosynthesis of tryptophan concludes with the reaction of indole glycerol phosphate with serine to produce tryptophan and glyceraldehyde-3-P (eq. 3-20), which is catalyzed by tryptophan synthase (EC 4.2.1.20).

Indole glycerol phosphate + Serine → Tryptophan + Glyceraldehyde-3-P   (3-20)

The enzyme consists of heterologous subunits α and β in a tetrameric structure $\alpha_2\beta_2$. Subunit α catalyzes the breakdown of indole glycerol phosphate to indole and glyceraldehyde-3-P, and subunit β catalyzes the production of tryptophan through the PLP-dependent condensation of indole with serine. The mechanism requires PLP to facilitate abstraction of the α-proton of serine and the α,β-elimination of water. Proton abstraction generates the α-carbanion, which provides the driving force for the elimination of water.

The mechanism also requires a coordination mechanism for the production of indole in the active site of subunit α and its use at the active site of subunit β, because indole is not a "free" intermediate; indole added to the reacting enzyme is not incorporated into tryptophan as efficiently as indole generated directly from indole glycerol phosphate. Indole produced from indole glycerol phosphate is "channeled" into the production of tryptophan (Creighton, 1970; DeMoss, 1962; Matchett, 1974; Yanofsky and Rachkeler, 1958). Structural analysis revealed the presence of a channel for the transfer of indole between sites (Miles, 2001; Miles et al., 1999).

A chemical mechanism is outlined in fig. 3-22, in which the actions of subunit α are shown in the upper half and those of the PLP-dependent subunit β in the lower half. Subunit α catalyzes the elimination of glyceraldehyde-3-P from indole glycerol phosphate to produce indole, which diffuses to the active site of subunit β. Indole is indicated as being transferred from subunit α to subunit β, as it must be because of kinetic data showing that free indole is not significantly incorporated into tryptophan. The structural basis for this is very beautiful and is illustrated in fig. 3-23. A "tunnel" connects the active sites of the two subunits, and it is through this passageway that indole diffuses between subunits.

The close contacts between indole glycerol phosphate in the active site of the α subunit and between the external aldimine of serine and PLP in the active site of the β subunit are depicted in fig. 3-24.

Role of Pyridoxal-5′-Phosphate in Radical Isomerizations

PLP has been implicated in reactions of organic radicals. Examples of enzymes that catalyze PLP-dependent radical reactions are aminomutases such as lysine 2,3-aminomutase and lysine 5,6-aminomutase (EC 5.4.3.2 and 3), which are required for lysine metabolism in anaerobic bacteria. β-Lysine is also a component of antibiotics produced in *Streptomyces* and *Nocardia*. Lysine 2,3-aminomutase contains PLP and [4Fe-4S] clusters and is activated by *S*-adenosylmethionine (SAM) (Chirpich et al., 1970; Moss and Frey, 1987; Petrovich et al., 1992; Song and Frey, 1991). Lysine 5,6-aminomutase is activated by PLP and the vitamin B_{12} coenzyme adenosylcobalamin (Baker et al., 1973).

EPR spectroscopic evidence implicates organic radicals as reaction intermediates in the case of lysine 2,3-aminomutase (Ballinger et al., 1992; Frey, 2001b). Radical initiation is brought about by a reaction between SAM and the [4Fe-4S] cluster that produces the 5′-deoxyadenosyl radical. In the reaction of lysine 5,6-aminomutase, the

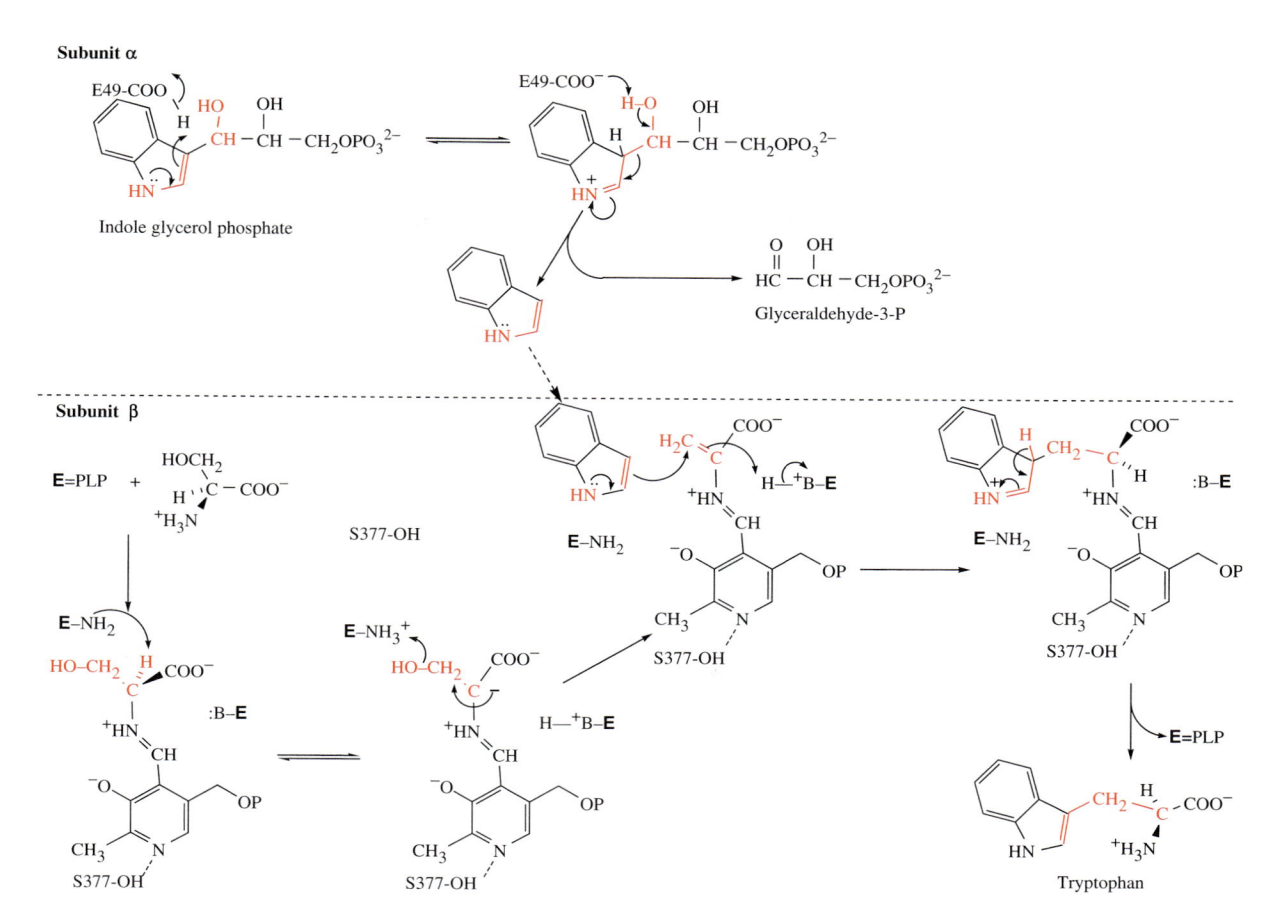

Fig. 3-22. A mechanism of the reaction of tryptophan synthase. The interface between the subunits symbolizes the tunnel shown in fig. 3-23. Indole is produced from indoleglycerol phosphate in the α subunit and migrates through the protein to the active site of the β subunit, where pyridoxal-5′-phosphate (PLP) catalyzes the dehydration of serine. Indole undergoes a β-replacement of the OH group of serine to form tryptophan. The ring nitrogen of PLP is not shown as protonated because the structure indicates that the nitrogen is hydrogen bonded to a serine residue, not an aspartate as in transaminases.

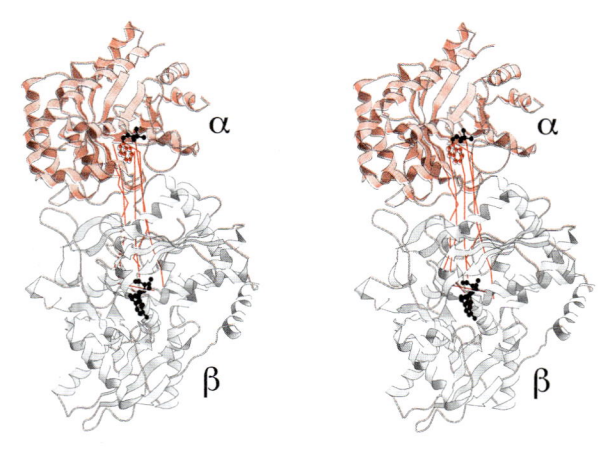

Fig. 3-23. A "tunnel" (red) connecting the active sites in tryptophan synthase runs between the two active sites in *Salmonella typhimurium* tryptophan synthase (EC 4.2.1.20), allowing indole generated in the α subunit to travel to the β subunit without being released into solution. Indole glycerol phosphate is shown as a ball-and-stick model in the α-subunit active site, with the indole moiety in red and the remainder in black. The pyridoxal-5′-phosphate internal aldimine is shown as a black ball-and-stick model in the active site of the β subunit. The illustration was generated using PDB 1QOQ (Weyand and Schlichting, 1999).

156   Enzymatic Reaction Mechanisms

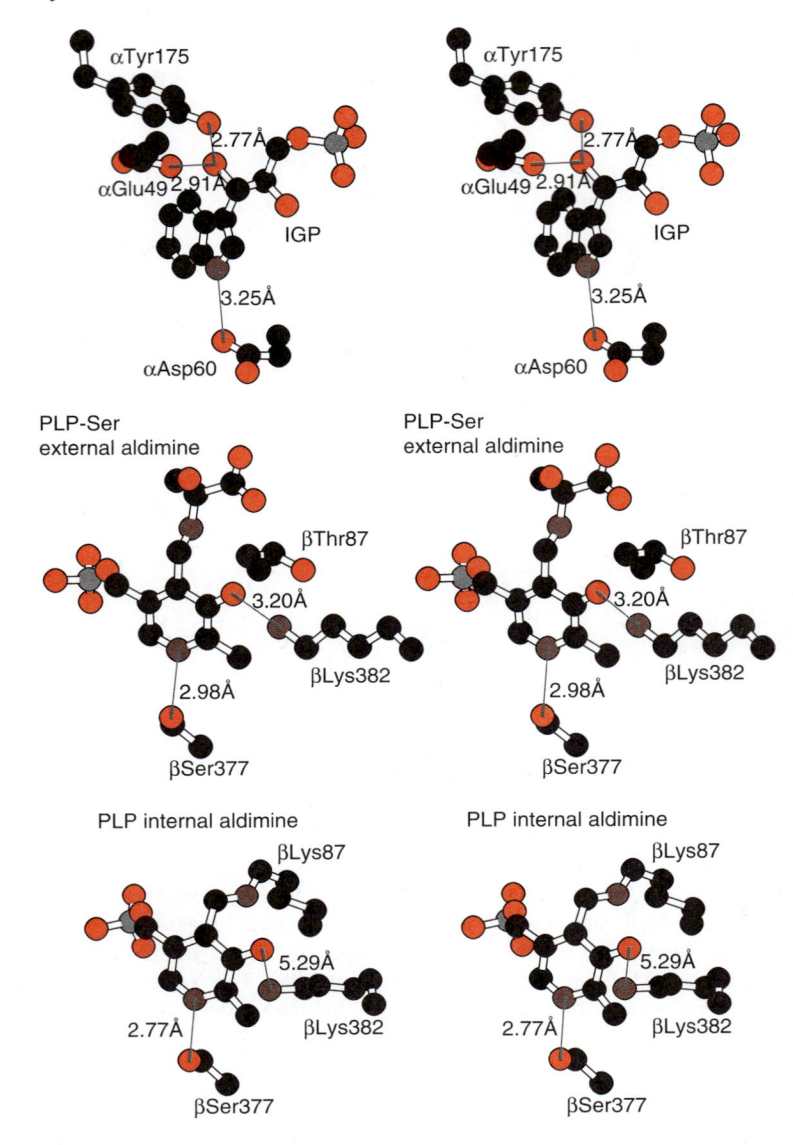

Fig. 3-24. Substrates in the active sites of the α and β subunits of tryptophan synthase. Three stereodiagrams show different substrate-intermediate complexes in the reaction of tryptophan synthase. The top stereopair shows indole glycerol phosphate bound in the α subunit. The middle and bottom stereoimages show the β subunit with the pyridoxal-5′-phosphate (PLP)–serine external aldimine and the PLP-enzyme lysine 87 internal aldimine, respectively. Observation of the external aldimine required altering the active-site residue β subunit lysine 87 to threonine. The top and bottom images were generated using PDB 1QOQ (Weyand and Schlichting, 1999), and the middle image used PDB 2TRS (Rhee et al., 1997).

5′-deoxyadenosyl radical arises from adenosylcobalamin. These processes are discussed in chapters 4 and 7. The amino group migration involves PLP and takes place by the mechanism in fig. 3-25.

The 5′-deoxyadenosyl radical initiates the process by abstracting a hydrogen atom from the side chain of lysine, which is bound to the active site as an external aldimine with PLP, to produce radical intermediate A and 5′-deoxyadenosine. PLP is postulated to facilitate the isomerization of radical A to radical C by way of the cyclic intermediate B.

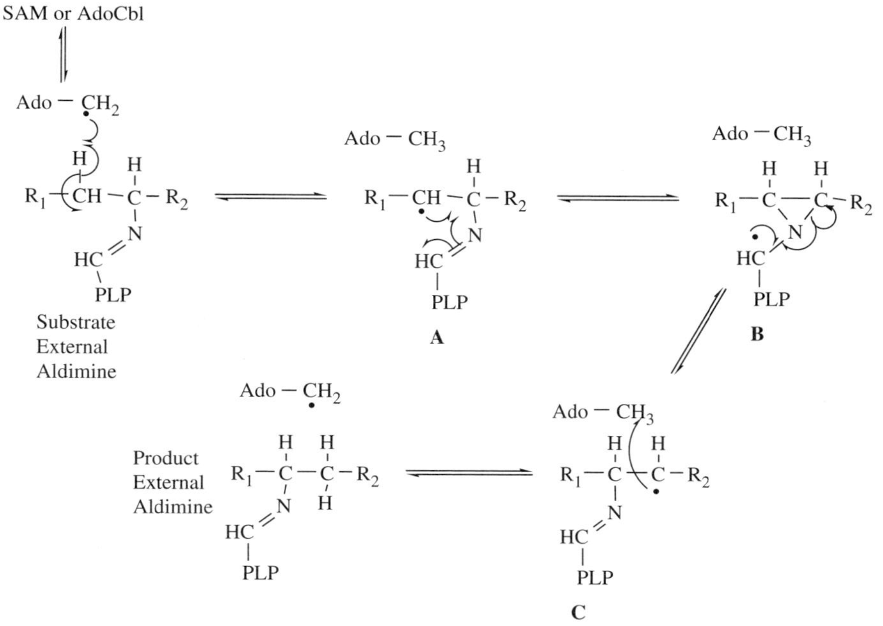

Fig. 3-25. The role of pyridoxal-5′-phosphate (PLP) in the radical-based rearrangement of aminomutases. PLP cannot facilitate the reaction of lysine 2,3-aminomutase by way of carbanionic intermediates. Instead, aminomutases induce radical formation through hydrogen abstraction by the 5′-deoxyadenosyl radical generated from S-adenosylmethionine or adenosylcobalamin (see chap. 4). PLP facilitates amino group migration in the external aldimine by means of radical isomerization.

Abstraction of a hydrogen atom from 5′-deoxyadenosine by C produces the external aldimine of the product and regenerates the 5′-deoxyadenosyl radical.

Another vitamin B_6-dependent reaction that involves a PLP-radical intermediate is the transformation of CDP-4-keto-6-deoxyglucopyranose into the corresponding CDP-4-keto-3,6-dideoxyglucopyranose, shown in fig. 3-26 (Rubinstein and Strominger, 1974a, 1974b; Weigel et al., 1992a, 1992b). The reaction is an overall reductive step on the biosynthetic pathway to ascarylose, a 3,6-dideoxysugar found in the lipopolysaccharide of cell walls in gram-negative bacteria. It is catalyzed by the actions of two enzymes, one designated E1 that contains pyridoxamine-5-phosphate (PMP), and one designated E3 that contains the redox cofactors flavin mononucleotide (following section) and iron-sulfur centers (see chap. 4). The redox cofactors of E_3 funnel reducing equivalents to the substrate-PMP intermediates generated at the active site of E1.

The transformations brought about by E1 are outlined in fig. 3-26 (Agnihotri and Liu, 2001). In step 1, PMP associated with E1·PMP forms an aldimine with CDP-4-keto-6-deoxyglucose, and an enzymatic base abstracts a 4′-hydrogen in step 2, to form a typical PLP α-carbanion shown in one resonance form. The carbanion eliminates the 3-OH group with acid catalysis in step 3, and the resulting conjugated anion is protonated at C3 in step 4. The reduction takes place in two one-electron steps beginning with step 5. The substrate intermediates remain bound to the complex E1·PMP, and the reducing equivalents supplied by NADH are processed by the iron-sulfur flavoprotein E3·FAD[2Fe–2S]. The putative radical produced in step 5 is consistent with the EPR spectrum of a kinetically competent radical observed as a transient intermediate. Reduction of this radical with an additional electron, followed by hydrolysis of the aldimine, produces the 3,6-dieoxysugar nucleotide.

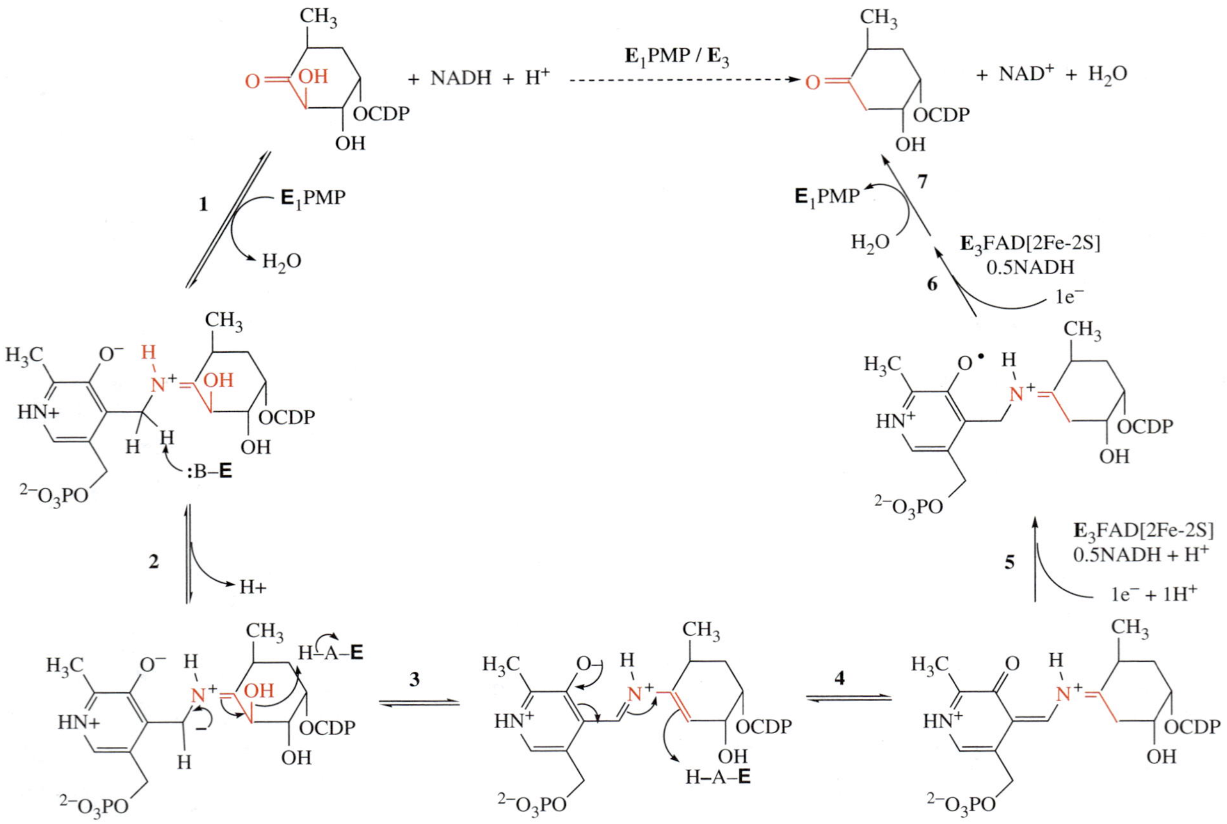

Fig. 3-26. Role of pyridoxal-5′-phosphate in deoxysugar formation. A hypothetical mechanism is shown for reduction of carbon 3 in 3,6-dideoxysugar biosynthesis.

Flavin Coenzymes

Structures of Flavin Coenzymes

The structures of the vitamin riboflavin and the flavin coenzymes are shown in fig. 3-27. Flavoproteins are yellow because FAD and FMN display characteristic absorption spectra with a band in the visible region centered at about 450 nm. The isoalloxazine ring is redox-active and exists in various oxidation states, including the dihydro forms shown in fig. 3-27, as well as semiquinone radicals (discussed later). The diverse redox properties of flavin coenzymes enable them to function in a broad array of enzymatic reactions. The 450-nm absorption bands of FAD and FMN are bleached in $FADH_2$ and $FMNH_2$, as well as in other forms that are transiently generated in enzymatic catalysis. Visible absorption spectrophotometry has been an important tool for analyzing the mechanisms of action of flavoproteins.

FAD and FMN are tightly but noncovalently bound to most flavoproteins. In certain cases, the flavin is not released on denaturation of the protein. In those instances, the flavin is covalently bonded to the protein. Covalent attachments from C-8α to histidine or cysteine or from C-6 to cysteine residues have been characterized and are depicted in fig. 3-28 (Singer and Edmondson, 1980). The covalent bonds to histidyl and cysteinyl side chains of proteins exert characteristic effects on the visible absorption spectra of the flavins.

Fig. 3-27. Structures of flavin coenzymes. The heterocyclic ring characteristic of flavin coenzymes is isoalloxazine. In riboflavin, N1 is carries the 1-ribityl substituent; in FMN, the 1-ribityl moiety is phosphorylated at the 5′ position; and in FAD, the phosphoryl group is transformed into an ADP substituent.

Mechanisms of Flavin Catalysis

Unlike most other coenzymes, a single chemical principle underlying the mechanism of flavin action cannot be offered. In general, flavins participate in redox reactions. In most but not all cases, a dihydroflavin is formed as an intermediate. When molecular oxygen is a substrate, it generally reacts with a dihydroflavin (FADH$_2$ or FMNH$_2$) that has been formed as an intermediate of substrate dehydrogenation. Dioxygen spontaneously oxidizes dihydroflavins to the corresponding flavin, with the production of hydrogen peroxide.

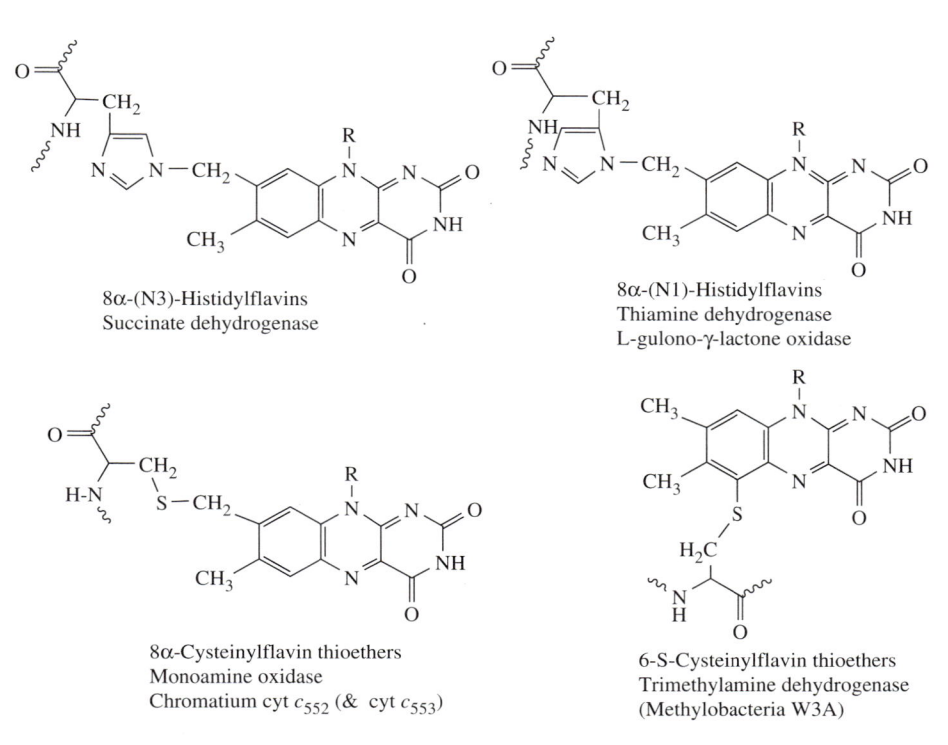

Fig. 3-28. Covalent linkages between the isoalloxazine rings of flavin coenzymes and certain enzymes. The covalent bonds to histidyl and cysteinyl side chains of proteins have characteristic effects on the visible absorption spectra of the flavins.

Enzymes that do not use O_2 as a substrate but produce dihydroflavin intermediates must protect the reduced flavin from reacting with O_2 that is adventitiously present. In some cases, such as the family of disulfide dehydrogenases, the dihydro-flavoprotein does not contain a dihydroflavin per se, but instead has a complex of a flavin and an enzymatic disulfide that share two electrons between them in a charge transfer that does not react with O_2.

Flavin mechanisms are complicated by the diverse chemistry of flavins, most of which is exploited in nature. An underlying theme is the ability of flavins to undergo redox reactions by either two-electron or one-electron transfer mechanisms, a property that allows flavin coenzymes to facilitate many biological redox reactions and to serve as a switch between two-electron and one-electron transfer systems. This property provides an interface between the two-electron redox chemistry in cytoplasmic metabolism and the one-electron redox chemistry in the membrane-bound electron transfer pathways. The switching capacity of flavin coenzymes also allows them to facilitate reactions between O_2 and two-electron donors in the actions of oxidases and oxygenases.

The structures of one-electron reduced forms of the isoalloxazine ring are shown in fig. 3-29 in two protonation states, which differ spectrophotometrically. The ability of flavins to undergo one- or two-electron redox reactions explains much about the biochemistry of flavins, including their participation in the membrane-bound electron transfer pathways, their ability to accept hydride from NADH and other substrates, and the ability of dihydroflavins to react with O_2, a paramagnetic molecule that is formally forbidden to react in two-electron processes.

Flavin-Dependent Oxidoreductases

Flavoproteins that catalyze two-electron redox reactions include glutathione reductase (EC 1.8.1.7) and transhydrogenase (EC 1.6.1.1) (eqs. 3-21 and 3-22).

$$\text{GSSG} + \text{NADPH} + \text{H}^+ \rightarrow 2\,\text{GSH} + \text{NADP}^+ \tag{3-21}$$

$$\text{NADH} + \text{NADP}^+ \rightarrow \text{NAD}^+ + \text{NADPH} \tag{3-22}$$

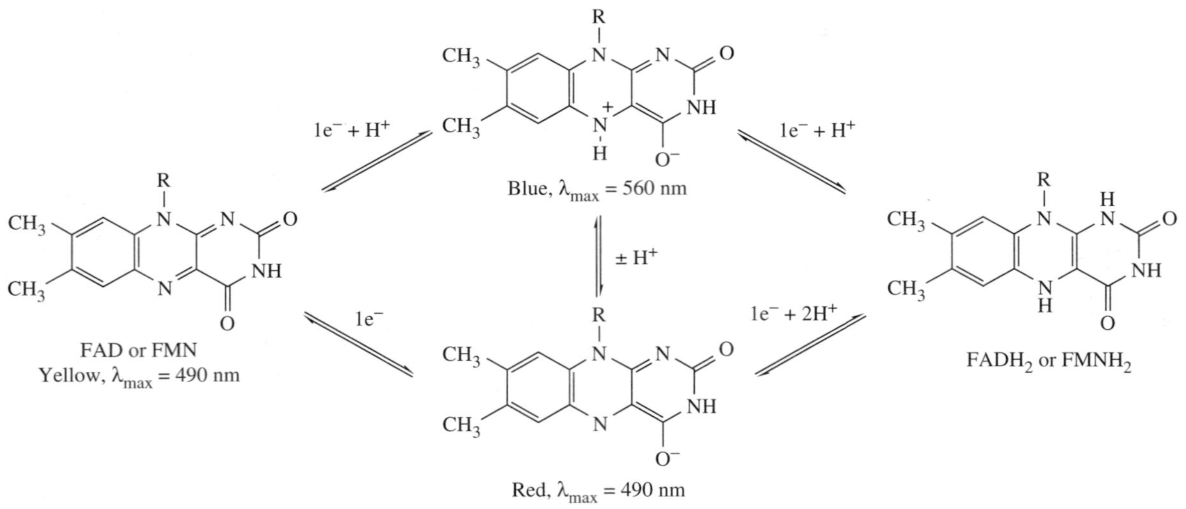

Fig. 3-29. Structures of one-electron, reduced semiquinone forms of flavins. One-electron redox reactions of flavins allow the central forms shown to participate in the redox biochemistry of flavoproteins. The dihydroflavins ($FADH_2$ and $FMNH_2$) are intermediates in most enzymatic reactions of flavin coenzymes.

Each reaction involves a two-electron transfer to produce a reduced flavoprotein followed by a two-electron reduction of a cosubstrate by the reduced flavoprotein. In the case of transhydrogenase, it is possible that the reaction proceeds by hydride transfer between the nicotinamide and flavin coenzymes according to eqs. 3-23a and 3-23b.

$$\text{E FAD} + \text{NADH} + \text{H}^+ \rightarrow \text{E FADH}_2 + \text{NAD}^+ \quad (3\text{-}23a)$$

$$\text{E FADH}_2 + \text{NADP}^+ \rightarrow \text{E FAD} + \text{NADPH} + \text{H}^+ \quad (3\text{-}23b)$$

In the case glutathione reductase and other members of the disulfide oxidoreductase family, including dihydrolipoyl dehydrogenase, glutathione reductase, and thioredoxin reductase the two-electron reduced form of the enzyme does not contain FADH$_2$ (Williams, 1995; Williams et al., 2000). Instead, it consists of a two-electron reduced enzymatic disulfide that interacts by charge-transfer complexation with FAD (see chap. 16). This is presented as *dihydro*-E FAD in the mechanism of eqs. 3-24a and 3-24b for glutathione reductase.

$$\text{E FAD} + \text{NADH} + \text{H}^+ \rightarrow \textit{dihydro-}\text{E FAD} + \text{NAD}^+ \quad (3\text{-}24a)$$

$$\textit{dihydro-}\text{E FAD} + \text{GS–SG} \rightarrow \text{E FAD} + 2\ \text{GSH} \quad (3\text{-}24b)$$

Flavoprotein Oxidases

Flavoprotein oxidases catalyze the oxidation of alcoholic or amino groups of substrates by dioxygen, generally to produce the carbonyl compound and hydrogen peroxide. An example is glucose oxidase, which catalyzes the reaction of glucose with O$_2$ to produce gluconolactone and hydrogen peroxide. The reaction proceeds in two steps (eqs. 3-25a and 3-25b). Other examples are D-amino acid oxidase and lactate oxidase (see chap. 17).

$$\text{E FAD} + \text{Glucose} \rightarrow \text{E FADH}_2 + \text{Gluconolactone} \quad (3\text{-}25a)$$

$$\text{E FADH}_2 + \text{O}_2 \rightarrow \text{E FAD} + \text{H}_2\text{O}_2 \quad (3\text{-}25b)$$

The mechanism by which O$_2$ reacts with FADH$_2$ is of interest. Molecular oxygen is paramagnetic and so should react by one-electron steps unless it is excited to its singlet state. Evidence indicates that a 4a-hydroperoxide intermediate is involved in a mechanism such as that depicted in fig. 3-30 (Bruice, 1984). The reaction of a dihydroflavin with O$_2$ to form a 4a-hydroperoxide is likely to proceed with electron transfer in the first step to produce a flavin semiquinone radical and the superoxide radical anion, which can then form a covalent bond at C4a by undergoing electron pairing. Elimination of hydrogen peroxide leads to the oxidized flavin.

Biopterin

Pterins are structurally related to flavins and function in similar enzymatic reactions. The best known pterin cofactor is tetrahydrobiopterin (structure **3-2**), which participates in the reaction of phenylalanine hydroxylase according to eq. 3-26.

3-2

Tetrahydrobiopterin (BH$_4$)

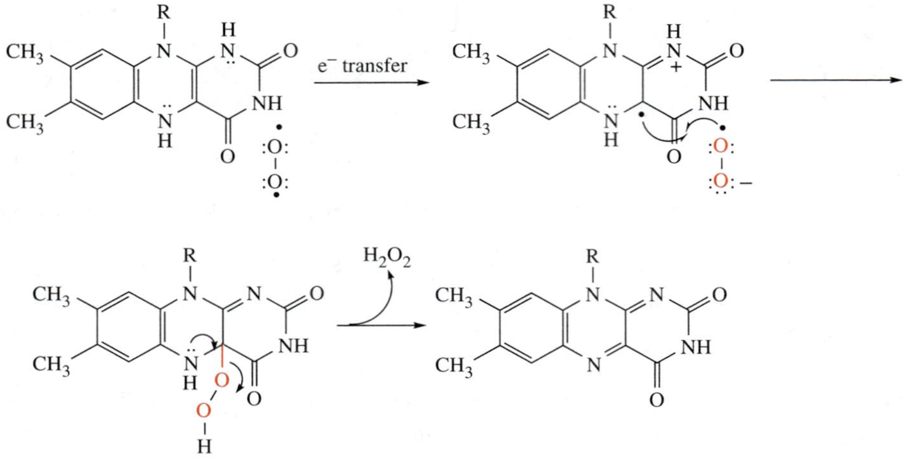

162   Enzymatic Reaction Mechanisms

Fig. 3-30. Flavin 4a-hydroperoxide as an intermediate in the oxidation of a dihydroflavin by O_2. The formation of a flavin 4a-hydroperoxide is likely to begin with a one-electron transfer from the dihydroflavin to oxygen to form the flavin semiquinone, shown in one resonance form, and superoxide radical anion. These two species react by electron pairing at C4a to form the 4a-hydroperoxide, which subsequently eliminates hydrogen peroxide to form the oxidized flavin.

The overall reaction is a monooxygenation, in which the oxygen incorporated into tyrosine is derived from O_2. Phenylalanine hydroxylase is an iron metalloprotein known as a mixed function monooxygenase; it requires an oxidizing and a reducing agent. Molecular oxygen is the oxidizing agent, and tetrahydrobiopterin serves as the reducing agent by reacting with O_2 and mediating the production of an oxygenating species. In a process that appears to be similar to the reaction of dihydroflavins with O_2, a hydroperoxybiopterin intermediate appears to be formed at the active site, and it potentiates the formation of an electrophilic oxy-iron species that hydroxylates phenylalanine. The reaction mechanism and structure of phenylalanine hydroxylase are discussed in greater detail in chapter 4.

Tetrahydrobiopterin is also one of the many cofactors of nitric oxide synthase, a topic of intense current interest because of the physiological role of nitric oxide as a second messenger. The role of tetrahydrobiopterin in nitric oxide formation has not been firmly established. The system is described in more detail in chapter 17.

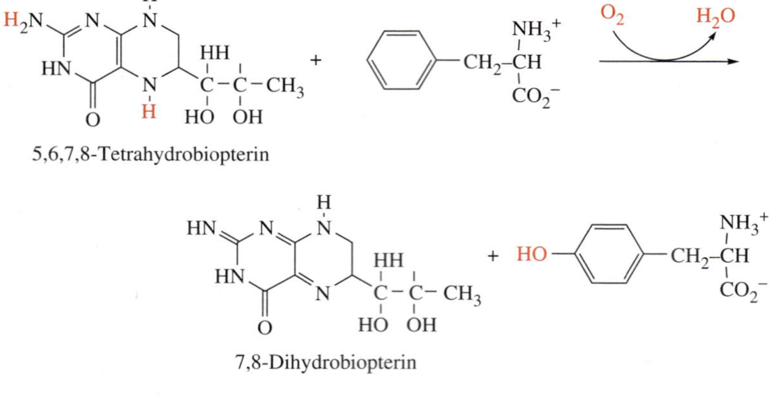

(3-26)

Biotin

Structure and Role as a Carboxyl Carrier

The vitamin biotin is covalently bonded to the ε-amino groups of lysine residues in biotin-dependent carboxylating enzymes, which catalyze reactions like that of acetyl CoA carboxylase (eq. 3-27).

$$CH_3-CO-SCoA + HOCO_3^- + ATP \rightarrow {}^-O_2C-CH_2-CO-SCoA + ADP + Pi \quad (3-27)$$

Other carboxylases catalyze the analogous carboxylations of propionyl CoA to methylmalonyl CoA, β-methylcrotonyl CoA to β-methylglutaconyl CoA, and pyruvate to oxaloacetate. The structures of biotin and the N^1-carboxybiotinyl group in a polypeptide chain are illustrated in fig. 3-31. The space-filling model of biotin shows that N3 is too sterically hindered by the biotin side chain to undergo carboxylation, whereas N1 is unhindered.

In biotin-dependent carboxylation reactions, the role of biotin is to accept a carboxyl group and transport it from one site to another within an enzyme complex (Knowles, 1989). Enzymes such as acetyl CoA carboxylase (eq. 3-27) consist of two types of catalytic subunits, a *biotin carboxylating* subunit that catalyzes the reaction of bicarbonate with ATP and biotin to form N^1-carboxybiotin according to eq. 3-28a, and a *substrate carboxylating* subunit that carries out the enolization of a substrate, usually an acyl CoA, and facilitates carboxylation of the enolate by N^1-carboxybiotin (eq. 3-28b). Biotin is the carrier of the carboxyl group between the two sites.

$$\text{Biotinyl-protein} + ATP + HC^{18}O_3^- \rightarrow {}^{-18}O_2C\text{–Biotinyl-}E + ADP + H^{18}OPO_3^{2-} \quad (3\text{-}28a)$$

$$^{-18}O_2C\text{–Biotinyl-protein} + CH_3\text{–CO–SCoA} \rightarrow \text{Biotinyl-}E \\ + {}^{-18}O_2C\text{–}CH_2\text{–CO–SCoA} \quad (3\text{-}28b)$$

Transcarboxylase (EC 2.1.3.1) catalyzes the reversible transfer of a carboxyl group from oxaloacetate to propionyl CoA to produce pyruvate and methylmalonyl CoA. This enzyme lacks the type of biotin carboxylating subunit that catalyzes the MgATP-dependent carboxylation of biotin; however, it consists of two subunits of the type that catalyze the

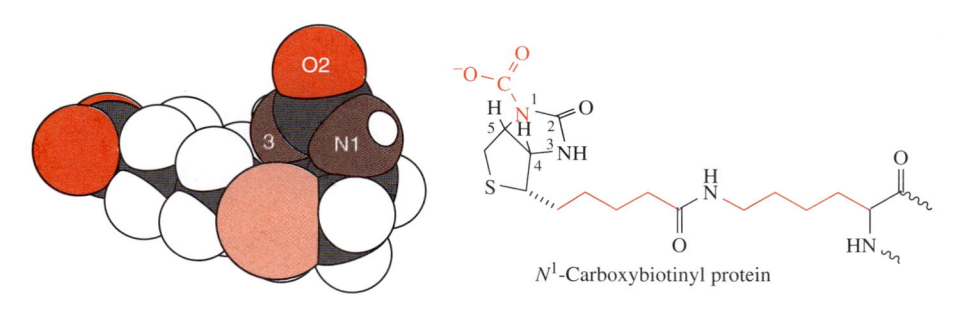

Fig. 3-31. Structures of biotin and a carboxybiotinyl protein. At the left is a space-filling model of biotin. Because of *cis*-fusion of the rings and the size of sulfur, the sulfur and carbonyl groups are sterically close. The stereochemistry of the side chain brings it within close contact to N3, so that only N1 is exposed to carboxylation. The structure of the N^1-carboxybiotinyllysyl group of a carboxylating protein is at the left. The highlighted bonds of the biotinyl and lysyl side chains are those about which rotation is allowed.

enolization and carboxylation of substrates by N^1-carboxybiotin, and biotin carries the carboxyl group from the active site of subunit to that of the other. Transcarboxylase was the first enzyme recognized to display two-site ping pong kinetics (see chap. 2).

The biotinyllysyl moiety of carboxylases and transcarboxylases is structurally well suited to carry out the function of transporting a carboxy group from one site to another through the conformational freedom allowed by rotation about the single bonds in the biotinyl and lysyl side chains (fig. 3-31).

Chemistry of Biotin and N^1-Carboxybiotin

The chemical properties of biotin and N^1-carboxybiotin can be explained by the *cis*-fused ring structure, which allows biotin to maintain the carboxyl group in a chemically poised state (Tipton and Cleland, 1988). It is a potential carboxylating agent within a carboxylating active site, but its structure makes it kinetically stable to nonenzymatic decarboxylation at pH 7. The steric clash between the *cis*-fused rings hinders formation of a tetrahedral transition state at N1.

A singular and informative feature of biotin chemistry is the kinetics of the exchange of the ureido protons with D_2O. The pH-rate profile for N1(H) exchange is concave, with acid and base catalyzed legs, a minimum (3 s^{-1}) at pH 6, and an exchange rate of 10 s^{-1} at pH 7. Exchange of the N3(H) is even slower at one-fifth the rate of N1(H), presumably because of the greater steric hindrance at N3 by the biotin side chain, as illustrated in the space-filling model in Fig 3-31. The exchange rates of biotin are much slower than of ordinary amides and ureas because of steric hindrance in the *cis*-fused structure (fig. 3-31). Both acid (H^+)– and base (HO^-)–catalyzed exchanges are slowed by steric hindrance.

Decarboxylation of carbamates is acid catalyzed but not base catalyzed, and acid catalyzed decarboxylation is slowed by steric hindrance in the *cis*-fused ring of N^1-carboxybiotin. This is exemplified by the decarboxylation of *N*-carboxyimidazolidone, a model compound lacking the fused ring, which undergoes acid catalyzed decarboxylation 30 times faster than N^1-carboxybiotin. Moreover, the pH-rate profile for decarboxylation of *N*-carboxyimidazolidone reaches its uncatalyzed minimum at pH 8.2, whereas that for N^1-carboxybiotin is minimal at pH 6.4. The pH-rate profiles show that at pH 7 decarboxylation of *N*-carboxyimidazolidone is predominantly acid catalyzed, whereas for N^1-carboxybiotin it is slow and uncatalyzed. Inasmuch as acid catalyzed decarboxylation requires protonation and is not observed at pH 7 for N^1-carboxybiotin, the *cis*-fused ring must interfere with protonation, as it does in proton exchange, and decarboxylation can only take place by the slow, uncatalyzed mechanism. This property of biotin makes it well suited as a carboxy-carrier at pH 7, where its stability against nonenzymatic acid catalyzed decarboxylation allows the N^1-carboxybiotinyl group to be transferred from one site to another in a stable form.

Mechanism of Biotin-Dependent Carboxylation

Two facts about the carboxylation of biotin by enzymes such as acetyl CoA carboxylase in eq. 3-28a are mechanistically revealing; MgATP is required, and ^{18}O from [^{18}O]bicarbonate is incorporated into inorganic phosphate (Knowles, 1989). These properties hold for all ATP-dependent biotin carboxylation reactions. The source of free energy for dehydrating bicarbonate is understood to be the hydrolysis of MgATP. Bicarbonate is unreactive as a carboxylating agent, whereas carbon dioxide is electrophilically reactive and can carboxylate biotin. The transfer of ^{18}O from bicarbonate into phosphate indicates the mechanism by which the hydrolysis of MgATP energizes the dehydration process in the active site of the biotin carboxylating subunit. MgATP phosphorylates an oxygen atom of bicarbonate to form "carboxyphosphate" according to eq. 3-29a. Carboxyphosphate is extremely reactive and should readily undergo decarboxylation to carbon dioxide and phosphate according to eq. 3-29b. [^{18}O]Carbon dioxide is immediately captured by biotin to form N^1-[^{18}O]carboxybiotin.

$$\text{MgATP} + \text{H}^{18}\text{O}-\overset{\overset{^{18}\text{O}}{\|}}{\text{C}}-^{18}\text{O}^- \rightleftharpoons \text{MgADP} + \left[^-\text{O}-\overset{\overset{\text{O}}{\|}}{\underset{\underset{\text{O}^-}{|}}{\text{P}}}-^{18}\text{O}-\overset{\overset{^{18}\text{O}}{\|}}{\text{C}}-^{18}\text{O}^- \right] \quad (3\text{-}29\text{a})$$

$$\left[^-\text{O}-\overset{\overset{\text{O}}{\|}}{\underset{\underset{\text{O}^-}{|}}{\text{P}}}-^{18}\text{O}-\overset{\overset{^{18}\text{O}}{\|}}{\text{C}}-^{18}\text{O}^- \right] \longrightarrow \ ^-\text{O}-\overset{\overset{\text{O}}{\|}}{\underset{\underset{\text{O}^-}{|}}{\text{P}}}-^{18}\text{O}^- + \left[\text{C}^{18}\text{O}_2 \right] \quad (3\text{-}29\text{b})$$

$$[\text{C}^{18}\text{O}_2] + \text{Biotinyl-E} \longrightarrow N^1\text{-}[^{18}\text{O}]\text{Carboxybiotinyl-E} \quad (3\text{-}29\text{c})$$

On being transported into the active site of the substrate carboxylating subunit, N^1-carboxybiotin becomes reactive as a carboxylating agent. Isotope effects indicate that N^1-carboxybiotin reacts by undergoing decarboxylation to generate carbon dioxide, which carboxylates the substrate enolate (see chap. 18). The kinetic isotope effect for the spontaneous, pH-independent decarboxylation of N^1-[^{13}C]carboxybiotin is 1.023 (Tipton and Cleland, 1988), and that for the enzymatic carboxylation of pyruvate is 1.024. These similar isotope effects are significantly different from those for alternative mechanisms. For example, the ^{13}C-kinetic isotope effect for acid catalyzed decarboxylation of biotin is 1.012. Carbon dioxide produced at the active site is immediately captured by the substrate enolate that is generated simultaneously (see chap. 18).

Phosphopantetheine Coenzymes

Structures of Phosphopantetheine Coenzymes

Coenzyme A (CoA) and the phosphopantetheine moieties of multienzyme complexes like fatty acid synthetase, nonribosomal polypeptide synthases, and polyketide synthases incorporate the vitamin pantothenic acid. The structures of these molecules are shown in fig. 3-32. The phosphopantetheine group is the acyl group-carrying moiety in coenzymatic action. CoA is the biological acyl group carrier in the form of acyl CoAs, and it functions in the capacity of an acyl substrate rather than as a true coenzyme. Phosphopantetheine bonded to a protein is more coenzyme-like in that, as in multienzyme synthases such as the fatty acid synthase, polyketide synthases, and polypeptide synthetases it works in concert with enzymes (see chap. 18).

Mechanism of Phosphopantetheine Action

The Chemical Role

The role of the sulfhydryl groups in CoA and phosphopantetheine is to activate the carboxylic acid groups of substrates for reactions they will undergo within the active sites

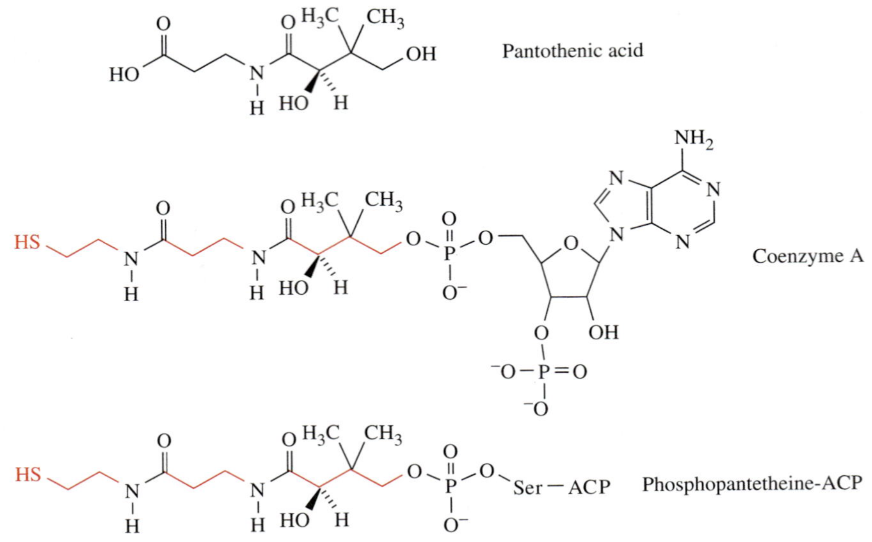

166   Enzymatic Reaction Mechanisms

Fig. 3-32. Structures of pantothenic acid and phosphopantetheine coenzymes. The structures of coenzyme A (CoA) and phosphopantetheine prosthetic groups of proteins are based on the vitamin pantothenic acid. The —SH group covalently binds acyl groups and chemically activates them. The phosphopantetheine moiety may be tightly bound and immobilized through protein interactions, especially in reactions of CoA, or it may be relatively freely mobile in the reactions of acyl carrier protein (ACP). The bonds that allow rotation are highlighted.

of cognate enzymes. Carboxylate groups are not very reactive in acyl group transfer and enolization. Acyl group transfer reactions require the initial addition of a nucleophile to the carbonyl of a carboxylic acid derivative to form a tetrahedral intermediate or transition state. Nucleophilic addition to a carboxylate ion leads to an accumulation of negative charge on the addition intermediate, an energetically and kinetically unfavorable process (structure **3-3**).

$$\text{3-3} \quad \text{R–COO}^- + \text{N:} \rightleftharpoons \text{R–C(O}^-\text{)(O}^-\text{)N} \quad << \quad \text{R–COS-CoA} + \text{N:} \rightleftharpoons \text{R–C(O}^-\text{)(S-CoA)N}$$

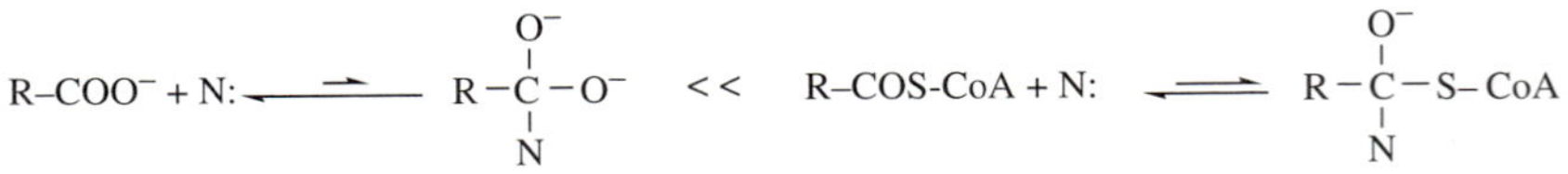

In the second step of an acyl group transfer, the leaving group from the tetrahedral intermediate would be an oxide ion (O^{2-}), a very poor leaving group. In thioesters such as CoA-esters, nucleophilic addition is much less disfavored because of little accumulation of negative charge, and there is a good leaving group. Thioesters are highly activated toward acyl transfer relative to carboxylate ions. The thioester carbonyl is also more electrophilic than the oxyester carbonyl because thioesters are more ketone-like than oxyesters. This difference can be attributed to the poor energy match between the p orbitals of sulfur and the π bond of the carbonyl group in a thioester, which allows less orbital overlap and less delocalization of nonbonding electrons from sulfur onto the carbonyl oxygen than in an oxyester.

The case for easier enolization of thioesters relative to carboxylate groups and oxyesters follows similar lines. The rate of base-catalyzed enolization is fastest with the species that has the most electrophilic carbonyl and produces the most stable enolate ion: the thioester (structure **3-4**).

$$\text{3-4} \quad \underset{|}{\overset{}{-}}\text{C}\overset{\text{O}^-}{\underset{}{=}}\text{C}-\text{O}^- \quad \ll \quad \underset{|}{\overset{}{-}}\text{C}\overset{\text{O}^-}{\underset{}{=}}\text{C}-\text{O}-\text{R} \quad < \quad \underset{|}{\overset{}{-}}\text{C}\overset{\text{O}^-}{\underset{}{=}}\text{C}-\text{S}-\text{R}$$

Increasing stability →

For these reasons, most biologically reactive carboxylic acid derivatives are CoA-esters. There are significant exceptions to this rule, such as the aminoacyl-tRNAs and certain metabolites, such as 2-phosphoglycerate, which contain carboxylate groups and undergo enzymatic enolization under the influence of coordinated divalent metal ions (see chap. 9).

The Physical Role

There appears to be a dichotomy in the physical role of phosphopantetheine in enzymatic reactions. In two well-studied cases of acyl CoA reactions the phosphopantetheinyl nucleotide functions as a binding anchor that induces important conformational effects on the cognate enzymes. Acetoacetyl CoA:succinate CoA transferase is discussed in chapter 1, and in that reaction, a detailed kinetic study shows that the phosphopantetheinyl nucleotide engages in important binding interactions with the enzyme that dramatically decrease the activation energy for CoA transfer. Methylmalonyl CoA mutase is discussed in chapter 7, and extensive structural evidence shows that the phosphopantetheinyl nucleotide induces an important conformational change that sequesters the acyl moiety from contact with the solution and allows it to undergo a radical isomerization in a protected environment. In these two cases, multiple noncovalent enzyme CoA binding interactions are very important in facilitating catalysis.

The role of phosphopantetheine in multienzyme complexes such as fatty acid synthases, polyketide synthases, and nonribosomal polypeptide synthetases is more complex. In these complexes, phosphopantetheine is covalently bonded to serine residues of proteins or folding domains that are components of the complexes. The thiol group of phosphopantetheine becomes acylated in these complexes and provides the chemical activation for the various processes that are catalyzed by these enzymes. The structural mobility of the phosphopantetheinyl moiety, because of rotation about the bonds highlighted in fig. 3-32, allows it to deliver the acyl groups to the various active sites of the other enzymatic subunits or domains within the complexes, where individual steps in the assembly of fatty acids, polyketides, and polypeptides take place (see chap. 18).

Folate Compounds

Folic acid is a vitamin and the precursor of folate compounds that are essential intermediates in the one-carbon metabolism required for the biosynthesis of nucleotides and nucleic acids, including DNA (Benkovic, 1980). Folate exists in three common oxidation states, the structures of which are shown in fig. 3-33. Folates consist of a pterin linked to *p*-aminobenzoate (PABA) through its amino group, with the carboxylate group glutamylated, generally with a polyglutamate tail in the most biologically active forms (Huang et al., 1998). The biologically important oxidation states are tetrahydrofolate (H$_4$folate) and dihydrofolate (H$_2$folate).

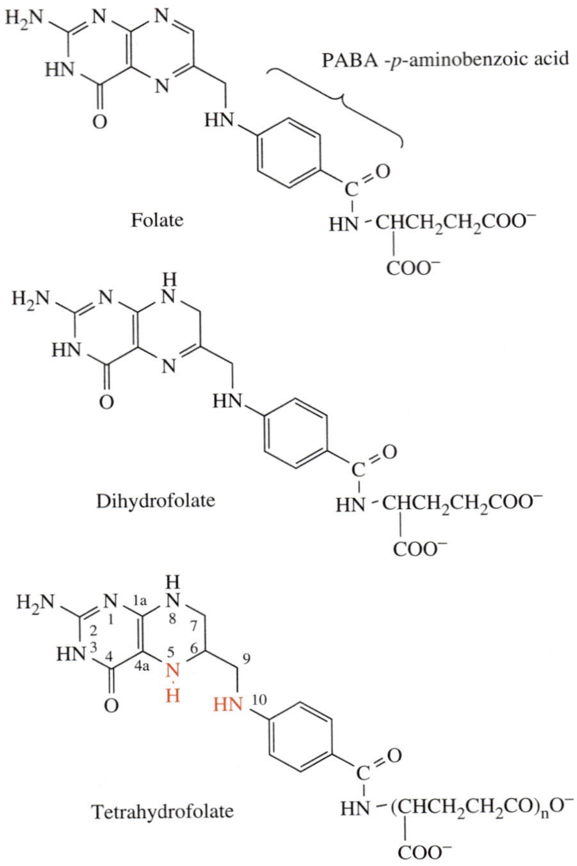

168   Enzymatic Reaction Mechanisms

Fig. 3-33. Structure and oxidation states of folic acid. The biologically important forms of the vitamin folic acid are dihydrofolate and tetrahydrofolate, shown at the bottom of the figure. p-Aminobenzoic acid (PABA), a nutritional factor, is incorporated in the biosynthesis of folate. The highlighted N^5 and N^{10} in tetrahydrofolate participate directly in biological reactions of folic acid. A polyglutamate is shown in the structure of tetrahydrofolate because the polyglutamate species with five of six glutamates are the most biologically active.

Folate Compounds of One-Carbon Metabolism

Tetrahydrofolate (H_4folate) compounds are intermediates in one-carbon metabolism at the three oxidation states of carbon corresponding to methanol, formaldehyde, and formate. Their structures and enzymatic interconversions are outlined in fig. 3-34, where 5-methyl-H_4folate is shown at the methanol level, 5,10-methylenetetrahydrofolate (methylene-H_4folate) is at the formaldehyde level, and 10-formyltetrahydrofolate (10-formyl-H_4folate) is at the formate level of oxidation. 5,10-Methenyltetrahydrofolate (methenyl-H_4folate) is at the formate oxidation level but is not known to be a primary formyl group donor in metabolism, although it indirectly donates formyl groups after being converted into 10-formyl-H_4folate. These compounds have come to be known as coenzymes because they were originally discovered through their apparent activation of enzymatic reactions. However, they do not function as co-catalysts but rather are methyl, methylene, or formyl donor substrates in biosynthetic reactions.

10-Formyl-H_4folate is produced by the action of formyltetrahydrofolate synthetase according to eq. 3-30. The kinetic mechanism has been reported to be random ter ter, with

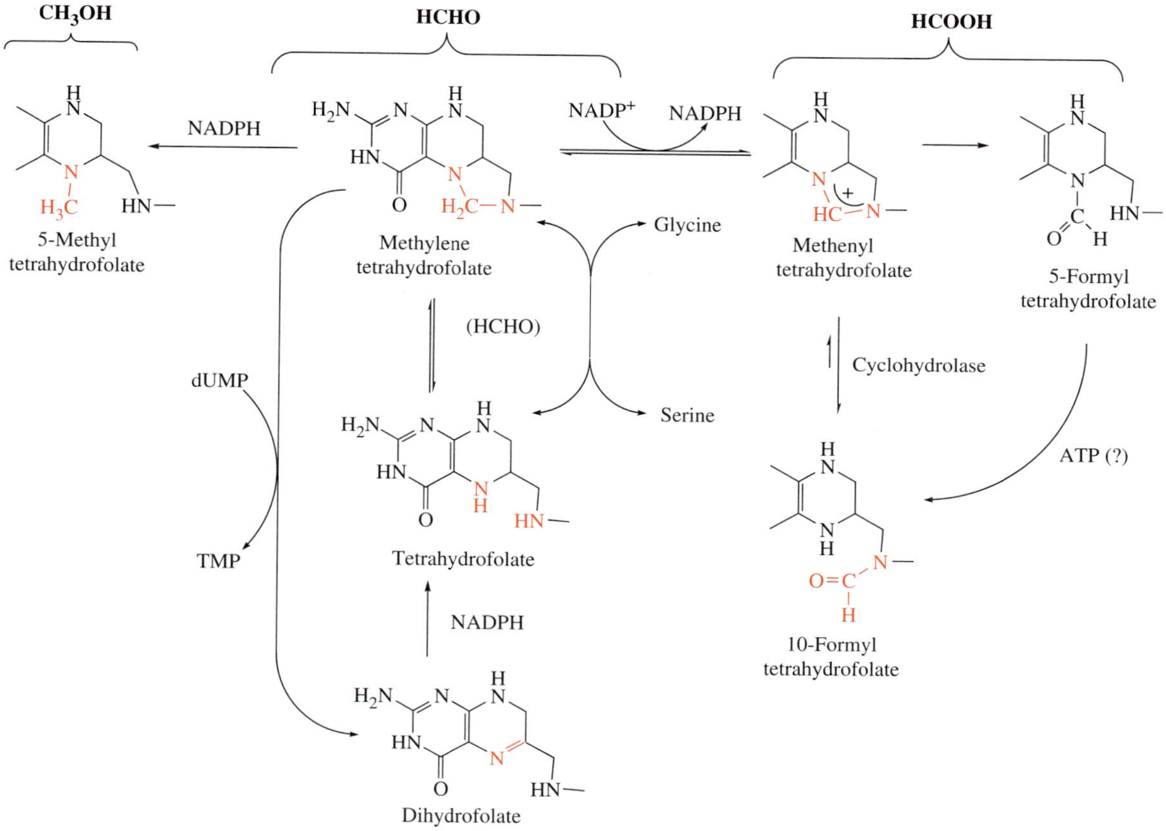

Fig. 3-34. Structures and interconversions of folate compounds in one-carbon metabolism.

a quaternary complex of enzyme, H$_4$folate, MgATP, and formate as a compulsory intermediate (Joyce and Himes, 1966).

$$H_4\text{folate} + \text{MgATP} + \text{HCOO}^- \rightarrow 10\text{-Formyl-H}_4\text{folate} + \text{MgADP} + P_i \quad (3\text{-}30)$$

The mechanism of formate activation and capture in the active site of the quaternary complex proceeds in a two-step mechanism shown in fig. 3-35. Formate is first phosphorylated to formyl phosphate by MgATP. Formyl phosphate is a highly reactive electrophile, so that the formyl group is immediately captured by the nearby nucleophilic N^{10} of H$_4$folate to release P$_i$ and produce 10-formyl-H$_4$folate (Buttlaire et al., 1976; Mejillano et al., 1989; Smithers et al., 1987).

Methylene-H$_4$folate is produced mainly by the actions of serine hydroxymethyltransferase and methenyl-H$_4$folate reductase. The action of cyclohydrolase in the dehydration of 10-formyl-H$_4$folate leads to methenyl-H$_4$folate. A reasonable chemical mechanism is sketched in scheme 3-4.

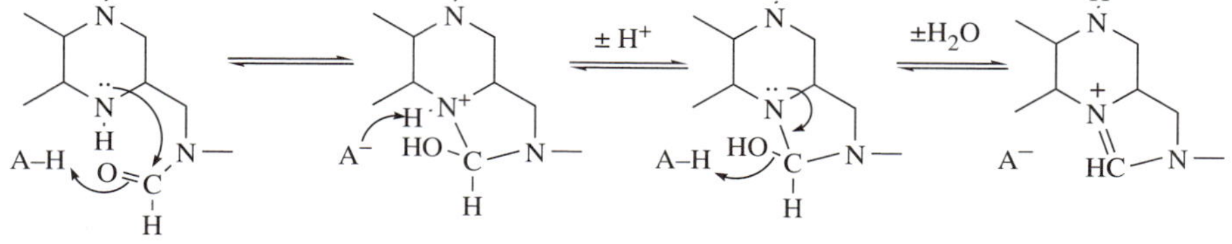

Scheme 3-4

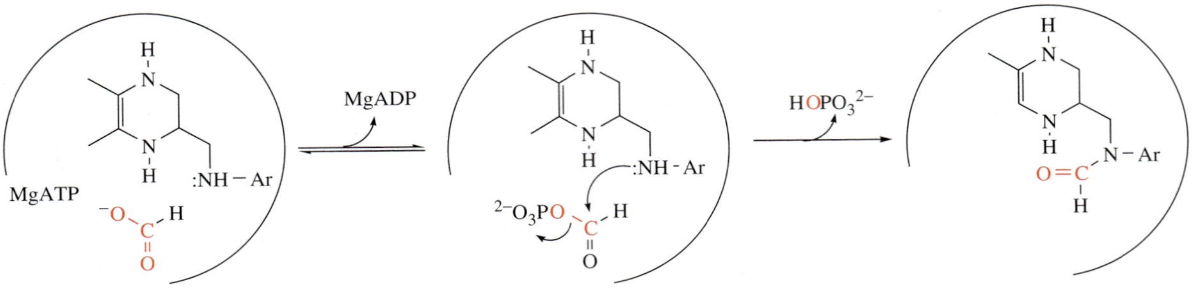

Fig. 3-35. Synthesis of 10-formyltetrahydrofolate by formyltetrahydrofolate synthetase.

Reduction of the iminium group by hydride transfer from NADPH is catalyzed by methenyl-H$_4$folate reductase to produce methylenetetrahydrofolate.

Enzymes in Tetrahydrofolate Metabolism

Serine Hydroxymethyltransferase

The other important source of methylene-H$_4$folate is the reaction of serine with H$_4$folate catalyzed by serine hydroxymethyltransferase (EC 2.1.2.1) according to eq. 3-31.

$$\text{Serine} + \text{H}_4\text{folate} \rightarrow \text{methylene-H}_4\text{folate} + \text{Glycine} \tag{3-31}$$

Serine hydroxymethyltransferase is a PLP-enzyme, the mechanism of which has attracted considerable interest (Schirch, 1980). This is a retroaldol-type reaction that is made possible by the stability of the observable and kinetically competent quinonoid α-carbanionic intermediate. We discuss this mechanism in further detail with other carbon-carbon bond forming and cleaving enzymes in chapter 14.

Thymidylate Synthase

Thymidylate (TMP) for DNA biosynthesis arises from the methylation of dUMP by methylene-H$_4$folate, which is catalyzed by thymidylate synthase according to the equation at the top of fig. 3-36. The mechanism of this reaction is of considerable interest for its chemical novelty, which serves as the basis for suicide inactivation of thymidylate synthase (EC 2.1.1.45) by 5-fluorodeoxyuridylate (FdUMP). In chapter 5, we discuss this mechanism and how it leads to inactivation by FdUMP. As shown in fig. 3-36, the formation of TMP leads to the conversion of methylene-H$_4$folate to H$_2$folate and not to H$_4$folate. The regeneration of H$_4$folate requires the action of dihydrofolate reductase, and serine hydroxymethyltransferase catalyzes the restoration of methylene-H$_4$folate. In chapter 5, we discuss the mechanism of action of dihydrofolate reductase and its inhibition by methotrexate, as well as the molecular basis for the actions of 5-fluorouracil and methotrexate as folate antagonists in cancer chemotherapy.

Formyl-Tetrahydrofolate–Dependent Formyl Transferases

Formyl transferases catalyze two steps in purine biosynthesis (Benkovic, 1980). Both enzymes require 10-formyl-H$_4$folate as the formyl donor. The first transfers the formyl

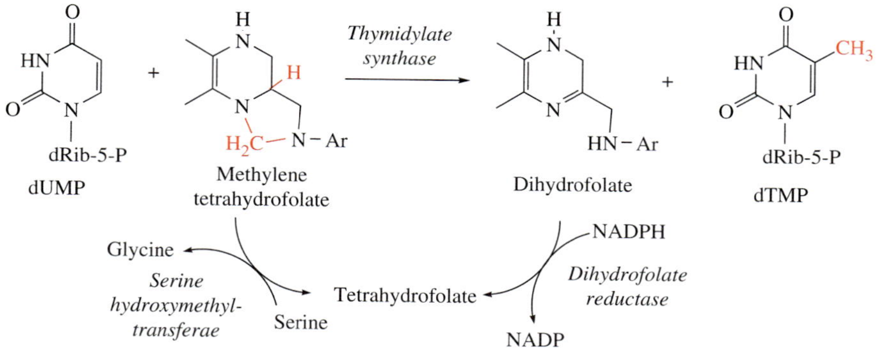

Fig. 3-36. The enzymatic conversion of dUMP into dTMP. Thymidylate synthase catalyzes the conversion of dUMP and methylenetetrahydrofolate into dTMP and dihydrofolate. The process would deplete the tetrahydrofolate pool were it not for dihydrofolate reductase, which catalyzes the reduction of dihydrofolate by NADPH. Serine hydroxymethyltransferase regenerates methylenetetrahydrofolate.

group to N^1-(5-phosphoribosyl)-glycinamide, an early intermediate to produce N^3-formyl-N^1-(5-phosphoribosyl)-glycinamide (eq. 3-32). The formyl group eventually becomes C8 in the purine ring. The other formyl transfer in purine biosynthesis takes place further along in the pathway, in which 5'-phosphoribosyl-4-carboxamide-5-aminoimidazole is transformed into 5'-phosphoribosyl-4-carboxamide-5-formamidoimidazole (eq. 3-33). The formyl group in this step eventually becomes C2 of the purine ring. In both of these reactions the formyl group is transferred from the N10-amino group of tetrahydrofolate to an amino group of an intermediate of purine biosynthesis.

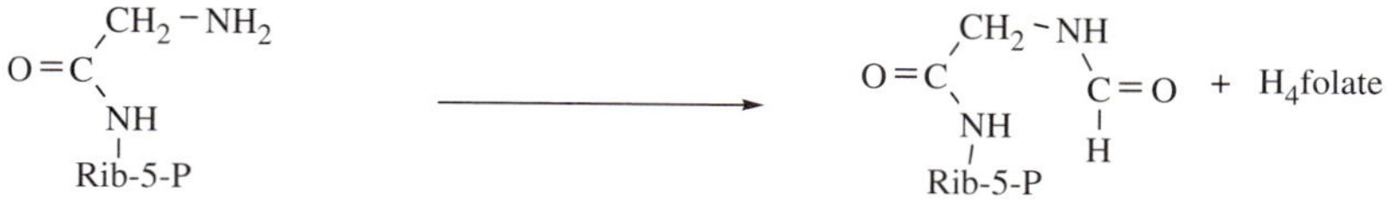

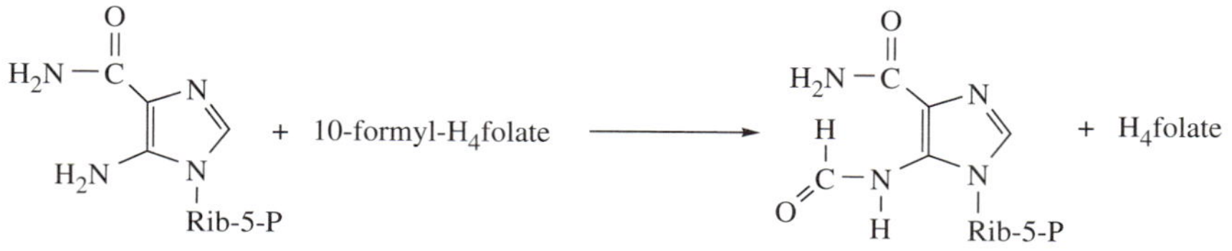

Biological Importance of Folate

Why should tetrahydrofolate derivatives of methanol, formaldehyde, and formate be necessary in biology? One-carbon units are useful in biosynthesis, and one can imagine other derivatives of methanol and formate that might fulfill this need. Formaldehyde presents significant problems to cells because it cross-links amino groups in proteins and nucleic acids and is toxic. Its very propensity to form cross-links between amino groups is put to use in its spontaneous reaction with H$_4$folate to form methylene-H$_4$folate, as shown in scheme 3-5.

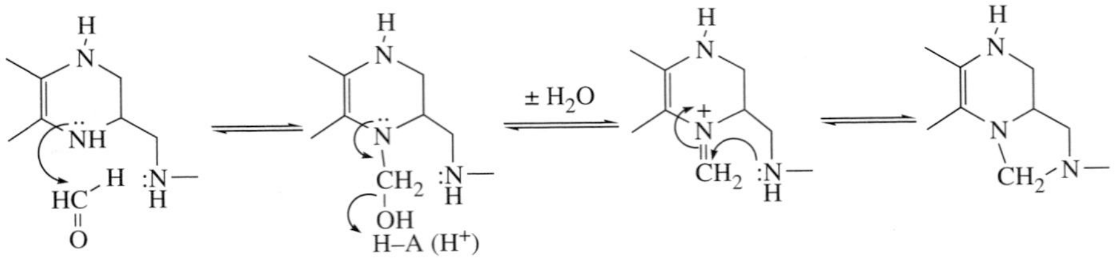

172 Enzymatic Reaction Mechanisms

Scheme 3-5

Methylene-H$_4$folate is stable in cells but can become reactive in the active sites of enzymes that use it as a source of formaldehyde. Moreover, it can be reduced to 5-methyl-H$_4$folate, which serves as a source of methyl groups for the production of methionine by methionine synthase (eq. 3-34).

$$\text{Homocysteine} + \text{5-methyl-H}_4\text{folate} \rightarrow \text{Methionine} + \text{H}_4\text{folate} \qquad (3\text{-}34)$$

Methionine synthase is a vitamin B$_{12}$–dependent enzyme that functions by a complex mechanism, which we discuss in chapter 15.

Formate is chemically unreactive and must be activated in some way to serve as a source of formyl groups in biosynthesis. Activation to formyl phosphate by reaction with ATP makes it excessively reactive and potentially toxic. Formyl-H$_4$folate synthetase produces formyl phosphate in its active site and then allows H$_4$folate to capture it. In the form of 10-formyl-H$_4$folate, the formyl group is not reactive enough to be toxic, but serves as a source of formyl groups in the active sites of enzymes. H$_4$folate balances maintains formaldehyde and formate in chemically poised states that are not toxic to cells but can become reactive in active sites of specific enzymes.

Amino Acid–Based Coenzymes

Pyruvoyl Decarboxylases

PLP-independent amino acid decarboxylases catalyze α-decarboxylations of histidine, S-adenosylmethionine, and phosphatidylserine to the corresponding amines (Li and Dowhan, 1988; Markham et al., 1982; Recsei and Snell, 1984; Satre and Kennedy, 1978; van Poelje and Snell, 1990). These enzymes incorporate a pyruvoyl residue, and the carbonyl group of the pyruvoyl moiety serves in the capacity of the aldehydic group of PLP in facilitating decarboxylation of an imine intermediate, as illustrated in scheme 3-6. The most thoroughly studied is the PLP-independent histidine decarboxylase (EC 4.1.1.22), which is discussed in more detail and compared with the PLP-dependent enzyme in chapter 8.

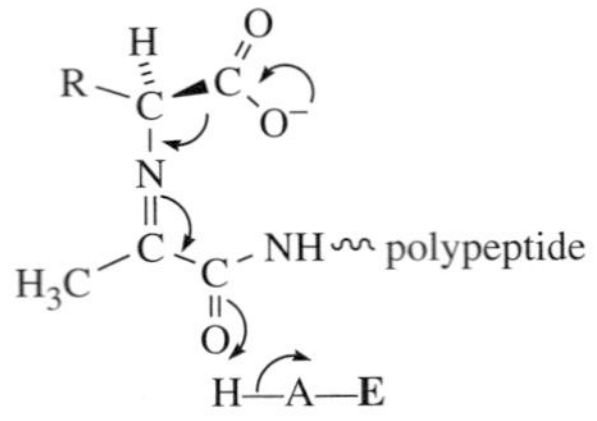

Scheme 3-6

The pyruvoyl moieties of PLP-independent α-amino acid decarboxylases are covalently linked to one of the polypeptide chains of the enzymes in place of one of the N-terminal amino acids. They arise through hydrolytic cleavage between adjacent serine residues of the primary biosynthetic translation products according to eq. 3-35.

$$\text{(3-35)}$$

The cleavage mechanism most likely entails nucleophilic participation by the neighboring Ser-β-OH in cleaving the peptide bond to form an intermediate ester, which undergoes β-elimination to produce an N-terminal dehydroalanyl peptide. Spontaneous hydrolysis of the N-terminal dehydroalanyl group would generate the pyruvoyl moiety. The cleaved polypeptides remain associated in the tertiary structure of the enzyme.

Pyruvoyl-enzymes are inactivated by carbonyl reagents such as sodium borohydride and phenylhydrazine. Reaction with phenylhydrazine gives protein chromophores similar to those of phenylhydrazones. Hydrolysis of a pyruvoyl-enzyme that has been reductively inactivated with borohydride yields lactate from the reduction of the pyruvoyl group. Reductive inactivation of a pyruvoyl-enzyme by sodium borohydride in the presence of the cognate [^{14}C]substrate results in the covalent incorporation of radioactivity from the ^{14}C-labeled substrate into the protein. These properties indicate that the pyruvoyl carbonyl group reacts with the α-amino group of the substrate to form a ketimine that is analogous to the external aldimine of PLP-dependent enzymes.

D-Proline reductase is a pyruvoyl-enzyme that catalyzes the reductive ring opening of proline to δ-aminovalerate (Hodgins and Abeles, 1969). The mechanism of this interesting reaction has not been investigated.

Methylidene Imidazolinone–Dependent Deaminases

Histidine ammonia-lyase (HAL; EC 4.3.1.3), also known as histidine deaminase, catalyzes the elimination of ammonia from histidine to form urocanate, and phenylalanine ammonia lyase (PAL) catalyzes the similar process on phenylalanine to form cinnamate (eq. 3-36).

$$\text{(3-36)}$$

These reactions are atypical in that they are β,α-eliminations in which a hydrogen in the β-position is lost rather than the more commonly eliminated α-hydrogen, as in α,β-eliminations. Both enzymes contain an amino acid-derived cofactor within their amino acid sequences. This cofactor is 3,5-dihydro-5-methylidene-4H-imidazol-4-one, or simply methylidene imidazolinone (MIO), and it is derived from a serine residue by internally facilitated cyclization with elimination of two molecules of water, as illustrated in scheme 3-7 (Langer et al., 2001; Rétey, 2003).

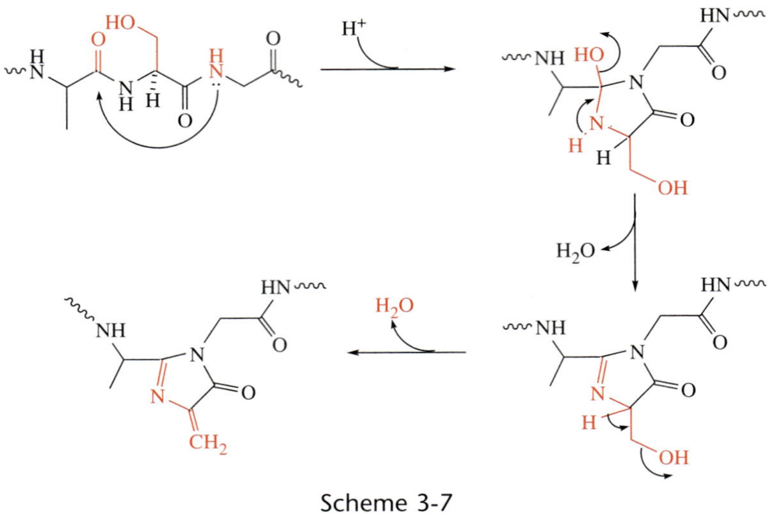

174 Enzymatic Reaction Mechanisms

Scheme 3-7

The cofactor displays chemical properties expected of a dehydroalanine, and for this reason it was long thought to be a dehydroalanyl residue. HAL and PAL are inactivated by sodium borohydride in the absence of substrates because of reduction of the methylidene group. Acid hydrolysis after reduction with NaB^3H_4 leads to tritiated alanine. HAL reacts with the nitromethane anion, which undergoes irreversible addition to the methylidene group (Givot et al., 1969). Cyanide undergoes nucleophilic addition to the methylidene group of PAL, and acid hydrolysis of PAL after inactivation with [^{14}C]cyanide produces [^{14}C]aspartic acid (Hodgins, 1971).

The posttranslational processing of a serine residue appears to be spontaneous; the dehydrations appear to be facilitated by the structures of the primary translation products and do not seem to require additional biological factors. Site-directed mutagenesis to replace serine with cysteine at the processing site leads to active enzyme, so that the elimination of hydrogen sulfide in place of water is facilitated by the enzyme (Langer et al., 1994). The mechanisms of the reactions of HAL and PAL are controversial and will be discussed in chapter 9.

Quinoproteins

Five quinone coenzymes, all derived from amino acids, have been characterized: pyrroloquinoline quinone (PQQ), tryptophan tryptophyl quinone (TTQ), cysteine tryptophyl quinone (CTQ), topaquinone (TPQ), and lysyltopaquinione (LTQ). PQQ quinone is found in bacterial alcohol dehydrogenases, and TTQ and CTQ are found in bacterial methylamine dehydrogenases. The quinoprotein dehydrogenases channel the electrons derived from the dehydrogenation of substrates into bacterial electron transport systems. TPQ is found in copper amine oxidases of eukaryotic and bacterial origin, and it transfers electrons to dioxygen in copper-dependent processes. LTQ is found in lysyl oxidase, which catalyzes the hydroxylation of lysyl residues in collagen.

Pyrroloquinoline Quinone

Methanol dehydrogenase (EC 1.1.99.8) catalyzes the reaction of methanol with one-electron acceptors such as phenazine methosulfate (PMS) to produce formaldehyde and the reduced acceptor. The enzyme is important in methanotrophic bacteria for the use of methane as the sole source of carbon. Methanol is produced from methane by the action

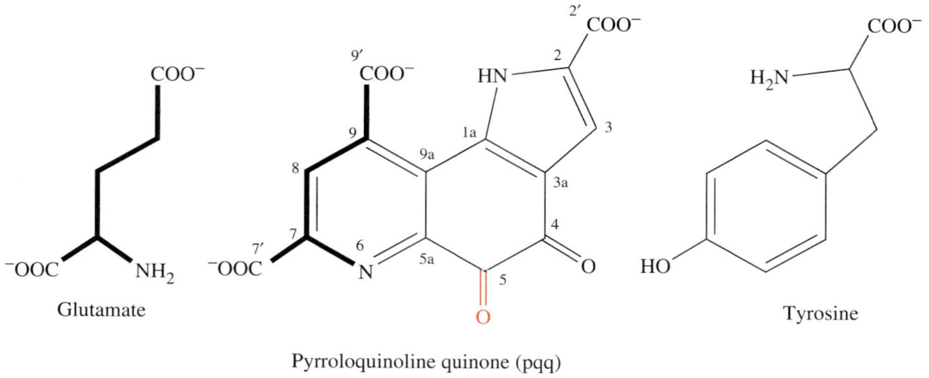

Fig. 3-37. The structure of pyrroloquinoline quinone (PQQ) is shown in the center. The molecule is derived from the carbon and nitrogen skeletons of glutamate and tyrosine in PQQ-producing strains of methylbacteria (Houck et al., 1988), and its biosynthesis requires the expression of at least seven genes in bacteria (Lidstrom, 1991).

of methane monooxygenase (see chap. 17) and further oxidized to formaldehyde by methanol dehydrogenase for use as a building block and fuel. The biological reaction is eq. 3-37, in which the electron acceptor is a cytochrome.

$$CH_3OH + 2\ cyt\ c_L(Fe^{3+}) \longrightarrow H_2C=O + 2\ cyt\ c_L(Fe^{2+}) + 2\ H^+ \quad (3\text{-}37)$$

The electrons enter the terminal electron transport pathway and are used to energize cellular processes, such as in ATP production, and formaldehyde enters metabolism as a one-carbon unit. Glucose dehydrogenase catalyzes a similar reaction.

The coenzyme of methanol and glucose dehydrogenases is PQQ, the structure of which is shown in fig. 3-37 (Duine and Frank, 1980; Duine et al., 1980, 1987). The molecule is derived biogenetically from glutamate and tyrosine by an oxidative process requiring the products of at least six genes (Houck et al., 1991; Toyama et al., 1997). The structure of purified methanol dehydrogenase shows PQQ bound as a calcium complex, and calcium is required for activity. This is a novel role for calcium, which normally participates in metabolic regulation rather than enzymatic catalysis.

Methylophilus methylotrophus methanol dehydrogenase has a molecular mass of 140 kDa and is a tetrameric protein with two types of subunits. It is formulated as $\alpha_2\beta_2$, and the molecular masses of the chains are 62 kDa and 8 kDa, respectively. The tetrasubunit protein contains two molecules of PQQ, as well as tightly bound calcium ions. The structure of methanol dehydrogenase is depicted in fig. 3-38A. The large subunit is composed largely of β-sheets arranged in a pseudo-eightfold symmetric β-propeller. In fig. 3-38B the close contacts of the calcium complex of PQQ with amino acids at the active site are delineated. Many hydrogen bonds connect the functional groups of PQQ to the protein, and Ca^{2+} serves as the nucleus for ligating the protein with the carboxylate-7′, N6, and the 5-carbonyl of PQQ. Asp 297 is near the 5-carbonyl, which is thought to be the functional group of PQQ that reacts chemically with methanol.

The hypothetical mechanism suggested here is reasonable and based on available evidence. The reaction with PMS as the electron acceptor follows ping pong kinetics according to eqs. 3-38a and 3-38b, suggesting that the oxidation of methanol takes place independently of the reduction of cytochrome (cyt) c_L (Duine and Frank, 1980).

176 Enzymatic Reaction Mechanisms

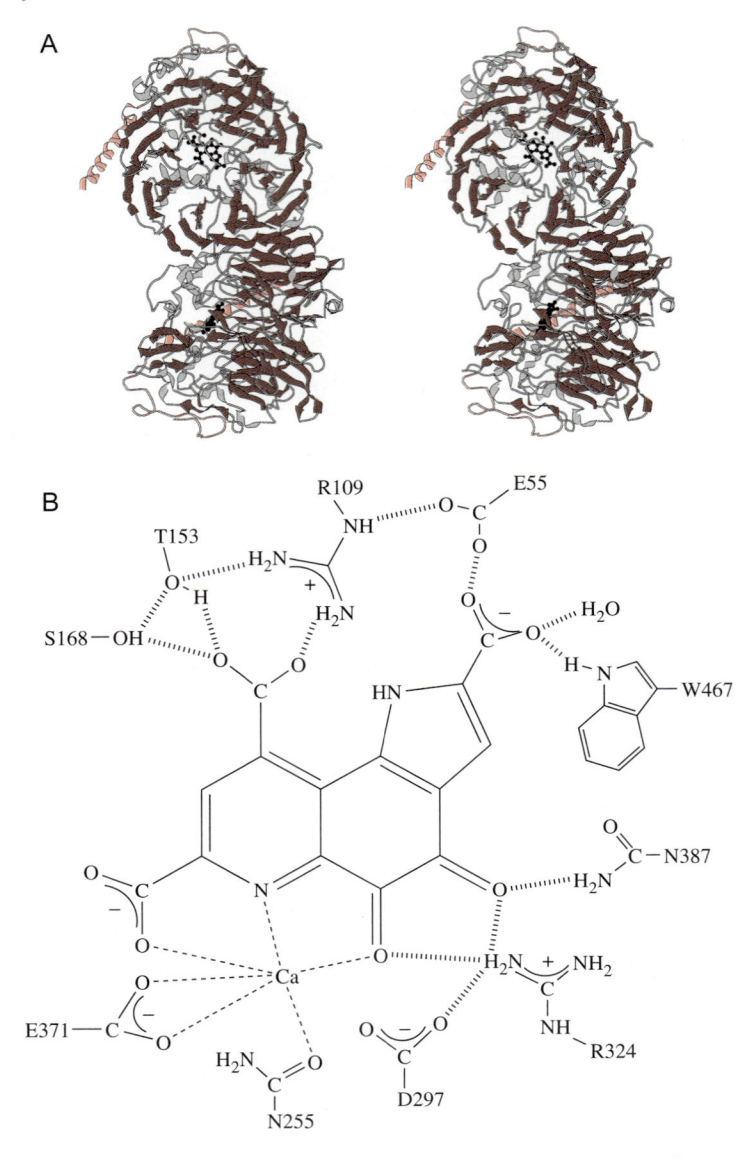

Fig. 3-38. Structure of methanol dehydrogenase. (A) Stereodiagram of the $\alpha_2\beta_2$ heterotetrameric methanol dehydrogenase (EC 1.1.99.8) from *Methylophilus methylotrophus* W3A1 with pyrroloquinoline quinone (PQQ) as a black ball-and-stick model. The small subunit (pink) is dwarfed by the large subunit, which consists of an eight-blade β-propeller fold surrounding the PQQ and calcium ion in the active site. The pseudo–eight-fold symmetry axis relating the eight β-leaflets of the large subunit is approximately perpendicular to the diagram. The figure was generated using the PDB coordinate file 1G72 for the crystallographic model at 1.9-Å resolution (Zheng et al., 2001). (B) Contacts between the enzyme and PQQ at the active site (Mathews, 1995).

$$CH_3OH + \mathbf{E} \cdot PQQ \rightleftharpoons H_2C=O + \mathbf{E} \cdot PQQH_2 \quad \text{Ping} \quad (3\text{-}38a)$$

$$\mathbf{E} \cdot PQQH_2 + 2\,PMS_{ox} \rightleftharpoons \mathbf{E} \cdot PQQ + 2\,PMS_{red} \quad \text{Pong} \quad (3\text{-}38b)$$

The dehydrogenation of CD_3OH proceeds with a kinetic isotope effect (k_H/k_D) of 4.3 with PMS as the electron acceptor, suggesting that C—H bond cleavage is rate limiting (Parkes and Abeles, 1984). Cyclopropanol inactivates the enzyme by forming an adduct with PQQ, and UV-spectroscopy indicates a C5-adduct of PQQ. Two reasonable mechanisms for the oxidation of methanol and reduction of PQQ are presented in fig. 3-39. In mechanism A, methanol undergoes nucleophilic addition to PQQ at C5 to form an adduct, which then loses a proton from the methyl group to form the hydroquinol of PQQ and formaldehyde. In mechanism B, hydride transfer from the methyl group of methanol

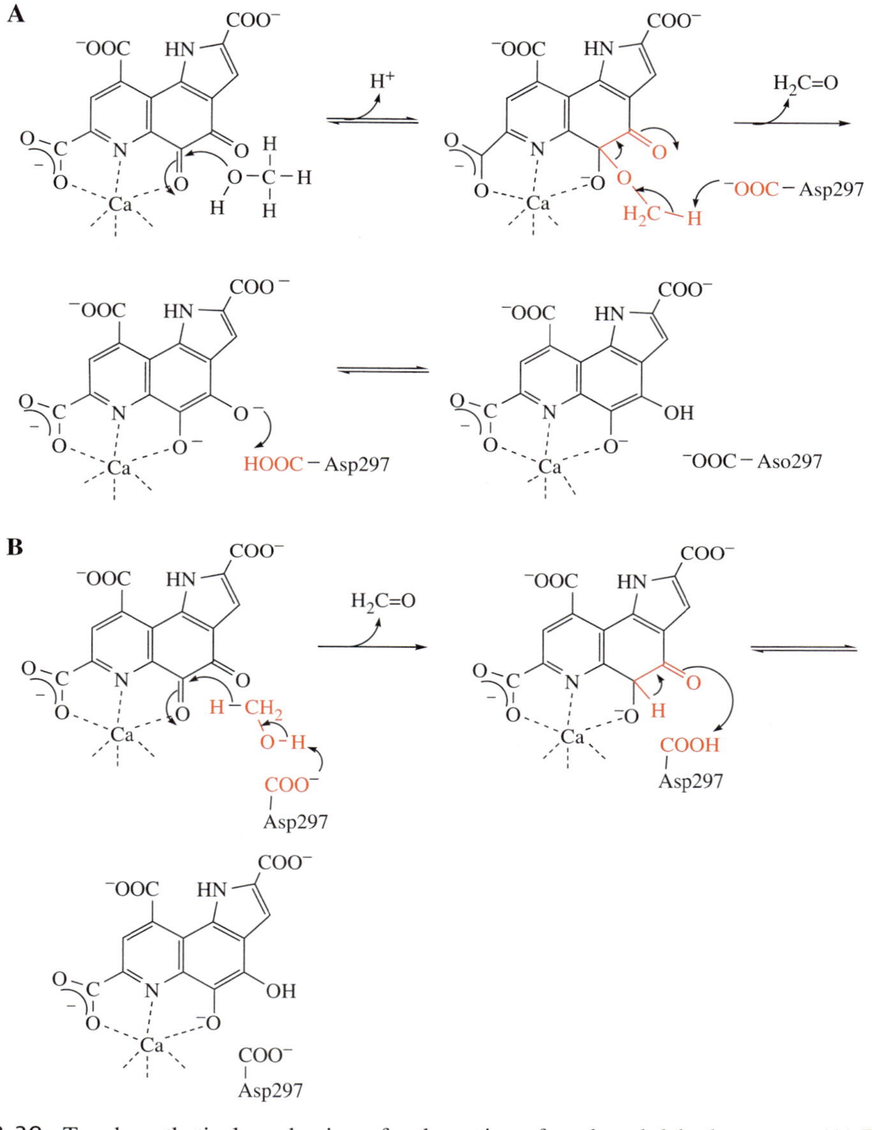

Fig. 3-39. Two hypothetical mechanisms for the action of methanol dehydrogenase. (A) The mechanism begins with the addition of the hydroxyl group of methanol to C5 of pyrroloquinoline quinone (PQQ), followed by the abstraction of a proton from the methyl group by Asp297, to generate formaldehyde and the hydroquinol of PQQ as its dianion. Protonation of the dianion by Asp297 completes the catalytic cycle. (B) In the hydride transfer mechanism, Asp297 abstracts the proton from the hydroxyl group of methanol in concert with hydride transfer from the methyl group to C5 of PQQ. This step generates formaldehyde directly and leads to the hydroxyketone tautomer of reduced PQQ. Proton transfer to the C4 carbonyl from Asp297 catalyzes tautomerization to the hydroquinol of PQQ. Coordination of the C5 carbonyl group in PQQ with Ca^{2+} polarizes the group and promotes nucleophilic addition in mechanism A or hydride transfer in mechanism B. Structural and molecular dynamics calculations favor hydride transfer and mechanism B (Xia et al., 1999; Zheng et al., 2001).

178   Enzymatic Reaction Mechanisms

to C5 of PQQ leads directly to formaldehyde and the ketol tautomer of reduced PQQ. Tautomerization to the hydroquinol of PQQ completes the catalytic cycle. In both mechanisms, Asp297 catalyzes the proton transfers required to drive the initial addition reactions and the formation of the hydroquinol form of PQQ. High-resolution structural analysis and molecular dynamics simulations support the hydride transfer mechanism B (Xia et al., 1999; Zheng et al., 2001).

The reoxidation of PQQH$_2$ in the cell is by cyt c_L, a one-electron acceptor. As a quinone, the reduced form of PQQ produced in the oxidation of methanol is well suited to donating its electrons in two one-electron steps by way of a semiquinone radical intermediate. Such a radical has been described briefly (Westerling et al., 1979).

Tryptophan Tryptophyl Quinone

Methylamine dehydrogenases (EC 1.4.99.3), which are periplasmic enzymes of methylobacteria that catalyze the oxidation of methylamine to formaldehyde by cyt c according to eq. 3-39, contain TTQ as part of their covalent structures.

$$CH_3NH_2 + 2\,cyt\,c\,(Fe^{3+}) + H_2O \longrightarrow H_2C=O + 2\,cyt\,c\,(Fe^{2+}) + NH_4^+ + H^+ \quad (3\text{-}39)$$

The enzymes are composed of two types of subunits, a large one designated α (40 to 48 kDa), and a small one designated β (8 to 16 kDa). There may be one of each (αβ) or two of each (α$_2$β$_2$) per molecule, depending on the species. Methylamine dehydrogenases differ from amine oxidases, which use dioxygen as the electron acceptor and are flavoproteins or topaquinone enzymes.

The structure of TTQ was determined by the application of every technique available to modern protein research (McIntire et al., 1991). Chemical derivatization by carbonyl reagents, including phenylhydrazines, semicarbazide, and hydroxylamine, coupled with UV-Vis spectrophotometric and resonance Raman spectroscopy indicated a quinone-like structure. Systematic degradation of the semicarbazide-derivatized protein, followed by amino acid sequencing and NMR and mass spectrometric analysis, showed that TTQ formed a cross-link of tryptophan residues within a peptide, with Trp 55 and Trp 57 in the cross-link. Simultaneously, x-ray crystallographic analysis implicated tryptophan residues in a novel arrangement and ultimately to formulation of the structure in fig. 3-40. A ribbon

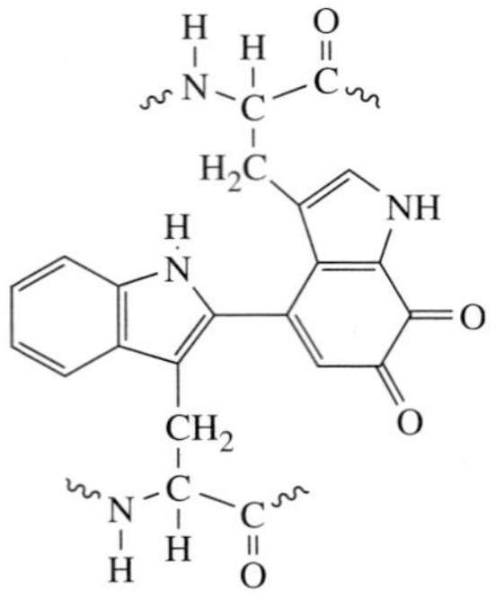

Tryptophan tryptophylquinone
(TTQ)

Fig. 3-40. The structure of tryptophan tryptophyl quinone (TTQ). Cross-linking and oxidation of two tryptophan residues to form TTQ was first discovered in methylamine dehydrogenase in a correlation of protein chemistry with x-ray crystallographic results (McIntire et al., 1991).

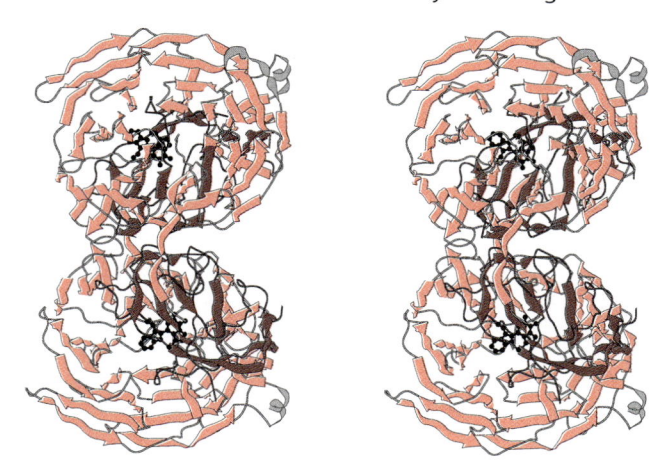

Fig. 3-41. The structure of the $\alpha_2\beta_2$ heterotetrameric *Paracocus denitrificans* methylamine dehydrogenase (EC 1.4.99.3) is shown in this stereodrawing. Both subunits largely consist of β-sheets. The larger subunit forms a seven-blade β-propeller fold (pink sheets with light gray loops or helices), reminiscent of the unrelated eight-bladed propeller of methanol dehydrogenase. The smaller subunit (brown sheets with dark gray loops or helices) provides the tryptophan residues that make up the tryptophan tryptophyl quinone (TTQ) moiety, which is situated at the center of one face of the β propeller of the other subunit. The image was generated from the 1.75-Å resolution model detailed in PDB 2BBK (Chen et al., 1998).

diagram of the structure of methylamine dehydrogenase showing the TTQ cross-link in the small subunit is shown in fig. 3-41.

Essential features of the mechanism have been delineated by kinetic and spectroscopic methods. Steady-state kinetic experiments indicated a ping pong mechanism that could be formulated as eqs. 3-40a and 3-40b, where the reduced form of the enzyme contains an amino group.

$$E_{ox} + CH_3NH_2 \rightleftharpoons E_{red}-NH_2 + H_2C=O \quad \text{Ping} \quad (3\text{-}40a)$$

$$E_{red}-NH_2 + 2\ \text{cyt}\ c\ (Fe^{3+}) + H_2O \rightleftharpoons E_{ox} + 2\ \text{cyt}\ c\ (Fe^{2+}) + NH_4^+ + H^+ \quad \text{Pong} \quad (3\text{-}40b)$$

Intermediates have been observed spectrophotometrically in transient kinetic studies that implicated a reduced TTQ in the reduced form of the enzyme. Spectrophotometric information implied that the reduced TTQ was an aminoquinol, and such a species was observed by ^{15}N NMR analysis of the reduced enzyme generated in the reaction of $CH_3^{15}NH_2$ in eq. 3-40a (Bishop et al., 1996). An ammonia adduct generated with a K_d of about 20 mM by reversal of eq.3-40a was also characterized as an inhibited form. A reasonable mechanism based on these facts and the three-dimensional structure is outlined in fig. 3-42.

Cysteine tryptophyl quinone is found in an amine dehydrogenase. In CTQ the cross-linking tryptophan of TTQ is replaced with cysteine linked through sulfur to the tryptophyl quinone moiety (Datta et al., 2001).

Topaquinone

Copper amine oxidases (CAOs; EC 1.4.3.6) incorporating TPQ derived from tyrosine within the amino acid sequences constitute a family (Klinman, 2003; Klinman and Mu, 1994).

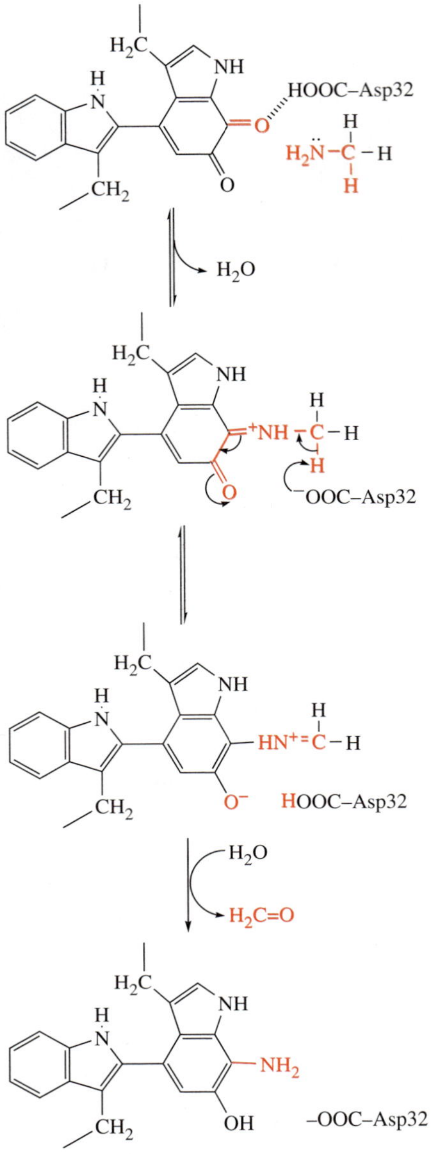

180 Enzymatic Reaction Mechanisms

Fig. 3-42. A mechanism for the dehydrogenation of methylamine by tryptophan tryptophyl quinone (TTQ) at the active site of methylamine dehydrogenase.

The coenzymatic group is the *para*-quinone derived from 2,4,5-trihydroxyphenylalanine, known as TPQ, shown in fig. 3-43. CAOs incorporating TPQ are found in serum, plants, and bacteria but not in archaea. A variant coenzymatic group is lysyltopaquinone (LTQ), which is found in lysyl oxidase (Tang and Klinman, 2001). The structures of TPQ and LTQ are shown in fig. 3-43.

Copper amine oxidases catalyze the oxidation of primary amines according to eq. 3-41.

$$RCH_2CH_2NH_2 + O_2 + H_2O \rightarrow RCH_2CHO + NH_3 + H_2O_2 \qquad (3\text{-}41)$$

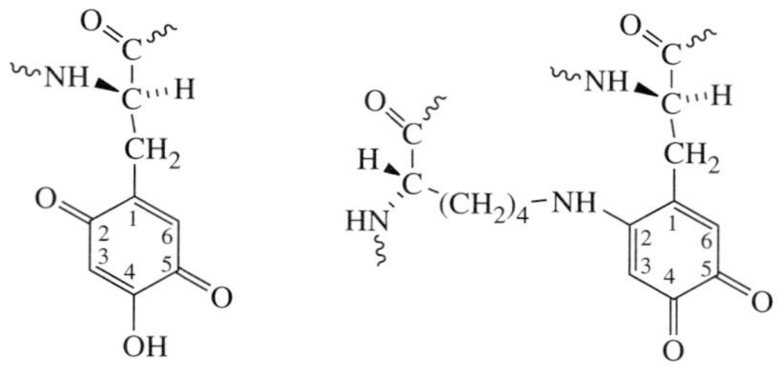

Fig. 3-43. The structures of topaquinone (TPQ) and lysyltopaquinone (LTQ).

The action of CAOs such as serum amine oxidase presented a large barrier to mechanistic analysis for many years. The main impediment was ignorance of the nature of the required coenzyme. The CAOs were inactivated by carbonyl reagents such as phenylhydrazine or *p*-nitrophenylhydrazine. They were subject to substrate-dependent inactivation by NaBH$_4$, which suggested that reaction of a substrate with the enzyme led to the formation of an iminium compound, analogous to the reactions of PLP. Reductive inactivation by NaBH$_4$ in the presence of [^{14}C]benzylamine led to the incorporation of radioactivity into the protein, further confirming iminium formation between benzylamine and the enzyme.

Purified CAOs were found to contain copper and a chromophoric group, which was not identified for many years. Identification of the chromophore as topaquinone spurred progress in the field (Janes et al., 1990). The chromophore was found to be a carbonyl species that reacted with phenylhydrazine to form a phenylhydrazone, and this was accompanied by inactivation of the enzyme. This chromophoric group could be labeled by reaction with [^{14}C]phenylhydrazine. Systematic proteolytic degradation of the ^{14}C-labeled protein led to a ^{14}C-peptide with spectral properties similar to those of the [^{14}C]phenylhydrazone form of the enzyme. Mass spectrometric and amino acid analysis of the peptide showed it to contain a quinone form of trihydroxyphenylalanine. Chemical synthesis of the possible isomers and correlation of their spectral properties with those of the ^{14}C-labeled peptide led to the identification of topaquinone. The x-ray crystallographic structure subsequently confirmed the structure of TPQ (Li et al., 1998).

The global and active site structures of the CAOs from *E. coli* and *Hansenula polymorpha* reveals mushroom-shaped dimeric molecules with identical active sites in each subunit, as shown in fig. 3-44. It proves to be significant that TPQ is not directly coordinated to Cu^{2+} in the structure of the mature enzyme. However, coordination is likely to be important in the biosynthesis of TPQ from the primary translation product, in which tyrosine occupies the position of TPQ in the mature enzyme.

The transformation of tyrosine in the primary translation product into TPQ in the mature enzyme depends on the presence of both Cu^{2+} and molecular oxygen (Klinman, 2003). This process takes place spontaneously and does not require the action of any other enzyme. The reaction proceeds with the transient and intermediate formation of a chromophore at 350 nm, which is thought to be a ligand to Cu^{2+} charge transfer complex. fig. 3-45 illustrates a mechanism for TPQ-formation in the active site that explains the roles of Cu^{2+} and oxygen. A minor resonance form of the charge transfer complex, with electron spin in the phenyl ring, may lower the barrier for the addition of molecular oxygen

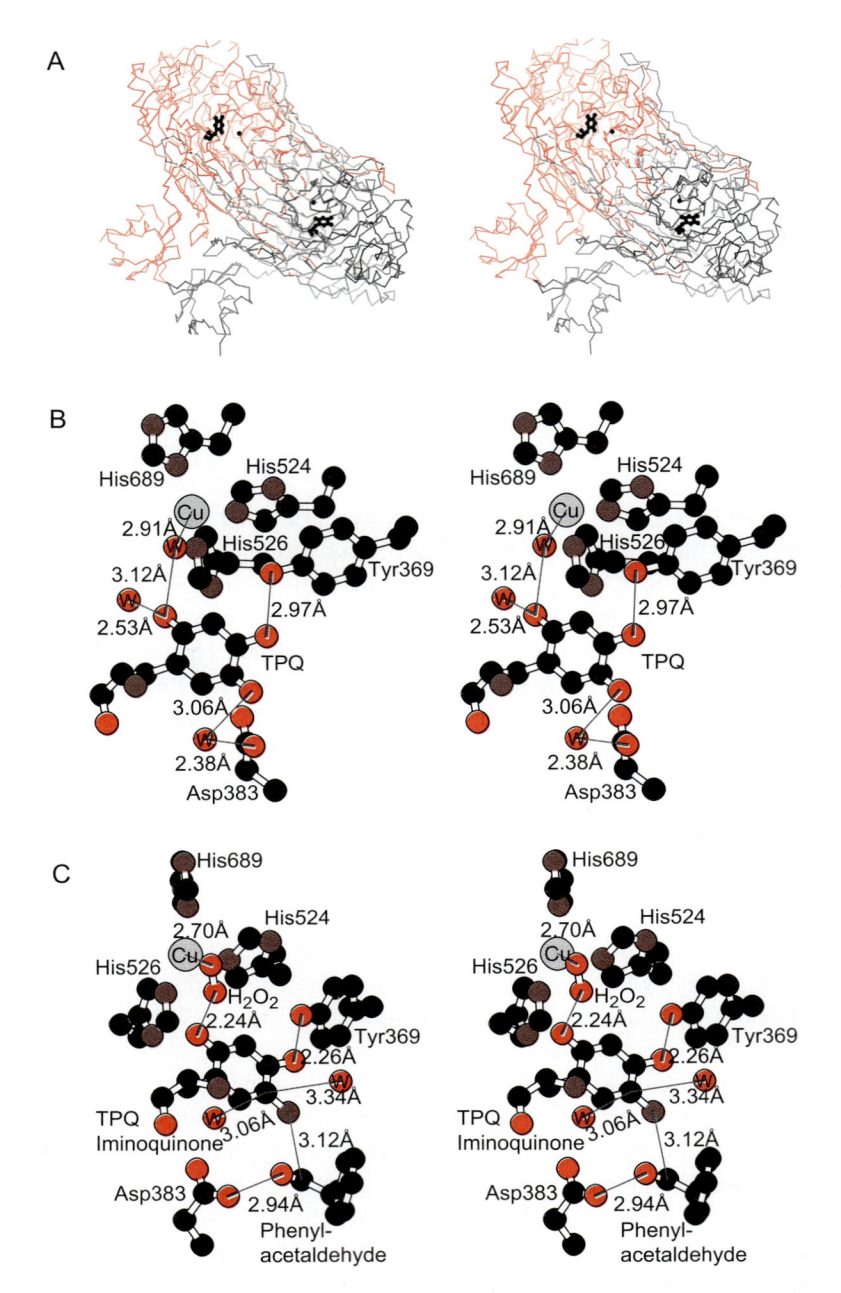

Fig. 3-44. The structure of copper amine oxidase from *E. coli*. (A) Cα trace stereodiagram of the mushroom-shaped *E. coli* homodimeric copper amine oxidase (EC 1.4.3.4) with the topaquinone (TPQ) moiety and the copper ion shown as black ball-and-stick models. (B) The active site with a water molecule situated between the TPQ and the copper atom. (C) The free-trapped active site product complex with hydrogen peroxide between the copper and the TPQ iminoquinone, which has accepted the nitrogen from phenylethylamine yielding phenyl-acetaldehyde.

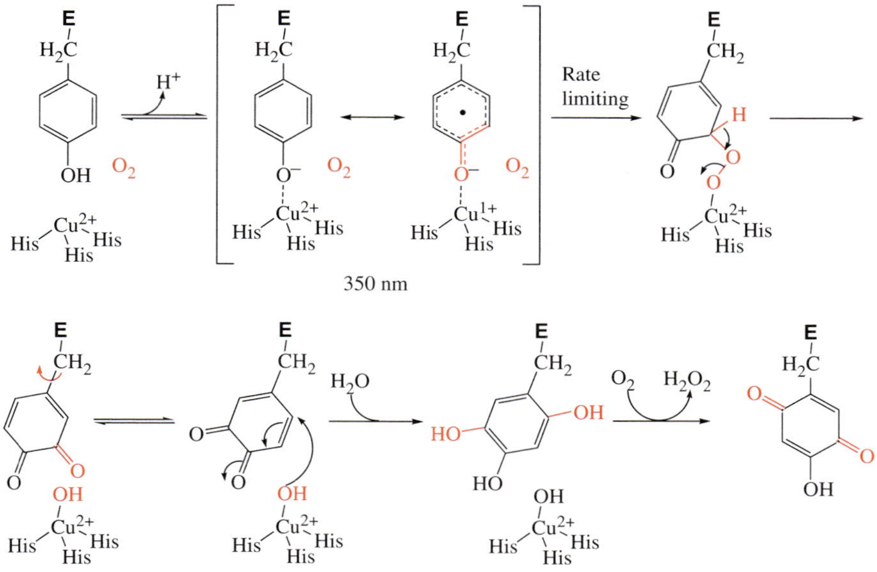

Fig. 3-45. A mechanism for topaquinone (TPQ) formation in a copper amine oxidase (CAO). Both Cu^{2+} and molecular oxygen are required to transform a tyrosine residue into TPQ at the active site of a CAO. This mechanism implicates copper as a means of lowering the barrier to the addition of the paramagnetic oxygen molecule to the phenyl ring of tyrosine. Formation of a ligand to copper charge transfer complex between tyrosine and Cu^{2+} increases the importance of a paramagnetic resonance form in the structure of the aromatic ring and lowers the barrier to formation of a bond to oxygen. Subsequent steps lead to oxygenation of the phenyl ring.

to the tyrosyl ring, and subsequent steps lead first to quinone formation and then to the addition of copper-bound hydroxide to generate 2.4.5-trihydroxyphenylalanine. Oxidation by molecular oxygen leads to TPQ.

Like the reactions of methanol dehydrogenase and methylamine dehydrogenase, the overall reaction of CAOs follow ping pong kinetics according to eqs. 3-42a and 3-42b.

$$\mathbf{E}_{ox} + PhCH_2NH_2 \rightleftharpoons \mathbf{E}_{red}\text{—}NH_2 + PhCHO \qquad \text{Ping} \qquad (3\text{-}42a)$$

$$\mathbf{E}_{red}\text{—}NH_2 + O_2 \rightleftharpoons \mathbf{E}_{ox} + NH_3 + H_2O_2 \qquad \text{Pong} \qquad (3\text{-}42b)$$

The first half reaction leads to the reduction of TPQ and concomitant transfer of the amino group to form an aminoquinol form of TPQ. A mechanism for this reaction is shown in fig. 3-46 (Mure et al., 2002), Reaction of benzylamine with TPQ leads to the iminium, or protonated Schiff base complex. Abstraction of a proton from the benzylic carbon potentiates electron transfer into TPQ and isomerization of the iminium group to form the benzaldiminium ion, which undergoes hydrolysis to benzaldehyde and the aminoquinol form of TPQ.

The mechanism of the oxidative half-reaction (eq. 3-42b) is less well understood. Two mechanisms are illustrated in fig. 3-47. They differ with respect to the role of copper. In mechanism A, copper participates in electron transfer between TPQ-aminoquinol and molecular oxygen. This begins with electron transfer from the aminoquinol to Cu^{2+} to form the aminoquinol semiquinone and Cu^{1+}, which transfers the electron to molecular oxygen. Reaction of the resulting superoxide and semiquinone leads to hydrogen peroxide and the

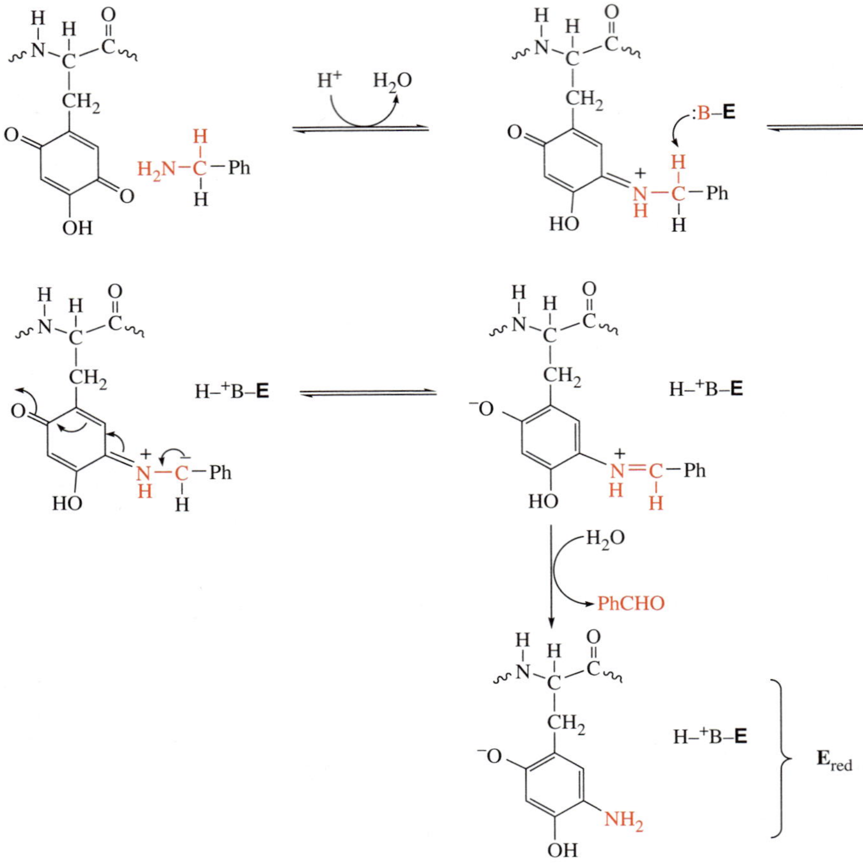

Fig. 3-46. A mechanism for the action of topaquinone in the reductive half-reaction of a copper amine oxidase.

iminoquinone of TPQ, which undergoes hydrolysis to ammonium ion and TPQ. In mechanism B, copper does not participate in electron transfer. The reaction begins with electron transfer from the aminoquinol directly to molecular oxygen. The resulting superoxide and aminosemiquinone react further to form hydrogen peroxide and the aminoquinone of TPQ. Hydrolysis of the aminoquinone leads to ammonium ion and TPQ.

The question of the possible involvement of copper in this process remains controversial. Mechanisms invoking Cu^{2+} in electron transfer such as mechanism A in fig. 3-47 are supported by the fact that the TPQ-aminoquinol/Cu^{2+} forms of reduced CAOs from a variety of species are in equilibrium with TPQ-aminosemiquinone/Cu^{1+} (Hartman and Dooley, 1995). This suggests that electron transfer from TPQ-aminoquinol and Cu^{2+} may be involved in the oxidative half reaction, as in mechanism A of fig. 3-47. However, in the case of the CAO from *H. polymorpha*, replacement of copper with cobalt does not inactivate the enzyme. The cobalt form is as active as the copper form, although replacement of copper by cobalt increases the K_m for molecular oxygen (Mills and Klinman, 2000; Mills et al., 2002). If electron transfer from the aminoquinol to Cu^{2+} is a required first step, as in the mechanism of fig. 3-47A, it would be difficult to imagine it occurring with Co^{2+} in place of Cu^{2+}, because the reduction potential for Co^{2+} is about 500 mV more negative than that for Cu^{2+}.

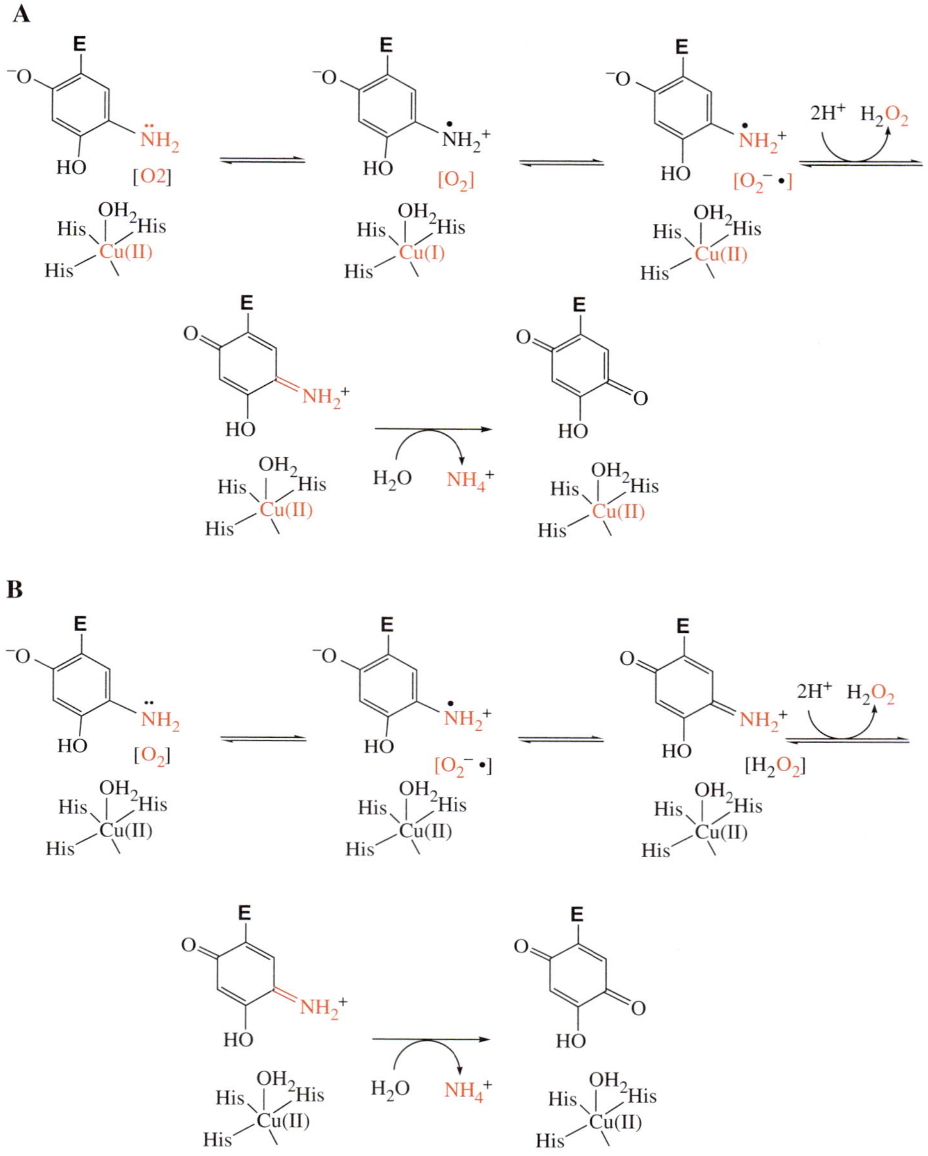

Fig. 3-47. Two mechanisms for the oxidative half-reaction of a copper amine oxidase (CAO). The essential difference between mechanisms A and B is that copper mediates electron transfer in A, whereas in mechanism B, direct electron transfer takes place from the aminoquinol and oxygen, and Cu^{2+} is a spectator. The fact that substitution of Co^{2+} for Cu^{2+} in CAO from *Hansenula polymorpha* allows the reaction to occur lends credence to mechanism B.

Copper mediates the oxygenation and oxidation steps in the biosynthesis of TPQ at the active sites of CAOs. Its role in catalysis is less clear.

References

Agnihotri G and HW Liu (2001) *Bioorg Chem* **29**, 234.
Baker JJ, C van der Drift, TC Stadtman (1973) *Biochemistry* **12**, 1054.
Ballinger MD, PA Frey, and GH Reed (1992) *Biochemistry* **31**, 10782.
Bartlett PA, KL McLaren, and MA Marx (1994) *J Org Chem* **59**, 2082.

Bartlett PA and K Satake (1988) *J Am Chem Soc* **110**,1628.
Bender SL, T Widlanski, and JR Knowles (1989) *Biochemistry* **28**, 7560.
Benkovic SJ (1980) *Annu Rev Biochem* **49**, 227.
Bishop GR, EJ Valente, TLWhitehead, KL Brown, RT Hicks, and VL Davidson (1996) *J Am Chem Soc* **118**, 12868.
Bishop GR, Z Zhu, TL Whitehead, RP Hicks, and VL Davidson (1998) *Biochem J* **330**, 1159.
Blake R, LP Hager, and RB Gennis (1978) *J Biol Chem* **253**, 1963.
Breslow R (1957) *J Am Chem Soc* **79**, 1762.
Breslow R (1958) *J Am Chem Soc* **80**, 3719.
Bruice TC (1984) *Isr J Chem* **24**, 54.
Buttlaire DH, RH Himes, and GH Reed (1976) *J Biol Chem* **251**, 4159.
Carpenter E, A Hawkins, J Frost, and K Brown (1998) *Nature* **394**, 299.
Carreras CW and DV Santi (1995) *Annu Rev Biochem* **64**, 721.
Chabriere E, X Vernede, B Guigliarelli, MH Charon, EC Hatchikian, and JC Fontecilla-Camps (2001) *Science* **294**, 2559.
Chen L, M Doi, RC Durley, AY Chistoserdov, ME Lidstrom, VL Davidson, and FS Mathews (1998) *J Mol Biol* **276**, 131.
Chirpich TP, V Zappia, RN Costilow, and HA Barker (1970) *J Biol Chem* **245**, 1778.
Creighton TE (1970) *Eur J Biochem* **13**, 1.
Datta S, Y Mori, K Takagi, K Kawaguchi, Z-W Chen, T Okajima, S Kuroda, T Ikeda, K Kano, K Tanizawa, and FS Mathews (2001) *Proc Natl Acad Sci U S A* **98**, 14268.
Davis L and DE Metzler (1972) In Boyer PD (ed): *The Enzymes*, vol 7, 3rd ed. Academic Press: New York, p 33.
DeMoss JA (1962) *Biochim Biophys Acta* **62**, 279.
Duine JA and J Frank Jr (1980) *Biochem J* **187**, 221.
Duine JA, Frank J, and JA Jongejan (1987) *Adv Enzymol Relat Areas Mol Biol* **59**, 169.
Duine JA, J Frank, and PE Verwiel.(1980) *Eur J Biochem* **108**,187.
Dunathan HC (1966) *Proc Natl Acad Sci U S A* **55**, 712.
Fasella P and GG Hammes (1967) *Biochemistry* **6**, 1798.
Fisher HF, EE Conn, B Vennesland, and FH Westheimer (1953) *J Biol Chem* **202**, 687.
Floss H and J Vederas (1982) Stereochemistry of pyridoxyl phosphate—catalyzed reactions. In Tamm C (ed): *New Comprehensive Biochemistry*, vol 3. Elsevier/North Holland Biomedical Press: Amsterdam, pp 161-199.
Frey PA (1989) *Biofactors* **2**, 1.
Frey PA (2001a) *Science* **294**, 2489.
Frey PA (2001b) *Annu Rev Biochem* **70**, 121.
Gerratana B, WW Cleland, and PA Frey (2001) *Biochemistry* **40**, 9187.
Givot IL, TA Smith, and RH Abeles (1969) *J Biol Chem* **244**, 6341.
Glaser L and H Zarkowsky (1973) In Boyer PD (ed): *The Enzymes*, vol 5, 3rd ed. Academic Press: New York, p 465.
Goldberger ML and E Racker (1962) *J Biol Chem* **237**, PC3841.
Goodman JL, S Wang, S Alam, FJ Ruzicka, PA Frey, and JE Wedekind (2004) *Biochemistry* **43**, 13883.
Gross JW, AD Hegeman, B Gerratana, and PA Frey (2001) *Biochemistry* **40**, 12497.
Gruys KJ, A Datta, PA Frey (1989) *Biochemistry* **28**, 9071.
Gruys KJ, CJ Halkides, and PA Frey (1987) *Biochemistry* **26**, 7575.
Hager LP and F Lipmann (1961) *Proc Natl Acad Sci U S A* **47**, 1768.
Hartman C and D Dooley (1995) *Methods Enzymol* **258**, 69.
Hegeman AD, JW Gross, and PA Frey (2001) *Biochemistry* **40**, 6598.
Heider J, X Mai, and MW Adams (1996) J Bacteriol. **178**, 780.
Hodgins DS (1971) *J Biol Chem* **246**, 2977.
Hodgins DS and RH Abeles (1969) *Arch Biochem Biophys* **130**, 274.
Holzer H and K Beaucamp (1958) *Angew Chem Int Ed* **71**, 776.
Houck DR, JL Hanners, and CJ Unkefer (1991) *J Am Chem Soc* **113**, 3162.
Huang T, C Wang, B Maras, D Barra, V Schirch (1998) *Biochemistry* **37**, 13536.
Inui H, K Ono, K Miyatake, Y Nakano, and S Kitaoka (1987) *J Biol Chem* **262**, 9130.
Janes SM, D Mu, D Wemmer, AJ Smith, S Kaur, D Maltby, AL Burlingame, and JP Klinman (1990) *Science* **248**, 981.
Jordan F (1999) *FEBS Lett* **457**, 298.

Jörnvall H, B Persson, M Krook, S Atrian, R Gonzalez-Duarte, J Jeffrey, and D Ghosh (1995) *Biochemistry* **34**, 6003.
Joyce BK and RH Himes (1966) *J Biol Chem* **241**, 5725.
Kern D, G Kern, H Neef, K Tittmann, M Killenberg-Jabs, C Wikner, G Schneider, and G Hubner (1997) *Science* **275**, 67.
Kessler D, J Rétey, GE Schulz (2004) *J Mol Biol* **342**, 183.
Klinman JP (2003) *Biochim Biophys Acta* **1647**, 131.
Klinman JP and D Mu (1994) *Annu Rev Biochem* **63**, 299.
Knowles JR (1989) *Annu Rev Biochem* **58**, 195.
Krampitz LO, G Gruell, CS Miller, KB Bicking, HR Skeggs, and JM Sprague (1958) *J Am Chem Soc* **80**, 5893.
Langer B, M Langer, and J Rétey (2001) *Adv Protein Chem* **58**, 175.
Langer M, A Lieber, and J Rétey (1994) *Biochemistry* **33**, 14034.
Li QX and W Dowhan (1988) *J Biol Chem* **263**, 11516.
Li R, JP Klinman, and FS Mathews (1998) *Structure* **6**, 293.
Mai X and MW Adams (1994) *J Biol Chem* **269**, 16726.
Mai X and MW Adams (1996) *J Bacteriol* **178**, 5890.
Maitra US and DB Sprinson (1978) *J Biol Chem* **253**, 5426.
Mansoorabadi SO, J Seravalli, C Furdui, V Krymov, GJ Gerfen, TP Begley, J Melnick, SW Ragsdale, and GH Reed (2006) *Biochemistry*, **45**, 7122.
Markham GD, CW Tabor, and H Tabor (1982) *J Biol Chem* **257**, 12063.
Matchett WH (1974) *J Biol Chem* **249**, 4041.
Mathews FS (1995) *Methods Enzymol* **258**, 191.
McIntire WS, DE Wemmer, AY Christoserdov, and ME Lidstrom (1991) *Science* **252**, 817.
Mejillano MR, H Jahansouz, TO Matsunaga, GL Kenyon, and RH Himes (1989) *Biochemistry* **28**, 5136.
Menon AL, H Hendrix, A Hutchins, MF Verhagen, and MW Adams (1998) *Biochemistry* **37**, 12838.
Menon S and SW Ragsdale (1997) *Biochemistry* **36**, 8484.
Miles EW (2001) *Chem Rec* **1**, 140.
Miles EW, S Rhee, and DR Davies (1999) *J Biol Chem* **274**, 12193.
Mills SA, Y Goto, Q Su, J Plastino, and JP Klinman (2002) *Biochemistry* **41**, 10577.
Mills SA and JP Klinman (2000) *J Am Chem Soc* **122**, 9897.
Moore JD, JR Coggins, R Virden, and AR Hawkins (1994) *Biochem J* **301**, 297.
Moss M and PA Frey (1987) *J Biol Chem* **262**, 14859.
Muller YA, G Schumacher, R Rudolph, and GE Schulz (1994) *J Mol Biol* **237**, 315.
Mure M, SA Mills, and JP Klinman (2002) *Biochemistry* **41**, 9269.
Murray JM, CG Saysell, CM Wilmot, WS Tambyrajah, J Jaeger, PF Knowles, SEV Phillips, and MJ McPherson (1999) *Biochemistry* **38**, 8217.
Muth WL and RN Costilow (1974) *J Biol Chem* **249**, 7463.
Nichols CE, J Ren, HK Lamb, AR Hawkins, and DK Stammers (2003) *J Mol Biol* **327**, 129.
Palmer J and RH Abeles (1976) *J Biol Chem* **251**, 5817.
Palmer J and RH Abeles (1979) *J Biol Chem* **254**, 1217.
Parkes C and RH Abeles (1984) *Biochemistry* **23**, 6355.
Petrovich RM, FJ Ruzicka, GH Reed, and PA Frey (1992) *Biochemistry* **31**, 10774.
Popják G (1970) In Boyer PD (ed): *The Enzymes*, vol 2, 3rd ed. Academic Press: New York, p 115.
Recsei PA and Snell EE (1984) *Annu Rev Biochem* **53**, 357.
Reed LJ (1960) In Boyer PD (ed): *The Enzymes*, vol 3, 3rd ed. Academic Press: New York, p 195.
Rétey J (1994) *Arch Biochem Biophys* **314**, 1.
Rétey J (2003) *Biochim Biophys Acta* **1647**, 179.
Rhee S, KD Parris, CC Hyde, SA Ahmed, EW Miles, and DR Davies (1997) *Biochemistry* **36**, 7664.
Rios A, J Crugeiras, TL Aymes, and JP Richard (2001) *J Am Chem Soc* **123**, 7949.
Rubinstein PA and JL Strominger (1974a) *J Biol Chem* **249**, 3776.
Rubinstein PA and JL Strominger (1974b) *J Biol Chem* **249**, 3782.
Satre M and EP Kennedy (1978) *J Biol Chem* **253**, 479.
Schirch L (1982) *Adv Enzymol Relat Areas Mol Biol* **53**, 83.

Sedewitz B, KH Schleifer, and F Gotz (1984) *J Bacteriol* **160**, 273.
Singer T and DE Edmondson (1980) *Methods Enzymol* **66**, 253.
Smith ET, Blamey JM, Adams MW (1994) *Biochemistry* **33**, 1008.
Smithers GW, H Jahansouz, JL Kofron, RH Himes, and GH Reed (1987) *Biochemistry* **26**, 3943.
Snell EE and SJ Di Mari (1970) In Boyer PD (ed): *The Enzymes*, vol 2, 3rd ed. Academic Press: New York, p 335.
Song KB and PA Frey (1991) *J Biol Chem* **266**, 7651.
Takata Y, T Yamada, Y Huang, J Komoto, T Gomi, H Ogawa, M Fujioka, and F Takusagawa (2002) *J Biol Chem* **277**, 22670.
Tang C and JP Klinman (2001) *J Biol Chem* **276**, 30575.
Tittmann K, R Golbik, K Uhlemann, L Khailova, G Schneider, M Patel, F Jordan, DM Chipman, RG Duggleby, and G Hubner (2003) *Biochemistry* **42**, 7885.
Tittmann K, M Vyazmensky, G Hubner, Z Barak, and DM Chipman (2005) *Proc Natl Acad Sci U S A* **102**, 553.
Toyama H, L Chistoserdova, and ME Lidstrom (1997) *Microbiology* **143**, 595.
van Poelje PD and EE Snell (1990) *Annu Rev Biochem* **59**, 29.
Wahl RC and WH Orme-Johnson (1987) *J Biol Chem* **262**, 10489.
Washabaugh M and WP Jencks (1988) *Biochemistry* **27**, 5044.
Weigel TM, LD Liu, and HW Liu (1992a) *Biochemistry* **31**, 2129.
Weigel TM, VP Miller, and HW Liu (1992b) *Biochemistry* **31**, 2140.
Westerling J, J Frank, and JA Duine (1979) *Biochem Biophys Res Commun* **87**, 719.
Westheimer FH, HF Fisher, EE Conn, and B Vennesland (1951) *J Am Chem Soc* **73**, 2403.
Weyand M and I Schlichting (1999) *Biochemistry* **38**, 16469.
Widlanski T, SL Bender, and JR Knowles (1989) *Biochemistry* **28**, 7572.
Williams CH Jr (1995) *FASEB J* **9**, 1267.
Williams CH, LD Arscott, S Muller, BW Lennon, ML Ludwig, PF Wang, DM Veine, K Becker, and RH Schirmer (2000) *Eur J Biochem* **267**, 6110.
Williams FR and LP Hager (1961) *J Biol Chem* **236**, PC36.
Wilmot CM, J Hajdu, MJ McPherson, PF Knowles, and SEV Phillips (1999) *Science* **286**, 1724.
Yanofsky C and M Rachmeler (1958) *Biochim Biophys Acta* **28**, 640.
Xia ZX, YN He, WW Dai, SA White, GD Boyd, and FS Mathews (1999) *Biochemistry* **38**, 1214.
Zheng YJ, Zx Xia, Zw Chen, FS Mathews, and TC Bruice (2001) *Proc Natl Acad Sci U S A* **98**, 432.

4
Coenzymes II: Metallic Coenzymes

The original coenzymes were small organic molecules that activated enzymes and participated directly in catalyzing enzymatic reactions. Most of them were derived from vitamins and were known as biologically "activated" forms of vitamins such as niacin, riboflavin, thiamine, and pyridoxal. Heme was in a separate category, perhaps because of its widespread biological role as an oxygen carrier, and because it was not a vitamin, it was not widely regarded as a coenzyme. However, heme was clearly an enzymatic prosthetic group in enzymes such as peroxidases and catalase, and it was known to participate in catalysis. Today, heme takes its place among the coenzymes. Other, more recently discovered metallic cofactors round out this chapter on metallocoenzymes.

Most of the detailed mechanisms of metallocoenzyme-dependent reactions are not known. Hypothetical mechanisms can often be written, and some of them are supported by a few experiments. Emerging principles are emphasized here for several of the more extensively studied metallocoenzymes. In other cases, the detailed mechanisms that we include in figures and schemes must be regarded as conjectural. We do not regard them as fanciful, but they have not been proved and are referred to as "a mechanism for" in recognition that other possible mechanisms have not been excluded. Space does not permit all conceivable mechanisms to be aired, and we hope that those shown here will stimulate discussion and experimentation.

Vitamin B_{12} coenzymes may be regarded as transitional from traditional coenzymes, in that the parent cyanocobalamin is a true vitamin, and its biologically activated forms adenosylcobalamin and methylcobalamin, with their covalent cobalt-carbon bonds, are organometallic compounds. For these reasons, we begin by discussing the vitamin B_{12} coenzymes.

Vitamin B$_{12}$ Coenzymes

Chemistry of B$_{12}$ Coenzymes

Structures

The structure in fig. 4-1 is that of *adenosylcobalamin*, the first B$_{12}$ coenzyme to be discovered. The molecule consists of the tetradentate corrin ring, cobalt in its 3+ oxidation state held within the corrin ring, the lower axial dimethylbenzimidazole α-ribotide ligand linked by a phosphodiester group to the corrin, and the 5′-deoxyadenosyl moiety covalently bonded to cobalt. The corrin ring is structurally and biosynthetically related to heme, but it differs in a number of respects, including that it is more highly reduced and incorporates extensive stereochemistry. The most unique and mechanistically important part of this molecule is the covalent Co—C5′ bond. This was the first stable organometallic compound of cobalt to be discovered. In the other B$_{12}$ coenzyme, *methylcobalamin*, the methyl group replaces the 5′-deoxyadenosyl group as the covalent substituent of cobalt. In vitamin B$_{12}$, the upper axial ligand of Co is cyanide (CN$^-$), which arises from the cyanide employed during purification. Cyanide cleaves adenosylcobalamin by displacing the 5′-deoxyadenosyl group, leading to cyanocobalamin, adenine, and Δ^4-2,3-dihydroxypentenal. The derivative

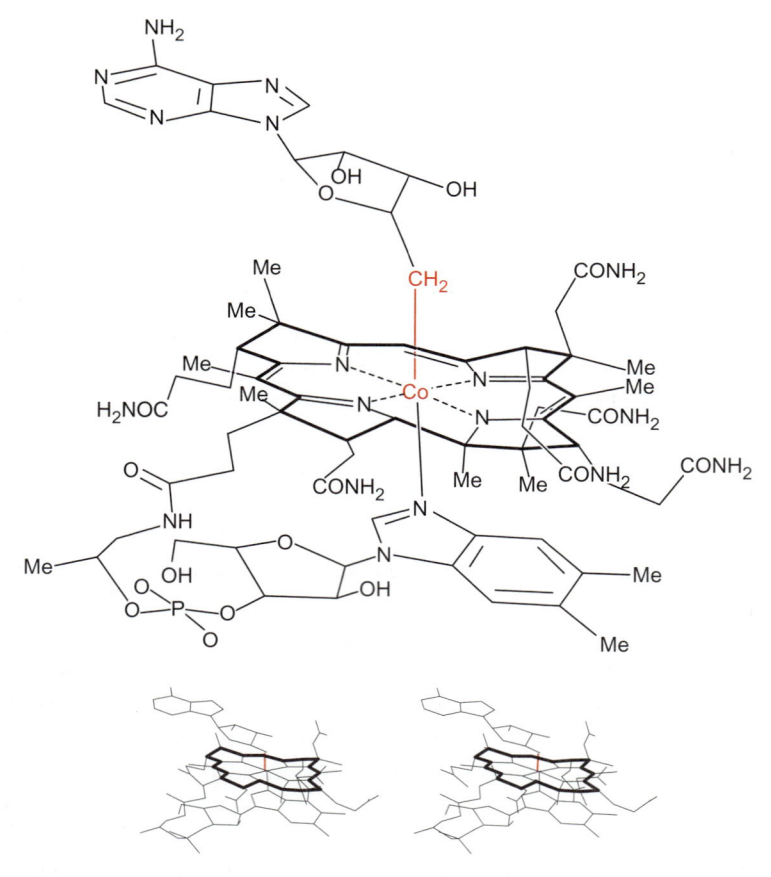

Fig. 4-1. In the structure of adenosylcobalamin, the corrin ring is biosynthetically related the porphyrin ring of heme. The pyrroline and pyrrolidine rings of corrin are reduced relative to the pyrrole rings of heme, and the substituents of these rings generate additional stereo centers. Cobalt is in the 3+ oxidation state and is diamagnetic. The chemically functional part of the molecule is the Co—C5′ bond (red).

of vitamin B_{12} in which a water molecule occupies the upper position is known as *aquocobalamin*, and when the upper axial ligand is hydroxide (HO⁻) it is known as *hydroxocobalamin*. The interconversion of aquocobalamin to hydroxocobalamin corresponds to the ionization of the coordinated water characterized by a pK_a of 8.

Adenosylcobalamin was discovered and characterized as an adenosyl derivative of vitamin B_{12} that activated glutamate mutase (Barker et alet al., 1960). The crystal structure showed the adenosyl group to be covalently bonded to cobalt through C5′ (Lenhart and Hodgkin, 1961). The discovery of several other enzymes requiring adenosylcobalamin or a close relative as a coenzyme soon followed (Abeles and Lee, 1961; Overath et al., 1962).

Chemical Properties of B_{12} Compounds

Cobalamin exists in the three oxidation states in scheme 4-1, where the corrin ring is symbolized by a square plane. Cob(III)alamin is red, cob(II)alamin is yellow, and cob(I)alamin is gray-green in color. Fig. 4-2 shows the visible absorption spectra of the three oxidation states. Cob(III)alamin and cob(I)alamin are diamagnetic, whereas cob(II)alamin is paramagnetic and displays a characteristic electron paramagnetic resonance (EPR) spectrum.

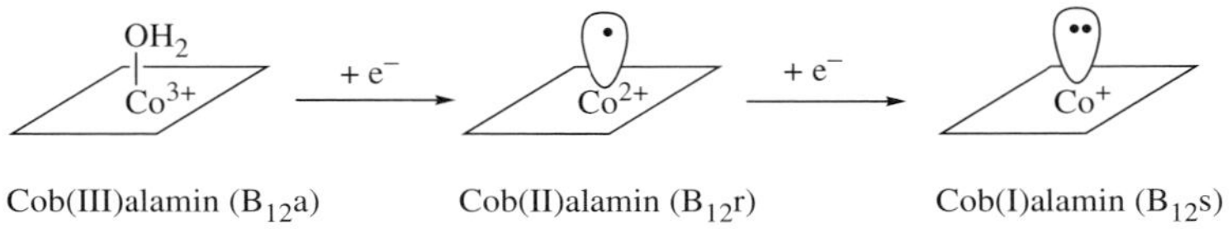

Cob(III)alamin ($B_{12}a$)   Cob(II)alamin ($B_{12}r$)   Cob(I)alamin ($B_{12}s$)

Scheme 4-1

Although the B_{12} coenzymes are at the 3+ oxidation state, forms observed spectroscopically at active sites as catalytic intermediates include cob(II)alamin and cob(I)alamin (Matthews,1999; Wagner et al., 1966). Cob(II)alamin appears as a transient intermediate in adenosylcobalamin-dependent reactions, and cob(I)alamin is an intermediate in alkyl group transfer reactions such as those catalyzed by methionine synthase, methyl coenzyme M synthase, and adenosylcobalamin synthetase. Cob(I)alamin is highly reactive as a nucleophile, being 40,000 times more reactive than the thiolate ion toward methyl iodide. As such, it is one of the most reactive nucleophiles in chemistry, and this property allows it to catalyze transalkylation reactions. Cob(I)alamin readily undergoes alkylation by alkyl halides, alkenes, and alkynes to form alkylcobalamins and alkenylcobalamins.

Adenosylcobalamin displays chemical properties that are relevant to its enzymatic reactions. We have alluded to its reactivity with cyanide to cleave away the adenosyl moiety. The Co—C5′ bond is also readily cleaved by acid hydrolysis and by photolysis (Hogenkamp, 1963; Hogenkamp et al., 1962). Acid hydrolysis proceeds with heterolytic cleavage of the Co—C5′ bond to produce cob(III)alamin, adenine, and Δ^4-2,3-dihydroxypentenal. Photolytic cleavage takes either of two courses depending on whether oxygen is present. Under anaerobic conditions, cob(II)alamin and 5′,8-*cyclo*-adenosine are the products of photolysis, as shown in fig. 4-3. In the air, the photolysis products are cob(III)alamin and adenosine-5′-aldehyde. The products imply that photolysis leads to homolytic cleavage of the Co—C5′ bond leading initially to cob(II)alamin and 5′-deoxyadenosine-5′-yl, the 5′-deoxyadenosyl free radical. In the absence of air, the free radical undergoes cyclization to 5′,8-*cyclo*-adenosine. This is a complex process, for which the mechanism is unknown. An analogous process has been observed in the radical-initiated cleavage of DNA. In the presence of air, the cob(II)alamin initially formed is oxidized to cob(III)alamin, and the 5′-deoxyadenosyl radical is captured by oxygen and transformed

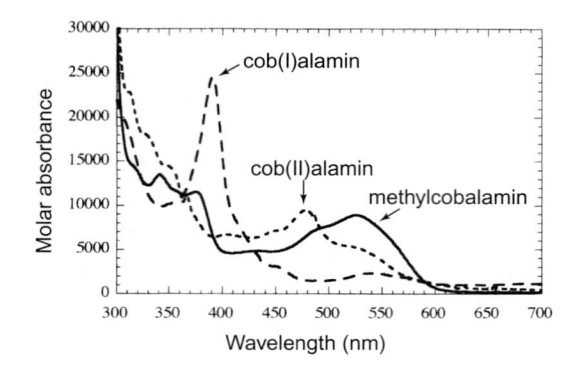

Fig. 4-2. Visible absorption spectra of cobalamin in three oxidation states. The cobalt in methylcobalamin is in the 3+ state, as it is in cob(III)alamin, which is aquocobalamin or hydroxocobalamin (B_{12a}), depending on pH. The absorption spectrum of B_{12a} is similar to but not identical with that of methylcobalamin. The visible absorption spectrum of adenosylcobalamin is very similar to that of methylcobalamin. Cobalt in cob(II)alamin (B_{12r}) is in the 2+ state, and it is paramagnetic. Cobalt in cob(I)alamin (B_{12s}) is in the 1+ oxidation state and is diamagnetic. (Spectra adapted from Matthews, 1999.)

into adenosine-5′-aldehyde. Presumably, the initially formed adenosine-5′-peroxy radical undergoes decomposition to the aldehyde.

In photolysis, the Co—C5′ bond of adenosylcobalamin breaks homolytically because it is the weakest bond in the molecule, and cleavage of this covalent bond allows the photoexcited states to lose energy. The weakness of this bond is further manifested in its thermal cleavage at higher than ambient temperatures. Being the weakest bond it is the first to break when adenosylcobalamin is heated. Systematic studies of the thermolytic cleavage have allowed the bond dissociation energy (BDE) for the Co—C5′ bond to be measured, and this has been found to be about 30 kcal mol^{-1} (Finke and Hay, 1984; Halpern et al., 1984). The bond is just stable enough to exist under ordinary physiological conditions.

Fig. 4-3. Hydrolytic and photolytic cleavage of the Co—C5′ bond in adenosylcobalamin. The products of acid hydrolysis arise from heterolytic cleavage of the Co—C5′ bond, and the products of photolysis arise from homolytic cleavage. Photolytic cleavage is brought about by light at all wavelengths in the absorption spectrum.

Adenosylcobalamin-Dependent Enzymes

Isomerization Reactions

Two classes of enzymes require adenosylcobalamin as the essential coenzyme. Members of the larger group catalyze molecular rearrangements following the pattern of eq. 4-1.

$$-\underset{X}{\overset{|}{C_\beta}}-\underset{H}{\overset{|}{C_\alpha}}- \quad \rightleftharpoons \quad -\underset{H}{\overset{|}{C_\beta}}-\underset{X}{\overset{|}{C_\alpha}}- \tag{4-1}$$

$$-X = -COSCoA, -CH(NH_2)CO_2^-, -OH, -NH_2, ...$$

In an isomerization typical of an adenosylcobalamin-dependent enzyme, a hydrogen atom and a group bonded to an adjacent carbon change positions; they undergo a cross-migration. Equation 4-1 shows that the hydrogen bonded to $C\alpha$ migrates to $C\beta$ and a group originally bonded to $C\beta$ migrates to $C\alpha$. The migrating hydrogen is generally chemically unreactive in the substrate or the product; it is not adjacent to a functional group that would increase its reactivity by an electronic effect such as electron withdrawal. The hydrogen migration takes place without exchange with solvent protons. The other migrating group, X, can be a carbon fragment, an hydroxyl group, or an amino group.

The best known adenosylcobalamin-dependent enzymatic reactions are described in eqs. 4-2 to 4-8 in fig. 4-4. Glutamate mutase was discovered in connection with the discovery of adenosylcobalamin. Methylmalonyl CoA mutase and dioldehydrase soon followed. The difficult chemical steps in reactions 4-2 to 4-8 characterize this class. The most obvious chemical problem is the migration of chemically unreactive hydrogen atoms. All of the reactions are in this respect unlike most enzymatic processes that take place by polar mechanisms involving nucleophilic, electrophilic, and acid-base catalysis. This occurs because coenzyme B_{12}-dependent reactions proceed by free radical mechanisms, the initiation of which is a special property of adenosylcobalamin.

For many years the pattern of eq. 4-1 pertained exclusively to adenosylcobalamin-dependent enzymes. However, it has come to light in recent years that lysine 2,3-aminomutase and related enzymes catalyze reactions analogous to that of D-lysine 5,6-aminomutase (eq. 4-6 in fig. 4-4), but use *S*-adenosyl-L-methionine (SAM) as the adenosyl coenzyme, not adenosylcobalamin (Baraniak et al., 1989; Moss and Frey, 1987). SAM is linked to many other enzymatic reactions in which difficult hydrogen-transfer processes occur. These are the radical SAM enzymes, which we discuss in a later section of this chapter.

Adenosylcobalamin and Hydrogen Transfer

The catalytic role of adenosylcobalamin was first discovered in studies of dioldehydrase. Research on this enzyme guided subsequent work on the other adenosylcobalamin-dependent enzymes for many years. For this reason, dioldehydrase serves as a case study for this group.

The most fundamental and essential element common to the reactions in fig. 4-4 is the transfer of unreactive hydrogen atoms. This fact inspired the first mechanistic work on hydrogen transfer in the action of dioldehydrase (EC 4.2.1.28). The role of adenosylcobalamin in this process was revealed in radiochemical experiments using tritium in place of hydrogen in the substrates and coenzyme. Dioldehydrase accepts ethylene glycol as a substrate as well as 1,2-propanediol, and reaction of a mixture of ethylene glycol and 1,2-[1-^{3}H]propanediol in the presence of dioldehydrase led to a mixture of [^{3}H]acetaldehyde

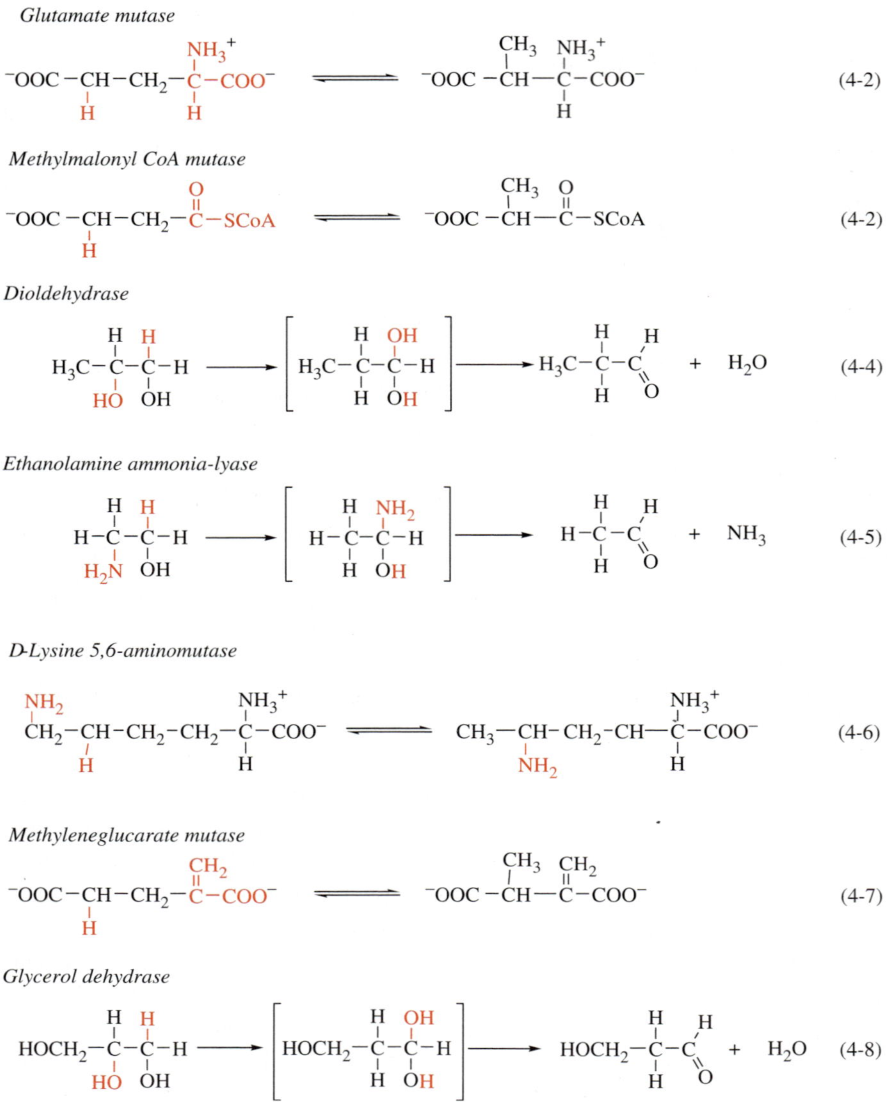

Glutamate mutase

$$^-OOC-CH-CH_2-\underset{H}{\underset{|}{\overset{NH_3^+}{\overset{|}{C}}}}-COO^- \rightleftharpoons {}^-OOC-CH-\underset{H}{\underset{|}{\overset{CH_3\;\;NH_3^+}{\overset{|\quad\;\;|}{C}}}}-COO^- \quad (4\text{-}2)$$

Methylmalonyl CoA mutase

$$^-OOC-CH-CH_2-\overset{O}{\overset{\|}{C}}-SCoA \rightleftharpoons {}^-OOC-CH-\overset{CH_3\;\;O}{\overset{|\quad\;\|}{C}}-SCoA \quad (4\text{-}2)$$

Dioldehydrase

$$H_3C-\underset{HO}{\underset{|}{\overset{H}{\overset{|}{C}}}}-\underset{OH}{\underset{|}{\overset{H}{\overset{|}{C}}}}-H \longrightarrow \left[H_3C-\underset{H}{\underset{|}{\overset{H}{\overset{|}{C}}}}-\underset{OH}{\underset{|}{\overset{OH}{\overset{|}{C}}}}-H\right] \longrightarrow H_3C-\overset{H}{\overset{|}{C}}-\overset{H}{\underset{O}{C\!\!\diagup\!\!\diagdown}} + H_2O \quad (4\text{-}4)$$

Ethanolamine ammonia-lyase

$$H-\underset{H_2N}{\underset{|}{\overset{H}{\overset{|}{C}}}}-\underset{OH}{\underset{|}{\overset{H}{\overset{|}{C}}}}-H \longrightarrow \left[H-\underset{H}{\underset{|}{\overset{H}{\overset{|}{C}}}}-\underset{OH}{\underset{|}{\overset{NH_2}{\overset{|}{C}}}}-H\right] \longrightarrow H-\overset{H}{\overset{|}{C}}-\overset{H}{\underset{O}{C\!\!\diagup\!\!\diagdown}} + NH_3 \quad (4\text{-}5)$$

D-Lysine 5,6-aminomutase

$$\underset{H}{\underset{|}{\overset{NH_2}{\overset{|}{CH_2}}}}-CH-CH_2-CH_2-\underset{H}{\underset{|}{\overset{NH_3^+}{\overset{|}{C}}}}-COO^- \rightleftharpoons CH_3-CH-CH_2-\underset{NH_2}{\underset{|}{CH}}-\underset{H}{\underset{|}{\overset{NH_3^+}{\overset{|}{C}}}}-COO^- \quad (4\text{-}6)$$

Methyleneglucarate mutase

$$^-OOC-CH-CH_2-\underset{H}{\underset{|}{\overset{CH_2}{\overset{\|}{C}}}}-COO^- \rightleftharpoons {}^-OOC-CH-\overset{CH_3\;\;CH_2}{\overset{|\quad\;\|}{C}}-COO^- \quad (4\text{-}7)$$

Glycerol dehydrase

$$HOCH_2-\underset{HO}{\underset{|}{\overset{H}{\overset{|}{C}}}}-\underset{OH}{\underset{|}{\overset{H}{\overset{|}{C}}}}-H \longrightarrow \left[HOCH_2-\underset{H}{\underset{|}{\overset{H}{\overset{|}{C}}}}-\underset{OH}{\underset{|}{\overset{OH}{\overset{|}{C}}}}-H\right] \longrightarrow HOCH_2-\overset{H}{\overset{|}{C}}-\overset{H}{\underset{O}{C\!\!\diagup\!\!\diagdown}} + H_2O \quad (4\text{-}8)$$

Fig. 4-4. Adenosylcobalamin-dependent rearrangement reactions. Shown are several but not all of the adenosylcobalamin-dependent rearrangement reactions catalyzed by enzymes. The adenosylcobalamin-dependent ribonucleotide reductase reaction is shown in eq. 4-9 and fig. 4-6.

and [^{3}H]propionaldehyde (Abeles and Zagalak, 1966). This experiment proved that hydrogen transfer could be intermolecular. The effect of varying the ratio of ethylene glycol to 1,2-[^{3}H]propanediol on the tritium content of the products proved that hydrogen transfer proceeded by a process that allowed both inter- and intramolecular transfer (Frey and Abeles, 1966; Frey et al., 1967). These experiments showed that either dioldehydrase or adenosylcobalamin mediated hydrogen transfer by a process that allowed tritium from the substrate to be exchanged with one or more hydrogen atoms within their structures.

Tritium transfer could be traced through the enzyme-coenzyme complex by using large amounts of enzyme and stoichiometrically equivalent amounts of adenosylcobalamin, the two of which form a tight complex. Catalysis by this complex could be interrupted by acid denaturation in the course of its reaction with 1,2-[1-^{3}H]propanediol. Repurification of the coenzyme led to [³H]adenosylcobalamin (Frey and Abeles, 1966). Activation of dioldehydrase by the [³H]adenosylcobalamin so produced, using unlabeled 1,2-propanediol as the substrate, led to [³H]propionaldehyde, which proved that adenosylcobalamin mediated hydrogen transfer. Chemical degradation of the enzymatically produced [³H]adenosylcobalamin proved that the tritium resided on C5′ of the deoxyadenosyl moiety, the carbon bonded to cobalt (Frey et al., 1967).

Activation of dioldehydrase with chemically synthesized [adenosyl-5′-³H]adenosylcobalamin in the reaction of unlabeled 1,2-propanediol led to [³H]propionaldehyde (Frey et al., 1967). These experiments proved that the 5′-methylene group of adenosylcobalamin mediated hydrogen transfer. Dioldehydrase also catalyzed the transfer of tritium from [³H]adenosylcobalamin to propionaldehyde, the product derived from 1,2-propanediol, proving that the reaction could be partially reversed. However, complete reversal of propionaldehyde to 1,2-proanediol could not be demonstrated (Frey et al., 1967). The role of adenosylcobalamin in mediating hydrogen transfer was verified in all of the other reactions in fig. 4-4.

Cleavage of the Co—C5′ bond was first observed in the suicide inactivation of dioldehydrase by glycolaldehyde (Wagner et al., 1966). The inactivated complex consisted of dioldehydrase, cob(II)alamin and 5′-deoxyadenosine as well as the carbon atoms derived from glycolaldehyde. The product derived from glycolaldehyde is now known to be the glycolaldehyde radical, the protonated form of the stable radical *cis*-ethanesemidione (Abend et al., 2000). Scheme 4-2 illustrates the chemical course of the cleavage of adenosylcobalamin in the inactivation of dioldehydrase and shows the structure of this radical.

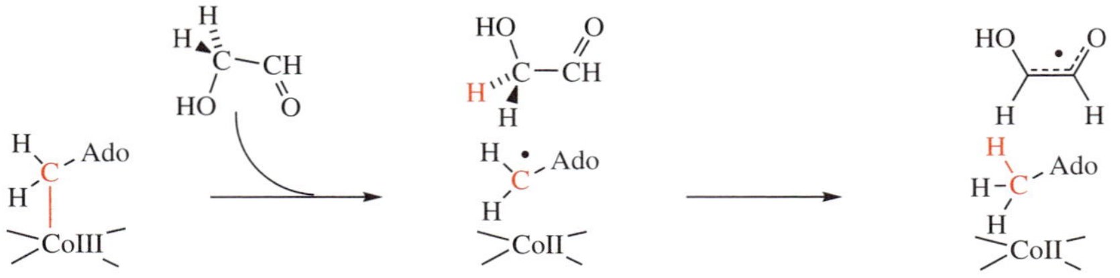

Scheme 4-2

The reaction of glycolaldehyde showed how adenosylcobalamin is cleaved to cob(II)alamin and 5′-deoxyadenosine. It also inspired experimentation showing that the substrate 1,2-propanediol induces the transient cleavage of adenosylcobalamin to cob(II)alamin at the active site (Wagner et al., 1966). Transient cleavage to cob(II)alamin also occurs in many of the other reactions in fig. 4-4, and substrate-derived radicals have been observed by EPR spectroscopy to be associated with the formation of cob(II)alamin (Cockle et al., 1972; Hamilton et al., 1972; Leutbecher et al., 1992; Michel et al., 1992; Padmukumar and Banerjee, 1995; Pilbrow, 1982; Valinsky et al., 1973; Zhao et al., 1992).

The inactivation of dioldehydrase by glycolaldehyde was documentation of a suicide inactivator. In this mode of irreversible inhibition, the inhibitor binds to the enzyme and undergoes initial steps of the catalytic mechanism. These steps lead to an analog of an intermediate that destroys enzyme activity. In the case of glycolaldehyde and dioldehydrase, the inactivator is an analog of proionaldehyde that reacts in the reverse direction but produces a radical that is so stable it cannot react further. The active site is then blocked and cannot catalyze the reaction of 1,2-propanediol. Many other examples of suicide inactivators, also known as mechanism-based inhibitors, are known. Some of them are in use as pharmaceutical agents, and most of them reveal significant facts about the enzymatic reaction mechanisms. Examples of suicide inactivators used in medical practice are given in chapter 5.

The role of adenosylcobalamin in mediating hydrogen transfer in the isomerization reactions may be described as in fig. 4-5. Substrate binding to the enzyme-coenzyme complex facilitates the cleavage of the Co—C5' bond, leading to cob(II)alamin and the 5'-deoxyadenosyl radical, which can exist only fleetingly because of its high reactivity. This property allows it to abstract unreactive hydrogen atoms from a substrate to form 5'-deoxyadenosine. The resulting substrate radical (S •) rearranges to a product radical (P •), which then abstracts a hydrogen atom from the methyl group of 5'-deoxyadenosine. The resulting 5'-deoxyadenosyl radical quickly recombines with cob(II)alamin to reconstitute adenosylcobalamin. The mechanisms of the various isomerization reactions must differ in the structures of the substrate and product related radicals and in the mode of radical isomerization. There are also geometric differences in the interactions of substrates with adenosylcobalamin in the active sites of these enzymes.

The question arises of the mechanism by which the Co—C5' breaks. It seems likely that binding interactions between an enzyme and adenosylcobalamin weakens the bond and allows it to undergo reversible homolytic scission. Such a process would require the use of perhaps 15 kcal mol^{-1} of binding energy to weaken the bond. Given the structure of adenosylcobalamin, it seems that substantial binding energy could potentially be available, and the use of 15 kcal mol^{-1} for weakening the Co—C5' bond would leave sufficient energy for maintaining adenosylcobalamin at the active site. A concept of the amount of binding energy that might be available for this process can be gained by considering the enthalpy of binding between a small molecule and a receptor protein. The enthalpy for the binding of the much smaller and structurally simpler molecule D-biotin to avidin is –23.4 kcal mol (Swamy, 1995). Avidin is a binding protein that uses little binding energy to do work such as conformational changes. It is not difficult to imagine that ≥30 kcal mol^{-1} of

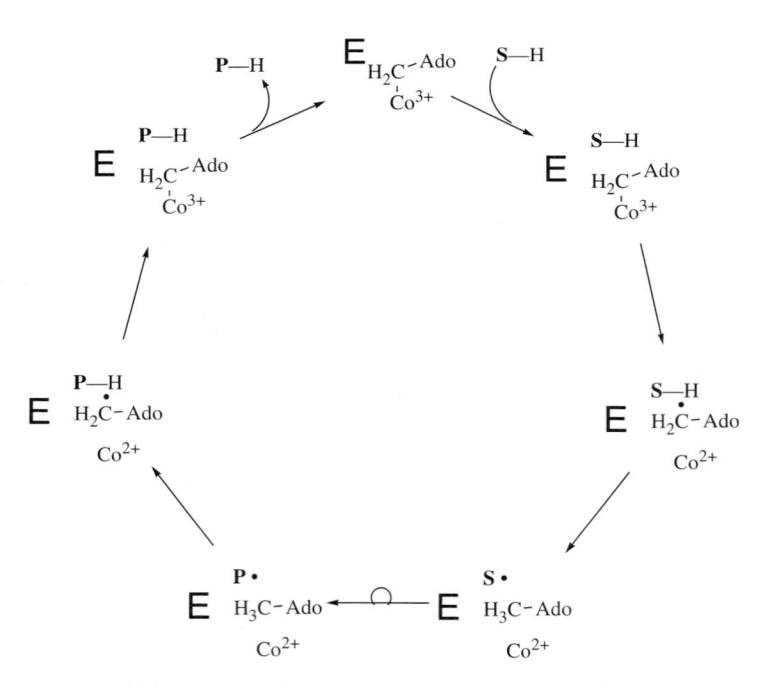

Fig. 4-5. Hydrogen transfer in adenosylcobalamin-dependent rearrangement reactions. The pathway of hydrogen transfer and radical formation is traced in the figure in a clockwise direction beginning at the top with the enzyme-adenosylcobalamin complex. The pathways of hydrogen transfer have been documented with tritium or deuterium-labeled substrate and adenosyl-5'-tritiated adenosylcobalamin in all of the enzymatic reactions.

enthalpy might be available from the binding of adenosylcobalamin to a cognate enzyme. The use of 15 kcal mol^{-1} for conformational work to weaken the Co—C5′ bond would leave ≥15 kcal mol^{-1} for binding the coenzyme, and this would correspond to a K_d of ≤10^{-7} M. Additional energy for this process is likely provided by the binding of a substrate, which stimulates Co—C5′ bond cleavage in most of the reactions of fig. 4-4.

Chemical Models for Radical Isomerizations

EPR experiments reveal the presence of radicals as transient intermediates at the active sites of most adenosylcobalamin-dependent enzymes. However, the exact mechanisms by which the initial substrate-related radicals undergo isomerization to product-related radicals are not known. Chemical modeling of radical rearrangements constitutes the available evidence for the mechanisms of radical isomerization in most of the adenosyl-cobalamin-dependent reactions. We present the chemical models in connection with the sections on the individual enzymes in chapters 7 and 8.

Although substrate- or product-related radical intermediates have been observed by EPR spectroscopy in most of the rearrangement reactions, the 5′-deoxyadenosyl radical has not been observed spectroscopically. However, an allylic analog of this radical with a double bond between ribosyl carbons 3′ and 4′ has been observed and characterized as an intermediate in the reaction of dioldehydrase activated with the coenzyme analog 3′,4′-anhydroadenosylcobalamin (Magnusson et al., in press).

Ribonucleotide Reductase

The ribonucleotide reductase (EC 1.17.4.2) from *Lactobacillus leichmanii* exemplifies the second type of adenosylcobalamin-dependent enzyme (Fontecave and Mulliez, 1999; Stubbe, 1990). This enzyme catalyzes the reduction of ribonucleoside triphosphates (NTPs) to deoxyribonucleoside triphosphates (dNTPs) according to eq. 4-9, in which Th is thioredoxin.

$$\text{Nucleoside triphosphate} + \text{Th}_{red} \rightarrow \text{Deoxynucleoside triphosphate} + \text{Th}_{ox} + H_2O \quad (4-9)$$

In the reaction of *Lactobacillus* ribonucleotide reductase the Co—C5′ bond of adenosylcobalamin plays the same role as in the isomerization reactions: initiation of radical formation. The process differs, however, in fundamental respects. Adenosylcobalamin initiates radical formation in the enzyme itself, and does not interact directly with the substrate (Stubbe, 1990). The enzymatic radical transfers the radical center to the substrate by abstracting the C3′(H) of the ribonucleoside triphosphate, thereby forming a substrate radical that undergoes subsequent dehydration and reduction processes.

Indirect evidence of enzyme radical formation is the reductase catalyzed exchange of protons from water with the methylene hydrogens of the Co—C5′ bond in the absence of a substrate (Abeles and Beck, 1967; Beck et al., 1966; Hogenkamp et al., 1968). This exchange is a consequence of reversible thiyl radical formation at Cys408, and it results from the exchangeability of the cysteine-thiol proton with water.

The mechanism in fig. 4-6 serves as a current working hypothesis for the reduction. The observation of a primary tritium kinetic isotope effect in the reduction of 3′-tritiated ribonucleoside diphosphates by the B$_{12}$-independent *Escherichia coli* ribonucleotide reductase implicates a mechanism involving C3′(H) and the substrate C3′-radical (Stubbe and Ackles, 1980). Observation of this isotope effect in the reduction by the *L. leichmannii* ribonucleotide reductase further supports the cleavage of C3′(H) in the mechanism (Stubbe et al., 1981).

198 Enzymatic Reaction Mechanisms

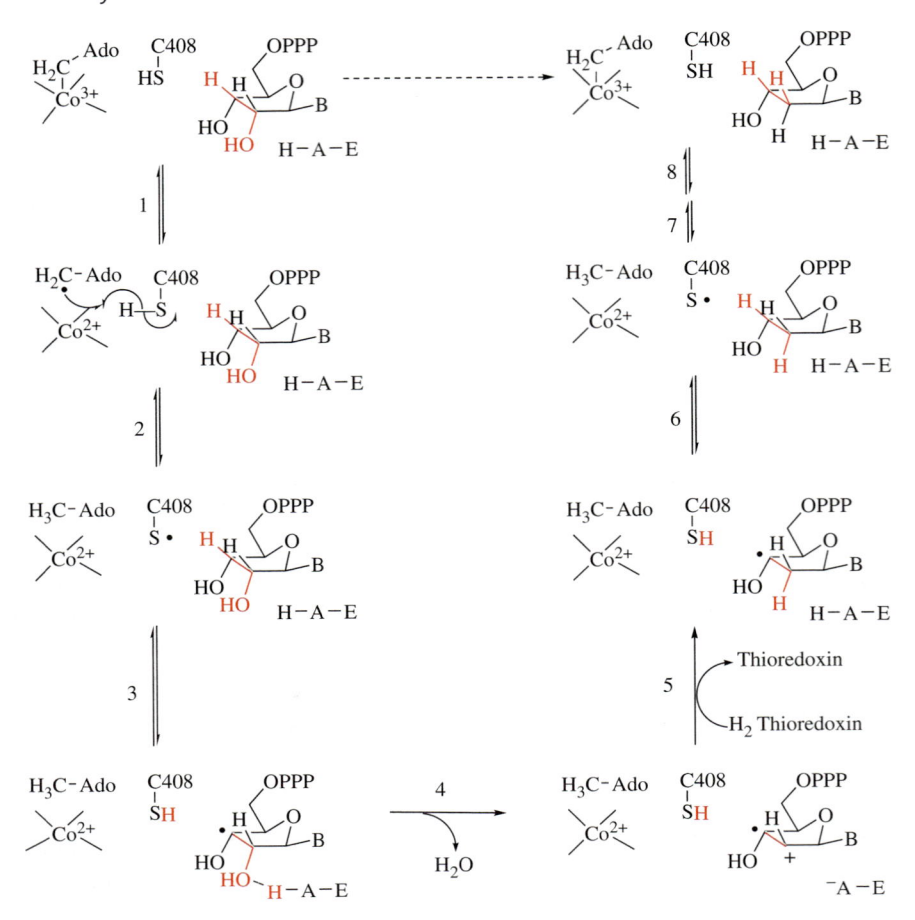

Fig. 4-6. A hypothetical mechanism of ribonucleotide reduction by B_{12}-dependent ribonucleotide reductases. Adenosylcobalamin initiates radical formation, presumably by way of the 5′-deoxyadenosyl radical, which abstracts a hydrogen atom from the sulfhydryl group of Cys408 to form the thiyl radical in step 2. In succeeding steps proceeding counterclockwise, the thiyl radical abstracts the C3′(H) from the substrate to form the substrate 3′-radical in step 3, which undergoes dehydration in step 4 and reduction in step 5 to generate the product-3′ radical. Abstraction of hydrogen from the SH group Cys408 by the product radical regenerates the thiyl radical in step 6, which reacts with 5′-deoxyadenosine and cob(II)alamin to complete the formation of product.

Before the discovery of C3′(H) isotope effects, no chemically credible mechanism for the reduction of a ribonucleotide had been written. The C3′(^{3}H)-kinetic isotope effect immediately brought mechanisms such as that in fig. 4-6 into play. The participation of a protein radical became clear from single turnover experiments showing that the C3′(^{3}H) was not transferred to adenosylcobalamin (Ashley et al., 1986). EPR studies implicated Cys408 as a thiyl radical by showing its formation and kinetic competence (Licht et al., 1996). The single turnover experiments had proved that adenosylcobalamin could not interact directly with the substrate (Ashley et al., 1986), but the thiyl radical form of Cys408 abstracted the C3′(H) of the substrate (see chap. 16).

The chemical mechanism of ribonucleotide reduction, as outlined in fig. 4-6, begins with cleavage of the Co—C5′ bond and thiyl radical formation at Cys408. This process does not require the presence of the substrate but does require an allosteric activator, which binds at the allosteric site some distance from the active site. Cleavage of the Co—C5′

bond and formation of the thiyl radical on Cys408 proceeds by a two-step mechanism, in which the Co—C5′ bond is first broken to form cob(II)alamin and the 5′-deoxyadenosyl radical, and the radical then abstracts the hydrogen atom from the sulfhydryl group of Cys408 to form 5′-deoxyadenosine and the thiyl radical (Chen et al., 2003a). As in other adenosylcobalamin-dependent reactions, 5′-deoxyadenosine is tightly bound as an intermediate and is never released from the enzyme. The process is reversible, and in $^{3}H_{2}O$ or $^{2}H_{2}O$ Cys408 becomes labeled with tritium or deuterium by solvent exchange. Because of this exchange and the mechanism of reversible Co—C5′ bond cleavage, the enzyme catalyzes the exchange of tritium or deuterium from the solvent into the C5′-methylene group of adenosylcobalamin. Solvent hydrogen exchange, like Co—C5′ bond cleavage, depends on the presence of an allosteric activator such as dGTP.

When a substrate binds to the active site, the thiyl radical abstracts C3′(H) in step 3 to form the substrate radical and a thiol group on Cys408. The proton of the thiol group in Cys408 is protected by the presence of the substrate from exchanging with the solvent. In step 4, the substrate radical eliminates the elements of water to form a new radical intermediate. The process is depicted in fig. 4-6 as an acid-catalyzed dehydration leading to a cation radical intermediate in the substrate, but this has not been proved. Reduction of the second radical intermediate by thioredoxin in step 5 leads to the 3′-radical of the deoxynucleotide, which abstracts the thiol hydrogen atom from Cys408 in step 6. The resulting thiyl radical at Cys408 then regenerates the Co—C5′ bond in adenosylcobalamin in steps 7 and 8, which represent reversal of steps 1 and 2. The structure of *Lactobacillus* ribonucleotide reductase and the relative locations of the coenzyme, substrate, and allosteric sites are discussed in chapter 16 together with B_{12}-independent ribonucleotide reductases.

Although the Cys408-thiyl radical has been observed by EPR spectroscopy and found to be kinetically competent, the other intermediates in fig. 4-6 have not been observed. Spectroscopic evidence for substrate-based radicals has been obtained by employing substrate analogs (van der Donk et al., 1998).

The adenosylcobalamin-dependent enzyme is a class II ribonucleotide reductase. The reaction mechanisms of other reductases (classes I and III) are thought to be the same from the point of thiyl radical formation, as demonstrated in steps 2 through 5 of fig. 4-6. Other reductases display different coenzyme requirements and chemistry for the generation of thiyl radicals.

We discuss two of these other coenzymes later in this chapter.

Methylcobalamin-Dependent Enzymes

Methionine Synthase

Most methyltransferases use SAM as the methyl donor substrate in the biosynthesis of hormones and neurotransmitters, the maturation and speciation of DNA, and in signal transduction. A few methyltransferases use 5-methyltetrahydrofolate and related compounds as methyl donors, and methylcobalamin is the coenzyme in those reactions. The most studied and important example is the B_{12}-dependent methionine synthase (EC 2.1.1.13), which catalyzes the eq. 4-10 (Matthews, 1999).

Homocysteine + 5-Methyltetrahydrofolate → Methionine + Tetrahydrofolate   (4-10)

Methionine synthase is a large, multidomain enzyme with a molecular mass of more than 200 kDa that contains cobalamin or methylcobalamin in its catalytic domain.

The steady-state kinetic mechanism is ping pong bi bi and proceeds in the two distinct chemical and kinetic steps of eqs. 4-11a and 4-11b.

200   Enzymatic Reaction Mechanisms

$$\text{E-MeB}_{12} + \text{Homocysteine} \rightarrow \text{E-B}_{12}\text{s} + \text{Methionine} \qquad (4\text{-}11\text{a})$$

$$\text{E-B}_{12}\text{s} + \text{Methyltetrahydrofolate} \rightarrow \text{E-MeB}_{12} + \text{Tetrahydrofolate} \qquad (4\text{-}11\text{b})$$

The chemical transformations illustrated in fig. 4-7 take place at the active site, and methylcobalamin mediates methyl group transfer while remaining tightly bound to the enzyme. In the methylation of homocysteine by methylcobalamin, the Co—CH$_3$ bond undergoes heterolytic cleavage with the bonding electrons remaining on cobalt in cob(I)alamin, B$_{12}$s in eq. 4-11a. Cob(I)alamin is a highly reactive nucleophile, and this reactivity may be a key to the efficiency of methyl group transfer in this system. 5-Methyltetrahydrofolate is a poor alkylating agent, and the high nucleophilic reactivity of cob(I)alamin does not overcome this barrier in solution. The question of the mechanism by which this barrier is overcome at the active site of methionine synthase remains open.

The ability of cob(I)alamin to accept the methyl group from 5-methyltetrahydrofolate in eq. 4-11b and of methylcobalamin to serve as a methylating agent for the thiol group of homocysteine form the chemical basis for catalysis in this system. We discuss the structure and function of methionine synthase in greater detail in chapter 15.

Methylcoenzyme M Synthase

The reduction of methylcoenzyme M by its reductase leads to methane production in methanogenesis. The enzymatic synthesis of methylcoenzyme M from coenzyme M follows the same course as methionine synthesis by methionine synthase. Methylcoenzyme M synthase is the B$_{12}$ enzyme that catalyzes methyl group transfer from an analog of 5-methyltetrahydrofolate to coenzyme M, analogous to the reaction in eq. 4-10. We describe this process in connection with the methanogenic coenzymes later in this chapter.

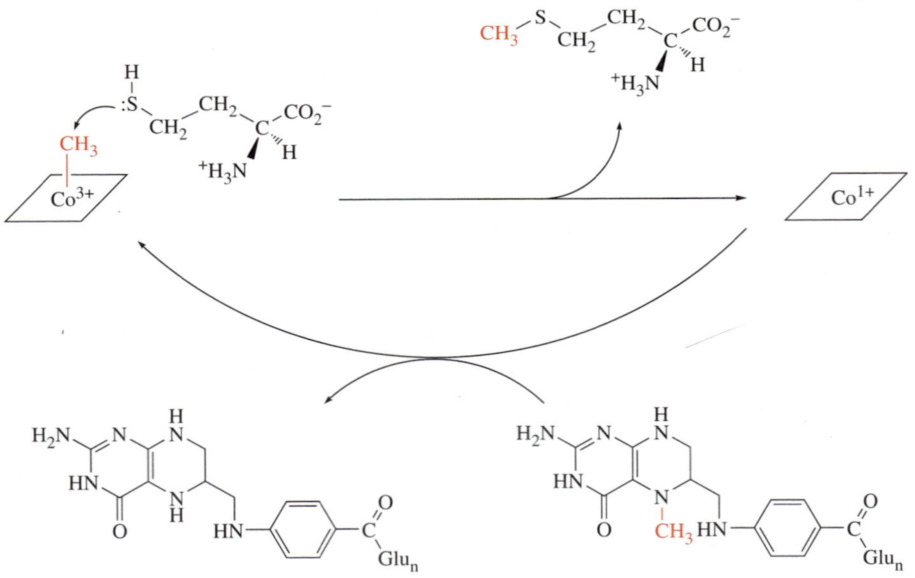

Fig. 4-7. Methyl group transfer in the reaction of methionine synthase. The reactions shown are catalyzed by the catalytic domain of methionine synthase. Other domains activate the enzyme and maintain the activity by maintaining methylcobalamin as the coenzyme. The domain structure and function of methionine synthases are discussed in chapter 15.

Heme Coenzymes

Many redox enzymes use oxygen or a peroxide as a substrate and have heme coenzymes that catalyze biochemical reactions. Heme in cytochromes also mediates terminal electron transfer pathways in all cells. Cytochromes *a*, *b*, and *c* typically link electron transfer complexes through their own redox cycling reactions. Heme coenzymes participate in catalysis by oxygen metabolizing enzymes such as catalase, peroxidases, and the cytochrome P450 monooxygenases. In discussing these reactions, we refer to diatomic oxygen as O_2 oxygen. In discussing the O atoms in molecules, we refer to them as oxo, oxide, or oxygen atoms, as appropriate. We focus on oxygen-metabolizing enzymes in the next section.

Chemistry of Oxygen and Heme

Oxygen Structure and Chemistry

Many enzymes that use oxygen as a substrate face a mechanistic barrier. Although oxygen is a moderately strong oxidizing agent, it is remarkably unreactive. The "low reactivity" of O_2 mainly refers to its reactions with organic compounds, which generally undergo oxidation by two-electron mechanisms. Only a few organic compounds react directly with oxygen, notably $FADH_2$ and $FMNH_2$ and hydroquinones. We discussed the likely mechanism of this reaction in chapter 3. In contrast, some transition metal ions readily react with O_2. We discuss the reasons for this before proceeding with heme enzymes.

The electronic structure of O_2 immediately explains its low reactivity with organic compounds. Molecular oxygen is a diradical, a paramagnetic molecule. The molecular orbital diagram of the 12 valence electrons in scheme 4-3 illustrates the electronic configuration of O_2.

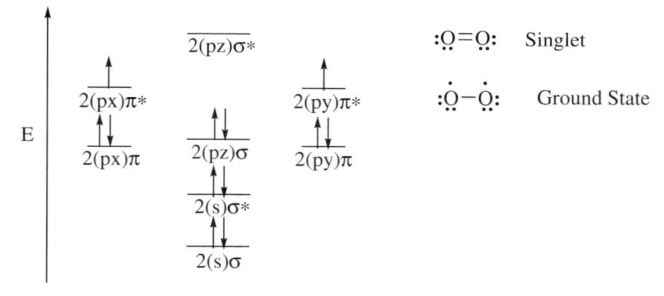

Scheme 4-3

Its ground-state, Lewis electronic structure appears to the right of the unpaired electrons. Each atom contributes 6 electrons from the 2s and 2p orbitals for a total of 12. They are used to fill the molecular orbitals following Hund's rules. The $2(px)p*$ and $2(py)p*$ energy levels lie well below that of the $2(pz)\sigma*$ orbital, so the last two electrons must occupy the lower energy levels singly as unpaired electrons. The Lewis structure shows them as dots in a diradical formulation.

The electronic structure of O_2 means that in its ground state it tends to react in one-electron processes, such as one-electron transfer or coupling with a free radical. Ordinary organic molecules do not tend to react in these modes. Exceptions include dihydroflavins, which can donate a single electron to O_2 to form the unusually stable flavin semiquinone radical and superoxide anion ($O_2^{\bullet-}$), which then undergoes coupling with electron pairing to form the flavin 4a-hydroperoxide (see chap. 3).

202 Enzymatic Reaction Mechanisms

$$O_2 \xrightarrow{-0.33\text{ v}} O_2^{-}\bullet \xrightarrow{0.87\text{ v}} H_2O_2 \xrightarrow{1.35\text{ v}} H_2O$$

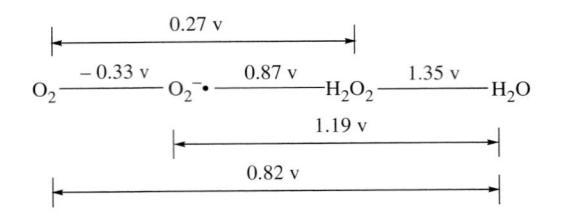

(0.27 v from O_2 to H_2O_2; 1.19 v from $O_2^{-}\bullet$ to H_2O; 0.82 v from O_2 to H_2O)

Fig. 4-8. Reduction potentials for species of oxygen.

The reduction potentials in fig. 4-8 reveal the fundamental energetics of oxygen reduction and the essential barrier to this process. The standard free energy change for a reaction can be calculated from the difference between the reduction potentials for the electron acceptor and donor by using eq. 4-12, in which $\Delta E° = E°$ (acceptor) $- E°$(donor), $F = 23{,}062$ cal mol^{-1} volt^{-1} (the Faraday constant), and $n =$ the number of electrons transferred.

$$\Delta G = -nF\Delta E° \tag{4-12}$$

Using this equation and the reduction potentials, $E°'$, referring to pH 7.0 and 25°C, we can calculate $\Delta E°' = +0.5$ V for the reduction of O_2 by NADH according to eqs. 4-13a and 4-13b.

$$0.5\ O_2 + 2e^- + 2\ H^+ \rightarrow H_2O \qquad E°' = 0.82\text{ V} \tag{4-13a}$$

$$NADH + H^+ \rightarrow NAD^+ + 2\ H^+ + 2e^- \qquad E°' = 0.32\text{ V} \tag{4-13b}$$

From eq. 4-12 and $\Delta E°'$, the value of $\Delta G°' = -26$ kcal mol^{-1}; the reduction of O_2 by NADH is energetically spontaneous and should take place. However, it does not occur in the absence of a catalyst. The reason is that NADH does not easily react in one-electron steps, while O_2 does not easily react in two-electron steps. The reaction is therefore kinetically blocked and occurs only in the presence of a catalyst.

NADH oxidases in cell extracts contain transition metal ions and flavins that can couple one- and two-electron transfer processes, and they catalyze the reduction of O_2 by NADH. A similar calculation for the reduction of O_2 by FADH$_2$ shows that the reaction is also energetically favored, and it actually occurs despite the fact that FADH$_2$ is an organic molecule. However, FADH$_2$ is special by virtue of the ease with which it undergoes electron transfer in one-electron steps, and this property allows it to react rapidly with oxygen (see chap. 3).

The unpaired electrons of O_2 become paired in the excited state known as *singlet oxygen*, in which one of the spins is flipped and the two highest energy electrons occupy a single 2pp* orbital. The consequent pairing requires an input of energy, and singlet oxygen is a chemically activated species that readily reacts with many organic compounds. In the past, this property of O_2 was put to use in the analysis of protein function. The photoactivated triplet state of methylene blue transfers energy to O_2 and leads to singlet oxygen, which reacts with the side chains of several amino acids, including histidine, cysteine, methionine, and tryptophan. The functional importance of these amino acids could be assessed by methylene blue–sensitized photo-oxidation under appropriate conditions. Singlet oxygen is appearing with increasing frequency in biological processes, such as the actions of neutrophils.

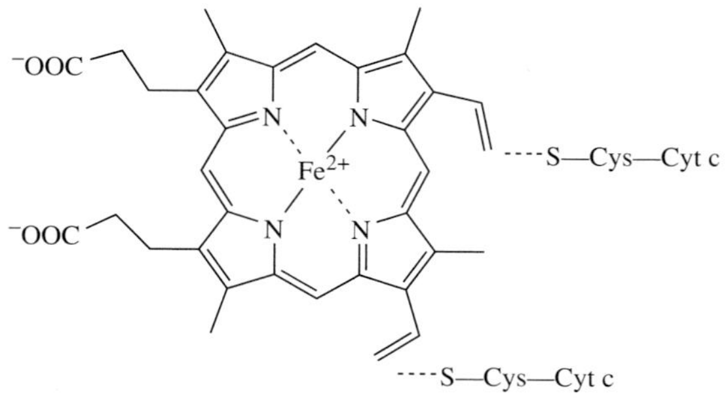

Heme (iron protoporphyrin IX)

Fig. 4-9. The structure of heme, iron protoporphyrin IX. Heme is the coenzyme of catalase, peroxidases, and cytochrome P450 monooxygenases. It is also the oxygen-binding prosthetic group of hemoglobin and the electron acceptor/donor in electron-transferring cytochromes. Heme is generally tightly but noncovalently bound to proteins. However, in cytochrome *c*, it is covalently bonded through its two vinyl groups to cysteine residues of the protein.

Structure and Chemistry of Heme

Figure 4-9 shows the structure of iron protoporphyrin IX, heme, which is found in hemoglobin, catalase, peroxidases, and cytochrome P450. In most cases, this molecule is tightly but noncovalently bound to its cognate enzyme. It is covalently bonded through a vinyl group to sulfur of cysteine residues in a few cases. Heme differs from the corrin system of B_{12} in several ways, including that it is more highly oxidized and more electronically conjugated.

Heme bound to hemoglobin, catalase, peroxidases, and cytochrome P450s displays markedly different chemical properties, and this is brought about by the different microenvironments of the heme binding sites. When bound to a protein, heme acquires new ligands to iron in the axial positions. Iron normally binds six ligands in an octahedral coordination sphere: the four nitrogens in the protoporphyrin ring and two axial ligands. Several functional groups in amino acids can serve as axial ligands to iron, including the side chains of histidine, cysteine, tyrosine, and methionine. The diverse chemical reactivities of heme enzymes depend to a significant degree on which amino acid side chains serve as axial ligands.

The extensive conjugation in heme allows it to participate in one-electron transfer processes through the formation of a highly delocalized cationic radical known as the π-cation radical. This property also allows heme to engage in fast electron transfer with the indole ring of nearby tryptophanyl residues. These properties of heme and tryptophan are put to use in heme enzymes.

Heme exists in three oxidation states of iron: ferrous (Fe^{2+}), ferric (Fe^{3+}), and ferrate (or ferryl, Fe^{4+}). Heme is in the ferrous state in hemoglobin and binds O_2 reversibly. Because of the diradical nature of O_2, its electronic interaction with hemoglobin allows its ligation to iron to be represented as the resonance state $Fe^{2+}-O_2 \leftrightarrow Fe^{3+}-O_2^{-}\bullet$, and it may be regarded as a ferric superoxide. Heme enzymes contain ferric heme in their resting states. The iron in heme bound to proteins resides in octahedral coordination environments and exists in high spin and low spin states in both the ferrous and ferric oxidation states. The electronic energy level diagrams in fig. 4-10 represent the spin states of iron in heme proteins. The d_{z^2} orbitals point toward the axial ligands, the $d_{x^2-y^2}$ orbital points toward the heme nitrogen ligands, and the d_{xy}, d_{yz}, and d_{xz} orbitals point into spaces between the six ligands. The ligand field splitting energy separates the e_g levels from the t_{2g} levels.

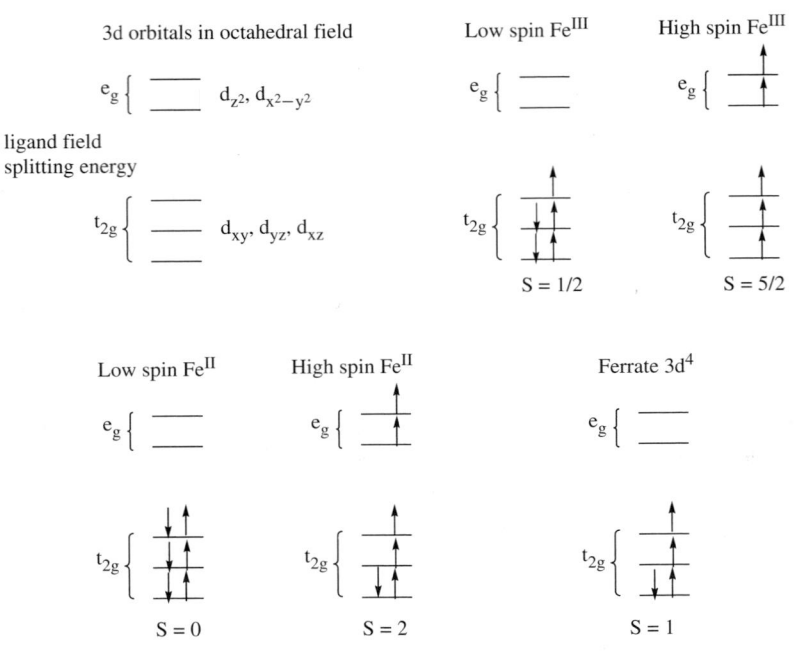

Fig. 4-10. Spin states of iron in heme.

The electrons are all paired in the low spin ferrous state, with six d electrons ($3d^6$), so that it has no spin and is diamagnetic and EPR-silent. The high spin ferrous state has four unpaired electrons, which because of being even in number often do not give an EPR signal. Both low and high spin ferric states give EPR signals, and these often give information about the axial ligands. The ferrate (Fe^{4+}) state often appears as an intermediate known as compound I in enzymatic oxygenation reactions. All known species have two unpaired electrons and $S = 1$.

The designation of formal positive charges on metal ions in coordination complexes can be confusing if they are viewed as point charges. To avoid confusion, the metal ions are often shown with oxidation states expressed as Roman numerals, such as Fe^{II} for ferrous heme and Fe^{III} for ferric heme. We adopt this notation in the following discussion.

Heme Enzymes

Catalase

Hydrogen peroxide produced in reactions of flavoprotein oxidases presents a detoxification challenge in cells. Catalase (EC 1.11.1.6) plays an important role in detoxifying hydrogen peroxide, and it does so by allowing for the recovery of one half of the oxygen that went into its production. Catalase catalyzes the dismutation of hydrogen peroxide according to eq. 4-14.

$$2 \, H_2O_2 \rightarrow O_2 + 2 \, H_2O \qquad (4\text{-}14)$$

In the structure of catalase, shown in fig. 4-11, Tyr358 contributes the lower axial ligand in heme, which is deeply buried within the enzyme structure. The sheltered environment above heme nevertheless allows access of water and hydrogen peroxide to the upper axial position. In the first step of the current working hypothesis for the mechanism, depicted in fig. 4-12, hydroperoxide replaces water as the upper axial ligand, and in step 2 an enzymatic

Coenzymes II: Metallic Coenzymes 205

base abstracts a proton. This strengthens the coordination bond to iron. In step 3, hydroperoxide undergoes the key step, dehydration of iron(III) peroxide to compound I. This requires acid catalysis by the enzyme and a two-electron transfer from heme because water departs with the two bonding electrons in hydroperoxide. One of the two electrons extracted from heme comes from iron and leads to the ferrate state and the second from the porphyrin

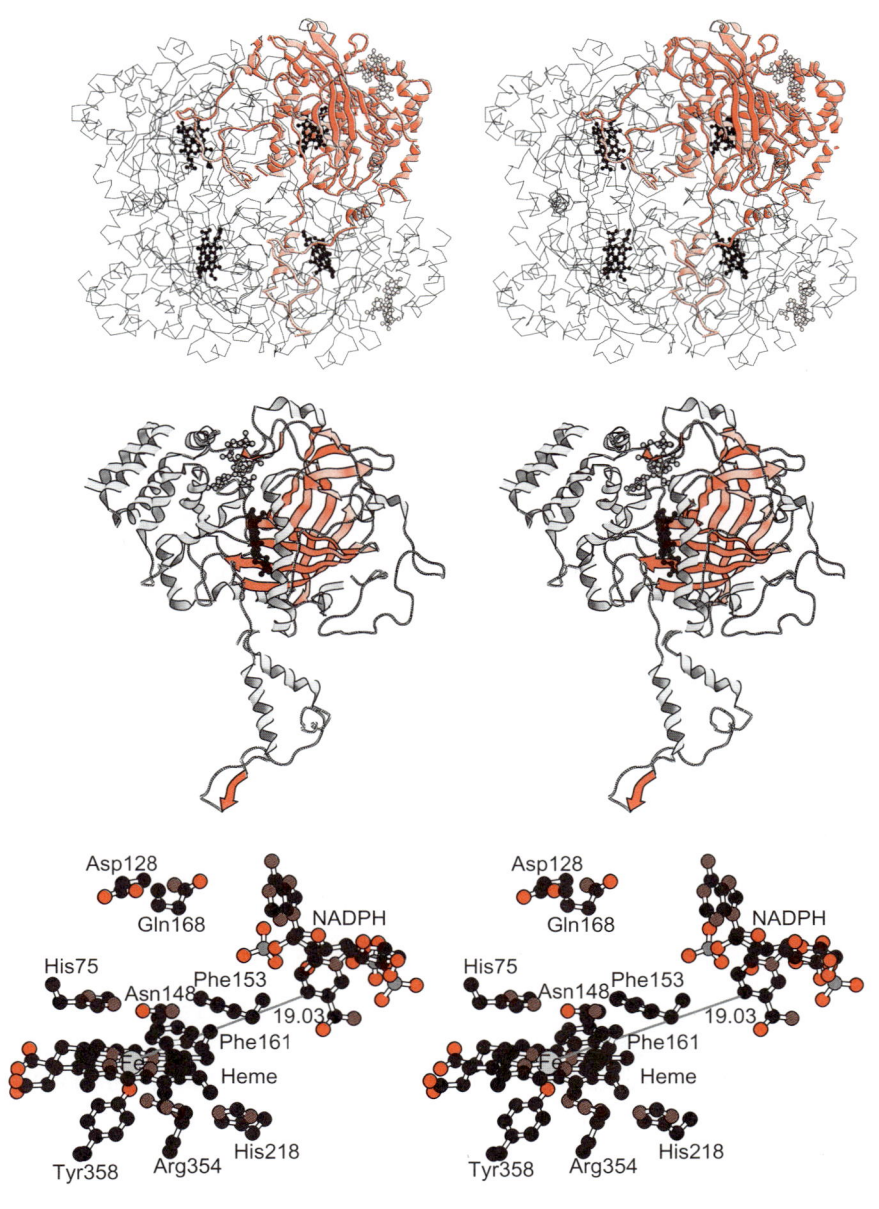

Fig. 4-11. One subunit of the tetrameric structure of catalase is shown as a ribbon diagram in red (top). The four black ball-and-stick models are hemes in each subunit. Two of the four subunits have NADH molecules indicated by pink ball-and-stick models. The structure of one subunit is shown in stereo (middle), with a view along the edge of the heme shown as a black ball-and-stick model. NADPH in this structure is shown as a gray-scale ball-and-stick model at the surface. Notice that heme is deeply buried in the protein. A stereoview (bottom) shows the active site as a ball-and-stick model. This view also shows the amino acid contacts with the heme and the 19-Å separation between the heme iron and C4 of NADH. The figures were generated using PDB 1DGF (Putnam et al., 2000).

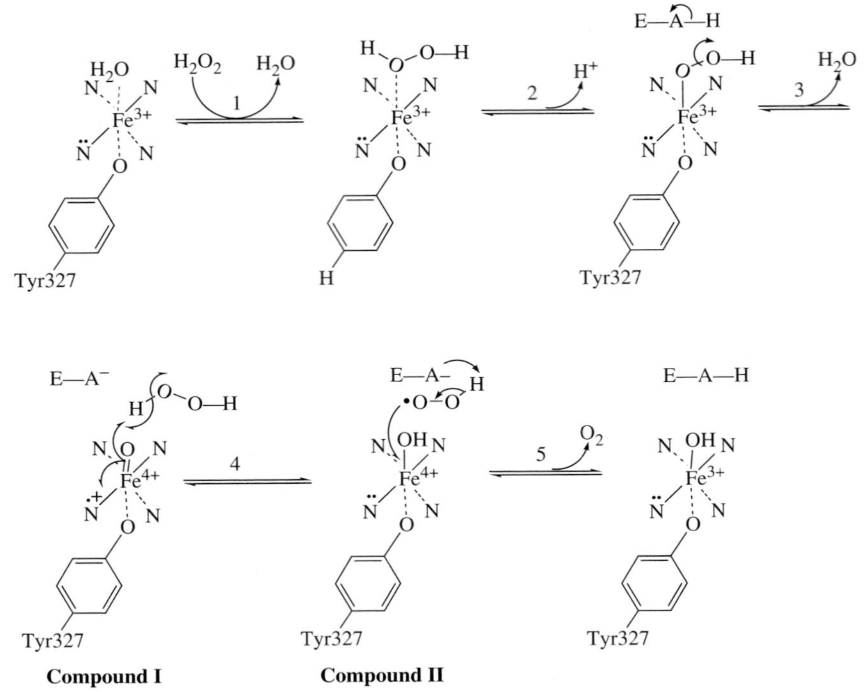

Fig. 4-12. The mechanism of dismutation of hydrogen peroxide by catalase begins with the exchange of hydrogen peroxide for water in the upper axial ligand and proceeds in five essential steps. Key process are the cleavage of the O—O bond in step 3 to form compound I, hydrogen abstraction from hydrogen peroxide by compound I in step 4, and electron transfer in step 5.

ring. Consequently, compound I consists of an Fe^{4+}-oxide and a π-cation radical in the porphyrin ring. This is the key species that also participates with alternative axial ligands in other heme enzyme reactions.

Compound I is a strong oxidant and may react in hydrogen atom abstraction. In step 4, compound I is postulated to abstract a hydrogen atom from the second molecule of hydrogen peroxide to produce the ferrate hydroxide and the hydroperoxide radical (hydrogen superoxide) and to discharge the porphyrin p-cation radical. Proton transfer to an enzymatic base and electron transfer to ferrate transforms the hydroperoxide radical to oxygen in step 5. The resulting ferric hydroxide can be restored to its starting aquo form by accepting a proton.

The absorption spectra of catalase in its resting and compound I and compound II states between 375 nm and 460 nm in fig. 4-13A show how the intermediates in the action of catalase can be distinguished as transient species. Reaction of catalase with one equivalent of hydrogen peroxide produces compound I, and reduction by one electron equivalent produces compound II. These species of catalase also display different features at longer wavelengths that further characterize them. The production of $^{18}O_2$ from $H_2^{18}O_2$ further supports the mechanistic hypothesis.

Peroxidases

Another way to detoxify peroxides is to reduce them. In the case of alkyl hydroperoxides, cytochrome *c* peroxidases catalyze their reduction to the corresponding alcohols and

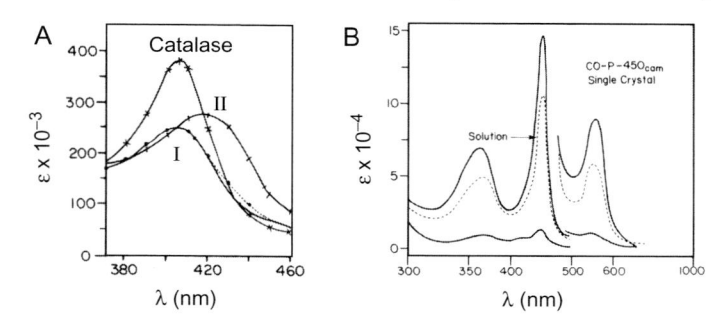

Fig. 4-13. The absorption spectra of catalase intermediates and cytochrome P450. (A) The absorption spectra of catalase between 375 nm and 460 nm in its resting form and two of its intermediates allow portions of the reaction mechanism to be traced. The curves are labeled for the resting enzyme (Catalase), and for compounds I and II. (B) The absorption spectrum of cytochrome P450 cam after reaction with CO. P450 in solution is shown with a continuous line, and in the crystal, it is shown in with a dashed line. (A, Adapted from Chance, 1952.)

water, using cytochrome c as the reducing agent. The overall reaction can be written as eq. 4-15, in which ferrocytochrome c is oxidized to ferricytochrome c.

$$ROOH + 2\ Cyt\ c(Fe^{2+}) + 2H^+ \rightarrow ROH + 2\ Cyt\ c(Fe^{3+}) + H_2O \qquad (4\text{-}15)$$

Histidine 174 occupies the lower axial ligand position of heme in cytochrome c peroxidase. The reaction mechanism proceeds initially as in catalase but diverges quickly. As shown in fig. 4-14, cleavage of the alkyl hydroperoxy heme ligand leads to a ferrate oxide as in compound I, but with the cation radical located in the indole ring of a nearby tryptophan residue instead of the heme ring. In a further divergence, this species undergoes reduction in two steps by ferrocytochrome c instead of reacting with a second molecule of peroxide. The overall reaction then consists of a net reduction of the alkyl hydroperoxide instead of a dismutation, as in the reaction of catalase.

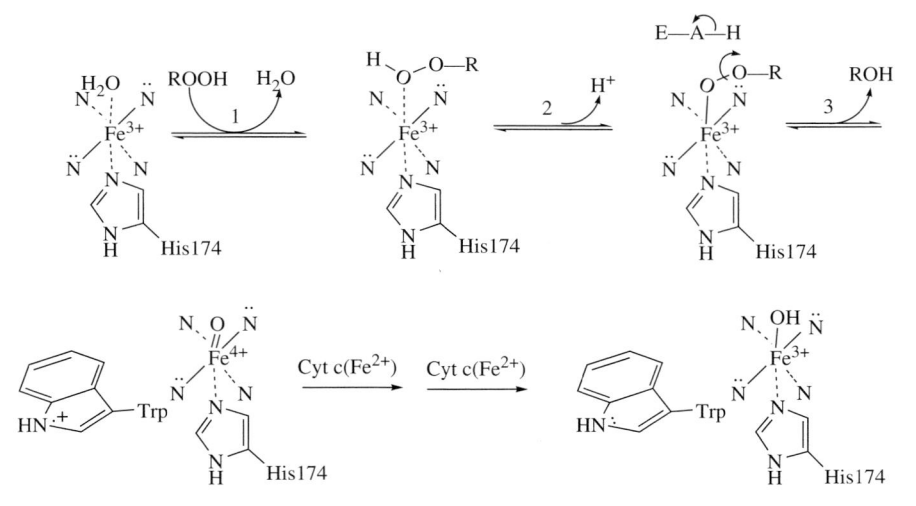

Fig. 4-14. A mechanism for alkyl hydroperoxide reduction by cytochrome c peroxidase is analogous to that of catalase to the formation of the ferryl oxy intermediate in step 3. This intermediate is then reduced in two steps by ferrocytochrome c.

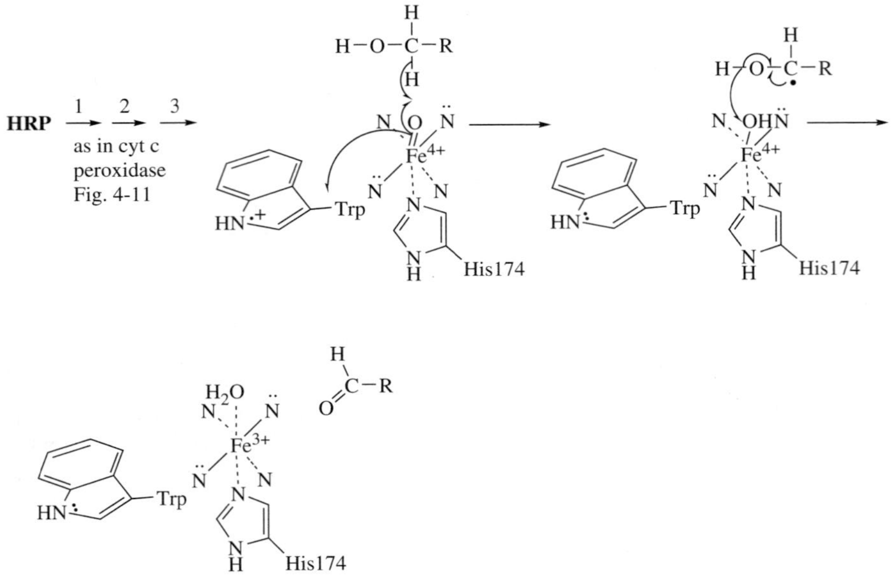

Fig. 4-15. A mechanism for the conversion of an alkyl hydroperoxide to the aldehyde by horseradish peroxidase. The reaction of horseradish peroxidase is analogous to those of catalase and cytochrome c peroxidase to the point of the ferryl oxy intermediate. It diverges in the reaction of this intermediate, which oxidizes a substrate alcohol to an aldehyde. The hypothetical mechanism shown is compatible with the oxygen labeling pattern.

Horseradish peroxidase (HRP) catalyzes the internal dismutation of alkyl hydroperoxides to aldehydes according to eq. 4-16.

$$RCH_2OOH \rightarrow RCHO + H_2O \qquad (4\text{-}16)$$

The mechanism follows a similar course as in cytochrome c peroxidase as far as the transformation to the alcohol and compound I. The alcohol is held within the site and undergoes oxidation to the aldehyde. A reasonable mechanism is that in fig. 4-15, where compound I abstracts a hydrogen atom to form the hydroxyalkyl radical and compound II, which then donates another hydrogen atom from the OH group of the hydroxyalkyl radical to form the aldehyde and heme.

Cytochrome P450s

The cytochrome P450 monooxygenases are the most thoroughly studied heme enzymes (Ortiz de Montellano, 1986; Sono et al., 1996). They derive their name from the 450-nm maximum in their absorption spectra when CO is added to the reduced protein and occupies the upper axial ligand of iron. Fig. 4-13b shows the spectrum of P450 cam. P450s function in coupling with cytochrome P450 reductases, which provide the reducing equivalents for the monooxygenation of substrates by oxygen. Equation 4-17 describes the overall stoichiometry in the oxygenation of a substrate by a cytochrome P450 and its associated reductase, in which R– is an organic group.

$$R\text{—}H + O_2 + NAD(P)H + H^+ \rightarrow R\text{—}OH + H_2O + NAD^+ \qquad (4\text{-}17)$$

P450s catalyze oxygenations in the biosynthesis of steroids, detoxification of xenobiotics, and catabolism of alkanes in bacteria. Most P450s are membrane-bound enzymes and difficult to study mechanistically. For this reason, the most intensive mechanistic work has been carried out on cytochrome P450cam (for camphor), a soluble bacterial enzyme, which is discussed in chapter 17. As in catalase and peroxidases, a ferrate cation radical species analogous to compound I plays a key mechanistic role.

Oxygen Binding and Electron Transfer

The most extensively studied heme proteins are not enzymes. The first three-dimensional structural model of a protein to be determined by x-ray crystallography was that of myoglobin, and the structure of hemoglobin followed soon afterward. Many structures of cytochrome *c* have also appeared. Each of these proteins binds heme in a unique way suited to its biological function.

Myoglobin and Hemoglobin

Myoglobin and hemoglobin regulate the supply of oxygen in tissues and blood by binding it reversibly. Hemoglobin delivers oxygen to tissues through the bloodstream, and myoglobin stores it within tissues and makes it available for metabolism as needed. Figure 4-16 shows the locations of heme in the structures of myoglobin and hemoglobin. Both proteins contain Fe^{2+}-heme and bind oxygen reversibly. Histidine residues ligate iron with the lower axial histidine tightly bound to iron and the upper axial histidine at long range, allowing for space to bind oxygen. Available structural evidence indicates that these proteins bind oxygen as a ferric superoxide complex. Although both ferric iron and superoxide are paramagnetic, the complex does not give an EPR signal, presumably because of antiferromagnetic coupling. The tetrameric structure of hemoglobin ($\alpha_2\beta_2$) allows it to bind oxygen cooperatively, allowing the binding and dissociation of oxygen to be maintained in a narrow range of oxygen partial pressure.

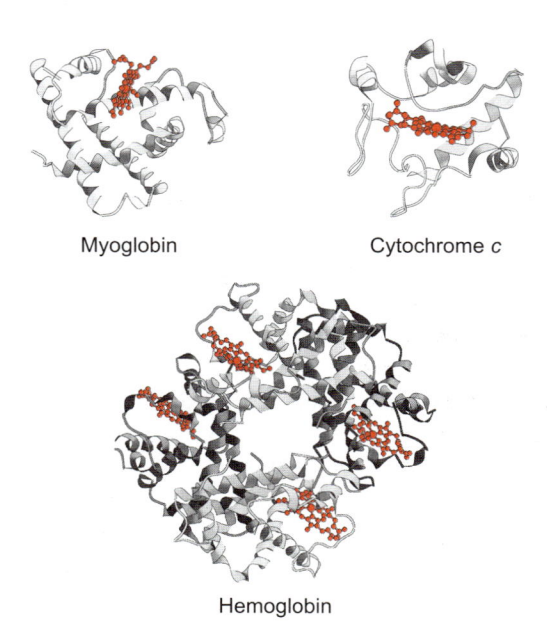

Fig. 4-16. The structures of sperm whale myoglobin (PDB 1A6M), albacore tuna heart cytochrome *c* (PDB 5CYT), and human hemoglobin A (PDB 1DSH) are shown with heme moieties as red ball-and-stick models (Kavanaugh et al., 2005; Takano and Dickerson, 1981; Vojtechovsky et al., 1999).

Cytochrome c

Cytochrome c is a small protein that facilitates electron transfer through its own reversible reduction and oxidation between the Fe^{3+}/Fe^{3+} states. Equation 4-15 for the reaction of cytochrome c peroxidase exemplifies the biological role of cytochrome c in one half of an electron transfer process. In keeping with its electron transfer function, heme in cytochrome c has upper and lower axial ligands: methionine in the upper and histidine in the lower position. The structure of a cytochrome c illustrated in fig. 4-16 shows the position of heme in the molecule.

Mononuclear Nonheme Iron

Many enzymes contain single ferrous or ferric iron at their active sites that participate directly in catalysis and provide the chemistry required for the biological transformation. The special catalytic properties of multi-iron coenzymatic sites are considered in later sections. The chemistry of these metalloenzymes may often be related to that of heme enzymes. However, the chemical frameworks of cobalamin and heme enzymes are substantially determined by their macrocyclic ligands, with modulation by axial protein ligands in heme enzymes. In mononuclear or dinuclear nonheme iron enzymes, amino acid side chains dominate the ligand sphere. The placement of protein ligands may be dictated or substantially influenced by the protein structure external to the metal ion itself. This allows protein-induced steric distortion of the geometry typical of ligand fields, and these distortions can significantly alter chemical properties, as for example the reduction potentials of metal ions. Structural distortions in metal ligand fields are known as *entatic* states (Vallee and Williams, 1968). These states are thought to enhance the chemistry of metal ions in biological catalysis.

Functional groups in amino acid side chains serving as ligands to iron can be chemically related to heme ligands, in particular the imidazole ring of histidine. Many enzymes contain nonheme iron in ligand environments somewhat analogous to heme enzymes, including the mononuclear iron oxygenases. The oxidation states, spin states, and electronic configurations of iron are similar or analogous to those of heme enzymes.

The primary experimental information on the structures of metal sites in enzymes arises from x-ray crystallography. The resolution of x-ray crystal structures allows the overall shape of a ligand sphere to be defined. However, details of bond lengths, exact angles, and oxidation states require higher resolution information, and this information usually comes from spectroscopy. A large number of optical and magnetic spectroscopies are typically employed to extract detailed information about metal ion sites in proteins. These include EPR and pulsed-EPR methods, such as electron nuclear double resonance (ENDOR) and electron spin-echo envelope modulation (ESEEM); circular dichroism (CD); magnetic circular dichroism (MCD); resonance Raman, Mössbauer, and x-ray absorption spectroscopies (XAS), including extended x-ray absorption fine structure (EXAFS) and x-ray absorption near-edge spectroscopies (XANES). Information obtained by these methods is mentioned in the following sections, but detailed explanations of the methods themselves are beyond the scope of this book. Brief explanations of the capabilities and applications of these methods can be found in the review literature (Holm et al., 1996).

Monooxygenases

Mixed-function oxygenases incorporate one atom of oxygen from O_2 into the product and require a reducing agent to reduce the other oxygen atom to water. A mononuclear iron enzyme of this type uses tetrahydrobiopterin or α-ketoglutarate as the reducing agent. We first consider a tetrahydrobiopterin-dependent monooxygenase.

Tetrahydrobiopterin (BH₄)

Phenylalanine Hydroxylase

Phenylalanine hydroxylase (PAH; EC 1.14.16.1) and the pterin-dependent tyrosine hydroxylase have been studied intensively for more than 40 years, and the mechanism of catalysis has been controversial during most of that time (Cappock and Caradonna, 1996). Tryptophan hydroxylase has not been studied as intensively but is of increasing interest in this field. These enzymes catalyze important biosynthetic reactions; the production of tyrosine by PAH; the production of 3,4-dihydroxyphenylalanine (L-DOPA) by tyrosine hydroxylase; and the production of 5-hydroxytryptophan, the precursor of serotonin, by tryptophan hydroxylase. Phenylketonuria (PKU) in humans is brought about by an inherited deficiency in PAH. L-DOPA and serotonin are important neurotransmitters, and L-DOPA is used in the treatment of Parkinson's disease. Early studies of phenylalanine hydroxylase led to the original discovery of the NIH shift (Guroff et al., 1967), which has since been observed in other hydroxylation reactions of aromatic rings. The isomerization in the NIH shift implies the participation of a carbocationic intermediate in the hydroxylation of an aromatic ring.

An early focus of controversy was the mechanistic role of tetrahydrobiopterin, and the participation of metal ions has also been controversial and sometimes uncertain. Advances have led to consensus views of the roles of tetrahydrobiopterin and iron in hydroxylation.

Eukaryotic phenylalanine hydroxylases display regulatory properties and have regulatory domains in their structures. Regulatory behavior introduces ambiguities and complexities into mechanistic studies (Kappock and Caradonna, 1996). The 32-kDa phenylalanine hydroxylase from *Chromobacterium violaceum* lacks the N-terminal regulatory domain found in liver PAHs, but the amino acid sequence is more than 35% identical to the C-terminal sequence of liver PAH, which includes the catalytic domain. The highly conserved His138 and His143 are essential for catalysis, as demonstrated by site-directed mutagenesis and kinetic assays (Balasubramanian et al., 1994). The structure of the homologous human PAH in fig. 4-17 shows a 2-His 1-carboxylate facial triad of iron ligands. This ligand motif is typical of mononuclear iron oxygenases (Bugg, 2001).

PAH catalyzes the hydroxylation of phenylalanine by oxygen and BH₄ as illustrated in fig. 4-18. The reaction leads to the cleavage of oxygen to incorporate one oxygen atom into phenylalanine to form tyrosine and the other into pterin 4a-carbinolamine, which is then recycled to BH₄ by the actions of a dehydratase and pteridine reductase. In vitro, the recycling of pterin can be short-circuited by use of 5,6-dimethyltetrahydrobiopterin (DMBH₄) in place of BH₄. Reaction of this pterin produces the quinonoid form of the dihydropterin, and it is rapidly reduced by DTT to regenerate DMBH₄, allowing for its recycling and catalytic function. The optical spectra of the pterin products can be exploited in assays of PAH activity. In the reaction of DMBH₄ the 4α-carbinolamine is dehydrated rapidly, and the formation of quinonoid-DMBH₂ can be observed as an increase in A_{347} and serves an accurate measure of the rate (Ayling et al., 1973). Alternatively, quinonoid-DMBH₂ can be recycled by reduction with DTT to DMBH₄, and the rate of tyrosine production can be

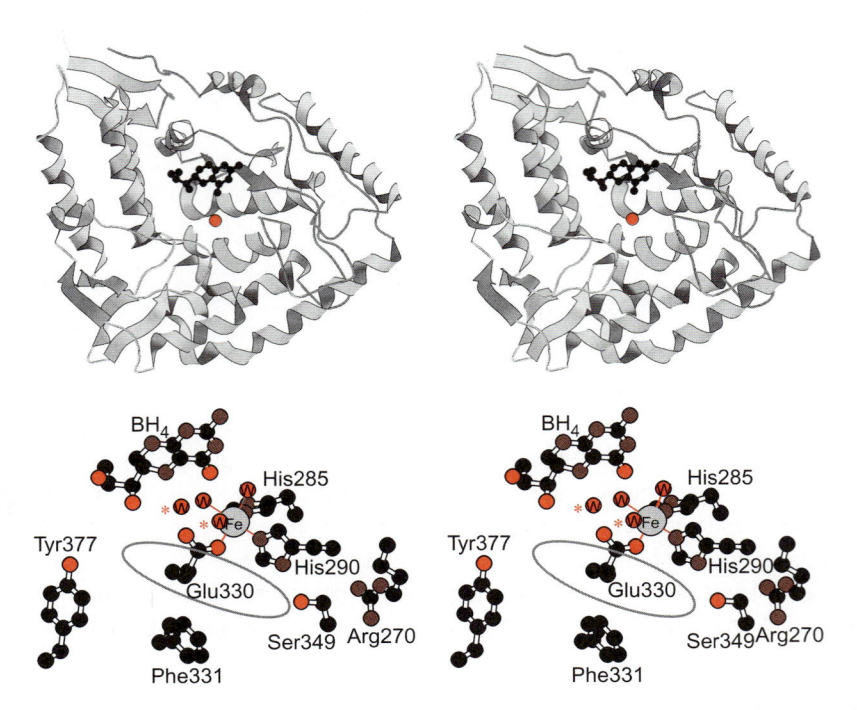

Fig. 4-17. The stereodiagram (top) shows the ribbon diagram of the structure of human phenylalanine hydroxylase, with the location of iron shown as a red ball and tetrahydrobiopterin (BH$_4$) as a black ball-and-stick model. The protein is largely helical, with three small β-leaflets, one near the BH$_4$ binding site. In the stereoview (bottom) of the active site, a ball-and-stick model has a gray circle indicating the vacant phenylalanine binding site. Mutations in residues Ser349, Tyr377, Phe331, or Arg270, adjacent to the Phe binding pocket, have all been implicated in phenylketonuria. The iron is bound in a 2-His 1-carboxylate facial triad by His285, His290, and Glu330. The tetrahydrobiopterin is near enough to iron to react with it directly at some point in the mechanism. The two water atoms marked by asterisks show the approximate dioxygen binding pocket predicted by modeling. This figure was prepared using PDB 1J8U (Anderson et al., 2001) and by referring to additional modeling (Teigen et al., 1999).

Fig. 4-18. Biopterin transformations in the hydroxylation of phenylalanine to tyrosine. The reaction of phenylalanine hydroxylase (PAH) leads to the oxidation of tetrahydrobiopterin (BH$_4$) to pterin 4a-carbinolamine according to the stoichiometry indicated at the top of the figure. Pterin 4a-carbinolamine is converted back into BH$_4$ by the actions of two enzymes, a dehydratase that produces the quinonoid dihydropterin and dihydropteridine reductase, a member of the short-chain dehydrogenase superfamily.

measured by any of a variety of methods. The choice of assays can reveal different aspects of catalysis, namely the oxidation of BH_4 and the oxygenation of phenylalanine. The steady-state kinetic mechanism of *C. violaceum* is reported to be sequential with partially ordered binding and oxygen binding first (Pember et al., 1989). Liver PAH also functions by a sequential kinetic mechanism (Kaufmann, 1962).

Detailed mechanistic studies of liver PAH have been plagued by observations of partially uncoupled hydroxylation and consequent hydrogen peroxide production (Kappock and Caradonna, 1996). Systematic studies of the *C. violaceum* PAH clarified issues related to uncoupling and the roles of tetrahydrobiopterin and iron in the hydroxylation of phenylalanine. *C. violaceum* PAH freed of iron and other transition metals catalyzes the phenylalanine-dependent oxidation of $DMBH_4$ by oxygen according to eq. 4-18 (Chen and Frey, 1998).

$$DMPH_4 + O_2 \; Quinonoid\text{-}DMBH_2 + H_2O_2 \qquad (4\text{-}18)$$

Although phenylalanine is absolutely required in the reaction of eq. 4-18, it is not hydroxylated or consumed by the reaction. The binding of phenylalanine at the active site presumably maintains the enzyme conformation in which $DMBH_4$ reacts with oxygen. Iron is not required for this reaction, showing that oxygen reacts directly with $DMBH_4$ at the active site. The value of k_{cat} for PAH in hydrogen peroxide production is about 1 s^{-1} with air as the source of oxygen.

Hydroxylation of phenylalanine by *C. violaceum* PAH requires ferrous iron in addition to $DMPH_4$ and proceeds by eq. 4-19 with no production of hydrogen peroxide.

$$Phenylalanine + DMPH_4 + O_2 \rightarrow Tyrosine + Quinonoid\text{-}DMPH_2 + H_2O \quad (4\text{-}19)$$

The turnover number for the fully coupled hydroxylation of phenylalanine in the air is $k_{cat} = 15 \text{ s}^{-1}$. Inasmuch as oxygen reacts directly with $DMPH_4$ at the active site in the absence of iron and is reduced to hydrogen peroxide, iron may play an indirect role in the oxidation of $DMPH_4$. However, the absolute requirement for iron in the hydroxylation of phenylalanine suggests that an iron-oxy species is the proximate oxygenating agent.

Hydrogen peroxide production in the absence of iron suggests that oxygen reacts with $DMPH_4$ to produce the 4a-hydroperoxide, which can be released from the active site and eliminate hydrogen peroxide. This reaction as outlined in scheme 4-4 would be similar and analogous to that of $FADH_2$ or of $FMNH_2$ with oxygen. The fact that this reaction is slower than the iron-dependent hydroxylation of phenylalanine proves that hydrogen peroxide production does not take place in the hydroxylation mechanism and is a side reaction, as expected for an uncoupled process. Iron is not required for this process, although it might facilitate it in some way, such as by initially binding oxygen in the active site. The structure of PAH in fig. 4-17 shows that iron is bound adjacent to the pterin site.

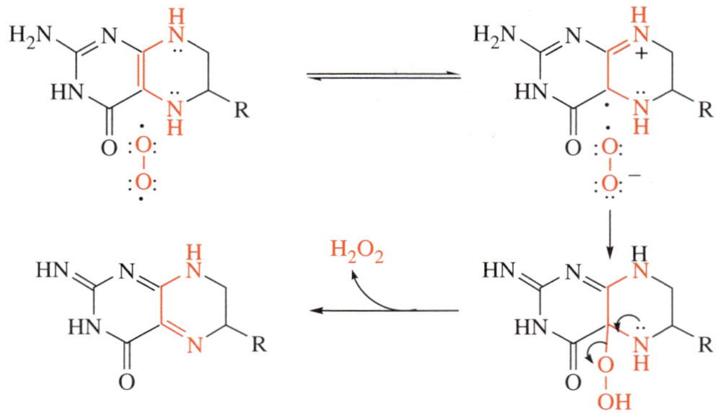

Scheme 4-4

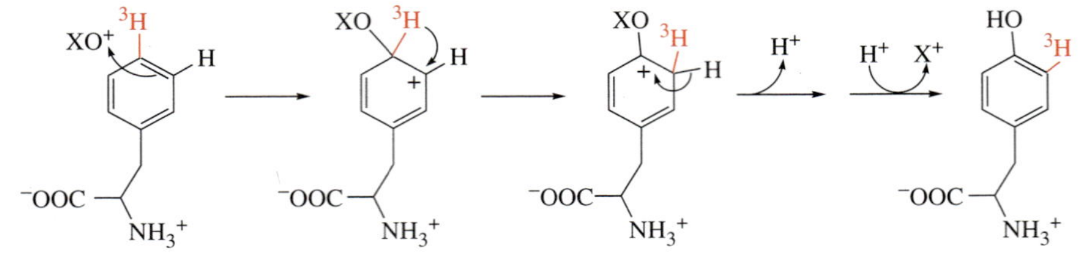

214   Enzymatic Reaction Mechanisms

Fig. 4-19. The NIH shift in the hydroxylation of phenylalanine by phenylalanine hydroxylase (PAH). The overall retention of tritium from [para-³H]phenylalanine in tyrosine resulting from hydroxylation by PAH is explained by a mechanism involving electrophilic aromatic substitution by an electrophilic oxygenating agent, represented by XO⁺. The true oxygenating agent is likely to be a ferryl oxy species at the active site. The initial carbocationic adduct in the electrophilic substitution is subject to the well-known 1,2-hydride shift in such species and leads to a second carbocation with the tritium in the *meta* position. Re-aromatization occurs by loss of ³H⁺ or H⁺ from the *meta* position, and H⁺ is preferentially lost because of the tritium kinetic isotope effect.

As in all PAHs, tetrahydropterin alone does not support hydroxylation of phenylalanine, and iron is required for this process. It is likely that iron reacts with some oxygenated species in an essential part of the hydroxylation mechanism. The sequence of events in hydroxylation is not known; however, it is possible to construct a logical mechanism based on available facts.

The NIH shift was an early observation in the reaction of PAH that gave fundamental information about the mechanism and influenced thinking about all enzymatic aromatic ring hydroxylation reactions. In the NIH shift experiment, the PAH-catalyzed hydroxylation of [*para*-³H]phenylalanine did not release the theoretical amount of tritium into water (Guroff et al., 1967). Instead, much of the tritium (>80%) was retained in the tyrosine purified from the reaction. This result implicated electrophilic aromatic substitution as the basic mechanism for hydroxylation of tyrosine by indicating the most obvious mechanism for tritium retention in the product. Tritium retention can occur through a 1,2-hydride shift in a carbocationic intermediate, as illustrated in fig. 4-19, where XO⁺ encodes the proximate oxygenating species as a highly electrophilic form of oxygen. 1,2-Hydride shifts are commonly observed in carbocationic intermediates in organic chemistry, and the observation of this shift in the action of PAH implicates such an intermediate.

The products of PAH-catalyzed hydroxylation of *p*-methylphenylalanine further support the participation of a carbocationic intermediate in aromatic ring hydroxylation. Both liver and *C. violaceum* PAH catalyze the hydroxylation of *p*-methylphenylalanine to a mixture of *m*-methyltyrosine and *p*-hydroxymethylphenylalanine, as shown in scheme 4-5 (Carr et al., 1995; Siegmund and Kaufmann, 1991). Migration of the methyl group to the *meta* position in 20% of the products implies a Meerwein-Pondorf 1,2-methyl migration characteristic of carbocationic intermediates.

6MPH₄ = 6-Methyltetrahydropterin

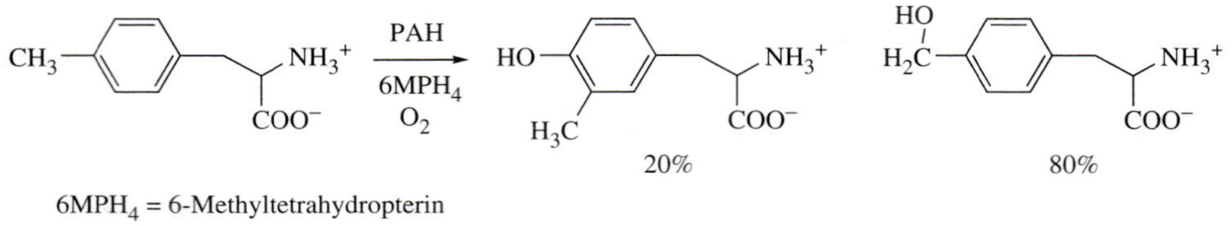

Scheme 4-5

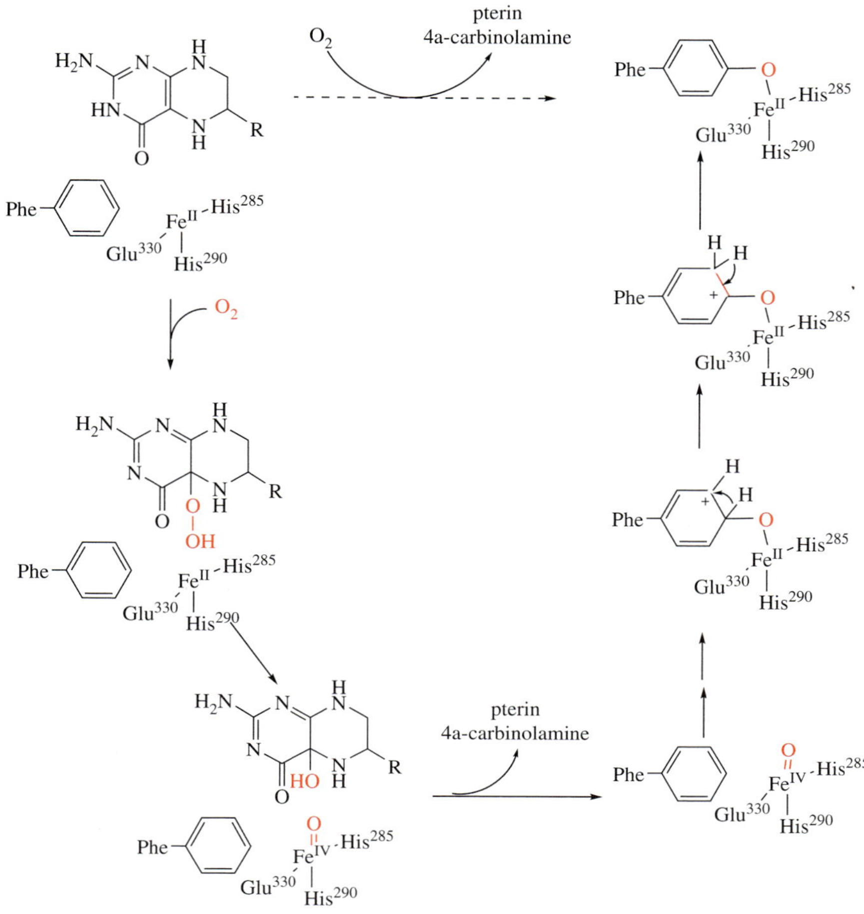

Fig. 4-20. A hypothetical mechanism for hydoxylation by phenylalanine hydroxylase (PAH). Experiments show that phenylalanine must be bound before tetrahydrobiopterin (BH$_4$) reacts with oxygen. The reaction mechanism proceeds counterclockwise, with the PAH/phenylalanine complex binding oxygen in the first step to undergo addition to BH$_4$ and form a bridged peroxy complex with ferrous iron. Heterolytic cleavage of the peroxy bridge generates the pterin 4a-carbinolamine and a ferryl oxy species. The nearby phenyl ring of phenylalanine undergoes electrophilic aromatic substitution with the oxygen of the ferryl oxy species to form a 3-carbenium ion, with the ferryl oxygen added to carbon 4. A 1,2-hydride shift of the 4-hydrogen atom to carbon 3 generates the 4-oxycarbenium ion, which can eliminate a proton from carbon 3 to form tyrosine. This is the same mechanism put forward for tyrosine hydroxylase (Fitzpatrick, 2003).

Hydroxylation of the methyl group in 80% of the products must follow a different mechanism similar to the hydroxylation of alkanes by cytochrome P450-dependent alkane monooxygenases (see chap.17). The ferryl oxy species in fig. 4-20 appears analogous to the oxygenating species in the alkane monooxygenases and evidently can carry out benzylic oxygenation, presumably by an analogous mechanism of hydrogen atom abstraction followed by oxygen rebound. The ferryl oxy species can abstract hydrogen atoms or undergo addition to double bonds.

Hydroxylation of *p*-[^{2}H$_3$]methylphenylalanine proceeds with a deuterium kinetic isotope effect of 8 for benzylic hydroxylation and a consequent reversal of the product ratio to 80% methyl migration and 20% benzylic hydroxylation (Siegmund and Kaufmann, 1991).

Much consideration has been given to the possible intermediate formation of arene oxide intermediates in enzymatic aromatic hydroxylation (Kappock and Caradonna, 1996). Consideration of arene oxides as possible intermediates was inspired by two experimental facts: the acid catalyzed hydrolysis of arene oxides was found to proceed with an NIH shift, and monooxygenases catalyze epoxide formation when presented with alkenes as substrates. However, no evidence for arene oxides as intermediates in enzymatic aromatic ring hydroxylation by tetrahydrobiopterin-dependent hydroxylases has been obtained. Moreover, it seems likely that epoxidation of alkenes represents an alternative, lower activation reaction for the highly electrophilic ferryl oxygenating species in the active sites of monooxygenases.

The PAH-catalyzed epoxidation of cyclohexadienylalanine (dihydrophenylalanine) to epoxycyclohexylalanine provides a clear example of the propensity for monooxygenases to effect epoxidation of alkenes. The illustration of a mechanism in scheme 4-6 shows how the option for aromatization of the key carbocationic intermediate by loss of a proton, as in fig. 4-20, is cut off by the dihydroaromatic ring, leaving epoxide formation as the best option. Re-aromatization in the last step of fig. 4-20 is a strong driving force for 1,2-hydride shift and loss of a proton in the carbocationic intermediate. However, aromatization is not possible for the intermediate carbocations in scheme 4-6. Epoxide formation is the remaining option in scheme 4-6.

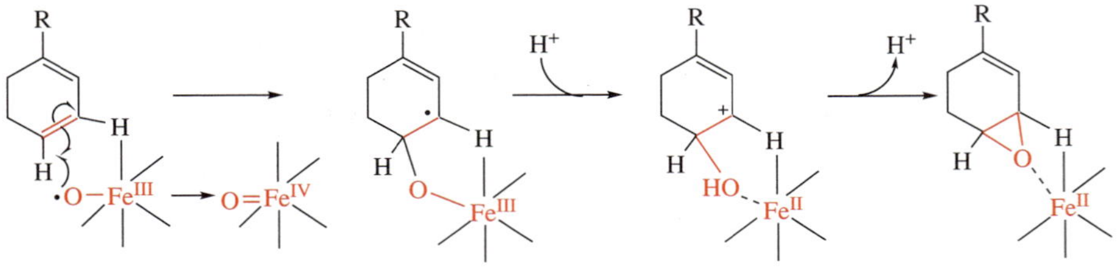

Scheme 4-6

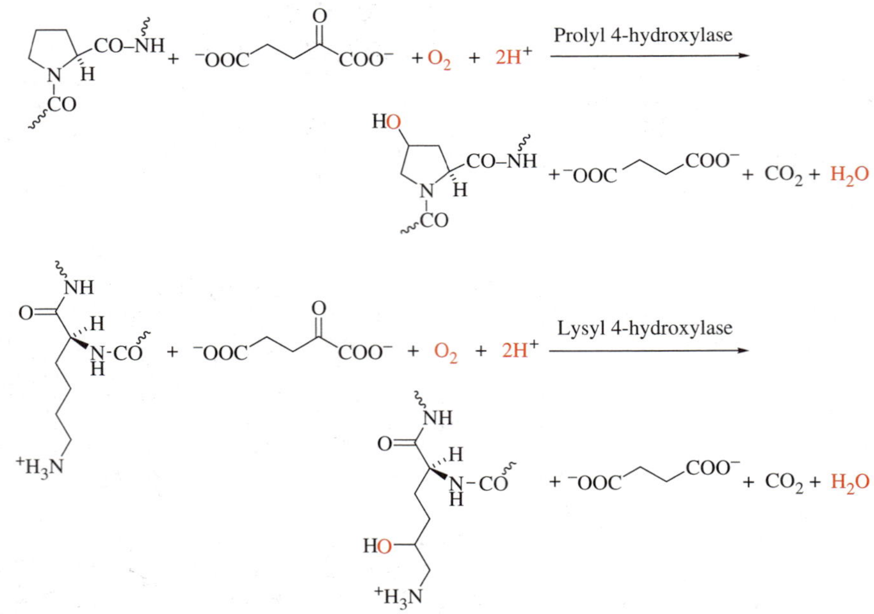

Fig. 4-21. Reactions of α-ketoglutarate–dependent monooxygenases.

α-Ketoglutarate as Reductant

Prolyl 4-hydroxylase and lysyl 5-hydroxylase are among the α-ketoglutarate-dependent, mononuclear iron monooxygenases. They catalyze the hydroxylation reactions shown in fig. 4-21, where α-ketoglutarate serves as the reducing agent and undergoes oxidation to CO_2 and succinate. Hydroxylations of prolyl and lysyl residues are important in collagen biosynthesis, and other α-ketoglutarate-dependent monooxygenases hydroxylate nucleotides and antibiotics. The active sites of the enzymes in this class bind iron in 2-His 1-carboxylate facial triad motifs. We discuss several examples of these reactions in chapter 17.

Dioxygenases

The descriptive term dioxygenase means that the enzyme incorporates both atoms of oxygen into the product. This stands in contrast to monooxygenases such as cytochrome P450s, which incorporate one oxygen atom into the product and reduce the other to water. The dioxygenases often but not always incorporate both atoms of oxygen into an aromatic ring, sometimes with cleavage of the ring. They include bacterial enzymes that function to degrade aromatic compounds to common metabolites.

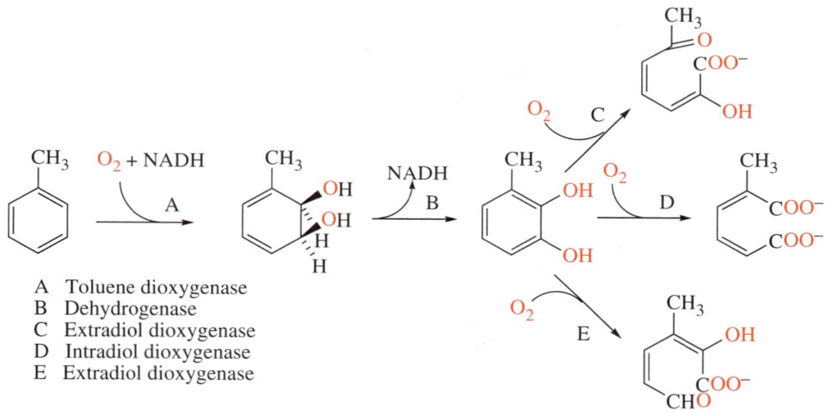

A Toluene dioxygenase
B Dehydrogenase
C Extradiol dioxygenase
D Intradiol dioxygenase
E Extradiol dioxygenase

Scheme 4-7

Scheme 4-7 illustrates the bacterial degradation of toluene by dioxygenases (Dagley, 1987). The process begins with the addition of oxygen to a double bond in an NADH-dependent reaction. Dehydrogenation of the product generates the catechol, which is a substrate for mononuclear iron dioxygenases. These enzymes add two additional oxygen atoms and cleave the aromatic ring on either side of the diol (extradiol cleavage) or between the hydroxyl groups (intradiol cleavage) (Lipscomb and Orville, 1992). Protocatechuate dioxygenases are discussed in chapter 17.

Oxo-Fe$_2$ Complexes

A number of enzymes carry out complex and difficult reactions with the aid of oxo di-iron cofactors. The well-known di-iron proteins include hemerythrin, which binds oxygen reversibly in worms, methane monooxygenase, fatty acyl desaturases, toluene monooxygenase,

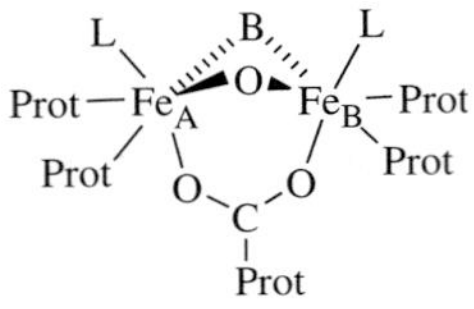

Oxo-Fe₂ complexes

Fig. 4-22. Structures of oxo-Fe₂ complexes in proteins. The generalized structure may include a bridging carboxyl group, a bridging oxygen atom or molecule, and protein (Prot) ligands (e.g., Asp, Glu, His, Tyr, Asn). External ligands (L) may include water, O_2, or HO^-.

and purple acid phosphatase. The similarities of many of these reactions to those of heme enzymes are striking, and mechanistic thought has often been inspired and guided by these similarities. However, the mechanism by which oxygen activates the di-iron complexes in preparation for interactions with substrates is unique to this class of coenzymes. Purple acid phosphatase differs in that it is not engaged in oxygen metabolism but uses a di-iron complex to hydrolyze phosphate esters.

Structures

The generalized structural formula for di-iron complexes in fig. 4-22 includes protein ligands, one of which is a bridging carboxyl group; a bridging oxygen atom or molecule; and other protein ligands, especially the carboxylate groups of glutamate and aspartate and imidazole ring of histidine. The complexes exist in the several oxidation states made possible by the presence of two irons, and the oxidation states vary with the function of the complexes. The oxidation states include diferric [Fe(III)Fe(III)], mixed valence [Fe(II)Fe(III)], and diferrous [Fe(II)Fe(II)]. The diferrous state activates oxygen in the oxygenases and desaturases and is not relevant to the purple acid phosphatase. The mixed-valence state is irrelevant to the activation of oxygen but is the functional state in purple acid phosphatases. Di-iron centers in their resting or diferric states display little structure in their optical spectra, with broad maxima in the 300 to 400 nm region due to ligand-metal charge transfer complexation. The purple acid phosphatase has a tyrosine phenolate ligand to ferric iron, which displays the typical 500 nm band and purple color due to the ligand metal charge transfer characteristic of this type of ferric-oxyanion interaction. All tyrosinate-ferric proteins are purple, as is the hemerythrin ferric peroxide. Peroxy-bridged di-iron complexes (Fe—O—O—Fe) are intermediate species in oxygenases and desaturases that display a blue-green color and absorb light at 600-700 nm.

In the diferric complexes the two iron centers are in the 3+ oxidation state with $S = 5/2$. However, they are antiferromagnetically coupled, so that $S = 0$ and there is no EPR signal. In the mixed-valence state, $S_1 = 5/2$ and $S_2 = 2$, so there is a characteristic EPR signal with $g_{ave} = 1.85$. The diferrous complexes have integer spins and should not display an EPR signal. However, a very weak signal is found far downfield in the case of methane monooxygenase because of very weak J coupling between ferrous irons through bridging water at nearly the same energy as the zero field splitting of ferrous iron (J/D mixing) (Fox et al., 1988). The various di-iron complexes display characteristic vibrational frequencies for the Fe—O ligands that are most conveniently observed by resonance Raman spectroscopy. The oxidation states of iron are most definitively determined by Mössbauer spectroscopy.

Reactions of Di-iron Enzymes

Proteins and enzymes incorporating di-iron complexes facilitate diverse biochemical reactions ranging from the reversible binding of oxygen by hemerythrin to the oxygenation of methane (see chap. 17), activation of ribonucleotide reductase (see chap. 16) and the desaturation of fatty acids. The first protein found to contain a di-iron complex was hemerythrin, the oxygen binding protein in marine worms and sea slugs. Hemerythrin binds oxygen reversibly in these animals in place of hemoglobin, which is absent. Because hemerythrin is not an enzyme, we limit discussion of this protein to the chemistry of oxygen binding. The diferrous state is functional and maintained by a cytochrome b_5–dependent reductase, which uses NADPH as the reducing agent. The diferrous hemerythrin binds oxygen reversibly to form a hydroperoxide complex according to eq. 4-20.

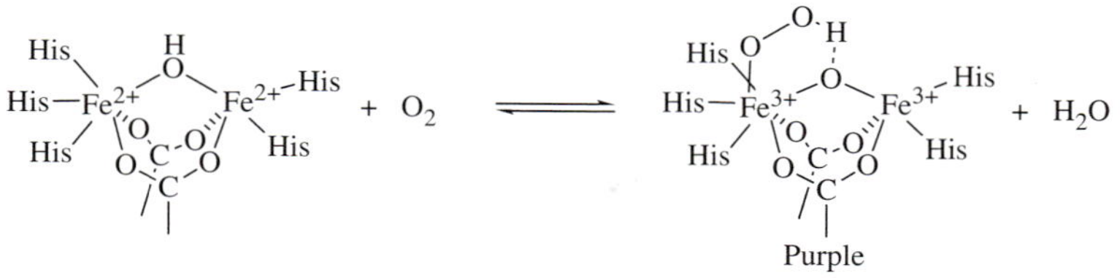

$$\text{(structure with } Fe^{2+}\text{)} + O_2 \rightleftharpoons \text{(Purple structure with } Fe^{3+}\text{)} + H_2O \quad (4\text{-}20)$$

Each ferrous site contributes one electron to reduce oxygen, first to superoxide anion ($O_2^{-\bullet}$) and then to hydroperoxide anion, which binds to one ferric site. The resultant ferric hydroperoxide complex is internally stabilized by hydrogen bonding to the bridging oxide group. The ferric hydroperoxide anion complex is purple because of ligand-to-metal charge transfer complexation.

Hydrogen Atom Abstraction

In the actions of di-iron monooxygenases and fatty acid desaturases, abstraction of hydrogen involves highly oxidized and oxygenated species of the di-iron complexes. The production of this species at an active site requires both a reducing agent and oxygen and follows the general course outlined in scheme 4-8.

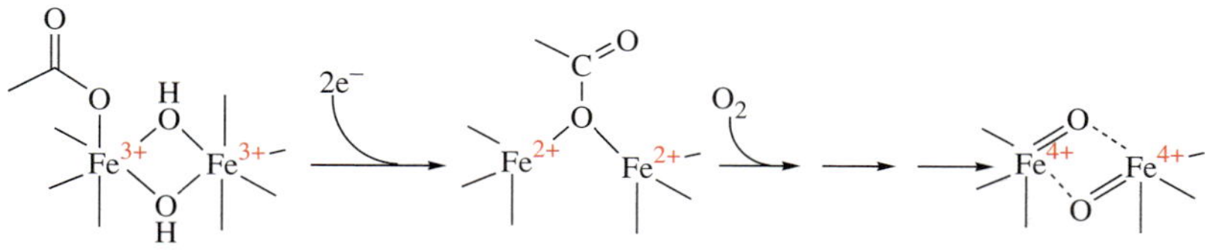

Scheme 4-8

The dioxo-Fe^{4+}_2 species in scheme 4-8 can abstract a hydrogen atom from an alkane or donate oxygen atom to reactive carbon, such as a carbon-carbon double bond. The dioxo-Fe^{4+}_2 species are key intermediates in the actions of methane monooxygenase (see chap. 17), toluene monooxygenase, and fatty acid desaturases.

Fatty Acid Desaturases

Double bonds at locations remote from the acyl groups of fatty acids are introduced by fatty acid desaturases. Desaturation of an alkyl chain is intrinsically oxidative, and in the cases of fatty acid desaturases oxygen is the oxidizing agent. The coenzymes are di-iron complexes analogous to those in methane monooxygenase (Broadwater et al., 1998a; Fox et al., 1993, 1994; Shanklin and Cahoon, 1998). Because of its solubility in water,

220 | Enzymatic Reaction Mechanisms

the most thoroughly studied desaturase is the plant Δ^9-stearoyl acyl carrier protein (ACP) desaturase.

Most desaturases are membrane bound, and less is known about them at the molecular level. However, amino acid sequence analysis shows that they contain motifs of eight histidines characteristic of di-iron enzymes: $HX_{(3 \text{ or } 4)}$—$HX_{(n)}$—$HX_{(2 \text{ or } 3)}$—HH—$X_{(m)}$—$(HX_{(2 \text{ or } 3)}HH)$, corresponding to three di-iron centers (Shanklin et al., 1994). The overall

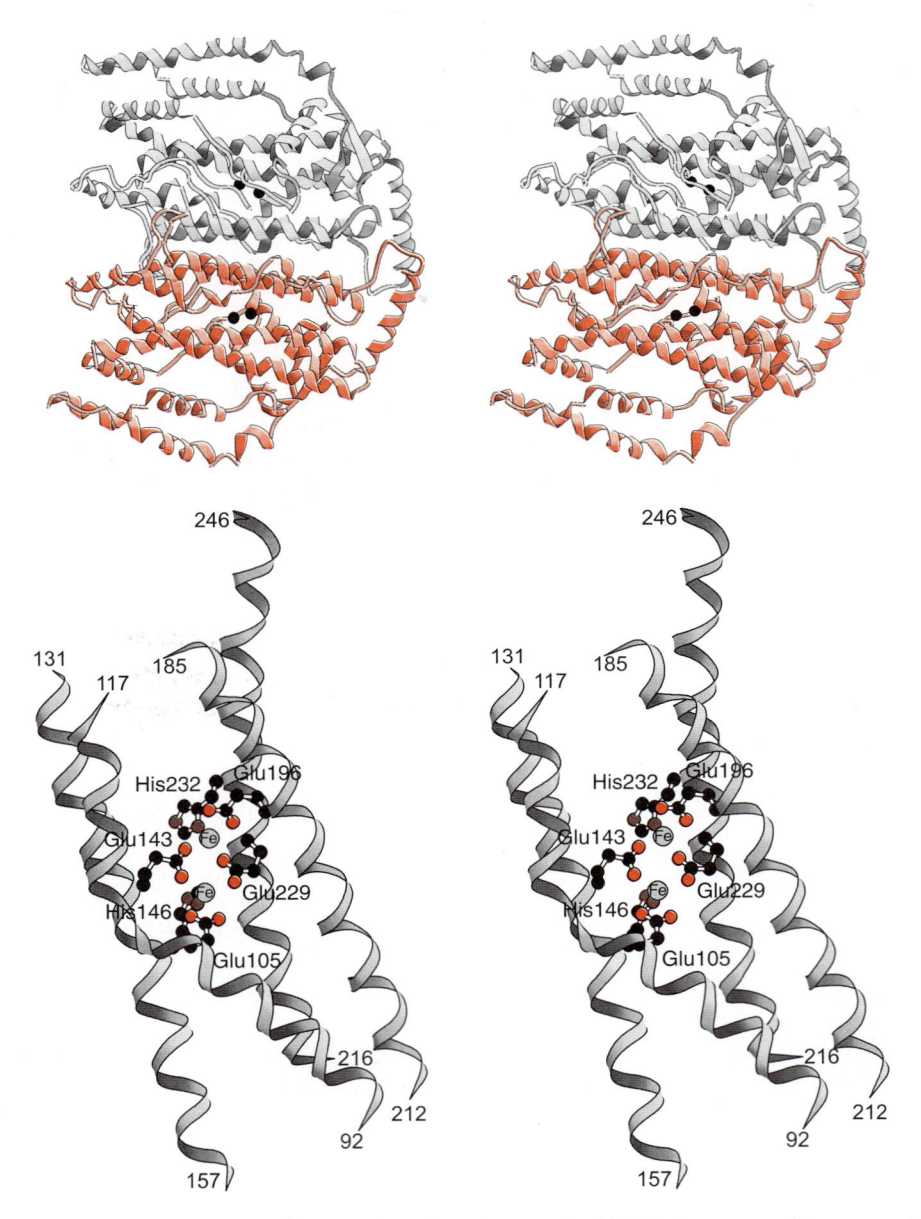

Fig. 4-23. The structure of Δ^9-stearoyl acyl carrier protein (ACP) desaturase. The stereoview (top) shows the dimeric enzyme, with one gray and one red subunit. The subunits are largely α-helical bundles, with the di-iron centers shown in black. A close-up stereoview (bottom) of the di-iron complex and its ligands in a ball-and-stick model show the iron ions represented by larger gray balls. The side chains of amino acids that contribute ligands project from the helices to trap the iron within a four-helix bundle. The illustration was prepared using PDB 1AFR (Lindqvist et al., 1996).

chemical process in the desaturation of the substrate stearoyl-acyl carrier protein by oxygen and NADPH proceeds according to eq. 4-21. Because of the requirement for both oxygen and NADPH, this is a mixed-function oxidase.

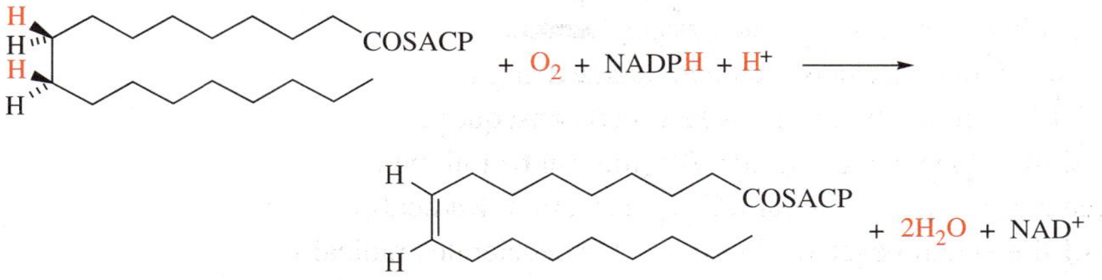

$$\text{stearoyl-COSACP} + O_2 + \text{NADPH} + H^+ \longrightarrow \text{oleoyl-COSACP} + 2H_2O + \text{NAD}^+ \quad (4\text{-}21)$$

The enzyme removes hydrogen from the 9- and 10-pro-R positions to give the cis- (Z-) geometric isomer of the unsaturated product (Behrouzian et al., 2002).

The structure of Δ^9-stearoyl ACP desaturase shown in fig. 4-23 reveals an association of essentially helical subunits (Lindqvist et al., 1996). The di-iron complexes are located within helix bundles that contribute ligands to the two iron ions.

Spectroscopic and crystallographic evidence implicates His179 and His265 as ligands to Fe$_A$ and Fe$_B$, respectively, Glu176 and Glu265 as bridging ligands, and Glu229 as a ligand to Fe$_B$ in the diferric complex. The residues numbers in fig. 4-23 differ because of species differences. Spectroscopic and kinetic evidence indicates that the reaction is initiated by reduction of the diferric complex to its diferrous state by the associated reductase (Broadwater et al., 1998b; Yang et al., 1999). Reaction of the diferrous complex with oxygen (Brazeau et al., 2001) is thought to lead to the dioxo-Fe$^{4+}_2$ species in scheme 4-8, as in the action of methane monooxygenase (see chap. 17). Abstraction of the C9(H) from the pro-R position by Q-desaturase would be the next likely step in the mechanism, and this is shown in the hypothetical mechanism outlined in scheme 4-9.

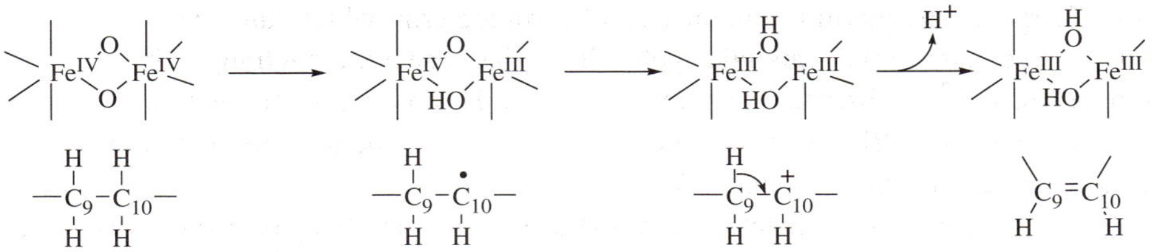

Scheme 4-9

A reasonable reaction of the radical intermediate is electron transfer to the second ferryl oxy iron to generate the carbocationic intermediate. This leaves the di-iron complex in its diferric hydroxide state, and the loss of a proton from the carbocation leads directly to the desaturated product.

The mechanism in scheme 4-9 may be related to one step in the hydroxylation of alkanes by methane monooxygenase (see chap. 17), which also involves abstraction of a hydrogen atom from an alkane. The question arises of why the desaturation substrate does not undergo oxygenation, as in the action of methane monooxygenase. It may be that the enzyme binding interactions hold the fatty acid side chain too far away from the di-iron center to allow the completion of an oxygenation rebound. Electron transfer, as in the second step of scheme 4-9, would then be favored because it can take place rapidly over a longer distance than oxygen transfer.

However, oxygenation does take place in the active site of Δ^9-desaturase when electron transfer is sufficiently slowed by use of a fluorinated substrate (Behrouzian et al., 2002). Reaction of (9R)-9-fluorostearoyl ACP as a substrate leads largely to desaturation products but also to almost 12% of 9-fluoro-10-hydroxystearoyl ACP. Formation of the hydroxylated product can be rationalized on the basis that electron transfer from the putative initial fluorinated radical intermediate would be slowed by fluorine. Fluorine is the most electronegative element and would decelerate electron transfer from the adjacent C10-radical intermediate. This would allow more time for the oxygen rebound mechanism at C10 to form the hydroxylated product. Other products included those arising from abstraction of C11(H) rather than the usual C10(H) in the first step, a presumed consequence of the presence of fluorine at C9 causing a shift in the reaction frame. This too can be rationalized on the basis of the effect of electronegative fluorine on initial radical formation. Radicals are electron deficient, and the presence of the electronegative fluorine at C9 decelerates radical formation through hydrogen abstraction at C10. Slowed hydrogen abstraction from C10 can enhance the contribution of occasional hydrogen abstraction at C11 and lead to corresponding products.

Purple Acid Phosphatases

The purple acid phosphatases contain mixed-valence (Fe^{II}/Fe^{III}) di-iron complexes or Zn^{II}/Fe^{III} complexes. They are purple because of ligand to metal charge transfer complexation between a tyrosine ligand and ferric iron. These enzymes are among a class of dinuclear metallic hydrolases in a later section on divalent metal ions.

Metallopterin Enzymes

Pterin cofactors defy classification in terms of their chemical action as coenzymes or as co-substrates for enzymes catalyzing complex reactions. Phenylalanine, tyrosine, and tryptophan hydroxylases require tetrahydrobiopterin as an integral participant in the chemistry by which they produce electrophilic hydroxyl groups from oxygen for insertion into aromatic rings. The pterin by-product is dihydrobiopterin, which must be reduced externally to potentiate further catalytic cycles. In the latter sense, the tetrahydrobiopterin may be regarded as a cosubstrate. However, its role in the production of electrophilic oxygen seems coenzymatic. Biopterin is also required as a coenzyme to support maximal activity of nitric oxide synthase (see chap. 17).

Pterin functions as a chelator of molybdenum in molybdopterin enzymes such as xanthine oxidase, dimethylsulfoxide (DMSO) reductase, and sulfite oxidase. In the molybdopterin enzymes, the pterin portion of the coenzyme does not participate chemically in catalysis apart from chelating the molybdenum. Perhaps the diverse roles of pterins in these different enzymes contributed to the confusion surrounding pterin cofactors for many years leading up to the present time. However, much has come to light within the past ten years, and these advances have allowed deeper insights into the mechanisms of action of a number of important and formerly mysterious enzymes.

Molybdopterin and Tungstopterin

Coenzyme Structures

A large group of oxidases and reductases have pterin-derived coenzymes, in which the dithiolene substituent serves as a chelator for molybdenum or tungsten (Hille, 1996). The structures of the molybdopterins as deduced from biochemical and crystallographic data are shown in fig. 4-24 (Johnson and Ragagopalan, 1982; Johnson et al., 1980, 1984; Kramer et al., 1987; Romao et al., 1995; Schindelin et al., 1996). In the crystal structure

Fig. 4-24. Structures of molybdopterin coenzymes.

of aldehyde:ferredoxin reductase from *Pyrococcus furiosus*, tungsten replaces molybdenum (Chan et al., 1995). The parent compound (R=H) is molybdopterin, and the nucleotide derivatives are known as molybdopterin guanosine diphosphate, molybdopterin cytosine diphosphate, and similar forms. The nucleoside in the molybdopterin nucleoside diphosphates tends to be species dependent in many bacteria, but different nucleotide derivatives have been found in single species (Hille, 1996). The ligands to molybdenum were determined initially by biochemical and spectroscopic methods and confirmed by x-ray crystallography.

The molybdopterin enzymes are oxidases or reductases, and they can be grouped in three classes based on coordination about molybdenum: the xanthine oxidase family, the sulfite oxidase family, and the DMSO reductase family (Hille, 1996). The molybdenum coordination patterns in fig. 4-25 distinguish these families. The xanthine oxidase family is characterized by a sulfo, an oxo, and a water ligand to molybdenum in addition to the dithiolene ring. The sulfite oxidase family has two oxo, a cysteinyl, the dithiolene, and possibly a water as ligands. The DMSO reductase family departs dramatically, having two pterindithiolenes giving four thiolate ligands and two other ligands; one may be oxo, sulfo, or seleno, and the other may be serine, cysteine, or selenocysteine.

Many molybdopterin enzymes contain other electron transfer cofactors such as heme, [2Fe–2S] centers, and FAD. The electron transfer cofactors serve the essential function of relaying electrons to or from the molybdenum center, which reacts either to extract electrons from or to insert electrons into the substrates. Extensive studies of the role of the pterin portion of the coenzyme have failed to provide any evidence that it undergoes chemical change in the reaction cycle (Hille, 1996). Certainly the pterin dithiolene chelates molybdenum (or tungsten), and the conformation of the pyran ring may modulate the reduction potential of the molybdenum ring. In some cases, the pterin ring may serve as a conduit for electron transfer; however, no evidence has been obtained for discrete changes in redox states or any direct catalytic participation of the pterin ring.

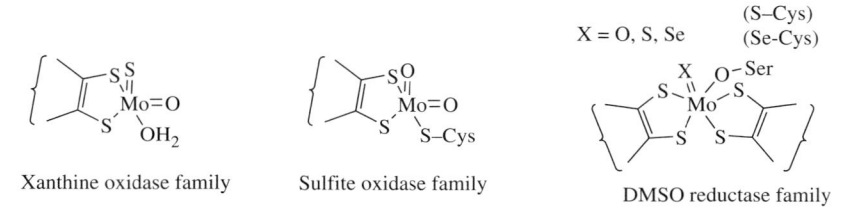

Fig. 4-25. Molybdenum coordination patterns can distinguish three families of molybdopterin enzymes: the xanthine oxidase family, the sulfite oxidase family, and the dimethylsulfoxide (DMSO) reductase family.

Xanthine Oxidase

The xanthine oxidase family catalyzes the oxidation of H—C=X groups in molecules and transforms them into HO—C=X groups, where X may be O or N. The oxygen atom incorporated into the product invariably originates with water and not O_2. The general chemical equation for these reactions may be written as eq. 4-22, in which the red-coded oxygen atoms represent the fate of ^{18}O.

$$\underset{X}{\overset{H}{\underset{|}{C}}} + H_2O \longrightarrow \underset{X}{\overset{OH}{\underset{|}{C}}} + 2H^+ + 2e^- \qquad (4\text{-}22)$$

$$\text{(xanthine)} + H_2O \longrightarrow \text{(uric acid tautomer)} + 2H^+ + 2e^- \qquad (4\text{-}23)$$

The family includes aldehyde oxidases and dehydrogenases, xanthine dehydrogenase, and a number of heterocyclic ring oxidases and dehydrogenases. In addition to the chemical similarities in the reactions they catalyze, the members of this family display extensive similarities in amino acid sequences (Hille, 1996). The title member of the family is xanthine oxidase from cow's milk, which catalyzes the oxidation according to eq. 4-23. In this case, the product is a tautomer of the compound resulting from insertion of an oxygen atom from water into the C8(H) bond. The two electrons extracted from the substrate are transferred to an external acceptor by the FAD and two [2Fe–2S] centers in the enzyme, which is a dimer with a molecular mass of 300 kDa. While xanthine oxidase is the most extensively studied enzyme in the family, the first structure to be obtained was that of aldehyde oxidase from *Desulfovibrio gigas* (Romano et al., 1995).

The structure of xanthine oxidase in fig. 4-26 shows the binding site of molybdopterin and its structural relationships with FAD and two [2Fe–2S] centers. Molybdenum in molybdopterin is in its 6+ oxidation state, Mo(VI), and oxidation of xanthine leaves it in its 4+, Mo(IV), oxidation state. Transfer of two electrons from Mo(IV) through the FAD and [2Fe–2S] centers to an external electron acceptor returns molybdenum to its Mo(VI) state for the next catalytic cycle.

The essential overall reaction of xanthine proceeds as outlined in fig. 4-27. No fewer than five mechanisms have been proposed for the transformation of the Michaelis complex to the product complex in fig. 4-27, and more are possible. Valuable information about inhibited complexes with substrate analogs has been obtained by XAS, ENDOR, ESEEM, and resonance Raman spectroscopies, and at least one intermediate at the 5+ oxidation state of molybdenum Mo(V) has been observed by EPR. However, the available information does not exclude any of the proposed mechanisms, some of which include proposals that the C8(H) of xanthine is removed as a proton or as a hydride ion. A major barrier to mechanistic analysis has been that the reaction proceeds with exchange of hydrogen with solvent protons and the incorporation of solvent oxygen into the product. All putative intermediates of molybdenum can exchange both protons and oxygen with water, making it impossible to trace these atoms definitively. The solution of this mechanism remains a challenge for the future.

It may prove to be simpler to solve the mechanism of the aldehyde oxidase reaction and use the revealed principles to understand xanthine oxidase. One of the mechanisms proposed for xanthine oxidase includes the initial ionization of the C8(H) from the purine ring ($pK_a = 14$). The corresponding ionization of the aldehyde substrate would be implausible

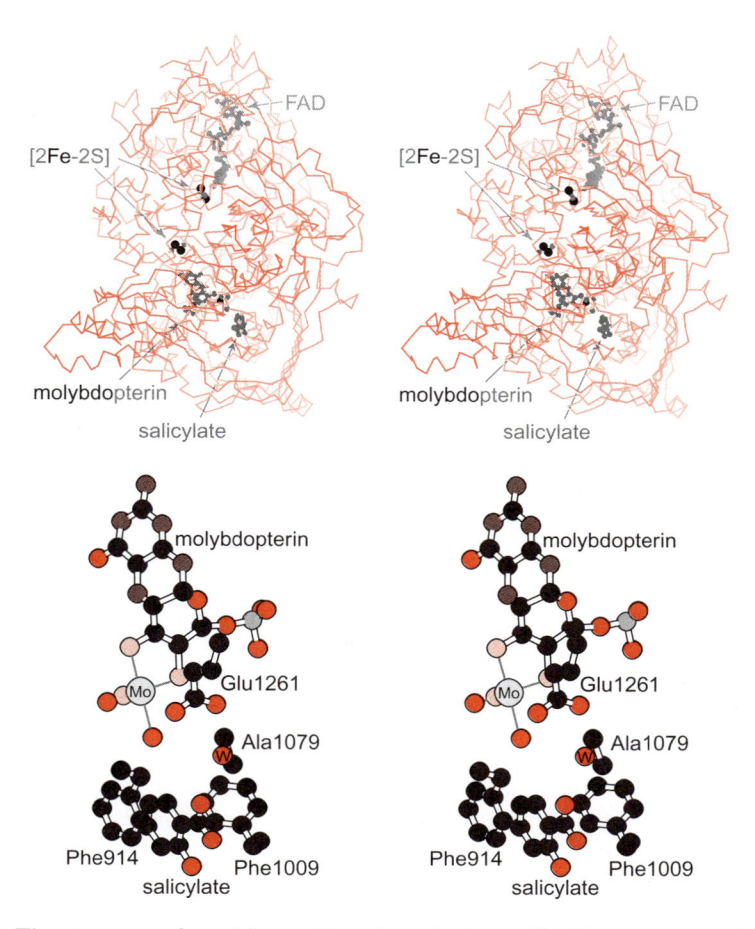

Fig. 4-26. The structure of xanthine oxidase from bovine milk. The stereoview (top) shows a Cα trace of the overall chain fold and the locations of the cofactors and salicylate bound to the substrate site. The gray-scale ball-and-stick models are FAD, the molybdopterin coenzyme, salicylate, and the [2Fe–2S] centers, with the Fe and Mo atoms emphasized in black. FAD is near the top, molybdopterin is in the center, and salicylate is near the molybdopterin at the lower right. Another stereoview (bottom) employs a ball-and-stick model of the molybdopterin site, showing the ligands to molybdenum, with salicylate resting between two phenylalanyl side chains. The FAD and [2Fe–2s] centers relay electrons to the external electron acceptor. The figure was prepared using PDB 1FIQ (Enroth et al., 2000).

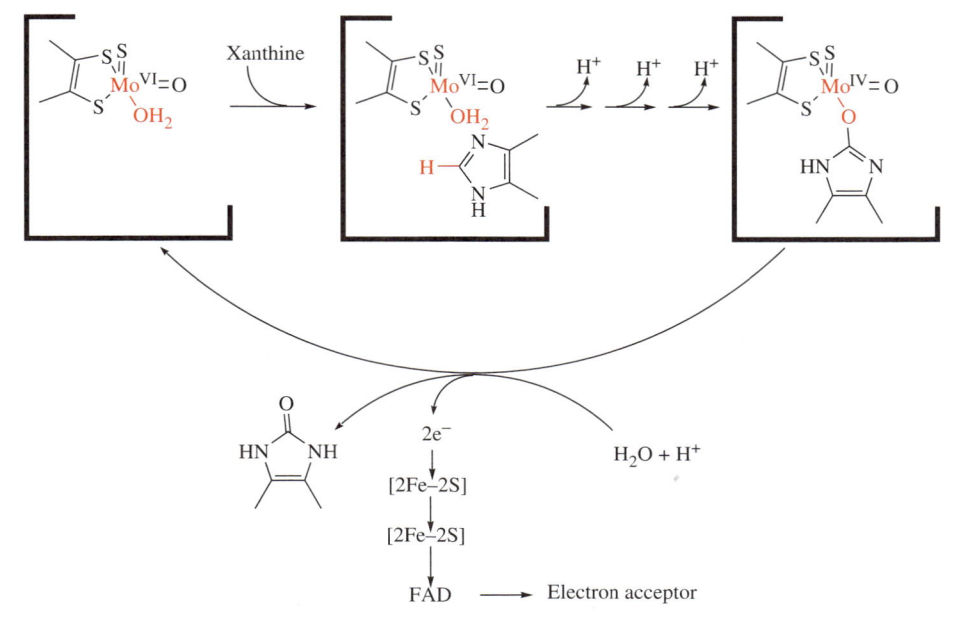

Fig. 4-27. The reaction cycle of xanthine oxidase.

226   Enzymatic Reaction Mechanisms

in the action of an aldehyde oxidase because the ionization of an aldehydic proton is unprecedented. If the members of this family function similarly, this one mechanism may be excluded.

Sulfite Oxidase Family

Oxidation of sulfite to sulfate and reduction of nitrate to nitrite are catalyzed according to eqs. 4-24 and 4-25 by sulfite oxidase (EC 1.4.2.1) and nitrate reductase (EC 1.7.1.1), respectively.

$$:SO_3^{2-} + H_2O + 2\text{Cyt } c^{3+} \rightarrow SO_4^{2-} + 2H^+ + 2\text{Cyt } c^{2+} \qquad (4\text{-}24)$$

$$NO_3^- + NADH + H^+ \rightarrow :NO_2^- + H_2O + NAD^+ \qquad (4\text{-}25)$$

Sulfite oxidase is a liver enzyme, and nitrate reductase is found in plants and yeast. Both contain molybdopterin as the coenzyme. Sulfite oxidase also contains heme for relaying electrons to cytochrome c, and nitrate reductase has heme and FAD for relaying electrons from NADH to the molybdopterin. The coordination of molybdenum in the sulfite oxidase family includes two oxo ligands and a cysteine, as illustrated in fig. 4-25. The ligands were identified largely in XAS studies of ligand to molybdenum distances in sulfite oxidases (Berg et al., 1979; Cramer et al., 1979, 1981; George et al., 1989) and nitrate reductase (Cramer et al., 1984).

Studies of chemical models for sulfite oxidase have clarified the mechanistic issues and led to an interesting working hypothesis. Because the reaction involves mainly the formation of a new bond to sulfite and not the cleavage of a preexisting bond, the mechanism is simpler than any possible mechanism for the xanthine oxidase family. Evidence from nonenzymatic reactions implicates the mechanism in fig. 4-28 for sulfite oxidase. Nucleophilic attack of sulfite through its non bonding electron pair on an oxo ligand of molybdenum provides a covalent conduit for fast electron transfer to molybdenum and leads directly to sulfate coordinated to Mo(IV). The nucleophilic properties of sulfite are well known from its nucleophilic addition reactivity with aldehydes. Exchange of sulfate with water leaves molybdenum in its reduced state, and electron transfer to cytochrome c is mediated by the heme cofactor.

An analogous sequence in the reverse direction, using FAD and heme to mediate electron transfer from NADH to molybdopterin, can be written for nitrate reductase.

Dimethylsulfoxide Reductase

DMSO reductase is a monomeric, 82-kDa, periplasmic, *Rhodobacter* enzyme that contains GDP-molybdopterin as its only coenzyme (Bastian et al., 1991; McEwan et al., 1991; Satoh et al., 1987). Unlike most molybdopterin enzymes, it lacks other electron

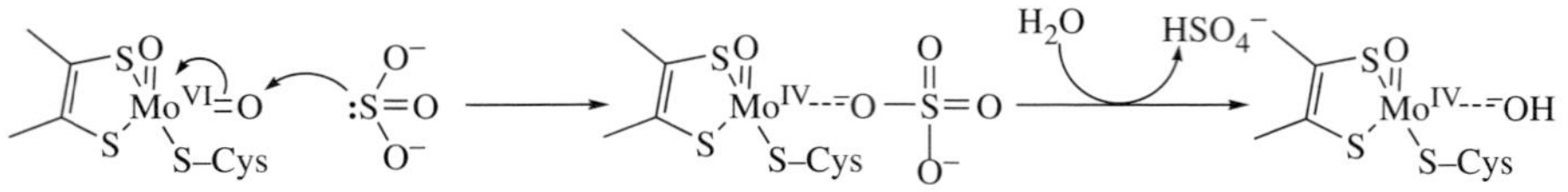

Fig. 4-28. A mechanism for the reduction of molybdopterin in sulfite oxidase by sulfite. Sulfite oxidase extracts a pair of electrons from sulfite while donating an oxygen to form sulfate. A likely mechanism takes advantage of the nucleophilic reactivity of sulfite. Nucleophilic attack on the oxo group of Mo^{VI}=O, accompanied by electron transfer to form Mo^{IV}, leads directly to sulfate formation.

transfer coenzymes. A homologous family of enzymes based on amino acid sequence identities includes biotin sulfoxide reductases, nitrate reductases, formate dehydrogenases, and polysulfide reductases (Hille, 1996; Yamamoto et al., 1995).

The crystal structure of DMSO reductase from *Rhodobacter capsulatus* is presented in fig. 4-29. The structure of the *bis*-GMP-molybdopterin coenzyme is consistent with biochemical and spectroscopic data, as reviewed in detail by Hille (1996). An oxo ligand and Ser147 constitute the two additional ligands to Mo(VI).

DMSO reductase catalyzes the reduction of DMSO to dimethylsulfide (eq. 4-26).

$$(CH_3)_2S\text{—}O + 2e^- + 2H^+ \rightarrow (CH_3)_2S: + H_2O \qquad (4\text{-}26)$$

An external electron donor is required to support the turnover of DMSO. DMSO reductase displays a tolerance for a broad range of dialkylsulfoxides and even oxygenates trialkylphosphines (Schultz et al., 1995). Oxygen-18 can be transferred to the reduced enzyme and then relayed to a trialkylphosphine to produce the ^{18}O-labeled trialkylphosphine oxide ($R_3P\text{—}^{18}O$). In the reduction of DMSO, the enzyme reacts in the form of its reduced molybdopterin, presumably as illustrated in fig. 4-30. On binding the substrate with oxygen weakly coordinated to Mo^{IV}, electron transfer through an oxygen atom leads to dimethylsulfide and $Mo^{VI}=O$. After release of dimethylsulfide, the oxo form of the enzyme must be reduced by an external electron source to regenerate the Mo^{IV}-molybdopterin.

Iron-Sulfur Centers

Iron-sulfur centers in enzymes and electron transfer proteins exist in various structural forms and carry out a variety of functions. By definition, these centers include sulfur ligands, often exclusively sulfur ligands. They range in structural complexity from the simplest mononuclear iron with four cysteine ligands to the elegant iron-molybdenum-sulfide center known as FeMoCo in nitrogenase. Their biochemical functions range from relaying electrons between other coenzymes to serving as Lewis acids, reducing elemental nitrogen, regulating genes (Beinert, 2000; Beinert et al., 1997) and even forming carbon radicals (Frey, 2001). In the following sections, we consider the structures and functions of iron-sulfur centers in enzymes.

Structures

Most iron-sulfur centers in proteins fall into one of the five main classes containing one to four irons shown in fig. 4-31. The simplest is the rubredoxin class, with one iron and four cysteine ligands in a tetrahedral array. From this simple structure, the iron-sulfur centers are built up into a hierarchy of structures with tetrahedral iron-sulfur apices. Multi–iron-sulfur centers include sulfide ions linking the irons to one another, with external cysteine or rarely with histidine ligands. The ferredoxin-type center is [2Fe–2S] with all cysteine ligands, and the Rieske-type center is [2Fe–2S] with two cysteine and two histidine ligands. The [4Fe–4S] centers have approximately cubic geometry in the iron-sulfide core and peripheral cysteine ligands, often four but sometimes only three. The [3Fe–4S] centers are structurally related to the four-iron centers but lack iron at one corner of the cube. The iron-molybdenum-coenzyme (FeMoCo) is the common name given to the nitrogen-reducing center in nitrogenase. Its structure is an elaboration of two four-iron centers with the insertion of a ninth sulfide and the replacement of one iron with molybdenum ligated to histidine and homocitrate.

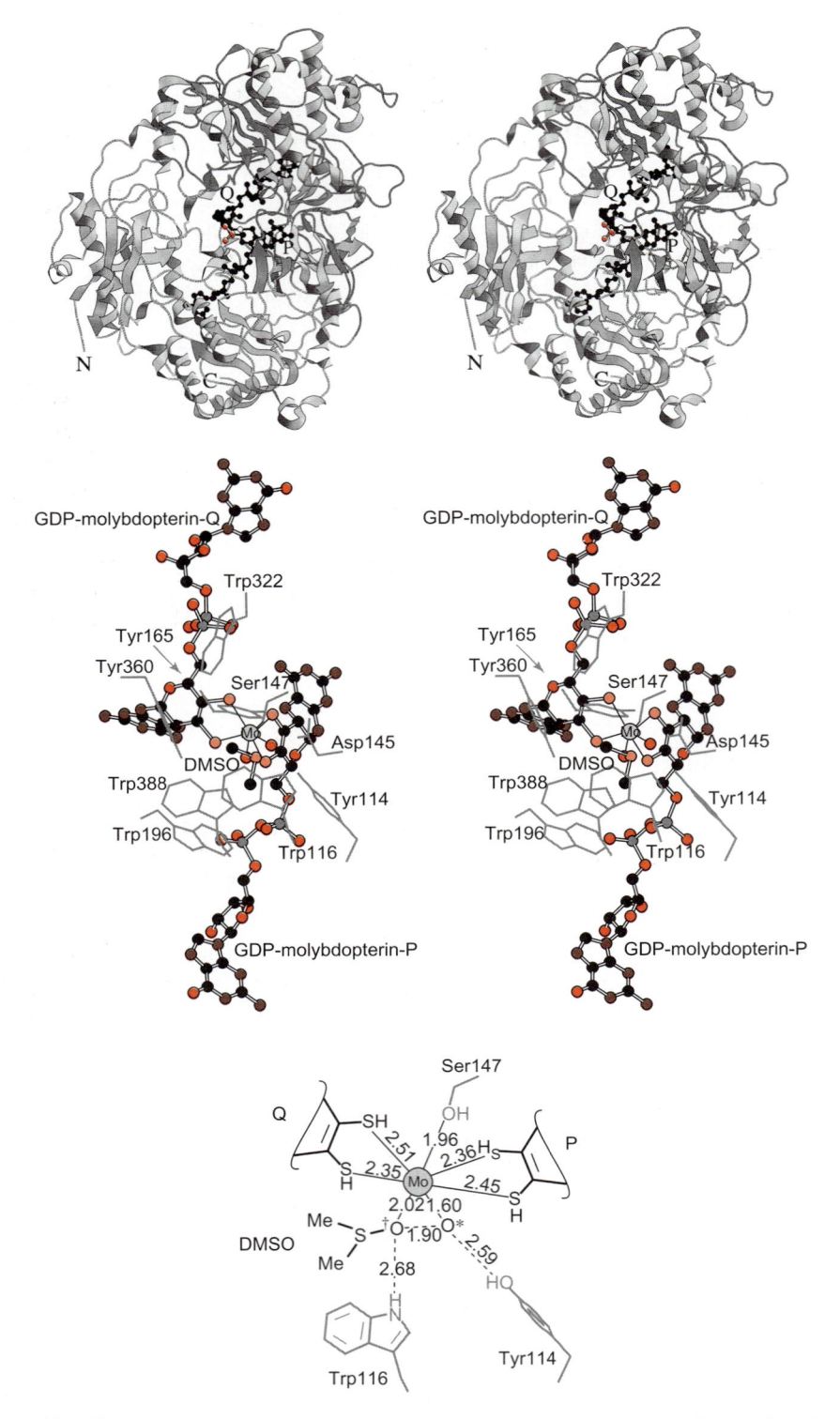

Fig. 4-29. The global structure of dimethylsulfoxide (DMSO) reductase from *Rhodobacter capsulatus* (top) shows the two GDP-molybdopterin moieties, P and Q, as black ball-and-stick models, with the DMSO highlighted in red adjacent to the Mo atom in the active site. A more detailed stereoimage (middle) of the active site shows the molecules colored by atom type and the protein side chains as gray stick figures. The interatomic distances among DMSO, Mo, and ligands are given (bottom). The DMSO oxygen (†) replaces an oxygen ligand to the Mo observed in structures of the oxidized enzyme. The protonation state of the other oxo ligand (*) is unknown, but it is probably OH$^-$, not H$_2$O, given the short distance to the metal. The figure was prepared using the PDB 4DMR of the 1.9-Å resolution model (McAlpine et al., 1998).

Fig. 4-30. A mechanism for the oxidation of reduced molybdopterin by dimethylsulfoxide (DMSO). The reduction of DMSO by its reductase begins with the reduced form of molybdopterin. Ligation of the oxo group in DMSO with Mo^{IV} creates a conduit for electron transfer, leading to the formation of dimethylsulfide and Mo^{VI}=O. Reduction of Mo^{VI}=O by a reducing agent returns the enzyme to its reduced form for another catalytic cycle. The electron donor is not known with certainty.

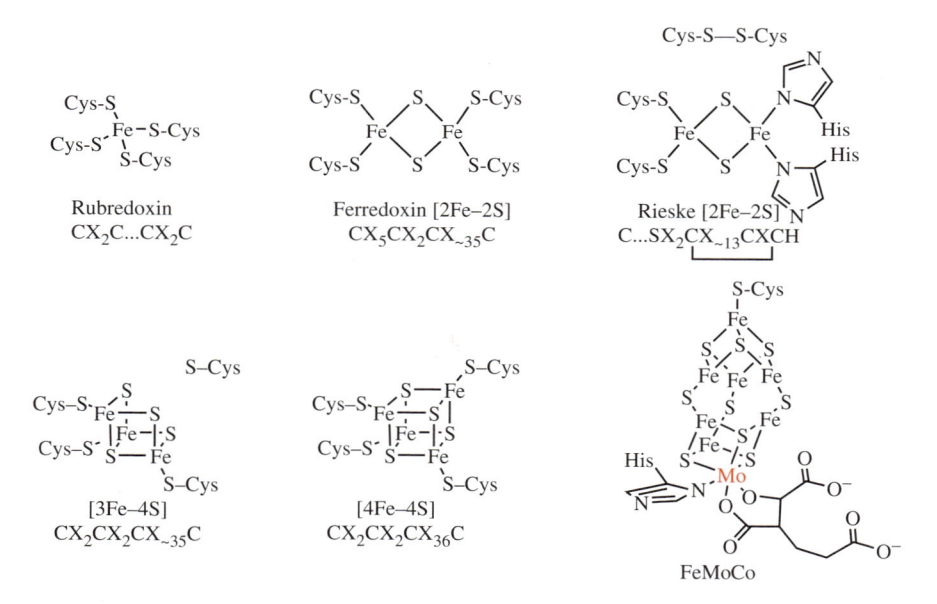

Fig. 4-31. Structures of the most frequently encountered iron-sulfur (Fe-S) centers. Most Fe-S centers are constructed with tetrahedral coordination of iron to four sulfur ligands, the simplest being the mononuclear iron in rubredoxin with four cysteine ligands. In multi-iron sites, the metal ions are bridged by sulfide ions, and the peripheral ligands usually are cysteine, although other ligands are sometimes found, most frequently histidine, as in the Rieske-type [2Fe–2S] center. Most [4Fe–4S] centers have four cysteines as peripheral ligands to the protein; however, in several cases, one cysteine is replaced by a ligand from a substrate or another coenzyme. In a few cases (not illustrated), an apical iron is linked by a peripheral sulfide bridge to another metal instead of to a protein ligand. FeMoCo is the coenzyme of nitrogenase. The reduction potentials of Fe-S centers are sensitive to small conformational effects, and the structures of Fe-S centers in enzymes are less symmetrical than implied by these drawings.

Iron-sulfur centers in electron transfer proteins are assembled within a diverse range of peptide chain folds, as illustrated in the six representative structures in fig. 4-32. Unlike di-iron centers, the structures of iron-sulfur centers are not associated with secondary structural motifs such as helix bundles.

Catalytic Functions

Electron Transfer

In the first known and most frequent function of iron-sulfur centers, they relay electrons between other redox active molecules, substrates, or other coenzymes. The observed oxidation states differ by one electron, indicating that the iron-sulfur centers engage in one-electron transfer processes. The reduction potentials vary over the ranges listed in table 4-1. No other biological redox molecules offer such a large range of reduction potentials. The iron-sulfur centers have often been regarded as biological relics from the origins of life, and they certainly appear in the earliest life forms. Their continued evolutionary survival may be attributable to their special redox properties and to their versatility as catalysts.

The range of reduction potentials observed, especially for [4Fe–4S] centers raises the question of how the variations in reduction properties are brought about. The structures of the [4Fe–4S] centers seem to be similar. The reduction potentials between different redox states naturally differ because of the internal electrostatic differences in the two redox states.

Pyrococcus furiosus
Rubredoxin
(1BRF)

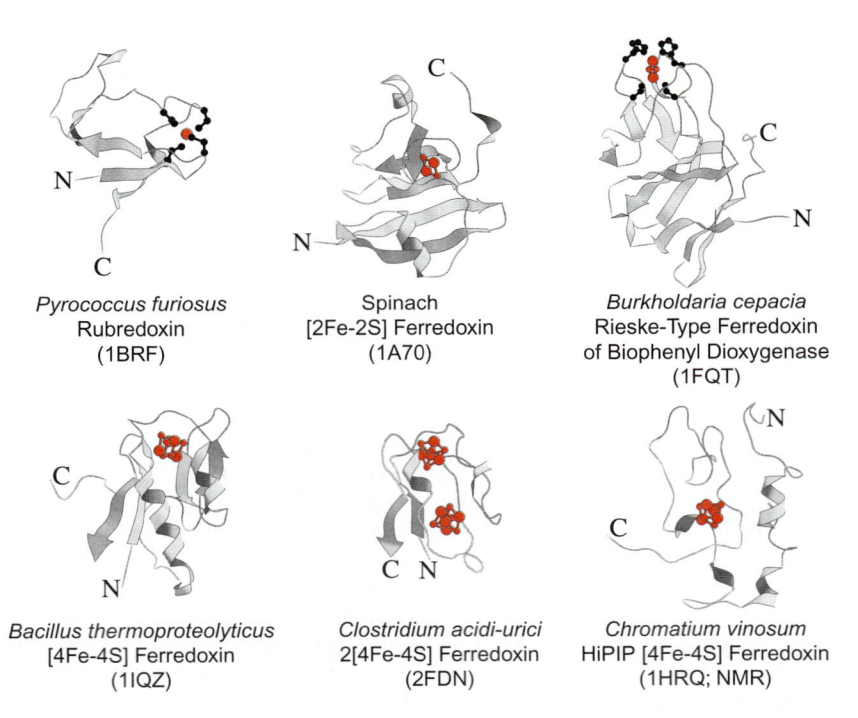

Spinach
[2Fe-2S] Ferredoxin
(1A70)

Burkholdaria cepacia
Rieske-Type Ferredoxin
of Biophenyl Dioxygenase
(1FQT)

Bacillus thermoproteolyticus
[4Fe-4S] Ferredoxin
(1IQZ)

Clostridium acidi-urici
2[4Fe-4S] Ferredoxin
(2FDN)

Chromatium vinosum
HiPIP [4Fe-4S] Ferredoxin
(1HRQ; NMR)

Fig. 4-32. Structures of typical iron-sulfur electron-transfer proteins. All of the iron-sulfur proteins included here are electron-transfer proteins. Notice the diversity of chain folds. The portions of the protein chains that donate ligands to iron tend to lie outside the secondary structural elements of the proteins. The figure was prepared using PDB files: 1BRF (Bau et al., 1998), 1A70 (Binda et al., 1998), 1FQT (Colbert et al., 2000), 1IQZ (Fukuyama et al., 2002), 2FDN (Dauter et al., 1997), and 1HRQ (Banci et al., 1995).

Table 4-1. Reduction Potentials for Iron-Sulfur Centers

Center Type	E_o (mV)
Rubredoxin	−60 to +5
Plant-type ferredoxins	−405 to −380
Rieske-type ferredoxins	−280 to −170
Four iron-sulfur centers	−650 to +360

The question of how the reduction potentials vary for apparently identical electrostatic states in different proteins must depend on microenvironmental effects imposed by the proteins. Electrostatic charges and electrical dipoles are likely to be important. In biological systems, the iron-sulfur centers are negatively charged overall because of the negative charges brought into the structures by the sulfide ions and thiolate groups of cysteine. The possible net charges on [4Fe–4S] centers with four cysteine ligands are accounted for in table 4-2, together with the oxidation and spin states observed in biological systems. The designated oxidation states differ from the net electrostatic charges and are assigned by convention as 1+, 2+, and so on for increasingly oxidized states.

Essentially all biological [4Fe–4S] centers are negatively charged overall, and this is also true for the rubredoxins and ferredoxins. The electrostatic properties of the microenvironments about iron-sulfur centers can be expected to modulate the ease with which they undergo reduction and oxidation (i.e., their reduction potentials). The variations in reduction potential have been attributed to electrostatic effects, including hydrogen bond donation to the sulfide or thiolate groups, the orientation of protein dipoles, and the proximity of electrostatically charged amino acid side chains. Distortion of the structures from their ideal geometries may also modulate reduction potentials.

Iron-sulfur centers participate in electron transfer processes in many biochemical systems, including the terminal electron transport pathways of cells and the actions of many complex enzymes. However, iron-sulfur centers turn out to be far more versatile in their biological functions, which now include action as Lewis acid catalysts and iron sensing functions in gene regulation, as well as unusual modes of electron transfer in complex mechanisms.

Aconitase and α,β-Dehydratases

A nonredox role for the iron-sulfur center in aconitase (EC 4.2.1.3) and α,β–hydro-lyases attests to the versatility of iron-sulfur centers in enzymatic catalysis (Beinert et al., 1996; Flint and Allen, 1996). Aconitase catalyzes the interconversion of citrate and isocitrate as illustrated in scheme 4-10.

Table 4-2. Net Charges and Observed Oxidation States in [4Fe–4S] Centers with Four Cysteine Ligands.

Fe^{3+}	Fe^{2+}	S^{2-}	Cys–S⁻	Net Charge	Oxidation State	Spin State
4	0	4	4	0	NA[a]	NA
3	1	4	4	1−	3+	$S = 1/2$
2	2	4	4	2−	2+	$S = 0$
1	3	4	4	3−	1+	$S = 1/2$
0	4	4	4	4−	NA	NA

[a] Not observed in biological systems.

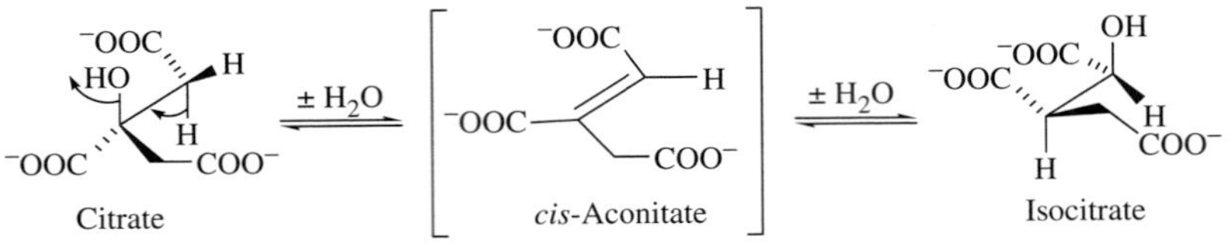

232 Enzymatic Reaction Mechanisms

Scheme 4-10

In the first step aconitase catalyzes the dehydration of citrate to *cis*-aconitate, which rarely dissociates from the active site but is immediately hydrated by addition of water to the adjacent carbon to form isocitrate. The structure of aconitase and several important aspects of the mechanism and stereochemistry of its action are presented in chapter 8. Here, we consider a single crucial aspect of the mechanism, the role of the iron-sulfur center.

For many years, a requirement for ferrous iron in the activity of aconitase was recognized but not explained. Divalent metal ions were required for the actions of many enzymes, and questions about their mechanistic functions have often been deferred pending more structural information. The structural information in the case of aconitase came when it was recognized that it contained a novel iron-sulfur center. Aconitase was found to contain acid-labile sulfide, an essentially definitive indicator of an iron-sulfur center (Kennedy et al., 1972). Subsequently, a mitochondrial protein that displayed a novel iron-sulfur EPR signal at $g_{ave} = 2.01$ proved to be aconitase (Ruzicka and Beinert, 1974). Careful biochemical and EPR spectroscopic studies traced the novelty in the iron-sulfur center to the fact that it existed as [3Fe–4S]$^+$ in its oxidized form. On addition of ferrous iron and a reducing agent, the novel center was transformed into a [4Fe–4S]$^{2+}$ center in the active enzyme. These results immediately explained the requirement of aconitase activity for ferrous iron and a reducing agent.

Further biochemical and spectroscopic experiments proved that the iron-sulfur center participated in catalysis as a coenzyme for dehydration. Aconitase with the [3Fe–4S]$^+$ center was inactive, and activation resulted from the incorporation of the fourth iron into the center. The most active aconitase contained the diamagnetic [4Fe–4S]$^{2+}$ center, but the enzyme displayed lower ($\approx$30%) activity after reduction of the center to its paramagnetic [4Fe–4S]$^+$ state. This was very important for the further characterization of the mechanism. The diamagnetic center could not give information about ligation to the fourth iron in EPR or ENDOR experiments, but the paramagnetic form gave a great deal of insight into this important aspect of the structure.

The results of Mössbauer and ENDOR spectroscopic experiments proved that a unique iron (Fe$_a$) in the [4Fe–4S] center formed coordination bonds with the substrate (Beinert et al., 1996, 1997; Emptage et al., 1983a, 1983b; Kennedy et al., 1987; Kent et al., 1985). The coordination of the departing OH$^-$ group by Fe$_a$ allowed the iron-sulfur cluster to serve as a Lewis acid in catalyzing the reversible dehydration of citrate to aconitate, by the mechanism illustrated in fig. 4-33.

A number of α,β-dehydrases and isomerases that involve α,β-dehydration contain iron-sulfur centers analogous to those of aconitase (Flint and Allen, 1996). These enzymes include some that are familiar members of nonmetalloenzyme families, such as serine dehydratase and fumarase. Mammalian fumarase is not a metalloenzyme, but some of the bacterial fumarases contain iron-sulfur centers analogous to aconitase. Serine dehydratases are generally PLP enzymes, but some of the bacterial serine dehydratases are iron-sulfur proteins. Bacterial isopropylmalate synthase is also in this group, as well as a number of other enzymes (Flint and Allen, 1996).

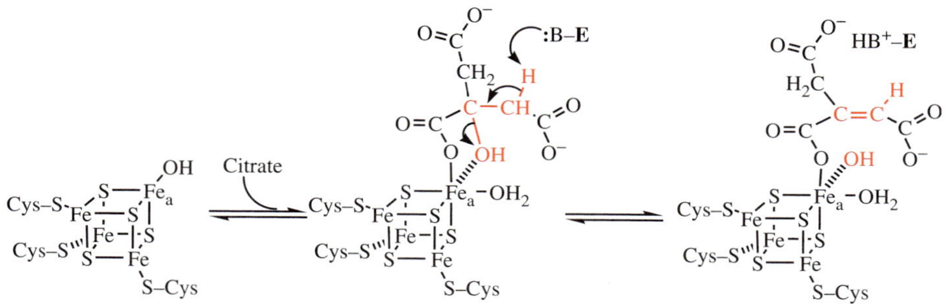

Fig. 4-33. The [4Fe–4S] center as a Lewis acid in the dehydration of citrate by aconitase. One iron in the [4Fe–4S] center of aconitase serves as a Lewis acid in two ways. Fe$_a$ binds the substrate by coordination of the carboxylate and hydroxyl groups, and it facilitates the departure of the hydroxyl group in the dehydration step by serving as a Lewis acid to facilitate the departure of the OH group from citrate.

β,α-Dehydratases

A few dehydratases catalyze the elimination of water in the reverse orientation from the usual α,β-elimination mode. The proton α-to an acyl or ketonic function is far more acidic (pK_a = 18 to 30) than that of an alkane C—H ($pK_a \geq 50$). The acidity of an α-proton can be further decreased by interactions of the acyl or ketonic group with electrophilic centers in a protein. The intrinsic and induced acidities of α-protons make it possible for them to be abstracted by bases in enzymes and allow α,β-dehydrations to proceed. The acidities of alkane C—H protons are too low for such reactions, and their acid strength cannot be increased by interactions with enzymatic sites. Dehydrations that do not follow the α,β-pattern are rare and proceed by novel mechanisms. One example is the reaction of dioldehydrase, which requires adenosylcobalamin and takes place by a radical mechanism.

A few dehydration reactions of molecules that include acyl groups occur in the reverse orientation from the α,β-elimination. The β-C—H bond is broken, and the hydroxyl group departs from the α-carbon. These reactions may be described as β,α-eliminations. They cannot proceed by abstraction of the β-hydrogens as protons because they are not acidic enough. Two examples are lactyl CoA dehydratase and 2-hydroxyglutaryl CoA dehydratase (Kuchta et al., 1986; Müller and Buckel, 1995). The β,α-orientations of these dehydrations are illustrated in eqs. 4-27 and 4-28.

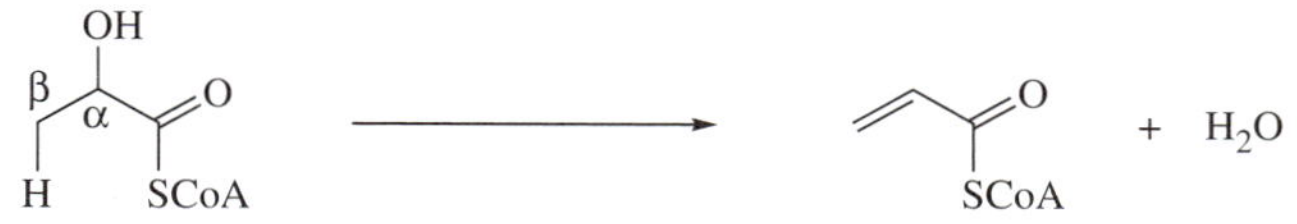

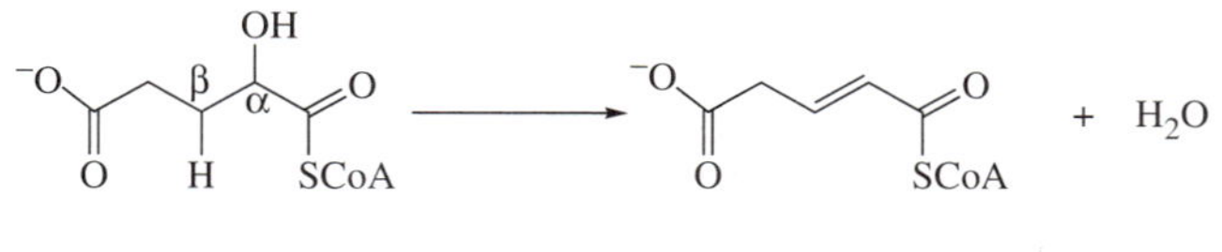

(2R)-Lactyl CoA dehydratase from *Clostridium propionicum* consists of two proteins that contain two types of iron-sulfur centers: a four-iron and a three-iron center. The enzyme also contains FMN and is activated by ATP (Flint and Allen, 1996). 2-(2R)-Hydroxyglutaryl CoA dehydratase from *Acidaminococcus fermentans* is a heterodimer (100 kDa) that contains reduced riboflavin, FMN, and approximately four Fe and four sulfide ions per dimer. It too is activated by ATP. The reaction mechanisms are not known, nor are the mechanisms of ATP activation. The reactions are thought to involve radical intermediates.

Molybdenum-Iron-Sulfur Center of Nitrogenase

Nitrogenase catalyzes the reduction of nitrogen gas to ammonia in a complex process that requires another iron-sulfur protein, as discussed in chapter 18. The actual reduction of nitrogen is brought about by the iron-molybdenum-sulfur coenzyme, the structure of which is illustrated in fig. 4-31.

S-Adenosylmethionine and Iron-Sulfur Centers

A rapidly growing number of enzymes have been found to require S-adenosylmethionine (SAM) and contain [4Fe–4S] centers that function together in catalysis. These enzymes may be grouped into two subclasses, those that use SAM as a coenzyme and those that use SAM as a substrate. The two subclasses both catalyze radical formation by reductive cleavage of SAM. The enzymes are genetically related and have become known as the *radical SAM enzymes* (Sofia et al., 2001).

The first two of these enzymes to be discovered represent the two subclasses. Lysine 2,3-aminomutase catalyzes the interconversion of α- and β-lysine and requires SAM as a coenzyme (Chirpich et al., 1970). Pyruvate formate-lyase activase (PFL activase) requires SAM as a substrate to activate PFL, and SAM is cleaved into methionine and 5′-deoxyadenosine in the activation process, which generates a glycyl radical in activated PFL (Knappe et al., 1984). The reactions catalyzed by several of these enzymes are illustrated in fig. 4-34 as eqs. 4-29 to 4-34. The SAM/[4Fe–4S] enzymes catalyze abstraction of hydrogen atoms from carbon atoms in substrates as a common feature. This is brought about by the 5′-deoxyadenosyl radical arising from reductive cleavage of SAM by the [4Fe–4S] centers.

Catalytic Action of S-Adenosylmethionine and [4Fe–4S] Centers

Lysine 2,3-Aminomutase

Research on lysine 2,3-aminomutase (EC 5.4.3.2) has clarified many aspects of the role of SAM and the 5′-deoxyadenosyl radical in these reactions. Lysine 2,3-aminomutase catalyzes reaction 4-29 in fig. 4-34. The role of PLP in the radical isomerization has been discussed in chapter 3 (fig. 3-25). The role of SAM is to mediate hydrogen transfer in the same way as adenosylcobalamin in coenzyme B_{12}-dependent reactions (Baraniak et al., 1989; Moss and Frey, 1987). Activation of the enzyme with [5′-^{3}H]SAM leads to [^{3}H]lysine and β-[^{3}H]lysine. In the mechanism, the 5′-deoxyadenosyl radical is generated reversibly in a reaction between SAM and [4Fe–4S]$^{1+}$ at the active site. This radical abstracts the 3-pro-R from the side chain of lysine, which is bound as its external aldimine with PLP, creating the lysyl=PLP radical and 5′-deoxyadenosine. The lysyl=PLP radical undergoes isomerization to the β-lysyl=PLP radical by the mechanism in fig. 3-25, and hydrogen abstraction from the methyl group of 5′-deoxyadenosine by the β-lysyl=PLP radical regenerates the 5′-deoxyadenosyl radical and produces the external aldimine of β-lysine.

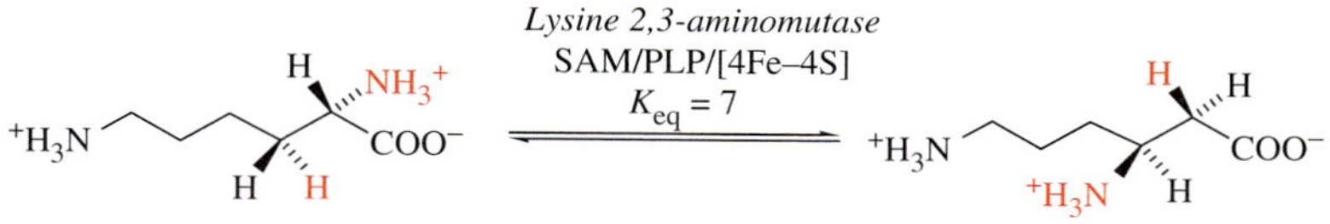

Coenzymes II: Metallic Coenzymes

$$\text{Lysine 2,3-aminomutase} \atop \text{SAM/PLP/[4Fe–4S]}$$

(4-29)

$$\text{PFL-Gly}_{734}\text{-H} + \text{SAM} \xrightarrow[\text{[4Fe–4S]}]{\text{PFL activase}} \text{PFL-Gly}_{734}\bullet + \text{Met} + 5'\text{-dAdo} \quad (4\text{-}30)$$

(inactive) → (active)

$$\text{ARR-Gly}_{680}\text{-H} + \text{SAM} \xrightarrow[\text{[4Fe–4S]}]{\text{Anaerobic ribonucleotide reductase (ARR)}} \text{ARR-Gly}_{680}\bullet + \text{Met} + 5'\text{-dAdo} \quad (4\text{-}31)$$

(inactive) → (active)

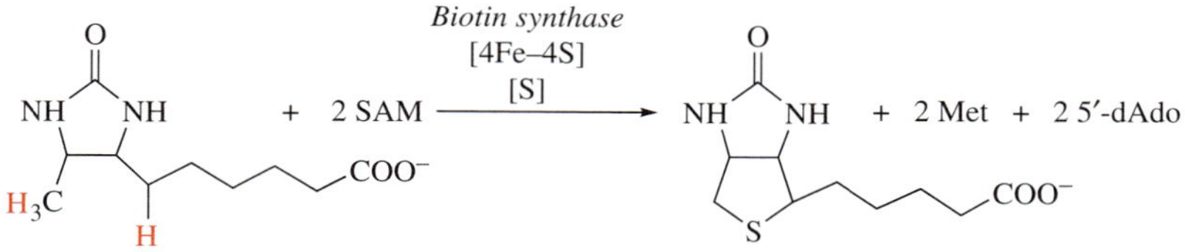

Biotin synthase [4Fe–4S] [S] + 2 SAM → + 2 Met + 2 5'-dAdo (4-32)

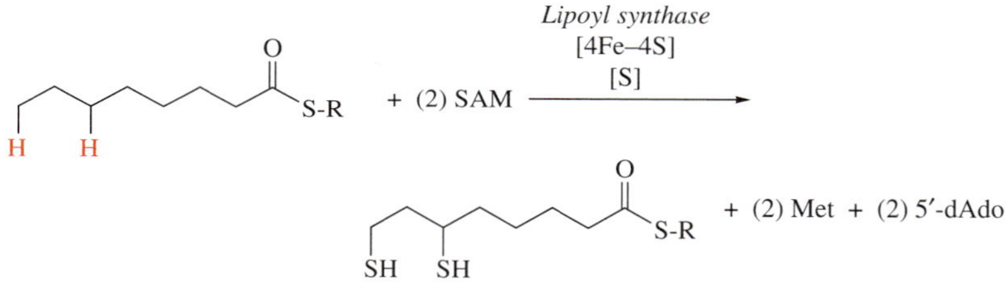

Lipoyl synthase [4Fe–4S] [S] + (2) SAM → + (2) Met + (2) 5'-dAdo (4-33)

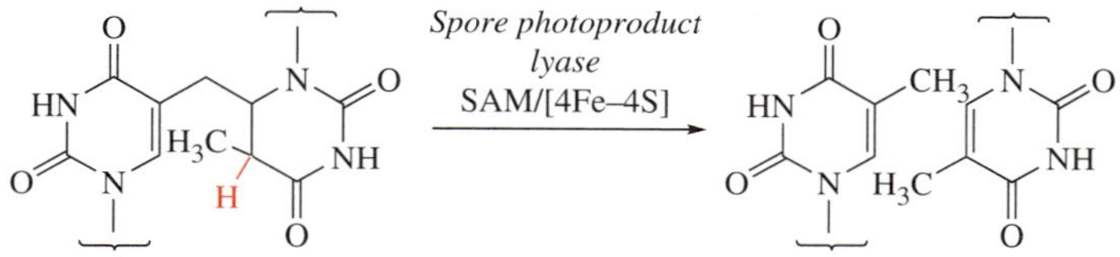

Spore photoproduct lyase SAM/[4Fe–4S] (4-34)

Fig. 4-34. Reactions catalyzed by enzymes that depend on *S*-adenosylmethionine (SAM) and [4Fe–4S] centers are of two types, those in which SAM and the [4Fe–4S] center function catalytically and those in which SAM functions stoichiometrically. In all of the reactions, SAM is cleaved to methionine and the 5'-deoxyadenosyl radical, which abstracts hydrogens (red) from the substrates. In the reactions of lysine 2,3-aminomutase and spore photoproduct lyase hydrogen, abstraction is catalytic because the hydrogen is returned the product. The pyruvate formate-lyase (PFL) and anaerobic ribonucleotide reductase (ARR) are active only when glycyl radicals are formed to initiate radical mechanisms. The sources of sulfur, [S], are not well defined but may be specialized iron sulfide centers.

An essential feature of this mechanism is the reversible cleavage of SAM to the 5'-deoxyadenosyl radical and methionine. This reaction cannot take place in the manner of the reversible homolytic scission of the Co—C5' bond in adenosylcobalamin because the S—C5' bond in SAM is too strong (>60 kcal mol^{-1}). The much weaker (30 kcal mol^{-1}) Co—C5' bond in adenosylcobalamin can be cleaved by the use of binding energy harvested from the interactions of an enzyme with remote parts of the coenzyme to strain the bond. In the case of SAM, the bond must be weakened before it can be cleaved.

Weakening of the S—C5′ bond is brought about by electron transfer into SAM from the [4Fe–4S]⁺ center (Frey, 1990; 2001).

The mechanism by which an electron is transferred to SAM and the S–C5′ bond is homolytically cleaved is a subject of intense investigation in this family of enzymes. Members of the radical SAM superfamily have a characteristic cysteine motif (CxxxCxxC) with no conserved fourth cysteine for binding iron in the [4Fe–4S] center. The ligands to the fourth iron have recently been discovered. Evidence from x-ray absorption spectroscopy (XAS) in the reaction of lysine 2,3-aminomutase indicates that the cleavage of *Se*-adenosylselenomethionine (*Se*SAM) leads to ligation of the [4Fe–4S] center with the selenoether group of selenomethionine (Cosper et al., 2001). *Se*SAM activates lysine 2,3-aminomutase nearly as well as SAM. ENDOR experiments with ¹⁷O and ¹⁵N prove that the carboxylate and amino groups of the methionyl moeity in SAM are the other ligands (Chen et al., 2003b; Walsby et al., 2002). The inner sphere electron transfer mechanism of eq. 4-35 is postulated to account for the reversible generation of the 5′-deoxyadenosyl radical in the reaction of SAM with the [4Fe–4S] center.

$$\text{(structure with [4Fe–4S]⁺/Ado-CH}_2\text{–SAM complex)} \rightleftharpoons \text{(structure with [4Fe–4S]}^{2+}\text{/·CH}_2\text{–Ado radical complex)} \tag{4-35}$$

Evidence by XAS spectroscopy strongly implicates ligation of the methionine side chain by iron. However, it has not been proved that the S—C5′ bond is cleaved in concert with electron transfer, as is implied by eq. 4-35, nor has the ligation of the thioether group of methionine by iron been proved in the actions of other reactions in this family.

Spore Photoproduct Lyase

The reaction of eq. 4-34 in fig. 4-34 describes the repair of methylene bridged thymine dimers in DNA. Spore photoproduct lyase, a member of the radical SAM superfamily, catalyzes this reaction. The enzyme requires SAM and contains a [4Fe–4S] center, but unlike lysine 2,3-aminomutase it does not require PLP. The enzyme cleaves SAM in the course of repairing DNA, and SAM mediates hydrogen transfer in this process (Cheek and Broderick, 2002; Rebeil and Nicholson, 2001). The elegant mechanism in fig. 4-35 accounts for the available facts.

Stoichiometric Reactions of *S*-Adenosylmethionine and [4Fe–4S] Centers

In reactions 4-30 through 4-33 of fig. 4-34, SAM undergoes irreversible reductive cleavage to 5′-deoxyadenosine and methionine. In the activation of PFL, benzylsuccinate synthase, and anaerobic ribonucleotide reductase glycyl residues are converted into radicals by hydrogen abstraction through the action of SAM/[4Fe–4S] centers. The glycyl radicals serve to initiate radical reactions of substrates. In the reactions of biotin synthase and lipoyl synthase, radicals have not been detected but may be involved as transient intermediates. We discuss PFL and PFL activase in chapter 8, biotin synthase in chapter 13, and anaerobic ribonucleotide reductase in chapter 16.

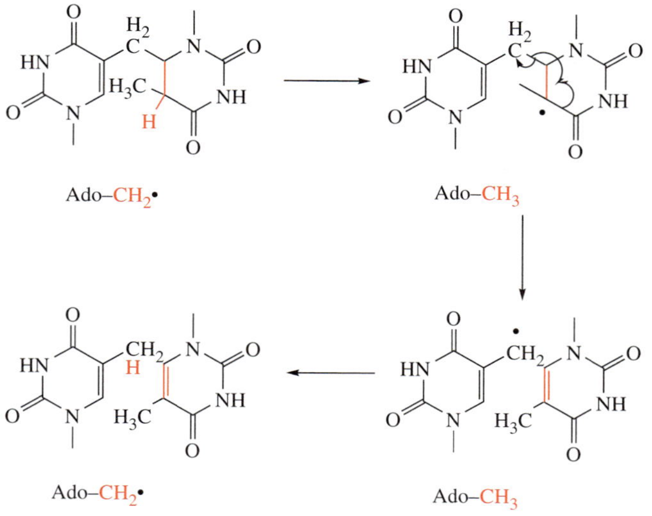

Fig. 4-35. A hypothetical mechanism for the repair of DNA by the spore photoproduct lyase. In this mechanism, the 5'-deoxyadenosyl radical initiates the reaction by abstracting hydrogen from C5 of the bridged dimer, and radical fragmentation then separates the rings, generates one thymine ring, and leads to the methylene radical of the other thymine. Hydrogen abstraction from 5'-deoxyadenosine completes the repair and regenerates the 5'-deoxyadenosyl radical.

Divalent Metal Ions

Divalent metal ions participate as cofactors in many enzymatic reactions. In the simplest and most obvious case, the magnesium ion in MgATP may be regarded as a cofactor in almost all reactions of ATP. In this role, Mg^{2+} almost always functions essentially to neutralize negative charges in ATP. In the following sections, we consider other roles of divalent metal ions in catalysis.

Electrostatic Activation of Coordinated Water

In general, the catalytic effect of divalent metal ions can be attributed to their electrostatic effects on the chemical properties of ligands to which they are coordinated. A fairly clear concept of the magnitude of this effect can be acquired by considering the acid strengthening effect of a divalent metal ion on a coordinated water molecule. Table 4-3 lists the range of effects of divalent metal ion coordination on the pK_a of water in a collection of divalent metal ions. The values of pK_a can be compared with the corresponding value of 15.7 for the ionization of a water molecule not coordinated to a metal. The values of pK_a in table 4-3 range from 2.5 to 11.3 log units below the value for water, corresponding to free energy differences of 3.5 to 16 kcal mol^{-1}. In general, the smaller ionic radius of the metal ion the lower the value of pK_a for coordinated water. Beryllium displays the largest effect, but this ion is biologically insignificant apart from its extreme toxicity. The biologically relevant ions are Ca^{2+}, Mg^{2+}, Zn^{2+}, Fe^{2+}, Mn^{2+}, and rarely Ni^{2+}.

Hydrolytic enzymes in which a single divalent metal ion coordinates a reacting water molecule include carboxypeptidase A and carbonic anhydrase. Note that these are zinc metalloenzymes with tetrahedrally coordinated zinc. The other ligands to zinc are three histidines in carbonic anhydrase and two histidines and a glutamate in carboxypeptidase A. The pK_a of water coordinated to tetrahedral zinc will be significantly lower than the

Table 4-3. Values of pK_a for Water Coordinated to Divalent Metal Ions

Metal	pK_a[a]	Ionic radius (Å)[b]	Ionic radius (Å)[c]
Ba^{2+}	13.1	1.49	—
Be^{2+}	4.3	0.59	0.41
Ca^{2+}	12.5	1.14	—
Cd^{2+}	9.8	1.09	0.92
Co^{2+}	9.4	0.89	0.72
Fe^{2+}	8.4	0.92	0.77
Mg^{2+}	11.4	0.86	0.63
Mn^{2+}	10.1	0.97	0.80
Ni^{2+}	9.0	0.83	0.69
Zn^{2+}	9.6	0.88	0.74

[a]K_a = [H$^+$][M(H$_2$O)$_5$OH]$^+$/[M(H$_2$O)$_6^{2+}$]
[b]From X-ray crystallographic measurements of hexacoordinate complexes (in high spin states where relevant).
[c]From X-ray crystallographic measurements of tetracoordinate complexes.
Data from Cotton and Wilkinson, 1988.

value of 9.6 in table 4-3 for hexacoordinated zinc because of the smaller ionic radius in tetrahedral coordination. This effect is likely to lead to a further lowering of perhaps 2.5 log units in the pK_a, so that ionization of coordinated water at pH 7 becomes possible. The importance of the lowered pK_a of zinc-coordinated water in the actions of carbonic anhydrase and carboxypeptidase A are discussed in chapters 9 and 6, respectively. In chapter 16, we discuss the function of zinc in lowering the pK_a of ethanol in the active site of alcohol dehydrogenase.

A single biologically relevant divalent metal ion in a hexacoordination sphere does not lower the pK_a of water enough to allow its ionization to hydroxide at pH 7. However, the effects of two divalent metal ions coordinated to a single water molecule can be expected to be twice as large as those listed in table 4-3, and would allow facile ionization of water to hydroxide at neutrality. This may explain the participation of two divalent metal ions in many enzymatic hydrolytic reactions. Hydrolytic enzymes in which two divalent metal ions participate include purple acid phosphatases, urease, alkaline phosphatase, and inorganic pyrophosphatase among others. Figure 4-36 shows the active site structures of urease, Fe/Fe purple acid phosphatase, leucine aminopeptidase, and arginase.

Electrostatic Activation of Enolization

The enolase superfamily of enzymes catalyze the enolization of substrates as steps in their mechanisms of action. Many members of this family depend on one or two divalent metal ions, usually Mg^{2+}, to facilitate enolization through the electrostatic stabilization afforded by coordination of a divalent metal ion to the enolate ions. This family is named for enolase, the most widely known member.

In chapter 1, we discuss the role of Mg^{2+} in facilitating enolization at the active site of enolase. A very high activation barrier must be negotiated in the dehydration of 2-phosphoglycerate to phosphoenolpyruvate catalyzed by enolase. Ionization of C2(H) constitutes the main barrier because of the high pK_a ($\approx$34) for this process. EPR experiments substituting Mn^{2+} for Mg^{2+} showed that two divalent metal ions coordinated the hydroxamate group of the specific inhibitor phosphonoacetohydroxamate (Poyner and Reed, 1992). X-ray crystallography confirmed the coordination of two magnesium ions to the carboxylate group of 3-phosphoglycerate and revealed an additional electrostatic interaction with

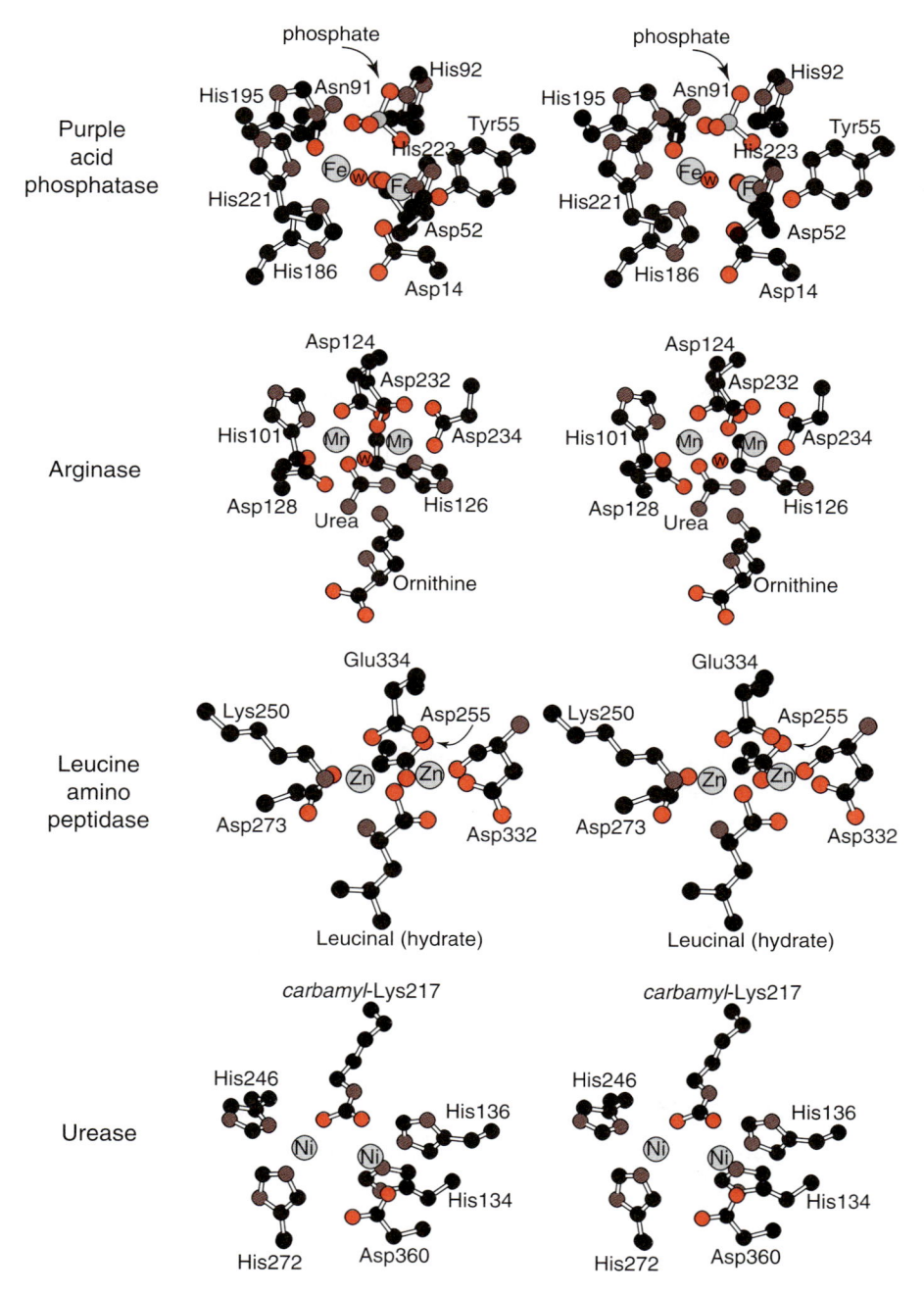

Fig. 4-36. Active site structures of hydrolytic enzymes with dimetallic centers. Stereoimages depict the active sites of four different hydrolytic enzymes that use iron, manganese, zinc, or nickel dimetalic centers for catalysis. The first is pig purple acid phosphatase (EC 3.1.3.2), which uses a di-iron center for phosphoester hydrolysis (PDB 1UTE; Guddat et al., 1999). The second is rat arginase (EC 3.5.3.1), which cleaves arginine to urea and ornithine using a dimanganese center (PDB 1HQG; Cox et al., 2001). The third is cow leucine aminopeptidase (EC 3.4.11.1), which catalyzes peptide bond hydrolysis using a dizinc center (PDB 1LAN; Strater and Lipscomb, 1995). The fourth is *Klebsiella aerogenes* urease (EC 3.5.1.5), which hydrolytically converts urea to carbon dioxide and ammonium using a dinickel center that is bridged by carbamylated lysine 217 (PDB 2KAU; Jabri et al., 1995).

a lysine-ε-aminium ion (Reed et al., 1996). Five electropositive charges are brought to bear on the carboxylate group of 2-phosphoglycerate, and the consequent electrostatic stabilization of the incipient enolized substrate, an *aci*-carbanion, presumably lowers the activation barrier to ionization. Enolase is discussed in more detail in chapters 1 and 9.

Copper as a Cofactor

Although copper is a trace nutrient, it is found in many proteins that carry out a variety of functions, including electron transfer, reversible oxygen binding, and catalysis of oxidase and oxygenase reactions. Copper is commonly in the 2+ oxidation state, which has one unpaired electron ($3d^9$) and $S = 1/2$ and gives a characteristic EPR signal at room temperature. The other oxidation state in biological systems is the diamagnetic 1+, with no unpaired electrons ($3d^{10}$). The 3+ oxidation state is known in chemistry but not in biological systems. Copper displays the full range of coordination possibilities, from linear 2 coordinate to 3, 4, 5, and 6 coordinate states.

Copper Proteins

Type 1 copper proteins are the blue copper proteins, which contain Cu^{2+} and display an intense blue color ($\lambda_{max} \approx 600$ nm, $\varepsilon \approx 1000$ to 7000 M^{-1} cm^{-1}). The extinction coefficients are about 100 times the normal values for copper complexes. Copper consists of two isotopes, ^{63}Cu (69%) and ^{65}Cu (31%), with spin 3/2 in the 2+ oxidation state and nearly equal magnetogyric ratios. Nuclear hyperfine splitting of the EPR signals lead to the complex pattern characteristic of Cu^{2+}. Type I copper proteins have at least one sulfur ligand, cysteine or methionine, and often contain both. Type II copper proteins in the 2+ oxidation state display normal optical and EPR spectra. Type II copper proteins almost always contain multinuclear copper centers. Type III copper proteins are thought to incorporate two antiferromagnetically coupled Cu^{2+} ions. Type III copper does not have an EPR signal because of antiferromagnetism, and the typical optical spectrum is a maximum at 330 nm.

Ascorbate Oxidase

The enzyme catalyzes the oxidation of ascorbic acid to dehydroascorbate in a four-electron reduction of oxygen to water according to eq. 4-36.

$$\text{ascorbate} + 1/2\, O_2 \longrightarrow \text{dehydroascorbate} + H_2O \qquad (4\text{-}36)$$

As the stoichiometry proves, a molecule of oxygen cannot be reduced in one turnover of ascorbate because only two electrons are available. Two molecules of ascorbate must be oxidized to reduce one molecule of oxygen. The enzyme must store reducing equivalents from ascorbate before reducing oxygen to two molecules of water.

The two copper centers in ascorbate oxidase may allow for the separate reactions of ascorbate and oxygen. Ascorbate oxidase contains copper types I, II, and III and a total of four moles of copper per mole of enzyme. All four coppers can be reduced to Cu^+ by ascorbate, and this process is accompanied by the transient appearance of EPR signals consistent with the presence of free radicals. Because of the structure of ascorbate, which allows for the extensive delocalization of an unpaired electron, the oxidation can proceed in one-electron steps corresponding to the two oxidation states of copper. The two copper centers are known from the x-ray structure to consist of a single type I copper site and a trinuclear copper site, in which the three coppers are separated by 12 Å to 15Å from the mononuclear copper site (Messerschmidt et al., 1992). It seems likely that the two copper sites interact separately with the two substrates, and it is thought that the ascorbate binding site is near the mononuclear copper, with the oxygen being reduced at the trinuclear copper site.

Galactose Oxidase

In fungi the 6-hydroxymethyl groups of D-galactosyl residues at the nonreducing termini of polysaccharides are oxidized to the corresponding aldehydes by the action of galactose oxidase (EC 1.1.3.9). Neither D-glucosyl nor L-galactosyl units are attacked, although the enzyme is remarkably unspecific when acting on small molecules such as simple alcohols and dihydroxyacetone according to eq. 4-37.

$$RCH_2OH + O_2 \rightarrow RCHO + H_2O_2 \tag{4-37}$$

The active site of galactose oxidase contains novel coenzymes consisting of a cysteine-bridged tyrosyl radical coordinated to a Cu^{2+} ion shown in fig. 4-37. The presence of Cu^{2+} and the tyrosyl radical implies that galactose oxidase should display an EPR spectrum; however, the resting enzyme is EPR silent. The nature of the site and an explanation for the EPR silence came about through careful experimentation and the application of logic (Whittaker and Whittaker, 1990). Active and inactive forms of galactose oxidase both proved by spectroscopy to contain only Cu^{2+}. Removal of copper and treatment with ferricyanide produced a free radical site on an amino acid, and the effect of biosynthetically incorporated tyrosine-3-d_2 on the spectrum proved it to be a tyrosyl residue. The absence of an EPR signal was attributed to antiferromagnetic coupling between the tyrosyl radical and Cu^{2+}. X-ray crystallography showed the thioether bridge between Cys228 and Tyr272 (Ito et al., 1994).

The mechanism of oxidation by galactose oxidase is not definitively known. However, a reasonable hypothetical mechanism is that in fig. 4-38. The unpaired electron on Tyr272 is delocalized, and significant spin on oxygen may allow it to serve in hydrogen abstraction from the substrate CH_2OH group to generate the substrate radical •CHOH. Electron transfer to Cu^{2+} produces Cu^+ and the substrate oxycation, which readily loses the hydroxy proton to form the aldehyde product. Reaction of the reduced enzyme with oxygen regenerates the Cu^{2+}—Tyr272 radical and hydrogen peroxide.

Other Copper Enzymes

Many other enzymes contain copper that participates in catalysis. These enzymes generally react in oxidation and oxygenation reactions. Examples include the quinoprotein copper amine oxidases (see chap. 3); a membrane-bound methane monooxygenase, about which little is known except that the reaction mechanism differs from that of the soluble methane monooxygenase; tyrosinase; laccase; cytochrome oxidase; and electron-transfer proteins such as the photosynthetic plastocyanins. One type of superoxide dismutase is a Cu/Zn enzyme.

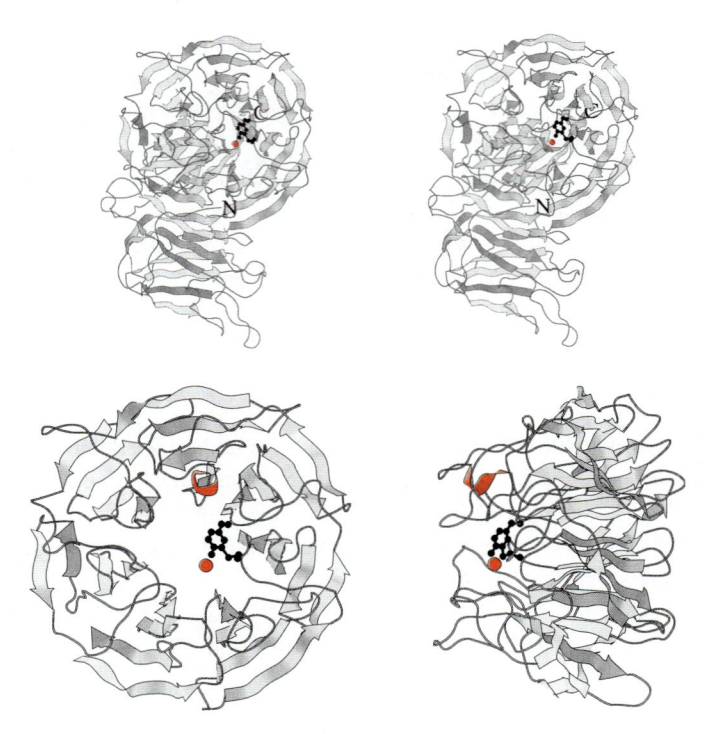

Fig. 4-37. Structure of galactose oxidase. A stereodiagram (top) shows the overall fold of galactose oxidase, with the copper ion highlighted in red and the thioether linked Tyr272 and Cys228 in black. Head-on and side views (bottom) of the 7-blade β-propeller domain show the copper ion and thioether-linked residues at the center. The entire propeller domain consists of β-sheet and loops, except for one small helix near the active site (red). The image was generated using PDB 1GOF (Ito et al., 1991).

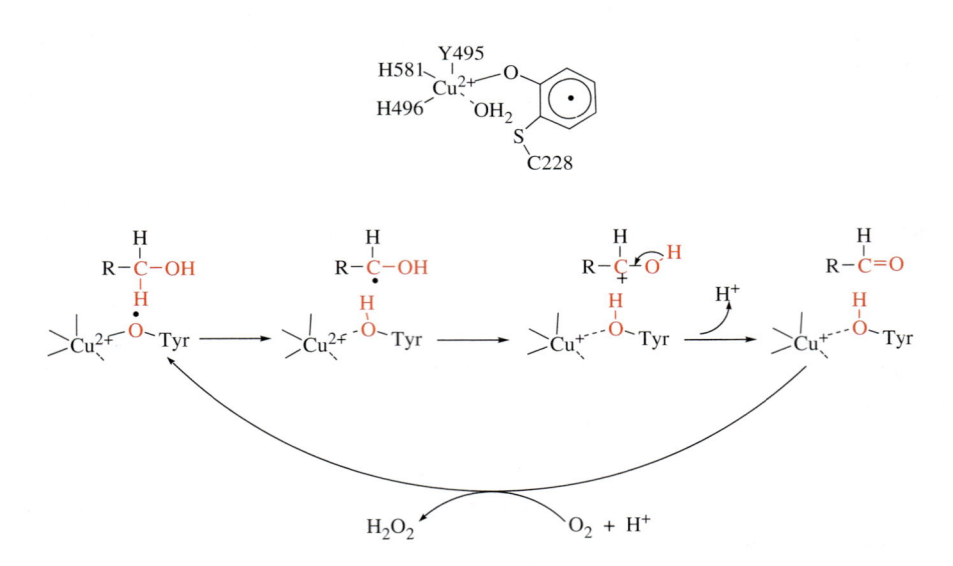

Fig. 4-38. Hypothetical mechanism of the reaction of galactose oxidase.

Nickel Coenzymes

Nickel was discussed in the section on divalent metal ions as the dinuclear metallocoenzyme of urease. Nickel has an interesting history in enzymology. Jack bean urease was the first enzyme to be crystallized in 1926 (Sumner, 1926), but the crystal structure of *Klebsiella* urease was not solved until 1995 (Jabri et al., 1995). In the intervening years, little was learned about the mechanism of action of urease, perhaps because the presence of nickel was not known until the crystal structure was solved. For almost 60 years, biochemists did not know that the first enzyme to be crystallized contained nickel. During that time, nickel turned up elsewhere, principally in enzymes of methanogenesis.

Nickel in Methanogenesis

Methanogenesis

The methanogenesis pathway as illustrated in fig. 4-39 reduces carbon dioxide to methane, and a nickel coenzyme plays an essential role in this process (DiMarco et al., 1990; Weiss and Thauer, 1993; Wolfe, 1991). In methanogenesis, tetrahydromethanopterin (H$_4$MPT) functions in a central role much like tetrahydrofolate, and factor F$_{420}$ is a 5-deazaflavin. These variants of familiar coenzymes are specially adapted by their reduction potentials to the chemistry required in the reduction of carbon dioxide to the level of methanol (Thauer et al., 1993; Weiss and Thauer, 1993). Novel coenzymes also play essential roles, including methanofuran (MFR), coenzyme M (CoM-SH), coenzyme B (CoB–SH), and factor F$_{430}$, the structures of which are shown in fig. 4-40. Factor F$_{430}$ is the nickel coenzyme in the last step of methane production by methyl coenzyme M reductase.

Methyl Coenzyme M Reductase

In methane production, methyl coenzyme M reductase catalyzes the reaction of methyl coenzyme M (CH$_3$-CoM) with coenzyme B (CoB-SH) to form methane and the disulfide CoM-S–S-CoB. Methyl coenzyme M reductase contains two copies of each of three subunits ($\alpha_2\beta_2\gamma_2$) (Thauer et al., 1993). The coenzyme F$_{430}$ of methyl coenzyme M reductase is related to corrin and heme but is more highly reduced and has Ni(II) as the coordinating metal (Diekert et al., 1980; Farber et al., 1991; Pfaltz et al., 1982). The location of F$_{430}$ in the structure of this enzyme is illustrated in fig. 4-41.

The mechanism of methane production in the reaction of methyl coenzyme M reductase is not known but is likely to be novel and of considerable interest for both theoretical and practical reasons. More than 10^9 tons of methane are produced annually in microbial processes, and this reaction may be a significant environmental factor. The chemistry of F$_{430}$ coupled with the binding properties of the reductase facilitates this remarkable reaction. The mechanism may involve the transient methylation of nickel in F$_{430}$. The reduction potential for the couple Ni(II)/Ni(I) of F$_{430}$ in solution is similar to that for Co(II)/Co(I) in cobalamin. Their similar chemical properties extend to the reactions of the reduced forms as nucleophiles with methyl iodide to form methyl-F$_{430}$ and methylcobalamin, respectively. However, methylcobalamin is stable in the absence of light, whereas methyl-F$_{430}$ decomposes to methane in aqueous solution. The transient formation of methyl-F$_{430}$ in the active site of the reductase can account for methane formation. This requires the one-electron reduction of F$_{430}$, presumably by CoB-SH, and methylation by methyl-CoM.

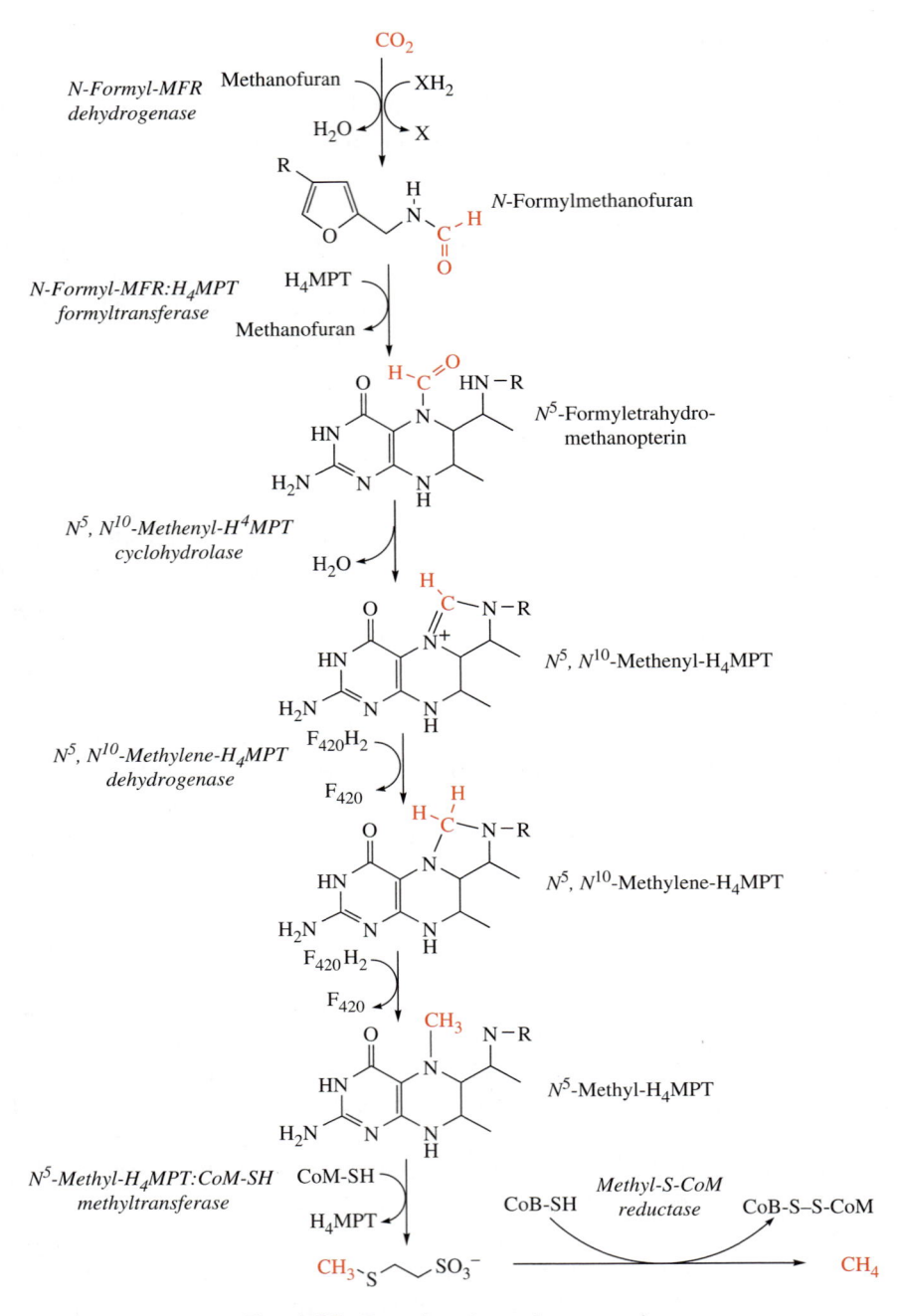

Fig. 4-39. Reactions in methanogenesis.

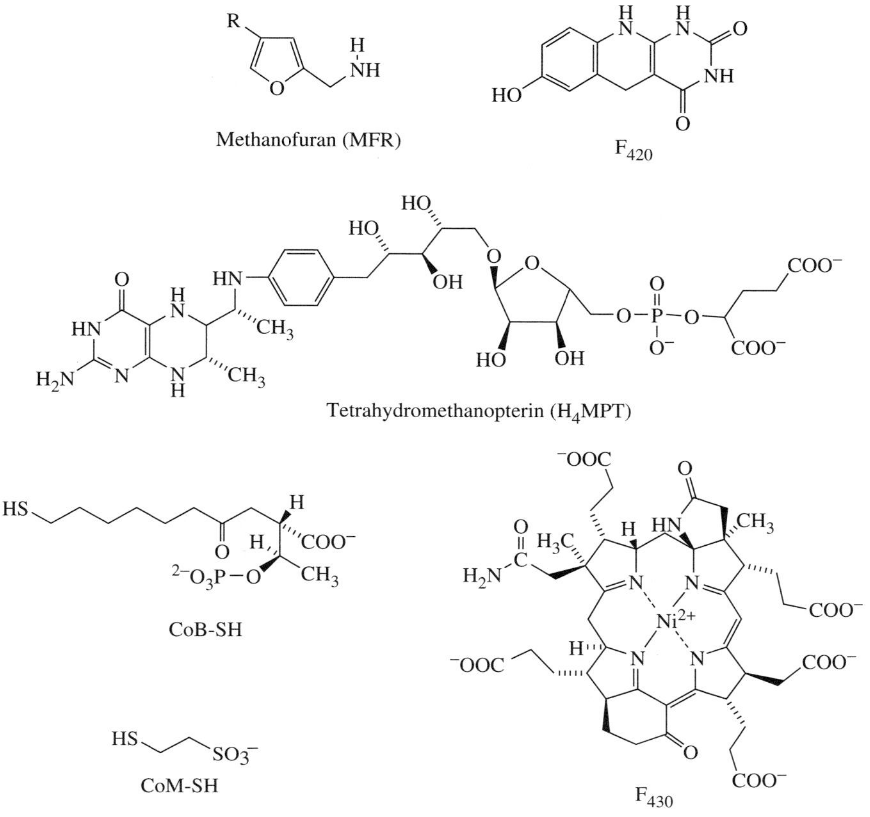

Fig. 4-40. Structures of specialized coenzymes in methanogenesis.

Spectroscopic and kinetic evidence indicates that the active form of methyl CoM reductase contains Ni(I)F$_{420}$ (Rospert et al., 1991). The chemical mechanism for the reduction of methyl-CoM remains to be elucidated definitively; the hypothetical mechanism in fig. 4-42 has been proposed (Berkessel, 1991).

Other Nickel Coenzymes

Hydrogenases catalyze the production of hydrogen gas in the reduction of hydrogen ions. The overall reaction may be expressed as $2H^+ + 2e^- = H_2$; $E°' = -0.414$ V. Most hydrogenases contain binuclear metal sulfur centers, including [NiFe] binuclear centers and [Fe$_2$] binuclear centers. In methanogenesis, hydrogen is the ultimate reducing agent in four of the steps, often as XH$_2$ in the first step of fig. 4-39, as the reducing agent in the production of H$_2$F$_{420}$, and in the reduction of HTP–S–S–CoM to HTP–SH and CoM–SH.

Carbon monoxide dehydrogenase/acetyl CoA synthase (CODH/ACS) contains Ni, Fe, and Cu (Doukov et al., 2002). It catalyzes two reactions, the reversible reaction of carbon dioxide with an electron donor to produce carbon monoxide and water and the reaction of carbon monoxide with CoA and methyltetrahydrofolate to produce acetyl CoA and tetrahydrofolate. The essential coenzyme is a [4Fe–4S] center linked through a bridging atom to nickel. Other nickel-containing proteins include a rare nickel superoxide dismutase (Youn et al., 1996) and nickel chaperonins (Hausinger, 1997).

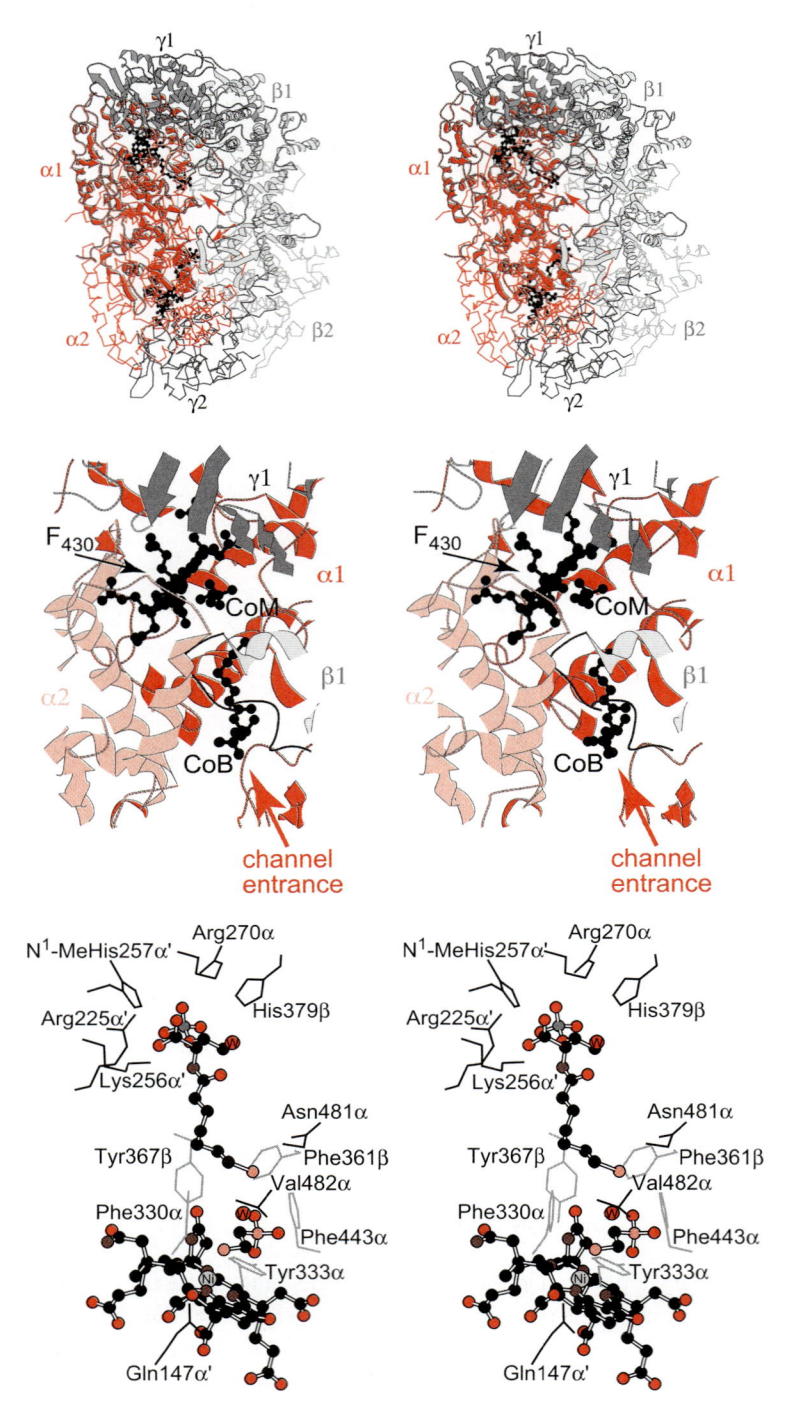

Fig. 4-41. Structure of methyl coenzyme M reductase. In the stereoview of the $\alpha_2\beta_2\gamma_2$ fold of the methyl coenzyme M reductase from *Methanobacterium thermoautotrophicum,* the α subunits are shown in red, the β subunits in light gray, and the γ subunits in darker gray, with one $\alpha\beta\gamma$ unit drawn using ribbon representation and the other $\alpha\beta\gamma$ unit as a Cα trace. The CoM, CoB, and F430 cofactors are all shown with black ball-and-stick models. Close-up stereoview (middle) shows the active site containing all three cofactors. The active site occurs at the junction of four subunits (1α, 1β, 1γ, and 2α), with access to the active site through the indicated channel entrance. A more detailed stereodiagram (bottom) of the active site shows ball-and-stick models of the cofactors colored by atom type, with the various active-site protein side chains shown as stick figures. The CoB and CoM thiol moieties are stacked above the Ni atom on one face of F430, with a single water molecule interceding. The illustration was generated using PDB 1MRO (Ermler et al., 1997).

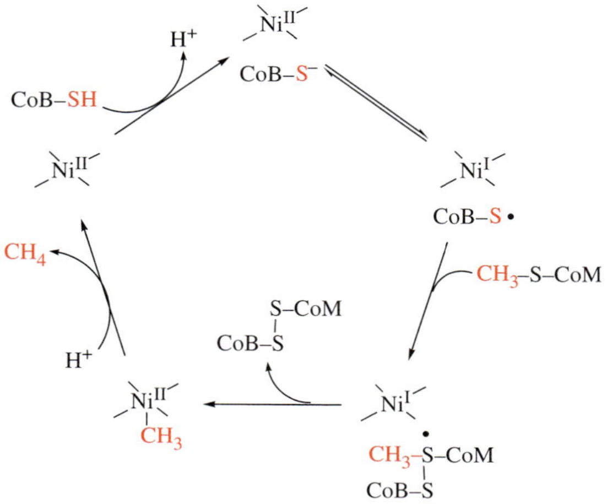

Fig. 4-42. A hypothetical mechanism of methane production by methyl CoM reductase was suggested by Berkessel (1991), but it has not been proven. It satisfies basic stoichiometric requirements, but it is not certain how the first step, reduction of Ni(II) to Ni(I), occurs and whether CoB is the reducing agent. There is evidence that Ni(I) and methyl-F_{430} participate as intermediates.

Long-Range Electron Transfer

Biological Electron Transfer

We have pointed out the likely importance of short-range electron transfer over distances of a few angstrom units within the confines of active sites in reaction mechanisms involving coenzymes, including heme, mononuclear iron, oxo-Fe_2, molybdopterins, copper, quinoproteins, flavoproteins, and tetrahydrobiopterin. Short-range electron transfer is very fast relative to enzymatic rates, occurring over distances corresponding to the lengths of a few covalent bonds (≈5 to 6 Å).

Electron transfer over significantly longer distances must occur in the actions of a few enzymes that we have already discussed as well as of many others, especially enzymes that participate in vectorial processes across membranes. Important examples of the latter are cytochrome oxidase and nitrogenase (see chap. 18). Long-range electron transfer in biology inevitably requires coenzymes, often metallocoenzymes.

In this chapter, we have presented molybdoproteins as enzymes that oxidize or reduce substrates, but we have not explained how the electrons involved are transferred to or from external electron acceptors or donors. The structure of xanthine oxidase includes a flavin binding site and two [2Fe–2S] centers intervening between the flavin and molybdopterin sites (fig. 4-26). In the reaction pathway (fig. 4-27), the substrate reduces the molybdopterin, and the catalytic cycle is completed by reoxidation of Mo(IV) to Mo(VI). This process consists of a sequence of long range electron transfers, in which the [2Fe–2S] centers relay electrons from molybdopterin to FAD, and electron transfer is extended from FAD to an external electron acceptor. The spatial relationships among the molybdopterin, the [2Fe–2S] centers, and FAD in the structure of xanthine oxidase are shown in fig. 4-26.

In chapter 3, we discussed the role of tryptophan tryptophyl quinone (TTQ) in the action of methylamine dehydrogenase. The dehydrogenation of methylamine leads to the aminoquinol form of TTQ, and its reoxidation requires electron transfer to cytochrome c. Electron transfer in this system is mediated by the copper protein amicyanin, which forms

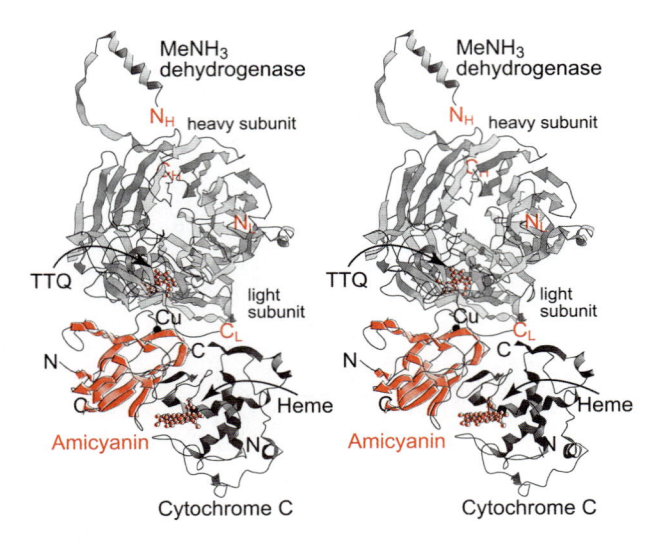

Fig. 4-43. Structure of a complex of methylamine dehydrogenase, amicyanin, and cytochrome c. The complex of methylamine dehydrogenase, amicyanin, and cytochrome c, shown here as a stereodiagram, indicates the necessity for long-range electron transfer in this multiprotein complex, as is required for the oxidative reactivation of the methylamine dehydrogenase tryptophan tryptophyl quinone (TTQ) cofactor. Electrons flow from TTQ to copper to heme. The image was generated using the PDB 2MTA (Chen et al., 1994).

a molecular complex with methylamine dehydrogenase and cytochrome c. The x-ray crystallographic structure of this complex in fig. 4-43 shows amicyanin intervening between methylamine dehydrogenase and cytochrome c, regardless of whether this structure accurately represents the functional complex, the reoxidation of the aminoquinol form of TTQ must proceed by long-range electron transfer.

Marcus Theory

The Marcus Equation

Marcus theory clarifies the interplay of physicochemical barriers in processes including electron transfer (Dutton and Moser, 1994; Marcus and Sutin, 1985; Moser et al., 1992; Page et al., 1999). The terms in the Marcus equation 4-38 are defined as follows.

$$\Delta G^{\ddagger} = \Delta G^{\ddagger}_{o}[1 + \Delta G^{o}/4\Delta G^{\ddagger}_{o}]^2 \qquad (4\text{-}38)$$

The experimental free energy of activation for electron transfer in a reaction under consideration is $\Delta G^{\ddagger}$. In the Marcus equation, $\Delta G^{\ddagger}_{o}$ is defined as the intrinsic activation energy for the reaction. The intrinsic activation energy pertains to a reference reaction that is identical in every way to the one under consideration except that it has no thermodynamic driving force; the standard free energy change is zero ($\Delta G^{o}_{o} = 0$). The true standard free energy change for the reaction under consideration is ΔG^{o} in the Marcus equation.

According to the Marcus equation, the free energy of activation is zero ($\Delta G^{\ddagger} = 0$) when $\Delta G^{o} = -4\Delta G^{\ddagger}_{o}$. Zero activation energy corresponds to no barrier for the reaction and confers a particular significance to the value of $4\Delta G^{\ddagger}_{o}$, which is known as the reorganizational energy and is often symbolized as λ ($\lambda = 4\Delta G^{\ddagger}_{o}$). Electron transfer imposes structural consequences on both the electron donor and acceptor. Simple loss of an electron by a donor will leave it in an unstable structure for its oxidized state. Similarly, acceptance of an electron will leave the acceptor in the wrong structure for its reduced state. Structural changes such as bond lengths and solvation must occur in connection with the transfer of an electron, and the reorganizational energy refers to these processes. The Marcus equation correlates the activation free energy $\Delta G^\ddagger$ for a reaction with the thermodynamic barrier ΔG° and the reorganizational free energy $4\Delta G^\ddagger_o$ (or λ).

Electron Tunneling

Being much smaller than atoms, approximately 1/1800 the size of a hydrogen atom, electrons readily undergo quantum mechanical tunneling, and this is important in biological electron transfer. Extensions of Marcus theory to include tunneling lead to simple equations that can be used to calculate rates of long-range electron transfer. These calculated rates in structurally defined systems compare very well with measured rates (Page et al., 1999). One parameter, $e^{-R\beta}$, relates the insulating effect of the electron transfer medium (β) and the edge-to-edge distance separating the electron donor and acceptor (R) to the electron transfer rate. In protein molecules, a value of 1.4 Å^{-1} for β correlates the rates very well. This value lies between those for intramolecular electron transfer (0.9 Å^{-1}) and electron transfer through a vacuum (estimated at 2.8 to 3.5 Å^{-1}).

Calculated electron transfer rates for many natural molecules with multiredox components agree very well with the measured rates and are much faster than typical values of k_{cat}. A survey of more than 30 proteins with multiredox centers showed separations of less than 14 Å for nearly all of the coupled redox components (Page et al., 1999). The redox centers in xanthine oxidase, shown in the structure in fig. 4-26, are an example of this spacing. The calculated and measured electron transfer rates were much faster than typical values of k_{cat}. Tunneling rates can be very fast, even when the electron transfer is thermodynamically uphill. Calculated rates for electron transfer against thermodynamic barriers of up to 0.5 eV remained at or above the value of k_{cat} (Page et al., 1999). Electron tunneling allows uphill electron transfer over distances of up to 14 Å at rates that are compatible with biochemical processes. The view that coupled redox centers must have similar midpoint potentials is superseded by this fact, which explains a number of apparently unorthodox biochemical reactions.

Natural Selection in Electron Transfer

It may seem reasonable to expect electron transfer pathways in redox proteins to be highly evolved to optimize the rates, especially for long-range electron transfer (Beratan et al., 1992). However, experimental and computational comparisons of the rates for naturally occurring electron transfers and artificially induced electron transfers in proteins showed the same correlations with distance. All the rates could be accounted for in theory, and all the rates were correlated on the same scale with the distances separating the electron transfer components (Page et al., 1999). Electron transfer through a protein medium is so fast that no survival advantage seems to exist for specialized electron transfer pathways. The many structures with redox components separated by 14 Å, the practical theoretical limit for fast electron tunneling, suggest that specialized pathways my not be found in the future.

References

Abeles RH and WS Beck (1967) *J Biol Chem* **242**, 3589.
Abeles RH and HA Lee Jr (1961) *J Biol Chem* **236**, 2347.
Abeles RH and B Zagalak (1966) *J Biol Chem* **241**, 1245.
Abend A, V Bandarian, GH Reed, and PA Frey (2000) *Biochemistry* **39**, 6250.

Andersen OA, T Flatmark, and E Hough (2001) *J Mol Biol* **314**, 279.
Ashley GW, G Harris, and J Stubbe (1986) *J Biol Chem* **261**, 3958.
Auclair K, Z Hu, DM Little, PR Ortiz de Montellano, and JT Groves (2002) *J Am Chem Soc* **124**, 6020.
Ayling J, R Pirson, W Pirson, and G Boehm (1973) *Anal Biochem* **51**, 80.
Balasubramanian S, RT Carr, CJ Bender, J Peisach, and SJ Benkovic (1994) *Biochemistry* **33**, 8532.
Banci L, I Bertini, A Dikiy, DH Kastrau, C Luchinat, and P Sompornpisut (1995) *Biochemistry* **34**, 206.
Baraniak J, ML Moss, and PA Frey (1989) *J Biol Chem* **264**, 1357.
Barker HA, RD Smyth, H Weissbach, JI Toohey, JN Ladd, and BE Volcani (1960) *J Biol Chem* **235**, 480.
Bastian NR, CJ Kay, MJ Barber and KV Rajagopalan (1991) *J Biol Chem* **266**, 45.
Bau R, DC Rees, DM Kurtz, RA Scott, HS Huang, MWW Adams, and MK Eidsness (1998) *J Biol Inorg Chem* **3**, 484.
Beck WS, RH Abeles, and WG Robinson (1966) *Biochem Biophys Res Commun* **25**, 421.
Behrouzian B, CK Savile, B Dawson, PH Buist, and J Shanklin (2002) *J Am Chem Soc* **124**, 3277.
Beinert H (2000) *J Biol Inorg Chem* **5**, 2.
Beinert H, RH Holm, and E Münck (1997) *Science* **277**, 653.
Beinert H, MC Kennedy, and CD Stout (1996) *Chem Rev* **96**, 2335.
Beratan DN, JN Onuchic, JR Winkler, and HB Gray (1992) *Science* **258**, 1740.
Berg JM, KO Hodgson, SP Cramer, JL Corbin, A Elsberry, N Periyadath, and EI Stiefel (1979) *J Am Chem Soc* **101**, 2774..
Berkessel A (1991) *Bioorg Chem* **19**, 101.
Binda C, A Coda, A Aliverti, G Zanetti, and A Mattevi (1998) *Acta Crystallogr D Biol Crystallogr* **54**, 1353.
Broadwater JA, J Ai, TM Loehr, J Sanders-Loehr, and BG Fox (1998b) *Biochemistry* **37**, 14644.
Broadwater JA, BG Fox, and JA Haas (1998a) *Fett Lipid* **100**, 103.
Bugg TH (2001) *Curr Opin Chem Biol* **5**, 550.
Carr RT, S Balasubramanian, PC Hawkins, and SJ Benkovic (1995) *Biochemistry* **34**, 7525.
Chan MK, S Mukund, A Kletzin, MWW Adams, and DC Rees (1995) *Science* **267**, 1463.
Chance B (1952) *Arch Biochem Biophys* **41**, 404.
Cheek J and JB Broderick (2002) *J Am Chem Soc* **124**, 2860.
Chen D, A Abend, J Stubbe, and PA Frey (2003a) *Biochemistry* **42**, 4578.
Chen D and PA Frey (1998) *J Biol Chem* **273**, 25594.
Chen D, C Walsby, BM Hoffman, and PA Frey (2003b) *J Am Chem Soc* **125**, 11788.
Chen L, RC Durley, FS Mathews, and VL Davidson (1994) *Science* **264**, 86.
Cockle SA, HA Hill, and RJP Williams (1972) *J Am Chem Soc* **94**, 275.
Colbert CL, MM-J Couture, LD Eltis, and JT Bolin (2000) *Structure* **8**, 1267.
Cosper NJ, SJ Booker, PA Frey, and RA Scott (2000) *Biochemistry* **39**, 15668.
Cotton FA and G Wilkinson (1988) *Advanced Inorganic Chemistry,* 5th ed, Wiley: New York.
Cox JD, E Cama, DM Colleluori, S Pethe, JL Boucher, DMansuy, DE Ash, and DW Christianson (2001) *Biochemistry* **40**, 2689.
Cramer SP, HB Gray, and KV Rajagopalan (1979) *J Am Chem Soc* **101**, 2772.
Cramer SP, JL Johnson, AA Ribiero, DS Millington, and KV Rajagopalan (1987) *J Biol Chem* **262**, 16357.
Cramer SP, LP Solomonson, MWW Adams, and LE Mortenson (1984) *J Am Chem Soc* **106**, 1467.
Cramer SP, R Wahl, and KV Rajagopalan (1981) *J Am Chem Soc* **103**, 7721.
Dagley S (1987) *Annu Rev Microbiol* **41**, 1.
Dauter Z, KS Wilson, LC Sieker, J Meyer, and JM Moulis (1997) *Biochemistry* **36**, 16065.
Diekert G, B Klee, and RK Thauer (1980) *Arch Microbiol* **124**, 103.
DiMarco AA, TA Bobik, and RS Wolfe (1990) *Annu Rev Biochem* **59**, 355.
Doukov TI, TM Iverson, J Seravalli, SW Ragsdale, and CL Drennan (2002) *Science* **298**, 567.
Dutton PL and CC Moser (1994) *Proc Natl Acad Sci U S A* **91**, 10247.
Emptage MH, J-L Dreyer, MC Kennedy, and H Beinert (1983a) *J Biol Chem* **258**, 11106.
Emptage MH, TA Kent, MC Kennedy, H Beinert, and E Münck (1983b) *Proc Natl Acad Sci U S A* **80**, 4674.

Enroth C, BT Eger, K Okamoto, T Nishino, T Nishino, and EF Pai (2000) *Proc Nat Acad Sci U S A* **97**, 10723.
Ermler U, W Grabarse, S Shima, M Goubeaud, and RK Thauer (1997) *Science* **278**, 1457.
Farber G, W Keller, C Kratky, B Jaun, A Pfaltz, C Spinner, A Kobelt, and A Eschenmoser (1991) *Helv Chim Acta* **74**, 697.
Finke RJ and BP Hay (1984) *Inorg Chem* **23**, 3041.
Finlay TH, J Valinsky, AS Mildvan, and RH Abeles (1973) *J Biol Chem* **248**, 1285.
Fitzpatrick PF (2003) *Biochemistry* **42**, 14083.
Fontecave M and E Mulliez (1999) In Banerjee R (ed): *Chemistry and Biochemistry of B_{12}*. Wiley: New York, p 731.
Fox BG, J Shanklin, J Ai, TM Loehr, and J Sanders-Loehr (1994) *Biochemistry* **33**, 12776.
Fox BG, J Shanklin, C Somerville, and E Münck (1993) *Proc Natl Acad Sci U S A* **90**, 2486.
Fox BG, KK Surerus, E Munck, JD Lipscomb (1988) *J Biol Chem* **263**,10553.
Frey PA (1997) *Curr Opin Chem Biol* **1**, 347.
Frey PA (2001) *Annu Rev Biochem* **70**, 121.
Frey PA and RH Abeles (1966) *J Biol Chem* **241**, 2732.
Frey PA, MK Essenberg, and RH Abeles (1967) *J Biol Chem* **242**, 5369.
Fukuyama K, T Okada, Y Kakuta, and Y Takahashi (2002) *J Mol Biol* **315**, 1155.
George GN, CA Kipke, RC Prince, RA Sunde, JH Enemark, and SP Cramer (1989) *Biochemistry* **28**, 5075.
Guddat LW, A Mcalpine, D Hume, S Hamilton, J De Jersey, and JL Martin (1999) *Structure Fold Des* **7**, 757.
Guroff G, JW Daly, DM Jerina, J Renson, B Witkop, and S Udenfriend (1967) *Science* **157**, 1524.
Halpern J, S-H Kim, and TW Leung (1984) *J Am Chem Soc* **106**,8317.
Hamilton JA, Y Tamao, RL Blakeley, and RE Coffman (1972) *Biochemistry* **11**, 4696.
Hanson L, WA Eaton, SG Sligar, IC Gunsalus, M Gouterman, and CR Connell (1976) *J Am Chem Soc* **98**, 2672.
Hausinger RP (1997) *J Biol Inorg Chem* **2**, 279.
Hille R (1996) *Chem Rev* **96**, 2757.
Hogenkamp HPC (1963) *J Biol Chem* **238**, 477.
Hogenkamp HPC, RK Ghambeer, C Brownson, RL Blakley, and E Vitols (1968) *J Biol Chem* **243**, 799.
Hogenkamp HPC, JN Ladd, and HA Barker (1962) *J Biol Chem* **237**, 1950.
Holm RH, P Kennepohl, and EI Solomon (1996) *Chem Rev* **96**, 2239.
Ito N, SE Phillips, C Stevens, ZB Ogel, MJ McPherson, JN Keen, KD Yadav, and PF Knowles (1991) *Nature* **350**, 87.
Ito N, SE Phillips, KD Yadav, and PF Knowles (1994) *J Mol Biol* **238**, 794.
Jabri E, MB Carr, RP Hausinger, and PA Karplus (1995) *Science* **268**, 998.
Johnson JL, BE Hainline, KV Ragalgopalan, and B Arison (1984) *J Biol Chem* **259**, 5414.
Johnson JL and KV Ragalgopalan (1982) *Proc Natl Acad Sci U S A* **79**, 6856.
Kappock TJ and JP Caradonna (1996) *Chem Rev* **96**, 2659.
Kaufmann S (1962) *Methods Enzymol* **5**, 809.
Kavanaugh JS, PH Rogers, and A Arnone (2005) *Biochemistry* **44**, 6101.
Kennedy C, R Rauner, and O Gawron (1972) *Biochem Biophys Res Commun* **47**, 740.
Kennedy MC, M Werst, J Telser, MH Emptage, H Beinert, and BM Hoffman (1987) *Proc Natl Acad Sci U S A* **84**, 8854.
Kent TA, MH Emptage, H Merkle, MC Kennedy, H Beinert, and E Münck (1985) *J Biol Chem* **260**, 6871.
Kuchta RD, GR Hanson, B Holmquist, and RH Abeles (1986) *Biochemistry* **25**, 7301.
Lee HA Jr and RH Abeles (1963) *J Biol Chem* **238**, 2367.
Lehert PG and DC Hodgkin (1961) *Nature* **192**, 937.
Leutbecher U, SPJ Albracht, and W Buckel (1992) *FEBS Lett* **307**, 144.
Licht S, GJ Gerfen, and J Stubbe (1996) *Science* **271**, 477.
Lindqvist Y, W Huang, G Schneider, and J Shanklin (1996) *EMBO J* **15**, 4081.
Lipscomb JD and AM Orville (1992) In Sigel H and A Sigel (eds): *Metal Ions in Biological Systems*, vol .8. Marcel Dekker: New York, p 243.
Magnusson OT, S Mansoorabadi, R Poyner, GH Reed, and PA Frey (in press) *Biochemistry*.
Marcus RA and N Sutin (1985) *Biochim Biophys Acta* **811**, 265.

Matthews RG (1999) In Banejee R (ed): *Chemistry and Biochemistry of B_{12}*. Wiley: New York, p 681.
McAlpine AS, AG McEwan, and S Bailey (1998) *J Mol Biol* **275**, 613.
McEwan AG, SJ Ferguson, and JB Jackson (1991) *Biochem J* **274**, 305.
Messerschmidt A, R Ladenstein, R Huber R, M Bolognesi, L Avigliano, R Petruzelli, A Rossi, and A Finazzi-Agro (1992) *J Mol Biol* **224**,179.
Michel C, SPJ Albracht, and W Buckel (1992) *Eur J Biochem* **205**, 767.
Miller RJ and SJ Benkovic (1988) *Biochemistry* **27**, 3658.
Moser CC, JM Keske, K Warncke, RS Farid, and PL Sutton (1992) *Nature* **355**, 796.
Moss ML and PA Frey (1987) *J Biol Chem* **262**, 14859.
Müller U and W Buckel (1995) *Eur J Biochem* **230**, 698.
Overath P, ER Stadtman, GM Kellerman, and F Lynen (1962) *Biochem Z* **336**, 77.
Padmakumar R and R Banerjee (1995) *J Biol Chem* **270**, 9295.
Page CC, CC Moser, X Chen, and PL Dutton (1999) *Nature* **402**, 47.
Pember SO, KA Johnson, JJ Villafranca, and SJ Benkovic (1989) *Biochemistry* **28**, 2124.
Pfaltz A, B Jaun, A Fassler, A Eschenmoser, R Jaenchen, HH Gilles, G Diekert, and RK Thauer (1982) *Helv Chim Acta* **65**, 828.
Pilbrow JR (1982) In Dolphin D (ed): *B_{12}*, vol 1. John Wiley & Sons: New York, p 431.
Poyner RR and GH Reed (1992) *Biochemistry* **31**, 7166.
Putnam CD, AS Arvai, Y Bourne, and JA Tainer (2000) *J Mol Biol* **296**, 295.
Rebeil R and WL Nicholson (2001) *Proc Natl Acad Sci U S A* **98**, 9038.
Reed GH, RR Poyner, TM Larsen, JE Wedekind, and I Rayment (1996) *Curr Opin Struct Biol* **6**, 736.
Romao MJ, M Archer, I Moura, JJG Moura, J Legall, R Engh, M Schneider, P Hof and R Huber (1995) *Science* **270**, 1150.
Rospert SR, R Böcher, SPJ Albracht, and RK Thauer (1991) *FEBS Lett* **291**, 371.
Ruzicka FJ and H Beinert (1974) *Biochem Biophys Res Commun* **58**, 556.
Satoh T and FN Kurihara (1987) *J Biochem (Tokyo)* **102**, 191.
Schindelin H, C Kisker, J Hilton, KV Rajagopalan, and DC Rees (1996) *Science* **272**, 1615.
Schultz BE, R Hille, and RH Holm (1995) *J Am Chem Soc* **117**, 827.
Shanklin J and EB Cahoon (1998) *Annu Rev Plant Physiol Plant Mol Biol* **49**, 611.
Shanklin J, E Whittle, and BG Fox (1994) *Biochemistry* **33**, 12787.
Siegmund HU and SJ Kaufmann (1991) *J Biol Chem* **266**, 2903.
Sofia HJ, G Chen, BG Hetzler, JF Reyes-Spindola, and NE Miller (2001) *Nucleic Acids Res* **29**, 1097.
Sono M, MP Roach, ED Coulter, and JH Dawson (1996) *Chem Rev* **96**, 2841.
Strater N and WN Lipscomb (1995) *Biochemistry* **34**, 14792.
Stubbe J (1990) *Adv Enzymol Rel Areas Mol Biol* **63**, 349.
Stubbe J and D Ackles (1980) *J Biol Chem* **255**, 8027.
Stubbe J, D Ackles, R Segal, and RL Blakely (1981) *J Biol Chem* **256**, 4843.
Sumner JB (1926) *J Biol Chem* **69**, 435.
Swamy MJ (1995) *Biochem Mol Biol Int* **36**, 219.
Takano T, RE Dickerson (1981) *J Mol Biol* **153**, 79.
Teigen K, NA Froystein, and A Martinez (1999) *J Mol Biol* **294**, 807.
Thauer RK, R Hedderich, and R Fischer (1993) In Ferry JG (ed): *Methanogenesis*. Chapman and Hall: New York, p 209.
Vallee BL and RJP Williams (1968) *Proc Natl Acad Sci U S A* **59**, 498.
van der Donk WA, G Yu, L Perez, RJ Sanchez, J Stubbe, V Samano, and MJ Robins (1998) *Biochemistry* **37**, 6419.
Vojtechovsky J, K Chu, J Berendzen, RM Sweet, and I Schlichting (1999) *Biophys J* **77**, 2153.
Wagner OW, HA Lee Jr, PA Frey, and RH Abeles (1966) *J Biol Chem* **241**, 1751.
Walsby CJ, D Ortillo, WE Broderick, JB Broderick, BM Hoffman (2002) *J Am Chem Soc* **124**, 11270.
Weiss DS and RK Thauer (1993) *Cell* **72**, 819.
Wolfe RS (1991) *Annu Rev Microbiol* **45**, 1.
Yamamoto I, N Wada, T Ujiye, M Tchibana, M Matsuzaki, H Kajiwara, Y Watanabe, H Hirano, A Okubo, T Satoh, and S Yamazaki (1995) *Biosci Biotechnol Biochem* **59**, 1850.
Yang Y-S, J Broadwater, SC Pulver, BG Fox, and EI Solomon (1999) *J Am Chem Soc* **121**, 2770.
Youn HD, EJ Kim, JH Roe, YC Hah, and SO Kang (1996) *Biochem J* **318**, 889.
Zhao Y, P Such, and J Rétey (1992) *Angew Chem Int Ed Engl* **31**, 215.

5
Enzyme Inhibition

One aspect of the importance of enzymes in biology can be appreciated by considering the attention that continues to be focused on the inhibition of enzymatic activity. Historically, inhibitors played important roles in the elucidation of metabolic pathways. An example was the use of malonate as a metabolic inhibitor, leading to the accumulation of succinate in metabolizing cell extracts. Malonate proved to be an inhibitor of succinate dehydrogenase, and its effect on metabolism revealed the importance of succinate as a metabolic intermediate and of succinate dehydrogenase in metabolism. In this way, malonate as an inhibitor played an important role in the elucidation of the tricarboxylic acid cycle. Countless competitive inhibitors have been applied in various ways to the characterization of active sites. Clues to the binding properties and specificities of active sites can be deduced by systematically varying the structures of competitive inhibitors and comparing the inhibition constants. In another application, competitive inhibitors are often used to shield active sites against the effects of group selective chemical modification. Because biological processes are catalyzed by enzymes, inhibitors can be used to manage biochemical dynamics to the advantage of humans. Examples include pharmaceutical agents and agricultural insecticides and herbicides, which are often inhibitors of specific enzymes. Many targets for ethical drugs are enzymes. For these reasons, the development of enzyme inhibitors is an important objective in pharmaceutical and agricultural research and development.

We explained the simplest kinetic properties of reversible inhibitors in chapter 2, and in chapter 1, we discussed the use of competitive reversible inhibitors and affinity-labeling agents for characterizing active sites. The importance of inhibitors in the pharmaceutical and agricultural industries has led to the discovery and invention of compounds that inhibit enzymes by special mechanisms and with very high binding affinities. These inhibitors often displayed special kinetic properties, which led to the development of kinetic paradigms, including slow-binding inhibition and tight-binding inhibition. In chapter 1, we introduced the theoretical significance of transition-state analogs as potent inhibitors of enzymes, and this theory has led to practical applications in the design of inhibitors. In this chapter, we consider some of the special applications of inhibitors in the pharmaceutical and agricultural industries.

Two-Substrate Analogs

Inhibition and Binding

Compounds that incorporate the structural features of two substrates of a multisubstrate enzyme are often good inhibitors. This occurs because a molecule that includes the major structural features of two substrates incorporates many more binding interactions to the active site than either substrate alone. Moreover, the binding of a two-substrate analog may be driven by a chelate effect, analogous to the binding of a multidentate ligand to a metal ion.

PALA and Aspartate Transcarbamylase

An early example of the binding of a two-substrate analog was the inhibition of aspartate transcarbamylase (ATCase) by N-(phosphonoacetyl)-L-aspartate (PALA), which is shown in structure **5-1**. ATCase is described in chapter 2.

ATCase catalyzes the reaction of aspartate with carbamyl-P to produce carbamylaspartate in the first step of pyrimidine biosynthesis. ATCase binds PALA very tightly at its active site in competition with its substrates (Collins and Stark, 1971). PALA is a very potent inhibitor of pyrimidine and DNA biosynthesis. PALA incorporates aspartate with an N-acyl substituent that is structurally related to carbamoyl-P but lacks the amido group and substitutes a methylene group in place of the oxygen linking the carbonyl group and phosphorus, making it a phosphonate rather than a phosphate.

Because the structure of PALA incorporates important features of both aspartate and carbamoyl-P, it can be regarded as a two-substrate analog of ATCase. The value of its inhibition constant, $K_I = 2.7 \times 10^{-8}$ M, is numerically within an order of magnitude of the product of the inhibition constants for succinate (9×10^{-4} M) and carbamoyl-P (2.7×10^{-2} M) (Collins and Stark, 1971). The carbon skeleton of succinate is similar to that of aspartate, so that this comparison seems reasonable. The product of two inhibition constants has units of M^2, which differs from that of an inhibition constant (M) and raises the question of how they can be compared. As an alternative, the free energy of binding PALA can be compared with the sum of free energies for binding succinate and carbamyl-P. The standard free energy ($\Delta G°$) for binding PALA is -10.6 kcal mol^{-1}, and the free energies for binding succinate and carbamoyl-P are -4.3 and -2.2 kcal mol^{-1}, respectively. The sum of free energies is -6.5 kcal mol^{-1}, significantly less than for PALA. Most likely the binding of PALA is further potentiated by the chelate effect.

PALA is less closely related in structure to the transition state for carbamoyl group transfer in ATCase than to the two substrates. In addition to lacking the amido group, PALA does not incorporate a tetrahedral carbon, which is the most characteristic structural feature of the intermediate that is likely to be similar in the transition state. However, the presence of the peripheral aspartate and phosphonate moieties relates PALA to both the transition state and the substrates.

Suicide Inactivation

In suicide inactivation of an enzyme, a molecule that is structurally related to the substrate (or product) binds to the active site and undergoes the first step or the first few steps of the usual catalytic process. At some point, the chemical difference from a substrate comes into play, and the reaction of the suicide inactivator takes a different course. The consequence for the enzyme can be disastrous when the inactivator becomes strongly bonded to the enzyme and blocks its further action. The term *suicide* refers to self-destruction by the enzyme, which recognizes the inactivator as if it is a substrate. In chapter 4, we encountered the earliest example of suicide inactivation in the reaction of glycolaldehyde with dioldehydrase, an adenosylcobalamin dependent enzyme (scheme 4-2). Glycolaldehyde reacts as if it were a substrate in the initial hydrogen abstraction function of adenosylcobalamin, leading to the formation of 5′-deoxyadenosine from the 5′-deoxyadenosyl radical, cob(II)alamin from the corrin portion of the coenzyme, and the stable *cis*-ethanesemidione radical from glycolaldehyde (Abend et al., 2000; Wagner et al., 1966). 5′-Deoxyadenosine, cob(II)alamin, and substrate and product related radicals normally participate in the catalytic mechanism, but the radicals are high-energy, metastable species that carry the reaction to product formation. In the reaction of glycolaldehyde, the enzyme produces a radical that is too stable to react further and blocks the active site.

Inhibitors that lead to self-destruction of enzymes have been known by various names, including mechanism-based inactivators (Silverman, 1988). Reference to these inhibitors as suicidal can be criticized as inaccurate because of the inability of an enzyme to form intent. We choose to accept suicide inactivation as put forward by early investigators in the field. Because English is a living language, many terms that come into general usage can be criticized as inaccurate. In this case, we choose the more colorful and succinct expression.

Thymidylate Synthase

5-Fluoro-dUMP

Certain drugs act as suicide inactivators of target enzymes, thereby disrupting disease-related metabolic processes. 5-Fluorouracil is an anticancer drug that can be converted into 5-fluoro-dUMP (FdUMP), which blocks thymidylate synthase (EC 2.1.1.45). FdUMP is a suicide inactivator of thymidylate synthase, an essential enzyme in the biosynthesis of dTTP for DNA biosynthesis. Rapidly proliferating cancer cells depend more on DNA biosynthesis than normal cells. 5-Fluorouracil is therefore more toxic to cancer cells than to normal cells. In this way, suicide inactivation of thymidylate synthase is a cancer chemotherapeutic mechanism.

Reaction and Molecular Properties

Thymine for DNA biosynthesis arises from the methylation of dUMP by methylene-H$_4$folate, which is catalyzed by thymidylate synthase according to eq. 5-1.

256　Enzymatic Reaction Mechanisms

$$\text{dUMP} + \text{methylene-H}_4\text{folate} \longrightarrow \text{H}_2\text{folate} + \text{dTMP} \qquad (5\text{-}1)$$

The methyl group of dTMP is derived from the methylene carbon and the 6-hydrogen of methylene-H$_4$folate, leading to H$_2$folate as the folate-product. The reaction is a reductive methylation, in which methylene-H$_4$folate serves as both the source of a methylene group and as a reducing agent for transforming it into a methyl group. Dihydrofolate is recycled by dihydrofolate reductase to H$_4$folate, and serine hydroxymethyltransferase converts the H$_4$folate into methylene-H$_4$folate, and this cycle fuels the continuous methylation of dUMP (see fig. 3-36). Thymidylate synthase in most cells is a dimer of identical subunits with molecular masses of 30 to 35 kDa, although it also appears in plants and protozoa as bifunctional proteins consisting of a domain for thymidylate synthase activity and one for dihydrofolate reductase (Carreras and Santi, 1995).

Reaction Mechanism

Several properties of thymidylate synthase provided clues to the mechanism of its reaction (Carreras and Santi, 1995). First, this enzyme catalyzes the exchange of tritium from pyrimidine-C5 of [5-^{3}H]dUMP into water. Second, thymidylate synthase is inactivated by 5-fluorodeoxyuridine-5′-phosphate (FdUMP) in the presence but not in the absence of methylene-H$_4$folate. Third, both FdUMP and methylene-H$_4$folate are covalently bonded to the inactivated enzyme. Fourth, covalent bonding of FdUMP in the inactivated *L. casei* thymidylate synthase is to Cys198 (Bellisario et al., 1976, 1979; Maley et al., 1979). These facts and the results of model nonenzymatic studies led to the postulated mechanism outlined in fig. 5-1 (Carreras and Santi, 1995; Danenberg et al., 1974; Lewis et al., 1981; Pogolloti and Santi, 1973; Santi and Brewer, 1968, 1974; Santi and Pogolotti, 1968; Santi et al., 1974). The reaction takes place within a ternary complex of thymidylate synthase, dUMP, and methylene-H$_4$folate. The steady-state kinetics indicate an ordered binding of substrates with dUMP leading when the monoglutamate of methylene-H$_4$folate is the methyl donor (Daron and Aull, 1978; Lorenson et al., 1967), but the binding mechanism becomes randomized with polyglutamyl substrates (Lu et al., 1984).

The chemical mechanism incorporates a methylene-bridged species as a key intermediate. A bridged intermediate had been proposed as early as 1959 (Friedkin, 1959). The current version in fig. 5-1 begins with the addition of the thiolate group of Cys198 to C6 of dUMP, a Michael type addition. The enolate-like adduct displays carbanion-like properties at C5, and in the second step, this carbanion-equivalent adds to the N^5-iminium ion derived from methylene-H$_4$folate after protonation of N^{10}. The C5-hydrogen of dUMP in the methylene-bridged intermediate is removed as a proton to produce an enolate-like form of the methylene-bridged intermediate, and protonation of N^5 on the H$_4$folate side of the methylene bridge allows the enolate to eliminate N^5 as H$_4$folate, forming a 5-methylidene group at C5 of the pyrimidine ring. Reduction of the methylidene group by direct hydride transfer from C6 of H$_4$folate generates the methyl group at C5 of the incipient dTMP in the form of an enolate, as well as dihydrofolate. The enolate of the dTMP adduct with Cys198 eliminates Cys198

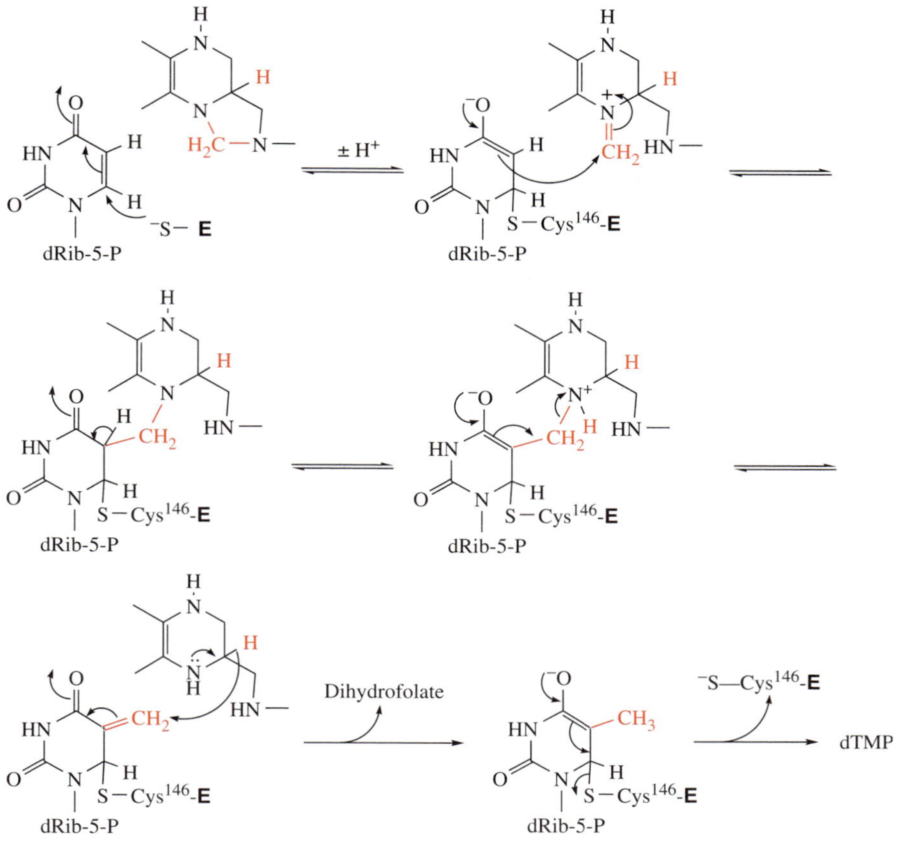

Fig. 5-1. A chemical mechanism for the reaction of thymidylate synthase. The structures shown are all enzyme-bound intermediates, none of which dissociates from the enzyme in the overall transformation of dUMP into dTMP.

in the last step to form dTMP. Mechanistic details have been thoroughly discussed in a review (Carreras and Santi, 1995).

The mechanism in fig. 5-1 accounts for the hydrogen exchange at C5 as in scheme 5-1 The initial addition of Cys198 to C6 is part of the normal reaction; however, protonation of C5, either by an enzymatic acid or an acid from the solvent, is adventitious and evidently not strictly stereospecific, so that tritium is eventually exchanged out. Because the exchange requires the adventitious and nonstereospecific deprotonation at C5, the exchange of tritium proceeds orders of magnitude slower than the overall reductive methylation to dTMP when methylene-H$_4$folate is present (Carreras and Santi, 1995).

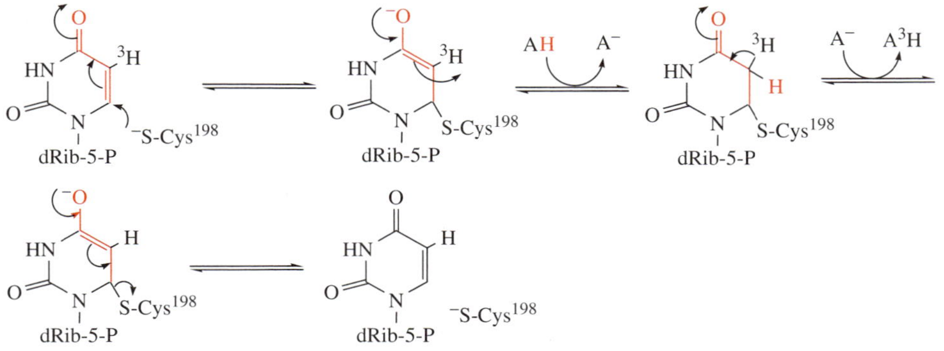

Scheme 5-1

Inactivation by FdUMP

The mechanism in fig. 5-1 offers a clear rationale for the inactivation of thymidylate synthase by FdUMP. This molecule is a suicide inactivator, which reacts as if it were a substrate for the first two steps of the normal mechanism, as shown in fig. 5-1. However, at the point of the elimination of the C5(H) in fig. 5-1, further steps are blocked by the presence of fluorine in place of hydrogen at C5 of FdUMP, and the reaction comes to rest. Fluorine differs fundamentally from hydrogen in the corresponding active complex. Hydrogen can depart as H^+, but fluorine can only depart as F^- and not F^+, so the reaction is frozen at the methylene-bridged species. The resulting ternary, covalent complex with FdUMP bonded both to Cys198 and methylene-H_4folate is stable. The covalent catalytic complex corresponding to the suicidal complex but generated with dUMP instead of FdUMP has been trapped by denaturation with TCA and shown to contain covalently bonded dUMP and methylene-H_4folate. Systematic degradation of the denatured ternary complex revealed the covalent linkage to Cys198 (Moore et al., 1986). With confirmation of the nucleophilic reaction of Cys198 with dUMP, as deduced from the reaction of FdUMP, the evidence in support of the mechanism in fig. 5-1 is compelling.

It seems reasonable that the cancer therapeutic effect of 5-fluoruracil in substantial part results from inactivation of thymidylate synthase, which is required for DNA biosynthesis in rapidly proliferating cells. However, other pyrimidine nucleotide methylating and hydroxymethylating enzymes that work by analogous mechanisms may also be inactivated by 5-fluorouracil derivatives.

Structure

Three-dimensional structures of thymidylate synthase with substrates and substrate analogs bound at the active site graphically support the catalytic mechanism in fig. 5-1 and the mechanism for inactivation by FdUMP in fig. 5-2. The original structure contained phosphate in the site for dUMP (Hardy et al., 1987). The structure of the enzyme inactivated by FdUMP and methylene-H_4folate clearly shows the methylene bridge between FdUMP and N^5 of H_4folate, as illustrated in part A of fig. 5-3 (Matthews et al., 1990).

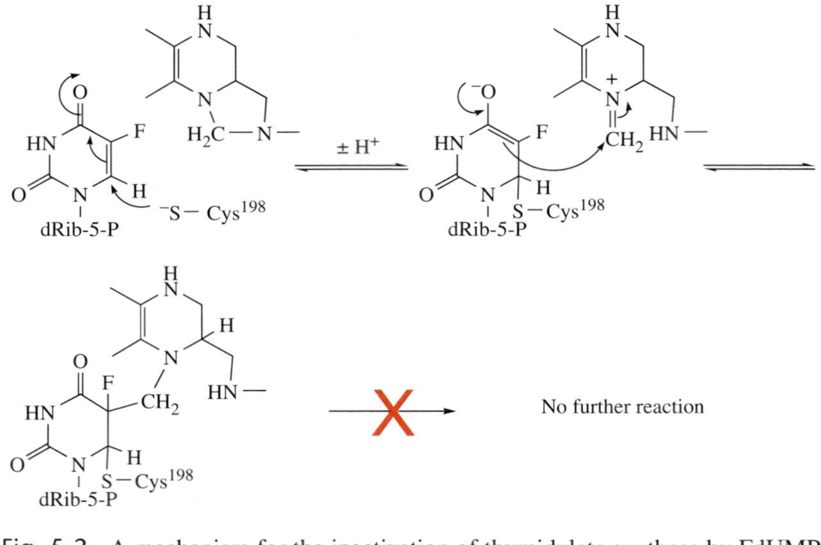

Fig. 5-2. A mechanism for the inactivation of thymidylate synthase by FdUMP.

Enzyme Inhibition   259

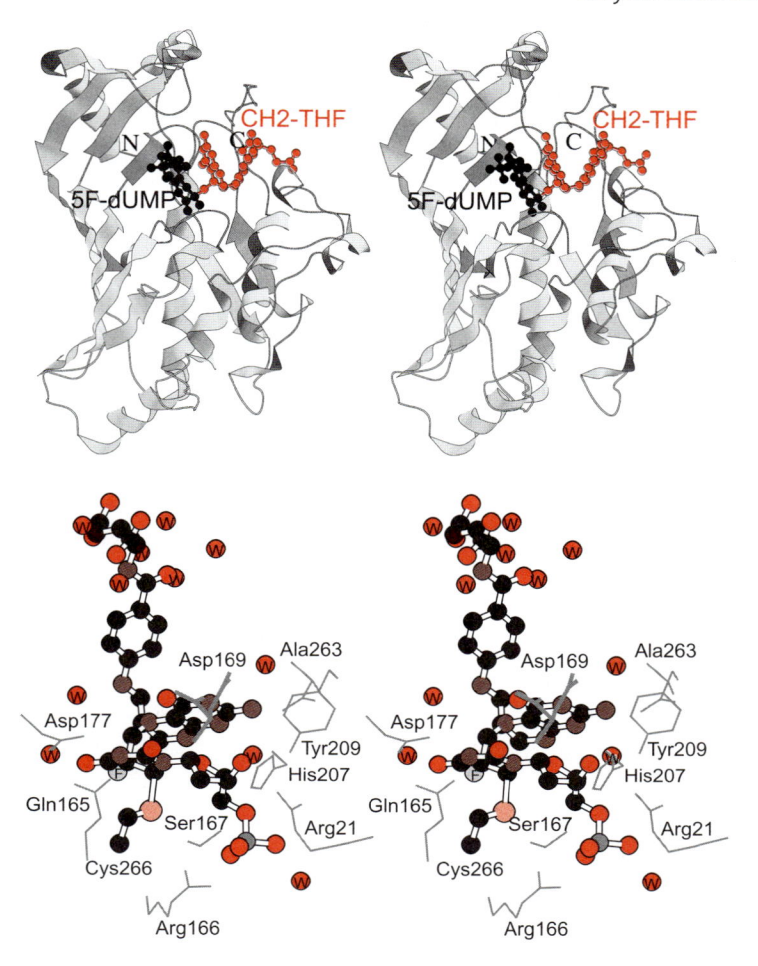

Fig. 5-3. Structure of thymidylate synthase. The stereo ribbon diagram (top) shows one subunit of *Escherichia coli* thymidylate synthase inactivated by FdUMP and methylene-H$_4$folate, with the methylene bridged ligands in the active site (Matthews et al., 1990). The active site interactions between amino acid side chains and substrates (bottom) are described in a review by Stroud and Finer-Moore (1993). The image was generated using PDB 1TSN (Hyatt et al., 1997).

The active site in thymidylate synthase is formed from portions of the two protomers of the dimeric molecule. Most of the close contacts between substrates and the active site involve one protomer; however, arginine residues from both promoters bind the phosphate group of dUMP at the base of the active site (Stroud and Finer-Moore, 1993).

The nearest contacts between amino acid side chains and the uracil ring of dUMP and methylene bridge of methylene-H$_4$folate are shown in part B of fig. 5-3. The adduct of FdUMP and methylene-H$_4$folate is shown in the site. Apart from the covalent bond linking Cys266 and C6 of dUMP, the active site is remarkable for the absence of acid-base side chains of amino acids interacting directly with the pyrimidine of dUMP or the N5 and N10 of methylene-H$_4$folate. The side chain of Gln165 is within hydrogen bonding distance of 4-oxo and N3(H) of the pyrimidine ring. Other amino acids may interact secondarily with substrates, but the structures show that fixed water molecules and a hydrogen-bonding network intervene between them and substrates. Extensive mutagenic analyses show that, except for cysteine in the active site, thymidylate synthase is remarkably resilient and can tolerate many changes in the amino acids that line the active site (Carreras and Santi, 1995). Mutation of cysteine to serine decreases the value of k_{cat} by 5000-fold, with little effect on K_m. Mutations of other amino acids peripheral to the active site produced variants of thymidylate synthase that displayed significant activity, although several mutations disrupted the hydrogen-bonding network (Sage et al., 1996).

By all indications, the addition of the thiolate group of cysteine to C6 of the uracil ring potentiates the reaction of thymidylate synthase, with the other steps being facilitated by the close proximity of the substrates within the active site and the hydrogen bonded network of water molecules in the active site. A report that mutations of a nearby serine decreased k_{cat}/K_m by 10^3 to 10^4-fold was attributed a role in maintaining the reactivity of the nucleophilic cysteine (Williams et al., 1998).

Suicide inactivation of thymidylate synthase by FdUMP gave important information about the catalytic mechanism. The hypothetical reaction of dUMP with methylene-H_4folate at the active site included a methylene bridged compound, a transient species that could not be observed directly or isolated (fig. 5-1). FdUMP arrested the reaction just at the point of the hypothetical methylene-bridged intermediate, which could then be observed by x-ray crystallography.

The mechanistic information derived from suicide inactivation is by no means limited to the observation of intermediates, although it almost invariably bears on the identification of intermediates. We shall encounter other examples of mechanistic information derived from suicide inactivation by substrates and inactivators.

β-Hydroxydecanoyl Thioester Dehydratase

Reaction and Mechanism

The oxygen-dependent desaturation of fatty acids discussed in chapter 4 in connection with fatty acyl carrier protein (ACP) desaturase cannot account for the biosynthesis of unsaturated fatty acids under anaerobic conditions. Little is known about this subject outside of the metabolism of *E. coli*. However, the oxygen-independent biosynthesis of palmitoleyl-ACP has been solved. This is brought about in part through the action of β-hydroxydecanoyl ACP dehydratase (Helmkamp et al., 1968), which participates in the biosynthesis of unsaturated fatty acids in *E. coli*. At the time of its discovery, the structure of the thioester portion of substrates had not been elucidated, but the enzyme accepted the cysteamine ester as a substrate. Today, the ACP is known to be the thiol portion of the thioester, and we refer to this enzyme as β-hydroxyldecanoyl thioester dehydratase (EC 4.2.1.60).

The conventional fatty acid biosynthesis by the two-carbon elongation process with acetyl CoA leads to the intermediate β-hydroxydecanoyl-ACP. In the next step of the biosynthesis of palmitoyl-ACP in *E. coli*, part of this intermediate is dehydrated to *trans*-α,β-decenoyl-ACP by the conventional β-hydroxyacyl-ACP dehydratase and then reduced to decanoyl-ACP. However, a fraction of β-hydroxydecanoyl-ACP is diverted to palmitoleyl-ACP biosynthesis by the action of β-hydroxydecanoyl thioester dehydratase, which catalyzes its dehydration to *cis*-β,γ-decenoyl-ACP by the novel mechanism illustrated in fig. 5-4.

The reaction proceeds initially through conventional α,β-dehydration to form the enzyme-bound α,β-decenoyl-ACP, and the enzyme then catalyzes isomerization of the double bond to β,γ-decenoyl-ACP, which is released from the active site. The isomerization step depends on the acidity of the C4(H), which can form a homoenolate on abstraction of C4(H) by a base at the active site.

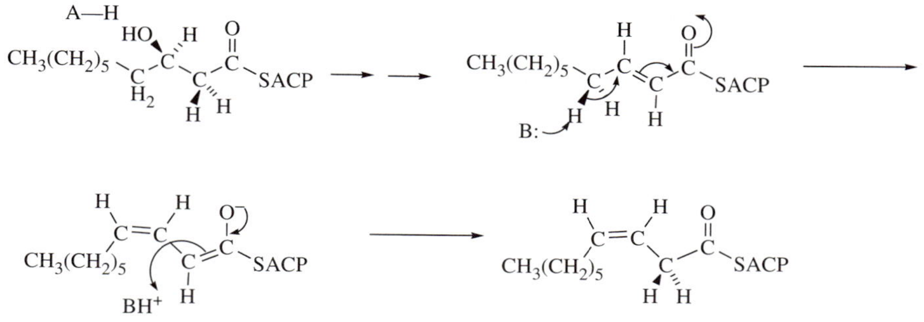

Fig. 5-4. Mechanism of β,γ-decenoyl acyl carrier protein (ACP) formation by β-hydroxydecenoyl thioester dehydratase. In the first step, the dehydration of β-hydroxydecanoyl-ACP proceeds to α,β-decenoyl-ACP, which does not dissociate from the active site. Initial dehydration may proceed in two steps, base-catalyzed enolization followed by acid-catalyzed elimination of water, or in a single, concerted acid-base–catalyzed dehydration. The isomerization of α,β-decenoyl-ACP to β,γ-decenoyl-ACP proceeds by base catalyzed homoenolization followed by protonation at the α-carbon.

Acetylenic Inactivator

Replacement of the double bond in β,γ-decenoyl cysteamine with a triple bond, the acetylenic group of β,γ-decynoyl cysteamine, leads to a molecule that binds at the active site in place of the product and is a suicide inactivator. The inactivation leads to covalent attachment of the inactivator to the enzyme molecule through alkylation of a histidine residue (Endo et al., 1970). The mechanism outlined in fig. 5-5 begins with the acetylenic analog of the product binding to the active site. The last step of the catalytic mechanism in fig. 5-4 is protonation of the homoenolate intermediate at the α-carbon to form β,γ-decenoyl cysteamine. With β,γ-decynoyl cysteamine in the site, abstraction of the α-proton C2(H), in essence the microscopic reverse of product formation, leads directly to the formation of an allene, a highly electrophilic alkylating group. The allene alkylates histidine in the active site, covalently binding the inhibitor to the site and inactivating the enzyme. The enzyme is the instrument of its own demise.

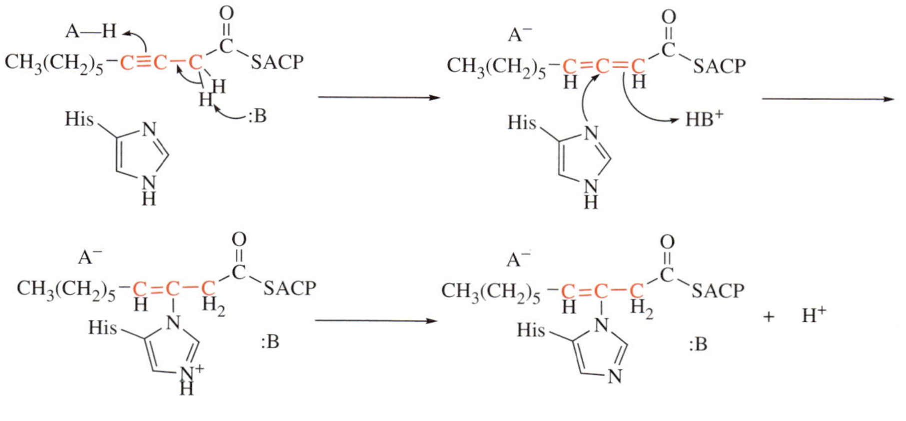

Fig. 5-5. Suicide inactivation of β-hydroxydecanoyl thioester dehydratase. β,γ-Decynoyl cysteamine binds at the active site in place of the normal product β,γ-decenoyl thioester. Abstraction of the α-proton leads to the electrophilic allene, which alkylates His70 at the active site. The covalently bonded inhibitor blocks the active site. Abstraction of the α-proton to form the allene corresponds to the microscopic reverse of the last step of the mechanism in fig. 5-4. Inactivation proceeds by way of a catalytic and suicidal process.

γ-Aminobutyrate Aminotransferase

γ-Aminobutyrate (GABA) aminotransferase (EC 2.6.1.19) catalyzes the reaction of GABA with α-ketoglutarate to form succinate semialdehyde and L-glutamate. The mechanism is typical of pyridoxal-5′-phosphate (PLP)–dependent transmanations and takes place in two overall processes: transfer of the amino group in GABA to PLP (eq. 5-2a) and transfer of the amino group from pyridoxamine phosphate to α-ketoglutarate (eq. 5-2b).

$$\text{}^{+}\text{H}_3\text{N-CH}_2\text{CH}_2\text{CH}_2\text{COO}^- + \text{E-NH}^+\text{=CH-PLP} \rightleftharpoons$$

$$\text{OHC-CH}_2\text{CH}_2\text{COO}^- + \text{E PLP-CH}_2\text{NH}_2 \quad (5\text{-}2a)$$

$$^-\text{OOC-CH}_2\text{CH}_2\text{-CO-COO}^- + \text{E-NH}_2\cdot\text{PLP-CH}_2\text{NH}_2 \rightleftharpoons$$

$$^-\text{OOC-CH}_2\text{CH}_2\text{-CH(NH}_3^+)\text{-COO}^- + \text{E-NH}^+\text{=CH-PLP} \quad (5\text{-}2b)$$

GABA and glutamate are major neurotransmitters, and for this reason GABA aminotransferase is a target for neurotherapeutic agents that can modulate the balance between these excitatory and inhibitory molecules.

The reaction mechanism is typical of aminotransferases, which invariably require PLP as the essential coenzyme. This background knowledge of PLP mechanisms, coupled with the importance of GABA aminotransferase as a drug target, has led to extensive efforts design suicide inactivators. These efforts have been successful in uncovering a large number of inactivators (Silverman, 1988). We have selected two for discussion here as suicide inactivators of GABA aminotransferase.

Gabaculine

Gabaculine is a natural product from *Streptomyces toyacaensis* that is a suicide inactivator of GABA aminotransferase. Its structure is shown in fig. 5-6 together with the mechanism by which it inactivates GABA aminotransferase. The portion of gabaculine shown in color resembles GABA, and the balance of the structure constitutes a dihydrobenzene ring.

Gabaculine undergoes the first steps of transamination in place of GABA, including transaldimination and the 1,3-prototrophic shift to the pyridoxamine imine. Then, a proton from the dihydrobenzene carbon of the ring is abstracted by an enzymatic base, with the consequence that the ring becomes aromatic, a far more stable structure. The aromatic stabilization energy of the resulting benzene ring presents a large barrier to reversal, so that complex is in a deep thermodynamic well and does not react further. All available evidence points to this mechanism, including the fact that inactivation proceeds with a deuterium kinetic isotope effect of 4.3 in the reaction of [2,3-^{2}H$_2$]gabaculine; inactivation by [2-^{3}H]gabaculine proceeds with incorporation of radioactivity into the undenatured enzyme; and the gabaculine-PLP complex is released when the inactive complex is denatured, proving that it is not covalently bonded to the protein (Rando, 1977; Rando and Bangerter, 1976; 1977). Gabaculine proved to be too toxic for use as a drug. This effect may result from its inactivation of other PLP-dependent enzymes (Soper and Manning, 1982; Wood et al., 1979).

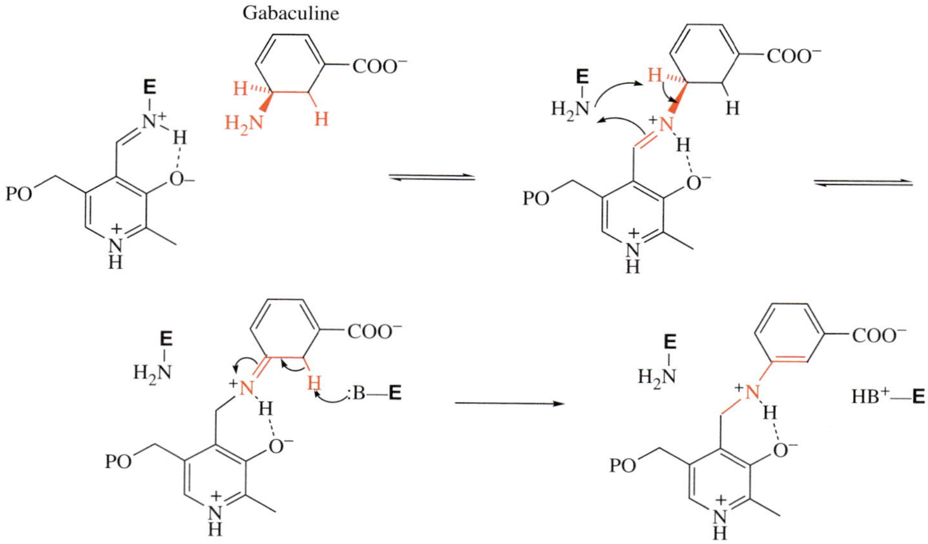

Fig. 5-6. Mechanisms of inactivation of GABA aminotransferase by gabaculine. Gabaculine is a natural product that incorporates the structural features of GABA (red) within a dihydrobenzene ring. On forming the external aldimine and undergoing the 1,3-prototropic shift characteristic of aminotransferases, abstraction of a β-proton from the dihydrobenzene ring leads to aromatization of the ring and a stable complex that does not react further.

γ-Vinyl GABA

Extensive developmental research on acetylenic and vinyl analogs of GABA as potential suicide inactivators of GABA aminotransferase led to the discovery of γ-vinyl GABA, shown in fig. 5-7 (Lippert et al., 1977). γ-Vinyl GABA is in use as the antiepileptic drug vigabatrin (Mumford and Cannon, 1994). The two mechanisms for suicide inactivation in fig. 5-7 can be written (Lippert et al., 1977; Metcalf, 1979). In Mech A, the C4(H) of the external aldimine is abstracted by a base at the active site and transferred to C6, which moves the double bond into conjugation with the π bond of the iminium group. The active site lysine then reacts as a nucleophile in a Michael-like addition to C5. Ultimate protonation at C4′ of PLP leads to a ketimine of PMP with C5 of the inactivator.

In Mech B, abstraction of C4(H) from the external aldimine as a proton and transfer to C4′ of PLP generates the ketimine analogous to the transamination intermediate in the reaction of GABA. This species is potentially reactive in alkylating the active site lysine or another nucleophile by Michael addition. Both mechanisms involve abstraction of C4(H), and the kinetic isotope effect of 3.9 in the reaction of γ-vinyl[4-^{2}H]GABA indicates that this step participates in rate limitation. In both mechanisms, denaturation of the inactivated complex would lead to the eventual hydrolysis of the bond linking PMP with the inhibitor and release of PMP, leaving the inactivator bound to the protein. Biochemical degradation of the inactive complex led to PMP and a peptide containing the active site lysine. The mass of this peptide was enhanced by 128, with the exact mass of the inhibitor with the amino group changed to a ketone (De Biase et al., 1991). This experiment ruled out mechanism B in fig. 5-7 (Likos et al., 1982; Ueno et al., 1982), which would have led to the isolation of a complex between PLP and γ-vinyl GABA. The enamine mechanism is discussed in a later section in connection with another suicide inactivator.

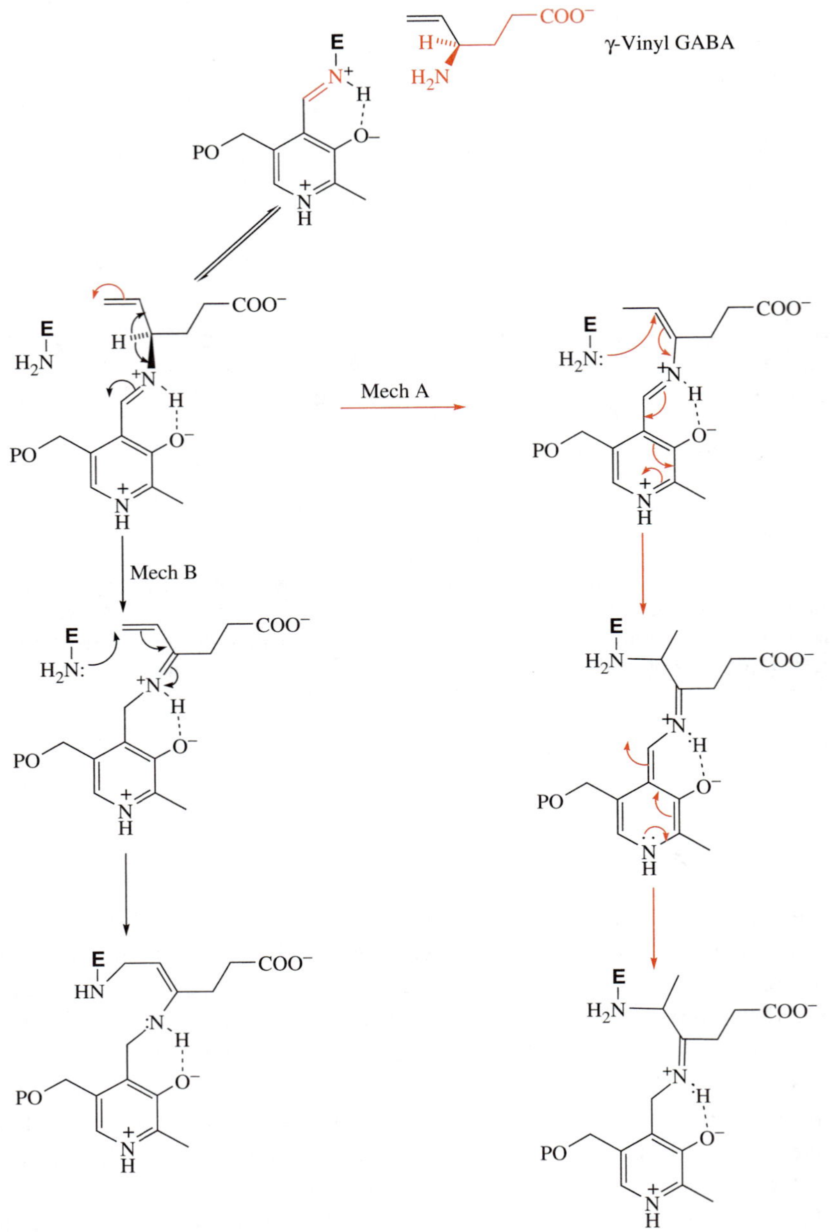

Fig. 5-7. Inactivation of GABA aminotransferase by γ-vinyl GABA. Two mechanisms can be considered, Mech A (red arrows) and Mech B (black arrows). γ-Vinyl GABA is the drug vigabatrin, which is used in the treatment of epilepsy.

Structure

The structure of GABA aminotransferase from pig liver is shown in fig. 5-8 (Storici et al., 2004). The pig liver enzyme is a dimer of identical 472-residue subunits, and it is 96% identical in amino acid sequence to the human brain enzyme. The overall chain fold is somewhat similar to other aminotransferases, and the structure was solved by molecular replacement with the structure of ornithine aminotransferase. The active site structure shown in the center of fig. 5-8 shows PLP bound as its internal aldimine with Lys329.

Enzyme Inhibition 265

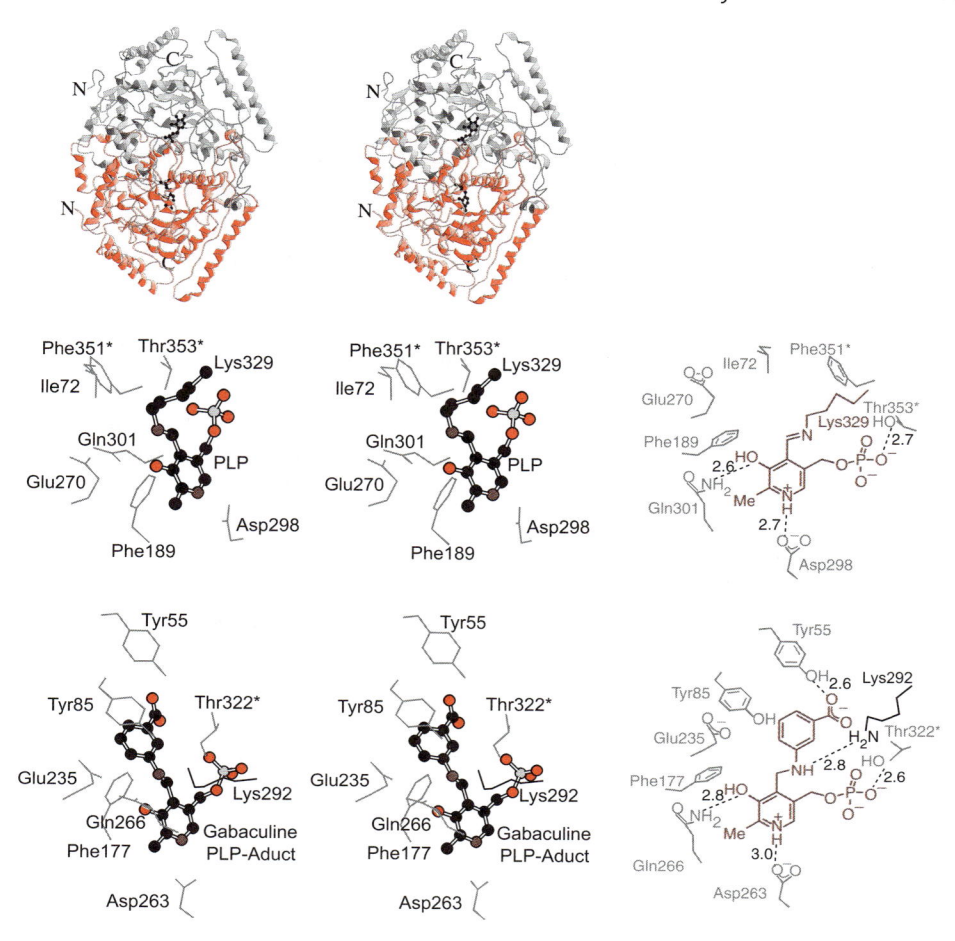

Fig. 5-8. Structure of GABA aminotransferase. The stereodrawing (top) of the overall chain fold in pig liver GABA aminotransferase has stick figures demonstrating pyridoxal-5′-phosphate (PLP) bound at the active sites of the subunits. A stereodrawing (center) of a ball-and-stick model of the active site shows PLP bound through Lys329 (PDB 1GTX; Storici et al., 2004). The close contacts are shown at center right. A stereodrawing (bottom) shows the active site of ornithine aminotransferase (EC 2.6.1.13) that has been inactivated with gabaculine (PDB 1GBN; Shah et al., 1997). The active site contacts are shown at the lower right in a two-dimensional drawing.

The active site of the inactive gabaculine complex of ornithine aminotransferase is shown at the bottom of fig. 5-8. This view shows the active site lysine, in ornithine aminotransferase.

In the ping pong bi bi kinetic mechanism typically found in the action of an aminotransferase, GABA binds to the active site in the first step producing the PMP-enzyme, and α-ketoglutarate binds in the second step to accept the amino group and form L-glutamate. This raises the question of how a single active site binds the monocarboxylate GABA and the dicarboxylates α-ketoglutarate and L-glutamate. Modeling studies indicate that Arg192 binds the α-carboxylate group of both substrates (Storici et al., 1999; Toney et al., 1995). The highly conserved Arg445 is shielded by a salt bridge with Glu270, unlike in most other aminotransferases, in which the corresponding residue binds the ω-carboxylate of substrates. The binding models for GABA aminotransferase include the postulate that the Arg445/Glu270 bridge is disrupted when α-ketoglutarate or L-glutamate bind, and Arg445 then binds the γ-carboxylate of those molecules.

Kinetics of Suicide Inactivation

Suicide inactivation often follows the same kinetic mechanism as affinity labeling in scheme 2-11, in which inactivation is time-dependent and displays saturation kinetics with respect to the concentration of inactivator. This rule applies in cases in which the inactivator is never converted into a dissociable product corresponding to a normal catalytic turnover. The time scale of inactivation is generally slower than the catalytic turnover of a true substrate, although in some cases it may be fast enough to require transient kinetic methods to measure. Two phenomena lead to complications in the kinetics. In studies in which a concerted effort is made to find a suicide inactivator that is extremely potent and fast acting, it can happen that the measurement of inactivation rates is convenient or possible only under conditions in which the enzyme and inactivator concentrations are comparable. In such cases, the steady-state approximation, on which the equations of scheme 2-11 are based, is not applicable. The kinetic profiles of those cases are complex, and they require special treatment on a case-by-case basis.

A second complication arises when the suicide substrate inactivates the target enzyme and is itself a substrate that is transformed into a dissociable product, analogous to the reaction of a true substrate. This phenomenon is described in the kinetic scheme 5-2, where I is the inactivator, I′ is an "activated" form of the inactivator that is generated by enzyme-action, E* is the suicide inactivated enzyme, and P_I is the product of transformation of the inactivator.

$$E + I \underset{k_2}{\overset{k_1}{\rightleftharpoons}} EI \xrightarrow{k_3} EI' \xrightarrow{k_4} E^*$$
$$EI' \xrightarrow{k_5} E + P_I$$

Scheme 5-2

The formation of P_I does not normally lead to inactivation, only the alternative reaction of EI′ to E* inactivates the enzyme. This means that more than one molecule of inactivator per enzyme molecule is required to inactivate a sample of enzyme. Kinetic treatments applicable to scheme 5-2 are in the literature (Galvez et al., 1981; Tatsunami et al., 1981; Tudela et al., 1987).

For practical purposes in the development of inactivators for applications, the partition ratio between inactivation and turnover of the inactivator (k_4/k_5 in scheme 5-2) can be determined in a simple experiment. A fixed amount of enzyme is mixed with the inactivator, at various ratios of enzyme to inactivator, and the reactions proceed to completion. The solutions are then freed of inactivator and products by dialysis or gel filtration, and the residual enzymatic activities are measured and plotted against the corresponding ratios of [inactivator]/[enzyme]. The intercept represents the partition ratio. Alternatively, the concentration of P_I produced in an experiment can be measured and divided by the enzyme concentration to obtain the partition ratio.

Fluorinated Analogs of GABA

The best suicide inactivators react with partition ratios of zero; that is, they do not undergo catalytic turnover to dissociable products. This is the case in the reactions of gabaculine and γ-vinyl GABA with GABA aminotransferase. The myriad chemical reactions of the external aldimines of PLP open varied lines of suicide inactivation mechanisms, and the fluorinated analogs of GABA exemplify some these mechanisms. We have seen two possible modes of Michael addition in the inactivation of GABA aminotransferase by γ-vinyl

GABA in fig. 5-7. Fluorinated derivatives of GABA constitute another class of inactivators that exploit the reactivity of fluorine as a leaving group to introduce reactive functional groups for suicide inactivation.

(*S*)-4-Amino-5-fluoropentanoic acid and (*S*)-4-Amino-5-fluoropentenoic acid inactivate GABA aminotransferase by the enamine mechanism, shown in fig. 5-9A (Silverman and George, 1986; Silverman and Invergo, 1986). The fluorine is in position to undergo β-elimination on carbanion formation at C4. Subsequent transimination with the active site lysine produces an enamine form of the inhibitor. Enamines are alkylating agents, and in this case the transiently formed enamine alkylates the internal aldimine at C4′, forming a stable carbon-carbon bond between PLP and C4 of the inactivator. The enamine mechanism was first discovered in suicide inactivation of other PLP enzymes (Likos et al., 1982; Ueno et al., 1982).

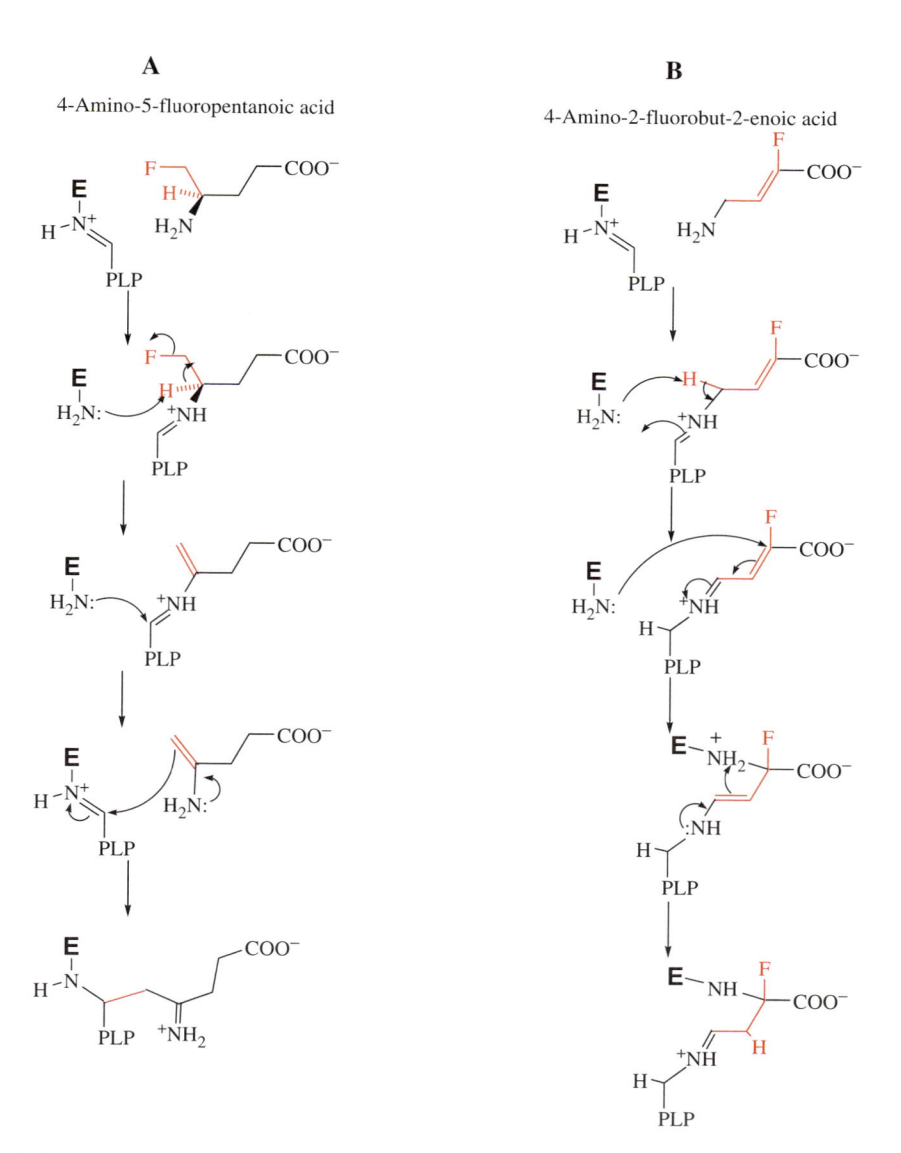

Fig. 5-9. Suicide inactivation of GABA aminotransferase by fluorinated analogs of GABA. (A) Inactivation by (*S*)-4-amino-5-fluoropentanoic acid follows the enamine mechanism. (B) Inactivation by Z-4-amino-2-fluorobut-2-enoic acid follows a Michael addition mechanism.

Z-4-Amino-2-fluorobut-2-enoic acid inactivates GABA aminotransferase inefficiently with a partition ratio of transamination to inactivation of about 750 (Silverman and George, 1988). This compound inactivates by a Michael addition mechanism, presumably according to the reactions shown in fig. 5-9B.

Kinetics of Slow-Binding and Tight-Binding Inhibition

Inactivation of an enzyme by a suicide inactivator is time dependent because the inactivating event is slower than the catalytic time scale. Suicide inactivators generally act irreversibly. However, inhibitors that bind reversibly often display time dependent inhibition for reasons other than covalent modification of the enzyme or coenzyme. Two general classes of time-dependent reversible inhibitors are recognized, those in which binding is a multistep process with one slow step and those that bind so tightly that the attainment of binding equilibrium is intrinsically slow. In this section, we consider kinetic treatments of slow-binding inhibition and tight-binding inhibition.

Slow Binding

Reversible inhibition by slow binding has been recognized for more than 20 years and has been reviewed (Morrison and Walsh, 1988). Two mechanisms are recognized, the most general and common of which is scheme 5-3, multistep inhibition in the reaction of a substrate A to product P.

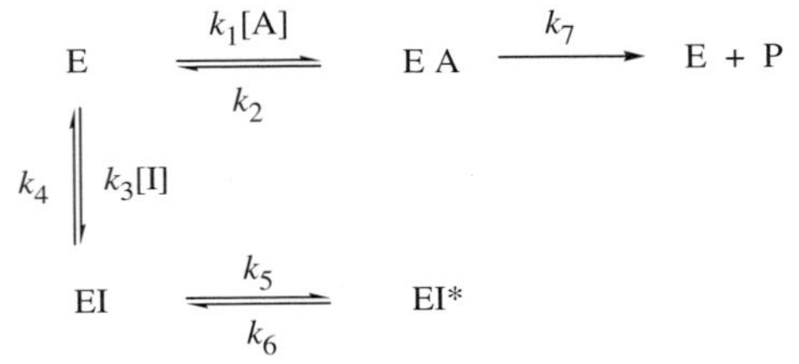

$$E \underset{k_2}{\overset{k_1[A]}{\rightleftharpoons}} EA \xrightarrow{k_7} E + P$$

$$k_4 \updownarrow k_3[I]$$

$$EI \underset{k_6}{\overset{k_5}{\rightleftharpoons}} EI^*$$

Scheme 5-3

In this mechanism, the initial rate in an assay of product formation changes with time at a given inhibitor concentration because the transformation of EI into EI* is slow relative to other steps. EI* formation represents a conformational isomerization of either the enzyme or the inhibitor within its binding site. The initial inhibitor binding steps governed by $k_3[I]$ and k_4 are generally fast relative to other steps and may be treated as equilibrium binding steps in classic reversible inhibition (i.e., $K_i = k_4/k_3$).

In most inhibition experiments, the enzyme is added to the mixture of substrate and inhibitor, and the initial rate, v, is measured as in an assay. In the mechanism of scheme 5-3, this may result in the formation of EI* being missed if k_5 is very small, and the inhibition can appear to be classic reversible competitive according to eq. 5-3, where v_o is the (observed) initial rate with the enzyme added last.

$$v_o = V[A]/\{K_m(1 + [I]/K_i) + [A]\} \tag{5-3}$$

In eq. 5-3, $K_i = k_4/k_3$. More often, the inhibition kinetics will be distorted by the partial transformation of EI into EI*, and the inhibition may appear to be noncompetitive. If the inhibitor is permitted to equilibrate with the enzyme and the reaction is then started by the addition of the substrate, the inhibition will still appear competitive, but the apparent value the inhibition constant will be much smaller, as shown by eq. 5-4, where v_{ss} is the steady-state initial velocity.

$$v_{ss} = V[A]/\{K_m(1 + K_I^*) + [A]\} \tag{5-4}$$

In eq. 5-4, $K_I^* = k_4 k_6/\{k_3(k_5 + k_6)\}$. The difference in inhibition depending on how the reaction is started can be dealt with by considering the time course for the change in the initial rate from v_o to v_{ss}. The time course depends on the rate of the transformation of EI into EI*, which is governed by k_5 and k_6 in scheme 5-3. In an experiment, the observed rate v changes from v_o to v_{ss}, and v at any time after the addition of enzyme is given by eq. 5-5.

$$v = v_{ss} - (v_o + v_{ss})e^{-kt} \tag{5-5}$$

The change from v_o to v_{ss} is governed by the rate constant k. The initial binding of I to form EI and the transformation of EI into EI* are kinetically linked, so that the value of k varies with the concentration of the inhibitor, as described by eq. 5-6.

$$k = k_6 + (k_5[I]/K_i)/(1 + [A]/K_m + [I]/K_i) \tag{5-6}$$

Because the value of k is hyperbolic with the concentration of inhibitor, its value ranges from k_6 at low inhibitor concentrations to $k_5 + k_6$ at high concentrations. Constants k_5 and k_6 can be evaluated by fitting values of k obtained at various inhibitor concentrations if K_m and K_i are known from separate experiments. The value of K_I^* can be obtained separately under conditions for eq. 5-4. Alternatively, K_I^* can be calculated from eq. 5-7 once k_5, k_6, and K_i have been evaluated.

$$k_5/k_6 = (K_i/K_i^*) - 1 \tag{5-7}$$

Infrequently, the initial inhibitor-binding step in scheme 5-3 may be so weak that it is missed in the kinetic experiments. Then the inhibitor may appear to be noninhibitory when the reaction is started by addition of enzyme but inhibitory when allowed to equilibrate with the enzyme and the reaction started by addition of the substrate.

Tight Binding

In the evaluation of inhibition, conditions are sought in which variable inhibition is observed experimentally, and the variation is employed to evaluate inhibitory potency, generally in the form of an inhibition constant K_I, or in pharmaceutical studies I_{50}, the concentration giving 50% inhibition. Very potent inhibitors introduce kinetic complications into these studies because of the underlying assumptions in classic kinetic treatments of inhibition. In the kinetics of reversible inhibition in chapter 2 and the foregoing kinetics of suicide inactivation and slow-binding inhibition, the equations depended on the assumption that the inhibitor concentration was much higher than that of the enzyme. The concentration of the inhibitor was not significantly changed by the addition of the enzyme. However, in studies of tight-binding inhibitors, the inhibition constant can be so low that it becomes necessary to use very low concentrations of inhibitor to observe any variations in inhibition. Then the concentrations of inhibitor and enzyme can be similar, in which case the assumptions on which conventional equations are based no longer apply.

An analogous situation can arise in site-directed mutagenesis experiments. A mutated enzyme may display such low activity that it must be assayed at a high concentration to measure activity, and when that concentration is within an order of magnitude of the concentration of a substrate or inhibitor, the standard kinetic equations are invalid.

Another complication arises in the evaluation of very potent inhibitors. Inhibition experiments must be carried out at inhibitor concentrations near K_i, which can be very low for a picomolar inhibitor. At very low concentrations of an inhibitor and an enzyme, the binding process does not come to equilibrium within the time frame of an initial rate measurement. Recall that the rate constants for ligand binding to an enzyme are in the range of 10^7 to 10^8 M^{-1} s^{-1} (see chap. 2), and the maximum is the diffusional limit for a macromolecule of about 10^9 M^{-1} s^{-1}. If the value of K_i for an inhibitor is 10^{-11} M, and $K_i = k_{off}/k_{on}$, the value of k_{off} cannot be larger than $10^9 \times 10^{-11}$ or 0.01 s^{-1}. At an inhibitor concentration equal to its K_i, the half-time for attaining binding equilibrium is 69 s. If the actual rate constant for binding (k_{on}) is the conventional 10^8 M^{-1} s^{-1}, the half-time will be 690 s. All ultra-high-affinity inhibitors are slow binding.

The kinetics of tight-binding inhibition is far more complex than we have heretofore encountered (Williams and Morrison, 1979). At the very low concentrations required for evaluation, the concentrations of inhibitor and enzyme are similar. The equations describing their behavior must take into account the conservation equations for the inhibitor, such as $[I_o] = [I] + [EI]$, as well as for the enzyme. The free inhibitor concentration is not the same as the total concentration. To illustrate the complications, consider the simplest possible kinetic model for a tight-binding inhibitor (scheme 5-4).

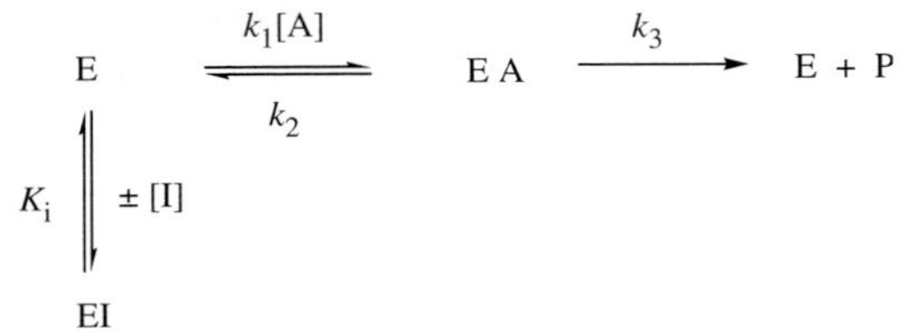

$$E \underset{k_2}{\overset{k_1[A]}{\rightleftharpoons}} EA \overset{k_3}{\longrightarrow} E + P$$

$$K_i \updownarrow \pm [I]$$

$$EI$$

Scheme 5-4

A rate equation can be obtained by considering the same equations for rate, steady state, and enzyme conservation written in chapter 2 for the first line of scheme 5-4 plus the conservation equation for the inhibitor: $[I]_o = [I] + [EI]$. It turns out to be quadratic eq. 5-8.

$$v_2 + vk_3[A]\{K_i/K_m + ([I_0] - [E_0])/(K_m + [A])\} - (k_3[A])^2 K_i[E_0]/\{(K_m + [A])K_m\} = 0 \tag{5-8}$$

No simple analysis method for extracting kinetic parameters using eq. 5-8 is available. It is advisable to collect initial rate data while varying $[E]_o$ and $[I]_o$; write the solution to the quadratic equation for eq. 5-8; and fit the data to the equation for the solution.

In pharmaceutical studies, values of I_{50} for various inhibitors are often the index of inhibitory efficiency. These values refer to the inhibitor concentrations at which 50% inhibition is observed under specified conditions. Equation 5-8 shows that the degree of inhibition in an experiment depends on both the enzyme and inhibitor concentrations. In reporting values of I_{50} for tight-binding inhibitors, it is necessary to include the enzyme concentration among the specified conditions.

Slow-Binding Inhibition

An inhibitor of chymotrypsin that we encountered in chapter 1, the transition-state analog inhibitor *N*-acetyl-L-leucyl-L-phenylalanine trifluoromethylketone, is a slow-binding inhibitor (Brady and Abeles, 1990). Among the many examples of slow-binding inhibition, cases from the pharmaceutical industry may be cited as the most widely known. In this section, we consider methotrexate, a well-known anticancer drug, as a classic slow-binding inhibitor of dihydrofolate reductase, a target for anticancer therapy (Huennekens, 1996).

Dihydrofolate Reductase

Function

Dihydrofolate reductase (DHFR; EC 1.5.1.3), a 20-kDa monomer, catalyzes the reduction of dihydrofolate by NADPH to form tetrahydrofolate and NADP⁺. The action of DHFR regenerates tetrahydrofolate from the dihydrofolate produced from methylenetetrahydrofolate in the thymidylate synthase catalyzed methylation of dUMP to dTMP. The thymine nucleotide is a building block for DNA, and for this reason, both DHFR and thymidylate synthase are essential for DNA biosynthesis.

Hydride transfer proceeds with migration of the pro-R hydrogen of NADPH to the si-face of C5 in dihydrofolate, as illustrated in scheme 5-5.

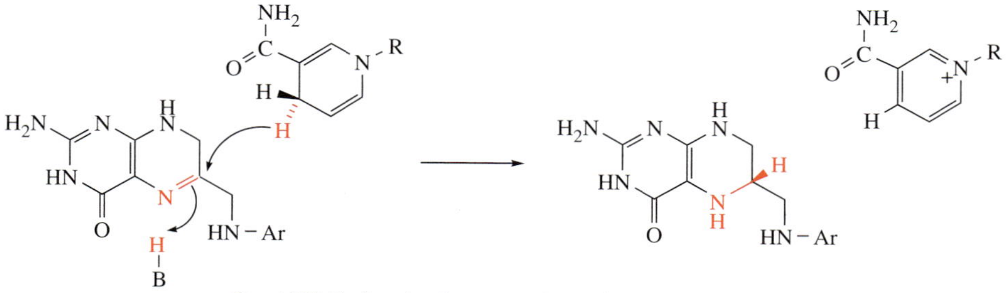

R = ADP-2′-phosphoribose   Ar = p-benzoyl polyglutamate

Scheme 5-5

Methotrexate

Aminopterin and amethopterin (methotrexate) are folate antagonists used in the treatment of leukemia and other malignancies. They are inhibitors of DHFR (Osborn et al., 1958; Peters and Greenberg, 1958). Structure **5-2** does not give clues about their potencies as inhibitors of DHFR. Nor did their inhibitory properties seem at first to be in accord with expectations that they would bind at the active site.

5-2

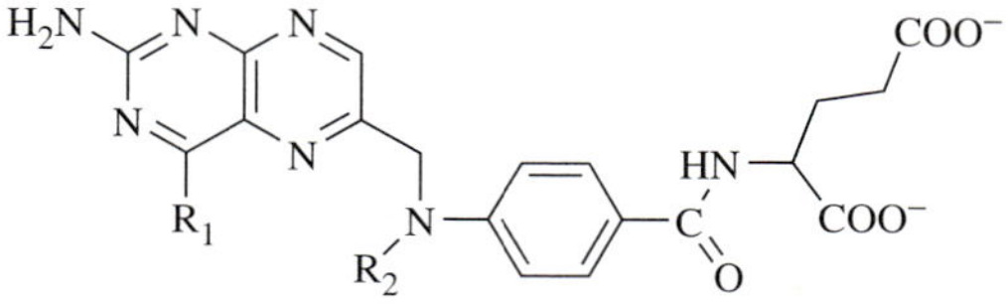

Methotrexate: R_1 = NH$_2$; R_2 = CH$_3$

Aminopterin: R_1 = NH$_2$; R_2 = H

Original inhibition data are shown graphically in fig. 5-10. The double reciprocal plot in fig. 5-10 shows that the lines do not meet on the ordinate, as would be expected for competitive inhibitors. It was at first concluded that the compounds were noncompetitive inhibitors of DHFR, with inhibition constants of approximately 10^{-9} M. Further studies showed the time course for inhibition to be slow. The half-time for dissociation, as estimated in dialysis studies, was 6 days in the absence of folate and 1 day in its presence, and the inhibition constants were revised to 10^{-11} M (Werkeiser, 1961). Based on the latter work, the compounds were described as *stoichiometric inhibitors*.

The slow time course acquired significance when these compounds were recognized as slow-binding inhibitors, so that full inhibition required significant time to reach equilibrium. The slow step in binding is attributed to a conformational change in DHFR, which is governed by a rate constant estimated at 0.03 s^{-1} (Cayley et al., 1981).

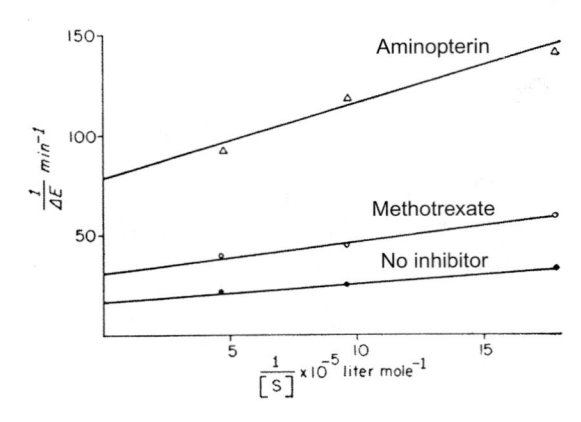

Fig. 5-10. Inhibition of dihydrofolate reductase (DHFR) by methotrexate. Apparent noncompetitive inhibition results from slow binding of methotrexate.

Structure

The crystal structure of DHFR shows an eight-stranded β-sheet encased within four helices and three loops, as illustrated in fig. 5-11 (Bolin et al., 1982; Filman et al., 1982; Reyes et al., 1995). NADPH is bound in an extended conformation adjacent to the folate binding site, with the dihydropyridine ring projected toward the pterin ring of the substrate. The structures of methotrexate and folate bound to DHFR differ, in that the pterin ring of methotrexate is rotated by 180 degrees relative to the orientation in folate. This results from a favorable hydrogen-bonded interaction of the pterin 1-amino group in methotrexate with two main chain carbonyl groups of DHFR, illustrated at the bottom right of fig. 5–11. Main chain carbonyl groups cannot form hydrogen bonds with the pterin 1-oxo group of folate.

At least five kinetically significant complexes participate in catalysis, and structures of five analogs of the *E. coli* FDHR are available (Sawaya and Kraut, 1997). The structures of the *E. coli* enzyme point to loop and subdomain movements in the overall mechanism. Loop1 between residues 9 and 24, also known as the Met20 loop, closes in on the active site in the ternary complex. This loop is not immobile, however, and appears in three distinct conformations, open, closed, and occluded. In the occluded conformation, the loop occupies a portion of the nicotinamide binding region and seems to occlude nicotinamide binding. The motions of loop 1 seem to be coupled to motions of the other loops. Structural information on the vertebrate DHFRs indicates that they are less conformationally mobile, and opening and closing of the substrate binding sites is not dominated by movements of loop 1 but rather by other conformational changes (Matthews et al., 1985a, 1985b; McTigue et al., 1992, 1993).

Structures of *E. coli* DHFR in complex with NADPH and methotrexate suggest that in the ternary complex nicotinamide C4 of NADPH is brought to within 2.8 Å of C6 of dihydrofolate (Sawaya and Kraut, 1997). This is in violation of the van der Waals contact and within 0.2 Å of the calculated transition-state distance (Wu and Houk, 1987). It seems clear that binding interactions force a sufficiently close interaction between NADPH and dihydrofolate to facilitate hydride transfer. Catalysis of proton transfer to N5 of dihydrofolate is less clear because of the absence of a nearby acid-base group. It may be that proton transfer to N5 is mediated by water molecules.

Enzyme Inhibition 273

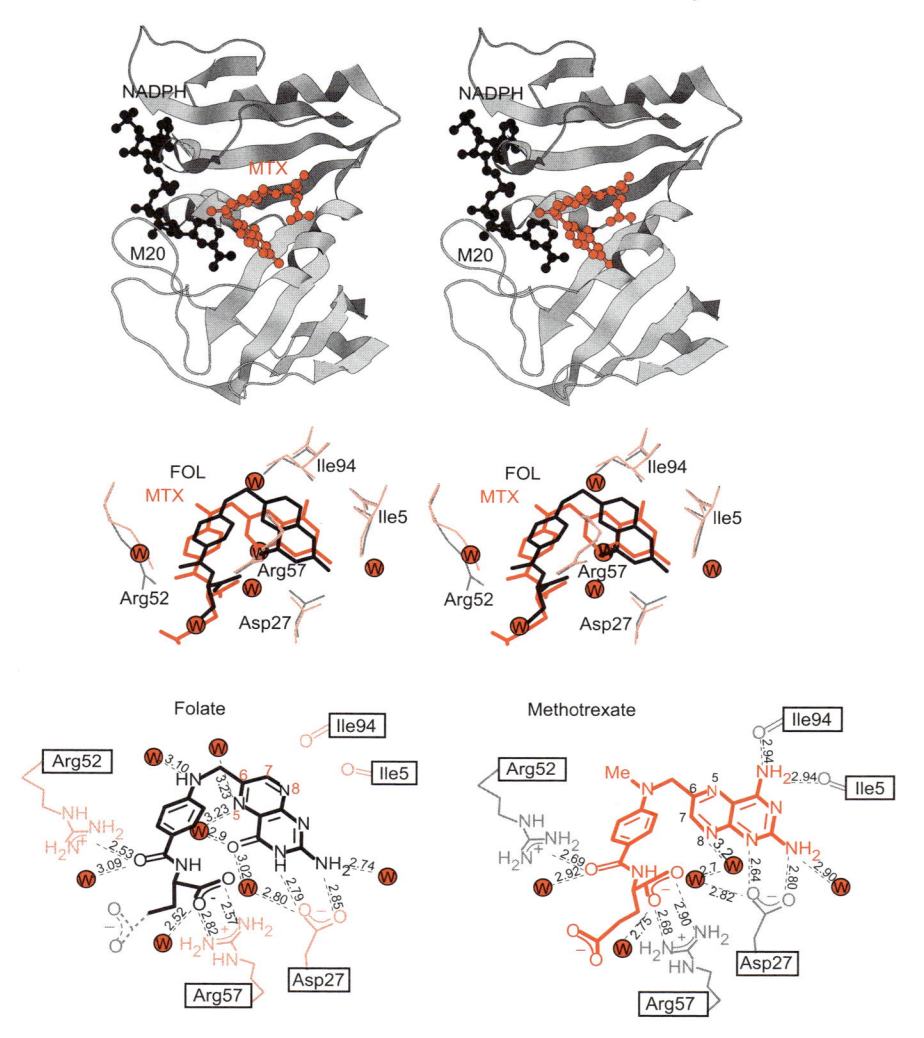

Fig. 5-11. Structure of dihydrofolate reductase (DHFR). A ribbon diagram (top) of *Lactobacillus casei* DHFR with NADPH shows methotrexate bound in the active site. Loop 1, the M20 loop, is in the closed conformation over NADPH. A stereodiagram (middle) of the V-shaped folate binding site in *E. coli* DHFR shows methotrexate in red and folate in black. Notice that the pterin ring of methotrexate is oriented with the 1-amino group projecting out and the 1-oxo group of folate projecting in. The two-dimensional diagrams (bottom) show the close contacts by folate (PDB 1DYI; Reyes et al., 1995) and methotrexate (PDB 4DFR; Bolin et al., 1982) at this site. Notice the hydrogen bonding between the 1-amino group of methotrexate with the main chain carbonyl groups of Ile94 and Ile5. Hydrogen bonding would not be possible between these carbonyl groups and the 1-oxo group of folate.

Kinetics

Based on transient kinetic studies, it seems that the high affinity of DHFR for NADPH and dihydrofolate makes it unlikely that free DHFR exists in vivo. The transient kinetics indicates that the product ternary complex undergoes dissociation of NADP$^+$, followed by binding of NADPH, then dissociation of tetrahydrofolate, and binding of dihydrofolate to form the substrate ternary complex (Fierke et al., 1987). This cycle may be represented by scheme 5-6, in which H$_2$fol is dihydrofolate and H$_4$fol is tetrahydrofolate. Structural analysis of several *E. coli* complexes indicates the presence of steric clashes between NADPH and

274 Enzymatic Reaction Mechanisms

tetrahydrofolate that may tend to drive dissociation of the tightly bound tetrahydrofolate (Sawaya and Kraut, 1997).

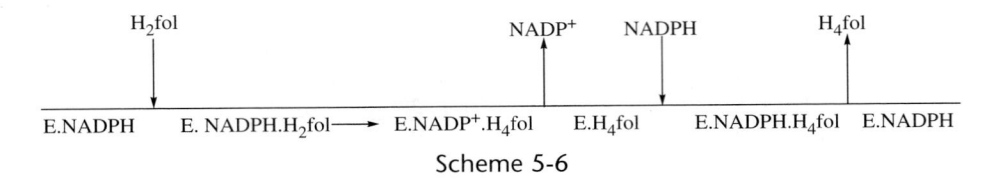

```
           H₂fol                          NADP⁺      NADPH           H₄fol
             │                              │          │                │
             ↓                              ↓          ↓                ↓
E.NADPH   E.NADPH.H₂fol ⟶ E.NADP⁺.H₄fol   E.H₄fol   E.NADPH.H₄fol   E.NADPH
```
Scheme 5-6

The transient kinetic studies revealed the overall pathway taken by DHFR, led to the measurement of many of the rate constants, and uncovered a kinetically significant nonchemical step assigned to a conformational change. The kinetic mechanism derived from the transient studies is shown in fig. 5-12. The dissociation of tetrahydrofolate (H₄fol in fig. 5-12) limits the rate in neutral solutions. Hydride transfer becomes rate limiting at high pH, as indicated by the appearance of a deuterium kinetic isotope effect in the reaction of (4R)-[4-²H]NADPH.

Structural Dynamics

NMR relaxation studies of the *E. coli* DHFR prove that loop movements, most directly the movements of the M20 loop, are fast enough to participate in the catalytic cycle, specifically the binding of substrates and release of products (Epstein et al., 1995; Falzone et al., 1994; Osborne et al., 1997). Movements of the loops that were implicated in binding in the crystal structures of enzyme substrate and analog complexes take place on the nanosecond time scale.

Prostaglandin H Synthase

Function

Prostaglandins, thromboxane, and prostacyclin modulate many important physiological processes, including the mediation of inflammation, allergy, and fever. Compounds that control their production are potentially important pharmaceutical agents (Marnett et al., 1999; Smith et al., 2000). In the first step of the biosynthesis of these lipids, arachidonic acid undergoes an oxygen-dependent transformation into prostaglandin G₂ (PGG₂), a hydroperoxide, and then its peroxidation to the corresponding alcohol prostaglandin H₂ (PGH₂).

Fig. 5-12. Kinetic mechanism of dihydrofolate reductase (DHFR). Transient kinetic analysis of *E. coli* DHFR led to the measurement of the rate constants for the transformations of ternary and binary complexes shown here (Fierke et al., 1987). Most of the second-order binding processes were too fast to observe by stopped-flow methods. Based on the available data, the main pathway for turnover is that defined by the complexes shown in boldface. The other processes take place at slower rates. H₂fol, dihydrofolate; H₄fol, tetrahydrofolate; N⁺, NADP⁺; NH, NADPH.

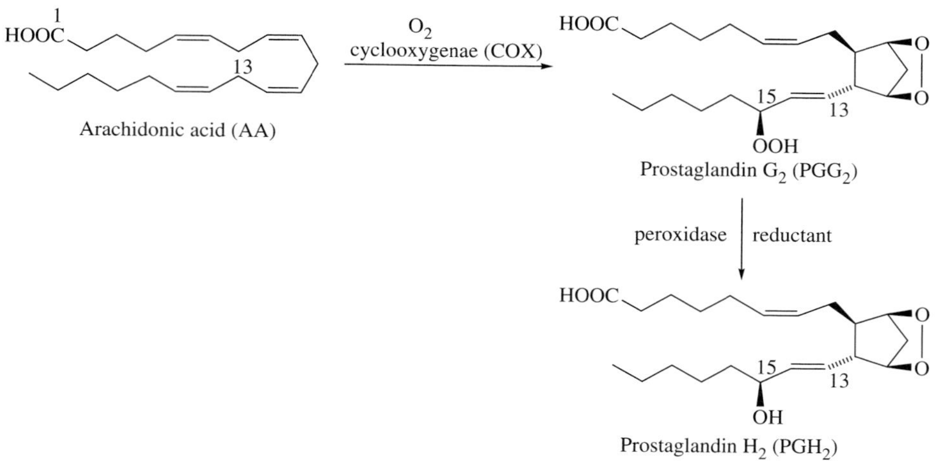

Fig. 5-13. Cyclooxygenase and peroxidase activities of prostaglandin H synthase (PGHS). PGHS is a heme enzyme that catalyzes two sequential reactions in the transformation of arachidonic acid (AA) into prostaglandin H_2 (PGH$_2$). Heme participates in both reactions, which take place at distinct active sites. The role of heme in cyclooxygenase (COX) activity is to generate the Tyr385 radical, which initiates the COX reaction to produce prostaglandin G$_2$ (PGG$_2$) (top line). The peroxidase activity of PGHS in converting PGG$_2$ into PGH$_2$ requires the direct action of heme in a typical peroxidase mechanism.

Prostaglandin H synthases (PGHS; EC 1.14.99.1) are heme proteins that catalyze both steps of this complex process, the cyclooxygenase and peroxidase reactions depicted in fig. 5-13. The cyclooxygenation of arachidonic acid to PGG$_2$ is an amazing process that includes the introduction of five optical centers and the formation of four new covalent bonds. The peroxidation of PGG$_2$ to PGH$_2$ appears to be a conventional peroxidase reaction of the type discussed in chapter 4.

Two PGHS enzymes have been examined in classic studies: a constitutive enzyme designated PGHS-1 and a closely related but different enzyme that appears in inflammation and the central nervous system and that is designated PGHS-2. In the literature, these enzymes are also referred to as COX-1 and COX-2 after their cyclooxygenase activities, which are the specific targets of pharmaceutical research. Inhibitors of COX-2 and COX-2 relieve pain and include such over-the-counter drugs as aspirin and ibuprofen. Because COX-2 appears in inflammation and is primarily responsible for the associated pain, it has become the favored target for the development of new drugs for pain treatment. In theory, a specific inhibitor of COX-2 can relieve pain while leaving COX-1 to its housekeeping functions. COX-2 inhibitors include Celebrex and drugs recently withdrawn because of side effects. Another PGHS has been reported as COX-3 (Chandrasekharan et al., 2002).

Structure

Many structures of COX-1 and COX-2 have been published, and in agreement with the amino acid sequence similarities between the two enzymes (>60% identities), they show very similar structures. COX-1 and COX-2 are membrane-associated dimers, with the active site buried deeply in the protein and accessible to membrane-bound substrates by passage through an apolar channel (Malkowski et al., 2000). Inhibitors bind in this channel and block access to the active site. The structure of sheep COX-1 in fig. 5-14 shows the overall chain fold, with ibuprofen at its binding site and blocking the channel. Arachidonic acid (AA) is also shown in its site.

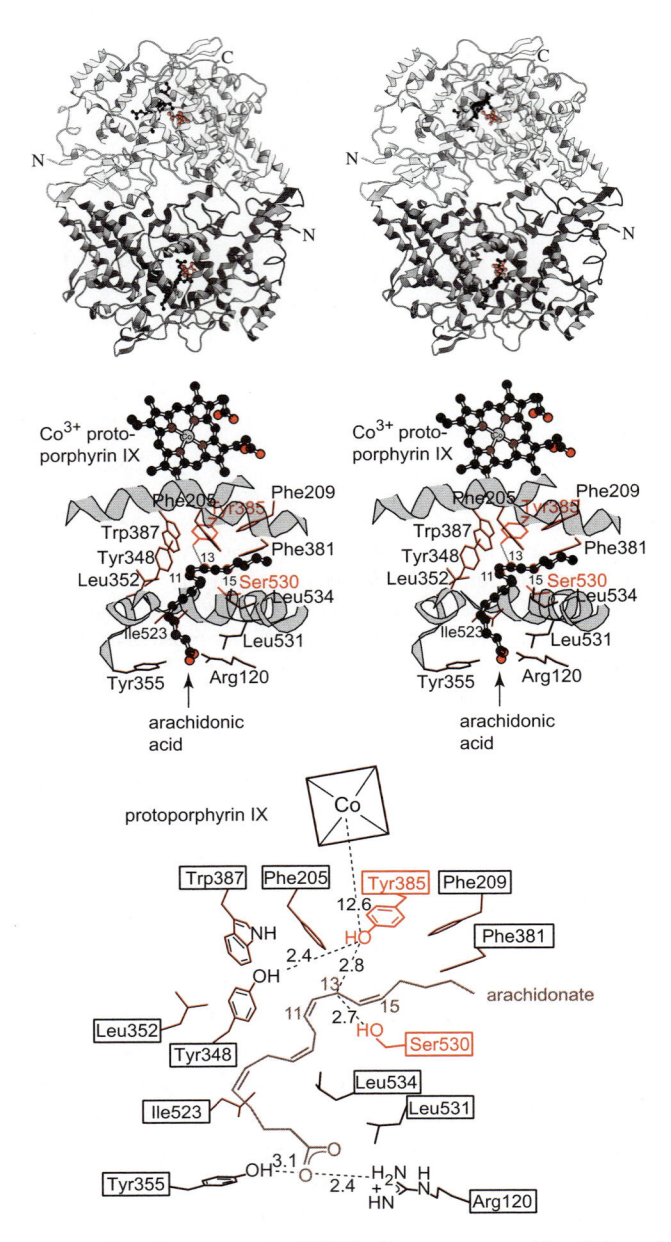

Fig. 5-14. Structure of cyclooxygenase (COX). The stereographic ribbon diagram (top) of homodimeric sheep prostaglandin H synthase-1 (top) depicts ball-and-stick models of heme (black) and ibuprofen (red) at their binding sites (PDB 1EQG; Selinsky et al., 2001). A stereo-drawing (middle) shows the active site, with arachidonic acid (AA) bound and the unreactive cobalt protoporphyrin IX in place of heme (PDB 1DIY; Malkowski et al., 2000). Notice that the carboxyl group of AA is in contact with Arg120, and AA is bound in a conformation that allows cyclooxygenation in the observed stereochemistry after the Tyr385 radical abstracts the C13(H). Ser530 and Tyr385 are on opposite sides of the site at a constriction that is thought to constrain AA in the correct conformation for the COX activity. The active site with Ser530 acetylated by aspirin is illustrated in fig. 5-16.

Aspirin alters the channels of COX-1 and COX-2 by covalent acetylation of Ser530 (COX-1 numbering). Ser530 and Tyr385 form a constriction in the channel. Other inhibitors bind noncovalently within the channel and at the same site in both types of COX.

Reaction Mechanism

The transformation of arachidonic acid into PGG_2 is a radical initiated, multistep process. PGH_2 is bound at the active site in a conformation that is compatible with the complex stereochemistry of the reaction mechanism (Kiefer et al., 2000). In the predicted conformation, the carboxyl group of AA should be associated with Arg120, and C13 should lie near Tyr385, which initiates catalysis as a tyrosyl radical (Goodwin et al., 1998; Shimokawa et al., 1990; Tsai et al., 1994). In a model that accounts for the stereochemistry, the pro-*S* hydrogen at C13 lies 2.5 Å from Tyr385. The binding model accommodates the stereochemistry of the postulated mechanism in fig. 5-15.

The initiating Tyr385 radical is formed by long-range electron transfer to the ferryl-heme. In step 1, the tyrosyl radical abstracts the 13-pro-*S* hydrogen from AA to form the C11-C13 allylic radical, shown with the unpaired electron on C11. Reaction with oxygen to form the 11-hydro-peroxide radical in step 2 is facile because of the diradical structure of oxygen. Ring closure in step 3 leads to the hypothetical C8 radical, which immediately undergoes ring closure in step 4 by electron pairing with a π-electron on C12, thereby generating the bicyclic C13-C15 allylic radical. Step 5 proceeds by coupling of the allylic

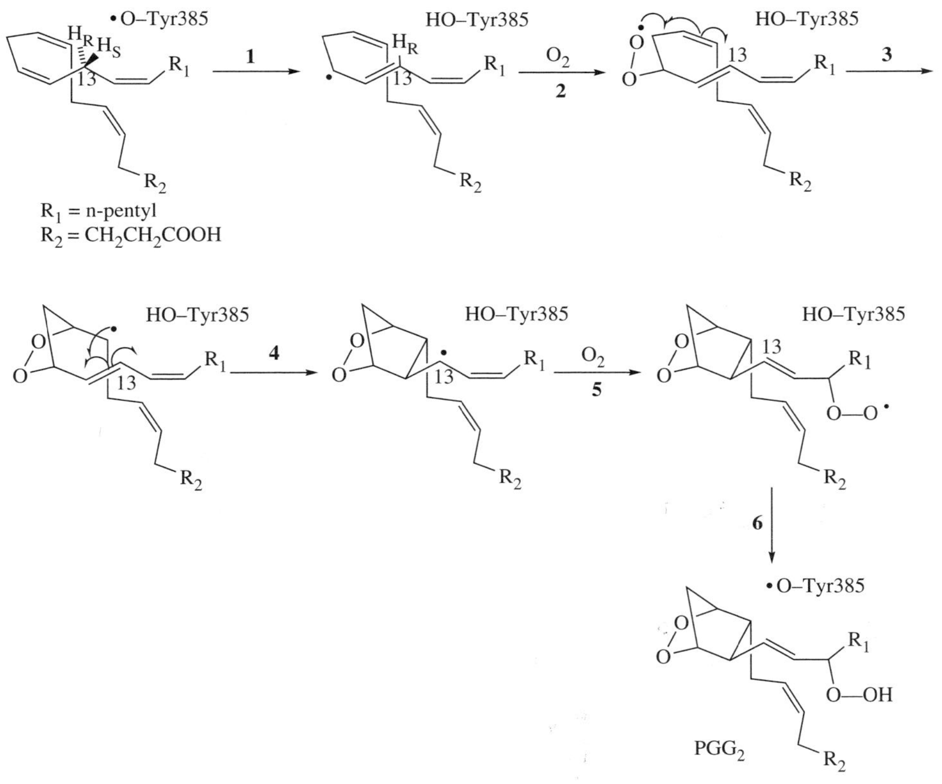

Fig. 5-15. Mechanism of the transformation of arachidonic acid (AA) into prostaglandin G_2 (PGG_2) by stepwise cyclooxygenation. Cyclooxygenase-1 (COX-1) and COX-2 catalyze this reaction by the 6-step, stereocontrolled radical mechanism shown. The initiating Tyr385 radical is formed by long-range electron transfer from Tyr385 to the ferryl heme.

278 Enzymatic Reaction Mechanisms

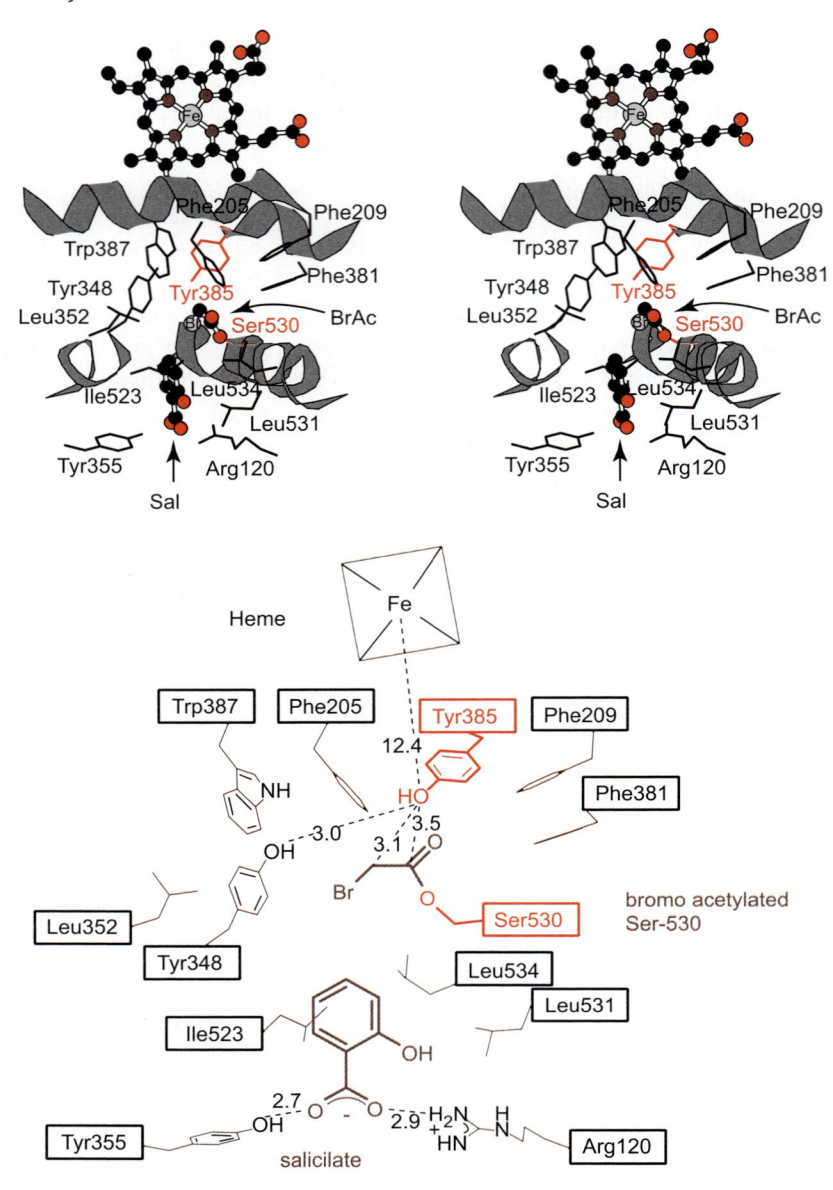

Fig. 5-16. The active site of prostaglandin H synthase is acetylated by aspirin. The stereodiagram (top) shows the cyclooxygenase-1 (COX-1) active site, Ser530, which is bromoacetylated by an aspirin analog, leaving no room for productive arachidonic acid (AA) binding (PDB 1PTH; Loll et al., 1995). Salicylate is reversibly bound in the AA binding tunnel (bottom).

radical with oxygen at C15 to form the 15-hydroperoxide radical, which is quenched in step 6 by Tyr385 to form PGG_2 and regenerate the Tyr385 radical.

The mechanism is supported by the identification of the tyrosyl radical by EPR analysis of samples freeze-quenched in the steady state (Tsai et al., 1999). In the Michaelis complex, the tyrosyl radical abstracts the C13 hydrogen from AA and leads to the C11-C15 pentadienyl radical formed in step 1. The EPR spectra of the radical generated with samples of AA synthesized with deuterium labeling at C11, C13(S), C15, or C16 allowed its structure to be assigned with certainty because of the differential nuclear hyperfine coupling constants for H and D (Peng et al., 2001, 2002; van der Donk et al., 2002).

Acetylation of Ser530 by aspirin blocks the COX activity but does not completely inhibit PGHS. The proximity of Ser530 to the AA binding site is clearly indicated in the stereodiagram of the active site in fig. 5-16, and this site in bromoaspirin-bromoacetylated PGHS is shown in fig. 5-16. AA is still a substrate for aspirin-acetylated PGHS-2, but the products are the noncyclic (15R)-hydroperoxy eicosatetraenoic acid and a lesser amount of the (11R)-isomer shown in fig. 5-17. Apparently, acetyl-Ser530 alters the conformation of AA in the active site in such a way as to disrupt the cyclization process in fig. 5-15, but it still allows the C11-C15 pentatienyl radical to react with oxygen, preferentially at C15 but also at C11 (Holtzman et al., 1992; Lecomte et al., 1994; Mancini et al., 1994).

Heme plays a dual role in the action of prostaglandin H synthase, in that it both potentiates the formation of PGG_2 from AA and also carries out the essential peroxidase chemistry in the subsequent transformation of PGG_2 into PGH_2. In initiating the COX reaction, the heme is first transformed into a ferryl-heme by reaction with the 15-hydroperoxide PGG_2 formed in step 6 of the mechanism in fig. 5-15. The ferryl-heme generates the Tyr385 radical by long-range electron transfer at a distance of 12.5 Å, and the tyrosyl radical can then initiate the COX reaction. The second role of heme is to facilitate peroxidase action on PGG_2, in which it is transformed into PGH_2 by the usual peroxidase mechanism (see chap. 4). This also lends a dual role to PGG_2 and leads to complications in the stoichiometry of the overall process (van der Donk et al., 2002).

Slow- and Fast-Binding COX Inhibitors

COX inhibitors have been classified as slow to bind or fast binding based on the kinetics of the inhibition. Fast-binding inhibitors include ibuprofen, which is not selective for COX-1 or COX-2. The kinetic mechanisms of eqs. 5-9 and 5-10 have been put forward to account for fast- and slow-binding inhibition, in which the conversion of EI to EI* is slow.

$$E + I \rightleftharpoons EI \tag{5-9}$$

$$E + I \rightleftharpoons EI \rightleftharpoons EI^* \tag{5-10}$$

In principle, the difference between EI and EI* could reside in the structure of the enzyme or of the inhibitor in its bound state.

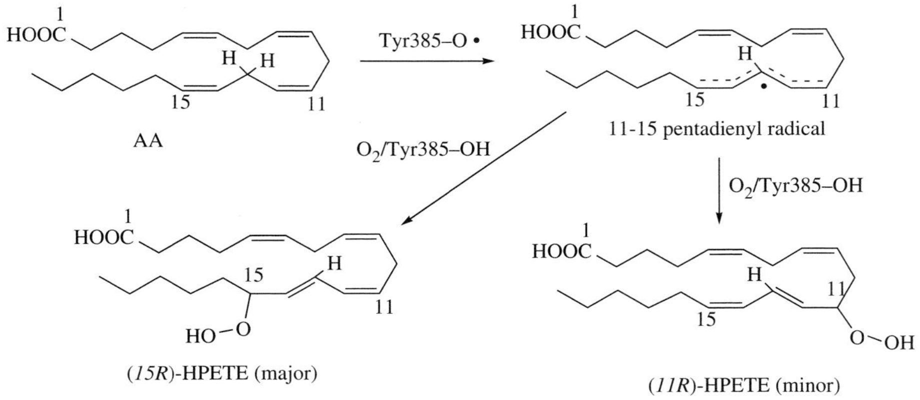

Fig. 5-17. Hydroperoxygenation of arachidonic acid by aspirin-acetylated prostaglandin H synthase (PGHS). The 11-15 pentadienyl radical in aspirin-inhibited PGHS does not undergo the cyclization processes seen in fig. 5-15, but it undergoes oxygenation at C15 and to some extent at C11 to form the two isomers of hydroperoxy-5Z,8Z,13E-eicosatetraenoic acid (HPETE).

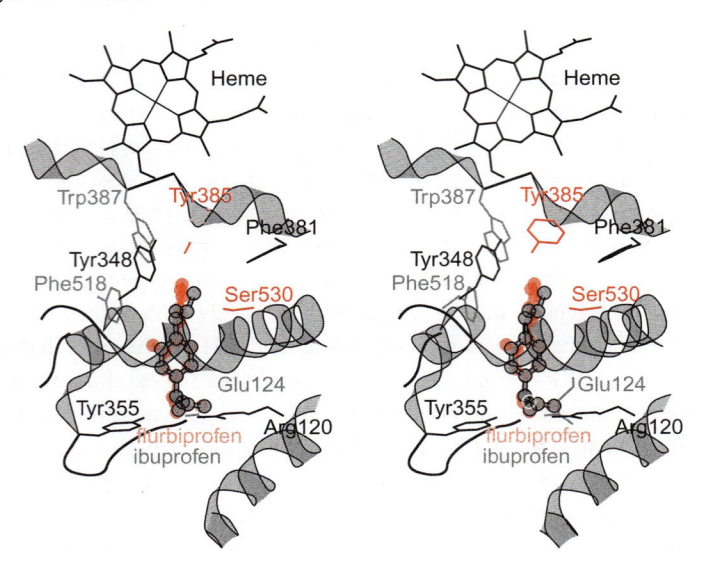

Fig. 5-18. Structures of ibuprofen and flurbiprofen bound to prostaglandin H synthase-2 (PGHS-2). Flurbiprophen (PDB 1EQH) and ibuprofen (PDB 1EQG) structures were aligned and are shown here bound in the arachidonic acid binding tunnel in almost identical conformations despite having very different binding kinetics. (From Selinsky et al., 2001, with permission.).

Slow binding cannot be attributed to differences in the structures of the enzyme or the bound inhibitors, as shown by the structures of complexes of PGHS-1 with slow- and fast-binding inhibitors (Selinsky et al., 2001). As illustrated in fig. 5-18, the structure of the active site of PGHS-1 with the fast-binding ibuprofen and slow-binding flurbiprofen does not reveal differences that could account for the kinetics of binding. The possibility that crystal forces could have driven both complexes into the EI* state was tested by dissolving the crystals and observing the dissociation rates for the inhibitors. Ibuprofen dissociated immediately, but flurbiprofen required 7 hours to dissociate (Selinsky et al., 2001). This experiment also proved that crystal forces could not have driven both complexes into the EI state. The structures simply do not explain the kinetics of binding. It may be that some inhibitors are kinetically hindered in negotiating the binding channel to the active site.

Tight-Binding Inhibition

Inhibitors that bind very tightly are generally also slow binding for the reasons given previously. However, tight-binding inhibition may be distinguished from the slow-binding inhibition of the type in the preceding section by the absence of biphasic kinetic behavior. The only kinetically significant species are the free enzyme E and the inhibited complex EI. Representative examples of tight-binding inhibitors developed in the pharmaceutical and agricultural sciences are presented in this section.

HMG-CoA Reductase

Properties and Function

HMG-CoA reductase (EC 1.1.1.34) catalyzes an essential step in the biosynthesis of cholesterol, the reduction of (R)-3-hydroxy-3-methylglutaryl coenzyme A (HMG-CoA) to mevalonate. In subsequent steps of cholesterol biosynthesis, mevalonate is phosphorylated in two steps by specific kinases to mevalonate-5-pyrophosphate and then decarboxylated in an ATP-dependent mechanism to isopentenyl pyrophosphate. This molecule and its isomer, dimethylallyl pyrophosphate, are the building blocks for squalene and ultimately cholesterol.

In hepatocytes, HMG-CoA reductase is a 97-kDa, membrane-bound protein comprising an integral membrane domain and a cytoplasmic catalytic domain (Parker et al., 1986). The catalytic domain is subject to inhibition by phosphorylation and can be released as a 53-kDa soluble protein by the proteolytic action of calpain. The human enzyme can also be separately expressed as a 58-kDa, soluble, dimeric enzyme (Mayer et al., 1988). Most bacterial HMG-CoA reductases are soluble enzymes, although their structures are not the same as the catalytic domains of the mammalian enzymes, and their active sites appear to differ (Istvan and Deisenhofer, 2000; Lawrence et al., 1995; Tabernero et al., 1999).

Inhibition of this enzyme in vivo by statins corrects overproduction of LDL-cholesterol and decreases susceptibility to cardiovascular disease. The reduction of HMG-CoA to mevalonate follows the course shown in eq. 5-11.

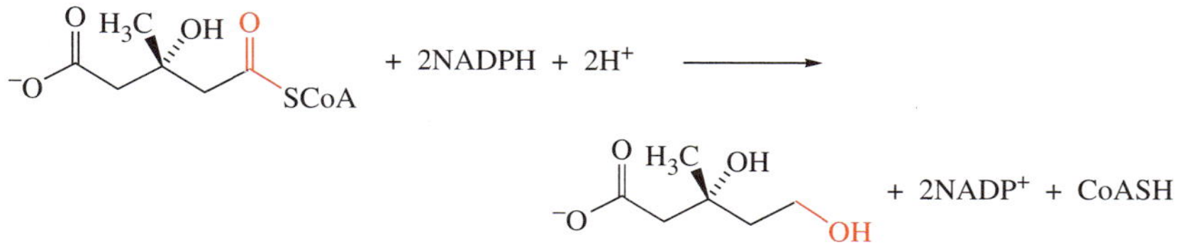

$$\text{HMG-CoA} + 2\text{NADPH} + 2\text{H}^+ \longrightarrow \text{mevalonate} + 2\text{NADP}^+ + \text{CoASH} \tag{5-11}$$

Reaction Mechanism

Reduction of HMG-CoA requires two equivalents of NADPH and takes place in several steps at the same active site. In the first step, HMG-CoA is reduced by NADPH to mevaldyl CoA, the hemiacetal of mevaldehyde, which eliminates CoASH (thiol form of CoA) to form mevaldehyde (scheme 5-7). After dissociation of NADP$^+$ and CoASH, the second molecule of NADPH binds and reduces mevaldehyde to mevalonate.

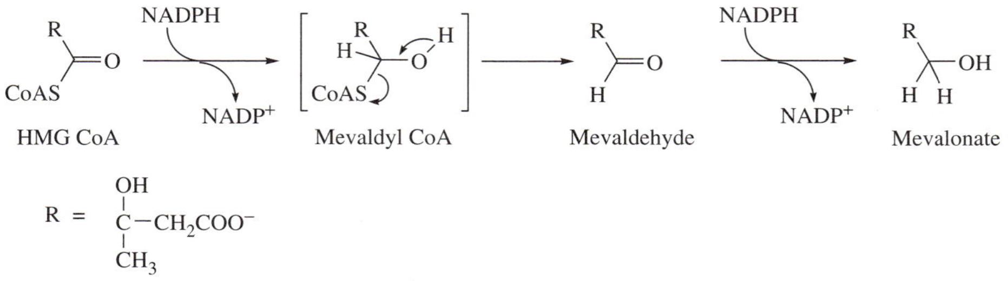

Scheme 5-7

The enzyme catalyzes partial reactions in addition to the overall forward and reverse reactions. They include the oxidation of mevaldyl CoA to HMG-CoA by NADP$^+$, the reduction of mevaldyl CoA to mevalonate by NADPH, the reduction of mevaldehyde to mevalonate by NADPH, and the oxidation of mevaldehyde in the presence of CoASH to HMG-CoA by NADP$^+$ (Frimpong and Rodwell, 1994a). These reactions represent steps or reversals of steps in scheme 5-7.

Results from chemical modification, pH dependence, multiple amino acid sequence alignments, and site-directed mutagenesis of conserved amino acids led to the assignments of Asp766, His865, and Glu558 of hamster reductase as amino acids likely to be engaged in catalysis (Darney and Rodwell, 1993; Frimpong and Rodwell, 1994b). Results of complementation experiments on coexpression of specifically mutated genes indicated that the active site would include Asp766 and Glu558 from different chains in the dimeric enzyme (Frimpong and Rodwell, 1994b). All of these mutated species displayed insignificant activities (≤0.3%) in the overall reaction, and D766N was inactive in all of the partial reactions. However, H865Q and E558Q displayed significant activity in all of the partial reactions. It was postulated that Asp766 participated in catalyzing hydride transfer steps 1 and 4 in the mechanism of fig. 5-19 because one of these steps is required in all of the reactions. His865 and Glu558 appeared to catalyze the elimination of CoASH from mevaldyl CoA, step 2 in fig. 5-19.

The structure of the catalytic domain of human HMG-CoA reductase in complex with HMG, CoASH and NADP$^+$ confirmed the presence of His866 (human numbering) in position to protonate coenzyme A and of Glu559 in the active site. As illustrated in the lower left of fig. 5-21, both Glu559 and Lys691 were hydrogen bonded with the acyl carbonyl group of the substrate, and Asp767 was hydrogen bonded to Lys691. The structure did not reveal the essential role of Asp767 (Asp766 in the hamster reductase), but Glu559 and Lys691 could catalyze proton transfer to and from the acyl carbonyl group in the hydride transfer steps 1 and 4 of fig. 5-19. Asp767 was found hydrogen bonded to Glu559 and Lys691.

Inhibition by Statins

A family of natural products and synthetic analogs known as statins bind very tightly to the active site of HMG-CoA reductase in place of HMG-CoA and inhibit the enzyme.

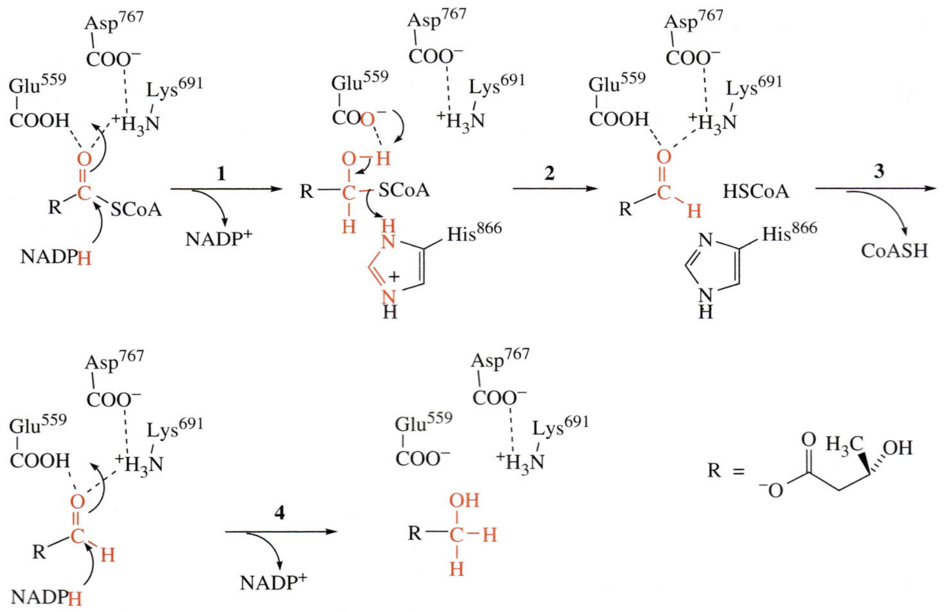

Fig. 5-19. A hypothetical mechanism for the catalytic roles of active site amino acids in HMG-CoASH reductase. The amino acid numbering is that of the human enzyme. Glu559, Asp767, and His866 have been implicated by site directed mutagenesis and kinetic assays in the overall and partial reactions (Frimpong and Rodwell, 1994). The structure revealed the presence of Glu559, His866, and Lys691 in the active site, with Asp767 hydrogen bonded to Lys691 (Istvan et al., 2000). The exact role of Asp767 is not known, and both Glu559 and Lys691 are close enough to the substrate acyl carbonyl to engage in proton transfer in the hydride transfer steps.

The statins are highly specific for inhibition of HMG-CoA reductase and are efficacious in decreasing LDL-cholesterol in hypercholesterolemia. A search of the PubMed database on Feb. 14, 2003 for reviews on this enzyme netted 897 hits, mainly medical reviews on the statins as drugs for the treatment of cardiovascular disease. The conclusion from double blind studies of more than 30,000 patients was that the statins uniformly and dramatically lowered LDL-cholesterol. Side effects appeared infrequently, in low single digit percentages, as myopathies. The statins are important drugs for the treatment and prevention of arteriosclerosis.

The naturally occurring statins include mevastatin, lovastatin, pravastatin, and simvastatin; synthetic analogs are fluvastatin and atorvastatin among others. Structures of representative statins are shown in fig. 5-20. The statins all include the essence of the mevalonate skeleton, excluding the 3-methyl group, and a large hydrophobic grouping. The specificity of statins for binding to HMG-CoA reductase is thought to reside in the mevalonate portion, and much of their high affinities is thought to result from the large hydrophobic moieties binding adjacent to the mevalonate subsite. The value of K_m for HMG-CoA is 4 μM, and the inhibition constants for statins are much lower (K_i = 0.1 to 2.3 nM). Structural studies indicate that binding of the hydrophobic portion leads to disorder in the C-terminal portion of the peptide chain, which includes His866 and other amino acids that bind CoASH (Istvan and Deisenhofer, 2001).

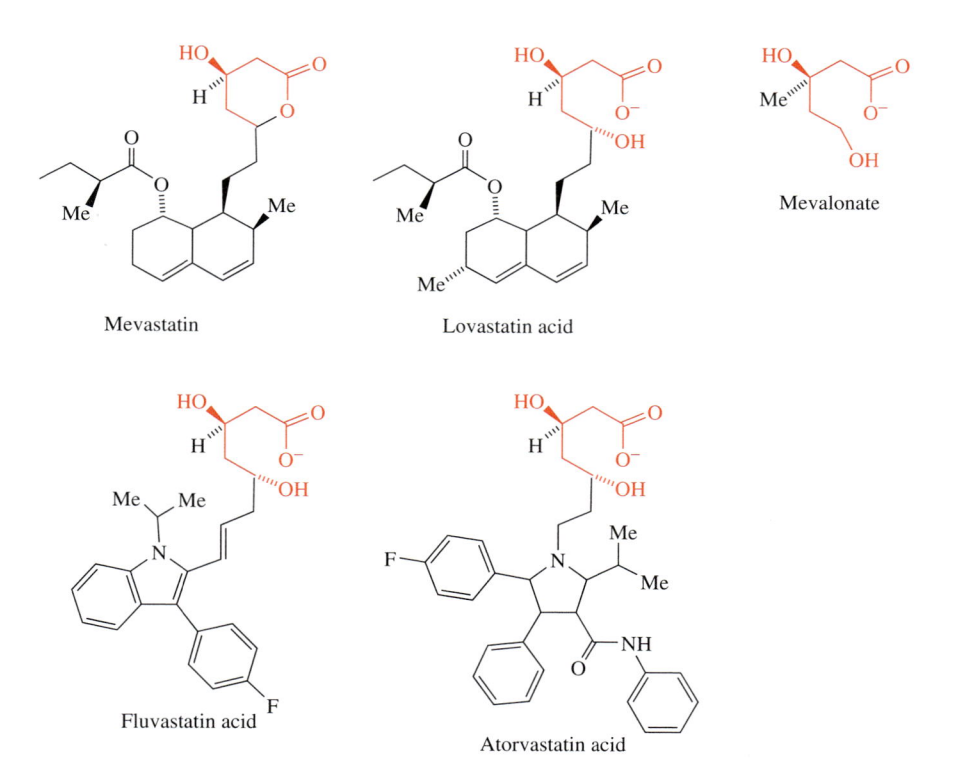

Fig. 5-20. Comparative structures of statins and mevalonic acid. Mevastatin and lovastatin are natural products that inhibit HMG-CoA reductase by binding tightly to the active site. An important specificity determinant appears to be the stereochemistry in the lactone ring. When the lactone ring is hydrolyzed, as in lovastatin acid, the resulting regiochemistry and stereochemistry are very similar to those of mevalonic acid, the product in the action of HMG-CoA reductase.

284   Enzymatic Reaction Mechanisms

Molecular Structure

The structure of the catalytic domain of human HMG-CoA reductase depicted in fig. 5-21 shows HMG, CoASH, and NADP bound at the active center in place of HMG-CoA and NADPH. In the crystal, the protein is a dimer of dimers. The structure shows that each subunit is constructed from the peptide chains of both subunits in the basic dimer. Amino acids in the active sites arise from both subunit chains, as had been inferred from complementation studies outlined earlier, although the active sites are not at the dimer interface.

As shown in the center of fig. 5-21, fluvastatin binds with the mevalonate portion in the mevalonate subsite, and the large hydrophobic portion in an adjacent site. The structures of complexes of six different statins with human HMG-CoA reductase indicate that the binding of the hydrophobic substituents of the statins imposes disorder on catalytic residues near the C-terminus of the enzyme (Istvan and Deisenhofer, 2001).

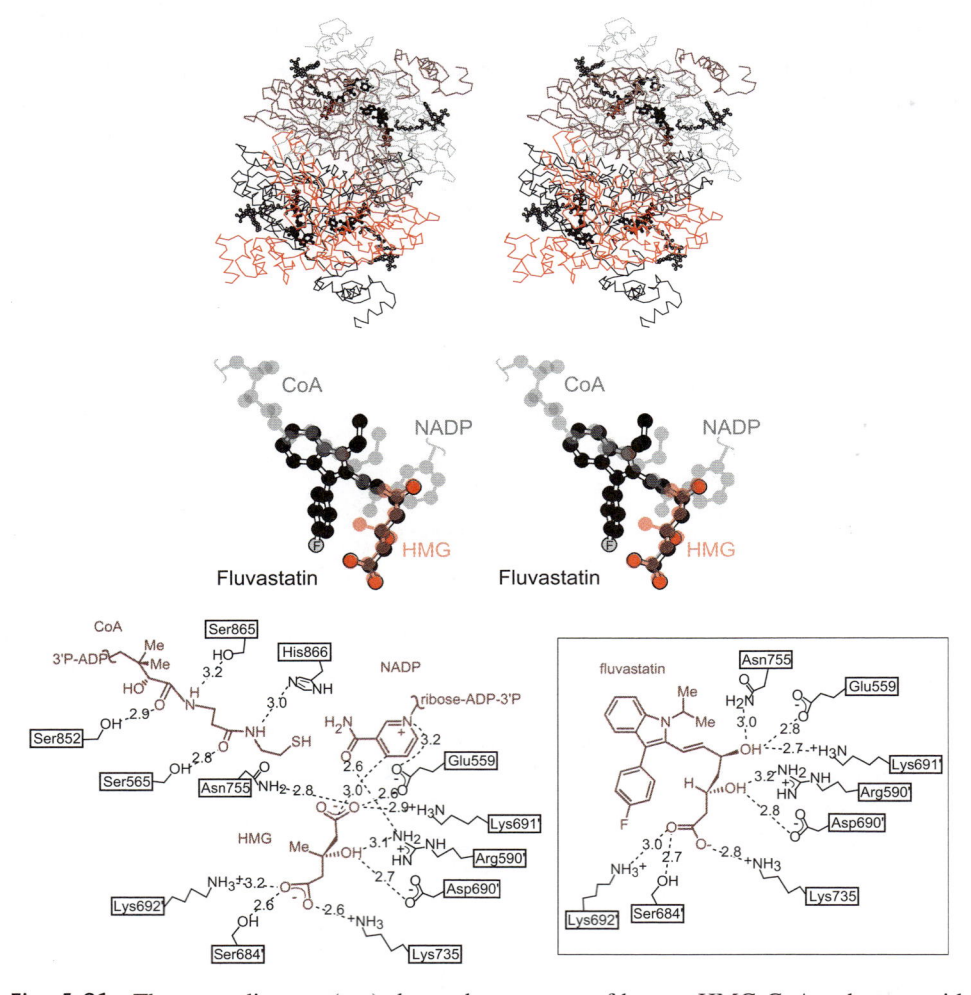

Fig. 5-21. The stereodiagram (top) shows the structure of human HMG-CoA reductase with HMG, CoASH, and NADP$^+$ bound at the substrate sites (PDB 1DQA; Istvan et al., 2000). The structure with fluvastatin bound at the active site is also available (PDB 1HWI; Istvan and Deisenhofer, 2001). The space-filling models (middle) of fluvastatin and HMG bound to the active site are shown in overlay. The two-dimensional maps (bottom) show the active site contacts for CoASH and HMG on the left and for fluvastatin on the right.

Enzyme Inhibition   285

The substrate and statin binding interactions shown at the bottom of fig. 5-21 illustrate the similarities in the binding of HMG and the mevalonate-like portion of fluvastatin. The interactions of the catalytic residues with HMG and CoASH are also shown at the bottom left of fig. 5-21. The structure shows the interactions of His866 with CoASH, consistent with its postulated role in donating a proton to the departing CoAS$^-$. The diagram also shows the proximity of Glu559 and Lys691 to the acyl carbonyl group of HMG, where they can catalyze proton transfer in the hydride transfer and CoASH elimination steps.

Alanine Racemase

Function and Properties

The conversion of L-alanine into D-alanine plays an essential role in gram-positive bacterial replication because cell wall biosynthesis requires UDP-*N*-acetylmuramyl pentapeptide, which contains D-Ala-D-Ala in the peptide, for insertion of the glycosyl peptide into the peptidoglycan. For this reason, alanine racemase (EC 5.1.1.1) is a target for the development of antibacterial pharmaceutical agents. Studies of alanine racemase from *E. coli* revealed it as a target for the naturally occurring antibiotics D-cycloserine and *O*-carbamoyl-D-serine (Kaczorowski et al., 1975a, 1975b; Wang and Walsh, 1978).

Bacterial alanine racemase form *B. stearothermophilus* is a homodimeric enzyme with a subunit molecular mass of 43 kDa. It requires PLP as the coenzyme and binds PLP as its internal aldimine with Lys39 (Shaw et al., 1997). As in other PLP-dependent enzymes, the coenzyme facilitates carbanion formation in the external aldimine with alanine (see chap. 3).

Reaction Mechanism

The first mechanistic information came from studies of suicide substrates directed toward the development of pharmaceutical agents against pathogenic bacteria. Suicide inactivators included L- and D-3-fluoroalanine, and L-and D-3-chloroalanine (Kollonitsch et al., 1978; Wang and Walsh, 1978). These compounds were found to be substrates that inactivated the racemase from *E. coli*, all with partition ratios of 790 to 920. Inactivation produced fluoride or chloride, proving that the halogen was eliminated in the inactivation process. The most obvious mechanism for elimination of halide involved the intermediate formation of the α-carbanion of the 3-haloalanine-PLP aldimine, which could either undergo protonation to the enantiomer or produce the halide by β-elimination according to the mechanism in scheme 5-8.

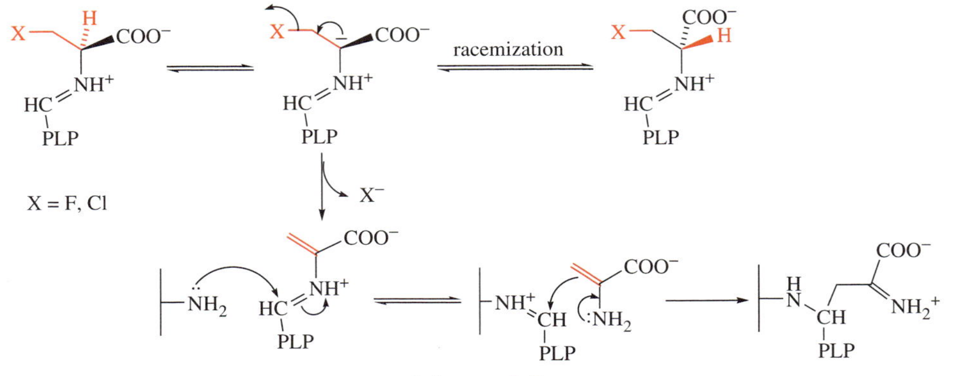

Scheme 5-8

The elimination of halide in this process was the first evidence for the formation of the α-carbanion as an intermediate in racemization. The similar partition coefficients among

286 Enzymatic Reaction Mechanisms

3-haloalanines, as well as O-carbamoyl-D-serine and O-acetyl-D-serine, suggested a common intermediate leading to inactivation. Most likely the partitioning of the α-carbanion in scheme 5-8 leading to either racemization or inactivation accounted for the similarities in partition ratios. The course of inactivation was verified by chemical degradation of the inactivated complex (Badet et al., 1984).

In an alternative, concerted racemization mechanism, proton transfer from the α-carbon to an enzymatic base would take place in concert with proton transfer to the opposite side of the α-carbon by an enzymatic acid. The concerted mechanism would not provide for the elimination of halide in the suicide inactivations by the 3-haloalanines. The simplest rationale accounting for both racemization and inactivation is that in scheme 5-8.

The detailed racemization mechanism in fig. 5-22 shows the participation of two acid-base groups in racemization. The importance of two catalytic groups rather than one has been proved in mutagenic and kinetic experiments (Sun and Toney, 1999; Watanabe et al., 1999).

The x-ray crystallographic structure (fig. 5-23) implicates Lys39 and Tyr265 in catalyzing proton transfer. Tyr265 catalyzes proton transfer between the α-carbanion and L-alanine, and Lys39 catalyzes the transfer between D-alanine and the α-carbanion.

The finding of two catalytic groups did not prove the intermediate formation of an α-carbanionic intermediate; instead, it raised the possibility of a concerted mechanism. Until recently, the best information implicating an α-carbanion was the suicide inactivation by 3-haloalanines (scheme 5-8). Multiple deuterium kinetic isotope effects on the racemization of alanine excluded a concerted mechanism and supported the stepwise process by way of the α-carbanion (Spies and Toney, 2003). The pH-k_{cat}/K_m profile for alanine racemase is bell-shaped and gives two pK_a values: 7.1 and 10.3. Ionization on the acid side is assigned to Tyr265 and on the basic side to amino group of alanine, which must be protonated to bind to the enzyme. The assignments are based on the temperature dependence of the kinetic pK_a values. On the acid side, the heat of ionization (ΔH) proved to be 6.5 kcal mol^{-1}, similar to the corresponding values of 6 and 5.6 kcal mol^{-1} for tyrosine and phenol, respectively. On the alkaline side the heat of ionization of 10.5 ± 0.7 kcal mol^{-1}

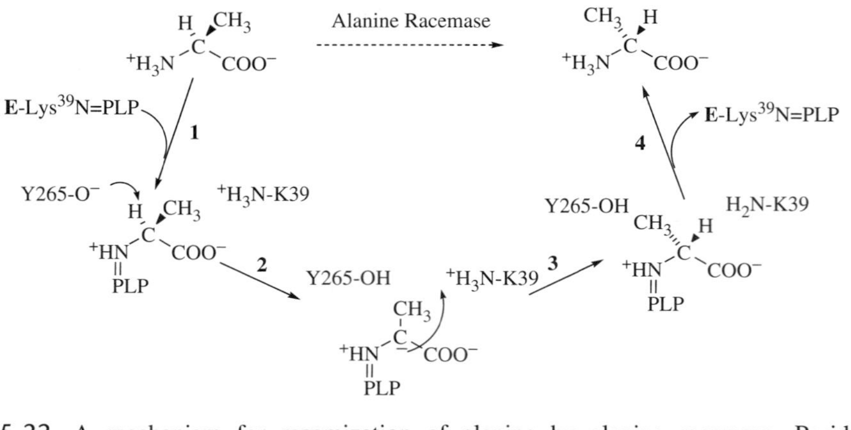

Fig. 5-22. A mechanism for racemization of alanine by alanine racemase. Pyridoxal-5′-phosphate (PLP) plays a central role in the mechanism, together with two acid-base groups. Reaction of the internal racemase-PLP aldimine with L-alanine produces the external L-alanine-PLP aldimine in step 1, and this is followed by abstraction of the α-proton by the first base, Tyr265, to form the α-carbanion. Protonation of the carbanion from the opposite side by the conjugate acid of the second catalytic group, Lys39, leads to the D-alanine-PLP aldimine in step 3. Transimination with Lys39 in step 4 produces free D-alanine and the internal racemase-PLP aldimine.

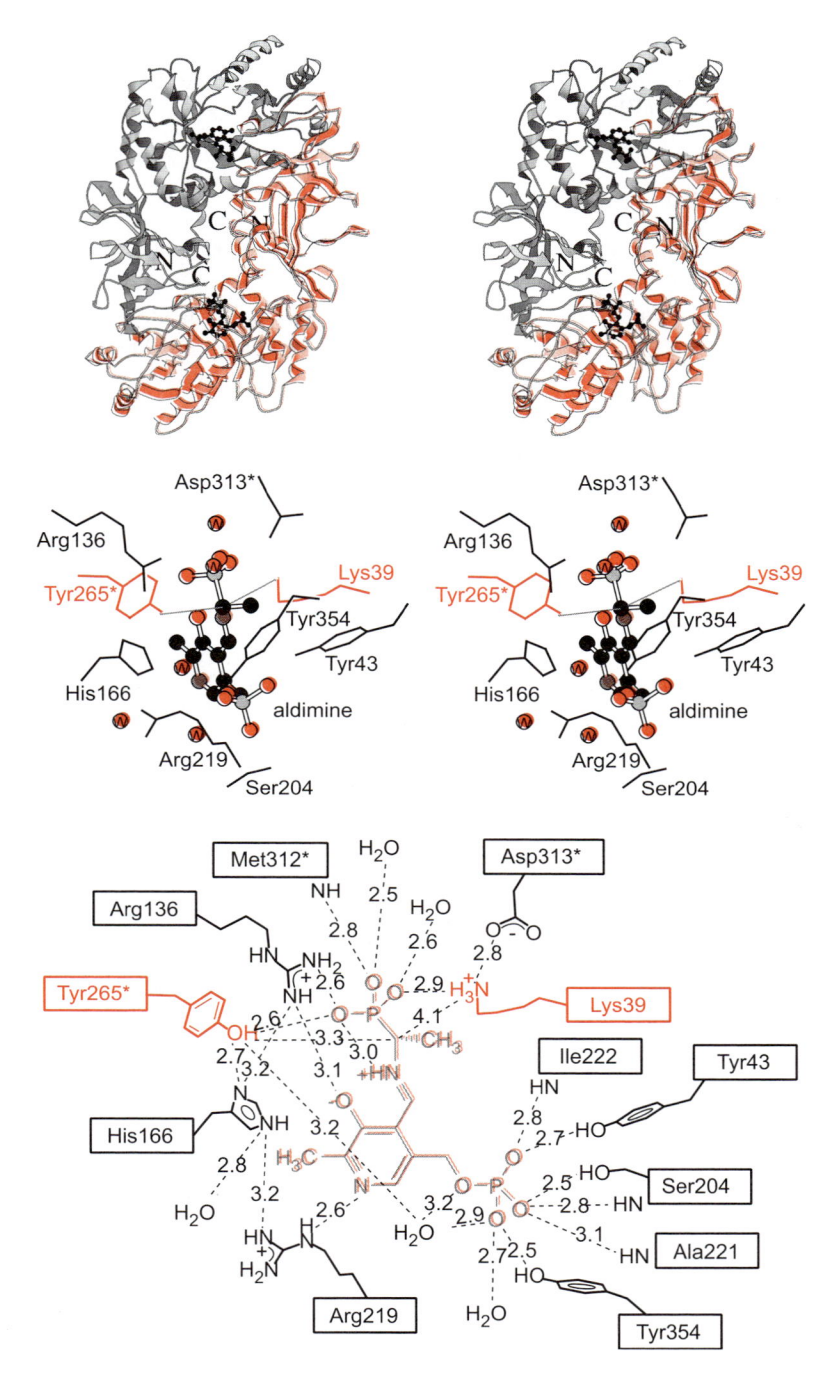

Fig. 5-23. Structure of alanine racemase from *Bacillus stearothermophilus*. The stereographic ribbon diagram (top) shows the dimeric enzyme. A stereodiagram (middle) highlights a ball-and-stick model of 1-aminoethylphosphonate bound at the active site, showing its contacts with amino acid side chains, including the acid-base catalytic groups Lys39 and Tyr265. A two-dimensional diagram (bottom) illustrates the hydrogen bonded contacts at the active site. The image was generated using PDB 1BD0 (Stamper et al., 1998) of the 1.6-Å resolution model of alanine racemase with 1-aminoethylphosphonate bound.

was the same as that of alanine and was assigned to alanine (Spies and Toney, 2003), although the kinetic value of pK_a was higher than that of alanine and more compatible with that of lysine, perhaps Lys39.

Inhibition by 1-Aminoethylphosphonate

The phosphonate analog of alanine, 1-aminoethylphosphonate binds tightly to alanine racemase and forms an inactive, external aldimine complex (Badet et al 1986). Inactivation is time dependent, and the *Bacillus stearothermophilus* racemase displays a value of 1 μM as the K_i for 1-aminoethylphosphonate in the initial binding step. Dissociation of the inhibitor is slow, with a half-time estimated at 25 days. The dianionic form of the inhibitor in its external aldimine form with alanine racemase may be an analog of the transition state for formation of the α-carbanionic external aldimine intermediate in catalysis (Badet et al., 1986). 1-Aminoethylphosphonate arises in vivo by hydrolysis from L-alanyl-L-1-aminoethylphosphonate (Alaphosphin). It inhibits alanine racemase from gram-positive and gram-negative bacteria.

Molecular Structure

The structures of alanine racemase and its complex with 1-aminoethylphosphonate revealed much about the catalytic mechanism (Stamper et al., 1998; Shaw et al., 1997). These structures also led to revisions in the conventional wisdom regarding the mechanism of action of PLP. The structure of the racemase-aminoethylphosphonate complex is shown in fig. 5-23. The structure with 1-aminoethylphosphonate bound differs at the active site in the placement of some of the amino acid side chains but is otherwise similar to the free enzyme. The placements of Lys39 and Tyr265 are compatible with their roles in acid-base catalysis, which had been proposed on the basis of the structure of the free enzyme and other data. The kinetically determined pK_a of 7.1 for Tyr265 from the pH-log k_{cat}/K_m profile requires some explanation for how the acidity of tyrosine could be increased by 3-pH units. The structure in fig. 5-23 does not provide a clear rationale. However, the pH dependence refers to the free enzyme, and that structure indicates an electrostatic and possible hydrogen bonding interaction (at 3.2 Å) between Tyr265 and Arg136 (Shaw et al., 1997). This could lower the pK_a of Tyr265 to 7.1 in the free enzyme.

The structure revealed two striking features. First, unlike other PLP enzymes the PLP domain is in an α,β-barrel. Second, unlike several other PLP enzymes, the PLP binding site does not include an acidic residue in contact with N1 of the pyridine ring. N1 is in contact with Arg219. The contact of PLP-N1 with an acidic residue had been expected on the basis that it would ensure that the pyridine ring was protonated, a feature that was thought to be generally important for resonance stabilization of the α-carbanionic intermediates in nearly all PLP-dependent reactions. This feature had been found in the structures of aspartate transaminase and dialkylglycine decarboxylase. However, the contact between PLP and Arg219 in alanine racemase virtually ensured that the pyridine ring of PLP must remain neutral, although hydrogen bonded to the presumably positively charged Arg219, throughout the catalytic process.

It had been thought that the protonated pyridinium ring of PLP would ensure that the α-hydrogen of the amino acid would be acidic enough to be abstracted. Delocalization of the α-carbanion into the pyridinium ring would stabilize the α-carbanion and lower the acidity of the α-hydrogen. A neutral pyridine ring would be much less effective in this role. In the same time frame, however, it was discovered that an iminium group in an α-amino acid was itself sufficient to lower the pK_a of the α-hydrogen by 7 pH units (see table 1-4 in chap. 1)

(Rios et al., 2001). These parallel discoveries permanently altered the standard paradigm for the role of PLP in catalyzing amino acid transformations. The protonation of N1 in PLP no longer seemed to be essential for α-carbanion formation. These findings also explained why the 495 nm chromophore of PLP-enzymes was so rarely seen, for it would be observed only for the α-carbanion with a pyridinium ring in the external aldimine.

5-Enolpyruvoylshikimate-3-Phosphate Synthase

Function and Mechanism of Action

5-Enolpyruvoylshikimate-3-phosphate (EPSP) synthase (EC 2.5.1.19) catalyzes an essential step in aromatic amino acid biosynthesis, the reaction of 3-phosphoshikimate with phosphoenolpyruvate to form EPSP according to eq. 5-12.

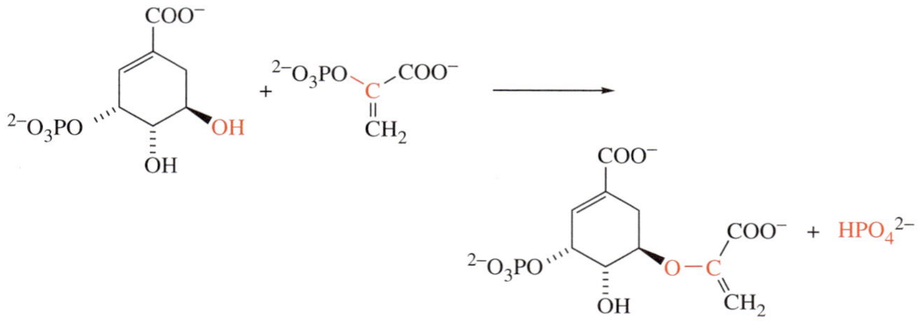

$$(5\text{-}12)$$

Further reactions convert EPSP into phenylalanine, tyrosine, or tryptophan. The reaction mechanism proceeds, as outlined in scheme 5-9, with the initial addition of the 5-OH group of 3-phosphoshikimate to C2 of phosphoenolpyruvate, followed by the elimination of phosphate.

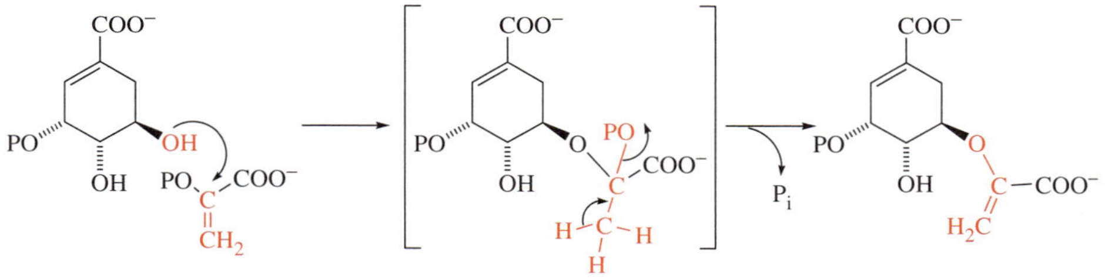

Scheme 5-9

When the reaction is conducted in ^3H_2O, the EPSP formed is labeled with tritium in the pyruvoyl group, as it would be if the tetrahedral adduct in scheme 5-9 were an intermediate (Anton et al., 1983). The intermediate can be isolated from the enzymatic reaction after quenching in the steady state, and transient-phase kinetic experiments show it to be kinetically competent as an intermediate (Anderson and Johnson, 1990). The intermediate was also characterized by ^{13}C NMR as an enzyme-bound species (Anderson et al., 1995).

Glyphosate

Humans and animals do not biosynthesize phenylalanine or tryptophan, which are essential amino acids that must be obtained from their diets. Plants cannot obtain aromatic amino acids from their environments and must biosynthesize them. EPSP synthase is an essential enzyme in the life of plants, and inhibitors of aromatic amino acid biosynthesis

290   Enzymatic Reaction Mechanisms

are herbicides. Glyphosate, or *N*-(methylphosphono)glycine, is such a molecule, and it is a very specific inhibitor of EPSP synthase. Glyphosate binds to EPSP synthase in a ternary complex with 3-phosphoshikimate (Anton et al., 1983; Steinrcken and Armhein, 1984). At one time, it was regarded as potentially a transition-state analog. Glyphosate is also an uncompetitive inhibitor with respect to EPSP in the reverse reaction and forms a ternary complex with EPSP and the synthase (Sammons et al., 1995). As an inhibitor with a value of $K_d = 54$ μM, glyphosate is not especially tight binding. However, it is specific for

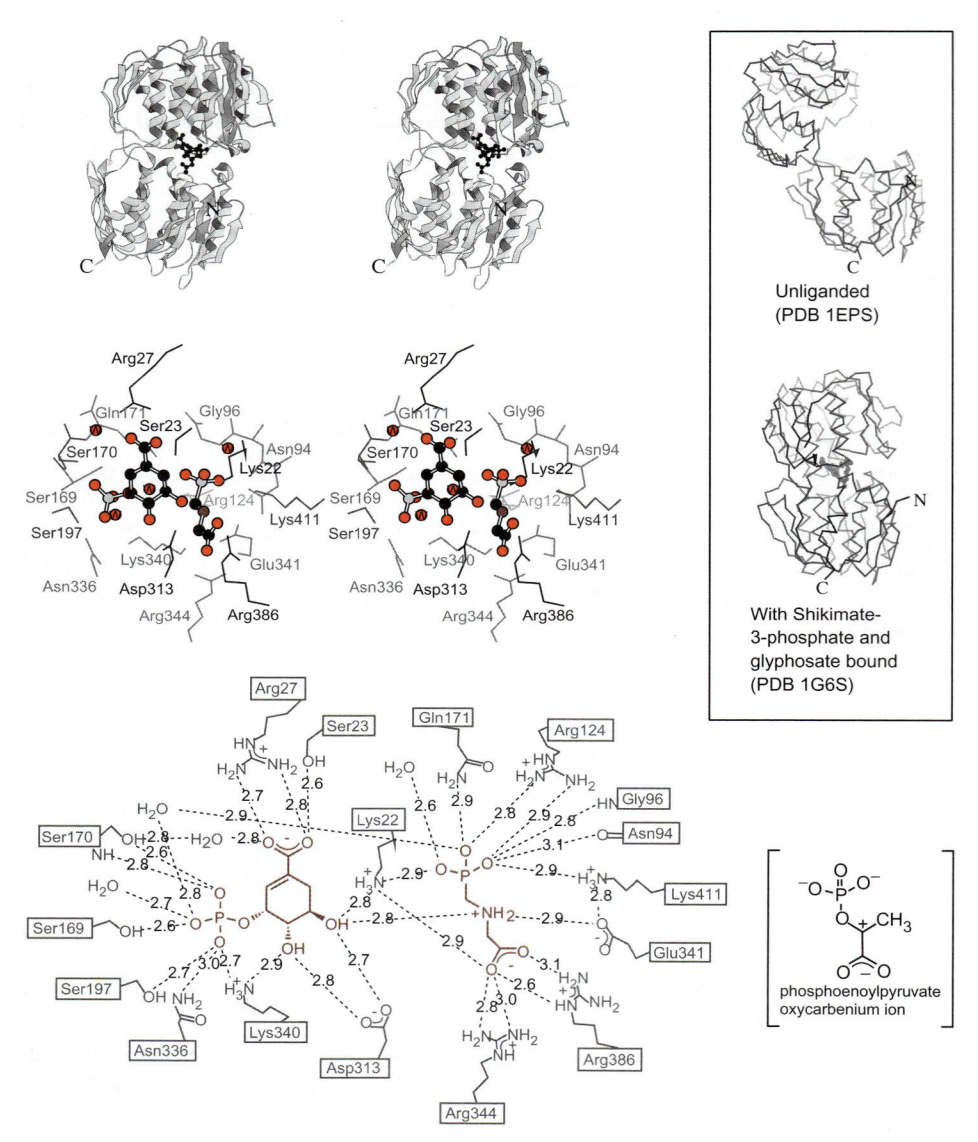

Fig. 5-24. Structure of 5-enolpyruvoylshikimate-3-phosphate (EPSP) synthase in complex with glyphosate and 3-phosphoshikimate. The stereodiagram (top) shows the overall fold of *E. coli* EPSP synthase with bound shikimate-3-phosphate and glyphosate (PDB 1G6S; Schonbrunn et al., 2001). The boxed Cα traces to the right show the large domain changes that accompany substrate binding (top is PDB 1ESP; Stallings et al., 1991). A closer view (middle) of the active site shows contacts in the shikimate-3-phosphate and glyphosate structure. A schematic representation (bottom) of the active site includes contact distances. The structure of phosphoenoylpyruvate oxycarbenium ion is included for direct comparison with glyphosate bound in its place.

EPSP synthase, in part perhaps because its binding depends on ternary complex formation with 3-phosphoshikimate and EPSP synthase.

The structure of the ternary complex with 3-phosphoshikimate is shown in fig. 5-24 (Schonbrunn et al., 2001). The structure shows how glyphosate seems to bind in place of phosphoenolpyruvate.

Glyphosate is a widely used herbicide that owes its success to several properties in addition to its binding affinity for EPSP synthase. It is a simple molecule that is inexpensive to produce. It has low toxicity for humans and animals. It kills most plants, although resistance has appeared. It is biodegradable and disappears from soil within a few weeks. The importance of its biodegradability cannot be overemphasized, because it is a safety feature that adds to its value as a short-term herbicide.

Acetylcholinesterase

Biological Function

Acetylcholinesterase (EC 3.1.1.7) catalyzes the hydrolysis of the neurotransmitter acetylcholine to choline and acetate in the central nervous systems of animals and humans. Ester hydrolysis is chemically more facile than amide hydrolysis, although the mechanism follows a similar course of tetrahedral adduct formation and decomposition. The enzymatic mechanism follows the same course as that of serine proteases such as chymotrypsin. Acetylcholine is released into nerve-nerve and neuromuscular junctions during neurotransmission, where it may interact reversibly with effector cell receptors. Acetylcholinesterase is present in high concentrations in association with the prejunctional and postjunctional cellular membranes, and it functions in the critical clearing of acetylcholine from the synaptic space. Acetylcholinesterase inhibitors slow the hydrolysis of synaptic acetylcholine, thereby hyperactivating cholinergic neurotransmission. Acetylcholinesterase inhibition can be lethal for animals because of respiratory paralysis or acute bradycardia, although other effects can include increased sweating, lacrimation, fasciculation (i.e., muscular twitching), and paralysis, depending on the organism and the extent of inhibition.

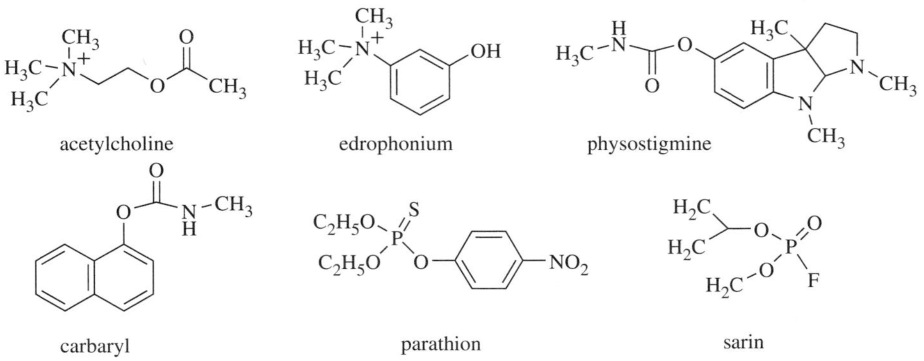

Inhibitors

Acetylcholinesterase inhibitors include man-made compounds such as insecticides, chemical warfare agents, and therapeutics, or as constituents of venoms. Three classes of small-molecule inhibitors may be described: cationic compounds, carbamates, and organophosphates.

292   Enzymatic Reaction Mechanisms

The cationic compounds mimic the quaternary ammonium group of acetylcholine and are noncovalent inhibitors such as the clinically useful compound, edrophonium. Carbamate inhibitors include physostigmine and the insecticide carbaryl; organophosphate inhibitors include the nerve gas Sarin and the insecticides Parathion and Malathion.

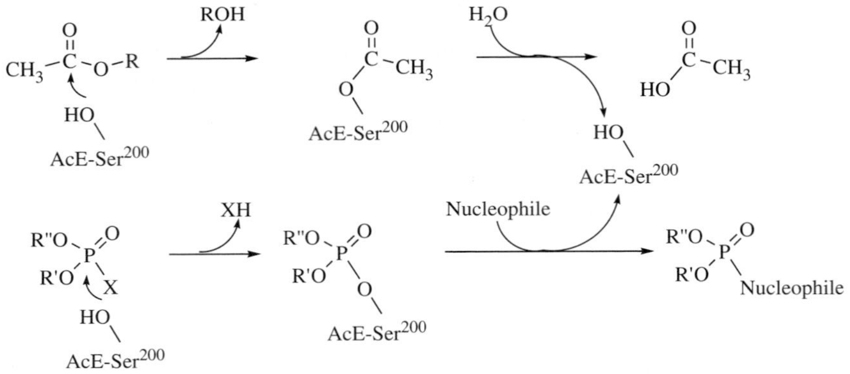

Scheme 5-10

The carbamates and organophosphates covalently modify the active site serine, as illustrated in the lower portion of scheme 5-10, to form either a carbamylated enzyme or an enzyme phosphotriester, both of which undergo hydrolysis so slowly as to effectively inactivate the enzyme. The upper portion of scheme 5-10 represents the enzymatic hydrolysis of acetylcholine, in which ROH is choline: $(CH_3)_3N^+CH_2CH_2OH$.

The covalently modified acetylcholinesterase can be reactivated by the addition of nucleophiles that react more rapidly with the phosphotriester or carbamyl groups than water in the imperceptively slow enzymatic hydrolysis. Pyridine aldoximine methiodide (PAM) is an effective antidote for poisoning by organophosphate compounds, because it combines an oxime nucleophile with the pyridinium cation that helps to target the drug to the active site much like the cationic inhibitors (Frode and Wilson, 1971).

Fasciculin

Another potent acetylcholinesterase inhibitor has been characterized from the venom of the green mamba *Dendroaspis angusticeps*. Two closely related 61–amino acid proteins, fasciculin 1 and 2, bind to acetylcholinesterase with fairly high affinities ($K_I \approx 10^{-10}$ M) (Rodriguez-Ithurralde et al., 1983). The faciculins belong to a family of "three-fingered" toxins found in cobra-family snake venoms that are 60 to 70 amino acids long and share a similar structure. This consists of a core domain with four disulfide bridges and three "finger" loops that interact with various target proteins depending on the toxin; targets include acetylcholinesterase, acetylcholine receptors, and K^+ and Ca^{2+} channels.

Figure 5-25 shows the complex of fasciculin 2 with *Torpedo californica* acetylcholinesterase. The active site catalytic triad (Ser200-His440-Glu327) is located in an active site gorge. Although the toxin does not interact directly with the active site, its large contact area (2000 Å) effectively occludes the active site by blocking the narrow active site gorge (Harel et al., 1995).

Reaction Mechanism

The mechanism of action of acetylcholinesterase is similar to that of chymotrypsin and serine proteases in many respects. The mechanism follows the course at the top of scheme 5-10. Nucleophilic catalysis by Ser200 in the first step leads to the acetyl-enzyme intermediate and choline, and hydrolysis of the intermediate leads to acetate.

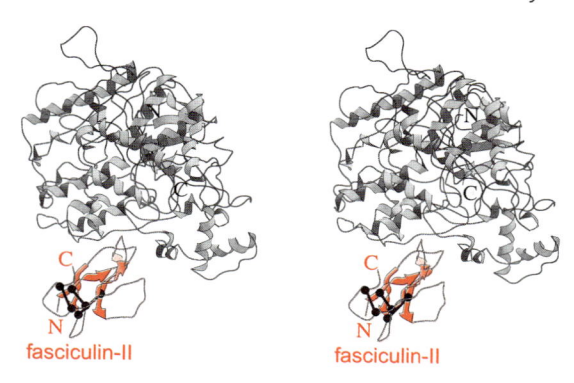

Fig. 5-25. The structure of acetylcholinesterase and the binding of fasciculin. The image was prepared using PDB 1FSS (Harel et al., 1995).

Diisopropylphosphofluoridate (DFP) inactivates acetylcholinesterase by phosphorylation of Ser200 to form the DFP-enzyme, just as it does chymotrypsin (see chap. 1). In the case of acetylcholinesterase, this reaction of DFP is the chemical basis for its toxicity as an early war gas. His440 and Glu327 in the active site function in the manner of His57 and Asp102 in chymotrypsin potentiate the nucleophilic reactivity of Ser200 by mediating proton transfer. Evidence of low barrier hydrogen bonding between His440 and Glu327 has been presented, and it is postulated to play the same role in catalysis by acetylcholinesterase as in chymotrypsin and other serine proteases (Massiah et al., 2001; Viragh et al., 2000).

Catalysis by acetylcholinesterase may vary from that of chymotrypsin with respect to stabilization of the tetrahedral addition intermediate. Acyl transfer generally involves the addition of a nucleophile to an acyl carbonyl group to form an anionic tetrahedral addition intermediate, as in the first step of scheme 5-11.

Scheme 5-11

Fig. 5-26. The diagram of the active site of acetylcholinesterase shows three hydrogen bonds in the tetrahedral anion binding site.

The oxyanionic group in the intermediate provides the driving force for expelling the leaving group in the second step. In the action of chymotrypsin, the structure of the active site provides two hydrogen bond donors, Gly193-NH and Ser195-NH, to stabilize the oxyanion and prevent its protonation. An alkoxide can accept up to three hydrogen bonds, as illustrated in scheme 5-11, but in chymotrypsin only two are provided by the active site. In the case of acetylcholinesterase, the active site provides three hydrogen bonds to stabilize the oxyanioninc intermediate. The three hydrogen bonds in the oxyanion binding site are shown in the active site structural map of acetylcholinesterase in fig. 5-26 (Sussman et al., 1991).

References

Abend A, V Bandarian, GH Reed, and PA Frey (2000) *Biochemistry* **39**, 6250.
Anderson KS and KA Johnson (1990) *J Biol Chem* **265**, 5567.
Anderson KS, RD Sammons, GC Leo, JA Sikorski, AJ Benesi, and KA Johnson (1990) *Biochemistry* **29**,1460.
Anton D, L Hedstrom, S Fish, and RH Abeles (1983) *Biochemistry* **22**, 5903.
Badet B, K Inagaki, K Soda, and CT Walsh (1986) *Biochemistry* **25**, 3275.
Badet B, D Roise, and CT Walsh (1984) *Biochemistry* **23**, 5188.
Bellisario RL, GF Maley, JH Galivan, and F Maley (1976) *Proc Natl Acad Sci U S A* **73**, 1848.
Bellisario RL, GF Maley, DU Guarino, and F Maley (1979) *J Biol Chem* **254**, 1296.
Bolin JT, DJ Filman, DA Matthews, RC Hamlin, and J Kraut (1982) *J Biol Chem* **257**, 13650.
Brady K and RH Abeles (1990) *Biochemistry* **29**, 7608.
Carreras CW and DV Santi (1995) *Annu Rev Biochem* **64**, 721.
Cayley PJ, SM Dunn, and RW King (1981) *Biochemistry* **20**.
Chandrasekharan NV, H Dai, KL Roos, NK Evanson, J Tomsik, TS Elton, and DL Simmons (2002) *Proc Natl Acad Sci U S A* **99**, 13926.
Collins KD and GR Stark (1971) *J Biol Chem* **246**, 6599.
Cook PF and WW Cleland (2006) *Enzyme Kinetics*. Garland: New York.
Danenberg PV, RJ Langenbach, and C Heidelberger (1974) *Biochemistry* **13**, 926.
Darnay BG and VW Rodwell (1993) *J Biol Chem* **268**, 8429.
Daron HH and JL Aull (1978). *J Biol Chem* **253**, 940.
DeBiase D, D Barra, F Bossa, P Pucci, and RA John (1991) *J Biol Chem* **266**, 20056.
Endo K, GM Helmkamp Jr, and K Bloch (1970) *J Biol Chem* **245**, 4293.
Epstein DM, DJ Benkovic, and PE Wright (1995) *Biochemistry* **34**, 11037.
Falzone CJ, PE Wright, and SJ Benkovic (1994) *Biochemistry* **33**, 439.
Fierke CA, KA Johnson, and SJ Benkovic (1987) *Biochemistry* **26**, 4085.
Friedkin M (1959) In Stohlman FJ (ed): *The Kinetics of Cellular Proliferation*. Grune and Stratton: New York, p 97.
Frimpong K and VW Rodwell (1994a) *J Biol Chem* **269**, 11478.
Frimpong K and VW Rodwell (1994b) *J Biol Chem* **269**, 1217.
Frode HC and IB Wilson (1971) In Boyer PD (ed): *The Enzymes*, vol 5, 3rd ed. Academic Press: New York, p 87.
Galvez J, R Varon, and F Garcia Carmona III (1981) *J Theor Biol* **89**, 37.
Goodwin DC, MR Gunther, LC Hsi, BC Crews, TE Eling, RP Mason, and LJ Marnett (1998) *J Biol Chem* **273**, 8903.
Hardy LW, JS Finer-Moore, WR Montfort, MO Jones, DV Santi, RM Stroud (1987) *Science* **235**, 448.
Harel M, GJ Kleywegt, RBG Ravelli, I Silman, and JL Sussman (1995) *Structure* **3**, 1355.
Helmkamp GM Jr, DJ Brock, and K Bloch (1968) *J Biol Chem* **243**, 3229.
Holtzman MJ, J Turk, and LP Shornick (1992) *J Biol Chem* **267**, 21438.
Huennekns FM (1996) *Protein Sci* **5**, 1201.
Hyatt DC, F Maley, and WR Montfort (1997) *Biochemistry* **36**, 4585.
Istvan ES and J Deisenhofer (2001) *Science* **292**, 1160.
Istvan ES, M Palnitkar, SK Buchanan, and J Deisenhofer (2000) *EMBO J* **19**, 819.
Kaczorowski G, L Shaw, M Fuentes, and CT Walsh (1975a) *J Biol Chem* **250**, 2855.
Kaczorowski G, L Shaw, R Laura, and CT Walsh (1975b) *J Biol Chem* **250**, 8951.

Kiefer JR, JL Pawlitz, KT Moreland, RA Stegeman, WF Hood, JA Gierse, AM Stevens, DC Goodwin, SW Rowlinson, LJ Marnett, WC Stallings, and RG Kurumball (2000) *Nature* **405**, 97.

Kollonitsch J, LM Perkins, AA Patchett, GA Doldouras, S Marburg, DE Duggan, AL Maycock, and SD Aster (1978) *Nature* **274**, 906.

Lawrence CM, VW Rodwell, and CV Stauffacher (995) *Science* **268**, 1758.

Lecomte M, O Laneuville, C Ji, DL CeWitt, and WL Smith (1994) *J Biol Chem* **269**, 13207.

Lewis CA Jr, PD Ellis, and RB Dunlap (1981) *Biochemistry* **20**, 2275.

Likos JJ, H Ueno, FW Feldhaus, and DE Metzler (1982) *Biochemistry* **21**, 4377.

Lippert B, BW Metcalf, MJ Jung, and P Casara (1977) *Eur J Biochem* **74**, 441.

Loll PJ, D Picot, and RM Garavito (1995) *Nat Struct Biol* **2**, 637.

Lorenson MY, GF Maley, and F Maley (1967) *J Biol Chem* **242**, 3332.

Lu YZ, PD Aiello, and RG Matthews (1984) *Biochemistry* **23**, 6870.

Maley GF, RL Bellisario, DU Guarino, and F Maley (1979) *J Biol Chem* **254**, 1301.

Malkowski MG, SL Ginell, WL Smith, and RM Garavito (2000) *Science* **289**, 1933.

Mancini JA, GP O'Neill, C Bayly, and PJ Vickers (1994) *FEBS Lett* **342**, 33.

Marnett LJ, DC Godwin, SW Rowlinson, AS Kalfutkar, and LM Landino (1999) In Barton DHR, K Nakanishi, and O Meth-Cohn (eds): *Comprehensive Natural Products Chemistry*. Pergamon Press: New York, p 225.

Massiah MA, C Viragh, PM Reddy, IM Kovach, J Johnson, TL Rosenberry, and AS Mildvan (2001) *Biochemistry* **40**, 5682.

Matthews DA, JT Bolin, JM Burridge, DJ Filman, KW Volz,, BT Kaufman, CR Beddell, JN Champness, DK Stammers, and J Kraut (1985a) *J Biol Chem* **260**, 381.

Matthews DA, JT Bolin, JM Burridge, DJ Filman, KW Volz,, and J Kraut (1985b) *J Biol Chem* **260**, 392.

Matthews DA, JE Villafranca, CA Janson, WW Smith, K Welsh, and S Freer (1990) *J Mol Biol* **214**, 937.

Mayer RJ, C Debouck, and BW Metcalf (1988) *Arch Biochem Biophys* **267**, 110.

McTigue MA, JF Davies, BT Kaufman and J Kraut (1992) *Biochemistry* **31**, 7264.

McTigue MA, JF Davies, BT Kaufman and J Kraut (1993) *Biochemistry* **32**, 6855.

Metcalf BW (1979) *Biochem Pharmacol* **28**, 1705.

Moore MA, F Ahmed, and RB Dunlap (1986) Biochemistry **25**, 3311.

Morrison JF and CT Walsh (1988) *Adv Enzymol* **61**, 201.

Mumford JP and DJ Cannon (1994) *Epilepsia* **35** (Suppl 5), 25.

Osborn MJ, M Freeman, and FM Huennekens (1958) *Proc Soc Exp Biol Med* **97**, 429.

Osborne MJ, J Schnell, SJ Benkovic, HJ Tyson, and PE Wright (1997) *Biochemistry* **40**, 9846.

Parker RA, SJ Miller, and DM Gibson (1986) *Adv Enzyme Regul* **25**, 329.

Peng S, NM Okeley, A-L Tsai, G Wu, RJ Kulmacz, and WA van der Donk (2001) *J Am Chem Soc* **123**, 3609.

Peng S, NM Okeley, A-L Tsai, G Wu, RJ Kulmacz, and WA van der Donk (2002) *J Am Chem Soc* **124**, 10785.

Peters JN and DM Greenberg (1958) *J Am Chem Soc* **80**, 6679.

Pogolotti AL and DV Santi (1974) *Biochemistry* **13**, 456.

Rando RR (1977) *Biochemistry* **16**, 4604.

Rando RR and FW Bangerter (1976) *J Am Chem Soc* **98**, 6762.

Rando RR and FW Bangerter (1977) *J Am Chem Soc* **99**, 5141.

Reyes VM, MR Sawaya, KA Brown, and J Kraut (1995) *Biochemistry* **34**, 2710.

Rios A, J Crugeiras, TL Aymes, and JP Richard (2001) *J Am Chem Soc* **123**, 7949.

Rodriguez-Ithurralde D, L Silveira, L Barbeito, F and Dajas (1983) *Neurochem Int* **5**, 267.

Roth BD (2002) *Prog Med Chem* **40**, 1.

Sage CR, EE Rutenber, TJ Stout, and RM Stroud (1996) *Biochemistry* **35**, 16270.

Sammons RD, KJ Gruys, KS Anderson, KA Johnson, and JA Sikorski (1995) *Biochemistry* **34**, 6433.

Santi DV and CF Brewer (1968) *J Am Chem Soc* **90**, 6236.

Santi DV and CF Brewer (1973) *Biochemistry* **12**, 2416.

Santi DV, CS McHenry, and H Sommer (1974) *Biochemistry* **13**, 471.

Santi DV and AL Pogolotti (1968) *Tetrahedron Lett* **59**, 6159.

Sawaya MR and J Kraut (1997) *Biochemistry* **36**, 586.

Schonbrunn E, S Eschenburg, WA Shuttleworth, JV Schloss, N Amrhein, JN Evans, and W Kabsch (2001) *Proc Natl Acad Sci U S A* **98**, 1376.
Selinsky BS, K Gupta, CT Sharkey, and PJ Loll (2001) *Biochemistry* **40**, 5172.
Shah SA, BW Shen, and AT Brunger (1997) *Structure* **5**, 1067.
Shaw JP, GA Petsko, and D Ringe (1997) *Biochemistry* **36**, 1329.
Shimokawa T, RJ Kulmacz, DL DeWitt, and WL Smith (1990) *J Biol Chem* **265**, 20073.
Silverman RB and BJ Invergo (1986) *Biochemistry* **25**, 6817.
Silverman RB and C George (1988) *Biochem Biophys Res Commun* **150**, 942.
Silverman RM (1988) *Mechanism-Based Enzyme Inactivation: Chemistry and Enzymology*, vols I and II. CRC Press: Boca Raton, 1988.
Smith WL, DL DeWitt, and RM Garavito (2000) *Annu Rev Biochem* **69**, 145.
Soper TS and JM Manning (1982) *J Biol Chem* **257**, 13930.
Spies MA and MD Toney (2003) *Biochemistry* **42**, 5099.
Stallings WC, SS Abdel-Meguid, LW Lim, HS Shieh, HE Dayringer, NK Leimgruber, RA Stegeman, KS Anderson, JA Sikorski, SR Padgette, and GM Kishore (1991) *Proc Natl Acad Sci U S A* **88**, 5046.
Stamper GF, AA Morollo, and D Ringe (1998) *Biochemistry* **37**, 10438.
Steinrucken HC and N Armhein (1984) *Eur J Biochem* **143**, 351.
Storici P, D De Biase, F Bossa, S Bruno, A Mozzarelli, C Peneff, RB Silverman, and T Schirmer (2004) *J Biol Chem* **279**, 363.
Stroud RM and JS Finer-Moore (1993) *FASEB J* **7**, 671.
Sun S and MD Toney (1999) *Biochemistry* **38**, 4058.
Sussman JL, M Harel, F Frolow, C Oefner, A Goldman, L Toker, and I Silman (1991) *Science* **253**, 872.
Tabermero L, DA Bochar, VW Rodbell, and CV Stauffacher (1999) *Proc Natl Acad Sci U S A* **96**, 7167.
Tatsunami S, N Yago, and M Hosoe (1981) *Biochim Biophys Acta* **662**, 226.
Toney MD, S Pascarella, and D De Biase (1995) *Protein Sci* **4**, 2366.
Tsai A, G Wu, G Palmer, B Bambai, JA Koehn, PJ Marshall, and RJ Kulmacz (1999) *J Biol Chem* **274**, 21695.
Tsai A-L, LC His, RJ Kulmacz, G Palmer, and WL Smith (1994) *J Biol Chem* **269**, 5085.
Tudela J, F García Cánovas, R Varón, F García Carmona, J Gálvez, and JA Lozano (1987) *Biochim Biophys Acta* **912**, 408.
Ueno H, JJ Likos, and DE Metzler (1982) *Biochemistry* **21**, 4377.
van der Donk WA, A-L Tsai, and RJ Kulmacz (2002) *Biochemistry* **41**, 15451.
Viragh C, TK Harris, PM Reddy, MA Massiah, AS Mildvan, and IM Kovach (2000) *Biochemistry* **39**,16200.
Wagner OW, HA Lee Jr, PA Frey, and RH Abeles (1966) *J Biol Chem* **241**, 1751.
Wang E and C Walsh (1978) *Biochemistry* **17**, 1313.
Watanabe A, Y Kurokawa, T Yoshimura, T Kurihara, K Soda, and N Esaki (1999) *J Biol Chem* **125**, 987.
Williams AW, RB Dunlap, and SH Berger (1998) *Biochemistry* **37**, 7096.
Williams JW and JF Morrison (1979) *Methods Enzymol* **63**, 437.
Wood JD, E Kurylo, and DSK Tsui (1979) *Neurosci Lett* **14**, 327.
Wu Y-D and KN Houk (1987) *J Am Chem Soc* **113**, 2353.

6
Acyl Group Transfer: Proteases and Esterases

Acyl group transfer processes are plentiful in enzymatic reactions. Examples may be found in ATP-dependent ligation in chapter 11, carbon-carbon bond formation in chapter 14, and fatty acid biosynthesis in chapter 18. In this chapter, we begin by presenting the basic chemistry of acyl group transfer. We then consider four major classes of proteases that catalyze acyl group transfer in the hydrolysis of peptide bonds.

Chemistry of Acyl Transfer

Acyl group transfer is so common in organic and biochemistry that the chemistry by which it occurs is often taken for granted. Early studies provided evidence for a mechanism initiated by nucleophilic addition of the acyl group acceptor to the carbonyl group to form a tetrahedral intermediate, analogous to the reversible addition of a nucleophilic molecule to the carbonyl group of an aldehyde or ketone. A mechanism of this type is shown in scheme 6-1 for acyl group transfer from a group :X to a nucleophile :G catalyzed by a general base.

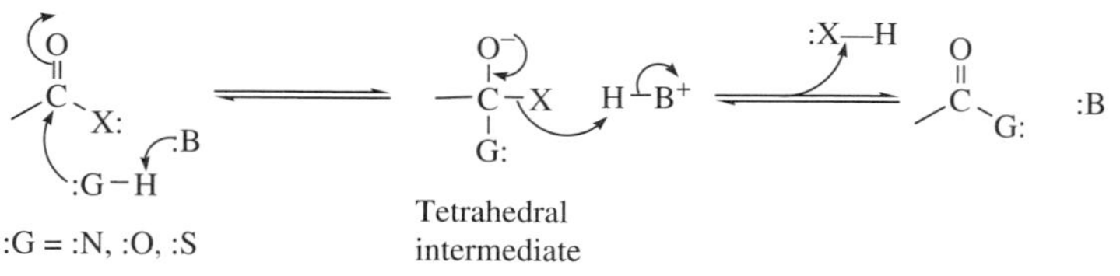

Scheme 6-1

This mechanism is drawn from a larger family of possible mechanisms involving specific acid-base, general acid, general base, or concerted general acid-base catalysis of nucleophilic addition to an acyl carbonyl group to form a tetrahedral intermediate, followed by the elimination of :X—H to produce the new acyl compound. In enzymatic reactions the nucleophilic atom G in scheme 6-1 is normally nitrogen, oxygen, sulfur, or a carbanionic species.

An acyl carbonyl group is less polar and correspondingly less reactive toward nucleophilic addition than an aldehyde or ketone. The reason is the effect on the heteroatom of nonbonding electrons, which reside in p orbitals that overlap the π orbital of the carbonyl group. The consequent delocalization of electrons stabilizes the carbonyl group and attenuates its reactivity with nucleophiles. Other factors being equal, the order of reactivity is thioester > ester > amide, which is the inverse of the degree of delocalization. Delocalization is least in thioesters because of the high energy of the sulfur p orbitals, which reside in the next higher principal quantum number relative to oxygen in the acyl carbonyl group. Delocalization is highest in the amide because nitrogen is better able to accommodate the positive charge than is oxygen in the charge-separated resonance forms shown in structure **6-1**, oxygen being more electronegative than nitrogen. In reaction with hydroxide ions, oxygen esters react about 10,000 times faster than amides.

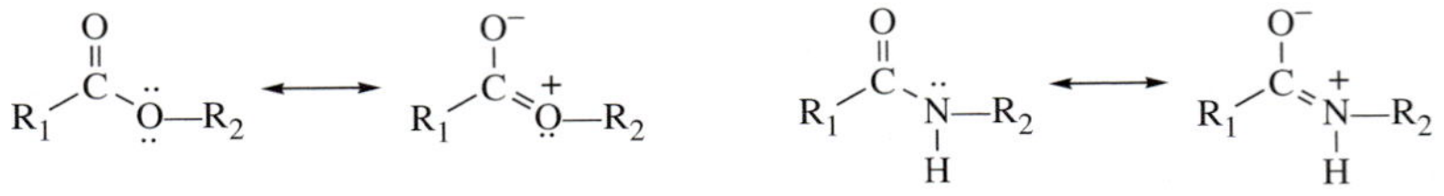

The fate of ^{18}O in the alkaline hydrolysis of esters in $H_2{}^{18}O$, or of ^{18}O-labeled esters in H_2O, provides evidence for the participation of tetrahedral intermediates. The nucleophile is hydroxide ion, and the leaving group is an alkoxide ion. Alkaline hydrolysis in $H_2{}^{18}O$ proceeds with the appearance of ^{18}O in the residual substrate during the course of the reaction (Bender, 1960). This observation is diagnostic of a mechanism involving the reversible addition of [^{18}O]hydroxide to form a tetrahedral intermediate. Such a mechanism for alkaline hydrolysis of a conventional ester is shown in fig. 6-1. This mechanism accounts for the appearance of ^{18}O in the residual substrate because proton transfer between the OH and O⁻ if the intermediate takes place at a significant rate relative to its breakdown in the forward and reverse directions.

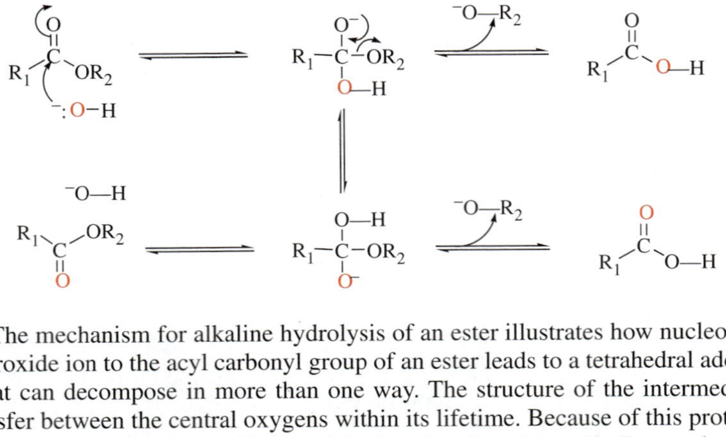

Fig. 6-1. The mechanism for alkaline hydrolysis of an ester illustrates how nucleophilic addition of hydroxide ion to the acyl carbonyl group of an ester leads to a tetrahedral addition intermediate that can decompose in more than one way. The structure of the intermediate allows proton transfer between the central oxygens within its lifetime. Because of this proton transfer, ^{18}O (red) can become equivalent with the original acyl carbonyl oxygen. Reversal of hydroxide addition then leads to ^{18}O-incorporation into the residual ester. Both intermediates can eliminate the alkoxide ion to form the ^{18}O-labeled carboxylic acid.

The ratio of rate constants for hydrolysis and ^{18}O-exchange, k_h/k_{ex}, is useful for describing the behavior of the tetrahedral intermediate. Values of k_h/k_{ex} are high (≥ 90) for aryl esters of benzoate and 12 to 27 for ethyl and methyl benzoate (Shain and Kirsch, 1968). These values indicate that proton transfer is much slower than breakdown of the tetrahedral intermediate when there is an excellent leaving group such as p-nitrophenol ($pK_a = 7$). When the leaving group is poor, as in methoxide or ethoxide ($pK_a = 15.5$ or 16.0), proton transfer within the tetrahedral intermediate competes with departure of the leaving group. Evidence from ^{18}O-kinetic isotope effects indicates that the lifetime of the tetrahedral intermediate is insignificant in the hydrolysis of p-nitrophenyl esters (Hess et al., 1998). Because of the low pK_a of p-nitrophenol, the leaving group departs without acquiring a proton, and these esters should be regarded as reacting by concerted displacement of p-nitrophenolate ion by hydroxide ion. p-Nitrophenyl esters react by a different mechanism than alkyl esters, one in which there is no discrete tetrahedral intermediate.

Tetrahedral addition intermediates in acyl transfer are not normally observed experimentally because they are often metastable. In a few cases, they can be observed as transient species. An example is the hydrolysis of acetyl imidazole, in the course of which a chromophore absorbing at 245 nm appears transiently. The kinetics of the formation and decay of this species implicates it as an intermediate, and the position of its chromophore is consistent with a tetrahedral adduct.

Evidence indicates that acyl transfer reactions of compounds with excellent leaving groups OR_2 in fig. 6-1 undergo concerted acyl transfer rather than stepwise transfer through tetrahedral intermediates. Alkaline hydrolysis of p-nitrophenyl or 2,4-dinitrophenyl esters in $H_2^{18}O$ is not accompanied by the appearance of the heavy isotope in the residual substrate (Shain and Kirsch, 1968). This could happen either if proton transfer in the mechanism of fig. 6-1 is not fast relative to the departure of an excellent leaving group, or if there is no tetrahedral intermediate when an excellent leaving group is available. The ^{18}O-kinetic isotope effects indicate that the mechanism does not include a discrete intermediate (Hess et al., 1998).

Enzymatic acyl transfer reactions include many in which the leaving groups require proton transfer to them as they depart, so that the mechanisms likely include tetrahedral intermediates. In these reactions, the leaving heteroatom is nitrogen or unactivated oxygen. In cases in which the leaving heteroatom is sulfur, the situation is less clear. The thiolate ion can be as good a leaving group as p-nitrophenolate if the pK_a of the conjugate acid is 7 or lower. In cases of cysteine and derived molecules such as glutathione, the normal pK_a is 8.5 and is subject to perturbation to 6 or 7 at an active site, as in glutathione S-transferases (Ladner et al., 2004). In chapter 1, we discuss the possibility of perturbation of the pK_a of coenzyme A from 11 to 7 in the active site of a coenzyme A transferase and give documented examples of perturbations on this scale in other active sites.

Acyl group transfer is subject to nucleophilic catalysis. In this mechanism, a highly reactive nucleophile initially accepts the acyl group from the acyl donor substrate to form a highly reactive acylating intermediate, which transfers the acyl group to an acceptor substrate. For example, imidazole catalyzes the hydrolysis of p-nitrophenylacetate according to the mechanism in eqs. 6-1 and 6-2.

$$R_1-\overset{O}{\underset{}{C}}-N\diagup^{+NH} + H_2O \longrightarrow R_1-\overset{O}{\underset{}{C}}-OH + HN^+\diagup NH \quad (6\text{-}2)$$

Imidazole reacts as the acetyl acceptor in eq. 6-1 to form acetyl imidazole and p-nitrophenol. Acetyl imidazole is a highly reactive acetylating agent because of its electrophilic reactivity, which is enhanced by the potential ability of the imidazole ring to exist as the imidazolium group. It reacts more rapidly with water in eq. 6-2 than p-nitrophenylacetate itself. Imidazole is a good catalyst in the hydrolysis.

Good acyl group donors have leaving groups with low values of pK_a. Acyl halides are excellent acylating agents because HCl is a strong acid. Acyl group transfer by donors with very good leaving groups are subject to nucleophilic catalysis by molecules such as imidazole and pyridine, which are reactive as nucleophiles in neutral aqueous solutions and form acyl intermediates that display high electrophilic reactivity. Acyl compounds with poor leaving groups do not display high reactivity with catalysts such as imidazole and pyridine. They tend to react at reasonable rates only with the most reactive nucleophiles such as thiolate, hydroxide, or alkoxide ions. Alternatively, they react under the influence of specific acid catalysis when the acyl oxygen in protonated by a very strong acid.

From a purely chemical standpoint, the most promising nucleophilic catalyst of acyl group transfer in an enzyme would be histidine. However, although histidine is a nucleophilic catalyst in enzymatic phosphotransfer (see chap. 10), it does not serve as a nucleophilic catalyst in enzymatic acyl group transfer. The most prevalent nucleophilic catalyst of acyl transfer by enzymes is serine, one of the least promising nucleophiles in proteins from a chemical standpoint. We discuss the mechanism by which serine functions in serine proteases and serine esterases. In cysteine proteases, the thiolate form of cysteine serves as the nucleophilic catalyst. In other classes of proteases, oxyanions are generated at the active sites to cleave the peptide bond.

Serine Proteases

The hydrolysis of peptide bonds underlies the processing of polypeptide hormones, removal of leader sequences from transported proteins, removal of N-terminal methionine from newly translated proteins, processing of polyproteins in viral infection, digestion of foreign proteins, digestion of cellular proteins in apoptosis, digestion of nutrient proteins, and other processes. The amide in a peptide bond is among the least reactive acyl groups, and its cleavage requires a highly reactive nucleophile.

In the action of serine proteases, serine functions as a nucleophilic catalyst, and the question of the means by which it acquires high reactivity is a central mechanistic issue. To function in this capacity, the hydroxyl group of serine should be ionized to the alkoxide, a process that is difficult at the neutral pH values in cells because of the pK_a of 13.4 for serine (Bruice et al., 1962). Nucleophilic reaction of serine should require a strong base to remove the proton from the hydroxyl group in the transition state. Serine proteases employ a histidine residue for this purpose. Histidine, with a pK_a of 7 would not seem to be strong enough as a base. The enhancement of basicity in the active site histidine is a subject of recent research. The most extensively studied serine proteases include chymotrypsin, trypsin, elastase, α-lytic protease, and subtilisin. All include active site triads similar to the Ser195, His57, and Asp102 of chymotrypsin. We emphasize mechanistic research on chymotrypsin (EC 3.4.21.1) because it has been the most extensively documented. Progress in the mechanistic analysis of the action of chymotrypsin has depended on the confluence of information

Acyl Group Transfer: Proteases and Esterases

obtained by biochemical, kinetic, crystallographic and spectroscopic methods. In several instances, key information about other serine proteases is cited.

The global chain folds of most serine proteases are similar; however, that of subtilisin is strikingly different. We compare the structures of subtilisin and chymotrypsin as an example of the convergent evolution of two protein families to similar active sites.

Chymotrypsin

In chapters 1 and 2, we discussed several aspects of catalysis by chymotrypsin as a case study for the following aspects of enzymatic action:

- The nature of enzyme structure
- Observation of a covalent intermediate by transient kinetics
- Identification of Ser195 as the nucleophilic catalyst by reaction with diisopropylphosphorofluoridate (DFP)
- Identification of His57 at the active site by affinity labeling with N-p-toluenesulfonyl-L-phenylalanine chloromethylketone (TPCK)
- Observation of a low-barrier hydrogen bond (LBHB) between His57 and Asp102

In this chapter, we explore the mechanism of action of chymotrypsin in greater detail.

The minimal kinetic mechanism in scheme 6-2 describes essential steps in the action of chymotrypsin. Although peptides are the natural substrates, esters, amides, and thioesters are also accepted, with preference for hydrophobicity in the acyl group, especially the aromatic groups of phenylalanine, tyrosine, and tryptophan, which are represented as $ArCH_2$. Equations 6-3 and 6-4 express k_{cat} and K_m for the mechanism in scheme 6-2, and $k_{cat}/K_m = (k_1 k_3)(k_2 + k_3)$.

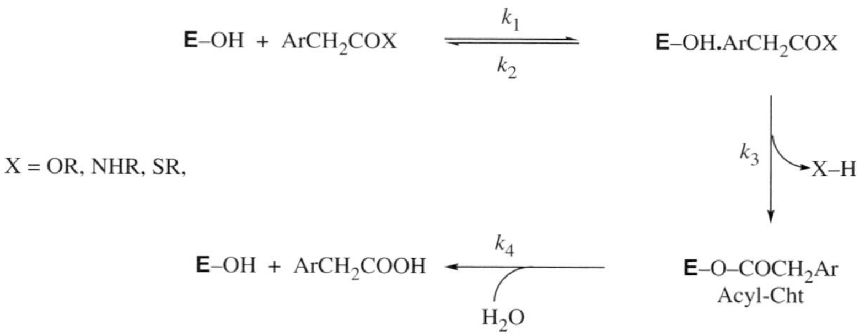

$$E\text{-OH} + ArCH_2COX \underset{k_2}{\overset{k_1}{\rightleftharpoons}} E\text{-OH} \cdot ArCH_2COX$$

$X = OR, NHR, SR,$

$$\downarrow k_3 \searrow X\text{-H}$$

$$E\text{-OH} + ArCH_2COOH \xleftarrow{k_4, H_2O} E\text{-O-COCH}_2Ar \quad \text{Acyl-Cht}$$

Scheme 6-2

$$k_{cat} = k_3 \left[\frac{k_4}{k_3 + k_4} \right] \tag{6-3}$$

$$K_m = \left[\frac{k_2 + k_3}{k_1} \right] \left[\frac{k_4}{k_3 + k_4} \right] \tag{6-4}$$

The steps governed by k_3 and k_4 are known as acylation and deacylation, respectively. Extensive evidence supports the kinetic mechanism (Hess, 1971).

Transient and Steady-State Kinetics

The reaction of the poor substrate p-nitrophenylacetate displayed the burst kinetics shown in fig. 2-13 in chapter 2 (Hartley and Kilby, 1954). This indicated a stepwise mechanism, in which the rapid acetylation of chymotrypsin was accompanied by release of a burst of yellow *p*-nitrophenolate ion, followed by the slower hydrolysis of the acetyl-chymotrypsin. Deacylation proved to be rate limiting in the hydrolysis of *p*-nitrophenylacetate. Because *p*-nitrophenylacetate was a poor substrate, burst kinetics could be observed in manual mixing experiments, as in fig. 2-13. As a highly reactive acetylating agent, *p*-nitrophenylacetate might have reacted differently from typical substrates, which reacted too fast to allow a burst to be observed by manual mixing. The experiment was repeated with the more specific aromatic amino acid esters, and they also displayed burst kinetics in rapid-mix, stopped-flow spectrophotometric experiments (Hess, 1971). These experiments strengthened the assignment of the mechanism.

The steady-state kinetic parameters in table 6-1 for the reactions of *N*-acetyl-L-tryptophan esters and the amide provided powerful indirect support for the mechanism. The fact that all the esters displayed the same value of k_{cat} could easily be explained by scheme 6-2 on the basis that $k_3 \gg k_4$, so that $k_{cat} = k_4$ for all of the esters. In the hydrolysis of esters, acylation is fast and deacylation is slower, and because the substrates all generate the same species of acyl-chymotrypsin, rate-limiting deacylation proceeds at the same rate for all the esters. Analogous results were obtained in the hydrolysis of the corresponding *N*-acetylphenylalanine esters (Zerner et al., 1964).

Partitioning of the Acyl-Chymotrypsin

The value of k_{cat} in the reaction of *N*-acetyl-L-tryptophan amide in table 6-1 is much lower than for the ester substrates. This raises the question of whether the amide reacts by a different mechanism. Alternatively, the mechanism can be the same with different relative values of rate constants. If $k_3 \ll k_4$ for amide substrates in eq. 6-3, then $k_{cat} = k_3$, and acylation would be rate limiting, thereby explaining the absence of a burst in the reactions of amide substrates. This rationalizes the behavior of amide substrates by the mechanism in scheme 6-2, but it is not evidence for an acyl-chymotrypsin intermediate.

The data in table 6-2 constitute affirmative, albeit indirect evidence for a common intermediate in the action of chymotrypsin, presumably an acyl-chymotrypsin. Chymotrypsin catalyzes transacylation as well as hydrolysis, a fact that itself suggests the participation of an acyl-chymotrypsin intermediate (Epand and Wilson, 1963). Particular support for an acyl-chymotrypsin is provided by the reactions of various amide and ester derivatives of *N*-acetyl-L-phenylalanine in the presence of chymotrypsin and alanine amide. Because alanine amide is a good acyl group acceptor in transacylation, chymotrypsin catalyzes both hydrolysis to *N*-acetyl-L-phenylalanine and acyl transfer to *N*-acetyl-L-phenylalanyl-L-alanine amide. As shown in table 6-2, the percent of acyl transfer product is the same for all derivatives of *N*-acetyl-L-phenylalanine-X; that is, it is independent of the leaving group. The most likely explanation is that there is a common intermediate, in this case the covalent *N*-acetyl-L-phenylalanyl-chymotrypsin. A constant percentage of transacylation for all amide and ester substrates would be very unlikely by any other mechanism.

Table 6-1. Kinetic Parameters for Chymotrypsin-Catalyzed Hydrolysis of Tryptophan Derivatives

N-Acetyl-L-tryptophan–X	$k_{cat}(s^{-1})$	K_m (mM)
X = OCH₃	28	0.095
OC₂H₅	27	0.097
O-*p*-NO₂Ph	31	0.002
NH₂	0.026	7.3

Data adapted from Schwert and Eisenberg, 1949.

Table 6-2. Partitioning of a Common Acyl-Chymotrypsin Intermediate between Water and Alanine Amide

N-Acetyl-L-Phenylalanine–X	N-AcPheAla-NH$_2$(%)
X = OCH$_3$	69
NH–p–N$^+$Me$_3$–Ph	67
NH–p–NMe$_2$–Ph	63
NHAlaNH$_2$	67

Data adapted from Fastrez and Fersht, 1973.

Acyl-Chymotrypsin and Ser195

In chapter 1, we presented biochemical experiments with the inactivator DFP, which implicated Ser195 as the nucleophile at the active site (Hartley and Kauffman, 1966; Hess, 1971). Covalent species of acyl-chymotrypsin could be isolated and even crystallized at low pH values when substrates with acyl groups displaying low reactivity at low pH were employed (Blow, 1971; Rossi and Bernhard, 1970). At low pH, the hydrolysis of indoleacryloyl-chymotrypsin was slow enough to allow its crystal structure to be determined (Henderson, 1970). All of these experiments implicated Ser195 as the site of acylation in chymotrypsin. Elastase, a member of the chymotrypsin family, reacted with a good substrate in crystals, which could be frozen to stop the reaction at the stage of the intermediate. Cryocrystallography revealed the structure of the acyl-elastase (Ding et al., 1994).

pH Dependence and His57

Plots of log(k_{cat}/K_m) for the action of chymotrypsin on specific substrates are bell-shaped, similar to the profile in fig. 2-18C in chapter 2. Complementary chemical and spectroscopic evidence indicate that the pK_a on the acid side should be assigned to the ionization of His57, which must be in its neutral form. Crystal structures show His57 in position to serve as the acid or base catalyst for acylation and deacylation of chymotrypsin. The pK_a on the alkaline side is assigned to the ionization of Ile16, the N-terminal amino acid, which must be in its protonated, cationic form. These assignments are based on evidence outlined later, allowing the pH behavior to be rationalized on the basis of structural and biochemical information.

The plot of log k_{cat} against pH shows an ascending limb with increasing pH breaking to a plateau with a pK_a near neutrality; this corresponds to the ascending limb in the profile of log(k_{cat}/K_m). The ionization represents a group either in a Michaelis complex or the acyl-chymotrypsin intermediate. This group is assigned as His57, and the pK_a of 7 for His57 in free chymotrypsin is directly confirmed by nuclear magnetic resonance (NMR) (Robillard and Shulman, 1974; Zhong et al., 1995). A plot of log(1/K_m) against pH describes a plateau at low pH that breaks downward on the alkaline side with a pK_a of 8.5; this corresponds to the descending limb in the profile of log(k_{cat}/K_m). Biochemical evidence indicates that the ionization is due to Ile16, the N-terminus of the B-chain, which must be in its cationic form for the substrate to bind. The kinetic pK_a of 8.5 is not, however, exactly that of Ile16 in the free enzyme, because the ionization is coupled to a conformational change.

Role of Ile16

To comprehend the identification of Ile16 as an ionizing group governing pH dependence, it is necessary to understand the origin of the B chain. Mature α-chymotrypsin consists of chains A, B, and C that arise from proteolytic processing of chymotrypsinogen, the inactive proenzyme form and primary translation product. Activation of chymotrypsinogen in vitro by the action of trypsin and activated chymotrypsin leads to the species depicted in fig. 6-2. Trypsin catalyzes peptide hydrolysis at sites of basic amino acids, and cleavage of the peptide bond between Arg15 and Ile16 by trypsin activates chymotrypsin. The resulting π-chymotrypsin is further processed into α-chymotrypsin by hydrolytic removal of the dipeptides Ser14Arg15 and Thr147Asn148 in several steps, with the intervening species being δ- and γ-chymotrypsin.

The central B-chain includes the active site, and its N-terminus is Ile16. In active chymotrypsin, Ile16 in its cationic, aminium form is paired with Asp194 (Blow, 1971), and in chymotrypsinogen, Asp194 is paired with His40 (Kraut, 1971a). The pairing of Ile16 with Asp194 leads to the burying of Ile16 into the interior structure and the displacement of the Asp194 side chain, among a number of other changes (Kraut, 1971a). The overall result is exposure of the active site for binding substrates. According to the current interpretations of the structures and pH-rate behavior, the loss of the N-terminal Ile16-amininium ion by dissociation of a proton at high pH releases Asp194 to pair with His40, with consequent occlusion of the active site. At high pH, the binding of substrates is inhibited, which explains the increase in K_m values at high pH values.

Biochemical experiments directly implicated Ile16 in the pH dependence (Hess, 1971; McConn et al., 1969). First, all of the amino groups in chymotrypsinogen were acetylated. Then, the peracetylated chymotrypsinogen was activated to polyacetylated δ-chymotrypsin, containing only a single amino group, the N-terminal amino group of Ile16. This polyacetylated enzyme was fully active and displayed the same alkaline pH dependence as ordinary δ-chymotrypsin. Further acetylation of polyacetylated δ-chymotrypsin inactivated the enzyme. The only amino group that could have been acetylated was that of Ile16. The aminium group of Ile16 was required for activity and must have been responsible for the alkaline limb of the pH-log(k_{cat}/K_m) profile.

A corollary of these experiments was that polyacetylated δ-chymotrypsin was fully active. All of the lysine-ε-amino groups and the N-terminus of the A-chain must have been acetylated,

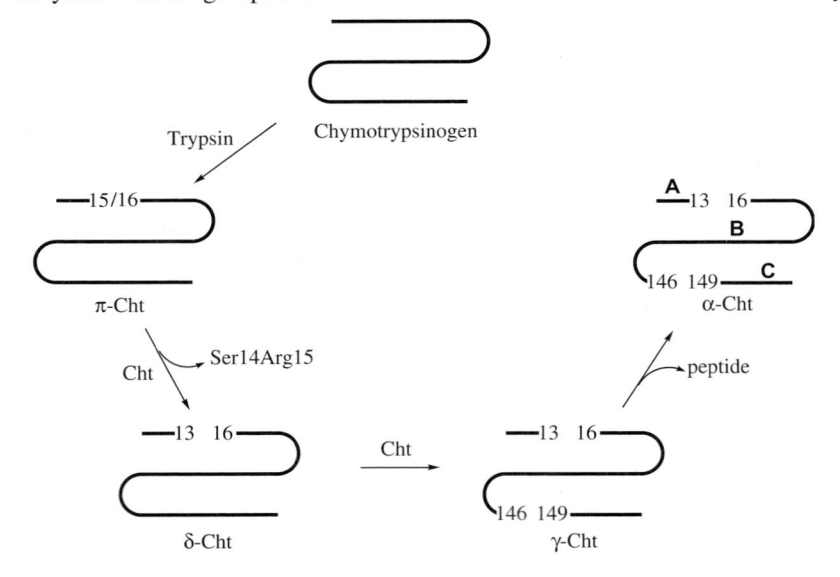

Fig. 6-2. Activation of chymotrypsinogen. Trypsin activates chymotrypsinogen by catalyzing hydrolysis of the peptide bond between Arg15 and Ile16 to produce π-chymotrypsin (π-Cht), a fully active form that participates in the further processing of the peptide chain (Wright et al., 1968). Further hydrolysis to excise Ser14Arg15 produces δ-Cht, and excision of Thr147Asn148 produces γ-Cht. γ-Cht contains a peptide bound to the active site that releases to generate α-Cht, consisting of three discrete polypeptides: A, B and C (Dixon and Matthews, 1989). The π-, δ-, γ-, and α-Cht forms are all active.

so that most of the positive charges were abolished. The net electrostatic charge of the enzyme must have been dramatically different from natural chymotrypsin, yet this did not alter enzymatic activity or even the pH-rate profile. We also encountered this phenomenon in chapter 1, in connection with chemical modification and neutralization of the surface carboxylate groups outside the active site of lysozyme.

Kinetic Isotope Effects

Typical kinetic isotope effects in the action of chymotrypsin are collected in table 6-3. Interpretation of isotope effects is necessarily guided by the established kinetic mechanism in scheme 6-1. However, they stand on their own and could in principle force revisions in the mechanism. However, they are compatible with the mechanism and can be interpreted to refine several aspects.

The chymotrypsin catalyzed hydrolysis of N-acetyl-L-tryptophan [^{15}N]amide proceeds with a significant kinetic isotope effect on $^{15}k_{cat}$. The value in table 6-4 (1.006) proves that C—N bond cleavage participates in rate limitation.

The value of $^{15}k_{cat}$ is low relative to the maximum $^{14}k/^{15}k$ for C—N cleavage (1.044 in table 2-4 in chap. 2) and may be explained in any of several ways. The transition state may be asymmetric, C—N cleavage may be kinetically coupled to another process, or 1.006 may be a secondary kinetic isotope effect. For example, in the acylation mechanism in scheme 6-3 (Ar = AcTrp R = H), nucleophilic attack on the amide carbonyl to form a tetrahedral intermediate proceeds with weakening of the C—N bond. This may lead to a secondary ^{15}N effect if formation of the tetrahedral intermediate governed by k_{31} limits the overall rate of acylation.

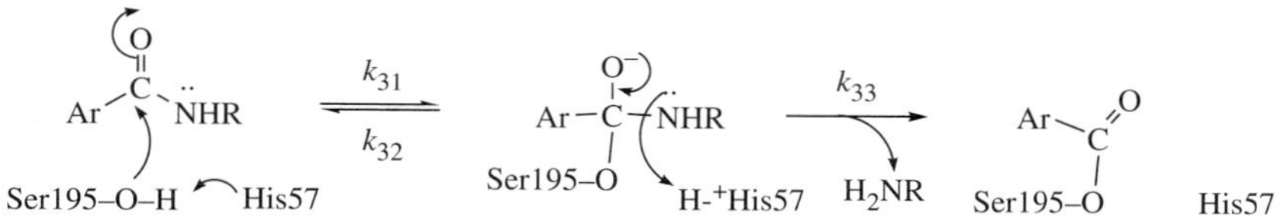

Scheme 6-3

The ^{15}N isotope effect for the hydrolysis of N-succinyl-L-phenylalanine p-nitroanilide (R = p-nitrophenyl in scheme 6-3) is 1.014, significantly smaller than the theoretical maximum but too large to be a secondary effect. In the reaction of this substrate, decomposition of the tetrahedral intermediate must limit the rate, and the lower value of the ^{15}N isotope effect indicates an asymmetric transition state or kinetic coupling with another process, such as a different chemical step or a conformational change. In scheme 6-3 the departure of the leaving group from the tetrahedral intermediate might occur at a similar rate as reversal to the Michaelis constant ($k_{33} \approx k_{32}$), and this would diminish the observed ^{15}N kinetic isotope effect. In any case, the ^{15}N kinetic isotope effects are compatible with the

Table 6-3. Kinetic Isotope Effects in the Action of Chymotrypsin

Substrate	Parameter	KIE	Study
N=AcTrp–^{15}NH$_2$[a]	$^{15}k_{cat}$	1.006	O'Leary and Kluetz, 1972
N-succinyl–Phe p-nitroanilide[a]	$^{15}k_{cat}$	1.014	Hengge and Stein, 2004
N-AcTyrOMe[a]	$^{D_2O}k_{cat}$	2.5	Bender et al, 1964

KIE, kinetic isotope effect.
[a] N-acetyltryptophan amide; N-succinylphenylalanine p-nitroanilide; N-acetyltyrosine methylester.

Table 6-4. Properties of the Low-Barrier Hydrogen Bond in Chymotrypsin and Transition-State Analog Complexes

		Inhibitor			
Physical Property[a]	Free Cht	AcF-CF$_3$	AcGF-CF$_3$	AcVF-CF$_3$	AcLF-CF$_3$
δ_H (ppm)	18.1	18.6	18.7	18.9	19.0
ϕ[b]	0.4	0.32	0.34	0.38	0.43
$\Delta(\delta_T - \delta_H)$ ppm[c]	—	0.8	0.8	—	0.8
k_{ex} (s^{-1}) 2[b]		282	123	—	12.4
$\Delta H^\ddagger_{ex}$ (kcal mol^{-1})[a]	10.4	15	16	—	19
pK_a[b]	7.0	10.7	11.1	11.8	12.1
K_I (μM)[c]	NA	17, 30 20, 40	18, 12	2.8, 4.5	1.2, 2.4

AcFCF$_3$, N-acetyl-L-phenylalanine trifluoromethylketone; AcGFCF$_3$, N-acetyl-glycyl-L-phenylalanine trifluoromethylketone; AcFCF$_3$, N-acetyl-L-valyl-L-phenyl-alanine trifluoromethylketone; AcFCF$_3$, N-acetyl-L-leucyl-L-phenylalanine trifluoromethylketone; Chy, chymotrypsin.
[a] Physical properties of His 57-Hδ1: δ_H, ^{1}H NMR chemical shift; ϕ, D/H fractionation factor; $\Delta(\delta_T - \delta_H)$, tritium isotope shift; k_{ex}, rate constant for exchange with solvent; ΔHá$_{ex}$, activation enthalpy for exchange at 25°C (Lin et al, 1998a, 1998b; Markley and Westler, 1996; Westler et al, 2002).
[b] The pK_a values for His57-Hε2 (Cassidy et al, 1997; Lin et al, 1998a).
[c] Inhibition constants reported (Brady and Abeles, 1990; Imperiali and Abeles, 1986; Liang and Abeles, 1987).

kinetics and prove that acylation of chymotrypsin limits the overall rate in the hydrolysis of amide substrates.

The solvent deuterium kinetic isotope effect ($^{D_2O}k_{cat}$) in table 6-3 is typical of nonenzymatic hydrolysis reactions (Schowen, 1972). Proton transfer is bound to be important in peptide hydrolysis and is likely to participate in rate limitation. Although a solvent kinetic isotope effect supports a role for proton transfer in a transition state, the assignment of the isotope effect to a particular proton in an enzymatic reaction is difficult. Problems arise from the fact that all acid-base functional groups and most amide hydrogens in an enzyme become labeled with deuterium in D$_2$O, and many of them participate in multiproton chains of hydrogen bonds that may impact catalysis. Other aspects of protein structure may be perturbed by the replacement of H$_2$O with D$_2$O. For example, hydrophobic interactions are stronger in D$_2$O than in H$_2$O (Kresheck et al., 1965; Schneider et al., 1965), and the rate of a conformational change that entails a hydrophobic interaction and participates in rate limitation may become a factor. Moreover, D$_2$O is more viscous than H$_2$O. This can lead to magnified normal solvent isotope effects when diffusion is rate limiting and even to inverse solvent isotope effects when a conformational change is rate limiting (Karsten et al., 1995).

Further complications are introduced by variations in H/D fractionation factors for protons that may be involved, even protons located at some distance from the reacting atoms. The possibility of multiproton catalysis can be investigated by the *proton inventory* method (Venkatasubban and Schowen, 1984). In this analysis, values of k_{cat} or k_{cat}/K_m are measured in various mixtures of H$_2$O and D$_2$O, so that the mole fraction of deuterium varies from 0.0 to 1.0. A Gross-Butler equation is fitted to the data to determine how many protons participate in catalysis. In the simplest case, the number of protons in flight in the transition state can be determined for reactions of simple molecules. The general Gross-Butler equation is eq. 6-5, where n is the fractional deuterium content, v_n is a rate parameter measured in the D$_2$O/H$_2$O at n fractional deuterium content, v_H is the same parameter measured in pure water, ϕ_I^T is the fractionation factor in the transition state, and ϕ_j^R is the fractionation factor in the reactant state. The numerators are the transition state, and the denominators are the reactant state contributions, respectively.

For enzymatic reactions, the rate parameters v_n are k_{cat} or k_{cat}/K_m. They are measured at various deuterium fractions n and plotted as v_n against n. A simplified Gross-Butler equation, such as 6-5a or 6-5b based on the principle that $k_D/k_H = \phi_D/\phi_H$, may be fitted to the data.

One proton
$$v_n = v_H[1 - n - n(k_D/k_H)] \tag{6-5a}$$

Two protons
$$v_n = v_H[1 - n - n(k_D/k_H)_1][1 - n - n(k_D/k_H)_2] \tag{6-5b}$$

A straight line is consistent with eq. 6-5a and may correspond in the ideal case to one proton experiencing a change in fractionation factor in the transition state. A plot with appropriate concavity may be consistent with the quadratic eq. 6-5b, ideally with two protons experiencing changes in fractionation in the transition state. A cubic equation would indicate three protons participating in the transition state. The value of k_D/k_H obtained by fitting the equations to experimental data is the inverse of the solvent kinetic isotope effect.

In nonenzymatic reactions of simple substrates, the proton inventory is valuable mechanistic information. In enzymatic reactions, the interpretation of proton inventory data may be more complex than the equations imply for the reasons stated earlier (Venkatasubban and Schowen, 1984). Moreover, Michaelis complexes can include hydrogen-bonding networks involving many protons that can affect the enzyme structure or its interactions with a substrate in ways that impact the experiment. The proton inventory does not contain information about which protons are involved. A straight line in the inventory can be ambiguous. It has been shown that the straight-line inventory in the hydrolysis of acetyl-chymotrypsin could result from compensatory interactions of a single proton in flight and several other protons displaying inverse isotope effects (Kresge, 1973; Pollack et al., 1972).

Proton inventories for hydrolysis by chymotrypsin vary depending on the substrate employed (Elrod et al., 1980; Pollack et al., 1970; Stein et al., 1983). In the hydrolysis of *p*-nitrophenylacetate, *N*-acetyltryptophan amide, and the *p*-nitrophenyl esters of Cbz-tryptophan and Cbz-phenylalanine the plots are linear. However, the plot is concave for the hydrolysis of the *p*-nitrophenyl ester of Cbz-glycine, which lacks the signature of specific substrates for chymotrypsin, an aryl methylene substituent on the α-carbon.

Ser195 Adducts

Two classes of compounds undergo simple addition to Ser195 to inhibit chymotrypsin. These inhibitors incorporate unsaturated, trigonal functional groups that form tetrahedral addition products with Ser195. Aryl boronic acids were the first to be discovered (Robillard and Shulman, 1974), and these early compounds inspired the development of peptide boronic acids as potent inhibitors (Zhong et al., 1995). The peptide boronic acids react with Ser195 as in eq. 6-6. His57 at the active site accepts the proton released in eq. 6-6.

$$\text{Ser195–OH} + \text{Peptide}-B\begin{smallmatrix}\text{OH}\\\text{OH}\end{smallmatrix} \longrightarrow \text{Peptide}-\underset{\underset{\text{Ser195}}{O}}{\overset{\text{OH}}{B^-}}-\text{OH} + H^+ \tag{6-6}$$

Peptide-boronates are tetrahedral, and in this respect, they are analogous to tetrahedral intermediates. However, they lack an oxyanionic group, and the negative charge resides on boron, so that the boronates are electrostatically unlike tetrahedral intermediates. The extremely high affinity of peptide boronic acids for chymotrypsin can be attributed in part to the peptide moiety. Moreover, the boron-oxygen bond to Ser195 is likely much stronger than the carbon-oxygen bond in a tetrahedral intermediate because of the greater electronegativity difference between boron and oxygen compared with carbon and oxygen. (It is mainly for this reason that C—F bonds and O—H bonds are so much stronger ($\approx$120 kcal mol^{-1}) than C—H or C—C bonds (90 – 100 kcal mol^{-1}.) The strength of boron-oxygen bonding is likely to contribute to the affinity of chymotrypsin for peptide boronic acids.

Another class of inhibitors with unsaturated, trigonal groups is the peptide trifluoromethylketones, which react with Ser195 according to eq. 6-7 (Imperiali and Abeles, 1986; Liang and Abeles, 1987). The proton is absorbed by His57 at the active site.

$$\text{Ser195-OH} + \text{Peptide}-\underset{CF_3}{\overset{O}{C}} \longrightarrow \underset{\underset{Ser195}{O}}{\overset{O^-}{\underset{|}{\underset{Peptide}{C}}}}CF_3 + H^+ \quad (6\text{-}7)$$

The trifluoromethylketone adducts incorporate an oxyanionic group and are sterically and electrostatically similar to the tetrahedral intermediates. The value of K_d for binding N-acetyl-L-leucine-L-phenylalanine trifluoromethylketone (AcLF-CF$_3$) is 0.2 nM (Brady and Abeles, 1990).

The related peptide aldehydes, such as N-acetyl-L-leucine-L-phenylalaninal (AcLP-CHO) incorporating H in place of CF$_3$, are much weaker inhibitors, and their reactions diverge in that the proton from Ser195 resides on the adduct itself, as in eq. 6-8, and not on His57 (Neidhart et al., 2001).

$$\text{Ser195-OH} + \text{Peptide}-\underset{H}{\overset{O}{C}} \longrightarrow \underset{\underset{Ser195}{O}}{\overset{OH}{\underset{|}{\underset{Peptide}{C}}}}H \quad (6\text{-}8)$$

The pK_a of the hemiacetal adduct is higher than 13.8. His57 cannot compete with the hemiacetal anion for the proton released from Ser195. The tetrahedral adducts of peptide aldehydes therefore lack an oxyanion, and His57 is not protonated at pH 7, unlike tetrahedral intermediates and the hemiketal adducts of peptide trifluoromethylketones.

The Oxyanion Site

In addition to revealing the tertiary fold of chymotrypsin, crystallographic structures unveiled essential features of the active site and its interactions with substrates. The most striking were the close contact of Asp102 with His57 and the presence of a specific binding site for the substrate-carbonyl and the oxyanionic group of the tetrahedral intermediate in scheme 6-3. The latter became known as the *oxyanion hole* and is referred to here as the *oxyanion site*. This site has received a great deal of attention in anticipation that it would provide the main driving force for stabilizing tetrahedral intermediates. As metastable species, the tetrahedral intermediates are regarded as analogs of the transition states for acylation and deacylation, and their stabilization would catalyze the reaction.

The oxyanion site in chymotrypsin is formed by the main chain peptide N–H groups of Gly193 and Ser195, which donate hydrogen bonds to the tetrahedral oxyanion. The interaction is shown in the structures of the tetrahedral adduct between Ser195 of chymotrypsin and the inhibitor N-acetyl-L-leucine-L-phenylalanine trifluoromethylketone (AcLF-TFMK) shown in fig. 6-3.

Vibrational spectroscopy has been exploited to investigate the interactions at the oxyanion site (Tonge and Carey, 1989, 1992). In studies of a series of substrates, the vibrational frequencies of substrate carbonyl groups at the active site were well correlated with reactivity, indicating that the better substrates engaged in stronger hydrogen bonding in the oxyanion site.

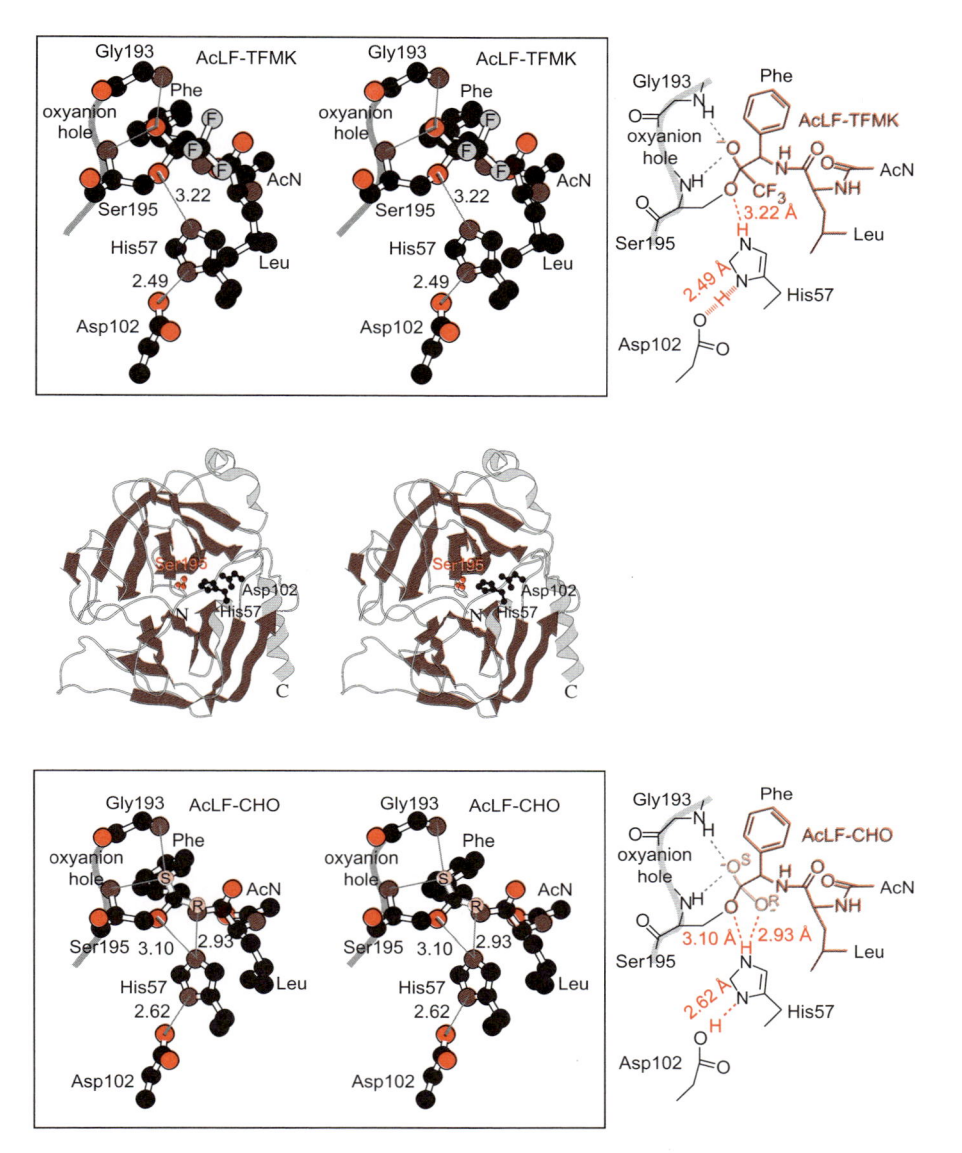

Fig. 6-3. AcLeuPhe-trifluromethylketone (AcLF-TFMK) and AcLeuPhe-CHO interact at the active site of chymotrypsin. AcLeuPhe-CF$_3$ (top) is bound to chymotrypsin (PDB 7GCH; Brady et al., 1990). Ac-LeuPhe-CF$_3$ binds as a hemiketal to Ser195 of chymotrypsin. The adduct is an analog of the tetrahedral intermediate in the hydrolysis of *N*-acetyl-L-leucine-L-phenylalanine esters and amides, with the CF$_3$ group in place of the leaving group. This picture of the active site is intended to highlight the close contact between His57-Nδ1 and Asp102-Oδ1 that characterizes the low-barrier hydrogen bond. The oxyanion binding site also consists of hydrogen bonds between the oxyanion of the adduct and the peptide amide groups of Gly193 and Ser195. The unliganded global fold of chymotrypsin (middle) is shows Ser195 in red and Asp102 and His57 in black ball-and-stick representations. The chymotrypsin active site complex (bottom) is displayed with AcLeuPhe-CHO (1GGD; Neidhart et al., 2001). The addition of Ser195 to this inhibitor results in an equilibrium mixture of *R*- and *S*-hemiacetal adducts, as partial density for the oxyanion in both conformations is observed.

Contrary to the prevailing perception, it is not clear that the oxyanion site is optimal for stabilizing oxyanions. In an optimized site, one might expect three hydrogen bond donors to satisfy the maximum hydrogen bonding potential of an alkoxide ion. However, there are only two donors in the oxyanion site of the chymotrypsin. There are three in the oxyanion site of acetylcholinesterase, which employs a triad of serine, histidine, and glutamate in catalyzing ester hydrolysis (Millard et al., 1999).

The Low-Barrier Hydrogen Bond

The LBHB between His57 and Asp102 has been implicated in the mechanism by which Ser195 reacts as a nucleophilic catalyst in the acylation of chymotrypsin (Cassidy et al., 1997, 2000; Frey et al., 1994; Lin et al., 1998a, 1998b; Neidhart et al., 2001; Westler et al., 2002). The proposed role of the LBHB is illustrated in fig. 6-4. The LBHB is formed in the complex of chymotrypsin with the tetrahedral intermediate and is postulated to stabilize this state, as well as the structurally related transition state.

The LBHB cannot be observed directly in catalysis because of the instability of tetrahedral intermediates. However, closely related analogs of tetrahedral intermediates display the LBHBs, which can be characterized spectroscopically and crystallographically. The close contact between His57 and Asp102 in transition-state analog complexes of chymotrypsin is consistent with the presence of an LBHB. The structure of the active site in this complex (AcLF-TFMK adduct) at 1.4-Å resolution is shown in fig. 6-3.

Chymotrypsin and other serine proteases share a property that may provide a glimpse into the transition state even in the absence of a transition-state analog. The protonated imidazole ring of His57 is a property in common between tetrahedral intermediates and free chymotrypsin at low pH. NMR experiments prove that at pH values below 7 His57-Hδ1, the proton-bridging His57 and Asp102, appears far downfield at 18 ppm (Markley, 1978; Robillard and Shulman, 1972). It is sufficiently downfield to be assigned as an LBHB (Frey et al., 1994). The downfield proton in chymotrypsin displays other spectroscopic and chemical properties of an LBHB.

The data in table 6-4 for complexes of chymotrypsin with peptide trifluoromethylketones and of free chymotrypsin at low pH values (<pH 6) characterize His57-Hδ1 as an LBHB by the criteria in table 1-5 in chapter 1. Typical values of chemical shift, fractionation factor, and the isotope effect on chemical shift for conventional hydrogen bonded protons listed in table 1-5 are very different from those in table 6-4, all of which are characteristic of LBHBs. The heats of chemical exchange of LBHBs with solvent protons are much higher than of

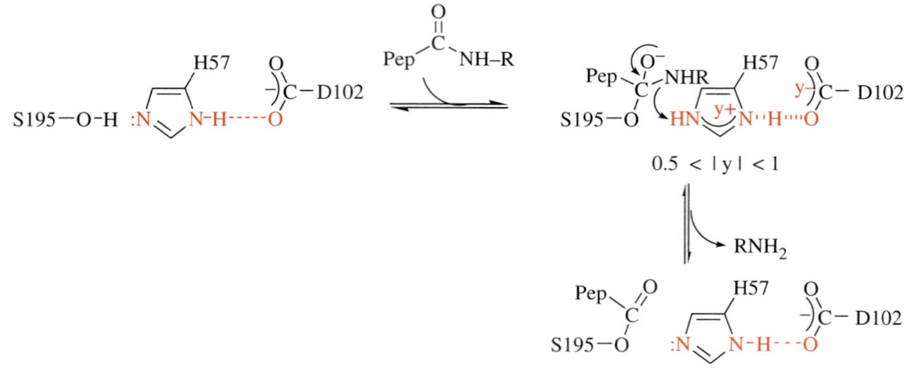

Fig. 6-4. Acylation of Ser195 of chymotrypsin and the role postulated for a low-barrier hydrogen bond.

conventionally hydrogen bonded protons, and the values in table 6-4 are among the highest that have been measured.

The LBHB lies at the center of a difference between free chymotrypsin and its complexes with transition-state analogs. His57 displays a normal pK_a in free chymotrypsin, whereas in transition-state analog complexes it displays an elevated pK_a. The high pK_a in a transition-state analog is regarded as the essential mechanistic role of the LBHB. The ideal base for abstracting a proton from Ser195 in the formation of the tetrahedral intermediate, and for donating a proton to the leaving group, should display a pK_a between that of Ser195 (13.4) and that of the leaving amino group (≈ 9). If the role of the LBHB is to increase the basicity of His57 in the transition state and intermediate, why does it not do so in free chymotrypsin?

Data in table 6-5 for several tetrahedral complexes and free chymotrypsin illuminate another factor governing the pK_a of His57. The pK_a of His57 is elevated only in tetrahedral complexes in which the adduct is negatively charged. As in free chymotrypsin, His57 ionizes "normally" in the diisopropylphosphoryl and AcLF-CHO adducts, which are neutral. The AcLF-CHO adduct is neutral because the proton from the addition of Ser195 resides on the hemiacetal adduct and not on His57. The negatively charged adduct with AcLF-TFMK displays another perturbed pK_a, that of the hemiketal oxygen. In aqueous solution, the hydroxyl group of a hemiketal of AcLF-TFMK is 9.5, but that of the Ser195 hemiketal in chymotrypsin is about 5. In the active site of this adduct, the pK_a of His57 is elevated by 4 to 5 units, and that of the hemiketal is depressed by 4.5 units. The two effects suggest a strong propensity toward preservation of a net charge of −1 at the active site. This microenvironmental effect has been observed in another context as a mechanism for expelling the anionic product from the active site (Johnson and Knowles, 1966).

The structure of the neutral AcLF-CHO adduct (fig. 6-3) is essentially that of the AcLF-TFMK adduct, with H in place of the CF_3 group. Yet the ionization of His57 in the AcLF-CHO complex appears to be normal, as in free and diisopropylphosphoryl-chymotrypsin. The ionization of His57 changes the net charge from −1 to 0 with little or no apparent effect on the pK_a of His57, which becomes engaged in the LBHB on protonation. If the same propensity for preserving the −1 net charge exists in free chymotrypsin and the neutral tetrahedral complexes, the normal pK_a for His57 would mean that in the absence of the LBHB, His57 would undergo ionization with a much lower pK_a of perhaps 2.

Subtilisin

Convergent Evolution

The tertiary fold and amino acid sequences of subtilisin (EC 3.4.21.62), named after the organism *Bacillus subtilis* from which subtilisin Carlsberg was first purified (Markland and Smith, 1971), are very different from serine proteases in the chymotrypsin family (Kraut, 1971b). However, subtilisin incorporates a catalytic triad (Ser221, His64, Asp32) that is very similar to other serine proteases. The structure of subtilisin in Fig. 6-5 is very different from that of chymotrypsin in Fig. 6-3. Apparently, these two enzymes evolved independently from separate precursors, and their evolution converged to structures with similar active site catalytic triads.

Early studies of the mechanism of action and specificity of subtilisin showed marked similarities with chymotrypsin, with differences in details (Markland and Smith, 1971). Both enzymes function by the kinetic mechanisms in schemes 6-2 and 6-3, and both are most active against substrates with hydrophobic amino acids in the scissile sites, although subtilisin is less stringent in its requirements. A significant difference is that subtilisin is less active on esters than chymotrypsin.

Table 6-5. Ionization Properties of Chymotypsin Complexes

Cht Complex		pk_a
AcLF^a-C(OH)(H,CF$_3$)-O- (tetrahedral adduct with S195, H57, D102)	—	His57
AcLF-C(O-)(CF$_3$)-O-S195; H57/D102 triad	<5	12
AcLF-C(OH)(H)-O-S195; H57/D102 triad	>13.5	6.5
iPr^{b}O-P(O-)(O-)-O-S195; H57/D102 triad	—	9.7
iPrO-P(=O)(OiPr)-O-S195; H57/D102 triad	—	7.4
S195-OH; H57/D102 triad	—	7.0

^{a}N-Acetyl-L-Leu-L-Phe.
^{b}Isopropyl.
Data summarized from Adebodun and Jordan et al, 1989; Neidhart et al, 2001; Lin et al, 1998; Liang and Abeles, 1987.

Chemical Mutagenesis

In the original site-specific mutagenesis experiment, Ser221 of subtilisin was changed into Cys221 by the chemical method outlined in fig. 6-6 (Neet and Koshland, 1966; Polgar and Bender, 1966). Reaction of subtilisin with phenylmethylsulfonyl fluoride, which inactivates serine proteases by sulfonylating the active site serine, led to phenylmethylsulfonyl subtilisin. Displacement of the phenylmethylsulfonyl group by thioacetate led to acetylthiol subtilisin, which underwent hydrolysis to generate cysteine at position 221. The resulting thiolsubtilisin displayed significant to high activity against p-nitrophenyl esters and acyl imidazoles, but no activity against peptides and other esters and amides (Neet and Koshland, 1968).

Engineered Subtilisin

Subtilisin has been adapted to alternative functions by systematic engineering. Selenosubtilisin, an analog of thiolsubtilisin, is a peptide ligase and links the N-terminus

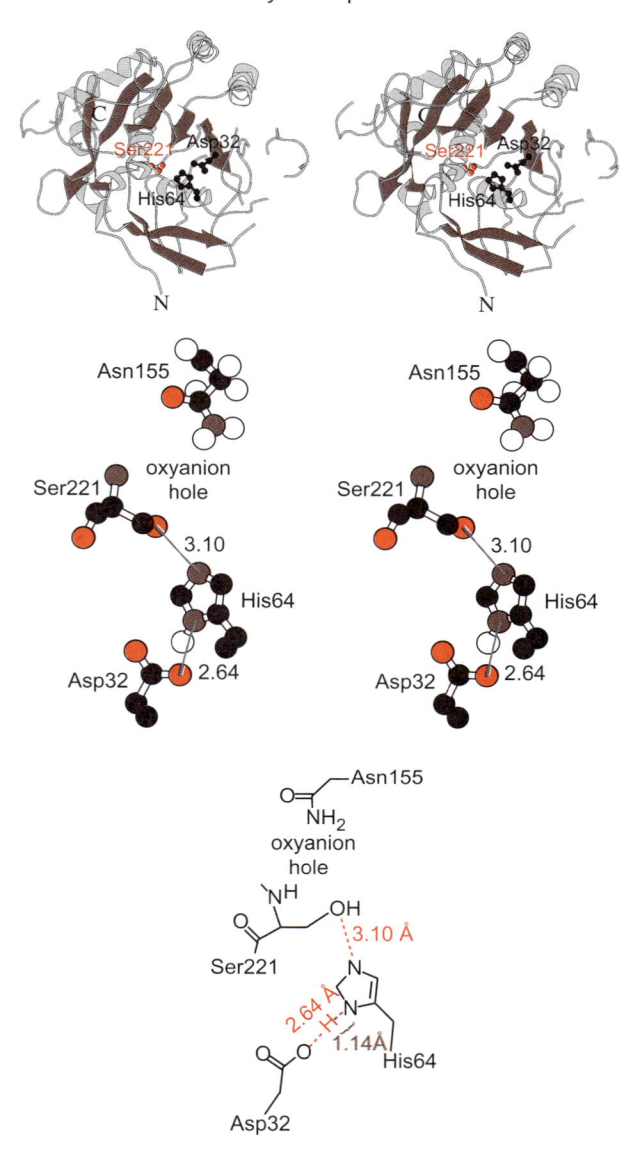

Fig. 6-5. The crystallographic structure of the active-site histidine and aspartate in *Bacillus lentus* subtilisin at 0.78-Å resolution shows the electron density for hydrogen in the low-barrier hydrogen bond (LBHB) (PDB 1GCI; Kuhn et al., 1998). Notice the very close contacts between His64 and Asp32 (2.6 Å), which are found in all structures of serine proteases. The high-resolution structure of subtilisin the N—H distance in the LBHB is 1.14 Å, substantially longer than the conventional distance of 0.9 Å in a weakly hydrogen-bonded NH group. The strong hydrogen bonding stretches the NH bond and leads to the downfield proton NMR signal and the isotope effects on the LBHB.

of a peptide with the C-terminal ester of another peptide under anaerobic conditions (Wu and Hilvert, 1989). Under aerobic conditions, selenosubtilisin is a peroxidase (Bell et al., 1993).

The peptide cleavage specificities of many proteases have been extended to preferred residues upstream and downstream from the cleavage site. The coding is as follows, reading left to right from the N-terminus: ... P3–P2–P1—P1′–P2′–P3′... . (The active site peptide is coded ... S1–S2–S3 * S1′–S2′–S3′..., with the active site at *.) For example, in the case of

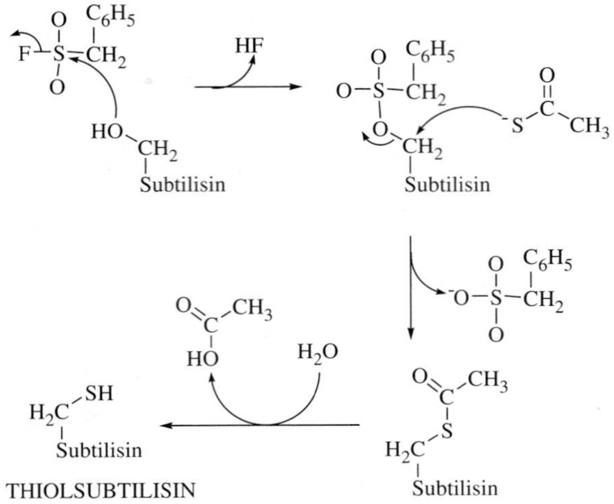

Fig. 6-6. Chemical mutagenesis of subtilisin into thiolsubtilisin. Phenylmethylsulfonyl fluoride (PMSF) inactivates serine proteases by sulfonylation of serine in active sites. Phenylsulfonate is an excellent leaving group and can be displaced by thiolate anions. Such an anion is thiolacetate, which displaces phenylmethylsulfonate under mild conditions, transforming the enzyme into acetyl thiolsubtilisin, which undergoes hydrolysis to thiolsubtilisin.

chymotrypsin leucine is preferred in the P2-site. This information for subtilisin is available (Perona and Craik, 1995). Mutation of His64 to alanine inactivates subtilisin. However, the variant displays activity against a peptide with histidine in the P2 or P1' positions. In these peptides, the histidine residue in the substrate substitutes for His64 and rescues the activity (Carter and Wells, 1987). This feature has been adapted to the specific cleavage of fusion proteins employed in high expression of enzymes (Forsberg et al., 1992).

The contribution of the oxyanion site in subtilisin has been probed by site-directed mutagenesis. In subtilisin, one hydrogen bond donor is the side chain amide of Asn155. Mutation of this residue to alanine lowered the activity by 200-fold, corresponding to an increase of about 3 kcal mol^{-1} in the activation energy (Bryan et al., 1986). This is on the high side of the kinetic effects observed for the loss of a single weak hydrogen bond in studies of the effects of hydrogen bonding on the action tyrosyl-tRNA synthetase (see chap. 2).

The Low-Barrier Hydrogen Bond

The structure of subtilisin at 0.78-Å resolution allows for the imaging of hydrogen atoms, and the proton bridging His64 and Asp32 at the active site has been imaged and modeled as an LBHB (Kuhn et al., 1998). The model of the active site structure of subtilisin at this resolution is shown in fig. 6-5. Similarly, the structure of pancreatic elastase at 0.95 Å reveals the presence of electron density between histidine and aspartate in the catalytic triad (Katona et al., 2002).

Cysteine Proteases

Families of peptide hydrolases incorporating cysteinyl residues as nucleophilic catalysts include the papain family and the caspases among others (Barrett et al., 1998). The plant enzyme papain is the most thoroughly studied from the mechanistic standpoint. Although papain is of plant origin, many related enzymes are found in animals.

Cathepsins, bleomycin hydrolase, and ubiquitin C-terminal hydrolase, among many other peptidases, are in the papain family (Barrett et al., 1998). Caspases are calcium-dependent cysteine proteases that cleave specifically at aspartate sites. They are currently under intensive investigation because of their key roles in apoptosis. The cysteine proteases have become important drug targets in the pharmaceutical industry (Lecaille et al., 2002). The mechanisms of action of all cysteine proteases are thought to be similar, and the extensive work on papain can serve as a case study for the group.

Papain

Papain (EC 3.4.22.2) is a monomeric enzyme isolated from papaya that is composed of 212 amino acids and is comparatively unspecific in its peptidase activity, relative to proteases such as chymotrypsin and trypsin, although it displays some preference for bulky, neutral aminoacyl sites in proteins (Glazer and Smith, 1971). Like chymotrypsin, this enzyme also catalyzes the hydrolysis of esters and thiolesters. Early observations focused attention on cysteine as essential for activity because of the sensitivity of papain to many alkylating agents, including iodoacetate. This property proved to be the first sign of the essential mechanistic feature of cysteine proteases, the importance of the Cys25-thiolate in catalysis. Cys25 is maintained in its thiolate form by the proximity of His159, which exists predominantly in its imidazolium form in the active site.

Kinetics and Isotope Effects

Extensive kinetic studies implicated the mechanism of scheme 6-3, with Cys25 in place of the Ser195 of chymotrypsin, as the overall kinetic mechanism for the action of papain (Glazer and Smith, 1971). This mechanism explains, among other things, the transesterification and transamidation activities of papain. The relative values of rate constants differ in the actions of chymotrypsin and papain, however. In the reaction of papain, the value of k_2 is much smaller than the value of k_3 (Theodoraou et al., 2001). Also unlike chymotrypsin, the value of the ^{15}N kinetic isotope effect in the hydrolysis of N-acetyl-L-arginine [^{15}N]amide is 1.02, about one half of the maximum effect of 1.044 (O'Leary et al., 1974). C—N cleavage is substantially rate limiting in the mechanism of scheme 6-3 for an amide substrate.

Structure

The crystal structure of papain shows it consists of two distinct domains folded around hydrophobic cores, with the active site at the interface (fig. 6-7). The N-terminal half of the molecule forms one domain with mainly α-helical secondary structures, and the C-terminal half forms a domain with mainly β-sheet secondary structure. At the interface, Cys25 in the N-terminal domain and His159 in the C-terminal domain form the active site, with Asn175 hydrogen bonded to His159. The active site contains a triad that is seemingly analogous to the serine, histidine, and aspartate triad in serine proteases. However, the chemical nature and action of the catalytic residues in papain vary significantly from the serine proteases. Figure 6-7 shows papain with an inhibitor bound to Cys25.

Roles of Cys25 and His159

The proximity of Cys25 and His159 in the structure raises questions about their interactions and the catalytic mechanism. Does the acylation of papain involve base catalysis by His159 to abstract the proton from the sulfhydryl group of Cys25, or do Cys25 and His159 exist

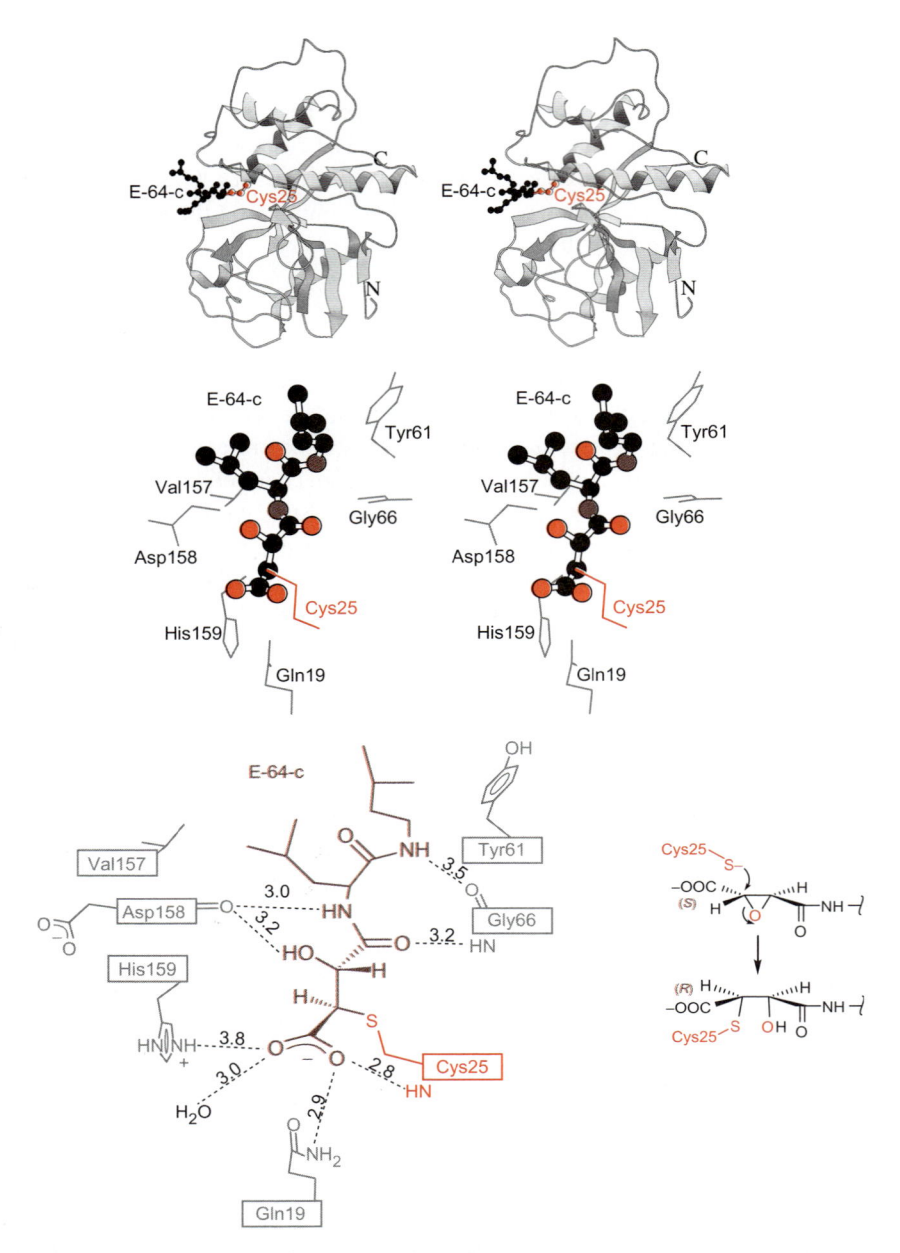

Fig. 6-7. The crystallographic structure (PDB 1PE6; Yamamoto et al., 1991) of papain shows that the molecule is folded into an N-terminal domain consisting of residues 15 to 111 and a C-terminal domain consisting of residues 112 to 207 as shown in stereo (top). The N-terminal domain is substantially helical, and the C-terminal portion is folded around a sheet. The two domains are bound together by crossover residues 1 to 14 interacting with the C-terminal domain and 208 to 212 interacting with the N-terminal domain, as well as the intervening residues 111 to 112. The active site, shown in stereo in more detail (middle) and schematically (bottom), is formed by Cys25 in the N-terminal lobe and His159 in the C-terminal lobe, with His159 hydrogen bonded to Asn175 (not shown). The enzyme is shown covalently modified by a nonspecific cysteine protease inhibitor E-64-c, which reacts by the mechanism shown in the right side of the bottom.

as a thiolate-imidazolium ion pair, so that acylation proceeds by simple addition of Cys25-thiolate to the acylcarbonyl group of a substrate? Alkylation of Cys25 by chloroacetamide and iodoacetamide suggested the latter (Polgar, 1973). Alkylation proved to be very fast, more than ten times faster than alkylation of cysteine. Moreover, alkylation proceeded with no solvent kinetic isotope effect, indicating that base catalysis of proton withdrawal from Cys25 did not accompany alkylation. On this basis, it was proposed that the thiolate-imidazolium ion pair of Cys25 and His159 reacted with the alkylating agents.

Potentiometric titrations of the pH dependence of proton release on removal of the methylthio group from methylthiol-papain (E-Cys25-S-SCH$_3$) by dithiothreitol (DTT) support the presence of the thiolate-imidazolium pair as the dominant species in the active site (Lewis et al., 1976). Similar results on methylthiolation by methanethiolsulfonate lead to the same conclusion. The results indicate that the ionizations of Cys25 and His159 are linked, such that the pK_a of the Cys25 shifts from 3.3 to 7.6 on deprotonation of His159 and the pKa of His159 shifts from 4.3 to 8.5 when Cys25 is deprotonated. Cys25 and His159 appear to exist predominantly as an ion pair, with $K_{eq} \approx 8$ to 12 (fig. 6-8). In microequilibria such as in fig. 6-8, the individual equilibrium constants cannot be rigorously assigned by measurements of proton release or uptake because of the absence of structural information about the origin of the protons. In the case of papain, additional information is available by NMR. The ionization of His159 in methylthiol-papain, as monitored by the NMR signal of imidazole-C2(H), displays a pK_a of 3.9 at 45°C, and it is completely ionized at pH 4 on demethylthiolation at pH 4 (Johnson et al., 1981). This behavior of methylthiol-papain as a model for papain-Cys25-SH supports the interpretation that Cys25 and His159 form an ion pair. NMR titration of His159 in fully active, succinylated papain indicated a pK_a of 8.6, further supporting the model (Lewis et al., 1981). Succinylation of papain maintained solubility for NMR titrations at high pH values. Notwithstanding that methylthiol protection of Cys25 constitutes an imperfect model for Cys25-SH (Migliorini and Creighton, 1986), the results compel the conclusion that Cys25 and His159 exist largely as an ion pair in the physiological pH range.

Unlike Ser195 in chymotrypsin, the Cys25-thiolate ion in papain is a strong nucleophile and can undergo nucleophilic addition to an acylcarbonyl group in the acylation of papain without base catalysis. His159 can serve as a proton donor to the leaving group in the decomposition of the tetrahedral intermediate. Such a mechanism for the action of papain is shown in fig. 6-9. This mechanism is supported by the proton inventory (Theodoru et al., 2001).

Caspases

The family of calcium-binding cysteine proteases have become known as *caspases*. They differ from papain in chain fold and are unrelated in amino acid sequence. Unlike the papain family, the active site histidine precedes the cysteine in the amino acid sequence, and there is no asparagine or glutamine hydrogen bonded to histidine in the active site. Available evidence indicates that the action of caspases follows the same mechanistic principles as papain. The active site cysteine and histidine functions as an ion pair, in which the highly nucleophilic cysteine-thiolate serves as the nucleophilic catalyst.

Aspartic Proteases

Several families of proteases have two aspartic acid residues in the active site that catalyze the hydrolysis of peptide bonds. The first aspartic protease to be recognized is the gastric enzyme pepsin (EC 3.4.23.1), which initiates the digestion of proteins in animals.

318   Enzymatic Reaction Mechanisms

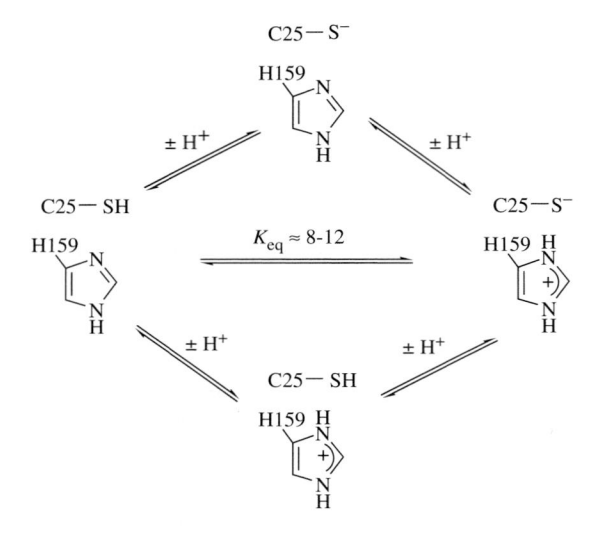

Fig. 6-8. Ionizations at the active site of papain. Potentiometric titrations and proton nuclear magnetic resonance analysis of the ionizations of His159 show that the dominant form of papain at physiological pH values is the ion pair at the right. The thiolate ion in this pair undergoes acylation in the mechanism, and His159 donates a proton to the leaving group in the breakdown of the tetrahedral intermediate (see fig. 6-9).

Homologs of pepsin are found in lysosomes, plants, and fungi. HIV protease (EC 3.4.23.16) is a member of another family of aspartic proteases that play essential roles in retroviral assembly. The structure of pepsin shows two major domains in the form of lobes, each of which contributes one of the aspartic acids in the active site at the domain interface. The secondary structures of the domains consist mainly of β-sheets, with a few short helical segments. Sequence homologies between the domains suggest that they are genetically related through gene duplication (Tang et al., 1978). HIV protease consists of identical 99-amino acid subunits that associate into a dimer, with each subunit contributing an aspartic acid to the active site. Much is known about the specificities and structures of these enzymes, but little is known about the details of the mechanism by which the two aspartic acid residues catalyze peptide hydrolysis.

Molecular Properties

Pepsin

Pepsin was the first enzyme to be identified and the second to be crystallized. The porcine enzyme is a 34.6-kDa protein composed of 324 amino acids. Pepsin is active from pH 1 to pH 6, with maximal activity at about pH 3.2 (Northrop, 2000). It displays a broadly based protease activity that cannot be characterized as specific, although certain sequence motifs are more susceptible than others. Pepsin acts very slowly on small peptide-like substrates but is very active on slightly longer substrates. Substrates with hydrophobic residues in the cleavage site (either side) seem to be preferred and are rapidly cleaved when flanked by two or more other residues (Fruton et al., 1976). A good chromogenic substrate is Lys–Pro–Ala–Glu–Phe–*p*-nitrophenylalanyl–Arg–Leu (Dunn et al., 1986).

The aspartic acids at the active site were identified by chemical modification, using affinity labeling agents. Asp32 underwent alkylation by the epoxy group of 1,2-epoxy-3-(*p*-nitrophenoxy)propane (Chen and Tang, 1972). Diazoacetylnorleucine ethyl ester

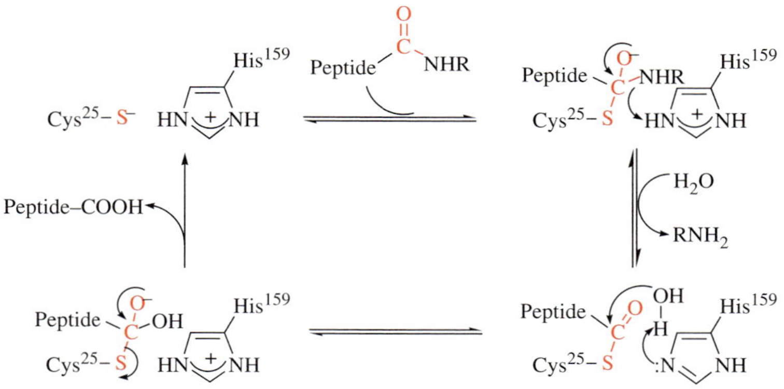

Fig. 6-9. A mechanism for the action of papain. In papain, Cys25 acts as the nucleophilic catalyst to cleave peptide bonds in substrates. Cys25 exists as its thiolate anion in an ionic interaction with the imidazolium group of His159, which is protonated at neutral pH. The thiolate ion is highly reactive and undergoes nucleophilic addition to the acylcarbonyl group of a peptide substrate to form a tetrahedral intermediate. Proton transfer from His159 to the leaving group facilitates its departure as RNH_2, with concomitant formation of the acyl-papain intermediate. His159 then abstracts a proton from a molecule of water as it undergoes nucleophilic addition to the acyl-papain. In the last step, the resulting tetrahedral intermediate decomposes to release the peptide product and regenerate active papain. This mechanism differs from that for the action of chymotrypsin, in that Cys25 is initially deprotonated in the resting enzyme, whereas in chymotrypsin a proton must be abstracted from Ser195 by His57 in the transition state for the formation of the tetrahedral intermediate.

reacted specifically with Asp215 in transesterification (Lundblad and Stein, 1969). Both chemical modifications were blocked by substrates.

Pepstatin, a potent inhibitor of pepsin and related enzymes, is isovaleryl–L-Val–L-Val–(3-hydroxy-4-amino-5-methyl)heptanoyl–L-Ala–(3-hydroxy-4-amino-5-methyl)heptanoic acid, shown in fig. 6-10. Pepstatin has been regarded as a transition-state analog (Marciniszyn et al., 1976). The inhibitor is engaged in a large number of protein contacts in its complex with pepsin, and these interactions may account for tight binding.

Human Immunodeficiency Virus Protease

In retroviral assembly, polyproteins are hydrolytically cleaved into viral proteins by action of viral proteases. Human immunodeficiency virus (HIV) protease performs this essential function in the assembly of the virus. It is essential for viral replication, and for that reason, it is an important drug target (Wlodawer, 1993).

Significant sequence homology with pepsin, together with the presence of the conserved motif Asp–Thr–Gly, suggested that retroviral proteases might be aspartic proteases (Toh et al., 1985). Site-directed mutagenesis of the conserved Asp25 in HIV protease confirmed its importance (Kohl et al., 1988; Seelmeier et al., 1988), and inhibition of retroviral proteases by pepstatin strengthened the analogy with pepsin. The structure of Rous sarcoma virus protease confirmed that the active site of the dimeric enzyme included the conserved aspartate residue from each subunit (Miller et al., 1989).

Limited supplies of purified HIV protease hampered structural studies. The 99–amino acid protein was chemically synthesized (Dawson and Kent, 2000) and crystallized to obtain the structure (Wlodawer et al., 1989), which proved to be the same as that of the recombinant enzyme (Lapatte et al., 1989; Navia et al., 1989). A structure of HIV protease shown in fig. 6-11 shows two aspartate residues at the active site. Many programs of inhibitor

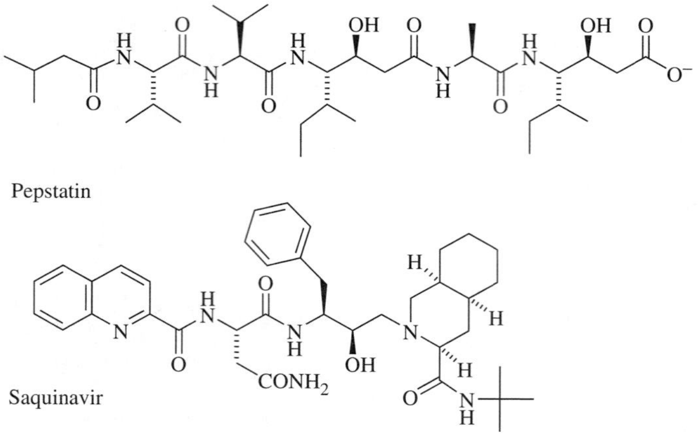

Pepstatin

Saquinavir

Fig. 6-10. Inhibitors of aspartic proteases. Pepstatin is a natural product from actinomycetes that inhibits pepsin and related enzymes, with inhibition constants in the nanomolar range. Saquinavir is a synthetic inhibitor of HIV protease (Roberts et al., 1990) that has been used in the treatment of HIV infection.

development have led to the determination of hundreds of structures of HIV protease with various compounds in the active site. The structures of HIV protease-inhibitor complexes have guided the development of new inhibitors (Wlodawer, 1993). An inhibitor that displays a subnanomolar IC_{50} is saquinavir, shown in fig. 6-10 (Roberts et al., 1990), has been used in the treatment of HIV disease.

Mechanism of Action

Despite pepsin being the first enzyme to be discovered and despite the all-out effort to develop inhibitors of HIV protease, remarkably little is known of the mechanism of action of aspartic proteases. In the conventional wisdom, the two aspartic acid side chains in the active site function together to catalyze nucleophilic attack by a water molecule on the carbonyl group of the scissile bond. Exactly how two aspartic acid residues can accomplish this feat is not known. The addition of a water molecule to a peptide carbonyl group requires the abstraction of a proton from the water molecule to generate an incipient hydroxide ion in the transition state. Water is a very weak acid (pK_a = 15), so that a strong base should be required to abstract a proton. How this is accomplished by an aspartate-β-carboxylate group, a very weak base (pK_a = 4), is a continuing puzzle.

Acid-Base Mechanism

The structures of aspartic proteases implied that the two aspartic acid side chains might engage in acid-base catalysis to facilitate peptide cleavage by water. In the mechanism of fig. 6-12A the two aspartic acid residues function separately as an acid and base, respectively (Davies, 1990). As in all mechanisms put forward for aspartic proteases, the question of how the weakly basic β-carboxylate group of aspartate abstracts a proton from water remains an enigma.

Iso-Mechanism

In enzymatic reaction mechanisms, the starting state of the free enzyme is regenerated at the conclusion of the reaction. The mechanism in fig. 6-12A is oversimplified because it

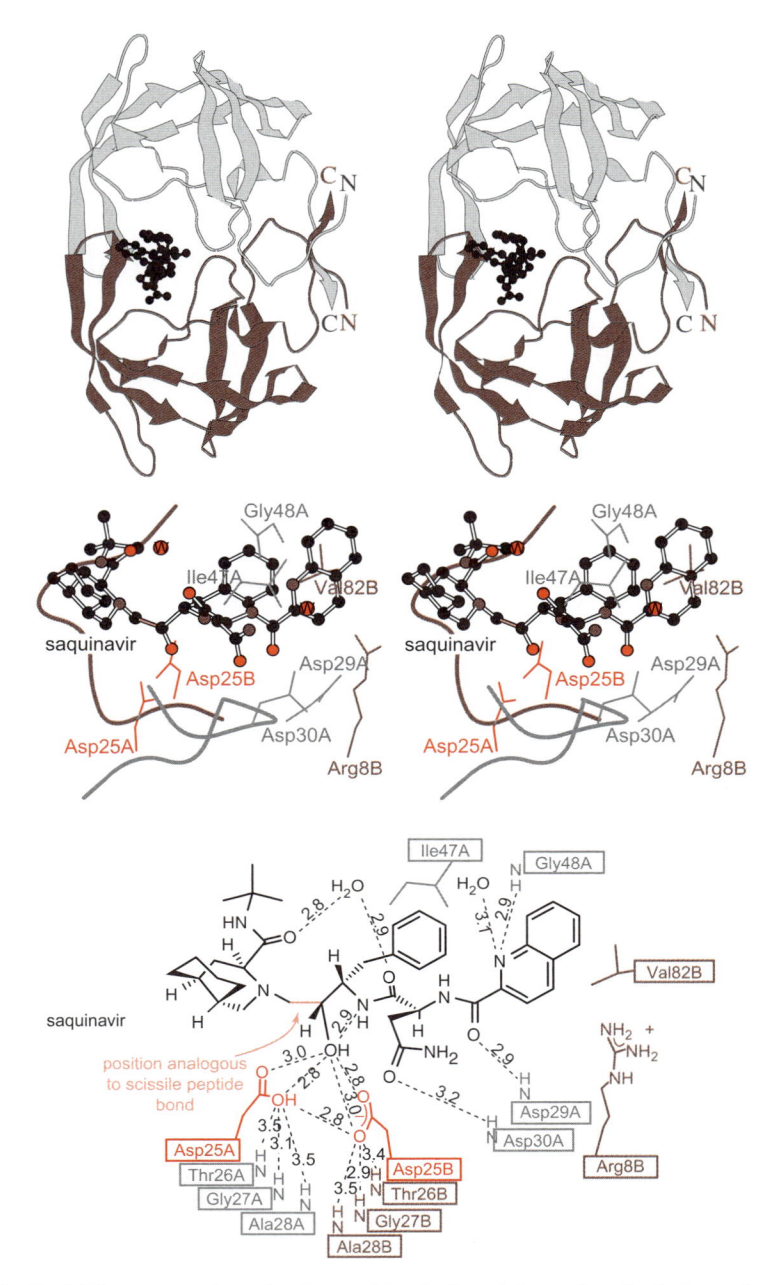

Fig. 6-11. In HIV protease, the subunits are identical, and the active site is formed at the interface so that the two essential aspartic acid residues are each part of a highly conserved Asp-Thr-Gly motif donated by each subunit. The aspartic acid residues are the same numbered residue from each subunit and are designated Asp25A from one subunit and Asp25B from the other. The inhibitor saquinavir is shown bound in the active site with the analogous bond to that which would be cleaved in a substrate highlighted (PDB 1HXB; Krohn et al., 1991).

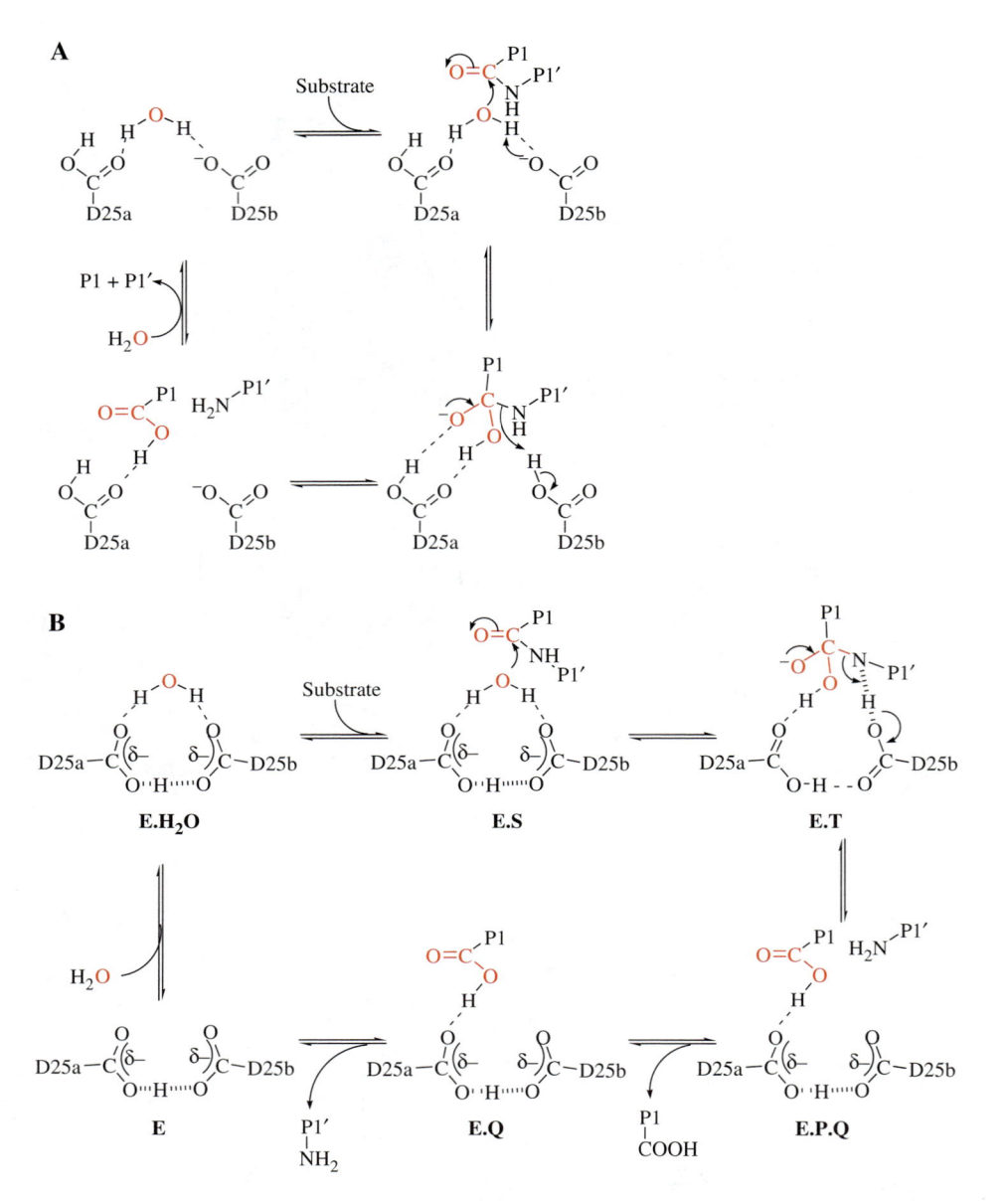

Fig. 6-12. In the two mechanisms for the action of two aspartic acids in the hydrolysis of a peptide bond by an aspartic protease, Asp25A and Asp25B are numbered as in HIV protease; the corresponding residues in pepsin are Asp32 and Asp215, respectively. (A) In this mechanism, Asp25B functions as a base to abstract a proton from water, and Asp25A donates a hydrogen bond to the tetrahedral intermediate resulting from the addition of water to the acylcarbonyl group of the peptide. Now protonated, Asp25B donates a proton to the leaving amino group of the tetrahedral intermediate. This mechanism is inspired by the observation of a water molecule between the aspartic acids in the structure of pepsin. (B) In this mechanism, the two aspartic acid side chains are linked by a low-barrier hydrogen bond (LBHB), and the two residues function through the LBHB to abstract a proton from water in the formation of the tetrahedral intermediate. The order of product release is unknown, and that shown here is arbitrary.

implies that the dissociation of the two products and the binding of a water molecule in preparation for the next catalytic cycle are trivial processes. These steps do not involve chemical changes, but they are kinetically significant. The actions of pepsin and HIV protease have been shown to follow iso-mechanisms; that is, the dissociation of the last product leads to a different state of the free enzyme than the one that initially binds a substrate molecule (Northrop, 2000). This means that the free enzyme at the end of a catalytic cycle must undergo an isomerization to the substrate-binding state before resuming the next catalytic cycle. The isomerization may entail processes such as a conformational change, a reorganization of protons among catalytic groups, or a change in solvation. Iso-mechanisms are increasingly recognized, and we discuss this in more detail in chapter 7 for the case of proline racemase.

Iso-mechanisms are manifested in various ways in different enzymes. In the case of HIV protease, a substrate is converted into two products. Product dissociation may be ordered or random, and if it leads directly to the substrate binding state of the enzyme at least one product must inhibit competitively with respect to the substrate according to Cleland's rules (see chap. 2). However, in the kinetics for pepsin and HIV protease, *both* products are noncompetitive inhibitors (Northrop, 2000). The free enzyme formed at the end of a catalytic cycle cannot be the same as the substrate binding form, and an isomerization process must intervene between catalytic cycles. It has been postulated that two isomerization processes take place between catalytic cycles in the action of aspartic proteases (Northrop, 2000).

Postulated Low-Barrier Hydrogen Bond

Simulations of the active site of HIV protease in molecular dynamics computations indicate the formation of an LBHB between the two aspartic acid side chains (Northrop, 2000; Piana and Carloni, 2000; Trylska et al., 2004). The LBHB is not short enough to be symmetrical, but the bond linking Asp25-Hδ1 is significantly elongated, and the contact distance between Asp25-Oδ1 and Asp125-Oδ2 is very short (2.5 Å). In one study of molecular dynamics, a very short hydrogen bond appeared between Asp-Oδ2 and the amine leaving group in the tetrahedral intermediate, compatible with LBHB formation (Trylska et al., 2004).

Experimental evidence for an LBHB is limited to the observation of low solvent D/H fractionation factors in proton inventory experiments (Northrop, 2000). One of the characteristics of an LBHB is a low D/H fractionation factor. A mechanism incorporating the LBHB in the aspartic acid dyad is shown in fig. 6-12B. This is a simplified mechanism that accounts for available information, in which the iso-step is a solvation change including the binding of a water molecule. Both Asp25 and Asp125 are protonated in the tetrahedral intermediate, which discharges the LBHB between them; however, this may be compensated by LBHB formation between Asp125 and the leaving amine, as suggested in molecular dynamics simulations (McCammon, 2004; Trylska et al., 2004). More complex mechanisms with additional steps have been put forward (Northrop, 2000; Trylska et al., 2004).

Metalloproteases

The largest family of proteolytic enzymes consists of metalloproteases (Barrett et al., 1998). The mononuclear zinc metalloproteases dominate this family, which also includes binuclear metalloproteases, and some of the latter enzymes contain manganese or cobalt. The metal ion at the active site of a metalloprotease is thought to bind the water molecule that will cleave the peptide bond by nucleophilic reaction with the peptide carbonyl group.

324 Enzymatic Reaction Mechanisms

Coordination of water with a metal ion lowers the value of its pK_a and eases the removal of a proton to generate a hydroxide ion in the transition state. In the coordination of water with Zn^{2+} in a tetrahedral complex, the pK_a of water is decreased from 15 to 9.6 (see table 4-3 in chap. 4). This significantly facilitates the abstraction of a proton from the coordinated water molecule by a weak base at the active site, to generate an incipient hydroxide ion in the transition state. Two extensively studied metalloproteases are the carboxypeptidase A and thermolysin. We discuss carboxypeptidase A in greater detail later. As in the case of chymotrypsin, the current theory of the action of carboxypeptidase A depends on the fusion of information obtained by biochemical, kinetic, crystallographic, and spectroscopic methods.

Carboxypeptidase A

An exopeptidase catalyzes the hydrolytic removal of the N-terminal or C-terminal amino acid from a peptide. Leucine aminopeptidase is a binuclear metallopeptidase that removes N-terminal amino acids (Barrett et al., 1998). Carboxypeptidase A (EC 3.4.17.1) is a mononuclear zinc metallopeptidase that removes C-terminal amino acids, with high selectivity for hydrophobic C-terminal amino acids. Carboxypeptidase A also catalyzes the hydrolysis of esters and thioesters.

Kinetics

The early efforts at steady-state kinetic analysis of the action of carboxypeptidase A were hampered by the selection of dipeptides as substrates. These molecules gave variable and complex results in kinetic studies (Hartsuck and Lipscomb, 1971). Tripeptide substrates had typical Michaelis-Menten behavior and were adopted in detailed steady-state and transient-phase kinetic studies.

Variation of pH affects the activity of carboxypeptidase A in a similar way as in the action of chymotrypsin. Plots of log k_{cat}/K_m against pH are bell-shaped curves, with the ascending leg demonstrating the pH dependence of k_{cat} and the descending leg demonstrating the pH dependence of K_m (Auld and Vallee, 1970). The value of pK_{EH_2} on the acid side is 6.2, and pK_{EH_2S} for the descending limb is 9.0. The pH dependence corresponds to the kinetic model of scheme 6-4.

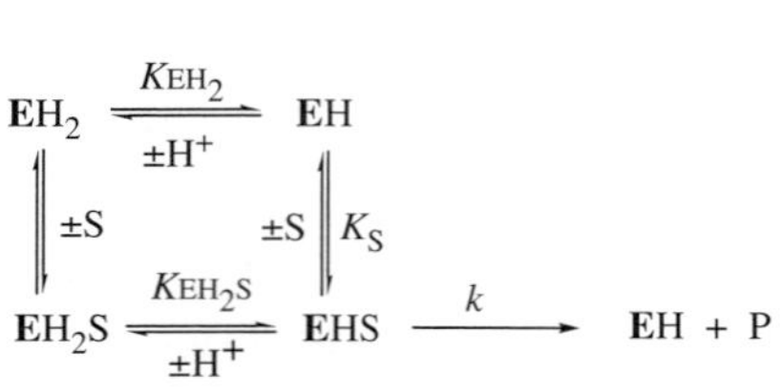

$$EH_2 \xrightleftharpoons[\pm H^+]{K_{EH_2}} EH$$
$$\updownarrow \pm S \qquad \pm S \updownarrow K_S$$
$$EH_2S \xrightleftharpoons[\pm H^+]{K_{EH_2S}} EHS \xrightarrow{k} EH + P$$

Scheme 6-4

The value of pK_{EH_2} varies with the metal ion in the active site, ranging from 5.3 for Co^{2+} to 6.4 for Mn^{2+}. Because Glu270 is in the active site and near the metal ion, it is regarded as the group ionizing with a pK_a of 6.2 in zinc-carboxypeptidase A.

The crystallographic structure indicated that the descending limb could be due to the ionization of Tyr248, which was thought to be the acid catalyst for breakdown of the tetrahedral intermediate (Hartsuck and Lipscomb, 1971). However, biochemical evidence showed that mutation of Tyr248 to Phe248 had very little effect on activity, and it did not abolish the descending limb of the pH-rate profile, ruling out Tyr248 as the ionizing group and as a catalytic group (Hilvert et al., 1986). Spectroscopic evidence shed some light on the pK_a of 9.0. The XAFS spectra between pH 7 and 11 revealed ionization in the Zn^{2+} coordination sphere, with a pK_a value exactly the same as the kinetic pK_a of 9.0 (pK_{EH2S}) (Zhang and Auld, 1993). This ionization may correspond to the group ionizing at the descending leg of the pH–log k_{cat}/K_m profile. The proposed ionizing group is the water molecule ligated to zinc. Ionization to a hydroxide ligand would alter the electrostatics at the active site and thereby interfere with substrate binding, which is consistent with the fact that the descending limb represents the effect of pH on K_m.

A detailed transient kinetic analysis conducted at low temperatures revealed the participation of two intermediates in the steady state, neither of which was a covalent acyl-enzyme (Galdes et al., 1983,1986). Kinetics at subzero temperatures was made possible by conducting stopped-flow kinetics at 4.5 M NaCl, with fluorescence detection of reaction steps. The kinetic mechanism in scheme 6-5 was identified for the hydrolysis of both peptides and esters.

$$E \underset{k_2}{\overset{k_1}{\rightleftharpoons}} ES_1 \underset{k_4}{\overset{k_3}{\rightleftharpoons}} ES_2 \overset{k_5}{\longrightarrow} E + P_1 + P_2$$

Scheme 6-5

Kinetic parameters and rate constants for the hydrolysis of a peptide and an ester substrate are given in table 6-6. The results show that the kinetics for paired peptide and ester substrates are different. Chemical analysis of the enzyme-bound intermediates revealed that at the steady state, the peptide substrate bound to the active site was not cleaved, so that both ES_1 and ES_2 were Michaelis complexes, presumably in different conformations. In contrast, the ester substrate was cleaved in the steady state, so that ES_2 had to be the product complex EP_1P_2. This is reasonable given the higher ($\approx 10^4$) chemical reactivity of esters with hydroxide. The two Michaelis complexes of the peptide substrate might have been stereoelectronic isomers, in which the peptide bond was twisted, forcing the nonbonding electrons of the leaving group out of the plane of the carbonyl group. This would decrease the resonance stabilization of the peptide bond and increase the electrophilic reactivity of the carbonyl group.

Peptide cleavage is thought to limit the rate in the hydrolysis of all peptides. In ester hydrolysis, cleavage precedes the rate-limiting step, and the substitution of Co^{2+} for Zn^{2+} affects the value of k_{cat}/K_m but not k_{cat} (Auld, 1987; Auld and Vallee, 1987).

Structure

Crystallographic and chemical modification experiments implicate Glu270 and Arg145 in the action of carboxypeptidase A (Christianson and Lipscomb, 1989; Hartsuck and Lipscomb, 1971; Lipscomb, 1973; Nau and Riordan, 1975; Quiocho and Lipscomb, 1971; Reeke et al., 1967; Riordan, 1973). The structure clearly defines Arg145 in contact with the substrate carboxylate, as shown in fig. 6-13 with a transition-state analog bound at the

Table 6-6. Kinetic Constants for Hydrolysis of Peptide and Ester Substrates by Carboxypeptidase A

Substrate	K_m^a (μM)	k_{cat}^a (s^{-1})	k_3 (s^{-1})	k_4 (s^{-1})	k_5 (s^{-1})	K_S^a (μM)
Dansyl–Ala–Ala–Phe[b]	13	1.2	40	3.5	1.3	102
Dansyl–Ala–Ala–Ophe[c]	1.6	0.06	53	0.5	0.06	129

[a]$K_S = k_2/k_1$; $K_m = K_S(k_4 + k_5)/(k_3 + k_4 + k_5)$; $k_{cat} = k_3k_5/(k_3 + k_4 + k_5)$.
[b]–20°C.
[c]–10°C.

326   Enzymatic Reaction Mechanisms

active site. The structure also shows close contacts between the inhibitor and Tyr248, Asn144, and Arg71. Biochemical evidence had indicated that the ligands to Zn^{2+} included a cysteine residue; however, the crystal structure showed the ligands to be His69, Glu72, His196, and a water molecule.

In the mechanism shown in fig. 6-14, Zn^{2+} binds a water molecule in position to undergo nucleophilic addition to the peptide carbonyl group. Water is not nucleophilic enough to react at a significant rate, but ligation to Zn^{2+} increases its acidity by shifting its pK_a from 15.5 to less than 10 (see table 4-3 in chap. 4). The nearby Glu270 assists the

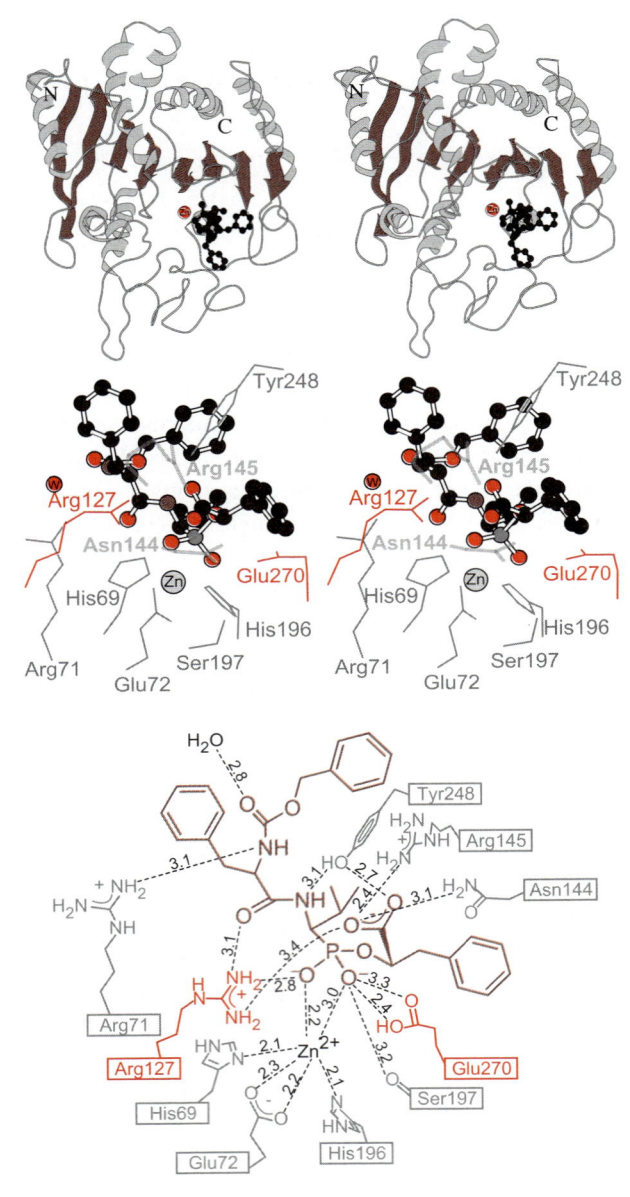

Fig. 6-13. The structure of carboxypeptidase A shows a phosphonate inhibitor at the active site. The overall fold of bovine carboxypeptidase A (top) is shown with the zinc atom in red and the phosphonate inhibitor (Bz-Phe-Val^P(O)-Phe) shown in black (PDB 7CPA; Kim and Lipscomb, 1991). A closer view of the active site (middle) highlights the bidentate interaction of the phosphonate inhibitor with the active site zinc. This is also shown schematically (bottom) with interatomic distances given in angstroms over dashed lines.

Acyl Group Transfer: Proteases and Esterases 327

ionization of the water ligand by serving as a base to generate hydroxide, a highly reactive nucleophile, in the transition state for addition to the peptide carbonyl. The resulting tetrahedral intermediate can revert to the substrate or eliminate the amine to generate the cleavage products.

Inhibition by Peptide Phosphonates

Analogs of substrates with a phosphonate in the scissile position are potent inhibitors of carboxypeptidase A (Phillips et al., 1992). These peptide phosphonates are regarded as analogs of the tetrahedral intermediate in the mechanism of fig. 6-14. The structures are compared in scheme 6-6.

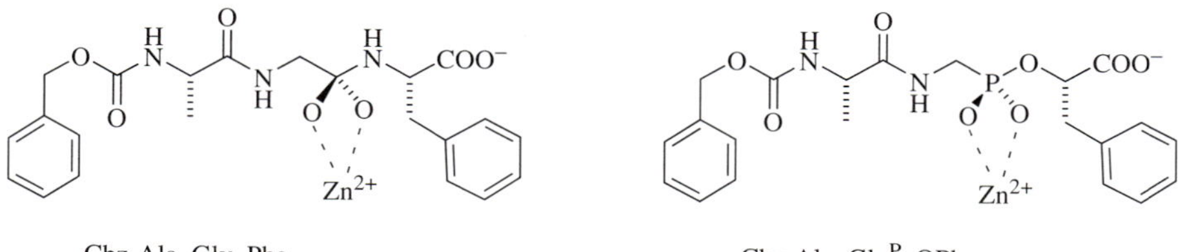

Cbz-Ala–Gly–Phe   Cbz-Ala–Gly^P–OPhe

Scheme 6-6

Mutation of Arg127 to lysine, methionine, or alanine decreases k_{cat}/K_m by factors of 1600 to 137000, depending on the substrate and mutation, implicating Arg127 in catalysis. These mutations also increase the values of K_i for the corresponding phosphonate inhibitors. For a given inhibitor, the degree to which the inhibition constant is increased by mutation of Arg127 corresponds to the degree to which the same mutation decreases the value of k_{cat}/K_m for the corresponding substrate. The graph in fig. 6-15 illustrates this effect for two inhibitor/substrate pairs. These results support the importance of Arg127 in binding the tetrahedral intermediate and, based on the Hammond postulate, the transition state.

Thermolysin

The zinc-endopeptidase thermolysin (34.6 kDa; EC 3.4.24.27) has an entirely different tertiary structure from that of carboxypeptidase A (Holden et al., 1987). The Zn^{2+} ligands

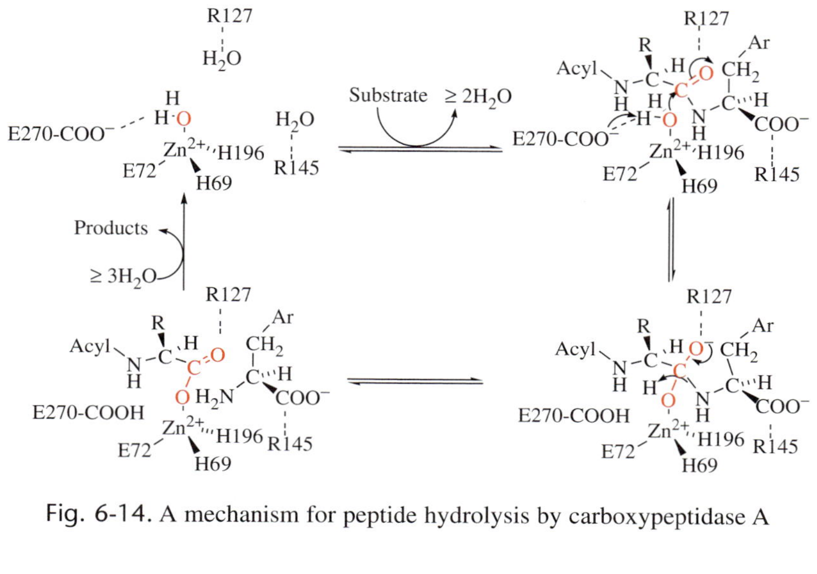

Fig. 6-14. A mechanism for peptide hydrolysis by carboxypeptidase A

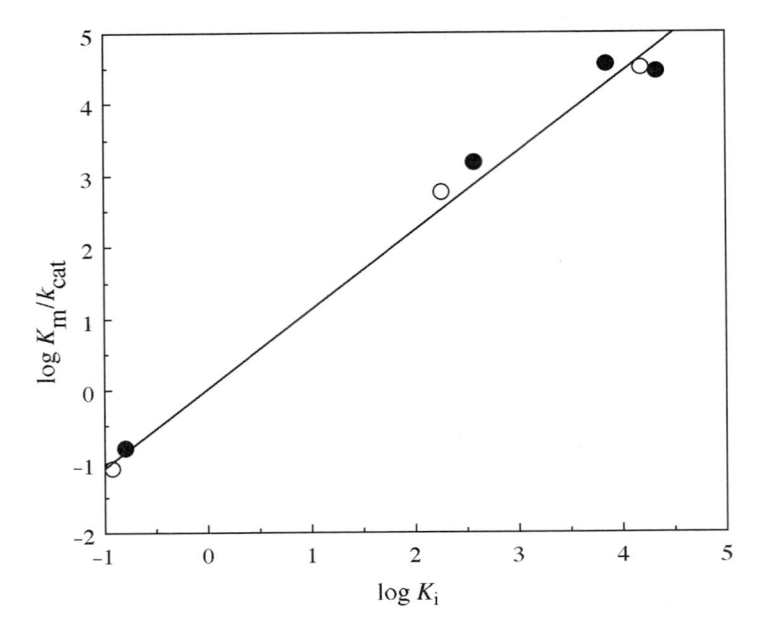

Fig. 6-15. Correlation of K_m/k_{cat} for two substrates of carboxypeptidase A with K_i for analogous phosphonate inhibitors. Values of $\log(K_m/k_{cat})$ for wild-type carboxypeptidase A and variants R127K, R127M, and R127A are plotted against $\log K_i$ for the corresponding phosphonate inhibitors or wild type and the same variants. The substrate and inhibitor (closed symbols) are CbzAlaGlyPhe, and CbzAlaGly^{P}Ophe, respectively, where Gly^P is the phosphonate analog of glycine. The substrate and inhibitor (open symbols) are CbzPheGlyPhe and CbzPheGly^{P}OPhe, respectively. (Data from Phillips et al., 1992).

are His142, His146, and Glu166, and the catalytic group is Glu143. The mechanism of action is similar to that of carboxypeptidase A, based on the structure of thermolysin and its interactions with transition-state analogs. Peptide phosphonates are potent inhibitors, and a linear relationship between $\log K_I$ and $\log(K_m/k_{cat})$ analogous to that in fig. 6-15 can be constructed, in which the structures of the inhibitors are varied (Bartlett and Marlowe, 1983).

Esterases

Structure and Function

In chapter 5, acetylcholinesterase is discussed in connection with inhibition and is related mechanistically to the serine proteases. Many esterases have serine residues that function as nucleophilic catalysts in the hydrolysis of esters. The hydrolysis of an ester is simpler and easier than the hydrolysis of an amide or peptide: an ester is intrinsically more reactive, by a factor of 10^4 for reaction with hydroxide ion. Perhaps for this reason, esterases depend less on an LBHB between histidine and glutamate or aspartate at the active site. Some esterases function without a catalytic triad analogous to that in chymotrypsin.

The active site of acetylcholinesterase provides three hydrogen bonds to the oxyanionic tetrahedral adduct between Ser200 and an ester substrate, unlike the two in chymotrypsin (see chap. 5). The LBHB in between His440 and Glu327 of acetylcholinesterase has been characterized and appears somewhat upfield from that in chymotrypsin (Massiah et al., 2001; Viragh et al., 2000). In the following section, we discuss an important esterase in which an LBHB unlike that in chymotrypsin has been observed in an inhibited complex.

Phospholipase A$_2$

Phospholipid metabolism includes hydrolysis of the *sn*-2 fatty acyl group to form the lysophospholipids (eq. 6-9). Phospholipase A$_2$s are 13- to 18-kDa, Ca^{2+}-dependent esterases.

$$\text{R}_2\text{-C(=O)-O-CH}_2\text{-CH(O-C(=O)-R}_1\text{)-CH}_2\text{-O-P(=O)(O}^-\text{)-OCH}_2\text{CH}_2\text{NH}_3^+ + \text{H}_2\text{O} \longrightarrow$$

$$\text{HO-CH}_2\text{-CH(O-C(=O)-R}_1\text{)-CH}_2\text{-O-P(=O)(O}^-\text{)-OCH}_2\text{CH}_2\text{NH}_3^+ + \text{R}_2\text{-C(=O)-OH}$$

(6-9)

The unique kinetics of the action of phopspholipases at phospholipid-water interfaces is reviewed in (Berg et al., 2001). Unlike serine esterases, there is no serine in the active site of a phospholipase A$_2$, and water reacts directly with the *sn*-2 ester group of the substrate. There is a histidine and aspartate pair in hydrogen bonded contact and in position to accept a proton from the attacking water molecule as it adds to the substrate ester. The His-Asp contact differs significantly from that in chymotrypsin, in that His48-N$^{\varepsilon 2}$ is hydrogen bonded to an *anti*-lone pair of Asp99 (structure **6-2**), unlike the contact of His57-N$^{\delta 1}$ with the *syn*-lone pair of Asp102 in chymotrypsin (see fig. 6-4). This is a significant difference because the *syn*- and *anti*- conformations of carboxylic acids display significantly different acidities, and relative proton affinity is an important determinant of low barrier hydrogen bonding.

6-2 H-O-H-----N$^{\delta}$(His48)N$^{\varepsilon}$H----O=C(Asp99)-O

syn-2-Phosphonate analogs of phospholipids inhibit phospholipase A$_2$ and are styled as transition-state analog inhibitors. The complex of the inhibitor MG14 with phospholipase A$_2$ has been structurally and spectroscopically characterized (Poi et al., 2003; Scott et al., 1990; Sekar et al., 1998). The complexes with bee venom and bovine phospholipases display downfield signals in the proton NMR spectra (17.6 and 18 ppm). Based on spectroscopic and H/D fractionation data, as well as exchange rates, an LBHB was assigned to the contact between a peripheral phosphonyl oxygen and His48-N$^{\varepsilon 2}$ (Poi et al., 2003). The structure at the active site is illustrated in fig. 6-16.

The significance of the LBHB in phospholipase A$_2$ remains uncertain. It certainly represents structural crowding in the active site. There could well be an LBHB between the attacking water molecule and His48 in the transition state of the reaction. On the approach to the transition state, the pK_a of water decreases, and that of His48 increases; if they cross at the transition-state, low-barrier hydrogen bonding is possible. Whether there would be an LBHB in the tetrahedral adduct is unclear. The LBHB in the phosphonate complex implies similarity between the pK_a values of His48 and the phosphonate. The pK_a of a phosphonate would be about 3 in water and higher in a less polar medium, so a pK_a match is likely in an apolar medium. The measured pKa for His48 in the complex is 9.0, three units higher than in the free enzyme (Poi et al., 2003). A tetrahedral adduct of water to an ester should display a pK_a of 11 or 12 in water, somewhat higher than that of His48 in the phosphonate complex.

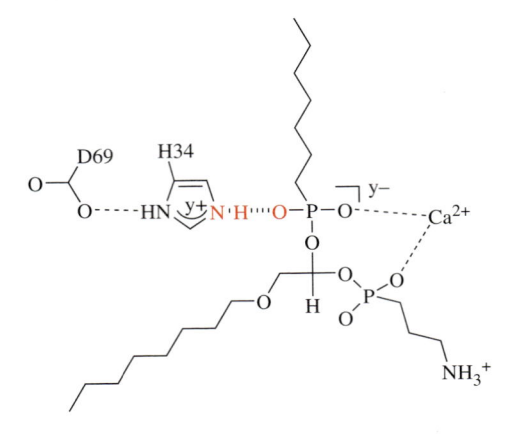

Fig. 6-16. Contacts of inhibitor MG14 at the active site of bee venom phospholipase A2. MG14 is regarded as a transition-state analog of the tetrahedral intermediate in the action of phospholipase A_2. Both phosphonate groups are coordinated to calcium, and the peripheral oxygen of the central phosphonate in the low-barrier hydrogen bond contact with His34 corresponds to the attacking water molecule in the tetrahedral intermediate.

Interactions of the tetrahedral intermediate with Ca^{2+} could well alter this picture and allow a lower value for the adduct, and this would tend to optimize hydrogen bonding.

References

Adebodun F and F Jordan (1989) *J Cell Biochem* **40**, 249.
Auld DS (1987) In Page M and A Williams (eds): *Enzyme Mechanisms*. Royal Society of Chemistry: London, p 240.
Auld DS and BL Vallee (1970) *Biochemistry* **9**, 4352.
Auld DS and BL Vallee (1987) In Neuberger A and K Brocklehurst (eds): *Hydrolytic Enzymes*. Elsevier: Amsterdam, p 201.
Barrett AJ, ND Rawlings, and JF Woessner, eds (1998). *Handbook of Proteolytic Enzymes*. Academic Press: San Diego, 1998.
Bartlett PA and CK Marlowe (1983) *Biochemistry* **22**, 4618.
Bell IM, ML Fisher, Z-P Wu, and D Hilvert (1993) *Biochemistry* **32**, 3754.
Bender ML (1960) *Chem Rev* **60**, 53.
Bender ML, GE Clement, FJ Kezdy, and H d'A Heck (1964) *J Am Chem Soc* **86**, 3680.
Berg OG, MH Gelb, MD Tsai, and MK Jain (2001) *Chem Rev* **101**, 2613.
Blow DM (1971) In Boyer PD (ed): *The Enzymes*, vol 3, 3rd ed. Academic Press: New York, p 185.
Brady K, A Wei, D Ringe, and RH Abeles (1990) *Biochemistry* **29**, 7600.
Bruice TC, TH Fife, JJ Bruno and NE Brandon (1962) *Biochemistry* **1**, 7.
Bryan P, MW Pantoliano, SG Quill, HY Hsiao, and T Poulos (1986) *Proc Natl Acad Sci U S A* **83**, 3743.
Carter P and JA Wells (1987) *Science* **237**, 394.
Cassidy CS, J Lin, and PA Frey (1997) *Biochemistry* **36**, 4576.
Cassidy CS, J Lin, and PA Frey (2000) *Biochem Biophys Res Commun* **273**, 789.
Chen KCS and J Tang (1972) *J Biol Chem* **247**, 2566.
Davies DR (1990) *Annu Rev Biophys Biophys Chem* **19**, 189.
Dawson PE and SBH Kent (2000) *Annu Rev Biochem* **69**, 923.
Ding X, BF Rasmussen, GA Petsko, and D Ringe (1994) *Biochemistry* **33**, 9285.
Dixon MM and BW Matthews (1989) *Biochemistry* **28**, 7033.
Dunn BM, M Jimenez, BF Parten, MJ Valler, CE Rolf, and J Kay (1986) *Biochem J* **237**, 899.
Epand RM and IB Wilson (1963) *J Biol Chem* **238**, 3138.
Fastrez J and AR Fersht (1973) *Biochemistry* **12**, 2025.
Forsberg G, B Baastrup, H Rondahl, E Holmgren, G Pohl, M Hartmanis, and M Lake (1992) *J Protein Chem* **11**, 201.
Frey PA, SA Whitt, and JB Tobin (1994) *Science* **264**, 1927.

Fruton JS (1976) *Adv Enzymol* **44**, 1.
Galdes A, DS Auld, and BL Vallee (1983) *Biochemistry* **22**, 1888.
Galdes A, DS Auld, and BL Vallee (1986) *Biochemistry* **25**, 646.
Hartley BS and DL Kauffman (1966) *Biochem J* **101**, 229.
Hartley BS and BA Kilby (1954) *Biochem J* **56**, 288.
Hartsuck JA and WN Lipscomb (1971) In Boyer PD (ed): *The Enzymes*, vol 3, 3rd ed. Academic Press: New York, p 1.
Hengge AC and RL Stein (2004) *Biochemistry* **43**, 742.
Hess GP (1971) In Boyer PD (ed): *The Enzymes*, vol 3, 3rd ed. Academic Press: New York, p 213.
Hess RA, AC Hengge, and WW Cleland (1998) *J Am Chem Soc* **120**, 2703.
Hilvert D, SJ Gardell, WJ Rutter, and ET Kaiser (1986) *J Am Chem Soc* **108**, 5298.
Holden HM, DE Tronrud, AF Mozingo, LH Weaver, and BW Matthews (1987) *Biochemistry* **26**, 8542.
Imperiali B and RH Abeles (1986) *Biochemistry* **25**, 3760.
Johnson CH and JR Knowles (1966) *Biochem J* **101**, 56.
Johnson FA, SD Lewis, and JA Shafer (1981) *Biochemistry* **20**, 44.
Karsten WE, C-J Lai, and PF Cook (1995) *J Am Chem Soc* **117**, 5914.
Katona G, RC Wilmouth, PA Wright, GI Berglund, J Hajdu, R Neutz, and CJ Schofield (2002) *J Biol Chem* **277**, 21962.
Kim H and WN Lipscomb (1991) *Biochemistry* **30**, 8171.
Kohl NE, EA Emini, WA Schleif, LJ Davis, JC Heimbach, RA Dixon, EM Scolnick, and IS Sigal (1988) *Proc Natl Acad Sci U S A* **85**, 4186.
Kraut J (1971a) In Boyer PD (ed): *The Enzymes*, vol 3, 3rd ed. Academic Press: New York, p 165.
Kraut J (1971b) In Boyer PD (ed): *The Enzymes*, vol 3, 3rd ed. Academic Press: New York, p 547.
Kresge AJ (1973) *J Am Chem Soc* **95**, 3065.
Kresheck GC, H Schneider, and HA Scheraga (1965) *J Phys Chem* **69**, 3132.
Krohn A, S Redshaw, JC Ritchie, BJ Graves, and MH Hatada (1991) *J Med Chem* **34**, 3340.
Kuhn P, M Knapp, SM Soltis, G Ganshaw, M Thoene, R Bott (1998) *Biochemistry* **37**, 13446.
Ladner JE, JF Parsons, CL Rife, GL Gilliland, and RN Armstrong (2004) *Biochemistry* **43**, 352.
Lapatto R, T Blundell, A Hemmings, J Overington, A Wilderspin, S Wood, JR Merson, PJ Whittle, DE Danley, and KF Geoghagen (1989) *Nature* **342**, 299.
Lecaille F, J Kaleta, and D Brömme (2002) *Chem Rev* **102**, 4459.
Lewis SD, FA Johnson, and JA Shafer (1976) *Biochemistry* **15**, 5009.
Lewis SD, FA Johnson, and JA Shafer (1981) *Biochemistry* **20**, 44.
Liang T-C and RH Abeles (1987) *Biochemistry* **26**, 7603.
Lin J, CS Cassidy, and PA Frey (1998a) *Biochemistry* **37**, 11940.
Lin J, WM Westler, WW Cleland, JL Markley, and PA Frey (1998b) *Proc Natl Acad Sci U S A* **95**, 14664.
Lundblad RL and WH Stein (1969) *J Biol Chem* **244**, 254.
Marciniszyn J Jr, JA Hartsuck, and J Tang (1976) *J Biol Chem* **251**, 7088.
Markland FS and EL Smith (1971) In Boyer PD (ed): *The Enzymes*, vol 3, 3rd ed. Academic Press: New York, p 561.
Markley JL (1978) *Biochemistry* **17**, 4648.
Markley JL and WM Westler (1996) *Biochemistry* **35**, 11092.
Massiah MA, C Viragh, PM Reddy, IM Kovach, J Johnson, TL Rosenberry, and AS Mildvan (2001) *Biochemistry* **40**, 5682.
McConn J, GD Fasman, and G Hess (1969) *J Mol Biol* **39**, 551.
Migliorini M and DJ Creighton (1986) *Eur J Biochem* **156**, 189.
Millard CB, G Kryger, A Ordentlich, HM Greenblatt, M Harel, ML Raves, Y Segall, D Barak, A Shafferman, I Silman, and JL Sussman (1999) *Biochemistry* **38**, 7032.
Miller M, M Jaskólski, JKM Rao, J Leis, and A Wlodawer (1989) *Nature* **337**, 576.
Navia MA, PMD Fitzgerald, BM Mckeever, C-T Leu, JC Heimbach, WK Herbert, IS Sigal, and PL Darke (1989) *Nature* **337**, 615.
Neet KE and DE Koshland (1966) *Proc Natl Acad Sci U S A* **56**, 1606.
Neet KE and DE Koshland (1968) *J Biol Chem* **243**, 6392.
Neidhart D, Y Wei, CS Cassidy, J Lin, WW Cleland, and PA Frey (2001) *Biochemistry* **40**, 2439.
Northrop DB (2000) *Acc Chem Res* **34**, 790.

O'Leary MH and MD Kluetz (1972) *J Am Chem Soc* **94**, 3585.
O'Leary MH, M Urberg, and AP Young (1974) *Biochemistry* **13**, 2077.
Perona JJ and CS Craik (1995) *Protein Sci* **4**, 337.
Phillips MA, AP Kaplan, WJ Rutter, and PA Bartlett (1992) *Biochemistry* **31**, 959.
Piana S and P Carloni (2000) *Proteins* **39**, 26.
Poi MJ, JW Tomaszewski, C Yuan, CA Dunlap, NH Andersen, MH Gelb, and MD Tsai (2003) *J Mol Biol* **329**, 997.
Polgár L (1973) *Eur J Biochem* **33**, 104.
Polgár L and ML Bender (1966) *J Am Chem Soc* **88**, 3153.
Roberts NA, JA Martin, D Kinchington, AV Broadhurst, JC Craig, IB Duncan, SA Galpin, BK Handa, J Kay, A Kröhn, RW Lambert, JH Merrett, JS Mills, KEB Parkes, S Redshaw, AJ Richie, DL Taylor, GJ Thomas, and PJ Machin (1990) *Science* **248**, 358.
Robillard G and RG Shulman (1972) *J Mol Biol* **71**, 507.
Robillard G and RG Shulman (1974) *J Mol Biol* **86**, 541.
Rossi GL and SA Bernhard (1970) *J Mol Biol* **49**, 85.
Schneider H, GC Krescheck and HA Scheraga (1965) *J Phys Chem* **69**, 1310.
Schowen RL (1972) *Prog Phys Org Chem* **9**, 275.
Schwert GW and MA Eisenberg (1949) *J Biol Chem* **179**, 665.
Scott DL, Z Otwinowski, MH Gelb, and PB Sigler (1990) *Science* **250**, 1563.
Seelmeier S, H Schmidt, V Turk, and K von der Helm (1988) *Proc Natl Acad Sci U S A* **85**, 6612.
Sekar K, A Kumar, X Liu, MD Tsai, MH Gelb, and M Sundaralingam (1998) *Acta Crystallogr D Biol Crystallogr* **54**, 334.
Shain SA and JF Kirsch (1968) *J Am Chem Soc* **90**, 5848.
Tang J, MNG James, IN Hsu, JA Jenkins, and TL Blundell (1978) *Nature* **271**, 618.
Toh H, M Ono, K Saigo, and T Mytata (1985) *Nature* **315**, 691.
Tonge PJ and PR Carey (1989) *Biochemistry* **28**, 6701.
Tonge PJ and PR Carey (1992) *Biochemistry* **31**, 9122.
Trylska J, P Grochowski, and A McCammon (2004) *Protein Sci* **13**, 513.
Venkatasubban KS and RL Schowen (1984) *CRC Crit Rev Biochem* **17**, 1.
Viragh C, TK Harris, PM Reddy, MA Massiah, AS Mildvan, and IM Kovach (2000) *Biochemistry* **39**, 16200.
Westler WM, PA Frey, J Lin, DE Wemmer, H Morimoto, PG Williams, and JL Markley (2002) *J Am Chem Soc* **124**, 4196.
Wlodawer A and JW Erickson (1993) *Annu Rev Biochem* **62**, 543.
Wlodawer A, M Miller, M Jaskolski, BK Sathyanarayana, E Baldwin, IT Weber, LM Selk, L Clauson, J Schneider, and SB Kent (1989) *Science* **245**, 616.
Wright HT, J Kraut, and PE Wright (1968) *J Mol Biol* **37**, 363.
Wu Z-P and D Hilvert (1989) *J Am Chem Soc* **111**, 4513.
Yamamoto D, K Matsumoto, H Ohishi, T Ishida, M Inoue, K Kitamura, and H Mizuno (1991) *J Biol Chem* **266**, 14771.
Zerner B, RPM Bond, and ML Bender (1964) *J Am Chem Soc* **86**, 3674.
Zhang K and DS Auld (1993) *Biochemistry* **32**, 13844.
Zhong S, K Haghjoo, C Kettner, and F Jordan (1995) *J Am Chem Soc* **117**, 7048.

7
Isomerization

Isomerization reactions are important in metabolism to potentiate further transformations that would otherwise be chemically impossible. A familiar example from glycolysis is phosphohexose isomerase, which catalyzes the interconversion of D-glusose-6-P and D-fructose-6-P. The formation of fructose-6-P makes it chemically feasible at a later step of glycolysis to cleave the six-carbon sugar into two three-carbon sugars, glyceraldehyde-3-P, and dihydroxyacetone-P by aldolase. No such cleavage of glucose-6-P into two three-carbon sugars is possible. The dihydroxyacetone-3-P is converted into glyceraldehyde-3-P by another isomerase, triosephosphate isomerase. In this way, glucose-6-P can be transformed into two molecules of glyceraldehyde-3-P, which can then be metabolized through glycolysis to pyruvate. Both reactions of phosphohexose and triosephosphate isomerases involve aldose/ketose interconversions and proceed by similar chemical mechanisms.

Other important isomerases include phosphomutases, epimerases, racemases, and carbon-skeleton mutases, all of which have their roles in metabolism. The chemical mechanisms vary with the classes of isomerases and include enolizations, hydride transfer, oxidation/reduction, phosphotransfer, and radical rearrangements. In this chapter, we consider the mechanisms by which enzymes catalyze isomerization reaction.

Aldose and Ketose Isomerases

Chemistry

The interconversions of glucose-6-P and fructose-6-P and of the triose phosphates can be formulated chemically, as in eq. 7-1.

334 Enzymatic Reaction Mechanisms

$$\text{HO-CH(H)-C(=O)-H} \rightleftharpoons \text{O=C(H)-CH(OH)-H} \quad (7\text{-}1)$$

The transformation in eq. 7-1 is an internal oxidation-reduction, in which the aldehyde group of the aldose is reduced and the neighboring alcoholic group is oxidized. This reaction can take place by either of two chemical mechanisms: an initial enolization at C2 to produce an enediolate intermediate that can be protonated at C1 to produce the product or a direct hydride transfer from C2 to C1. These mechanisms are outlined in scheme 7-1.

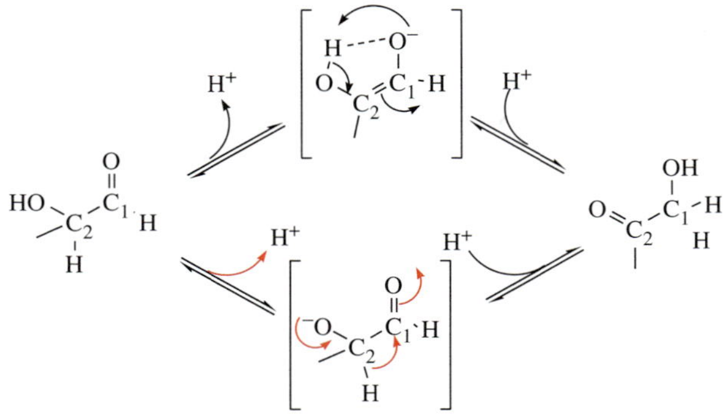

Scheme 7-1

Loss of the proton C2(H) by enolization in the upper pathway leads to the enediolate intermediate, and return of the proton to C1 (black arrows in scheme 7-1) leads to the ketose product. The hydride transfer mechanism in the lower pathway begins with the dissociation of the alcoholic proton to form the alcoholate intermediate. The alcoholate provides the driving force for the 1,2-hydride transfer (colored arrows in scheme 7-1) accompanied by protonation of the oxygen at C1. The two mechanisms require different hydrogen transfer regimes. The 1,2-hydride shift is favored in specific-base catalysis, that is, in hydroxide ion catalysis at high pH values (Nagorski and Richard, 1996). These conditions lead to alkoxide formation, which can undergo the 1,2-hydride shift. General base catalysis at neutral pH values promotes the enolization mechanism. Both mechanisms are found in enzymatic aldose/ketose isomerizations, the enolization mechanism in those enzymes that do not depend on a divalent metal ion, and the 1,2-hydride shift in enzymes that require a divalent metal ion to stabilize the alkoxide ion required for the hydride shift.

Phosphoglucose Isomerase

Evidence bearing on the distinction between the mechanisms in scheme 7-1 came in studies of phosphogluocose isomerase (EC 5.3.1.9), which catalyzes the interconversion of D-glucose-6-P and D-fructose-6-P (Rose and O'Connell, 1961). Although these phosphohexoses exist in solution in their cyclic pyranose and furanose forms, respectively, the enzyme catalyzes ring opening, and the aldose/ketose reaction takes place at the active site according to the pattern in eq. 7-1. Carefully executed quantitative studies showed that reaction of [2-^{3}H]glucose-6-P takes place with fractional loss of tritium into the medium and fractional transfer to C1 of the product. At low temperatures, most of the tritium was retained in the product, but at higher temperatures, up to 80% was exchanged into the medium. These results were incompatible with a mechanism involving the 1,2-hydride shift and implicated the enolization mechanism.

Isomerization

The stereochemistry of tritium transfer from C1 to C2 in the actions of several aldose/ketose isomerases showed that tritium was abstracted from C1 and added back to C2 from the same face of the molecule, so that the intermediates were likely to be *cis*-enediolates, not *trans*-enediolates (Rose, 1970). The chemical mechanism of enzymatic aldose/ketose isomerization could then be formulated as in scheme 7-2, assuming a single enzymatic base to abstract the proton from C1 and relay it to C2. Because most of the tritium in the substrate was retained in the product, this base had to be derived from a monoprotic acid and did not contain a proton in its conjugate base form. Among enzymatic bases, it could be a carboxylate or an imidazole ring. Tyrosine was not excluded at the time, but it was not considered to be an effective base at neutral pH because of its pK_a of 10.2. It has been found to ionize in the physiological pH range in the special active-site environments of UDP-galactose 4-epimerase and alanine racemase (Berger et al., 2001; Liu et al., 1997; Spies and Toney, 2003). The x-ray crystallographic structures of aldose/ketose isomerases that act by the enolization mechanism show that the base is a glutamate residue.

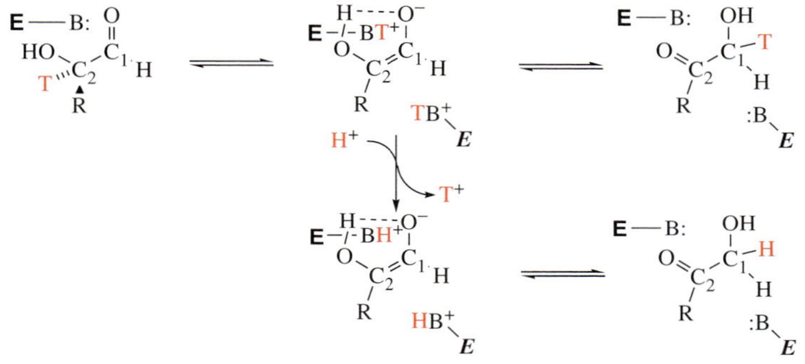

Scheme 7-2

Most aldose/ketose isomerases, including phosphoglucose isomerase and triosephosphate isomerase, catalyze enolization of the substrate, and this raises a number of still unsettled questions regarding details of the mechanism. For example, the value of the carbon pK_a for an α-hydroxy aldehyde such as glyceraldehyde-3-P is about 17, and the mechanism of ionization to an enolate through the action of a weak base, a carboxylate, must be explained. A few aldose/ketose isomerases act by the lower mechanism in scheme 7-1, the 1,2-hydride shift. Catalysis by these enzymes, such as xylose isomerase, depends on the function of a divalent metal ion, which seems to potentiate hydride transfer. Triosephosphate isomerase and xylose isomerase represent the two classes.

Triosephosphate Isomerase

The interconversion of glyceraldehyde-3-P (GAP) and dihydroxyacetone phosphate (DHAP) by triosephosphate isomerase (TIM; EC 5.3.1.1) follows the stereochemistry shown in scheme 7-2, where $R = CH_2OPO_3^{2-}$ (Rose, 1970). Unlike phosphoglucose isomerase, the reactions of (R)-[2-^{3}H]GAP or (R)-[1-^{3}H]DHAP proceed largely with exchange of tritium into the solvent and only a small retention in the product. TIM from most sources is a dimeric protein with a molecular mass of about 54 kDa and subunit molecular mass of about 27 kDa, depending on the source. Affinity-labeling studies with glycidol phosphate (1,2-epoxypropanol-3-P) and bromohydroxyacetone phosphate as active-site–directed alkylating agents led to the alkylation of Glu165, which was presumed to be the base in the enolization mechanism of scheme 7-2 (Coulson et al., 1970; Waley et al., 1970).

336 Enzymatic Reaction Mechanisms

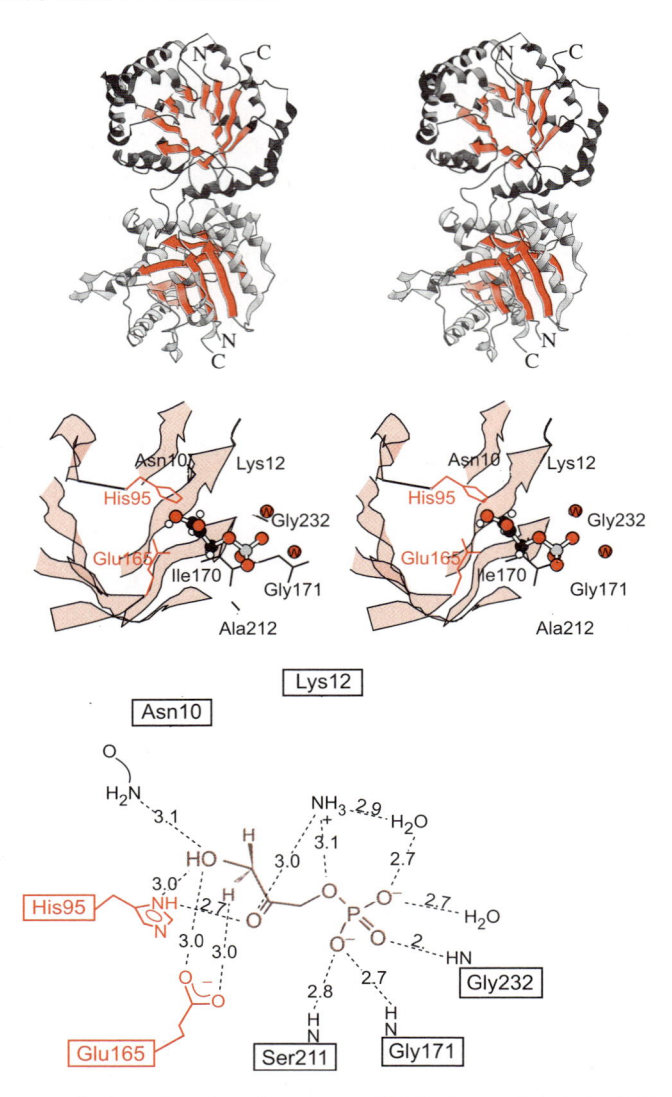

Fig. 7-1. Structure of triosephosphate isomerase (TIM) shows the chain fold of the original TIM barrel, or β-barrel, a frequently encountered structure (PDB 1YPI; Lolis et al., 1990). The TIM from yeast has dihydroxyacetone phosphate (DHAP) bound at the active site (PDB 1NEY; Jogl et al., 2003). The stereodrawing (center) of a ball-and-stick model shows DHAP at the active site and stick diagrams of nearby amino acid side chains. The catalytic side chains of Glu165 and His95 are shown in red. A two-dimensional map (bottom) of the active site shows the protein contacts with DHAP.

The structure of TIM gave the first view of the famous and most prevalent protein fold among enzymes, the TIM barrel or α,β barrel, shown at the top of fig. 7-1 (Alber et al., 1981; Banner et al., 1975; Lolis et al., 1990; Wierenga et al., 1987). This fold appears in about 10% of enzyme structures. The active site of TIM includes three amino acids in close contact with substrates or inhibitors, Lys12 that binds the phosphate group, and Glu165 and His95, which catalyze the aldose/ketose transformation. The ball and stick diagram of the active site in the center of fig. 7-1 suggests that Glu165 catalyzes proton transfer between C1 and C2, and His95 interacts closely by hydrogen bonding with the *cis*-enediolate intermediate. The close contacts are mapped in two dimensions at the bottom of fig. 7-1. In the highest resolution structure of the complex of TIM with DHAP at 1.2 Å, the contact between His95 and O2 in the substrate is extraordinarily close, 2.6 Å, for a hydrogen bond of the type O—-H–N (Jogl et al., 2003). We return to this topic later in the discussion of the reaction mechanism.

Mutation of Glu165 to aspartate decreased the value of k_{cat} 1500-fold for GAP and 240-fold for DHAP, with small decreases in K_m (Straus et al., 1985). Mutation of His95 to glutamine decreased the activity about 400-fold relative to wild-type TIM (Nickberg et al., 1988). The mutations did not completely inactivate TIM but lowered the value of k_{cat}/K_m two to three orders of magnitude. The residual activities in the mutated variants could be attributed to changes in rate limitation in the multistep mechanism and to takeover of the essential acid-base functions of Glu165 and His95 by other enzymatic groups or buffer components. Because the chemical steps did not limit the rate for wild-type TIM (discussed later), the mutations of catalytic groups did not lower the overall activity to the maximum degree expected for loss of function in acid-base catalysis. The mutations led to changes in rate limitation from diffusion to the chemical steps affected by the mutations (Blacklow et al., 1988).

Nonenzymatic, base-catalyzed enolization of GAP led to the *cis*-enediolate intermediate in solution (Richard, 1984). The *cis*-enediolate underwent elimination of phosphate to form methyl glyoxal, as in scheme 7-3, much more rapidly than isomerization to DHAP or reversion to GAP.

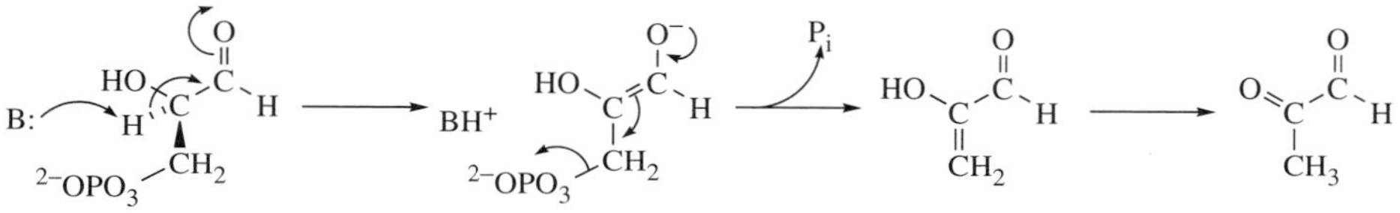

Scheme 7-3

The elimination of phosphate does not occur at the active site of TIM, presumably because the *cis*-enediolate is held in a conformation in which elimination is stereoelectronically forbidden. Such a conformation would place the phosphate group in the plane of the *cis*-enediolate p-bonding system. Methylglyoxal synthase catalyzes the transformation of GAP into methylglyoxal by the mechanism in scheme 7-3 (Cui and Karplus, 2003; Marks et al., 2004).

A careful analysis of the hydrogen bonding about Nδ1 and Nε2 of His95 indicated that the imidazole ring of His95 in free TIM is not in its conjugate acid form. It is not an imidazolium ring in the physiological pH range (Lolis et al., 1990). An ^{15}N nuclear magnetic resonance (NMR) study of TIM confirmed this by clearly proving the neutrality of His95 in the free enzyme (Lodi and Knowles, 1991). Stabilization of the *cis*-enediolate intermediate appears to involve hydrogen bond donation from the neutral, conjugate base form of His95. Given the presumed role of His95 as an electrophilic catalyst, its conjugate acid had been expected to be its functional form. We return to this topic later in the discussion of the reaction mechanism.

Detailed kinetic analysis of the overall reaction and the exchange kinetics, as well as transient kinetic measurements of binding and enolization, allowed several of the rate constants in the mechanism to be assigned and the free energy profile in fig. 7-2 to be constructed (Albery and Knowles, 1976). The graph showed that the free energy barriers were remarkably similar, and in particular that diffusion to form Michaelis complexes could be rate limiting. No distinct rate-limiting step could be assigned, and this explained the low values of deuterium kinetic isotope effects in the reactions of deuterated substrates, which were 1.0 in the forward reaction of [2-^{2}H]GAP and 2.9 in the reverse reaction of [(1R)-1-^{2}H]DHAP (Leadlay et al., 1976). Moreover, the value of k_{cat}/K_m (10^8 M^{-1} s^{-1}) was

338   Enzymatic Reaction Mechanisms

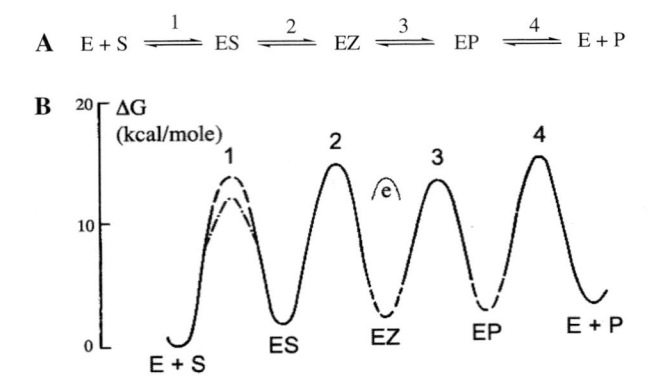

$$A \quad E + S \underset{}{\overset{1}{\rightleftharpoons}} ES \underset{}{\overset{2}{\rightleftharpoons}} EZ \underset{}{\overset{3}{\rightleftharpoons}} EP \underset{}{\overset{4}{\rightleftharpoons}} E + P$$

Fig. 7-2. (A) Free energy profile for the reaction of triosephosphate isomerase (TIM). (B) The free energy profile was calculated from kinetic data on the overall and exchange kinetics for the action of TIM. The profile refers to the mechanism shown in A, in which ES is the Michaelis complex of GAP, EP is the product complex with DHAP, and EZ is the complex of TIM with the *cis*-enediolate intermediate. (B, From Albery and Knowles, 1976, with permission.)

typical of binding rate constants for enzymes (see chap. 2), suggesting that substrate and product diffusion could be rate limiting.

Diffusion as the rate-limiting process means that the enzyme is as highly evolved as it can be because further improvements in the chemistry of the process cannot increase the rate. This is consistent with a theory of the evolution of enzymes, which holds that incremental evolution lowering the activation barriers for individual steps will eventually lead to similar barriers for all steps and rate limitation by diffusion. In this theory, the diffusional barrier eventually sets the chemical barriers. The graph in fig. 7-2 supports the theory in the case of TIM.

Rate limitation by diffusion can be tested experimentally by studying the effects of the viscosity of the solvent medium on the value of k_{cat}/K_m. If a rate is limited by diffusion, it will be sensitive to changes in the diffusion rates of the reactants. The coefficients of diffusion are determined by the molecular weights of the molecules, the medium viscosity, and temperature. The temperature dependence is similar for most molecules, and little can be done about molecular weights in enzymatic processes. However, the solvent viscosity can be manipulated by addition of viscogens (Brouwer and Kirsch, 1982). In enzymatic work, the viscogen must not interfere with the interactions of the enzyme and substrate; it must not be an inhibitor. To be useful for studying molecular diffusion, it must alter the microviscosity of the medium. This means that the molecular weight of the viscogen must be on the order of that of the substrates. Polymers such as Ficoll and polyethylene glycol (PEG) can increase the macroviscosity but not the microviscosity, and they do not slow the diffusion of small molecules. They can serve as control viscogens in viscosity studies. Molecules such as glycerol and sucrose are often used as viscogens in enzymatic work, but it is essential to prove that they do not inhibit by binding to the enzyme or substrate at the concentrations required to increase the viscosity.

In studies of the effects of viscosity on the kinetic parameters for TIM, the results convincingly implicated diffusion as the rate-limiting process in the mechanism (Blacklow et al., 1988). In both forward and reverse directions, increasing the viscosity (sucrose or glycerol) decreased the values of k_{cat}/K_m by the maximum amount and not more; the rate decreases could be attributed to decreased diffusion rates. The value of k_{cat}/K_m in this case (4×10^8 M^{-1} s^{-1}) was similar to the maximum values of the second order rate constants for binding that had been found for other enzymes (Albery and Knowles, 1976) (see chap. 2). The effects of viscosity confirmed that substrate diffusion and binding limited the rate.

In addition to proving that His95 is unprotonated in the physiological pH range, spectroscopy provided valuable information about the polarization of the carbonyl group of substrates at the active site. Experiments employing Fourier transform infrared spectroscopy (FTIR) showed that the carbonyl stretching frequency of DHAP is decreased by about 19 cm^{-1} in the Michaelis complexes, strongly implicating an interaction of the carbonyl oxygen with an electrophilic group (Belasco and Knowles, 1980; Komives et al., 1991). This corresponds to a weakening of the carbonyl bond by about 10%, or 12 kcal mol^{-1}, due to polarization (Chittock et al., 1999). FTIR has been similarly employed in analogous studies of other enzymes (Wharton, 2000).

Crystal structures implicate His95 as the only group interacting with the carbonyl oxygen, presumably as a hydrogen bond donor. In conventional hydrogen bonding, one might expect His95 to be most effective as an electrophilic donor to the anionic *cis*-enediolate if it were in its electropositive, protonated form. However, in TIM it is in its neutral form. This raises the question of whether hydrogen bond between the neutral imidazole ring of His95 and the *cis*-enediolate is a low-barrier hydrogen bond (LBHB). The pK_a values may be similar or even matched, which is a condition for LBHB formation. Such hydrogen bonding has been postulated to account for the stabilization of enolate ions at enzymatic active sites (Gerlt et al., 1997). We discussed LBHBs in chapters 1 and 5 in connection with serine proteases and acetylcholinesterase.

Enolates at active sites must be stabilized by interactions with electrophilic groups. In the case of enolase, two magnesium ions provide sufficient electrostatic stabilization to facilitate the enolization of the extraordinarily weak carbon acid 2-phosphoglycerate (see chap. 1). GAP and DHAP are much stronger carbon acids (p$K_a \approx 17$) and so are much less difficult to enolize; however, their carbon acidities must be increased at the active site of TIM to allow formation of the *cis*-enediolate. This is most likely brought about by the electrophilic interaction that polarizes the carbonyl group. The absence of an obvious alternative electrophilic species, such as a metal ion, focuses attention on the hydrogen bond donated by His95 to the carbonyl group of the substrate.

A correlation of available information leads to the chemical mechanism shown in fig. 7-3 for the action of TIM. Base catalysis by Glu165 leads to abstraction from GAP to

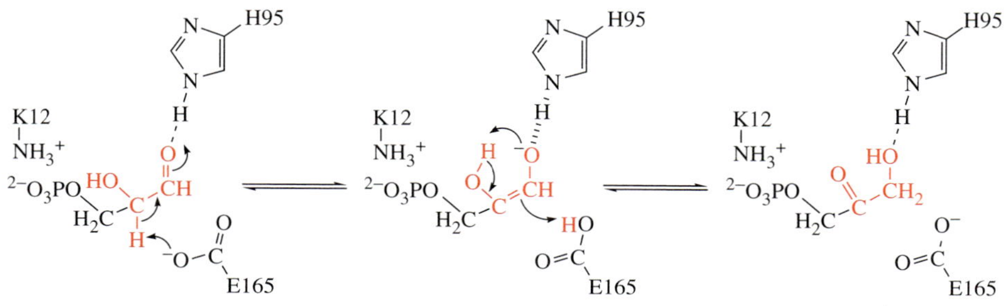

Fig. 7-3. A mechanism of catalysis at the active site of triosephosphate isomerase (TIM). In the accepted mechanism, Glu165 serves as the base that abstracts the α-proton from glyceraldehyde-3-P (GAP), and His95 serves as the electrophilic catalyst that stabilizes the enediolate intermediate. The conjugate acid form of Glu65 donates a proton to the enediolate to form dihydroxyacetone phosphate (DHAP).

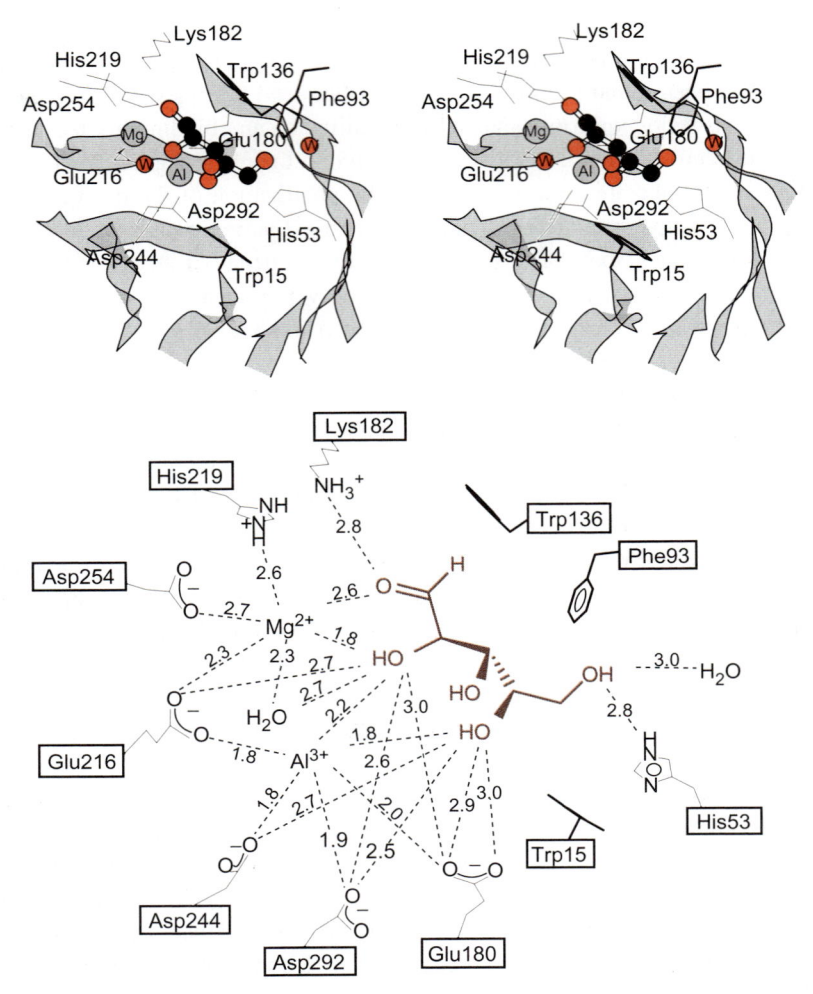

Fig. 7-4. Structure of the active site of xylose isomerase. The global structure of *Athrobacter* (strain B3728) xylose isomerase (PDB 1XLG; Collyer et al., 1990) is based on the triosephosphate isomerase (TIM) barrel chain fold, although the amino acid sequence is not homologous to that of TIM. The active site is unlike that of TIM, but as in TIM, the active site is located inside the C-terminal end of the β-barrel. This is common but not universal for enzymes with this fold. The structure of the active site is shown with Mg^{2+} and Al^{3+} interacting with the substrate. The aluminum was included for the structural characterization because it does not support catalysis, which requires a second divalent metal ion, but it does allow the formation of a stable substrate active site complex amenable to crystallographic characterization. Coordination of Mg^{2+} is thought to lower the pK_a of the α-hydroxyl group (see chap. 4). The Mg^{2+}-alkoxide resulting from ionization can undergo isomerization to the alternative Mg^{2+}-alkoxide for xylulose by the 1,2-hydride shift shown at the bottom of scheme 7-1.

form the enediolate intermediate, which is stabilized by hydrogen bonding from His95. Proton transfer from the conjugate acid of Glu165 to C1 leads to DHAP. The hydrogen bond linking His95 and the enediolate intermediate has not been definitively characterized and is a subject of continuing interest.

The structure of TIM includes a flap or lid (residues 166 to 176) that closes over the active site and seals in the substrate and intermediate. Flap closure does not depend on the presence of the substrate, but dynamics experiments by NMR spectroscopy show that the rate of flap opening and closing is on the order of the turnover number (k_{cat})

(Rozovsky et al., 2001; Rozovsky and McDermott, 2001). Flap dynamics may limit the rate, and mutations in the flap greatly decrease the catalytic rate (Xiang et al., 2004).

Xylose Isomerase

In contrast to TIM and phosphoglucose isomerase, exchange of substrate hydrogen with solvent has never been observed in the reaction of xylose isomerase (EC 5.3.1.5), which catalyzes the interconversion of D-xylose and D-xylulose (eq. 7-2).

$$\text{D-xylose} \rightleftharpoons \text{D-xylulose} \tag{7-2}$$

The isomerization takes place in the noncyclic aldose/ketose forms of the sugars according to eq. 7-1. The complete absence of exchange suggested that the reaction might occur by the 1,2-hydride shift mechanism along the lower pathway of scheme 7-1. This concept drew support from the requirement of xylose isomerase for a divalent metal ion. Coordination of Mg^{2+} with the α-hydroxycarbonyl groups of the xylose and xylulose could lower the pK_a of the hydroxyl group and allow its ionization to the alkoxide (see chap. 4). In accord with the chemical requirements of the hydride shift mechanism (Nagorski and Richard, 1996), formation of the alkoxide would provide the driving force for the 1,2-hydride shift.

The structure of xylose isomerase, shown in fig. 7-4, supports the assignment of the 1,2-hydride shift mechanism by showing that two metal ions form inner sphere complexes bridging the substrate and active site. Magnesium ion ligates C1(O) and C2(O) that undergo aldose-ketose isomerization, and aluminum (Al^{3+}) ligates C2(O). The two metal ions and Lys182 contribute six positive charges in the active site, and five acidic residues contribute five negative charges, making the active site positive overall. Direct ligation of Mg^{2+} and Al^{3+} should drive the ionization of C2(OH) and lead to overall neutrality in the site. The C2-alkolide so formed would undergo the 1,2-hydride shift by the lower mechanism in scheme 7-1. The structure does not reveal a base catalyst in position to abstract a proton from C1 or C2, as would be required in an enolization mechanism. All of the conditions for alkoxide formation are met, and all of the experimental evidence supports the assignment of the 1,2-hydride shift as the mechanism in the case of xylose isomerase.

Phosphomutases

α-Phosphoglucomutase

Extensive mechanistic evidence on the action of rabbit muscle phosphoglucomutase (EC 5.4.2.2) establishes the principles underlying catalysis by cofactor-dependent phosphomutases (Knowles, 1980; Liu et al., 1997b; Mildvan, 1974; Rhyu et al., 1984). The interconversion of D-glucose-6-P and α-D-glucose-1-P by this enzyme is an essential step in the metabolism of glycogen, which is a source of α-D-glucose-1-P through the action of glycogen phosphorylase. Active phosphoglucomutase is a phosphoenzyme, with a residue of phosphoserine at the active site and a requirement for magnesium ion (Ray and Roscelli, 1964; Ray, 1969). The phosphoserine participates in the catalytic cycle, which can be represented by eqs. 7-3a through 7-3d.

$$E\text{-Ser}^{116}\text{-OP} + \text{Glc-1-P} \rightarrow E\text{-Ser}^{116}\text{-OP.Glc-1-P} \quad (7\text{-}3a)$$

$$E\text{-Ser}^{116}\text{-OP.Glc-1-P} \rightarrow E\text{-Ser}^{116}\text{-OH.Glc-1P,6P} \quad (7\text{-}3b)$$

$$E\text{-Ser}^{116}\text{-OH.Glc-1P,6-P} \rightarrow E\text{-Ser}^{116}\text{-OP.Glc-6-P} \quad (7\text{-}3c)$$

$$E\text{-Ser}^{116}\text{-OP.Glc-6-P} \rightarrow E\text{-Ser}^{116}\text{-OP} + \text{Glc-6-P} \quad (7\text{-}3d)$$

The phosphoenzyme binds glucose-1-P to form the Michaelis complex in eq. 7-3a, and phosphoryl transfer from Ser116 to C6(OH) of glucose-1-P forms the dephosphoenzyme and enzyme-bound glucose-1,6-bisphosphate in eq. 7-3b. The noncovalently bound species Glc-1P,6-P in eqs. 7-3b and 7-3c symbolizes glucose-1,6-bisphosphate, the compulsory intermediate. In eq. 7-3c, phosphoryl transfer from glucose-1,6-bisphosphate to Ser116 regenerates the phosphoenzyme and forms the product glucose-6-P, which dissociates in eq. 7-3d. Notice that phosphotransfer is intermolecular; the phosphoryl group of the product is that of the initial phosphoenzyme, and the phosphoryl group of the substrate becomes that of the phosphoenzyme produced in the reaction cycle.

Occasionally, the noncovalently bound glucose-1,6-bisphosphate adventitiously dissociates from the enzyme according to eq. 7-3e, leaving the enzyme dephosphorylated.

$$E\text{-Ser}^{116}\text{-OH.Glc-1P,6-P} \rightarrow E\text{-Ser}^{116}\text{-OH} + \text{Glc-1P,6-P} \quad (7\text{-}3e)$$

The dephosphoenzyme is inactive, but it can be reactivated by addition of glucose-1,6-bisphosphate and reversal of eq. 7-3e. Purification of phosphoglucomutase leads to significant dephosphorylation, and the purified enzyme too is activated by glucose-1,6-bisphosphate. For these reasons, glucose-1,6-bisphosphate is regarded a cofactor of phosphoglucomutase; it maintains enzyme activity by restoring the dephosphoenzyme to the phosphoenzyme and must be present for the enzyme to display maximal activity. Because glucose-1,6-bisphosphate is required for maximum activity, this enzyme is sometimes known as a cofactor-dependent phosphoglucomutase.

Evidence leading to the foregoing description of the mechanism is compelling and includes the tracing of the pathway of phosphoryl group transfer by use of ^{32}P. The adventitious dissociation of glucose-1,6-bisphosphate allowed a number of informative experiments to unveil the mechanism to be designed.

In some drawings of the structure of glucose, C1(OH) and C6(OH) seem to be well separated and unlikely to interact with a single phosphoserine residue. However, glucose-6-P and glucose-1-P exist in cyclic forms that tend to bring the two hydroxyl groups into reasonable proximity for interaction with a single phosphoserine. Scheme 7-4 illustrates this principle, showing how a phosphoserine can interact alternately with both C1(OH) and C6(OH).

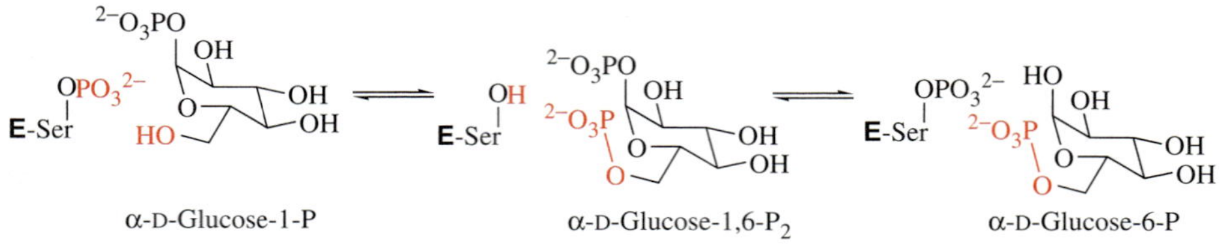

Scheme 7-4

Rabbit muscle phosphoglucomutase is a 62-kDa enzyme with the multidomain structure shown in fig. 7-5 at 2.4-Å resolution. Three of the domains are topologically similar and the fourth is distinct. Ser116 lies at the base of the large cleft. A higher resolution structure will be of considerable interest for deducing details of interactions at the active site.

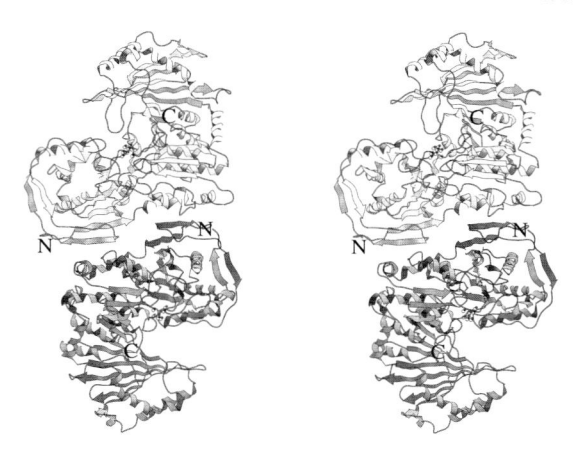

Fig. 7-5. The structure of rabbit muscle phosphoglucomutase. α-Phosphoglucomutase (cofactor-dependent phosphoglucomutase) is a large, homodimeric protein with a complex multidomain structure (PDB 3PMG; Liu et al., 1997). In this structure, phosphoserine is shown in red at the active site.

β-Phosphoglucomutase

Bacterial phosphoglucomutases differ from the muscle enzyme α-phosphoglucomutase (EC 5.4.2.6), in that they are smaller proteins and use aspartate as the nucleophilic catalyst instead of serine. Because glucose-1,6-bisphosphate rarely dissociates, these enzymes are known as cofactor-independent phosphoglucomutases. Otherwise, a similar catalytic cycle is followed, in which phosphoaspartate and Mg^{2+} at the active site mediate phosporyl group transfer. The role of phosphoaspartate is remarkable because acylphosphates are high-energy phosphorylating agents ($\Delta G^{\circ\prime}$ = 10 kcal mol^{-1}), much higher than the 3 kcal mol^{-1} for glucose-6-P and 5 kcal mol^{-1} for glucose-1-P. A significant energetic barrier must be overcome for phosphoryl group transfer from the sugar phosphates to aspartate at the active site.

High-resolution structures are available for β-phosphoglucomutases, and one of these is illustrated in fig. 7-6. In this structure, obtained under cryogenic conditions, the trivalent molecule simulating a phosphoryl group in flight is believed to be MgF_3^-, based on ^{19}F NMR spectroscopy (Baxter et al., 2006). The structure appears to simulate the transition state for PO_3^- transfer in the mechanism. The structure is consistent with a metastable PO_3^- between and partially bonded to the anomeric C1(O) and C6(O) in the transition state.

In the detailed mechanistic analysis of phosphoryl group transfer, the transition state has been characterized as one with a low degree of bonding to the entering and departing oxygen atoms at the transition state. The observation of this structure focuses attention on the fact that any degree of bonding, which implies the presence of a restoring force, must be regarded as evidence for a metastable intermediate. We return to this subject in chapter 10.

Phosphoglycerate Mutases

The interconversion of 3-phosphoglycerate and 2-phosphoglycerate is an essential step in glycolysis on the way to phosphoenolpyruvate and pyruvate. The transformation is catalyzed by phosphglycerate mutase (EC 5.4.2.1). Most phosphoglycerate mutases catalyze this reaction by a process that is analogous to that of phosphoglucomutases, in which the phosphoenzyme is active, and 2,3-diphosphoglycerate is an intermediate and a cofactor. These are known as the cofactor-dependent phosphoglycerate mutases.

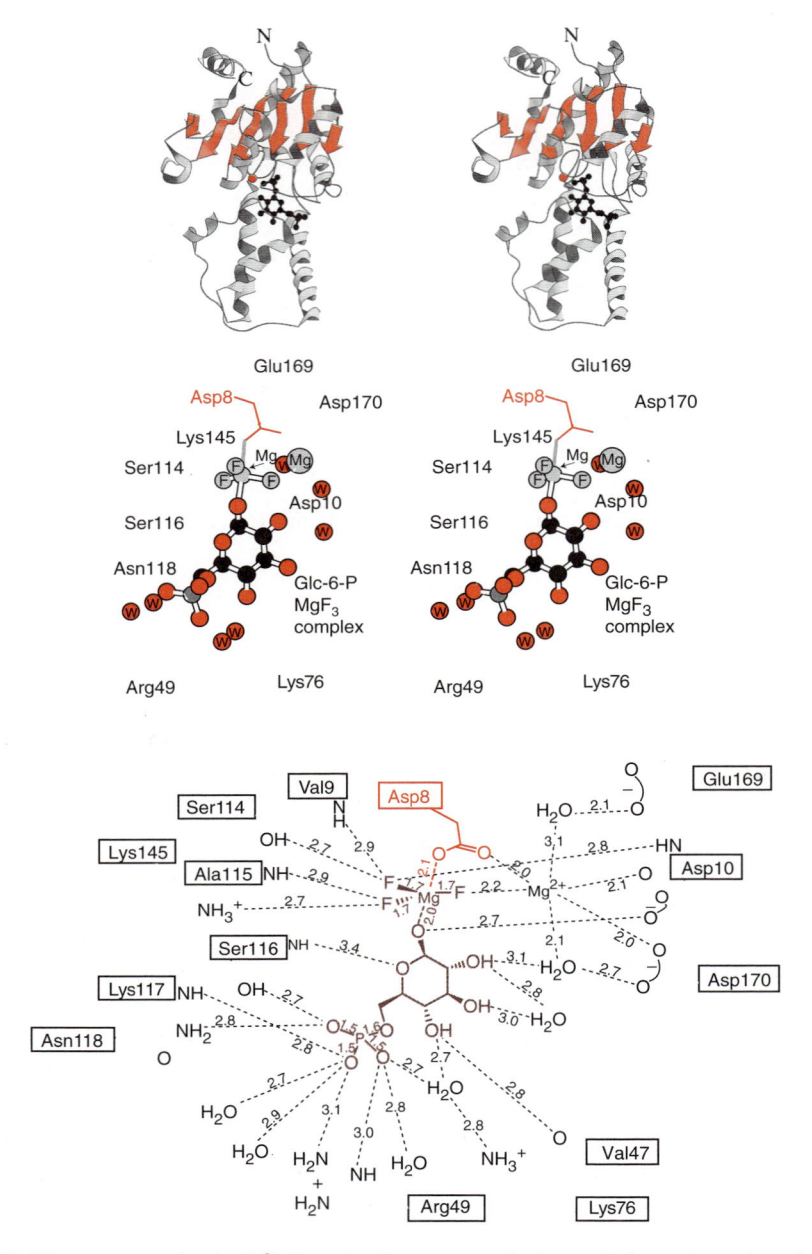

Fig. 7-6. The structure (top) of β-phosphoglucomutase (cofactor-independent phosphoglucomutase) is smaller than that of muscle phosphoglucomutase, and Asp8 serves as the nucleophilic catalyst for mediating phosphoryl group transfer instead of serine. This structure shows glucose-6-P or glucose-1,6-bisphosphate at the active site (PDB 1O08; Lahiri et al., 2003). A stereodrawing (middle) of a ball-and-stick model shows glucose-6-P bound at the active site and its contacts with amino acid side chains, which are drawn as stick models. Several fixed water molecules are shown as red balls, and Mg^{2+} is shown as a gray ball. The structure shown as MgF_3 simulates a phosphoryl group in flight.

Isomerization 345

They use histidine residues to mediate phosphoryl group transfer, and the active enzymes contain phosphohistidine residues at the active sites. The catalytic cycle is represented by eqs. 7-4a through 7-4d.

$$E\text{-His-P} + \text{Glycerate-3-P} \rightarrow E\text{-His-P.Glycerate-3-P} \quad (7\text{-}4a)$$

$$E\text{-His-P.Glycerate-3-P} \rightarrow E\text{-His.Glycerate-3P,2P} \quad (7\text{-}4b)$$

$$E\text{-His.Glycerate-3P,2P} \rightarrow E\text{-His-P.Glycerate-2-P} \quad (7\text{-}4c)$$

$$E\text{-His-P.Glycerate-2-P} \rightarrow E\text{-His-P} + \text{Glycerate-2-P} \quad (7\text{-}4d)$$

Analogous to phosphoglucomutase and glucose-1,6-bisphosphate, 2,3-diphosphoglycerate is a compulsory intermediate that can adventitiously dissociate, and for this reason it is also a cofactor that maintains enzymatic activity by keeping the active site histidine phosphorylated.

Phosphoglycerate mutase has two histidine residues at the active site (Bond et al., 2001). This raises the question whether the phosphoryl group is shuttled between histidine residues, with each histidine relaying the group to C3(OH) or C2(OH) of glycerate, respectively. The fact that only a single phosphoenzyme can be isolated does not support the concept of both histidine residues mediating phosphotransfer. Moreover, stereochemical evidence essentially rules out a two-histidine relay mechanism.

The stereochemical course of phosphoryl group transfer by cofactor-dependent phosphoglycerate mutase is illustrated in fig. 7-7 (Blättler and Knowles, 1980). Chirality at phosphorus is brought about by stereospecific placement of the isotopes of oxygen in the nonbridging, peripheral positions through chemical synthesis. The figure shows how the overall conversion of 3-phosphoglycerate to 2-phosphoglycerate proceeds with

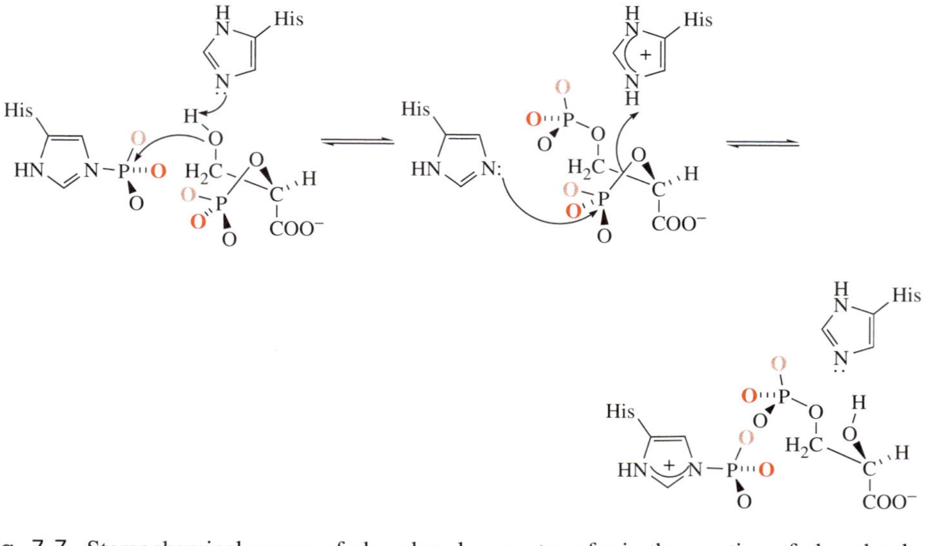

Fig. 7-7. Stereochemical course of phosphoryl group transfer in the reaction of phosphoglycerate mutase. The symbols O, **O**, and **O** represent the isotopes of oxygen, ^{16}O, ^{17}O, and ^{18}O, respectively. The catalytic phosphohistidine on the left was generated in the preceding turnover and possesses the opposite configuration at the phosphorus from that in the substrate. The configuration is again inverted on transfer to C2(OH) of the substrate to form the intermediate glycerate-2,3-bisphosphate (diphosphoglycerate). Transfer of the 2-phosphoryl group to His completes the isomerization to 3-phosphoglycerate. The overall process proceeds with retention of configuration at the phosphorus, but inversion takes place at each step.

retention of configuration at phosphorus. Inversion takes place at each of the two transfer steps required in the overall reaction, so that the original configuration is restored in the product. The configuration in the phosphoenzyme has not been determined but has been inferred from the empirical observations of inversion at each step in enzymatic phosphoryl group transfer processes.

Retention of configuration at phosphorus is inconsistent with a two-histidine relay mechanism for the phosphoglycerate mutase reaction. Such a mechanism would require three phosphoryl group transfer steps in each enzymatic turnover, and this would lead to overall inversion of configuration at phosphorus, contrary to the observed retention.

The reaction mechanism in the case of the wheat germ phosphoglycerate mutase differs fundamentally from the 2,3-diphosphoglucerate-dependent enzymes. A phosphoenzyme has never been observed, 2,3-diphosphoglycerate is not an intermediate, and the activity does not depend on the presence of 2,3-diphosphoglycerate. This enzyme is known as a cofactor-independent phosphoglycerate mutase. The energetically unlikely formation of glycerate-2,3-cyclic phosphate as a transient intermediate is ruled out by the fact that the reaction in $H_2^{18}O$ does not lead to the incorporation of ^{18}O into 2- or 3-phosphoglycerate (Breathnach and Knowles, 1977). The weight of evidence has been interpreted to suggest that the wheat germ enzyme functions through a reaction cycle that includes a phosphoenzyme intermediate along the lines of scheme 7-5, in which the nucleophile XH has not been assigned.

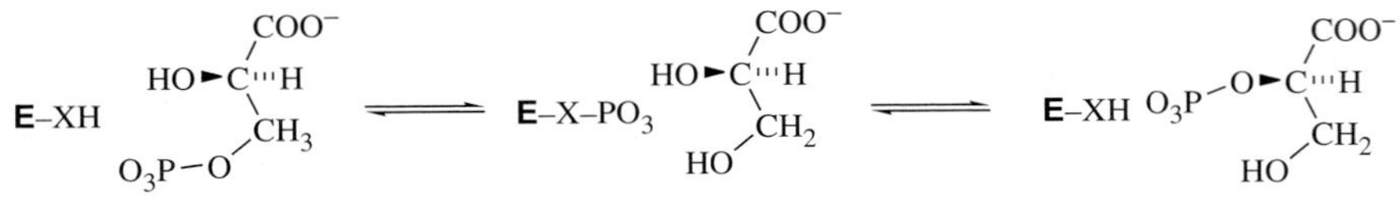

Scheme 7-5

Evidence in support of scheme 7-5 includes retention of configuration at phosphorus (Blättler and Knowles, 1980) and evidence from radiochemical experiments that phosphotransfer from 2-phosphoglycolate and other 2-phosphohydroxyacids to [^{14}C]glycerate can be detected (Breathnach and Knowles, 1977).

Racemases and Epimerases

Because enzymatic reactions are generally stereospecific, the observation of a nonstereospecific process at an active site raises the question of the mechanism by which nonstereospecificity is attained. This question arises for all racemases and epimerases, where nonstereospecificity is the sole biological raison d'être for the enzyme. We have already considered the case of alanine racemase, a paradigm for pyridoxal-5'-phosphate (PLP)–dependent amino acid racemases, and found that nonstereospecific proton transfer is brought about by the actions of two acid-base catalysts placed on opposite sides of the α-carbon (see chap. 5). In this section, we consider two PLP-independent amino acid racemases and the mechanism by which they carry out nonstereospecific proton transfer. We also discuss the actions of other racemases and an epimerase.

Proline Racemase

In early work on clostridial proline racemase, it was found to be PLP-independent (Cardinale and Abeles, 1968). The racemization of proline is chemically interesting because it is the only α-amino acid that cannot be racemized nonenzymatically by heating its copper complex with base. Proline racemase (EC 5.1.1.4) is a homodimeric enzyme with a subunit molecular mass of 38 kDa (Rudnick and Abeles, 1975).

Enzymatic racemization of α-deuteroproline proceeds with a primary kinetic isotope effect, and racemization in D_2O leads to the incorporation of deuterium into the product. These facts explained the original observation of "overshoot" in the racemization of proline in D_2O (Cardinale and Abeles, 1968). As illustrated in fig. 7-8, the optical rotation of L-proline in D_2O decreases in the course of racemization and becomes negative before rising to zero at equilibrium. This occurs because the approach to equilibrium depends on the rates in both forward and reverse reactions. Because the α-deueroproline produced in D_2O reacts in the reverse direction with a primary kinetic isotope effect, slower than the reaction of L-proline in the forward direction, the forward rate remains faster than the reverse rate until both L- and D-proline become labeled at C2 with deuterium. This leads to a transient excess of D-deuteroproline and a negative optical rotation.

The overshoot implies that proline racemase catalyzes the incorporation of deuterium into the racemization product faster than it catalyzes exchange of deuterium into the unracemized substrate. If this is true, the enzyme may not catalyze the exchange of deuterium into either D- or L-proline independently of conversion into its enantiomorph. The mechanism in fig. 7-9 accounts for racemization, the deuterium kinetic isotope effect, and the overshoot. It has become known as an *iso*-mechanism for the fact that the free enzyme exists in isomeric states, with one of two catalytic acid-base groups in its conjugate acid and the other in its conjugate base form.

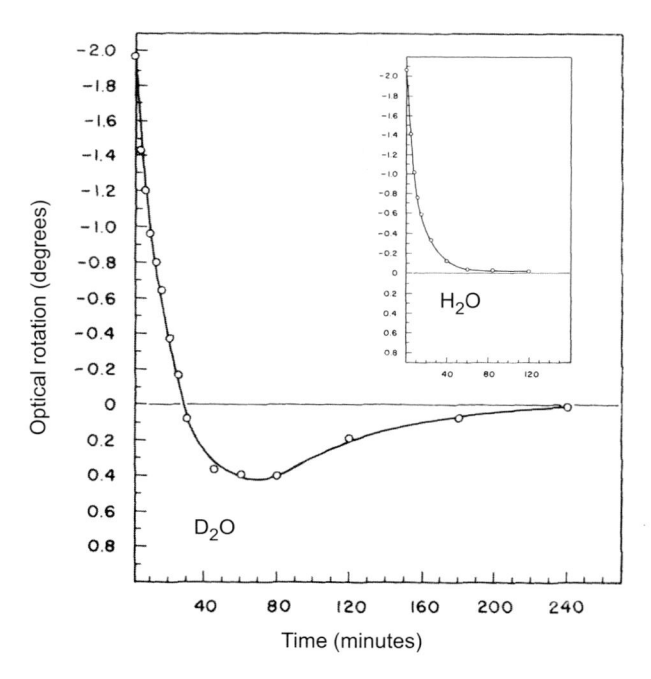

Fig. 7-8. Overshoot phenomenon in the enzymatic racemization of L-proline in D_2O. The optical rotation in water increases to zero (inset) as the racemization of proline approaches equilibrium. Because the optical rotations of L- and D-proline are exactly opposite, plus and minus, respectively, and the equilibrium constant is exactly 1.0, this is ideal behavior. However, in D_2O, the optical rotation first becomes positive and then rises to zero as the reaction approaches equilibrium. This overshoot phenomenon results from the kinetic isotope effect for the reverse reaction of 2-deuteroproline. Adapted from fig. 3 of Cardinale and Abeles (1968).

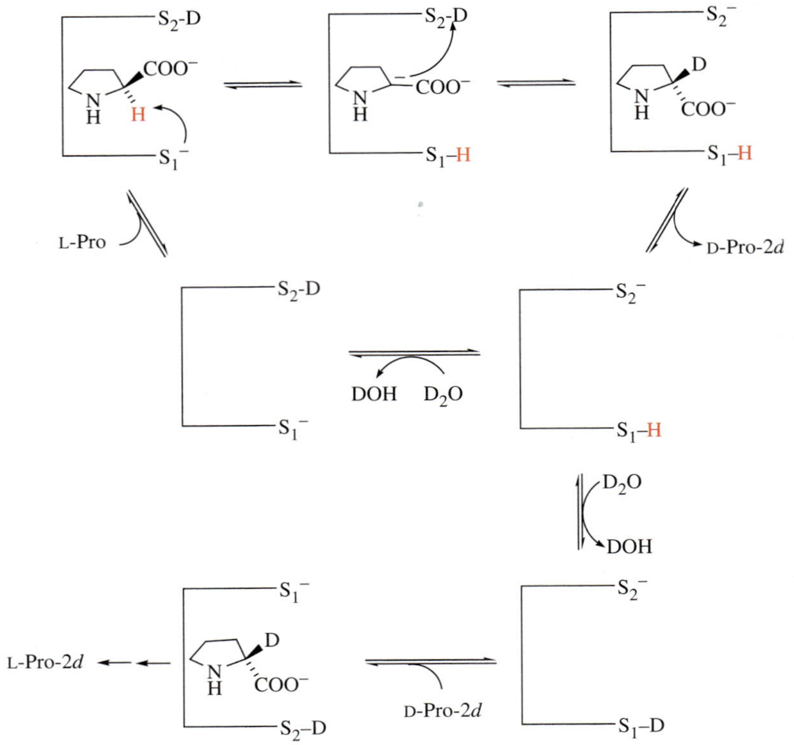

Fig. 7-9. An *iso*-mechanism for racemization by proline racemase. Notice that S_1^- can react only on the bottom face of L-proline, and S_2–D can react only on the top face. In the two-base mechanism, the transformation of L-proline into D-proline-2*d* leaves S_1–H and S_2^-, that is, an isomeric form of the enzyme. The enzyme must undergo isomerization before it can carry out another catalytic cycle, and for this reason, the mechanism is known as an *iso-mechanism*.

In the case of proline racemase, biochemical evidence indicates that the two catalytic groups are the thiol groups of cysteine residues, S_1–H and S_2–H in fig. 7-9. An oxidized form of the enzyme is inactive and can be activated by addition of thiols, and iodoacetate inactivates the enzyme by alkylating a single cysteine in the amino acid sequence (Rudnick and Abeles, 1975). Substrate binding experiments by equilibrium dialysis indicate one binding site per molecule of dimeric enzyme. It has been proposed that the active site is formed at the interface of the two subunits, with one cysteine from each subunit contributing the two cysteine thiol groups S_1–H and S_2–H (Rudnick and Abeles, 1975).

In the hypothetical mechanism of fig. 7-9 the free enzyme exists as two isomers, with one of the two cysteine residues bearing a proton and the other ionized. One of the isomeric enzyme forms interacts productively with L-proline and the other with D-proline. The isomeric forms of the enzyme can undergo interconversion only as the free enzyme and not within the substrate and product complexes.

In tests of the *iso*-mechanism, a few percent racemization of L-proline in D_2O, followed by quenching of the enzyme, led to the production of D-proline-2*d* and *no exchange* of deuterium into the recovered L-proline. The converse experiment led to L-proline-2*d* and no exchange of deuterium into D-proline (Cardinale and Abeles, 1968). This is in accord with the catalytic cycle of the *iso*-mechanism at the top of fig. 7-9. Moreover, the experiments showed that in an iso-mechanism both catalytic residues must be monoprotic acid-base groups, a condition satisfied by two cysteine thiol groups. A polyprotic catalyst such as lysine would allow at least one of the enantiomers to undergo exchange without first undergoing racemization. The mechanism allows exchange of deuterium from D$_2$O into a substrate only by isomerization of one form of the free enzyme into the other form, as illustrated at the bottom of fig. 7-9 with exchange of H for D, and then overall reversal from the deuterated product.

Racemization is a one-substrate-product process that proceeds to equilibrium with a value of 1.00 for K_{eq}. Such a reaction can be monitored under reversible and even equilibrium conditions by measuring optical rotation, a method that is not especially sensitive but is very accurate. Such studies of the action of proline racemase illuminated features of the reaction that explicitly confirmed the mechanism in fig. 7-9, while revealing the complexities of reversible enzymatic catalysis. In chapter 2, we explained the advantages of initial rate measurements in kinetic analysis of reactions in steady states. Under initial rate conditions, an enzymatic reaction is irreversible, and this simplifies the kinetics. The advantages lie wholly in the realm of experimental mechanistic analysis in vitro and do not pertain to conditions in vivo. However, kinetic information may be uniquely accessed in studies carried out under reversible conditions. Such studies are rare in mechanistic enzymology; however, a detailed kinetic analysis of proline racemase under reversible conditions revealed a phenomenon that proved to be valuable in testing the mechanism in fig. 7-9.

Observation of the action of proline racemase by optical rotation throughout the approach to equilibrium as a function of substrate concentration revealed the appearance of product inhibition at later stages of the reaction at very high concentrations of substrate (Fisher et al., 1986a). When studied under conditions of reversibility, which pertain to monitoring the approach to equilibrium, the kinetics becomes complex, as set forth in the original reference (Fisher et al., 1986a). However, it will be clear that at high concentrations of substrate, the product may well accumulate to inhibitory concentrations in the course of approach to equilibrium. The reaction has been analyzed under conditions of three regimes of substrate concentration, undersaturation, saturation, and oversaturation for the simplified kinetic mechanism in scheme 7-6.

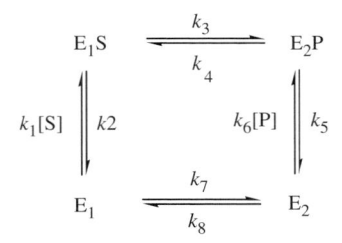

$$\begin{array}{ccc} E_1S & \underset{k_4}{\overset{k_3}{\rightleftharpoons}} & E_2P \\ k_1[S] \updownarrow k_2 & & k_6[P] \updownarrow k_5 \\ E_1 & \underset{k_8}{\overset{k_7}{\rightleftharpoons}} & E_2 \end{array}$$

Scheme 7-6

At very high concentrations of substrate, the enzyme quickly generates sufficient product to slow down product dissociation by virtue of the k_6[P] process, and this results in oversaturation behavior. Under oversaturation conditions, the reaction flux in the forward direction decreases with increasing substrate concentration because of noncompetitive product inhibition. In extreme oversaturation, hardly any E_2 and E_1 are present, and the rate becomes wholly limited by the conversion of E_2 into E_1.

The oversaturation phenomenon made it possible to study the mechanism of the interconversion of the two enzyme forms, with the following results. The deuterium fractionation factor (D/H) for the rate of conversion of one enzyme form into the other proved to be about 0.5, a value that was consistent only with cysteine-thiol groups in proteins (Belasco et al., 1986a,b). This supported the original assignment of cysteine residues as the acid-base catalysts. The interconversion of enzyme forms proved to be catalyzed by buffers, proving that the two cysteine residues could not directly interact by proton transfer between them.

350   Enzymatic Reaction Mechanisms

The D/H fractionation factors for racemization in the two directions under saturating (non-oversaturating) conditions, which refers to the catalytic processes, also were less than 0.5, consistent with catalysis by thiol groups (Fisher et al., 1986b). It was further found that the deuterium kinetic isotope effect in the reactions of L- or D-2-deuteroproline decreased at very high substrate concentrations, consistent with isomerization of E_2 into E_1 becoming rate limiting under oversaturating conditions.

Glutamate Racemase

Glutamate racemase (EC 5.1.1.3) catalyzes the racemization of glutamate by a two-base mechanism driven by two cysteine residues similar to that of proline racemase. This enzyme plays an important role in the biosynthesis of peptidoglycans for bacterial cell walls. Glutamate racemase is activated by UDP-*N*-acetylmuramyl-L-alanine, a peptidyoglycan precursor (Ho et al., 1995; Doublet et al., 1994).

Glutamate racemase from *Lactobacillus fermenti* is monomeric in solution, with a molecular mass of 28.3 kDa, and it displays values of 0.3 mM and 70 s^{-1} for K_m and k_{cat} in both reaction directions (Gallo and Knowles, 1993). The enzyme from *Aquifex* crystallizes as a dimer, which seems to be the functional form. It displays an overshoot in D_2O and the other mechanistic properties of proline racemase (Gallo et al., 1993; Tanner et al., 1993). The observation of significantly different deuterium kinetic isotope effects in the forward and reverse directions (2.5 and 3.4) was consistent with asymmetry in the action of the two cysteine residues.

Mutations of Cys73 and Cys184 to alanine inactivate the enzyme as a racemase. However, the C73A-enzyme catalyzes elimination of HCl from *threo*-3-chloro-L-glutamate and the C184A-enzyme catalyzes elimination from *threo*-3-chloro-D-glutamate according to scheme 7-7 (Tanner et al., 1993).

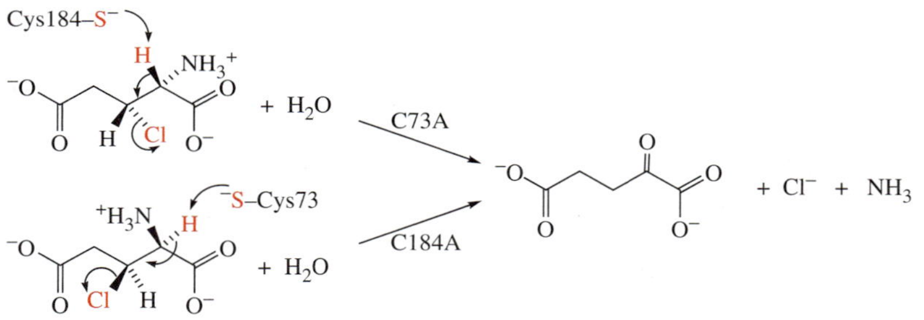

Scheme 7-7

The stereochemistry suggests that Cys73 abstracts the C2(H) from D- and Cys184 from L-glutamate in the racemization reaction. This assignment is supported by the stereochemistry with which the mutated enzymes catalyze the dehydrations of D- and L-*N*-hydroxyglutamate (Glavas and Tanner, 1999).

Although the structure of proline racemase is not known, that of glutamate racemase from *Aquifex pyrophilus* is available (Hwang et al., 1999). The structure in fig. 7-10 illustrates how the active site is constructed, with Cys70 and Cys178 (*Aquifex* numbering) on opposite sides of glutamine bound at the active site. Figure 7-10 shows the overall chain fold and an expanded view of the active site. In addition to the two cysteine residues, Asp7, Glu147, and His180 in the active site play important roles in catalysis, as revealed by the impairment in activities (k_{cat} and k_{cat}/K_m) on mutation of these residues (Glavas and Tanner, 2001; Hwang et al., 1999).

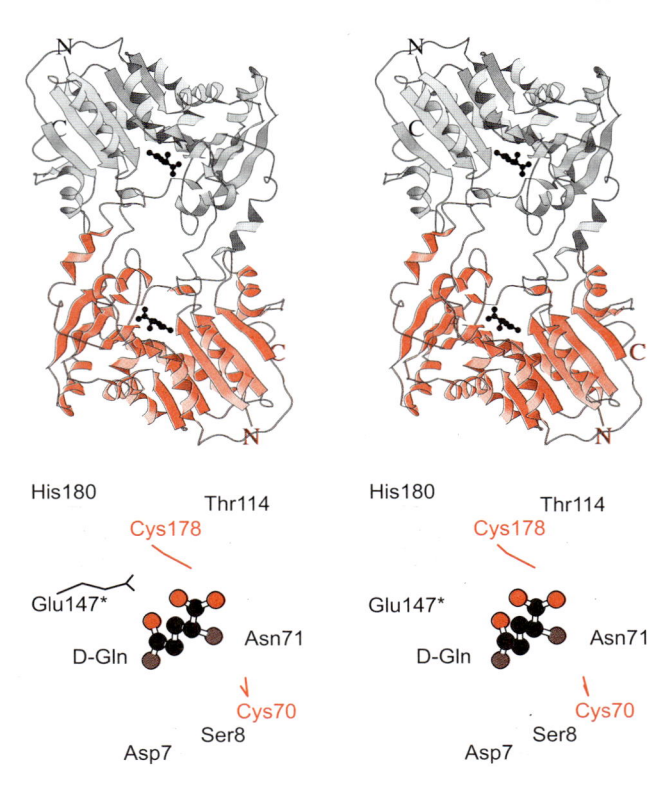

Fig. 7-10. Structure of glutamate racemase from *Aquifex pyrophilus*. Like proline racemase, glutamate racemase functions by a two-base iso-mechanism. The stereodrawing (bottom) shows a ball-and-stick model of glutamine at the active site, with Cys70 and Cys178 on opposite sides of the α-carbon of substrate analog D-glutamine. Residue Glu147* is from the other subunit (PDB 1B74; Hwang et al., 1999).

Ionization of Amino Acid α-Protons

A fundamental issue in the mechanisms of the reactions of proline and glutamate racemases is the abstraction of the C2(H) proton of an α-amino acid by a cysteine-thiolate. The value of pK_a for cysteine is 8.5, and the pH-rate profile for the action of proline racemase is in accord with this value. The value of pK_a for C2(H) of an amino acid is about 29 (Rios et al., 2000) (see chap. 1). A thiolate would not seem to be sufficiently basic to abstract a C2(H) proton from an amino acid. Presumably the basicities of the cysteine thiolates in the active sites of these racemases may be increased, and the C2(H)-acidities of the amino acid substrates may be increased in the Michaelis complexes. However, the structure of glutamate racemase in fig. 7-10 does not offer any insights into how these effects may be brought about in the active site. The absence of a metal cofactor to coordinate with the substrates and increase the acidity of C2(H) in the substrates leaves open the question of how the thiolates can carry out racemization by abstracting the α-protons.

Several other enzymes that catalyze the ionization of α-protons, including enolase (see chap. 1), mandelate racemase in the following section, and aldose/ketose isomerases in preceding sections, the acidities of α-protons seem to be increased by metal ion complexation or strong hydrogen bonding.

Mandelate Racemase

A two-base racemization mechanism is also adopted in a variant fashion by mandelate racemase in the interconversion of L- and D-mandelate (eq. 7-5). Mandelate racemase (EC 5.1.2.2) is a dimer of dimers in solution with a subunit molecular mass of 38 kDa.

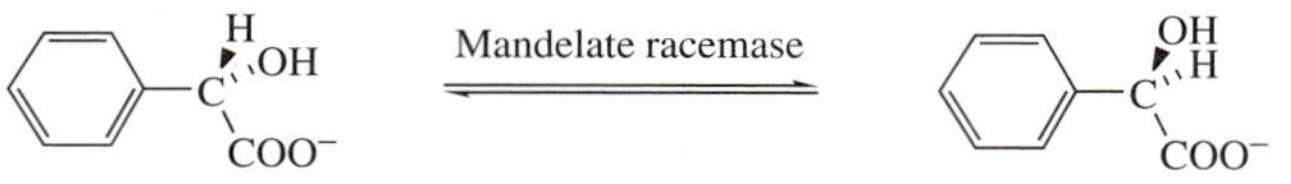

$$\text{L-mandelate} \xrightleftharpoons{\text{Mandelate racemase}} \text{D-mandelate} \tag{7-5}$$

As an α-hydroxyacid, the pK_a of mandelic acid is thought to have a value of about 30. Unlike glutamate and proline racemases, mandelate racemase requires a divalent metal ion (Mg^{2+}) for activity. The likely role of Mg^{2+} is to lower the pK_a for ionization of the α-proton, C2(H), to a value in the physiological pH range. The active site also includes a lysine and a histidine residue, one of which is always positively charged and contributes to the positive electrostatic field that locally neutralizes the carboxylate group of mandelate. The possibility that low-barrier hydrogen bonding with Glu317 contributes to stabilizing the enolate has also been suggested (Gerlt et al., 1997).

Unlike proline and glutamate racemases, the acid-base catalysts are not cysteine-thiol groups (Landro et al., 1991; Neidhart et al., 1991; Powers et al., 1991). The first indication of this came in studies modeled after those conducted with proline racemase. The racemization of α-deuteromandelate proceeds with a primary deuterium kinetic isotope effect, and the racemization of unlabeled mandelate in D_2O displays an overshoot. However, unlike proline racemase, the enzyme catalyzes the exchange of deuterium into unreacted L-mandelate from the start of its racemization. Like proline racemase, it does not catalyze exchange of deuterium into unreacted D-mandelate during racemization in D_2O. This can be explained on the basis that one of the acid-base groups is polyprotic and the other is monoprotic, as in scheme 7-8, in which L-mandelate (S-mandelate) is pictured interacting with two acid-base groups at the active site.

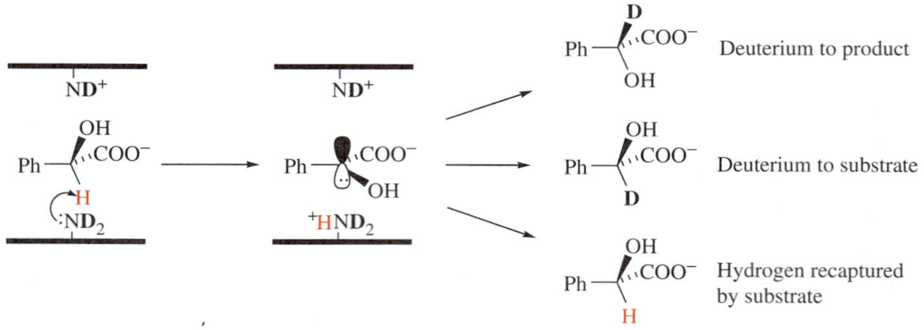

Scheme 7-8

The base :ND_2 can abstract the α-proton to form the carbanionic intermediate. The intermediate is flanked by two acid groups, $^+HND_2$ at the si-face and ND^+ at the re-face. The intermediate can be protonated on its si-face either by the same proton it had donated or by one of the deuterons originally bonded to :ND_2; the polyprotic nature of this base explains how L-mandelate can undergo deuterium exchange from the beginning of its racemization. The intermediate can also be deuterated by ND^+ to form α-deutero-D-mandelate. The exercise in scheme 7-8 carried out with D-mandelate gives the opposite result, no incorporation of deuterium into the unreacted substrate because of the monoprotic acid-base catalyst.

The structural framework of mandelate racemase is an α,β-barrel (TIM barrel) shown in fig. 7-11 for the variant K116R, and it is a member of the enolase superfamily. These enzymes include enolase (see chap. 1), galactonate dehydratase, glucarate dehydratase, muconate-lactonizing enzymes, *N*-acylamino acid racemase, β-methylaspartate ammonia-lyase, and *O*-succinylbenzoate synthase (Babbitt et al., 1996). All of the members of this superfamily catalyze the removal of an α-proton from a carboxylic acid. The structure at the active site shows how Mg^{2+} is coordinated to the α-hydroxyl and carboxylate groups of mandelate, as well as to acidic residues of the enzyme. The acid-base catalysts Lys166 and His297 in the wild type enzyme flank the substrate on opposite sides, with Lys166 in position to interact with the *si*-face and His297 on the *re*-face of the putative carbanionic intermediate. In scheme 7-8, ND_2 is Lys166, and ND^+ is His297.

Mutations of Lys166 and His297 inactivate mandelate racemase. However, mutation of His297 does not disable Lys166 as a catalyst. This is shown by the fact that the H297N-racemase, although inactive in racemization, still catalyzes the exchange of deuterium from D_2O into L-mandelate and the elimination of bromide from L-*p*-(bromomethyl) mandelate (Landro et al., 1991). The pH-rate profile for this exchange is instructive, in that it displays a break at pH 6.4, corresponding to the pK_a of an enzymatic group, presumably Lys166. The low pK_a of Lys166 is explained, as in the case of acetoacetate decarboxylase (see chap. 1) by the proximity of Lys164 in the structure (fig. 7-11). The positive ε-aminium group of Lys164 placed near Lys166 raises the barrier to adding a proton to Lys166 and decreases its basicity.

The proximity of His297 to the positive electrostatic field created by Lys164 and Mg^{2+} raises the question of how His297, which catalyzes prototrophy on the *re*-side of the enolic-intermediate (Landro et al., 1991), can display a "normal" value of pK_a. The structure shows His297 hydrogen bonded to Asp270, and mutation of the latter to Asn270 decreases enzymatic activity about 10,000-fold relative to the wild type, although no significant global structural change accompanies the mutation (Schafer et al., 1996). The mutated enzyme catalyzes exchange of the α-proton of L- but not of D-mandelate, and it catalyzes the elimination of bromide from L-*p*-(bromomethyl)mandelate but not from D-*p*-(bromomethyl)mandelate. Asp270 seems as important to catalysis of proton transfer from D-mandelate as is His297 itself. The mutation D270N alters the alkaline limb of the pH-rate profile in the L to D direction to a pK_a of 10. All evidence indicates that the "normal" pK_a exhibited by His297 in wild-type mandelate racemase is due to its hydrogen-bonded interaction with Asp270.

Enolization of mandelate requires stabilization of the *aci*-carbanion intermediate. Chelation by Mg^{2+} can stabilize this intermediate; however, Glu317 is also in position to interact with the carboxylate of the *aci*-carbanion. Structural information suggests that Glu317 may contribute a hydrogen bond, perhaps a strong hydrogen bond, to the *aci*-carboxylate group. Mutation to Gln317 does not bring about a significant change in the global structure but decreases activity 4500-fold in the D-to L-direction and 29,000-fold in the L-to D-direction (Mitra et al., 1995).

The reaction of mandelate racemase from *Pseudomonas putida* is partially diffusion controlled (St Maurice and Bearne, 2002). An assessment of the activation parameters, after correction for partial diffusion limitation, indicates that the principal contribution to the free energy of transition-state stabilization for *aci*-carbanion formation is enthalpic ($\Delta H^\ddagger = 23$ kcal mol^{-1}), with a significant entropic component ($T\Delta S^\ddagger = +2$ kcal mol^{-1}). The principal stabilization of the transition state may be attributed to hydrogen bonded, electrostatic, and hydrophobic interactions between the transition state and enzymatic side chains and Mg^{2+}.

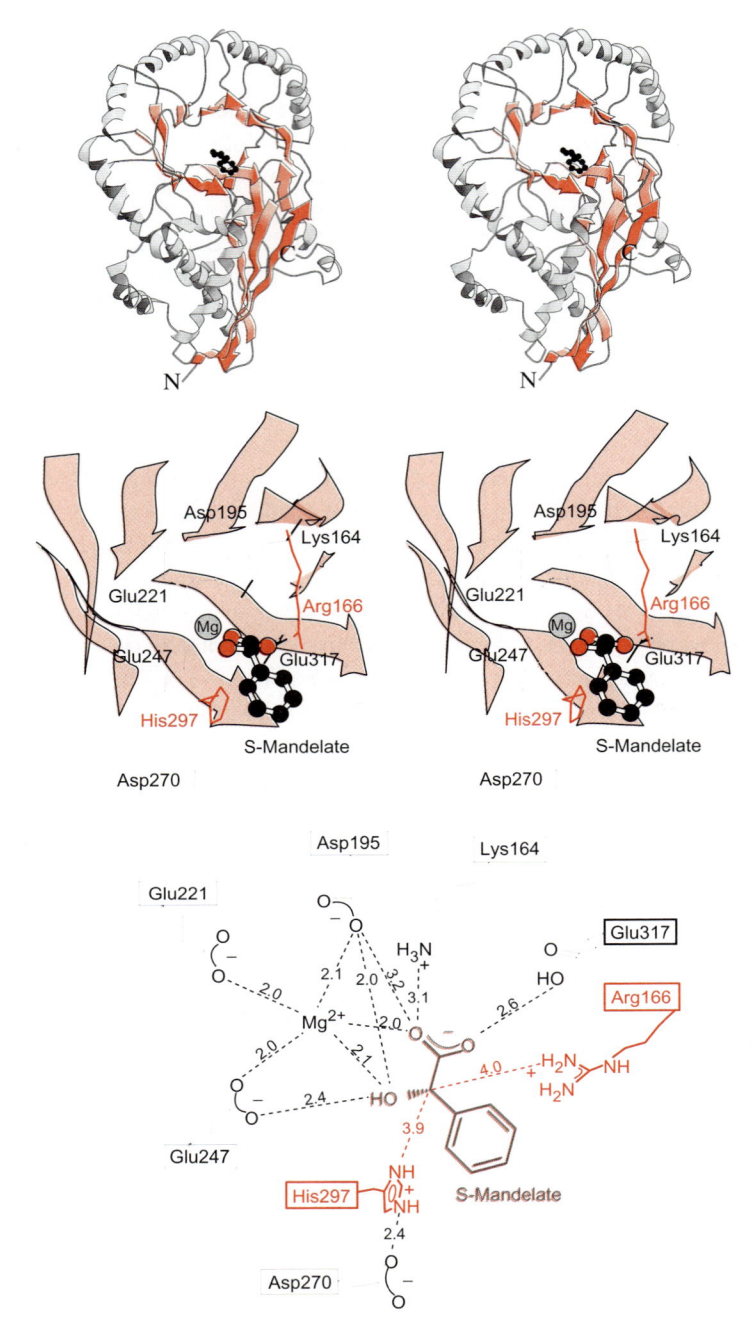

Fig. 7-11. Structure of mandelate racemase. Mandelate racemase from *Pseudomonas aeruginosa* is another triosephosphate isomerase (TIM) barrel enzyme, and the structure of the K166R variant is shown (top) with *S*-mandelate bound at the active site (PDB 1MDL; Kallarakal et al., 1995). A ball-and-stick model (middle) of *S*-mandelate in stereo at the active site shows its proximity to amino acid side chains. His297 and Lys166 are the acid-base catalysts in the wild-type enzyme. In this variant enzyme, Lys166 has been changed to an arginine to remove its function as an acid-base catalytic residue. The two-dimensional map (bottom) of the active site shows the close contacts between *S*-mandelate and the variant.

UDP-Galactose 4-Epimerase

All racemases discussed to this point and others encountered later in this volume function by two-base mechanisms. UDP-galactose 4-epimerase (GalE; EC 5.1.3.2) catalyzes the epimerization of glycosyl-C4 in UDP-galactose 4-epimerase by a radically different mechanism. The reaction is the interconversion of UDP-galactose and UDP-glucose, eq. 7-6.

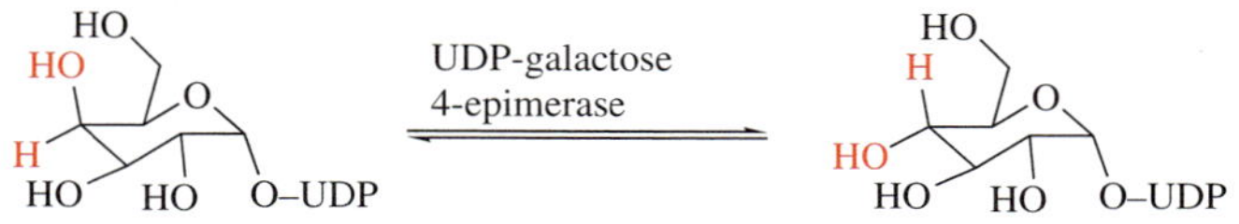

$$\text{UDP-galactose} \underset{}{\overset{\text{UDP-galactose 4-epimerase}}{\rightleftharpoons}} \text{UDP-glucose} \qquad (7\text{-}6)$$

The enzyme is a dimer of identical subunits and plays an essential role in the production galactosyl units and the metabolism of galactose as an energy source in all cells. The most thoroughly studied is GalE from *Escherichia coli*, which has a subunit molecular mass of 39 kDa. GalE is a product of the *gal* operon in *E. coli*, which encodes the enzymes of galactose metabolism. The other products of the gal operon are galactose mutarotase (GalM), galactokinase (GalK), and galactose-1-P uridylyltransferase (GalT).

GalE and all other 4-epimerases are NAD$^+$ dependent, and the coenzyme plays a central role in the mechanism. The enzyme displays a moderately high turnover number ($k_{cat} = 760$ s^{-1}) and the value of k_{cat}/K_m is 3.4×10^6 M^{-1} s^{-1} (Liu et al., 1997). The overall mechanism in fig. 4-12 is extensively documented and generally accepted. Reversible dehydrogenation of either UDP-galactose or UDP-glucose by NAD$^+$ to form NADH and UDP-4-ketoglucose as tightly bound intermediates can be regarded as a minimal description of the mechanism. A more complete description is that in fig. 7-12, which encompasses the foregoing description but includes a rationale for nonstereospecificity in the dehydrogenation by NAD$^+$. We discuss four aspects of the mechanism: dehydrogenation at glycosyl-C4 by NAD$^+$, the molecular basis for nonstereospecificity in hydride transfer by NAD$^+$, the mechanism of hydride transfer, and the enhancement of the reactivity of NAD$^+$ by uridine nucleotides. The accumulated evidence bearing on these processes compels acceptance of the transformations in fig. 7-12.

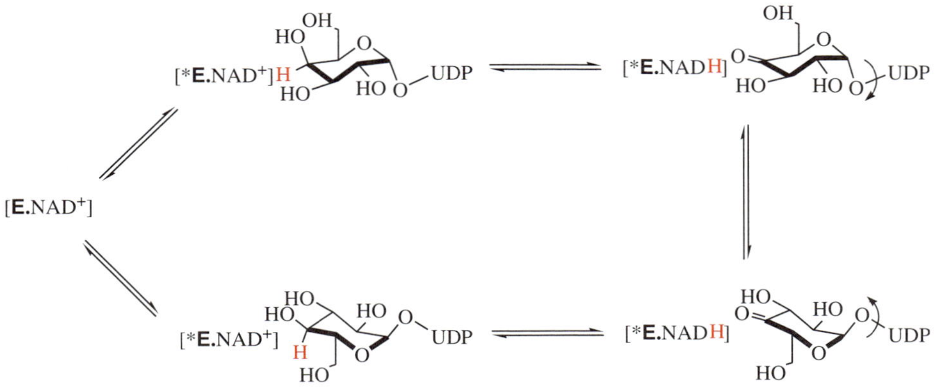

Fig. 7-12. Epimerization mechanism for the action of UDP-galactose 4-epimerase. Extensive evidence proves that NAD$^+$ tightly bound at the active site reversibly dehydrogenates glycosyl-C4 of UDP-glucose or UDP-galactose. The nonstereospecificity of this process leads to epimerization at glycosyl-C4.

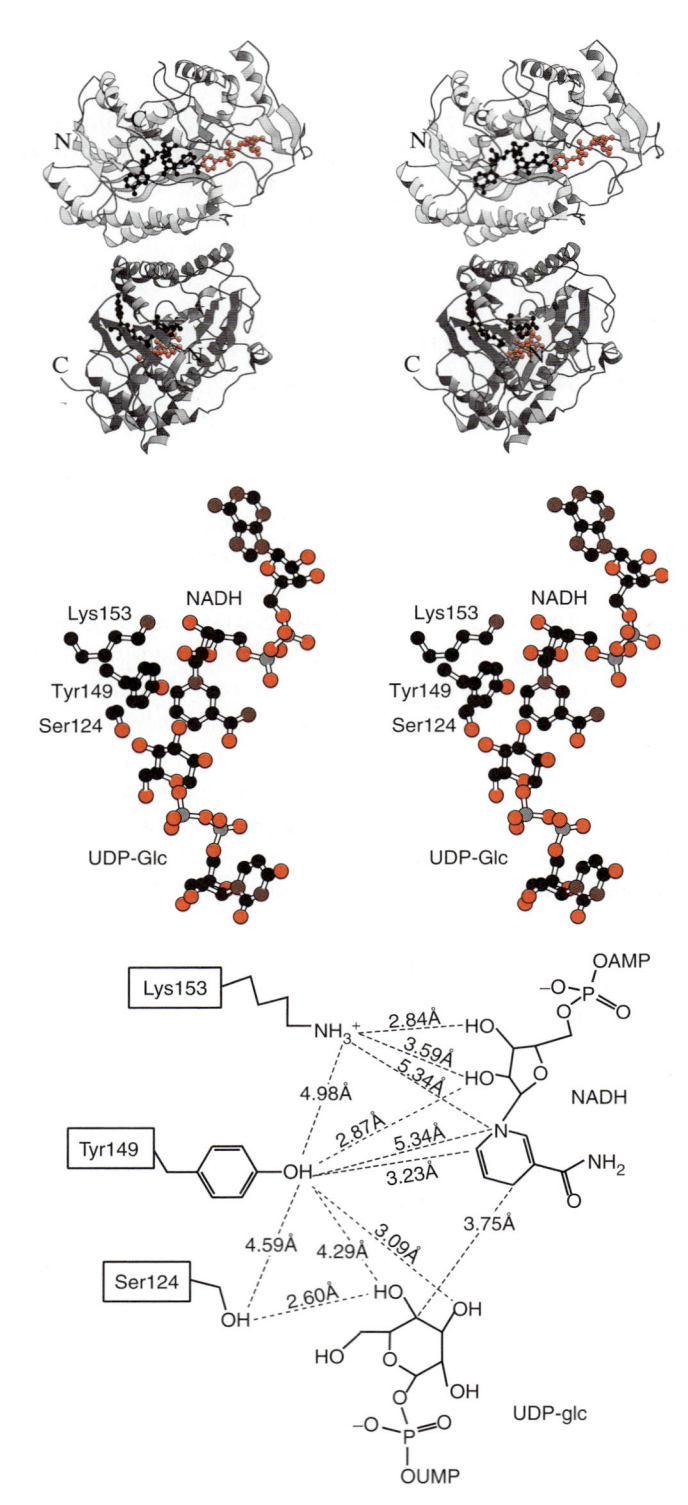

Fig. 7-13. The structure of UDP-galactose 4-epimerase (GalE) from *E. coli* consists of a dinucleotide fold, which binds NAD$^+$, and a smaller domain that binds UDP-glucose or UDP-galactose, which is shown (top) with NAD+ in black and a substrate analog, UDP-phenol, in red (PDB 2UDP; Thoden et al., 1996a). The two domains create a cleft, and the binding site for the substrate fixes carbon-4 of the substrate pyranosyl ring near carbon-4 of the nicotinamide ring of NAD$^+$. A ball-and-stick model in stereo (middle) shows UDP-glucose bound to the active site of GalE with NADH in an abortive complex (PDB 1XEL; Thoden et al., 1996c). This model shows the amino acid side chains of Tyr149, Ser124, and Lys153 near the substrate glycopyranosyl ring. A two-dimensional map (bottom) illustrates the interactions of NAD$^+$ and UDP-glucose with GalE.

Dehydrogenation at Glycosyl-C4

GalE contains one tightly bound molecule of NAD$^+$ per subunit when purified to homogeneity, and it is transiently reduced to NADH when substrate is added (Liu et al., 1997; Nelsestuen and Kirkwood, 1971; Wilson and Hogness, 1964). The C4(H) and C4(O) of UDP-glucose are retained in the product; that is, they are not exchanged with water, and tritium is transferred between NADH and UDP-4-ketoglucose at the active site (Adair et al., 1973; Anderson et al., 1956; Kowalsky and Koshland, 1956; Maitra and Ankel, 1971; Nelsestuen and Kirkwood, 1971, Wee and Frey, 1973). Epimerization of substrates labeled with tritium at glycosyl-C4 proceed with very small tritium kinetic isotope effects that are slightly higher than secondary effects and suggest that the C4(H) bond is cleaved in the reaction. Tritium is transferred from substrates or NaB^{3}H$_4$ to the B side (pro-S) of NAD$^+$ (Wee and Frey, 1973). These facts implicate NAD$^+$ in transient dehydrogenation of glycosyl-C4 and rule out plausible alternative mechanisms proceeding by transient dehydrogenation at C2, C3, or C6. The observation of compulsory intramolecular transfer of deuterium from glucosyl-C4 to galactosyl-C4 proves that the transformation takes place at a single active site and is brought about by a single NAD$^+$ (Glaser and Ward, 1970).

The x-ray crystallographic structure of GalE, shown in fig. 7-13, reveals a single active site in each subunit, and the two sites in the dimeric molecule are not in contact or connected in any way (Bauer et al., 1992; Thoden et al., 1996a). A novel aspect of GalE.NAD$^+$ is the tight binding between the enzyme and coenzyme, which are purified as a complex. Exchange of radiochemically labeled NAD$^+$ with GalE.NAD$^+$ can be induced by some sort of structural perturbant, such as a low concentration of urea (Burke and Frey, 1993; Vanhooke, 1993). The NAD$^+$ contacts to GalE include eight direct hydrogen bonds between heteroatoms of the adenosyl moiety and enzymatic groups, including main chain and side chain hydrogen bonds, and no fixed water molecules in contact with the riboside (Thoden et al., 1996b). The structures of ordinary pyridine nucleotide-dependent dehydrogenases show very few direct contacts between protein groups and the adenosyl moiety of NAD$^+$, as well as showing the presence of fixed water molecules surrounding the adenosyl moiety. It seems that the hydrogen-bonding network ensures the tight binding of NAD$^+$ to GalE.

Nonstereospecificity

Binding and kinetic data prove that strong binding between the active site and the UDP moiety anchors the substrate to the active site, and there is very little binding of the sugar moiety (Kang et al., 1975; Wong and Frey, 1977). Binding information on GalE.NAD$^+$ shows that $\Delta G° = 0$ for binding glucose, and studies of GalE.NADH indicate only -2 kcal mol^{-1} for binding a glycosyl moiety. In contrast, GalE.NAD$^+$ and GalE.NADH bind the UDP moiety with standard free energies of -5 and -7 kcal mol^{-1}, respectively. Weak sugar binding inspired the conceptualization of the conformational change in the UDP-4-ketoglucose intermediate by torsion about the bond connecting the anomeric oxygen of the sugar with the β-phosphorus of the UDP moiety, as illustrated in fig. 7-12, to explain nonstereospecific hydride transfer at C4. Molecular models verify that this torsion would lead to the projection of C4(H), C4(OH), and C3(OH) of C4-epimers in similar arrays toward NAD$^+$, as illustrated in fig. 7-14 for UDP-L-arabinose and UDP-D-xylose, which are epimeric at C4 and interconverted by GalE almost as efficiently as UDP-glucose and UDP-galactose. The models show that NAD$^+$ in a single location within the active site can access C4(H) of the epimeric sugars and abstract it if a base catalyst is placed between C4(OH) and C3(OH) groups.

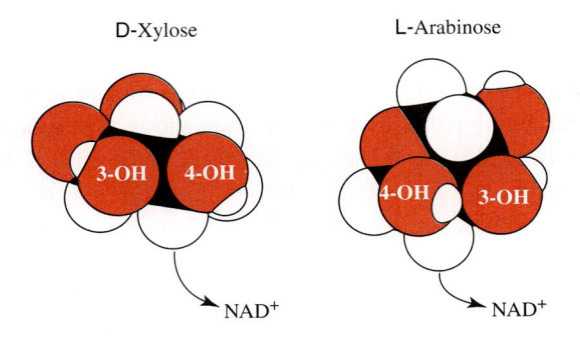

Fig. 7-14. In a model for nonsterospecific hydride transfer by UDP-galactose 4-epimerase (GalE), the stereochemical relationship between the pyranosyl rings of the GalE substrates UDP-D-xylose and UDP-L-arabinose are viewed from the pyranosyl-C4,C3 end. The similarities extend to the reversed positions of the 4-OH and 3-OH and similar positions of the C4-H in the two models, which differ in torsional orientation about the bond linking C_1–O and P_β of UDP.

The binding data and torsional model afford a rationale for the role of the UDP moiety as the binding anchor. Nonstereospecificity in hydride transfer at glycosyl-C4 would not be possible if the glycosyl moiety were rigidly held through many contacts with the active site. Minimal interactions with the glycosyl moiety allow the mobility required for nonstereospecificity in hydride transfer, but the price that must be paid for mobility is weakness in glycosyl binding. The UDP anchor overcomes this, and in its absence 4-ketoglucose would escape from the active site and leave the enzyme inactive in the form of Gal.NADH. fig. 7-15 shows the binding conformations of UDP-glucose, UDP-6-deoxy-6-fluoroglucose, and UDP-4-deoxy-6-fluorogalactose in the active site (Thoden et al., 1997). The glycosyl groups are bound in various orientations differing mainly by torsion about the bond linking the anomeric oxygen and the β-phosphorus of UDP, which is held

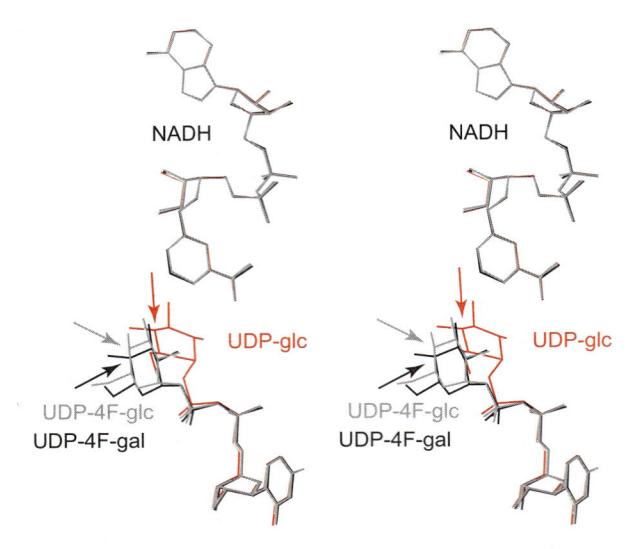

Fig. 7-15. The binding conformations of UDP-sugars at the active site of UDP-galactose 4-epimerase (GalE) are shown for UDP-glucose, UDP-4-deoxy-4-fluoroglucose, and UDP-4-deoxy-4-fluorogalactose. Notice that the pyranosyl moieties differ in torsional orientation about the pyranosyl-C_1–O and P_β of UDP, confirming the proposed mechanism for nonstereospecific action.

in essentially the same conformation. The structure with UDP-glucose and UDP-galactose bound to an inactive doubly mutated GalE further supports the binding model (Thoden and Holden, 1998). This structural evidence proves the binding flexibility in the glycosyl subsite and further supports the anchoring role of UDP and the model in figs. 7-12 and 7-14 for nonstereospecific hydride transfer.

Mechanism of Hydride Transfer

Affinity labeling experiments failed to identify an acid-base catalyst in the active site. Such a catalyst is required to support proton transfer accompanying hydride transfer in the dehydrogenation of an alcoholic group. The crystal structure revealed the basis of acid-base catalysis in the action of GalE (Liu et al., 1997; Thoden et al., 1996c). In the structure, Tyr149 and Ser124 interact with the C4(OH) group of the substrate through hydrogen bonding, and Lys153 binds the 2- and 3-OH groups of the nicotinamide riboside moiety of NAD$^+$ through a bifurcated hydrogen bond. Mutation of Tyr149 to Phe decreases the value of k_{cat} to 1/10,000 that of wild-type enzyme, and mutation of Ser124 to Ala decreases the activity to 1/3000 of wild-type, both with modest changes in the value of K_m. The value of pK_a for Tyr149 is 6.1, as determined spectrophotometrically (Liu et al., 1997), so that it exists as the phenolate ion at neutral pH and can serve as the base catalyst for dehydrogenation, as illustrated in scheme 7-9.

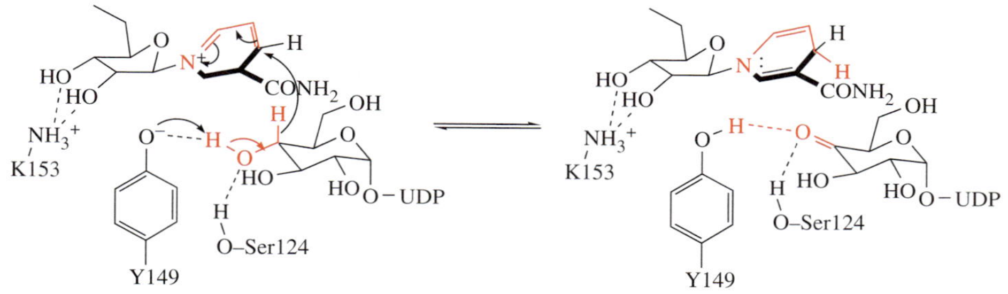

Scheme 7-9

Epimerization by wild-type GalE displays no pH dependence and no deuterium kinetic isotope effect in the reaction of UDP-D-glucose-d_7 (Berger et al., 2001). The absence of a deuterium kinetic isotope effect and the very small value of the tritium kinetic isotope effect prove that hydride transfer does not limit the rate and suggest that a conformational change controls the rate. The conformational change may include rotation about the bond linking the anomeric oxygen to the β-phosphorus of the UDP moiety. Mutation of Ser124 to Ala124 changes the rate-limiting step to hydride transfer, as shown by the deuterium kinetic isotope effect of 2.5 (Berger et al., 2001). The spectrophotometrically measured pK_a of Tyr149 in S124-GalE is 6.3, and that for epimerization by the mutated enzyme (log k_{cat}/K_m versus pH) is also 6.3, proving that tyrosine functions as the acid-base catalyst in hydride transfer (Berger et al., 2001).

Uridine Nucleotide–Induced Reductive Inactivation

The 4-epimerases from yeast and *E. coli* display the property of UMP-enhanced reduction of NAD$^+$ to NADH by sugars, including galactose, glucose, arabinose, and xylose among others (Kalckar et al., 1970). UMP and UDP also enhance the rate of reductive inactivation by NaBH$_3$CN (Davis et al., 1974; Liu et al., 2000). Mutations of Lys153 or Tyr149 dramatically decrease and essentially abolish phenomenon (Liu et al., 1997; Swanson et al., 1993).

The binding of UDP to GalE.NAD⁺ leads to significant polarization of the π-electron system in the nicotinamide ring, as shown by the downfield perturbation of the ^{13}C NMR signal for [*nicotinamide*-4-^{13}C]NAD⁺ and the upfield perturbation of the ^{15}N NMR signal for [*nicotinamide*-1-^{15}N]NAD⁺ bound to GalE (Burke and Frey, 1993). The effects of nucleotides on the NMR signals are abolished by mutation of Tyr149 to Phe149 and of Lys153 to Met153 (Wei et al., 2001). The magnitude of the ^{13}C NMR perturbation corresponds to a 150 mV more positive reduction potential for NAD⁺, and this correlates very well with the magnitude of the rate enhancement (≈3000-fold) brought about by UDP (Burke and Frey, 1993; Liu et al., 2000).

All the kinetic and spectroscopic results suggest that the binding of uridine nucleotides activates NAD⁺ through an electrostatic effect. As shown in fig. 7-13 and scheme 7-9, the Tyr149-phenolate ion interacts with the positive electrostatic field created by the ε-iminium group of Lys153 and the quaternary N1 of NAD⁺. A lateral motion of Tyr149 can lead to a polarization of the p-electron system of NAD⁺, increasing its reduction potential and reactivity.

Ribulose-5-P 4-Epimerase

The C4(H) of ribulose-5-P is not activated for abstraction of a proton, but epimerization at C4 is catalyzed by ribulose-5-P 4-epimerase (EC 5.1.3.4). The enzyme does not contain NAD(P)⁺ and is not activated by a pyridine nucleotide, ruling out a dehydrogenation mechanism. In the open-chain sugar, the 2-keto group makes C3(H) acidic enough to be abstracted in an enolization, and in principle, this could allow dehydration between C3 and C4. Hydration of the resulting double bond in the opposite stereo sense could allow for epimerization at C4. However, this mechanism is not sustained by the experimental results; for example, there is no exchange of C3(H) with solvent hydrogen (McDonough and Wood, 1961).

As shown in fig. 7-16, the ability of the 2-keto group to induce enolization can be brought into play in another mechanism, a retro-aldol process with cleavage between C3 and C4. This creates two-substrate fragments as intermediates, glycolaldehyde-P from C4 and C5 and the enol of dihydroxyacetone-P from C1, C2, and C3, both of which remain bound to the active site. Torsional reorientation of the aldehydic carbon in glycolaldehyde-P followed by reformation of the C3–C4 bond generates the epimer. The mechanism requires an acid-base catalytic group to abstract the proton from C4(OH) in the first step and return it in the last step, and it requires a means of stabilizing the enolate intermediate. This mechanism is experimentally supported.

Epimerization of [3-^{13}C]- and [4-^{13}C]ribulose-5-P proceed with primary ^{13}C primary kinetic isotope effects of 1.85% and 1.5%, respectively, at pH 7 and increasing to 2.5% and 2.1% at pH 5.5, showing that that the C3–C4 bond is cleaved in the reaction (Lee et al., 2000a). The low values at pH 7 are attributed to C3–C4 cleavage not being solely rate limiting. At pH 5.5, the cleavage is more highly rate limiting, with the result that the isotope effects are larger. The mutations H97N and Y229F increase the isotope effects further. Reactions of [3-^{2}H]- and [4-^{2}H]ribulose-5P proceed with secondary deuterium kinetic isotope effects, which also increase at low pH and in the reactions of the mutated epimerase.

The amino acid sequence of ribulose-5-P 4-epimerase is related to class II aldolases, the divalent metal ion–dependent aldolases, and the activity of 4-epimerase is Zn^{2+} dependent, consistent with the involvement of an aldol-type mechanism (Johnson and Tanner, 1998; Lee et al., 2000b). Ribulose-5-P 4-epimerase carrying the mutation H97N can catalyze the condensation of dihydroxyacetone-P and glycolaldehyde-P to a mixture of ribulose-5-P and xylulose-5-P at a slow rate.

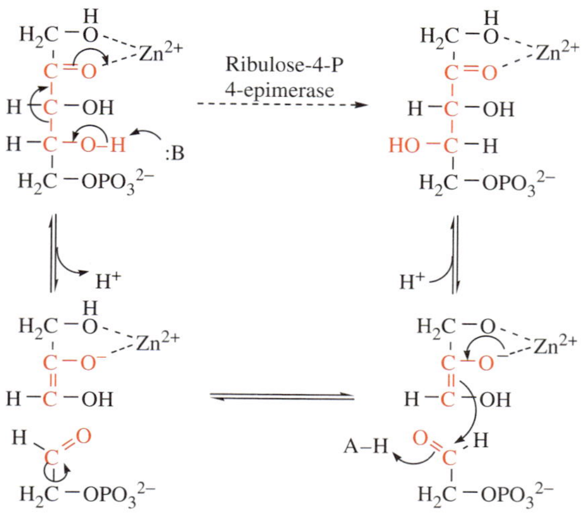

Fig. 7-16. Mechanism of epimerization for ribulose-5-P 4-epimerase. An enzymatic base :B abstracts the proton from C4(OH), promoting cleavage of the C4–C3 bond and forming glycolaldehyde-P and the enolate of dihydroxyacetone-P, both of which remain bound to the active site. The enolate is stabilized by the coordinated Zn^{2+}. Rotation about the aldehydic C–C bond of glycolaldehyde-P reorients the aldehyde group, and addition of the enolate to the aldehyde to form the epimer is facilitated by proton transfer to the aldehyde oxygen by the enzymatic acid A–H.

The structure of ribulose-4-P 4-epimerase is consistent with the mechanism in fig. 7-16 (Luo et al., 2001). The global structure and the active site, which is formed at the interface of subunits, are mapped in fig. 7-17. The structure and mutagenesis data indicate that the acid-base catalysts are Asp120′ and Tyr229′ from the subunit adjacent to the binding site for ribulose-5-P and Zn^{2+} (Lee et al., 2000b; Samuel et al., 2001).

UDP-*N*-Acetylglucosamine-2-Epimerase

Epimerization by reversible elimination may be added to epimerization by reversible dehydrogenation and reversible aldol cleavage discussed in the preceding sections. Elimination is the means by which bacterial UDP-*N*-acetylglucosamine-2-epimerase (EC 5.1.3.14) produces UDP-*N*-acetylmannosamine, as illustrated in scheme 7-10.

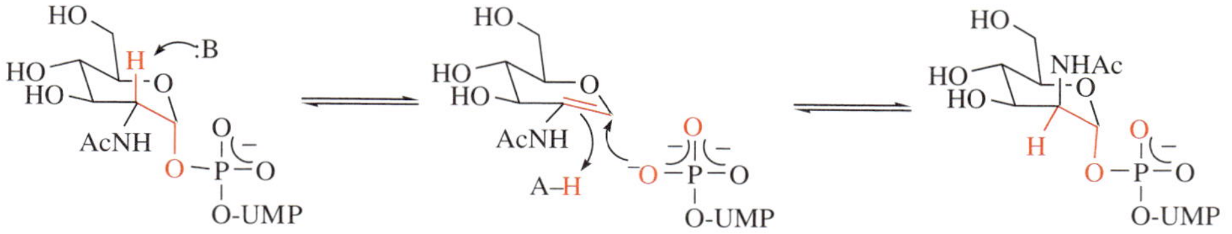

Scheme 7-10

The mechanism requires *anti*-elimination of UDP to form the glucal followed by *syn*-readdition of UDP to C1 of the glucal. It is a two-base mechanism with the two acid-base catalysts engaged in proton transfer to and from C2 at opposite faces of the sugar.

The most incisive evidence implicating elimination is cleavage of the bond linking C1 with the anomeric oxygen in the course of the reaction. Proof of cleavage is the fact that the anomeric oxygen undergoes positional exchange with the two peripheral

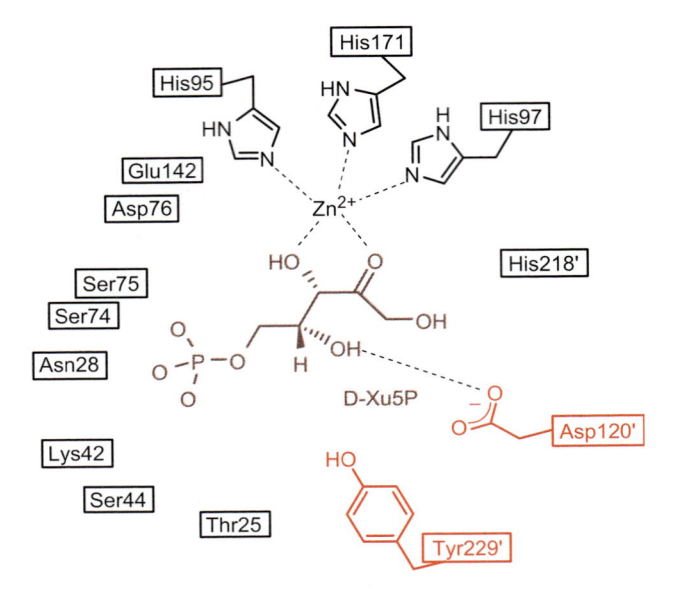

Fig. 7-17. The schematic structure of ribulose-5-P 4-epimerase shows D-xylulose-5-phosphate placed in the active site, interacting with the zinc ion. The only structure available of ribulose-5-P 4-epimerase lacks substrate, product, or analogs, but it clearly shows the coordinated zinc ion and the various active site residues that are donated by two adjacent subunits (PDB 1K0W; Lou et al., 2001). The exact contacts between substrates and enzyme are unknown. A few active-site residues have been shown to be important for catalysis using site-directed mutagenesis and steady-state kinetic analysis. The core protein fold (not shown) and active site are conserved between ribulose-5-P 4-epimerase and L-fuculose-1-P aldolase, which share common mechanistic elements.

oxygens bonded to the β-phosphorus of the UDP moiety (Morgan et al., 1997; Sala et al., 1996). We illustrate this in scheme 7-10 by tracing the anomeric oxygen in color and showing that it does not necessarily return to the anomeric position in the product. In the elimination of UDP from UDP-*N*-acetylglucosamine, the anomeric oxygen becomes equivalent with the other two peripheral oxygens bonded to the β-phosphate of UDP. On reformation of the bond to C1, any one of these three oxygens becomes the anomeric oxygen.

^{31}P NMR experiments with ^{18}O-labeled substrate reveal positional isotope exchange of ^{18}O. Oxygen-18 bonded to phosphorus shifts the ^{31}P NMR signal slightly upfield to varying degrees depending on the P–O bond order. With ^{18}O in the anomeric position, the P–^{18}O bond order is 1.0 (scheme 7-10) giving an isotope shift of about 0.017 ppm. With ^{18}O in a nonbridging position of the product, the P–O bond order is 1.5 and an isotope shift about 0.025 ppm. The ^{31}P NMR signals for the β-phosphorus of the UDP moiety in the substrate and 2-epimeric product, starting with ^{18}O in the anomeric position of UDP-*N*-acetylglucosamine, clearly show positional isotope exchange of ^{18}O, proving cleavage of the C1(O) bond in the reaction.

A structure of the bacterial 2-epimerase·UDP complex at 2.5 Å (fig. 7-18) reveals the chain fold as an αβα-sandwich, with the active site on the surface (Campbell, 2000). The fold is related to that of phosphorylase, which also catalyzes C–O cleavage at α-D-aldose-1-phospho sites (see chap. 12). The active site includes basic amino acids that may be involved in proton transfer.

The mammalian 2-epimerase is related but different in that it is a bifunctional protein, UDP-*N*-acetylglucosamine/mannosamine kinase, and catalyzes cleavage of the substrate

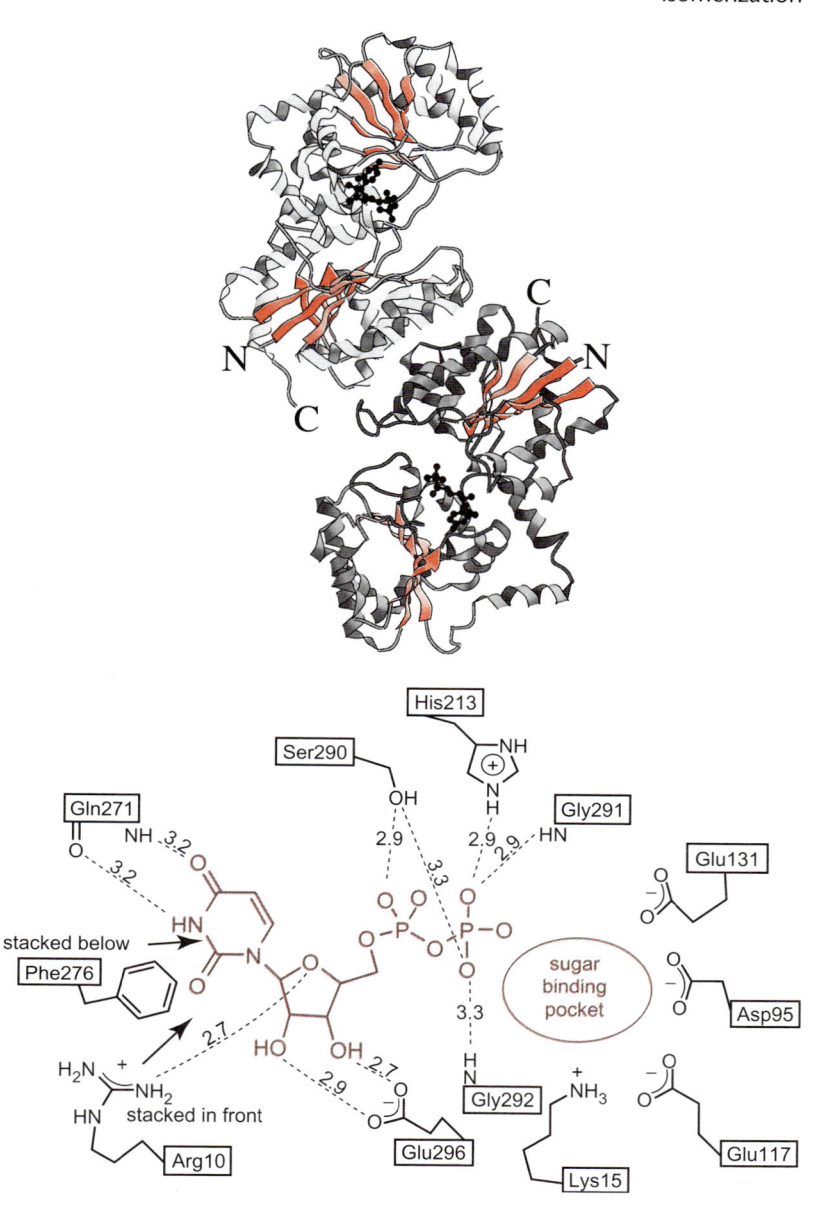

Fig. 7-18. Structure of UDP *N*-acetylglucosamine 2-epimerase at its active site. The *E. coli* UDPglcNac 2-epimerase is a homodimeric enzyme with two subdomains sandwiched together to form a large active-site cleft, shown containing UDP (black) (PDB 1F6D; Campbell et al., 2000). At the core of each subdomain are parallel β-sheets (red). A schematic of the active site is shown below the enzyme fold. The absence of the hexose moiety in the structure makes it impossible to show the exact contacts, but several likely candidates are included based on proximity to the β-phosphate of UDP.

to α-mannosamine and its phosphorylation by ATP to mannosamine phosphate. *anti*-Elimination of UDP is similar to the bacterial enzyme, and this is followed by *syn*-addition of water to the glycal (Chou et al., 2003). Experiments similar to those described for proline racemase implicate a two-base mechanism in abstraction and return of a proton at C2, as implied in scheme 7-10. In D$_2$O deuterium is incorporated at C2 of the product but not into the residual substrate.

Chorismate Mutase

Chorismate mutase (EC 5.4.99.5) catalyzes the isomerization of chorismate into prephenate, an essential step in the biosynthesis of aromatic amino acids, as illustrated in fig. 7-19. Aromatic amino acids are essential nutrients for mammals, which lack the enzymes for their biosynthesis, including chorismate mutase and enolpyruvoylshikimate phosphate synthase (see chap. 5).

The reaction of chorismate mutase is interesting for the apparently simple chemistry involved and the apparently simple requirements for catalysis. There is no obvious need for acid-base or nucleophilic catalysis, so a direct chemical interaction between chorismate and enzymatic side chains is not expected. Biochemical and structural studies of chorismate mutase bear out these expectations. The reaction may be regarded as a biological Claisen rearrangement, which is moderately well understood in organic chemistry.

The isomerization of chorismate is presented in fig. 7-19 as a concerted process, with cleavage of the C5–O bond and formation of the new bond to C1 in the transition state. Heavy-atom kinetic isotope effects show that cleavage of the C5–O bond is well advanced in the transition state, whereas the bond linking C1 with the methylene carbon of the pyruvoyl moiety is only slightly formed (Gustin et al., 1999). That is, the mechanism may be described as concerted and asynchronous. The measurements were by the remote label technique on the enzyme from *Bacillus subtilis*, which was found to be only slightly limited by diffusion, and on a mutated variant that was not at all diffusion limited. The reaction of the C5–^{18}O substrate displayed 4.5% and 5.3% kinetic isotope effects in reactions of the wild-type and variant enzymes, respectively. Reactions of [1-^{13}C]chorismate proceeded with 0.43% and 0.57% isotope effects, showing only slight bond formation in the transition state.

The isotope effects are accentuated compared with those of the nonenzymatic reaction, suggesting that the enzymatic process may involve a more dissociative mechanism, in which the C5–O bond is more cleaved in the transition state than in the nonenzymatic reaction.

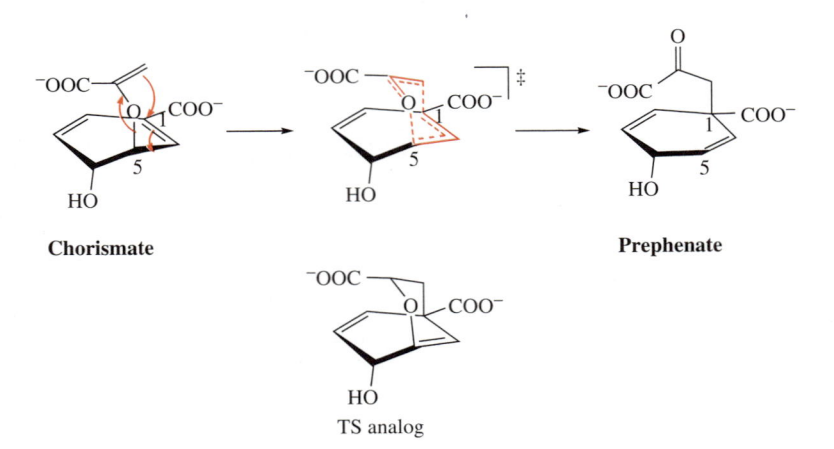

Fig. 7-19. Transformation of chorismate into prephenate by chorismate mutase is an intramolecular electrocyclic reaction, in which the C5(O) bond of chorismate is broken, a new bond between C1 of chorismate and the methylene group of the pyruvyl moiety is formed, and the double bond of C1=C6 migrates to C5=C6. The reaction is analogous to the Claisen rearrangement in organic chemistry and is not subject to catalysis by acids or bases. The nature of the transition state led to the synthesis of a transition-state analog (TS) analog, which proved to be an inhibitor and served as the hapten for generating a monoclonal antibody to catalyze the chorismate mutase reaction.

In this case, the accumulation of negative charge on the enolpyruvoyl group would be enhanced at the active site relative to the reaction in solution. The active site, illustrated in fig. 7-20, is lined with basic amino acids on one side (Chook et al., 1994). This should provide a positive electrostatic field that can be expected to stabilize the accumulation of negative charge on the enolpyruvoyl group in the transition state (Chook et al., 1994; Kast et al., 1996).

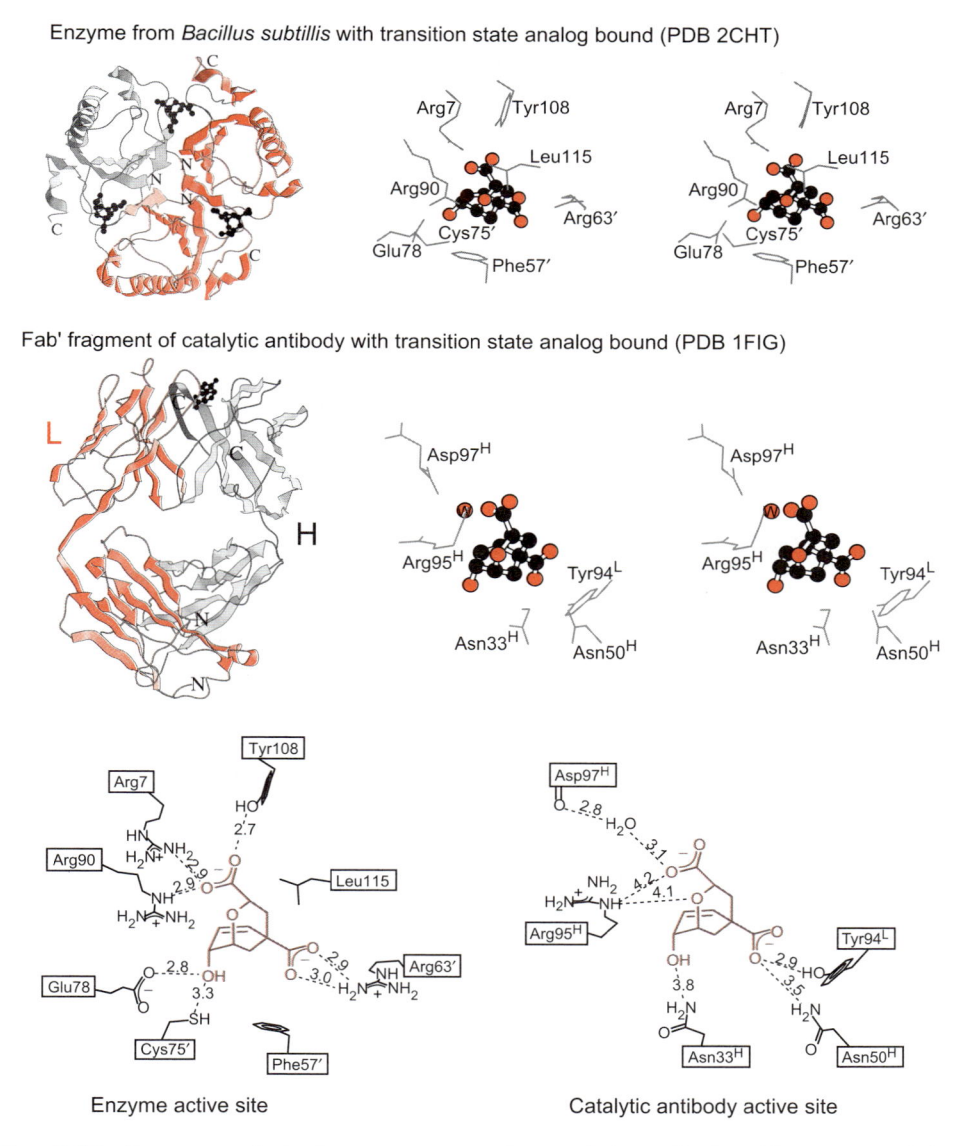

Fig. 7-20. Structures of chorismate mutase and a catalytic antibody with mutase activity. At the top are a ribbon diagram of the trimeric chorismate mutase from *Bacillus subtilis* and a stereodrawing of a ball-and-stick model of the transition-state analog in fig. 7-19 at the active site (PDB 2CHT; Chook et al., 1993). In the middle are a ribbon diagram of a catalytic antibody (Fab' fragment) that catalyzes the mutase reaction and a stereodiagram of a ball-and-stick model of the transition-state analog at its active site (PDB 1FIG; Haynes et al., 1994). In the enzyme and the catalytic antibody, one side of the transition-state analog is lined with basic amino acids. At the bottom are two-dimensional maps of the contacts between the transition-state analog and the active sites.

A stepwise mechanism, in which cleavage of the C5–O bond takes place in the transition state to form a transient ion-paired intermediate, with no significant bond formation to C1 can be ruled out. Secondary tritium kinetic isotope effects suggest 40% cleavage of the C5–O bond (Addadi et al., 1983). This mechanism too would be promoted by the positive electrostatic field in the active site.

The simplicity of this reaction, and the stability of the transition-state analog in fig. 7-19 stimulated interest in the generation of a catalytic antibody for chorismate mutation. A hapten based on the structure of the transition-state analog elicited monoclonal antibody formation, and two catalytic antibodies were obtained. The Fab fragments catalyzed the reaction with k_{cat} values of 0.07 min^{-1} and 2.7 min^{-1}, respectively (Hilvert et al., 1988; Jackson et al., 1988). These values were 10^2- to 10^4-fold higher than the rate constant for the uncatalyzed reaction. The enzyme catalyzes this reaction with a rate enhancement of 10^6, only 10^2- to 10^4-fold higher than values of k_{cat} for the catalytic antibodies. The relatively low enzymatic rate enhancement arises mainly from the relatively fast rate of the nonenzymatic reaction. One side of the binding site for the transition-state analog in the catalytic antibody incorporates basic amino acids, shown in fig. 7-20, albeit fewer than in the active site of chorismate mutase.

Δ^5-3-Ketosteroid Isomerase

Migration of the carbon-carbon double bond in 5-androstene-3,17-dione to 4-androstene-3,17-dione is brought about by Δ^5-3-ketosteroid isomerase (EC 5.3.3.1) (fig. 7-21). The double bond migrates essentially as the result of a 1,3-proton transfer from C4 to C6. Ionization of a C4(H) from the substrate forms the homoenolate ion in fig. 7-21, which can be protonated on oxygen to the enol on C4 to regenerate the substrate or on C6 to form the product, in which the C=C bond is in conjugation with the ketone group.

Proton transfer proceeds with partial exchange of the proton in flight with solvent hydrogen, and the process is not entirely stereospecific (Viger et al., 1981; Zawrotny et al., 1996). Isomerization in D$_2$O results in the incorporation of about 5% deuterium into the products. Isomerization of a substrate labeled with deuterium in the 4β-posistion leads to 65% retention of deuterium in the products, with one third remaining at C4. The same reaction catalyzed by the enzyme with Asp38 mutated to Glu38 leads to retention of 60% of the deuterium at C4. Solvent exchange of the proton in flight proves that the reaction cannot result from a 1,3-hydride shift mechanism. Retention of substantial deuterium at C4, even more than at C6 in the reaction catalyzed by the D38E variant, proves that much of the proton transfer takes place faster than solvent exchange and that it is not absolutely stereospecific.

The value of pK_a for Tyr14 is greater than 10.9, so that it functions as its conjugate acid and donates or shares the phenolic proton with the enolic intermediate (Kuliopulos et al., 1991). The intermediate displays a characteristic absorption band, which appears in the steady state of the reaction (Xue et al., 1991).

The inhibitor, dihydroequilenin, shown in fig. 7-21 binds to the active site, and the complex displays a downfield ^{1}H NMR signal that represents a slowly exchanging low-barrier hydrogen bonded proton (Zhao et al., 1996). The LBHB appears at 18.2 ppm, and its D/H fractionation factor is 0.4, consistent with an LBHB. Moreover, the activation enthalpy for exchange of this proton is 10.4 kcal mol^{-1}, also consistent with an LBHB. The LBHB most likely bridges the phenolic oxygens of Tyr14 and the inhibitor, as illustrated in fig. 7-21.

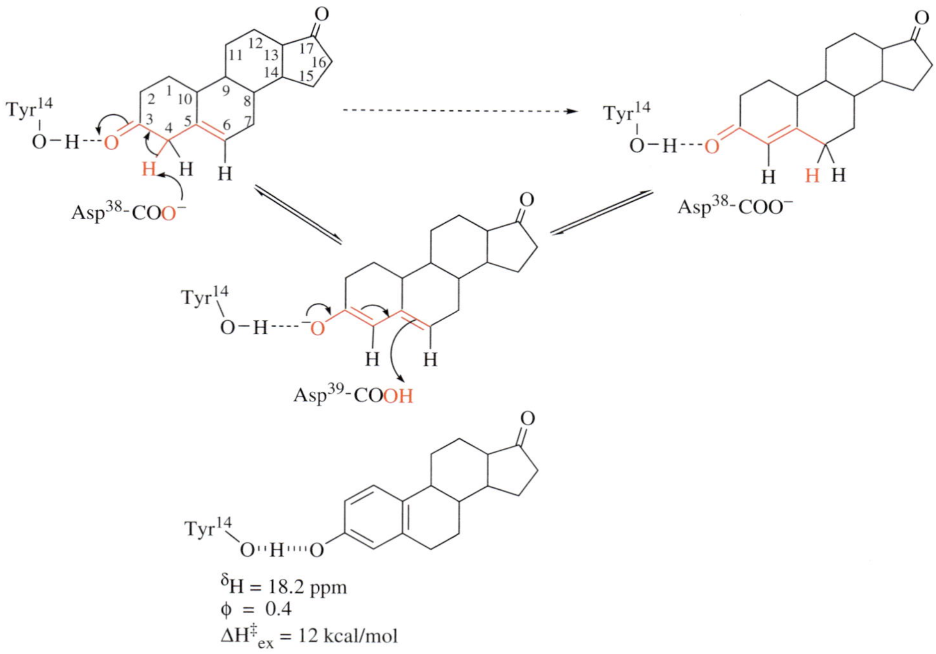

δ_H = 18.2 ppm
ϕ = 0.4
$\Delta H^{\ddagger}_{ex}$ = 12 kcal/mol

Fig. 7-21. Isomerization by Δ^5-3-ketosteroid isomerase. A Δ^5-3-ketosteroid can undergo homoenolate formation by dissociation of a proton from C4. The homoenolate may be protonated at any of three loci: the oxyanion, C4, or C7. Protonation at C7 leads to isomerization to the conjugated Δ^4-3-ketosteroid. The homoenolate intermediate is readily detected spectrophotometrically at 240 nm; the reaction is typically monitored at about 248 nm, which is the isosbestic point for the homoenolate and the Δ^4-3-ketosteroid. The inhibitor dihydroequilenin, shown below the homoenolate, incorporates a phenolic hydroxyl group and is almost isosteric with the substrate. The pK_a of the phenolic group is similar to that of the homoenolic intermediate and to that of Tyr14 in the active site. When bound to the active site, dihydroequilenin forms a low-barrier hydrogen bond with Tyr14. The intervening proton displays a downfield nuclear magnetic resonance signal at 18.2 ppm, a low value for its D/H fractionation factor and a high value for it activation enthalpy for exchange with protons of water. It has been postulated that low-barrier hydrogen bonding between Tyr14 and the homoenolate intermediate facilitates the reaction.

Structures of Δ^5-3-ketosteroid isomerase have been obtained by NMR and x-ray crystallography (Wu et al., 1997; Massiah et al., 1998; Ha et al., 2001; Nam et al., 2001). The highest resolution structure, shown in fig. 7-22, is of the D40N-variant of the Pseudomonas enzyme, which has Tyr16 and Asp40 in place of Tyr14 and Asp38. The structure is consistent with the assignment of an LBHB between Tyr16 and dihydroequilenin, as in fig. 7-21. The similar pK_a values of Tyr16 and the phenolic group in dihydroequilenin, and the close contact between these groups in the active site imposed by the structure of the enzyme, may facilitate formation of the LBHB.

Strong hydrogen bonding with Tyr14 may be important in stabilizing the homoenolate intermediate in fig. 7-21 and in facilitating its formation. The pK_a of the corresponding homoenol, the conjugate acid of the intermediate in fig. 7-21, may be similar to that of Tyr14, and this would create a favorable circumstance for strong hydrogen bonding in the intermediate and in the transition state for its formation.

368   Enzymatic Reaction Mechanisms

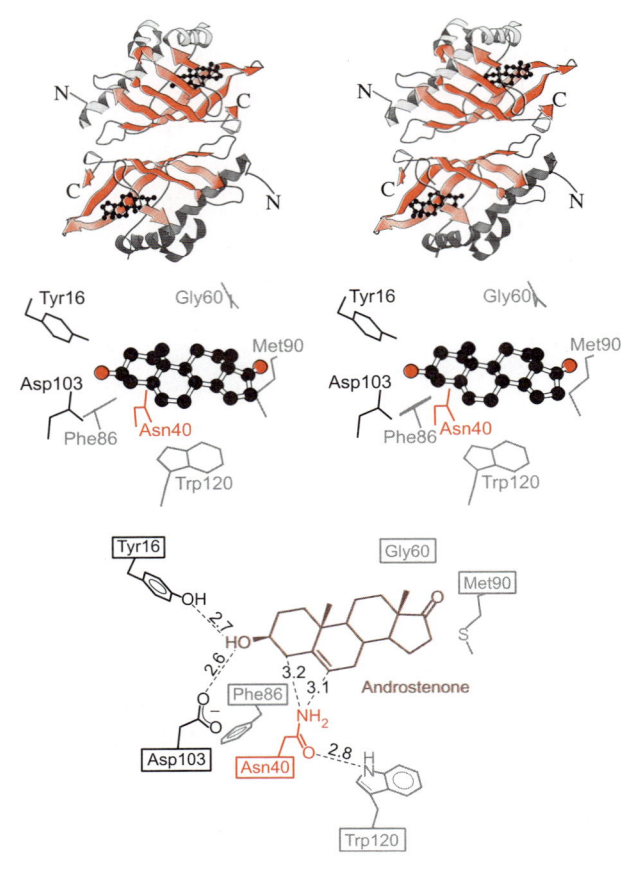

Fig. 7-22. Structure of *Pseudomonas putida* D40N-Δ^5-3-ketosteroid isomerase in complex with a substrate analog. The acid-base catalysts are Tyr16 and Asp40, and mutation of Asp40 to asparagine greatly decreases the activity. Androsten-3β-ol-17-one is an unreactive substrate analog bound at the active site. The ribbon stereodiagram at the top shows the overall chain fold and location of the active sites in the two subunits of the homodimer (PDB 1E3R; Ha et al., 2000). At the center is a ball-and-stick model of the substrate analog in stereo, showing the locations of amino acid side chains, including Tyr16 and Asn40 (in place of Asp40). At the bottom is a two-dimensional diagram of the contacts between the substrate analog and amino acid side chains at the active site.

Radical Isomerizations

Enzymatic isomerization reactions of the type defined in scheme 7-11 are catalyzed by two classes of enzymes, those that require adenosylcobalamin as the radical initiating coenzyme, and those that require *S*-adenosylmethionine (SAM) and a [4Fe–4S] cluster as the radical initiating system. The latter are members of the radical SAM superfamily (Sofia et al., 2001).

$$-\underset{X\ \ H}{C_\beta}-\underset{}{C_\alpha}- \quad \rightleftharpoons \quad -\underset{H\ \ X}{C_\beta}-\underset{}{C_\alpha}-$$

$-X = -COSCoA, -CH(NH_2)COO^-, -OH, -NH_2, ...$

Scheme 7-11

Both classes of enzymes generate the 5'-deoxyadenosyl radical as the initiator of radical mechanisms. We consider two examples from the adenosylcobalamin class and one radical SAM isomerase.

Glutamate Mutase

The discovery of vitamin B_{12} coenzymes, the first biologically functional forms of the vitamin to be found, took place in connection with research on the enzymology of glutamate metabolism in *Clostridia* (Barker et al., 1958). Glutamate mutase (EC 5.4.99.1) catalyzes the interconversion of L-glutamate and L-β-methylaspartate and requires a vitamin B_{12} coenzyme, the most widely studied of which is adenosylcobalamin (see chap. 4). Glutamate mutase consists of two components, a small protein (14.7 kDa) designated S or ε and a dimeric protein (2 × 50 kDa) designated E or σ (Buckel et al., 1999). The functional protein consists of two copies of each subunit, or $ε_2σ_2$. Neither component displays activity in the absence of the other, and the two do not form a tight complex. A fivefold excess of component S relative to E is required for maximum activity. A fusion protein of E and S forms a dimer that displays 80% of the activity of glutamate mutase (Chen and Marsh, 1997).

Equation 7-7 illustrates the overall transformation catalyzed by glutamate mutase, showing how the reaction follows the pattern of scheme 7-11, with $K_{eq} = 11$, favoring β-methylaspartate.

$$\text{L-glutamate} \xrightleftharpoons{\text{Glutamate mutase}} \text{L-β-methylaspartate} \qquad (7\text{-}7)$$

Notice that hydrogen transfer proceeds with inversion of configuration at the terminus of migration $C_β$, or C4 of glutamate (Sprecher et al., 1966b). In the next section, we see that methylmalonyl CoA mutase catalyzes an apparently analogous hydrogen transfer with retention of configuration.

Although the large subunit E is pink when purified from *Clostridia*, indicative of the presence of cobalamin, the pure subunit expressed in *E. coli* from the cloned gene does not itself bind adenosylcobalamin. Binding requires the small subunit S, which incorporates the motif DxHxxG. As in methionine synthase (see chapter 15), methylmalonyl CoA mutase, methylene glutarate mutase, and lysine 5,6-aminomutase, this motif contributes histidine to displace dimethylbenzimidazole from the lower axial position and coordinate the cobalt of adenosylcobalamin. The dimethylbenzimidazole ribotide of the coenzyme is bound in a pocket of the small subunit and may serve as an anchor. The small subunit is not completely folded in solution in the absence of a cobalamin (Tollinger et al., 1998). The large subunit E forms a complex with S in the crystal structure, in which adenosylcobalamin resides at the interface (Reitzer et al., 1998). The structure is shown in fig. 7-23 with glutamate in the active site. The substrate binds to a cluster of three arginine residues known as an "arginine claw" at a distance of 6.5 Å between Co and the nearest OH, the same distance estimated in simulating the electron paramagnetic resonance (EPR) spectrum of the radical intermediate (discussed later).

As in all adenosylcobalamin-dependent reactions, the coenzyme participates by mediating hydrogen transfer and initiating free radical formation (Buckel et al 1999; Marsh and Drennan, 2001) (see chap. 4). Adenosylcobalamin and all substrate- and product-related intermediates remain bound to the enzyme throughout the reaction. Homolytic scission of the Co–C5′ bond generates the 5′-deoxyadenosyl radical, which abstracts a hydrogen atom from the substrate, either the 4-pro-*R* hydrogen of L-glutamate or a methyl hydrogen of L-β-methylaspartate, depending on the direction of reaction, to form 5′-deoxyadenosine and a substrate-related radical. The reaction proceeds by a radical isomerization to the product-related free radical, which abstracts a methyl-hydrogen from 5′-adenosine to

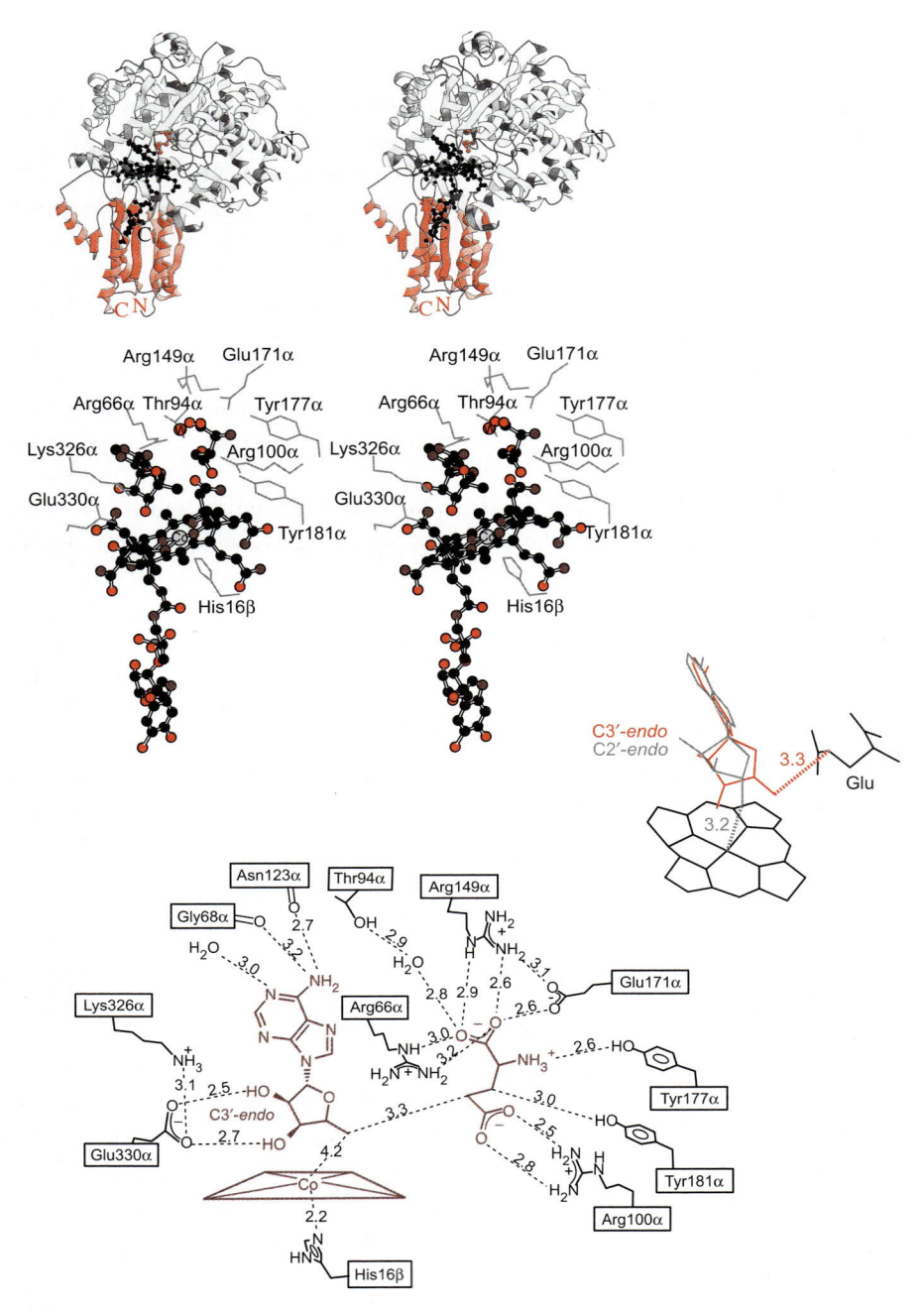

Fig. 7-23. Structure of glutamate mutase. At the top is a stereodrawing of a ribbon diagram for glutamate mutase from *Clostridium cochlearium*, showing the large α subunit shaded in gray and the small β subunit in red (PDB 1I9C; Gruber et al., 2002). Adenosylcobalamin is bound at the interface, with the lower axial cobalamin base and its tether projecting into the β subunit and replaced in the axial cobalamin position with His16β. At the center is a ball-and-stick model in stereo of adenosylcobalamin and glutamate at the active site. At the bottom is a two-dimensional diagram of the protein contacts of the adenosyl moiety and glutamate. To the side are overlaid line drawings of the C3′-endo (red) and C2′-endo (gray) forms of the ADP-5-deoxyribose that show the distance of the C5′ to the cobalt atom (in the C2′-endo conformation) or to the substrate (in the C3′-endo).

370

regenerate the 5-deoxyadenosyl radical. Recombination with cob(II)alamin forms adenosylcobalamin, and the product dissociates.

A central question has been the exact nature of transient intermediates and the mechanism of isomerization. The glutamate-related radical, arising from the abstraction of a hydrogen atom from C4 of glutamate, appears in the EPR spectrum observed in the steady state after mixing glutamate mutase with glutamate and freezing at 50 K (Bothe et al., 1998). The spectrum consists of that for cob(II)alamin interacting by exchange and dipolar coupling with the radical derived by hydrogen abstraction from C4 of glutamate. Simulations indicate that the cobalt of cob(II)alamin and the unpaired electron in the substrate-C4 radical are 6.6 Å apart A deuterium label at C4 of glutamate sharpens the radical features in the spectrum, and a ^{13}C label at C4 broadens the radical features, identifying C4 as the locus of spin. These effects arise from the spitting constant for nuclear hyperfine coupling of deuterium with an unpaired electron in a π-radical being smaller (1/6) than for hydrogen, and the nuclear hyperfine coupling of ^{13}C being large (Weil et al., 1994). Labeling at C3 or C2 in glutamate has little or no effect on the radical component of the spectrum. The predominant radical at the active site in the steady state is that with the unpaired electron at glutamate-C4, the substrate-radical illustrated in fig. 7-24.

Unlike other adenosylcobalamin-dependent isomerizations, no nonenzymatic model is available for the interconversion of the glutamate and β-methylaspartate radicals. The elimination/addition mechanism in fig. 7-24 is regarded as the most likely route. It would be difficult to simulate in a chemical model because of diffusion and radical combination. At the active site, diffusion is prevented. Support for this mechanism includes the fact that inhibition by acrylate and glycine is highly synergistic, and the combination elicits radical formation at the active site (Beatrix et al., 1995; Buckel et al., 1999). Moreover, glycine and acrylate are detectable as intermediates (Chih and Marsh, 1999, 2000).

Methylmalonyl CoA Mutase

The interconversion of methylmalonyl CoA and succinyl CoA is a step in the metabolism of valine in bacteria and higher animals. The enzyme is a homodimer (α_2) in most

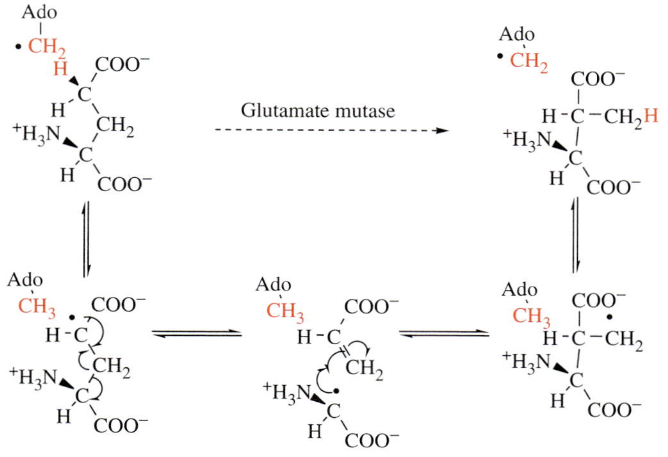

Fig. 7-24. Mechanism for radical isomerization by glutamate mutase. The substrate radical arising from hydrogen abstraction by the 5-deoxyadenosyl radical is undergoing elimination of acrylate, with concomitant formation of the glycine radical, as enzyme-bound intermediates. Addition of the glycine radical to C2 of acrylate produces the product radical, which abstracts a hydrogen atom from 5-deoxyadenosine to form β-methylaspartate.

eukaryotes and a heterodimer ($\alpha_2\beta_2$) in bacteria (Banerjee and Chowdhury, 1999). The β subunit in the bacterial enzymes appears homologous with the α subunit. The reaction appears similar to that of glutamate mutase, and it follows the same overall course as the other adenosylcobamin-dependent isomerases. However, the mechanism of the radical isomerization very likely differs from the fragmentation process in the action of glutamate mutase.

The reaction of methylmalonyl CoA mutase (EC 5.4.99.2) can be written to show the stereochemistry and exchange of the O=C–SCoA moiety with an adjacent hydrogen as in eq. 7-8.

$$\text{(reaction scheme)} \tag{7-8}$$

That COSCoA migrates, not COO⁻, is proved by the fact that [2-^{14}C]methylmalonyl CoA is transformed into [3-^{14}C]succinyl CoA and not [2-^{14}C]succinyl CoA (Eggerer et al., 1960a). The 1,2-migration of hydrogen proceeds with retention of configuration, unlike the reaction of glutamate mutase (Rétey and Lynen, 1964; Sprecher and Sprinson, 1964, Sprecher et al., 1966a). This difference arises from a difference in processing the radical intermediates and not from a fundamental difference in the mechanism of action of adenosylcobalamin. As in all coenzyme B_{12} reactions, adenosylcobalamin mediates hydrogen transfer (Frey and Abeles, 1966; Rétey and Arigoni, 1966) and initiates radical formation by undergoing homolytic fission of the Co–C5′ bond to form cob(II)alamin. The resulting 5′-deoxyadenosyl radical abstracts a hydrogen atom from the substrate bound nearby and forms 5′-deoxyadenosine. The resultant substrate-radical undergoes isomerization to the product radical, which abstracts a hydrogen atom from 5-deoxyadenosine to form the product and regenerate the 5′-deoxyadeosyl radical.

Crystal structures of methylmalonyl CoA mutase from *Propionibacterium shermanii* yielded valuable information about the global structure, as well as about interactions between the enzyme and adenosylcobalamin and substrates (Mancia and Evans, 1998; Mancia et al., 1996). Because the active site is located in the α subunit, which is similar in sequence to the eukaryotic enzyme, we show structures of the α subunit in fig. 7-25. The enzyme exists in an open conformation when crystallized in the absence of a substrate and in a closed conformation when crystallized in the presence of a substrate or CoA analog. In the closed conformation, the two main domains are a TIM barrel and a B_{12}-binding domain similar to that in methionine synthase and the S subunit of glutamate mutase. Adenosylcobalamin binds with coordination of cobalt to His610 in the triad Asp610-Gly609-His610 and its dimethylbenzimidazole ribotide side chain buried in the B12 binding domain. The upper face and deoxyadenosyl ligand are brought into contact with the TIM barrel. In the absence of a substrate or CoA analog the TIM barrel is open. The barrel closes on binding the substrate or CoA-analog. In the closed conformation, the methylmalonyl or succinyl moiety of the substrate is deeply buried near the upper face of the coenzyme, the ADP moiety is near the surface of the TIM barrel, and the phosphopantetheine moiety lies within the barrel. The TIM barrel provides a tunnel or groove that binds the substrate, projects the reactive portion into the catalytic site, and seals it from contact with the solution. The closing of the TIM barrel attending the binding of a substrate or CoA analog is illustrated in fig. 7-25. The adenosyl binding site includes Tyr89, which has been linked with the homolytic cleavage of the Co–C5′ bond (Vlasie and Banerjee, 2003).

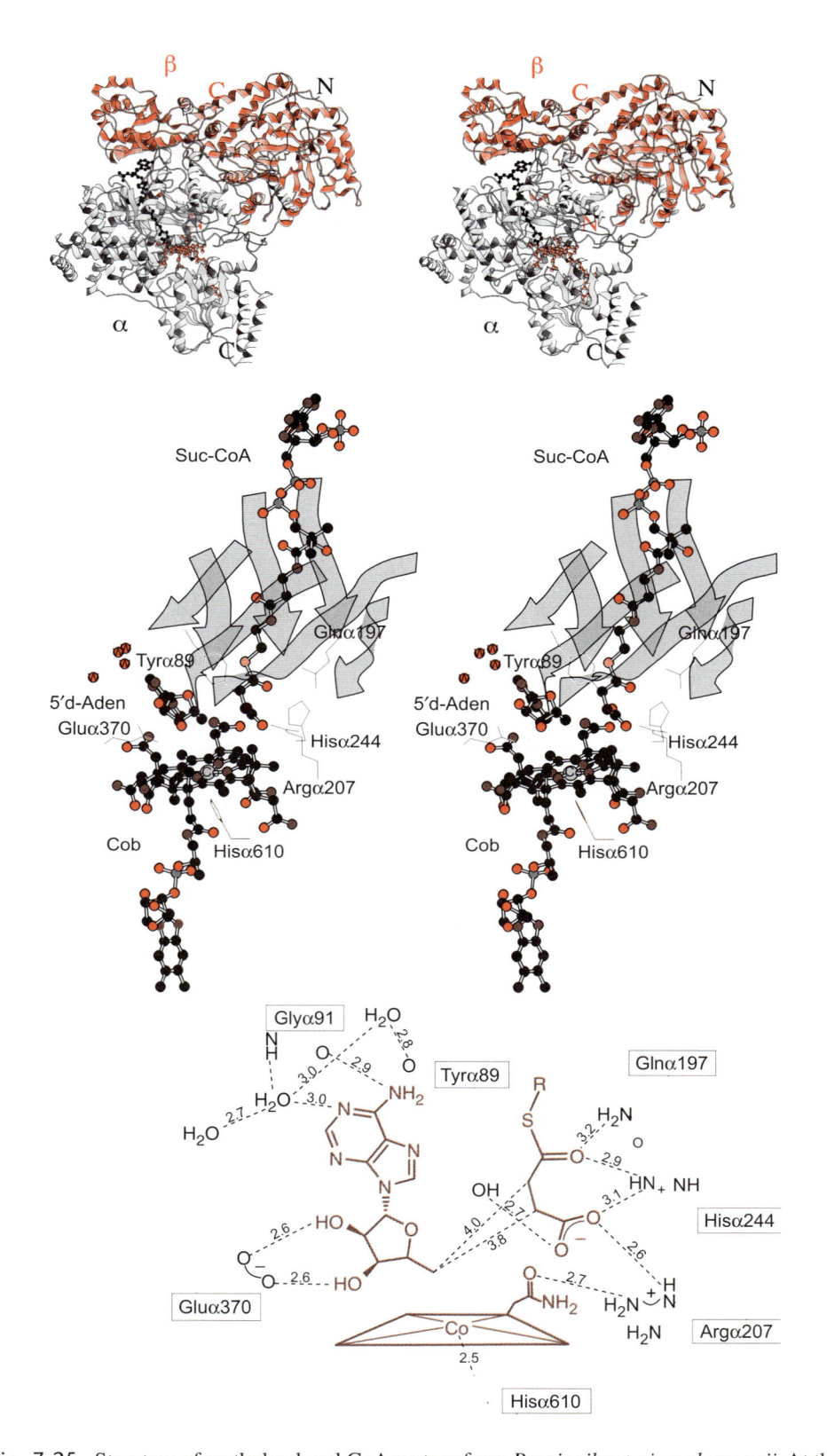

Fig. 7-25. Structure of methylmalonyl CoA mutase from *Propionibacterium shermanii*. At the top is a ribbon diagram in stereo of the molecule, showing the α subunit in gray and the β subunit in red. Cobalamin and succinyl CoA are shown as a black ball-and-stick model (PDB 4REQ; Mancia and Evans, 1998). The cobalamin is deeply buried inside the α subunit, and the CoA molecule traverses a tunnel from the surface, with the adenosyl moiety at the surface and the succinyl moiety in the interior near the cobalamin. In the center is a ball-and-stick diagram in stereo of the cobalamin, 5′-deoxyadenosine, and succinyl moiety of succinyl CoA. At the bottom is a two-dimensional map of contacts between the enzyme and 5′-deoxyadenosine and succinyl CoA, respectively.

374 Enzymatic Reaction Mechanisms

Unlike glutamate mutase, a relevant nonenzymatic model for the reaction of methylmalonyl CoA mutase is available. The reaction depends on the generation of a free radical analogous to the substrate-radical in the enzymatic reaction. This is possible by application of standard radical chemistry. Heating 2,2'-azobisisobutyronitrile (ABIN) extrudes N_2 and releases isobutyryl radicals, which abstract hydrogen atoms from tributyltin hydride. The resultant tributyltin radicals abstract halogen atoms from alkyl halides to form alkyl radicals. When this chemistry is applied to the 2-bromomethylene analog of a methylmalonate diester, the outcome is illustrated in scheme 7-12 (Halpern, 1985; Wollowitz and Halpern, 1984, 1988).

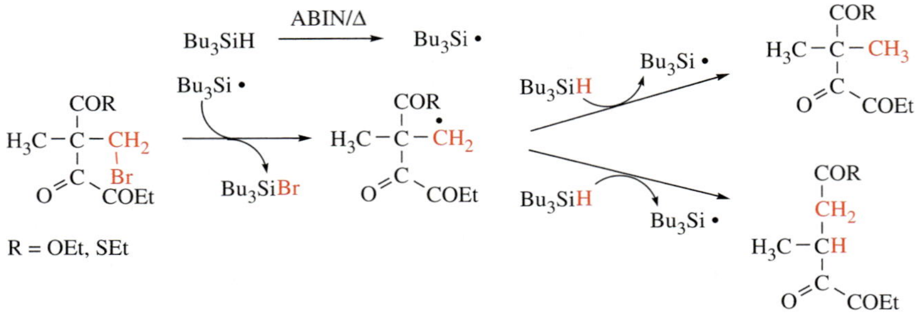

Scheme 7-12

The rearrangement is analogous to that brought about by methylmalonyl CoA mutase. Isomerization cannot result from fragmentation of the initial radical to an acrylate and radical, as in glutamate mutase, because this would lead to radical combination products, and they are not observed. The mechanism is likely to be similar to that shown for the enzymatic reaction illustrated in fig. 7-26. The radicals are quenched by abstraction of hydrogen atoms from tributyltin hydride.

The most likely isomerization mechanism in the enzymatic and model reactions is illustrated in fig. 7-26. In the active site, the 5-deoxyadenosyl radical from homolytic cleavage of the Co–C5' bond abstracts a hydrogen atom from the methyl group of methylmalonyl

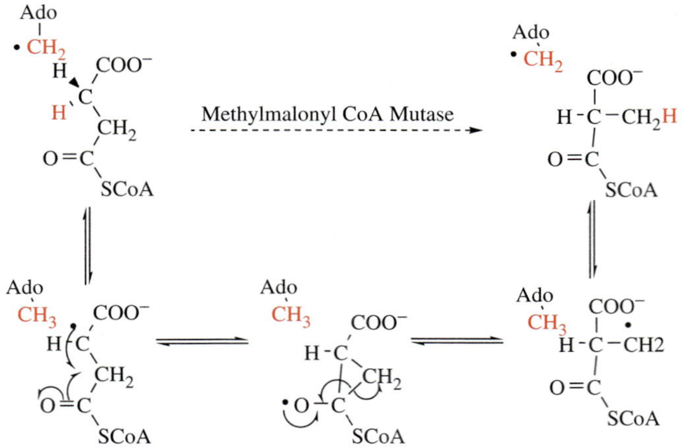

Fig. 7-26. Mechanism for radical isomerization in methylmalonyl CoA mutase. The substrate-radical arising from abstraction of a hydrogen atom from C2 of succinyl CoA by the 5'-deoxyadenosyl radical is shown as undergoing internal cyclization by addition of C2 to the acyl carbonyl carbon to form a cyclopropyl oxyl radical intermediate. The quasi-symmetric intermediate can open to the product-radical, which abstracts a hydrogen atom from the methyl group of 5'-deoxyadenosine to form methylmalonyl CoA and regenerate the 5'-deoxyadenosyl radical.

CoA to form 5-deoxyadenosine and a transient methylene radical. Addition of the π-radical center to the carbonyl group of the COSCoA moiety forms the cyclopropyl oxyl radical in fig. 7-26, which can open to the product-radical or revert to the substrate-radical. The product-radical abstracts a methyl hydrogen from 5-deoxyadenosine to regenerate the coenzyme-radical and produce succinyl CoA. Radical isomerization was first proposed in 1960 (Eggerer et al., 1960b), before the hydrogen transfer function of adenosylcobalamin and homolytic scission of the Co–C5′ bond were discovered.

Extensive studies of adenosylcobalamin bound to methylmalonyl CoA mutase, and parallel nonenzymatic studies of analogs of adenosylcobalamin, have been directed to the question of whether the replacement of the dimethylbenzimidazole moiety of the coenzyme by the histidine imidazole would weaken the Co–C5′ bond through a *trans* effect. No evidence of a substantial effect has been found (Banerjee and Chowdhury, 1999; Dong et al., 1998; Finke et al). Analysis of the enzyme structure suggests that substrate binding energy is mobilized to destabilize the Co–C5′ bond, presumably by forcing the adenosyl group into a conformation that weakens the bond. Substrate binding accelerates the cleavage of this bond by 10^{12}-fold (Padmakumar et al., 1997).

Cleavage of the Co–C5′ bond is kinetically coupled to hydrogen transfer from the substrate to the 5′-deoxyadenosyl moiety. This was first revealed by the observation of a large deuterium kinetic isotope effect in the cleavage of the bond by [*methyl*-^{2}H$_3$]methylmalonyl CoA (Padmakumar et al., 1997). When the complex of mutase and adenosylcobalamin is mixed with methylmalonyl CoA in a stopped-flow spectrophotometer, a fraction (≈20%) of the coenzyme is cleaved in the steady state to cob(II)alamin, as monitored by the decrease in A_{525}. This takes place too fast to measure accurately at ambient temperatures (>600 s^{-1}). With [*methyl*-^{2}H$_3$]methylmalonyl CoA the rate is much slower, 28 s^{-1}, corresponding to a deuterium kinetic isotope effect of more than 20. The isotope effect arises from kinetic coupling between Co–C5′ cleavage and hydrogen abstraction from the substrate by the intermediate 5′-deoxyadenosyl radical. The large deuterium isotope effect in hydrogen abstraction step slows the overall cleavage of the Co–C5 bond and cob(II)alamin formation.

The turnover number (k_{cat}) for the overall reaction is only 60 s^{-1}, whereas cob(II)alamin formation coupled with hydrogen transfer and substrate radical formation is much faster (>600 s^{-1}). The overall reactions of deuterated substrates display modest deuterium kinetic isotope effects of 3.5 to 4.2, depending on the source of the enzyme (Michenfelder et al., 1987; Miller and Richards, 1969). Mutation of Tyr89 to Phe lowers the value of k_{cat} to 1000th the wild-type value, abolishes the overall deuterium kinetic isotope effect, and suppresses the formation of observable cob(II)alamin in the steady state (Vlasie and Banerjee, 2003). The mutation presumably makes Co–C5′ bond formation fully rate limiting. The contact between the adenosyl moiety and Tyr89 may be important in the cleavage of this bond. Cob(II)alamin formation on addition of substrate to the complex of enzyme and adenosylcobalamin is accompanied by the appearance of a radical signal. The envelope of the EPR spectrum arises from cob(II)alamin and the radical with the unpaired electron on the carbon skeleton of the substrate (Banerjee and Chowdhury, 1999). The two paramagnetic species, Co(II) and radical, are spatially close and spin coupled in a triplet system. A detailed analysis of the triplet EPR spectrum observed with deuterated and ^{13}C-labeled substrates allowed the structure to be modeled (Mansoorabadi et al., 2005). The radical proved to be the succinyl CoA–related species in fig. 7-26, which is far less unstable than the methylene radical in the methylmalonyl CoA–related species. Analysis of the interaction of this radical with cob(II)alamin revealed the relative orientations of the two paramagnetic species and fitted into the active site. The modeled structure showed a distance of 6.0 Å between the unpaired electron in the succinyl CoA radical and Co(II).

Lysine 2,3-Aminomutase

In chapter 3, we introduced the enzymatic interconversion of L-lysine and L-β-lysine in connection in connection with PLP, and in chapter 4, we discussed the role of SAM and a [4Fe–4S] center. The enzyme appears in many bacteria and functions either in lysine metabolism or antibiotic biosynthesis. The enzyme from *Clostridium subterminale* is oligomeric and composed of 47-kDa subunits (Chirpich et al., 1970; Ruzicka et al., 2000), and the enzyme from *Bacillus subtilis* appears to be a homotetramer (Chen et al., 2000). All species require PLP, SAM, and a unique [4Fe–4S] center for activity.

The reaction is analogous to B_{12}-dependent isomerization reactions because it follows the pattern of scheme 7-11. The 3-pro-*R* hydrogen of L-lysine is transferred to the 2-pro-*R* position of L-β-lysine, and the α-amino group of L-lysine undergoes cross migration to the β-carbon in L-β-lysine (Aberhart et al., 1983).

Like adenosylcobalamin-dependent reactions, the 5′-deoxyadenosyl radical plays a central role (Frey, 1990). However, in lysine 2,3-aminomutase (EC 5.4.3.2) this radical arises from the homolytic cleavage of the S–C5′ bond in SAM instead of the Co–5′ bond in adenosylcobalamin. In this reaction, SAM has been regarded as "a poor man's adenosylcobalamin" (Frey, 1993). Later experiments implicated the mechanism in scheme 7-13 for the reversible cleavage of SAM (Chen et al., 2003; Cosper et al., 2000; Lieder et al., 1998).

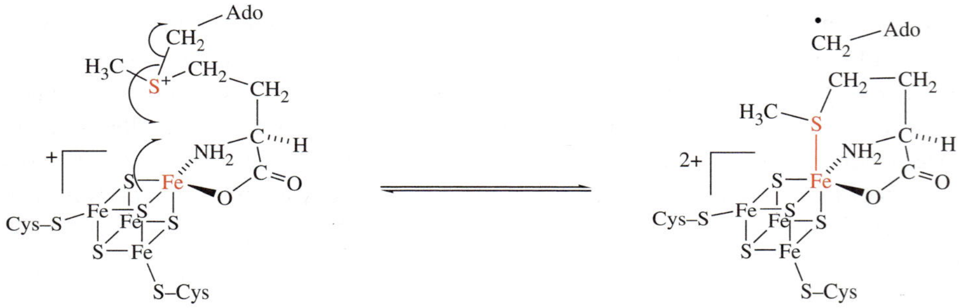

Scheme 7-13

Direct ligation of SAM to Fe in the [4Fe–4S] center through the carboxylate and a-amino groups of SAM has been proved by ^{17}O-ENDOR and ^{15}N-ENDOR spectroscopy. *Se*-adenosylselenomethionine (SeSAM) functions well in place of SAM, and direct ligation of selenium in this coenzyme analog to Fe in the cleaved state has been proved by selenium x-ray absorption spectroscopy.

The overall mechanism of aminomutation and the roles of the 5′-deoxyadenosyl radical and PLP are illustrated in fig. 7-27 (Baraniak, 1989; Moss and Frey, 1987, 1990). The α-amino group of lysine reacts with the internal Lys^{337}-PLP aldimine to form the external aldimine. The 5-deoxyadenosyl radical from SAM abstracts the 3-pro-*R* hydrogen atom from the side chain to form the substrate-radical intermediate A, which undergoes isomerization to the product-radical C by way of the azacyclopropylcarbinyl radical B. Abstraction of a methyl hydrogen from 5-deoxyadenosine by the product radical C produces the β-lysyl-PLP external aldimine, which undergoes transaldimination with Lys337 to release the product and regenerate the internal aldimine. Evidence supporting this mechanism includes the characterization of the product-radical C as the dominant radical in the steady state by EPR and electron spin-echo envelope modulation (ESEEM) spectroscopy and in the transient state by rapid mix-freeze-quench EPR (Ballinger et al., 1992, 1995; Chang et al., 1996). L-4-Thialysine is a substrate, and the 4-thialysyl analog of the substrate-radical A has also been characterized by EPR as the dominant radical in the

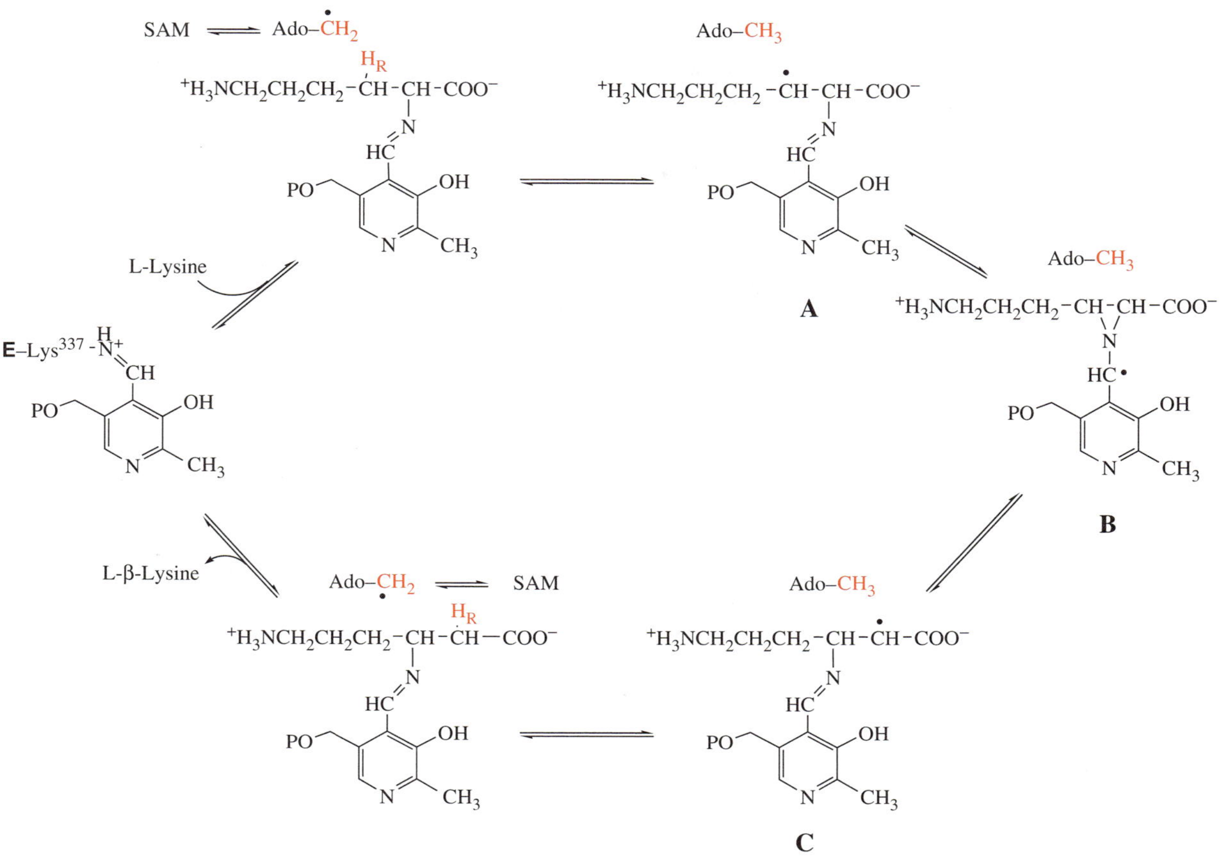

Fig. 7-27. A mechanism of amino group migration in lysine 2,3-aminomutase. Lysine undergoes transaldimination with the internal aldimine formed between pyridoxal-5′-phosphate (PLP) and Lys337 of the enzyme. The 5-deoxyadenosyl radical generated from the S-adenosyl-L-methionine (SAM)/[4Fe–4S] reaction in scheme 7-13 abstracts the 3-pro-R hydrogen from the side chain of the substrate to form the substrate radical A, which undergoes internal cyclization to the azacyclopropyl carbinyl radical B. This quasi-symmetric intermediate can open to the product-radical, which abstracts a hydrogen atom from the methyl group of 5′-deoxyadenosine to form the internal aldimine of β-lysine with PLP and regenerate the 5′-deoxyadenosyl radical.

steady state (Wu et al., 1995). S-3′,4′-Anhydroadenosylmethionine activates the enzyme in place of SAM, and the 5′-deoxy-3′,4′-anhydroadenosyl radical has been identified as an intermediate in the steady state by EPR spectroscopy and in the transient state by rapid mix-freeze-quench EPR (Magnusson et al., 2000).

The structure of lysine 2,3-aminomutase in complex with SeSAM, PLP, lysine, and the cluster [4Fe–4S]$^{2+}$ shown in fig. 7-28 demonstrates how the three coenzymes interact with one another and the substrate. This complex differs from the active complex by one electron: the active Michaelis complex has [4Fe–4S]$^{1+}$. The external aldimine of PLP and L-lysine is adjacent to SeSAM, with the lysyl side chain between the two coenzymes, anchored at the end by contacts with Asp293 and Asp330, and the pro-R C3(H) within 4 Å of carbon-5′ of SAM. SAM itself is ligated through it's α-amino and carboxylate groups to one Fe in the cluster, and the selenium atom in the selenomethionine moiety is poised above the same Fe at a distance of 3.1 Å. The selenium is in excellent position to react with Fe in the SeSAM-cleavage step, illustrated for SAM in scheme 7-13. The C5′ of SAM is

378   Enzymatic Reaction Mechanisms

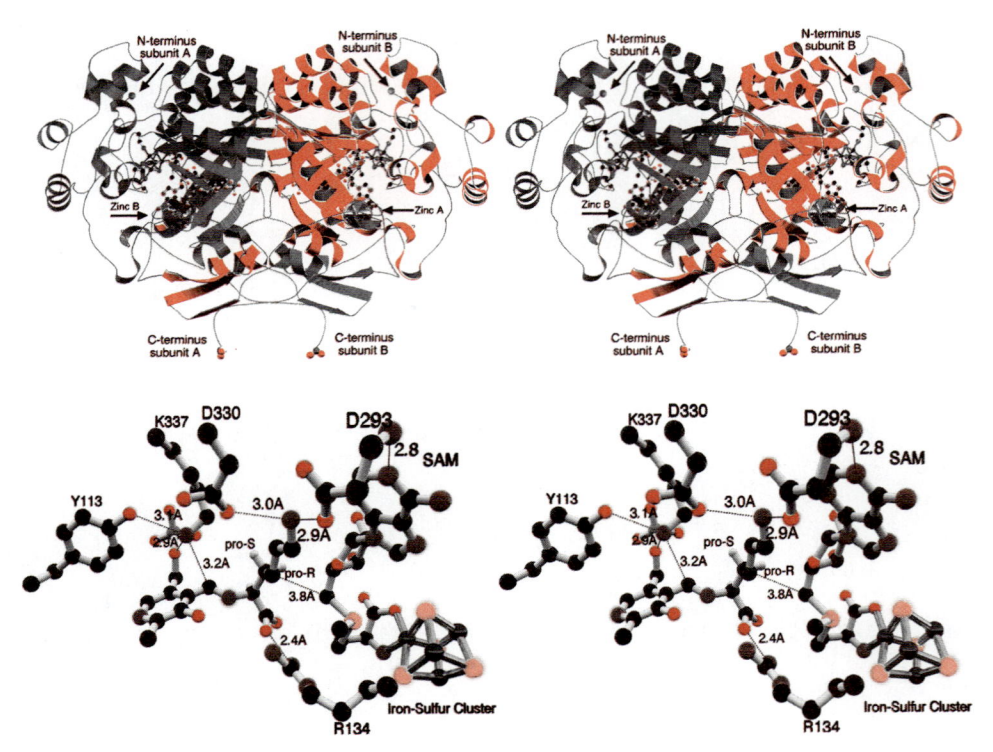

Fig. 7-28. Structure of a complex of lysine 2,3-aminomutase with S-adenosyl-L-methionine (SAM), pyridoxal-5'-phosphate (PLP), and lysine. Two adjacent subunits of the homohexameric lysine 2,3-aminomutase from *Clostridium subterminale*, one gray and the other red, are shown in stereo in the top panel (PDB 2A5H; Lepore et al., 2005). Each subunit contains one noncatalytic zinc ion, as well as the PLP-lysine external aldimine, SAM, and the [4Fe-4S] cluster. The bottom panel shows the arrangement of the catalytically important components at the active site in stereo. The [4Fe-4S] cluster is liganded by the three cysteine residues that are a common motif (CxxCxxxC, not shown) of the radical SAM family (see chap. 4). The Fe atom of the cluster that does not make direct contact with the enzyme has an octahedral coordination sphere with three contacts to sulfur atoms in the cluster, two close contacts with the SAM methionine carboxylate and amino groups, and one open or long-distance interaction (3.2 Å) with the SAM sulfur atom. (Courtesy of Bryan Lepore, Brandeis University, Waltham, MA.)

in excellent position to abstract the 3-pro-*R* hydrogen from lysine on cleavage to the 5'-deoxyadenosyl radical.

The lysine 5,6-aminomutase, also found in bacteria, including *Clostridia*, is a heterodimeric enzyme that catalyzes the analogous isomerization of the ε-amino group of L-β-lysine to produce 3,5-diaminohexanoate (Stadtman, 1973). The enzyme also accepts D-lysine as a substrate. Unlike the 2,3-aminomutase, this enzyme requires adenosylcobalamin instead of the SAM/[4Fe–4S] system, and like the 2,3-aminomutase it also requires PLP. The mechanism is thought to be analogous to that in fig. 7-27, involving migration of the ε-amino group instead of the α-amino group, with adenosylcobalamin as the source of the 5'-deoxyadenosyl radical. A structure of the internal aldimine form of this enzyme showed a TIM barrel in one subunit and a dinucleotide fold in the other, with the cobalamin bound to the TIM barrel and PLP in the other subunit bonded through the aldimine linkage to Lys144 in the TIM barrel (Berkovitch et al., 2004).

Newer Isomerases

The great variety of chemical mechanisms in the foregoing examples of isomerization demonstrates the versatility of enzymes as catalysts. The reaction types presented in preceding pages exemplify the chemistry available to enzymes, but they do not exhaust the repertoire. Neither do the following examples of enzymatic isomerization.

UDP-Galactopyranose Mutase

UDP-galactopyranose mutase (EC 5.4.99.9) catalyzes the interconversion of the pyranose and furanose forms of UDP-galactose, that is the reversible transformation of UDP-galactopyranose into UDP-galactofuranose, an intermediate in bacterial cell wall biosynthesis. The overall reaction is illustrated (fig. 7-29) within the context of a biochemically unusual mechanism.

The enzyme contains FAD and is activated by NADH or by other reducing agents such as dithionite, which reduce FAD to $FADH_2$. It is clear that the active enzyme contains $FADH_2$. A number of mechanisms have been suggested, including two radical mechanisms. However, the simple mechanism in fig. 7-29 is supported by the available evidence.

In this mechanism, N5 of $FADH_2$ functions as a nucleophile to displace the ADP moiety of UDP-galactopyranose in step 1, leading to glycosylation of the dihydroalloxazine ring. Iminium formation accompanies opening of the galactopyranosyl ring in step 2, and the ring closes again by nucleophilic addition of the glycosyl 4-hydroxyl group to the iminium center in step 3. Reformation of the bond between ADP and the furanosyl ring generates UDP-galactofuranose and regenerates $FADH_2$ in step 4. Powerful evidence in support of this mechanism is provided by the appearance of a reduced adduct between a hexose and $FADH_2$ when the enzyme is treated with $NaBH_3CN$ in the presence of UDP-galactopyranose (Soltero-Higgin et al., 2004). This would represent hydride reduction of the iminium intermediate in the mechanism of fig. 7-29. Other evidence in support of fig. 7-29 includes the fact that 1-deaza-FAD activates the enzyme, whereas 5-deaza-FAD does not (Huang et al., 2003). 5-Deaza-FAD lacks N5, which is required for every step of the mechanism in fig. 7-29. Moreover, ^{18}O-positional isotope exchange experiments prove that the anomeric oxygen of UDP-galactopyranose undergoes exchange with nonbridging oxygen bonded to the β-phosphorus of the ADP moiety at the same rate as the overall reaction (Barlow et al., 1999). This experiment proves that the bond between galactopyranose-C1 and the anomeric oxygen must break in the course of the reaction, as it does in steps 1 and 4 of fig. 7-29. Radical mechanisms are inspired by the observation of the blue semiquinone form of FAD on reduction, showing that the semiquinone is a stable species at the active site (Fullerton et al., 2003). However, the radical mechanisms do not explain the substrate-dependent inactivation by $NaBH_3CN$, with formation of a hexose-$FADH_2$-adduct. The structure (Beis et al., 2005) is compatible with the mechanisms so far proposed.

Pseudouridine Synthase

Many species of RNA, including tRNA, rRNA, and snRNA, but not mRNA, contain pseudouridine (Ψ) in specific locations within their nucleotide sequences. Pseudouridine arises from posttranscriptional modification of specific uridine sites by action of pseudouridine synthases (Ψ synthases; EC 5.4.99.12). The overall transformation is illustrated at the top of fig. 7-30.

The reaction must proceed by reversible cleavage of the N-ribosyl bond of the uridine residue, reorientation of the uracil ring about an axis bisecting N2 and C6, and reformation of the ribosyl linkage to C5 to produce Ψ, a C-nucleoside.

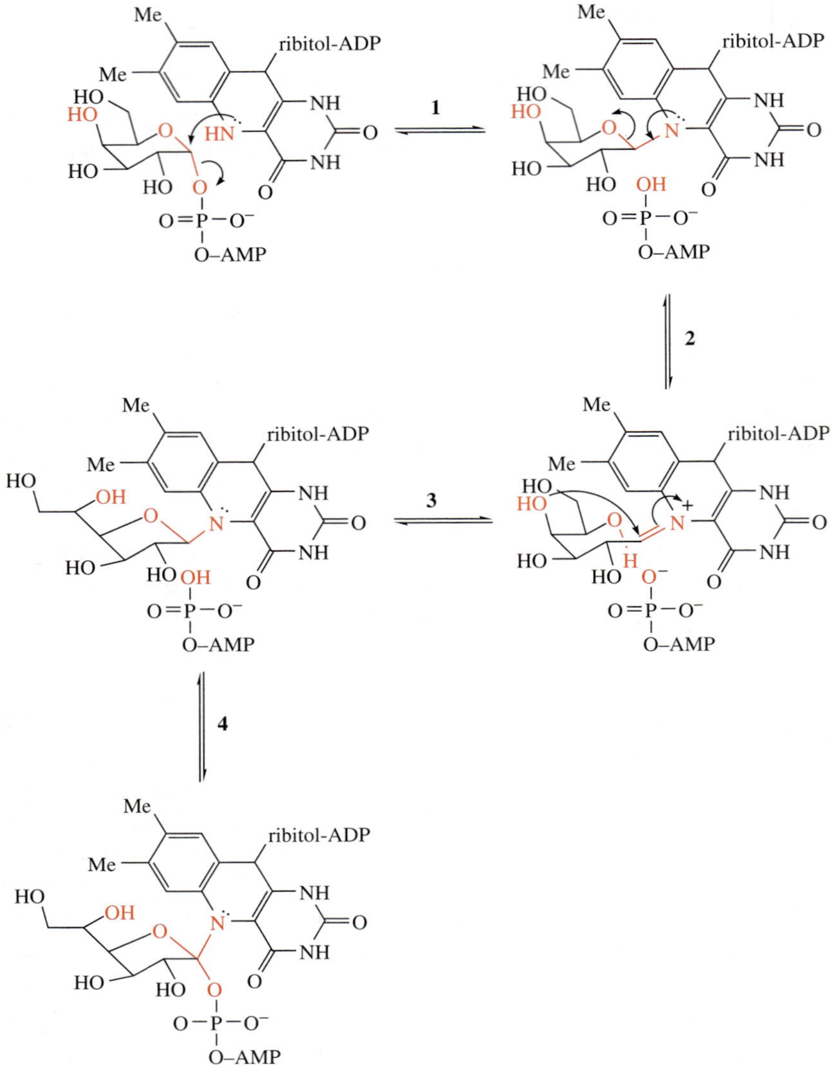

380 Enzymatic Reaction Mechanisms

Fig. 7-29. In a mechanism for the action of UDP-galactopyranose mutase, FADH$_2$ functions as a nucleophile to displace the ADP moiety of UDP-galactopyranose and glycosylate the dihydroalloxazine ring. Iminium formation accompanies opening of the galactopyranosyl ring, which re-closes by nucleophilic addition of the 4-hydroxyl group to the iminium center. Transglycosylation to ADP generates UDP-galactofuranose and regenerates FADH$_2$. The exact mechanisms of the proton transfer steps are not specified in this illustration.

Five families of Ψ synthases are known, based on amino acid sequence comparisons. No two families display significant sequence homologies; however, all five families share similarities in chain fold, and each one has a conserved aspartate residue in the active site (Foster et al., 2000; Hamilton et al., 2005; Hoang and Ferré-D'Amaré, 2004; Huang et al., 1998; Kaya et al., 2004). Each Ψ synthase modifies a specific species of RNA at defined uridine sites. For example, Ψ synthase PsiSI from *E. coli* modifies U38, U39, or U40 of tRNA, whereas Ψ synthase TruB modifies U55 in the TΨC loop of tRNA (Foster et al., 2000; Hamilton et al., 2005).

Two reaction mechanisms are under consideration for the action of Ψ synthases. In both mechanisms, the conserved aspartate serves as a nucleophilic catalyst in reversibly cleaving

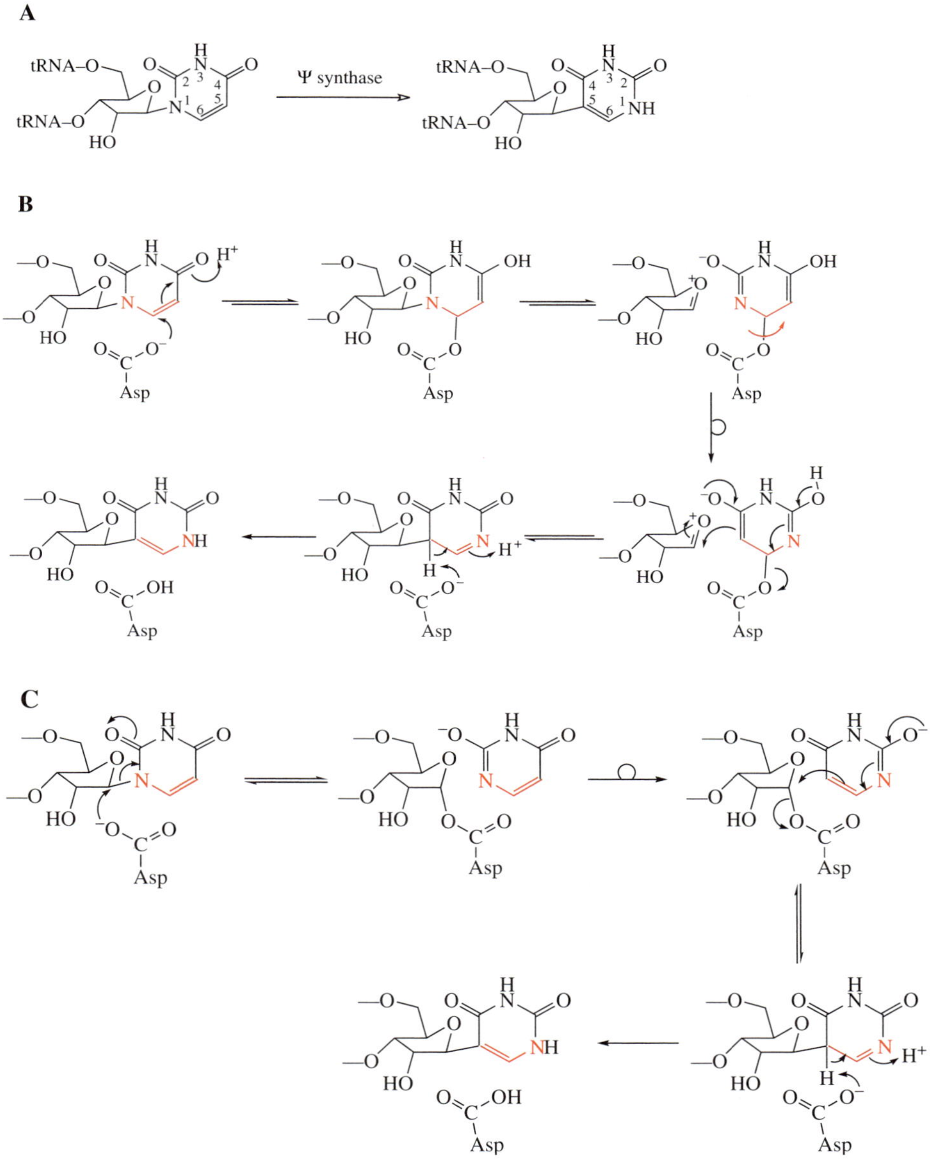

Fig. 7-30. Two mechanisms are under consideration for the action of pseudouridine synthase. (A) Overall transformation brought about by the action of Ψ synthase on a uridine residue in RNA. (B) In one of the mechanisms under consideration for the action of Ψ synthase, the conserved aspartate residue undergoes nucleophilic Michael addition to C6 of the uracil ring, and the resulting adduct undergoes C–N cleavage to the ribosyl carbocation and the uracil-Ψ adduct. A rotation of the uracil ring brings C5 into position to be ribosylated. Proton transfer from C5 to N1 completes the reaction. (C) An alternative mechanism involves C–N cleavage by ribosyl transfer to the conserved aspartate. The freed uracil remains bound at the active site and can rotate 180 degrees to bring C5 into position for ribosylation. Subsequent steps are as those in B.

the N-ribosyl bond (Huang et al., 1998). Mutation of the conserved aspartate residue in all Ψ synthases so far studied abolishes the activity. In the mechanism outlined in fig. 7-30B, the conserved aspartate functions after the fashion of the active site cysteine in thymidylate synthase by initially undergoing Michael addition to the uracil ring at C6. The resulting 5,6-dihydrouridine-like structure is labilized to N-ribosyl cleavage, just as 5,6-dihyrouridine is labile to hydrolysis. On cleavage to the ribosyl-C1' carbocation the uracil ring is free to rotate about the C6-N3 axis, bringing C5 into position for ribosylation to the C-nucleoside. The initial C-ribosyl intermediate undergoes proton transfer from C5 to N1, in a process likely mediated by the conserved aspartate residue (Hamilton et al., 2005). This attractive mechanism has the advantage of precedent for Michael addition to C6. It also provides for the sequestration of the uracil ring in the active site through its covalent bond to aspartate. The mechanism does not provide for stabilization of the ribosyl carbocation during the changeover of uracil orientations. Glycosyl carbocations are extremely reactive species (see chap. 12).

In the alternative mechanism shown in fig. 7-30C, the conserved aspartate reacts directly with the ribosyl moiety by nucleophilic substitution at ribosyl-C1', releasing the uracil ring. Uracil must be allowed to rotate 180 degrees about the C6-N3 axis but not to dissociate from the active site (Huang et al., 1998). When oriented with C6 adjacent to ribosyl-C1' in the covalent acylal intermediate, ribosyl transfer to C6 generates the C-nucleoside bond. Proton transfer from C6 to N1 by the same mechanism as in fig. 7-30B, presumably with base catalysis by aspartate, leads to Ψ.

Evidence supporting the first mechanism (fig. 7-30B) includes experiments carried out with a substrate analog, FUra-tRNA^{Phe}, containing 5-fluorouracil in place of uracil at the reaction site (Gu et al., 1999; Huang et al., 1998). This molecule reacted with Ψ synthase as an apparent suicide substrate. It appeared to react in the early steps, albeit slowly, but it did not complete the catalytic cycle. The inactivation product proved to be a covalent conjugate of FUra-tRNA^{Phe} and Ψ synthase. Careful biochemical characterization indicated that the conjugate included a covalent bond between a group on the protein and C6 of the fluorouracil ring in FUra-tRNA^{Phe}. The experiments proved that N-ribosyl linkage remained intact in the conjugate, even after denaturation. The exact linkage between Ψ synthase and FUra-tRNA^{Phe} was not identified, but the degradation product, 5-hydroxy-5,6-dihydro-5-fluorouracil, was consistent with a bond to a nucleophilic group of the enzyme.

Both mechanisms in fig. 7-30 have strong and weak points. The pH-rate profiles are bell-shaped curves for both k_{cat} and k_{cat}/K_m, with the ascending limb attributed to the ionization of the conserved aspartate, $pK_a = 6.7$ for k_{cat}/K_m (Hamilton et al., 2005). The descending limb has not been assigned. An important fact relating to the mechanism is the absence of acid-base groups in the active site other than the conserved aspartate. The evidence for fig. 7-30B is currently more extensive than for fig. 7-30C, but further work is required to derive a definitive picture.

References

Aberhart DJ, SJ Gould, H-J Lin, TK Thiuruvengadam, and BH Weiller (1983) *J Am Chem Soc* **105**, 5461.
Adair WL, O Gabriel, D Ullrey, and HM Kalckar (1973) *J Biol Chem* **248**, 4635.
Addadi L, EK Jaffe, and JR Knowles (1983) *Biochemistry* **22**, 4494.
Albery WJ and JR Knowles (1976) *Biochemistry* **15**, 5627.
Anderson L, DF Diedrich, and AM Landel (1956) *Biochim Biophys Acta* **22**, 573.
Babbitt PC, MS Hasson, JE Wedekind, DR Palmer, WC Barrett, GH Reed, I Rayment, D Ringe, GL Kenyon, and JA Gerlt (1996) *Biochemistry* **35**, 16489.
Ballinger MD, PA Frey, and GH Reed (1992) *Biochemistry* **31**, 10782.
Ballinger MD, PA Frey, GH Reed, and R LoBrutto (1995) *Biochemistry* **34**, 10086.

Banerjee R and S Chowdhury (1999) In Banerjee R (ed): *Chemistry and Biochemistry of B$_{12}$.* John Wiley & Sons: New York, p 707.
Banner DW, AC Bloomer, GA Petsko, DC Phillips, CI Pogson, and IA Wilson (1975) *Nature* **255**, 609.
Baraniak J, ML Moss, and PA Frey (1989) *J Biol Chem* **264**, 1357.
Barker HA, H Weissbach, and RD Smyth (1958) *Proc Natl Acad Sci U S A* **44**, 1093.
Barlow JN, ME Girvin, and JS Blanchard (1999) *J Am Chem Soc* **121**, 6968.
Baxter N, L Olguin, M Golicnik, G Feng, A Hounslow, W Bermel, M Blackburn, F Hollfelder, J Waltho, and N Williams (2006) *Proc Natl Acad Sci U S A,* in press.
Beatrix B, O Zelder, FK Kroll, G Örlygsson, BT Golding, and W Buckel (1995) *Angew Chem Int Ed Eng* **34**, 2398.
Beis K, V Srikannathasan, H Liu, SW Fullerton, VA Bamford, DA Sanders, C Whitfield, MR McNeil, and JH Naismith (2005) *J Mol Biol* **348**, 971.
Belasco JG, TW Bruice, WJ Albery, and JR Knowles (1986a) *Biochemistry* **25**, 2558.
Belasco JG, TW Bruice, LM Fisher LM, WJ Albery, and JR Knowles (1986b) *Biochemistry* **25**, 2564.
Belasco JG and JR Knowles (1980) *Biochemistry* **19**, 472.
Berger E, A Arabshahi, Y Wei, JF Schilling, and PA Frey (2001) *Biochemistry* **40**, 6699.
Berkovitch F, E Behshad, KH Tang, EA Enns, PA Frey, and CL Drennan (2004) *Proc Natl Acad Sci U S A* **101**, 15870.
Blackburn GM, NH Williams, SJ Gamblin, and SJS Smerdon (2003) *Science* **301**, 1184.
Blacklow SC, RT Raines, WA Lim, PD Zamore, and JR Knowles (1988) *Biochemistry* **27**, 1158.
Blättler WA and JR Knowles (1980) *Biochemistry* **19**, 738.
Bond CS, MF White, and WN Hunter (2001) *J Biol Chem* **276**, 3247.
Bothe H, DJ Darley, SP Albracht, GJ Gerfen, BT Golding, and W Buckel (1998) *Biochemistry* **37**, 4105.
Breathnach R and JR Knowles (1977) *Biochemistry* **16**, 3054.
Brouwer AC and JF Kirsch (1982) *Biochemistry* **21**, 1302.
Buckel W, G Bröker, H Bothe, and A Pierek (1999) In Banerjee R (ed): *Chemistry and Biochemistry of B$_{12}$.* John Wiley & Sons: New York, p 757.
Burke JR and PA Frey (1993) *Biochemistry* **32**, 13220.
Campbell RE, SC Mosimann, ME Tanner, NC Strynadka (2000) *Biochemistry* **39**, 14993.
Cardinale GJ and RH Abeles (1968) *Biochemistry* **7**, 3970.
Chang CH, MD Ballinger, GH Reed, and PA Frey (1996) *Biochemistry* **35**, 11081.
Chen D, FJ Ruzicka, and PA Frey (2000) *Biochem J* **348**, 539.
Chen H-P and EN Marsh (1997) *Biochemistry* **36**, 14939.
Chih HW and EN Marsh (1999) *Biochemistry* **38**, 13684.
Chih HW and ENG Marsh (2000) *J Am Chem Soc* **122**, 10732.
Chirpich TP, V Zappia, RN Costilow, and HA Barker (1970) *J Biol Chem* **245**, 1778.
Chittock RS, S Ward, A-S Wilkinson, P Caspers, B Mensch, MGP Page, and CW Wharton (1999) *Biochem J* **338**, 153.
Chook YM, JV Gray, H Ke, and WM Lipscomb (1994) *J Mol Biol* **240**, 476.
Chou WK, S Hinderlich, W Reutter, and ME Tanner (2003) *J Am Chem Soc* **125**, 2455.
Collyer CA, K Henrick, and DM Blow (1990) *J Mol Biol* **212**, 211.
Cosper NJ, SJ Booker, F Ruzicka, PA Frey, and RA Scott (2000) *Biochemistry* **39**, 15668.
Coulson AFW, JR Knowles, JD Priddle, and RE Offord (1970) *Nature* **227**, 180.
Cui Q and M Karplus (2003) *Adv Protein Chem* **66**, 315.
Davis JE, LD Nolan, and PA Frey (1974) *Biochim Biophys Acta* **334**, 442.
Dong S, R Padmakumar, N Maiti, R Banerjee, and TG Spiro (1998) *J Am Chem Soc* **120**, 9947.
Doublet P, J van Heijenoort, and D Mengin-Lecreulx (1994) *Biochemistry* **33**, 5285.
Eggerer H, P Overath, F Lynen, and ER Stadtman (1960a) *J Am Chem Soc* **82**, 2643.
Eggerer H, ER Stadtman, P Overath, and F Lynen (1960b) *Biochem Z* **333**, 1.
Fisher LM, WJ Albery and JR Knowles (1986a) *Biochemistry* **25**, 2529.
Fisher LM, JG Belasco, TW Bruice, WJ Albery and JR Knowles (1986b) *Biochemistry* **25**, 2543.
Foster PG, L Huang, DV Santi, and RM Stroud (2000) *Nat Struct Biol* **7**, 23.
Frey PA (1990) *Chem Rev* **90**, 1343.
Frey PA (1993) *FASEB J* **7**, 662.
Frey PA and RH Abeles (1966) *J Biol Chem* **241**, 2732.
Fullerton SW, S Daff, DA Sanders, WJ Ingledew, C Whitfield, SK Chapman, JH Naismith (2003) *Biochemistry* **42**, 2104.

Gallo KA and JR Knowles (1993) *Biochemistry* **32**, 3981.
Gallo KA, ME Tanner, and JR Knowles (1993) *Biochemistry* **32**, 3991.
Gerlt JA, MM Kreevoy, WW Cleland, and PA Frey (1997) *Chem Biol* **4**, 259.
Glaser L and L Ward (1970) *Biochim Biophys Acta* **198**, 613.
Glavas S and ME Tanner (1999) *Biochemistry* **38**, 4106.
Gruber K, R Reitzer, and C Kratky (2002) *Angew Chem Int Ed Eng* **40**, 3377.
Gu X, Y Liu, and DV Santi (1999) *Proc Natl Acad Sci U S A* **96**, 14270.
Gustin DJ, P Mattei, P Kast, O Wiest, L Lee, WW Cleland, and D Hilvert (1999) *J Am Chem Soc* **121**, 1756.
Ha NC, Choi G, Choi KY, Oh BH (2001) *Curr Opin Struct Biol* **11**, 674.
Ha NC, MS Kim, W Lee, KY Choi, and B-H Oh (2000) *J Biol Chem* **275**, 41100.Halpern J (1985) *Science* **227**, 869.
Hamilton CS, CJ Spedaliere, JM Ginter, MV Johnston, and EG Mueller (2005) *Arch Biochem Biophys* **433**, 322.
Haynes MR, EA Stura, D Hilvert, and IA Wilson (1994) *Science* **263**, 646.
Hilvert D, SH Carpenter, KD Nared, and MT Auditor (1988) *Proc Natl Acad Sc U S A* **85**, 4953.
Ho HT, PJ Falk, KM Ervin, BS Krishnan, LF Discotto, TJ Dougherty, and MJ Pucci (1995) *Biochemistry* **34**, 2464.
Hoang C and AR Ferré-D´Amaré (2004) *RNA* **10**, 1026.
Huang L, M Pookanjanatavip, X Gu, and DV Santi (1998) *Biochemistry* **37**, 344.
Huang Z, Q Zhang, and HW Liu (2003) *Bioorg Chem* **31**, 494.
Hwang KY, CS Cho, SS Kim, HC Sung, YG Yu, and Y Cho (1999) *Nat Struct Biol* **6**, 422.
Jackson DY, JW Jacobs, R Sugasawara, SH Reich, PA Bartlett, and PG Schultz (1988) *J Am Chem Soc* **110**, 4841.
Jogl G, S Rozovsky, AE McDermott, and L Tong (2003) *Proc Natl Acad Sci U S A* **100**, 50.
Johnson AE and ME Tanner (1998) *Biochemistry* **37**, 5746.
Jörnvall H, B Persson, M Krook, S Atrian, R González-Duarte, J Jeffrey, and D Ghosh (1995) *Biochemistry* **34**, 6003.
Kalckar HM, AU Bertland, and B Bugge (1970) *Proc Natl Acad Sci U S A* **65**, 1113.
Kallarakal AT, B Mitra, JW Kozarich, JA Gerlt, JG Clifton, GA Petsko, and GL Kenyon (1995) *Biochemistry* **34**, 2788.
Kang UG, LD Nolan, and PA Frey (1975) *J Biol Chem* **250**, 7099.
Kast P, M Asif-Ullah, N Jiang, and D Hilvert (1996) *Proc Natl Acad Sci U S A* **93**, 5043.
Kaya Y, M Del Campo, J Ofengand, and A Malhotra (2004) *J Biol Chem* **279**, 18107.
Knowles JR (1980) *Annu Rev Biochem* **49**, 877.
Komives EA, LC Chang, E Lolis, RF Tilton, G Petsko, and JR Knowles (1991) *Biochemistry* **30**, 3011.
Koshland DE and A Kowalsky (1956) *Biochim Biophys Acta* **22**, 575.
Kuliopulos A, GP Mullen, L Xue, AS Mildvan (1991)*Biochemistry* **30**, 3169.
Lahiri SD, G Zhang, D Dunaway-Mariano, and KN Allen (2003) *Science* **299**, 2067.
Landro JA, AT Kallarakal, SC Ransom, JA Gerlt, JW Kozarich, DJ Neidhart, and GL Kenyon (1991b) *Biochemistry* **30**, 9274.
Leadlay PF, WJ Albery, and JR Knowles (1976) *Biochemistry* **15**, 5617.
Lee LV, RR Poyner, MV Vu, and WW Cleland (2000b) *Biochemistry* **39**, 4821.
Lee LV, MV Vu, and WW Cleland (2000a) *Biochemistry* **39**, 4808.
Lepore BW, FJ Ruzicka, PA Frey, and D Ringe (2005) Proc Natl Acad Sci *U S A* **102**, 13819.
Lieder KW, S Booker, FJ Ruzicka, H Beinert, GH Reed, and PA Frey (1998) *Biochemistry* **37**, 2578.
Liu Y, A Arabshahi, and PA Frey (2000) *Bioorganic Chem* **8**, 29.
Liu Y, WJ Ray Jr, S Baranidharan (1997b) Acta Crystallogr D Biol Crystallogr **53** (Pt 4),392.
Liu Y, JB Thoden, J Kim, E Berger, AM Gulick, FJ Ruzicka, HM Holden, and PA Frey (1997a) *Biochemistry* **36**, 10675.
Lodi PJ and JR Knowles (1991) *Biochemistry* **30**, 6948.
Lolis E, T Alber, RC Davenport, D Rose, FC Hartman, and GA Petsko (1990) *Biochemistry* **29**, 6609.
Luo Y, J Samuel, SC Mosimann, JE Lee, ME Tanner, and NC Strynadka (2001) *Biochemistry* **40**, 14763.
Magnusson O Th, GH Reed, and PA Frey (2001) *Biochemistry* **40**, 7773.

Maitra US and H Ankel (1971) *Proc Natl Acad Sci U S A* **68**, 2660.
Mancia F and PR Evans (1998) *Structure* **6**, 711.
Mancia F, NH Keep, A Nakagawa, PF Leadlay, S McSweeney, B Rasmussen, P Bösecke, O Diat, and PR Evans (1996) *Structure* **4**, 339.
Mansoorabadi S, R Padmakumar, N Fazliddinova, M Vlasie, R Banerjee, and GH Reed (2005) *Biochemistry* **44**, 3153.
Marks GT, M Susler, and DH Harrison (2004) *Biochemistry* **43**, 3802.
Marsh EN, and CL Drennan (2001) *Curr Opin Chem Biol* **5**, 499.
Massiah MA, Abeygunawardana C, Gittis AG, Mildvan AS (1998) *Biochemistry* **37**, 14701.
McDonough MW and WA Wood (1961) *J Biol Chem* **236**, 1220.
Michenfelder M, WE Hull, and J Rétey (1987) *Eur J Biochem* **168**, 659.
Mildvan AS (1974) *Annu Rev Biochem* **43**, 357.
Miller WW and JH Richards (1969) *J Am Chem Soc* **91**, 1498.
Mitra B, AT Kallarakal, JW Kozarich, JA Gerlt, JG Clifton, GA Petsko, and GL Kenyon (1995) *Biochemistry* **34**, 2788.
Morgan PM, RF Sala, and ME Tanner (1997) *J Am Chem Soc* **119**, 10269.
Moss ML, and PA Frey (1987) *J Biol Chem* **262**, 14859.
Moss ML and PA Frey (1990) *J Biol Chem* **265**, 18112.
Nagorski RW and JP Richard (1996) *J Am Chem Soc* **118**, 7432.
Nam GH, Jang DS, Cha SS, Lee TH, Kim DH, Hong BH, Yun YS, Oh BH, Choi KY (2001) *Biochemistry* **40**, 13529.
Neidhart DJ, PL Howell, GA Petsko, VM Powers, R Li, GL Kenyon, and JA Gerlt (1991) *Biochemistry* **30**, 9264.
Nelsestuen GL and S Kirkwood (1971) *J Biol Chem* **246**, 7533.
Nickbarg EB, RC Davenport, GA Petsko, and JR Knowles (1988) *Biochemistry* **27**, 5948.
Padmakumar R, R Padmakumar, and R Banerjee (1997) *Biochemistry* **36**, 3713.
Powers VM, CW Koo, GL Kenyon, JA Gerlt, and JW Kozarich (1991) *Biochemistry* **30**, 9255.
Ray WJ (1969) *J Biol Chem* **244**, 3740.
Ray WJ Jr and GA Roscelli (1964) *J Biol Chem* **239**, 1228.
Reitzer R, M Krasser, G Jogl, W Buckel, and C Kratky (1998) *Acta Crystallogr D Biol Crystallogr* **D54**, 1039.
Rétey J and D Arigoni (1966) *Experientia* **22**, 783.
Rétey J and F Lynen (1964) *Biochem Biophys Res Commun* **16**, 358.
Rhyu GI, WJ Ray Jr, and JL Markley (1984) *Biochemistry* **23**, 252.
Richard JP (1984) *J Am Chem Soc* **106**, 4926.
Rios A, TL Aymes, and JP Richard (2000) *J Am Chem Soc* **122**, 9373.
Rose IA (1970) In Boyer PD (ed): *The Enzymes*, vol 2, 3rd ed. Academic Press: New York, p 281.
Rose IA and EJ O'Connell (1961) *J Biol Chem* **236**, 3086.
Rozovsky S, G Jogl, L Tong, and AE McDermott (2001) *J Mol Biol* **310**, 271.
Rozovsky S and AE McDermott (2001) *J Mol Biol* **310**, 259.
Ruzicka FJ, KW Lieder, and PA Frey (2000) *J Bacteriol* **182**, 469.
Sala RF, PM Morgan, and ME Tanner (1996) *J Am Chem Soc* **118**, 3033.
Samuel J, Y Luo, PM Morgan, NC Strynadka, and ME Tanner (2001) *Biochemistry* **40**, 14772.
Schafer SL, WC Barrett, AT Kallarakal, B Mitra, JW Kozarich, JA Gerlt, JG Clifton, GA Petsko, and GL Kenyon (1996) *Biochemistry* **35**, 5662.
Sofia HJ, G Chen, BG Hetzler, JF Reyes-Spindola, and NE Miller (2001) *Nucleic Acids Res* **29**, 1097.
Soltero-Higgin M, EE Carlson, TD Gruber, and LL Kiessling (2004) *Nat Struct Mol Biol* **11**, 539.
Spies MA and MD Toney (2003) *Biochemistry* **42**, 5099.
Sprecher M, MY Clark, and DB Sprinson (1966a) *J Biol Chem* **241**, 872.
Sprecher M and DB Sprinson (1964) *Ann N Y Acad Sci* **112**, 655.
Sprecher M, RL Switzer, and DB Sprinson (1966b) *J Biol Chem* **241**, 864.
St Maurice M and SL Bearne (2002) *Biochemistry* **41**, 4048.
Stadtman TC (1973) *Adv Enzymol Relat Areas Mol Biol* **38**, 413.
Straus D, R Raines, E Kawashima, JR Knowles, and W Gilbert (1985) *Proc Natl Acad Sci U S A* **82**, 2272.
Swanson BA and PA Frey (1993) *Biochemistry* **32**, 13231.
Tanner ME, KA Gallo, JR Knowles (1993) *Biochemistry* **32**, 3998.
Thoden JB, PA Frey, and HM Holden (1996a) *Protein Sci* **5**, 2149.

Thoden JB, PA Frey, and HM Holden (1996b) *Biochemistry* **35**, 2557.
Thoden JB, PA Frey, and HM Holden (1996c) *Biochemistry* **35**, 5137.
Thoden JB, AD Hegeman, G Wessenberg, MC Chapeau, PA Frey, and HM Holden (1997) *Biochemistry* **36**, 6294.
Thoden JB and HM Holden (1998) *Biochemistry* **37**, 11469.
Tollinger M, R Konrat, BH Hilbert, ENG Marsh, and B Kräutler (1998) *Structure* **6**, 1021.
Vanhooke JL (1993) Ph.D. Dissertation, University of Wisconsin-Madison.
Viger A, S Coustal, and A Marquet (1981) *J Am Chem Soc* **103**, 451.
Vlasie MD and R Banerjee (2003) *J Am Chem Soc* **125**, 5431.
Waley SG, JC Miller, IA Rose, and EF O'Connell (1970) *Nature* **227**, 181.
Wee TG and PA Frey (1973) *J Biol Chem* **248**, 33.
Wei Y, J Lin, and PA Frey (2001) *Biochemistry* **40**, 11279.
Weil JA, JR Bolton Jr, and JA Wertz (1994) *Electron Paramagnetic Resonance*. Wiley: New York.
Wharton CW (2000) *Nat Prod Rep* **17**, 447.
Wierenga RK, KH Kalk, and WJG Hol (1987) *J Mol Biol* **198**, 109.
Wilson DB and DS Hogness (1964) *J Biol Chem* **239**, 2469.
Wollowitz S and J Halpern (1984) *J Am Chem Soc* **106**, 8319.
Wollowitz S and J Halpern (1988) *J Am Chem Soc* **110**, 3112.
Wong SS and PA Frey (1977) *Biochemistry* **16**, 298.
Wu W, KW Lieder, GH Reed, and PA Frey (1995) *Biochemistry* **34**, 10532..
Wu ZR, Ebrahimian S, Zawrotny ME, Thornburg LD, Perez-Alvarado GC, Brothers P, Pollack RM, Summers MF (1997) *Science* **276**, 415.
Xiang J, JY Jung, and NS Sampson (2004) *Biochemistry* **43**, 11436
Xue LA, A Kuliopulos, AS Mildvan, and P Talalay (1991) *Biochemistry* **30**, 4991.
Zawrotny ME, DC Hawkinson, G Blotny, and RM Pollack (1996) *Biochemistry* **35**, 6438.
Zhao Q, C Abeygunawardana, P Talalay, and AS Mildvan (1996) *Proc Natl Acad Sci U S A* **93**, 8220.

8
Decarboxylation and Carboxylation

Decarboxylation is an essential process in catabolic metabolism of essentially all nutrients that serve as sources of energy in biological cells and organisms. The most widely known biological process leading to decarboxylation is the metabolism of glucose, in which all of the carbon in the molecule is oxidized to carbon dioxide by way of the glycolytic pathway, the pyruvate dehydrogenase complex, and the tricarboxylic acid cycle. The decarboxylation steps take place in thiamine pyrophosphate (TPP)–dependent α-ketoacid dehydrogenase complexes and isocitrate dehydrogenase. The latter enzyme does not require a coenzyme, other than the cosubstrate NAD^+. Many other decarboxylations require coenzymes such as pyridoxal-5′-phosphate (PLP) or a pyruvoyl moiety in the peptide chain.

Biological carboxylation is the essential process in the fixation of carbon dioxide by plants and of bicarbonate by animals, plants, and bacteria. Carboxylation by enzymes requires the action of biotin or a divalent metal cofactor, and it requires ATP when the carboxylating agent is the bicarbonate ion. The most prevalent enzymatic carboxylation is that of ribulose bisphosphate carboxylase (rubisco), which is responsible for carbon dioxide fixation in plants.

Chemistry of Decarboxylation and Carboxylation

The basic chemistry of decarboxylation is illustrated by mechanisms A to D in fig. 8-1. The mechanisms all require some means of accommodation for the electrons from the cleavage of the bond linking the carboxylate group to the α-carbon. In mechanism A, an electron sink at the β-carbon provides a haven for two electrons. Acetoacetate decarboxylase functions by this mechanism (see chap. 1), as well as PLP- and TPP-dependent

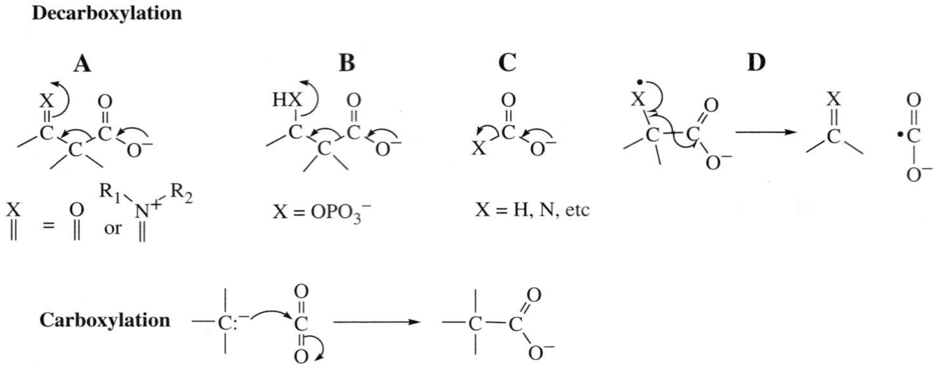

388 Enzymatic Reaction Mechanisms

Decarboxylation

A   **B**   **C**   **D**

X = O or R₁N⁺R₂   X = OPO₃⁻   X = H, N, etc

Carboxylation

Fig. 8-1. Mechanisms of decarboxylation in enzymatic reactions. (A) The most common decarboxylation mechanism is decarboxylation of a β-ketoacid or β-iminium acid. The β-iminium decarboxylation takes place in thiamine pyrophosphate–, pyridoxal-5′-phosphate–, and pyruvoyl-dependent decarboxylation. (B) Biological decarboxylation with departure of a leaving group rarely occurs. One example is mevalonate pyrophosphate decarboxylase. (C) The mechanism applies to formate dehydrogenase and decarboxylation of carbamates. (D) The radical fragmentation is seen with increasing frequency and always involves secondary electron transfer events to generate a radical or to oxidize the formyl radical produced in the C—C cleavage. The carboxylation of a carbanion by carbon dioxide is the chemical process in enzymatic carbon dioxide fixation.

decarboxylases (see chap. 3). In mechanism B, a leaving group at the β-carbon departs with two electrons. Mevalonate-5-diphosphate decarboxylate functions by mechanism B and is discussed in a later section. In mechanism C, a leaving group replaces the α-carbon and departs with a pair of electrons. A biological example is formate dehydrogenase, in which the leaving group is a hydride that is transferred to NAD⁺. In mechanism D, a free radical center is created adjacent to the α-carbon and potentiates the homolytic scission of the bond to the carboxylate group. Mechanism D requires secondary electron transfer processes to create the radical center and quench the formyl radical. Mechanism D is appearing with increasing frequency but is still uncommon in enzymatic processes. In this chapter, we discuss examples of mechanisms A, B, and D. Mechanism C occurs in biotin-dependent carboxylases within complex, multienzyme systems and is discussed in chapter 18.

All biological carboxylation reactions follow at some level the course sketched in fig. 8-1. Those in which bicarbonate is the carboxylating agent require ATP and are preceded by the ATP-dependent dehydration of bicarbonate to carbon dioxide, which then reacts with a carbanion, as illustrated in fig. 8-1. In this chapter, we discuss carbon dioxide fixation by ribulose bisphosphate carboxylation, which uses carbon dioxide as the carboxylating agent, and phosphoenolpyruvate carboxylase, which uses bicarbonate and ATP to generate carbon dioxide in the active site.

Decarboxylases

Most decarboxylases require the action of a coenzyme, usually TPP for α-ketoacid decarboxylation, PLP for amino acid decarboxylation, or an enzymatic pyruvoyl group for certain α-amino acid decarboxylations. β-Ketoacid decarboxylations do not require a coenzyme, and we discuss the role of iminium intermediates in the action of acetoacetate decarboxylase (see chap. 1). The NAD⁺-dependent isocitrate dehydrogenase and malic enzyme generate β-ketoacids and then catalyze their decarboxylation, presumably by mechanism A in fig. 8-1, where X = O. We consider NAD⁺-dependent enzymes in chapter 16.

An increasing number of decarboxylases function by radical mechanisms, presumably mechanism D in fig. 8-1. The radical decarboxylations require metallocoenzymes to facilitate radical formation. We next consider coenzyme-dependent decarboxylases.

Pyruvate Decarboxylase

The most thoroughly studied TPP-dependent decarboxylation is that catalyzed by pyruvate decarboxylase (PDC; EC 4.1.1.1) from yeast (Jordan, 2003). Several other TPP-dependent decarboxylation processes have been studied in varying detail, including the E1 components of the pyruvate dehydrogenase and α-ketoglutarate dehydrogenase complexes (see chap. 18), benzoylformate decarboxylase (Hasson et al., 1998), and acetolactate synthase. In chapter 3, we discussed the role of TPP in forming the lactyl-TPP adduct, which sets up the thiazolium ring as an electron sink for decarboxylation by mechanism A in fig. 8-1. The overall mechanism for the action of PDC in fig. 8-2 is supported by numerous

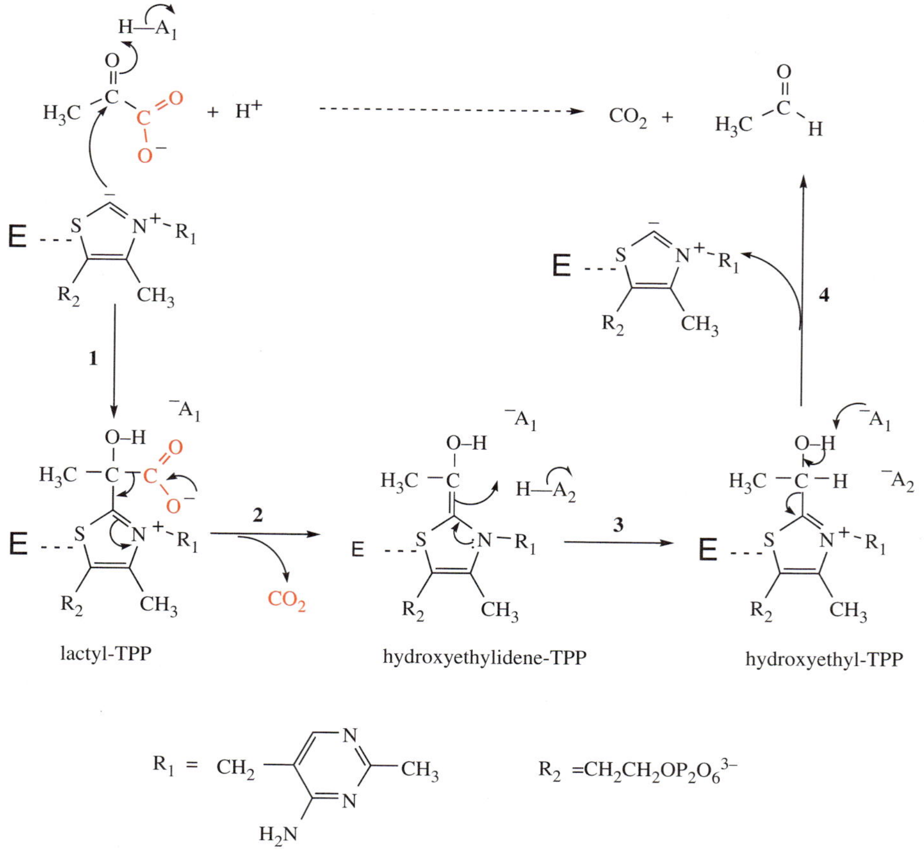

Fig. 8-2. The role of thiamine pyrophosphate (TPP) in the decarboxylation of pyruvate by pyruvate decarboxylase. The hypothetical role of the pyrimidine ring in TPP is outlined in scheme 8-1, and the mechanism by which the ylide carbanion of TPP is generated is discussed in the text, which also provides evidence for the participation of the intermediates lactyl-TPP, hydroxyethylidene-TPP, and hydroxyethyl-TPP.

experiments that implicate lactyl-TPP, hydroxyethylidene-TPP, and hydroxyethyl-TPP as intermediates. Details of the catalysis of proton transfer remain somewhat unclear, despite extensive structural information and many experiments employing site-directed mutagenesis and kinetic analysis. The most illuminating results derive from biochemical experiments with the wild-type and specifically mutated enzyme.

Thiamine Pyrophosphate Intermediates

Four types of evidence support the participation of the TPP-intermediates shown in fig. 8-2. Biochemical experiments on the role of TPP are complicated by the fact that PDC is purified with tightly bound TPP in the active site, and this fact places certain limitations on the experiments that can be conducted. Acid-quenched samples of PDC.TPP in the steady state of its reaction with ^{14}C-pyruvate contained a radioactive compound that proved to be chromatographically indistinguishable from synthetic hydroxyethyl-TPP (Carlson and Brown, 1961; Holzer and Beaucamp, 1958; Krampitz et al., 1958).

A yeast apo-PDC can be constructed by mutation of Glu91 to Asp91 (E91D-PDC). The mutated enzyme remains active but binds TPP weakly (Jordan, 2003). Synthetic lactyl-TPP is accepted by apo-E91D-PDC as an intermediate, proving its chemical competence. The reaction leads to the partitioning of lactyl-TPP into either pyruvate in the back reaction or into acetaldehyde plus carbon dioxide in the forward direction. When the enzymatic reaction of wild-type PDC.TPP with pyruvate is quenched by denaturation of the protein in the steady state, both lactyl-TPP and hydroxyethyl-TPP appear in the quenched solution. Quantitative data obtained by ^{1}H NMR analysis of rapid mix-acid quenched samples in transient kinetic studies show that they are kinetically competent as intermediates (Tittman et al., 2003).

Spectrophotometric and crystallographic evidence with substrate analogs implicate the central species in fig. 8-2, hydroxyethylidene-TPP, as an intermediate. Derivatives of pyruvate, with an aromatic substituent in place of C3, behave as suicide inactivators because of electronic stabilization of the enamine analogous to hydroxyethylidene-TPP. Compounds such as X–C$_6$H$_4$–COCOOH and X–C$_6$H$_4$–CH=CHCOCOOH, where X is a substituent on the aromatic ring, react with TPP rapidly at the active site and undergo decarboxylation, but the resulting enamines are too stable to react further (Kuo and Jordan, 1983; Menon-Rudolph et al., 1992). They seem to be too stable as enamines to undergo the protonation step to the hydroxyethyl analogs. These enamine analogs of hydoxyethylidene-TPP incorporate strong chromophores because of the conjugated double bonding shown in color in structure **8-1,** and these prominent optical properties allowed them to be identified as the decarboxylation products at the active site.

8-1

2-(*p*-Nitrobenzylidene)-1-hydroxyethylidene-TPP

In recent crystallographic work, 3-hydroxypyruvate co-crystallized with PDC·TPP gave a complex whose structure proved to be compatible with dihydroxyethylidene-TPP, an analog of hydroxyethylidene-TPP (Fiedler et al., 2002).

Structure of Yeast Pyruvate Decarboxylase

Several structures of PDC from yeast and bacteria reveal much about the chemical environment at the active site and the structure of TPP at this site (Arjunan et al., 1996; Dobritzsch et al., 1998; Dyda et al., 1993). The structure of yeast PDC in fig. 8-3 reveals the complexities of a tetrameric multidomain protein with several binding sites. In addition

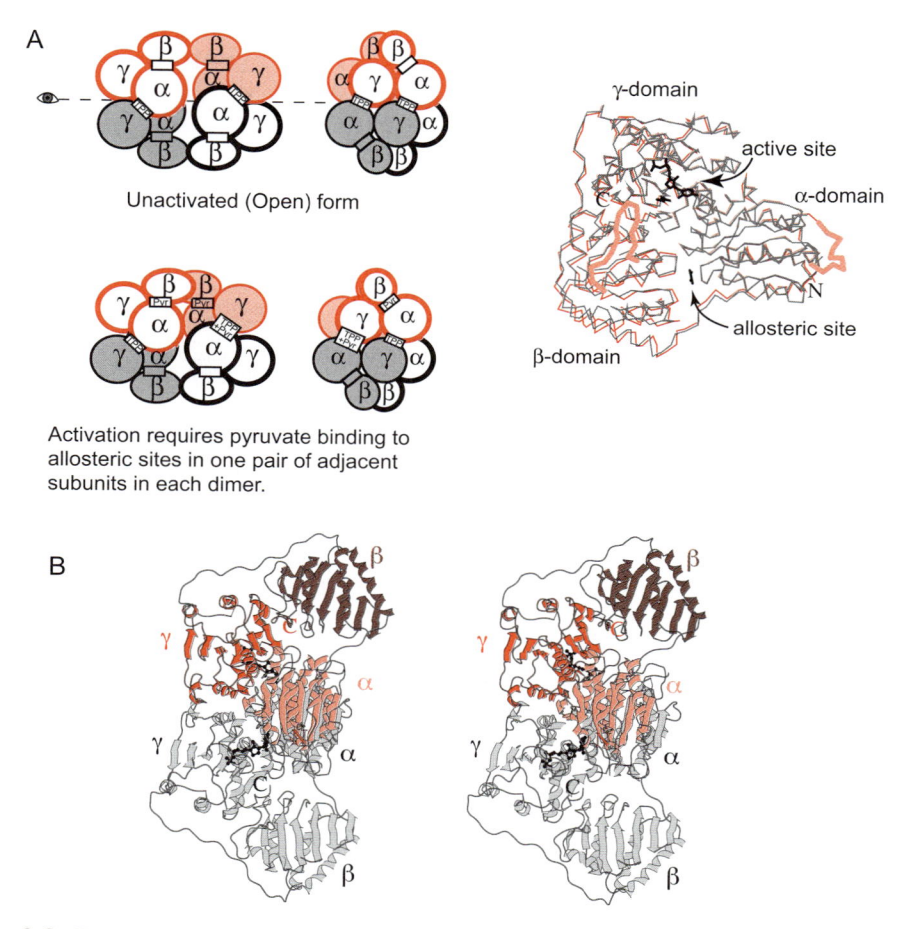

Fig. 8-3. Pyruvate decarboxylase from yeast is a homotetrameric protein that is subject to allosteric regulation. The four subunits of the structure are assembled as a dimer of dimers, with each subunit composed of three distinct domains (α, β, and γ). (A) Binding of pyruvate to the allosteric site (located between domains α and β in each subunit) in one subunit per dimer drives the holoenzyme from the inactive, open form to the active, closed conformation. Thiamine pyrophosphate (TPP) is retained in the active sites located between the α and γ domains of adjacent subunits. To the right, overlaid Cα traces of the open (in gray with TPP bound; 2.4-Å resolution; PDB 1PYD; Dyda et al., 1993) and closed (in red with pyruvamide and thiamine pyrophosphate (TPP) in black; 2.4-Å resolution; PDB 1QPB; Lu et al., 2000) forms of the enzyme are shown. (B) Stereoimage of a ribbon representation of an open-form dimer is displayed with the top subunit colored by domain.

Continued

392 Enzymatic Reaction Mechanisms

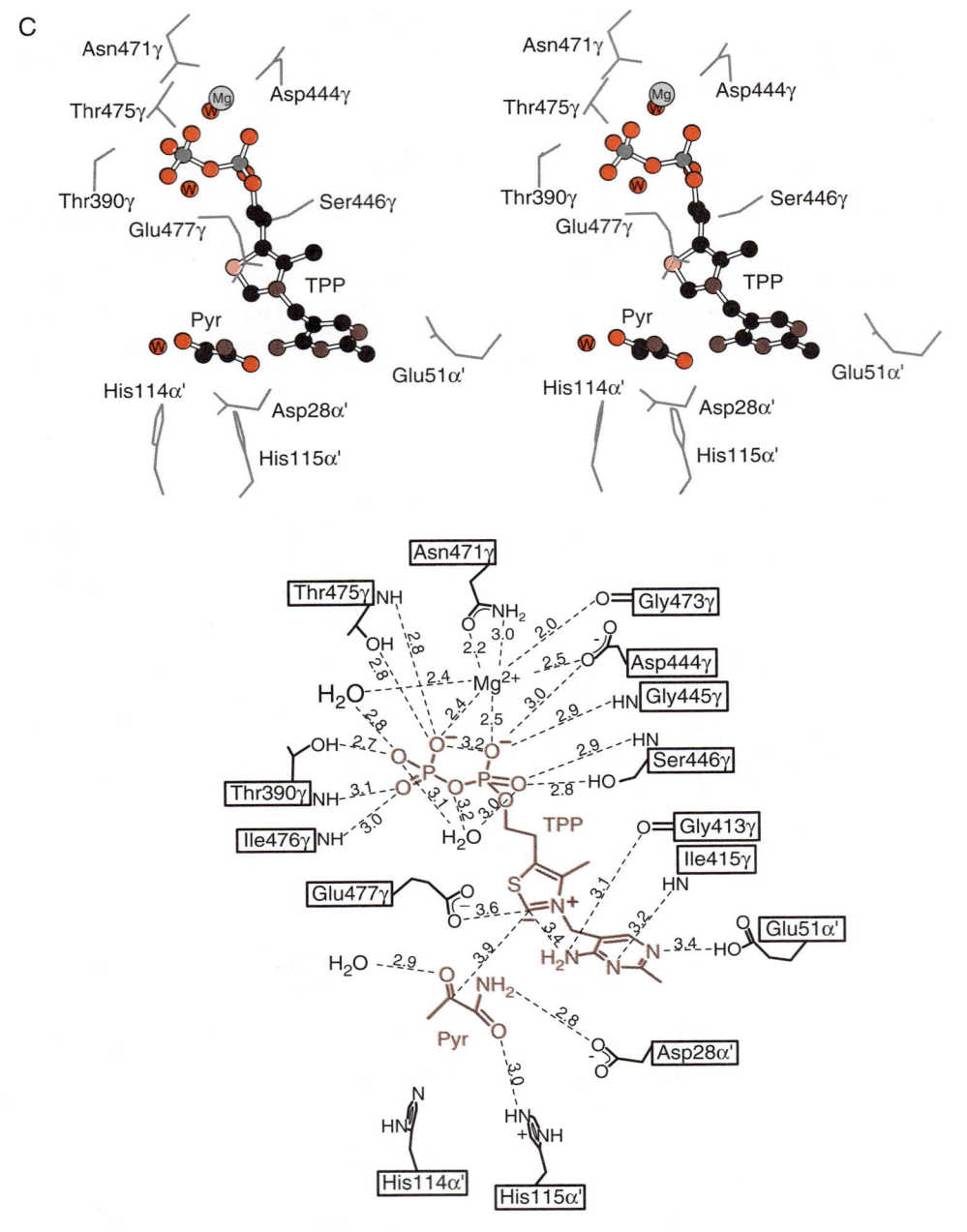

Fig. 8-3. cont'd (C) The active site of the closed form of the enzyme is shown in stereo with pyruvamide and MgTPP bound. A schematic drawing of the stereoimage is included beneath it.

to the TPP and pyruvate sites in the active center, there is an allosteric pyruvate binding site. Nonlinear double reciprocal plots of activity against pyruvate concentration show substrate activation. Pyruvamide in place of pyruvate also activates by binding to the allosteric site, and inclusion of pyruvamide in assays can eliminate the nonlinear kinetics.

The structure of the PDC·TPP complex shows the conformation of TPP as V-shaped centered on the methylene bridge between the thiazolium and pyrimidine rings, and this structure has mechanistic implications. The V-shaped structure is found in all TPP-dependent enzymes studied to date. The structure of PDC further confirms the role assigned to an amino acid sequence motif found in all TPP-dependent enzymes, GDGX$_{26}$N(C)N (Hawkins et al., 1989). This motif provides the protein ligands for Mg^{2+} engaged in binding the pyrophosphate moiety of TPP. In PDC, the Mg^{2+} ligands are Asp444, and Asn471. Amino acid residues in the active site that appear to be functionally important in catalysis include Glu477, Asp28, and Glu51; the latter forms a hydrogen bonded contact with N1 of the pyrimidine ring. The contact is important in binding TPP but may also play a mechanistic role.

Mechanism of Catalysis

The role of TPP in catalysis shown in fig. 8-2 is well established. However, the roles of the enzyme and of amino acid residues at the active site are less certain. We discuss three aspects of catalysis: ionization of TPP to its C2-anion, decarboxylation of lactyl-TPP, and proton transfer.

Experiments implicate Glu51 and the pyrimidine ring of TPP in the initial ionization of the thiazolium-C2(H) preceding its addition to pyruvate in step 1 of fig. 8-2. The V structure of TPP places the exocyclic amino group of the pyrimidine ring within 3.0 to 3.4 Å of C2 in the thiazolium ring, making the amino group a candidate base to abstract C2(H), as in scheme 8-1.

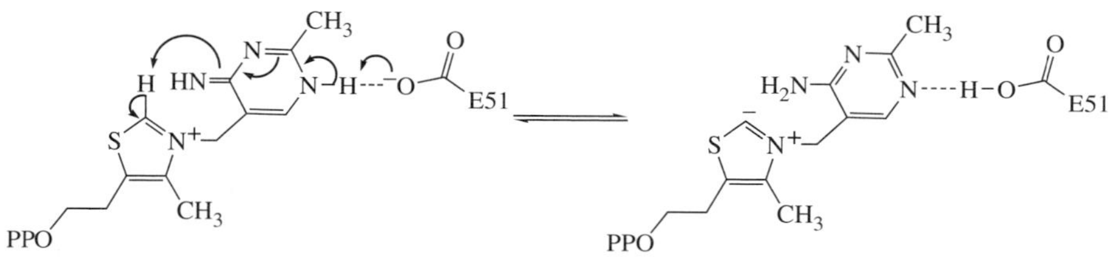

Scheme 8-1

The amino group is not basic enough to abstract the C2(H) from TPP. However, as the iminopyrimidine tautomer, it is postulated to mediate proton transfer to Glu51.

Evidence in support of this mechanism is the observation of a chromophore centered between 300 and 310 nm and a circular dichroism band at 310 to 320 nm (Jordan et al., 2002). This band is similar to those for chemical models of the iminopyrimidine ring on the right in scheme 8-1. The mechanism implicates Glu51 in promoting the ionization through proton transfer mediated by the iminopyrimidine ring.

2-Lactylthiazolium salts undergo decarboxylation sluggishly in water but very fast in ethanol (Crosby et al., 1970). This process corresponds to step 2 in fig. 8-2, which illustrates the destruction of electrostatic charge on decarboxylation of lactyl-TPP to the enamine hydroxyethylidene-TPP. Because the transition state is less polar than the substrate, decarboxylation is promoted by a medium of low dielectric constant such as ethanol. The effective dielectric constant in an enzymatic active site is generally lower than that of water, and this is true of PDC. Evidence of the apolar nature of the active site is afforded by the fluorescence properties of the TPP-analog thiochrome pyrophosphate, the alkaline oxidation product of TPP. The emission wavelength of free thiochrome pyrophosphate is well correlated with the solvent dielectric constant, and correlation of the fluorescence emission spectrum of the same compound in the active site of PDC indicates that the effective dielectric constant is 13 to 15 (Jordan et al., 1999). The effective dielectric constant is much lower than 80, that of water, and even lower than that of ethanol.

Another measure of the effect of medium polarity in promoting the decarboxylation of lactyl-TPP is its effect on the ionization of hydroxyethyl-TPP, which also leads to hydroxyethylidene-TPP. The process is illustrated for hydroxybenzyl-TPP in eq. 8-1. The value of pK_a for the ionization of hydroxybenzyl-TPP in water is about 15 (Barletta et al., 1997). However, at the active site of PDC this compound exists as the enamine, hydroxybenzylidene-TPP at pH values as low as pH 6, as determined spectrophotometrically.

$$\text{hydroxybenzyl-TPP} \rightleftharpoons \text{hydroxybenzylidene-TPP} + H^+ \quad (8\text{-}1)$$

The value of pK_a for hydroxybenzylidene-TPP at the active site must be lower than 6, and this corresponds to a stabilization of the enamine by a factor of 10^9 relative to water (Jordan et al., 1999). The low pK_a at the active site is well correlated with the low effective dielectric constant, and this accounts for a large part of the rate enhancement for decarboxylation brought about by PDC.

The low pK_a of hydroxybenzyl-TPP at the active site of PDC would make it difficult to protonate hydroxybenzylidene-TPP generated in the decarboxylation of benzoylformate. This problem may not arise in the decarboxylation of pyruvate because of the higher intrinsic pK_a of hydroxyethyl-TPP, which is estimated to be about 18 (Jordan, 2003), almost 3 units higher than that of hydroxybenzyl-TPP. The microenvironment at the active site of PDC may lower the pK_a to the neutral range, allowing hydroxyethylidene-TPP to be efficiently protonated to hydroxyethyl-TPP and then decomposed to acetaldehyde and TPP.

Steps 1, 3, and 4 in fig. 8-2 involve acid-base catalysis by PDC. The catalytic groups given generic designations in fig. 8-2 are not definitively known but have been tentatively assigned as Asp28 and Glu477 (Jordan, 2003). There appears to be no general pattern in catalysis of proton transfer among TPP-dependent enzymes. For example, benzoylformate decarboxylase is similar in many ways to PDC, but the amino acid residues participating in acid-base catalysis, although not definitively assigned, seem to be different from those in PDC (Jordan, 2003).

Amino Acid Decarboxylases

The decarboxylation of amino acids by enzymes also require a coenzyme: PLP or a pyruvoyl moiety of the enzyme. Numerous PLP-dependent amino acid decarboxylases are known, and PLP facilitates both α-decarboxylation of α-amino acids and the β-decarboxylation of aspartate. A growing number of α-decarboxylases incorporate a pyruvoyl moiety within their polypeptide chains, and the pyruvoyl-carbonyl group facilitates α-decarboxylation in place of PLP. The pyruvoyl enzymes include S-adenosylmethionine decarboxylase (EC 4.1.1.50) (Tolbert et al., 2001), phosphatidyl serine decarboxylase (EC 4.1.1.65) (Dowhan, 1997), aspartate α-decarboxylase (EC 4.1.1.11) (Albert et al., 1998), and the original member of this family, the bacterial histidine decarboxylase. In this section, we consider PLP-dependent and pyruvoyl α-decarboxylases and aspartate β-decarboxylase.

The general mechanism for PLP-dependent α-decarboxylation is outlined in fig. 8-4. The principle of external aldimine formation and decarboxylation to form the α-carbanion, shown in its quinonoid resonance form, is followed in all cases. Protonation of the α-carbanion and transaldimination to the internal aldimine leads to the amine product. Several PLP-dependent α-decarboxylases have been studied. In the following section, we consider a hybrid α-decarboxylase that also catalyzes transamination.

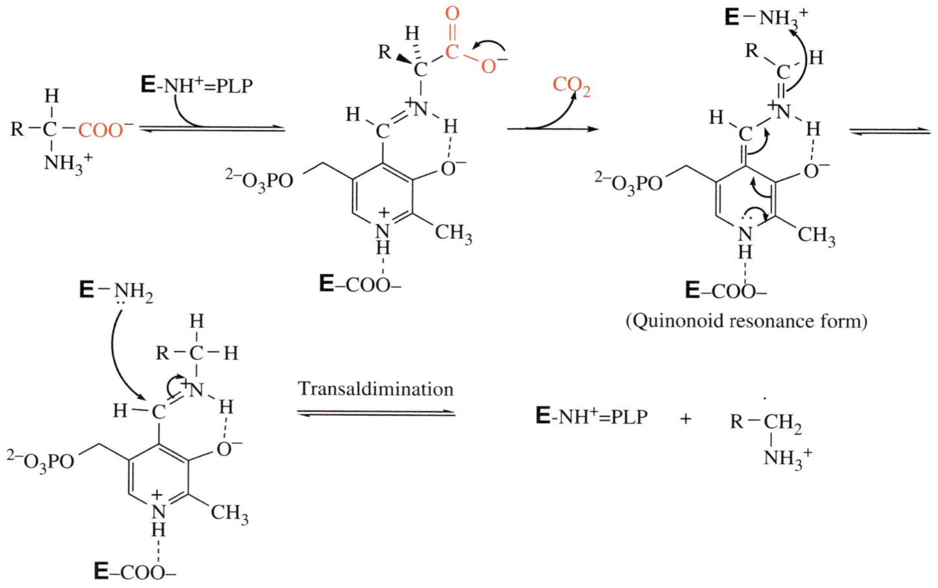

Fig. 8-4. A minimal mechanism for the action of a pyridoxal-5′-phosphate (PLP)–dependent α-decarboxylase. α-Decarboxylation begins with reaction of the internal aldimine **E**-NH⁺=PLP with the α-amino acid to form the external aldimine. Decarboxylation leads to the corresponding α-carbanionic PLP-species shown in its quinonoid resonance form. Protonation of the α-carbon in the next step leads to the external aldimine of the decarboxylated amino acid. Transaldimination produces the amine and regenerates the internal aldimine **E**-NH⁺=PLP (see chap. 3).

Dialkylglycine Decarboxylase

Dialkylglycine decarboxylase (EC 4.1.1,64), a tetrameric PLP-enzyme (47-kDa subunit) accepts a variety of substrates and requires a monovalent cation for maximal activity. The reaction catalyzed includes both α-decarboxylation and transamination as defined in eq. 8-2.

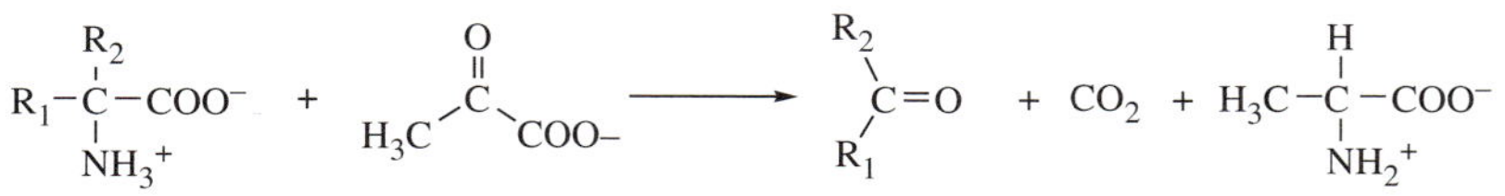

$$R_1-\underset{\underset{NH_3^+}{|}}{\overset{\overset{R_2}{|}}{C}}-COO^- + H_3C-\overset{\overset{O}{\|}}{C}-COO^- \longrightarrow \underset{R_1}{\overset{R_2}{\diagdown}}C=O + CO_2 + H_3C-\underset{\underset{NH_2^+}{|}}{\overset{\overset{H}{|}}{C}}-COO^- \qquad (8\text{-}2)$$

2,2-Dialkylglycines associated with peptide antibiotics are metabolized in bacteria and fungi through the action of this enzyme. The reaction involves α-decarboxylation of the substrate followed by transamination with pyruvate to form a ketone, CO_2, and alanine.

The mechanism in fig. 8-5 describes the chemical process in steps.

The overall reaction follows ping pong kinetics with the formation of the **E.PMP** intermediate. The first half-reaction to form the **E.PMP** is known as decarboxylation, and the second to transfer the amine group to pyruvate is known as transamination. Transient kinetics show the two half-reactions proceeding at comparable rates, with k_{cat} values in the range of 25 s⁻¹ and 75 s⁻¹ for decarboxylation and transamination, respectively. Transaldimination seems to be rate limiting in the decarboxylation half-reaction, so that the quinonoid-carbanion is not observed; however, the quinonoid is observed transiently in the transamination half-reaction (Zhou et al., 2001).

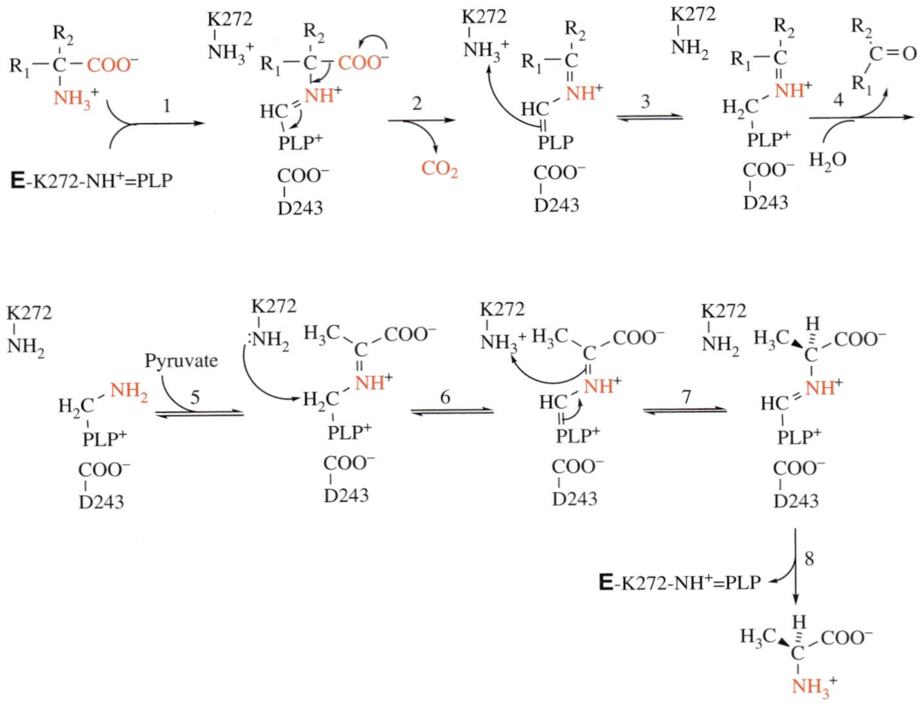

396 Enzymatic Reaction Mechanisms

Fig. 8-5. A mechanism for the action of dialkylglycine decarboxylase. In the upper line, the dialkylglycine and enzyme–pyridoxal-5′-phosphate (PLP) aldimine react in step 1 to form the external aldimine, which undergoes decarboxylation in step 2, analogous to the mechanism in fig. 8-4. However, unlike the mechanism in fig. 8-4, the resulting ketimine intermediate is protonated on C4′ of PLP in step 3 and in step 4 suffers hydrolysis to the ketone product $R_1R_2C=O$ and pyridoxamine 5′-phosphate (PMP). The lower line outlines the transamination part of the mechanism, in which PMP reacts with pyruvate in steps 5 to 8 to form alanine and regenerate the internal aldimine of the enzyme with PLP.

The mechanism for α-decarboxylation and transamination in fig. 8-5 includes the α-carbanionic intermediate shown as its quinonoid resonance form. Transient kinetic analysis revealed the formation of a long-wavelength chromophore characteristic of the quinonoid intermediate in the transamination half-reactions of L-alanine and L-aminobutyrate, but not in the decarboxylation. The results suggest that decarboxylation is rate limiting in the overall reactions of these substrates. Reactions of slower substrates may be kinetically limited by transaldimination. Computations indicate that the imine linkage of the external aldimine contributes more to stabilizing the carbanion resulting from decarboxylation than does the pyridinium ring (Toney, 2001).

When interacting with substrates in which R_1 or R_2 in eq. 8-2 is hydrogen and not an alkyl group, dialkylglycine decarboxylase can also catalyze an overall transamination reaction. In overall transamination, the transamination half-reaction in fig. 8-5 operates in both directions, leading to the classic reaction of an amino acid with a ketoacid to transfer the amino group and to form the alternative ketoacid and amino acid, respectively.

The structure of dialkylglycine decarboxylase illustrated in fig. 8-6 shows the binding site for PLP. The monovalent cation binding site lies near the active site. Larger cations such as Rb^+ and K^+ activate the enzyme, and smaller cations such as Na^+ and Li^+ inhibit the enzyme. Crystallographic evidence indicates conformational differences that depend on the size of the cation (Hohenester et al., 1994). Kinetic evidence indicates that this

Decarboxylation and Carboxylation 397

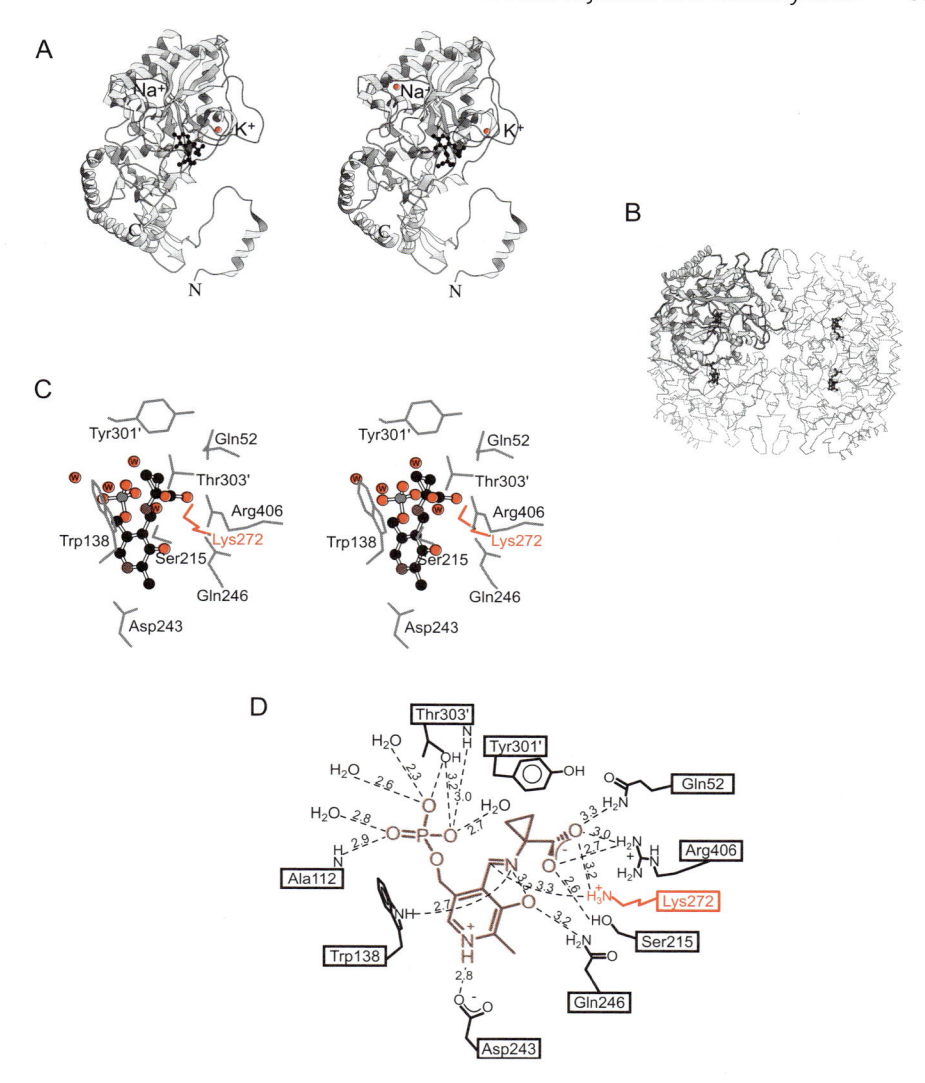

Fig. 8-6. Structure of dialkylglycine decarboxylase. One subunit of the homotetrameric 2,2-dialkylglycine decarboxylase from *Pseudomonas cepacia* is shown with the pyridoxal-5′-phosphate (PLP)-1-amino-1-cyclopropane carboxylate (ACC) external aldimine (black) bound at the active site (2.0-Å resolution; PDB 1D7R; Malashkevich et al., 1999). Two noncatalytic, monovalent metal ions (Na$^+$ and K$^+$) are also shown; different-sized monovalent metal ions have been shown to mediate active site conformational changes such that Li$^+$ and Na$^+$ are inhibitory and K$^+$ and Rb$^+$ are activating (Hohenester et al., 1994). (B) The homotetrameric holoenzyme is shown. (C) The active site with the ACC substrate analog inhibitor is reversibly bound to the PLP. (D) The structure is shown as a two-dimensional scheme.

enzyme undergoes conformational changes in solution with time that involve the cation binding site (Zhou and Toney, 1998).

The dual reaction specificity and broad substrate specificity of dialkylglycine decarboxylase suit it to studies of the hypothesis of stereoelectronic control in PLP reactions by enzyme-substrate interactions in three subsites (Dunathan, 1966). This hypothesis is discussed in chapter 3, where it was supported by the observation of suprafacial intramolecular proton transfer in transamination reactions. The hypothesis is further supported by

398 Enzymatic Reaction Mechanisms

studies of dialkylglycine decarboxylase, and scheme 8-2 illustrates part of the evidence (Sun et al., 1998).

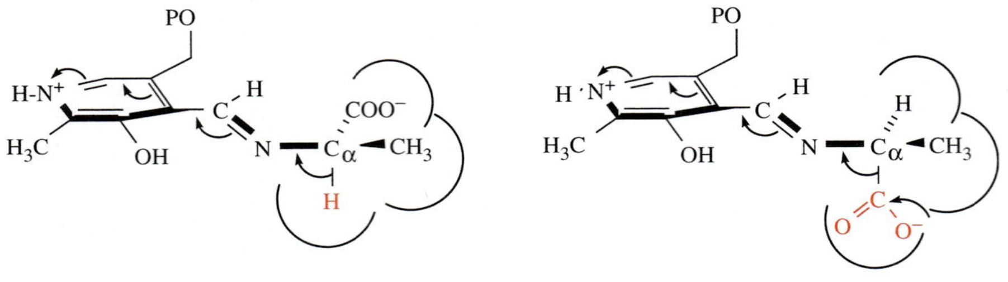

D-Alanine   L-Alanine

Scheme 8-2

The enzyme catalyzes the decarboxylation of aminomalonate, proving that it can accommodate two carboxylate groups, one in the decarboxylation subsite and a second in another subsite. It also catalyzes the decarboxylation of L-alanine but not of D-alanine, supporting the hypothesis that the reacting carboxylate group must be held orthogonal to the π-bonding system of PLP in the reactive conformation. The carboxylate of D-alanine does not react, presumably because it cannot occupy the reactive decarboxylation site. If the methyl groups of D- and L-alanine occupy the same subsite, this would place the carboxylate of D-alanine in the alternative, nondecarboxylating subsite and the α-hydrogen orthogonal to the p-bonding system, in position for removal in transamination. This explains how the enzyme catalyzes the transamination of D-alanine and not its decarboxylation.

Aspartate β-Decarboxylase

The decarboxylation of aspartate to alanine is a β-decarboxylation reaction (eq. 8-3), as distinguished from the α-decarboxylations that are most typical of amino acid decarboxylases (Tate and Meister, 1971).

$$H^+ + {}^-OOC-\overset{H}{\underset{{}^+H_3N}{C}}-CH_2COO^- \longrightarrow CO_2 + {}^-OOC-\overset{H}{\underset{{}^+H_3N}{C}}-CH_3 \quad (8\text{-}3)$$

The bacterial enzyme is PLP dependent, and the coenzyme stabilizes the β-carbanion of the alanine skeleton resulting from decarboxylation. PLP has the capacity to stabilize β-carbanions and α-carbanions of amino acids (see chap. 3). The chemical mechanism is shown in fig. 8-7. In this mechanism, the reaction proceeds initially along the lines of an aminotransferase through steps 1 to 3. β-Decarboxylation analogous to reaction A in fig. 8-1 then takes place in step 4, and the resulting resonance stabilized β-carbanion is protonated in step 5. Steps 6 to 8 to form the product are analogous to the reversal of steps 1 to 3. The mammalian aspartate β-decarboxylase (EC 4.1.1.12) is not PLP-dependent (Rathod and Fellman, 1985).

The ^{13}C kinetic isotope effect in the reaction of the *Alcaligenes* β-decarboxylase is small, 1.0099 relative to the maximum of 1.05, showing that decarboxylation is not entirely rate limiting (Rosenberg and O'Leary, 1985). The protonation of the β-carbanion in step 5 of fig. 8-7 proceeds with inversion of configuration (Chang et al., 1982).

Aspartate β-decarboxylase gradually loses activity during turnover but can be rescued by addition of pyruvate, which is itself converted into alanine. Loss of activity is brought about a side reaction, in which the decarboxylation intermediate occasionally undergoes hydrolysis to pyruvate and **E.PMP**, which is inactive against aspartate, as shown in the

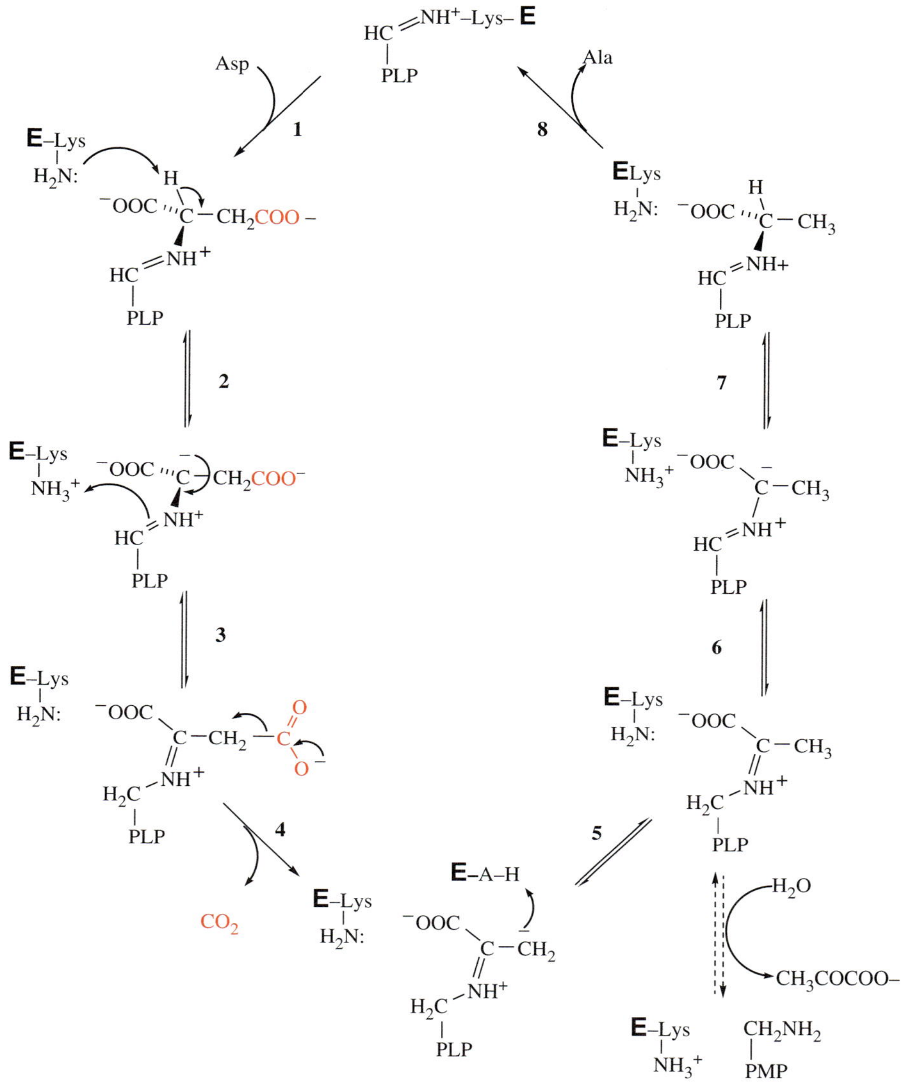

Fig. 8-7. A mechanism for decarboxylation by aspartate β-decarboxylase. Aspartate β-decarboxylase functions in the first few steps much as an aminotransferase, in that the α-proton is labilized in the isomerization of the external aldimine through steps 2 and 3 to the ketimine. Decarboxylation then occurs in step 4, and protonation of the β-carbanion occurs in step 5. Isomerization of the ketimine to the external pyridoxal-5′-phosphate (PLP)–aldimine of alanine in steps 6 and 7 is followed by transaldimination and release of the product alanine. The side reaction in the lower right is an aminotransferase-like hydrolysis to pyruvate and pyridoxamine 5′-phosphate (PMP).

lower right of fig. 8-7. (This would naturally happen in the case of an aminotransferase.) The addition of free pyruvate brings its concentration high enough to reverse the inactivation and regenerate **E**-PLP.

Histidine Decarboxylase

PLP-dependent and PLP-independent histidine decarboxylases (EC 4.1.1.22) catalyze the same reaction. A PLP-independent histidine decarboxylase was one of the first

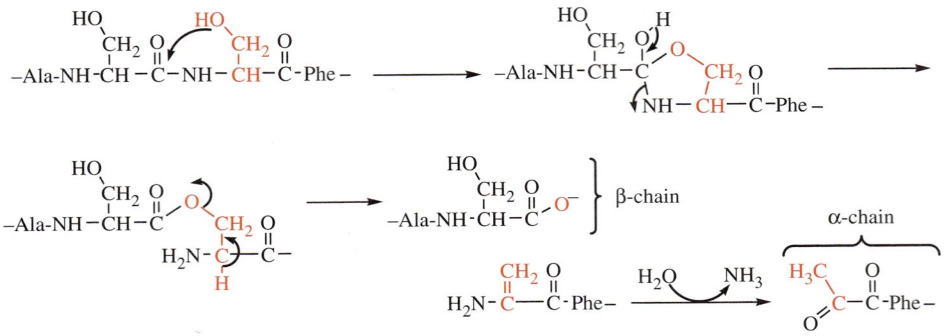

Fig. 8-8. A mechanism for the cleavage of pro-histidine decarboxylase into α and β subunits. Spontaneous cleavage between serine residues in the -Thr-Ala-Ser-Ser-Phe- segment of pro-histidine decarboxylase may follow the course shown here, which is suggested by ^{18}O-labeling experiments (Recsei and Snell, 1985). The figure demonstrates how the 3-hydroxyl group of serine is transferred to the C-terminal carboxyl group of serine in the β subunit.

pyruvoyl-enzymes to be discovered (Recsei and Snell, 1984; van Poelje and Snell, 1990). Pyruvoyl-enzymes catalyze α-decarboxylation of amino acids, similar to many PLP-dependent enzymes, and they include histidine, S-adenosylmethionine, phosphatidyl serine, and aspartate α-decarboxylases. Pyruvoyl-enzymes are found in both prokaryotes and eukaryotes. D-Proline reductase is also a pyruvoyl-enzyme (Hodgins and Abeles, 1967).

The pyruvoyl moiety in histidine decarboxylase arises in a self-cleavage reaction of the newly translated proenzyme (Recsei and Snell, 1984). The 35- to 37-kDa proenzyme includes a Ser-Ser sequence that undergoes cleavage to generate two peptide fragments, the 25- to 28-kDa α subunit, which includes the pyruvoyl moiety blocking its N-terminus, and the 9- to 11-kDa β subunit (Recsei et al., 1983). The exact lengths of the proenzyme and subunits differ slightly among histidine decarboxylases from different species, but the known proenzymes include the sequence -Thr-Ala-Ser-Ser-Phe- at the cleavage site (Huynh and Snell, 1985). Internal cleavage involves the β-hydroxyl group of a seryl residue reacting as a nucleophile with the neighboring acylcarbonyl group, as illustrated in fig. 8-8. Evidence supporting serinolysis consists essentially of the fate of ^{18}O in the proenzyme labeled with [3-^{18}O]seryl residues. After processing, the β-chain contained ^{18}O in the carboxyl and β-hydroxyl groups (Recsei and Snell, 1985). The processed enzyme is dodecameric $(\alpha\beta)_6$ and about 208 kDa.

Pyruvoyl-enzymes display characteristic chemical properties. They react with and are inactivated by carbonyl reagents such as NaBH$_4$ and phenylhydrazine. PLP-enzymes also react with these reagents, but pyruvoyl-enzymes do not contain PLP. Hydrolysis of a pyruvoyl-enzyme that has been reduced with borohydride yields a mole of lactate per mole of protein. Reaction with phenylhydrazine generates protein chromophores similar to those of phenylhydrazones.

A reasonable decarboxylation mechanism is depicted in fig. 8-9. Support for this mechanism includes the trapping of an iminium intermediate by reduction of the enzyme with NaBH$_4$ in the presence of ^{14}C-labeled histidine or histidine methyl ester. Histidine decarboxylase is inhibited by histidine methyl ester, which forms an imine with the keto group of the pyruvoyl moiety (Alston and Abeles, 1987).

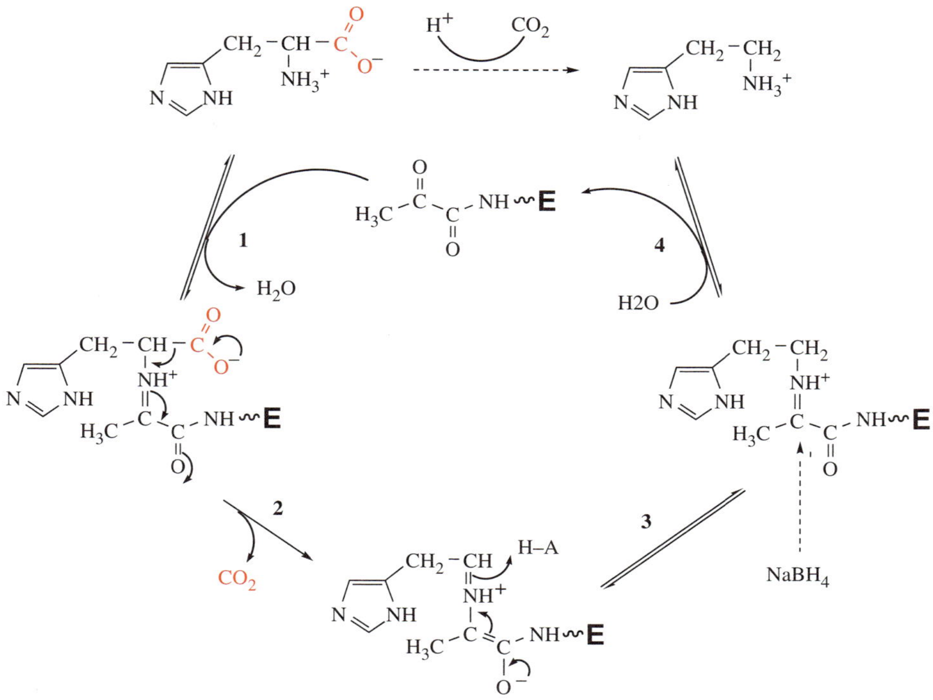

Fig. 8-9. A mechanism for decarboxylation of histidine by pyruvoyl-dependent histidine decarboxylase. The keto group of the pyruvoyl moiety fills the role played by the aldehydo group in pyridoxal-5′-phosphate (PLP)–dependent decarboxylation reactions. In the first step, the pyruvoyl moiety reacts with the amino group of histidine to form an imine linkage between the substrate and enzyme. The imine includes a potential electron sink for decarboxylation in the form of the acyl-carbonyl group of the pyruvoyl moiety, and decarboxylation proceeds in step 2 by a process homologous to mechanism A in fig. 8-1. Protonation at the α-carbon in step 3 forms the product imine, which undergoes hydrolysis in step 4 to histamine and the pyruvoyl-enzyme. Evidence for imine formation is discussed in the text and by the structure shown in fig. 8-9.

Decarboxylation of histidine by the *Lactobacillus* enzyme proceeds with a ^{13}C isotope effect ($^{12}k/^{13}k$) measured under essentially saturating (k_{cat}) conditions of 1.0334 ± 0.0005 and an α-^{15}N isotope effect ($^{14}k/^{15}k$) of 0.9799 ± 0.0006. Both isotope effects are smaller than the maximum effects for decarboxylation and ketimine formation, and the results indicate that both processes, steps 1 and 2 in fig. 8-9, contribute to rate limitation (Abell and O'Leary, 1988a). Decarboxylation by the PLP-dependent histidine decarboxylase from *Morganella morganii* proceeds with similar isotope effects, while indicating that the transition state is slightly more product-like for the PLP-dependent reaction (Abell and O'Leary, 1988). Solvent deuterium kinetic isotope effects indicate that the imine intermediates are normally protonated; they are the iminium ions shown in fig. 8-9. The isotope effects and kinetic parameters do not reveal a chemical advantage of either PLP or the pyruvoyl moiety in promoting enzymatic decarboxylation of histidine.

The structure of histidine decarboxylase with histidine methyl ester bound as an imine to the active site pyruvoyl group is shown in fig. 8-10 (Gallagher et al., 1989). The ester cannot undergo decarboxylation and allows the imine to be observed as a structural analog of the true intermediate.

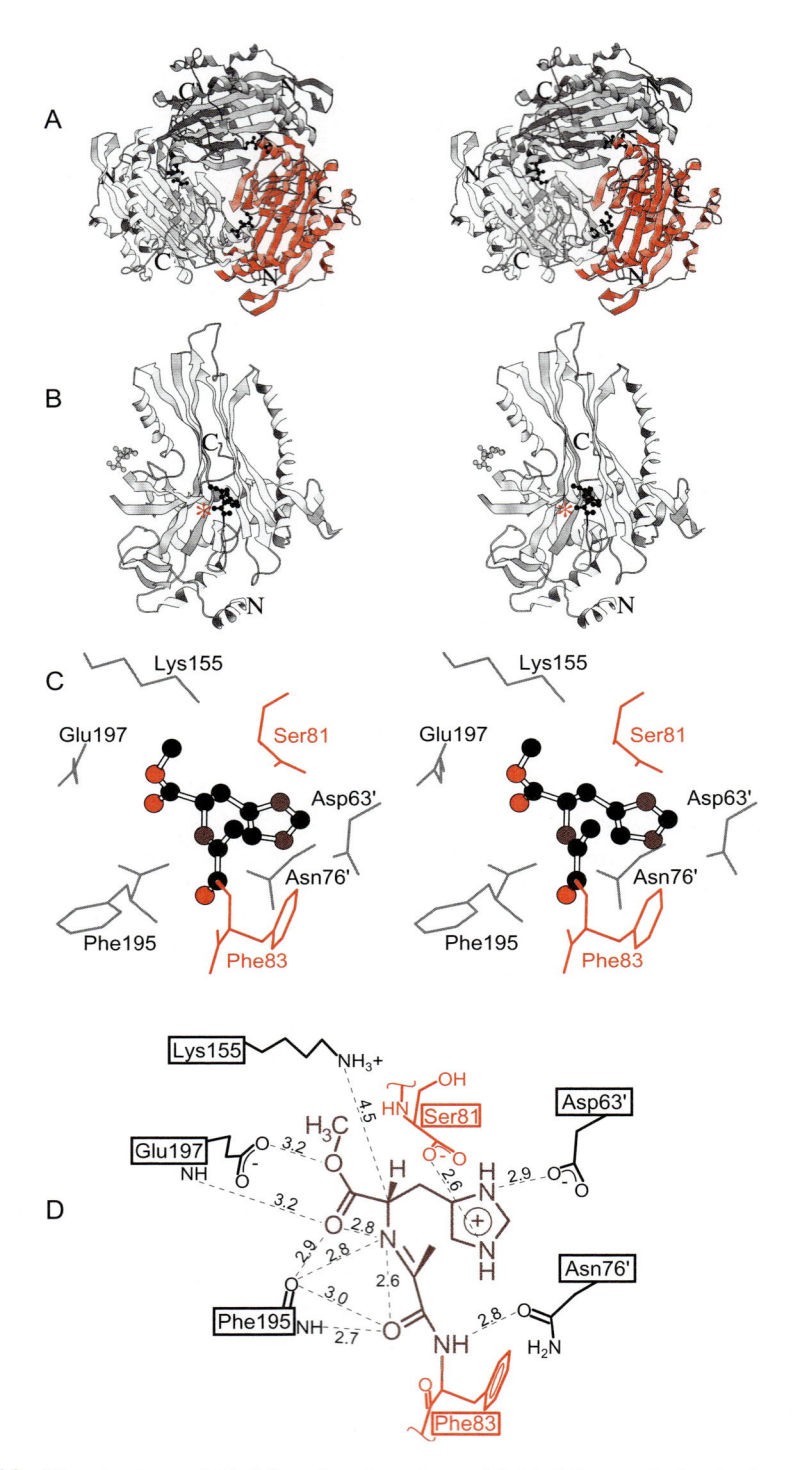

Fig. 8-10. The structure of histidine decarboxylase with histidine methyl ester bound to the active site. (A) The overall fold of the homotrimeric histidine decarboxylase from *Lactobacillus* 30a is shown in stereo, with one subunit highlighted in red and the histidine methyl ester pyruvoyl adduct in black (2.5-Å resolution; PDB 1IBV; Worley et al., 2002). The enzyme displays cooperativity with histidine, and the D53N/D54N double mutant, which locks the enzyme in the T state, was used for this structure. (B) A single subunit is shown with the active site pyruvoyl cofactor located in the cleft of a β-sandwich, close to the subunit interface. The cofactor is derived by serinolysis of the intact proenzyme to generate two chains (residues 1 to 81 and 83 to 311), the longer of which has the pyruvoyl moiety at its N-terminus (red asterisk in B). The pyruvamide His-methylester adduct is bound at the active site and shown in stereo (C) and in two dimensions (D).

Acetoacetate Decarboxylase

A few decarboxylases do not require the assistance of a coenzyme. A well-studied example is the decarboxylation of acetoacetate (eq. 8-4) by acetoacetate decarboxylase (AAD; EC 4.1.1.4), a dodecamer of identical 40-kDa subunits from *Clostridium acetobutylicum*.

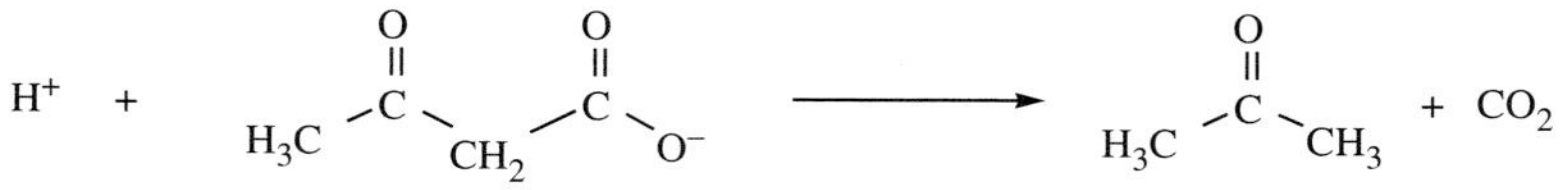

$$H^+ + H_3C-\underset{\underset{O}{\|}}{C}-CH_2-\underset{\underset{O}{\|}}{C}-O^- \longrightarrow H_3C-\underset{\underset{O}{\|}}{C}-CH_3 + CO_2 \quad (8\text{-}4)$$

As a β-ketoacid, acetoacetate undergoes nonenzymatic decarboxylation at a significant rate, especially under acidic conditions by the mechanism in scheme 8-3, analogous to mechanism A in fig. 8-1.

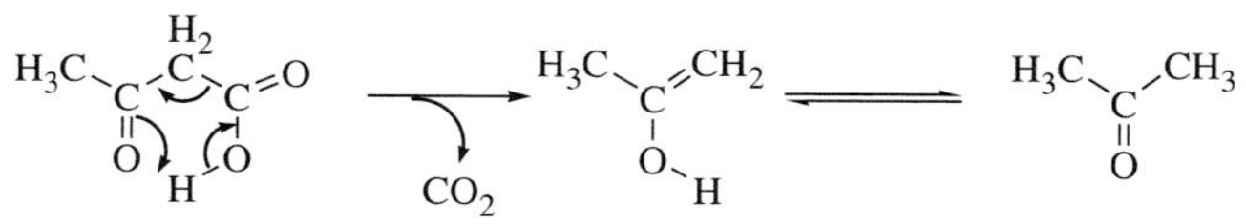

Scheme 8-3

Acetoacetic acid incorporates within its structure the intrinsic chemical properties required for facile decarboxylation, an internal electron sink in the β-carbonyl group and an internal acid catalyst to donate a proton to the nascent enolate ion in the transition state. Chemical catalysis of this reaction would require improving some aspect of the mechanism, and this is possible by increasing the reactivity of the internal electron sink.

Transformation of the β-keto group to a β-iminium ion increases its polarity and facilitates decarboxylation, and this is the essential mechanism by which AAD catalyzes the reaction (Fridovich and Westheimer, 1962; Hamilton and Westheimer, 1959; Warren et al., 1966). AAD is inactivated by NaBH$_4$ in the presence of acetone or acetoacetate, but not in their absence. This is the property exhibited by enzymes that form iminium ions between carbonyl groups of substrates and the ε-amino group of a lysyl residue at the active site (see chap. 1). AAD displays this property; accordingly, reductive inactivation by NaBH$_4$ in presence of [3-^{14}C]acetoacetate followed by acid hydrolysis of the inactive protein to amino acids leads to the formation of N^ε-[^{14}C]isopropyl-Lys115, shown in structure **8-2**.

8-2

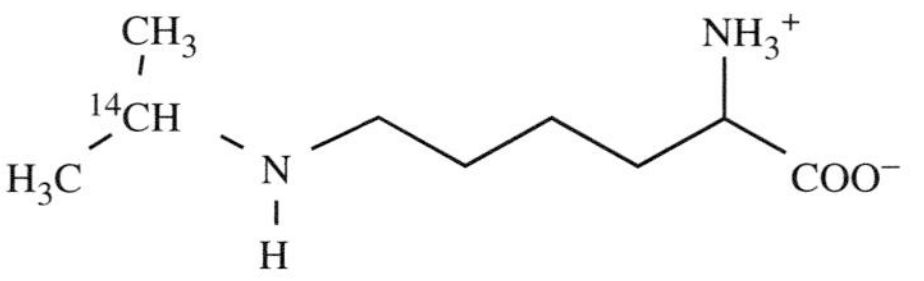

Based on the foregoing evidence, the mechanism in fig. 8-11 is generally accepted for the action of AAD. The reaction of acetoacetate with Lys115 at the active site generates an iminium linkage at the β-carbon of acetoacetate, thereby providing a superior electron sink for decarboxylation by mechanism A in fig. 8-1. The resulting enamine undergoes tautomerization to the iminium form of acetone, which undergoes hydrolysis to acetone.

The optimal pH for the action of AAD in the decarboxylation of acetoacetate is pH 6, more than four units below the normal pK_a for a lysine residue. To catalyze decarboxylation by iminium formation with acetoacetate, Lys115 in the active site must function in its neutral, conjugate base form. The pH dependence might be explained if some process other than the chemistry of iminium formation and decarboxylation controlled the rate. Alternatively, Lys115 might display a low value of pK_a. The fact that it could be acylated selectively by acetic anhydride in neutral solution suggested a low pK_a (O'Leary and

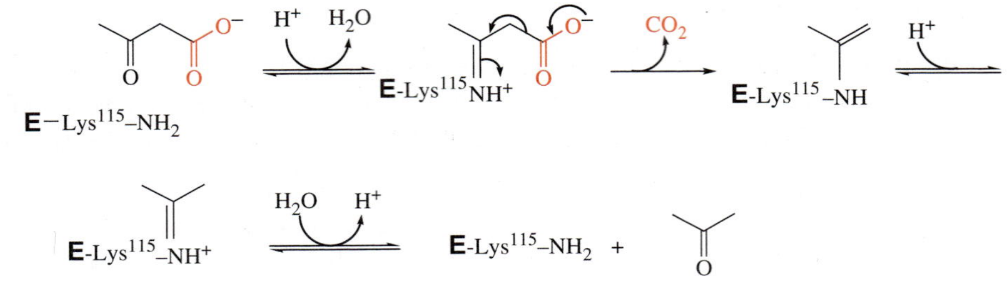

Fig. 8-11. A mechanism for decarboxylation of acetoacetate by acetoacetate decarboxylase.

Westheimer, 1968). The pH dependence of acylation by dinitrophenyl propionate indicated a pK_a of 6 for the active site of lysine (Schmidt and Westheimer, 1971).

The ambiguities in kinetic measurements of pK_a were overcome by the use of a reporter group to measure the thermodynamic pK_a of the alkylated Lys115 and unmask the microenvironmental effects of the active site on ionization (Frey et al., 1971; Kokesh et al., 1971). 5-Nitrosalicylaladehyde inhibits AAD reversibly with a low value of K_i by forming an aldimine complex with Lys115. This aldimine is reduced by $NaBH_4$, irreversibly inactivating the enzyme and introducing the chromophoric p-nitrophenyl group into the active site. The ionization of the alkylated Lys115 and of the p-nitrophenol group can be observed as a function of pH by spectrophotometry. The same chemistry and optical spectroscopy could be carried out on the model compound derived from methylamine in place of Lys115 in AAD, as illustrated in scheme 8-4, with the results indicated.

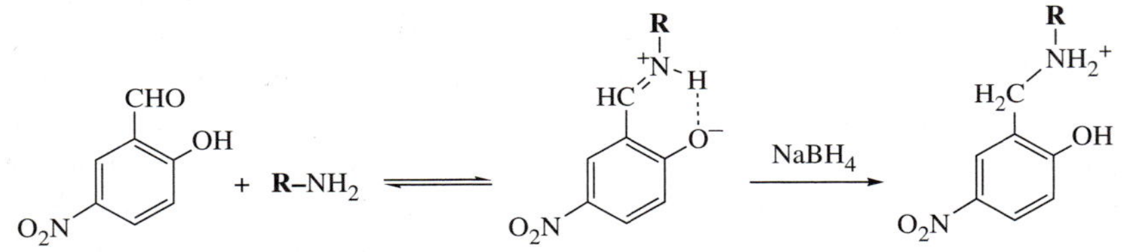

R =	CH_3	AAD–Lys
$pK(NH_3^+)$	10.7	6
$pK(OH)$	5.9	2.4

Scheme 8-4

The group $R\text{-}NH_2^+-$ in scheme 8-4 is reductively alkylated Lys115 when the reporter is reduced onto AAD, and it is the methyl group when methylamine is reductively alkylated. The OH group is p-nitrophenol in both complexes. The results show that both values of pK_a are dramatically decreased in the active site relative to the model compound. The active site is acid strengthening for neutral and positively charged acids. This particular result strongly implicates an electrostatic effect in the microenvironment. A nonpolar microenvironment would be acid weakening for the phenolic group (see chap. 2). The active site must contain a positive charge in addition to that potentially contributed by Lys115. That positive charge stabilizes the negative charge on the p-nitrophenolate oxygen and destabilizes the positive charge on the ε-aminium group of Lys115, thereby lowering both pK_a values. This effect apparently causes the ε-aminium ion of Lys115 to display a pK_a of 6. The low pK_a explains the pH-rate profiles for decarboxylation of acetoacetate and the acylation of Lys115 by acylating agents in neutral solutions.

The positive charge in the active site was hypothesized to originate with an electrostatic interaction between Lys116 and Lys115 in AAD (Frey et al., 1971; Kokesh and Westheimer, 1971). In a test this hypothesis, the consequences of specific mutations at

Lys115 and Lys116 were evaluated with respect to enzymatic activity and the pK_a of the reporter group (Highbarger et al., 1996). Mutation of Lys115 to either cysteine or glutamine inactivated the enzyme. The inactive K115C-AAD could be reactivated by alkylation of Cys115 with 2-bromoethylamine, converting it into the 4-thialysyl residue, which is nearly isosteric with a lysyl residue. Mutation of Lys116 to cysteine, arginine, or asparagine significantly decreased the enzymatic activity. The Lys116-variants gave useful information about the effect of Lys116 on the ionization of the reporter group. Studies of the reporter labeled variants indicated that the pK_a of Lys115 in K116R-AAD was similar to that of wild-type AAD but elevated to above 9.2 in K116C- and K116N-AAD. Aminoethylation of K116C-AAD to the 4-thialysyl-16 derivative restored the pK_a of 5.9 to Lys115 and much of the enzymatic activity. Lys116 must have been responsible for lowering the pK_a of Lys115.

The pK_a of Lys115 accounts for the pH dependence but also raises the question of how the iminium group formed between Lys115 and acetoacetate remains protonated and positively charged in the presence of the ε-aminium group of Lys116. The normal pK_a for an iminium group is between 7 and 8, and the acid-strengthening effect of Lys116 might be expected to lower the pK_a to 3 of 4 and lead to loss to of the proton and formation of the relatively unreactive imine. One rationale is that the closed active site offers no chemical route for ionization in the intermediates. In this scenario, the ε-iminium intermediates would be kinetically trapped in a destabilizing environment created by Lys116. This would generate a driving force for decarboxylation, which destroys the β-iminium group to form an enamine fig. 8-11. The resolution of this question awaits a crystal structure for AAD in an iminium complex with acetone or an analog of acetoacetate. Such an analog might be the inhibitor acetopyruvate (Tagaki et al., 1968).

Mevalonate Pyrophosphate Decarboxylase

Decarboxylation mechanism B in fig. 8-1 appears to be followed in the action of mevalonate pyrophosphate decarboxylase (MPD; EC 4.1.1.33). The conversion of mevalonate pyrophosphate (MevPP) into isopentenyl pyrophosphate is a step in the biosynthesis of cholesterol and is an ATP-dependent decarboxylation (eq. 8-5).

$$^{3-}O_6P_2OCH_2-CH_2-\underset{CH_2}{\overset{HO\ CH_3}{C}}-CH_2-\overset{O}{\underset{}{C}}-O^- + ATP \longrightarrow$$

(8-5)

$$^{3-}O_6P_2OCH_2-CH_2-\underset{CH_2}{\overset{CH_3}{C}}=CH_2 + CO_2 + ADP + P_i$$

MPD is homodimeric, with 43- to 45-kDa subunits in rats, mice, and humans (Michihara et al., 1997, 2002; Toth and Huwyler, 1996). Steady-state kinetic analysis of the action of MPD from chicken liver indicated an ordered binding mechanism, with MevPP binding first and then MgATP (Jabalquinto and Cardemil (1989). Stereochemical analysis of the reaction using P-chiral [γ-^{18}O]ATPγS showed that the overall reaction proceeds with inversion of configuration at the γ-phosphate, indicating a single substitution at phosphorus in the mechanism (Iyengar et al., 1985). The mechanism in fig. 8-12 is consistent with the available facts. Initial phosphorylation of MevPP by MgATP to the hypothetical intermediate, 3-phospho-MevPP, facilitates decarboxylation in the next step by introducing phosphate as an excellent leaving group β-to the carboxylate.

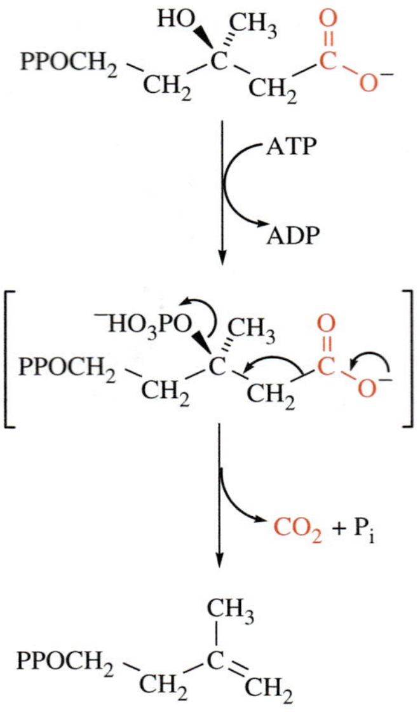

Fig. 8-12. Proposed mechanism for the action of mevalonate pyrophosphate decarboxylase. Phosphorylation of mevalonate pyrophosphate to 3-phosphomevalonate pyrophosphate (bracketed) would introduce a good leaving group (phosphate) in the β-position. Decarboxylation could then proceed by mechanism B in fig. 8-1, coupled with the departure of phosphate. The intermediate would not necessarily dissociate from the enzyme and could remain tightly bound or undergo decarboxylation too rapidly to undergo dissociation.

Phosphate leaves with the bridging oxygen atom and without cleavage of any bonds to phosphorus, consistent with the stereochemistry. In this way, phosphorylation by ATP facilitates mechanism B in fig. 8-1.

Substitution of one methyl-hydrogen in MevPP with fluorine to 3'-fluoro-MevPP (structure **8-3**) dramatically slows the rate of decarboxylation. The compound is a reversible competitive inhibitor, $K_i = 0.01$ μM (Nave et al., 1985; Reardon and Abeles, 1987), and also a substrate that undergoes decarboxylation at 1/2500 the rate of MevPP (Dhe-Paganon et al., 1994).

8-3

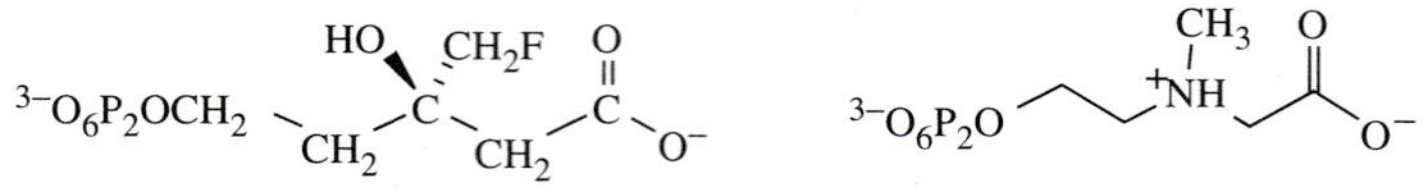

3'-flluoro-MevPP

N-methyl-N-carboxymethyl-2-pyrophosphoethanolamine

Fluorine is small and unlikely to present a steric barrier to the reaction; however, it is strongly electron withdrawing and would be likely to exert an electronic effect on the transition state for decarboxylation. In particular, if the transition state included the accumulation of positive charge at C3 because of the departure of the phosphate group in the intermediate 3-phospho-MevPP, the fluoro substituent in 3'-fluoro-MevPP would raise the energy of the transition state and decrease the rate. The slow rate at which 3'-fluoro-MevPP reacts suggests that the transition state is electropositive at C3. Moreover, MPD catalyzes the phosphorylation of 3'-fluoro-MevPP by MgATP, in accord with the mechanism. This interpretation is supported by the interaction of N-methyl-N-carboxymethyl-2-pyrophosphoethanolamine (see structure **8-3**) with MPD.

This compound incorporates a positive charge on nitrogen in neutral solutions and is a good inhibitor of MPD ($K_i = 0.75$ μM) (Dhe-Paganon et al., 1994). The positively charged nitrogen is in the same position as C3 of the transition state for decarboxylation.

MPD appears to incorporate arginine residues in the active site (Jabalquinto et al., 1983). A structure is available (PDB 1F14); however, it does not include ligands such as substrates or inhibitors bound to the active site, so there is no information about the location of the active site.

Radical-Based Decarboxylases

It has become apparent in recent years that enzymatic decarboxylations can proceed by radical fragmentation according to mechanism D in fig. 8-1. Enzymatic decarboxylation by this mechanism is far less common than the polar mechanisms A or C, but they are being recognized with increasing frequency. We briefly consider three well-established examples.

Pyruvate-Formate Lyase

When *Escherichia coli* cells are grown anaerobically, the glycolytic pathway degrades glucose to pyruvate, as in aerobic growth, and pyruvate serves as the main source of acetyl CoA. However, under anaerobic conditions, the cells do not use the pyruvate dehydrogenase complex to metabolize pyruvate, and although acetyl CoA is produced from pyruvate, formate is produced in place of NADH and CO_2. This metabolism of pyruvate arises from the action of pyruvate-formate lyase (PFL; EC 2.3.1.54) according to eq. 8-6.

$$\text{Pyruvate} + \text{CoASH} \rightarrow \text{Acetyl CoA} + \text{Formate} \qquad (8\text{-}6)$$

Under anaerobic conditions, pyruvate is metabolized by this enzyme instead of by the pyruvate dehydrogenase complex (Becker et al., 1999; Knappe and Sawere, 1990; Knappe and Wagner, 1995; Knappe et al., 1993; Frey et al., 1994; Parast et al., 1995a; Volker-Wagner et al., 1992). PFL is a 170-kDa homodimeric (α_2) free radical enzyme, with a glycyl radical at position 734 in the polypeptide chain.

The glycyl radical serves as a coenzymatic prosthetic group that facilitates the cleavage of the carboxylate group from pyruvate. It is very interesting from a structural standpoint because it is an exceptionally stable radical in the absence of oxygen. Its stability may be attributed to captodative delocalization of the unpaired electron, as illustrated by the three resonance forms in scheme 8-5, which represent a larger number of forms.

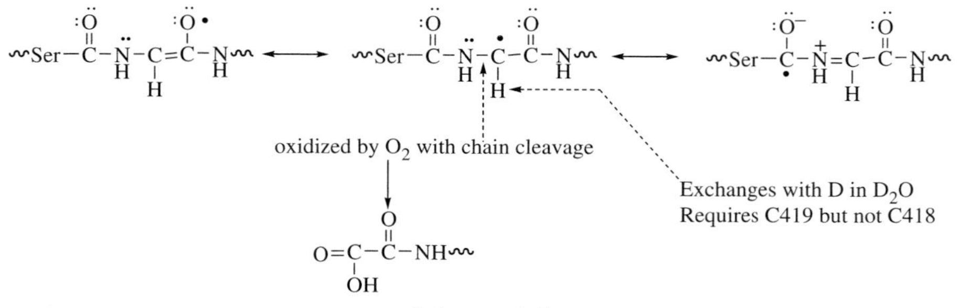

Scheme 8-5

Chemical properties of the Gly734 radical, also noted in scheme 8-5, include its sensitivity to oxygen, with cleavage of the peptide chain to form an oxalyl-residue blocking the N-terminal peptide, and the Cys419-dependent exchange of the α-hydrogen with D_2O.

408 Enzymatic Reaction Mechanisms

Chain cleavage by oxygen facilitated the location of the position of the glycyl radical in the amino acid sequence. Exchange of the α-hydrogen with deuterium simplified the characterization of the radical by EPR. Exchange of this hydrogen with deuterium led to a dramatic narrowing of the EPR signal because of the smaller nuclear hyperfine coupling constant of deuterium with the unpaired electron relative to hydrogen (Volker-Wagner et al., 1992). Substitution of ^{13}C at the α-carbon significantly broadened the EPR signal, as did ^{13}C at the acyl carbonyl and even at the neighboring upstream acyl carbonyl because of the larger nuclear hyperfine coupling constant for ^{13}C relative to ^{12}C (Weil et al., 1994). The ^{13}C effects could be attributed to delocalization of the unpaired electron (see scheme 8-5) and further supported the assignment of the glycyl radical (Volker-Wagner et al., 1992).

The kinetic mechanism is ping pong bi bi, with an acetyl-*S*-enzyme intermediate, as described by scheme 8-6, in which •E represents the active, glycyl-radical form of PFL.

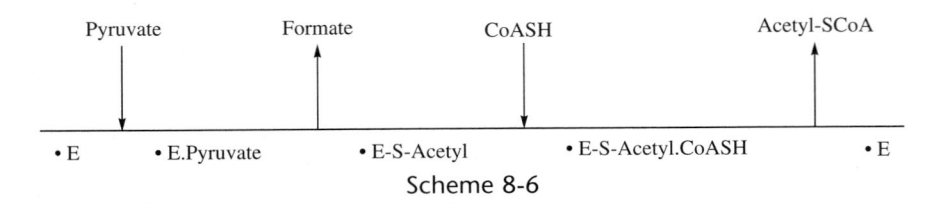

```
        Pyruvate        Formate         CoASH                    Acetyl-SCoA
           │               ↑              │                          ↑
           ↓               │              ↓                          │
  •E         •E.Pyruvate        •E-S-Acetyl      •E-S-Acetyl.CoASH        •E
```
Scheme 8-6

The free enzyme and the covalent intermediate •E–*S*-acetyl are both radicals, although the radical sites may not be identical. The reaction is reversible, and in the reverse direction the reaction of deuteroformate proceeds with a deuterium kinetic isotope effect of 3.6 on k_{cat}/K_m (Brush et al., 1988). Hydrogen abstraction from formate is at least partially rate limiting in the reverse direction.

Inasmuch as PFL is a radical enzyme and formate is the decarboxylation product, a decarboxylation mechanism similar to A in fig. 8-1 seemed to be unlikely, and a radical-based mechanism appeared likely. As a nonenzymatic model, the reaction of hydrogen peroxide and Fe^{2+} with pyruvoyl esters to form acetate and formyl esters (Bernardi et al., 1973) may be relevant to the radical-based decarboxylation of pyruvate. A decarboxylation mechanism such as D in fig. 8-1 and scheme 8-7 could be relevant to the action of PFL.

Scheme 8-7

Cys418 and Cys419 are remote from Gly734 in the amino acid sequence, yet they are required for the activity of PFL. Cys418 becomes acetylated in the reaction, and Cys419 is required for the solvent-exchange of the α-hydrogen of the glycyl radical (Parast et al., 1995b). The structure of PFL in fig. 8-13 reveals the spacing and potential interactions among Gly734, Cys418, Cys419, and the substrate. In the structure, Cys418 lies nearest the substrate binding site, whereas Gly734 is remote, with Cys419 intervening between them and in potential contact with either or both of them. Cys419 could facilitate solvent exchange of the α-hydrogen of Gly734 by a radical interchange process such as in scheme 8-8.

Decarboxylation and Carboxylation 409

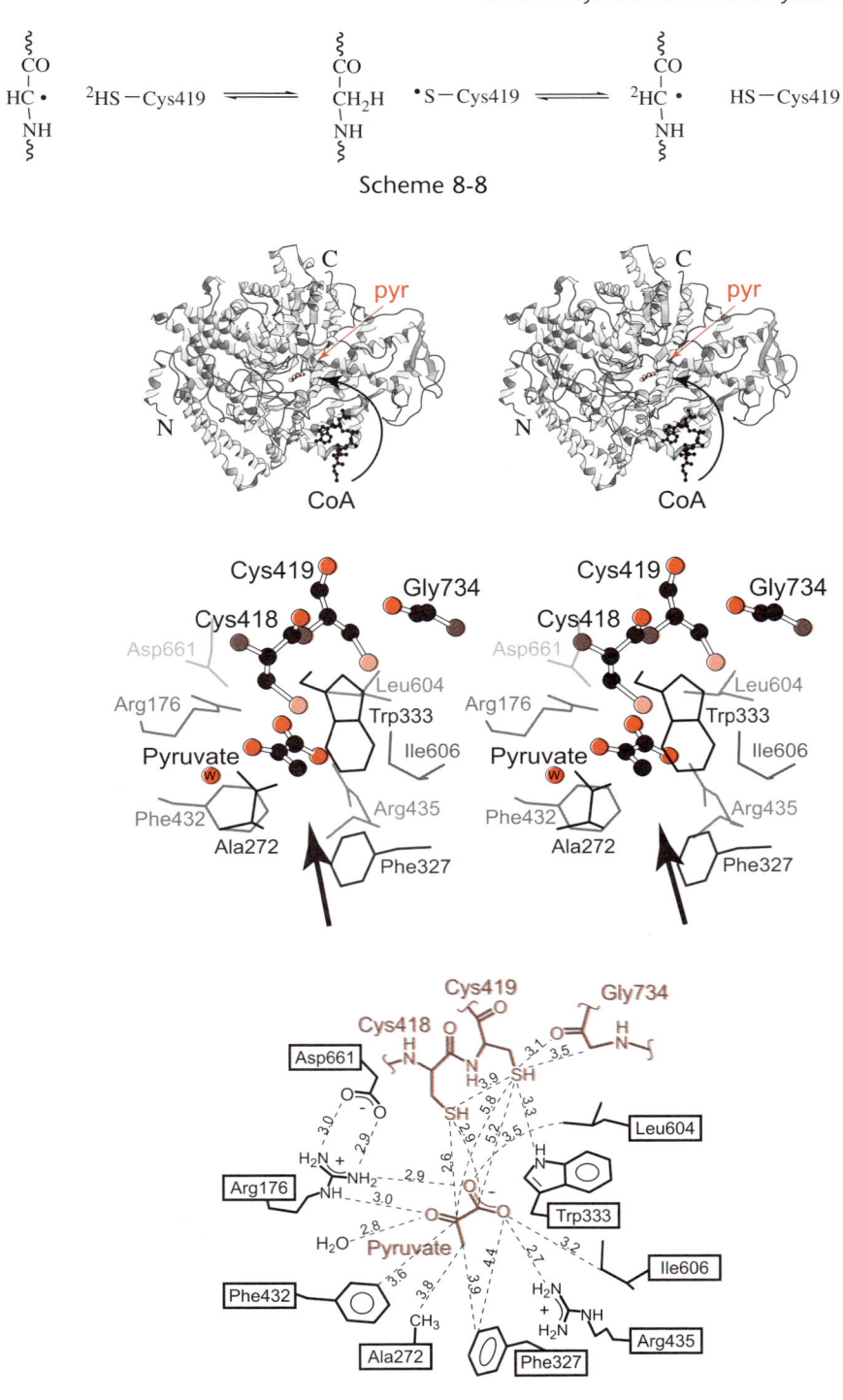

$$\overset{\xi}{\underset{\xi}{\overset{CO}{\underset{NH}{|}}}}\overset{}{\underset{}{HC\cdot}} \quad {}^2HS-Cys419 \quad \rightleftarrows \quad \overset{\xi}{\underset{\xi}{\overset{CO}{\underset{NH}{|}}}}\overset{}{\underset{}{CH_2H}} \quad {}^{\cdot}S-Cys419 \quad \rightleftarrows \quad \overset{\xi}{\underset{\xi}{\overset{CO}{\underset{NH}{|}}}}\overset{}{\underset{}{{}^2HC\cdot}} \quad HS-Cys419$$

Scheme 8-8

Fig. 8-13. The structure of pyruvate formate lyase from *E. coli*. Pyruvate (red) and CoA (black) are shown bound to *E coli* pyruvate formate lyase (1.53-Å resolution; PDB 1H16; Becker and Kabsch, 2002). A black arrow shows the large conformational change required (including an *anti* to *syn* glycosidic bond rotation) to bring the CoA sulfhydryl functionality, which is shown close to the "o" in CoA up to the active site. It has been proposed that this large, sweeping motion occurs with each round of catalysis, and it has been called the *fishing model*. The active site with pyruvate bound adjacent to the catalytic cysteine and glycine residues is shown in the middle panel in stereo and shown in the bottom panel in two dimensions. Access to the active site from the solvent is indicated by the arrow in the middle panel.

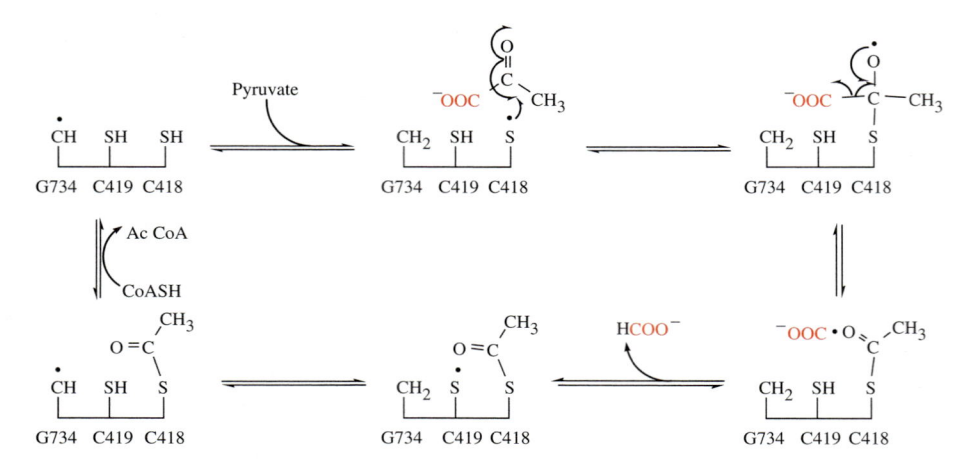

Fig. 8-14. A mechanism for the action of pyruvate-formate lyase (PFL). The unpaired electron resides on Gly734 in the resting enzyme but may be relayed to Cys419 or Cys418 by hydrogen transfer. The pyruvate binding site is near Cys418, and it is postulated that the thiyl radical at that site undergoes addition to the pyruvate carbonyl group. Radical fragmentation produces the formyl radical and the acetylthioester on Cys418. The formyl radical is quenched to formate by hydrogen abstraction, presumably from Cys419, which abstracts a hydrogen atom from Gly734. The S-acetyl-Cys418 reacts by acyl group exchange with CoASH to form acetyl CoA (Becker et al., 1999).

Solvent exchange depends on nonstereospecificity in hydrogen transfer, in contrast to the stereospecific hydrogen abstraction from the α-carbon of Gly734 in the posttranslational modification of PFL (discussed later). Thiyl radical exchange between Cys419 and Cys418 could take place by process of hydrogen transfer between them. This would allow the radical center to oscillate between Gly734 and Cys418 by way of Cys419. The Cys418-thiyl radical could react directly with the substrate, and Gly734 would serve as a stabilizing and protective haven for the unpaired electron whenever the active site is vacant.

The mechanism in fig. 8-14 draws on available information about the chemical processes that have been described for PFL, the structure, and the hypothetical mechanism of decarboxylation in the chemical model of scheme 8-7 (Becker et al., 1999). The mechanism shown is not proven in every respect. It is based on the hypothesis that the unpaired electron on Gly734 of the resting enzyme can be translocated to Cys419 or Cys418 by hydrogen transfer among these residues. Support for thiyl radical formation is provided by EPR studies of the reactions of substrate analogs with PFL, in particular the observation of sulfur-based radicals in the suicide inactivation of PFL by mercaptopyruvate (Parast et al., 1995a) and in the reaction of molecular oxygen with PFL (Reddy et al., 1998). The mechanism is further based on the hypothesis of thiyl radical addition to the carbonyl group of pyruvate. The radical fragmentation of the thiyl adduct to the formyl radical is based on the nonenzymatic model in scheme 8-7. The mechanisms by which the formyl radical is quenched and the acetyl group is transferred to CoASH are not proven, although those shown in fig. 8-14 are reasonable.

Pyruvate-Formate Lyase Activase

The radical center on Gly734 in PFL arises by posttranslational modification of Gly734 catalyzed by PFL activase, a radical SAM enzyme that catalyzes hydrogen abstraction from Gly734. The activase, a 85-kDa, homodimeric enzyme in *E. coli*, contains a [4Fe–4S] center analogous to that of lysine 2,3-aminomutase (Cheek and Broderick, 2001; Walsby et al., 2002a, 2002b; Wong et al., 1993;). The iron-sulfur cluster carries out the reductive cleavage

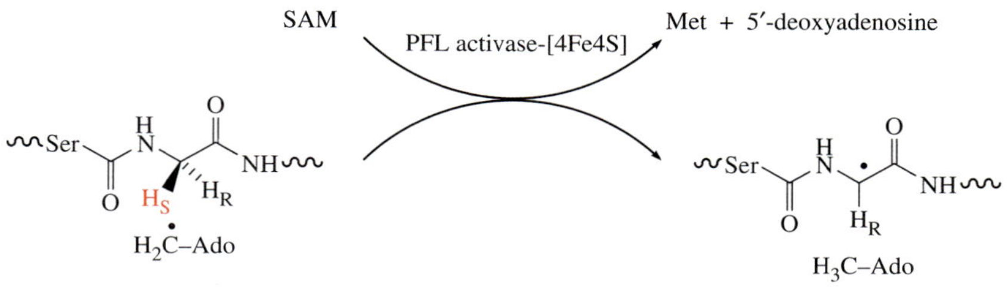

Fig. 8-15. Activation of pyruvate formate-lyase (PFL) by PFL activase. PFL activase introduces the glycyl radical at Gly734 of PFL by a radical mechanism. PFL activase is a radical SAM enzyme that contains a [Fe4–S4] cluster typical of this enzyme family. The iron- sulfur center reductively cleaves SAM into methionine and the 5′ deoxyadenosyl radical, which abstracts the pro-S hydrogen from Gly734 to produce 5′ deoxyadenosine and the Gly734 radical.

of SAM into methionine and the 5′-deoxyadenosyl radical. Abstraction of the pro-S hydrogen from Gly734 by the 5′-deoxyadenosyl radical forms the Gly734 radical and 5′-deoxyadenosine (Frey et al., 1994). The reaction follows the course outlined in fig. 8-15.

Coproporphyrinogen Oxidases

The transformation of propionate side chains of coproporphyrinogen III into the vinyl side chains of protoporphyrinogen IX in the biosynthesis is an oxidative decarboxylation catalyzed by coproporphyrinogen oxidases (EC 1.3.3.3). The overall reaction is illustrated in fig. 8-16, where the electron acceptor is molecular oxygen in aerobes and is as yet unidentified in anaerobes. The enzyme catalyzes the transformation of two propionate side chains into vinyl groups. This is a difficult reaction that appears to defy the mechanistic patterns in fig. 8-1. However, there is reason to believe that the C—C cleavage may proceed by mechanism D in fig. 8-1.

The enzyme in aerobes contains manganese and uses molecular oxygen as the electron acceptor. The enzyme from anaerobes differs, in that it cannot use molecular oxygen and is a radical SAM enzyme, with the [4Fe–4S] clusters characteristic of this family of enzymes (Layer et al., 2002). The anaerobic enzyme seems to function initially to cleave SAM into methionine and the 5-deoxyadenosyl radical, which then abstracts the β-hydrogen

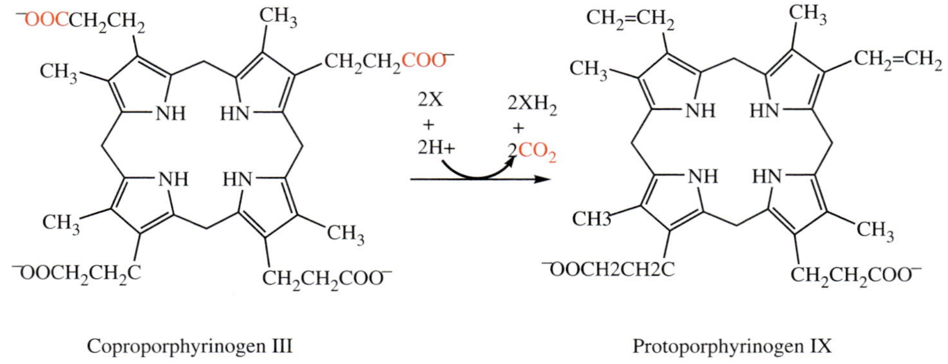

Fig. 8-16. Reaction of coproporphyrinogen oxidase. In the biosynthesis of heme, the oxidative decarboxylation of coproporphyrinogen III to introduce the vinyl groups in protoporphyrinogen IX is catalyzed by the title oxidase. Oxidase forms part of the name of this enzyme because the oxidizing agent is molecular oxygen in aerobic organisms. In anaerobes, another unidentified electron acceptor is required. In aerobes, the enzyme is a manganese protein, and in anaerobes, it is a radical SAM enzyme. Both reactions are thought to proceed by a radical mechanism, in which the carboxyl group is cleaved by a process analogous to mechanism D in fig. 8-1.

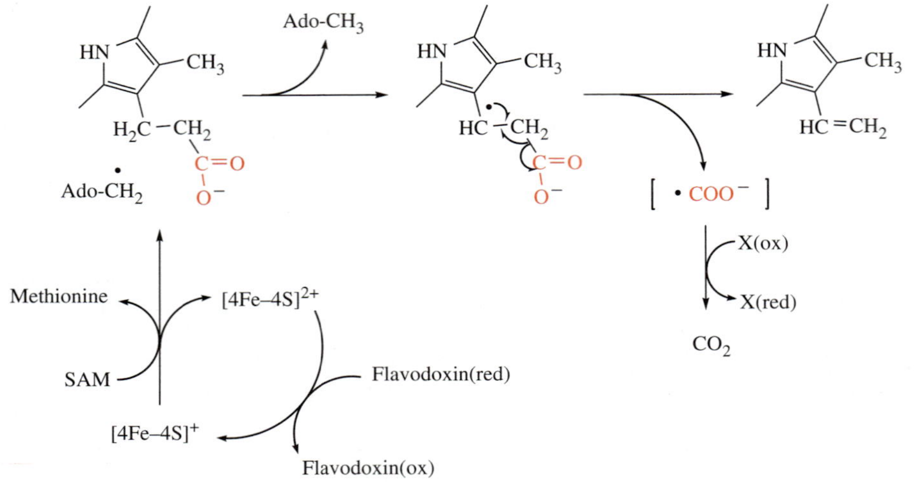

Fig. 8-17. Events in the oxygen-independent oxidative decarboxylation of coproporphyrinogen were postulated by Layer and colleagues (2002). In anaerobes, the transformation of propionate side chains in coproporphyrinogen into vinyl groups is oxygen independent and catalyzed by a radical SAM enzyme. The enzyme has been characterized, and the electron transfer proteins are being investigated. The scheme is subject to modification as the electron transfer pathway is elucidated. The key step in the decarboxylation is the abstraction of a hydrogen atom from the β-carbon of the propionate side chain by the 5-deoxyadenosyl radical derived from SAM. Radical fragmentation leads to cleavage of the C—C bond to generate the vinyl group and putative formyl radical, which undergoes oxidation by an unknown electron acceptor.

from a propionate side chain, to form an allylic radical intermediate, as depicted in fig. 8-17. This intermediate could undergo radical fragmentation to produce the formyl radical anion by mechanism D in fig. 8-1. The further electron transfer steps to oxidize the formyl radical anion to CO_2 and regenerate SAM are under investigation.

Oxalate Decarboxylase

Oxalate undergoes at least three types of metabolism in various organisms, all of which lead to decarboxylation but by very different mechanisms and with the formation of very different products. Oxalate decarboxylase (OxDC; EC 4.1.1.23) is found in fungi and bacteria. It catalyzes the decarboxylation to form carbon dioxide and formate according to eq. 8-7.

$$HOOC\text{—}COO^- \rightarrow CO_2 + HCOO^- \tag{8-7}$$

Oxalate oxidase (EC 1.2.3.4) catalyzes the reaction of oxalate with molecular oxygen to form two moles of carbon dioxide and a mole of hydrogen peroxide. Oxalyl CoA decarboxylase (4.1.1.8) is a TPP-dependent enzyme that catalyzes the conversion of oxalyl CoA to carbon dioxide and formyl CoA.

OxDC is a manganese-enzyme that requires molecular oxygen to activate it but does not use it as a substrate. The main mechanistic information comes from heavy atom kinetic isotope effects and the crystal structure. The structure of the enzyme from *Bacillus subtilis* is shown in fig. 8-18 (Anand et al., 2002). The enzyme is a 264-kDa hexamer, in which each subunit is composed of two similar but not identical domains. In original structure shown in fig. 8-18, each domain incorporated a manganese binding site consisting of three histidine residues and a glutamate. Site b in the C-terminal domain included a second

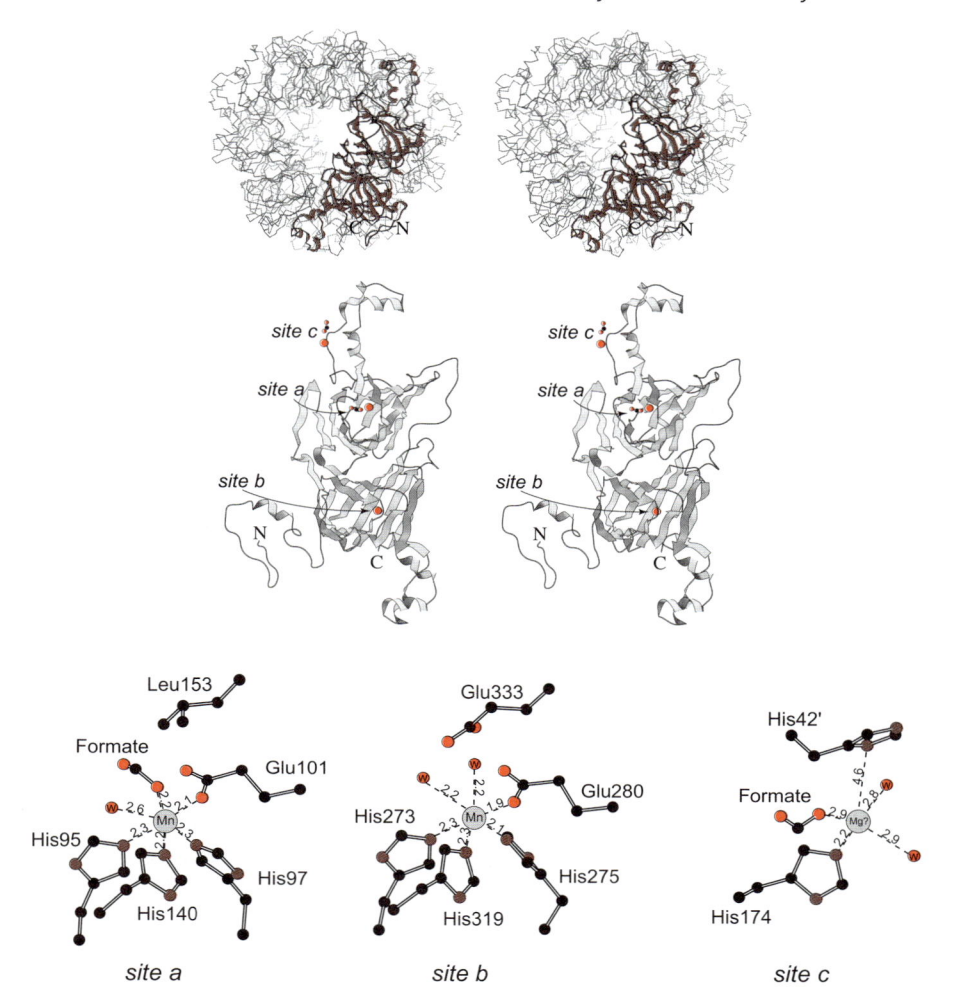

Fig. 8-18. *Bacillus subtilis* oxalate decarboxylase is a homohexameric enzyme and is depicted in the top panel in stereo, with one subunit drawn in brown ribbons and the other five subunits drawn as gray Cα traces. The middle panel shows a single subunit with its bicupin fold (one cupin fold surrounds site a, and the other cupin fold surrounds site b). Three divalent metal ions are bound to each subunit in three distinct sites: a, b, and c. Manganese is bound to sites a and b and magnesium to site c. Formate was also observed bound to the metal ions in sites a and c. Each site is shown schematically in the bottom panel. Site b has been proposed to support oxalate decarboxylase activity because mutation of Glu333 results in a 25-fold activity drop (1.9-Å resolution; PDB 1L3J; Anand et al., 2002).

glutamate residue in position to function as a catalyst (Glu333), and this site was thought to be the functional active site. Site a in the N-terminal domain included Leu153 in the position corresponding to Glu333 in the b site, and site a was thought not to be functional. In a later structure, sites a and b were found to be identical, with a catalytic glutamate in each site (Just et al., 2004). Magnesium is bound in a third site, c, that is unlike sites a and b.

The pH dependence for log V/K shows a single break downward with increasing pH, corresponding to a pK_a of 4.2, exactly the value for the first ionization of oxalate, and leading to the conclusion that the substrate is hydrogen oxalate, HOOC–COO⁻ (Reinhardt et al., 2003). The reaction proceeds without significant exchange of solvent oxygen into the unreacted oxalate or the formate produced.

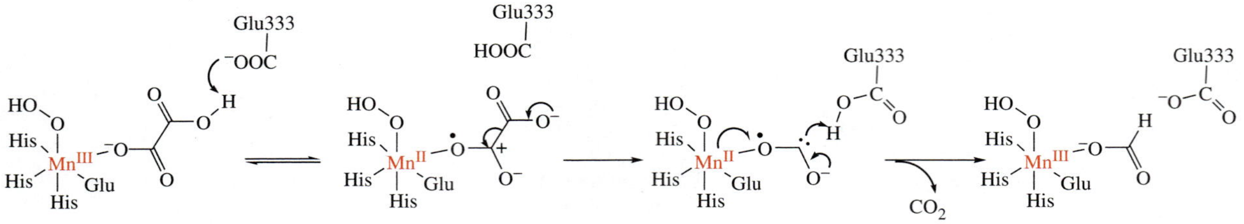

Fig. 8-19. A mechanism for the action of oxalate decarboxylase.

Because the action of OxDC produces two products that can be separately analyzed for ^{13}C and ^{18}O by isotope ratio mass spectrometry, formate after oxidation to carbon dioxide, two ^{13}C and two ^{18}O kinetic isotope effects can be measured in each experiment (Reinhardt et al., 2003). Because at natural abundance a given molecule of oxalate rarely contains two heavy isotopes, the ^{13}C and ^{18}O isotope effects on formate production interrogate a different part of the reaction from that of the isotope effects on carbon dioxide production. At pH 4.2, the ^{13}C and ^{18}O kinetic isotope effects for the production of formate and for carbon dioxide formation are smaller than the maximum values but normal; the heavier isotope slows the reaction by on the order of 1%. At pH 5.7, both ^{13}C isotope effects and the ^{18}O isotope effect on formate production are normal; however, the ^{18}O isotope effect on the production of carbon dioxide is the inverse.

The crystallographic and biochemical results are consistent with the mechanism in fig. 8-19 (Reinhardt et al., 2003). The isotope effects require a mechanism in which a reversible step precedes decarboxylation. In fig. 8-19, the equivalent of hydrogen atom transfer occurs, in which an electron is transferred to Mn^{III} and a proton is transferred to Glu333, generating the oxalyl radical coordinated to Mn^{II} as the first step. Decarboxylation of the radical leads to carbon dioxide and a formyl radical anion. In the last step, the formyl radical anion acquires the equivalent of a hydrogen atom by process of electron transfer from Mn^{II} and proton transfer from Glu333 to complete the formation of formate and return both manganese and Glu333 to their original states. Notice that the reaction takes place with the Mn^{III} hydroperoxide complex, which explains the activation of this enzyme by molecular oxygen. The oxidation states of manganese in this mechanism are the presumptive states and have not been confirmed spectroscopically.

Orotidine Monophosphate Decarboxylase

An essential decarboxylation in nucleic acid biosynthesis does not fit any of the mechanistic patterns in fig. 8-1. Because of the absence of an obvious mechanism, many have been proposed. Orotidine monophosphate decarboxylase (ODC; EC 4.1.1.23) is a homodimeric enzyme and catalyzes the decarboxylation of OMP to UMP according to eq. 8-8.

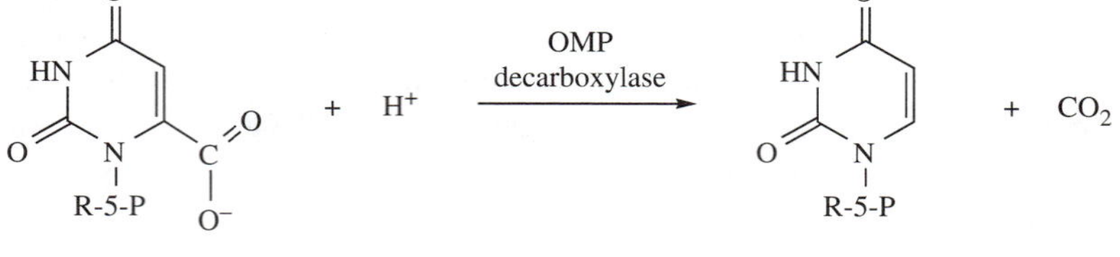

The enzyme has the distinction of displaying the largest rate enhancement factor (10^{17}) that has been measured (see chap. 1). The large factor owes more to the slow nonenzymatic rate than to the turnover number, which is about 30 s^{-1}. Because the substrate has no built-in electron sink for decarboxylation, the molecule is exceedingly unreactive in decarboxylation.

In most enzymatic reactions, the problem of unreactive substrates is overcome by some sort of enzyme-catalyzed preliminary chemical alteration in the substrate, either in a reaction with an amino acid side chain or a reaction with a coenzyme or metal ion to increase its reactivity. One such process, at one time suggested to explain the decarboxylation by ODC, was the covalent addition mechanism shown in fig. 8-20, among five mechanisms that have been proposed.

In the covalent addition mechanism, an enzymatic nucleophile adds in the Michael sense to C5, and leads to protonation at C6. This chemically reasonable process has the virtue of simultaneously introducing a leaving group at C5 and the hydrogen at C6 that will be needed in the product. Then, decarboxylation can proceed by mechanism B of fig. 8-1 to eliminate CO_2 and the enzymatic nucleophile. However, the values of heavy atom kinetic isotope effects appear to rule out the covalent addition mechanism. The ^{13}C kinetic isotope effect for enzymatic decarboxylation of OMP is 1.025 at pH 6.8 and 1.05 at pH 4.0, and decarboxylation is essentially rate limiting at pH 4.0 but only partially rate limiting at pH 6.8. The isotope effect reaches a plateau of about 1.035 at high pH (Smiley et al., 1991). The pH dependence is the reverse of expectations for a covalent addition mechanism. The reaction of [5-^{2}H]OMP proceeds with no detectable secondary deuterium kinetic isotope effect, indicating no change in hybridization at C5 in the transition state or preceding steps (Acheson et al., 1990).

Decarboxylation could be understood if the ring nitrogen were positively charged to stabilize the developing negative charge on C6 on decarboxylation to an ylid. As in the nonenzymatic decarboxylation of model compounds, preliminary protonation of the 2-oxo group by the enzyme could generate the quaternary ring nitrogen, as illustrated in fig. 8-20. Decarboxylation to the ylid would then proceed, followed by protonation to UMP (Beak and Siegel, 1976). An analogous mechanism with preliminary protonation of the 4-oxo group would also lead to quaternization of the ring nitrogen and decarboxylation to an ylid, as shown in fig. 8-20, and the 4-oxo group is more basic than the 2-oxo group (Phillips and Lee, 2001). However, the structure of ODC does not reveal the presence of an acidic group near either the 2-oxo or the 4-oxo group of UMP bound at the active site. Moreover, the ^{15}N kinetic isotope effect for enzymatic decarboxylation of OMP (1.0036) is similar to that for nonenzymatic decarboxylation of *N*-methyl picolinate (1.0053), in which bond order at nitrogen cannot change (Rishavy and Cleland, 2000). The results were interpreted as inconsistent with protonation at the 2-oxo or 4-oxo groups in the enzymatic reaction.

Four crystal structures of ODC from different species and with different ligands at the active site appeared nearly simultaneously (Appleby et al., 2000; Harris et al., 2000; Miller et al., 2000; Wu et al., 2000). The structures did not reveal the presence of acidic amino acid residues that could protonate the 2-oxo or 4-oxo groups. All the structures show that the side chains of a conserved aspartate and a conserved lysine, Asp70 and Lys72 in ODC from *Methanobacterium thermoautotrophicum*, lie near the site that would be occupied by the carboxylate group of OMP. These two amino acids are further associated in an alternating cluster with another aspartate and another lysine residue, Lys42 and Asp75B (B subunit) in *M. thermoautotrophicum* ODC. The structure in fig. 8-21 of this enzyme shows Asp70 and Lys72 residing in the active site and proximal to the C6 of UMP.

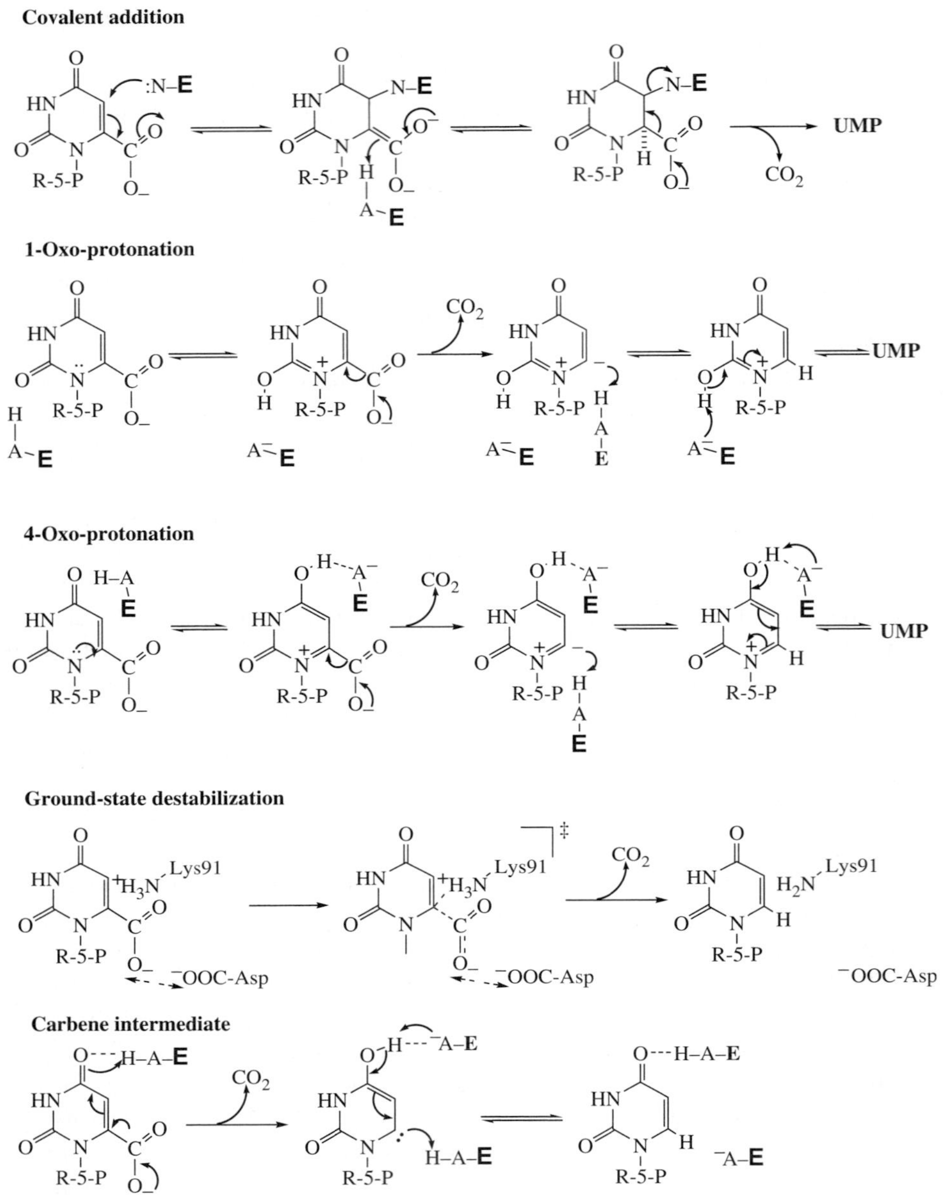

Covalent addition

1-Oxo-protonation

4-Oxo-protonation

Ground-state destabilization

Carbene intermediate

Fig. 8-20. The five mechanisms for the decarboxylation of orotidine-5′-phosphate (OMP) are taken from a larger number that have been postulated. The first mechanism, covalent addition, appears to have been ruled out by the heavy atom kinetic isotope effects (Smiley et al., 1991). The 2-oxo protonation and 4-oxo protonation mechanisms are not supported by the structure in the active site. The ground-state destabilization mechanism appears compatible with the structure and is supported by quantum mechanical calculations (Lee and Houk, 1997) and heavy atom isotope effects, as is the carbene mechanism (Rishavy and Cleland, 2000).

The proximity of Asp70 to the carboxylate of OMP led to the proposition of ground state destabilization as the mechanism of decarboxylation. Lys72 is invoked to facilitate the process. In this mechanism, it is postulated that in the reactive conformation the carboxylate groups of OMP and Asp70 are brought very close together, and the consequent charge repulsion energizes the decarboxylation to the incipient C6-anion, which is

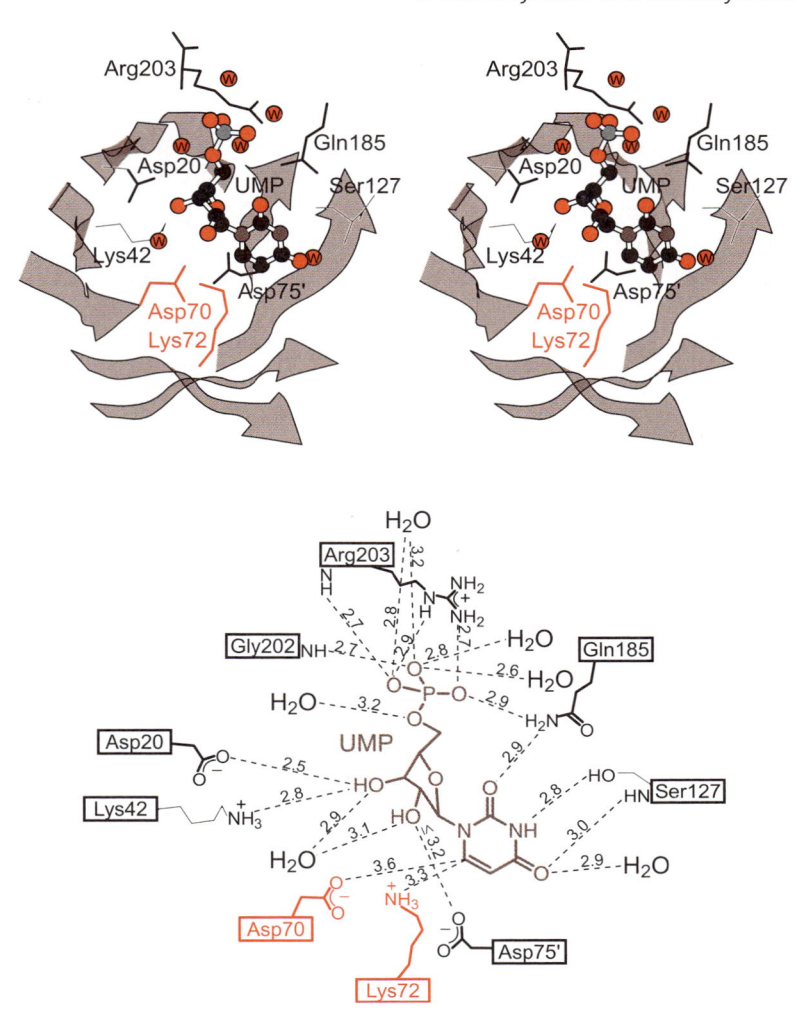

Fig. 8-21. Structure of orotidine-5-phosphate decarboxylase with UMP at the active site. Orotidine-5-phosphate decarboxylase from *Methanobacterium thermoautotrophicum* is a homodimeric enzyme, and the monomeric unit has the common $(\alpha\beta)_8$-barrel (TIM-barrel) fold. The active site is shown in stereo with product UMP bound and catalytic residues Lys72 and Asp70 highlighted in red (1.5-Å resolution; PDB 1LOQ; Wu and Pai, 2002). The active site is shown in a two-dimensional scheme in the bottom panel.

stabilized by the aminium group of Lys72 in the transition state, which is shown in fig. 8-20. On decarboxylation, C6 is protonated by Lys72. In agreement with the experimental results, quantum mechanical and molecular dynamics calculations support this mechanism and the carbene mechanism, the lower entries in fig. 8-20 (Hur and Bruice, 2002; Lee and Houk, 1997; Wu et al., 2000).

Another structural aspect that supports the ground state destabilization mechanism is the conformation of OMP when bound to the active site. Naturally, the structure cannot be determined directly because of the decarboxylation of OMP to UMP prior to the collection of structural data. Even variants of ODC with "essential" residues mutated retain sufficient activity to decarboxylate OMP to UMP by the time crystals have formed. However, the doubly mutated D70A/K72A-ODC does form a complex with OMP, and in the structure the carboxylate group is found to be thirty degrees out-of-plane relative to the pyrimidine ring (Wu et al., 2002). All structures of free OMP show the carboxylate group in plane, as expected for conjugation of the carboxylate and pyrimidine π-systems. The twist of the carboxylate out of plane is thought to be imposed by steric crowding in the active site, and it is likely to be more severe in the wild-type enzyme. The twist in the Michaelis complex would destabilize the carboxylate group of OMP, while at the same time placing it in the conformation required for generating the electron pair resulting from decarboxylation in an orbital orthogonal to the π-system of the pyrimidine ring, as required for proper protonation by Lys72. The twist may constitute a stereoelectronic contribution to catalysis by ODC.

If this mechanism proves to be correct, it seems that the enzyme must prevent the protonation of the two carboxylate groups as they are brought together. Data on dicarboxylic acids show that two carboxylate groups can become quite basic when they are brought together in the same molecule. For example the second pK_a of *cis*-caronate is 8.3 in water and would be much higher in a medium of low dielectric constant, perhaps 13 in ethanol. The ground-state destabilization mechanism could work for ODC if the carboxylate groups were held at a distance that allowed their basicity to be similar to or slightly higher than that of Lys72, so that a significant fraction would be in the appropriate ionization state shown in figs. 8-20 and 8-21. The proper ionization states of Asp70 and Lys72 may be maintained by the electrostatic interactions in the cluster of residues Lys42, Asp70, Lys72, and Asp75B.

ODC binding interactions to the 5-phosphoribosyl moiety of OMP are important in catalysis. Mutations of groups in ODC that bind the 5′-phosphate group dramatically decrease activity. The results indicate that remote binding interactions are critical to the structural integrity of the transition state.

Carboxylases

From a purely theoretical chemical standpoint, carboxylation may be regarded as the reverse of decarboxylation. However, in the functioning world of the biosphere, it is a different matter. Decarboxylation reactions usually produce carbon dioxide (CO_2), a gas that either evaporates or undergoes hydration to bicarbonate ($HOCO_2^-$). A major problem in carboxylation by a carboxylase becomes the acquisition of carbon dioxide, either by the absorption of the tiny amount of the gas dissolved in the cytosol of a cell or by the dehydration of bicarbonate, the major form in which carbon dioxide exists in a cell. Absorption and use of dissolved carbon dioxide is problematic, although it does occur in the action of ribulose-1,5-bisphosphate carboxylase (rubisco), an important enzyme in carbon dioxide fixation in plants. We discuss this enzyme as a case study of a carbon dioxide–fixing carboxylase.

Carboxylases more often use bicarbonate as the source of carbon dioxide. To release carbon dioxide, bicarbonate must be dehydrated, and this requires energy to shift the equilibrium toward carbon dioxide. Carboxylases that extract carbon dioxide from bicarbonate use ATP to drive the reaction. Most such enzymes are biotin carboxylases and use biotin to trap the carbon dioxide formed in the ATP-dependent step and to release it in the carboxylation step (see chap. 3). Most biotin-dependent carboxylases are complex enzymes, and acetyl CoA carboxylase is discussed in chapter 18. Phosphoenolpyruvate carboxylase (PEPC) catalyzes the phosphorylation-dependent dehydration of bicarbonate in a biotin-independent process, and the carbon dioxide produced in the active site carboxylates the substrate. PEPC also is important in carbon dioxide fixation in plants, and we discuss this enzyme as a case study of a phosphorylation-dependent carboxylase.

Ribulose-1,5-Bisphosphate Carboxylase

Reaction and Properties

The carboxylation of ribulose-1,5-bisphosphate (RuBP) leads ultimately to its cleavage into two molecules of 3-phosphoglycerate (3-PGA), as shown in the upper pathway of fig. 8-22. Ribulose-1,5-bisphosphate carboxylase (rubisco; EC 4.1.1.39) also catalyzes another reaction of RuBP, oxygenation and cleavage into one molecule of 3-PGA and one molecule of phosphoglycolate. For this reason, the enzyme is often known as RuBP carboxylase/oxygenase, but it is more frequently called *rubisco*. The oxygenase activity arises from the intrinsic reactivity of the enediolate intermediate with oxygen. In the early evolution of plants, O_2 was not present, so the oxygenase activity was not a side reaction. Because oxygen is ubiquitous in the aerobic world of today—it is produced by plants— the oxygenase activity of rubisco can be suppressed only by increasing the rate of carboxylation. Plant rubisco seems to have evolved over millennia to its optimal carboxylase

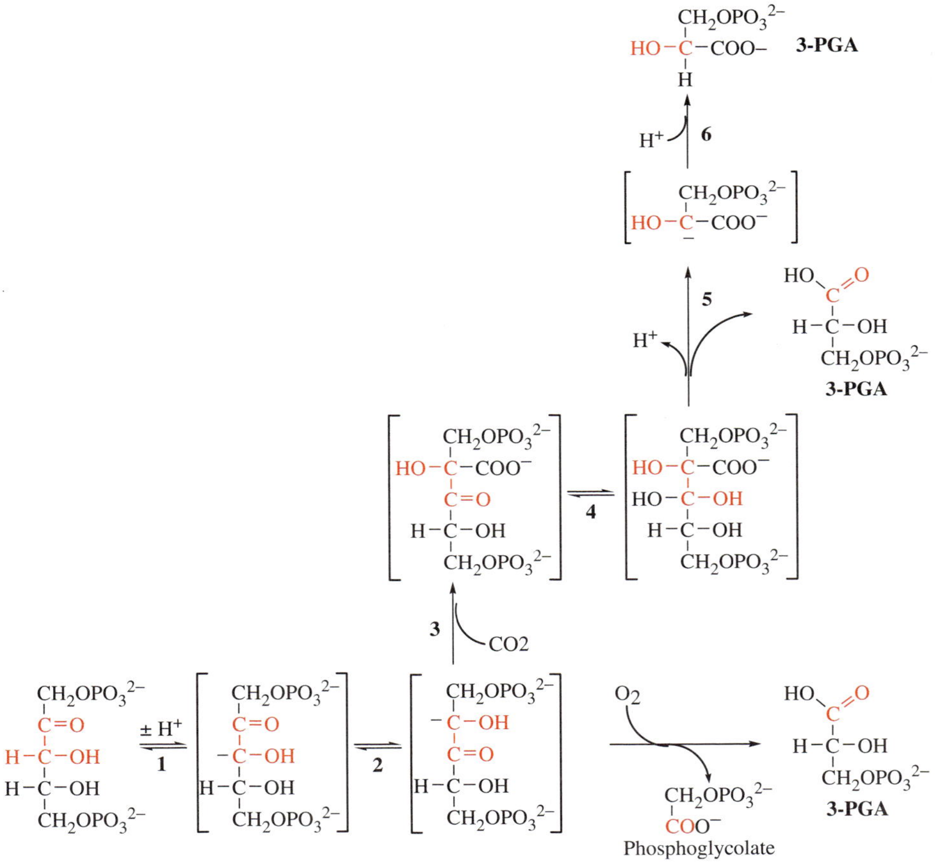

Fig. 8-22. Reactions catalyzed by ribulose-1,5-bisphosphate carboxylase. Rubisco catalyzes two reactions of RuBP: carboxylation and cleavage into two molecules of 3-phosphoglycerate and oxygenation and cleavage into one molecule of 3-phosphoglycerate and one molecule of phosphoglycolate. Both reactions begin with the ionization of RuBP at C3 to the C3-carbanion and its isomerization to the C2-carbanion (steps 1 and 2). Then, they diverge, with carboxylation following the upper pathway (steps 3 to 6) and oxygenation following the lower pathway. The two reactions take place simultaneously, so that RuBP is partitioned between them. The bracketed species in the figure represent enzyme-bound intermediates.

activity in terms of k_{cat}, about 10 s^{-1}, so that the only other means of suppressing the oxygenase reaction is by increasing the concentration of carbon dioxide, an impractical and undesirable measure in the biosphere. The ratio of k_{cat}/K_m for carboxylase and oxygenase activities for rubisco from plants is about 80, and the ratios for bacterial enzymes range from 10 to 20. The oxygenase reaction decreases the potential yield of 3-PGA, but part of this is recovered in the further metabolism of phosphoglycolate.

The 3-PGA from the action of rubisco enters the metabolic pathways and serves as the source of carbon in the biosynthesis of every other molecule in the plant. The rubisco reaction is one of the most important in the biosphere, because it is the first step in the production of most of the biomass on Earth. Rubisco is also plentiful in the leaves of plants (≈50% of soluble protein) which, given the low activity of rubisco and low carbon dioxide concentration in leaves, require a great deal of enzyme to sustain their needs for growth.

Rubisco from photosynthetic bacteria such as *Rhodospirillum rubrum* is a 100-kDa homodimer (L2), and from plants and cyanobacteria, it is a 550-kDa hexadecamer composed of eight large and eight small subunits (L8S8). The large subunits incorporate the active site, and the amino acid sequences of the catalytic subunits in the two classes are 30% identical (Hartman and Harpel, 1994). The small subunits can be removed from the L8S8 species, leaving an octamer of the large subunits, and these do not display activity in the overall reaction or in any of the partial reactions catalyzed by the intact enzyme (Andrews et al., 1986).

Activation

Rubisco exists in active and inactive states, and activation of the resting enzyme follows an interesting and for many years novel course. Activation requires a minimum of two equivalents of carbon dioxide, which bind at distinct sites (Lorimer et al., 1976; Miziorko, 1979). The first molecule binds in a site that does not permit its exchange with free carbon dioxide, and this activates the enzyme. The activation site is the ε-amino group of Lys201 spinach enzyme (Lys191 in *R. rubrum*), which binds carbon dioxide as an N-carbamate and forms the active enzyme.

Rubisco also requires magnesium ions for activity, with magnesium binding after carbamylation of Lys201. Structural information indicates that the *N*-carbamyl group of Lys201 is a ligand to magnesium in certain complexes.

Mechanism of Action

The basic chemistry of the carboxylation and cleavage of RuBP is well understood and outlined in fig. 8-22. Carboxylation is a multistep process that begins with the ionization of the only enolizable site in the molecule, abstraction of C3(H) by the carbamate to produce the C3-carbanion (Lane and Miziorko, 1978). Carboxylation takes place on C2, so that the C3-cabanion must undergo isomerization to the C2-carbanion, as shown in step 2 of fig. 8-22. Isomerization itself proceeds following the course in scheme 8-9, which shows that it essentially consists of tautomerization, or proton transfer from C3(OH) to C2(O), a process that is catalyzed by the enzyme.

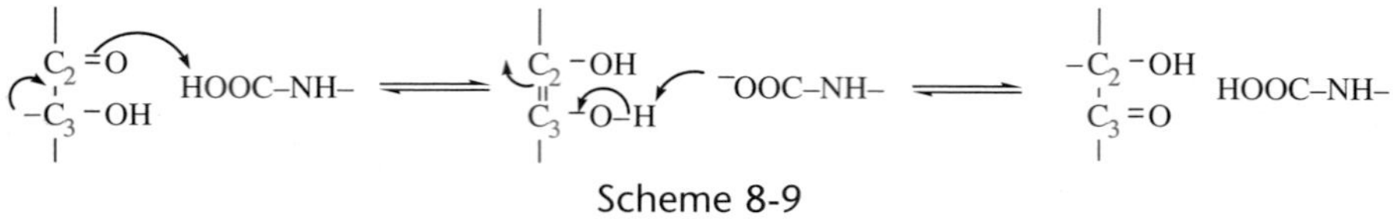

Scheme 8-9

Table 8-1. Catalytic Residues in Rubisco from Spinach and *R. rubrum*

Spinach[a]	*Rhodospirillum rubrum*[a]
Lys175	Lys166
Lys201 (N$^\varepsilon$-cabamate)	Lys191 (N$^\varepsilon$-cabamate)
Lys334	Lys329
His294	His287

[a]Active-site residues were identified by active-site affinity labeling or/sitedirected mutation of conserved residues and confirmed in crystal structures (Soper et al, 1988; Hartman et al, 1987; Lorimer and Hartman, 1988; Hartman et al, 1985; Lundqvist and Schneider, 1991; Newman and Gutteridge, 1993; Knight et al, 1990; Andersson, 1996)

Carboxylation of the C2-carbanion in step 3 is essentially the reverse of decarboxylation mechanism A in fig. 8-1. Cleavage of the primary carboxylated intermediate begins with hydration of the carbonyl group in step 4 of fig. 8-22, and cleavage of the C2-C3 bond in step 5 produces one molecule of 3-PGA and the enolate of a second molecule, which on protonation in step 6 is released as 3-PGA.

The active sites of the dimeric and hexadecameric enzymes are very similar, and the catalytic residues have been identified by affinity labeling, by site-directed mutagenesis of conserved residues and biochemical analysis of variants, and by x-ray crystallography. In this chapter, we employ the numbering of amino acid residues in the spinach enzyme, unless otherwise indicated. The sequence numbers of the catalytic residues in spinach and *R. rubrum* rubisco are listed in table 8-1.

An analog of the intermediate 3-keto-2-carboxy-D-arabinitol-1,5-bisphosphate (structure **8-4**) binds very tightly to rubisco and is regarded as a transition-state analog. This compound is 2-carboxyarabinitol-1,5-bisphosphate (CABP), and the value of $K_d \leq 10^{-11}$ M (Pierce et al., 1980). The D-*ribo*-epimer is a much weaker inhibitor, with $K_d = 1.5 \times 10^{-6}$ M.

8-4

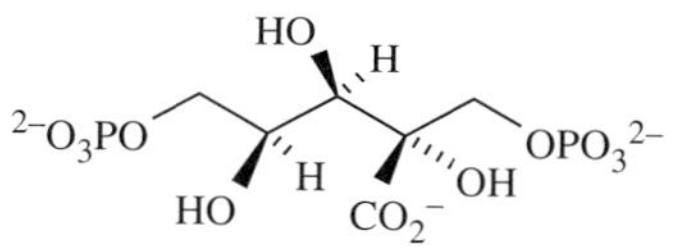

2-Carboxy-D-arabinitol-1,5-bisphosphate, CABP
$Kd \leq 10^{-11}$ M

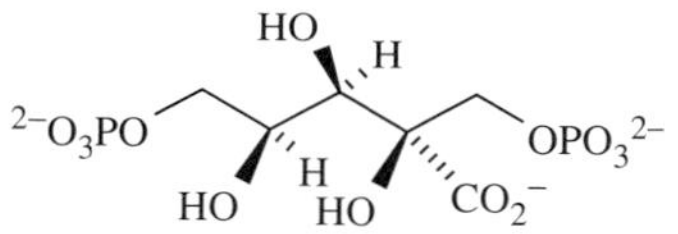

2-Carboxy-D-ribitol-1,5-bisphosphate
$Kd \leq 10^{-6}$ M

The differential binding free energy arising from the difference in configuration at C2 is 7 kcal mol^{-1} at 25°C. CABP has been a useful ligand for rubisco in crystal structures that have provided valuable information about the interactions of amino acid side chains with the carboxylated intermediate. Figure 8-23 illustrates the contacts between rubisco and CABP.

Calcium ion binds to rubisco but does not activate the enzyme. The structure of spinach rubisco with RuBP bound to the active site could be obtained by substitution of Ca^{2+} for Mg^{2+} in the crystallization of the complex (Taylor and Andersson, 1997). This structure is shown in fig. 8-24 and is thought to be related to that of the active magnesium complex, which cannot be determined with RuBP in the active site. The structure with calcium provides information regarding the identification of the base that abstracts C3(H) from the substrate in step 1 of the mechanism in fig. 8-22. The structure shows that the Lys201-*N*-carbamoyl group is in position to perform this function. Consensus holds that the *N*-carbamoyl moiety functions as the base to remove C3(H) (Cleland et al., 1998; Mauser et al., 2001; Newman and Gutteridge, 1993; Taylor and Andersson, 1997).

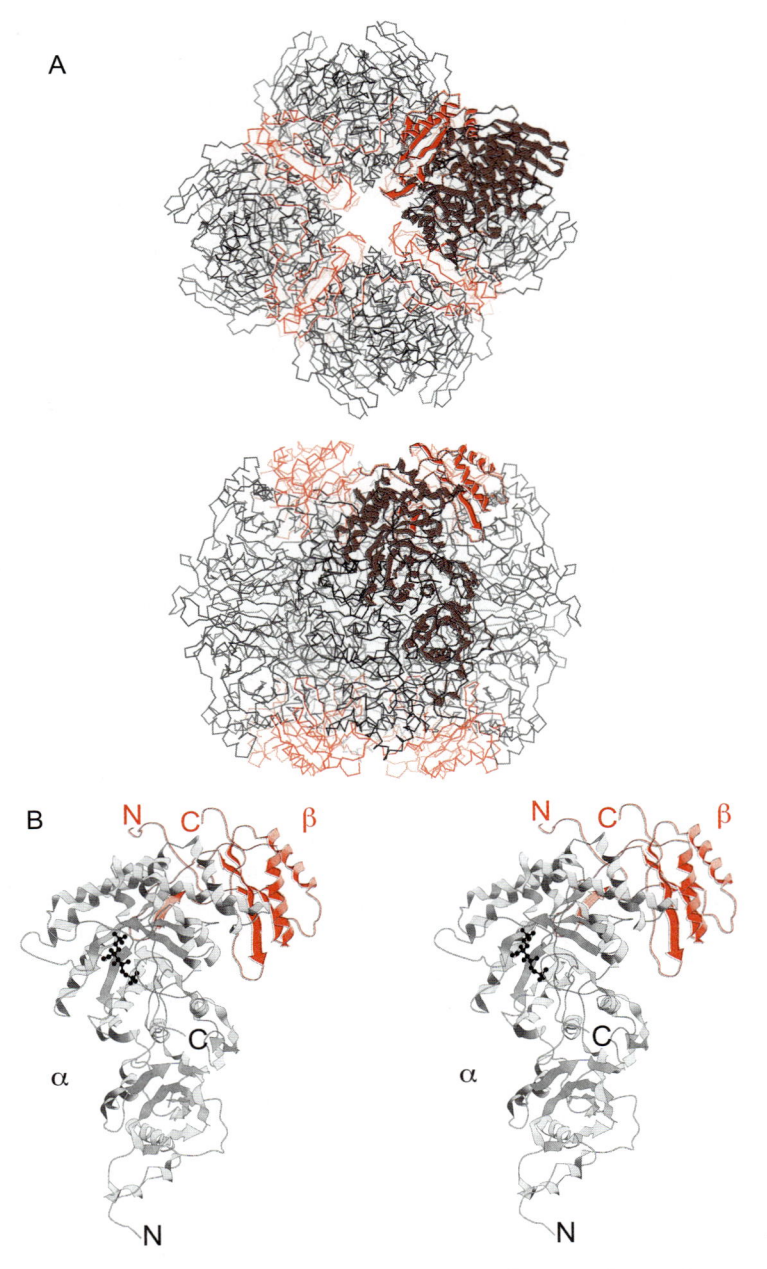

Fig. 8-23. Spinach rubisco with CABP and Mg^{2+} bound to the active site. (A) Heterohexadecameric $(\alpha\beta)_8$ *Spinacia oleracea* rubisco is shown from two perspectives: looking down the fourfold axis of symmetry and rotated 90 degrees, looking down a twofold axis of symmetry. In both images, one $\alpha\beta$ unit is shown as a ribbon drawing (with the large subunit in brown and the small subunit in red) and the others as Cα traces (with the large subunits in gray and the small subunit in red). (B) A large image in stereo of a large and small subunits with the transition-state analog 2-carboxyarabinitol-bisphosphate (CABP) bound at the active site in the large subunit (1.8-Å resolution; PDB 1IR1; Mizohata et al., 2002).

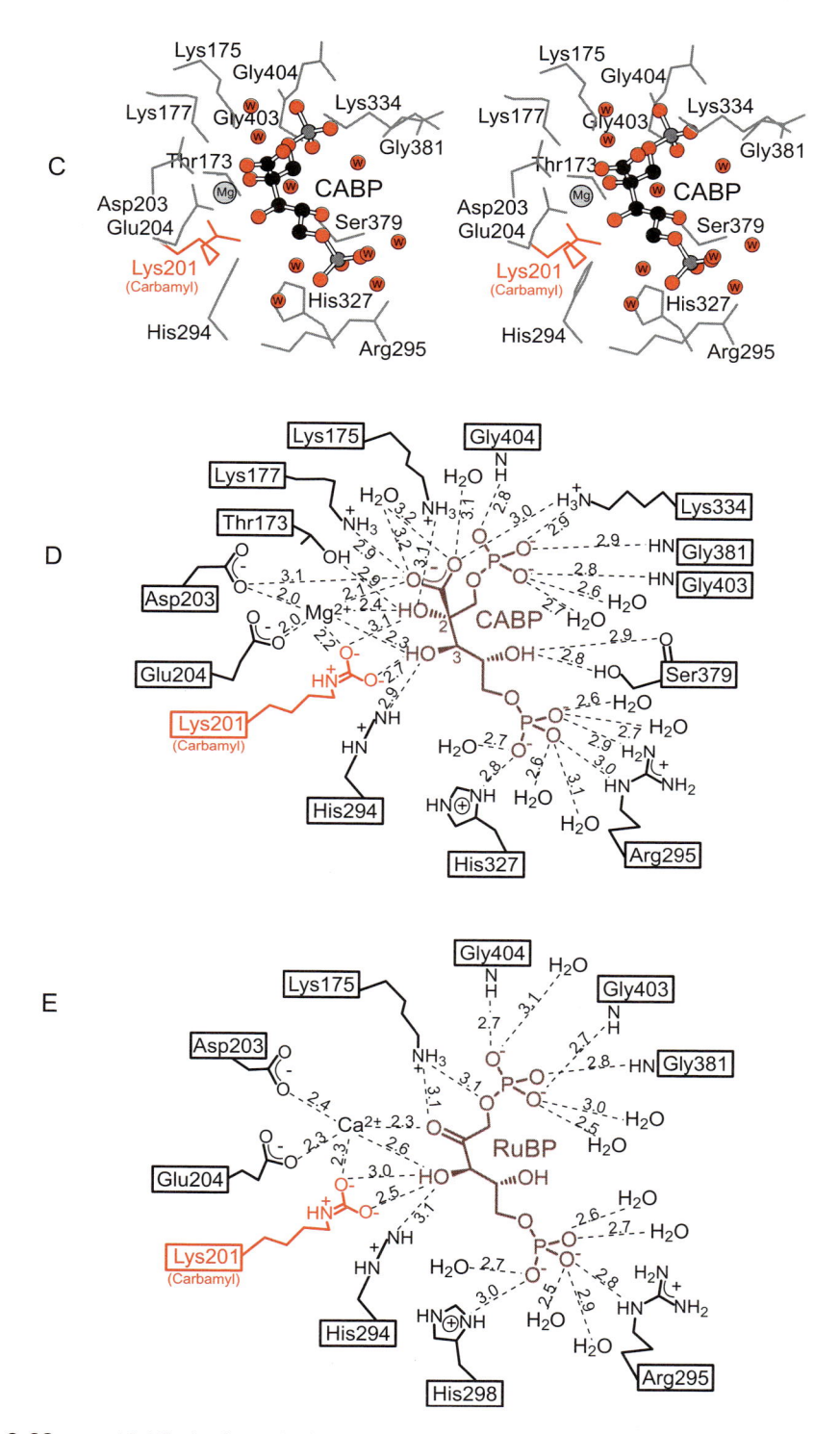

Fig. 8-23. cont'd (C) A closer look at the active site complex shows Mg^{2+}, CABP, and all of the active site contacts, including the carbamyl-Lys201. (D) The same information is shown in two dimensions, with interatomic contact distances given in angstroms. (E) Similar information is given for a complex of spinach rubisco with Ca^{2+} and ribulose 1,5-bisphosphate (2.2-Å resolution; PDB 1RXO; Taylor and Andersson, 1997). This substrate complex is observable because calcium does not support catalysis.

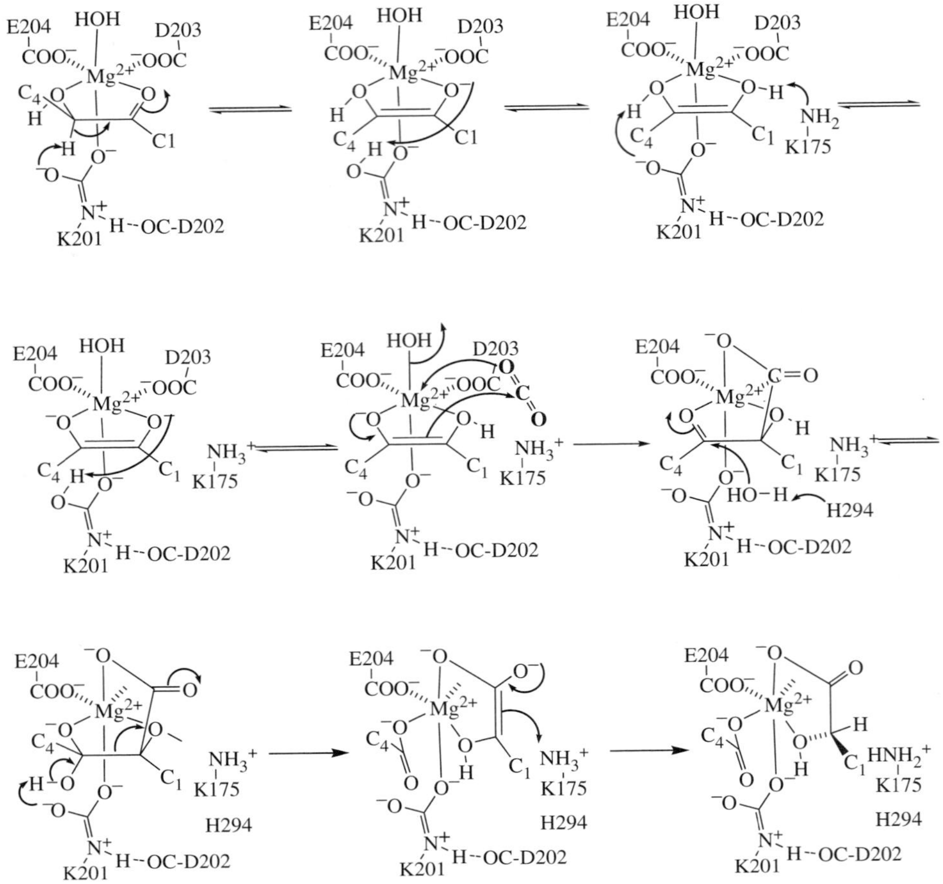

424 Enzymatic Reaction Mechanisms

Fig. 8-24. The mechanism for the action of rubisco in carboxylation and cleavage of RuBP accounts for the stereochemistry of carboxylation on the *si*-face of the C2-carbanion in fig. 8-22, for the stereochemistry of the post-cleavage proton transfer to the second molecule of 3-PGA, and for the role of the *N*-carbamoyl group of Lys201 in the enolization of RuBP and other proton transfer steps (Cleland et al., 1998).

The mechanism includes a minimum of six chemical steps, some of which can be studied separately from the overall reaction. For example, step 1 in fig.8-22, abstraction of C3(H) from the substrate, is a reversible process, and in D_2O or ^3H_2O, it leads to the incorporation of deuterium or tritium into RuBP. Next, the central carboxylated intermediate 3-keto-2-carboxy-D-arabinitol-1,5-bisphosphate can be obtained as a sufficiently stable compound to add to rubisco. The active enzyme catalyzes its hydrolytic cleavage to 3-PGA.

The partial reactions have been exploited for evaluating the catalytic properties of specifically mutated variants of rubisco. The assignment of Lys334 to promoting carboxylation of the enediolate at C2 in step 3 resulted from site-directed mutation and analysis of the competency of the variants in partial reactions (Hartman and Lee, 1989; Lorimer et al., 1993). Mutation of this residue abolished overall carboxylation activity, but not the C3(H)-exchange or the cleavage of the intermediate, nor was oxygenase activity abolished, although its chemistry was altered (Harpel et al., 1995).

Mutation of Lys166 (*R. rubrum*) greatly impairs but does not abolish C3(H) exchange, and it alters the course of the reaction of the enediolate. Most of the enediolate undergoes β-elimination of phosphate, and the part that undergoes carboxylation is improperly processed to the formation of pyruvate (Harpel et al., 2002). This residue appears to play a role in the carboxylation and processing of the carboxylated intermediate. In particular, it may interact with the intermediate to maintain the C2(OH) group and prevent β-elimination of phosphate. The exact function of His294 is uncertain. It may serve as a base to mediate proton transfer or play a more general role in stabilizing one or another transition state.

The overall mechanism in fig. 8-24 assigns an expanded catalytic role to the *N*-carbamoyl group of Lys201 and explains the stereochemistry of carboxylation and proton transfer to the C2 carbanion of 3-PGA after cleavage of the carboxylated intermediate (Cleland et al., 1998). In this mechanism, the resonance enhancement of the base properties of the *N*-carbamoyl group coordinated to magnesium ion is emphasized by showing this group in its zwitterionic resonance form.

Phosphoenolpyruvate Carboxylase

Carboxylation of phosphoenolpyruvate (PEP) by bicarbonate through the action of PEP carboxylase (PEPC; EC 4.1.1.31) produces phosphate and oxaloactetate according to eq. 8-9.

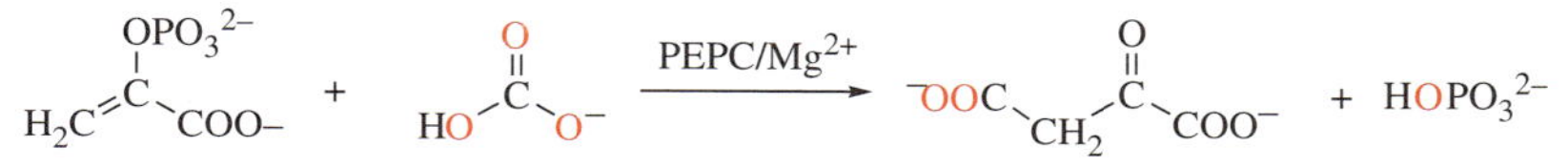

$$\text{H}_2\text{C}=\overset{\text{OPO}_3^{2-}}{\underset{}{\text{C}}}-\text{COO}^- \; + \; \text{HO}-\overset{\text{O}}{\underset{}{\text{C}}}-\text{O}^- \xrightarrow{\text{PEPC/Mg}^{2+}} \; ^-\text{OOC}-\text{CH}_2-\overset{\text{O}}{\underset{}{\text{C}}}-\text{COO}^- \; + \; \text{HOPO}_3^{2-} \quad (8\text{-}9)$$

The reaction is Mg^{2+} dependent and follows a random sequential kinetic mechanism with highly synergistic binding, in which Mg^{2+} binds first in equilibrium step, followed by PEP and then bicarbonate (Janc et al., 1992b). Oxygen-18 in bicarbonate appears in the phosphate produced, proving that the mechanism involves a chemical linkage between bicarbonate and phosphate. The reaction of [(*S*)-^{16}O,^{17}O]thiophospho-enolpyruvate with [^{18}O]bicarbonate (in $H_2^{18}O$) and configurational analysis of the [^{16}O,^{17}O^{18}O]thiophosphate produced proves that the overall reaction proceeds with inversion of configuration at phosphorus (Hanson and Knowles, 1982). The mechanism must involve an uneven number of substitution steps at phosphorus, most likely one.

The proposed chemical mechanism is outlined in fig. 8-25. Evidence supporting this mechanism includes the transfer of ^{18}O from bicarbonate to phosphate, which suggests the formation of an intermediate in which bicarbonate and phosphate are linked as carbonic

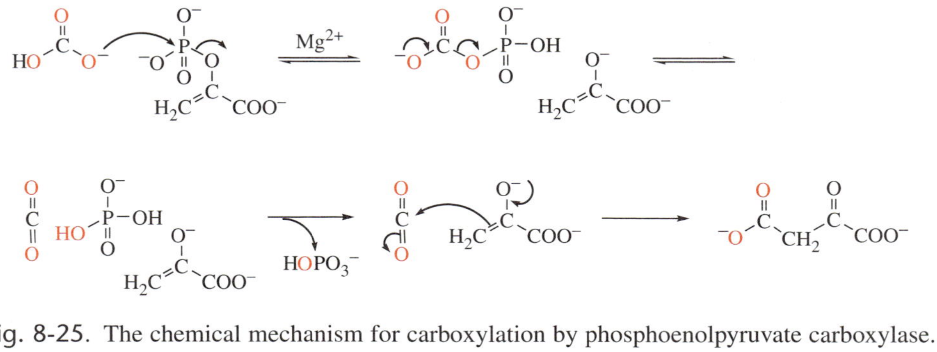

Fig. 8-25. The chemical mechanism for carboxylation by phosphoenolpyruvate carboxylase.

phosphoric anhydride, commonly known as carboxyphosphate, as shown in fig. 8-25. This also leads to the intermediate formation of enolpyruvate, and after decarboxylation of carboxyphosphate, the carbon dioxide formed carboxylates the enolpyruvate to oxaloacetate. The carboxylation step is essentially the reverse of decarboxylation mechanism A in fig. 8-1.

Further evidence supporting the mechanism is provided by experiments with analogs of bicarbonate or PEP. Formate reacts slowly in place of bicarbonate for the first part of the mechanism, the upper line in fig. 8-25, to form formyl phosphate and pyruvate (Janc et al., 1992a). PEPC also catalyzes the hydrolysis of formyl phosphate to formate and phosphate, and reaction of [^{18}O]formate leads ultimately to the incorporation of one ^{18}O into phosphate. Formate reacts slowly (1%) in place of bicarbonate and with a high value of K_m, 25 mM compared with 0.18 mM for bicarbonate. However, formyl phosphate is chemically analogous to carboxyphosphate, which is too unstable to isolate, and the production of formyl phosphate supports the mechanistic role of carboxyphosphate in the mechanism.

The geometric isomers (Z)- and (E)-3-fluorphosphoenolpyruvate (F-PEP) react in place of PEP (Janc et al., 1992c). Both are partitioned between carboxylation and hydrolysis to fluoropyruvate. (E)-F-PEP is partitioned 86% to carboxylation to (S)-3-fluorooxalacetate, and (Z)-F-PEP is partitioned 35% to (R)-3-fluoropyruvate. The stereochemistry corresponds to phosphorylation on the 2-si-face of PEP. Carboxylation of F-PEP with [^{18}O]bicarbonate leads to incorporation of more than one ^{18}O into phosphate, an important observation supporting the participation of carboxyphosphate as an intermediate. Apparently, the slow reaction of F-PEP allows the cleavage of carboxyphosphate to carbon dioxide and phosphate to be reversible in the second step of the mechanism in fig. 8-25, and this allows positional isotope exchange (PIX) of ^{18}O in phosphate to precede the carboxylation of enol-3-fluoropyruvate. In the course of the reaction of F-PEP with [^{18}O]bicarbonate, ^{18}O is incorporated into residual F-PEP.

Carboxylation of F-PEP by [^{13}C]bicarbonate proceeds with ^{13}V/K isotope effects of 1.049 for the (Z)-isomer and 1.009 for the (E)-isomer (Janc et al., 1992c). These isotope effects are those expected for carboxylation by carbon dioxide and not by carboxy phosphate, with carbon dioxide formed in a reversible step preceding carboxylation.

All things considered, the mechanism in fig. 8-25 appears well supported by all the experimental evidence. The details of the interactions between the metal ion and substrates and the active site of the enzyme are not revealed by the biochemical data. The structure in fig. 8-26 is of *Zea mays* PEPC with 3,3-dichloropyruvate bound at the active site in place of PEP.

Vitamin K–Dependent Carboxylase

Vitamin K originally derived its name from one of its biologic functions, the coagulation of blood, as a "Koagulation" factor. A biochemical description of its role in blood coagulation appeared in the mid-1970s (Esmon et al., 1975; Nelsestuen et al., 1974; Stenflo et al., 1974). Glutamyl residues in active prothrombin were found to be carboxylated on the γ-carbons, and this posttranslational modification was traced to the effects of vitamin K. The carboxylated glutamyl residues readily chelate Ca^{2+}, a messenger in the blood-clotting cascade. Two classes of vitamin K are phylloquinone (vitamin K_1), a plant pigment, and menaquinone (vitamin K_2). Menaquinone (structure **8-5**) is the coenzyme for vitamin K–dependent carboxylase.

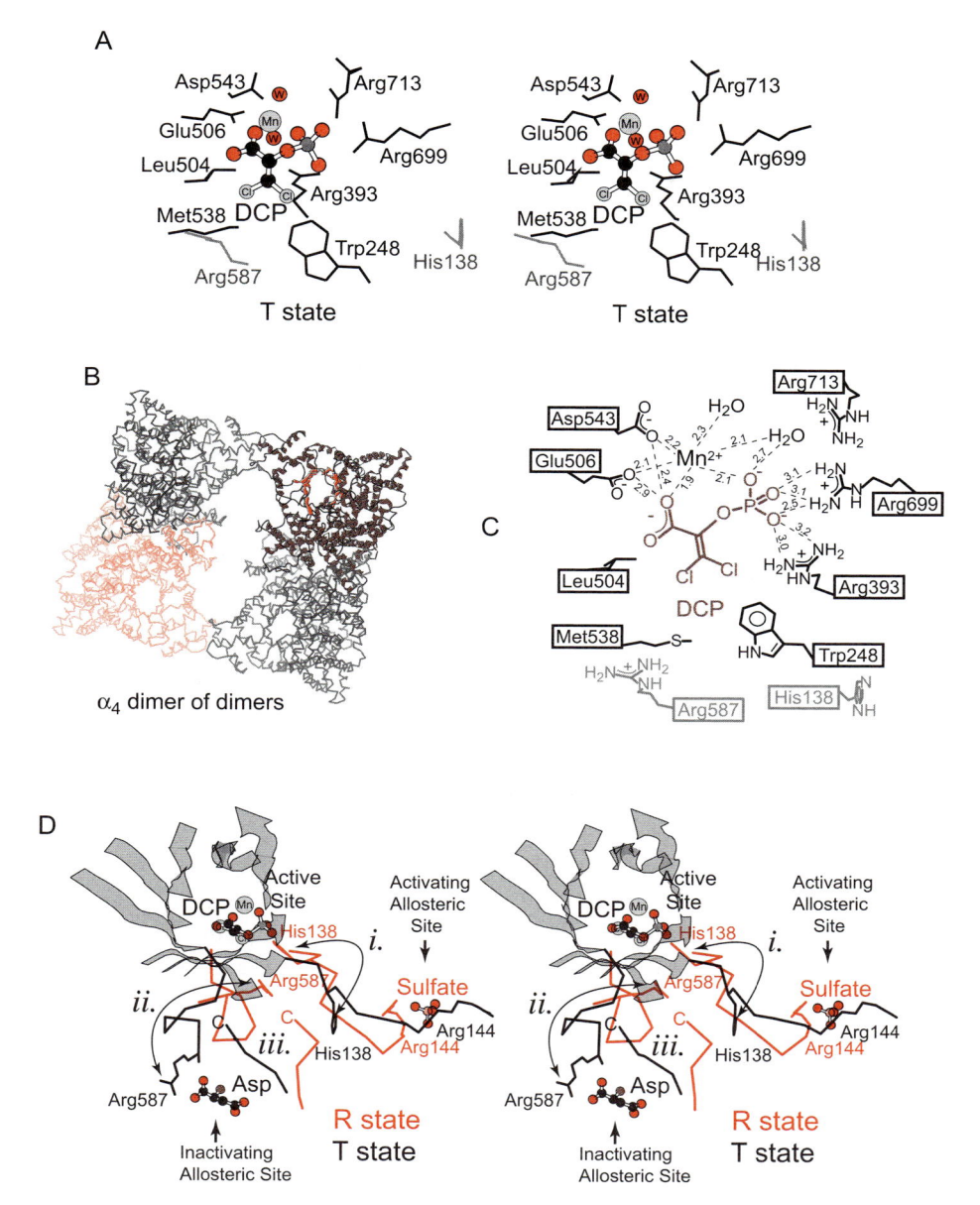

Fig. 8-26. Structure of phosphoenolpyruvate carboxylase with 3,3-dichlorophosphoenolpyruvate bound to the active site. (A) The *E. coli* phosphoenolpyruvate (PEP)carboxylase has 3,3-dichlorophosphoenolpyruvate and Mn^{2+} bound at the active site. This structure also has aspartate bound in an adjacent, inactivating allosteric site such that the active-site conformation is that of a T state (2.35-Å resolution; PDB 1JQN; Matsumura et al., 2002). The R state has Arg587 and His138 (gray) close to the active site, as shown in the similar structure from *Zea mays* with sulfate bound in an activating allosteric site (3.0-Å resolution; PDB 1JQO; Matsumura et al., 2002). (B) The dimer of dimers subunit arrangement of the homotetrameric *E. coli* PEP carboxylase shows the β-barrel highlighted in red in one subunit. (C) The active site contacts to the substrate analog and metal ion. (D) The *E. coli* enzyme T state (black) is compared with the corn enzyme R state (red), showing the large-scale loop movements (*iii.*) that transport the His138 (*i.*) and Arg587 (*ii.*) residues in and out of the active site.

428   Enzymatic Reaction Mechanisms

8-5   Menaquinone

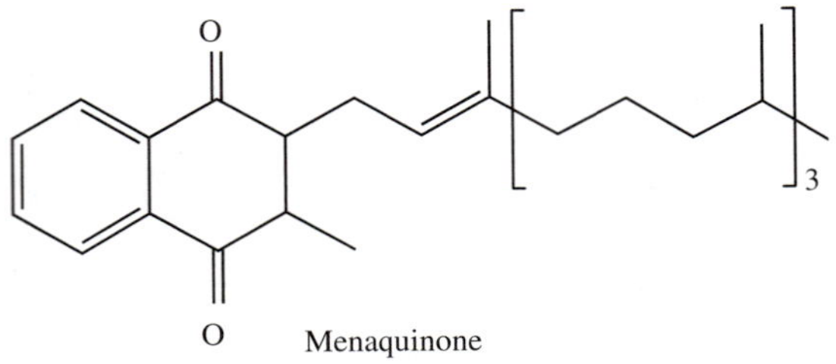

Vitamin K–dependent carboxylase is an integral membrane enzyme that catalyzes the O_2-dependent carboxylation of glutamyl residues (Glu) in vitamin K–dependent proteins to γ-carboxylglutamyl (Gla) residues (Berkner, 2001; Dowd et al., 1995; Suttie, 1993). Prothrombin and other blood coagulation factors are substrates for vitamin K–dependent carboxylase, and carboxylation of these proteins is required for blood clotting. In recent years, other proteins have been found to be γ-glutamyl carboxylated as well (Berkner, 2000). The carboxylation domains are homologous and rich in glutamyl residues, and carboxylation of all the glutamyl residues in a domain proceeds processively (Berkner, 2000; Benton et al., 1995; Morris et al., 1995).

Vitamin K carboxylase has presented barriers to detailed mechanistic analysis. Purification has been difficult, although highly purified preparations derived from expression of the human cDNA have been described (Wu et al., 1991a, 1991b). Information about propeptide and vitamin K binding loci has been obtained in photoaffinity labeling and mutagenic experiments (Kuliopulos et al., 1994; Roth et al., 1995; Sugiura et al., 1996; Yamada et al., 1995). Vitamin K interacts at a site in a carboxy terminal domain, and pro-peptides that undergo carboxylation interact in a domain in the N-terminal part of the protein (Berkner, 2000).

The mechanism of O_2-dependent carboxylation is poorly understood. Certain steps in the process are reasonably well defined and shown in fig. 8-27. Both vitamin K–dependent carboxylase and vitamin K epoxide reductase are required to maintain activity, because the epoxide is a proximal product of the carboxylation process and must be reduced to the hydroquinone form of vitamin K for a new cycle of carboxylation. Inhibition of the vitamin K epoxide reductase by warfarin is the basis for its anticoagulation effect.

Removal of the C4(H) from the glutamyl side chain (γ-H) is the most difficult and least understood step in the mechanism. Current thought holds that it is removed as a proton by

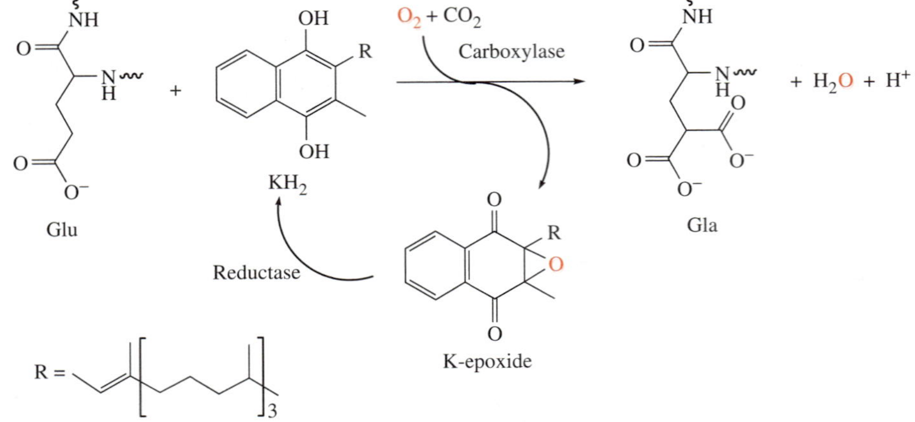

Fig. 8-27. Carboxylation of glutamyl side chains by vitamin K–dependent carboxylase.

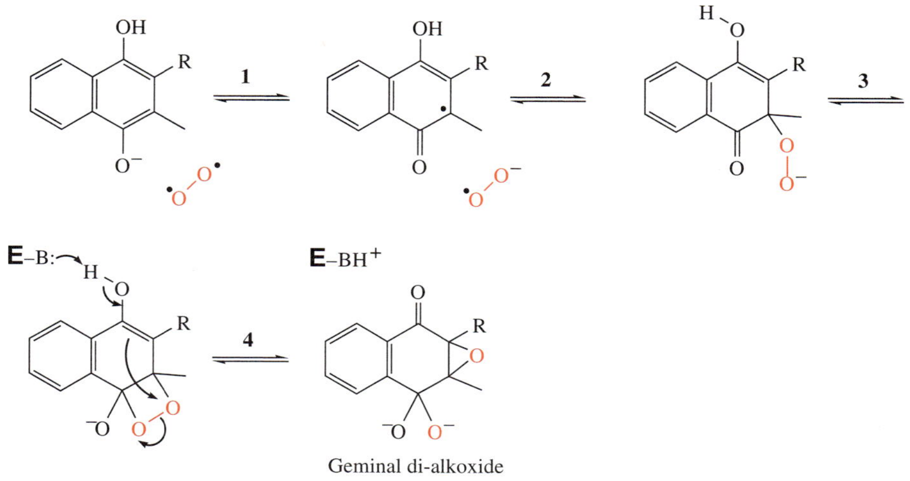

Fig. 8-28. A mechanism for oxygenation of vitamin K to a geminal di-alkoxide of vitamin K epoxide. The geminal di-alkoxide form of vitamin K epoxide is postulated to serve as the strong base for abstracting a γ-proton from a Glu side chain in its carboxylation to Gla by vitamin K–dependent carboxylase (Dowd et al., 1995).

a strong base to form a carbanion, which is carboxylated by CO_2 bound in an adjacent site by the carboxylation mechanism in fig. 8-1. The role of O_2 is thought to be to energize the formation of a strong base through epoxidation of vitamin K (Dowd et al., 1995). The strong base would abstract C4(H) from a glutamyl side chain to allow carbanion formation and carboxylation. Carboxylase-catalyzed, CO_2-independent and O_2-dependent exchange of tritium from ^3H_2O into glutamyl side chains is cited as evidence for carbanion formation. The carbon pK_a value for C4(H) should be similar to that for acetate ion, which is 33.5 (see table 1-4 in chap. 1), unless the γ-carboxylate of the side chain is coordinated to a metal ion or engaged in low barrier hydrogen bonding. In any case, a very strong base is required.

A geminal di-alkoxide form of vitamin K epoxide, shown in fig. 8-28, has been suggested to serve as a strong base to abstract the C4(H) from the glutamyl side chain (Dowd et al., 1995). Reaction of O_2 with the hydroquinone form of vitamin K is postulated to lead to a peroxy anion that reacts further to an oxetane. This might occur through electron transfer in step 1 of fig. 8-28 to form a vitamin K radical and superoxide ion. Radical coupling in step 2 would generate a peroxy anion that could undergo intramolecular addition to the neighboring carbonyl in step 3 to form the postulated oxetane. This mechanism differs from the one suggested (Dowd et al., 1995) but allows for the addition of the paramagnetic O_2^- in a spin-allowed process. The oxetane can undergo a base-catalyzed internal electrocyclic process to form the di-alkoxide of vitamin K epoxide in step 4 of fig. 8-28. A suitable enzymatic base for epoxide formation in step 4 appears to be a highly basic amine (Rishavy et al., 2004).

Many issues remain regarding the action of vitamin K–dependent carboxylase. Is the di-alkoxide vitamin K epoxide a strong enough base? The pK_a of a ketone hydrate is about 14. That of the hydrate anion should be five units higher at 19, based on the rule that successive pK_a values for mononuclear polyhydroxylic acids differ by five units. For example, the values for H_3PO_4 are 2, 7, and 12. On this basis, the di-alkoxide, if it could be sustained in an enzymatic site, would not appear to be strong enough to abstract a γ-proton from a glutamyl residue. However, if the γ-carboxylate of the glutamyl side chain should be in its conjugate acid state (COOH), or if it should be ligated to a metal ion or in a strongly hydrogen bonded state, the pK_a of the γ-hydrogen would be lower than 33, the expected value in aqueous solution. These matters may be settled in future research.

References

Abell LM and MH O'Leary (1988a) *Biochemistry* **27**, 5933.
Abell LM and MH O'Leary (1988b) *Biochemistry* **27**, 5927.
Acheson SA, JB Bell, ME Jones, and R Wolfenden (1990) *Biochemistry* **29**, 3198.
Albert A, V Dhanaraj, U Genschel, G Khan, MK Ramjee, R Pulido, BL Sibanda, F von Delft, M Witty, TL Blundell, AG Smith, and C Abell (1998) *Nat Struct Biol* **5**, 289.
Alston TA and RH Abeles (1987) *Biochemistry* **26**,4082.
Anand R, PC Dorrestein, C Kinsland, TP Begley, and SE Ealick (2002) *Biochemistry* **41**, 7659.
Andersson I (1996) *J Mol Biol* **259**, 160.
Andrews TJ, GH Lorimer, and J Pierce (1986) *J Biol Chem* **261**, 12184.
Appleby TC, C Kinsland, TP Begley, and SE Ealick (2000) *Proc Natl Acad Sci* **97**, 2005.
Arjunan P, T Umland, F Dyda, S Swaminathan, W Furey, M Sax, B Farrenkopf, Y Gao, D Zhang, and F Jordan (1996) *J Mol Biol* **256**, 590.
Barletta G, Y Zou, WP Huskey, and F Jordan (1997) *J Am Chem Soc* **119**, 2356.
Beak P and B Siegel (1976) *J Am Chem Soc* **98**, 3601.
Becker A, K Fritz-Wolf, W Kabsch, J Knappe, S Schultz, and AF Volker Wagner AF (1999) *Nat Struct Biol* **6**, 969.
Becker A and W Kabsch (2002) *J Biol Chem* **277**, 40036.
Benton ME, PA Price, and JW Suttie (1995) *Biochemistry* **34**, 9541.
Berkner KL (2000) *J Nutr* **130**, 1877.
Bernardi R, T Caronna, R Galli, F Minisci, and M Perchinunno (1973) *Tetrahedron Lett* **9**, 645.
Carlson GL and GM Brown (1961) *J Biol Chem* **236**, 2099.
Chang CC, A Laghai, MH O'Leary, and HG Floss (1982) *J Biol Chem* **257**, 3564.
Cleland WW, TJ Andrews, S Gutteridge, FC Hartman, and GH Lorimer (1998) *Chem Rev* **98**, 549.
Crosby J, R Stone, and GE Lienhard (1970) *J Am Chem Soc* **92**, 2891.
Dhe-Paganon S, J Magrath, and RH Abeles (1994) *Biochemistry* **33**,13355.
Dobritzsch D, S Konig, G Schneider, and G Lu (1998) *J Biol Chem* **273**, 20196.
Dowd P, SW Ham, S Naganathan, and R Hershline (1995) *Annu Rev Nutr* **15**, 419.
Dowhan W (1997) *Methods Enzymol* **280**, 81.
Dunathan HC (1966) *Proc Natl Acad Sci U S A* **55**, 712.
Dyda F, W Furey, S Swaminathan, M Sax, B Farrenkopf, and F Jordan (1993) *Biochemistry* **32**, 6165.
Esmon CT, JA Dadowski, and JW Suttie (1975) *J Biol Chem* **250**, 4744.
Fiedler E, S Thorell, T Sandalova, R Golbik, S Konig, and G Schneider (2002) *Proc Natl Acad Sci U S A* **99**, 591.
Frey PA, F Kokesh, and FH Westheimer (1971) *J Am Chem Soc* **93**, 7266.
Frey M, M Rothe, AF Wagner, and J Knappe (1994) *J Biol Chem* **269**,12432.
Fridovich I and FH Westheimer (1962) *J Am Chem Soc* **84**, 3208.
Gallagher T, EE Snell, and ML Hackert (1989) *J Biol Chem* **264**, 12737.
Hamilton GA and FH Westheimer (1959) *J Am Chem Soc* **81**, 6332.
Hansen DE and JR Knowles (1982) *J Biol Chem* **257**,14795.
Harpel MR, FW Larimer, and FC Hartman (2002) *Biochemistry* **41**, 1390.
Harpel MR, EH Serpersu, JA Lamerdin, ZH Huang, DA Gage, and FC Hartman (1995) *Biochemistry* **34**, 11296.
Harris P, JC Navarro Poulsen, KF Jensen, and S Larsen (2000) *Biochemistry* **39**, 4217.
Hartman FC and MR Harpel (1994) *Annu Rev Biochem* **63**, 197.
Hartman FC and EH Lee (1989) *J Biol Chem* **264**, 11784.
Hartman FC, S Milanez, and EH Lee (1985) *J Biol Chem* **260**, 13968.
Hartman FC, TS Soper, SK Niyogi, RJ Mural, RS Foote, S Mitra, EH Lee, R Machanoff, and FW Larimer (1987) *J Biol Chem* **262**, 3496.

Hasson MS, A Muscate, MJ McLeish, LS Polovnikova, JA Gerlt, GL Kenyon, GA Petsko, and D Ringe (1998) *Biochemistry* **37**, 9918.
Hawkins CF, A Borges, and RN Perham (1989) *FEBS Lett* **255**, 77.
Highbarger LA, JA Gerlt, and GL Kenyon (1996) *Biochemistry* **35**, 41.
Hodgins D and RH Abeles (1967) *J Biol Chem* **242**, 5158.
Hohenester E, JW Keller, and JN Jansonius (1994) *Biochemistry* **33**, 13561.
Holzer H and K Beaucamp (1958) *Angew Chem* **71**, 1776.
Hur S and TC Bruice (2002) *Proc Natl Acad Sci U S A* **99**, 9668.
Huynh QK and EE Snell (1985) *J Biol Chem* **260**, 2798.
Iyengar R, E Cardemil, and PA Frey (1986) *Biochemistry* **25**, 4693.
Jabalquinto AM and E Cardemil (1989) *Biochim Biophys Acta* **996**, 257.
Jabalquinto AM, J Eyzaguirre, and E Cardemil (1983) *Arch Biochem Biophys* **225**, 338.
Janc JW, WW Cleland, and MH O'Leary (1992a) *Biochemistry* **31**, 6441.
Janc JW, MH O'Leary, and WW Cleland (1992b) *Biochemistry* **31**, 6421.
Janc JW, JL Urbauer, MH O'Leary, and WW Cleland (1992c) *Biochemistry* **31**, 6432.
Jordan F (2003) *Nat Prod Rep* **20**,184.
Jordan F, H Li, and A Brown (1999) *Biochemistry* **38**, 6369.
Jordan F, Z Zhang, and E Sergienko (2002) *Bioorg Chem* **30**, 188.
Just VJ, CEM Stevenson, L B owater, A Tanner, DM Lawson, and S Bornemann (2004) *J Biol Chem* **279**, 19867.
Knappe J, S Elbert, M Frey, and AF Wagner (1993) *Biochem Soc Trans* **21**, 731.
Knappe J and G. Sawers (1990) *FEMS Microbiol Rev* **6**, 383.
Knappe J and AF Wagner (1995) *Methods Enzymol* **258**, 343.
Knight S, I Andersson, and C-L Brändén (1990) *J Mol Biol* **215**, 113.
Kokesh F and FH Westheimer (1971*) J Am Chem Soc* **93**, 7270.
Krampitz LO, G Gruell, CS Miller, KB Bicking, HR Skeggs, and JM Sprague (1958) *J Am Chem Soc* **80**, 5893.
Kuliopulos A, NP Nelson, M Yamada, CT Walsh, B Furie, BC Furie, and DA Roth (1994) *J Biol Chem* **269**, 21364.
Kuo DJ and F Jordan (1983) *J Biol Chem* **258**, 13415.
Lane MD and HM Miziorko (1978) *Basic Life Sci* **11**, 19.
Layer G, K Verfurth, E Mahlitz, and D Jahn (2002) *J Biol Chem* **277**, 34136.
Lee JK and KN Houk (1997) *Science* **276**, 942.
Lorimer GH, MR Badger, and RJ Andrews (1976) *Biochemistry* **15**, 529.
Lorimer GH, YR Chen, and FC Hartman (1993) *Biochemistry* **32**, 9018.
Lu G, D Dobritzsch, S Baumann, G Schneider, and S Konig (2000) *Eur J Biochem* **267**, 861.
Lundqvist T and G Schneider (1991) *Biochemistry* **30**, 904.
Malashkevich VN, P Strop, JW Keller, JT Jansonius, and MD Toney (1999) *J Mol Biol* **294**, 193.
Matsumura H, Y Xie, S Shirakata, T Inoue, T Yoshinaga, Y Ueno, K Izui, and Y Kai (2002) *Structure* **10**, 1721.
Mauser H, WA King, JE Gready, and TJ Andrews (2001) *J Am Chem Soc* **123**, 10821.
Menon-Rudolph S, S Nishikawa, X Zeng, and F Jordan (1992) *J Am Chem Soc* **114**, 10110.
Michihara A, K Akasaki, Y Yamori, and H Tsuji (2002) *Biol Pharm Bull* **25**, 302.
Michihara A, M Sawamura, Y Nara, K Ikeda, and Y Yamori (1997) *J Biochem (Tokyo)* **122**, 647.
Miller BG, AM Hassell, R Wolfenden, MV Milburn, and SA Short (2000*) Proc Natl Acad Sci U S A* **97**, 2011.
Miziorko HM (1979) *J Biol Chem* **254**, 270.
Mizohata E, H Matsumura, Y Okano, M Kumei, H Takuma, J Onodera, K Kato, N Shibata, T Inoue, A Yokota, and Y Kai (2002) *J Mol Biol* **316**, 679.
Morris DP, RD Stevens, DJ Wright, and DW Stafford (1995) *J Biol Chem* **270**, 30491.
Nave JF, H d'Orchymont, JB Ducep, F Piriou, and MJ Jung (1985) *Biochem J* **227**, 247.
Nelsestuen GL, TH Zytkovicz, and JB Howard (1974) *J Biol Chem* **249**, 6347.
Newman J and S Gutteridge (1993) *J Biol Chem* **268**, 25876.
O'Leary MH and FH Westheimer (1968) *Biochemistry* **7**, 913.
Palekar AG, SS Tate, and A Meister (1970) *Biochemistry* **26**, 2310.
Parast CV, KK Wong, JW Kozarich, J Peisach, and RS Magliozzo (1995a) *Biochemistry* **34**, 5712.
Parast CV, KK Wong, SA Lewisch, JW Kozarich, J Peisach, and RS Magliozzo (1995b) *Biochemistry* **34**, 2393.

Phillips LM and JK Lee (2001) *J Am Chem Soc* **123**, 12067.
Pierce J, NE Tolbert, and R Barker (1980) *Biochemistry* **19**, 934.
Rathod PK and JH Fellman (1985) *Arch Biochem Biophys* **238**, 435.
Reardon JE and RH Abeles (1987) *Biochemistry* **26**, 4717.
Recsei PA, QK Huynh, and EE Snell (1983) *Proc Natl Acad Sci U S A* **80**, 973.
Recsei PA and EE Snell (1984) *Annu Rev Biochem* **53**, 357.
Recsei PA and EE Snell (1985) *J Biol Chem* **260**, 2804.
Reddy SG, KK Wong, CV Parast, J Peisach, RS Magliozzo, and JW Kozarich (1998) *Biochemistry* **37**, 558.
Reinhardt LA, D Svedruzic, CH Chang, WW Cleland, and NG Richards (2003) *J Am Chem Soc* **125**,1244.
Rishavy MA and WW Cleland (2000) *Biochemistry* **39**, 4569.
Rishavy MA, BN Pudota, KW Hallgren, W Qian, AV Yakubenko, JH Song, KW Runge, and KL Berkner (2004) *Proc Natl Acad Sci U S A* **101**, 13732.
Rosenberg RM and MH O'Leary (1985) *Biochemistry* **24**, 1598.
Roth DA, ML Whirl, LJ Velazquez-Estades, CT Walsh, B Furie, and BC Furie (1995) *J Biol Chem* **270**, 5305.
Schmidt DE Jr and FH Westheimer (1971) *Biochemistry* **10**, 1249.
Smiley JA, P Paneth, MH O'Leary, JB Bell, and ME Jones (1991) *Biochemistry* **30**, 6216.
Soper TS, RJ Mural, FW Larimer, EH Lee, R Machanoff, and FC Hartman (1988) *Protein Eng* **2**, 39.
Stenflo J, P Fernlund, W Egan, and P Roepstorff (1974) *Proc Natl Acad Sci U S A* **71**, 2730.
Sugiura I, B Furie, CT Walsh, and BC Furie (1996) *J Biol Chem* **271**, 17837.
Sun S, CK Bagdassarian, and MD Toney (1998) *Biochemistry* **37**, 3876.
Suttie JW (1993) *FASEB J* **5**, 445.
Tagaki W, JP Guthrie, and FH Westheimer FH (1968) *Biochemistry* **7**, 905.
Tate SS and A Meister (1971) *Adv Enzymol* **35**, 503.
Taylor TC and I Andersson (1997) *J Mol Biol* **265**, 432.
Tittman K, R Golbik, K Uhlemann, L Khailora, G Schneider, M Patel, F Jordan, DM Chipman, RG Duggleby, and G Hübner (2003) *Biochemistry* **42**, 7885.
Tolbert WD, JL Ekstrom, II Mathews, JA Secrist III, P Kapoor, AE Pegg, and SE Ealick (2001) *Biochemistry* **40**, 9484.
Toney MD (2001) *Biochemistry* **40**,1378.
Toth ML and L Huwyler (1996) *J Biol Chem* **271**, 7895.
van Poelje PD and EE Snell (1990) *Annu Rev Biochem* **59**, 29.
Volker-Wagner AF, M Frey, FA Neugebauer, W Schafer, and J Knappe (1992) *Proc Natl Acad Sci U S A* **89**, 996.
Warren S, B Zerner, and FH Westheimer (1966) *Biochemistry* **5**, 817.
Weil JA, JR Bolton Jr, and JA Wertz (1994) *Electron Paramagnetic Resonance*. Wiley: New York.
Wong KK, BW Murray, SA Lewisch, MK Baxter, TW Ridky, L Ulissi-DeMario, and JW Kozarich (1993). *Biochemistry* **32**, 14102.
Worley S, E Schelp, AF Monzingo, S Ernst, and JD Robertus (2002) *Proteins Struct Funct Genet* **46**, 321.
Wu N, W Gillon, and EF Pai (2002) *Biochemistry* **41**, 4002.
Wu N, Y Mo, J Gao, and EF Pai (2000) *Proc Natl Acad Sci U S A* **97**, 2017.
Wu N and EF Pai (2002) *J Biol Chem* **277**, 28080.
Wu SM, DP Morris, and DW Stafford (1991b) *Proc Natl Acad Sci U S A* **88**, 2236.
Wu SM, WF Cheung, D Frazier, and DW Stafford (1991a) *Science* **254**, 1634.
Yamada M, A Kuliopulos, NP Nelson, DA Roth, B Furie, BC Furie, and CT Walsh (1995) *Biochemistry* **34**, 481.
Zhou X, X Jin, R Medhekar, X Chen, T Dieckmann, and MD Toney (2001) *Biochemistry* **40**,1367.
Zhou X, S Kay, and MD Toney MD (1998) *Biochemistry* **37**, 5761.

9
Addition and Elimination

α,β-Elimination/Addition Reactions

Most elimination and addition reactions in biochemistry proceed by α,β-elimination/addition mechanisms. In the case of elimination, the leaving group is β to an activating functional group in the substrate. The activating group may be the carbonyl group of a ketone or aldehyde, the iminium group derived from an aldehyde or ketone, or the acyl-carbonyl of a carboxylic acid or ester, and the proton is α to the activating group. Addition reactions in this class are the same reactions in reverse, and they follow the course of the Michael addition in organic chemistry. The generic process is illustrated in scheme 9-1.

X = O, N   Y = O, N, S   Z = H, C, SCoA, OR, O⁻

Scheme 9-1

Substituents among the activating and leaving groups are diverse and are presumed to account for the significant variations among enzymes in the class. A few enzymes in this class catalyze elimination/addition without the assistance of a coenzyme or cofactor. They presumably incorporate sufficiently acidic (A—H) or basic (:B) amino acid side chains to catalyze the proton transfer processes, or they may stabilize carbanionic intermediates by low-barrier hydrogen bonding. Others employ divalent metal ions, pyridoxal-5′-phosphate (PLP), [4Fe–4S] centers, or NAD⁺ to facilitate the reactions. Cofactors and coenzymes increase the acidity of C_α—H or improve the propensity of the leaving group Y to depart.

In most cases, the major barrier consists of increasing the acidity of the C_α—H group, which decreases the pK_a. In a few cases, as when the leaving group is a carboxylic acid or a phosphate, no catalysis is required for it to depart.

Cofactor-Independent α,β-Elimination/Addition Reactions

Limited space prevents discussion of the many enzymes that catalyze cofactor-independent α,β-eliminations. We address the actions of fumarase and crotonase because of the historic emphasis on the biochemical significance of these enzymes. Many other dehydratases and ammonia lyases also belong in this group.

Fumarase

In the tricarboxylic acid cycle, fumarate arises from the action of succinate dehydrogenase, and fumarase (EC 4.2.1.2) catalyzes the addition of water to form S-malate (eq. 9-1).

$$\begin{array}{c}\text{H}\\ \\ \text{}^-\text{OOC}\end{array}\!\!\!\!C\!=\!C\!\!\!\!\begin{array}{c}\text{COO}^-\\ \\ \text{H}\end{array} + H_2O \xrightleftharpoons{\text{Fumarase}} \begin{array}{c}H_R\\ H_S\,\text{C}\\ \text{}^-\text{OOC}\end{array}\!\!-\!\!\begin{array}{c}\text{COO}^-\\ \text{C}\,\text{H}\\ \text{OH}\end{array} \quad (9\text{-}1)$$

The reaction can be monitored in either direction, and in various studies, the kinetic parameters may be quoted as such (e.g., fumarate formation, or malate formation). The body of knowledge about the action of fumarase is surprisingly incomplete, given the importance of the enzyme in metabolism. The reaction itself raises deep mechanistic problems. In particular, in the direction of fumarate formation the value of pK_a for H_R (eq. 9-1) is very high, about 30, but the action of a class II fumarase does not involve any coenzyme or cofactor that may increase its acidity. The situation is further complicated by the fact that key studies are of fumarase from diverse species such as yeast, *Escherichia coli*, and hog. In this connection, we consider cofactor-independent fumarases. FumC in *E. coli* is a class II, cofactor-independent, homotetrameric, 200-kDa, thermally stable enzyme analogous to the mammalian and yeast fumarases. Cofactor-dependent fumarases exist, such as the *E. coli* class I fumarases FumA and FumB, which are dimeric, 120-kDa enzymes that are iron dependent and superoxide sensitive. We discuss iron-dependent dehydration and hydration in a later section on aconitase.

An early study on the kinetics of isotope exchange at equilibrium led to the suggestion that the action of fumarase involved the initial dehydration of malate to a carbocationic intermediate, which lost the β-proton to form fumarate (Hansen et al., 1969). In the reverse direction, this would amount to protonation of fumarate to the carbocation followed by the addition of a hydroxyl group from water. Further kinetic and inhibition studies showed that the rates were not limited by the chemical transformation and indicated an acid-base catalyzed elimination mechanism by way of a carbanionic intermediate.

Deep mechanistic issues in the action of fumarase remain to be resolved. Kinetic studies are hampered by complications, one of which is the fact that the reaction follows an iso-mechanism; that is, one in which different forms of the enzyme react with fumarate in the forward direction and with malate in the reverse direction. The two forms are different both in the protonation states of acid-base groups and in their conformations (Rebholz and Northrop, 1994; Rose, 1998). An iso-mechanism first appears in this volume in connection with the actions of HIV protease, proline and glutamate racemases. In the case of fumarase, acid-base catalysis involves two catalytic groups, which can be expected to differ in ionization state in the substrate and product Michaelis complexes and the related free enzyme forms, as illustrated in fig. 9-1. Interconversion of the isomeric enzyme forms is a step in the catalytic cycle that appears, in a detailed analysis, to be about 30% rate limiting (Rebholz and Northrop, 1994).

In the malate to fumarate direction, the rate is not limited by the elimination of water in neutral solutions, but diffusion, as shown by the kinetic isotope effects and viscosity studies (Blanchard and Cleland, 1980; Sweet and Blanchard, 1990). The pH-rate profile is a bell-shaped curve with a maximum at neutral pH and at $pK_1 = 5.5$ and $pK_2 = 7.8$. The profile represents changes in rate-limiting step in response to ionizations of enzymatic groups, as indicated by the pH dependence of isotope effects. The ^{18}O isotope effect for the reaction of (S)-[2-^{18}O]malate is near unity above pH 7, but at pH, 5 it is 1.08, including 1.03 for the equilibrium isotope effect. Similarly, the primary deuterium kinetic isotope effect in the reaction of (2S,3R)-[3-^2H]malate is near unity at low and high pH values (5 and 9) but inverse, 0.915, at pH 7. The ^{18}O kinetic effects show that a step or steps that do not involve the dehydration of malate limit the rate in neutral solutions. The absence of a primary deuterium kinetic isotope effect under the same conditions as the observation of an ^{18}O-isotope effect rules out a concerted proton abstraction and water elimination. The mechanism must be stepwise, with abstraction of the α-proton preceding elimination of the β-hydroxyl group, to produce an intermediate aci-carbanion, as illustrated in scheme 9-2.

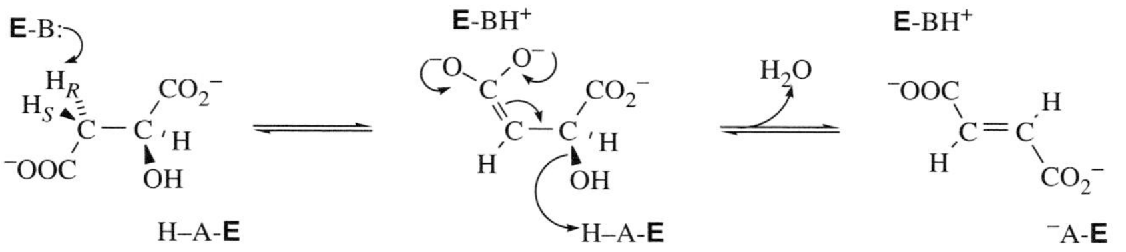

Scheme 9-2

The *aci*-carbanionic intermediate in scheme 9-2 incorporates planar and tetrahedral carbons at positions 3 and 2 and a distinct pattern of functional groups. The structure of the ionized form of 3-nitrolactate (structure **9-1**) is very similar, and it is a potent inhibitor (K_i = 27 nM) (Porter and Bright, 1980). 3-Nitrolactate is structurally similar to the *aci*-carbanionic intermediate.

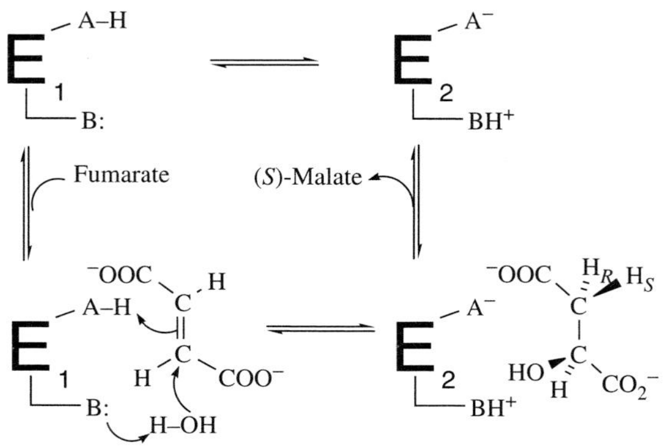

Fig. 9-1. Iso-mechanism in the action of fumarase. Regardless of the detail in the action of fumarase, two acid-base groups catalyze proton transfer, and the ionization states of these groups in part define two forms of the enzyme, E_1 and E_2. In E_1, the groups exist as A–H/B:, and in E_2, they occur as A-/BH+. E_1 binds fumarate and facilitates its transformation into malate, and E_2 binds malate and facilitates its transformation into fumarate. The two forms must undergo isomerization with each catalytic turnover.

9-1 3-Nitrolactate

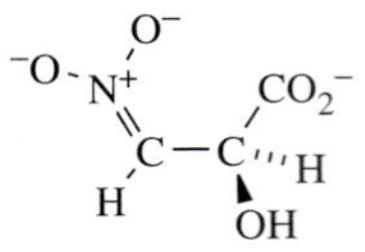

Several structures of the class II fumarase from *E. coli* are available. The structure of FumC from *E. coli* is shown in fig. 9-2 in complex with malate bound to one site (B) and pyromellitic acid (PMA to the other site (A) (Weaver and Banaszak, 1996). The structure shows the active site composed of amino acid residues from three subunits of the tetrameric enzyme. Potential acid-base catalytic residues include His188 and Lys324. Mutation of His188 to asparagine severely decreases the activity, and His188 is likely to be an acid-base catalyst for dehydration. A second catalytic group remains to be identified. There is much amino acid sequence identity among the class II fumarases, especially in the active site region, so that the structures of yeast and mammalian fumarases are likely to be similar to FumC from *E. coli* (Estévez et al., 2002).

The outstanding questions about the mechanism of action of fumarase include the identity of a second acid-base catalyst and the means by which the aci-carbanionic intermediate is stabilized. The second catalytic group may be identified by further structural, mutagenic and kinetic analysis. The stabilization of the aci-carbanion is a major question. The class II fumarases do not function with the aid of coenzymes or cofactors such as PLP, divalent metal ions, or iron-sulfur centers. The acidity of C3(H) in malate cannot be increased by a cofactor and this excludes a cofactor. The abstraction of this proton must be facilitated by the stabilization of the aci-carbanionic intermediate at the active site by some unknown means. The possibility of stabilization through low barrier hydrogen bonding has not been ruled out.

Enoyl CoA Hydratase

Crotonase, or enoyl CoA hydratase (ECH; EC 4.2.1.17), catalyzes the elimination of water from (3S)-3-hydroxybutyryl CoA to form *trans*-2-crotonyl CoA according to eq. 9-2.

$$(9\text{-}2)$$

Like fumarase, the action of ECH does not require a cofactor or coenzyme; however, the chemical barrier to α,β-elimination is lower in the case of ECH because the α-hydrogen of the substrate is more acidic ($pK_a \approx 21$) (see table 1-4 in chap. 1).

Intensive research on ECH from rat liver has led to a clear description of the active site in terms of its structure and interactions with substrates and water, the effects of binding interactions on physicochemical properties of substrates, the stereochemistry of the addition of water, and the implication of Glu144 and Glu164 in the mechanism of hydration. Much evidence based on kinetic isotope effects could be interpreted to indicate a concerted mechanism. However, the results of structure/function analyses implicated a stepwise mechanism of dehydration and hydration in reverse.

Site-directed mutagenesis of Glu144 and Glu164 implicates these residues in the action of ECH (D'Ordine et al., 1994a; Hofstein et al., 1999). ECH catalyzes the exchange of the 2-pro-*S* hydrogen of butyryl CoA with D_2O at a rate enhancement of 3×10^8 relative to

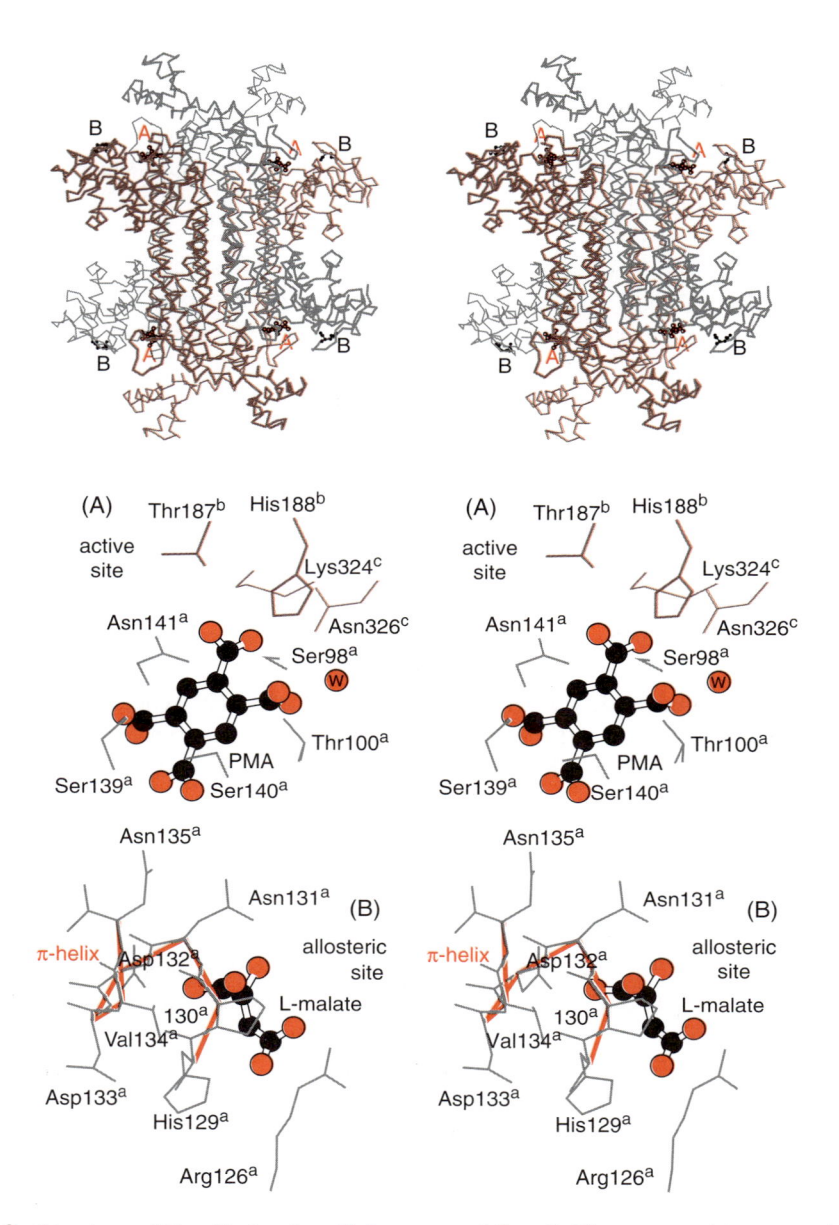

Fig. 9-2. Structure of FumC, the class II fumarase of *E. coli*. The top panel is a stereoimage of a Cα trace of the homotetrameric fold of *E. coli* class II fumarase C (2.0-Å resolution; PDB 1FUP; Weaver and Banaszak, 1996). Two multicarboxylate binding sites exist per monomer. L-Malate (black), which is an activator and a substrate, is bound at the B site, and pyromellitic acid (PMA; red), a potent inhibitor of FumC, is bound at the active site (A site). The lower panel shows one set of adjacent binding sites with the A site (PMA) above the B site (L-malate). The A site is at the interface of two monomers, and the B site is largely composed of contacts to a π-helix, which is a fairly unusual secondary structural feature of this enzyme.

the nonenzymatic process. The 2-pro-S hydrogen is diastereotopically opposite the 2-pro-R hydrogen abstracted in the dehydration of (3S)-3-hydroxyburyryl CoA. The rates of both the exchange of butyryl CoA and dehydration of (3S)-3-hydroxyburyryl CoA by ECH are decreased more than 100-fold by the mutation of Glu164 to glutamine, suggesting that this residue functions to abstract the α-proton in both reactions. Mutation of Glu164 or Glu144 severely decreases k_{cat} in the reaction of crotonyl 3′-dephosphoCoA, 7700-fold and 630,000-fold in the cases of E144Q-ECH and E164Q-ECH, respectively, with little effect on K_m. Like E164Q-ECH, the E144Q-ECH is also inactive in catalyzing the exchange of the 2-pro-S hydrogen in butyryl CoA, despite the presence of Glu164. The stereochemistry of the exchange of the 2-pro-S hydrogen in butyryl CoA, and the consequences of glutamate to glutamine mutations at the active site emphasize the importance of the interactions between the active site and the substrate-3-hydroxyl group. Glu144 and Glu164 must carry out essential catalytic functions in the overall reaction.

The structure of ECH focused attention on the likely functions of Glu144 and Glu164, shown in fig. 9-3 in direct contact with a water molecule poised to undergo addition to the double bond of 4-(N,N-dimethylamino)cinnamoyl CoA (Bahnson et al., 2002). The hydrogen bonded contacts of the two γ-carboxyl groups suggested that both were ionized as carboxylate groups, which made little chemical sense for a pretransition state or Michaelis complex. It was suggested, based on this structure, that one proton and the hydroxyl group from a single water molecule would be added to the double bond in a concerted or a stepwise mechanism (Bahnson et al., 2002). The apparent absence of an acidic residue in the active site raised the possibility that the observed structure might represent a dead end complex that formed a stable crystal and was not a true intermediate. In any case, the structure gave important information about the substrate binding interactions and the involvement of Glu144 and Glu164 in catalysis.

The primary ^{18}O and deuterium kinetic isotope effects, measured by the equilibrium perturbation method, in the dehydration of (3S)-3-hydroxybutyryl CoA are 1.05 for $^{18}k_{cat}/K_m$ and 1.6 for $^D k_{cat}/K_m$ (Bahnson and Anderson, 1989). The secondary deuterium isotope effects are 1.12 and 1.13 at C2 and C3, respectively, and 1.00 at C4, which is not involved in the chemical change. The magnitude of the ^{18}O effect proves that the C–O cleavage is rate limiting, and the small primary deuterium effect, as well as the secondary effect at C2, indicate that proton transfer is also at least partially rate limiting. This can happen in a concerted mechanism or when two steps have similar activation energies and both are partially rate limiting. The double-isotope method in theory allows such ambiguities to be resolved (Hermes et al., 1982). The ratio of values for the α-secondary $^D k_{cat}/K_m$ in the hydration of the alternative substrate [2-^{2}H]crotonylpantetheine in H$_2$O and D$_2$O, determined to be 1.00, indicates a concerted mechanism or equal partitioning of an α-deuterocarbanionic intermediate in H$_2$O and D$_2$O (Bahnson and Anderson, 1991).

Structure-function experiments are most consistent with a stepwise mechanism for ECH (Fang et al., 2002). The stereoselectivity for addition of water to crotonyl CoA to form (3S)-3-hydroxybutyryl CoA is $K_S K_r = 400,000$ (Wu et al., 2000). The rates at which E164Q- or E164D-ECH produce (3S)-3-hydroxybutyryl CoA are dramatically decreased without effect on the rate at which the (3R)-epimer is produced. For E164Q-ECH $K_S K_r = 0.33$ and for E164D-ECH $K_S K_r = 1000$. Mutation of Glu144 decreases the rates of formation of both epimers. Glu144 is required to produce the (3R)-epimer but Glu164 is not. The carbon acidity of 3-hydroxybutyryl-oxyCoA, in which the sulfur is replaced by oxygen, is lower 10,000-fold lower than that of the SCOI-ester. The exchange rates and hydration rate for crotonyl-oxo CoA are consistent with the stepwise mechanism in scheme 9-3 for the action of ECH.

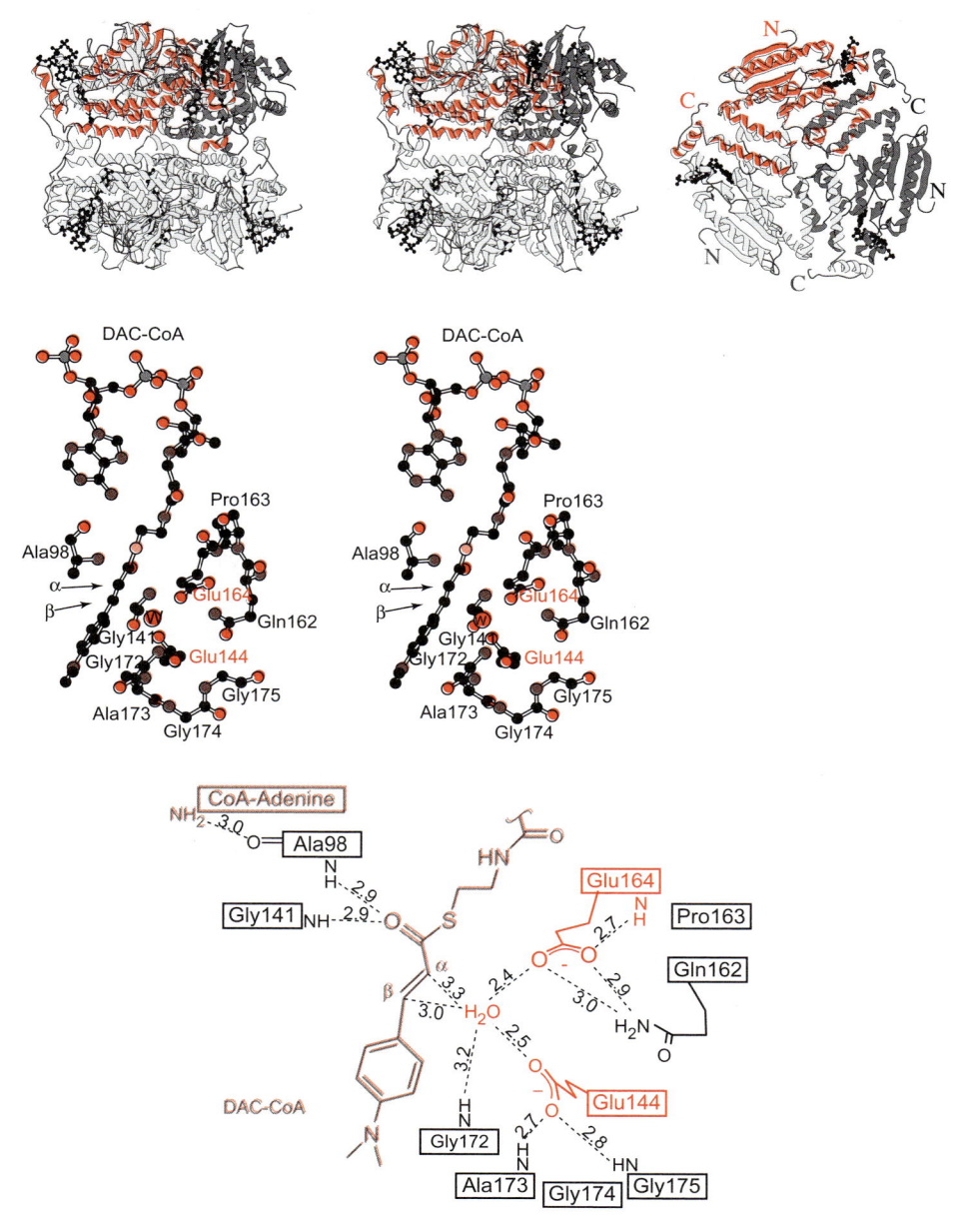

Fig. 9-3. Structure of enoyl CoA hydratase. The top panel shows an edge-on view of the dimer of trimers homohexameric assembly of *Rattus norvegicus* enoyl CoA hydratase (2.3-Å resolution; PDB 1EY3; Bahnson et al., 2002). To the right of the stereopair is a view of one trimeric half of the holoenzyme viewed from the perspective of the top down in the stereoimage. Each subunit has one molecule of 4-dimethylaminocinnamoyl CoA (DAC-CoA) bound per active site (black). DAC-CoA is as active a substrate as crotonyl CoA, but equilibrium heavily favors the dehydrated substrate over hydrated product making this an ideal compound for characterizing a catalytically relevant enzyme substrate complex. The DAC-CoA carbons α and β to the CoA carbonyl are indicated adjacent to a bound water molecule and the two active-site glutamate residues (Glu164 and Glu144). The active site is depicted in a two-dimensional scheme in the bottom panel.

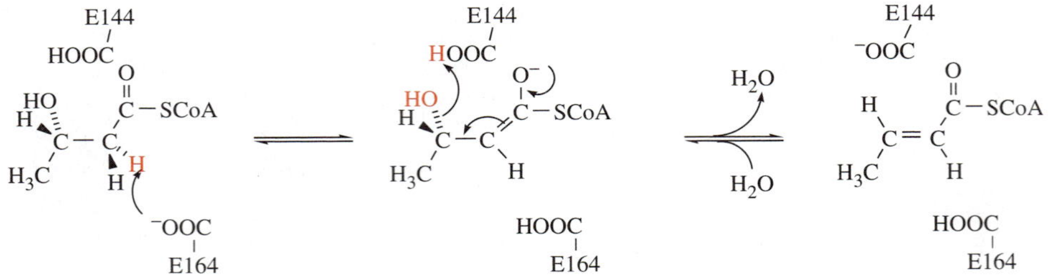

440 Enzymatic Reaction Mechanisms

Scheme 9-3

Although the α-carbon pK_a for a thioester is about 21 (see table 1-4 in chap. 1), and perhaps 20 in the 3-hydroxyacyl CoA-substrates for ECH, the enolate intermediate in scheme 9-3 must be stabilized by interactions with the enzyme to some degree if it is to be a viable intermediate in catalysis. The stabilization manifests itself in the form of electronic polarization of α,β-unsaturated acyl CoAs bound to ECH (D'Ordine et al., 1994b). Polarization is detected in the perturbations of the UV-Vis and ^{13}C NMR spectra of the bound substrates relative to the same molecules in solution. For the purpose of UV spectral analysis, the chromophoric phenyl-substituted cinnamoyl CoAs signal the electronic polarization as red shifts in their spectra on binding to ECH. Linear free energy relationships of the effects of electron donating and withdrawing substituents X on the spectra indicate that binding to ECH increases the contribution of the charge-separated resonance form to structure **9-2**.

9-2

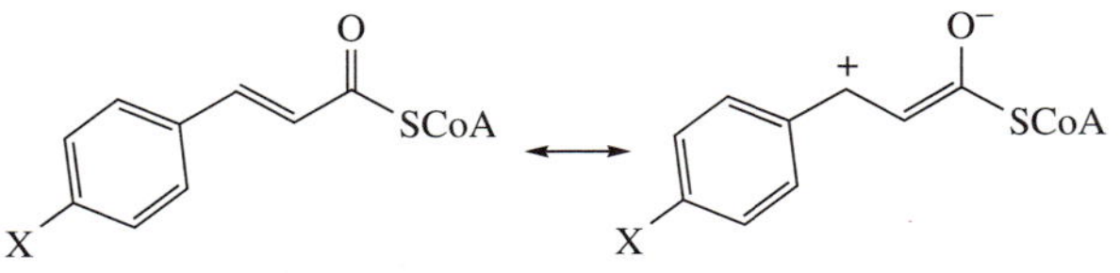

p-Substituted cinnamoyl CoA and a charge sepatated resonance form

This polarization is confirmed by the ^{13}C NMR spectra of the parent [3-^{13}C]cinnamoyl CoA, which reveal a 3.2 ppm downfield perturbation of the ^{13}C NMR signal on binding to ECH, indicating electronic deshielding at C2. Analogous experiments with [2-^{13}C]acyl CoA show upfield perturbations, with increased shielding at C2 (D'Ordine et al., 2002). Raman spectra further reinforce the conclusion that binding to ECH polarizes the π-electrons of α,β-unsaturated acyl CoAs toward the acyl oxygen. A quantitative analysis of the spectra indicates that binding to ECH perturbs the electronic structure by about 3.2 kcal mol^{-1} toward that of the charge separated structures. The enolate, with enhanced negative charge on the acyl oxygen, should experience even greater stabilization in its interaction with a stabilizing group on at the active site, perhaps a hydrogen bond donor such as Glu164 in scheme 9-3.

Cofactor-Dependent α,β-Elimination/ Addition Reactions

The chemical barriers to α,β-elimination are often overcome by the actions of cofactors such as divalent metal ions or coenzymes such as NAD$^+$ or PLP with the substrates. A divalent metal ion in coordination with a carboxylate ligand significantly increases the acidity of the α-hydrogens and so facilitates ionization to an enolate. A divalent or trivalent metal ion in coordination with the β-leaving group functions as a Lewis acid in facilitating its departure with a pair of electrons. A PLP-dependent enzyme forms an external aldimine with an α-amino acid, thereby increasing the acidity of the α-proton by about 7 pK_a units (see table 1-4 in chap. 1) and facilitating the elimination of a leaving group from the β-carbon.

Examples of cofactor-dependent enzymatic elimination are considered in the following sections.

Enolase and the Role of Mg^{2+}

Electrophilic catalysis is nowhere more clearly established than in the action of enolase in catalyzing the dehydration of 2-phosphoglycerate to phosphoenolpyruvate. The structure of enolase (EC 4.2.1.1) is shown in fig. 9-4. Enolase from rabbit muscle is a homodimeric TIM- barrel enzyme, with the active site at the N-terminal end of the barrel as in triose phosphate isomerase and other TIM barrels. Enolase may be regarded as the parent of a family of enzymes known as the enolase superfamily in which enolization is a central step in catalysis (Babbitt et al., 1996).

The enolization mechanisms in this superfamily vary over the entire range, from a metal ion–independent mechanism in the action of triose phosphate isomerase (see chap. 7) to the single Mg^{2+}–catalyzed enolization by mandelate racemase (see chap. 7) to the two Mg^{2+}–catalyzed enolization by enolase (see chap. 1). There seems to be a correlation between the carbon acid pK_a of the substrate and the number of Mg^{2+} ions involved in enolization within the enolase superfamily.

In chapter 1, we explained the role of two Mg^{2+} ions in facilitating the enolization of 2-phosphoglycerate. The active site of enolase with 2-phosphoglycerate and Mg^{2+} bound at the active site is illustrated in fig. 1-13 of chapter 1, and it clearly shows two Mg^{2+}-ions in coordination with the two carboxylate-oxygens of 2-phosphoglycerate. The crystal structure and Mn^{2+}– electron paramagnetic resonance (EPR) spectra of enolase-inhibitor complexes both reveal this interaction (Poyner and Reed, 1992; Reed et al., 1996). The crystal structure also shows a direct contact between the carboxylate group of 2-phosphoglycerate and the Lys395-ε-aminium group. The substrate-carboxylate group is directly coordinated to electrophilic species contributing five positive charges to the active site. The active site is electrostatically neutral overall because of the contributions of acidic amino acid side chains coordinated to the two Mg^{2+} ions and the three negative charges on the substrate.

The dehydration of 2-phosphoglycerate by enolase seems to follow the course outlined in fig. 9-5. Electrostatic polarization of the substrate carboxylate by the two Mg^{2+}-ions and Lys395 facilitates the abstraction of C2(H) as a proton by Lys345 to form the *aci*-carbanion.

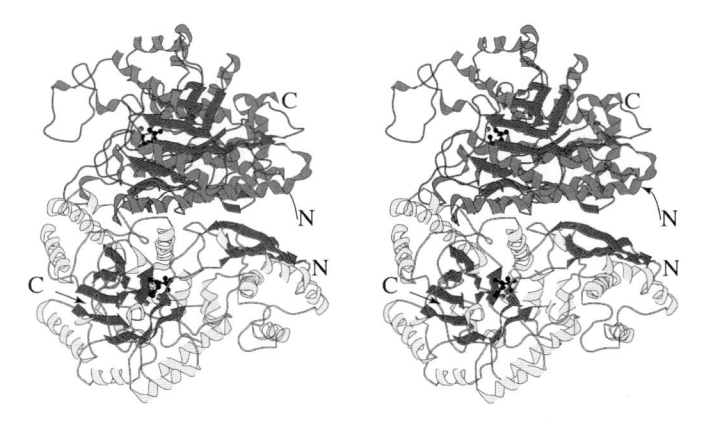

Fig. 9-4. Structure of yeast enolase. Homodimeric yeast enolase is shown as a stereoimage with 2-phosphglycerate (black) and two magnesium ions (red) at each active site (1.8-Å resolution; PDB 1ONE; Larsen et al., 1996). The β-sheet portions of each fold (brown) highlight the β-barrel domain that cradles each active site. More detailed images of the active site of this complex may be viewed in fig. 1-13.

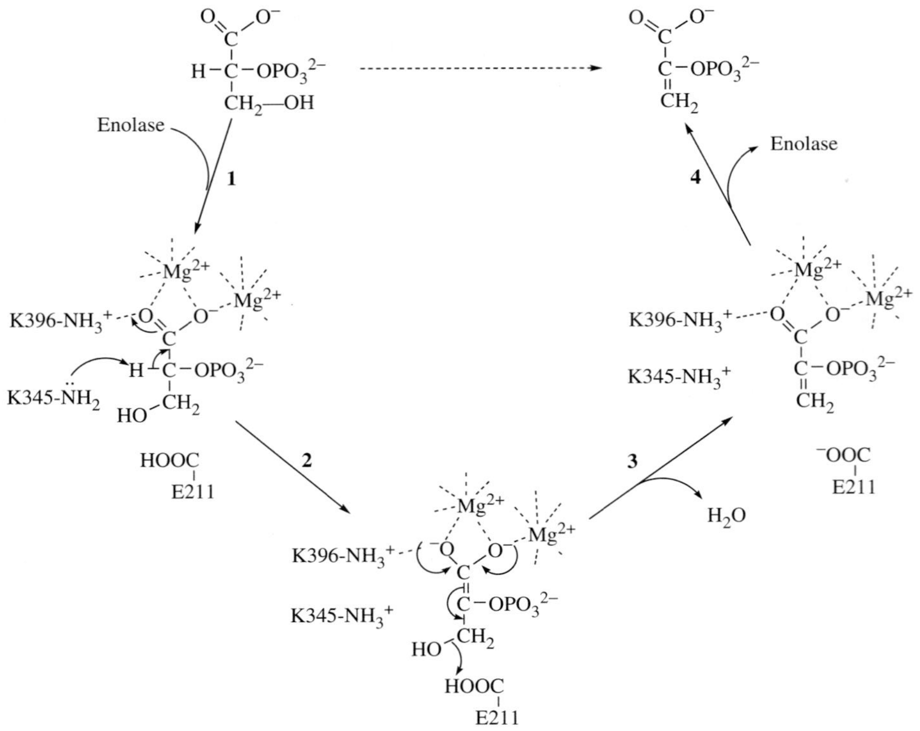

Fig. 9-5. A mechanism for the dehydration of 2-phosphoglycerate (2-PGA) in the action of enolase is extensively supported by structural and kinetic results and by structure-function studies conducted by site-directed mutagenesis. 2-Phosphoglycerate (2-PGA) binds to the site with its carboxylate group coordinated to two magnesium ions and, through hydrogen-bonded ion pairing, to Lys396. Lys345 functions as a base to abstract C2(H) as a proton from 2-PGA to form the *aci*-carbanion. The elimination of water is driven by the carbanion and further catalyzed by the carboxylic acid group of Glu211, which donates a proton to the departing hydroxyl group and forms water. The side chains of Lys345 and Glu211 are neutral in the Michaelis complex formed in step 1. They normally are ionized ε-aminium and γ-carboxylate in neutral solutions because of the values of pK_a for these groups, but the forward action of enolase involves the phenomenon of reverse protonation.

Elimination of water is driven by the relief of negative charge on the *aci*-carbanion and acid catalysis by Glu211. As shown in fig. 9-5, the side chains of Lys345 and Glu211 are both neutral (not ionized) in the Michaelis complex formed in step 1. They would normally be ionized ε-aminium and γ-carboxylate in neutral solutions because of the values of pK_a for these groups. However, the forward action of enolase involves the phenomenon of *reverse protonation* (Sims et al 2003), which is explained in chapter 2. The microscopic equilibrium in enolase includes the reverse protonated species, and in the action of enolase these are the most catalytically efficient forms. The basicity of the ε-amino group and acidity of the γ-carboxylic acid provide driving force for proton abstraction in step 2 and protonation of the leaving group in step 3. This phenomenon appears, as shown in fig. 9-5, to lead to an iso-mechanism, in which the enzyme must undergo isomerization to the reverse protonated form between turnovers.

Aconitase and the Role of [4Fe–4S] Centers

Aconitase (EC 4.2.1.3) catalyzes the conversion of citrate into isocitrate, an essential step in the tricarboxylic acid cycle of aerobic metabolism. The reaction proceeds by the elimination of the elements of water to form *cis*-aconitate as an intermediate, followed by the re-addition of the elements of water in reverse orientation to the double bond of *cis*-aconitate. Figure 9-6 describes the overall process and stereochemical consequences in the action of aconitase. The reaction is complex, and the discovery of the mechanism by which it takes place is an interesting story in mechanistic enzymology. For example, the process by which the elements of water are added to either end of the double bond of *cis*-aconitate does not involve the oversimplified pivotal motion illustrated in fig. 9-6, which is shown for the purpose of defining the mechanistic issue.

A central fact of the action of aconitase is that although the intermediate *cis*-aconitate does not normally dissociate from the active site before being converted into isocitrate, it occasionally escapes, as illustrated in fig. 9-6, and it can be found free in solution. It readily returns to the active site and is converted into either isocitrate or citrate.

Second, as in the action of aldose/ketose isomerases (see chap. 7), the tritium abstracted from the substrate as a proton by a basic group can be incorporated into the product faster than it undergoes exchange with the solvent (Rose and O'Connell, 1967) Reaction of [3-^{3}H]isocitrate leads to [2-^{3}H]citrate in the reverse reaction. Experiments show that tritium abstracted from a given molecule of a substrate can be returned to a product molecule derived from a different substrate molecule. Tritium transfer can be intermolecular as well as intramolecular. Proof of intermolecular transfer is that the aconitase-catalyzed transformation of mixtures of [3-^{3}H]isocitrate and 4-methylisocitrate produces both 4-methyl[^{3}H]isocitrate and 4-methyl[^{3}H]citrate in addition to [2-^{3}H]citrate. The simplest rationale for the label in 4-methylisocitrate and 4-methylcitrate depends on the transient dissociation of *cis*-aconitate and 4-methyl-*cis*-aconitate. Tritium abstracted from [3-^{3}H]isocitrate is retained by aconitase on dissociation of *cis*-aconitate, and it is captured by free 4-methyl-*cis*-aconitate derived from 4-methylisocitrate. Intermolecular transfer of

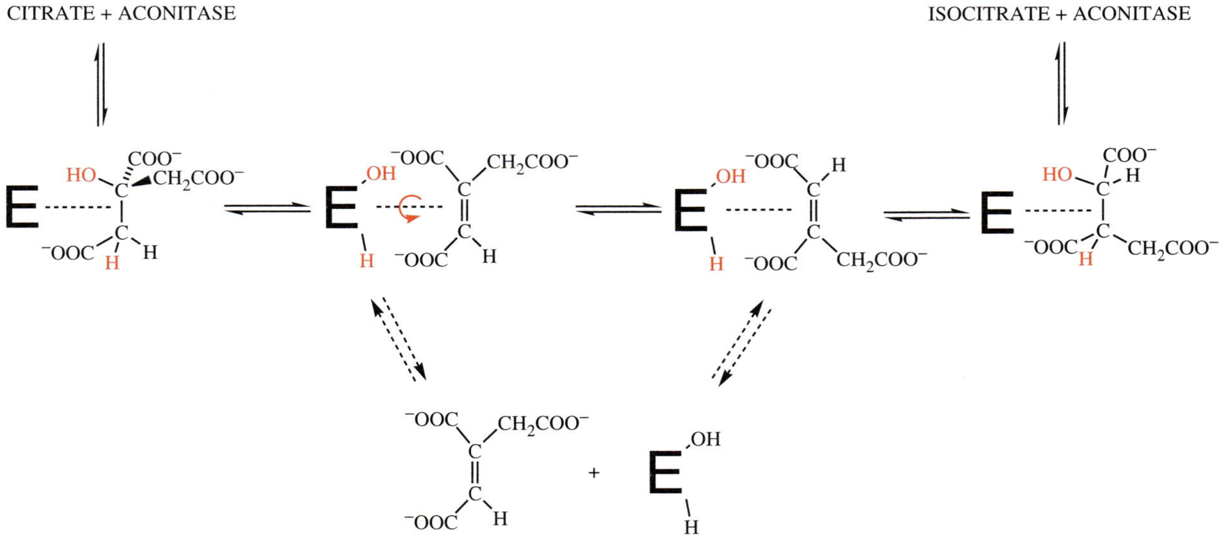

Fig. 9-6. The actions of aconitase in the transformation of citrate into isocitrate. After binding to aconitase to form the Michaelis complex, citrate undergoes dehydration to *cis*-aconitate in the active site. Water seems to be added back to *cis*-aconitate in the reverse orientation, that is, to the opposite carbon of the double bond. It seems as if aconitate could undergo a pivotal reorientation (red arrow). Addition of the elements of water after reorientation leads to the formation and dissociation of isocitrate. A further complication is the occasional, reversible dissociation of *cis*-aconitate into solution. The actual process involves the action of a [4Fe–4S] center as a Lewis acid, as illustrated in fig. 9-7.

tritium by aconitase indicates that the abstracted tritium is sterically sequestered from solvent exchange or solvent exchange is intrinsically slow.

Third, the hydroxyl group eliminated in the formation of the intermediate *cis*-aconitate is not retained in the product; reaction of [3-^{18}O]citrate produces unlabeled isocitrate. The hydroxyl group is eliminated into the solvent as water, or it is transferred to a group that undergoes oxygen exchange with water faster than the formation of product.

Aconitase is activated by ferrous ions (Fe^{2+}), so that the reaction mechanism may be expected to be analogous to that of enolase, β-methylaspartase, and aspartase. However, the function of iron in the action of aconitase is quite different. Purified pig heart and beef heart mitochondrial aconitases are iron-sulfur proteins containing the [3Fe–4S] center (Beinert et al., 1996, 1997; Kennedy et al., 1972; Ruzicka and Beinert, 1978). Activation by Fe^{2+} converts the cluster into [4Fe–4S]$^{2+}$, which participates directly in the dehydration/hydration process by the mechanism outlined in fig. 9-7. Briefly, the iron-sulfur cluster serves as a Lewis acid in facilitating the elimination of the hydroxyl group to form *cis*-aconitate. A hydroxyl group ligated to iron in the cluster is added to *cis*-aconitate in product formation. The activating Fe^{2+} occupies the unique site in the iron-sulfur cluster, as illustrated in fig. 9-7. In this site, iron is not ligated to a cysteine residue of the enzyme, and because of the absence of a cysteine ligand the iron is subject to dissociation and exchange with free ferrous ions in solution. The unique iron participates as the Lewis acid in catalysis.

Mössbauer spectroscopy, employing ^{57}Fe to discover the oxidation state of iron, first revealed the uniqueness of the fourth iron site in aconitase (Beinert et al., 1997). Biochemical studies in combination with Mössbauer analysis revealed that the iron sites in aconitase could be distinguished, both in the [3Fe–4S]-center and in the [4Fe–4S]-center. The three irons designated Fe$_b$ in the [3Fe–4S]$^+$ center could be replaced by ^{57}Fe only after all the iron had been removed from aconitase and it was then reconstituted with ^{57}Fe (Emptage et al., 1983a). In contrast, ^{57}Fe could be incorporated into the fourth site (Fe$_a$) simply by activating the enzyme with ^{57}Fe and a reducing agent. Mössbauer spectroscopy showed that Fe$_a$ (ΔE_Q of 0.83 mm/s and δ of 0.44 mm/s) and Fe$_b$ (ΔE_Q of 1.30 mm/s and δ of 0.44 mm/s) within the center were distinct and did not exchange with one another. The Mössbauer parameters for Fe$_a$ were perturbed differentially by addition of citrate or isocitrate, indicating increased ferrous character and possible direct and unique interactions with the substrate and product (Beinert et al., 1996; Emptage et al., 1983b; Kent et al., 1985).

Proof that substrates donated ligands to the fourth iron (Fe$_a$) of aconitase came from electron nuclear double resonance (ENDOR) spectroscopy and subsequently from x-ray crystallography. Hyperfine coupling between a nucleus with spin and a paramagnetic center can be detected by ENDOR spectroscopy. By using the reduced form of aconitase with the [4Fe–4S]$^+$ center as the paramagnetic probe, ^{17}O-ENDOR signals could be observed and characterized when the ligands to Fe$_a$ were labeled with oxygen-17. These signals were observed with H$_2$^{17}O and free aconitase, showing that the solvent donated a ligand to iron. Further studies with ^{17}O-labeled substrates and inhibitors showed that the β-carboxylate and HO group of the substrates also donated ligands to Fe$_a$ (Beinert et al., 1996; Kennedy et al., 1987). The ligand to Fe$_a$ derived from the solvent turned out to vary with the ligation state of iron. Experiments with ^{1}H- and ^{2}H-ENDOR spectroscopy showed that in the absence of substrate the ligand was HO$^-$ and in the presence of a substrate it was H$_2$O.

Integration and correlation of all the biochemical and spectroscopic information on aconitase made it possible to write a detailed mechanism for the role of the [4Fe–4S]$^{2+}$ center in the dehydration of steps of the aconitase reaction. The mechanism in fig. 9-7 accounts for the function of the iron-sulfur center in the dehydration of citrate and hydration of *cis*-aconitate at the active site of aconitase. In this mechanism, Fe$_a$ functions as a Lewis acid catalyst by

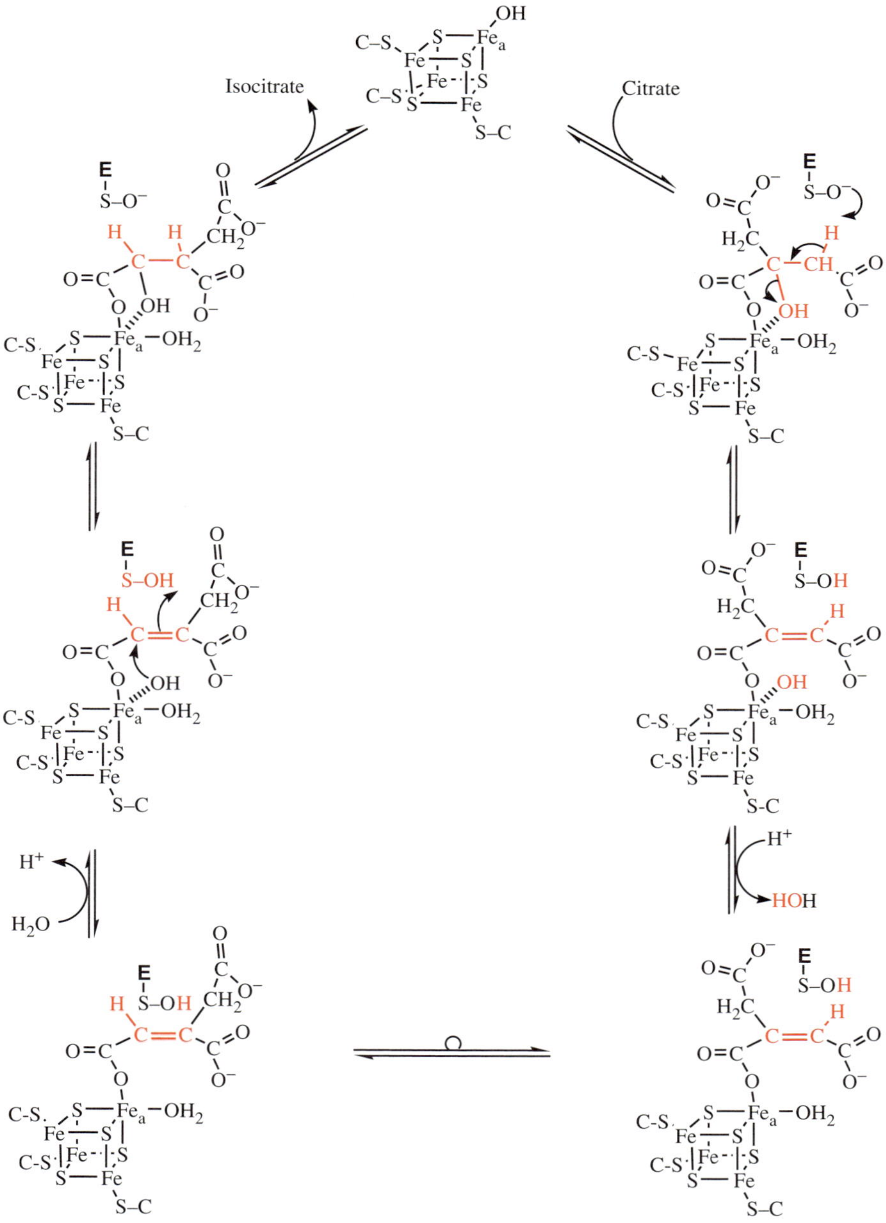

Fig. 9-7. The [4Fe–4S] center as a Lewis acid in the action of aconitase. One iron in the [4Fe–4S] center of aconitase serves as a Lewis acid in two ways. It binds the substrate by forming coordination bonds to the carboxylate and hydroxyl groups, and it facilitates the departure of the hydroxyl group in the dehydration step by serving as a Lewis acid. In many other dehydration reactions, the Brønsted acid catalysts for dehydration donate a proton to the hydroxyl group. Coordination of the carboxylate and hydroxyl groups to iron has been proven in the action of aconitase by ^{17}O-ENDOR spectroscopy of complexes of aconitase with ^{17}O-labeled substrates (Werst et al., 1990).

accepting and stabilizing the HO⁻ group in the dehydration step and facilitating its addition to cis-aconitate in the second step. In any dehydration reaction, as in the production of cis-aconitate from citrate or isocitrate, an acid is required to facilitate the departure of the OH group. A Brønsted acid can in principle fill this role by donating a proton. A Lewis acid can also serve by coordinating an electron pair of hydroxide ion (HO⁻). In the aconitase mechanism, iron (Fe$_a$) serves as a Lewis acid in facilitating the departure of the hydroxide ion.

A base is also required to abstract the β-proton and is shown as E–S–O⁻ in fig. 9-7. This base in aconitase has been shown by x-ray crystallography to be the ionized 3-hydroxyl group of Ser642. The structure of S642A-aconitase in fig. 9-8 shows the overall chain fold and the [4Fe–4S] cluster at the active site, as well as citrate coordinated with Fe$_a$. The nearest amino acid side chain to the proton abstracted in the formation of cis-aconitate would be Ser642 in the position of Ala642 in fig. 9-8. The pK_a of serine is 13.4, far too high for serine to serve as a base catalyst in neutral solution. Ser642 is in close contact with the side chain of Arg447, which presumably lowers the pK_a of the β-OH group and allows it to function as a base. The overall charge on an iron-sulfur cluster is negative (see chap. 4), and the substrate contributes three negative charges, so that the basic groups in the active site do not contribute sufficient positive charge to create an electropositive microenvironment. The only means of by which Ser642 can mediate proton transfer is through the acid strengthening effect of its hydrogen bonded interaction with Arg447. Then in the cis-aconitate intermediate state, the proton abstracted by Ser642 would be shielded from the solvent and only slowly exchangeable on the time scale of catalysis.

Cytosolic aconitase is similar in that it contains the transformable iron-sulfur center and may be catalytically active. However, in eukaryotes the tricarboxylic acid cycle takes place in the mitochondria, and there is little need for aconitase in the cytosol. Cytosolic aconitase functions principally as an iron sensor as the iron regulatory element in iron metabolism (Beinert et al., 1997).

dTDP-Glucose 4,6-Dehydratase and the Role of NAD⁺

The dehydration of dTDP-glucose to dTDP-4-keto-6-deoxyglucose by dTDP-glucose 4,6-dehydratase (4,6-dehydratase; EC 4.2.1.46) is the first step in deoxysugar biosynthesis in E. coli, and other bacteria. This is a unique dehydration. Although not an α,β-elimination overall, the step of the mechanism in which water is eliminated is an α,β-elimination. NAD⁺ plays an essential role in facilitating the α,β-elimination process, in addition to catalyzing hydride transfer from C4 of the substrate to C6 of the product. The steps of the reaction are illustrated in fig. 3-4, which emphasizes the function of NAD⁺. In the mechanism, dTDP-glucose is transiently converted into two consecutive, enzyme-bound intermediates, dTDP-4-ketoglucose and dTDP-4-ketoglucose-5,6-ene. As proved by rapid mix-quench kinetics, with detection by matrix-assisted laser desorption/ionization–time-of-flight (MALDI TOF) mass spectrometry, dTDP-4-ketoglucose-5,6-ene is a true intermediate and is kinetically competent (Gross et al., 2000). dTDP-4-ketoglucose cannot be detected in the steady state of the catalytic reaction. However, dTDP-xylose, which cannot undergo dehydration, reacts in place of dTDP-glucose with E.NAD⁺ to form the complex E.NADH.dTDP-4-ketoxylose, as determined spectrophotometrically and by x-ray crystallography (Beis et al., 2003; Gerratana et al., 2001; Hegeman et al., 2001).

Like UDP-galactose 4-epimerase, 4,6-dehydratase is homodimeric with a subunit molecular mass of 39 kDa (Glaser and Zarkowsky, 1973), and the purified enzyme contains one molecule of NAD⁺ per subunit. Like 4-epimerase, it is a member of the short-chain dehydrogenase/reductase superfamily of NAD⁺-dependent enzymes (Jornvall et al., 1995).

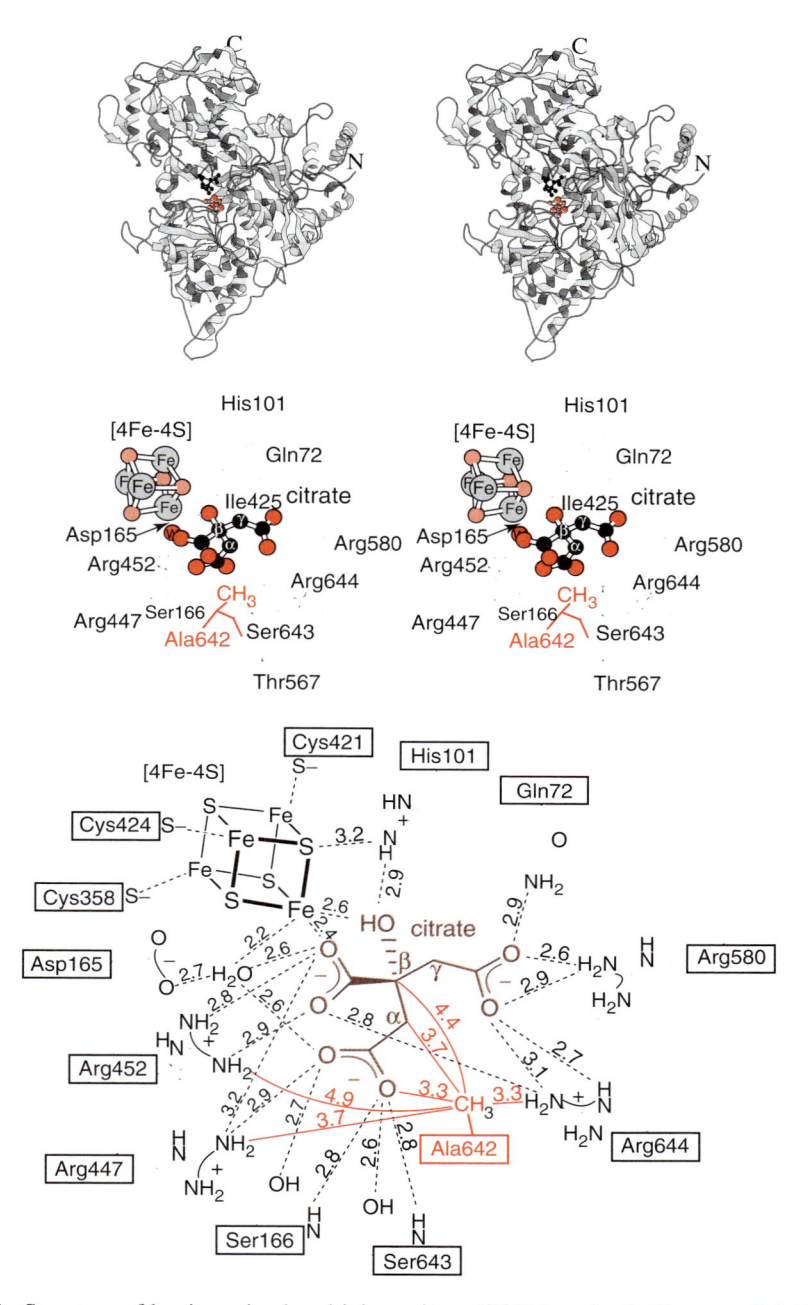

Fig. 9-8. Structure of bovine mitochondrial aconitase (S642A variant). The overall fold of the bovine mitochondrial aconitase is shown in stereo in the top image, with the [4Fe-4S] cluster in red and citrate in black (1.81-Å resolution; PDB 1E96; Lloyd et al., 1999). Details of the active site contacts are given in the second stereopair, including the site of the S642A mutation (red) adjacent to the citrate, which is ligating one iron atom of the cluster. The active site is depicted in two dimensions in the bottom panel, with interatomic distances shown in angstroms.

448   Enzymatic Reaction Mechanisms

All of these enzymes include essential tyrosine and lysine residues in a YxxxK motif, and most of them also have a serine or threonine 25 residues upstream. In UDP-galactose 4-epimerase, these residues are Ser124, Tyr149, and Lys153 (see chap. 7). In the *E. coli* 4,6-dehydratase, they are Thr134, Tyr160, and Lys164 (Hegeman et al., 2001). These conserved residues catalyze the hydride transfer steps in the mechanism. The pK_a of the conserved tyrosine is lowered to 6.4 by the positive electrostatic field created by the conserved lysyl residue and the quaternary N1 of NAD$^+$ (Gerratana et al., 2001). The resultant phenolate group of tyrosine serves as the base for abstracting the proton from the C4-hydroxyl group on hydride transfer from glucosyl-C4 to NAD$^+$. We discuss this mechanism in chapter 7 in the section on the 4-epimerase, and it is similar in the reaction of the 4,6-dehydratase (Gerratana et al., 2001).

A dehydration mechanism, as shown in fig. 9-9, involves the action of two catalytic residues acting on dTDP-4-ketoglucose at the active site. The groups are Asp135 and Glu136 and are conserved in all 4,6-dehydratases. Mutation of either one severely decreases catalytic activity (Hegeman et al., 2001a). α,β-Elimination of water between glucosyl-C5 and C6 of the substrate is potentiated by dehydrogenation to dTDP-4-ketoglucose, which makes C5(H) sufficiently acidic (pK_a ≈ 18) to be abstracted by Glu136 acting as a base. Proton transfer to the C6(OH) group by Asp135 acting as an acid facilitates dehydration to dTDP-4-ketoglucose-5,6-ene, and its reduction by 1,4-hydride transfer from NADH leads to the 4-keto-6-deoxyglucose moiety in the product.

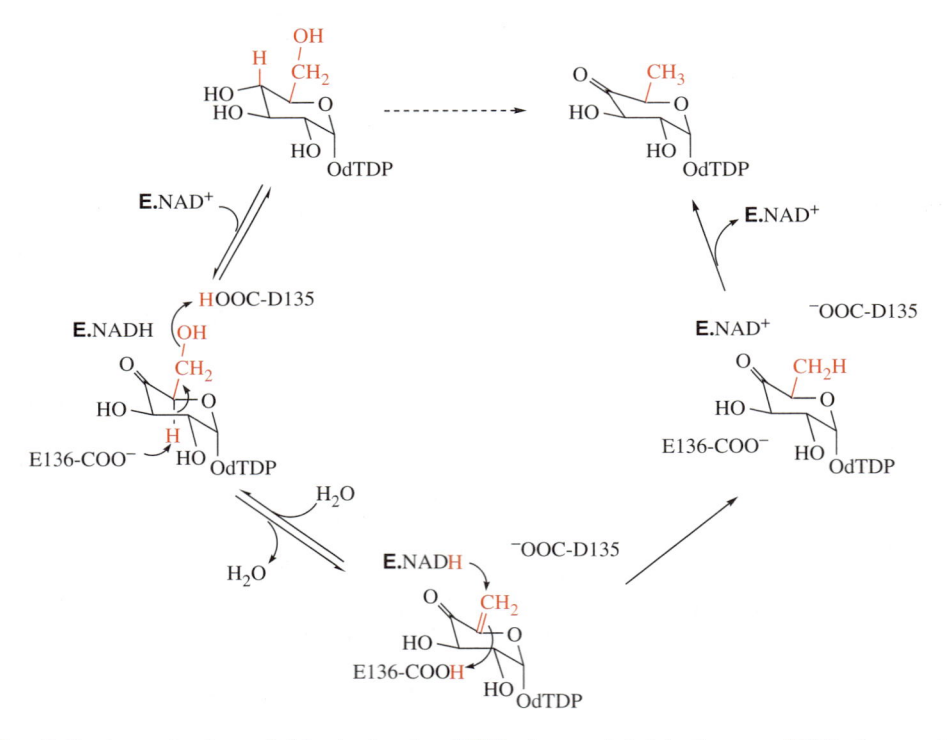

Fig. 9-9. A mechanism of dehydration by dTDP-glucose 4,6-dehydratase. dTDP-glucose 4,6-dehydratase contains tightly bound NAD$^+$ that potentiates the elimination of water by dehydrogenation at glucosyl-C4. The overall transformation of dTDP-glucose into dTDP-4-keto-6-deoxyglucose (dashed arrow) proceeds through dehydrogenation of the substrate to dTDP-4-ketoglucose (lower left in the first step). Glu136 and Asp135 functioning as acid and base, respectively, then catalyze dehydration. Hydride transfer from **E**.NADH to C6 of dTDP-4-ketoglucose-5,6-ene completes the process. Experimental evidence supports this mechanism.

The functions of Asp135 and Glu136 in fig. 9-9 are supported by unique catalytic properties of the specifically mutated enzyme (Gross et al., 2001). Wild-type 4,6-dehydratase accepts dTDP-6-deoxy-6-fluoroglucose as a substrate, in which the elements of hydrogen fluoride are eliminated instead of water. Fluoride is an excellent leaving group and does not require acid catalysis. Its reaction as a substrate should not require the action of the acid catalyst, shown as Asp135 in fig. 9-9. Accordingly, mutation of Asp135 to alanine or asparagine essentially abolishes activity toward dTDP-glucose but does not alter the activity toward dTDP-6-deoxy-6-fluoroglucose. Mutation of Glu136 to alanine or glutamine, however, severely decreases activity toward dTDP-6-deoxy-6-fluoroglucose. Another reaction catalyzed by the wild-type 4,6-dehydratase is exchange of the C5(H) in dTDP-4-ketoglucose with solvent protons. Mutation of Glu136 decreases the exchange activity by 70- to 700-fold, whereas mutation of Asp135 has very little effect on the exchange activity. These results allow the assignment of Asp135 as the acid catalyst and Glu136 as the base catalyst in dehydration.

The transient kinetic analysis with MALDI TOF detection revealed that 4,6-dehydratase catalyzed the reversal of all steps leading to dTDP-4-ketoglucos-5,6-ene (Gross et al., 2000). The exchange of C5(H) and C6(OH) of the substrate with solvent protons and oxygen could be observed by mass spectrometry. This technique was employed in a search for an enolization intermediate in the dehydration of dTDP-4-ketoglucose (Hegeman et al., 2002). This intermediate would be unmasked by the observation of C5(H)-exchange in the absence of the exchange of C6(O)—actually by the observation of faster exchange of C5(H) than of C6(O). Quantitative mass spectrometry of the residual substrate during the overall reaction in isotopically labeled water would provide information. These experiments were conducted with deuterium-labeled substrate in $H_2^{18}O$ in a search for the loss of C5(D) and incorporation of C6(^{18}O) in the residual substrate, with the results illustrated in scheme 9-4.

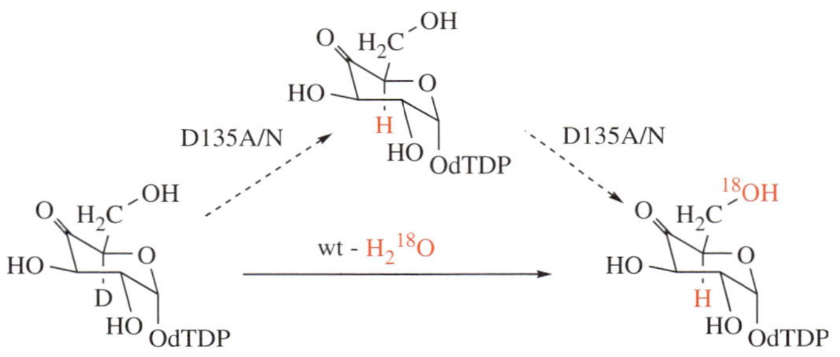

Scheme 9-4

The loss of deuterium and incorporation of ^{18}O catalyzed by the wild-type enzyme took place at the same rate, with no detectable intermediate. The same experiments conducted with D135A- or D135N-dehydratase, which were handicapped for catalyzing dehydration, revealed the exchange of C5(D) slightly faster than C6(^{18}O) incorporation (at rates several orders of magnitude slower than the wild-type enzyme). The results indicated that the dehydration step in fig. 9-9 was concerted, or the second step was several orders of magnitude faster than enolization.

The structure of dTDP-glucose 4,6-dehydratase with dTDP-xylose bound to the active site is compatible with the current thought regarding the mechanism and is shown in fig. 9-10. dTDP-xylose undergoes the dehydrogenation to dTDP-4-ketoxylose but cannot undergo dehydration (Gerratana et al., 2001; Hegeman et al., 2001). The overall structure is very similar to that of UDP-galactose 4-epimerase, a dinucleotide fold for binding the pyridine nucleotide and a smaller substrate-binding domain. The active site tyrosine, lysine, and threonine involved in hydride transfer occupy the same locations as the corresponding

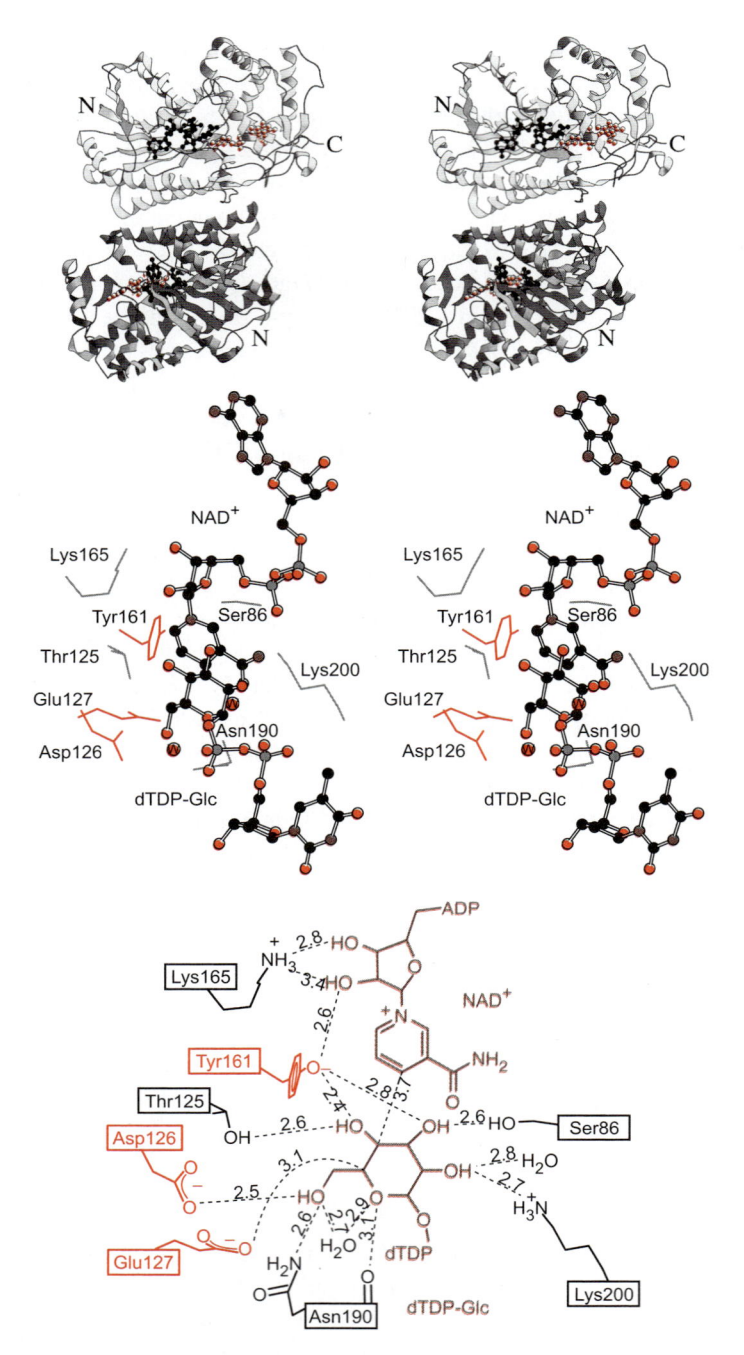

Fig. 9-10. The structure and active site of *Streptomyces suis* dTDP-glucose 4,6-dehydratase. The homodimeric *S. suis* dTDP-glucose 4,6-dehydratase is shown in the upper panel with dTDP-xylose (black) and NADH (red) bound in an abortive complex (1.5-Å resolution; PDB 1OC2; Beis et al., 2003). A complex of *S. suis* dTDP-glucose 4,6-dehydratase with substrate dTDP-glucose and cofactor NAD$^+$ bound are shown in active site in the middle panel (2.2-Å resolution; PDB 1KER; Allard et al., 2002). This complex was observed in crystals grown at low pH (pH 5.4), preventing turnover. Three residues, Tyr160, Glu127, and Asp126 (red), are involved directly in acid-base catalysis in the course of this multistep reaction. The active site is shown schematically in the bottom panel.

residues in the 4-epimerase. The conserved Glu127 and Asp136residues appear to be in position to carry out the functions assigned by the biochemical properties of specifically mutated variants. The structure at high resolution reveals additional detail (Beis et al., 2003). The nucleotide sugar is in the form of dTDP-4-ketoxylose and the pyridine nucleotide is NADH, as indicated by spectrophotometric analysis (Gerratana et al., 2001). The dihydronicotinamide ring is in a boat conformation, with the *si*-face of C4 and the 4-pro-*S* hydrogen projecting toward the keto group of dTDP-4-ketoxylose.

Pyridoxal-5′-Phosphate–Dependent α,β-Elimination

Serine and threonine dehydratases (EC 4.3.1.17 and 4.3.1.19) catalyze PLP-dependent eliminations of water from threonine and serine, respectively. The products are α-ketobutyrate

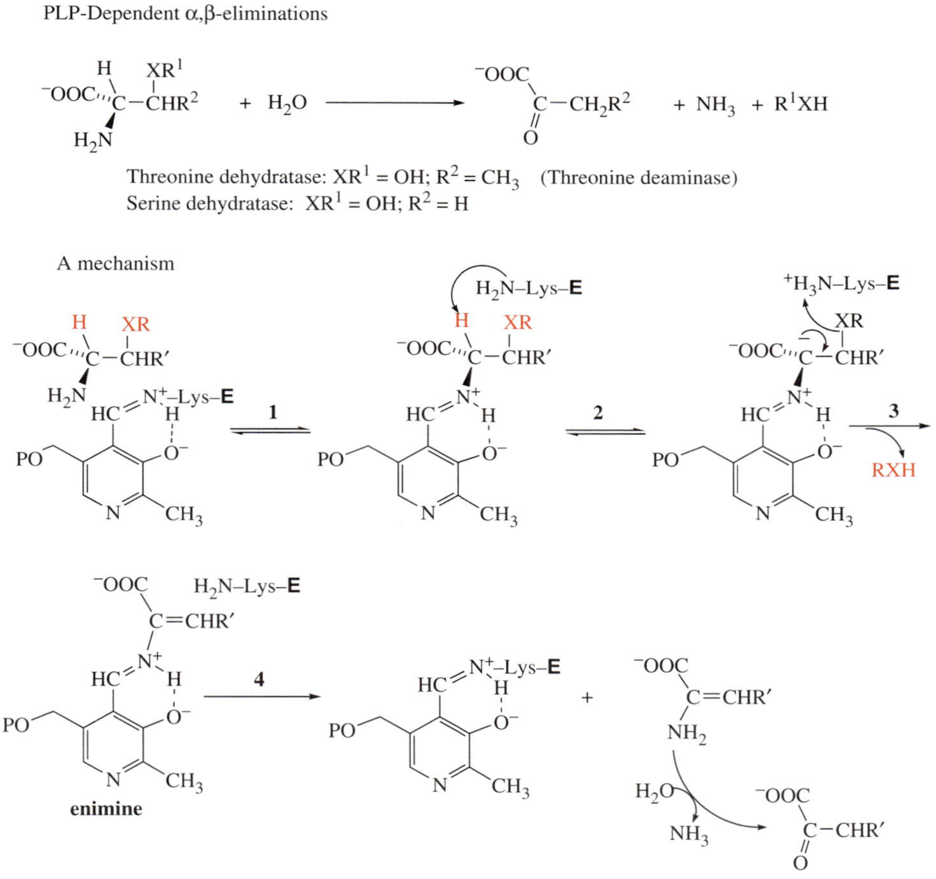

Fig. 9-11. Mechanism for catalysis of α,β-elimination by pyridoxal-5′-phosphate (PLP), At the top is the generic equation for α,β-elimination reactions of amino acids with potential leaving groups at C3, the β-carbon. Two examples are threonine dehydratase and serine dehydratase. Most enzymatic processes of this type are PLP dependent, and the chemical mechanisms are as shown, beginning with transaldimination to form the external aldimine in step 1. Ionization of C2(H) in step 2 involves the action of a base, which may be the active site lysyl residue or another base at the active site. The resulting carbanion undergoes elimination of the leaving group, which is subject to acid catalysis. Transaldimination in step 4 returns the PLP to the internal aldimine and releases the product enamine, which undergoes hydrolysis to the α-ketoacid and ammonia. Steps 1 to 3 also take place in enzymatic β-replacement reactions, such as tryptophan synthase, cystathionine β-synthase, and *O*-acetylserine sulfhydrylase (fig. 9-12).

and pyruvate, respectively, as shown in fig. 9-11. In the first step of the chemical mechanism, the internal enzyme-PLP aldimine reacts by transaldimination with the α-amino group of the substrate to form the external aldimine (see chap. 3). In the external aldimine, the pK_a of the substrate α-proton is decreased by at least 7 units (see table 1-4 in chap. 1), and in step 2 a base at the active site abstract the proton to form the a-carbanion. Step 3 is the elimination of the β-hydroxyl group by the driving force of the α-carbanion. The departure of the hydroxyl group is catalyzed by an acid at the active site. The resulting enimine (ene-imine) undergoes transaldimination in step 4 to regenerate the internal aldimine and release the amino acid enamine, which undergoes hydrolysis in solution to the α-ketoacid and ammonia. The identities of the acid and base catalysts are not certain; the active site lysyl residue could in principle fulfill both roles in the separate steps, but acid-base remains to be characterized. A structure for the internal aldimine of threonine dehydratase is available, with no substrate or inhibitor bound to the active site. No structure is available for a serine dehydratase.

Pyridoxal-5'-Phosphate–Dependent β-Replacement

In β-replacement, the β-oxygen of serine or an *O*-acylserine is eliminated by α,β-elimination and then replaced by some other group to form a new amino acid. The best known examples are tryptophan synthase (EC 4.2.1.20), cystathionine synthase (EC 2.5.1.48), and *O*-acetylserine sulfhydrylase (EC 4.2.99.8). These reactions are illustrated in fig. 9-12. Tryptophan synthase is required for tryptophan biosynthesis, and we discuss the mechanism

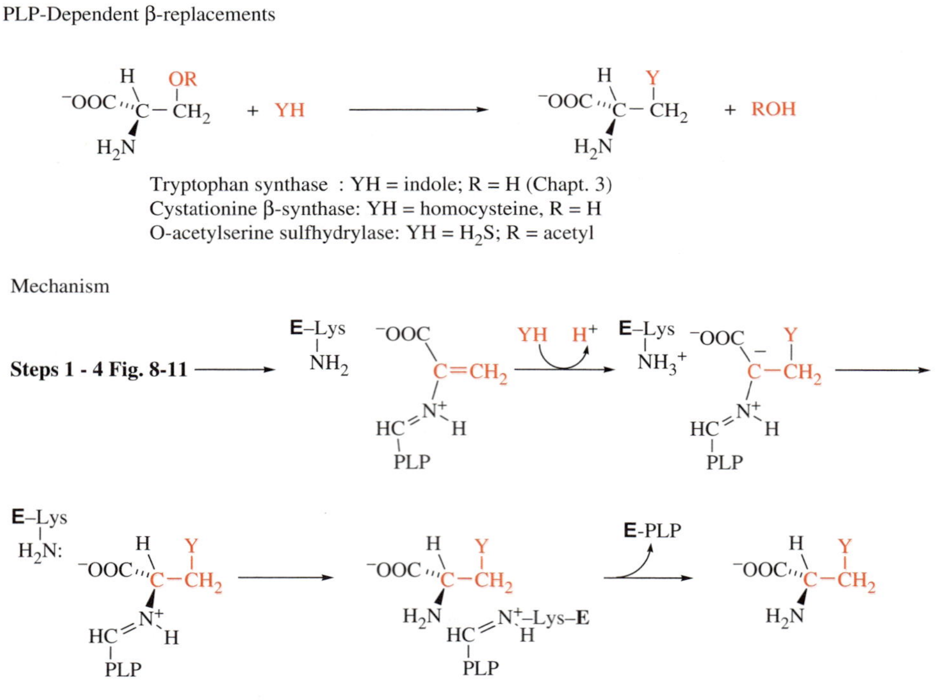

Fig. 9-12. Mechanism for pyridoxal-5'-phosphate (PLP) catalysis of β-replacement reactions. β-Replacement enzymes include tryptophan synthase, cystathionine β-synthase, and *O*-acetylserine sulfhydrylase. Serine or an *O*-acyl derivative undergoes the first four steps of α,β-elimination, as in fig. 9-11. Then, the replacement molecule YH undergoes nucleophilic addition to the intermediate from step 4 of fig. 9-11 in a series of steps that are analogous to the reversal of α,β-elimination. The resulting amino acid ligated to Y is the β-replacement product, tryptophan, cystathionine, or cysteine.

and the role of a tunnel through the protein in chapter 3. Cystathionine β-synthase and O-acetylserine sulfhydrylase are involved in cysteine biosynthesis. Cystathionine produced by cystathionine synthase can be cleaved by β,γ-elimination to cysteine, as described in the next section. O-Acetylserine sulfhydrylase produces cysteine by replacement of the O-acetyl group of the substrate with hydrosulfide (HS⁻).

β-Replacement reactions proceed by essentially the same mechanism as the α,β-elimination in fig. 9-11 forward to the enimine intermediate in step 3, followed by reversal of the preceding steps with the replacement of the original leaving group by the new group. The replacement group is indole for tryptophan synthase, homocysteine for cystathionine β-synthase, and hydrosulfide (HS⁻) for O-acetylserine sulfhydrylase. The mechanism of β-replacement is illustrated in fig. 9-12. After the first three steps of α,β-elimination, the replacement molecule reacts by addition to the enimine intermediate, in principle by the reverse of the preceding elimination step. Other steps are also analogous to the reversal of corresponding steps in α,β-elimination.

The mechanism in fig. 9-12 should mandate a ping pong bi bi kinetic mechanism for the overall process, where the central, covalently modified intermediate, designated F in chapter 2, is the enimine resulting from the elimination of the leaving group. This is the case for O-acetylserine sulfhydrylase (Tai and Cook, 2000). The steady-state kinetics of the action of tryptophan synthase is sequential and not ping pong, because the replacement group, indole, is generated from indole glyerol phosphate by the α subunit concomitant with the dehydration of serine by the β?subunit. Indole then passes through a tunnel in the enzyme to the enimine intermediate bound to PLP in the β subunit (see chap. 3).

In the reaction of O-acetylserine sulfhydrylase, hydrosulfide replaces acetate to produce cysteine. The basic steady-state kinetic mechanism is scheme 9-5, although additional interactions leading to sequential components can be observed under some conditions (Tai et al., 1993).

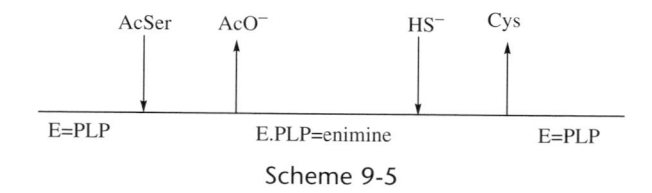

```
         AcSer    AcO⁻           HS⁻     Cys
           │       ↑              │       ↑
           ↓       │              ↓       │
   ─────────────────────────────────────────────
    E=PLP        E.PLP=enimine          E=PLP
```

Scheme 9-5

Substrate inhibition due to complexation of HS⁻ with E=PLP and AcSer with E.PLP-ene-imine is observed at very high concentrations of substrates, as is typical of ping pong mechanisms (see chap. 2). The enzyme is not specific for HS⁻ and accepts a large variety of other nucleophiles (Tai and Cook, 2000). The first half-reaction limits the rate, which displays a primary deuterium kinetic isotope effect of 2.8 on the value of k_{cat}/K_m in the reaction of O-acetyl-[2-²H]serine (Tai and Cook, 2000). Stopped-flow transient kinetic analysis, with spectrophotometric detection of intermediates, confirms that the first half-reaction limits the rate (Tai and Cook, 2000). The enimine (or aminoacrylate) intermediate can be observed in the presence of O-acetylserine in the absence of HS⁻, and it displays a spectrophotometric absorption maximum at 470 nm (Tai and Cook, 2000). Although the elimination proceeds in two steps in the generic mechanism of fig. 9-11, the departure of acetate and abstraction of the α-proton in the action of O-acetylserine sulfhydrylase are postulated to be concerted (Tai and Cook, 2000, 2001). Acetate is an excellent leaving group, and its elimination likely does not require acid catalysis. No intermediate was detected in the kinetic analysis.

The structure of resting O-acetylserine desulfurylase shows PLP bound as an internal aldimine to Lys41 (Burkhardt et al., 1998). A structure with a substrate or product bound in the active site is not available. However, a structure is available with Lys41 deleted and methionine bound to PLP as an analog of the external aldimine intermediate (Burkhardt et al., 1999), and this is shown in fig. 9-13. This structure revealed a substantial conformational difference when compared with the resting enzyme.

The structure shows the pyridine nitrogen atom of PLP within hydrogen-bonding distance of Ser272. The structures of tryptophan synthase and cystathionine β-synthase similarly show N1 of PLP hydrogen bonded to a serine residues (Meier et al., 2001). It is difficult to imagine that N1 of PLP would carry a proton and be positively charged in these structures. As in alanine racemase (see chap. 7) it appears that PLP is not a pyridinium compound in these active sites. The structures contradict expectations of a pyridinium ring in PLP that functions as an electron sink in the action of PLP, at least in the cases of alanine racemase and β-replacement enzymes. The nitrogen atom in the pyridine ring is inductively electron withdrawing, and this would increase the effect of the iminium linkage in facilitating α-carbanion formation in the external aldimine intermediates of PLP catalysis. It seems clear that the main driving force for ionization of the α-proton in the external aldimine is the iminium group itself, as shown by the effect of this group in table 1-4 in chapter 1. However, a neutral pyridine ring in PLP does not favor delocalization of negative charge in a quinonoid resonance form of the α-carbanionic intermediates. Quinonoid intermediates absorbing at about 500 nm are not observed as intermediates. In aminotransferases and related PLP-enzymes, the pyridine nitrogen of PLP is engaged in hydrogen bonded ion pairing with an aspartate β-carboxylate group, so that PLP carries a pyridinium ring that can function as a resonance electron sink to stabilize a quinonoid-type α-carbanion (see chaps. 3 and 13).

Pyridoxal-5'-Phosphate–Dependent β,γ-Elimination

Cystathionase catalyzes the hydrolysis of cystathionine to cysteine and homoserine by a β,γ-elimination mechanism facilitated by PLP. The enzyme is more formally known as cystathionine γ-lyase (EC 4.4.1.1). Cystathionine is produced by cystathionine β-synthase (EC 4.2.1.22), a PLP-dependent heme-enzyme that catalyzes the reaction of serine with homocysteine to form cystathionine. PLP in cystathionine β-synthase catalyzes the dehydration of serine and ligation of homocysteine, and heme serves as a redox sensor in regulating enzyme activity (Taoka et al., 2002). The combined actions of these two enzymes produce cysteine and homoserine from serine and homocysteine.

The production of cysteine by β,γ-elimination from cystathionine can be described by the mechanism in fig. 9-14. Tautomerization of the external PLP-cystathionine aldimine in two steps by ionization of the α-proton and its transfer to C4' of PLP moves the iminium group to the α-carbon of the homocysteine moiety. The resulting ketiminium group increases the acidity of a β-proton and allows it to be abstracted in the elimination of cysteine. Proton transfer steps lead to the external PLP-aldimine of 2-amino-2-butenoate, which is released into solution on transaldimination of PLP to the active site lysine. Hydrolysis of this enamine produces α-ketobutyrate.

Other cystathionine processing enzymes catalyze analogous reactions. Cystathionine γ-synthase (EC 2.5.1.48) catalyzes the reaction of *O*-succinylhomoserine with cysteine to produce succinate and cystathionine (Clausen et al., 1998). Unlike β-synthases that use serine, the γ-synthase binds PLP with Asp173 in a hydrogen bonded ion pair with the pyridine nitrogen of PLP. Lys198 is engaged in aldimine linkage with PLP. Cystathionine β-lyase (EC 4.4.1.8) catalyzes the hydrolysis of cystathionine to homocysteine, pyruvate, and ammonia (Clausen et al., 1996). The β-lyase binds PLP as an aldimine with Lys210 and with the pyridine nitrogen in a hydrogen bonded ion pair with Asp185.

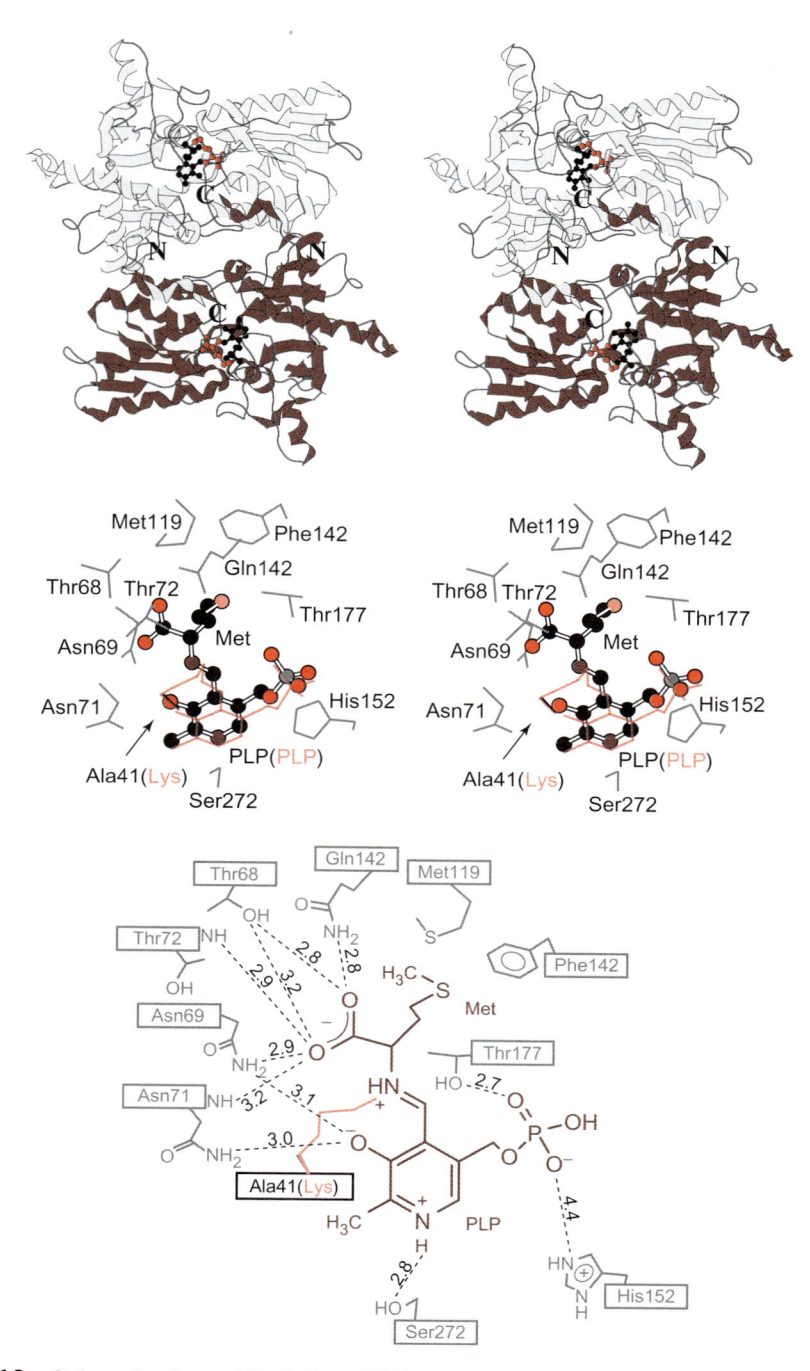

Fig. 9-13. *O*-Acetylserine sulfhydrylase (K41A variant) from *Salmonella typhimurium* is shown in the top stereoimage with pyridoxal-5'-phosphate (PLP) (black) and methionine (red) external aldimine in the active site (2.3-Å resolution; PDB 1D6S; Burkhard et al., 1999). The middle panel shows more active site details and includes a red stick model of the PLP internal aldimine structure bound to the enzyme through Lys41 align (2.2-Å resolution; PDB 1OAS; Burkhard et al., 1998). Lys41 was converted to Ala41 in the enzyme used for the external aldimine structure. A comparison of the two structure shows that the PLP tilts 13 degrees so that the C4' move toward the active site entrance (to the back in the figure). The motion accompanies changes in the protein structure, most noticeably in the "asparagine loop" (Pro67-Thr68-Asn69-Gly70), which results in a constriction of the active site entrance allowing only small molecules, such as the second substrate HS$^-$, to enter. The active site is depicted schematically in two dimensions in the lower panel.

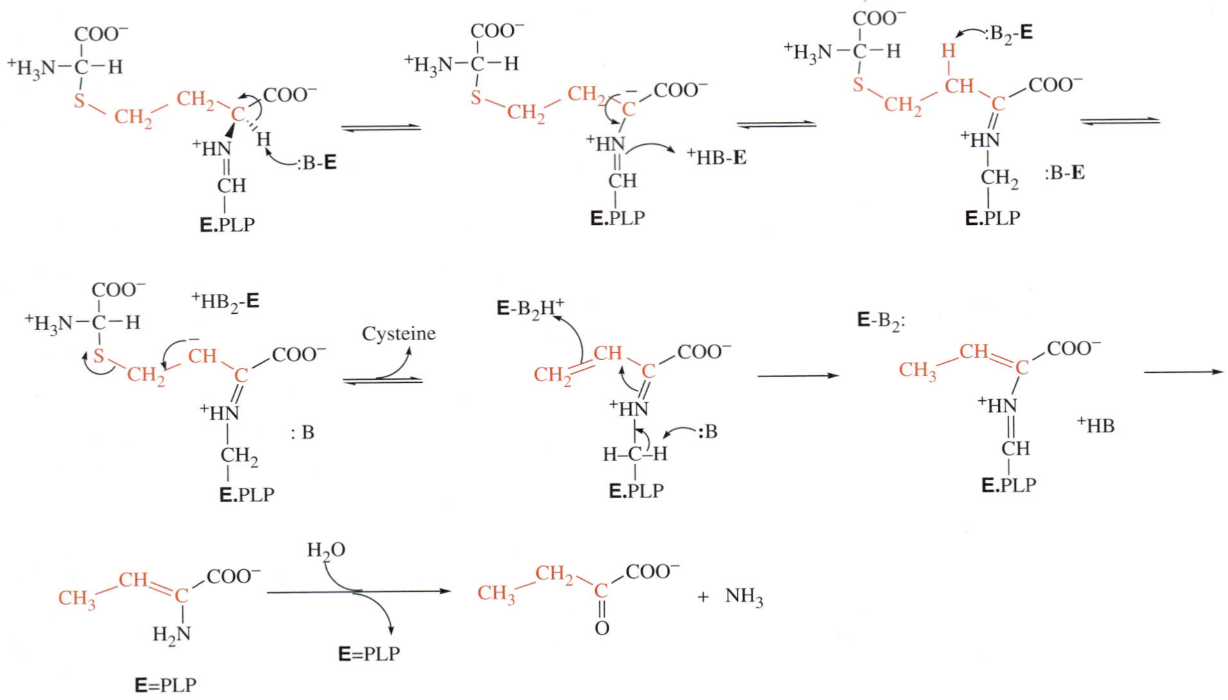

Fig. 9-14. A mechanism for pyridoxal-5′-phosphate (PLP)–dependent β,γ-elimination in the action of cystathionine γ-lyase. The substrate cystathionine forms an external PLP-aldimine with the enzyme on the homocysteinyl side of the molecule. The elimination of cysteine and subsequent production of α-ketobutyrate proceeds in several steps. **E=PLP** designates the internal aldimine, and **E.PLP** designates the external aldimine complexes.

β,α-Elimination/Addition Reactions

A few enzymes catalyze elimination/addition with abstraction of a proton β-to an activating functional group and the leaving group in the α-position. Histidine ammonia-lyase (HAL) and phenylalanine ammonia-lyase (PAL) catalyze the elimination of the α-amino groups of histidine or phenylalanine (Langer et al., 2001; Rétey, 2003). These enzymes incorporate a novel coenzymatic prosthetic group, methylidene imidazolinone, within the polypeptide chains (see chap. 3).

Methylidene Imidazolone–Dependent Elimination and Addition

Histidine Ammonia-Lyase

In the first step of the catabolic metabolism of histidine, histidine ammonia-lyase (HAL; EC 4.3.1.3) catalyzes the elimination of ammonia to form *trans*-urocanic acid according to eq. 9-3.

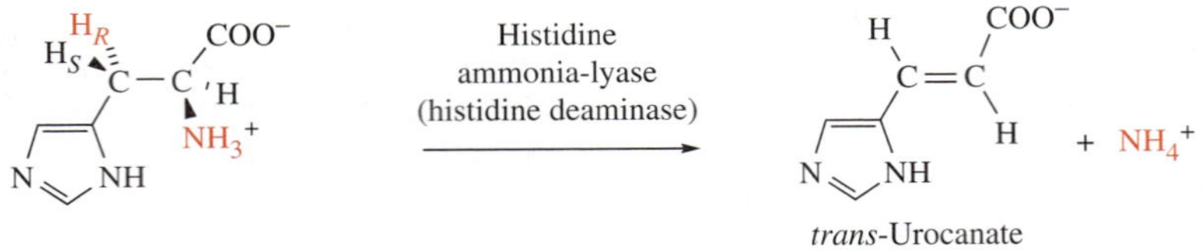

HAL is also known as histidine deaminase and histidase. Treatment with sodium borohydride inactivates HAL because of the reduction of the coenzymatic prosthetic group, which is derived from Ser143 (Langer et al., 1994).

For many years the coenzymatic group was thought to be dehydroalanine; however, the crystal structure showed it to be 3,5-dihydro-5-methylidene-4H-imidazol-4-one, or simply methylidine imidazolone, shown in fig. 9-15 with the structure of HAL (Schwede et al., 1999). Imidazolone formation arises from cyclization of the polypeptide chain through dehydration between the acyl carbonyl of Ala142 and the peptide amide of Gly144. The intervening Ser143 undergoes a further dehydration by β-elimination of its 3-hydroxyl group to generate the methylidene group. The methylidene imidazolone is subject to borohydride reduction leading to inactivation, and acid hydrolysis of the reduced enzyme generates alanine from position 143.

The most essential mechanistic problem in the action of HAL is abstraction of the β-proton, with maintenance of the α-amino group in its aminium or protonated form. A nearby base must abstract the β-proton, while being prevented from accepting a proton from the aminium group. The base must be strong to abstract the β-proton, which is not very acidic. Electron withdrawal by the sp^2-hydridized ring carbon makes the β-proton more acidic than an ordinary alkyl proton, but it is much less acidic than a proton activated by a ketone or ester group.

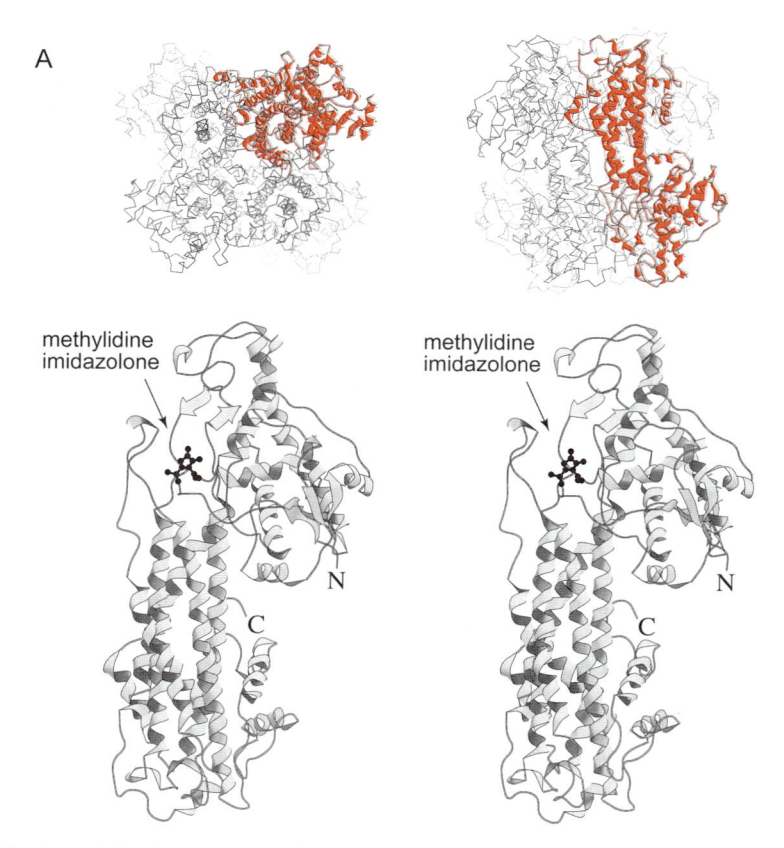

Fig. 9-15. A and B. Structure of histidine ammonia-lyase and its coenzyme methylidene imidazolone. (A) At the top of the panel are two views of the overall fold of the homotetrameric *Pseudomonas putida* histidine ammonia-lyase, with one of the monomers in red (2.1-Å resolution; PDB 1B8F; Schwede, 1999). Below those images is a stereopair showing a ribbon representation of one of the monomers, with the methylidene imidazolone moiety indicated.

Continued

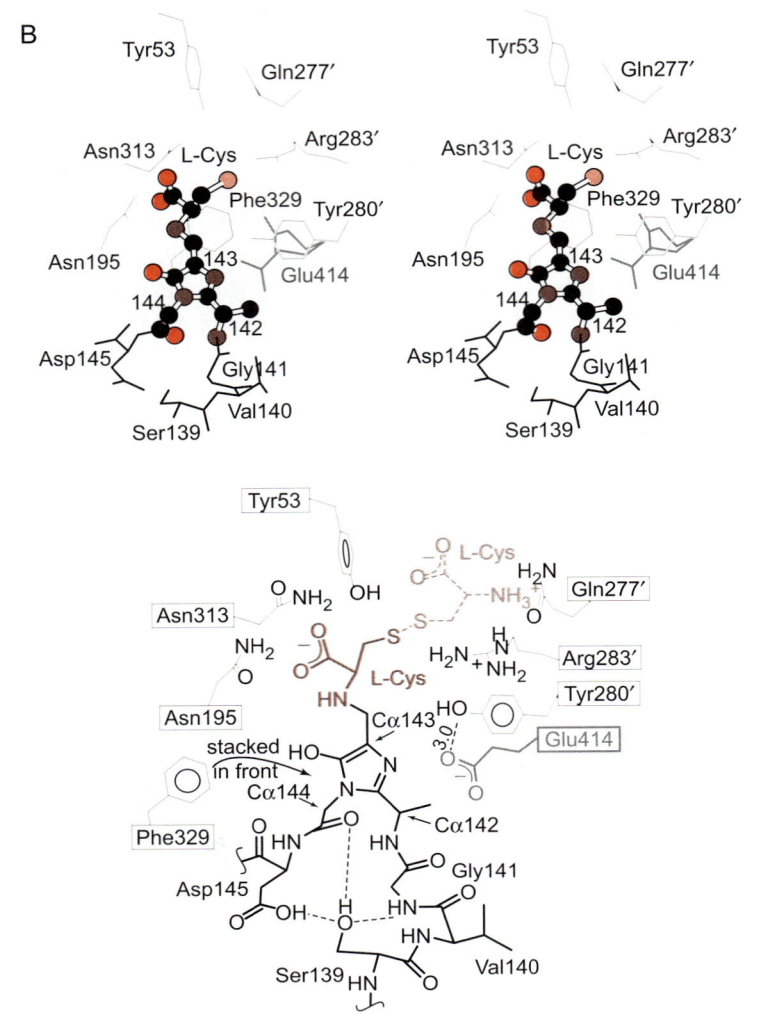

Fig. 9-15, cont'd (B) A more detailed stereodiagram of the active site shows L-cysteine covalently attached to the methylidene imidazolone (1.0-Å resolution; PDB 1GKM; Baedeker and Schulz, 2002). The methylidene imidazolone is labeled with numbers in three places, indicating the α-carbons of the amino acid residues from which the cofactor was derived. Glu414 showed partial density in two discrete conformations. The bottom panel shows the active site schematically in two dimensions. Some density adjacent to the cysteine moiety may indicate fractional occupancy by the disulfide cystine.

The kinetic mechanism is characterized by ordered release of urocanate and then ammonia. The two hypothetical mechanisms in fig. 9-16 have been suggested. In mechanism 1, the histidine-amino group undergoes nucleophilic addition to the methylidene moiety of the coenzyme. The amino group is thought to become a better leaving group, so that in the second step, the β-proton may be more easily abstracted by an enzymatic base at the same time that the amino group departs to form urocanate, which dissociates from the enzyme. The amino group is then eliminated from its adduct with the dehydroalanyl group.

In mechanism 2 of fig. 9-16, the histidine ring undergoes nucleophilic addition to the electrophilic dehydroalanyl moiety. In the adduct, the positive charge on the imidazolium ring activates the β-proton in the side chain of histidine for abstraction by an enzymatic base in concert with the elimination of ammonia. Urocanate then dissociates followed by ammonia.

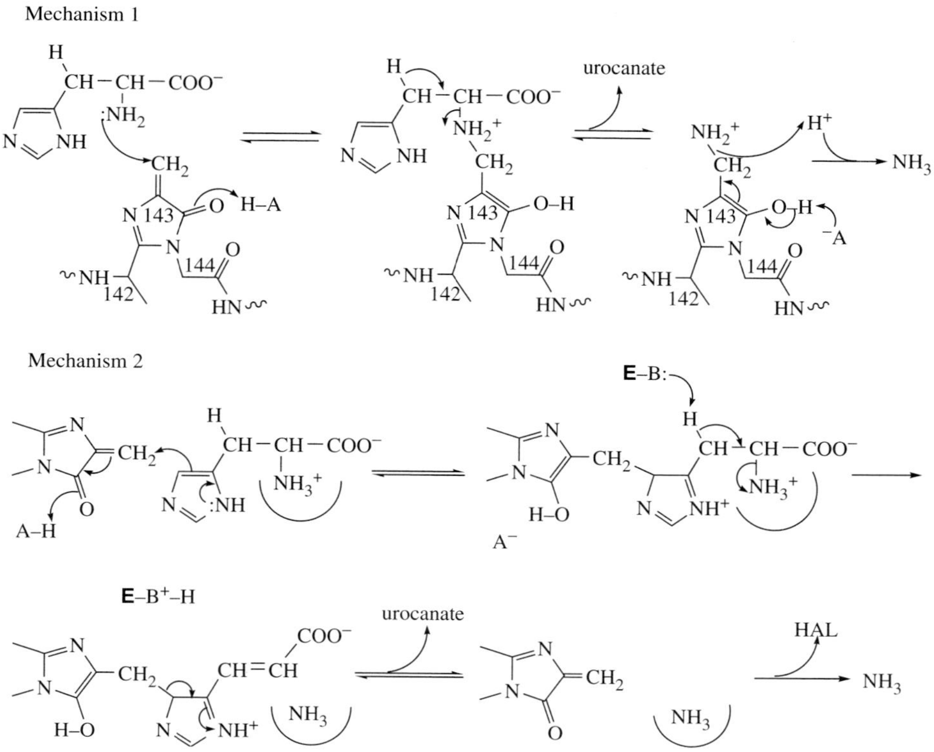

Mechanism 1

Mechanism 2

Fig. 9-16. Hypothetical mechanisms for the action of histidine ammonia-lyase. The mechanisms differ in the function assigned to methylidene imidazolone. In mechanism 1, the amino group of histidine undergoes nucleophilic addition, and its reactivity as a leaving group in the resulting complex drives its elimination. An enzymatic base simultaneously accepts the benzyl-like β-proton. In mechanism 2, the imidazole group of histidine undergoes nucleophilic addition to the coenzyme, and this process increases the acidity of the β-proton in histidine. The increased ease of β-proton abstraction becomes the driving force for elimination of the amino group.

The question of which mechanism more accurately describes the enzymatic process has been controversial. In addition to the structural differences in the role of methylidene imidazolone, the mechanistic principles underlying its action in the two mechanisms are different. In mechanism 1, it is postulated to improve the leaving group reactivity of the amino group, and this is the proposed driving force for elimination. In mechanism 2, the methylidene imidazolone increases the acidity of the β-proton in the side chain of histidine. The driving force is then the ease with which the β-proton may be abstracted in the elimination of ammonia. The fact that 2-nitrohistidine is a very reactive substrate for the borohydride-reduced HAL provides indirect support for mechanism 2. The nitro group bonded to the imidazole ring in place of C2(H) increases the acidity of the β-proton in the side chain and facilitates the elimination of ammonia. In this interpretation, the nitro group replaces the role of the methylidene imidazolone in the mechanism. In mechanism 1 the nitro group should have the effect of increasing the value of k_{cat}, although perhaps not of k_{cat}/K_m, but it does not. The situation becomes ambiguous if the dissociation of ammonia is rate limiting.

Mutation of any residue in the active site, including Arg283, Tyr53, Tyr280, Glu414, Gln277, Phe329, Asn195, and His83, significantly decreases the kinetic parameters (Rother et al., 2001). The most sensitive residue appears to be Glu414, which is postulated to carry out an essential function as a base.

Phenylalanine Ammonia-Lyase

The reaction of the plant enzyme phenylalanine ammonia-lyase (PAL; EC 4.3.1.5) is similar to that of HAL in that ammonia is eliminated to form cinnamate (eq. 9-4).

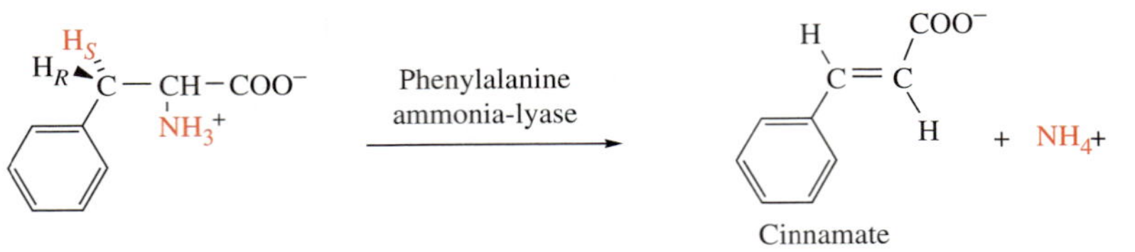

$$\text{Phenylalanine} \xrightarrow{\text{Phenylalanine ammonia-lyase}} \text{Cinnamate} + NH_4^+ \qquad (9\text{-}4)$$

Methylidene imidazolone is the coenzymatic moiety fashioned by posttranslational, autocatalytic modification of Ser202 and surrounding residues (Langer et al., 1997). The activity of PAL toward phenylalanine is abolished either by reduction of methylidene imidazolone with sodium borohydride or by site-directed mutation of Ser202 to Ala202. Chemical modification by phenylglyoxal, with substrate protection, and site-directed mutagenesis implicate Arg174 in catalytic function.

Two mechanisms similar to those in fig. 9-16 for HAL have been suggested for the action of PAL. An adaptation of mechanism 2 in fig. 9-16 to the reaction of PAL is shown in fig. 9-17. Two lines of evidence support the mechanism in fig. 9-17 (Schuster and Rétey, 1995). First, borohydride-inactivated PAL and S202A-PAL catalyze the deamination of 4-nitro-L-phenylalanine at a rate two orders of magnitude faster than the deamination of phenylalanine. Second, although L-tyrosine is not a substrate for wild-type PAL, 3-hydroxy-L-phenylalanine is a good substrate. The 4-nitro group would electronically activate the β-proton of phenylalanine for abstraction, obviating the need for the methylidene imidazolone moiety. Inductive electron withdrawal by the phenolic group in tyrosine would deactivate ring carbons 2 and 6 for electrophilic substitution by methylidene imidazolone; however, resonance electron donation by the phenolic group at ring position 3 of 3-hydroxy-L-phenylalanine would activate ring positions 2 and 6 toward electrophilic substitution. The electronic effects of ring substituents support the mechanism in fig. 9-17.

Standing against the foregoing analysis is the fact that 4-nitro-L-phenylalanine is also a substrate for wild-type PAL, and the 4-nitro group would increase the reactivity of the β-proton by either mechanism. It is difficult to reconcile the behavior of tyrosine and 3-hydroxy-L-phenylalanine with wild-type PAL in the context of a mechanism such as 1 in fig. 9-16.

The fact that 3-(1,4-cyclohexadienyl)-L-alanine is a reasonably good substrate for PAL (reacting 14-fold to 40-fold slower than phenylalanine, depending on source) raises further questions about the mechanism in fig. 9-17 (Hanson et al., 1979). In the case of the cyclohexadienyl substrate, electrophilic addition of methylidene imidazolone would lead to a localized, tertiary carbenium ion on C1 of the ring (scheme 9-6). It seems questionable whether this would be an intermediate.

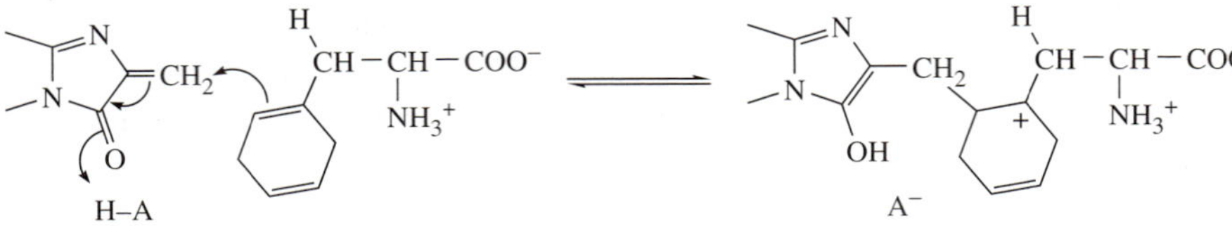

Scheme 9-6

The pH dependence in the reaction of PAL is more compatible with mechanism 1 of fig. 9-16 for the actions of PAL and HAL than mechanisms 2 in figs. 9-16 and 9-17.

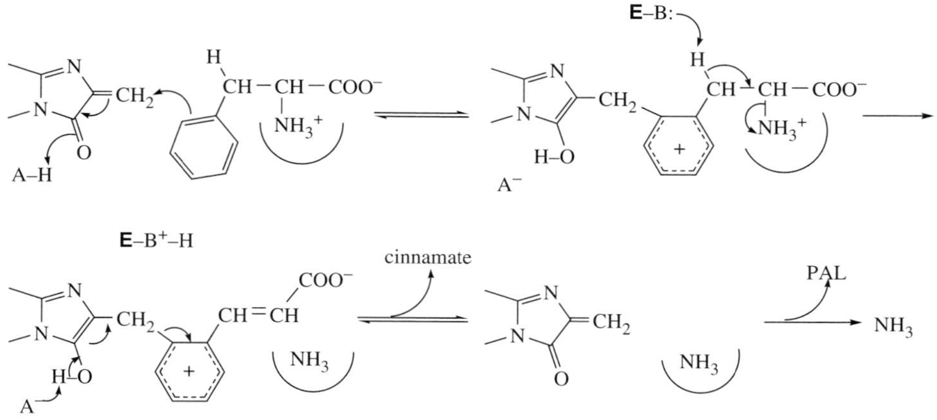

Fig. 9-17. A mechanism for the action of phenylalanine ammonia-lyase involves electrophilic attack on the aromatic ring of phenylalanine. This mechanism is analogous to mechanism 2 for histidine ammonia-lyase (fig. 9-16), in which the coenzymatic methylidene imidazolone moiety undergoes electrophilic addition to the phenyl ring of the substrate to generate a resonance delocalized carbenium ion at position 1, adjacent to the β-carbon of phenylalanine. The carbenium ion increases the acidity of the β-proton for transfer to a base, providing the driving force for the elimination of ammonia. A competing mechanism is similar to mechanism 1 in fig. 9-16 for the action of histidine ammonia-lyase.

The pH dependence for the reactions of phenylalanine and 3-(1,4-cyclohexadienyl)-L-alanine indicates that substrates react with their α-amino groups neutral, not as aminium ions (Hermes et al., 1985). The pH dependence for inhibition by cyclohexylalanine also indicates that PAL binds the monoanion. The pH dependencies are consistent with a mechanism in which the amino group undergoes addition to the electrophilic methylidene group of the coenzyme. This conclusion also follows from the pH dependence of ^{15}N isotope effects (Hermes et al., 1985).

The ^{15}N isotope effect in the reaction of the slow substrate 3-(1,4-cyclohexadienyl)-L-alanine is 0.9921 when the substrate is deuterated and 1.0047 with the undeuterated substrate. The deuterated substrate reacts with a primary kinetic isotope effect $^Dk_{cat}$ value of 2.0, which corresponds to an intrinsic effect of 4 to 6 after correction for the forward commitment (Hermes et al., 1985). The isotope effects show that the base-catalyzed elimination of ammonia is substantially rate limiting in the reaction of 3-(1,4-cyclohexadienyl)-L-alanine. The most probable mechanism is stepwise, with abstraction of the β-proton to form a discrete carbanionic intermediate. The isotope effects rule out the concerted mechanism. The isotope effects are much smaller in the reaction of phenylalanine because this step is not rate limiting.

The kinetic evidence does not support the mechanism in fig. 9-17 for PAL Moreover, the structure of a crystal formed in the presence of cinnamate and ammonium sulfate shows cinnamate at the active site and electron density attached to methylidene imidazolone that is compatible with an amino group (Calabrese et al., 2004). The crystal appears to be the product complex, in which the aminomethylidene imidazolone complex is held together by the presence of cinnamate, which according to the kinetics must dissociate before ammonia can be eliminated. The structure also suggests that the β-proton of phenylalanine may be activated for abstraction by a positive electrostatic field created by three helix dipoles oriented toward the β-carbon. All things considered, the weight of evidence favors mechanism 1 in fig. 9-16 for the actions of both PAL and HAL.

Carbonic Anhydrase

The hydration of carbon dioxide by the action of carbonic anhydrase (EC 4.2.1.1) defies assignment to a chemical category because it may be regarded as an addition of water to carbon dioxide or as a carboxylation of water. We choose to regard it as an addition, while recognizing that it is also a carboxylation. It is an apparently simple reaction that takes place spontaneously in acidic solutions to form carbonic acid, H_2CO_3 (pK_a = 6.1). In neutral solutions, carbonic acid exists as bicarbonate, and the overall reaction of carbonic anhydrase may be written as in eq. 9-5.

$$CO_2 + H_2O \underset{}{\overset{\text{Carbonic anhydrase}}{\rightleftharpoons}} \underset{HO}{\overset{O}{\underset{}{C}}}{-}O^- + H^+ \qquad (9\text{-}5)$$

In a chemical sense, the proton released can be regarded as originating with carbonic acid resulting from the addition of water to carbon dioxide, and this is true for the nonenzymatic reaction. However, in the enzymatic process, this proton arises in a different way, and its release from the enzyme constitutes a key step in the mechanism.

Carbonic anhydrases (CAs) are zinc metalloenzymes with the important function of maintaining bicarbonate and carbon dioxide at equilibrium in all cells and in the bloodstream. Carbonic anhydrase exists in animals, plants, and bacteria in distinct forms or isozymes, such as CA I through CA VII. There is no reason to expect differences in the basic chemical mechanism in the actions of the various isozymes, although they exhibit somewhat different values of kinetic parameters. Isozyme II, the most active species, is also the most thoroughly studied, and the human enzyme is known as HCA II. The kinetic parameters for HCA II are k_{cat} = 10^6 s^{-1} and k_{cat}/K_m = 1.2×10^8 M^{-1} s^{-1} (Steiner et al., 1975). The lower activities of other isozymes have inspired experimental designs to generate clues to the mechanism of action of HCA II. The value of k_{cat}/K_m places HCA II in the category of enzymes that function at the diffusional limit for enzymes in water (see chap. 2), so that one can regard it as maximally efficient.

Detailed studies showed that the reaction proceeds through eqs. 9-6 and 9-7, which can be distinguished kinetically (Lindskog and Silverman, 2000; Silverman and Lindskog, 1988).

$$\text{E-ZnOH}^- + CO_2 \rightleftharpoons \text{E-ZnOH}^-\cdot CO_2 \rightleftharpoons \text{E-ZnHCO}_3^- \rightleftharpoons \\ \text{E-ZnH}_2O + HCO_3^- \qquad (9\text{-}6)$$

$$\text{E-ZnH}_2O \rightleftharpoons H^+\text{E-ZnOH}^- \overset{B}{\rightleftharpoons} \text{E-ZnOH}^- + BH^+ \qquad (9\text{-}7)$$

The buffer components of the medium are symbolized by B and BH$^+$ in eq. 9-7. The steps in eq. 9-6 never limit the rate. The rate is limited by the first step in eq. 9-7, proton transfer within the enzyme with water ligated to zinc in place of OH$^-$ or HCO$_3^-$ at high buffer concentrations or by proton transfer to the buffer in the second step of eq. 9-7 at low buffer concentrations.

In HCA II, His94, His96, and His119 ligate zinc, and His64 serves as an acid-base catalyst. The structure of HCA II and its active site is shown in fig. 9-18. In the active site, zinc is coordinated to hydroxide, in addition to the histidine residues. An extensive hydrogen-bonding network, including Thr199, Thr200, Glu106, and a water molecule bridging it to

His64 (not shown in fig. 9-18), fixes the structure and likely participates in stabilizing and orienting the hydroxide ligand.

Extensive research on the action of HCA II and on the structures and functions of specifically mutated variants is consistent with the mechanism illustrated in fig. 9-19. Zinc in the resting enzyme has Zn^{2+}—OH^- poised to undergo nucleophilic addition to a polar double bond. An adjacent hydrophobic pocket, one side of which is defined by the carbon skeleton of Thr199, forms the carbon dioxide binding site. Water occupies this site in the resting enzyme and is displaced by carbon dioxide in the first step. The hydroxide ligand of zinc then undergoes nucleophilic addition to carbon dioxide to form bicarbonate as a ligand to zinc, with the OH group as the zinc ligand. Then a water molecule enters the site and ligates zinc, distorting the complex and moving bicarbonate aside. The dissociation of bicarbonate in the next step is accompanied by the binding of a water molecule at the carbon dioxide binding site, and the zinc-ligated water molecule enters the axial position. An internal proton transfer through the hydrogen-bonding network converts the zinc-ligated water into a hydroxide and transfers the proton to His64. In the last step, proton transfer to the basic component of the buffer completes the cycle.

Much evidence supports the mechanism in fig. 9-19. The reaction rate displays two ionizing groups with pK_a values near 7 (Steiner et al., 1975). One of the groups is assigned as the zinc-water ligand, and it accounts for the existence of the zinc-hydroxide in neutral solution. This pK_a is perturbed upward by altering the hydrogen-bonding network by site-directed mutagenesis (Krebs et al., 1993; Liang et al., 1993). Replacing zinc with other divalent metal ions also perturbs it. A linear free energy correlation of the zinc-water pK_a with activity proves the importance of the zinc-hydroxide to the mechanism (Kiefer et al., 1995).

The higher pK_a is that of His64, which is required for the internal proton transfer after the release of bicarbonate (Tu et al., 1989). This transfer is required to regenerate the zinc-hydroxide. Mutation of His64 to alanine dramatically decreases the activity, part of which can be rescued by imidazole buffers (Tu et al., 1989). The significance of His64 can be appreciated by the fact that it is replaced by Lys64 in HCA III, which is less active than CA II. Mutation of Lys64 in CA III to His64 increases the activity about 10-fold, further documenting the importance of His64 at this position (Jewell et al., 1991). However, His64 does not remove the proton from the active site, a process that requires the action of a buffer or water molecule.

Carbon dioxide binds weakly to a hydrophobic site, and molecular dynamics studies indicate it is adjacent to zinc-hydroxide and bordered by the carbon skeleton of Thr199 and hydrophobic amino acid side chains (Liang and Lipscomb, 1990; Merz, 1991). Mutation of Thr199 to enlarge the side chain decreases the size of the hydrophobic site and dramatically decreases activity (Fierke et al., 1991).

The basis of the slow release of a proton to the medium has been studied. In pursuit of the reasons for this, a detailed study of internal proton transfer in HCA III, altered by specific amino acid mutations at positions that altered the pK_a of zinc-water, gave sufficient data to construct a linear free energy correlation. Treatment of the data by Marcus theory revealed an intrinsic barrier of only 1.5 kcal mol^{-1} but the reorganization energy or work function of 10 kcal mol^{-1} (Silverman et al., 1993) (see chap. 4 for the equations of the Marcus theory). This makes good sense in terms of the mechanism in fig. 9-19, because the addition of hydroxide to carbon dioxide must be an easy reaction, whereas the binding of water and management of proton transfers must involve a good deal of reorganization.

Much structural information on carbonic anhydrases and specific mutagenic variants, as well as divalent metal ion variants, is available. Carbonic anhydrases are among the most intensively studied enzymes. The hydration of carbon dioxide is far from the only reaction catalyzed by carbonic anhydrases, which also display esterase activity. Other specialized

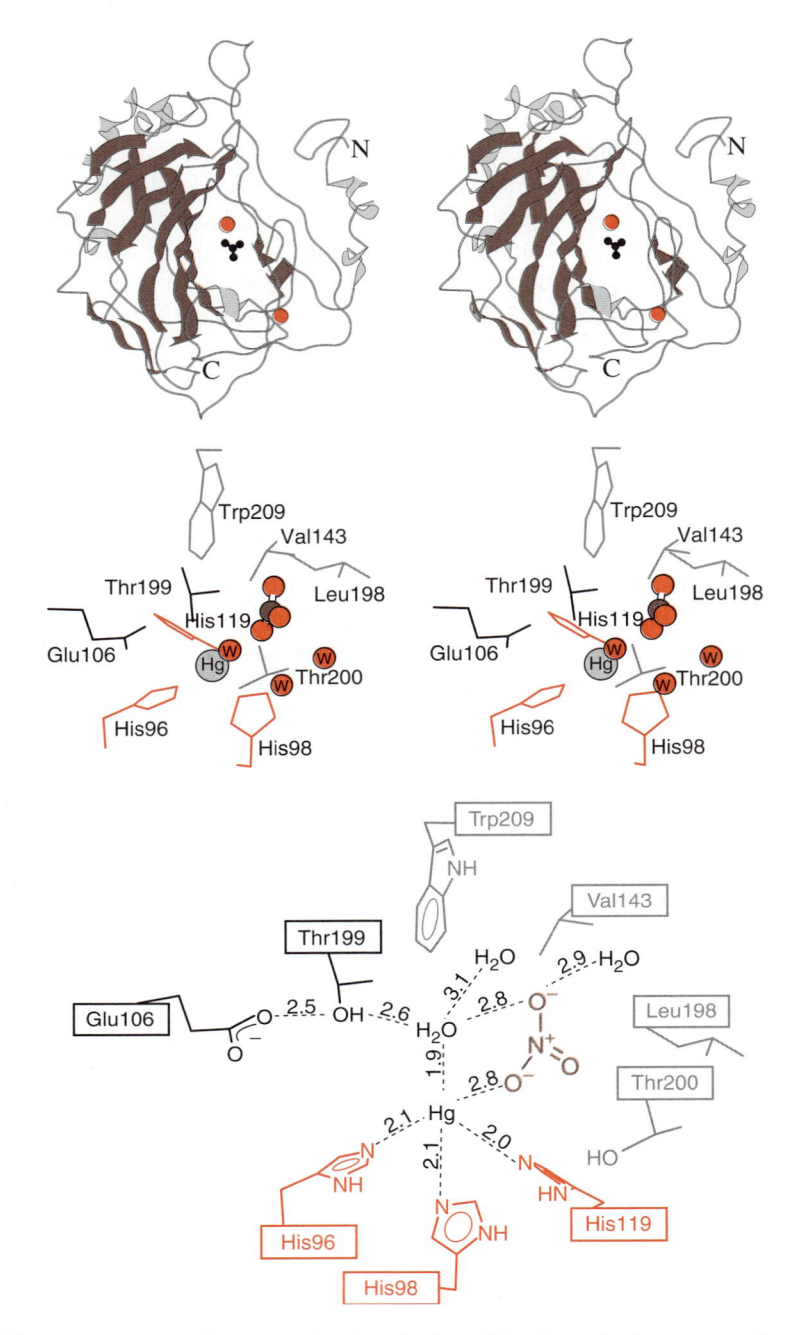

Fig. 9-18. The structure of human carbonic anhydrase II is shown in the to stereodiagram (top) with two mercury ions in red and a nitrate in black (1.9-Å resolution; PDB 1CAN; Mangani and Hakansson, 1992). The middle panel shows the active site with mercury, which does not support catalysis, and a nonreactive substrate analog that is nitrate bound. The active-site complex observed, although totally incapable of turnover, is a good approximation of the complex of product bicarbonate with the Zn^{2+}. The active-site complex is shown schematically in two dimensions in the lower panel, with distances given in angstroms.

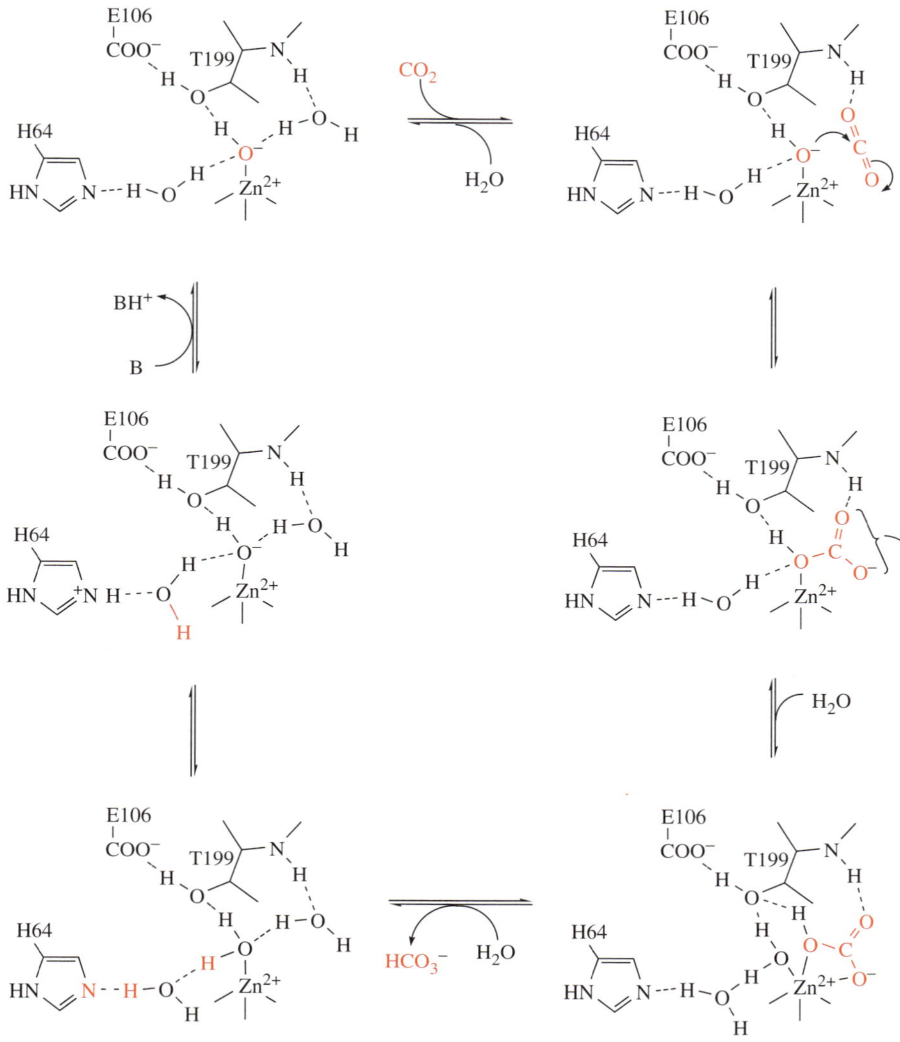

Fig. 9-19. A mechanism for the action of human carbonic anhydrase II. In the resting enzyme, zinc is coordinated to three histidine residues and hydroxide ion. His64 and Thr199 are linked to the hydroxide ligand through a hydrogen-bonding network. The reaction cycle begins in the upper left and proceeds clockwise, with the reacting species shown in red at each step. His64 and the hydrogen-bonding network participate in catalysis.

aspects of the structure and function of carbonic anhydrase are dealt with in more detail in reviews (Christianson and Cox, 1999; Lindskog and Silverman, 2000; Pocker et al., 1988).

Isomerization and Elimination

Catalytic Process

A few enzymes catalyze both isomerization and elimination in the same catalytic cycle. In the dehydration of glycols and deamination of ethanolamine, ordinary elimination is not chemically feasible because of the absence of electron withdrawing groups in the substrates that would facilitate the required proton abstraction. In these cases, the enzymes catalyze the initial isomerization of the substrate and place it into a structure that easily undergoes elimination. Although the isomerization process is the more chemically difficult, we consider these reactions as eliminations because of the overall consequences of the process.

Coenzyme B$_{12}$–Dependent Elimination

Dioldehydrase

In bacteria that metabolize glycols such as 1,2-propanediol or glycerol, the process begins with the dehydration of these molecules to the propionaldehyde and 3-hydroxypropionaldehyde, respectively. Dioldehydrase (DDH; EC 4.2.1.28) accepts ethylene glycol and 1,2-propanediol as substrates. Both (R)- and (S)-1,2-propanediol are substrates, with the ratio of k_{cat} values of (R)/(S) = 1.8. The apparent absence of stereo preference for enantiomeric substrates is exceptional among enzymes, but it is compensated in this case by the novel stereospecificities for hydrogen and hydroxyl group transfer brought about by DDH, as shown in eqs. 9-8 and 9-9.

$$\text{(9-8)}$$

$$\text{(9-9)}$$

In the reaction of (S)-1,2-propanediol, eq. 9-8, the pro-S hydrogen at C1 is transferred to the pro-R position at C2 of propionaldehyde, and the C2(OH) of the substrate is eliminated as water. The opposite takes place in the reaction of (S)-1,2-propanediol, eq. 9-9. The pro-R hydrogen at C1 of the substrate is transferred to the pro-S position at C2 of propionaldehyde, and the C2(OH) is transferred to and retained at C1 of propionaldehyde. All of this takes place by virtue of the initial formation of propionaldehyde hydrate at the active site, with stereospecific hydrogen transfers controlled by the initial configuration at substrate C2, followed by stereospecific dehydration of the hydrates (Rétey et al., 1966; Zagalak et al., 1966).

The action of DDH is adenosylcobalamin-dependent, and the transfer of C2(OH) in substrates to C1 and converse migration of a hydrogen from substrate-C1 to product-C2 is typical of the action of adenosylcobalamin (see chap. 4). The evidence implicating adenosylcobalamin as the mediator of hydrogen transfer is presented in chapter 4, where the essential properties of adenosylcobalamin and its weak Co—C5′ bond are introduced. These facts allow the minimal mechanism in fig. 9-20 to be written. Cleavage of the Co—C5′ bond and the binding of substrate lead to the complex of cob(II)alamin, the 5′-deoxyadenosyl radical, and the substrate in the active site.

The extremely reactive 5′-deoxyadenosyl radical abstracts a hydrogen atom from C1 of 1,2-propanediol to form 5′-deoxyadenosine and the substrate-related C1-radical. The initial radical undergoes an isomerization, in which the C2(OH) migrates to C1 and the unpaired electron to C2. The product-related C2-radical abstracts a hydrogen atom from 5′-deoxyadenosine to regenerate the coenzyme and propionaldehyde hydrate, which undergoes dehydration release as propionaldehyde by the action of DDH. Very large deuterium

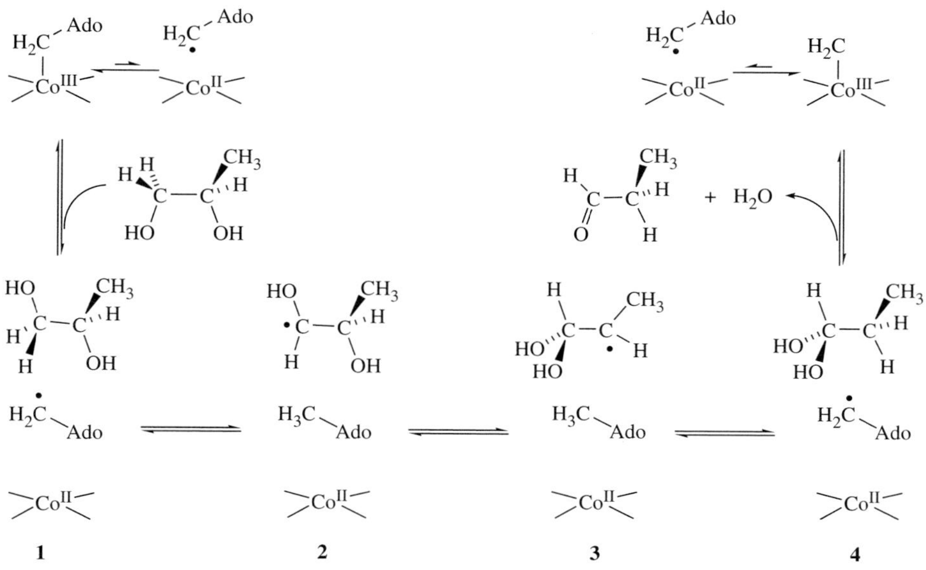

Fig. 9-20. In the mechanism for the function of adenosylcobalamin in the reaction of dioldehydrase, the abbreviated structure of adenosylcobalamin at the top represents the complex of dioldehydrase (DDH) with the coenzyme, and the Co—C5' bond is strained by enzyme-coenzyme binding interactions. The strain may be increased to a small degree when 1,2-propanediol binds to the active site. The 5-deoxyadenosyl radical generated transiently by homolytic cleavage of the Co—C5' bond in complex 1 abstracts a hydrogen atom from C1 of the substrate to form 5-deoxyadenosine and the substrate radical in complex 2. Rearrangement of the radical leads to the product-related radical in complex 3, which abstracts a hydrogen atom from 5-deoxyadenosine to form propionaldehyde hydrate in complex 4. DDH catalyzes dehydration of propionaldehyde hydrate, and propionaldehyde undergoes dissociation as the Co—C5' bond is restored. There is evidence for the presence of complexes 1 and 2 in the steady state.

and tritium kinetic isotope effects characterize the hydrogen transfer processes. The kinetic isotope effect for the overall reaction of 1,2-[1-^{2}H]propanediol with deuterium in the transferring position is $^Dk_{cat} = 12$ (Frey et al., 1965). The kinetic isotope effect for the transfer of tritium from [5'-^{3}H]adenosylcobalamin to propionaldehyde is 83 (Essenberg et al., 1971). The large tritium isotope effect is regarded as a composite of secondary effects and hydrogen tunneling.

Evidence in support of this mechanism, in addition to the hydrogen transfer function of adenosylcobalamin, includes the observation by EPR of low-spin cob(II)alamin and a substrate derived radical as intermediates in the steady state (Schepler et al., 1975; Valinsky et al., 1974). The EPR spectra of cob(II)alamin at the active site of DDH and of the substrate-based radical and cob(II)alamin at the active site of DDH in the steady state of the reaction of 1,2-propanediol are shown in fig. 9-21. The structure of the radical intermediate is shown in complex 2 of fig. 9-20, as proved by the line broadening effect of 1,2-[1-^{13}C]propanediol and the absence of such an effect with 1,2-[2-^{13}C]propanediol (Toraya, 2003). Analysis of the weak spin coupling between low-spin cob(II)alamin and the radical in complex 2 indicates a separation of about 9 Å (Schepler et al., 1975).

The preceding complex 1, with cob(II)alamin and the 5'-deoxyadenosyl radical, cannot be observed in the steady state, presumably because of the high energy and extreme reactivity of the primary radical. However, an allylic analog of this radical is observed at the active site of DDH with 3',4'-anhydroadenosylcobalamin as the coenzyme (Reed and Mansoorabadi, 2003). The allylic analog is the 3',4'-anhydro-5'-adenosyl radical shown on the right side

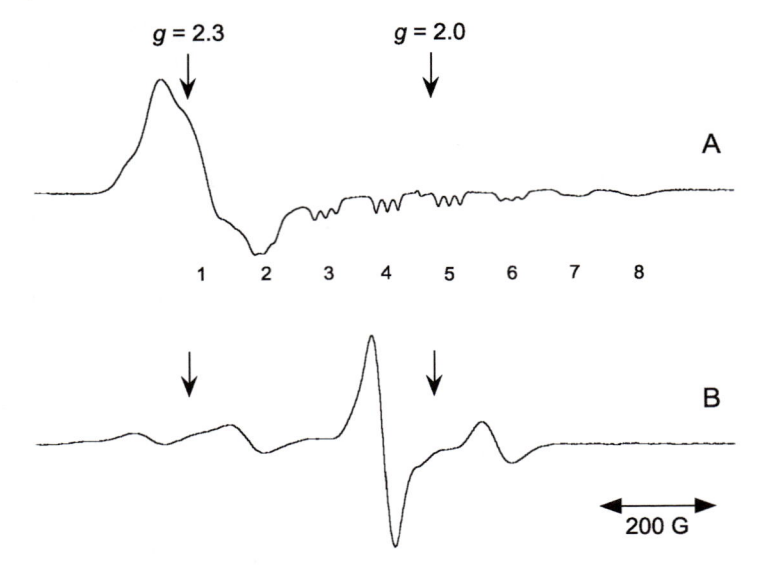

Fig. 9-21. Electron paramagnetic resonance (EPR) spectra of cob(II)alamin and the substrate radical at the active site of dioldehydrase. (A) The EPR spectrum is that of cob(II)alamin at the active site of dioldehydrase and recorded at 4°K. Notice the eight triplets due to nuclear hyperfine coupling of the unpaired electron of cobalt with ^{14}N in the nitrogen ligands. The eight triplets become doublets when the lower axial ligand of adenosylcobalamin is ^{15}N from [$^{15}N_2$]dimethylbenzimidazole, proving that dimethylbenzimidazole remains the lower ligand at the active site (Abend et al., 1999), unlike the cases of methylmalonyl CoA and glutamate mutases (see chap. 7). (B) The EPR spectrum is that of a complex of dioldehydrase, adenosylcobalamin, and 1,2-propanediol frozen in the steady state at 4°K. The signal at g = 2 is that of a radical derived from the substrate, and that at g = 2.3 is derived largely from that of cob(II)alamin. The unpaired electrons of the organic radical and cob(II)alamin are spin coupled, resulting in the attenuation of their individual features.

of scheme 9-7. The allylic radical reacts at only 0.1% the rate of adenosylcobalamin because of its stability, so that it is the predominant intermediate in the steady state (Magnusson and Frey; 2002).

Scheme 9-7

Coupling between the allylic radical and low-spin cob(II)alamin led to a complex EPR spectrum and indicated a very close contact. Analysis of the spectra obtained in the presence and absence of 1,2-propanediol by simulation led to the structural models in fig. 9-22 (Reed and Mansoorabadi, 2003). The fact that the analog underwent cleavage to cob(II)alamin and the allylic nucleoside radical in the absence of the substrate meant that the binding interactions between DDH and 3′,4′-anhydroadenosylcobalamin were sufficient to cleave the Co—C5′ bond (24 kcal mol^{-1}). The distances separating the unpaired electron in the allylic radical and cobalt in low-spin cob(II)alamin were 3.5 Å or 3.4 Å in the presence or absence of 1,2-propanediol, respectively.

Substrate-independent cleavage of the unsaturated coenzyme analog by DDH stands in contrast to the substrate-dependent cleavage of adenosylcobalamin (Wagner et al., 1967). Apparently, the substrate dependence for cleaving the Co—C5′ bond in adenosylcobalamin represents capture of an undetectable equilibrium concentration of the 5′-deoxyadenosyl radical by hydrogen abstraction from the substrate to form the more stable substrate-C1 radical, in which the p-radical orbital overlaps the doubly occupied nonbonding orbitals of the C1-hydroxyl group. In the case of 3′,4′-anhydroadenosylcobalamin, allylic stabilization of the 3′,4′-anydro-5′-adenosyl radical shifts the equilibrium toward Co—C5′ cleavage, so that it is observed spectroscopically irrespective of the presence of 1,2-propanediol.

DDH is a dimer of trimeric units composed of three types of subunits—α, β, and γ—and may be formulated as (αβγ)$_2$. The structure of DDH in fig. 9-23 shows the binding site for cobalamin at the interface of subunits, with the nucleoside binding to the TIM barrel of the α subunit and the lower dimethylbenzimidazole ligand interacting with the β subunit (Toraya, 2003). The structure confirms the conclusion, based on EPR spectroscopy, that the lower axial ligand of adenosylcobalamin is the dimethylbenzimidazole moiety and not a protein ligand (Abend et al., 1998; Yamanishi et al., 1998).

The structure of DDH places C1 of the substrate at a distance of 9 Å from cobalt in cob(II)alamin, the same distance as reported for the substrate related radical. This distance, considered together with the placement of the adenine ring and the structural constraints within the ribosyl moiety, imply the simple model shown in fig. 9-24 for chemical communication between the Co—C5′ bond and the substrate. According to this model, cleavage of the Co—C5′ bond is followed by a rotation of 94 degrees about the N-ribosyl bond of 5′-deoxyadenosyl radical and places the unpaired electron at carbon-5′ into contact with

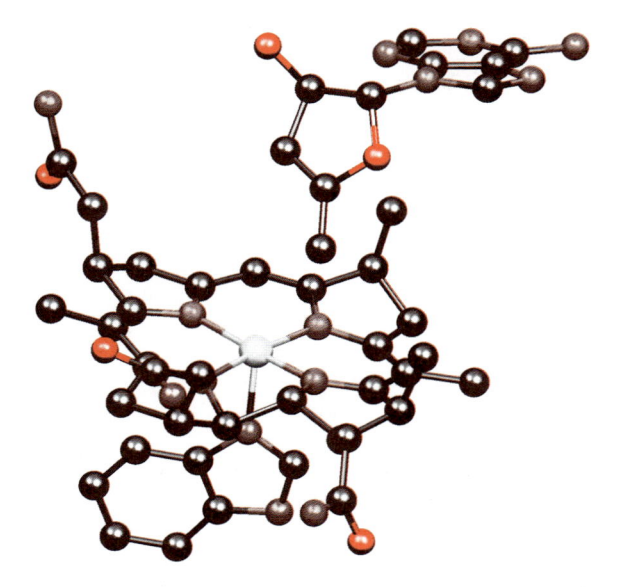

Fig. 9-22. Structures of the allylic 5′-deoxyadenosyl radical and cob(II)alamin at the active site of dioldehydrase.

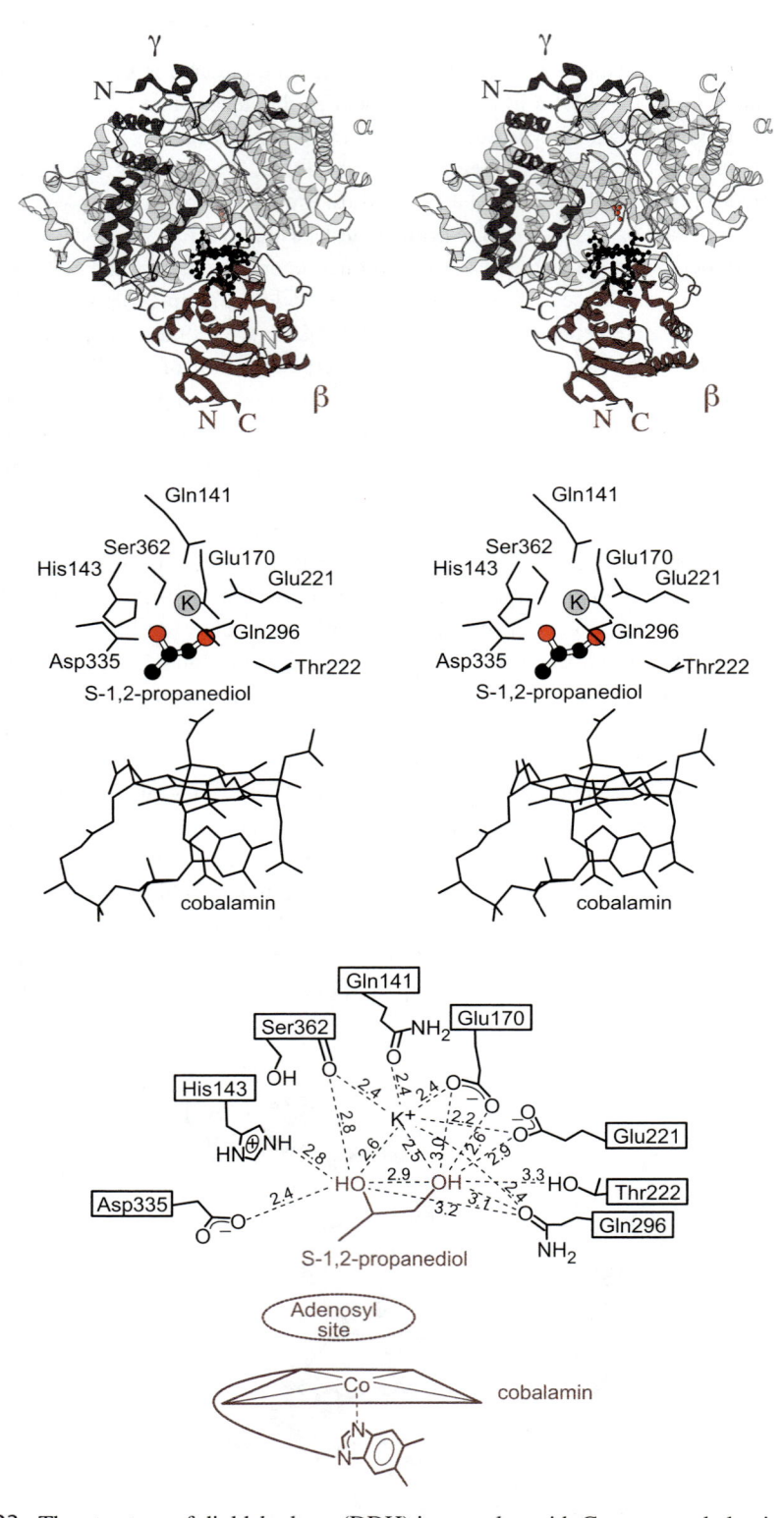

Fig. 9-23. The structure of dioldehydrase (DDH) in complex with Co-cyano cobalamin and S-1, 2-propanediol. *Klebsiella oxytoca* DDH is shown in the top stereodiagram (1.8-Å resolution; PDB 1UC4; Shibata et al., 2003). The enzyme is composed of α, β, and γ subunits with the adenosyl cobalamin (cyanocobalamin in black) bound at the α/β subunit interface and the diol binding site sequestered within the α subunit (propane diol in red) above the adenosyl cobalamin. The middle stereodiagram shows the active site in more detail, with the monovalent metal binding site (occupied here by K+) adjacent to the propane diol. The space between the diol and the corrin is presumably the adenosine binding pocket, as is shown in the two-dimensional active-site scheme (bottom).

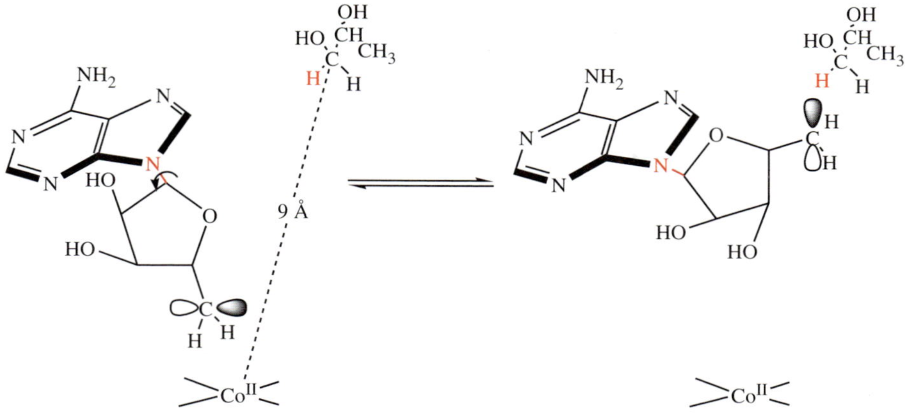

Fig. 9-24. Mechanism for translocation of the 5-deoxyadenosyl radical in dioldehydrase. The analysis of weak coupling between the substrate-related radical and low-spin cob(II)alamin within complex 2 of the mechanism in fig. 9-20 indicated a separation of about 9 Å, and the same distance was found between cobalt and 1,2-propanediol in the structure of dioldehydrase (DDH) (see fig. 9-23). A mechanism for translocating C5' of the 5'-deoxyadenosyl radical to the substrate must therefore exist. The simplest mechanism is that shown here, a 94-degree rotation about the N-ribosyl bond. This motion does not violate van der Waals contacts in the active site and allows perfect placement of C5' to abstract the hydrogen atom from the substrate. The same motion in reverse would allow the Co—C5' bond to be regenerated in complex 4 of fig. 9-19 after isomerization of the substrate radical and hydrogen transfer from 5-deoxyadenosine to the product-related radical.

the C1(H) of the substrate (Toraya, 2003). This mode of chemical communication nicely explains many of the biochemical and structural results.

The question of the mechanism for hydroxyl group transfer in the interconversion of complexes 2 and 3 in fig. 9-20 has been considered. The two main issues have been whether the migration of the hydroxyl group is acid catalyzed or base catalyzed and whether potassium ion plays a chemical role. A chemical model was available for the acid-catalyzed elimination of the hydroxyl group from the ethylene glycol radical (Walling and Johnson, 1975). In this model, the radical was initially generated by hydrogen atom abstraction from ethylene glycol by the hydroxyl radical (HO•). In acetic acid, the elimination of water eventually led to acetaldehyde through a radical chain mechanism, presumably by way of an oxycation radical intermediate. Adaptation of this model to the action of DDH for the generation of the ethylene glycol radical led to mechanism A in fig. 9-25.

In a theoretical treatment, it was concluded that isomerization at the active site of DDH proceeds through a cyclic, hydrogen bonded transition state in a concerted mechanism, as illustrated in mechanism B of fig. 9-25 (Smith et al., 1999). Mechanisms A and B in fig. 9-25 differ in the interactions of the migrating water molecule with the carbon radical skeleton. However, they become similar when it is recognized that in the transition state for mechanism B, the migrating oxygen is separated from C1 and C2 by van der Waals contact distances and consists of the elements of water in a hydrogen bonded state. This mechanism has been characterized as being facilitated by partial proton transfer (Smith et al., 1999).

The ethylene glycol radical can also be converted into acetaldehyde nonenzymatically by base catalysis, in a radical chain process in which hydroxide ion is eliminated (Bansal et al., 1973). However, this mechanism is not adaptable to the re-addition of hydroxide to form acetaldehyde hydrate, a process that is required to explain the stereochemistry in the action of DDH.

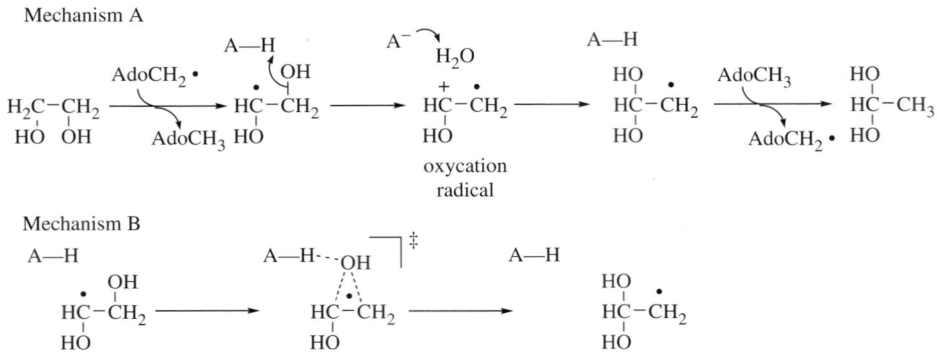

Fig. 9-25. Mechanisms for the isomerization of substrate radicals in the action of dioldehydrase (DDH). The transformation of the radical in complex 2 of fig. 9-20 to that in complex 3 involves the intramolecular migration of an hydroxyl group. The process requires acid catalysis, and two mechanisms are illustrated. Mechanism A is adapted to the action of DDH from that proposed for the dehydration of ethylene glycol radical to acetaldehyde in acetic acid (Walling and Johnson, 1975). Mechanism B requires partial proton donation to the migrating hydroxyl group in a concerted process (Smith et al., 1999). In the transition state for mechanism B, the migrating species is separated from the carbon skeleton by van der Waals distances, as in an eliminated molecule of water that does not escape.

A further mechanistic question is the role of potassium ion. The action of DDH depends on the presence of a monovalent cation, preferentially potassium ion among the alkali metal ions. Other monovalent cations having the same ionic radius as potassium, including ammonium and thallous ions (Tl$^+$), similarly activate DDH. In the structure of DDH with 1,2-propanediol bound to the active site, potassium ion is associated with the two hydroxyl groups of the substrate, which appear to serve as solvation for the potassium ion (see fig. 9-23). Given that hydroxyl group migration requires electrophilic catalysis, the potential participation of potassium ion in this process naturally arises.

A mechanistic function for potassium in catalysis has been suggested (Toraya, 2003; Toraya et al., 1999). The presence of potassium in contact with 1,2-propanediol in the structure supports a role for the cation. If this contact persists in the radical intermediate, magnetic interactions between the nuclei of ^{203}Tl and ^{205}Tl and the unpaired electron of the radical would be expected when the reaction is activated by Tl$^+$. However, no interactions appear in the EPR, Tl-ENDOR, or electron spin-echo envelope modulation (ESEEM) spectra (Schwartz et al., 2003). The role of the monovalent cation is unknown. It is possible to imagine a structural function, but direct participation in catalysis may be unlikely.

Amino acid side chains in the active site of DDH must provide acid catalysis of hydroxyl group migration from C1 to C2 of the 1,2-propanediol-radical, as illustrated in fig. 9-25. As shown in fig. 9-23, the active site includes Glu170 and His143 close to 1,2-propanediol. Mutation of His143 to alanine decreased the values of k_{cat} by 77-fold and of k_{cat}/K_m by 400-fold relative to the wild-typ, respectively. Mutation of Glu170 to Ala decreased the value of k_{cat} by 38000-fold relative to the wild type (Toraya, 2003).

Ethanolamine Ammonia-Lyase

The adenosylcobalamin-dependent elimination of ammonia from ethanolamine catalyzed by ethanolamine ammonia-lyase (EAL; EC 4.3.1.7) appears related to the reaction of dioldehydrase, with the NH$_3^+$ group in place of an OH group (eq. 9-10).

$$^+H_3N-CH_2CH_2-OH \rightarrow CH_3CHO + NH_4^+ \qquad (9\text{-}10)$$

The EAL molecule consists of six α subunits and six β subunits and may be designated as $\alpha_6\beta_6$. Much mechanistic information is available (Bandarian and Reed, 1999); however, there is no crystal structure of EAL.

Mechanistic information on the action of EAL is in many ways similar to that for dioldehydrase, and in a few ways more extensive. There are several similarities. First, adenosylcobalamin mediates hydrogen transfer from C1 of ethanolamine to C2 of acetaldehyde. Second, in the steady state of the reaction with 2-aminopropanol, adenosylcobalamin is cleaved at the active site to cob(II)alamin and a substrate-derived C1-radical that is weakly spin coupled to low-spin cob(II)alamin. Third, analysis of the spin coupling indicates that the substrate-radical and cobalt in cob(II)alamin are separated by a distance of about 11 Å (Schepler et al., 1975; Bandarian and Reed, 2002). Fourth, the reaction of EAL proceeds with very large deuterium and tritium kinetic isotope effects. Fifth, EAL reacts with the coenzyme analog 3′,4′-anhydroadenosylcobalamin to cleave the Co—C5 bond and form the allylic 5′-deoxyadenosyl radical, as in scheme 9-7. Sixth, EAL undergoes suicide inactivation by reaction with glycolaldehyde to form cob(II)alamin and the *cis*-ethanesemidione radical, as in the suicide inactivation of dioldehydrase (Abend et al., 2000) (see chap. 4).

Spectroscopic analysis of the action of EAL exceeds that for dioldehydrase in several respects. Pulsed EPR experiments show that in the substrate-radical derived from 2-aminopropanol the unpaired electron lies within 3.3 Å of the methyl group of 5′-deoxyadenosine (Warncke and Utada, 2001). This EAL intermediate would correspond to complex 2 of DDH in fig. 9-20. The product-related radical is also in van der Waals contact with the methyl group of 5′-deoxyadenosine (Warncke, 2005).

The action of EAL is expected to differ in a few respects from that of DDH. The elimination of the amino group from the substrate-radical should not require acid catalysis because it exists as an aminium group under physiological conditions and should depart as NH_3. It cannot be proved that the NH_3 is added back to C1 to form an hydroxyaminal radical analogous to the aldehyde hydrate radical in complex 3 of DDH in fig. 9-20. The formation of this hydrate in the reaction of DDH is necessary to explain the stereochemistry. However, no such stereochemistry is available for the reaction of EAL. Moreover, the EPR results in the steady state of the reaction of ethanolamine with EAL indicate the presence of the acetaldehyde radical and not either the substrate radical or a hydroxyaminal radical analogous to complexes 2 and 3, respectively, in fig. 9-20. The reaction of EAL may follow the course in scheme 9-8. Much remains to be learned about the mechanism of the reaction of EAL.

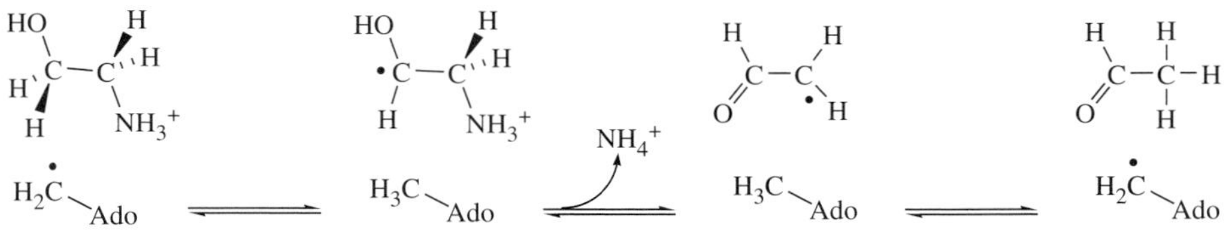

Scheme 9-8

References

Abend A, V Bandarian, GH Reed, and PA Frey (2000) *Biochemistry* **39**, 6250.
Abend A, R Nitsche, V Bandarian, E Stupperich, and J Rétey (1999) *Angew Chem Int Ed* **37**, 625.
Allard ST, K Beis, MF Giraud, AD Hegeman, JW Gross, RC Wilmouth, C Whitfield, M Graninger, P Messner, AG Allen, DJ Maskell, and JH Naismith (2002) *Structure* **10**, 81.
Babbitt PC, MS Hasson, JE Wedekind, DR Palmer, WC Barrett, GH Reed, I Rayment, D Ringe, GL Kenyon, and JA Gerlt (1996) *Biochemistry* **35**, 16489.
Baedeker M and G Schulz (2002) *Eur J Biochem* **269**, 1790.
Bahnson BJ and VE Anderson (1989) *Biochemistry* **28**, 4173.

Bahnson BJ and VE Anderson (1991) *Biochemistry* **30**, 5894.
Bahnson BJ, VE Anderson, and GA Petsko (2002) *Biochemistry* **41**, 2621.
Bandarian V and GH Reed (2002) *Biochemistry* 41, 8580.
Bansal KM, M Gratzel, A Henglein, and E Janata (1973) *J Phys Chem* **77**, 16.
Beinert H, RH Holm, and E Münck (1997) *Science* **277**, 653.
Beinert H, MC Kennedy, and CD Stout (1996) *Chem Rev* **96**, 2335.
Beis K, ST Allard, AD Hegeman, G Murshudov, D Philp, and JH Naismith (2003) *J Am Chem Soc* **125**, 11872.
Blanchard JS and WW Cleland (1980) *Biochemistry* **19**, 4506.
Burkhard P, GSJ Rao, E Hohenester, KD Schnackerz, PF Cook, and JN Jansonius (1998) *J Mol Biol* **283**, 121.
Burkhard P, CH Tai, CM Ristroph, PF Cook, and JN Jansonius (1999) *J Mol Biol* **291**, 941.
Calabrese JC, DB Jordan, A Boodhoo, S Sariaslani, and T Vannelli (2004) *Biochemistry* **43**, 11403.
Christianson DW and JD Cox (1999) *Annu Rev Biochem* **68**, 33.
Clausen T, R Huber, B Laber, HD Pohlenz, and A Messerschmidt (1996) *J Mol Biol* **262**, 202.
Clausen T, R Huber, L Prade, MC Wahl, and A Messerschmidt (1998).*EMBO J* **17**, 6827.
D'Ordine RL, BJ Bahnson, PJ Tonge, and VE Anderson (1994a) *Biochemistry* **33**, 14733.
D'Ordine RL, PJ Tonge, PR Carey, and VE Anderson (1994b) *Biochemistry* **33**, 12635.
Emptage MH, J-L Dreyer, MC Kennedy, and H Beinert (1983a) *J Biol Chem* **258**, 11106.
Emptage MH, TA Kent, MC Kennedy, H Beinert, and E Münck (1983b) *Proc Natl Acad Sci U S A* **80**, 4674.
Essenberg MK, PA Frey, and RH Abeles (1971) *J Am Chem Soc* **93**, 1242.
Estévez M, J Skarda, J Spencer, L Banaszak, and TM Weaver (2002) *Protein Sci* **11**, 1552.
Feng Y, HA Hofstein, J Zwahlen, and PJ Tonge (2002) *Biochemistry* **41**, 12883.
Fierke CA, TL Calderone, and JF Krebs (1991) *Biochemistry* **30**, 11054.
Frey PA, GL Karabatsos, and RH Abeles (1965) *Biochem Biophys Res Commun* **18**, 551.
Gerratana B, WW Cleland, and PA Frey (2001) *Biochemistry* **40**, 9187.
Glaser L and H Zarkowsky (1973). In Boyer PD (ed): *The Enzymes*, vol 5, 3rd ed. Academic Press: New York, pp 465-480.
Gross JW, AD Hegeman, B Gerratana, and PA Frey (2001) *Biochemistry* **40**, 12497.
Gross JW, AD Hegeman, MM Vestling, and PA Frey (2000) *Biochemistry* **39**,13633.
Hansen JN, EC Dinovo, and PD Boyer (1969) *J Biol Chem* **244**, 6270.
Hanson KR, EA Havir, and C Ressler (1979) *Biochemistry* **18**, 1431.
Hegeman AD, JW Gross, and PA Frey (2001) *Biochemistry* **40**, 6598.
Hegeman AD, JW Gross, and PA Frey (2002) *Biochemistry* **41**, 2797.
Hermes JD, CA Roeske, MH O'Leary, and WW Cleland (1982) *Biochemistry* **21**, 5106.
Hermes JD, PM Weiss, and WW Cleland (1985) *Biochemistry* **24**, 2959.
Hofstein HA, Y Feng, VE Anderson, and PJ Tonge (1999) *Biochemistry* **38**, 9508.
Jewell DA, C Tu, SR Paranawithana SM Tanhauser, PV LoGrasso, PJ Laipis, and DN Silverman (1991) *Biochemistry* **30**, 1484.
Jornvall H, B Persson, M Krook, S Atrian, R Gonzalez-Duarte, J Jeffery, and D Ghosh (1995) *Biochemistry* **34**, 6003.
Kennedy C, R Rauner, and O Gawron (1972) *Biochem Biophys Res Commun* **47**, 740.
Kennedy MC, M Werst, J Telser, MH Emptage, H Beinert, and BM Hoffman (1987) *Proc Natl Acad Sci U S A* **84**, 8854.
Kent TA, MH Emptage, H Merkle, MC Kennedy, H Beinert, and E Münck (1985) *J Biol Chem* **260**, 6871.
Kiefer LL, SA Paterno, and CA Fierke (1995) *J Am Chem Soc* **117**, 6831.
Krebs JF, JA Ippolito, DW Christianson, and CA (1993) *J Biol Chem* **268**, 27458.
Langer B, M Langer, and J Rétey J (2001) *Adv Protein Chem* **58**, 175.
Langer B, D Rother, and J Rétey (1997) *Biochemistry* **36**,10867.
Langer M, G Reck, J Reed, and J Rétey (1994) *Biochemistry* **33**, 6462.
Larsen TM, JE Wedekind, I Rayment, and GH Reed (1996) *Biochemistry* **35**, 4349.
Liang J-Y and WN Lipscomb (1990) *Proc Natl Acad Sci U S A* **87**, 3675.
Liang Z, Y Xue, G Behravan, B-H Jonsson, and S Lindskog (1993) *Eur J Biochem* **211**, 821.
Lindskog S and DN Silverman (2000) In Chegwidden WR, ND Carter, and YJ Edwards (eds): *The Carbonic Anhydrases. New Horizons.* Berkhäuser Verlag: Basel, pp 175-195.
Lloyd SJ, H Lauble, GS Prasad, and CD Stout (1999) *Protein Sci* **8**, 2655.
Magnusson OT and PA Frey (2002) *Biochemistry* **41**, 1695.

Maier M, M Janosik, V Kery, JP Kraus, and P Burkhard (2001) *EMBO* **20**, 3910.
Mangani S and K Hakansson (1992) *Eur J Biochem* **210**, 867.
Merz KM (1991) *J Am Chem Soc* **113**, 406.
Porter DJT and HJ Bright (1980) *J Biol Chem* **255**, 4772.
Poyner RR and GH Reed (1992) *Biochemistry* **31**, 7166.
Rebholz KL and DB Northrop (1994) *Arch Biochem Biophys* **312**, 227.
Reed GH and SO Mansoorabadi (2003) *Curr Opin Struct Biol* **13**, 716.
Reed GH, RR Poyner, RM Larsen, JE Wedekind, and I Rayment (1996) *Curr Opin Struct Biol* **6**, 736.
Rétey J (2003) *Biochim Biophys Acta* **1647**, 179.
Rétey J, A Umani-Rouchi, J Seibl, and D Arigoni (1966) *Experientia* **22**, 502.
Rose IA (1998) *Biochemistry* **37**, 17651.
Rose IA and EL O'Connell (1967) *J Biol Chem* **242**, 1870.
Rother D, L Poppe, S Viergutz, B Langer, and J Rétey (2001) *J Eur J Biochem* **268**, 6011.
Ruzicka FJ and H Beinert (1978) *J Biol Chem* **253**, 2514.
Schepler KL, WR Dunham, RH Sands, JA Fee, and RH Abeles (1975) *Biochim Biophys Acta* **397**, 510.
Schuster B and J Rétey (1995) *Proc Natl Acad Sci U S A* **92**, 8433.
Schwartz P, R LoBrutto, GH Reed, and PA Frey (2003) *Helv Chim Acta* **86**, 3764. .
Schwede TF, J Rétey, and GE Schulz (1999).*Biochemistry* **38**, 5355.
Shibata N, Y Nakanishi, M Fukuoka, M Yamanishi, N Yasuoka, and T Toraya (2003) *J Biol Chem* **278**, 22717.
Silverman DN and S Lindskog (1988) *Acc Chem Res* **21**, 30.
Silverman DN, C Tu, X Chen, SM Tanhauser, AJ Kresge, and PJ Laipis (1993) *Biochemistry* **32**, 10757.
Sims PA, TM Larsen, RR Poyner, WW Cleland, and GH Reed (2003) *Biochemistry* **42**, 8298.
Smith DM, BT Golding, and L Radom (1999) *J Am Chem Soc* **121**, 5700.
Steiner H, BH Jonsson, and S Lindskog (1975) *Eur J Biochem* **59**, 253.
Sweet WL and JS Blanchard (1990) *Arch Biochem Biophys* **277**,196.
Tai C-H and PF Cook (2000) *Adv Enzymol Rel Areas Mol Biol* **74B**, 185.
Tai C-H and PF Cook (2001) *Acc Chem Res* **34**, 49.
Tai C-H, SR Nalabolu, TM Jacobson, DE Minter, and PF Cook (1993) *Biochemistry* **32**, 6433.
Taoka S, BW Lepore, O Kabil, S Ojha, D Ringe, and R Banerjee (2002) *Biochemistry* **41**, 10454.
Toraya T (2003) *Chem Rev* **103**, 2095.
Toraya T, K Yoshizawa, M Eda, and T Yamabe (1999) *J Biochem* **126**, 650.
Tu CK, DN Silverman, C Forsman, B-H Jonsson, and S Lindskog (1989) *Biochemistry* **28**, 7913.
Valinsky JE, RH Abeles, and JA Fee (1974) *J Am Chem Soc* **96**, 4709.
Walling C and RA Johnson (1975) *J Am Chem Soc* **97**, 2405.
Warncke K (2005) *Biochemistry* **44**, 3184.
Warncke K and Utada AS (2001) *J Am Chem Soc* **123**, 8564.
Weaver TM and L Banaszak (1996) *Biochemistry* **35**, 13955.
Werst MM, MC Kennedy, H Beinert, and BM Hoffman (1990) Biochemistry 29, 10526.
Wu WJ, WY Feng, X He, HS Hofstein, DP Raleigh, and PJ Tonge (2000) *J Am Chem Soc* **122**, 3987.
Yamanishi M, S Yamade, H Muguruma, Y Murakami, T Tobimatsu, A Ishida, J Yamauchi and Toraya T (1998) *Biochemistry* **37**, 4799.
Zagalak B, PA Frey, GL Karabatsos, and RH Abeles (1966) *J Biol Chem* **241**, 3028.

10
Phosphotransfer and Nucleotidyltransfer

Chemistry of Phosphoryl Group Transfer

Phosphotransferases, phosphatases, and nucleotidyltransferases catalyze nucleophilic substitution at phosphorus. They constitute a dominant class of enzymes in intermediary metabolism, energy transduction, nucleic acid biosynthesis and processing, and regulation of many cellular processes, including replication, cellular development, and apoptosis. The mechanisms of the action of these enzymes have been studied intensively at several levels, ranging from the biosynthesis of metabolites and nucleic acids to unmasking signaling networks to elucidating the molecular mechanisms of catalysis. We focus on the chemical mechanisms of the reactions of biological phosphates. More than 40 years of research on this chemistry reveals that the mechanisms can be grouped into two classes: the phosphoryl group (PO_3^-) transfer mechanisms and the nucleotidyl or alkylphosphoryl group ($ROPO_2^-$) transfer mechanisms. Because the fundamental chemical mechanisms of these reactions are not treated in textbooks, we begin by considering this chemistry and then move on to the enzymatic reaction mechanisms.

Phosphomonoesters

Transition State for Phosphoryl Group Transfer

Phosphomonoesters, phosphoanhydrides, and phosphoramidates undergo substitution at phosphorus by transfer of the phosphoryl (PO_3^-) group, that is, by P—O and P—N cleavage. The current description of a typical phosphoryl group transfer mechanism is one in which the phosphoryl donor and acceptor interact weakly with the phosphoryl group in flight in a transition state in which the total bonding to phosphorus is decreased relative to the ground state (eq. 10-1).

$$\begin{array}{c} O \\ \parallel \\ R-X-P-O^- \\ | \\ O^- \end{array} + :Y \longrightarrow \left[\begin{array}{c} O \\ \parallel \\ R-X----P----Y \\ / \quad \backslash \\ O^- \quad O^- \end{array} \right]^{\ddagger} \longrightarrow \begin{array}{c} O \\ \parallel \\ ^-O-P-Y \\ | \\ O^- \end{array} + RX: \quad (10\text{-}1)$$

The bonding is weak between phosphorus and the leaving group R–X and between phosphorus and the accepting group Y in the transition state of eq. 10-1. Because of decreased bonding to phosphorus, this is a loose transition state that has been described as *dissociative*. The latter should not be confused with the *dissociative mechanism*, which is considered later. To avoid confusion, we use the term *loose transition state*. Detailed studies indicate that the bonding denoted by the dashed lines in eq. 10-1 represents partial covalency on the order of 10% to 20% of the strength of a full covalent bond, or a bond order of 0.1 to 0.2.

In certain cases, in which R—X is an extremely good leaving group, evidence indicates that the reaction takes place in steps by a dissociative mechanism in which the phosphoryl group is released by the donor as discrete monomeric metaphosphate $[PO_3]^-$ in the first step of eq. 10-2.

$$\begin{array}{c} O \\ \parallel \\ R-X-P-O^- \\ | \\ O^- \end{array} \xrightarrow[\text{state 1}]{\text{Transition}\;\ddagger} \xrightarrow{RX:} \left[\begin{array}{c} O \\ \parallel \\ P \\ // \backslash \\ O \quad O \end{array} \right] \xrightarrow{Y:} \xrightarrow[\text{state 2}]{\text{Transition}\;\ddagger} \begin{array}{c} O \\ \parallel \\ ^-O-P-Y \\ | \\ O^- \end{array} \quad (10\text{-}2)$$

In the second step, a nucleophile captures the highly electrophilic metaphosphate. Because of its extreme reactivity, metaphosphate does not diffuse freely in solvents such as water or simple alcohols but reacts with nucleophiles within its solvation sphere.

pH Dependence of Hydrolysis

The pH dependence for the hydrolysis of methyl phosphate, and alkyl phosphates in general, is illustrated in fig. 10-1 (Benkovic and Schray, 1973). The inflections in the profile arise from the unique reactions of the four ionic species. In the physiological pH range, the most reactive form by more than four orders of magnitude is the monoanion ($CH_3OPO_3H^-$), which undergoes hydrolysis by P—O bond cleavage. The neutral form reacts much more slowly and by C—O bond cleavage, and the dianion is essentially unreactive. Strong acids catalyze the hydrolysis of the neutral form by both P—O and C—O cleavage.

Several lines of evidence indicate that the monoanion reacts as if the mechanism were stepwise as in the upper line of scheme 10-1, where $v = k_1[ROPO_3H^-]$ (Butcher and Westheimer, 1954; Kirby and Varvoglis, 1967).

478   Enzymatic Reaction Mechanisms

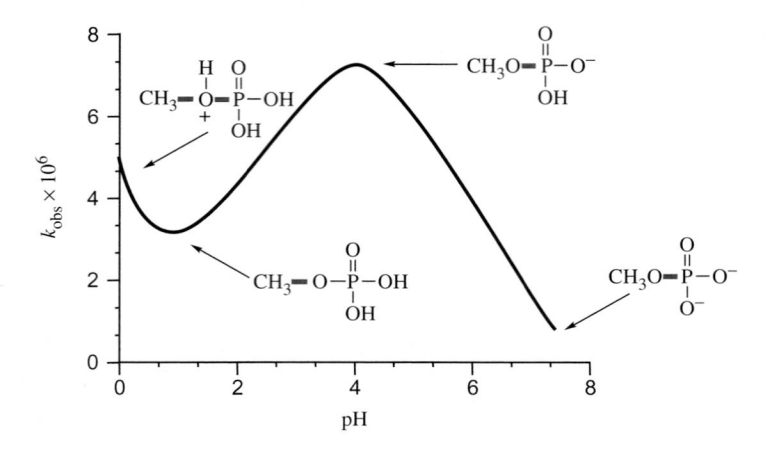

Fig. 10-1. The pH-rate profile for the nonenzymatic hydrolysis of methylphosphate. The pH dependence and rate constants for methylphosphate are typical of alkyl phosphates in general. Notice that the rate constants are very small, in keeping with the stability of phosphate esters. They had to be measured by extrapolation of data obtained at higher temperatures. The red bonds are those cleaved by hydrolysis of the different ionic species; the monoanion is the most reactive by a factor of more than 10^4.

Scheme 10-1

In the mechanism on the upper line, initial proton transfer to the bridging oxygen is a fast, but highly unfavorable process because of the weak basicity of the bridging oxygen. However, the bridge-protonated, zwitterionic species is highly reactive, with an excellent leaving group, and the two negatively charged oxygens provide a strong driving force to expel the alcohol. The expulsion of ROH is rate limiting. The zwitterionic species is not a free intermediate and exists only transiently, is in contact with solvent, and is quenched by water within the solvation sphere.

The kinetically equivalent lower pathway in scheme 10-1 can be ruled out as follows. The rate law is $v = k_2 K_b [\text{ROPO}_3 \text{H}^-]$, where K_b is the known hydrolysis constant for phosphate monoanion. From the experimentally measured rate constant k_{obs} ($= k_2 K_b$), and the value of K_b, 10^{-7} M^{-1} s^{-1} can be calculated for k_2. However, the rate constant for the electronically similar and sterically more hindered reaction of hydroxide with trimethylphosphate is 3.3×10^{-2} M^{-1} s^{-1}, more than four orders of magnitude larger than the calculated value. Steric and electronic effects of the methyl groups in trimethylphosphate would slow the reaction relative to ROPO$_3$H$_2$, not accelerate it. The reaction of hydroxide with ROPO$_3$H$^-$ is unreasonable for the mechanism (Benkovic and Schray, 1973; Bruice and Benkovic, 1966).

Electronic Effects on Hydrolysis

Electron withdrawing R-groups in scheme 10-1 increase the stability of the leaving group (ROH) but decrease the basicity of the bridging oxygen. Because of these compensating effects, the hydrolytic rate is only modestly sensitive toward electron withdrawal by the substituent R. Conversely, electron-donating groups decrease the stability of the leaving group but increase the basicity of the bridging oxygen. The compensatory effects lead to little sensitivity of the rate to electron donation by the substituent. A plot of log k_{obs} against pK_a of the leaving group for phosphomonoester monoanions displays a slope of −0.3. Weak inductive effects are the expected behavior for the mechanism in the upper line of scheme 10-1.

Activation Parameters and Solvent Partitioning

Because the rate-limiting step in the hydrolytic mechanism in scheme 10-1 is unimolecular, the entropy of activation ($\Delta S^\ddagger$) should be near zero, and experiments show that it is within 3 cal mol^{-1} deg^{-1} of zero in the reactions of many phosphomonoesters (Benkovic and Schray, 1973). In a bimolecular reaction, two molecules collide to form the transition state, and this leads to a negative value of $\Delta S^\ddagger$, on the order of −20 cal mol^{-1} deg^{-1}. The hydrolysis of phosphomonoesters behaves as if the transition state is formed by the ester within its solvation sphere. The apparent monomolecularity in this case does not exclude the participation of water in forming the transition state, because water is a solvating molecule that is present in both the ground state and transition state. Although the low value of $\Delta S^\ddagger$ is consistent with scheme 10-1, it does not prove the mechanism.

An alternative to scheme 10-1 might be a mechanism in which the internal proton transfer to the leaving group takes place in the transition state in concert with the departure of the leaving group and the entry of the acceptor. In such a mechanism, the substitution of deuterium for protium should lead to a significant deuterium kinetic isotope effect. However, the hydrolysis of methyl phosphate in D$_2$O proceeds with no primary deuterium kinetic isotope effect. $k_{H/D}$ = 0.87 (Bunton et al., 1958), consistent with the absence of hydrogen transfer in the transition state. An exception in the case of 2,4-dinitrophenyl phosphate monoanion is discussed later.

Experiments in mixed alcohol/water solvents show that the product mixture is nearly the same as the ratio of alcohol to water, as it would be if discrete [PO$_3$]$^-$ were randomly captured by either water or the alcohol. Both water and alcohol can solvate phosphate anions, so that both can be available to enter the transition state in the mechanism of eq. 10-1.

Special Cases of Hydrolysis

The two oxyanionic substituents on phosphorus in the reactive species in scheme 10-1 seem to be generally important as a driving force in phosphoryl group transfer. A few special cases reinforce this point. Phosphoanhydrides are important phosphoryl group donors in biochemistry, the most important example being ATP. The hydrolysis of γ-phenylpropyl-diphosphate dianion is instructive (Miller and Westheimer, 1966). This compound exists in solution as the two tautomers A and B (structure **10-1**), and the tautomeric mixture undergoes hydrolysis 2000 times faster than P^1,P^2-*bis*-γ-phenylpropyldiphosphate dianion.

10-1

Phenylpropyldiphosphate A   Phenylpropyldiphosphate B

P^1,P^2-*bis*-Phenylpropyldiphosphate   P^1,P^1-*bis*-Phenylpropyldiphosphate

The difference is likely due to the high reactivity of tautomer B of γ-phenylpropyldiphosphate. This is verified by the reactivity of P¹,P¹-*bis*-γ-phenylpropyldiphosphate dianion, which undergoes hydrolysis at nearly the same rate as that estimated for tautomer B of phenylpropyldiphosphate (Miller and Ukena, 1969). These relationships focus attention on the importance of two oxyanionic substituents as the driving force for phosphoryl group transfer.

The hydrolytic rates of 2,4-dinitrophenylphosphate monoanions and dianions further accentuate the impact of the dianionic driving force and the importance of proton transfer in the monoanion. The relevant structures are shown in structure **10-2**.

The relative rates for the monoanion and dianion are inverted in this case because of the resonance and inductive electron withdrawal by the two nitro groups. These effects draw electron density away from the bridging oxygen, dramatically decreasing its basicity, so that a pre-equilibrium proton transfer is ineffective in generating the zwitterionic form in scheme 10-1. At the same time, strong electron withdrawal lends the leaving group a pK_a of 4 for dinitrophenol, much lower than the value of 15 for a simple alcohol, and making it such a good leaving group that protonation is not required in the reaction of the dianion. The monoanion, on the other hand, does not have the required dianionic driving force, and it can react only by transferring the proton. Because of the low basicity of oxygen in the leaving group, the proton is transferred in the transition state, which is confirmed by the observation in this case of a primary solvent deuterium kinetic isotope effect of 1.45 (Kirby and Varvoglis, 1967).

^{18}O Kinetic Isotope Effects

Three isotope effects must be taken into consideration in the interpretation of ^{18}O kinetic isotope effects in the hydrolysis of phosphomonoester monoanions (Cleland and Hengge, 1995; Hengge, 2002). They are the primary ^{18}O-kinetic isotope effect on P—^{18}O bond cleavage when the isotope is bridging, that is, part of the leaving group; the secondary ^{18}O-kinetic isotope effect when the isotope is a peripheral, nonbridging, substituent of phosphorus; and the equilibrium ^{18}O isotope effect for protonation of ^{18}O. A normal, primary ^{18}O kinetic isotope effect for P—O bond cleavage is 1.03 (Hengge, 2002), and the equilibrium isotope effect for protonation of ^{18}O is 1.015 (Weiss et al., 1986). The expected secondary ^{18}O-kinetic isotope effect depends on the mechanism of P—O bond cleavage. If the mechanism can be described by scheme 10-1 or by eqs. 10-1 or 10-2, the P—O bond order to peripheral oxygens does not decrease and may increase slightly in the transition state. In this case, the secondary ^{18}O-kinetic isotope effect is unity or slightly inverse. If the mechanism involves an increase in the total bond order to phosphorus in the transition state, as in the formation of a pentavalent oxyphosphorane, the P—O bond order to peripheral oxygens will decrease, and the secondary kinetic isotope effect will be normal, up to 1.025 (Hengge, 2002).

Available results support the mechanism of scheme 10-1 for the hydrolysis of alkyl phosphate monoanions. The remote label method (see chap. 2) allows accurate ^{18}O-kinetic isotope effects to be measured for the reactions of molecules such as glucose-6-phosphate monoanion and *m*-nitrobenzylphosphate monoanion. *m*-Nitrobenzylphosphate reacts as an alkyl phosphate with a highly basic leaving group ($pK_a = 14$). The secondary ^{18}O-kinetic isotope effects in the reactions of both glucose-6-phosphate and *m*-nitrobenzylphosphate are near unity or slightly inverse, and P—O bond cleavage is rate limiting, consistent with the mechanism in scheme 10-1 (Grzyska et al., 2003; Weiss et al., 1986).

Measurements of secondary ^{18}O kinetic isotope effects in the reactions of alkaline phosphatase and hexokinase led to results similar to those found in nonenzymatic reactions (Jones et al., 1991; Weiss and Cleland, 1989). In each case, slightly inverse effects indicate a loose transition state for phosphoryl group transfer in the enzymatic mechanisms, similar to the nonenzymatic reactions.

Phosphoryl Group Transfer to Nucleophiles Other Than Solvents

In the mechanism of eq. 10-1, :Y may be a nucleophile other than water. As formulated in eq. 10-1, the complications of proton transfer in the mechanism can be avoided by the choice of leaving group, as we saw in the case of 2,4-dinitrophenylphosphate. The nucleophilic and leaving group reactivities of :Y and R—X, respectively, can be varied and exploited to clarify finer mechanistic details. When R—X in eq. 10-1 is a substituted pyridine, that is X = N$^+$, and :Y is another substituted pyridine or an oxygen nucleophile such as a carboxylic acid, the reactivities of both the leaving group and nucleophile can be varied over large ranges. This allows structure-function tests of the mechanisms in eqs. 10-1 and 10-2.

Suppose that the two-step mechanism of eq. 10-2 were the general case for phosphoryl group transfer. When the leaving group is poor and the nucleophile is reactive, the first step should be rate limiting. Conversely, when the leaving group is excellent and the nucleophile is relatively unreactive the second step should be rate limiting. These relationships should lead to breaks in linear free energy plots according to the Brønsted catalysis law ($\log k = \beta\, pK_a + C$) (see chap 1). In eq. 10-2, when R—X is held constant and :Y is varied over a large range of reactivities, the plot of $\log k$ against pK_a of the nucleophile should display a break at the point where the rate-limiting step changes. Conversely, when :Y is held constant and R—X is varied, a plot of $\log k$ against pK_a of the leaving group should show a break at the point the rate-limiting step changes. Such plots obtained in the reactions of large ranges of nucleophilic and leaving group reactivities do not display breaks (Bourne and Williams, 1984; Herschlag and Jencks, 1989a, 1989b; Skoog and Jencks, 1984). It seems that the mechanism of eq. 10-1 best describes the mechanism of phosphoryl group transfer when proton transfer to the leaving group is not involved. In this mechanism, the nucleophile participates in but does not dominate the transition state.

Evidence of nucleophilic reactivity in the transition state is augmented by the observation of an α-effect (see chap. 1) when the acceptor is an α-effect nucleophile, such as hydroperoxide or hydroxylamine (Herschlag and Jencks, 1990). The results indicate that many nonenzymatic phosphoryl group transfer reactions are not stepwise as in eq. 10-2 but may be concerted as in eq. 10-1.

When the leaving group is varied the value of the slope (β_{lg}) is large and negative, showing that bond cleavage to the leaving group is important in the transition state. (Notice the difference from the hydrolysis of phosphomonoester monoanions, where proton transfer is part of the mechanism.) In contrast, when the nucleophile is varied, the value of β_{nuc} is very small, showing that bond formation to the nucleophile is small. In the emerging picture, both the nucleophile and leaving group participate in the transition state, but there is little bonding between phosphorus and either the nucleophile or the leaving group. Secondary linear free energy plots indicate that the nucleophile and leaving group sense the presence of each other in the transition state.

Stereochemistry

If a phosporyl group transfer takes place by the mechanism of eq. 10-1, it must proceed with inversion of stereochemical configuration in the reaction of a P-chiral substrate. Accordingly, the alcoholysis of arylphosphates with chiral phosphorus, by virtue of ^{16}O, ^{17}O, and ^{18}O as peripheral oxygen substituents, proceeds with inversion of configuration (Buchwald et al., 1984). This is illustrated in eq. 10-3 for phenylphosphate.

$$\text{Ph-O-P(}^{16}O\text{)(}^{17}O\text{)(}^{18}O\text{)} + ROH \longrightarrow \text{R-O-P(}^{16}O\text{)(}^{17}O\text{)(}^{18}O\text{)} + \text{Ph-OH} \qquad (10\text{-}3)$$

(S_P) \qquad\qquad (R_P)

In *t*-butanol, phosphoryl transfer by the same molecules proceeds with racemization at phosphorus (Freeman et al., 1987). The stereochemistry indicates that inversion takes place when the nucleophilic acceptor is a solvating molecule, but racemization occurs when the acceptor is not a solvating molecule. The results suggest that the planar metaphosphate may be an intermediate in the absence of a proximal acceptor. In enzymatic reactions, an acceptor would always be present in the active site.

Discrete Metaphosphate Anion?

Foregoing evidence indicates that metaphosphate anion $[PO_3]^-$ does not normally exist as a discrete species except when the leaving group is very good, as for ROH of scheme 10-1. In any case, it is not a freely diffusible species in most hydroxylic solvents, which trap it very rapidly. A positional isotope exchange (PIX) experiment provides further evidence that metaphosphate anion can exist transiently (Cullis and Nichols, 1987). In one version of a PIX experiment, the P—^{18}O bond order measured by ^{31}P NMR spectroscopy changes on interconversion of bridging and nonbridging ^{18}O. The bonding of ^{18}O to P induces an upfield perturbation ($\Delta\delta_P$) on the ^{31}P NMR signal, the magnitude of which is proportional to the P—O bond order. For a single bond, $\Delta\delta_P = 0.017$ ppm, and for a double bond, $\Delta\delta_P = 0.034$ ppm, as long as all other substituents are oxygen. For a bond order of 1.5, the upfield perturbation is intermediate. The PIX rearrangement shown in eq. 10-4 takes place in *t*-butanol or acetonitrile.

$$\text{Ado-O-P(O)(O}^-\text{)-O-PO}_3^{2-} \rightleftharpoons \text{Ado-O-P(O)(O}^-\text{)-O-PO}_3^{2-} \qquad (10\text{-}4)$$

The oxygen in color represents ^{18}O. The simplest explanation for this rearrangement is that AMP is a sufficiently good leaving group to allow the β-phosphoryl group to be released transiently as metaphosphate anion and AMP. The three peripheral oxygen atoms of the phosphate group in AMP are tortionally equivalent, so that the recapture of metaphosphate leads to ADP with bridging or nonbridging ^{18}O.

μ-Monothiopyrophosphate trianion [HO$_3$P–S–PO$_3$]$^{3-}$ undergoes hydrolysis at a rate 10^7 times that of pyrophosphate trianion, presumably because of the weakness of the P—S bond and the importance of bond cleavage in the transition state (Halkides and Frey, 1991). Tris, (HOCH$_2$)$_3$C–NH$_2$, at molar concentrations is a highly reactive nucleophile that *does not increase the rate* at which μ-monothiopyrophosphate reacts. However, Tris captures the phosphoryl group to form both *O*- and *N*-phospho-Tris (Lightcap and Frey, 1992). The results indicate that in this special case, discrete monomeric metaphosphate is an intermediate in the mechanism of eq. 10-2, where water or Tris can occupy the solvation sphere and react with [PO$_3$]$^-$. The weakness of the P—S bond presumably allows HO$_3$P–S$^-$ to leave fast enough to generate metaphosphate transiently in advance of its reaction with solvating water or Tris.

Phosphodiesters

The mechanistic properties of P—O bond cleavage in reactions of phospodiesters are, in most respects, unlike those in reactions of phosphomonoesters (Schray and Benkovic, 1973). The hydrolysis of phosphodiester monoanions is very much slower ($k_{monoester}/k_{diester} \approx 10^4$) for the same R groups. The rates of hydrolysis are more sensitive to the stability of the leaving group than in phosphomonoesters; the more stable the leaving group the faster the hydrolysis. Linear free energy plots of log k against pK_{lg} give slopes of about −1.2. Unlike phosphomonoesters, the rates are equally sensitive to the reactivity of the nucleophile. In further contrast, the entropy of activation ($\Delta S^\ddagger$) in the hydrolysis of a phosphodiester is large and negative (about −25 cal mol^{-1} deg^{-1}) consistent with a bimolecular reaction. Hydrolysis proceeds in D$_2$O with a primary kinetic isotope effect of about 1.6, consistent with proton transfer in the transition state. The mechanism in scheme 10-2, with a tight transition state, is consistent with the evidence.

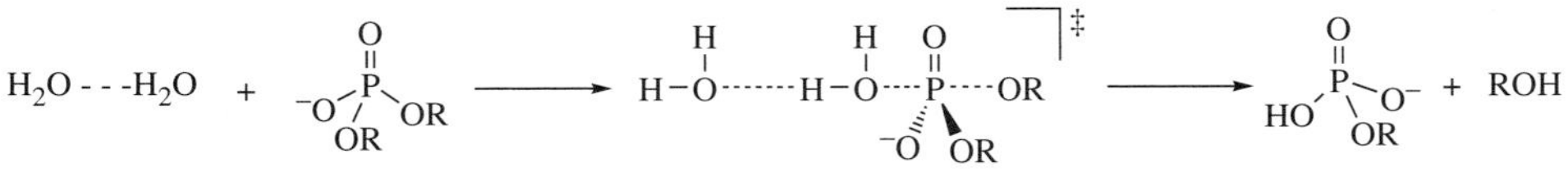

Scheme 10-2

In a tight transition state, the bond orders between phosphorus and both the leaving group and acceptor are about 0.5, and the net bond order to phosphorus in the transition state is equal to or larger than in the ground state. The tight transition state has been described as *associative*, but it should not be confused with an associative mechanism, in which a discrete pentavalent adduct is formed as an intermediate. In this extreme case, where the bond order is unity to both the entering and leaving group, there is a pentavalent intermediate. Pentavalent intermediates are rare but do occur, especially in reactions of five-member ring cyclic triesters and phosphodiesters under acidic conditions (Westheimer, 1971).

Available information on the secondary ^{18}O-kinetic isotope effects for reactions of phosphodiesters is consistent with a tight transition state: increased bond order to phosphorus and decreased bond order to the peripheral oxygens in the transition state.

Phosphodiesterases or nucleotidyltransferases, a few of which are discussed in later sections, catalyze most enzymatic alkylphosphoryl group transfer reactions.

Phosphotriesters

Because of the absence of negative charge on phosphotriesters, they are far more reactive toward nucleophiles than phosphodiester monoanions. They are especially reactive toward

Table 10-1. Relative Reactivities of p-Nitrophenyl Esters of Phosphate

Compound	k_{H_2O} (M^{-1}min^{-1})	[k_{HO^-}(M^{-1}min^{-1})]
p-O$_2$NPhOPO$_3^{2-}$	2×10^{-4}	—
p-O$_2$NPhO—PO$_2^-$ (OCH$_3$)	3×10^{-7}	3×10^{-2}
p-O$_2$NPhO—P(=O)(O—)(O—) (five-membered ring)	8×10^{-6}	60

negatively charged nucleophiles such as hydroxide ion. The reactivities of phosphotriesters toward nucleophiles relative to phosphomonoester dianions and phosphodiester monoanions are brought clearly into focus by the data in table 10-1 (Schray and Benkovic, 1973). The phosphotriester is 2000 times as reactive as the phosphodiester monoanion toward hydroxide ion. The transition states for the reactions of phosphotriesters are very tight.

Table 10-1 also shows the greater reactivity of the phosphomonoester dianion toward water relative to the phosphodiester monoanion and even the neutral phosphotriester. In this case, the leaving group is p-nitrophenol, an excellent leaving group (pK_a = 7) that does not require protonation to depart. The reactivity of the phosphomonoester dianion is due to the special mechanism by which it reacts through a loose transition state, which is not available to the phosphodiester and phosphotriester. The alkyl substituents prevent bond delocalization in a hypothetical alkylmetaphosphate [ROPO$_2$]. Extensive bond and charge delocalization is essential to the stability of metaphosphate monoanion PO$_3^-$, whether as an incipient species in a transition state or as a discrete intermediate.

The rule in nonenzymatic reactions is that transfer of the PO$_3^-$ group proceeds through a loose transition state, and transfer of a ROPO$_2^-$ group proceeds through a tight transition state. Available evidence indicates that the rule holds for enzymatic reactions as well.

Five-Member Ring Phosphoesters

Five-member ring cyclic phosphodiesters occupy a special place in biochemistry as intermediates in the actions of ribonucleases and ribozymes. They also display special chemical reactivities that are important in the chemistry of RNA and in enzymatic and ribozymatic reactions of RNA. Six-member ring cyclic phosphodiesters such as 3′,5′cyclic AMP are also important in biochemistry; however, they do not display special kinetic properties. In this section, we consider the exceptional kinetic reactivities of five-member ring cyclic phosphoesters.

The rates of acid and base-catalyzed hydrolytic reactivities of ethylene phosphate and diethylphosphate are multimillion-fold different (structure **10-3**) (Westheimer, 1968).

10-3

Relative hydrolytic rates   ~10^6–10^7 1

The five-membered ring is strained, and strain is relieved by hydrolysis to hydroxyethyl phosphate. Relief of strain in the overall reaction can explain the rate difference only if the mechanism includes relief of strain in a rate-limiting step.

Normally, the hydrolysis of a phosphodiester proceeds through a tight transition state. In the case of five-membered ring cyclic phosphates, the relief of strain on formation of the transition state is sufficient to allow the formation of trigonal bipyramidal oxyphosphoranes as discrete intermediates in an associative mechanism. Formation of trigonal bipyramids in five-membered ring cyclic phosphorus compounds follows preference rules (Westheimer, 1968):

1. Apical positions are occupied by the more electronegative groups.
2. Anionic substituents must be equatorial.
3. In the five-member ring, one substituent must be apical and the other equatorial.

The formation of a trigonal bipyramid in a cyclic phosphate requires apical attack of a nucleophile. By the principal of microscopic reversibility, the leaving group departs from an apical position. Strain in the O—P—O bond angle is relieved in the O—P—O 90-degree angle of the trigonal bipyramical cyclic phosphorane, and this accounts for rate enhancement in the reactions of five-member ring cyclic phosphates. As a final point of chemical fact, pentavalent phosphorus compounds are subject to shape changes, in which two apical groups become equatorial and two equatorial groups become apical. This process is known as pseudorotation, because the transformation of trigonal bipyramids appears is as if the molecule had rotated by 90 degrees. Pseudorotation does not entail bond cleavages, only the shortening of bonds to apical groups coupled with lengthening of bonds to equatorial groups, together with the associated bond angle transitions.

Consider the mechanism of acid-catalyzed hydrolysis of a five-member ring cyclic triester in fig. 10-2. Methyl ethylene phosphate undergoes either endocyclic or exocyclic cleavage on hydrolysis to a phosphodiester. As shown in table 10-2, both exocyclic and endocyclic cleavages take place at comparable rates, 10 million times faster than the hydrolysis of trimethylphosphate.

The figure shows the mechanism by which both exocyclic and endocyclic cleavage take place through trigonal pyramidal oxyphosphorane intermediates in the acid catalyzed reaction. The leaving group is always protonated and always departs from an apical position. In a trigonal bipyramid with a five-membered ring, one substituent is equatorial and the other apical; both cannot be apical for steric reasons. The incoming nucleophile must enter an apical position. The methoxyl group cannot be in apical position after addition of the nucleophile, which is water in this case. The only way that methanol can leave in exocyclic cleavage is for the trigonal pyramidal intermediate to change shape through pseudorotation, Because the most polar groups must be apical, pseudorotation is most facile for the neutral form, and this is illustrated in fig. 10-2; notice that it leads to an apical position for the methoxyl group and its eventual departure. Because pseudorotation is normally not rate-limiting, exocyclic cleavage can be as fast as endocylcic cleavage.

The mechanism in fig. 10-2 explains the results in table 10-2. Exocyclic and endocyclic cleavage of methyl ethylene phosphate are equally fast because pseudorotation does not limit the rates. Methyl propylene phosphonate undergoes fast ring opening because ring strain is relieved on formation of the trigonal bipyramidal intermediate. However, exocyclic cleavage is slow because the intermediate cannot undergo pseudorotation rapidly without violating the preference rule requiring the polar oxy groups to occupy apical positions. In the lower entry, the cyclic phostonate, trigonal bipyramid formation is subject to a large barrier because of the absence of an oxy group in the ring to occupy an apical position and the rule that one ring substituent must be apical. No relief of ring strain can occur in the hydrolytic mechanism.

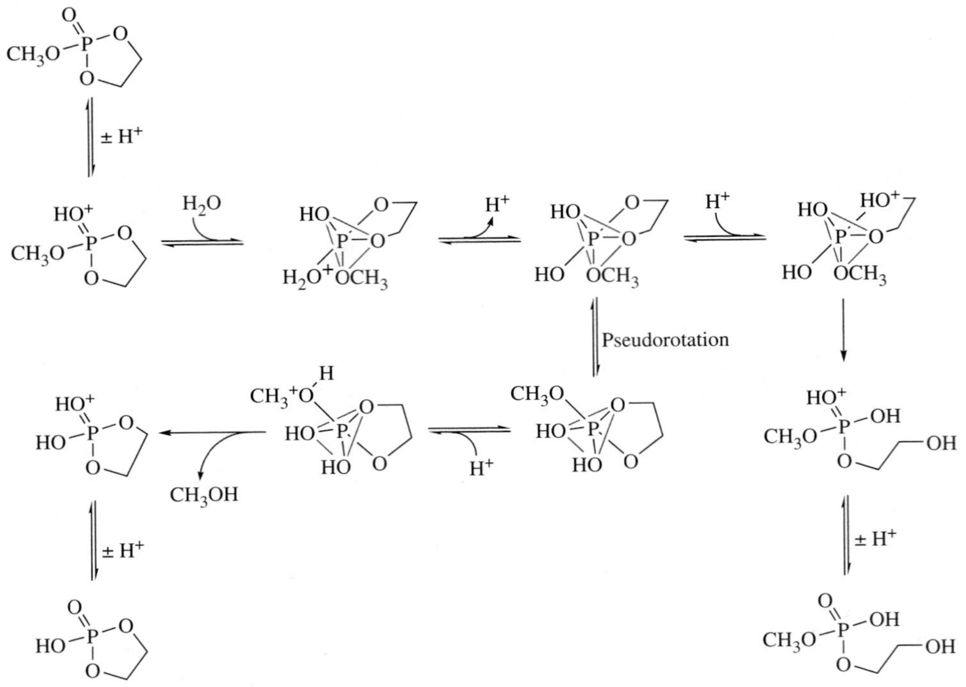

486 Enzymatic Reaction Mechanisms

Fig. 10-2. Acid-catalyzed hydrolysis of a methyl ethylene phosphate. Methyl ethylene phosphate is a five-membered-ring cyclic phosphotriester that can undergo endocyclic or exocyclic cleavage on hydrolysis to a phosphodiester. The mechanism is shown by which both exocyclic and endocyclic cleavage take place through trigonal pyramidal oxyphosphorane intermediates in the acid-catalyzed reaction. Initial protonation (upper left) is followed by addition of water to form a cationic trigonal bipyramid. The red lines define the plane of the trigonal bipyramid. Proton transfer to the apical ring oxygen (upper horizontal line) makes the ring oxygen a good leaving group. On ring cleavage, the cationic phosphodiester loses a proton to form the product methyl hydroxyethyl phosphate (lower right). Exocyclic cleavage requires elimination of methanol and proceeds by pseudorotation of the neutral trigonal bipyramid, leading to another neutral trigonal bipyramid with the apical methoxyl group. Protonation of the apical methoxyl group allows it to leave from an apical position to form ethylene phosphate.

Table 10-2. Relatives Rates of Phosphotriester Hydrolysis

Phosphotriester	Relative Cleavage Rate	
	Exocyclic	Endocyclic
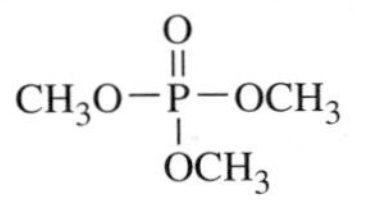 $CH_3O-P(=O)(OCH_3)-OCH_3$	(1)	(1)
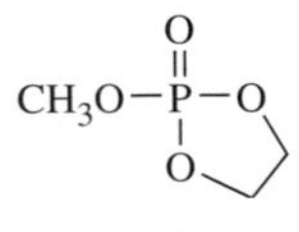 $CH_3O-P(=O)(O-)-O$ (5-ring)	10^7	10^7
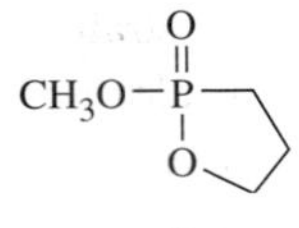 $CH_3O-P(=O)-$ (6-ring)	1	10^7
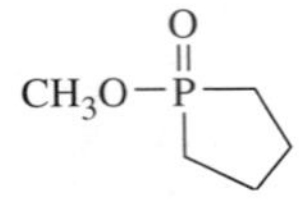 $CH_3O-P(=O)-$ (acyclic pentyl)	1	—

No examples of pseuorotation have been found in enzymatic mechanisms. However, five-member ring cyclic phosphates are intermediates in a number of enzymatic reactions, and relief of strain is likely important in these mechanisms.

Enzymatic Phosphoryl Group Transfer

Single and Double Displacements

The question of the possible intermediate formation of a covalent phosphoenzyme intermediate inevitably arises in every study of enzymatic substitution at phosphorus. This question is less difficult to settle today than in the past, partly because of the development of biochemical and physical methods to characterize phosphoenzymes and partly because of an increased awareness of the existence of artifacts resulting from adventitious phosphotransfer processes. In this section, we consider the methods in use to distinguish single-displacement and double-displacement mechanisms.

Steady-State Kinetics

The kinetics of phosphoryl group transfer reactions can diagnose the involvement of a covalent phosphoenzyme whenever ping pong kinetics mandates a covalent intermediate, and the only chemically reasonable intermediate is a phosphoenzyme. The observation of sequential kinetics does not give information about the participation of a phosphoenzyme. The kinetic mechanisms in scheme 10-3 clarify the contributions and limitations of steady-state kinetics to the assignment of single- and double-displacement mechanisms in bisubstrate group transfer reactions. The generic reaction is of a phosphodonor A–P reacting with a phosphoacceptor B to form products A and B–P.

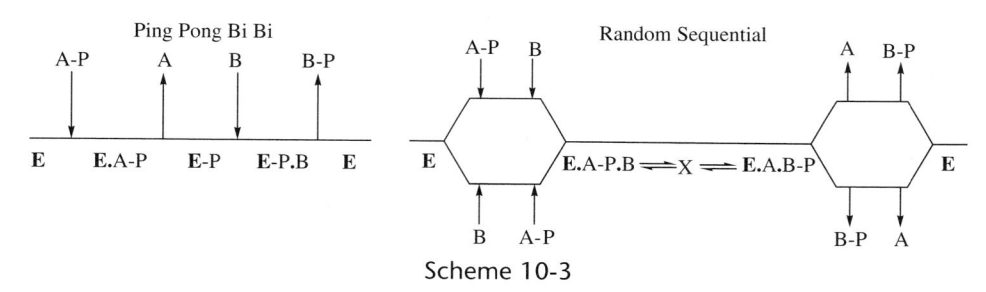

Scheme 10-3

The ping pong mechanism on the left side of scheme 10-3 is definitive in diagnosing a phosphoenzyme intermediate only when the kinetics is unequivocal. In chapter 2, we discuss the pitfalls in assigning this mechanism and the methods of verification, in particular the observation of the relevant exchange reactions and their rates. Note that the mechanism mandates the formation of the E-P on mixing the enzyme with the phophodonor (A-P) irrespective of the presence of the acceptor (B). This fact simultaneously mandates the exchange of labeled A (A*) with A-P at a kinetically competent rate.

The sequential mechanism on the right side of scheme 10-3 does not provide information about the possible participation of a phosphoenzyme because steady-state kinetics cannot determine the number of steps in the interconversion of the ternary complexes. If it is a single step phosphoryl group transfer, then the possible intermediate X in scheme 10-3 does not exist. Alternatively, if there are two steps, X may include a covalent phosphoenzyme (E-P.A.B). Because of the requirement for ternary complex formation before any chemical step, the sequential mechanism does not allow the exchange reactions typical of the ping pong mechanism. Analogous ambiguities exist in the kinetics of phosphatases and ATP-dependent synthetases. Other methods are required to test for the possible participation of a phosphoenzyme in a sequential mechanism.

Stereochemistry

The tetrahedral array of ligands in phosphates, and the stability of phosphate ligands against spontaneous exchange, introduces the same potential for chirality at phosphorus that is found in carbon compounds. Most phosphate anions are not chiral because of the presence of two or three oxygens among the substituents. These may be made different by substitution of sulfur for one of the oxygens. This substitution alters the delocalization of electrostatic charge in such a way that one negative charge is localized on sulfur, which is always singly bonded to phosphorus. In contrast, charge is equally delocalized among oxygens, and the P—O bond order may be higher than 1.0, for example 1.5 or 1.25. The structural formulations for phosphate and phosphorothioate monoanions and dianions are shown in structure **10-4** (Frey and Sammons, 1985).

In stereochemical analysis of substitution at phosphorus in biological molecules, the phosphate center may be made chiral by substitution of oxygen with sulfur or heavy isotopes of oxygen (structure **10-5**).

$$a > b > c \ : \ S > OR > {}^{18}O > {}^{17}O > O$$

The configurational symbols are assigned by discounting bond orders and electrostatic charges and ordering the substituents according to the priority rules indicated (structure **10-5**). Viewing the molecule from the side opposite the lowest priority group, usually O for ^{16}O, R_P is assigned to the isomer in which the substituents a, b, and c appear with decreasing priority in a clockwise array. The isomer with the same substituents in counterclockwise array is assigned as S_P. Any biological phosphate compound can be synthesized with chiral phosphorus; and enzymatic, mass spectrometric, and NMR methods for determining the configuration at phosphorus are also available (Buchwald et al., 1982; Frey, 1982, 1989, 1992; Gerlt et al., 1983; Knowles, 1980).

Stereochemical analysis of a large number of phosphotransferase and nucleotidyltransferase reactions sustains the following rule: Every enzymatic substitution at phosphorus proceeds with inversion of configuration at phosphorus. It follows that a single-displacement mechanism proceeds with overall inversion of configuration, and a double-displacement proceeds with overall retention of configuration at phosphorus. Stereochemical results have settled many controversies and ambiguities regarding the involvement or absence of covalent phosphoenzymes in enzymatic reaction mechanisms.

Characterization of Covalent Phosphoenzymes

When evidence indicates the participation of a covalent phosphoenzyme, its characterization becomes essential to understanding the reaction mechanism. This can be straightforward when the phosphoenzyme can be isolated, crystallized, and its structure determined by x-ray crystallography or NMR. However, these are not routine procedures, and in any case phosphoenzymes are not always chemically stable enough to survive global structural analysis. Biochemical methods are available for the identification of a phosphorylated amino acid residue in a phosphoprotein.

Often, the first step in the isolation of a covalent intermediate is gel permeation chromatography of an enzyme that has been treated with a radiochemically labeled substrate. Gel filtration of a solution containing nucleoside diphosphate kinase and [γ-^{32}P]ATP can separate the [^{32}P]phosphoenzyme from unreacted [γ-^{32}P]ATP. Two control experiments can verify the [^{32}P]phosphoenzyme. The gel filtration experiment with [^{14}C]ATP does not yield a radioactive protein, and the denatured [^{32}P]phosphoenzyme retains the ^{32}P. The importance of these control experiments cannot be overstated. Their omission has led to mistaken conclusions in several studies of phosphotransferases and other group transferases, in which the proteins were found to bind substrates or products noncovalently but tightly enough to survive gel filtration.

The chemical properties of phosphoamino acids guide the biochemical characterization of the phosphoamino acid residue in a protein. Phosphoserine and phosphothreonine are acid-stable but labile to β-elimination of phosphate in 0.1 M base at 60°C. The resulting dehydroamino acid residue can be reduced by NaB^{3}H$_4$ to generate tritiated alanine or α-aminobutyrate at the phosphorylation site in the protein. Phosphoserine and phosphothreonine, or peptides containing them, can be isolated from partial acid digests of phosphoproteins and characterized. Phosphotyrosine is stable in dilute acid and base. Phosphohistidine, phospholysine, and phosphoarginine are very labile to hydrolysis at low pH and significantly labile even in neutral solutions, but they are stable to alkaline hydrolysis. They can be isolated from alkaline digests of phosphoproteins and chemically characterized. Phosphoaspartate and phosphoglutamate are labile to hydrolysis at both low and high pH and are reasonably stable at pH 7. As acyl phosphates, they are subject to reduction by sodium borohydride, which displaces phosphate and further reduces the resulting aldehyde to the alcohol. Reduction of a phosphoaspartate residue in Ca/Mg-dependent ATPase with NaB^{3}H$_4$ leads to a residue of 4-hydroxy-[4-^{3}H]aminobutyrate in place of phosphoaspartate (Degani and Boyer, 1973). Reaction of an acyl phosphate with hydroxylamine displaces phosphate and converts the acyl group into a hydroxamate.

The actions of a few nucleotidyltransferases and ATP-dependent ligases involve covalent nucleotidyl enzymes, such as AMP-enzymes or UMP-enzymes. Two methods are available for characterizing them. Mass spectral analysis can identify the nucleotide. The nucleotidyl-amino acid can easily be converted into the corresponding phosphoamino acid by chemical removal of the nucleoside through periodate cleavage between the 2'- and 3'-OH groups followed by α,β-elimination of the phosphate group from C5' (Yang and Frey, 1979), The resulting phosphoprotein can be characterized as described earlier.

Phosphotransferases

Adenylate Kinase and Nucleoside Diphosphate Kinase

These two phosphotransferases are well studied and discussed in chapter 2. Although the reactions are chemically similar, the mechanisms are quite different. Adenylate kinase acts by a random sequential kinetic mechanism, and group transfer takes place with inversion of configuration at phosphorus (see chap. 2). Nucleoside diphosphate kinase acts by a ping pong bi bi kinetic mechanism and a double-displacement chemical mechanism through a phosphoenzyme intermediate (P-His122), and group transfer proceeds with retention of configuration at phosphorus (see chap. 2). The mechanism is thought to arise on the principle of "economy in the evolution of binding sites" (see chap. 2).

Creatine Kinase

Creatine phosphate serves as a cellular storage species of high-energy phosphate that is readily available for the formation of ATP from ADP by the action of creatine kinase (EC 2.7.3.2). Creatine phosphate is a high-energy phosphate ($\Delta G'^\circ = -10$ kcal mol^{-1} for hydrolysis) by virtue of being a phosphoramidate. The reversibility of the reaction of creatine kinase (eq. 10-5), allows creatine phosphate to be formed whenever the concentration of ATP is high, and it makes creatine phosphate an efficient reserve of high-energy phosphate to generate ATP under conditions of high-energy demand.

$$\text{creatine phosphate} + \text{MgADP} \rightleftharpoons \text{creatine} + \text{MgATP} \quad (10\text{-}5)$$

The reaction proceeds by an equilibrium random sequential kinetic mechanism (Morrison and Cleland, 1966), and phosphoryl transfer takes place with inversion of configuration at phosphorus, indicating a single-displacement mechanism (Hansen and Knowles, 1981). A review summarizes early biochemical studies of the reaction mechanism (Kenyon and Reed, 1983).

The action of creatine kinase requires a divalent metal ion, either Mg^{2+} or Mn^{2+} in complex with ADP. The more prevalent Mg^{2+} is the likely activating metal ion in cells. Structurally and stereochemically distinct complexes of divalent metal ions with ADP or ATP exist in solution, and species that bind to a particular enzyme can be determined in various ways. For example, the following two structures of MgADP (structure **10-6**) differ in configuration at P$_\alpha$. In solution, the two Mg^{2+} complexes are in equilibrium, but only the Δ-isomer binds to creatine kinase.

10-6 Λ-MgADP Δ-MgADP

Many more chemically and stereochemically distinct structures of MgATP are possible because of β,γ; α,β; and α,β,γ coordination and the possibility of P-chiral centers at both P$_\alpha$ and P$_\beta$. Studies of coordination exchange-inert complexes, such as Cr(III) and Co(III) complexes of ADP and ATP, which can be purified and structurally characterized, allows their interactions with enzymes to be studied and the configurations Λ or Δ to be determined as the species binding to a particular enzyme (Cleland, 1982). By extension, the corresponding structures of the magnesium complexes would be expected to bind. Alternatively, thionucleotides can be synthesized with sulfur stereospecifically substituted for oxygen at P$_\alpha$ or P$_\beta$. The differential divalent metal binding properties of S and O lead to patterns of enzyme activation by different metals. In favorable cases, configurational assignments of metal-nucleotides at active sites can be made (Cohn, 1982).

The structures of Mn-nucleotides bound to the active sites of enzymes can also be determined by synthesis of chirally labeled ^{17}O-nucleotides and exploitation of the paramagnetism of manganese. The EPR signal due to unpaired electrons of Mn^{2+} is broadened by nuclear hyperfine coupling with ^{17}O when it is a ligand. For example, when ADP is labeled with ^{17}O in the α-pro-R position, the Λ-Mn complex displays broadening of the Mn^{2+} EPR signal. The Mn-EPR spectrum obtained in $H_2^{17}O$, with MnATP bound at an active site, reports the number of water molecules bound to Mn in the enzyme-MnATP complex. On the basis of data of this type obtained with ADP and ATP stereospecifically labeled with ^{17}O, the structure of MnATP in the active site of creatine kinase is known to involve the Δ configuration at P_α and Λ at P_β, and Mn^{2+} is coordinated to three molecules of water. Mn^{2+} is hexavalent, with three ligands from ATP, three from water molecules, and no ligands from the enzyme itself (Burgers and Eckstein, 1980; Leyh et al., 1982).

A crystal structure of the creatine kinase from *Torpedo californicus* in a complex with ADP, nitrate, and creatine reveals much more about the active site (Lahiri et al., 2002). Nitrate, NO_3^-, is sterically and electrostatically similar to metaphosphate monoanion, PO_3^-, and is regarded as an analog. The complex is thought to be analogous to the transition state for phosphoryl transfer (Reed and Cohn, 1972).

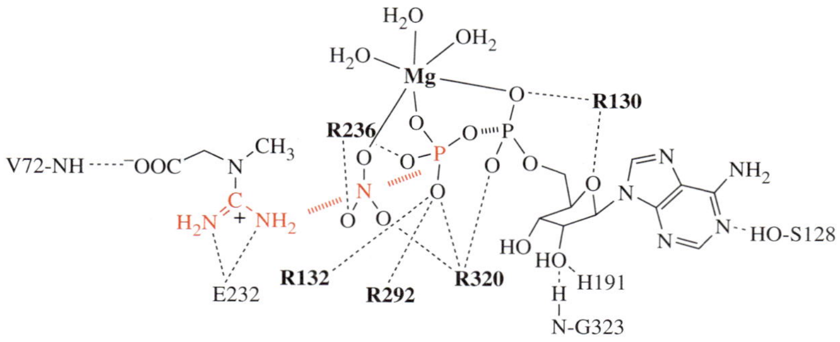

Scheme 10-4

Scheme 10-4 illustrates the steric orientations of creatine and nitrate with respect to MgADP and also the principal interactions with amino acid side chains and hydrogen bonding peptide groups. A striking feature is the positive charge in the active site. The five arginine residues, presumably carrying positive charges, form ion-paired hydrogen bonds with ADP^{3-} and NO_3^-. Moreover, Mg^{2+} contributes two more positive charges, for an excess of three. Glutamate 232 presumably partially compensates with an additional negative charge; it is thought to absorb the proton released by phosphorylation of the guanidino group in creatine. Mutation of Glu232 to aspartate, a conservative change, decreases the activity by 500-fold (Cantwell et al., 2001). Other peptide interactions with the substrates are mediated by bridging fixed water molecules (Lahiri et al., 2002). Arginine kinase is similar to creatine kinase (Yousef et al., 2002).

Creatine kinase is a dimer of chemically identical, approximately 40-kDa subunits, depending on species. However, the subunits are not structurally identical in the complex with MgADP, nitrate, and creatine (Lahiri et al., 2002). Creatine and nitrate are bound with MgADP to only one of the subunits; the other subunit has MgADP at the active site, and there are associated differences in the polypeptide chains. Given the concentrations of nitrate and creatine in the crystallization experiment, the enzyme should have been saturated. The differential structures of the subunits suggest negative cooperativity in creatine binding for this enzyme (Lahiri et al., 2002).

Acetate Kinase

Acetyl phosphate arises in bacteria through the action of phosphotransacetylase, which catalyzes the displacement of CoASH from acetyl CoA by phosphate. Acetate kinase (EC 2.7.2.1) catalyzes phosphotransfer from acetyl phosphate to ADP to form ATP according to eq. 10-6.

$$\underset{H_3C}{}\overset{O}{\underset{}{C}}-OPO_3^{2-} + MgADP^- \xrightleftharpoons{\text{Acetic kinase}} \underset{H_3C}{}\overset{O}{\underset{}{C}}-O^- + MgATP^{2-} \quad (10\text{-}6)$$

The mechanism of this reaction was controversial because of apparent conflicts between kinetic and biochemical results. The enzyme could be phosphorylated by either ATP or acetyl phosphate (Anthony and Spector, 1972; Purich, 1982), and the phosphoenzyme could be dephosphorylated by ADP, but the steady-state kinetics was found to be sequential (Janson and Cleland, 1974). It seemed that a phosphoenzyme intermediate and double-displacement mechanism might be taking place in the interconversion of ternary complexes in this case. However, stereochemical analysis showed that phosphotransfer in the reverse of eq. 10-6 proceeded with inversion of configuration at phosphorus, indicating a single-displacement mechanism (Blättler and Knowles, 1979a).

The covalent phosphoryl-acetate kinase might be regarded as a side product in the interaction of acetate kinase with the reactive phosphorylating agents ATP or acetyl phosphate. However, it turns out to be a manifestation of the dual biological function of acetate kinase. Phospho-acetate kinase is now known to be a phosphodonor to the sugar phosphotransferase transport system in bacteria. In this system, sugars bound to specific receptors are phosphorylated to 6-phosphosugars on transport into the cells. A cytosolic phosphodonor for this system is phosphoenolpyruvate. However, phospho-acetate kinase also functions in this capacity (Fox et al., 1986). Acetate kinase has two distinct biological functions, and the participation of the phosphoenzyme in the sugar phosphotransfer system introduces complications when interfaced with mechanistic analysis of the acetate kinase function.

The structure of acetate kinase with ADP bound to the active site, shows the overall chain fold and the active site contacts with the product (Buss et al., 2001). The phosphorylation site is near the nucleotide binding site. ATP can generate phospho-acetate kinase whenever it is needed for sugar transport, although its formation and reaction with acetate are not fast enough to aaccount for the reaction of eq. 10-6.

Phosphoglycerate Kinase

A reaction chemically similar to that of acetate kinase is the dephosphorylation of 1,3-diphosphoglycerate (1,3DPG) with ADP to form 3-phosphoglycerate (3PGA) and ATP catalyzed by phosphoglycerate kinase (eq. 10-7).

$$\begin{array}{c} O\underset{}{\diagdown}\underset{}{C}\diagup OPO_3^{2-} \\ | \\ H-C-OH \\ | \\ CH_2OPO_3^{2-} \end{array} + MgADP^- \xrightleftharpoons{\text{Phosphoglycerate kinase}} \begin{array}{c} O\underset{}{\diagdown}\underset{}{C}\diagup O^- \\ | \\ H-C-OH \\ | \\ CH_2OPO_3^{2-} \end{array} + MgATP^{2-} \quad (10\text{-}7)$$

1,3-Diphosphoglycerate arises from the action of glyceraldehyde-3-phosphate dehydrogenase in glycolysis, so that phosphoglycerate kinase (EC 2.7.2.3) plays an essential role in glycolysis. It is also important in CO_2 fixation in plants, so that it is essentially ubiquitous in nature.

Misleading evidence seemed to implicate a covalent phosphoenzyme in the mechanism. The observations were explained by the presence of a tightly but noncovalently bound substrate in the enzyme preparation (Johnson et al., 1976; Walsh and Spector, 1971). The kinetics proved to be sequential (Janson and Cleland, 1974; Larsson-Raznikiewicz, 1967), and inversion of configuration at phosphorus was found in a stereochemical study (Webb and Trentham, 1980). All evidence pointed to a single-displacement mechanism in the interconversion of the ternary complexes E.1,3DPG.ADP and E.3PGA.ATP.

Phosphoglycerate kinase from all species is monomeric and has a molecular mass of 48 to 50 kDa, depending on species variation. It is very active, with a turnover number of about 1000 s^{-1} for the yeast enzyme. The structure is shown in fig. 10-3. The large cleft is open in the free enzyme and closed in the ternary complexes. It is evident that a large conformational change accompanies the binding of substrates.

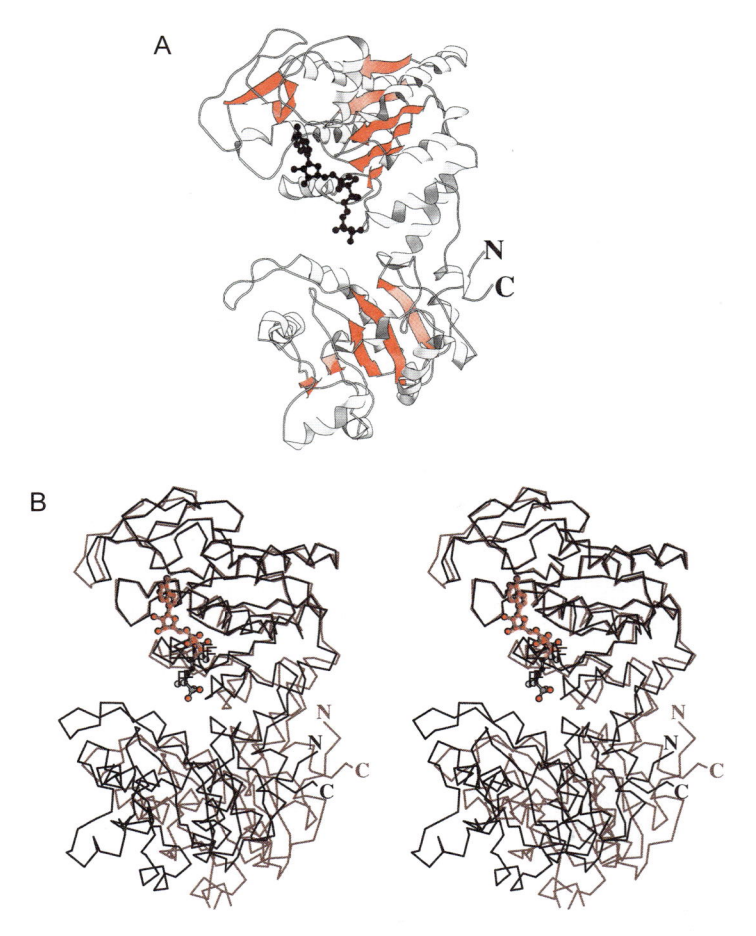

Fig. 10-3. Structure of phosphoglycerate kinase. (A) Ribbon diagram of phosphoglycerate kinase from *Trypanosoma brucei* with a bi-substrate analog, adenylyl 1,1,5,5-tetrafluoropentane-1,5-bisphosphate bound (black) (1.6-Å resolution; PDB 16PK; Bernstein et al., 1998). The enzyme has two domains with β-sheets (red) at their cores. (B) A Cα trace alignment of the bi-substrate analog structure with a related complex of ADP and phosphoglycerate kinase from *Bacillus stearothermophilus* (1.65-Å resolution; PDB 1PHP; Davies et al., 1994). The two domains are drawn in together in the bi-substrate analog complex compared with the ADP complex.

Continued

494   Enzymatic Reaction Mechanisms

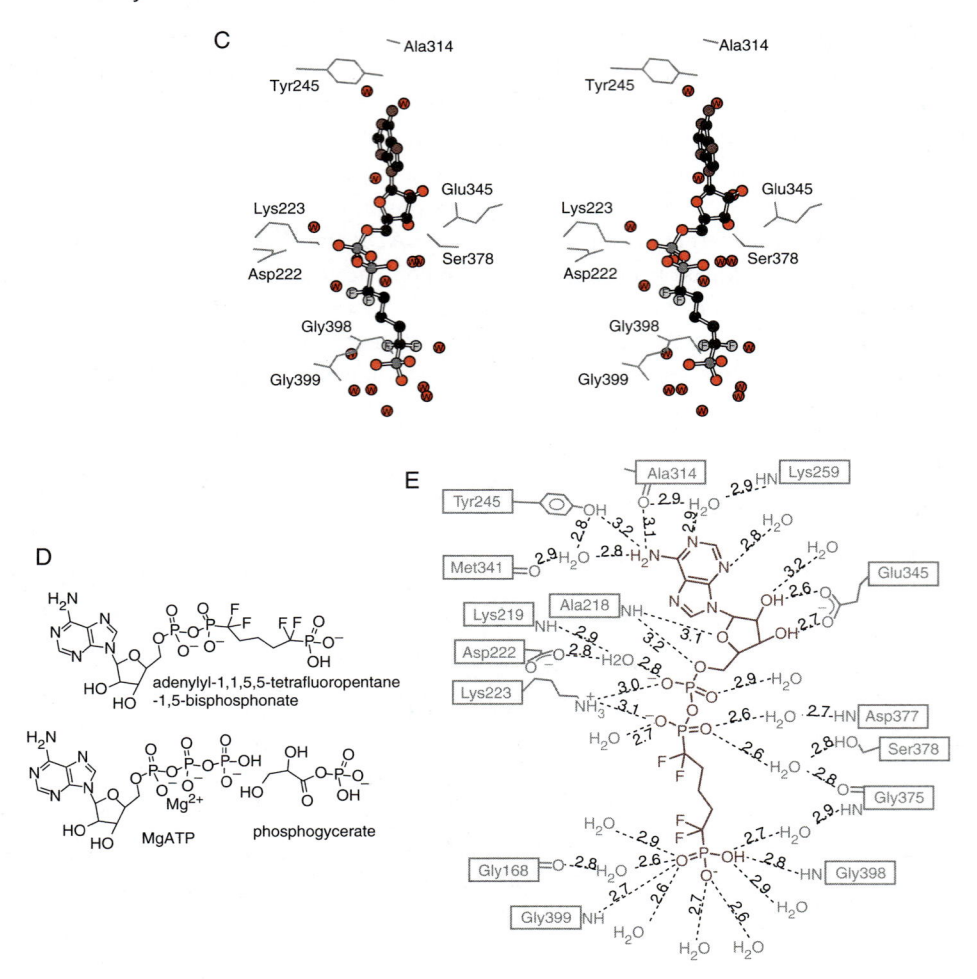

Fig. 10-3, cont'd. (C) The active site of the bi-substrate analog complex in stereo. (D) The structures of the bi-substrate analog and the substrates are compared. (E) Information from the image in C is presented schematically in two dimensions, with distances given in angstroms.

The preceding sections demonstrate that the characterization of enzymatic phosphotransfer mechanisms is not a simple matter. The examples exemplify the complications that arise in the analysis of these and any other enzymatic reaction mechanisms. All possibilities must be considered, with appropriate attention to details.

Pyruvate Kinase

The action of pyruvate kinase (EC 2.7.1.40), in addition to its importance in glycolysis, is interesting for its unique mechanistic aspects and relationships with other phosphokinases such as phosphoenolpyruvate carboxykinase (EC 4.1.1.49). Also of interest are the fundamental mechanistic differences between the actions of pyruvate kinase and pyruvate phosphate dikinase (EC 2.7.9.1). Pyruvate kinase catalyzes phosphoryl group transfer between phosphoenolpyruvate and ADP according to eq. 10-8.

$$\text{phosphoenolpyruvate} + \text{MgADP}^- + \text{H}^+ \xrightleftharpoons{\text{Pyruvate kinase}} \text{pyruvate} + \text{MgATP}^{2-} \tag{10-8}$$

The enzyme from many species is tetrameric and composed of identical 40- to 50-kDa subunits. Pyruvate kinase in many species displays allosteric behavior, usually homotropic activation by phosphoenolpyruvate and heterotropic activation by fructose-1,6-bisphosphate (Kayne, 1973).

Unlike most other phosphotransferases, pyruvate kinase catalyzes the enolization of pyruvate and the ketonization of enolpyruvate, in addition to phosphoryl group transfer (Kuo et al., 1979; Robinson and Rose, 1972). There is no evidence of a phosphoenzyme intermediate, and phosphoryl group transfer proceeds with inversion of configuration at phosphorus, consistent with a single-displacement mechanism (Blättler and Knowles, 1979b; Pliura et al., 1980).

The mechanism in fig. 10-4 explains the available facts. Two divalent metal ion binding sites with different affinities for Mg^{2+} can be detected in complexes of the enzyme with ATP and substrate analogs (Baek and Nowak, 1982). Ligation of one site with Mn^{2+} and the other with Mg^{2+} allows ligands for the two sites to be assigned by Mn^{2+}-EPR (Lodato and Reed, 1987). The Mn^{2+}-complex of pyruvate kinase with ATP and oxalate, a structural analog of enolpyruvate, shows that oxalate and the γ-phosphate of ATP coordinate Mn^{2+}. Nuclear hyperfine broadening of the Mn^{2+} EPR spectrum by ^{17}O-labeled oxalate and ATP demonstrate these interactions. One Mg^{2+} is shown bridging enolpyruvate and ATP in fig. 10-4. The other Mg^{2+} site is the tridentate complex with ATP.

The equilibrium constant for the reaction of pyruvate kinase is of interest. Most phosphotransferase reactions of ATP operate in the direction of phosphorylation of a substrate by ATP. For example, $\Delta G°'$ for the phosphorylation of glucose by ATP in the reaction of hexokinase is -5 kcal mol^{-1}. Although pyruvate kinase does catalyze the phosphorylation of pyruvate, the equilibrium lies in favor of ATP-formation, and $\Delta G°'$ for reaction 10-8 is -6 kcal mol^{-1}. This is a consequence of the high phosphoryl group transfer potential of phosphoenolpyruvate; $\Delta G°'$ for its hydrolysis is -14 kcal mol^{-1}. Phosphoenolpyruvate is a high-energy phosphodonor because enolpyruvate initially formed in step 2 of fig. 10-4 undergoes ketonization in step 3, an energetically downhill reaction. The same phenomenon accompanies the hydrolysis of phosphoenolpyruvate. It is likely that the phosphotransfer

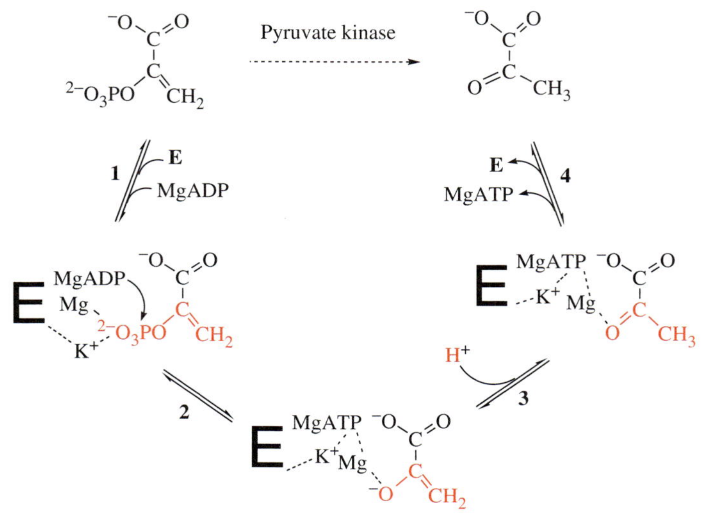

Fig. 10-4. A mechanism for the action of pyruvate kinase.

between ATP and enolpyruvate is energetically similar to that between ATP and other alcohols such as glucose. However the enolization/ketonization process in the reaction of phosphoenolpyruvate strongly influences the overall energetics. Based on the standard free energy of hydrolysis of phosphoenolpyruvate, it seems that the value of $\Delta G^{\circ\prime}$ for the enolization of pyruvate is about +10 kcal mol^{-1}, essentially the difference between the values of $\Delta G^{\circ\prime}$ for the hydrolysis of phosphoenolpyruvate and a hexose phosphate.

Notice the potassium ion coordinated to the phosphoryl group in flight in the mechanism of Fig 10-4. Muscle pyruvate kinase was the first enzyme found to require potassium ion for activity (Boyer et al., 1942). More than 40 other enzymes have subsequently been found to require potassium ion for activity (Larsen and Reed, 2001). The structure with L-pholactate in place of phosphoenolpyruvate showed potassium coordinated to the phosphate group and enzymatic side chains Asn74, Ser76, Asp112, and Thr113 (Larsen et al., 1994). The structure of the complex with oxalate and ATP showed potassium coordinated to the γ-phosphate of ATP (Larsen et al., 1997), as shown in fig. 10-5. Note the tridentate coordination of one Mg^{2+} by ATP and the coordination of the second Mg^{2+} by the γ-phosphate of ATP and oxalate. As a structural analog of enolpyruvate, oxalate in its complex with pyruvate kinase and ATP has been regarded as an analog of the ternary complex with enolpyruvate formed in step 2 of fig. 10-4. This analogy has been supported by the reactivity of oxalyl phosphate as a substrate in place of phosphoenolpyruvate (Kofron and Reed, 1990).

Although they are homologous to the eukaryotic enzymes, bacterial pyruvate kinases do not require potassium ion. In the aligned amino acid sequences, Glu117 in the active site of the muscle enzyme is replaced by lysine in the pyruvate kinases from *Corynebacterium glutamicum* and *Escherichia coli*. According to one hypothesis, the ε-aminium group of this lysine in a bacterial enzyme satisfies the role of potassium ion in the muscle enzyme. This interpretation is supported by the consequences of mutating Glu117 in the muscle enzyme to Lys117, which leads to an active pyruvate kinase that does not require potassium ion (Laughlin and Reed, 1997).

The mechanism by which pyruvate kinase incorporates the proton in step 3 of fig. 10-4 into the methyl group of pyruvate is not known in detail. The proton is stereospecifically incorporated on the *si*-face of enolpyruvate (Kuo et al., 1979). The proton donor seems to be a monoprotic acid with a high pK_a (Rose et al., 1991). The structure of muscle pyruvate kinase shows Lys269 near the re-face of pyruvate and Thr327 near the *si*-face of pyruvate bound to the active site (Larsen et al., 1994). A similar orientation of Lys240 and Thr298 exists in the phosphoglycolate complex of yeast pyruvate kinase (Jurica et al., 1998). Lys269 (240 yeast) is not in position to donate a proton to the *si*-face of enolpyruvate; it plays a role in stabilizing the transition state for phosphoryl group transfer (Bollenbach et al., 1999). Mutation of Thr298 to alanine in yeast pyruvate kinase modestly decreases the values of k_{cat} and k_{cat}/K_m ($\approx$10-fold). This mutation causes a similar decrease in catalysis of the exchange of tritium from [^{3}H]pyruvate with water (Susan-Resign and Nowak, 2003). The effect is insufficient to implicate Thr298 as the driving force for protonation of enolpyruvate, although Thr298 may mediate the process and serve as the proximal source of the proton. Evidence points to water as the source of the proton in step 3 of fig. 10-4 (Susan-Resign and Nowak, 2003). This would be in accord with the assignment of a high pK_a, monoprotic acid (Rose et al., 1991).

Pyruvate kinase catalyzes a number of reactions in addition to eq. 10-8. These include the phosphorylation of fluoride by ATP, the phosphorylation of hydroxylamine by ATP, and the decarboxylation of oxaloacetate. Kinetics and pH dependencies of these reactions have been reported (Dougherty and Cleland, 1985).

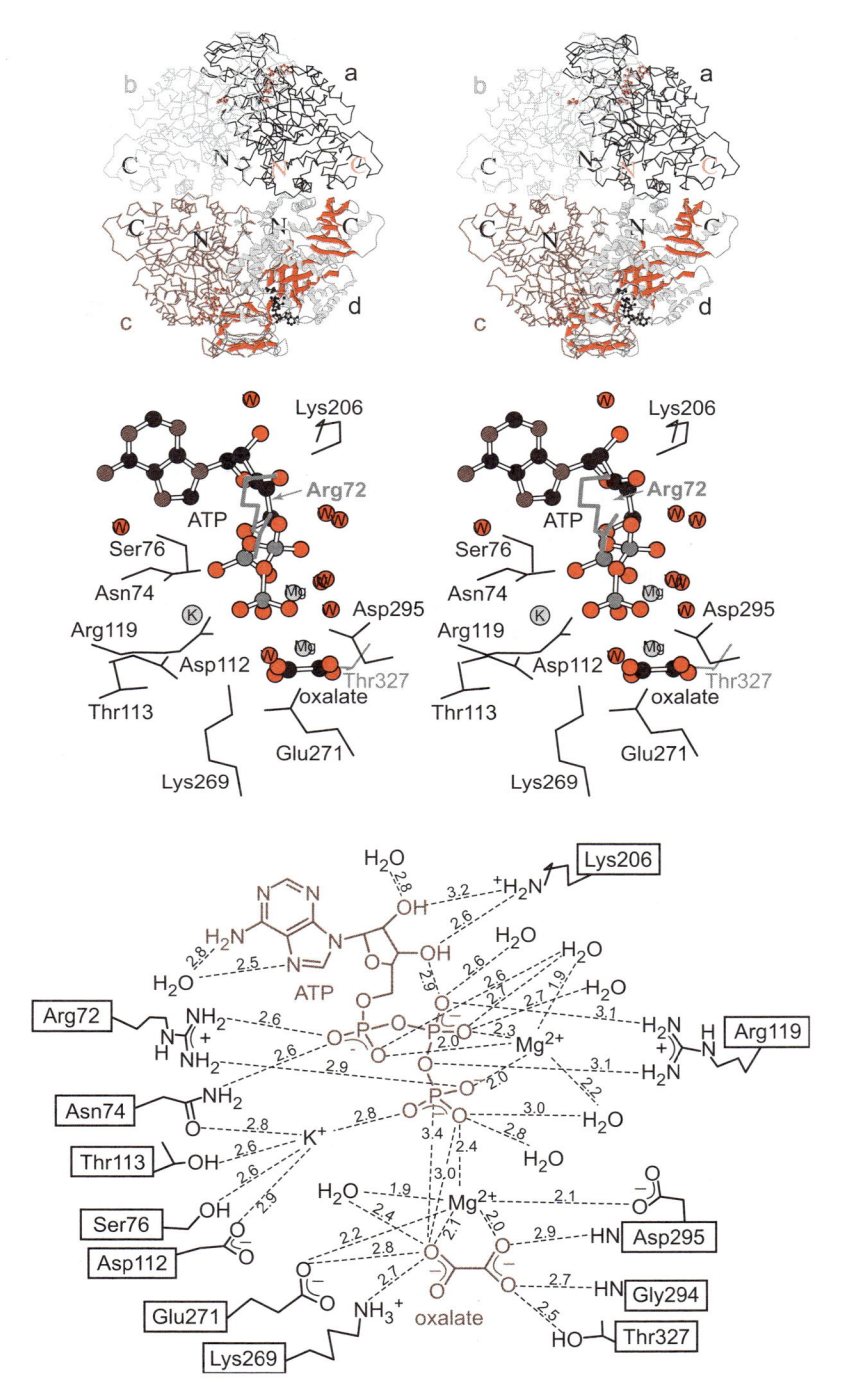

Fig. 10-5. The structure of pyruvate kinase in complex with ATP and oxalate. The top stereoview shows the *bis*-Mg^{2+}-ATP, oxalate, K$^+$ complex of rabbit muscle pyruvate kinase (2.10-Å resolution; PDB 1A49; Larsen et al., 1998). The enzyme is a homotetramer, with subunits a to c drawn as different-colored Cα traces and subunit d presented as a ribbon drawing. The active site is at the N-terminal face of a β-barrel (TIM-barrel) subdomain, which can be seen in subunit d, with the β-sheets in red and the ligands in black. The active site is shown in more detail in the center panel in stereo and is presented in two dimensions in the bottom panel.

497

Pyruvate Phosphate Dikinase

The energetically difficult production of phosphoenolpyruvate from pyruvate and ATP is overcome by use of both high-energy phosphoanhydride groups in ATP. The value of $\Delta G°'$ for hydrolysis of MgATP to MgADP and P_I is -7.8 kcal mol^{-1}, and for hydrolysis to AMP and MgPP$_I$, it is -10.3 kcal mol^{-1} (Arabshahi and Frey, 1998). The energy of both phosphoanhydride bonds is used in the action of pyruvate phosphate dikinase (EC 2.7.9.1) to produce phosphoenolpyruvate from pyruvate and ATP in bacteria and plants. The overall reaction is described by eq. 10-9.

$$\text{Pyruvate}^- + \text{MgATP}^{2-} + P_i^{2-} \rightleftharpoons \text{Phosphoenolpyruvate}^{3-} + \text{MgPP}_i^{2-} + \text{AMP}^{2-} + 2\text{H}^+ \quad (10\text{-}9)$$

As the reaction stands in eq. 10-9, the free energy available from the cleavage of ATP into AMP and PP$_i$ is $\Delta G'° = -10.3$ kcal mol^{-1} (Frey and Arabshahi, 1998), whereas -14 kcal mol^{-1} are required to phosphorylate pyruvate. On this basis, the reaction is not spontaneous as written ($\Delta G°' = +3$ kcal mol^{-1}). However, the hydrolysis of PP$_i$ by inorganic pyrophosphatase will generate an additional -4.6 kcal mol^{-1} (Jencks and Regenstein, 1970), making the production of phosphoenolpyruvate spontaneous in vivo.

The enzymatic mechanism follows a stepwise course according to eqs. 10-10a to 10-10c (Wood et al., 1977). Because the three steps (a, b, and c) take place independently, the covalent intermediates **E–PPMg** and **E–P** can be isolated.

$$\mathbf{E} + \text{MgATP} \rightleftharpoons \mathbf{E-PPMg.AMP} \rightleftharpoons \mathbf{E-PPMg} + \text{AMP} \quad (10\text{-}10a)$$

$$\mathbf{E-PPMg} + P_i \rightleftharpoons \mathbf{E-PPMg.}P_i \rightleftharpoons \mathbf{E-P.}\text{MgPP}_i \rightleftharpoons \mathbf{E-P} + \text{MgPP}_i \quad (10\text{-}10b)$$

$$\mathbf{E-P} + \text{Pyruvate} \rightleftharpoons \mathbf{E-P.}\text{Pyruvate} \rightleftharpoons \mathbf{E} + \text{Phosphoenolpyruvate} \quad (10\text{-}10c)$$

Moreover, the enzyme catalyzes the exchange of [^{14}C]AMP into ATP, by virtue of eq. 10-10a, and of ^{32}PP$_I$ into ATP, by virtue of eqs. 10-10a and 10-10b. Reaction of the enzyme with MgATP produces **E–PPMg**; reaction of **E–PPMg** with P_I produces **E–P**; and reaction of **E–P** with pyruvate produces phosphoenolpyruvate. The pyrophosphoryl and phosphoryl groups are bonded to a histidine residue, and the steady-state kinetics is hexa-uni ping pong, scheme 10-5 (Thrall et al., 1993).

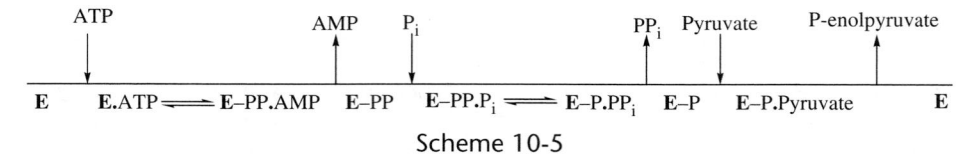

```
       ATP              AMP   P_i              PP_i  Pyruvate   P-enolpyruvate
        ↓                ↑     ↓                ↑      ↓            ↑
  E    E.ATP ⇌ E–PP.AMP  E–PP  E–PP.P_i ⇌ E–P.PP_i  E–P  E–P.Pyruvate   E
```
Scheme 10-5

The chemical mechanism in fig. 10-6 traces the fates of the and γ-phosphoryl groups in the reaction, in which the β-phosphoryl of ATP becomes the phosphate group in phosphoenolpyruvate, and the γ-phosphoryl group is transferred to phosphate in the formation of pyrophosphate. This is one of the few examples of enzymatic nucleophilic substitution at the β-phosphorus of ATP. The chemistry of the process is outlined in fig. 10-6, which shows how the β-phosphate of ATP becomes covalently bonded to the active site histidine

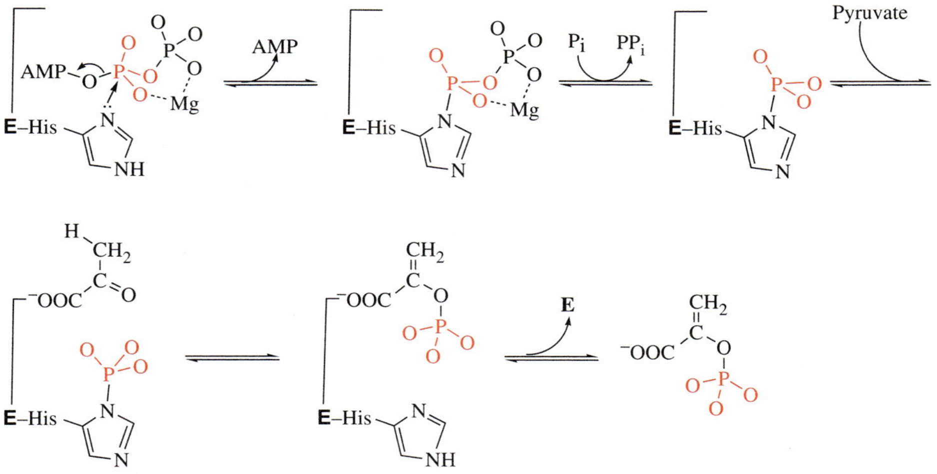

Fig. 10-6. Chemistry of phosphoenolpyruvate formation by pyruvate phosphate dikinase. Within the complex of enzyme with ATP, a histidine residue in the ATP binding domain reacts as a nucleophile with the β-phosphate of ATP to displace AMP and form the pyrophosphoryl enzyme (**E–PP**). Inorganic phosphate then reacts with the terminal phosphate to form PP$_i$ and the phosphoryl enzyme (**E–P**). Then, pyruvate binds to the pyruvate binding domain and undergoes phosphorylation by phosphohistidine in the ATP binding domain.

in the ATP binding domain to form the covalent **E-PP** intermediate and AMP. Reaction of E-PP with phosphate produces PP$_I$ and the covalent **E-P**, which then phosphorylates pyruvate to phosphoenolpyruvate.

Pyruvate and ATP bind to different domains of pyruvate phosphate dikinase and react independently. An N-terminal domain produced by truncation through recombinant DNA methodology catalyzes the partial reactions of eqs. 10-10a and 10-10b, and a C-terminal domain catalyzes the partial reaction of eq. 10-10c (Xu et al., 1995). The domains reside in physically separate locations in the three-dimensional structure and must interact through conformational transitions (Hertzberg et al., 2002).

Phosphofructokinase

An essential and the first committed step in glycolysis is the phosphorylation of β-D-fructose-6-P (F6P) to fructose-1,6-bisphosphate (FBP) by ATP catalyzed by phosphofructokinase (PFK; EC 2.7.1.11) according to eq. 10-11.

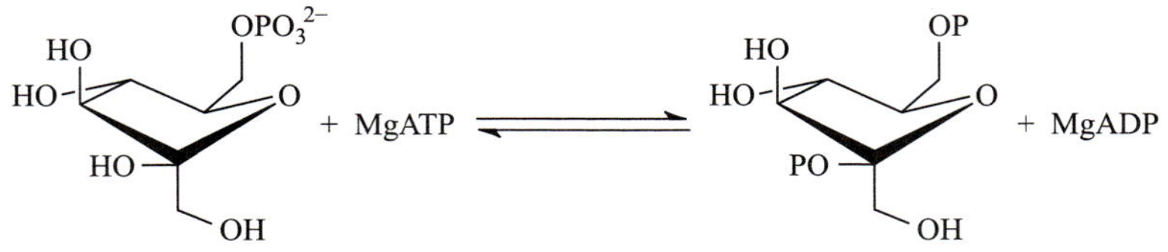

$$\text{F6P} + \text{MgATP} \rightleftharpoons \text{FBP} + \text{MgADP} \quad (10\text{-}11)$$

PFK is found in animals, plants, and bacteria. It displays complex behavior in studies of subunit composition because of aggregation at high concentrations (Bloxham and Lardy, 1973). The basic functional unit in most species appears to be tetrameric with identical subunits ranging in size from 35 to 53 kDa in various species. PFK is specific for F6P as the phosphoryl acceptor but accepts phosphoryl donors other than ATP.

Like many phosphotransferases, PFK catalyzes the exchange reactions typical of ping pong mechanisms at very low rates, much slower than overall catalysis (Bloxham and Lardy, 1973). As a further complication, double-reciprocal plots of activity often appear parallel. However, at low concentrations of substrates the double-reciprocal plots converge to the left of the ordinate, and inhibition patterns are most consistent with a random sequential kinetic mechanism. Phosphoryl group transfer proceeds with inversion of configuration at phosphorus (Lowe et al., 1981). The mechanism appears to involve ternary complex formation and a single step, in-line transfer of a phosphoryl group from ATP to F6P, which are bound at adjacent sites. PFK requires both monovalent and divalent cations for activity. The most favored cations are potassium and magnesium ions.

The structure of PFK from *E. coli* crystallized in the presence of substrates shows the products bound at adjacent sites in fig. 10-7. The crystal contains two conformations of the product complex, one in which ADP and FBP are both coordinated to Mg^{2+} and a second in which the product molecules are about 1.5 Å further apart and Mg^{2+} is coordinated only to ADP. The two conformations may represent two steps in the overall mechanism of product formation and dissociation.

Site-directed mutagenesis experiments on PFK from *E. coli* show that Asp127 is particularly important in the mechanism (Hellinga and Evans, 1987). Mutation of Asp127 to serine decreases the value of k_{cat} by 18,000. Asp127 is thought to function as the base that removes the proton from the 1-hydroxyl group of F6P in the transfer of the phosphoryl group from ATP, as illustrated in scheme 10-6.

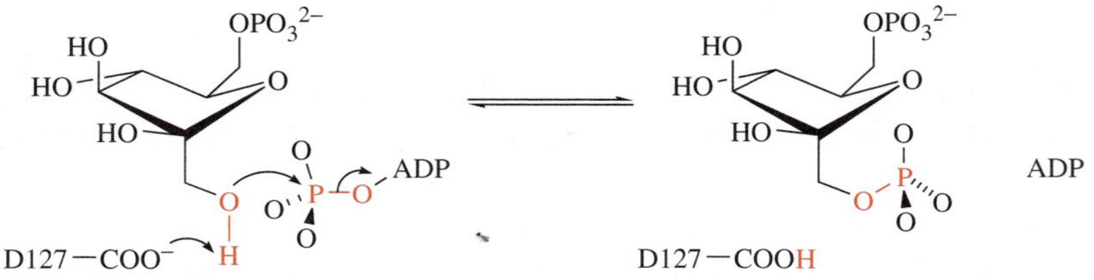

Scheme 10-6

As a gatekeeper for glycolysis, PFK displays regulatory properties. ATP at high concentrations inhibits the mammalian enzymes, as does citrate (Bloxham and Lardy, 1973). At inhibitory concentrations of ATP, plots of initial rate against F6P-concentration are sigmoidal. Allosteric activators include AMP, cAMP, FBP, and P_i. Metabolic studies in mammals show that the essential crossover in the overall regulation of glycolysis is at the point of PFK. Whenever glycolysis is active, as under anaerobic conditions or hormonal activation, PFK is active. Under conditions in which glycolysis activity is attenuated, PFK activity is inhibited by ATP or citrate.

PFK from *E. coli* and *B. stearothermophilus* are tetrameric and display both homotropic and heterotropic allosteric effects. Plots of initial rate against concentration of F6P are sigmoidal, showing homotropic activation by the substrate binding at the active site. In heterotropic allosteric regulation, the bacterial enzymes are activated by ADP and inhibited by phosphoenolpyruvate binding at the allosteric site. In-depth studies show that mutation of active site residues can differentially alter heterotropic allosteric effects. Certain mutations alter activation by ADP and others alter inhibition by PFK, indicating that the heterotropic effects involve different conformational transitions (Fenton et al., 2003). Mutations at the active site also alter the cooperativity of F6P binding (Berger and Evans, 1990). Arg162 and Arg243 from adjacent subunits interact with F6P, and mutation of these residues to serine decreases the binding of F6P as well as cooperativity with minimal effect on the value of k_{cat}. Mutation of Arg72, which bridges F6P and ATP, to serine decreases both activity and cooperativity.

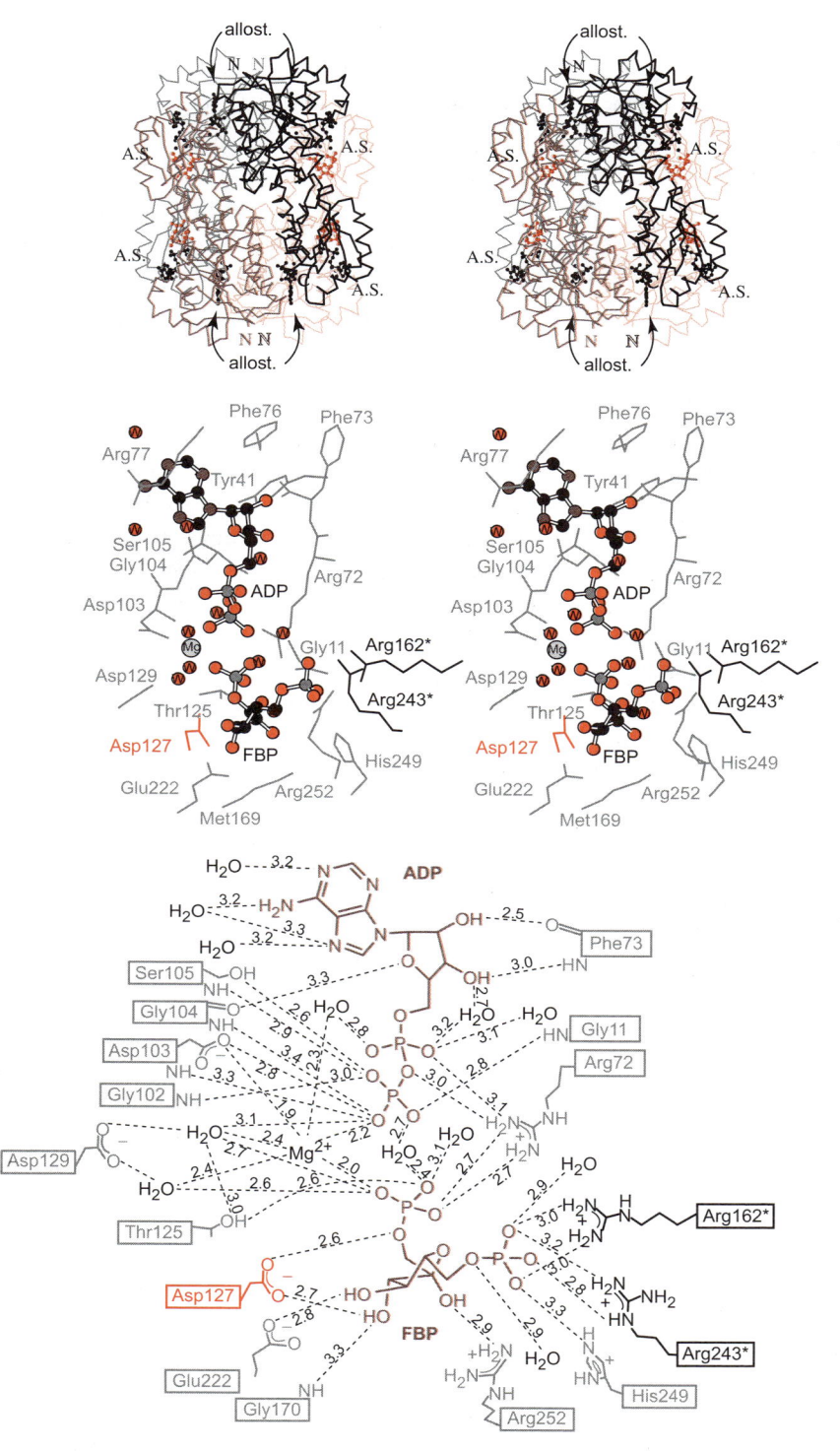

Fig. 10-7. Structure of phosphofructokinase. The top stereoimage shows the homotetrameric phosphofructokinase from *E. coli* with its reaction products of Mg^{2+} ADP (black) and fructose 1,6-bisphosphate (FBP; red) bound to each active site labeled A.S. (2.4-Å resolution; PDB 1PFK; Shirakihara et al., 1988). This enzyme is allosterically regulated, and a second ADP molecule is bound to each subunit in the sites labeled allost., with this particular complex representing the activated or R state. A detailed stereoimage of the active site is shown in the middle. Two arginine residues (asterisks) from an adjacent subunit interact with one of the FBP phosphate moieties. The catalytically important aspartate residue is shown in red. The bottom panel shows the active site in two dimensions, with distances given in angstroms.

502   Enzymatic Reaction Mechanisms

Most discussion of the regulatory consequences of mutating residues in the active or allosteric sites are based on the concerted transition model (Berger, and Evans, 1991; Kundrot and Evans, 1991). In one study, a hybrid species of *E. coli* PFK was assembled to contain just one native active site and one native allosteric site, with other active and allosteric sites desensitized through mutagenesis (Fenton and Reinhart, 2002). The free energy of activation by ADP proved to be about 20% of that for the all wild-type PFK. The results indicated that the regulatory behavior of the hybrid enzyme was not consistent with the concerted transition model.

Protein Phosphorylation: Protein Kinase A

Enzyme phosphorylation and dephosphorylation reactions underlie most signaling strategies for controlling eukaryotic cellular processes. Protein kinases and phosphatases play essential roles in the first signaling system to be discovered, the glycogen phosphorylase cascade, in which glycogen phosphorylase becomes activated through the phosphorylation of a serine residue (Fischer and Krebs, 1955; Krebs et al., 1959; Walsh et al., 1968). The activation process is much more complex than a single protein phosphorylation; it is a series of individually controllable reactions originating with a hormonal signal. The essential features of the cascade are outlined in fig. 10-8, showing that hormonal stimulation at the cell membrane leads to cAMP-induced stimulation of the activation of glycogen phosphorylase. Epinephrine activates membrane-bound adenylyl cyclase to produce cAMP,

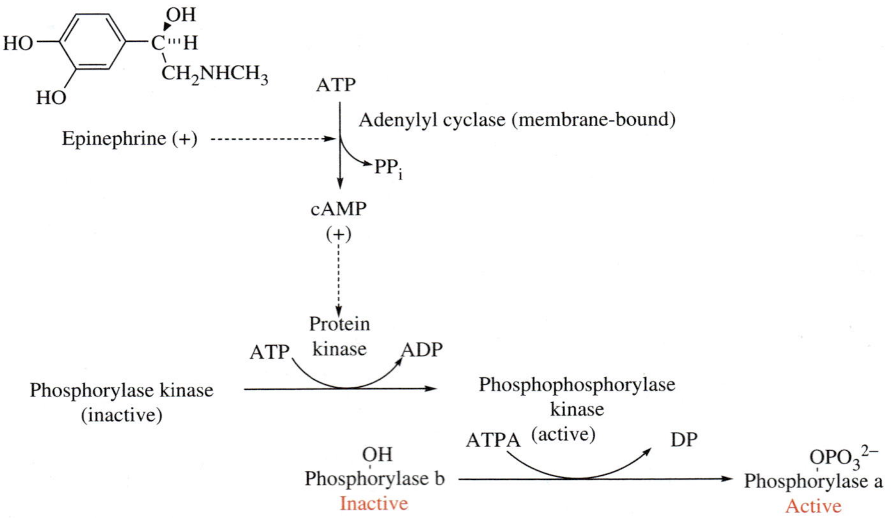

Fig. 10-8. The glycogen phosphorylase activation cascade. Phosphorylase b is the inactive dephospho form of glycogen phosphorylase, and phosphorylase a is active glycogen phosphorylase, which catalyzes the cleavage of glycogen by phosphate to form α-D-glucose-1-phosphate. The transformation cascade for the activation of phosphorylase begins with the hormonal activation of adenylyl cyclase by epinephrine, which catalyzes the formation of cAMP from ATP. cAMP then stimulates the cAMP-dependent protein kinase, which catalyzes the phosphorylation of phosphorylase kinase to its active form. The active phosphophosphorylase kinase catalyzes the phosphorylation of phosphorylase b. A recurrent theme in signaling is represented in this cascade; the cAMP-dependent protein kinase catalyzes protein phosphorylation in this and other activation cascades, whereas the phosphorylase kinase is specific for the activation of phosphorylase. In this way, cAMP-dependent protein kinase is a central element in the interaction of complementary phosphorylation cascades that regulate cellular processes.

which activates cAMP-dependent protein kinase. One of the targets of cAMP-dependent protein kinase is phosphorylase kinase, which in its active, phosphorylated form catalyzes the phosphorylation of glycogen phosphorylase b to its active form, glycogen phosphorylase a. Because cAMP relays the hormonal signal to the cAMP-dependent protein kinase, it is known as a second messenger.

cAMP-dependent protein kinase also catalyzes the phosphorylation of other enzymes in the regulation of their activities. For example, its action inhibits glycogen synthase, preventing the incorporation of glucose phosphates into glycogen. The complementary actions of cAMP-dependent protein kinase on the enzymes of glycogen metabolism ensure the availability of glucose phosphates as energy sources when needed.

The protein kinases as a group are among the most numerous of all enzymes. There are hundreds of families and thousands of kinases in eukaryotes. At least 518 protein kinase genes have been identified in the humane genome alone (Manning et al., 2002). They constitute about 1.7% of the proteins encoded in the genome. The great number of protein kinases arises from the many cellular processes that they regulate and the operating principle of the regulatory cascade. In the glycogen phosphorylase cascade, the posthormonal steps include two protein phosphorylations and two protein kinases. Even more protein kinases participate in other activation cascades. For example, in mitogen activation there is a MAP kinase kinase kinase kinase (MAP4K). Mitogen activation is a four-stage protein kinase cascade. An activation cascade amplifies the original signal, which may be initiated by only one or a few molecules. Each stage of a cascade amplifies the signal by activating an enzyme that can activate many more enzymes through turnover. The importance of amplification cascades in cellular processes cannot be overstated.

In the glycogen phosphorylase cascade, the phosphorylation sites are serine residues in phosphorylase kinase and phosphorylase b. Serine or threonine residues are often phosphorylated in other signaling systems as well. However, the discovery that the transforming protein kinase in Rous sarcoma virus (Collett et al., 1978) phosphorylates tyrosine residues (Hunter and Sefton, 1980) ushered in a new family of tyrosine kinases. Most phosphorylation sites in signaling cascades are serine, threonine, or tyrosine. These are relatively low-energy phosphoamino acids, unlike phosphohistidine, phospholysine, and phosphoaspartate.

The individual protein kinases in regulatory cascades are tightly regulated by diverse mechanisms. In most cases, a protein kinase consists of catalytic components and regulatory components. The cAMP-dependent protein kinase consists of two catalytic and two regulatory subunits. Other protein kinases may consist of catalytic domains and regulatory domains. Membrane-bound protein kinases also contain membrane-spanning domains that anchor them to membranes. Each subunit or domain of a protein kinase has its function in the catalytic event or in controlling or localizing the protein kinase activity. Most research on protein kinases concerns the widely divergent regulatory properties. In this chapter, we focus on the catalytic mechanism of protein phosphorylation.

The catalytic subunits and catalytic domains of protein kinases have much in common with one another (Taylor, 1990). They have similar core structures and active sites (Madhusudan et al., 1994; Sicheri et al., 1997; Taylor et al., 1995). The cAMP-dependent protein kinase, also known as protein kinase A (PKA), is a thoroughly studied example of a protein kinase and is involved in the first signaling cascade to be discovered. We discuss the mechanism of action of this enzyme as a case study for catalysis of protein phosphorylation, keeping in mind that the regulatory properties are unique and do not extend to other protein kinases.

The action of PKA serves as a paradigm for phosphorylation of enzymes and other proteins. The resting, inactive PKA is tetrameric and composed of two identical catalytic subunits (C subunits) and two identical regulatory subunits (R subunits) arranged in the

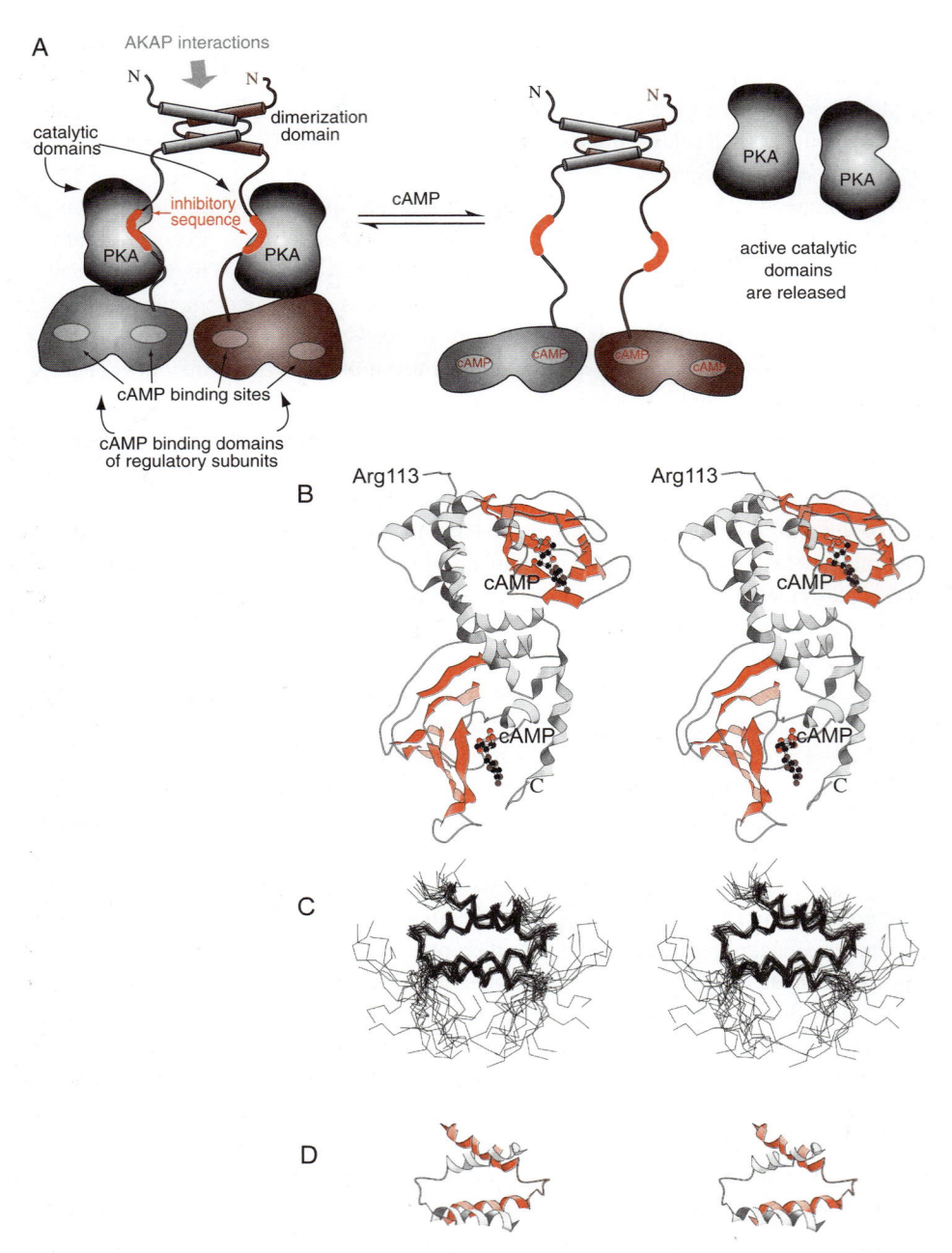

Fig. 10-9. Structure of the regulatory subunit of cAMP-dependent protein kinase A (PKA) and the cAMP sites. (A) The global structure of PKA regulatory complex is shown with the inactive catalytic subunits associated with the dimeric inhibitory subunits. cAMP binding releases active PKA from the regulatory complex. (B) The ribbon stereodiagram shows the cAMP binding domain of the bovine Riα regulatory subunit in stereo, with cAMP bound to both sites (2.8-Å resolution; PDB 1RGS; Su et al., 1995). (C) The mouse Riiα N-terminal dimerization interface is shown as 24 conformers resolved by nuclear magnetic resonance (NMR). (D) Two pairs of ordered helices are demonstrated by NMR (PDB 1L6E; Morikis et al., 2002).

configuration shown in fig. 10-9. The binding of cAMP to the regulatory subunits leads to their dissociation from the catalytic subunits, according to eq. 10-12, and brings about activation of PKA (Gill and Garren, 1970).

$$C_2R_2 + 4cAMP \rightleftharpoons C_2 + R_2(cAMP)_4 \qquad (10\text{-}12)$$
Inactive $\qquad\qquad\qquad$ Active

The C subunits function as the active kinase, and the cAMP dependence arises from the dissociation of R subunits.

Action of the Catalytic Subunit

The amino acid sequences of the C subunits in diverse eukaryotic cells display very little species variation (Lee et al., 1983; Showers and Mauerer, 1986). The subunit molecular mass is 41 kDa. Extensive chemical modification experiments employing affinity labeling and group selective chemical agents led to the identification of contact points in the binding of ATP to the catalytic subunits (Taylor et al., 1990). These experiments identified Lys72, Asp184, and Glu170 as essential and conserved residues that were protected by ATP from affinity labeling or chemical modification. Analogous experiments on other protein kinases implicated homologous residues in those enzymes as well.

Lys72 is sulfonylated by the affinity label *p*-fluorosulfonylbenzoyl 5′-adenosine (FSBA) (Zoller et al., 1981). FSBA inactivates the catalytic subunit, and ATP protects the enzyme from inactivation and labeling by FSBA. Lys72 is also readily acetylated by acetic anhydride, as are Lys76 and Lys47, and ATP protects against these chemical modifications (Buechler et al., 1989). Lys72 is conserved in all species of PKA and in other protein kinases as well, indicating an essential function for this residue (Taylor, 1990).

Dicyclohexylcarbodiimide (DCCD) inactivates the catalytic subunit and ATP protects the enzyme. Radiolabeled DCCD is not incorporated into the protein because of a secondary reaction of the DCCD-activated Asp184, which initially reacts. In chemical modification by carbodiimides, amines such as glycine methyl ester are often included to react with the activated carboxyl groups (see table 1-8 in chap 1). In the case of the catalytic subunit of PKA, Lys72 in the active site reacts with the DCCD-activated Asp184, leading to an amide crosslink in the enzyme (Buechler and Taylor, 1989). Asp184 is conserved in protein kinases.

The water soluble carbodiimide 1-ethyl-3(3-dimethylaminopropyl)-carbodiimide (EDC) also inactivates the catalytic subunit. Both ATP and an inhibitory peptide, an analog of the phosphorylation site in a protein, are required to protect against inactivation by EDC (Buechler and Taylor, 1990). Inclusion of [^{14}C]glycine methyl ester with EDC labels Glu170, and this is blocked by the combination of ATP and inhibitory peptide. This residue is also conserved in protein kinases. Acidic residues in the carboxyl terminus also react with EDC (Taylor et al., 1990).

The C subunit of PKA catalyzes the phosphorylation of specific serine or threonine residues in proteins. Structural and kinetic analyses show that important recognition factors in synthetic peptide substrates include arginine residues at positions P-6, P-3, and P-2 and a large hydrophobic residue at P+1 relative to serine in the phosphorylation site (Moore et al., 2003). For example, the peptide GRTGRRNSI incorporates all of the recognition features and is an excellent substrate. A smaller substrate is LRRASGL, also known as Kemptide. LRRAAGL is not phosphorylated, and it is a good inhibitor. The arginine residues at P-6, P-3, and P-2 are recognition features not shared with the sequences recognized by other protein kinases. The arginine side chains make contacts with glutamate residues in the active site.

The structure of the C subunit is shown in fig. 10-10 with an unreactive analog of ATP and an inhibitory peptide bound at the active site. The structure consists of a large core formed in a typical nucleotide fold from residues 40 to 300, flanked at the amino end by a 39-residue helix and at the carboxyl end by a 50-residue tail. The kinase core structure consists of an N-terminal lobe that binds the nucleotide and a larger C-terminal lobe that conveys peptide substrate specificity (Knighton et al., 1991). The figure shows the placement of the substrate analogs and the conformation of key loop elements that constiue the active site (catalytic loop, Mg binding loop, activation loop and the P+1 loop). In the most active form, Thr197 in an activation loop is phosphorylated (Adams et al., 1995). This and Ser338 are known as the autophosphorylation site a in the C subunit because they are phosphorylated in the purified protein and in the protein expressed in *E. coli*. Interactions of phospho-Thr197 with Arg165 and Lys189 hold the C subunit in the correct conformation for substrate recognition, catalysis, and inhibition by the R subunit (Johnson et al., 2001).

The kinetics of the action of PKA on small peptide substrates can be described in outline by scheme 10-7. The details of the kinetics vary with the structure of the peptide substrate.

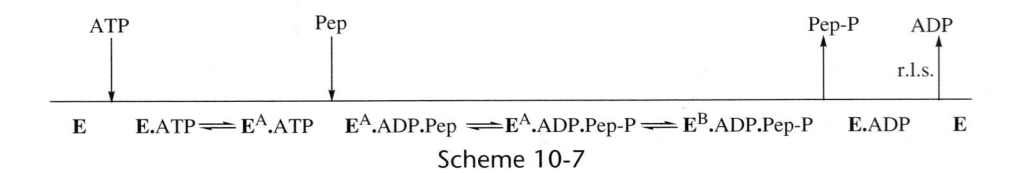

$$E \quad E.ATP \rightleftharpoons E^A.ATP \quad E^A.ADP.Pep \rightleftharpoons E^A.ADP.Pep\text{-}P \rightleftharpoons E^B.ADP.Pep\text{-}P \quad E.ADP \quad E$$

with arrows showing inputs ATP, Pep and outputs Pep-P, ADP (r.l.s.)

Scheme 10-7

Under many conditions, the dissociation of ADP limits the rate. For example, in transient studies of the phosphorylation of Kemptide (LRRASGL) or of GRTGRRNSI, burst kinetics is observed in the formation of the phosphopeptides. The burst amplitudes are on the order of enzyme concentrations (Grant and Adams, 1996; Zhou and Adams, 1997). With the peptide GRTGRRNSI as substrate, the burst rate constant is 150 s^{-1} and the rate constant for ADP dissociation is 16 s^{-1}, near the turnover number. ATP-linked protein conformational changes and post-phosphorylation protein conformational changes are implicated in the mechanism (Lew et al., 1997; Shaffer and Adams, 1999). Such conformational states are indicated by the symbols E^A and E^B in scheme 10-7.

Although scheme 10-7 indicates two protein conformational changes, there is evidence of multiple conformational changes, and the reaction of a protein substrate is likely to be still more complex (Johnson et al., 2001). In a major conformational change attending the binding of ATP, the smaller N-terminal lobe closed down on the larger C-terminal lobe (Johnson et al., 2001). The small lobe is conformationally more mobile than the larger lobe. The locus of motion is the glycine-rich, phosphate-binding loop, which includes Gly50, Gly52, and Gly55. This loop is mobile in the free enzyme and wound about the triphosphate of ATP in ternary complexes, with the amide nitrogen of Ser53 contacting the γ-phosphate in a hydrogen bond (Johnson et al., 2001).

Structure-Function Relationship in a Protein Tyrosine Kinase

The question of the nature of the transition state in the action of protein kinase A is difficult to address because it is not possible to vary the electronic properties of serine in substrates, and protein kinases are not well suited to measurements of secondary ^{18}O kinetic isotope effects. The protein tyrosine kinases offer greater possibilities for structure-function analysis.

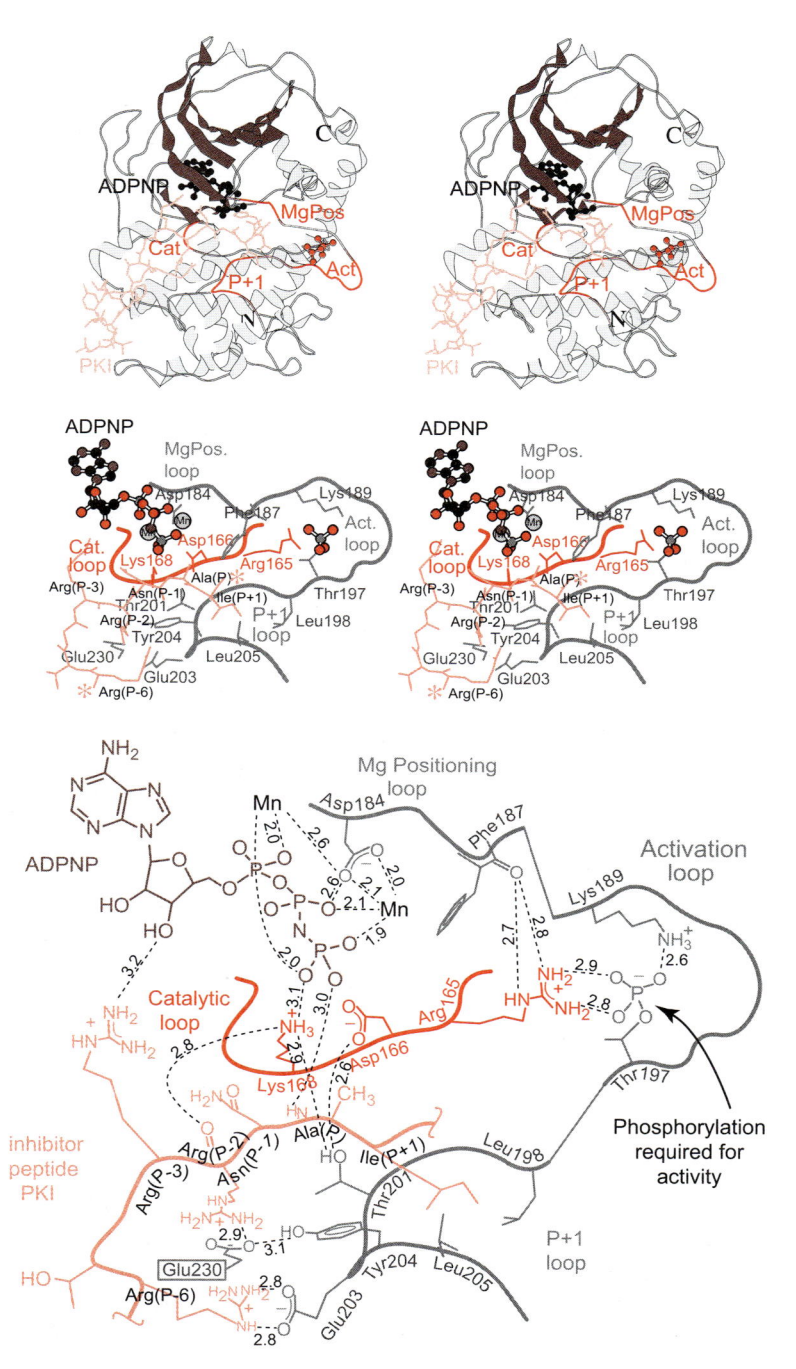

Fig. 10-10. Structure of the catalytic subunit of mouse cAMP-dependent protein kinase A (PKA). The global structure of the catalytic subunit in stereo at the top shows the locations of the ATP-analog *bis*-Mn^{2+} AMPPNP and an inhibitory peptide (2.2-Å resolution; PDB 1ATP; Zheng et al., 1993). In the middle, a stereodiagram shows the catalytic loop in red and the activation loop in gray. The inhibitory peptide (pink) is shown with Ile in the P+1 pocket and Ala at the P position. The termini of the peptide are labeled (asterisks). Phosphorylation of Thr197 in the activation loop is required for assembly of a functional active site, as mediated by the interaction of phospho-Thr with Lys189 (affecting the Mg positioning loop) and Arg165 (affecting the activation loop). At the bottom is a two-dimensional map of the ATP and peptide binding site, showing the contacts between the enzyme and ligands.

Because the structures of protein serine kinases and protein tyrosine kinases are very similar, apart from differential recognition of the hydroxymethyl and phenolic groups (Madhusudan et al., 1994; Sicheri et al., 1997), there is reason to expect the essential features of the transition states to be similar. For that reason, we cite a linear free energy structure-function study of a tyrosine kinase, Csk, which catalyzes the reaction of eq. 10-13.

$$\text{Tyr-OH} + \text{MgATP} \underset{}{\overset{\text{CsK}}{\rightleftharpoons}} \text{Tyr-OPO}_3^{2-} + \text{MgADP} \quad (10\text{-}13)$$

Structure-function analysis of Csk implicates a loose transition state with little bonding of phosphorus to either the donor or acceptor (Kim and Cole, 1998).

Csk accepts fluorotyrosyl substrates, in which the extent and orientation of ring fluorination controls the value of pK_a for the phenolic group. The reactions of Csk with the peptides NGXTA, where **X** is tyrosine or a mono-, di-, tri-, or tetrafluorotyrosyl residue, give significant information about the mechanism and transition state. The plot of pH against log (k_{cat}/K_m) for phosphorylation of the tyrosine peptide breaks downward at lower pH values corresponding to a pK_{a1} near 7, much lower than that of tyrosine. This must be an ionization in Csk, perhaps similar to one observed in the action of PKA (Yoon and Cook, 1987), and the tyrosyl moiety in the peptide substrate must be in its un-ionized, phenolic state. The fluorinated peptides give bell-shaped pH-log(k_{cat}/K_m) profiles, with additional downward breaks at higher pH values. The lower break is the same as with the tyrosyl peptide, but the higher values of pK_{a2} are similar to those of the fluorotyrosyl peptides. The tyrosyl and fluorotyrosyl peptides react as neutral phenolic groups, not as phenolate ions.

To construct a linear free energy correlation, kinetic studies were focused on the reverse reaction of eq. 10-13, which displayed an equilibrium constant of 10 in the forward direction. Initial rates of the reverse reaction of the phosphorylated peptides could be conveniently measured. The plot of log (k_{cat}/K_m) against pK_{a2} of the tyrosylpeptide or fluorotyrosylpeptide was linear and gave a value of 0.33 for β_{lg}. This was much smaller than the values of β_{lg} for transferring the PO_3^- group to oxyanionic or nitrogen acceptors, and close to the value of 0.27 for the nonenzymatic hydrolysis of phosphomonoester monoanions (Kirby and Varvoglis, 1967). The values of β_{lg} and pK_{a2} in the structure-function study, as well as the implication of an enzymatic group ionizing with a $pK_{a1} = 7$, implicated proton transfer in the mechanism. The enzymatic group may have been Asp314. Mutation of this residue decreased the rate of the phosphotransfer step by 10,100-fold (Grace et al., 1997).

Action of the Regulatory Subunit

The R subunit of PKA incorporates two cAMP binding sites within tandem, gene duplication sequences in the C-terminal region of the sequence. The sites are similar but not identical, and both must be occupied to induce the dissociation of PKA into catalytic and regulatory subunits (Smith et al., 1981; Takio et al., 1984; Titani et al., 1984). The N-terminal region encompasses the dimer interaction site, and the intervening sequence includes a peptide inhibitory site that occupies the peptide binding site of the catalytic subunit in the inactive, holo-PKA. The peptide inhibitory sequence is similar to the sequence recognized by the catalytic subunit and includes the ArgArg motif. In a particularly significant difference between two classes of regulatory subunits, R^I and R^{II}, the peptide inhibitory sequences in the class of R^{II} include serine in the phosphorylation position, and this serine is autophosphorylated by the catalytic subunit. In the class R^I subunits, this serine is replaced by another amino acid, usually alanine or glycine, which cannot be phosphorylated. In PKAs with class R^I regulatory subunits, the regulatory subunit is not phosphorylated, but the holo-PKA binds MgATP.

All evidence indicates that the autophosphorylation and pseudophosphorylation sequences in the regulatory subunits inhibit the action of the catalytic subunit toward protein substrates (Taylor, 1990). Dissociation of the regulatory subunits on binding cAMP at both sites activates the catalytic subunits toward glycogen phosphorylase kinase or other enzyme substrates.

The interaction of the inhibitory peptide sequence of the regulatory subunit with the active site of the catalytic subunit is a recurrent theme in signaling systems. In variations on this, a protein kinase may consist of catalytic and regulatory domains within a single peptide chain, in which the inhibitory sequence resides in the regulatory domain and occupies the catalytic site of the catalytic domain. Release of the inhibitory peptide may be brought about by another interaction of the regulatory domain, leading to activation of the catalytic domain. The secondary activating interaction may be the binding of another molecule or a phosphorylation event.

The structure of a PKA-regulatory dimer of class R^I is shown in fig. 10-9 with cAMP bound at both sites in each subunit. The basic subunit structure consists of a TIM barrel, unlike most other nucleotide binding protein. The N-terminal sequences constitute the dimer interface. The pseudophosphorylation sequence of the inhibitory peptide region is shown in color.

Protein Kinase Inhibitor and A-Kinase Anchoring Proteins

The protein kinase inhibitor (PKI) is a 75–amino acid polypeptide that contains an inhibitory sequence similar to the consensus sequence for substrates, with alanine in place of serine. The binding of its consensus sequence to PKA is augmented by additional contacts between PKA and three arginine residues and a phenylalanine in PKI, making PKI a potent inhibitor (Walsh et al., 1990). Crystal structures of ternary complexes comprising PKA, ATP, and PKI reveal important features of the interactions of peptides with PKA. The biological roles of PKI are under investigation.

Although PKA is a soluble protein, it can be found in membrane-associated states (Pawson and Scott, 1997). Membrane association is mediated by A-kinase anchoring proteins (AKAPs), which bind the R subunits of PKA. The AKAPs incorporate targeting motifs that direct PKA to specific subcellular locations (Johnson et al., 2001).

Phosphomonoesterases

Phosphomonoesterases catalyze the hydrolysis of phosphate monoesters to the corresponding alcohol and orthophosphate. They are commonly known as phosphatases. The reactions proceed by P—O bond cleavage and phosphoryl group transfer to water. This view of their action is accentuated by the fact that many of these enzymes catalyze phosphoryl group transfer to acceptors other than water when presented with high concentrations of alcohols. Phosphatases carry out important steps in many biological processes, including metabolism, nucleic acid processing, energy transduction, and signaling. They function by comparably diverse mechanisms, including nucleophilic catalysis by histidine, cysteine, or serine, or by electrophilic catalysis by divalent metal ions.

The classifications of phosphatases have been nearly as diverse as their biological functions and catalytic mechanisms. Clinical chemists referred to *alkaline phosphatase* and *acid phosphatase*, after the assay conditions using nonphysiological substrates at set concentrations. These names did not correspond to the pH dependence of the kinetic parameters for the enzymes. A classification into families based on their substrate specificities and molecular properties included alkaline phosphatases, purple acid phosphatases, low-molecular-weight acid phosphatases, high-molecular-weight acid phosphatases, and protein phosphatases (Vincent et al., 1992). A simple and less cumbersome classification has been suggested that places the enzymes into families defined by reaction mechanisms and active-site catalytic groups, as follows: serine phosphatases, histidine phosphatases, cysteine phosphatases, and metallophosphatases (Van Etten, 1982). The serine, histidine, and cysteine phosphatases function by nucleophilic catalysis, with covalent phosphoserine, phosphohistidine, or phosphocysteine intermediates. The metallophosphatases employ divalent metal ions in electrophilic catalysis, and they include purple acid phosphatase and inorganic pyrophosphatase. We adopt the latter classification.

Scheme 10-8 illustrates the general mechanism of action of the serine, histidine, and cysteine phosphatases. In this mechanism, the phosphomonoester binds to the active site to form a Michaelis complex.

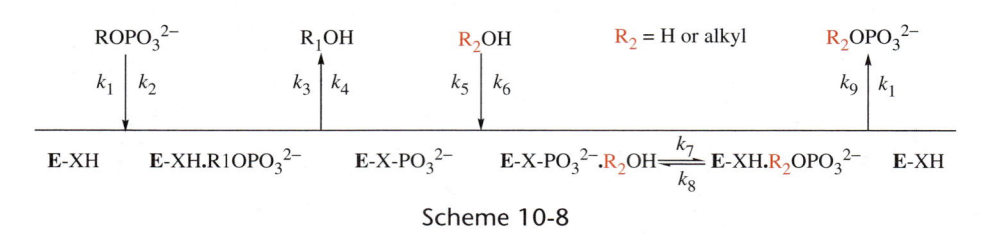

$$\text{E-XH} \xrightleftharpoons[k_2]{k_1 \, \text{ROPO}_3^{2-}} \text{E-XH·R1OPO}_3^{2-} \xrightleftharpoons[k_4]{k_3 \, R_1\text{OH}} \text{E-X-PO}_3^{2-} \xrightleftharpoons[k_6]{k_5 \, R_2\text{OH}} \text{E-X-PO}_3^{2-}\cdot R_2\text{OH} \xrightleftharpoons[k_8]{k_7} \text{E-XH·R}_2\text{OPO}_3^{2-} \xrightleftharpoons[k_{1}]{k_9 \, R_2\text{OPO}_3^{2-}} \text{E-XH}$$

R_2 = H or alkyl

Scheme 10-8

A nucleophilic amino acid then accepts the phosphoryl group to form a covalent phosphoenzyme and release the alcohol product. The covalent phosphoenzyme binds water, or another acceptor, and transfers the phosphoryl group. Phosphate or the new phosphomonoester dissociates from the active site. The three families that function by this mechanism employ serine, histidine, or cysteine as the nucleophilic catalyst XH in scheme 10-8. Compelling evidence implicates this mechanism in the action of each of these families of phosphatases.

Histidine Phosphatases

Phosphatases displaying full activity at pH 2.5, with subunit molecular masses of 40 to 60 kDa and dimeric structures, constitute a distinct family. The human lysosomal and prostatic acid phosphatases employ histidine as the nucleophilic catalyst (Van Etten, 1982). Trapping experiments with the substrate *p*-nitrophenyl [^{32}P]phosphate and alkaline denaturation in the steady state leads to a ^{32}P-labeled protein, which on alkaline hydrolysis produces to δ^1-phosphohistidine (Van Etten and Hickey, 1977). Histidine phosphatases catalyze transphosphorylation, in which an added alcohol reacts with the covalent phosphoryl-enzyme, E–X–P in scheme 10-8, to accept the phosphoryl group. The transphosphorylation reaction proceeds with overall retention of configuration at phosphorus when a P-chiral substrate is employed (Saini et al., 1981). The stereochemistry is consistent with two inversions at phosphorus and supports the double-displacement mechanism in scheme 10-8 for the action of histidine phosphatase.

The active site includes histidine, arginine, and aspartate residues, as indicated by chemical modification (Van Etten, 1982), site-directed mutagenesis based on amino acid sequence alignments (Ostantin et al., 1994), and x-ray crystallography of the rat enzyme (Lindqvist et al., 1993). As shown in fig. 10-11, the conserved and catalytically important residues Arg11, His12, Arg15, Arg79, His257, and Asp258 are found in the active site. His12 is the nucleophilic catalyst, and Arg11, Arg15, and Arg79 engage in binding the phosphomonoester. Kinetic evidence obtained on the enzyme from *E. coli*, and less complete kinetic information on the human enzyme implicates Asp258 in catalysis of the phosphorylation of His12.

The corresponding residues in the enzyme from *E. coli* are Arg16, His17, Arg20, Arg92, His303, and Asp304 (Ostantin and Van Etten, 1993; Ostantin et al., 1992). Mutation of any

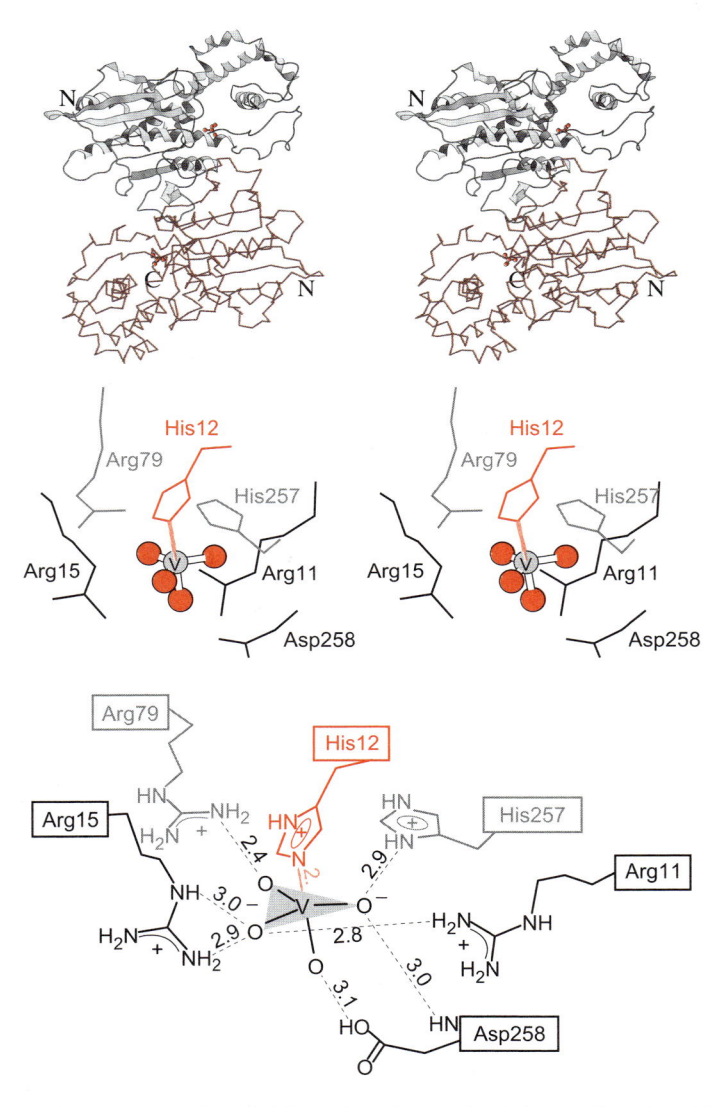

Fig. 10-11. In the structure of a histidine phosphatase (top), homodimeric acid phosphatase (from rat) is shown in complex with the transition-state analog vanadate (red) at the active site (3.0-Å resolution; PDB 1RPT; Lindqvist et al., 1994). A more detailed image of the active site (middle stereoimage) reveals the trigonal bipyrimidal vanadyl His12 adduct. The vanadyl His12 and surrounding enzyme residues are shown in two dimensions in the lower panel, with distances given in angstroms.

of these residues to alanine abolishes or severely decreases activity, the main effect being on the maximum velocity. Kinetic evidence implicates Asp304 as the acid-base catalyst in phosphoryl group transfer. The wild-type enzyme displays very similar maximal velocities for a large number of aryl and alkyl phosphates, with leaving groups ranging in acidity from pK_a values of 7 to 10. The wild-type enzyme catalyzes transphosphorylation from p-nitrophenylphosphate to ethylene glycol, and the rate of formation of p-nitrophenol increases in proportion to the concentration of ethylene glycol. This behavior implicates the mechanism of scheme 10-8, with hydrolysis of the phosphoryl-enzyme being rate limiting. Phosphotransfer to glycerol represents the interception of the phosphoryl-enzyme by glycerol.

Mutation of Asp304 to alanine decreases the maximum velocity by 30- to 1600-fold, depending on the substrate. The plot of log V_m against pK_a for the leaving group displays a negative slope of –0.51, illustrated in fig. 10-12, strongly implicating phosphoryl-enzyme-formation as the rate-limiting step in the mechanism of scheme 10-8. Mutation of Asp304 changes the rate-limiting step and implicates Asp304 as the acid-base catalyst in phosphoryl transfer. Mutation of His303 to alanine similarly decreases the activity, but the variant does not display the linear free energy correlation with substrates that vary in leaving group ability, ruling out His303 as the acid-base catalyst. The slope of –0.5 in the linear free energy correlation in fig. 10-12 for the D304A-phosphatase implies that bond cleavage to the leaving group is important in the transition state, as it would be in a loose transition state, in which there is decreased bonding to phosphorus relative to the ground state.

One of the characterizing features of the histidine phosphatases is their activity at acidic pH values, a property that inspires their widespread designation as acid phosphatases.

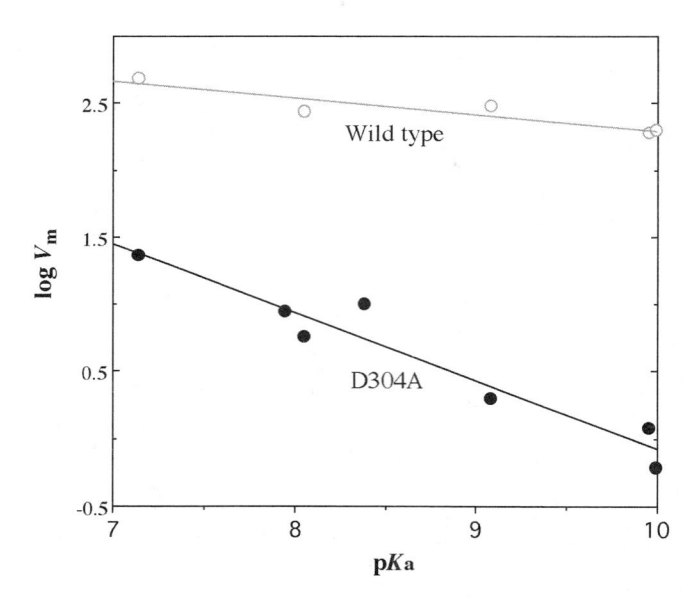

Fig. 10-12. Linear free energy correlations in the action of *E. coli* histidine phosphatase. The plot of log V_m for the D304A-phosphatase against pK_a of the leaving group for the hydrolysis of a series of ring-substituted phenyl phosphates appears linear, with a slope of –0.5. This plot indicates that the rate is limited by the leaving group ability of the phenolic groups and with phosphorylation of the enzyme being rate limiting in the mechanism of scheme 10-8. The wild-type phosphatase displays much higher values of V_m for the same substrates and a much smaller dependence on the leaving group (slope of –0.08), consistent with rate-limiting dephosphorylation in the mechanism of scheme 10-8. (Data from Ostantin et al., 1994).

The pH dependencies of the human and *E. coli* histidine phosphatases are typical (Ostantin and Van Etten, 1993; Van Etten, 1982). Plots of log V_m or log V_m/K_m against pH describe high plateaus that do not break downward at pH values as low as 2 but turn downward at pH values higher than 5. Although the pH dependence is not perfectly understood, it raises questions about the function of His17 as the nucleophilic catalyst (His12 in the human enzyme). Such a pH dependence would seem to require a very low value of pK_a for the nucleophilic histidine, which must function in the form of its conjugate base to become acylated in the mechanism of scheme 10-8. The structure shows the active site composed of three arginine residues in addition to the nucleophilic histidine and catalytic aspartate (see fig. 10-11). Any reasonable interpretation of the charge states of these residues reveals that the nucleophilic histidine resides in a positive electrostatic field. This would present an electrostatic barrier to protonation of histidine and depress its pK_a, just as described in chapter 1 for the active site-lysine in acetoacetate decarboxylase.

The mechanism in fig. 10-13 is supported by available information. A phosphomonoester is bound at the active site primarily by electrostatic attraction to the three arginine side chains, which maintain His12 in its conjugate base form. Phosphorylation of His12 is catalyzed by proton transfer from Asp304 to the leaving group. Hydrolysis of the phosphoryl-enzyme follows the reverse course, with water in place of the alcohol leaving group.

Cysteine Phosphatases

A large superfamily of signaling phosphatases employs cysteine as the nucleophilic catalyst for the hydrolysis of phosphoproteins (Guan and Dixon, 1991; Zhang, 2003; Zhang and Dixon, 1994). All members of the superfamily include a small catalytic domain that appears similar, as well as variable regulatory and localization domains (Zhang, 2003). Most members catalyze the hydrolysis of phosphotyrosine residues in phosphoproteins, and they have become known as protein tyrosine phosphatases (PTPs). However the superfamily includes enzymes with different specificities; for example, PTEN and myotubularin catalyze the hydrolysis of phosphoinositides, the dual specific VH1-like and Cdc14-like phosphatases catalyze the hydrolysis of phosphoserine/phosphothreonine in addition to phosphotyrosine residues, and the superfamily includes mRNA triphosphatases. The human genome encodes more than 100 PTPs. The smallest member of the superfamily is the low-molecular-weight PTP (LMW PTP), which comprises a catalytic domain with a molecular mass of about 18 kDa (Van Etten, 2003). The superfamily includes intracellular soluble

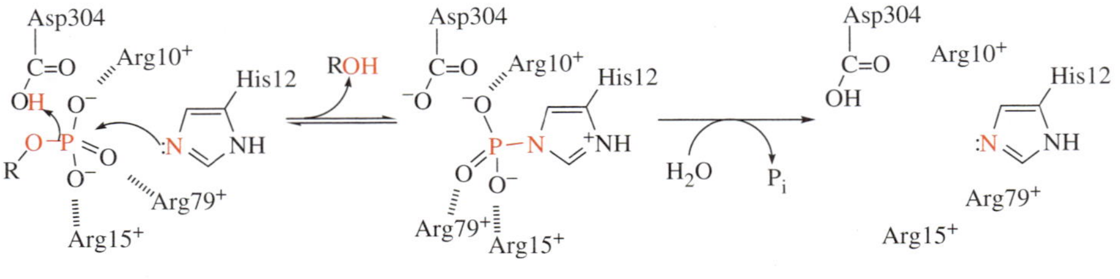

Fig. 10-13. In a mechanism for the action of human acid phosphatase, a histidine phosphatase, the phosphomonoester is bound to the active site mainly by electrostatic attraction to Arg10, Arg15, and Arg79 near His12, which is in its conjugate base and reactive state. Departure of the alcohol ROH is catalyzed by proton donation from Asp304, concomitant with transfer of the phosphoryl group to His12, to form the phosphoryl-enzyme intermediate. Hydrolysis of the phosphoryl-enzyme follows the reverse mechanism, with water in place of the alcohol.

PTPs and receptor-type, often membrane-bound PTPs. All PTPs include the active-site amino acid sequence motif C(X)$_5$R(S/T) in their catalytic domains, which are similar in structure.

The PTPs function as switches in cell signaling. Dephosphorylation of a phosphotyrosine or phosphoinositide can turn a biological function on or off, and this is the most essential function in signaling. Signals must include both on and off switches to be effective, and PTPs provide a biological mechanism for these functions. A large research effort is directed toward the elucidation of the regulatory properties of these enzymes in controlling cell functions. The biological roles of many PTPs are implied by the biological phenomena leading to their identification and cloning, and they include processes such as differentiation, cell-cell communication, regulation of the cell cycle, immune response, apoptosis, and metabolism. Impaired regulation of PTPs can lead to diseases such as cancer and diabetes (Zhang, 2001).

The biological role of LMW PTP is less specifically known. The LMW PTP is found in many animal organs, including liver, heart, placenta, and red cells and also in bacteria (Van Etten, 2003). Studies in vitro show that the mammalian enzymes display activity toward phosphoproteins such as phosphotyrosyl angiotensin, tyrosine kinase P^{40}, phosphotyrosyl IgG, phosphotyrosyl casein, phosphotyrosyl peptides from the insulin receptor, epidermal growth factor (EGF) receptor, and platelet-derived growth factor (PDGF) receptor (Van Etten, 2003). Biological studies in cells indicate a variety of consequences associated with overexpression of this enzyme.

Most of the available mechanistic information about PTPs refers to the LMW PTP or the slightly more complex *Yersinia* PTP, which is found in the plague causing bacteria and is required for virulence. These PTPs are the simplest members of the superfamily and the most convenient for mechanistic analysis. The mechanisms of action of the catalytic domains in other PTPs are likely to be similar to the action of the small PTPs.

Transient-phase kinetic studies support the kinetic mechanism of scheme 10-8 for the LMW and *Yersinia* PTPs (Zhang and Van Etten, 1991; Wu and Zhang, 1996). Burst kinetics and steady-state data indicate fast phosphorylation and rate-limiting dephosphorylation of these enzymes. The overall reaction proceeds with retention of configuration at phosphorus, consistent with the double-displacement in the mechanism of scheme 10-8 (Saini et al., 1981). The downfield ^{31}P NMR signal (+16.2 ppm relative to phosphoric acid) of the phosphoryl-PTP (**E–X–P** in scheme 10-8) is characteristic of a phosphorothiolate ester and not of a phosphate ester or phosphoramidate (Davis et al., 1994; Wo et al., 1992). Specific mutation of the cysteine residue in the conserved motif **C**(X)$_5$**R**(S/T) of each enzyme, Cys403 in *Yersinia* and Cys12 in LMW, completely eliminates activity (Davis et al., 1994; Guan and Dixon, 1991).

The dependence of activity on pH suggested the importance of an acidic catalytic group in the action of both *Yersinia* and LMW PTPs, and mutagenesis of conserved acidic residues implicated Asp356 of *Yersinia* and Asp127 of LMW PTPs as essential residues (Zhang et al., 1994a, 1994b). Mutation of Asp356 in the *Yersinia* PTP both lowered the activity by at least three orders of magnitude and eliminated the pH dependence. Transient-phase kinetic analysis indicated that Asp127 in the LMW PTP functioned as a base in the dephosphorylation step of the mechanism, presumably to remove a proton from water in the dephosphorylation step of scheme 10-8 (Wu and Zhang, 1996). Mutagenesis experiments also proved the importance of Arg409, showing that mutation to alanine decreased the value of V_m/K_m by a factor of 2×10^5 (Zhang et al., 1994c).

Crystal structures of *Yersinia* and LMW PTPs are available as free enzymes and with transition-state analogs such as tungstate or vanadate bound at the active sites (Denu et al., 1996; Jia et al., 1995; Stuckey et al., 1994; Zhang et al., 1997, 1998). The structure of the

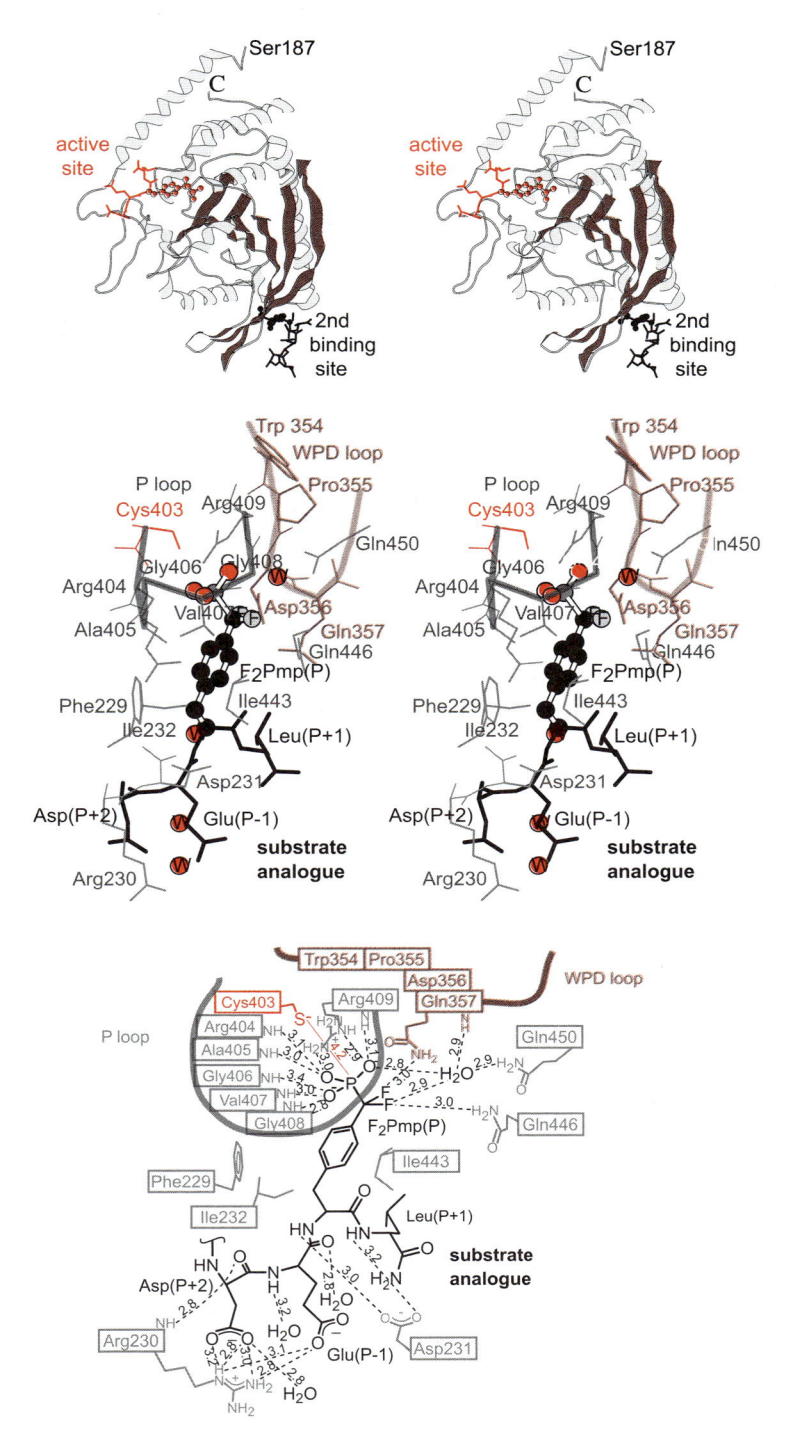

Fig. 10-14. Protein tyrosine phosphatase from *Yersinia pestis* (YopH) is shown (top) with a nonhydrolyzable phosphotyrosine hexapeptide mimetic bound in the active site and in a second nonreactive site (1.5-Å resolution; PDB 1QZ0; Phan et al., 2003). The middle panel provides a closer view of the active site, highlighting the mimetic bound in the active site flanked by the active-site Cys403 (red)–containing P loop and the adjacent WPD loop. Interactions between the mimetic and the enzyme are shown in two dimensions (bottom), with distances given in angstroms.

516 Enzymatic Reaction Mechanisms

Yersinia PTP in fig. 10-14 shows the chain fold of the catalytic domain and the spatial relationships among Cys403, Asp356, and Arg409. The overall structure consists of a twisted β-sheet flanked by helices, similar to a dinucleotide binding domain. Cysteine 403 is placed in position to accept the phosphoryl group and is hydrogen bonded to the nearby Thr410. The phosphate is held in place by a hydrogen bonded ionic attachment of two oxygens to Arg409 and by three main chain hydrogen bonds to the third peripheral oxygen. In the structures of vanadate or tungstate complexes, the trigonal bipyramidal inhibitors are bonded to Cys403 of *Yersinia* or Cys12 of LMW PTPs, and the catalytic Asp356 in *Yersinia* or Asp127 in LMW PTPs are in position to serve as acid catalysts for the leaving alcoholic group.

Biochemical and computational evidence indicates that the nucleophilic cysteinyl residue in the free enzyme is maintained as a thiolate ion by the compensating positive electrostatic field created by helix dipoles and Arg409 (Dillet et al., 2000). It should be noted that with Cys–S$^-$, Asp–COOH, and Arg–NHC(NH$_2$)$_2$$^+$, the active site is electrostatically neutral, overall, with positive dipoles contributed by main chain N–H ends of helices in position to bind a phosphate group. With these features in mind, and the available kinetic information, the mechanism can be formulated as in fig. 10-15. The active site

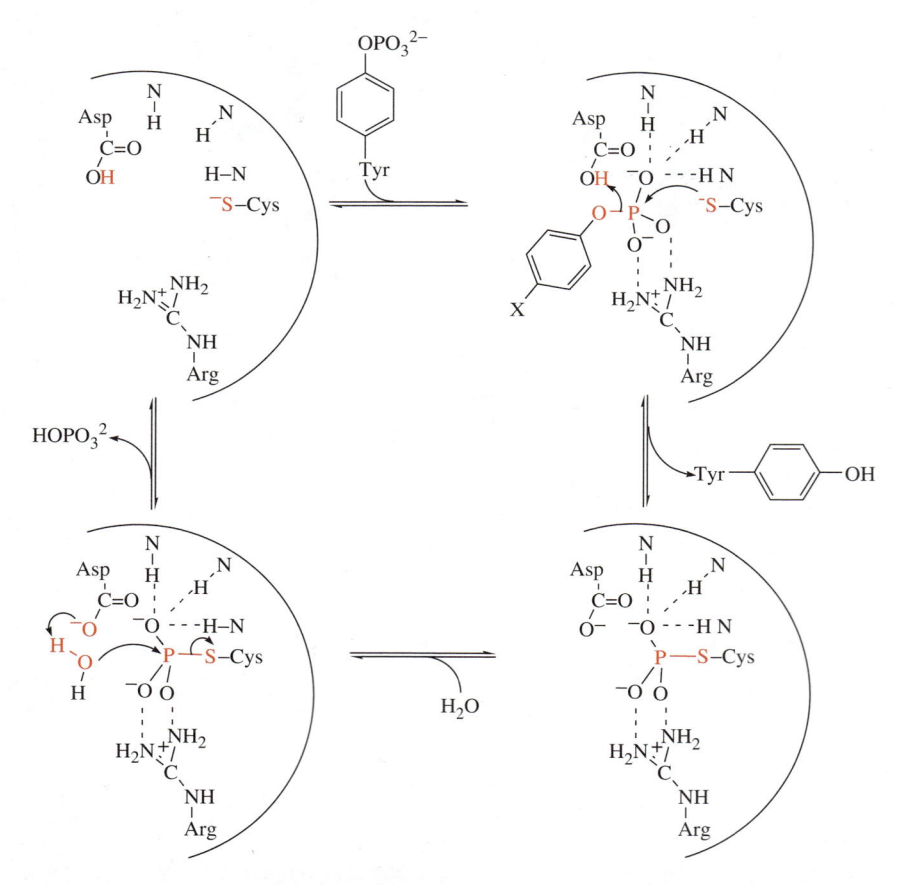

Fig. 10-15. A mechanism for the action of a protein tyrosine phosphatase. In a low-molecular-weight PTPase, the nucleophilic residue is Cys12, and in *Yersinia* protein tyrosine phosphatase (PTP), it is Cys405. The substrate binds through its phosphate group to the active-site arginine and three hydrogen bonds to main chain N–H groups. Transfer of the phosphoryl group to Cys–S$^-$ is catalyzed by Asp–COOH, which donates a proton to the leaving phenolic group. The phosphoryl-enzyme undergoes hydrolysis by essentially the reverse of phosphorylation, with water in place of the tyrosine.

Table 10-3. Oxygen-18 kinetic Isotope Effects in the Nonenzymatic and Protein Tyrosine Phosphatase–Catalyzed Hydrolysis of p-Nitrophenyl Phosphate

Catalyst[a]	$^{18}(V_m/K_m)_{bridge}$	$^{18}(V_m/K_m)_{nonbridge}$
Yersinia PTP (wild type)	1.0160	1.0001
D356A-PTP	1.0275	1.0007
Nonenzymatic (dianion)	1.0230	0.9993
Nonenzymatic (monoanion)	1.0106	1.0224

[a]All isotope effects are measured by the remote label method with ^{15}N in the nitro group. Data from (Zhang, 2003).

cysteine is in position to capture the phosphoryl group in flight to form the covalent phosphoryl-enzyme intermediate. This is accompanied by the donation of a proton from the aspartic acid side chain to the leaving group. The phosphoryl-enzyme then reacts with water to form phosphate by a mechanism that is essentially the reverse of the phosphorylation step.

The nature of the transition states in the phosphoryl transfer steps is revealed by the ^{18}O-kinetic isotope effects in table 10-3. The use of *p*-nitrophenyl phosphate as a substrate facilitates the application of the remote label method (see chap. 2), with ^{15}N in the nitro group as the remote label. When ^{18}O is in the position bridging phosphorus and the *p*-nitrophenyl group, the observation of an isotope effect proves that P—O bond cleavage is rate limiting for k_{cat}/K_m. The secondary ^{18}O-isotope effects observed when ^{18}O is nonbridging report on changes in bond order between phosphorus and peripheral oxygen atoms in the transition state. The data in table 10-3 show that P—O cleavage is rate limiting in the PTP-catalyzed reaction. Moreover, the nonbridge-^{18}O isotope effect is minimal, consistent with little or no change in bonding to the peripheral oxygens. Mutation of Asp to Ala increases the bridge-^{18}O isotope effect to the value for the nonenzymatic hydrolysis of dianionic *p*-nitrophenyl phosphate, and the nonbridge-^{18}O effects are also similar. The results indicate that, as in the nonenzymatic reaction, the transition state is loose, with decreased bonding to the leaving group and little bonding to the entering nucleophile, as in the nonenzymatic reaction. The isotope effects also confirm that the substrate reacts as the dianion.

Serine Phosphatases

The most widely studied serine phosphatases, the alkaline phosphatases (EC 3.1.3.1), are found in most organisms and catalyze both hydrolysis and transphosphorylation of phosphate esters according to the mechanism in scheme 10-8, where X is the β-oxygen of serine (Coleman, 1992; Schwartz and Lippman, 1961; Wilson et al., 1964). They are metalloenzymes that require three divalent metal ions, 2 Zn^{2+} and 1 Mg^{2+} (or another Zn^{2+}), for activity. The metal ions appear to increase the nucleophilic reactivity of serine through coordination with the β-hydroxyl group, decreasing its basicity and facilitating removal of the hydroxyl proton. The metal ions also bind the substrate-phosphate group.

The most thoroughly studied member of this family is alkaline phosphatase from *E. coli*. High-resolution structures are available for the enzyme/phosphate complex, the covalent phosphoryl-enzyme, and a vanadate/enzyme complex (Holtz et al., 1999; Kim and Wyckoff, 1991; Stec et al., 2000). The trigonal bipyramidal vanadate is regarded as a transition-state analog. The structure shown in fig. 10-16 is of the *E. coli* phosphatase, showing the overall structure of the dimeric enzyme. Each subunit consists of 448 amino acid residues organized in a ten-strand β-sheet flanked on each face by helices, and the β-sheets are fused into a single sheet at the subunit interface.

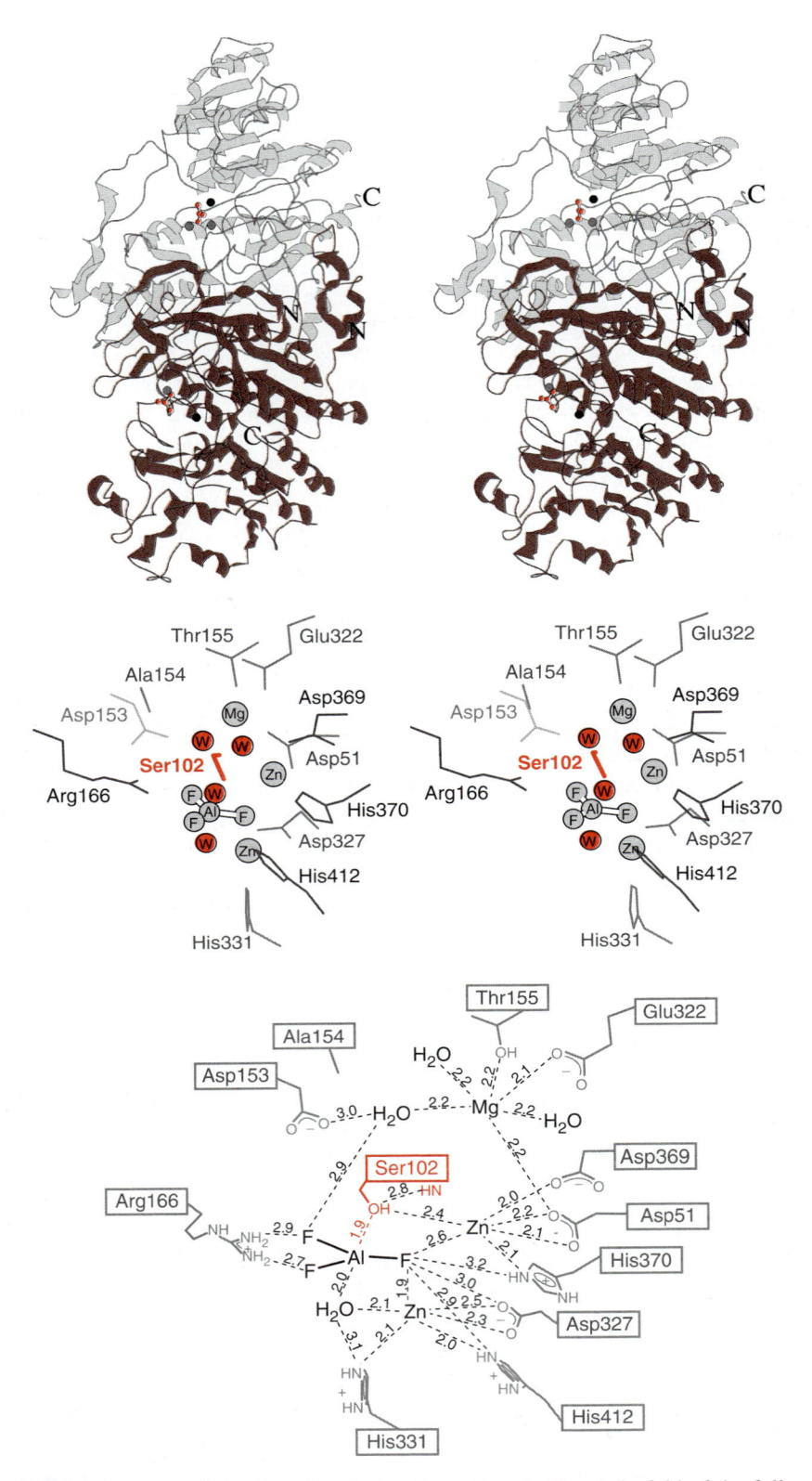

Fig. 10-16. Structure of alkaline phosphatase from *E. coli*. The chain fold of the fully active D330N, dimeric alkaline phosphatase variant is shown with transition-state analog AlF$_3$ (red), two Zn^{2+} (gray), and one Mg^{2+} (black) bound in each subunit (2.0-Å resolution; PDB 1KH5; Le Du et al., 2002). The active site is shown in stereo (middle) and in two dimensions (bottom), with the nucleophilic serine in red.

Each active site contains a zinc ion coordinated to Asp327, His331, His412 and phosphate; a zinc ion coordinated to Asp51, Asp369, His370, and Ser102; and a magnesium ion coordinated to Asp51, Thr155, Glu322, and phosphate. The structure does not show a specific binding site for the leaving group, which may explain how the enzyme displays activity against phosphomonoesters in general. In addition to the metal ions, Arg166 forms part of the binding site for phosphate. An analysis of the electrostatics at the active site leads to the conclusion that in the free enzyme the site carries a charge of +3 if the β-oxygen of Ser102 is a hydroxyl group or +2 if it is an alkoxide ion. It seems likely to be an alkoxide, and this would explain its reactivity as a nucleophilic catalyst.

Alkaline phosphatase from *E. coli* is very active against *p*-nitrophenyl phosphate, with a turnover number (k_{cat}) of at least 45 s^{-1}. Mutation of the Ser102, the active site nucleophilic catalyst, to glycine, alanine, or cysteine dramatically decreases the activity (Stec et al., 1998). These mutations do not perturb the overall structure in important ways, as determined by high-resolution x-ray crystallography, so that the kinetic parameters may be regarded as significant. The value of k_{cat} for the S102G- and S102A-phosphatases are about 1/37,000 that for the wild-type phosphatase, and the values of k_{cat}/K_m are about 1/200,000 that of the wild-type enzyme. The S102C-variant is somewhat more active, but still 1/250 as active in terms of k_{cat} and 1/7000 as active in terms of k_{cat}/K_m. The glycine and alanine variants are thought to function by a mechanism in which a water molecule or hydroxide react directly with the bound substrate in place of Ser102. Despite their low activities, the turnover numbers of the glycine and alanine variants are still on the order of 2×10^4 the rate of nonenzymatic hydrolysis, showing that the disabled enzymes retain significant activity (Holtz et al., 1999).

Limited information is consistent with a transition state for phosphoryl transfer in which P—O bond cleavage is well along the reaction coordinate, that is, with a loose (or dissociative) transition state. Secondary ^{19}O-kinetic isotope effects are consistent with a loose transition state (Weiss and Cleland, 1989). Because of its high activity against substrates with a broad range of leaving groups, *E. coli* alkaline phosphatase would seem to be a good candidate for studies of linear free energy relationships, that is for linear correlations of kinetic parameters with pK_a values of leaving groups. Such studies are often defeated in enzymatic research by binding selectivity at active sites. However, alkaline phosphatase is extremely sensitive to product inhibition by phosphate ($K_i < \mu M$), so that the spectrophotometric assays usually used are not sensitive enough to allow accurate evaluation of kinetic parameters for all substrates. Aryl phosphorothioates are poor substrates that clearly react with phosphorylation of the enzyme rate limiting, and the thiophosphate produced is less inhibitory than phosphate. A linear free energy plot of log (k_{cat}/K_m) against the pK_a of the leaving group for a series of aryl phosphorothioates is linear, with a slope of –0.8 for β_{lg} (Hollander and Herschlag, 1995). This slope is in the range with that for the nonenzymatic hydrolysis (–1.1) for the same compounds, showing that P—O bond cleavage characterizes the transition state.

Product inhibition can be overcome in the kinetic analysis by use of a highly sensitive radiochemical assay and ^{32}P-labeled substrates. Application of this assay allows accurate initial rates for alkaline phosphatase to be determined without interference by product inhibition. A linear free energy correlation of log (k_{cat}/K_m) against the pK_{alg} for a series of aryl phosphates gives a straight line with a slope of –0.85 for β_{lg} (O'Brien and Herschlag, 2002). The results indicate that the phosphorylation step is rate limiting and that the transition state is loose (or dissociative). Further studies of the pH dependence for k_{cat}/K_m indicate that the value of p$K_a \leq 5.5$ for a group required for activity, presumably Ser102, the nucleophilic catalyst. There is no need for a basic amino acid side chain to remove the proton from Ser102; it is acidic enough to be ionized in the physiological pH range, presumably through the electrostatic effects of the two divalent metal ions to which it is coordinated. Bis-coordination of a hydroxyl group by two divalent metal ions is a recurrent theme, which we have seen in the case of urease in chapter 4 and will see again in the next section on metallophosphatases.

The question of the biological function of alkaline phosphatases often stimulates discussion. There is reason to believe that it harvests phosphate in metabolism. However, transphosphorylation may be significant in certain settings. In *E. coli*, alkaline phosphatase is periplasmic, where phosphorylated compounds can be found in the absence of ATP-dependent phosphotransferases. It may be that alkaline phosphatase mediates transphosphorylation in the periplasm. It seems that this enzyme may have more than one, in some cases perhaps several biological roles in metabolism. Genetic and biochemical evidence indicates that *E. coli* alkaline phosphatase participates in phosphite metabolism and displays phosphate hydrogenase activity according to eq. 10-14 (Yang and Metcalf, 2004).

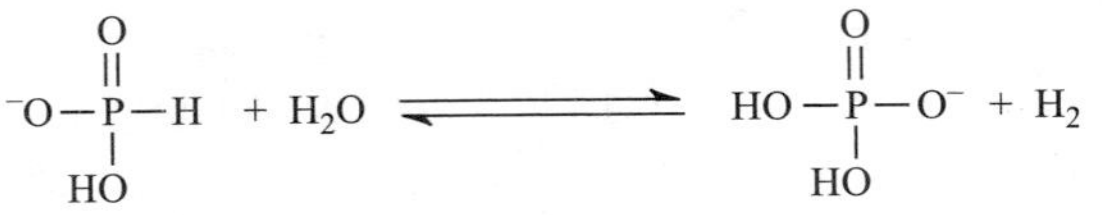

$$^-O-\underset{\underset{HO}{|}}{\overset{\overset{O}{\|}}{P}}-H + H_2O \rightleftharpoons HO-\underset{\underset{HO}{|}}{\overset{\overset{O}{\|}}{P}}-O^- + H_2 \qquad (10\text{-}14)$$

The hydrogenase reaction could proceed by a hydride transfer mechanism through a metaphosphate-like transition state, as in the other reactions catalyzed by alkaline phosphatase.

Metallophosphatases and Pyrophosphatases

A large family of phosphatases uses *bis*-divalent metal centers to catalyze the hydrolysis of phosphomonoesters and phosphoanhydrides. Purple acid phosphatases have diiron centers and are purple because of ligand-to-iron charge transfer (see chap. 4). Other members of this family have zinc, manganese, or magnesium in *bis*-metallo centers and do not display color. They include protein serine/threonine phosphatases, which are Mg^{2+} or Mn^{2+} dependent and are important in the dephosphorylation of signaling phosphoproteins (Das et al., 1996; Goldberg et al., 1995). Other members of this family include inorganic pyrophosphatase, in which the requirement for multiple Mg^{2+} ions includes the assembly of a *bis*-Mg^{2+} center (Heikinheimo et al., 2001). The *bis*-metallo centers serve the important function of increasing the acidity of water coordinated between the two divalent metal ions. By decreasing the acid dissociation constant of water into the neutral range, they facilitate the formation of a *bis*-metallo-hydroxide ion, which can accept a phosphoryl group in flight without the need for base catalysis.

Inorganic pyrophosphatase is activated by Mg^{2+} or Mn^{2+}, but Mg^{2+} is most likely the physiological activating ion. Extensive kinetic analysis shows that four divalent metal ions participate in the enzymatic hydrolysis of pyrophosphate (Belogurov et al., 2000). The structure reveals the assembly of two Mn^{2+} ions into a *bis*-manganese center in the active site (Heikinheimo et al., 1996). Two other divalent metal ions are in coordination with the substrate, pyrophosphate. A kinetic model supported by available evidence is provided in scheme 10-9 (Baykov et al., 2000; Heikinheimo et al., 2001).

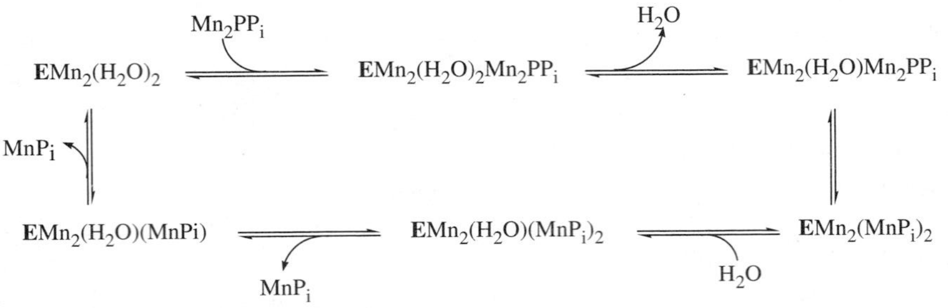

$$\begin{array}{ccccc}
& \xrightarrow{Mn_2PP_i} & & \xrightarrow{H_2O} & \\
EMn_2(H_2O)_2 & \rightleftharpoons & EMn_2(H_2O)_2Mn_2PP_i & \rightleftharpoons & EMn_2(H_2O)Mn_2PP_i \\
\updownarrow {\scriptstyle MnP_i} & & & & \updownarrow \\
EMn_2(H_2O)(MnP_i) & \rightleftharpoons & EMn_2(H_2O)(MnP_i)_2 & \rightleftharpoons & EMn_2(MnP_i)_2 \\
& \xrightarrow{MnP_i} & & \xrightarrow{H_2O} &
\end{array}$$

Scheme 10-9

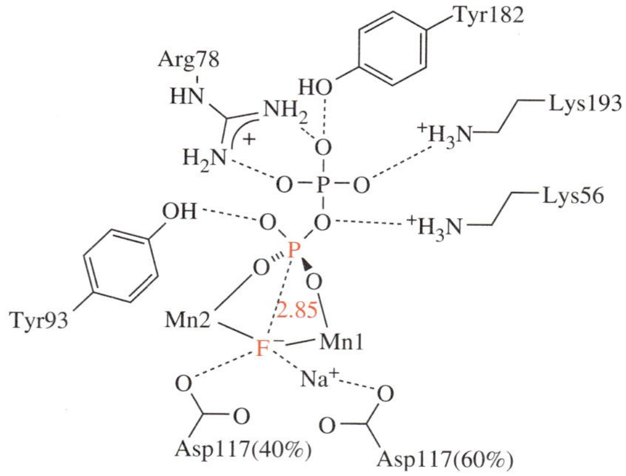

Fig. 10-17. Structure of inorganic pyrophosphatase from yeast. In the active site, the enzyme with fluoride is bound in place of a water molecule or hydroxide ion between the two divalent metal ions in the *bis*-Mn^{2+} center (Heikinheimo et al., 2001). The other two Mn^{2+} ions are coordinated to pyrophosphate.

Structures of several complexes in scheme 10-9 are available. Shown in fig. 10-17 are models from x-ray crystallographic structural models of the yeast pyrophosphatase. The bis-manganese complex is shown with Mn^{2+}-pyrophosphate and fluoride in place of hydroxide or water in the bridging position to prevent cleavage of pyrophosphate. A hydroxide in this location would be in position to accept the phosphoryl group in flight as Mn$_2$-pyrophosphate is cleaved into two Mn-phosphates.

Enzymatic Nucleotidyl Group Transfer

Nucleotidyltransferases

Most enzymes that catalyze transfer of a phosphoryl ester group, ROPO$_2^-$ as distinguished from PO$_3^-$, are nucleotidyltransferases. These are enzymes like DNA and RNA polymerases, adenylyl cyclase, nucleotide sugar pyrophosphorylases, and NAD$^+$ pyrophosphorylase. Nucleotidyl group transfer also takes place in the actions of certain ATP-dependent ligases or synthetases, which are presented in chapter 11. In the next section, we consider the nucleotidyltransferases.

Chemistry of Nucleotidyl Transfer

In the sections on phosphoryl group transfer, we encountered single-displacement mechanisms with no covalent phosphoryl-enzyme intermediate and double-displacement mechanisms involving covalent intermediates. Nucleotidyltransferases also function by single- or double-displacement mechanisms. However, very little information about the nature of transition states is available for nucleotidyltransferases. Constructions of linear free energy correlations are impractical because nucleotidyltransferases display high selectivity for substrates. In most cases, the only guide to the nature of the transition state is the basic chemistry, which seems to mandate a tight transition state, with substantial P—O bond formation to the entering nucleophile and significant bond cleavage to the leaving group. One study of the primary and secondary ^{18}O-kinetic isotope effects in the action of a nucleotidyltransferase supports the expectation of a tight transition state.

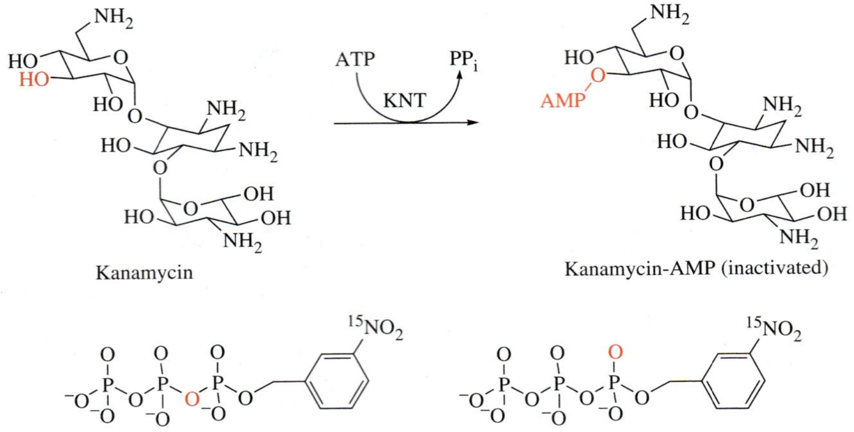

522 Enzymatic Reaction Mechanisms

Fig. 10-18. Inactivation of kanamycin by the action of kanamycin nucleotidyltransferase (KNT). Aminoglycoside nucleotidyltransferases inactivate the aminoglycoside antibiotics by catalyzing the chemical modification of a hydroxyl group. For example, KNT catalyzes transfer of the adenosine-5′-phosphoryl group from ATP to the 4′-OH group of kanamycin. Like other aminoglycoside nucleotidyltransferases, KNT displays little selectivity for ATP and will accept any nucleoside-5′-triphosphate. It even accepts m-nitrobenzyl triphosphate as a slow subtrate and catalyzes transfer of the m-nitrobenzylphosphoryl group to kanamycin. The latter substrate facilitated the measurement of ^{18}O-kinetic isotope effects by the remote label method, using ^{15}N as the remote label. The labeling patterns are illustrated, with ^{18}O shown in red.

Kanamycin nucleotidyltransferase (KNT) catalyzes the transfer of an AMP group from ATP to the 4′-OH group of a kanamycin. This reaction inactivates the aminoglycoside as an antibiotic. The structures of kanamycin and its inactive, adenylylated form are shown in fig. 10-18. Like other aminoglycoside nucleotidyltransferases, KNT displays little selectivity for ATP and will accept any nucleoside-5′-triphosphate, and the reaction follows a sequential kinetic mechanism (Van Pelt and Northrop, 1984). The structure of KNT does not show extensive contact with the nucleoside portion of the nucleotide (Pedersen et al., 1995). In the case of KNT, this lack of specificity allows m-nitrobenzyl triphosphate to function as a poor substrate in place of ATP. Being a poor substrate, m-nitrobenzyl triphosphate reacts slowly enough to allow the chemical step to be rate limiting (Gerratana et al., 2001). The m-nitrobenzyl group allows the remote label method to be employed in the measurement of ^{18}O-kinetic isotope effects, with ^{15}N as the remote label. The structures of remotely ^{15}N-labeled substrates employed in measurements of ^{18}O-kinetic isotope effects are shown in fig. 10-18. The normal primary ^{18}O-kinetic isotope effect of 1.016 ± 0.0003 with ^{18}O in the bridging position of the bond undergoing cleavage in m-nitrobenzyl triphosphate proves that P—O bond cleavage is rate limiting with this substrate. The observation of a normal secondary ^{18}O kinetic isotope effect of 1.0033 ± 0.0004 with ^{18}O in the nonbridging positions proves that the P—O bond orders in the transferring group are slightly decreased in the transition state relative to the ground state, consistent with a tight transition state. The secondary kinetic isotope effect implicates substantial bonding to the 4′-OH group of kanamycin in an oxyphosphorane-like transition state, as in nonenzymatic alkylphosphoryl group transfer.

Adenylyl Cyclase

In a central reaction of hormonal signaling, adenylyl cyclase catalyzes the transformation of ATP into 3′,5′-cyclic-AMP (cAMP). This reaction proceeds by internal displacement of pyrophosphate through nucleophilic attack by the 3′-OH group of ATP on P_α, as illustrated in eq. 10-15. The stereochemical consequence is overall inversion of configuration at P_α of ATP, implicating a single-displacement mechanism (Gerlt et al., 1980).

$$\text{ATP} \cdot Mg^{2+} \xrightleftharpoons{\text{Adenylyl cyclase}} \text{cAMP} + MgPP_i \qquad (10\text{-}15)$$

Adenylyl cyclase (EC 4.6.1.1) is membrane bound and composed of two cytoplasmic domains separated by membrane-spanning domains. Although adenylyl cyclase is the catalytic component of cAMP-mediated hormonal signaling, communications from the hormone receptor to the catalytic component is a dominant theme. GTP-binding proteins, known as G proteins, transmit information from hormonal receptors to adenylyl cyclase, and the signals transmitted by G proteins are controlled by a switching mechanism involving the hydrolysis of GTP (Gilman, 1987; Casey and Gilman, 1988). The diagram in fig. 10-19 illustrates a generic G-protein signaling scheme, one that may represent the activation of adenylyl cyclase by epinephrine in touching off the glycogen phosphorylase activation cascade in fig. 10-8.

G proteins are heterotrimeric and represented in fig. 10-19 as $G_{\alpha\beta\gamma}$. Hormonal stimulation in G protein mediated activation of adenylyl cyclase operates by use of binding energy to transmit a signal. This is a recurrent theme in the scheme in fig. 10-19. A hormone H binds to its receptor **R**, and the binding energy released increases the binding affinity of the receptor, presumably through a conformational change, for the G protein complex $G_{\alpha\beta\gamma}$·GDP. Binding to the receptor weakens the affinity of the G protein for GDP, presumably through a conformational change, stimulating the release of GDP and giving GTP access to the nucleotide site. On binding GTP, the structure of the G protein is destabilized,

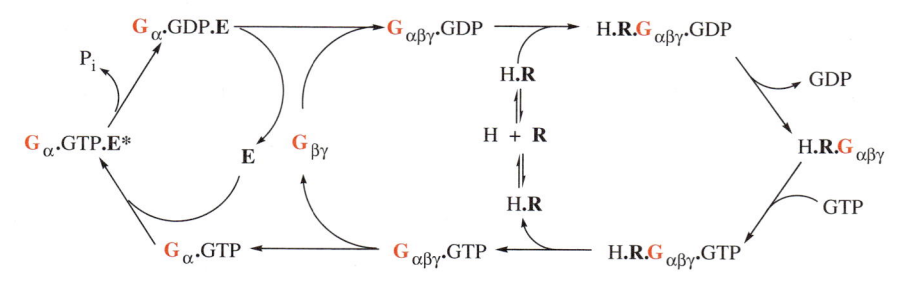

Fig. 10-19. Mediation of hormonal activation of adenylate cyclase by a G protein. A G protein comprises three subunits, α, β, and γ, and normally binds GDP (top center). On binding to a receptor-hormone complex H·**R**, the binding of GDP is weakened, leading to its dissociation. Binding of GTP weakens the interaction with the receptor and leads to dissociation from the receptor. The β,γ portion of the G protein dissociates to generate the complex of G_α with GTP (lower left), and this form binds and activates adenylyl cyclase. **E*** designates active adenylyl cyclase, and **E** designates unactivated cyclase. The GTP bound to G_α and **E*** in the activated complex (left center) slowly undergoes hydrolysis to GDP, switching off the activating signal. G proteins interact with a number of receptors and serve to link the activation of adenylyl cyclase with the hormonal signal.

presumably through a conformational change, inducing the dissociation of the β and γ subunits as the heterodimer $G_{βγ}$ This stabilizes the resulting complex of $G_α$·GTP, which possesses the unique property of binding and activating adenylyl cyclase. So long as this complex persists, adenylyl cyclase is active and produces cAMP. However, the GTP is not perfectly stable in the complex of adenylyl cyclase and $G_α$·GTP, and it undergoes hydrolysis to GDP on a scale that is slow relative to enzymatic turnover. The complex of $G_α$·GDP does not retain the capacity to bind and activate adenylyl cyclase very well, presumably because of a conformation change induced by the hydrolysis of GTP, and the activation signal is switched off. Dissociation from adenylyl cyclase ensues, accompanied by re-binding of $G_{βγ}$ to complete the cycle. These events take place on the surface of the membrane within which the hormone receptor and adenylyl cyclase remain anchored. The cycle allows the activity of adenylyl cyclase to be switched on so long as sufficient hormone is present. However, whenever the hormone is absent, the hydrolysis of GTP silences adenylyl cyclase. This brief discussion is limited to activation, but G proteins can also mediate inhibition of adenylyl cyclase in extended regulatory phenomena. The principle of G protein mediation extends to other signaling systems, including the ras P21 GTPase, which is implicated in cell proliferation and cancer.

The mechanism of action of adenylyl cyclase is under continuing investigation. Kinetic information indicates that cAMP dissociates from the product complex before $MgPP_i$ (Dessauer and Gilman, 1997). Membrane binding hampers progress on the reaction mechanism; however, the cytoplasmic domains can be expressed in E. coli. Soluble forms of activating G protein can also be expressed, and the three components undergo association to produce active adenylyl cyclase. Kinetic evidence implicates two divalent metal ions in the reaction mechanism (Garbers and Johnson, 1975). X-ray crystallographic analysis of complexes of the two cytoplasmic domains with $G_α$, ATP analogs, and Mg^{2+}, Mn^{2+} or Zn^{2+} show the locations of two divalent metal ions at the active site (Tesmer et al., 1999). Shown in fig. 10-20 is the active site, with β-L-2′,5′-dideoxy adenosine 3′–monophosphate bound with triphosphate and Mg^{2+}. The two metal ions are linked together by mutual ligation to the conserved residues Asp396 and Asp440.

Consideration of the structure in fig. 10-20, together with other structures in which ATPαS is the ATP analog and Zn^{2+} is the divalent metal, leads to a proposal for the structure of a Michaelis complex between adenylyl cyclase and ATP (Tesmer et al., 1999). In this model, the 3′-OH group of ATP is ligated to Me_A, lowering its pK_a value and thereby increasing its nucleophilic reactivity for the internal displacement of pyrophosphate to form cAMP. Simultaneous coordination of Me_A by $P_α$ can be expected to increase the electrophilic reactivity of $P_α$ toward nucleophilic attack by the 3′-oxygen. The mechanistic concept is outlined in fig. 10-21.

The postulated roles of the two metal ions in the mechanism of fig. 10-21 are similar to those proposed in the action of DNA polymerase. The biological roles of adenylyl cyclase and nucleic acid polymerases are different, but the underlying chemistry is similar. Moreover, the secondary structural motif of the βαββαβ fold defining the active site of adenylyl cyclase is similar to the palm-like domains in polymerases. Like the polymerases, a conformational closing of the active site accompanies substrate binding (Tesmer et al., 1999).

Intervention in activation of adenylyl cyclase is an objective in pharmaceutical research. G protein mediated activation offers the possibility of blocking any of several steps in fig. 10-19 through the action of a drug. Compounds that inhibit adenylyl cyclase itself must directly block its action. The P-site inhibitors include naturally occurring compounds that bind to the active site, many with high affinity (Johnson et al., 1985; Dessauer and Gilman, 1996). These inhibitors are substrate analogs that cannot undergo cyclization. One of many such compounds is adenosine-2′,5′-dideoxy-3′-triphosphate. This and other P-site

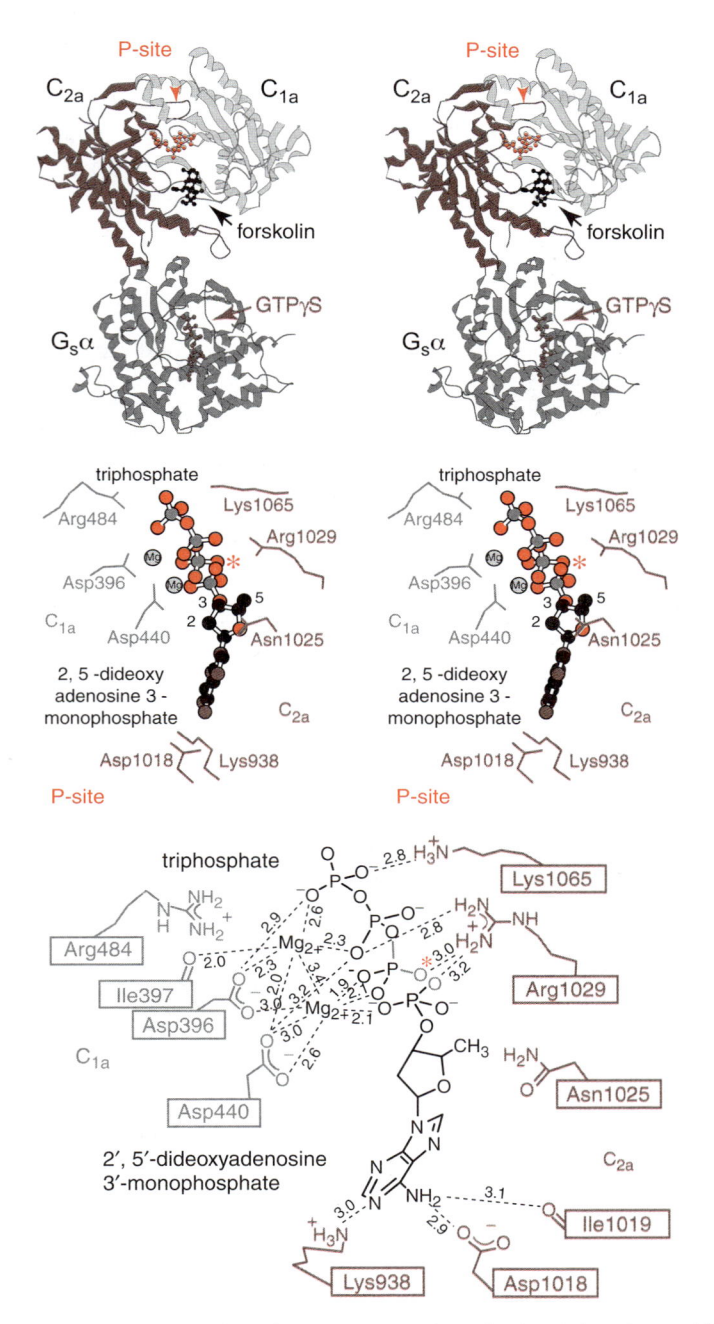

Fig. 10-20. Structure of a complex of catalytic domains of adenylyl cyclase with the activating domain of a G protein. The complex of canine VC$_{1a}$, and rat IIC$_{2a}$ adenylyl cyclase subunits with bovine G$_s\alpha$ G protein is shown in the top stereoimage (2.4-Å resolution; PDB 1CUL; Tesmer et al., 2000). GTPγS is bound to the G protein, forskolin, which stabilizes the C$_{1a}$C$_{2a}$ complex, which is bound in the adenylyl cyclase ventral cleft, and 2′,5′-dideoxyadenosine 3′ monophosphate and triphosphate are bound in the P site. The P site is shown in stereo in the middle panel and in two dimensions in the bottom panel. Occupancy by both molecules in this site is partial, because they are mutually exclusive due to overlap (asterisk).

525

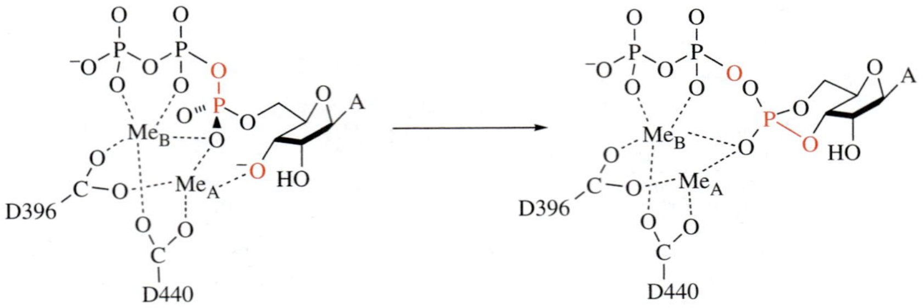

526 Enzymatic Reaction Mechanisms

Fig. 10-21. A mechanism for the action of adenylyl cyclase. Two divalent metal ions participate in the reaction of adenylyl cyclase. In this mechanism, Me$_A$ coordinates the 3-OH group of ATP and the pro-R oxygen on Pα of ATP, as well as an aspartate residue (Tesmer et al., 1999). Me$_B$ coordinates the pyrophosphate and pro-R oxygen on Pα of ATP, as well as an aspartate residue. Me$_A$ facilitates nucleophilic attack by lowering the pK_a of the 3'-OH group, allowing nucleophilic attack by the 3'-oxyanion on P$_\alpha$.

inhibitors bind to the complex of adenylyl cyclase and MgPP$_I$ (Tesmer et al., 2000). The P-site inhibitors are thought to intercept the cyclase·PP$_I$ complex that arises from the dissociation of cAMP in the kinetic mechanism.

Uridylyltransferases

Galactose metabolism is essential in all cells and depends on the presence of the enzymes of the Leloir pathway. Human galactosemia is most often due to a deficiency in galactose-1-phosphate uridylyltransferase (EC 2.7.7.12), which catalyzes the reaction of UDP-glucose with galactose-1-P to produce UDP-galactose and glucose-1-P according to eq. 10-16.

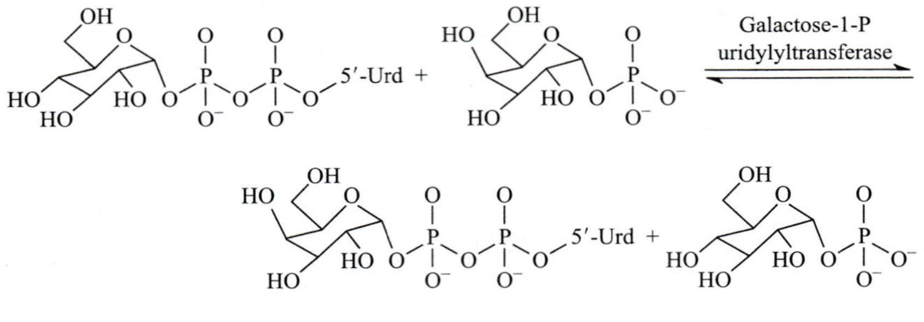

(10-16)

Note the cleavage of the bond P$_\alpha$—O shown in color. The transferred group is uridine-5'-phosphoryl (uridylyl) and not the galactosyl or glucosyl group. The overall reaction is nearly isoenergetic, and the mechanism involves a slightly higher energy intermediate. This enzyme is GalT in *E. coli* and GALT in yeast. We use GalT because most mechanistic work is on the enzyme from *E. coli*.

The uridylyl acceptors on the two sides of eq. 10-16 are sterically and electrostatically similar, as are the uridylyl donors. A single binding site could accommodate both substrates. A mechanism that allows joint use of a single binding site is based on the principle of economy in the evolution of binding sites. This is a double-displacement mechanism with a covalent uridylyl-enzyme intermediate. Such a mechanism would allow a single binding site for a nucleotide sugar, provided that a nucleophilic group in the site could accept the uridylyl group transiently during the changeover of acceptor molecules.

Accordingly, the kinetics in the action of *E. coli* GalT is ping pong bi bi as in scheme 10-10, where the enzyme nucleophile is **E-N** (Wong and Frey, 1974a). Moreover, the covalent intermediate is readily isolated (Wong and Frey, 1974b).

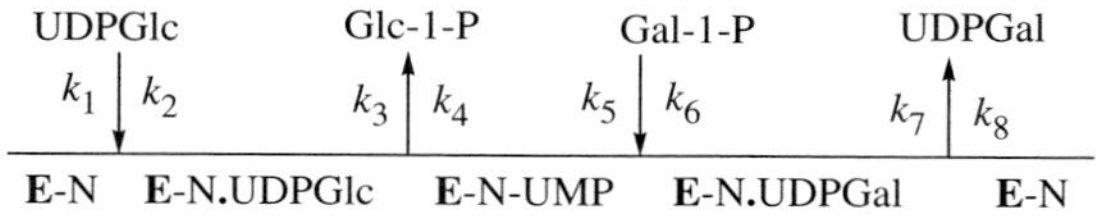

$$\begin{array}{ccccc}
\text{UDPGlc} & \text{Glc-1-P} & \text{Gal-1-P} & & \text{UDPGal} \\
k_1 \downarrow k_2 & k_3 \uparrow k_4 & k_5 \downarrow k_6 & & k_7 \uparrow k_8 \\
\hline
\text{E-N} \quad \text{E-N·UDPGlc} & \text{E-N-UMP} & \text{E-N·UDPGal} & & \text{E-N}
\end{array}$$

Scheme 10-10

As in all ping pong mechanisms, the enzyme catalyzes characteristic exchange reactions. It catalyzes the exchange of [^{14}C]glucose-1-P into UDP-glucose through the reversibility of steps k_1 through k_4 in the absence of galactose, and it catalyzes the exchange of [^{14}C]galactose-1-P into UDP-galactose through the reversibility of steps k_5 through k_8 in the absence of glucose. The relationship $1/V^{X1} + 1/V^{X2} = 1/V^F + 1/V^R$, relating maximum exchange rates and maximum forward and reverse rates, is satisfied (see chap. 2).

The covalent intermediate can be produced by reaction of the enzyme with [α-^{32}P] UDP-glucose followed by gel filtration, and chemical degradation of this intermediate proves that the enzyme-nucleophile is a histidine residue (Yang and Frey, 1979). The only essential histidine residues are His164 and His166, as proven by site-directed mutagenesis (Field et al., 1989). H166G-GalT is chemically rescued by imidazole. Moreover, H166G-GalT catalyzes the reaction of glucose-1-P with uridine-5′-phosphoimidazolide to form UDP-glucose, and galactose-1-P reacts similarly (Kim et al., 1990; Ruzicka et al., 1998). H164G-GalT does not catalyze these reactions, proving that the nucleophilic catalyst in the active site is His166.

GalT accepts (R_P)-UDPαS-glucose as a good substrate. (R_P)-UDPαS-glucose incorporates a chiral center at P$_\alpha$, which undergoes P—O cleavage, and the reaction of this substrate proceeds with overall retention of configuration to produce (R_P)UDPαS-galactose, as demonstrated by ^{31}P NMR spectroscopy (Sheu et al., 1978). Overall retention is required by the mechanism in scheme 10-10. Moreover, there is proof that the P-configuration of the uridine-5′-thiophosphryl-GalT is inverted relative to (R_P)UDPαS-glucose, verifying that each step proceeds with inversion at P$_\alpha$, leading to overall retention in the two steps according to scheme 10-11 (Arabshahi et al., 1986). Retention of configuration at each step would also lead to overall retention, but the reaction actually proceeds with inversion at each step.

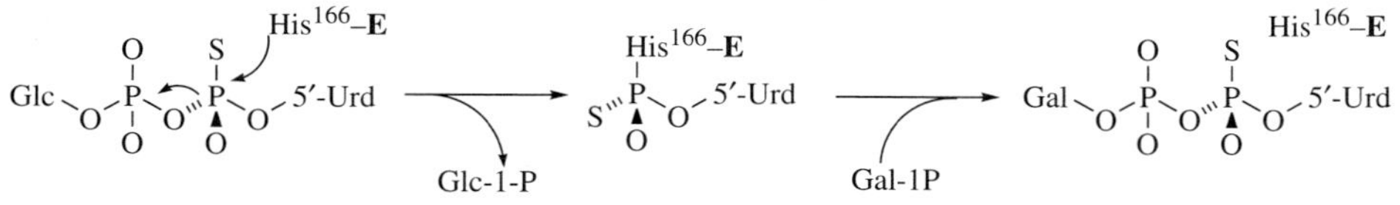

Scheme 10-11

GalT is a dimer of identical 39-kDa subunits. Although it catalyzes nucleotidyltransfer and is a metalloenzyme (Ruzicka et al., 1995), the metal ions do not participate in the chemistry of nucleotidyltransfer. Each subunit contains one Zn^{2+} and one Fe^{2+}. The locations of the metal ions are shown in the structure of the uridylyl-GalT in fig. 10-22 (Wedekind et al., 1996). The structure shows how several residues in the highly conserved active-site sequence motif $_{160}$CSNPHPHGQ$_{168}$ interact with the uridylyl group. The essential residues are His164, His166, and Gln168 (Field et al., 1989; Geeganage et al., 2000). His166 is covalently bonded to the uridylyl group in transit; the imidazole ring of His164 is one of four ligands to Zn^{2+}; the main chain carbonyl group of His164 forms a hydrogen bond to imidazo-N$^{\delta 1}$ of His166; and the side chain of Gln168 is hydrogen bonded to the phosphoryl group.

528   Enzymatic Reaction Mechanisms

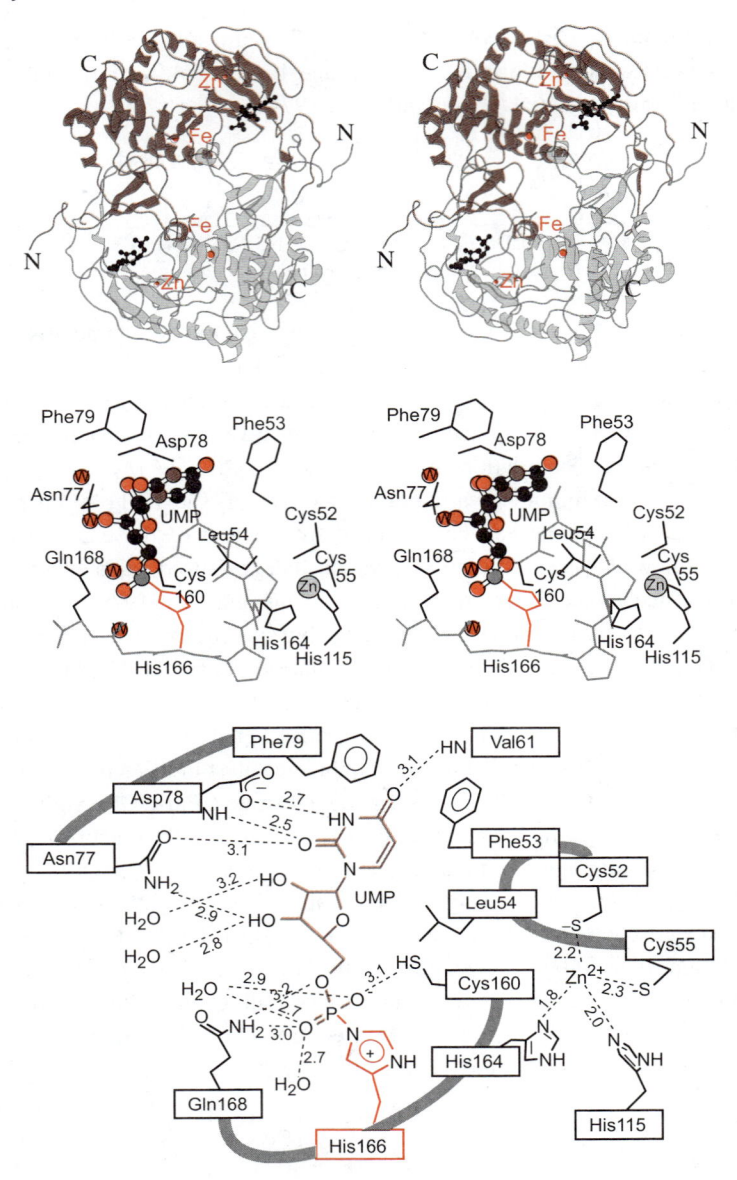

Fig. 10-22. Structure of GalT from *E. coli* in its uridylylated form. Galactose-1-phosphate uridylyltransferase from *E. coli* (GalT) is a homodimeric enzyme with two noncatalytic metal binding sites (containing Zn^{2+} and Fe^{2+}, both in red) in each half-barrel fold (1.8-Å resolution; PDB 1HXP; Wedekind et al., 1995). The uridylyl moiety is shown in black. The iron site is just outside of the half-barrel, near the dimer interface, which it helps to stabilize. The zinc site is adjacent to the active site, and its occupancy is required for the formation of a functional active site; a variety of divalent metal ions (Zn^{2+}, Fe^{2+}, Cd^{2+}, Co^{2+}, and Mn^{2+}) in this site can serve this structural requisite. The middle panel shows uridylylated His166 and the adjacent structural divalent metal binding site. A two-dimensional map (bottom) shows the active site, with distances given in angstroms.

The major lesion in human galactosemia is the substitution of Gln188 (Gln168 in the *E. coli* enzyme) to arginine. Mutation of Gln168 to arginine in *E. coli* GalT decreases the overall activity by 10^5, and both the formation and reaction of the uridylyl-GalT intermediate are comparably compromised (Geeganage and Frey, 1998).

The chemically analogous reaction of MgUTP with glucose-1-P to form UDP-glucose and $MgPP_I$ is catalyzed by UDP-glucose pyrophosphorylase (EC 2.7.7.10) by a different mechanism. The kinetics is ordered bi bi sequential, with MgUTP binding first (Tsuboi et al., 1969). The reaction proceeds with inversion of configuration at P_α of UTPαS (Sheu and Frey, 1978). The mechanism likely involves the binding of MgUTP and glucose-1-P at adjacent sites and direct transfer of the uridylyl group from MgUTP to glucose-1-P, with no covalent intermediate.

DNA Polymerases

Four families of DNA polymerases (EC 2.7.7.7) catalyze strand elongation in a variety of DNA processing functions, including the replication and repair of DNA. The diverse functional specificities of DNA polymerases notwithstanding, the mechanism of 3′,5′-phosphodiester bond formation is thought to be similar in the actions of all nucleic acid polymerases. The most thoroughly studied DNA polymerase I from *E. coli* participates in the repair of DNA and serves as a case study of the mechanism of action in this family. DNA replication requires a complex of enzymes assembled at an origin of

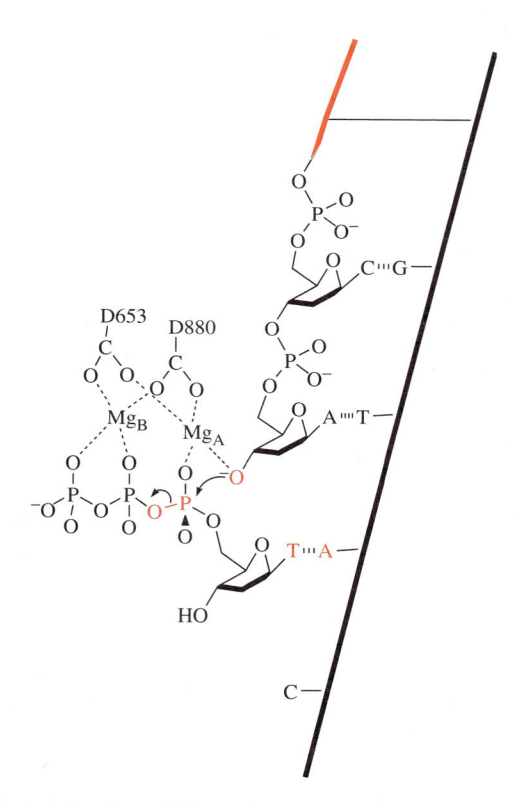

Fig. 10-23. A mechanism for the addition of a nucleotide by the action of DNA polymerase. Elongation of a strand of DNA by DNA polymerase requires the presence of dATP, dGTP, dTTP, dCTP, and a segment of primer/template DNA as substrates, as well as magnesium ion. The primer strand (red) of the primer-template is required because DNA polymerase elongates strands but does not initiate new ones. The template strand (black) is required to select the dNTP to be added in each elongation step through hydrogen-bonded base pairing, as illustrated above in the pairing of dTTP with A in the template strand. The chemical mechanism is facilitated by two Mg^{2+} ions, in which each metal is coordinated to two aspartate residues of the enzyme and to the substrates. The numbering of aspartate residues is correlated with the numbering in fig. 10-24. Coordination of Mg_A with P_α and the terminal 3′-OH group of the primer strand catalyzes nucleophilic attack. Coordination of Mg_B with P_β and P_γ facilitates departure as $MgPP_i$.

replication that carry out many processes in addition to the formation of 3′,5′-phosphodiester bond formation between deoxynucleotides (Waga and Stillman, 1998). We limit this discussion to the chemical mechanism of deoxynucleotide addition by DNA polymerases.

DNA polymerases do not initiate strand formation but catalyze the ordered addition of deoxynucleotides to an incomplete strand in duplex DNA under the direction of a complementary strand. The basic chemical process and mechanism shown in fig. 10-23 is supported by available biochemical evidence and the structures of DNA polymerases (Beese and Steitz, 1991; Steitz, 1993). The observation of DNA polymerase activity in vitro requires the presence of a primer/template DNA, all four deoxynucleoside triphosphates, Mg^{2+}, and a DNA polymerase. The primer or template serves the dual role of providing a strand for elongation and a complementary strand to serve as the template for ordering the addition of nucleotides to the primer. At each elongation cycle, the deoxynucleotide chosen is specified by the Watson-Crick base pairing rules. The hydrogen bonding patterns in the allowed A:T and C:G base pairs are shown in structure **10-7**.

10-7

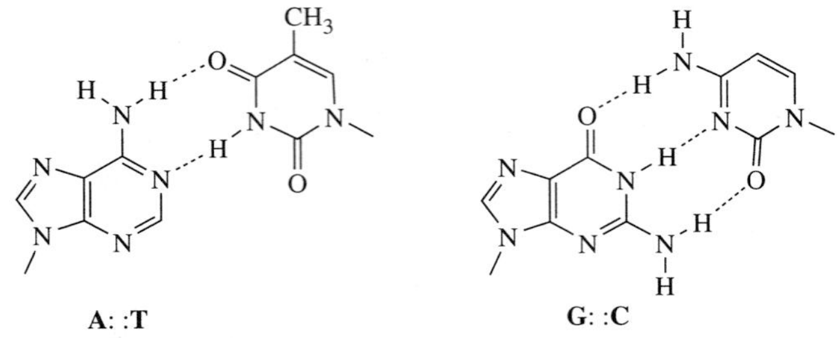

A: :T   G: :C

The structure shown in fig. 10-24 is typical of three of four families of DNA polymerases, those that include the βαββαβ fold alluded to in the section on adenylyl cyclase. This hand-like motif contains two aspartate residues that are conserved in this family of DNA polymerases and that bind two divalent metal ions that participate in catalysis. The primer/template is bound along a groove adjacent to the divalent metal ions.

The structure in the active site shows the two Mg^{2+} ions coordinated to the triphosphate of a dNTP and in position to coordinate the 3′-OH group at the growing end of the primer. In fig. 10-23, this group is shown with the proton dissociated to generate the 3′-oxyanion for nucleophilic attack on P_α of dATP to displace $MgPP_i$ and form the 3′,5′-phosphodiester bond. The role of Mg_A^{2+} in the mechanism presumably includes decreasing the pK_a value of the 3′-OH group through coordination, thereby facilitating loss of the proton to form the nucleophilic oxyanion. In another role, Mg_A^{2+} is coordinated to the P_α group of the entering dNTP, which should increase the electrophilic reactivity of P_α and further promote nucleophilic attack. This mechanism for the action of DNA polymerase (Beese and Steitz, 1991; Steitz, 1993), and the structural similarities in the active sites of DNA polymerase and adenylyl cyclase are the inspiration for the similar mechanism put forward for the action of adenylyl cyclase in fig. 10-21 (Tesmer et al., 1999).

The structure of DNA polymerase shown in fig. 10-24 is compatible with the mechanism in fig. 10-23. The two metal ions are shown in position to coordinate the triphosphate moiety of dNTP and the 3′-OH end of a primer strand. The two aspartate residues (Asp830 and Asp653) that bind the metals are conserved in DNA polymerases.

The detailed kinetic mechanism for the action of DNA polymerase I from *E. coli* in fig. 10-25 was derived from a series of transient kinetic and isotope exchange experiments (Joyce and Benkovic, 2004). The rate constants for individual steps, interpreted in the light of the rate constant for a single turnover, require that a conformational change be included as an essential step in the mechanism. This conformational change may represent a frame-shift

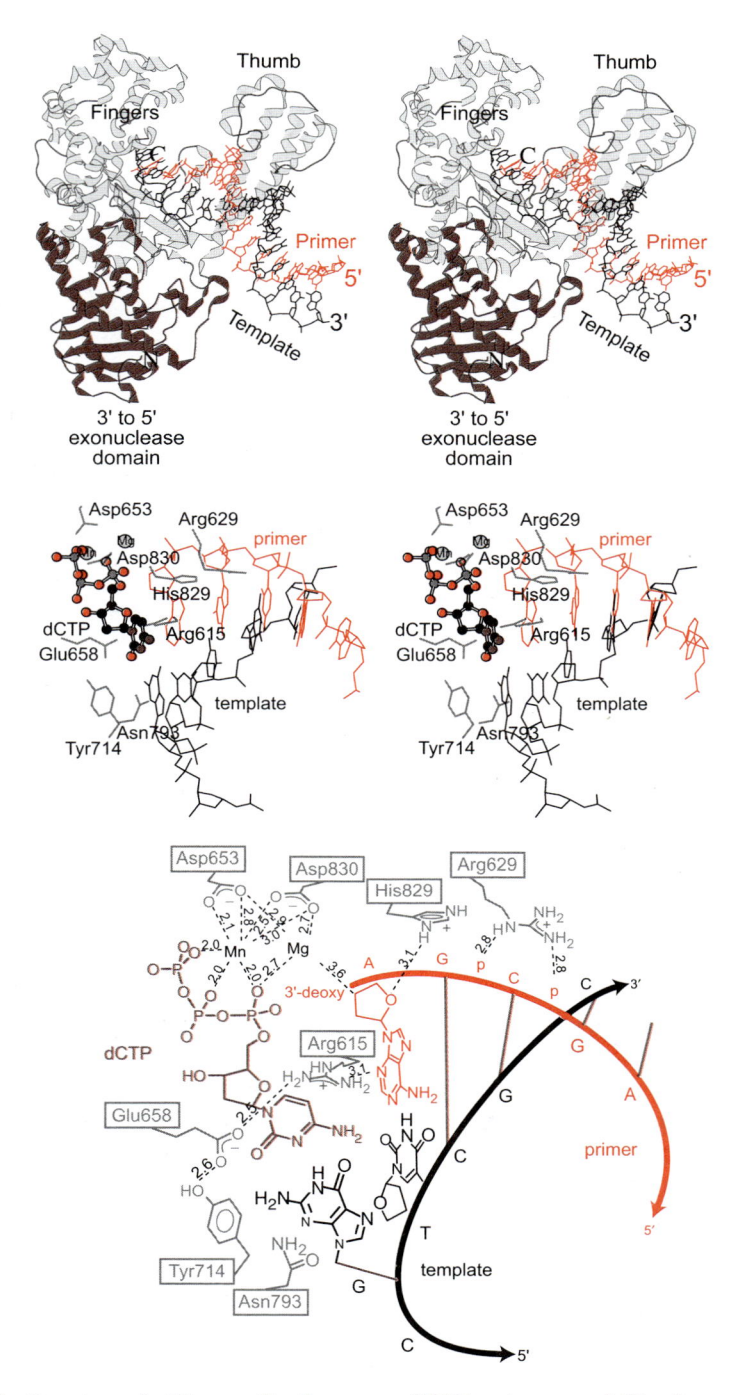

Fig. 10-24. Structure of a Klenow-like fragment of DNA polymerase I. The fingers and thumb of the handlike active site of this Klenow-like fragment of DNA polymerase I from *Bacillus stearothermophilus* are indicated in the global structure at the top (1.7-Å resolution; PDB 1L3S; Johnson et al., 2003). The template DNA is shown in black and the priming strand in red. The active site in the center shows the two metal ions and aspartate ligands with dCTP base paired with the template and base stacked with the terminal nucleotide on the primer strand. The primer strand terminates with a 2'3'-dideoxy residue, and elongation is halted in the complex shown. The two-dimensional map of the active site at the bottom shows the contact distances.

532   Enzymatic Reaction Mechanisms

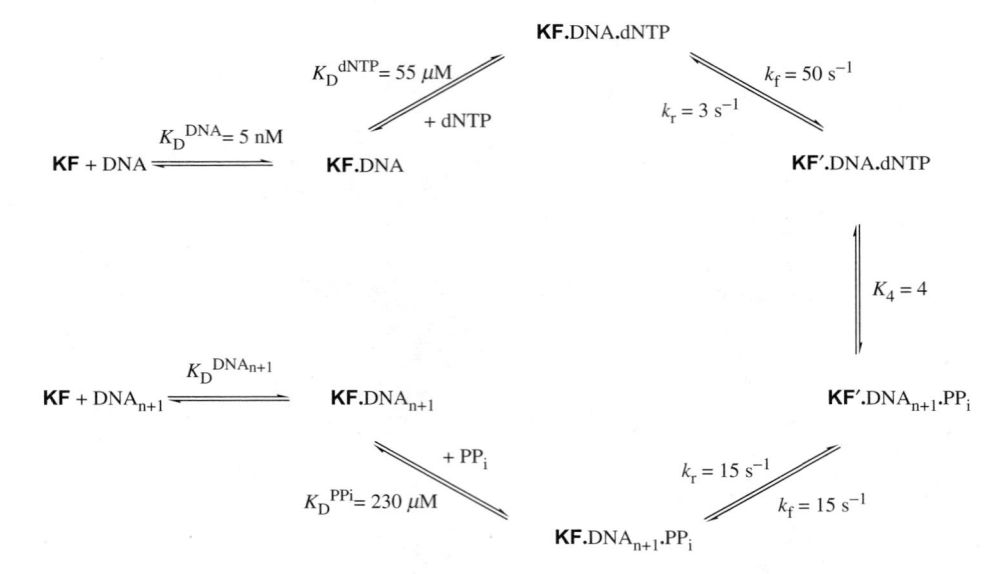

$$\mathbf{KF} + \mathbf{DNA} \underset{K_D^{DNA}=\,5\,\text{nM}}{\rightleftharpoons} \mathbf{KF.DNA} \underset{K_D^{dNTP}=\,55\,\mu M}{\xrightarrow{+\,dNTP}} \mathbf{KF.DNA.dNTP} \underset{k_r\,=\,3\,s^{-1}}{\overset{k_f\,=\,50\,s^{-1}}{\rightleftharpoons}} \mathbf{KF'.DNA.dNTP}$$

$$K_4 = 4$$

$$\mathbf{KF} + \mathbf{DNA}_{n+1} \underset{K_D^{DNA_{n+1}}}{\rightleftharpoons} \mathbf{KF.DNA}_{n+1} \underset{K_D^{PPi}=\,230\,\mu M}{\xrightarrow{+\,PP_i}} \mathbf{KF.DNA}_{n+1}\mathbf{.PP_i} \underset{k_r\,=\,15\,s^{-1}}{\overset{k_f\,=\,15\,s^{-1}}{\rightleftharpoons}} \mathbf{KF'.DNA}_{n+1}\mathbf{.PP_i}$$

Fig. 10-25. Kinetic mechanism in the action of DNA polymerase I from *E. coli*.

within the enzyme-primer/template complex, moving the enzyme along the template. The internal equilibrium constants are small (e.g., $K_4 = 4$), so that the free energies of intermediates are comparable.

DNA polymerases catalyze chemically crucial steps in DNA replication, the stepwise addition of deoxynucleotides in correct order. A unique mechanistic aspect of enzymatic nucleotide polymerization is that the enzymes themselves do not determine the substrate specificity at each cycle of nucleotide addition. The nucleotide sequence in the template strand of the primer/template, the cosubstrate, determines the order of nucleotide addition.

DNA polymerase I catalyzes a second process in *E. coli*. The fidelity of deoxynucleotide incorporation is very high but not sufficient to ensure the retention of genetic information. The occasional mis-incorporated deoxynucleotide introduces a lesion in the growing structure. Recognition of mis-incorporation is an important auxiliary function of DNA polymerase I, which contains 3′,5′-exonuclease activity that removes offending lesions when they occur. This activity is catalyzed by a specialized domain that functions to "edit" the growing chain and correct adventitious mistakes as they occur.

In DNA replication, enzymes other than DNA polymerase catalyze such essential steps as priming, unwinding, and ligation. The mechanisms of these processes are considered in later chapters.

Topoisomerases

Supercoiling in DNA is important in the transmission of genetic information because the recognition of nucleotide sequences requires the disruption of base pairing through local unwinding of the helix. Negative supercoiling in DNA strains the helix and promotes unwinding, a process that is required in transcription and replication. Topoisomerases (EC 5.99.1.2 and EC 5.99.1.3) control supercoiling in DNA. Type I topoisomerases catalyze the relaxation of supercoiling, and type II topoisomerases catalyze ATP-dependent negative supercoiling.

Any alteration in the topology of DNA must entail strand cleavage, the induction of a topological change, and then strand religation. Strand cleavage and religation are the chemical processes carried out by topoisomerases, which also guide the intervening topological changes. The chemical problem of cleavage and religation is elegantly solved through covalent catalysis by tyrosine residues in topoisomerases. Reaction of a tyrosine residue with a 3′,5′-phosphodiester linkage in DNA to form a tyrosyl-3′-phosphodiester or tyrosyl-5′-phosphodiester linkage between DNA and the topoisomerase maintains the phosphodiester bond energy during the time required for the guided topological change. Religation consists of reversing the cleavage reaction to restore the 3′,5′-phosphodiester. The chemistry of cleavage and religation is isoenergetic and does not require complex chemical machinery.

A focus of research on topoisomerases is the physical mechanism of topological change (Redinbo et al., 1999; Wang, 1995, 1998, 2002). In the action of a type I topoisomerase, one of the two strands is cleaved by nucleophilic action of a tyrosine residue to bind the 3′- or 5′-end covalently as a phosphodiester and create a 3′-OH or 5′-OH end. Type I topoisomerases are approximately 100-kDa enzymes comprising subtypes IA and IB that are unrelated in amino acid sequence and act by different mechanisms. Type IA forms a tyrosine-5′-phosphodiester as the intermediate, and the topological change seems to take place by a strand passage mechanism, as illustrated in fig. 10-26. Type IB topoisomerase cleaves the strand in the opposite orientation to form tyrosine-3′-phosphodiester and a 5′-OH end. The topological rearrangement takes place by a controlled rotation mechanism, and the 5′-OH group then displaces the tyrosine residue to religate the chain.

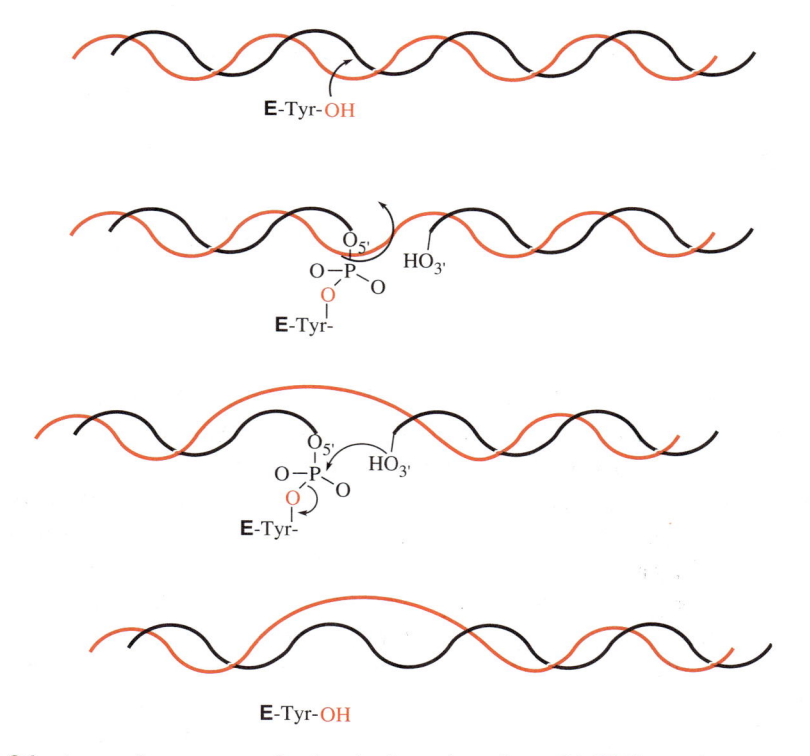

Fig. 10-26. A strand passage mechanism in the action of type IA DNA topoisomerase. A tyrosine residue of topoisomerase displaces the 3′-OH end at a phosphodiester linkage in negatively supercoiled DNA, forming a covalent phosphodiester linkage to the 5′-phospho end. The complementary strand passes through the nick in the cleaved strand, partially relaxing the helix. The 3′-OH group then displaces the tyrosine residue to religate the transiently cleaved strand and regenerate the active site tyrosine. This cycle relaxes the DNA helix by one turn.

534 Enzymatic Reaction Mechanisms

Type II topoisomerases cleave both DNA strands of a double helix by forming tyrosine phosphodiesters with both strands. The cleavage points are separated by four nucleotides, resulting in sticky ends that can facilitate strand dissociation for the topological transition and reassociation for the religation process. In the dissociated state, a separate segment of double helix is captured, and in an ATP hydrolysis-dependent process, the double helix is passed through the separated segments. This increases the negative coiling and requires the energy of ATP-hydrolysis. Then, the sticky ends of the cleaved double helix guide the religation of the two cleaved strands.

The structures of DNA topoisomerases include holes that can admit a DNA double helix. The chemical cleavages by tyrosine residues take place within the complexes of topoisomerase and DNA. The structure of one topoisomerase is shown in fig. 10-27 in complex with a segment of double-stranded DNA. The DNA binding surface is lined with basic amino acids that form ionic contacts with the phosphodiester linkages and hold the DNA in position for cleavage by the active site tyrosine residue.

Topoisomerase action is essential to the regulation of transcription and replication. Because of this, several antibiotics and drugs inhibit the actions of topoisomerases. The antibacterial drug ciprofloxacin (Cipro) blocks bacterial DNA replication and RNA transcription by inhibiting a topoisomerase. Doxorubicin (Adriamycin) inhibits eukaryotic topoisomerase II and is used in cancer chemotherapy to kill rapidly dividing cells (structure **10-8**).

10-8 Ciprofloxacin   Doxorubicin

Ribozymes

Newly transcribed species of RNA undergo processing to mature RNA, and several of the required steps proceed in diverse species by self-processing (Doudna and Cech, 2002). For example, introns can be spliced out by action of the pre-RNA itself to form rRNA, tRNA, or mRNA, without the intervention of a protein enzyme. Moreover, replication of certain viral RNAs includes phosphodiester cleavage by the action of RNA domains. Self-processing is enzyme-like activity and can be truly enzymatic when multiple turnovers occur. For this reason, such species of RNA are known as ribozymes (Doudna and Cech, 2002; Kruger et al., 1982). Most naturally occurring ribozymes facilitate transesterification, an isoenergetic reaction. However, RNase P is a ribonucleoprotein, in which RNA functions as the catalyst in the hydrolysis of phosphodiester linkages. The ultimate ribozyme is probably the ribosome, which catalyzes peptide bond formation in protein biosynthesis (see chap. 18).

The current repertoire of natural ribozymes is small and expected to grow. The simple, nonribosomal ribozymes may be classified as those that do not require divalent metal ions and those in which divalent metal cations participate in catalysis. The hammerhead, the hepatitis delta virus (HDV), and the hairpin ribozymes, the crystal structures of which are shown in fig. 10-28, represent the first group.

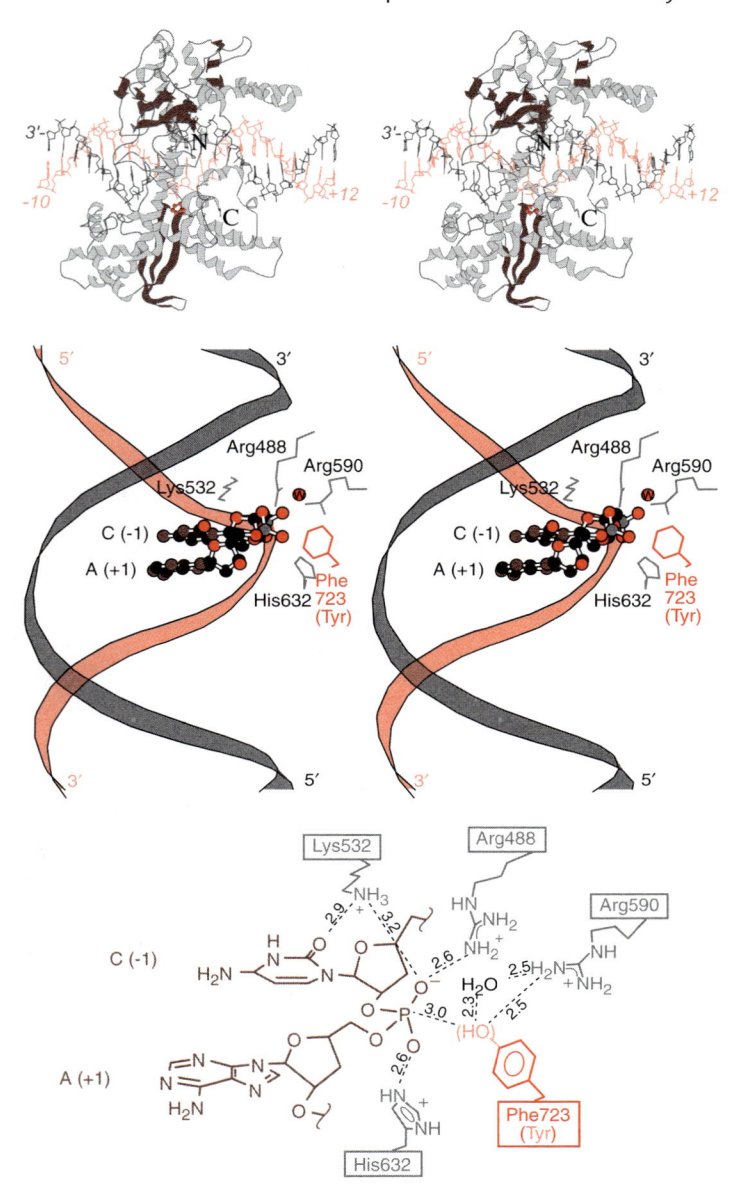

Fig. 10-27. Structure of human topoisomerase I B. Human topoisomerase I Y723F variant is shown in complex with double-stranded DNA, with the double helix passing directly through a large hole in the center of the enzyme (2.6-Å resolution; PDB 1EJ9; Redinbo et al., 2000). The strand that is cleaved and religated is shown in pink. The active site tyrosine nucleophile (Y723) has been removed by conversion to phenylalanine shown in red in the middle panel. The nucleotide flanking the site of phosphodiester bond cleavage are shown as ball-and-stick drawings within the pink and gray helices representing the DNA. The active site is shown schematically in the lower panel, with distances given in angstroms.

X-ray crystallography of ribozymes has been difficult, presumably because of structural mobility in RNA. To enhance the chances of initiating and growing ordered crystals, the HDV and hairpin ribozymes were co-crystallized with an RNA binding protein of known structure, and the structures of these complexes were determined (Ferre-D'Amare et al., 1998; Rupert and Ferre-D'Amare, 2001). This strategy succeeded with the HDV and hairpin ribozymes, as shown in fig. 10-28. The structure of the hammerhead ribozyme was determined in the absence of a protein (Pley et al., 1994).

536 Enzymatic Reaction Mechanisms

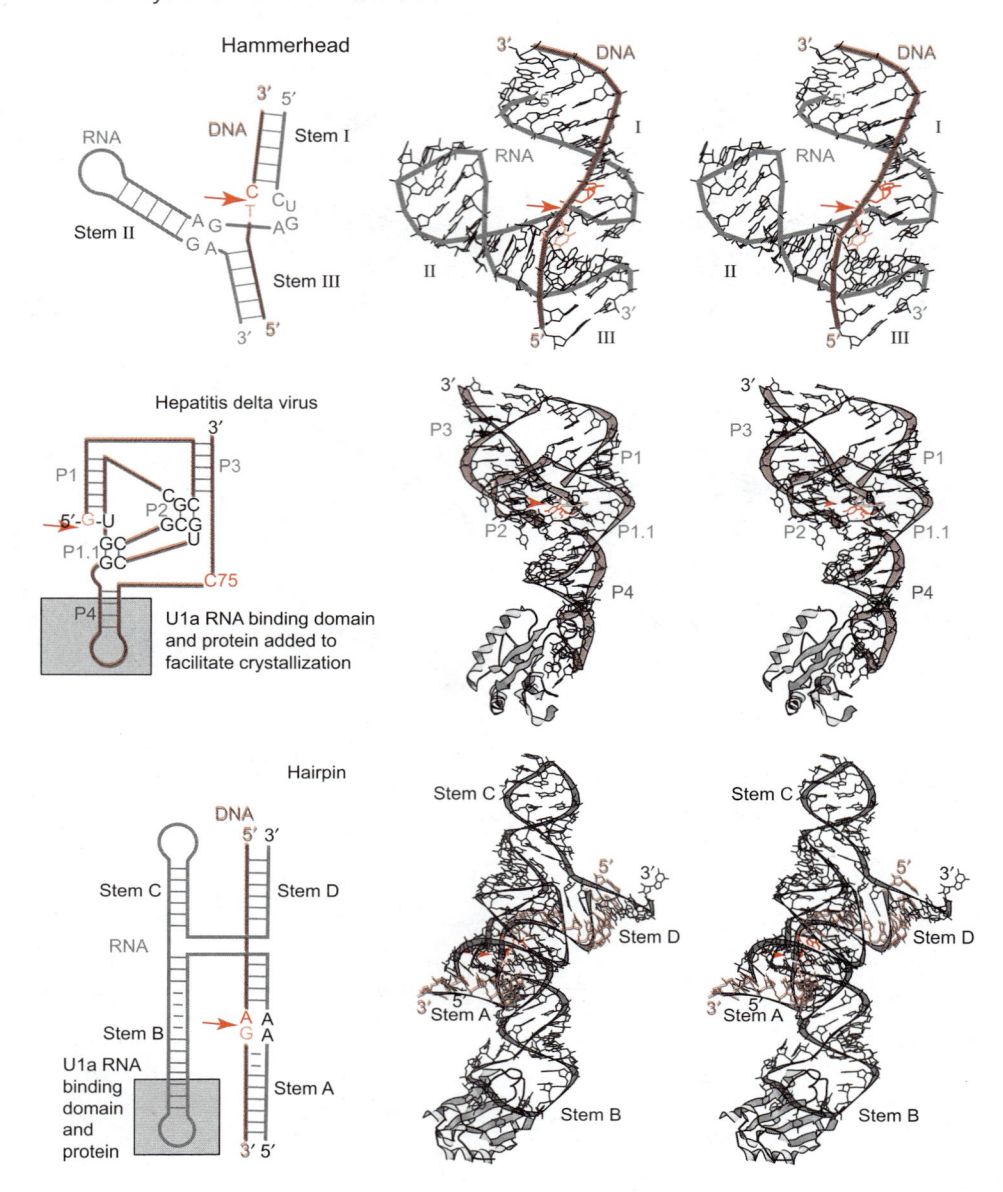

Fig. 10-28. Structures of hammerhead, hepatitis delta virus (HDV), and hairpin ribozymes. Three catalytic RNA structures are shown with secondary structural schemes to the right and with stereoimages of the three-dimensional structures to the left: 1, the hammerhead ribozyme (2.6-Å resolution; PDB 1HMH; Pley et al., 1994); 2, the HDV ribozyme (2.3-Å resolution; PDB 1CX0; Ferre-D'Amare et al., 1998); and 3, the hairpin ribozyme (2.4-Å resolution; PDB 1M5K; Rupert et al., 2002). Many fixed metal ions are associated with these structures but are not included for clarity. Alteration of the last two ribozymes to include the U1a RNA binding domain and the addition of U1a RNA binding protein to the crystallization trials were required for obtaining ordered structures by x-ray crystallography.

Available information indicates that the hairpin, HDV, and hairpin ribozymes function by a mechanism that exploits the most unique chemical property of RNA, base catalyzed hydrolysis with anchimeric participation by the 2′-OH group, as illustrated in fig. 10-29. This mechanism does not involve or require a divalent metal cation. The available structures may not represent the active conformations of the ribozymes, so that the acid and base catalysts are difficult to identify (Doudna and Cech, 2002). Candidate acid-base

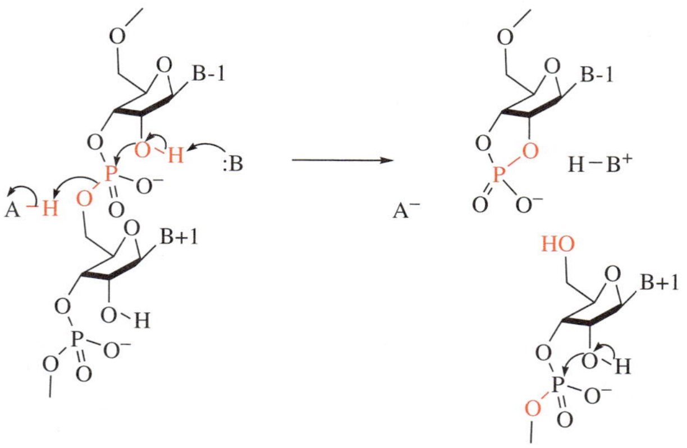

Fig. 10-29. A mechanism for 3′,5′-phosphodiester cleavage by hammerhead, hepatitis delta virus, and hairpin ribozymes.

catalysts are the heterocyclic bases A, G, and C. The ionizing properties of the bases may not seem appropriate for acid-base catalysis; the pK_a values are well below or above the physiological pH. However, as we point out in chapter 1, microenvironmental effects can perturb pK_a values by four or more units, and such perturbations could shift the pK_a values of the heterocyclic bases into the neutral range. A well-known example of such an effect in a nucleic acid is the elevation of the pK_a of the cytosine ring in a DNA triplex (Sklenar and Feigon, 1990). Perturbations in pK_a also occur in RNA (Legault and Pardi, 1994). The chemical steps may not be rate limiting, so that pK_a values on the acid or alkaline side of neutrality would not be problematic.

Powerful evidence implicates cytidine 75 (C-75) as a catalyst in the self-cleavage of HDV. The structure ties this residue to the cleavage site (Ferre-D'Amare et al., 1998). Mutation of C-75 abolishes ribozymatic activity, and added imidazole partially rescues the activity (Perrotta et al., 1999). Other heterocyclic bases partially rescue activity with pH dependencies corresponding to their pK_a values (Shih and Been, 2001). Variants with replacement of C-75 by other heterocyclic bases of varying pK_a display activities with pH dependencies corresponding to the pK_a of the substituted heterocyclic base (Nakano et al., 2000). The assignment of C-75 as a catalyst is as secure as the assignment of an acid-base catalyst in a protein enzyme.

Structures of HDV in two conformations, precleavage and postcleavage, reveal new aspects of the reaction mechanism. In the precleavage structure, a divalent metal ion chelated by backbone phosphate oxygens and uracil-oxo groups is in contact with the 5′-oxygen of the scissile phosphodiester through an intervening water molecule, and C-75 is positioned near the 2′-OH group that attacks the scissile phosphate. The divalent metal ion is expelled in the postcleavage structure. In the current mechanism, C-75 functions as the base to abstract the proton from the 2′-OH group, and the divalent metal ion facilitates the departure of the 5′-OH group, acting as a Lewis acid through the intervening water molecule (Ke et al., 2004).

The first ribozyme to be discovered is the group I intron from *Tetrahymena* (Kruger et al., 1982). Group I introns splice out intervening sequences in pre-RNA by the mechanism in fig. 10-30. The process depends on mediation of transesterification by guanosine, GMP, GDP, or GTP. The RNA cleavage sequence binds by base pairing to a guide sequence in a process that forms the helix P1, and the guanosine nucleotide binds to the G site. The 3′-OH group of the guanosine nucleotide cleaves the phosphodiester in the cleavage site

538 Enzymatic Reaction Mechanisms

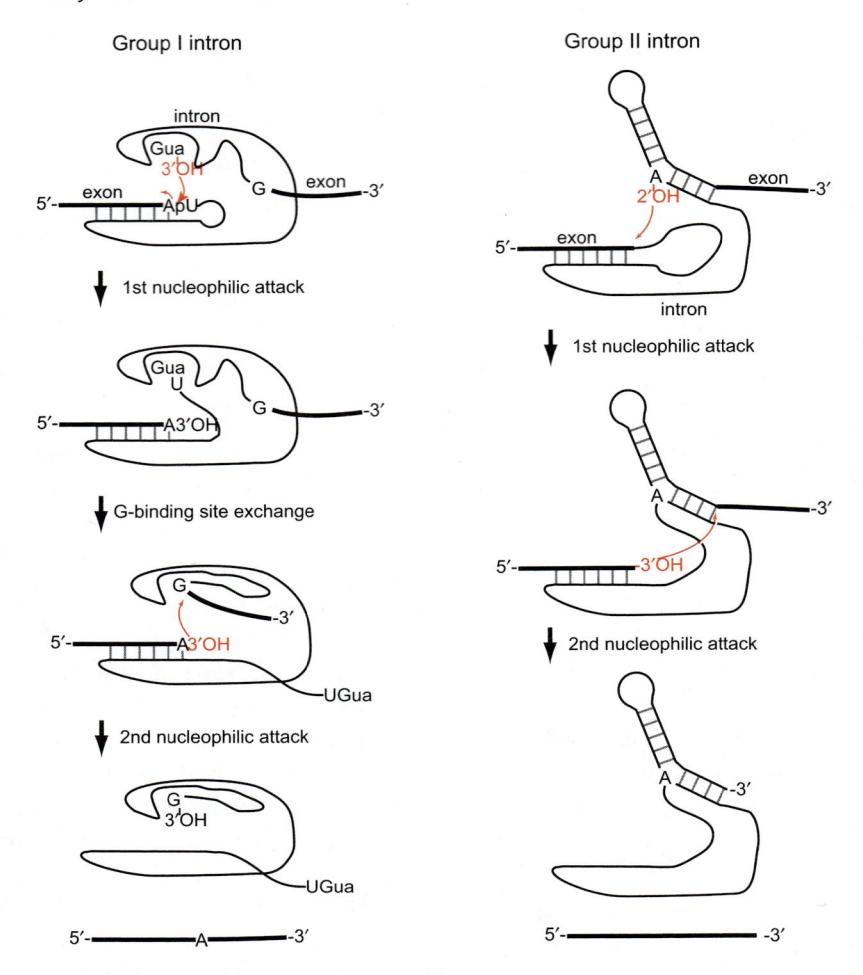

Fig. 10-30. Self-splicing by group I and group II introns. The group I introns undergo self-splicing in a process mediated by an external guanosine or guanosine nucleotide, which initially cleaves at the splice site. The group II introns employ the 2′-OH group of an internal adenosine nucleotide for this function. (Adapted from Doudna and Cech, 2002.)

to liberate its 3′-OH end and form a phosphodiester linkage to the guanosine nucleotide at the 3′-end of the intron. Then, the guanosine nucleotide end dissociates from the G site and is replaced by an internal guanosine nucleotide at the splice site. The liberated 3′-OH group in the active site then cleaves the phosphodiester at the internal G to liberate it as the 5′-end of the intron and generate spliced RNA. Group II introns function in an analogous sequence but use the 2′-OH group of an internal adenosine nucleotide in place of the external guanosine nucleotide employed by group I introns.

Divalent metal ions participate directly in the action of group I introns (Narlikar and Herschlag, 1997; Shan et al., 2001). The roles of divalent metal ions are diverse, and as many as three seem to be involved. Substitution of sulfur or amino groups in various locations of the cleavage site, and correlations with consequent divalent metal ion selectivities, indicates a mechanism such as that in fig. 10-31. The metal coordination serves to organize the reacting groups in alignment for nucleophilic attack by the 3′-OH group of guanosine. Two divalent metal ions are postulated to catalyze the reaction by decreasing the pK_a of the 3′-OH group of guanosine (Me$_B$), increasing its nucleophilic reactivity, and by increasing the electrophilic reactivity of the phosphodiester group (Me$_A$).

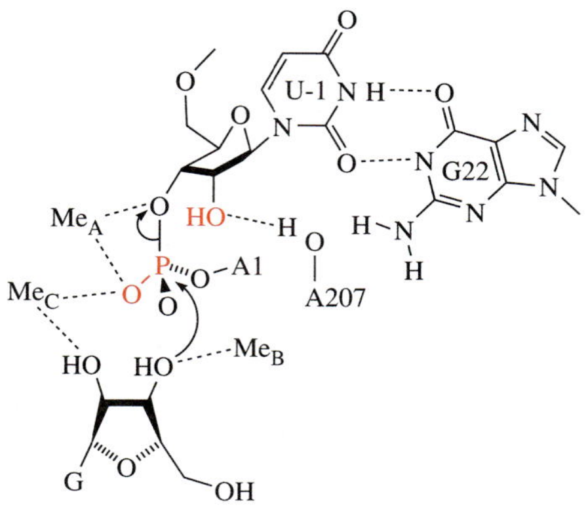

Fig. 10-31. Mechanism for divalent metal ion-assisted transesterification by group I intron from *Tetrahymena*.

Phosphodiesterases

Ribonuclease A

Pancreatic ribonuclease, one of the most intensively studied proteins, has been named ribonuclease A (RNase A; EC 3.1.27.5). Classic experiments on RNase established important principles of protein structure and function, including the fact that tertiary structure is determined by the amino acid sequence. This information appears in most textbooks of biochemistry. The mechanism of action of RNase has also been studied intensively, and we focus on this aspect.

RNase A is a small enzyme of 124 amino acids and one of the early successes in protein structural analysis by x-ray crystallography. Limited proteolytic cleavage by subtilisin produces a large fragment named RNase S and a small fragment, the S-peptide. RNase S is inactive, but addition of the S-peptide restores activity. The S-peptide contains His12, which is part of the active site.

Early chemical modification experiments with iodoacetate led to the identification of His12 and His119 as essential for activity. After alkylation of RNase A by iodoacetate, systematic analysis showed that part of the inactive protein contained carboxymethyl-His12 and part contained carboxymethyl-His119 (Lennette and Plapp, 1979). Both residues were shielded from alkylation by competitive inhibitors or substrates. The structure

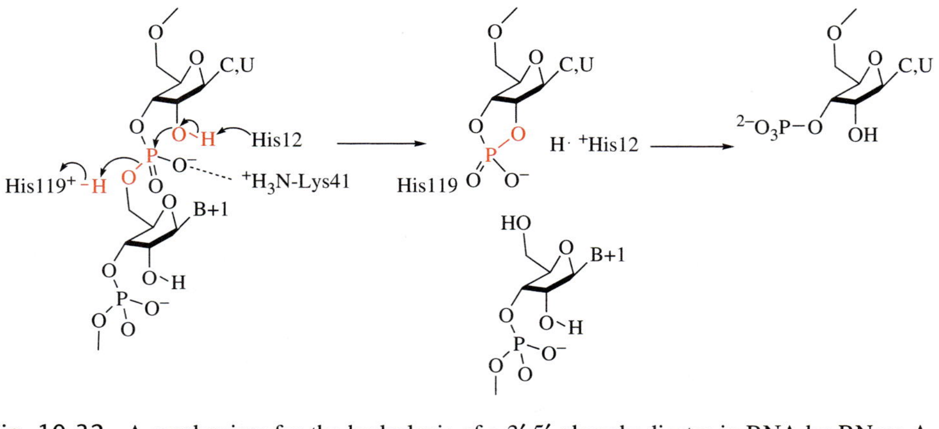

Fig. 10-32. A mechanism for the hydrolysis of a 3′,5′-phosphodiester in RNA by RNase A.

540 Enzymatic Reaction Mechanisms

confirmed the presence of His12 and His119 in the active site, and site-directed mutagenesis confirmed their importance for activity (Thompson and Raines, 1994). The pH dependence for k_{cat}/K_m showed a bell-shaped profile with pK_a values of 5.4 and 6.4 (Rosario and Hammes, 1969).

Success in the chemical modification of RNase A by iodoacetate can be attributed to two factors. This reagent is most reactive with sulfhydryl groups in proteins and is widely used to alkylate them (see chap. 1). RNase contains four disulfide bonds in its structure,

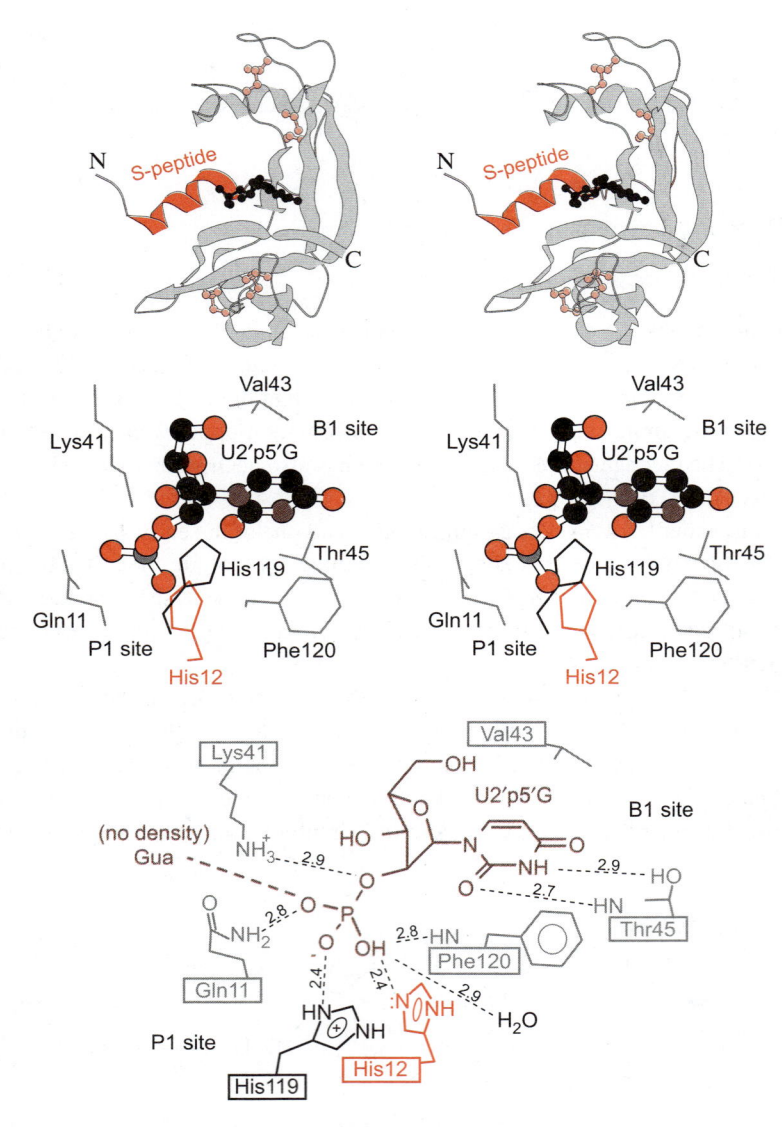

Fig. 10-33. The structure of RNase A. Bovine ribonuclease A isolated from the pancreas is shown with uridylyl-2′,5′-guanosine bound (2.0-Å resolution; PDB 1EOS; Vitagliano et al., 2000). Eight cysteine residues participate in four disulfide bounds that help to stabilize this excreted enzyme. There is too little electron density observed for the guanosine moiety for it to be included in the model. The active site appears to contain uridylyl-2′-phosphate (middle), with the active-site His12 shown in red. This histidine residue is part of the S-peptide (red in top panel). A two-dimensional map of the active site is included at the bottom, with the approximate position of the substrate guanosine indicated; interatomic distances are given in angstroms.

but there are no sulfhydryl groups, leaving the histidine residues as targets for alkylation. The active site binds negatively charged species, and iodoacetate carries a negative charge, so that it is attracted to the active site. For this reason, His12 and His119 in the active site react preferentially with iodoacetate.

RNase A catalyzes the endonucleolytic hydrolysis of RNA at pyrimidine sites. The primary products are the 5'-OH end and a 2',3'-cyclic phosphate end of two RNA fragments. In a secondary process, RNase catalyzes the hydrolysis of the 2',3'-cyclic phosphate end. A mechanism of acid-base catalyzed hydrolysis leading to these products is illustrated in fig. 10-32. In this mechanism, His12 serves as the base that abstracts the proton from the 2'-OH group as it engages in nucleophilic attack on the phosphodiester, and His119 serves as the acid in donating a proton to the leaving 5'-OH group in the formation of the 2',3'-cyclic phosphate. This mechanism is supported by the structure of RNase shown in fig. 10-33.

The mechanism in fig. 10-32 is written as a single-step displacement, implying a trigonal bipyramidal transition state. This is consistent with the fact that the reaction proceeds with inversion of configuration at phosphorus (Usher et al., 1970). The possibility that the mechanism might involve a discrete trigonal pyramidal oxyphosphorane intermediate could not be excluded on the basis of the stereochemistry. RNase accepts the *m*-nitrobenzyl ester of 3'-UMP as substrates. This allows the primary and secondary ^{18}O kinetic isotope effects to be measured by the remote label, internal competition method, with $^{15}NO_2$ as the remote label (Sowa et al., 1997). The chemical transformation limits the rate in this case. The ^{18}O-kinetic isotope effect of 1.016 for bridging ^{18}O shows that P—O cleavage to the leaving group is a property of the transition state. The secondary ^{18}O-kinetic isotope effect of 1.005 for nonbridging ^{18}O shows that the P—O bond order for the nonbridging oxygens decrease in the transition state. The secondary isotope effects do not support the formation of a discrete oxyphosphorane intermediate, and they indicate a concerted mechanism by way of a trigonal bipyramidal transition state. The sum of P—O bond orders for the attacking 2'-OH and the leaving m-nitrobenzyl alcohol is 1.13 to 1.20, corresponding to a tight transition state. This is in contrast to the loose transition state for phosphoryl group transfer. Evidence indicates that Lys41 stabilizes the transition state by donating a hydrogen bond to a nonbridging oxygen (Messmore et al., 1995).

References

Adams JA, ML McGlone, R Gibson, and SS Taylor (1995) *Biochemistry* **34**, 2447.
Anthony RS and LB Spector (1972) *J Biol Chem* **247**, 212.
Arabshahi A, RS Brody, A Smallwood, TC Tsai, and PA Frey (1986) *Biochemistry* **25**, 5583.
Baek YH and T Nowak (1982) *Arch Biochem Biophys* **217**, 491.
Baykov AA, IP Fabrichniy, P Pohjanjoki, AB Zyryanov, and R Lahti (2000) *Biochemistry* **39**, 11939.
Beese LS and TA Steitz (1991) *EMBO J* **10**, 25.
Belogurov GA, IP Fabrichniy, P Pohjanjoki, VN Kasho, E Lehtihuhta, MV Turkina, BS Cooperman, A Goldman, AA Baykov, and R Lahti (2000) *Biochemistry* **39**, 13931.
Benkovic SJ and K Schray (1973) In Boyer PD (ed): *The Enzymes*, vol 8, 3rd ed. Academic Press: New York, p 201.
Berger SA and PR Evans (1990) *Nature* **343**, 575.
Berger SA and PR Evans (1991) *Biochemistry* **30**, 8477.
Bernstein BE, DM Williams, JC Bressi, P Kuhn, MH Gelb, GM Blackburn, and WG Hol (1998) *J Mol Biol* **279**, 1137.
Blättler WA and JR Knowles (1979a) *Biochemistry* **18**, 3927.
Blättler WA and JR Knowles (1979b) *J Am Chem Soc* **101**, 510.
Bloxham DP and HA Lardy (1973) In Boyer PD (ed): *The Enzymes*, vol 8, 3rd ed. Academic Press: New York, p 239.

Bollenbach TJ, AD Mesecar, and T Nowack (1999) *Biochemistry* **38**, 9137.
Bourne N and A Williams (1984) *J Am Chem Soc* **106**, 7591.
Boyer PD, HA Lardy, and PH Philipp (1942) *J Biol Chem* **146**, 673.
Bruice TC and SJ Benkovic (1966) *Bioorganic Mechanisms*, vol 2. Benjamin: New York, pp 1-30.
Buchwald SL, JM Friedman, and JR Knowles (1984) *J Am Chem Soc* **106**, 4911.
Buchwald SL, DE Hansen, A Hassett, and JR Knowles (1982) *Methods Enzymol* **87**, 279.
Buechler JA and SS Taylor (1989) *Biochemistry* **28**, 2065.
Buechler JA and SS Taylor (1990) *Biochemistry* **29**, 1937.
Buechler JA, TA Vedvick, and SS Taylor (1989) *Biochemistry* **28**, 3018.
Burgers PMJ and F Eckstein (1980) *J Biol Chem* **255**, 8229.
Buss KA, DR Cooper, C Ingram-Smith, JG Ferry, DA Sanders, and MS Hasson (2001) *J Bacteriol* **183**, 680.
Butcher WW and FH Westheimer (1954) *J Am Chem Soc* **77**, 2420.
Cantewell JS, WR Novack, PF Wang, MJ McLeish, GL Kenyon, and PC Babbitt (2001) *Biochemistry* **40**, 3056.
Casey PJ and AG Gilman (1988) *J Biol Chem* **263**, 2577.
Cleland WW (1982) *Methods Enzymol* **87**, 159.
Cleland WW and AC Hengge (1995) *FASEB J* **9**, 1585.
Cohn M (1982) *Annu Rev Biophys Bioeng* **11**, 23.
Coleman JE (1992) *Annu Rev Biophys Biomol Struct* **21**, 441.
Collett MS and RL Erikson (1978) *Proc Natl Acad Sci U S A* **75**, 2021.
Cullis PM and D Nicholls *Chem Comm* **783** (1987).
Dahlberg ME and SJ Benkovic (1991) *Biochemistry* **30**, 4835.
Das AK, NR Helps, PT Cohen, and D Barford (1996) *EMBO J* **15**, 6798.
Davies GJ, SJ Gamblin, JA Littlechild, Z Dauter, KS Wilson, and HC Watson (1994) *Acta Crystallogr D Biol Crystallogr* **50**, 202.
Davis JP, M-M Zhou, and RL Van Etten (1994) *J Biol Chem* **269**, 8734.
Degani C and PD Boyer (1973) *J Biol Chem* **248**, 8222.
Denu JM, DL Lohse J Vijayalskshmi, MA Saper, and JE Dixon (1996) *Proc Natl Acad Sci U S A* **93**, 2493.
Dessauer CW and AG Gilman (1996) *J Biol Chem* **271**, 16967.
Dessauer CW and AG Gilman (1997) *J Biol Chem* **272**, 27787.
Dillet V, RL Van Etten, and D Bashford (2000) *J Phys Chem B* **104**, 11321.
Doudna JA and TR Cech (2002) *Nature* **418**, 222.
Dougherty TM and WW Cleland (1985) *Biochemistry* **24**, 5870.
Fenton AW, NM Paricharttanakul, and GD Reinhart (2003) *Biochemistry* **42**, 6453.
Ferre-D'Amare AR, K Zhou, and JA Doudna (1998) *Nature* **395**, 567.
Field TL, WS Reznikoff, and PA Frey (1989) *Biochemistry* **28**, 2094.
Fischer EH and EG Krebs (1955) *J Biol Chem* **216**, 121.
Fox DK, ND Meadow, and S Roseman (1986) *J Biol Chem* **261**, 13498.
Freeman S, JM Friedman, and JR Knowles (1987) *J Am Chem Soc* **109**, 3166.
Frey PA (1982) *Tetrahedron* **38**, 1541.
Frey PA(1989) *Adv Enzymol* **62**, 119.
Frey, PA (1992) In Boyer, PD and Sigman, DS (eds): *The Enzymes*, 3rd ed., Academic Press: New York, pp. 141–186.
Frey PA and RD Sammons (1985) *Science* **228**, 541.
Geeganage S and PA Frey (1998) *Biochemistry* **37**, 14500.
Geeganage S, VWK Ling, and PA Frey (2000) *Biochemistry* **39**, 5397.
Gerlt JA, JA Coderre, and S Mehdi (1983) *Adv Enzymol* **55**, 291.
Gerlt JA, JA Coderre, and MS Wolin (1980) *J Biol Chem* **255**, 331.
Gerratana B, PA Frey, and WW Cleland (2001) *Biochemistry* **40**, 2972.
Gill GN and LD Garren (1970) *Biochem Biophys Res Commun* **39**, 335.
Gilman AG (1987) *Annu Rev Biochem* **56**, 615.
Goldberg J, H Huang, Y Kwon, P Greengard, A Nairn, and J Kuriyan (1995) *Nature* **376**, 745.
Grace MR, CT Walsh, and PA Cole (1997) *Biochemistry* **36**, 1874.
Grant BD and JA Adams (1996) *Biochemistry* **35**, 2022.
Grzyska PK, PG Czyryca, J Purcell, and AC Hengge (2003) *J Am Chem Soc* **125**, 13106.
Guan KE and JE Dixon (1991) *J Biol Chem* **266**, 17026.

Halkides CJ and PA Frey (1991) *J Am Chem Soc* **113**, 9843.
Hansen DE and JR Knowles (1981) *J Biol Chem* **256**, 5967.
Heikinheimo P, J Lehtonen, A Baykov, R Lahti, BS Cooperman, and A Goldman (1996) *Structure* **4**, 1491.
Heikinheimo P, V Tuominen, AK Ahonen, A Teplyakov, BS Cooperman, AA Baykov, R Lahti, and A Goldman (2001) *Proc Natl Acad Sci U S A* **98**, 3121.
Hellinga HW and PR Evans (1987) *Nature* **327**, 437.
Hengge AC (2002) *Acc Chem Res* **35**, 105.
Hengge AC, WA Edens, and WW Cleland (1994) *J Am Chem Soc* **116**, 5045.
Herschlag D and WP Jencks (1989a) *J Am Chem Soc* **111**, 7579.
Herschlag D and WP Jencks (1989b) *J Am Chem Soc* **111**, 7587.
Herschlag D and WP Jencks (1990) *J Am Chem Soc* **112**, 1951.
Herzberg O, CC Chen, S Liu, A Tempczyk, A Howard, M Wei, D Ye, and D Dunaway-Mariano (2002) *Biochemistry* **41**, 780.
Hoff RH, L Wu, B Zhou, Z-Y Zhang, and AC Hengge (1999) *J Am Chem Soc* **121**, 9514.
Hollfelder F and D Herschlag (1995) *Biochemistry* **34**, 12255.
Holtz KM, B Stec, and ER Kantrowitz (1999) *J Biol Chem* **274**, 8351.
Hunter T and BM Sefton (1980) *Proc Natl Acad Sci U S A* **77**, 1311.
Janson CA and WW Cleland (1974) *J Biol Chem* **249**, 2567.
Jencks, WP, and J Regenstein (1970) In Sober HA (ed): *Handbook of Biochemistry and Molecular Biology*, CRC Press: Cleveland, p J187.
Jia Z, D Barford, AJ Flint, and NK Tonks (1995) *Science* **268**, 1754.
Johnson DA, P Akamine, E Radzio-Andzelm, M Madhusan, and SS Taylor (2001) *Chem Rev* **101**, 2243.
Johnson PE, SJ Abbott, and JR Knowles (1976) *Biochemistry* **15**, 2893.
Johnson RA, KH Jacobs, and G Schultz (1985) *J Biol Chem* **260**, 114.
Johnson SJ, JS Taylor, and LS Beese (2003) *Proc Nat Acad Sci U S A* **100**, 3895.
Jones JP, PM Weiss, and WW Cleland (1990) *Biochemistry* **30**, 3634.
Joyce CN and SJ Benkovic (2004) *Biochemistry* **43**, 14317.
Jurica MS, A Mesecar, PJ Heath, W Shi, T Nowak, and BL Stoddard (1998) *Structure* **6**, 195.
Kayne FJ (1973) In Boyer PD (ed): *The Enzymes*, vol 8, 3rd ed. Academic Press: New York, p 353.
Ke A, K Zhou, F Ding, JH Cate, JA Doudna (2004) *Nature* **429**, 201.
Kenyon GL and GH Reed (1983) *Adv Enzymol Relat Areas Mol Biol* **54**, 367.
Kim EE and HW Wyckoff (1991) *J Mol Biol* **218**, 449.
Kim J, FJ Ruzicka, and PA Frey (1990) *Biochemistry* **29**, 10590.
Kim KK and PA Cole (1998) *J Am Chem Soc* **120**, 6851.
Kirby AJ and AG Varvoglis (1967) *J Am Chem Soc* **89**, 415.
Knighton DR, JH Zheng, LF Ten Eyck, VA Ashford, NH Xhong, SS Taylor, and JM Sowadski (1991) *Science* **253**, 407.
Knowles JR (1980) *Annu Rev Biochem* **49**, 877.
Kofron JL and GH Reed (1990) *Arch Biochem Biophys* **280**, 40.
Krebs EG, DJ Graves, and EH Fischer (1959) *J Biol Chem* **234**, 2867.
Kruger K, PJ Grabowski, AJ Zaug, J Sands, DE Gottschling, and TR Cech (1982) *Cell* **31**, 147.
Kundrot CE and PR Evans (1991) *Biochemistry* **30**,1478.
Kuo DJ, EL O'Connell, and IA Rose (1979) *J Am Chem Soc* **101**, 5025.
Lahiri SD, P-F Wang, PC Babbitt, MJ McLeish, GL Kenyon, and KN Allen (2002) *Biochemistry* **41**, 13861.
Larsen TM, MM Benning, I Rayment, and GH Reed (1998) *Biochemistry* **37**, 6247.
Larsen TM, MM Benning, GE Wesenberg, I Rayment, and GH Reed (1997) *Arch Biochem Biohphys* **345**, 199.
Larsen TM, LT Laughlin, HM Holden, I Rayment, and GH Reed (1994) *Biochemistry* **33**, 6301.
Larsen TM and GH Reed (2001) In Bertino I, A Sigel, and H Sigel (eds): *Handbook on Metalloproteins*. Marcel Dekker: New York, p 9.
Larsson-Raznikeiwicz M (1967) *Biochim Biophys Acta* **132**, 33.
Laughlin LT and GH Reed (1997) *Arch Biochem Biophys* **348**, 262.
LeDu MH, C Lamoure, BH Muller, OV Bulgakov, E Lajeunesse, A Menez, and JC Boulain (2002) *J Mol Biol* **316**, 941.

Lee DC, DF Carmichael, EG Krebs, and EG McKnight (1983) *Proc Natl Acad Sci U S A* **80**, 3608.
Legault P and A Pardi (1994) *J Am Chem Soc* **116**, 8390.
Lennette EP and BV Plapp (1979) *Biochemistry* **18**, 3938.
Lew J, SS Taylor, and JA Adams (1997) *Biochemistry* **36**, 6717.
Leyh TS, RD Sammons, PA Frey, and GH Reed (1982) *J Biol Chem* **257**, 15047.
Lightcap E and PA Frey (1992) *J Am Chem Soc* **114**, 9750.
Lindqvist Y, G Schneider, and P Vihko (1993) *J Biol Chem* **268**, 20744.
Lindqvist Y, G Schneider, and P Vihko (1994) *Eur J Biochem* **221**, 139.
Lodato DT and GH Reed (1987) *Biochemistry* **26**, 2243.
Lowe G, PM Cullis, RL Jarvest, BVL Potter, and BS Sproat (1981) *Philosophical Transactions of the Royal Society of London. Series B, Biological Sciences* **293**, 75.
Madhusudan S, EA Trafny, N-H Xhong, JA Adams, LF Ten Eyck, SS Taylor, and JM Sowadski (1994) *Protein Sci* **3**, 176.
Manning G, DB Whyte, R Martinez, T Hunter, and S Sudarsanam (2002) *Science* **298**, 1912.
Messmore JM, DN Fuchs, and RT Raines (1995) *J Am Chem Soc* **117**, 8057.
Miller DL and FH Westheimer (1966) *J Am Chem Soc* **88**, 1507.
Miller DL and T Ukena (1969) *J Am Chem Soc* **91**, 3050.
Moore MJ, JA Adams, and SS Taylor (2003) *J Biol Chem* **278**, 10613.
Morikis D, M Roy, MG Newlon, JD Scott, and PA Jennings (2002) *Eur J Biochem* **269**, 2040.
Morrison JF and WW Cleland (1966) *J Biol Chem* **241**, 673.
Nakano S, DM Chadalavada, PC Bevilacqua (2000) *Science* **287**, 1493.
Narlikar GJ and D Herschlag (1997) *Annu Rev Biochem* **66**, 19.
O'Brien PJ and D Herschlag (2002) *Biochemistry* **41**, 3207.
Ostantin K, A Saeed, and RL Van Etten (1994) *J Biol Chem* **269**, 8971.
Pawson T and JD Scott (1997) *Science* **278**, 2075.
Pedersen LC, MM Benning, and HM Holden (1995) *Biochemistry* **34**, 13305.
Perrotta AT, I Shih, and MD Been (1999) *Science* **286**, 123.
Phan J, K Lee, S Cherry, JE Tropea, TR Burke Jr, and DS Waugh (2003) *Biochemistry* **42**, 13113.
Pley HW, KM Flaherty, and DB McKay (1994) *Nature* **372**, 68.
Pliura DH, D Schomburg, JP Richard, PA Frey, and JR Knowles (1980) *Biochemistry* **19**, 325.
Purich D (1982) *Methods Enzymol* **87** (C), 1.
Redinbo MR, JJ Champoux, and WGJ Hol (1999) *Curr Opin Struct Biol* **9**, 29.
Redinbo MR, JJ Champoux, and WG Hol (2000) *Biochemistry* **39**, 6832.
Reed GH and M Cohn (1972) *J Biol Chem* **247**, 3073.
Robinson JL and IA Rose (1972) *J Biol Chem* **247**, 1096.
Rosario EJ and GG Hammes (1969) *Biochemistry* **8**, 1884.
Rose IA, DJ Kuo, and JVB Warms (1991) *Biochemistry* **30**, 722.
Rupert PB, AP Massey, ST Sigurdsson, and AR Ferre-D'Amare (2002) *Science* **298**, 1421.
Ruzicka FJ, S Geeganage, and PA Frey (1998) *Biochemistry* **37**, 11385.
Ruzicka FJ, JE Wedekind, J Kim, I Rayment, and PA Frey (1995) *Biochemistry* **34**, 5610.
Saini MS, SL Buchwald, RL Van Etten, and JR Knowles (1981) *J Biol Chem* **256**, 10453.
Schwartz JH and F Lippman (1961) *Proc Natl Acad Sci U S A* **47**, 1996.
Shaffer J and JA Adams (1999) *Biochemistry* **38**, 5572.
Shan S, AV Kravchuk, JA Piccirilli, and D Herschlag (2001) *Biochemistry* **40**, 5161.
Sheu KFR and PA Frey (1978) *J Biol Chem* **253**, 3378.
Sheu KF, JP Richard, and PA Frey (1979) *Biochemistry* **18**, 5548.
Shih IH and MD Been (2001) *Proc Natl Acad Sci U S A* **98**, 1489.
Shirakihara Y and PR Evans (1988) *J Mol Biol* **204**, 973.
Showers MO and RA Mauerer (1986) *J Biol Chem* **261**, 16288.
Sicheri F, I Moarefi, and J Kuriyan (1997) *Nature* **385**, 602.
Sklenar V and J Feigon (1990) *Nature* **345**, 836.
Skoog MT and WP Jencks (1984) *J Am Chem Soc* **106**, 7597.
Smith SB, HD White, JB Siegel, and EG Krebs (1981) *Proc Natl Acad Sci U S A* **78**, 1591.
Sowa GA, AC Hengge, and WW Cleland (1997) *J Am Chem Soc* **119**, 2319.
Stec B, MJ Hehir, C Brennan, M Nolte, and ER Kantrowitz (1998) *J Mol Biol* **277**, 647.
Stec B, KM Holtz, and ER Kantrowitz (2000) *J Mol Biol* **299**, 1303.
Steitz TA (1993) *Curr Opin Struct Biol* **3**, 31.

Stuckey JA, HL Schubert, EB Fauman, Z-Y Zhang, JE Dixon, and MA Saper (1994) *Nature* **370**, 571.
Su Y, WR Dostmann, FW Herberg, K Durick, NH Xuong, L Ten Eyck, SS Taylor, and KI Varughese (1995) *Science* **269**, 807.
Susan-Resign D and T Nowak (2003) *J Biol Chem* **278**, 12660.
Takio K, SB Smith, EG Krebs, KA Walsh, and K Titani (1984) *Biochemistry* **23**, 4200.
Taylor SS, JA Buechler, and W Yonemoto (1990) *Annu Rev Biochem* **59**, 971.
Taylor SS, E Radzio-Andzelm, and T Hunter (1995) *FASEB J* **9**, 1255.
Tesmer JJG, CW Dessauer, RK Sunahara, LD Murray, RA Johnson, AG Gilman, and SR Sprang (2000) *Biochemistry* **39**, 14464.
Tesmer JJG, RK Sunahara, RA Johnson, G Gosselin, AG Gilman, and SR Sprang (1999) *Science* **285**, 756.
Thompson JE and RT Raines (1994) *J Am Chem Soc* **116**, 5467.
Thrall SH, AF Mehl, LJ Carroll, and D Dunaway-Mariano (1993) *Biochemistry* **32**, 1803.
Titani K, T Sasagawa, LH Ericsson, S Kumar, SB Smith, EG Krebs, and KA Walsh (1984) *Biochemistry* **23**, 4193.
Tsuboi KK, K Fukunaga, and JC Petricciani (1969) *J Biol Chem* **244**, 1008.
Usher DA, DI Richardson, and F Eckstein (1970) *Nature* **228**, 663.
Van Etten RL (1982) *Ann N Y Acad Sci* **390**, 27.
Van Etten RL (2003) In Bradshaw R and E Dennis (eds): *Handbook of Cell Signaling*. Elsevier: New York, 2003, p 733.
Van Etten RL and ME Hickey (1977) *Arch Biochem Biophys* **183**, 250.
Vincent JB, MW Crowder, and BA Averill (1992) *Trends Biochem Sci* **17**, 105.
Vitagliano L, A Merlino, A Zagari, and L Mazzarella (2000) *Protein Sci* **6**, 1217.
Waga S, and B Stillman (1998) *Annu Rev Biochem* **67**, 721.
Walsh CT and L Spector (1971) *J Biol Chem* **246**, 1255.
Walsh DA, KL Angelos, SM Van Patten, DB Glass, and LP Garetto (1990) In Kemp BE (ed): *Peptides and Protein Phosphorylation*. CRC Press: Boca Raton, FL.
Walsh DA, JP Perkins, and EG Krebs (1968) *J Biol Chem* **243**, 3763.
Wang JC (1996) *Annu Rev Biochem* **65**, 635.
Wang JC (1998) *Q Rev Biophys* **31**, 107.
Wang JC (2002) *Nature Rev Mol Cell Biol* **3**, 430.
Webb MR and DR Trentham (1980) *J Biol Chem* **255**, 1775.
Wedekind JE, PA Frey, and I Rayment (1995) *Biochemistry* **34**, 11049.
Wedekind JE, PA Frey, and I Rayment (1996) *Biochemistry* **35**, 11560.
Weiss PM and WW Cleland (1989) *J Am Chem Soc* **111**, 1928.
Weiss PM, WB Knight and WW Cleland (1986) *J Am Chem Soc* **108**, 2761.
Westheimer FH (1968) *Acc Chem Res* **1**, 70.
Wilson IB, J Dayan, and K Cyr (1964) *J Biol Chem* **239**, 4182.
Wo Y-Y, M-M Zhou, P Stevis, JP Davis, Z-Y Zhang, and RL Van Etten (1992) *Biochemistry* **31**, 1712.
Wong LJ and PA Frey (1974a) *Biochemistry* **13**, 3889.
Wong LJ and PA Frey (1974b) *J Biol Chem* **249**, 2322.
Wood HG, WE O'brien, and G Micheales (1977) *Adv Enzymol Relat Areas Mol Biol* **45**, 85.
Wu L and Z-Y Zhang (1996) *Biochemistry* **35**, 5426.
Xu Y, M McGuire, D Dunaway-Mariano, and BM Martin (1995) *Biochemistry* **34**, 2195.
Yang SL and PA Frey (1979) *Biochemistry* **18**, 2980.
Yang SL and WW Metcalf (2004) *Proc Natl Acad Sci U S A* **101**, 7919,
Yoon MY and PF Cook (1987) *Biochemistry* **26**, 4118.
Yousef MS, F Fabiola, JL Gattis, T Somasundaram, and MS Chapman (2002) *Acta Crystallogr D Biol Crystallogr* **58**, 2009.
Zhang M, CV Stauffacher, D Lin, and RL Van Etten (1998) *J Biol Chem* **273**, 21714.
Zhang M, M Zhou, RL Van Etten, and CV Stauffacher (1997) *Biochemistry* **36**, 15.
Zhang Z, E Harms, and RL Van Etten (1994a) *J Biol Chem* **269**, 25947.
Zhang Z-Y (2001) *Curr Opin Chem Biol* **5**, 416.
Zhang Z-Y (2003) *Acc Chem Res* **36**, 385.
Zhang Z-Y and JE Dixon (1994) *Adv Enzymol Relat Areas Mol Biol* **68**, 1.
Zhang Z-Y and RL Van Etten (1991) *J Biol Chem* **266**, 1516.
Zhang Z-Y, Y Wang, and JE Dixon (1994b) *Proc Natl Acad Sci U S A* **91**, 1624.

Zhang Z-Y, Y Wang, L Wu, E Fauman, JA Stuckey, ML Schubert, MA Saper, and JE Dixon (1994c) *Biochemistry* **33**, 15226.
Zheng J, DR Knighton, LF ten Eyck, R Karlsson, N Xhong, SS Taylor, and JM Sowadski (1993) *Biochemistry* **22**, 2154.
Zheng J, EA Trafny, DR Knighton, NH Xuong, SS Taylor, LF Teneyck, and JM Sowadski (1993) *Acta Crystallogr D Biol Crystallogr* **49**, 362.
Zhou J and JA Adams (1997) *Biochemistry* **36**, 15733.
Zoller MJ, NC Nelson, and SS Taylor (1981) *J Biol Chem* **256**, 10387.

11
ATP-Dependent Synthetases and Ligases

Ligation and the Energy of ATP

The joining of two molecules is energetically unfavorable in an aqueous medium when the substrates correspond to hydrolysis products. In biochemistry, such ligations are driven by the free energy released by the hydrolysis of MgATP or an energetically equivalent molecule. The ATP-dependent synthetases and ligases catalyze reactions in which water is extracted from two molecules that become joined. The amount of free energy available depends on the site at which the ATP molecule is cleaved. The most common cleavage modes and the free energy change under standard conditions, which are pH = 7.0, 25°C, and 1 mM free Mg^{2+}, are given in eqs. 11-1 and 11-2 (Alberty, 1994; Arabshahi and Frey, 1995). Hydrolysis of the α,β-phosphoanhydride linkage to form AMP and PP_i releases 3.2 kcal mol^{-1} more free energy than hydrolysis of the β,γ-linkage.

$$\text{MgATP} + H_2O \rightleftharpoons \text{AMP} + \text{MgPP}_i \qquad \Delta G^{\circ\prime} = -10.9 \text{ kcal mol}^{-1} \qquad (11\text{-}1)$$

$$\text{MgATP} + H_2O \rightleftharpoons \text{MgADP} + P_i \qquad \Delta G^{\circ\prime} = -7.7 \text{ kcal mol}^{-1} \qquad (11\text{-}2)$$

In the actions of ATP-dependent ligases and synthetases, the free energy released in the hydrolysis of MgATP is used to overcome the energetic barrier to the elimination of water. The general principle is exemplified by the free energy barrier for the formation of ethyl acetate from acetate and ethanol under standard conditions, which is $\Delta G' = +4$ kcal mol^{-1} (Jencks and Regenstein, 1970) The free energy change in the hydrolysis of MgATP to MgADP and P_i is $\Delta G' = -7.7$ kcal mol^{-1} under the same conditions (Alberty, 1994). If these two reactions can be made to be interdependent, or coupled, the overall process would be the reaction of acetate, ethanol, and ATP to produce ethyl acetate, MgADP, and

P_i, and the overall standard free energy change would be $\Delta G' = -3.7$ kcal mol^{-1}, making it a spontaneous or energetically downhill process.

In the action of an ATP-dependent synthetase or ligase, the enzyme links the hydrolysis of MgATP with the ligation of the molecules by catalyzing the phosphorylation or adenylylation of one substrate and then the displacement of phosphate or AMP by the other substrate.

Activation by Phosphorylation

Glutamine Synthetase

Two types of glutamine synthetases are found in bacteria and eukaryotes. The bacterial glutamine synthetases, designated GS I (EC 6.3.1.2), are the most thoroughly studied. All species of GS I are dodecameric, 600- to 640-kDa enzymes assembled as two layers of hexameric rings associated face to face (Eisenberg et al., 2000; Stadtman and Ginsburg, 1974). Eukaryotic synthetases, designated GS II, are less understood, but essential aspects of their reaction mechanisms appear to be similar to that of GS I (Eisenberg et al., 2000; Meister, 1974a). Both GS I and GS II can be found in bacteria, although GS I is predominant. Eukaryotes contain only GS II. In this chapter, we discuss the reaction mechanism of GS I and GS II and the structure of GS I.

Reaction and Energetics

The reaction of glutamate, ammonium ion, and MgATP, catalyzed by glutamine synthetase, produce glutamine, MgADP, and P_i, as shown in eq. 11-3c. The overall reaction proceeds in two chemical steps, in which glutamate is first phosphorylated to γ-glutamyl phosphate in eq. 11-3a. γ-Glutamyl phosphate is shown in brackets to indicate its status as an intermediate. In the reaction of eq. 11-3b, phosphate is displaced from γ-glutamyl phosphate by ammonia derived from ammonium ion, the third substrate.

$$\text{glutamate} + \text{ATP} \rightleftharpoons [\gamma\text{-glutamyl phosphate}] + \text{ADP} \quad (11\text{-}3a)$$

$$[\gamma\text{-glutamyl phosphate}] + \text{NH}_4^+ \rightleftharpoons \text{glutamine} + \text{HPO}_4^{2-} + \text{H}^+ \quad (11\text{-}3b)$$

$$\text{glutamate} + \text{ATP} + \text{NH}_4^+ \rightleftharpoons \text{glutamine} + \text{ADP} + \text{HPO}_4^{2-} + \text{H}^+ \quad (11\text{-}3c)$$

The standard free energy change for the hydrolysis of glutamine is -3.4 kcal mol^{-1} (Jencks and Regenstein, 1970). The hydrolysis of MgATP to MgADP and P_i provides more than enough energy to produce glutamine from glutamate and ammonium ion, and the value of $\Delta G'^\circ$ for the overall reaction of glutamine synthetase is -4.3 kcal mol^{-1}.

Asparagine synthetase catalyzes an analogous reaction in which the products derived from MgATP are AMP and MgPP$_i$ (Meister, 1974b). In this case, the intermediate is β-aspartyl adenylate, which reacts with ammonium ion to form asparagines and AMP. The value of $\Delta G'^{\circ}$ for the hydrolysis of asparagines is -3.6 kcal mol^{-1} (Jencks and Regenstein, 1970), so that with the cleavage of MgATP to AMP and MgPP$_i$, the overall standard free energy change for the synthesis of asparagines is -7.6 kcal mol^{-1}. This mode of MgATP cleavage would not appear to be necessary from a purely chemical standpoint. Asparagine synthetases are less well understood than glutamine synthetases, although a structure is available (Larsen et al., 1999).

Kinetic Mechanism

Kinetic studies have been hampered by complexities in metal ion activation and biological regulation of GS I. The enzyme binds two divalent metal ions, Mg^{2+} or Mn^{2+}, or both, which differentially activate. Moreover, the GS I is generally purified in variant states of partial posttranslational modification in which a specific tyrosine residue has been adenylylated (Ginsburg and Stadtman, 1974). Adenylyl-GS I is less active than the unadenylylated enzyme, and a typical sample purified from bacteria is partially adenylylated. Negative cooperative effects may further complicate studies (Rhee et al., 1981; Shrake et al., 1980). The kinetics of isotope exchange at equilibrium shows that random binding of substrates occurs (Wedler, 1974; Wedler and Boyer, 1972). The rates of exchange at equilibrium are not equal for substrate/product pairs; for example, the exchange rates for glutamate/glutamine are faster than for MgATP/MgADP,P$_i$. However, isotope exchange at equilibrium indicates random sequential binding.

Chemical Mechanism

γ-Glutamyl phosphate cannot be isolated and characterized as the intermediate because in solution it undergoes internal cyclization of the α-amino group on the γ-acyl phosphate to produce pyrrolidine-5-carboxylate. The isolation of pyrrolidine-5-carboxylate from reactions of glutamine synthetases with glutamate and MgATP provided indirect evidence for the formation of γ-glutamyl phosphate (Krishnaswamy et al., 1962).

Positional isotope exchange (PIX) experiments with [^{18}O]ATP provided further evidence by proving the glutamate-dependent reversible cleavage of MgATP to MgADP at the active site (Midelfort and Rose, 1976). PIX for this case is illustrated in fig. 11-1, where ^{18}O appears in color. The fast, glutamate-dependent migration of β,γ-bridging ^{18}O to β-nonbridging ^{18}O proved that MgATP is cleaved rapidly and reversibly in a reaction with glutamate.

Detection of ^{18}O-PIX in GS I depended on biochemical and chemical degradation and mass spectral analysis (Midelfort and Rose, 1976). Positional exchange of ^{18}O in ATP can also be observed by ^{31}P NMR spectroscopy because of the ^{18}O-induced isotope shifts in the ^{31}P signals and their bond order dependence (Cohn and Hu, 1978). The upfield perturbation induced by ^{18}O in a P—O single bond is 0.017 ppm and in a P=O double bond 0.034 ppm. The bond order proportionality extends to the fractional bond orders found in phosphate anions. The ^{31}P NMR spectrum of partially ^{18}O-labeled inorganic phosphate is shown in fig. 11-2.

GS I requires two divalent metal ions bound at two distinct sites, n_1 and n_2, for maximal activity (Hunt and Ginsburg, 1980). The metal ion in site n_2 interacts directly with MgATP, and the divalent metal at site n_2 maintains the active conformation. The site n_2 displays a higher affinity for the divalent metal ion, with K_d in the μM range, whereas the n_1 site displays weaker affinity, with K_d in the mM range.

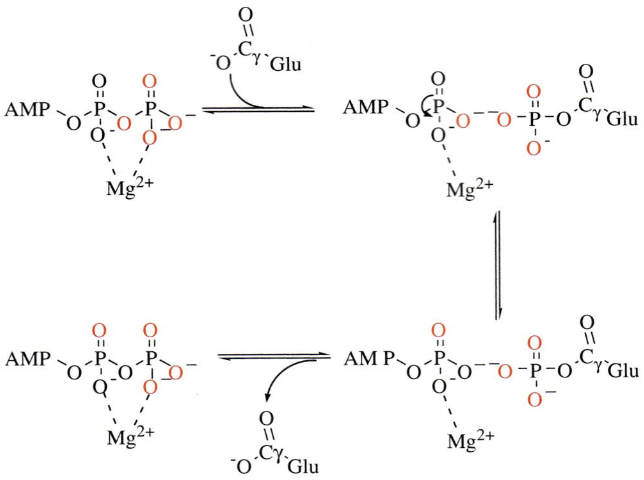

Fig. 11-1. Positional isotope exchange catalyzed by glutamine synthetase. Oxygen 18 in the β,γ-bridging position of ATP undergoes a glutamate-dependent exchange into a nonbridging position at the active site of glutamine synthetase. The rate of this exchange is compatible with the rate of the overall reaction; the exchange is kinetically competent to be brought about by a catalytic step in the mechanism. The exchange requires cleavage of the β,γ-phosphoanhydride bond, and the simplest mechanism involves the formation of the γ-glutamyl phosphate intermediate. The exchange involves torsional motion of the β-phosphate of MgADP, and this motion is not restricted by Mg^{2+} coordination on the time scale of catalysis.

Regulation in Bacteria

The *Escherichia coli* GS I is a dodecameric enzyme that displays complex regulatory properties that support the central role of glutamine in nitrogen metabolism. A number of molecules that derive nitrogen from glutamine inhibit glutamine synthetase by feedback inhibition. These include CTP, carbamoyl phosphate, histidine, tryptophan, and glucosamine-6-P. Other nitrogen-containing compounds that also inhibit include alanine, serine, and glycine. CTP and AMP inhibit by competing with ATP and glutamine.

At another level, GS I is subject to covalent modification by adenylylation of a tyrosine residue (Tyr397 in *S. typhimurium*), as illustrated in fig. 11-3. Adenylylation by ATP decreases the activity and is catalyzed by an adenylyltransferase (P_{II} in fig. 11-3), which is itself subject to covalent modification by uridylylation of a tyrosine residue. The uridylyl-adenylyltransferase catalyzes the complementary de-adenylylation and reactivation of adenylyl-GS I by reaction of P_i with the adenylyl group to release ADP. The uridylyltransferase that catalyzes uridylylation of the adenylyltransferase is itself subject to regulation. The products of the action of GS I, glutamine and P_i, inhibit the uridylyltransferase, tending to keep GS I in its less-active adenylyl form. α-Ketoglutarate, a precursor of glutamine, and ATP activate the uridylyltransferase, facilitating de-adenylylation of GS I to its more active form.

Inhibition

Glutamine synthetases are subject to inhibition by naturally occurring analogs of glutamate that bind at the active site and are or become analogs of γ-glutamyl

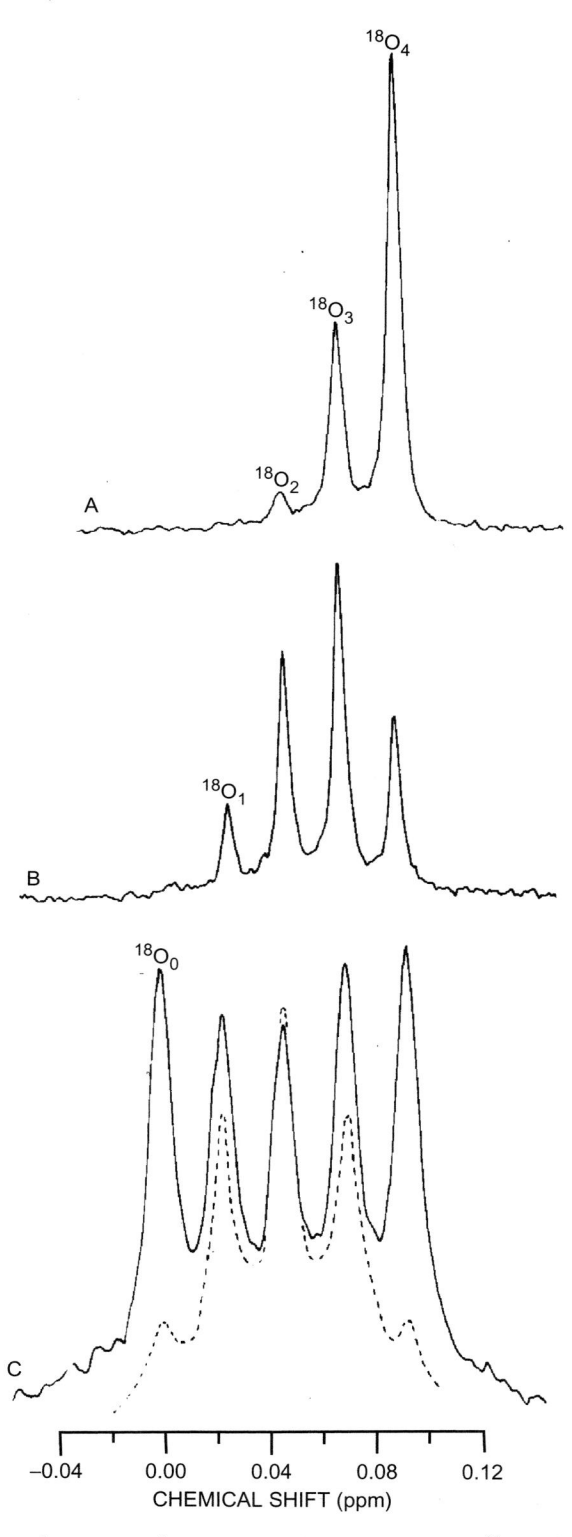

Fig. 11-2. The ^{31}P nuclear magnetic resonance (NMR) spectra of [^{18}O]phosphate with various degrees of ^{18}O enrichment. (A and B) ^{31}P NMR spectra of phosphate show extensive ^{18}O exchange. Each peak position is labeled, ^{18}O$_n$, according to the number, n, of ^{18}O atoms exchanged in each phosphate isotopomer. (C) An evenly populated exchange distribution (solid line) is superimposed with a 50% exchanged distribution (dashed line). (Adapted from Bock and Cohn, 1978.)

552 Enzymatic Reaction Mechanisms

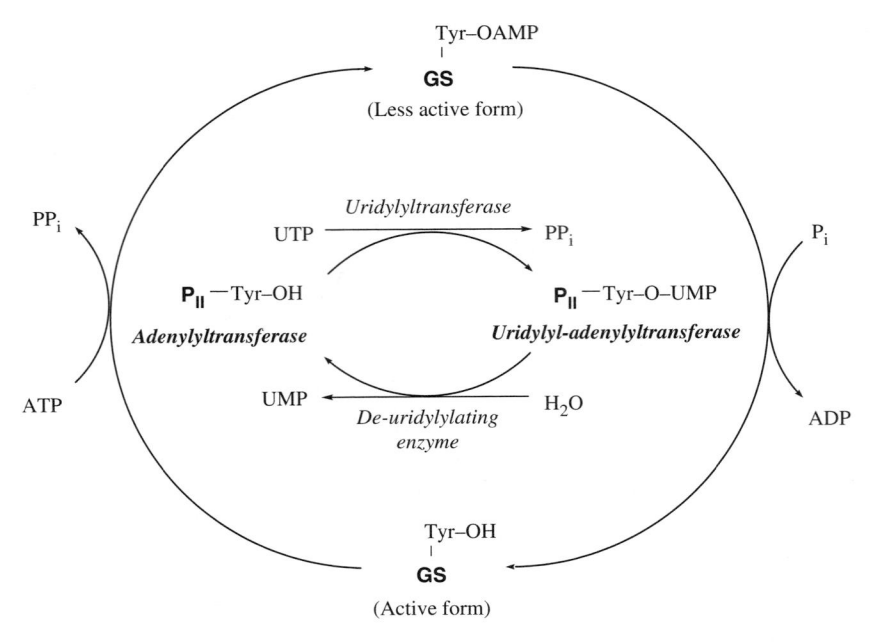

Fig. 11-3. Covalent regulation of bacterial glutamine synthetase (GS I).

phosphate (structure **11-1**). Methionine sulfoximine and phosphinothricin are well known and inhibit with K_i values of 0.1 mM and 25 µM, respectively.

11-1 Methionine sulfoximine, Methionine sulfoximine phosphate, Phosphinothricin

Inhibition by methionine sulfoximine in the presence of ATP is accompanied by time-dependent phosphorylation to methionine sulfoximine phosphate, which is an essentially irreversible inhibitor (Ronzio and Meister, 1968). Methionine sulfoximine is a toxin that causes convulsions and epileptic seizures in animals. Many analogs of these molecules have been synthesized and evaluated as inhibitors, Essential inhibitory properties of 46 compounds are compiled in a review (Eisenberg et al., 2000).

Structure

The structure of GS I in fig. 11-4 shows the hexameric structure and intersubunit interactions of an ADP and phoshinithricin complex. The twelve active sites are formed at the interfaces of opposing subunits and define an hourglass shape open at both ends, with the metal sites n_1 and n_2 near the passageway and the MgATP and glutamate sites at the top and bottom, respectively. Two flaps guard the passageway and entrance to the bottom. The Glu327 flap is at the bottom, and the Asp50′ flap is at the passageway adjacent to the substrate binding site. His271 and Arg344 bind the α- and β-phosphoryl groups of ATP. Glu131 and Asn264 bind the α-amino group of glutamate, Arg321 binds the α-carboxylate, and Arg359 binds the γ-carboxylate of glutamate. His269, Glu357, and Glu129 coordinate metal n_2, and Glu129 also forms a hydrogen bond with His271 in contact with the α-phosphoryl group of ATP. Glu212, Glu220, and Glu131 coordinate metal $n1$; Glu212 contacts the ammonium ion. Asp50′ of the opposite subunit undergoes conformational changes and participates in catalysis by accepting a proton from NH_4^+ and allowing the formation of NH_3 at the active site. Ammonia has the nucleophilic properties required for reaction with γ-glutamyl phosphate.

A mechanism incorporating the chemical and structural changes in the action of GS I is shown in fig. 11-5 (Eisenberg, 2000). Bearing in mind that substrate-binding order is not compulsory, fig. 11-5 depicts events beginning with ATP binding in the top of the hourglass, accompanied by a shift of the Asp50′ flap and coordination of the γ-phosphate with metal n_2. The binding of glutamate proceeds with a major relocation of the Glu327 loop, positioning Glu327 near the site of tetrahedral intermediate formation. The ammonium ion binds in a cluster of carboxylate groups. The γ-carboxylate of glutamate reversibly accepts the γ-phosphoryl group of ATP to form γ-glutamyl phosphate as a transient intermediate.

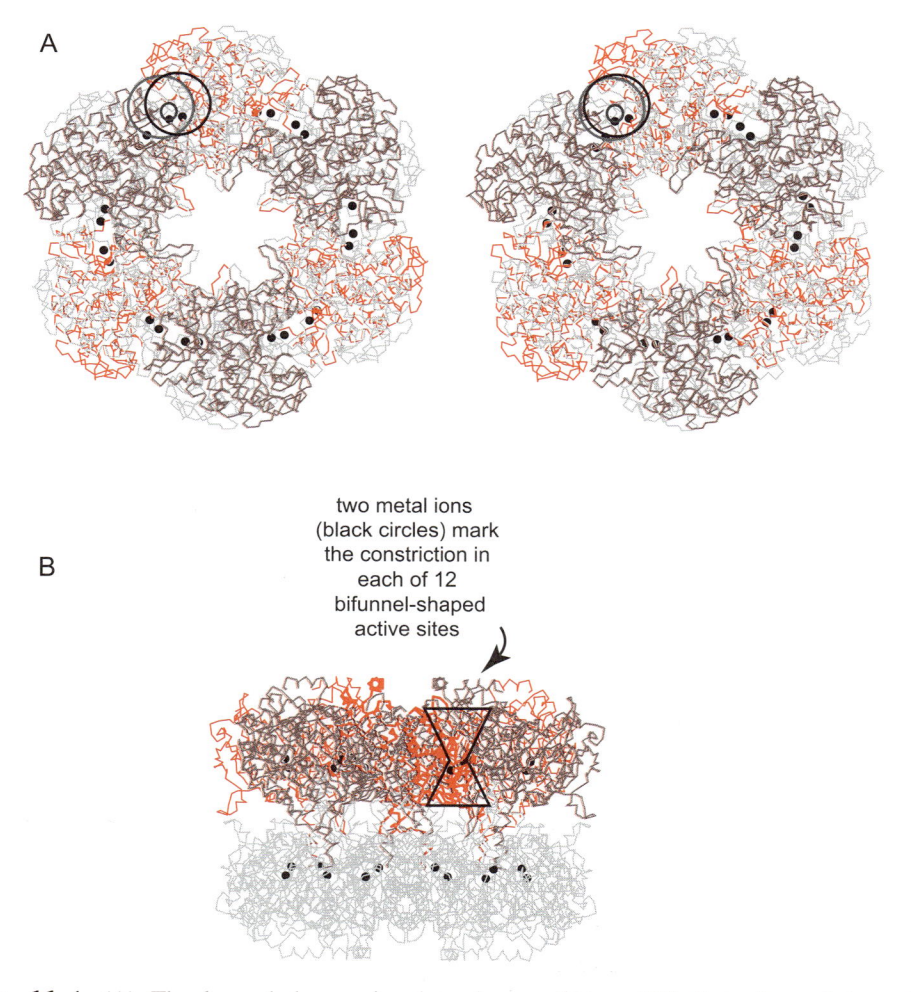

Fig. 11-4. (A) The homododecameric glutamine synthetase (GS I) is from *Salmonella typhimurium*, with Mn^{2+} ions shown in black (2.89-Å resolution; PDB 1FPY; Gill and Eisenberg, 2001). (B) The global structure consists of a stacked dimer of hexameric rings shown from the side. There is one subunit per active site formed at each subunit interface in the two rings. Each active site is shaped like an hourglass (often referred to in this context as a *bifunnel*), with two divalent metal ions at the central constriction; the hourglass is represented in A and B for one active site as large circles for the top and bottom and a small circle for the constriction.

Continued

554 Enzymatic Reaction Mechanisms

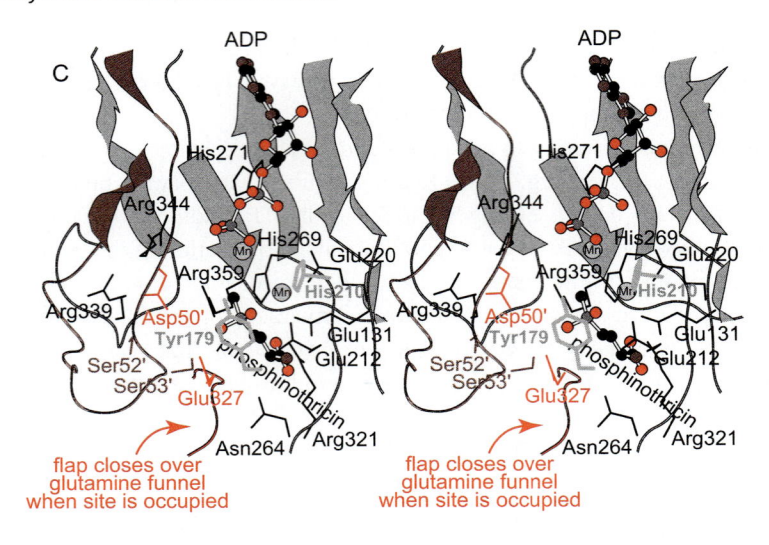

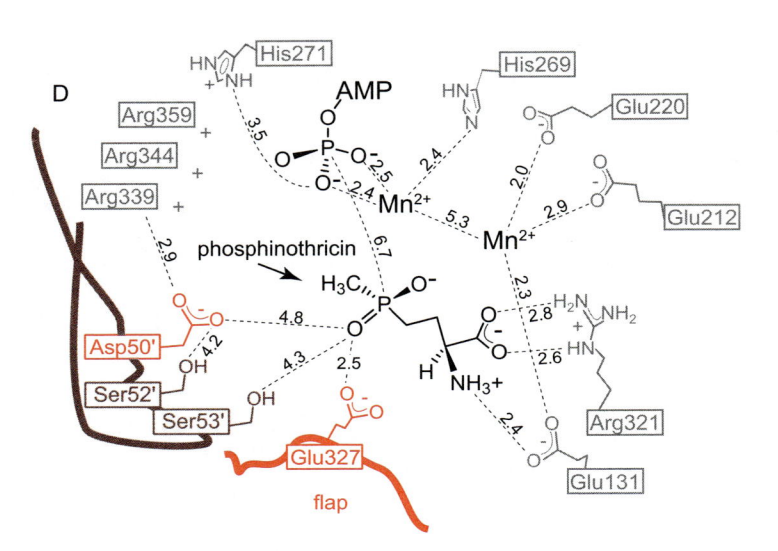

Fig. 11-4. cont'd (C) A closer view is provided of a single active site with product ADP and intermediate analog phosphinothricin bound. Two catalytically important residues, Asp50' and Glu327 (from different subunits) are shown in red, as is a flap that closes over the glutamine funnel. (D) A two-dimensional drawing shows the active site, with distances given in angstroms.

At the same time, Asp50' reversibly abstracts a proton from NH_4^+, and the NH_3 so-formed engages in nucleophilic addition to the γ-carbonyl group of γ-glutamyl phosphate to form the tetrahedral addition intermediate. Glu327 accepts a proton from the aminium (NH_3^+) group of the intermediate before or in concert with the elimination of phosphate ion. The resulting glutamine dissociates from the site. This mechanism accommodates the biochemical and structural facts and does not violate any known principles.

Carbamoyl Phosphate Synthetase

Pyrimidine biosynthesis begins with the formation of carbamyl phosphate (carbamoyl-P) through the action of carbamoyl-P synthetase (CPS). Arginine biosynthesis and the urea

ATP-Dependent Synthetases and Ligases   555

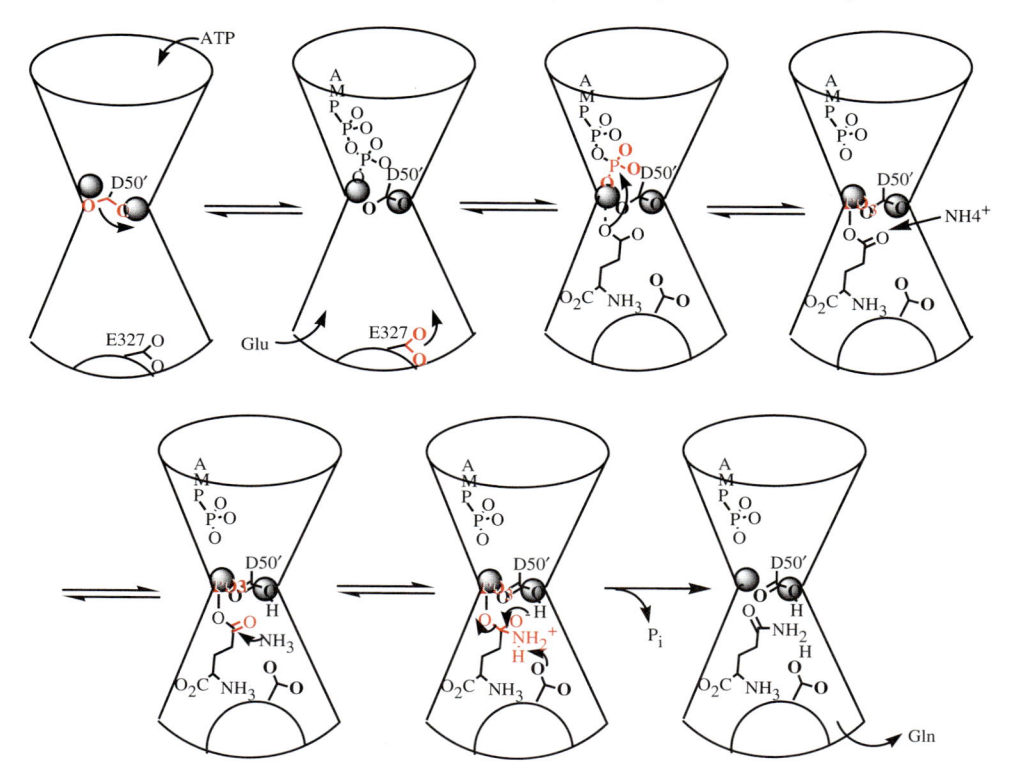

Fig. 11-5. A mechanism for the action of glutamine synthetase (GS I). Adapted from Eisenberg et al. (2000) with permission.

cycle also depend on CPS and the biosynthesis of carbamoyl-P. The three known types of CPS are designated as I, II, and III (Anderson, 1995; Meister, 1989). CPS I is engaged in arginine biosynthesis and the urea cycle in the liver, uses ammonia as the source of nitrogen in carbamoyl-P, and is activated by *N*-acetylglutamate. CPS II uses glutamine as the source of nitrogen and is not activated by *N*-acetylglutamate. CPS III uses glutamine for nitrogen and is activated by *N*-acetylglutamate. Prokaryotes contain only one CPS, generally a CPS II.

The CPS in *E. coli* (EC 6.3.5.5) is dimeric and consists of small (α) and large (β) subunits of 42 and 119 kDa, respectively. The reaction requires glutamine for nitrogen and proceeds according to eq. 11-4.

$$\text{HO-CO-O}^- + \text{H}_2\text{N-CO-CH}_2\text{-CH}_2\text{-CH(NH}_3^+\text{)-COO}^- + 2\text{ATP} + \text{H}_2\text{O} \xrightarrow{\text{CPS II/Mg}^{2+}}$$

$$\text{H}_2\text{N-CO-OPO}_3^{2-} + {}^-\text{O-CO-CH}_2\text{-CH}_2\text{-CH(NH}_3^+\text{)-COO}^- + 2\text{ADP} + \text{HOPO}_3^- + \text{H}^+ \qquad (11\text{-}4)$$

Ammonia can also be used as a source of nitrogen, albeit at a slower rate. Carbamoyl-P is high in energy, with $\Delta G'^\circ$ of about 12.7 kcal mol^{-1}, so that two moles of MgATP are required.

Chemical Mechanism

The chemical sequence of steps captures the energy of MgATP while employing its chemical properties mechanistically in facilitating the reaction. One MgATP participates directly in the dehydration of bicarbonate, as shown in eq. 11-4 by the appearance of ^{18}O (color) from [$^{18}O_3$]bicarbonate in the $H^{18}OPO_3^-$ produced (Jones and Spector, 1960). The amido group of glutamine serves as the source of nitrogen for carbamoyl-P, and the phosphate is derived from the second MgATP. The mechanisms by which the amido group of glutamine is extracted and the free energy of MgATP is utilized are kinetically and chemically complex (Meister, 1989). The partial reactions of eqs. 11-5, 11-6, and 11-7 catalyzed by CPS II constitute clues to the chemical steps in the mechanism.

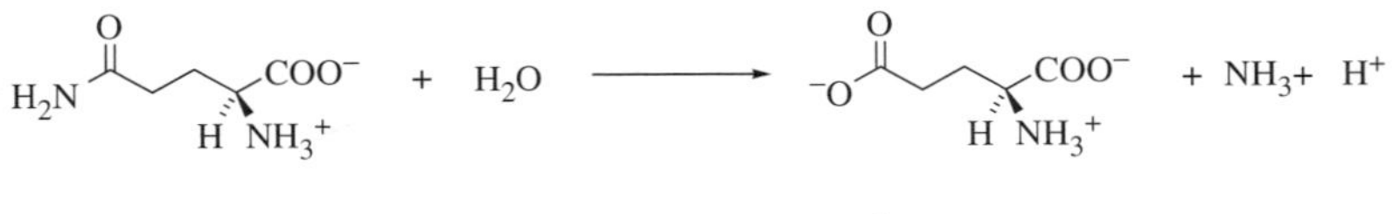

$$H_2N\text{-CO-CH}_2\text{CH}_2\text{-CH(NH}_3^+)\text{-COO}^- + H_2O \longrightarrow {}^-O\text{-CO-CH}_2\text{CH}_2\text{-CH(NH}_3^+)\text{-COO}^- + NH_3 + H^+ \quad (11\text{-}5)$$

$$ATP + H_2O \xrightarrow{HCO_3^-/Mg^{2+}} ADP + P_i \quad (11\text{-}6)$$

$$H_2N\text{-C(=O)-OPO}_3^{2-} + ADP \xrightleftharpoons{Mg^{2+}} H_2N\text{-C(=O)-O}^- + ATP \quad (11\text{-}7)$$

Reaction 11-5 implicates the hydrolysis of glutamine to form ammonia as the source of nitrogen, Bicarbonate-dependence in the ATPase reaction of eq. 11-6 implicates carboxyphosphate as a possible intermediate in the formation of carbamate (Anderson and Meister, 1965). The reversible reaction of carbamoyl-P with MgADP to produce carbamate and MgATP in eq. 11-7 implies the intermediate formation of carbamate and its phosphorylation by the second molecule of MgATP to carbamoyl-P. The sequence of chemical steps in the overall reaction of eq. 11-4 can be formulated as a sequence of reactions. The process begins with the production of ammonia, as in eq. 11-5, followed by the incorporation of ammonia into carbamoyl-P in the overall sequence of steps shown in fig. 11-6.

MgATP reversibly phosphorylates bicarbonate to carboxyphosphate, a highly reactive and transient species. Chemical evidence for carboxyphosphate as an intermediate includes trapping experiments with $NaBH_4$, which lead to ATP-dependent production of formate (Meister, 1989). The ammonia generated in eq. 11-5 captures the carboxyphosphate in a nucleophilic displacement of phosphate to form carbamate. Phosphorylation of carbamate by MgATP generates carbamoyl-P, which is finally released from the enzyme.

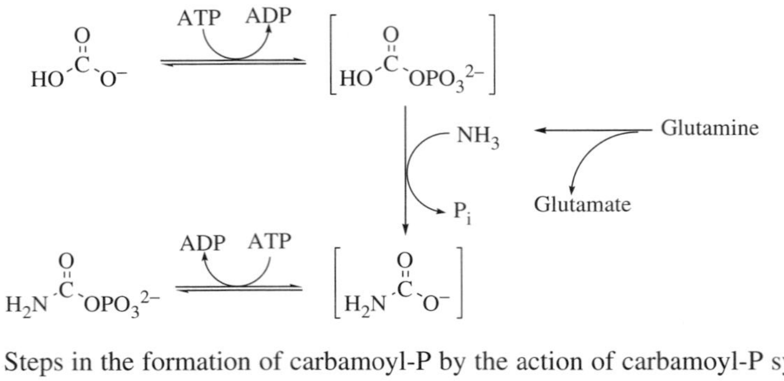

Fig. 11-6. Steps in the formation of carbamoyl-P by the action of carbamoyl-P synthetase.

The overall reaction proceeds in the four distinct chemical steps shown in fig. 11-6, all of which must be catalyzed by CPS II. As shown by kinetic experiments, none of the intermediates dissociates from the enzyme before it undergoes further reactions to the ultimate production and release of carbamoyl-P.

Kinetics

Bicarbonate dependence in the ATPase activity of CPS II implicates carboxyphosphate as a transient intermediate. Intermediate formation of carboxyphosphate also explains the transfer of ^{18}O from bicarbonate to inorganic phosphate (eq.11-4). The ATPase reaction is much slower than the overall reaction, raising the question of the kinetic competence of carboxyphosphate. PIX experiments monitoring the bridge to nonbridge isomerization of β,γ-bridging ^{18}O in MgATP in rapid mix-quench transient kinetic experiments further support carboxyphosphate as an intermediate and prove its kinetic competence (Mullins et al., 1991; Raushel and Villafranca, 1979, 1980; Wimmer et al., 1979).). The observation of ^{18}O-PIX implicates a reversible step in the mechanism of the bicarbonate-dependent ATPase reaction of eq. 11-6. An ATPase mechanism such as in scheme 11-1, where carboxyphosphate and MgADP remain enzyme-bound until after the hydrolysis of carboxyphosphate, can account for the ^{18}O-PIX on the basis of the fast reversal of carboxyphosphate formation and its slower hydrolysis to bicarbonate and inorganic phosphate.

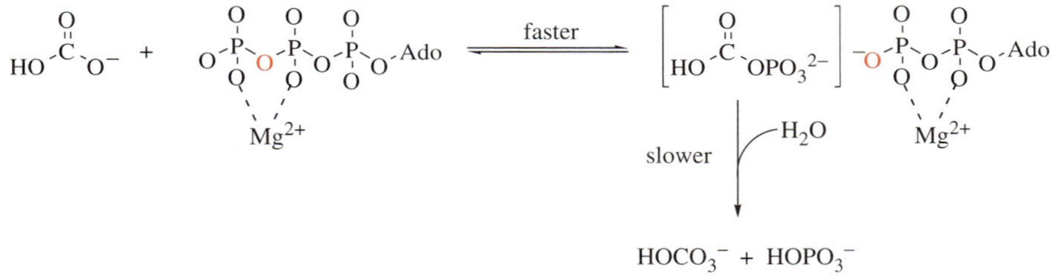

Scheme 11-1

The glutaminase activity of CPS II suggests that ammonia from glutamine may be a discrete intermediate. However, the glutaminase reaction of eq. 11-5 is slower by orders of magnitude than the overall reaction of eq. 11-4. Free ammonia (NH_3) serves as a substrate in place of glutamine, further supporting the proposition of its participation (Huang and Raushel, 1999). Ammonium ion is not, however, a substrate, so that glutamine is a much better substrate at physiological pH. Experiments with mixed $^{15}NH_4Cl$ and unlabeled glutamate show that less than 5% of the carbamoyl-P formed could arise from the reaction of free ammonia, the balance deriving nitrogen from glutamine (Mullins and Raushel, 1999). This experiment proves that ammonia derived from glutamine cannot be freely dissociable, and as an intermediate it must react faster with enzyme-bound carboxyphosphate faster than it dissociates from the enzyme.

The glutaminase activity can be separated from carbamoyl-P synthesis by active site-directed alkylation. Reaction of CPS II from *E. coli* with a chloroketone analog of glutamine, L-2-amino-4-oxo-7-chloro[^{14}C]pentanoate selectively abolishes activity with glutamine as the nitrogen donor, but it has no effect on the reaction of ammonia (Meister, 1989). The chemical modification abolishes the glutaminase activity and is inhibited by glutamine. Chemical degradation of the ^{14}C-labeled enzyme led to S-[^{14}C]carboxymethylcysteine, which was identified by biochemical methods as Cys269 in the small subunit. Mutation of Cys269 inactivates the glutaminase activity, as well as overall carbamoyl-P synthesis from glutamine (Meister, 1989). A conserved Cys/His pair in the small subunit is implicated in

558 Enzymatic Reaction Mechanisms

the glutaminase action of the small subunit, which presumably acts by a mechanism analogous to the action of cysteine proteases described in chapter 6. Adapted to CPS II, the thiolate of Cys269 undergoes nucleophilic addition to the γ-carboxamido group of glutamine, and ammonia is then eliminated from the tetrahedral adduct to form the Cys269-S-γ-glutamylthioester intermediate. Hydrolysis of the S-γ-glutamyl intermediate to glutamate regenerates Cys269.

The CPS-catalyzed MgATP formation from carbamoyl-P and MgADP implicates carbamate as a potential intermediate. As further evidence, the observation of ADP-dependent bridge/nonbridge PIX of ^{18}O in carbamoyl-P implicates carbamate as an intermediate, and the fact that PIX is faster than the overall reaction verifies the kinetic competence of carbamate (Raushel and Villafranca, 1980). Carbamate as an intermediate must be strictly enzyme bound, taking into consideration that free carbamate decomposes to ammonia and carbon dioxide with a half-time of 70 ms (Wang et al., 1972). Carbamate released from the enzyme would not be recaptured by diffusion fast enough to prevent its nonenzymatic decomposition.

Structure

The chemical steps in the action of CPS are mechanistically diverse, and it would be difficult to imagine them taking place at a single active site. CPS II from *E. coli* contains three distinct active sites that are widely separated and connected by tunnels through the protein, analogous to the indole tunnel in tryptophan synthase (see chap. 3). Figure 11-7 shows the structure of CPS from *E. coli* (Thoden et al., 1999). A glutaminase site is in the small subunit, and the bicarbonate and carbamyl-P binding sites are in the large subunit. The glutaminase site is about 45 Å from the bicarbonate site, and the bicarbonate site is about 35 Å from the carbamoyl-P site. Two tunnels through the protein connect the widely separated sites and allow the intermediates to flow from one site to another without

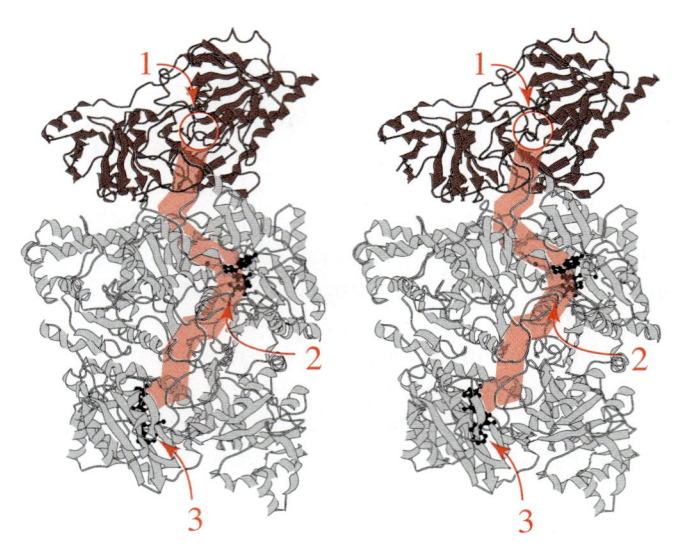

Fig. 11-7. Structure of carbamoyl phosphate synthetase (CPS) from *E. coli*. The structure of *E. coli* CPS shows the tunnel (pink) linking the three active sites, which are labeled with red numbers and arrows (2.1-Å resolution; PDB 1BXR; Thoden et al., 1999). The sites are numbered in order so they correspond with the reactions occurring in eqs. 11-5, 11-6, and 11-7 respectively.

dissociating from the protein. Ammonia generated at the glutaminase site in the small subunit diffuses through the ammonia tunnel to the bicarbonate binding site, where carboxyphosphate is generated by reaction of bicarbonate with MgATP. Carbamate from the reaction of ammonia with carboxyphosphate at this site diffuses through the carbamate tunnel to the carbamoyl-P binding site, where it is phosphorylated by MgATP. There are distinct MgATP binding sites for the two molecules consumed in the overall reaction.

Results of site-directed mutagenesis to modify the ammonia tunnel verifies its function (Huang and Raushel, 2000a, 2000b). Mutations to constrict passage through the tunnel decouple the hydrolysis of glutamine from carbamoyl-P synthesis and inhibit CPS II.

Activation by Adenylylation

Many ligases cleave the α,β-phosphoanhydride bridge of ATP, producing AMP and PP_i instead of ADP and P_i. In some cases, the higher free energy of the α,β-phosphoanhydride is clearly required or beneficial for driving the reaction (Frey and Arabshahi, 1999). In other cases, such as that of asparagines synthetase, the reason is obscure. We consider two examples: DNA ligases and aminoacyl tRNA synthetases. In the action of DNA ligase, a phosphodiester bond is formed between phosphomonoester and alcoholic groups in nicked DNA, and this proceeds with the cleavage of the α,β-phosphoanhydride linkage in either NAD or MgATP. The value of $\Delta G'^\circ$ for the hydrolysis of the phosphodiester in $2',3'$-cyclic-CMP is -5 kcal mol^{-1}. If that for a phosphodiester linkage in DNA is similar, it would seem that the energy of the β,γ-phosphoanhydride linkage would suffice, but the α,β-linkage in ATP or NAD is cleaved instead. The free energy of a phosphodiester bridge in DNA might be higher and so require the higher energy for DNA repair.

In the action of aminoacyl tRNA synthetases, an ester is formed between tRNA and an amino acid, a process that can be expected to require at least $+8.4$ kcal mol^{-1} based on the standard free energy change for the hydrolysis of valyl tRNA (Jencks and Regenstein, 1970). Here, the -7.7 kcal mol^{-1} of the β,γ-phosphoanhydride linkage could not allow a favorable equilibrium constant, and the higher free energy of the α,β-bridge is brought into play.

DNA Ligase

Nicks in DNA arise in the discontinuous synthesis of the lagging strand during replication. They also arise in the repair of DNA strands from which segments have been excised to remove lesions and then partially re-synthesized by DNA polymerase. The nicks are repaired by the action of DNA ligases (Lehman, 1974). The nicks formally correspond to hydrolytic cleavage of phosphodiester groups, and DNA ligase catalyzes the "restoration" of phosphodiester linkages.

DNA ligases repair nicks with $3'$-OH and $5'$-phosphate ends in double stranded DNA. They couple the repair to the hydrolysis of α,β–phosphoanhydride linkages in ATP or NAD^+. Bacteria have NAD^+-dependent DNA ligases (EC 6.5.1.2) and a few have ATP-dependent ligases. Eukaryotes, viruses, and archaea have ATP-dependent DNA ligases (EC 6.5.1.1).

Chemical Mechanism

The chemistry of the *E. coli* DNA polymerase revealed the coupling mechanism linking NAD^+ hydrolysis to phosphodiester bond formation in nicked DNA (Lehman, 1974).

560 Enzymatic Reaction Mechanisms

Reaction of the ligase with NAD⁺ produced ligase-adenylate (ligase-AMP) and NMN. The modified enzyme displayed a marked blue shift in its ultraviolet absorption spectrum because of the AMP moiety covalently linked to the protein, with stoichiometry exceeding 0.7 AMP per molecule of ligase. Biochemical degradation of the ligase-AMP concluding with alkaline hydrolysis led to lysine-N^ε-IMP, with alkaline hydrolysis removing the 6-amino group from the adenine ring of AMP (Gumport and Lehman, 1971).

Ligase-adenylate is also produced in the reaction of T4-DNA ligase with MgATP (Weiss et al., 1968). The pH dependence for adenylylation of T4 DNA ligase indicates that the pK_a of Lys159 at the active site is 8.4. The standard free energy change for the hydrolysis of the ligase-adenylate ($\Delta G'^\circ$) is -13.4 kcal mol^{-1} (Arabshahi and Frey, 1999).

Reaction of the ligase-adenylate with nicked DNA under controlled conditions leads to small amounts of adenosine-5′-diphosphoryl-DNA (Harvey et al., 1971; Olivera et al., 1968). As a transient intermediate, this species does not accumulate. The ligase-adenylate can be made to accumulate because of being produced in the absence of the cosubstrate. The chemical steps in nick repair by NAD⁺-dependent DNA ligase can be formulated as in fig. 11-8, which shows adenylylation of the ligase by NAD⁺, adenylyltransfer to the phosphate group in the nick, and displacement of AMP from the ADP moiety of adenosine-5′-diphosphoryl-DNA.

Kinetic Mechanism

Inasmuch as a ligase-adenylate can be produced in the absence of nicked DNA, the reaction may follow ping pong bi bi kinetics. Double reciprocal plots of initial rates against either of the substrates are families of parallel lines (Modrich and Lehman, 1973). The ligase from *E. coli* catalyzes the exchange of NMN with [*nicotinamide*-4-³H]NAD⁺ in the absence of nicked DNA. The appearance of [4-³H]NMN in a pool of NMN takes place at

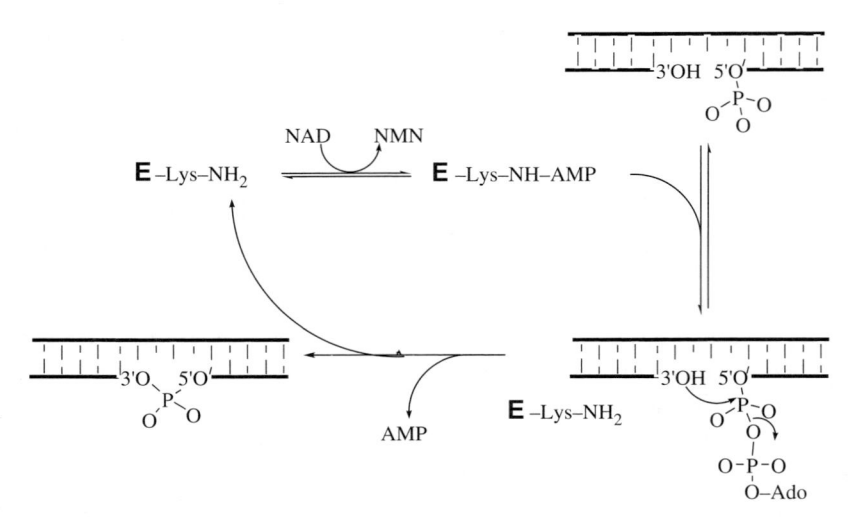

Fig. 11-8. Chemical steps in the action of NAD-dependent DNA ligase. Sealing of nicks with 3′-OH and 5′-phosphate begins with the adenylylation of the active-site lysine-ε-amino group by NAD⁺ to form the ligase-adenylate intermediate and release nicotinamide mononucleotide (NMN). This step requires MgATP in the case of ATP-dependent DNA ligases. The ligase adenylate then transfers the adenylate group to the 5′-phosphate group at a nick in double-stranded DNA to form the intermediate adenosine-5′-diphosphoryl-DNA. The ligase then catalyzes the displacement of AMP from the adenosine-5′-diphosphoryl moiety by the 3′-OH group in the nick, thereby forming the phosphodiester bridge and releasing AMP.

a maximum rate about twice that for the overall joining reaction, proving that, given the ping pong kinetics, adenylylation of the enzyme is fast enough to participate in the mechanism. The kinetics for the *E. coli* DNA ligase can be formulated as in scheme 11-2.

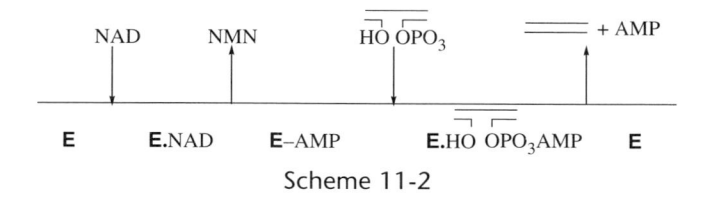

```
           NAD     NMN         HO OPO₃              + AMP
            │       │            │                    │
            ↓       ↑            ↓                    ↑
    E       E.NAD       E–AMP        E.HO OPO₃AMP        E
```
Scheme 11-2

Because the kinetics is ping pong, the adenylyltransfer steps can be isolated and their rates measured, and because each step requires Mg^{2+}, the rate of phosphodiester bond formation can be measured as well. The rates of all of the steps can be compared with the overall rate of strand joining. Some of the results for two bacterial DNA ligases are summarized in table 11-1; one ligase is NAD-dependent and the other is ATP dependent. The rate constants are typical of those for bacteriophage T4, chlorella virus, and archaeal DNA ligases (Hall and Lehman, 1969; Ho et al., 1997; Nakatani et al., 2002). Those for the NAD-dependent ligase are much higher in the presence of ammonium ion. As a rule in enzymology, potassium and ammonium ions are essentially interchangeable in activating enzymes that depend on monovalent cations. The ligase activities measured in vitro are sufficient to meet cellular needs for DNA replication and repair (Lehman, 1974).

Structure of ATP-Dependent Ligase

Few structures of DNA ligases are available, none with DNA bound at the active site. The structure of chlorella virus DNA ligase in its adenylate form is shown in fig. 11-9. The large cleft is found in other DNA ligases and may represent a DNA binding locus. Several amino acid side chains form close contacts with the adenylate moiety.

Aminoacyl-tRNA Synthetases

Translation

Translation of the genetic code requires that a connection be made between the ordered triplet coding in an mRNA-ribosome complex and the mixture of amino acids that must be incorporated in a definite order into a protein. This translation is carried out by the

Table 11-1. Rate Constants for Steps in the Action of DNA Ligases

DNA Ligase	Reaction	Rate Constant (min⁻¹)
E. coli NAD-dependent (+NH₄⁺)[a]	DNA joining	28
	NAD/NMN exchange	50
	Adenylylation of DNA	10
E. coli NAD-dependent (−NH₄⁺)	DNA joining	1.4
	NAD/NMN exchange	50
	Adenylylation of DNA	9
Niseria meningitides (ATP-dependent)[b]	DNA joining	0.48
	Adenylylation of ligase	1.6
	Adenylylation of DNA	1.7
	Phosphodiester formation	1.2

[a]Modrich and Lehman, 1973.
[b]Magnet and Blanchard, 2004.

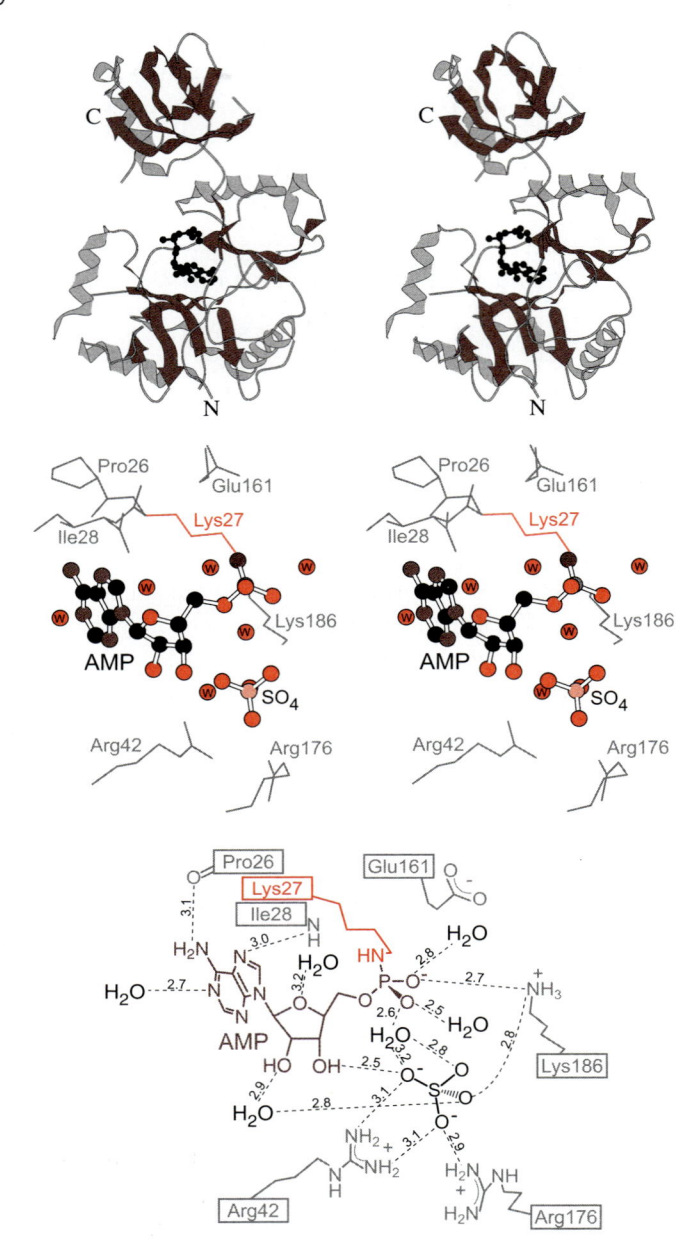

Fig. 11-9. Structure of chlorella virus DNA ligase-adenylate. DNA ligase from the chlorella virus is shown in the top panel with an active-site lysine residue adenylated (2.0-Å resolution; PDB 1FVI; Odell et al., 2000). The middle stereoimage shows the active site with the adenylated Lys27 (red) and sulfate. The bottom panel shows the active site in two dimensions, with distances given in angstroms.

20 specific aminoacyl-tRNA synthetases (aaRSs), one for each amino acid. Each aaRS catalyzes the ligation of an amino acid to the 3'-end of its cognate tRNA^{aa}. For each amino acid, there is also a specific tRNA^{aa}. All of the tRNAs have similar secondary and tertiary structures, such as that illustrated in fig. 11-10, and each has structural features that allow it to be recognized by a specific aaRS. The anti-codon loop of a tRNA^{aa} complements the mRNA-codon for that amino acid. Other structural features of a tRNA^{aa} constitute recognition elements for one aaRS, allowing only that amino acid to be ligated to the 2'- or 3'-OH

ATP-Dependent Synthetases and Ligases 563

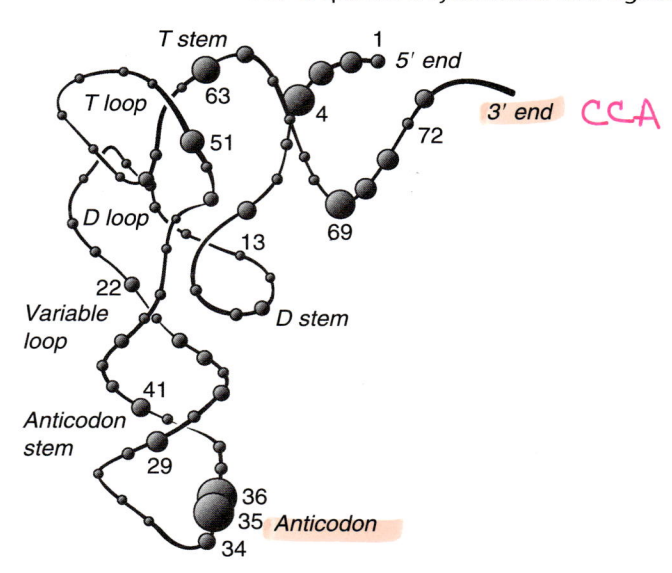

Fig. 11-10. The generic secondary and tertiary structure of tRNA is illustrated. The aminoacyl accepting group is the 3′-CCA end. The spheres represent nucleotide sites that have been found to play important roles in recognition by interacting directly with cognate aminoacyl tRNA synthetases. The larger the sphere, the greater the frequency with which the nucleotide engages in recognition contacts. Adapted with permission from a review (Ibba and Söll, 2000).

group of adenosine at the 3′-CCA end. Only TyrRS produces Tyr-tRNA^{Tyr}, and only AlaRS produces Ala-tRNA^{Ala}. The ligated amino acid, the aminoacyl-tRNA^{aa}, is in its chemically activated state for incorporation into a protein when the anti-codon binds to the codon of mRNA. The anti-codon loop (fig. 11-10) of a specific aminoacyl-tRNA^{aa} binds to the complementary coding triplet of mRNA on the ribosome. The anti-codon in Ala-tRNA^{Ala} is complementary only to the codon for alanine and binds only to that codon in mRNA, so that alanine is incorporated only in its turn into a protein. In this way Ala-RS translates the alanine part of the genetic code, Tyr-RS translates the tyrosine part, and so forth.

Classes of Synthetases

Aminoacyl tRNA synthetases fall into two classes with respect to both amino acid sequences and tertiary structures (Carter, 1993; Ibba and Söll, 2000). Class I and class II synthetases are listed in table 11-2. The class I synthetases aminoacylate the 2′-OH groups

Table 11-2. Classes of Aminoacyl-tRNA Synthetases

Class I	Class II
ArgRS	AlaRS
CysRS	AsnRS
GlnRS	AspRS
GluRS	GlyRS
IleRS	HisRS
LeuRS	LysRS
MetRS	PheRS
TrpRS	ProRS
TyrRS	SerRS
ValRS	ThrRS
	SeCysRS[a]

[a] Activation of selenocysteine with a specialized tRNA^{Ser} for incorporation into E. coli proteins such as glutathione peroxidase.

of cognate tRNAs and, with the exception of PheRS, the class II synthetases aminoacylate the 3′-OH groups. The differences between the two classes extend to editing mechanisms, but both classes function by the same basic chemical mechanism apart from their specificities for 2′- and 3′-OH groups. The two classes display distinctly characteristic amino acid sequence motifs and chain folds.

Mechanism of Action

The mechanism of action of TyrRS is discussed in chapter 2, and other aminoacyl tRNA synthetases act by the same general mechanism. Scheme 11-3 illustrates the kinetic and chemical mechanism for the adenylylation of an amino acid to its aminoacyl adenylate.

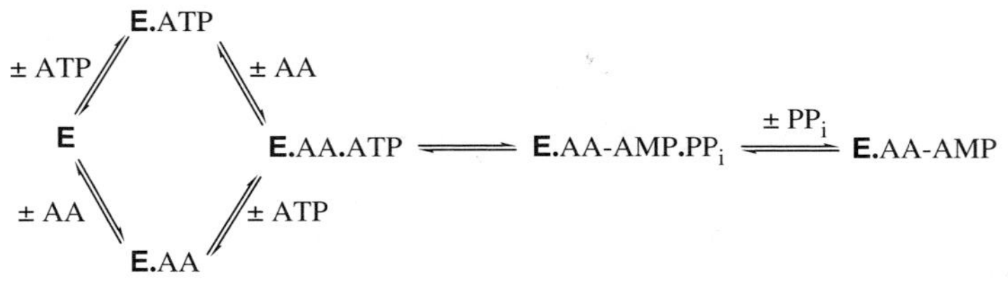

$$E \underset{\pm AA}{\overset{\pm ATP}{\rightleftarrows}} \begin{matrix} E \cdot ATP \\ E \cdot AA \end{matrix} \underset{\pm ATP}{\overset{\pm AA}{\rightleftarrows}} E \cdot AA \cdot ATP \rightleftarrows E \cdot AA\text{-}AMP \cdot PP_i \underset{}{\overset{\pm PP_i}{\rightleftarrows}} E \cdot AA\text{-}AMP$$

Scheme 11-3

The amino acid and MgATP bind sequentially to the active site, with either substrate leading, to form a ternary complex. Reaction of the amino acid with MgATP forms the aminoacyl adenylate and $MgPP_i$, which dissociates from the enzyme. The aminoacyl adenylate does not dissociate but reacts directly with the 3′-OH end of its cognate $tRNA^{aa}$ in subsequent steps to form the aminoacyl-tRNA and AMP. Because all of the steps are reversible, The rate of approach to equilibrium is conveniently measured by the rate of exchange of $[^{31}P]PP_i$ into ATP.

The aminoacyl adenylate in scheme 11-3 is kinetically competent for at least some aaRSs and likely for all, as demonstrated by rapid mix-quench kinetic experiments measuring the rate of its formation in the case of IleRS, which is faster than the overall reaction. Kinetic experiments measuring the rate at which the complex of IleRS and the isoleucyl adenylate ($E \cdot AA\text{-}AMP$) reacts with $tRNA^{Ile}$ to form the Ile-$tRNA^{Ile}$ show that it corresponds to the overall rate. That the aminoacyl adenylate is a compulsory intermediate is proved by the fact that its formation, detected by release of $^{31}PP_i$ from $[\gamma\text{-}^{31}P]ATP$, displays burst kinetics, whereas the formation of Ile-$tRNA^{Ile}$ does not (Fersht and Kaethner, 1976a, 1976b).

This mechanism of amino acid activation is a general but incomplete story. In some parts of the biological world, certain aminoacyl-tRNAs arise through postactivation biochemical transformations. For example, Gln-$tRNA^{Gln}$ and Asn-$tRNA^{Asn}$ can be produced from Glu-$tRNA^{Gln}$ and Asp-$tRNA^{Asn}$, respectively. These and other examples are discussed in a review (Ibba and Söll, 2000).

Fidelity of Action

Because the mRNA-ribosome complexes recognize the anti-codon loops of aminoacyl-tRNAs, and not the aminoacyl groups themselves, it is extremely important for the aaRSs to act with very high fidelity. One mistake in about 3000 amino acids incorporated into a protein is regarded as a threshold above which an organism cannot thrive and may not survive. Most aminoacyl-tRNA synthetases function with higher fidelity, on the order of one mistake in 10,000. Subjects of ongoing research include the means by which the synthetases recognize specific amino acids and tRNAs and the mechanisms by which fidelity is maintained and occasional mistaken events are corrected.

In a widely held view, editing by an aaRS operates by a two-sieve system. In the coarse sieve, any amino acid with a side chain too large for the active site is sterically excluded, whereas an amino acid with a side chain of similar or smaller size may be activated. The fine sieve comes into play when mis-activation occurs, and the mistake is corrected by hydrolysis to the free amino acid and AMP (Fersht, 1977; Schimmel and Schmidt, 1995; Schmidt and Schimmel, 1994).

Mistaken events in the action of an aaRS can include the mis-activation of an amino acid and the mis-acylation of a tRNA. A few of the aaRSs that activate sterically similar amino acids are prone to mis-activation or mis-acylation, or both, and they incorporate machinery to correct them. For example, IleRS can accept valine for activation at a frequency of about 1/150 that of isoleucine (Baldwin and Berg, 1966). Valine is activated to valyl adenylate, but no Val-tRNA^{Ile} is formed. Instead, in the presence of tRNA^{Ile}, hydrolysis occurs and valine is released with AMP in a process known as *pretransfer editing*. Valine is smaller by one methylene group than isoleucine and can fit into the site well enough to react too often. The values of k_{cat} for reactions of the two amino acids are comparable, but the values of k_{cat}/K_m differ by a factor of 150 (Schimmel). In the case of ValRS, tRNA^{Val} is occasionally mis-acylated with threonine, which is isosteric, but the Thr-tRNA^{Val} is quickly hydrolyzed, preventing the incorporation of threonine in place of valine into proteins (Fersth and Kaether, 1976). This process is known as *posttransfer editing*.

Pretransfer editing has been associated with aaRS-catalyzed, tRNA-dependent exchange of ^{31}PP into ATP, which occurs by reversal of the mechanism in scheme 11-3. In posttransfer editing, this exchange is not observed, presumably because of the rapid formation and hydrolysis of mis-acylated tRNA.

The mechanism of the hydrolysis of mis-activated amino acids by aaRSs involves complex machinery. In the case of IleRS, both valyl adenylate and valyl-tRNA^{Ile} mistakenly formed undergo hydrolysis at a site about 30 Å from the site of aminoacyl adenylate formation (Lin et al., 1997; Nurecki et al., 1998). This editing site is contained within a peptide termed the CP1, for connective peptide, which in its isolated form catalyzes the hydrolysis of mis-acylated tRNA (Lin et al., 1997). Moreover, the hydrolysis of valyl adenylate requires tRNA^{Ile}, and variants of tRNA^{Ile} lacking a 3'-OH on adenosine at the 3'-CCA end do not fulfill the tRNA requirement (von der Haar and Cramer, 1975; 1976; Nordin and Schimmel, 2003). tRNA^{Ile} with 3'-deoxy, 3'-amino, or 3'-fluoro ends does not support pretransfer or posttransfer editing. IleRS does not catalyze the hydrolysis of mis-charged Val-tRNA^{Ile} with 3'-deoxy, 3'-amino, or 3'-fluoro ends. An intact 3'-OH end is strictly required for pretransfer (aminoacyl adenylate) and posttransfer (aminoacyl-tRNA) hydrolysis of mis-activated amino acids.

The exact relation between pretransfer and posttransfer editing remains unclear. Misacylation of tRNA may take place at least transiently in both editing mechanisms. If so, the ^{31}PP$_i$/ATP exchange characteristic of pretransfer editing may represent the steady-state accumulation of the mis-activated aminoacyl adenylate, analogous to **E**.AA-AMP in scheme 11-3, allowing it to undergo rapid reversal by reaction with ^{31}PP$_i$, thereby accounting for the ^{31}PP$_i$/ATP exchange. In posttransfer editing, mis-acylated tRNA may be formed too rapidly to allow the accumulation of the mis-activated aminoacyl adenylate, so that the exchange becomes very slow or cannot be detected.

ValRS as a class I synthetase incorporates a similar editing site as IleRS. In the class I enzymes, the hydrolytic site in CP1 excludes amino acid side chains larger than or chemically different from the cognate amino acid. Class II synthetases also display editing functions that differ in detail from the class I enzymes (Beebe et al., 2003a, 2003b). Still another editing mechanism operates to exclude homocysteine and homoserine from activation by LysRS, a class II synthetase. Mis-activation of homocysteine or homoserine by LysRS leads to internal cyclization and lactone or thiolactone, as shown in scheme 11-4 (Jakubowski, 1997).

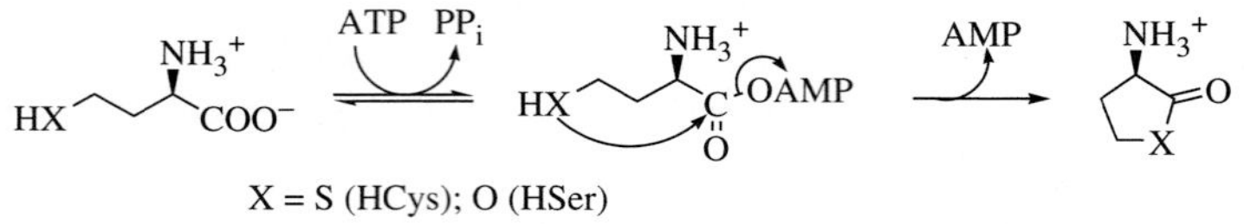

X = S (HCys); O (HSer)

Scheme 11-4

Fidelity of aaRSs for tRNA appears to be brought about by specific interactions of synthetases with their cognate tRNAs. The synthetases interact specifically with several segments of tRNA structure, each synthetase in a unique way with its cognate tRNA. Parts of the tRNA structure frequently engaged in these interactions include N73, the anti-codon, certain base pairs in the acceptor stem, the D stem, the phosphate backbone, and other loci. The drawing of tRNA in fig. 11-10 illustrates the frequencies with which specific nucleotides in the sequence are brought into play in binding to the twenty synthetases. The nucleotides most frequently found in contact with one of the synthetases are the larger spheres, those that interact less frequently appear as smaller spheres, and those that never interact do not appear as spheres.

Ubiquitin

ATP-dependent proteolysis plays many important roles in cells, and ubiquitin is one molecule that confers ATP-dependence in controlled proteolysis. Ubiquitin is a small protein (74 amino acids) that serves as a marker for proteins, in guiding them for cell functions and identifying them for proteolysis by the proteasome (Pickart, 2001, 2004; Wilkinson, 1999).

Marking of a protein by ubiquitin consists of ligating it through its C-terminal glycine to a lysine-ε-amino group in an "iso-peptide" linkage. Ligation of ubiquitin is brought about by three proteins, and it includes adenylylation of the C-terminal glycine as the initial chemical activation event (Hershko et al., 1980, 1984). The proteins include an ubiquitin-activating enzyme, designated E1, that catalyzes its own ubiquitylation in the two-step process outlined scheme 11-5. Adenylylation of ubiquitin at the C-terminal glycine residue forms the ubiquitin-glycyl adenylate, and this then reacts with a cysteine residue of E1 to form an ubiquityl thioester of E1.

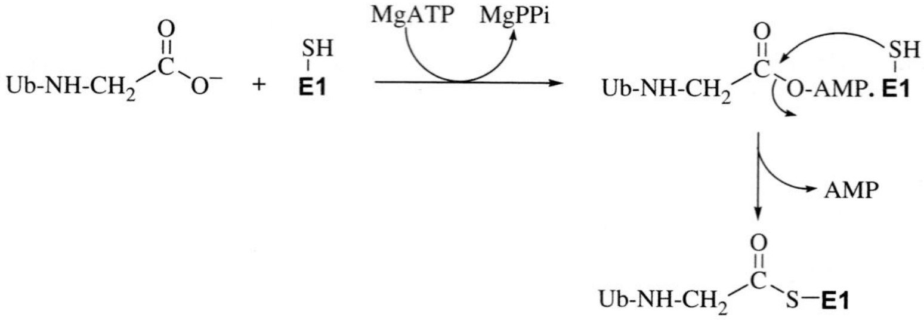

Scheme 11-5

Transesterification then occurs to the sulfhydryl group of a cysteine residue of a second enzyme designated E2 to form an ubiquityl-E2, as in fig. 11-11. A third enzyme E3 accepts the ubiquityl moiety and catalyzes its transfer it to a lysine-ε-amino group of a protein molecule, the ultimate substrate and destination of ubiquitin.

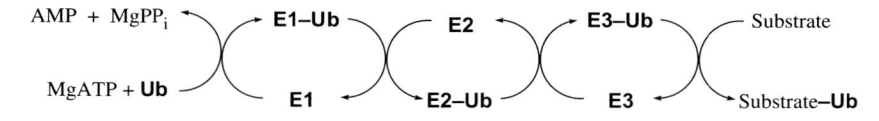

```
AMP + MgPP_i              E1–Ub              E2              E3–Ub              Substrate
              ⟩⟨              ⟩⟨              ⟩⟨              ⟩⟨
MgATP + Ub         E1              E2–Ub              E3              Substrate–Ub
```

Fig. 11-11. The ubiquitination cascade.

The ubiquitin activating enzyme E1 specifically reacts with ubiquitin and MgATP but is less selective about which of several species of E2 to which it will donate the ubiquityl moiety. The ubiquityl-E2s catalyze ubiquitylation of a large variety of E3s, which determine the ultimate destination of ubiquitin. Proteins that are mono-ubiquitinated are directed to a variety of cellular regulatory processes (Pickart, 2001, 2004;Wilkinson, 1999). Proteins that are poly-ubiquitylated, in which ubiquityl moieties themselves undergo ubiquitin-polymerization, often are coded for proteolytic degradation and are directed to the proteasome. Most proteins that undergo rapid turnover in cells undergo poly-ubiquitination and proteolysis.

The adenylylation step is typical of ATP-dependent activation and makes use of the free energy associated with cleaving the α,β-phosphoanhydride bond of ATP. The resultant glycyl-adenylate is an excellent acylating agent for thioester formation to initiate the cascade.

References

Alberty RA (1994) *Pure Appl Chem* **66**, 1641.
Anderson PM (1995) In Walsh PJ and P Wright (eds): *Nitrogen Metabolism and Excretion.* CRC Press: New York, p 33.
Anderson PM and A Meister (1965) *Biochemistry* **4**, 2803.
Arabshahi A and PA Frey (1999) *J Biol Chem* **274**, 8586.
Baldwin AN and P Berg (1966) *J Biol Chem* **241**, 831.
Beebe K, E Merriman, and P Schimmel (2003a) *J Biol Chem* **278**, 45056.
Beebe K, L Ribas de Pouplana, and P Schimmel (2003b) *EMBO J* **22**, 668.
Bock JL and M Cohn (1978) *J Biol Chem* **253**, 4082.
Carter CW Jr (1993) *Annu Rev Biochem* **62**, 715.
Cohn M and A Hu (1978) *Proc Natl Acad Sci U S A* **75**, 200.
Eisenberg D, HS Gill, GMU Pfluegl, and SH Rotstein (2000) *Biochim Biophys Acta* **1477**, 122.
Fersht AR (1977) *Biochemistry* **16**, 1025.
Fersht AR and MM Kaethner (1976a) *Biochemistry* **15**, 818.
Fersht AR and MM Kaethner (1976b) *Biochemistry* **15**, 3342.
Frey PA and A Arabshahi (1995) *Biochemistry* **34**, 11307.
Gill HS and D Eisenberg (2001) *Biochemistry* **40**, 1903.
Gumport RI and IR Lehman (1971) *Proc Natl Acad Sci U S A* **68**, 2559.
Hall ZW and IR Lehman (1969) *J Biol Chem* **244**, 43.
Harvey CL, TF Gabriel, EM Wilt, and CC Richardson (1971) *J Biol Chem* **246**, 4523.
Hershko A, A Ciechanover, H Heller, AL Haas, and IA Rose (1980) *Proc Natl Acad Sci U S A* **77**, 1783.
Hershko A, H Heller, E Eytan, G Kaklij, and A Ciechanover (1983) *J Biol Chem* **258**, 8206.
Ho CK, JL Van Etten, and S Shuman (1997) *J Virol* **71**, 1931.
Huang X, and FM Raushel (1999) *Biochemistry* **38**, 15909.
Huang X, and FM Raushel (2000a) *Biochemistry* **39**, 3240.
Huang X, and FM Raushel (2000b) *J Biol Chem* **275**, 23233.
Hunt JB and A Ginsburg (1980) *J Biol Chem* **255**, 590.
Ibba M and D Söll (2000) *Annu Rev Biochem* **69**, 617.
Jakubowski H (1997) *Biochemistry* **36**, 11077.
Jencks WP and W Regenstein (1970) In Sober HA (ed): *Handbook of Biochemistry and Molecular Biology.* CRC Press: Cleveland, OH, pp J183-J185.

Jones ME and L Spector (1960) *J Biol Chem* **235**, 2897.
Khedouri E, PM Anderson, and A Meister (1966) *Biochemistry* **5**, 3552.
Krishnaswamy PR, V Pamilians, and A Meister (1962) *J Biol Chem* **237**, 2932.
Larsen TM, SK Boehlein, SM Schuster, NG Richards, JB Thoden, HM Holden, and I Rayment (1999) *Biochemistry* **38**, 16146.
Lin L, SP Hale, and P Schimmel (1997) *Nature* **384**, 33.
Magnet S and JS Blanchard (2004) *Biochemistry* **43**, 710.
McClain WH and HB Nicholas (1987) *J Mol Biol* **194**, 635.
Meister A (1974a) In Boyer PD (ed): *The Enzymes*, vol 10, 3rd ed. Academic Press: New York, p 561.
Meister A (1974b) In Boyer PD (ed): *The Enzymes*, vol 10, 3rd ed. Academic Press: New York, p 699.
Meister A (1989) *Adv Enzymol Relat Areas Mol Biol* **62**, 315.
Modrich P and IR Lehman (1973) *J Biol Chem* **248**, 7502.
Mullins LS, CJ Lusty, and FM Raushel (1991) *J Biol Chem* **266**, 8236.
Mullins LS, and FM Raushel (1999) *J Am Chem Soc* **121**, 3803.
Nakatani M, S Ezaki, H Atomi, and T Imanaka (2002) *Eur J Biochem* **269**, 650.
Nureki O, DG Vassylyev, M Tateno, A Shimada, T Nakama, S Fukui, M Konno, TL Hendrickson, P Schimmel, and S Yokoyama (1998) *Science* **280**, 578.
Odell M, V Sriskanda, S Shuman, and DB Nikolov (2000) *Mol Cell* **6**, 1183.
Olivera BM, ZW Hall, and IR Lehman (1968) *Proc Natl Acad Sci U S A* **61**, 237.
Pickart CM (2001) *Mol Cell* **8**, 499.
Pickart CM (2004) *Cell* **116**, 181.
Raushel FM and JJ Villafranca (1979) *Biochemistry* **18**, 3424.
Raushel FM and JJ Villafranca (1980) *Biochemistry* **19**, 3170.
Rhee S, FC Wedler, Y Sugiyama, and P Chock (1981) *J Biol Chem* **256**, 644.
Ronzio R and A Meister (1968) *Proc Natl Acad Sci U S A* **59**, 164.
Schimmel P and E Schmidt (1995) *Trends Biochem Sci* **20**, 1.
Schmidt E and P Schimmel (1994) *Science* **264**, 265.
Shapiro BM and ER Stadtman (1968) *J Biol Chem* **243**, 3769.
Shrake A, E Whitely, and A Ginsburg (1980) *J Biol Chem* **255**, 581.
Stadtman ER and A Ginsburg (1974) In Boyer PD (ed): *The Enzymes*, vol 10, 3rd ed. Academic Press: New York, p 755.
Thoden JB, G Wesenberg, FM Raushel, HM Holden (1999) *Biochemistry* **38**, 2347.
von der Haar F and F Cramer (1975) *FEBS Lett* **56**, 215.
von der Haar F and F Cramer (1976) *Biochemistry* **15**, 4131.
Wang TT, SH Bishop, and A Himoe (1972) *J Biol Chem* **247**, 4437.
Wedler FC (1974) *J Biol Chem* **249**, 5080.
Wedler FC and PD Boyer (1972) *J Biol Chem* **247**, 984.
Weiss B, A Thompson, and CC Richardson (1968) *J Biol Chem* **243**, 4556.
Wilkinson KD (1999) *J Nutr* **129**, 1933.
Wimmer MJ, IA Rose, SG Powers, and A Meister (1979) *J Biol Chem* **254**, 1854.

12
Glycosyl Group Transferases

Glycosyl group transfer underlies the biosynthesis and breakdown of all nucleotides, polysaccharides, glycoproteins, glycolipids, and glycosylated nucleic acids, as well as certain DNA repair processes. Glycosyl transfer consists of the transfer of the anomeric carbon of a sugar derivative from one acceptor to another, as in eq. 12-1, which describes the transfer of a generic pyranosyl ring between nucleophilic atoms :X and :Y of acceptor molecules.

$$\text{(pyranosyl)–X–R}_1 + \text{:Y–R}_2 \rightleftharpoons \text{(pyranosyl)–Y–R}_2 + \text{:X–R}_1 \qquad (12\text{-}1)$$

The stereochemistry at the anomeric carbon is not specified in eq. 12-1, but the leaving group occupies the axial position in an α-anomer or the equatorial position in a β-anomer. The overall transfer can proceed with either retention or inversion of configuration. In biochemistry, the acceptor atoms can be oxygen, nitrogen, sulfur, or in the biosynthesis of C-nucleosides even carbon. The great majority of biological glycosyl transfer reactions involve transfer between oxygen atoms of different acceptor molecules.

Enzymes catalyzing glycosyl transfer are broadly grouped according to whether the acceptor :Y–R$_2$ in eq. 12-1 is water or another molecule. In the actions of glycosidases, the acceptor is water, and glycosyl transfer results in hydrolysis of a glycoside, a practically irreversible process in dilute aqueous solutions. In the action of glycosyltransferases, the acceptors are molecules with hydroxyl, amide, amine, sulfhydryl, or phosphate groups.

Chemical Mechanisms

Chemistry of Glycoside Hydrolysis

Acid Catalysis

The simplest nonenzymatic glycosyl transfer reaction is the hydrolysis of a glycoside, and early studies revealed the fundamental fact that glycosides are much less reactive toward hydrolysis in basic solutions than in acidic solutions. This fact underlies much that is known about the mechanism of glycosyl transfer; that is, the anomeric carbon of a glycoside is remarkably unreactive toward direct nucleophilic attack, but it becomes reactive when one of the oxygens is protonated by an acid, as illustrated in fig. 12-1 for the acid-catalyzed hydrolysis of a generic glycoside. The reaction by both mechanisms in fig. 12-1 proceeds by pre-equilibrium protonation of the glycoside to form oxonium ion intermediates, which are subject to hydrolysis by water.

The two mechanisms in fig. 12-1 are of interest. The mechanism proceeding through exocyclic cleavage of the glycoside has historically been regarded as the more likely, and for this reason, the route through endocyclic cleavage has received little consideration. A wealth of information in nonaqueous solvation experiments, such as acetolysis (McPhail et al., 1992), has implicated endoocyclic cleavage as a universal and in some cases dominant mechanism. Endocyclic cleavage is significant in the methanolysis and hydrolysis of a β-glycoside (Liras and Anslyn, 1994). Exocyclic cleavage has not been implicated in an enzymatic glycosyl transfer reaction, although such a mechanism has been proposed in the reaction of lysozyme. The mechanism would have stereochemical implications undermining the rules that have come into play for analyzing the mechanisms of enzymatic glycosyl transfer.

The mechanism of the hydrolytic step is of some interest. Does it take place by a nucleophilic displacement (S_N2), or does the C—O bond break to form an oxacarbenium ion intermediate that is rapidly quenched by water? The two mechanisms are illustrated in fig. 12-2. The oxacarbenium ion is stabilized by resonance electron donation from oxygen, which delocalizes the positive charge ($-O^+=C< \leftrightarrow -O-C^+<$), but the lifetime of oxacarbenium ions in aqueous solutions is very short, on the order of nanoseconds to picoseconds (Amyes and Jencks, 1989). The lifetimes are within one to two orders of magnitude of low-frequency covalent bond vibrations, raising questions of whether they could be discrete intermediates. Certainly they could not escape from an aqueous solvation sphere before

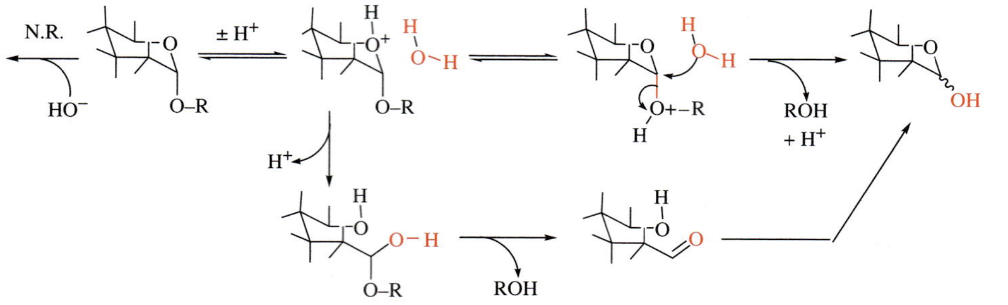

Fig. 12-1. Two mechanisms for the acid-catalyzed hydrolysis of an *O*-glycoside. *O*-Glycosides display very low reactivity toward hydroxide ions, but they are susceptible to acid-catalyzed hydrolysis. Protonation of the glycoside on either ether oxygen can lead to hydrolysis, most directly through exocyclic cleavage in the upper mechanism. Endocyclic cleavage leads to the ring-opened hemiacetal. Elimination of the alcohol to the ring-opened saccharide followed by ring closure leads to the pyranose.

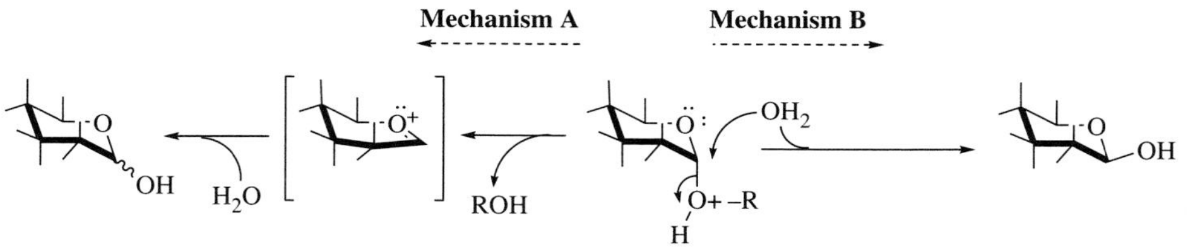

Fig. 12-2. Two mechanistic concepts for acid-catalyzed hydrolysis of a pyranoside. In mechanism A, the oxonium alcohol is eliminated to form a glycosyl oxacarbenium ion as a transient intermediate that reacts rapidly with water to form the pyranose sugar. This mechanism is analogous to a classical S_N1 mechanism. In mechanism B, a water molecule displaces the oxonium alcoholic group in a classical S_N2 mechanism through a single transition state, in which the entering water and departing alcohol are both partially bonded to the glycosyl-C1.

reacting with water molecules. Detailed studies indicate that in water as the solvent, the acceptor nucleophile participates in the transition state, but there is little bonding between the glycosyl-C1 and the attacking nucleophile or the leaving group (Davies et al., 1997; Knier and Jencks, 1980). The lifetime of an oxacarbenium ion may be extended in the nonaqueous environment of an active site through in-plane electrostatic stabilization by an anion.

2-Acetamido Participation: Anchimeric Assistance

The 2-acetamido group in polysaccharides containing *N*-acetylglucosamine, *N*-acetylgalactosamine, N-acetylmannosamine, or *N*-acetylmuramic acid can in principle facilitate glycosyl transfer. This group participates in the acid catalyzed hydrolysis of *N*-acetylglucosamine glycosides by internal nucleophilic covalent catalysis (Piszkiewicz and Bruice, 1968). The process is illustrated in fig. 12-3. In 2-acetamide–assisted hydrolysis the carbonyl oxygen internally displaces the protonated leaving group to form a cyclic, amidinium intermediate. The amidinium group is an excellent leaving group that is readily displaced by a water molecule to form the pyranose product. Note that this mechanism involves compulsory retention of configuration at the anomeric carbon generated by hydrolysis. 2-Acetamido participation is thought to take place in the actions of a number of glycosaminidases (Davies et al., 1997).

Transition State

Structure-function analysis of the rates of glycosyl transfer, with variations in nucleophilic reactivities of the acceptors and in the stabilities of the leaving groups has given information about the nature of the transition state (Craze et al., 1978; Knier and Jencks, 1980). The linear free energy correlations in free solutions show little sensitivity of the rate to nucleophilic reactivity of the acceptor nucleophile. Log plots of rate constant against pK_a of the acceptor show slopes on the order of –0.1, indicating little covalent bonding to the leaving group and the attacking nucleophile in the transition state. In contrast, the rates are very responsive to leaving group ability. Log plots of rate constant against the pK_a of the leaving group show high sensitivity to the stability of the leaving group, with slopes of $\beta_{lg} \approx 1.0$, indicating that the bond to the leaving group is largely broken in the transition state. The linear free energy correlations show that glycosyl transfer in free solution proceeds through a loose transition state, with little bonding between the anomeric carbon and both the attacking nucleophile and leaving group. In the oxacarbenium-like transition state, substantial positive charge on the anomeric center is generated, but there is detectable

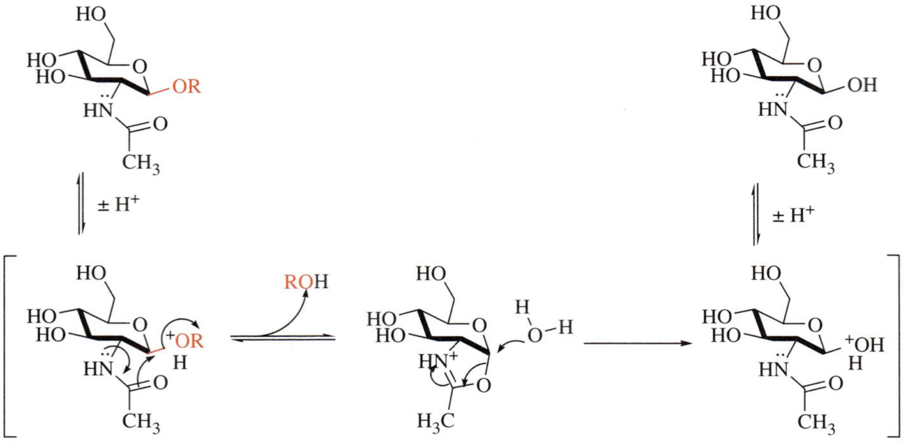

572   Enzymatic Reaction Mechanisms

Fig. 12-3. Participation of the 2-acetamido group in acid-catalyzed hydrolysis of a glycoside. In acid-catalyzed hydrolysis of *N*-acetylglycosamine, glycosides the acetamido group assists the process by neighboring group participation in the expulsion of the leaving group. The protonated alcohol substituent is an oxonium ion, an excellent leaving group, which is displaced by the weakly nucleophilic but neighboring *N*-acetyl group. The intermediate also incorporates an excellent leaving group and is readily hydrolyzed by water.

interaction with the entering and leaving groups. The linear free energy correlations are analogous to those for phosphoryl transfer, which proceeds through a metaphosphate-like transition state (see chap. 10).

The secondary deuterium kinetic isotope effects for glycosyl transfer with deuterium at the anomeric carbon are in the range of 1.1 to 1.2 (Bennett and Sinnott, 1986; Davies et al., 1997). These values are in the range expected for oxacarbenium ion formation at the transition state. They are compatible with the loose transition state, indicated by the linear free energy correlations.

The transition state for glycosyl transfer in free solution indicates oxacarbenium-like character that includes long-range interactions with the acceptor and leaving group, as illustrated in scheme 12-1.

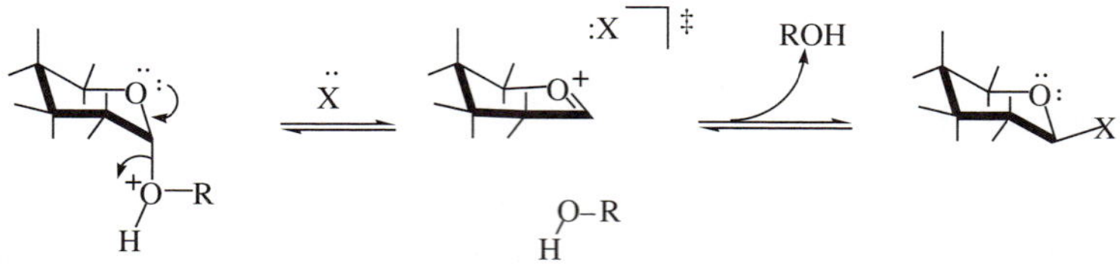

Scheme 12-1

In a solvating solution, there is always a potential acceptor in the solvation sphere. An acceptor is generally present in an active site, either in the form of the co-substrate or a nucleophilic catalytic group. In the case of nonenzymatic hydrolysis of the nicotinamide riboside moiety of NAD$^+$, where there is an excellent leaving group that does not require acid catalysis to depart, analysis of kinetic isotope effects indicate only a trace of interaction between a practically discrete ribo-oxacarbenium ion intermediate and the leaving nicotinamide and an entering water molecule (Berti and Schramm, 1997). The ribo-oxacarbenium ion is for practical purposes a discrete, solvated species separated from the leaving nicotinamide-N1 by 2.7 Å and from a solvating water molecule by 3.0 Å, and the solvent captures the intermediate.

The acid hydrolyses of α- and β-methylglucosides proceed with small, primary 1-^{13}C kinetic isotope effects of 1.007 and 1.011, respectively, and substantial secondary 1-^{2}H isotope effects of 1.137 and 1.089 (Bennett and Sinnott, 1986). The transition states computed from these effects indicate that both reactions proceed through oxacarbenium ion intermediates in half-chair conformations, as illustrated in scheme 12-1 for an α-glycoside (Lee et al., 2004).

The oxacarbenium ion is short-lived, cannot diffuse freely, and is captured by water within its solvation sphere. Isotope effects for the α-glucosidase-catalyzed hydrolysis of α-methylglucoside also indicate an intermediate oxacarbenium ion-intermediate. However, the much larger 1-^{13}C isotope effect of 1.032 in the β-glucosidase-catalyzed hydrolysis of β-methylglucoside indicates an S_N2-type transition state with oxacarbenium ion character (Lee et al., 2004). Such a mechanism is illustrated in scheme 12-2, where weak bonding to the leaving group and attacking nucleophile contribute significantly to the transition state. Larger primary 1-^{13}C and smaller secondary 1-^{2}H isotope effects distinguish this mechanism from that in scheme 12-1.

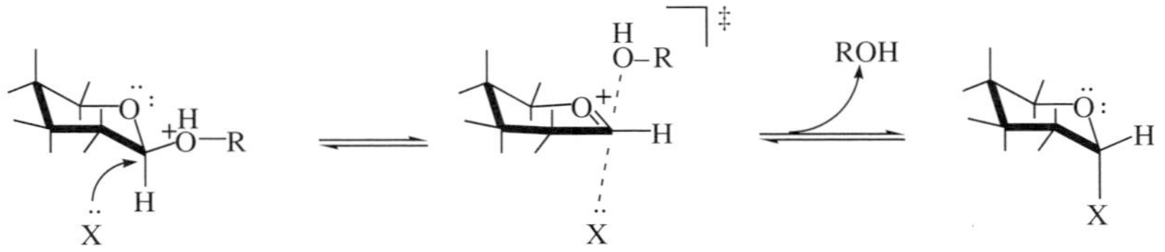

Scheme 12-2

In this connection, consideration should be given to a fundamental difference between α- and β-pyranosides. Backside attack on carbon-1 of a β-pyranoside is sterically hindered and appears to require a significant conformational change toward a half-chair (Davies et al., 1997). In contrast, nucleophilic attack on carbon-1 of an α-pyranoside takes place on the open face and is unhindered. Because of the steric effect, the conformational change required for the reaction of a β-glucoside could be coupled to glucosyl transfer and contributes toward rate limitation, directly implicating it in a transition state that includes the attacking nucleophile.

Enzymatic Glycosyl Transfer

Retaining and Inverting Enzymes

Enzymatic glycosyl transfers take place with either inversion or retention of configuration at the anomeric carbon and never with epimerization. As a general rule, retaining enzymes catalyze the reaction by a double-displacement mechanism and nucleophilic catalysis. The nucleophilic group is a carboxylate in the active site or a 2-acetamido group in the glycosyl donor substrate. Inverting enzymes catalyze the reaction with direct transfer of the glycosyl group to the acceptor substrate. In either case, mechanistic evidence indicates that glycosyl transfer proceeds through a loose, oxacarbenium ion–like transition state.

As a general but not inviolable rule, glycosyl transferring enzymes contain two acidic residues in their active sites (Davies et al., 1997). The acidic residues are separated by 9 to 10 Å in the inverting enzymes and about 5 to 6 Å in the retaining enzymes. These separations are compatible with the catalytic functions of the residues. Exceptions to this rule include enzymes that catalyze reactions of N-ribosyl transfer or of S-glycosyl transfer, in which the leaving groups require alternative catalytic strategies. We discuss enzymes that fit the pattern and those that do not.

Generic Mechanisms

The mechanisms in fig. 12-4 illustrate the functions of the acidic residues in glycosyl transfer. In a retaining enzyme, the carboxylate group of one residue acts as the nucleophilic catalyst and is transiently glycosylated, and the other is the acid catalyst in facilitating the departure of the leaving group. The second acidic residue then becomes a base catalyst in assisting glycosyl transfer to the acceptor substrate. In an inverting enzyme, one residue acts as an acid catalyst in promoting the departure of the leaving group, and the other acts as a base catalyst to abstract a proton from the acceptor substrate.

Although the mechanisms in fig. 12-4 apply generally to glycosidases and glycosyltransferases, exceptions exist among well-studied enzymes, some of which are discussed in the following sections. In certain cases, an inverting enzyme may not need a catalyst to assist the departure of a good leaving group. In other cases, reasons for exceptions are not obvious, and deep mechanistic issues remain.

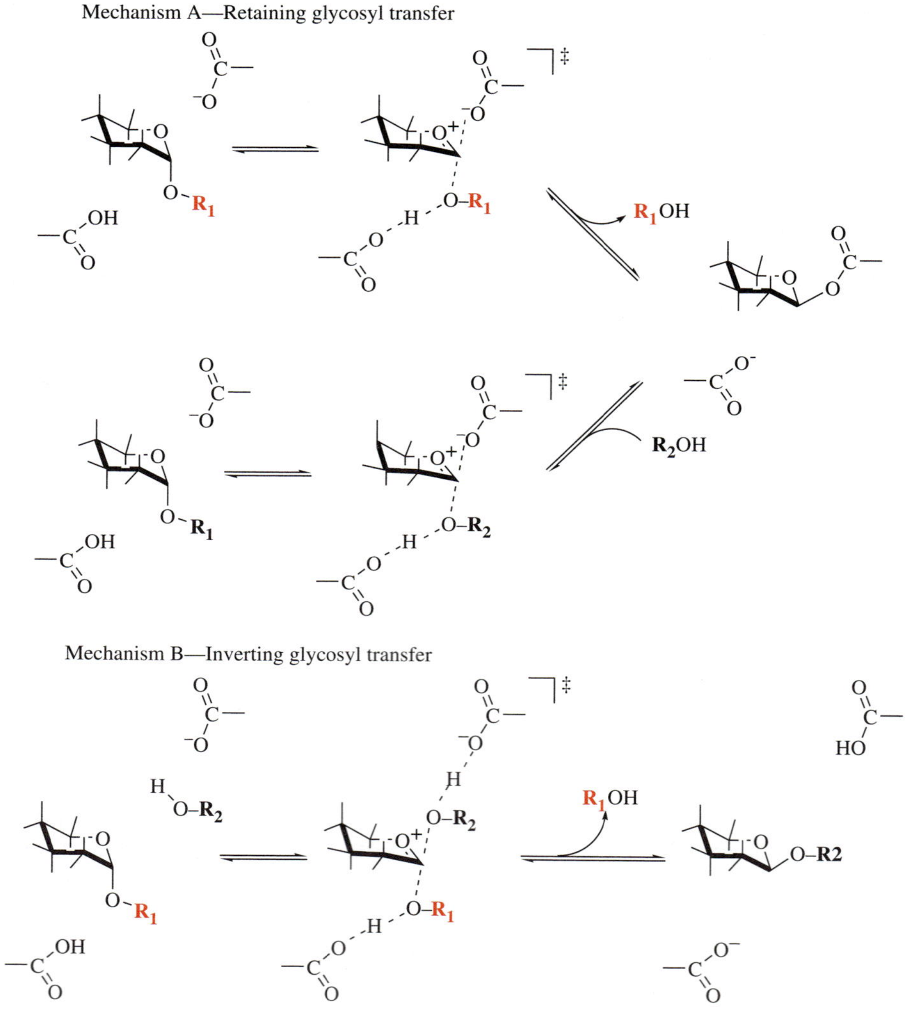

Fig. 12-4. Mechanisms for retaining and inverting glycosyl transferases. Most glycosyltransferases and glycosidases have two acidic residues at their active sites. In the retaining enzymes, one residue acts as a nucleophilic catalyst and accepts the transferring glycosyl group in a covalent glycosyl-enzyme intermediate. The other residue functions as an acid-base catalyst. In the inverting enzymes, both acidic residues act as acid-base catalysts.

Glycosyltransferases

Among retaining glycosyltransferases, the minimal studies of sucrose phosphorylase have yielded the clearest mechanistic information, and the intensive studies of glycogen phosphorylase have failed to solve the mechanism of glycosyl transfer. Among inverting glycosyltransferases, the intensively studied purine nucleoside phosphorylase presents a clear picture of the transition state, and transition-state analog inhibitors may have medical applications in the future. We focus on these enzymes in this chapter, while also emphasizing that they represent special subjects. Special aspects of the actions of other glycosyltransferases may emerge as they are subjected to penetrating analysis. A major class of glycosyltransferases use nucleotide sugars such as UDP-glucose as the glycosyl donors for the assembly of complex carbohydrates. Mechanistic analysis of these enzymes is in its infancy and growing rapidly (Hu et al., 2003).

Glycosyltransferases constitute a huge class of enzymes with highly diverse amino acid sequences. They have been organized into sequence-based families, currently 65 and rapidly increasing with the avalanche of genomic information (Coutinho et al., 2003; Davies and Henrissat, 2002).

Sucrose Phosphorylase

Reaction, Stereochemistry, and Kinetics

In the reaction of sucrose phosphorylase (EC 2.4.1.7), phosphate displaces α-D-fructose from sucrose to form α-D-glucose-1-P, as illustrated at the top of fig. 12-5. The reaction involves the transfer of the glucosyl group between α-D-fructose and phosphate, with overall retention of configuration at the anomeric carbon of glucose. The enzyme also catalyzes the exchange of radiochemically labeled fructose into sucrose in the absence of phosphate (Fitting and Doudoroff, 1952). The stereochemistry suggests a double-displacement mechanism, with the intermediate formation of a covalent glucosyl–enzyme as an intermediate. Such a mechanism is likely to obey ping pong kinetics, and it does in the case of the enzyme from *Pseudomonas saccharohila*, as illustrated at the bottom of fig. 12-5 (Silverstein et al., 1967).

The ping pong mechanism explains both the overall retention of configuration at the anomeric carbon of glucose and the exchange of fructose into glucose in the absence of phosphate on the basis of the formation of a covalent glucosyl-enzyme intermediate. Note that in the kinetic mechanism (fig. 12-5) sucrose reacts with the enzyme to form a covalent glucosyl-enzyme with the release of fructose into solution in the absence of phosphate. Reversibility of this process allows the exchange to occur. Furthermore, the formation of a covalent glucosyl-enzyme is likely to occur by backside attack of an enzymatic nucleophile, with inversion of configuration. The subsequent glucosyl transfer to phosphate from the glucosyl-enzyme is also likely to occur with inversion at the anomeric carbon, resulting in overall retention of configuration. Experiments explicitly prove each and every one of the foregoing statements for the reaction of sucrose phosphorylase.

Glucosyl-Enzyme Intermediate

An explicit test of the chemical and kinetic mechanism in fig. 12-5 would be the isolation and characterization of the hypothetical covalent glucosyl-enzyme intermediate. Failure to isolate a covalent intermediate could not disprove the mechanism, but its actual isolation and chemical characterization provided the first convincing test of the retaining mechanism

576   Enzymatic Reaction Mechanisms

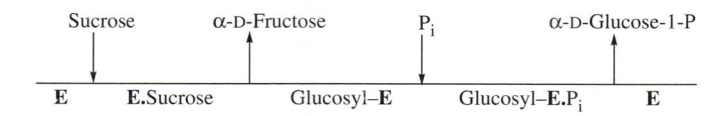

Fig. 12-5. The reaction and kinetic mechanism of sucrose phosphorylase. Transfer of the glucosyl group in sucrose to phosphate by sucrose phosphorylase produces α-D-glucose-1-P and α-D-fructose. The kinetics follows the ping pong kinetic mechanism shown at the bottom of the figure.

in enzymatic glycosyl transfer (Silverstein et al., 1967; Voet and Abeles, 1970). The experiments outlined in fig. 12-6 led to the first proof of the existence of a covalent glucosyl-enzyme. According to the kinetic mechanism in fig. 12-5, sucrose must react with the enzyme in the absence of phosphate to form the glucosyl-enzyme. In the case of sucrose phosphorylase, the isolation of a covalent intermediate is complicated by the hydrolytic activity in the absence of phosphate. This arises from the hydrolysis of the glucosyl-enzyme intermediate with a half-time on the order of seconds. The glucosyl-enzyme had to be isolated at first in a denatured form by mixing the enzyme with radioactive sucrose and then quenching the reaction with hot methanol. The precipitated protein was radioactive when [glucose-^{14}C]sucrose was the substrate but not with [fructose-^{14}C]sucrose. Precipitation of the glucosyl-enzyme with concentrated ammonium formate at pH 3 also led to a chemically stable glucosyl-enzyme, which retained its integrity after solubilization in 7 M urea. In retrospect, referring to mechanism A in fig. 12-4, stabilization at low pH probably resulted from protonation of the acidic residue in the intermediate, which must be in its conjugate base form to catalyze glucosyl transfer from the covalent intermediate.

The denatured glucosyl-enzyme and glucosyl-peptides derived from it display a property that allows a key test of the mechanism. They are remarkably labile to hydrolysis in weakly alkaline solutions (pH 8). Lability to alkaline hydrolysis is not a general property of glycosides. In the case of sucrose phosphorylase, sensitivity to alkaline hydrolysis arises from the chemistry of the enzymatic group to which the glucosyl moiety is bonded, a carboxyl group. Attack of hydroxide ion on the carbonyl group of the glucosyl ester, as illustrated in fig. 12-6, directly liberates glucose through esterolytic cleavage and not through glucosyl transfer. Silylation of the glucose released stabilized the anomeric configuration, and the silylated product proved to be the β-anomer. Alkaline hydrolysis releases β-D-glucose, proving that the β-anomeric configuration for the glucosyl-enzyme. Overall retention of configuration therefore must result from inversion at each step of the double-displacement mechanism. Additional evidence confirms that the active-site nucleophile is an acidic residue (Mieyal and Abeles, 1972; Voet and Abeles, 1970).

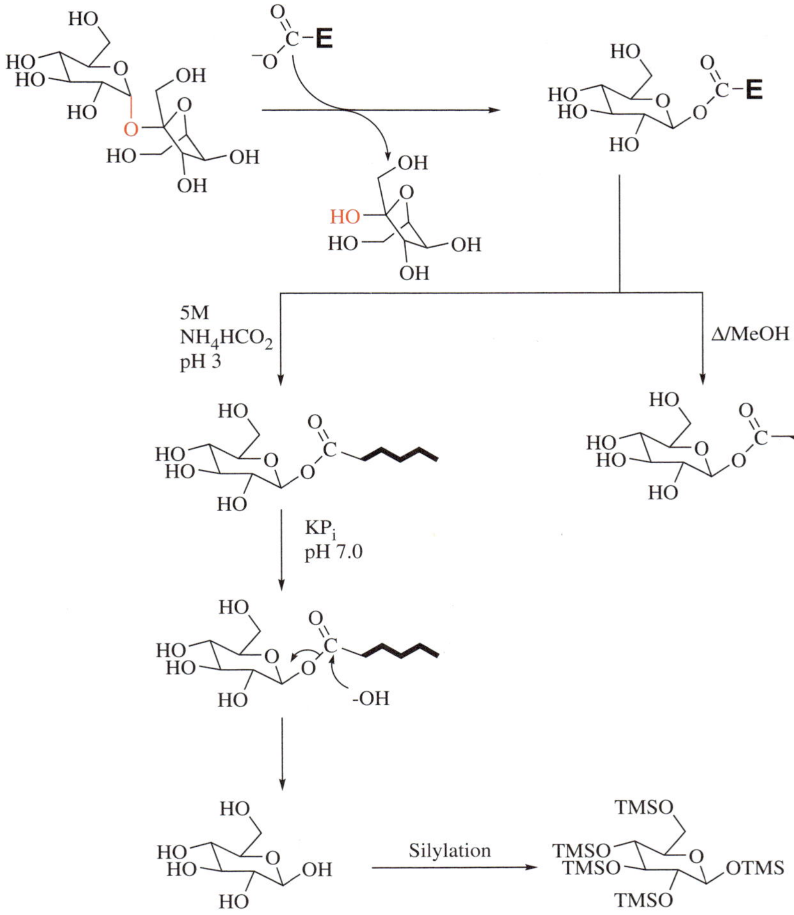

Fig. 12-6. Isolation of the glucosyl-enzyme intermediate of sucrose phosphorylase and degradation to β-D-glucose. The glucosyl-enzyme resulting from reaction of sucrose with *Pseudomonas saccharophila* sucrose phosphorylase was chemically stabilized by denaturation in hot methanol and by precipitation with ammonium formate at pH 3.0. The precipitated glucosyl-enzyme redissolved at pH 7 underwent esterolytic cleavage by hydroxide ion to release β-D-glucose, which was identified as its trimethylsilyl derivative. This proved that the formation of the glucosyl-enzyme proceeded with inversion of configuration at the anomeric carbon.

The structure of sucrose phosphorylase from *Bifidobacterium adolescentis* confirms the presence of the two acidic residues at the active site, Asp192 as the nucleophilic catalyst and Glu322 as the acid-base catalyst (Sprogøe et al., 2004). The two residues are separated by 5.5 Å, consistent with the spacing in other retaining glycosyl transferases.

Glycogen Phosphorylase

Reaction and Stereochemistry

α-Glucan phosphorylases that catalyze phosphorolysis at the nonreducing ends of α-1,4-glucans with retention of configuration at the anomeric carbon constitute a large family of very similar enzymes. The most thoroughly studied member is glycogen phosphorylase (phosphorylase; EC 2.4.1.1). This enzyme catalyzes the reversible glucosylation of the nonreducing end of glycogen by α-D-glucose-1-P, as illustrated in fig. 12-7 in the direction of glycogen elongation. The equilibrium slightly favors glucosylation of glycogen, with an equilibrium value of 3.6 for the ratio of [P$_i$]/[glucose-1-P] at pH 6.8. However, the biologically important process is the reverse, the phosphorolytic cleavage of glycogen to α-D-glucose-1-P. Arsenate also reacts in place of phosphate in an arsenolysis reaction that leads to the degradation of glycogen to glucose because of the hydrolysis of the arsenate ester. The most intensively studied phosphorylase is that from rabbit muscle, which is a homodimeric enzyme composed of 97-kDa subunits, Historical developments in research on phosphorylases are given in reviews (Graves and Wang, 1972; Davies et al., 1997).

Despite the fact that carbanions and free radicals play no part in the action of phosphorylase, it is a pyridoxal-5′-phosphate (PLP) enzyme, and PLP is absolutely required for activity. The most intriguing aspects of the reaction mechanism are the role of PLP and the mechanism by which retention of anomeric configuration is brought about. Neither of these has been satisfactorily explained to the date of this writing, despite several decades of intensive study in a number of laboratories.

Phosphorylase catalyzes several other reactions that bear on the mechanism of glucosyl transfer. The reactions in fig. 12-8 all proceed stereospecifically and practically irreversibly at the active site of phosphorylase, and all require phosphate as a co-substrate or as a cofactor. D-Glucal and 1-heptenitol have sp^2 carbons at the potentially anomeric positions and undergo addition reactions that require protonation at the adjacent carbon. Addition of the nonreducing end of glycogen to D-glucal in fig. 12-8B incorporates a 2-deoxyglucosyl group to the end and requires protonation at C2. Addition of phosphate to 1-heptenitol in fig. 12-8C

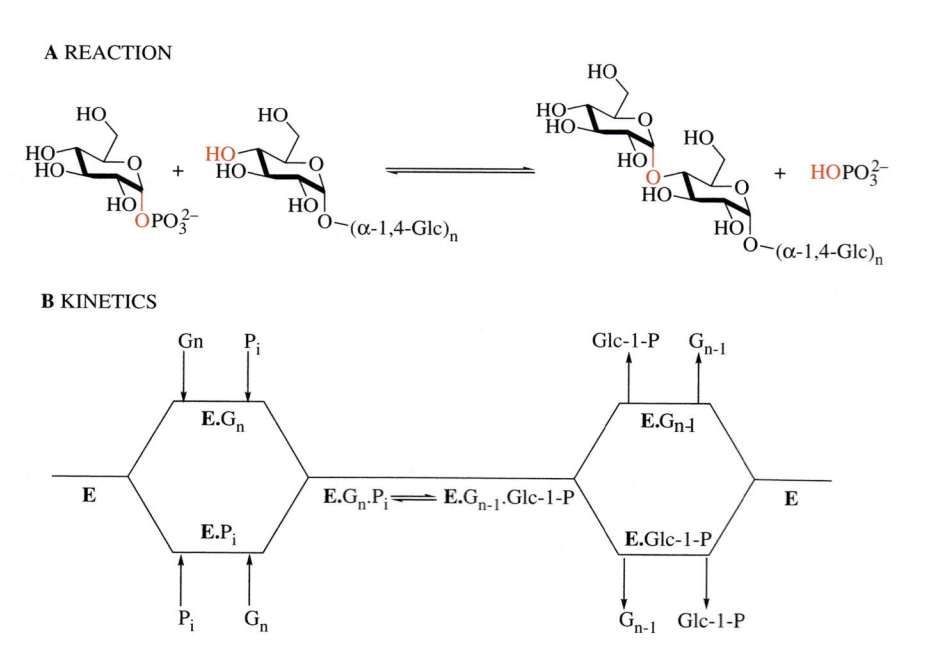

Fig. 12-7. The reaction of glycogen phosphorylase. (A) Glycogen phosphorylase catalyzes the reversible glucosylation of the nonreducing end of glycogen by α-D-glucose-1-P. The cleaved bond and the atoms participating in bond cleavage and formation are shown in red. (B) The kinetic mechanism of phosphorylase action is equilibrium random sequential, in which both substrates bind to phosphorylase to form a ternary complex before any product dissociates from the enzyme. Product dissociation is likewise random. Glycogen phosphorylase is a retaining glucosyltransferase and is in this respect like sucrose phosphorylase. Unlike sucrose phosphorylase, the kinetic mechanism is not ping pong bi bi.

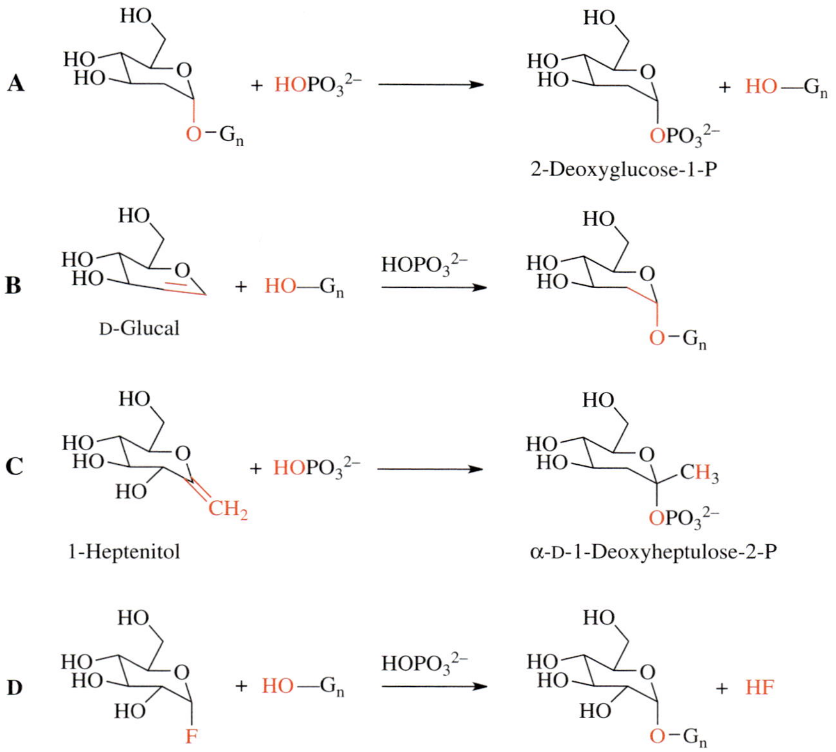

Fig. 12-8. Nonphysiological reactions catalyzed by phosphorylase. These reactions are catalyzed unidirectionally by phosphorylase at rates well below that for the physiologic reaction. However, each reaction is related in some way to the normal action of phosphorylase. Notice that all of the reactions require the presence of inorganic phosphate, either as a cosubstrate or as an activator. The 2-deoxyglucose glycogen analog supports a transformation similar to the physiological reaction (A). The reactions of D-glucal (B) and 1-heptenitol (C) are related to reactions of corresponding oxacarbenium ions resulting from protonation at C2 of D-glucal and the methylene group of 1-heptenitol. Elimination of fluoride drives oligimerization with 1-deoxy-1-fluoroglucose and a growing glycogen chain (D). The atoms rendered in red undergo chemical changes.

leads to α-D-1-deoxyheptulose-2-P and requires protonation at C1. The irreversibility of this reaction has been put to use in the crystallization of a product ternary complex of phosphorylase (Johnson et al., 1990). The addition of the nonreducing end of glycogen to α-1-fluoro-D-glucose in fig. 12-8D requires phosphate and extends the glycogen chain by one glucosyl residue. It is not known whether phosphate reacts chemically with α-1-fluoro-D-glucose to form α-D-glucose-1-P at the active site, which then undergoes glucosyl transfer to glycogen.

Regulation

Phosphorylase is subject to regulation by two mechanisms, allosteric activation by AMP and covalent activation through phosphorylation (Davies et al., 1997; Graves and Wang, 1972). Phosphorylase exists in two forms, phosphorylase b and phosphorylase a. The dephospho form, phosphorylase b, is inactive in its resting form but is activated by AMP, an allosteric effector. Phosphorylase b is subject to allosteric inhibition by glucose, which binds to the active site.

580   Enzymatic Reaction Mechanisms

The hormonal regulation of phosphorylase and the associated phosphorylation cascade are outlined in chapter 10. The activation cascade in fig. 10-8 begins with epinephrine-activation of adenylyl cyclase to produce cAMP, which activates protein kinase A to catalyze the phosphorylation of phosphorylase kinase to its active phospho-phosphorylase kinase, and this phosphorylates Ser14 of phosphorylase b to the active phosphorylase a. Dephosphorylation by phosphorylase phosphatase converts the enzyme back to phosphorylase b.

Phosphorylase a and AMP-activated phosphorylase b a display similar activities. The allosteric, AMP binding site in phosphorylase b is near Ser14, the phosphorylated site in phosphorylase a. By different mechanisms, AMP binding and phosphorylation of Ser14 induce similar conformational changes in phosphorylase.

Kinetics

Stereochemical retention in the action of phosphorylase suggests that the mechanism should follow the course outlined in fig. 12-4A, as in the reactions of sucrose phosphorylase and many retaining glycosidases. Such double-displacement mechanisms are most often associated with ping pong kinetics. However, unlike sucrose phosphorylase, the steady-state kinetics for the actions of muscle phosphorylase a and b and phosphorylases from a variety of other tissues and organisms are consistent only with the equilibrium random binding mechanism illustrated at the bottom of fig. 12-7 (Engers et al., 1969, 1970a, 1970b). In this mechanism, all substrates must be present at the active site before any chemical transformation takes place. This mechanism erects a barrier to the observation and characterization of either a covalent or noncovalent glucosyl-enzyme intermediate. A noncovalent glucosyl-enzyme would be one in which the intermediate is a discrete glucosyl-oxacarbenium ion. There are other, perhaps more formidable barriers in the case of phosphorylase.

An important and original experiment employing ^{18}O-positional isotope exchange (PIX) accentuated the importance of the presence of an oligosaccharide in the glucosyl-acceptor binding site, while simultaneously providing indirect evidence for a glucosyl-enzyme intermediate (Kokesh and Kakuda, 1977). Phosphorylase catalyzes the ^{18}O-PIX in scheme 12-3, in which the anomeric oxygen in α-D-glucose-1-P undergoes exchange with nonbridging oxygens of phosphate.

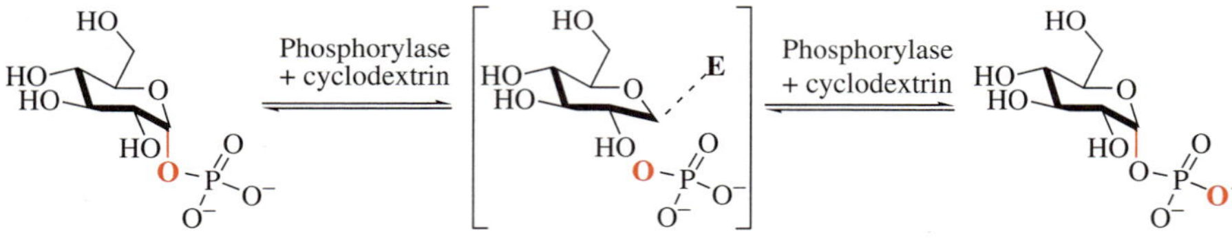

Scheme 12-3

Potato phosphorylase catalyzes the cyclodextrin-dependent ^{18}O-PIX in α-D-[1-^{18}O]glucose-1-P. As cyclic oligoglucosides, cyclodextrins lack nonreducing ends and cannot act as glucosyl acceptors. They can, however, bind to the glycogen binding site in place of glycogen and provide the structural signal guiding the enzyme into its active conformation. The occurrence of ^{18}O-PIX proves that the anomeric ^{18}O of -[1-^{18}O]glucose-1-P suffers cleavage from the glucosyl moiety and is then replaced by a peripheral oxygen of the phosphate group. There must be an intermediate form of the glucosyl moiety; it is covalently bonded to a nucleophilic atom of phosphorylase itself, or it exists transiently as a stabilized oxacarbenium ion. The nature of the glucosyl phosphorylase intermediate in the ^{18}O-PIX experiment is unknown.

Inhibition of phosphorylase by glucono-δ-lactone is enhanced by phosphate, as is the more potent inhibition by nojirimycin tetrazole (structure **12-1**) (Gold et al., 1971; Mitchell et al., 1996).

12-1   Glucono-δ-lactone   Nojirimycin tetrazole

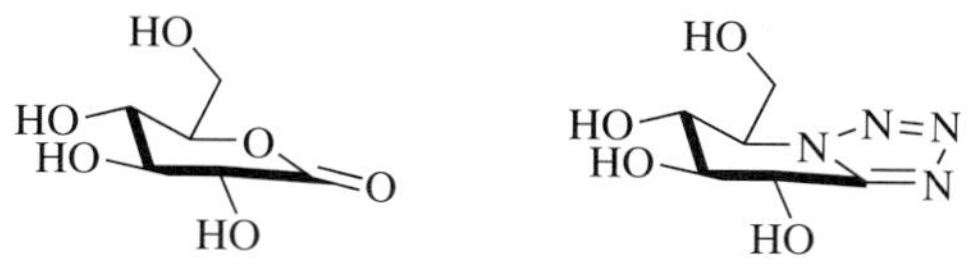

These compounds incorporate sp^2 hybridization at C1, similar to an oxacarbenium ion. The interaction of these compounds with phosphorylase is consistent with a transition state in which the anomeric carbon is in-plane with C2 and C5 and may be oxacarbenium ion-like.

Formation of an oxacarbenium-like transition state should lead to a secondary deuterium isotope effect if glycosyl transfer is rate limiting. However, no secondary α-deuterium kinetic isotope effect could be detected in the reaction of α-D-[^{2}H]glucose-1-P (Firsov et al., 1974). A step other than glycosyl transfer may limit the rate.

Structure

The many structures available for phosphorylase, determined beginning in 1976 (Fletterick et al., 1976a, 1976b; Johnson et al., 1977) convey a clear picture of regulatory aspects, and they show large structural transitions accompanying many ligand-binding processes. The structures show the AMP binding site near the phosphorylation site at Ser14 at the interface of subunits. Comparisons of the relaxed R states show the similarities between phosphorylase a and complexes of phosphorylase b with AMP, as well as the differences from the taut T state of phosphorylase b.

The structures have been less informative about the mechanism of glucosyl transfer (Davies et al., 1997). Certain structures show that the 5'-phosphate of PLP is near the phosphate binding site, an important fact in considering the role of PLP. An important structure is that of phosphorylase b with heptulose-2-P and maltoheptaose at the active site and AMP at the allosteric site (Johnson et al., 1990). This structure shows a hydrogen-bonded contact between the 5'-phosphate of PLP and the phosphate group of α-D-1-deoxyheptulose-2-P, which arises from the addition of phosphate to heptenitol (fig. 12-8C). This complex of phosphorylase may be regarded as that of an analog of a ternary product complex in the cleavage of an α-glucan by phosphate.

All structures have failed to reveal the presence of a nucleophilic amino acid side chain as a candidate for mediating glucosyl transfer by a double-displacement mechanism such as that in fig. 12-4A. The absence of a candidate nucleophilic catalyst in phosphorylase prompted the proposition of an S_Ni mechanism, in which an intermediate oxacarbenium ion is stabilized by in-plane contact with the phosphate anion and serves as a transient intermediate (Palm et al., 1990).

A structure of phosphorylase in complex with nojirimycin tetrazole and phosphate is shown in fig. 12-9 (Mitchell et al., 1996). This structure lacks an oligosaccharide in the glycogen binding site, but it should be otherwise related to the putative transition state. As in all other structures, no candidate nucleophilic group of the enzyme is found near the β-face of the inhibitor. However, it has been noted that the peptide carbonyl group of His377 is in position to interact with C1 of the inhibitor on its β-face. Catalysis of glucosyl transfer by a peptide carbonyl could follow the course outlined in fig. 12-3 for neighboring group participation in glycosyl transfer by the 2-acetamido group. This would be an instance of a main chain group oriented through binding interactions to function by a mechanism established for nonenzymatic intramolecular catalysis.

Pyridoxal-5'-Phosphate and the Mechanism of Glycosyl Transfer

PLP is bonded as an internal aldimine to Lys680 of rabbit muscle phosphorylase and is an essential cofactor (Madsen and Withers, 1986; Palm et al., 1990). Extensive biochemical and structural work proves that PLP does not act to stabilize a carbanion or labilize an acidic proton, but the phosphate group is placed near the glycosyl transfer site and is essential for activity. The following facts lead to this conclusion:

- Reduction of the internal PLP-aldimine by borohydride does not inactivate phosphorylase (Fischer et al., 1958).
- The 4'-CHO group of PLP is required for activation, presumably to bond with Lys680 (Graves and Wang, 1970).
- Pyridoxal alone cannot activate apo-phosphorylase, but it activates in the presence of fluorophosphate or phosphite ions (Parrish et al., 1977).
- Pyridoxal-5'-diphosphoryl-α-1-D-glucose binds to apo-phosphorylase and reacts to donate the α-D-glucosyl moiety to the nonreducing end of glycogen (Withers et al., 1981).
- The structures of phosphorylase show the phosphate group of PLP near the glycosyl transfer site.
- The phosphonate analogs of PLP shown in fig. 12-9 activate phosphorylase (Yan et al., 1979 ; Stirtan and Withers, 1996).

In-depth studies show that the 5'-phosphate must be dianionic in active phosphorylase. Evidence includes the fact that pyridoxal-5'-sulfate, with monoanionic sulfate, does not activate phosphorylase. Furthermore, ^{31}P-NMR data indicate that the 5'-phosphate of PLP is monoprotonated and therefore monoanionic in unactivated phosphorylase b, whereas in phosphorylase a and AMP-activated phosphorylase b it is dianionic (Feldman and Hull, 1977; Höri et al., 1979).

Mechanism of Action

Despite many in-depth studies over more than fifty years, a consensus has not emerged regarding the mechanism of action of phosphorylase. The outstanding issues include the nature of the transition state, the mechanism by which glucosyl transfer proceeds with retention of configuration at the anomeric carbon, and the exact role of the 5'-phosphate of PLP. A loose transition state with oxacarbenium ion character is generally assumed to be involved, but this is based on expectations inspired by knowledge of other glycosyltransferases and the nonenzymatic mechanism of glycosyl transfer, not on evidence from studies of phosphorylases. There are competing theories of the mechanism of glucosyl transfer with stereochemical retention and of the role of PLP.

In one theory, phosphate plays three roles, as acceptor, as acid-base catalyst, and as electrostatic, in-plane stabilizer of an oxacarbenium ion-intermediate (Palm et al., 1990). In this mechanism, the 5'-phosphate of PLP serves to shuttle protons to phosphate; as such, the 5'-phosphate functions in both its monoprotonated and unprotonated forms. This theory has merit in that it explains everything and is consistent with the structure of the product complex of phosphorylase b, α-D-1-deoxyheptulose-2-P, maltoheptaose, and AMP (Johnson et al., 1990). It has been criticized as inconsistent with the fact that phosphorylase activated by pyridoxal and fluorophosphates or phosphite and by the pyridoxal-5'-phosphonates in

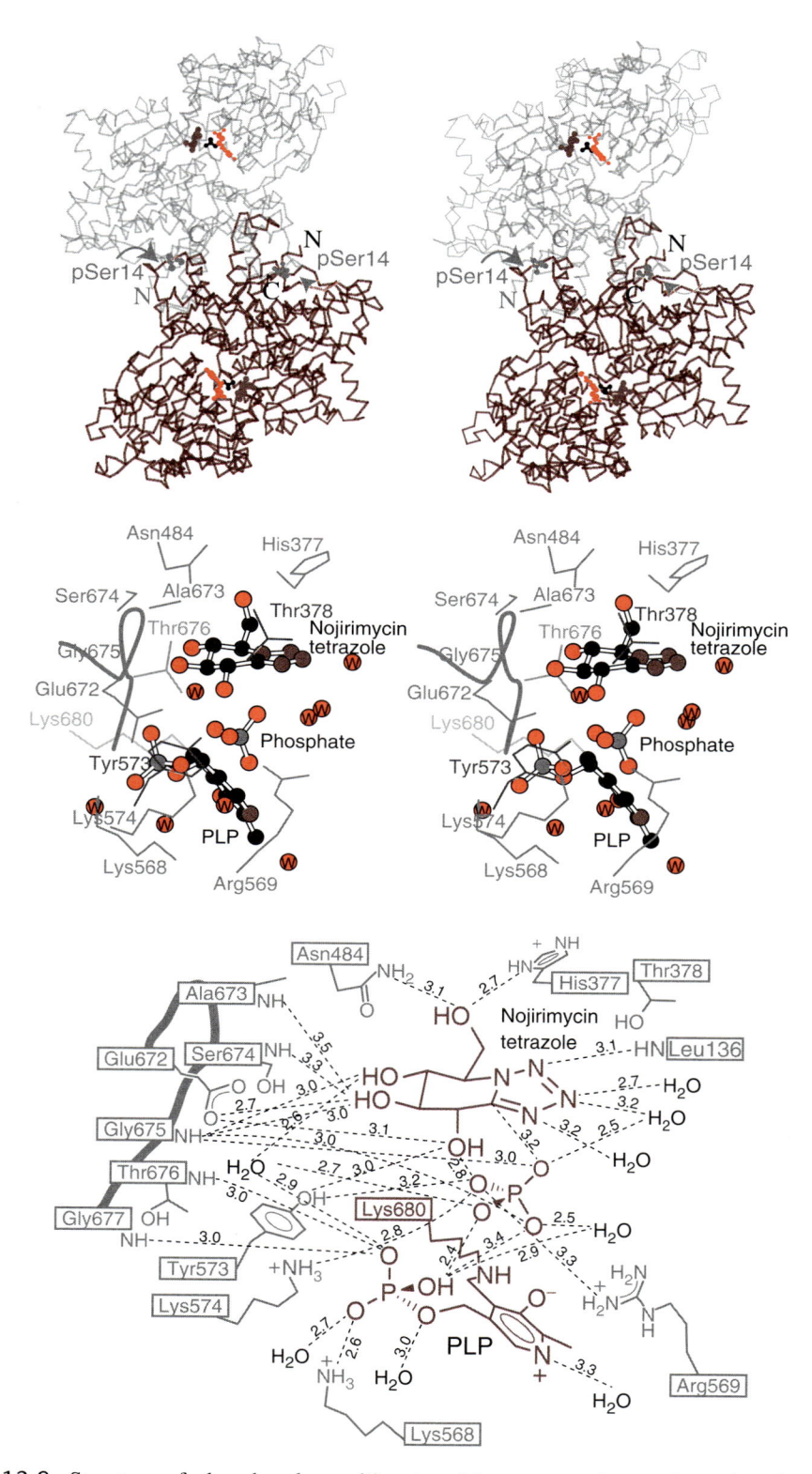

Fig. 12-9. Structure of phosphorylase with a transition-state analog at the active site. The x-ray crystallographic structure of rabbit glycogen phosphorylase with nojirimycin tetrazole (brown), pyridoxal-5′-phosphate (PLP) (red), and phosphate (black) bound at the active site (2.5-Å resolution; PDB 1NOI; Mitchell et al., 1996). This is allosterically regulated by various mechanisms, including activation by phosphorylation of Ser14 (shown phosphorylated in gray in the top panel). The middle panel shows the active site in stereo, and the bottom panel shows the active site in two dimensions, with interatomic distances given in angstroms.

584 Enzymatic Reaction Mechanisms

fig. 12-10 display the same pH-rate behavior despite the great differences in the pK_a values (range, 4.2 to 7.2) (Davies et al., 1997; Stirtan and Withers, 1996). These objections would be muted if proton transfer is not rate limiting.

An alternative theory holds that stereochemical retention must arise through covalent bonding between the glucosyl group and an enzymatic nucleophile yet to be identified, and that the unprotonated, dianionic 5'-phosphate of PLP functions by ground state destabilization to facilitate the action of phosphate in cleaving the glucosidic bond (Davies et al., 1997). This theory has merit in that only the dianionic 5'-phosphate is observed in active forms of phosphorylase, and all other retaining glycosyltransferases function by the double-displacement mechanism in fig. 12-4A.

The mechanism of action of phosphorylase is not understood. Much more work will be required to unveil this mechanism.

Purine Nucleoside Phosphorylase

Many enzymes catalyze ribosyl transfer in the hydrolysis, phosphorolysis, and biosynthesis of nucleotides and nucleosides. Those subjected to intensive mechanistic analysis include AMP nucleosidase (Mentch et al., 1987), nucleoside hydrolase, orotate phosphoribosyl transferase, NAD-dependent ADP-ribosyl transferases, and purine nucleoside phosphorylase (EC 2.4.2.1) (Schramm, 1998). The transition states for ribosyl transfer deduced from the magnitudes of intrinsic kinetic isotope effects are similar, albeit not identical. However, they indicate oxacarbenium ion–like character at the transition states, with a variable degree of bond breaking in the leaving group. In this chapter, we focus on the purine nucleoside phosphorylase (PNP) as a case study.

Reaction, Stereochemistry, and Kinetics

PNP in mammals participates in the maintenance of balance among purine nucleosides and deoxynucleosides. PNP deficiency is associated with T-cell deficiency. T-cell action is implicated in T-cell lymphoma, rheumatoid arthritis, lupus, psoriasis, and tissue rejection in transplantation. Research on PNP has been stimulated by the need for specific inhibitors that might be useful in controlling these disorders. The focus of research is on defining the transition state and designing and synthesizing transition-state analogs for possible use in therapy. Mammalian PNP is a homodimeric enzyme with a subunit molecular mass of 40 kDa.

PNP accepts inosine, guanosine, and 2'-deoxyguanosine as substrates and catalyzes the phosphorolysis of inosine according to fig. 12-11. Phosphorolysis takes place with inversion

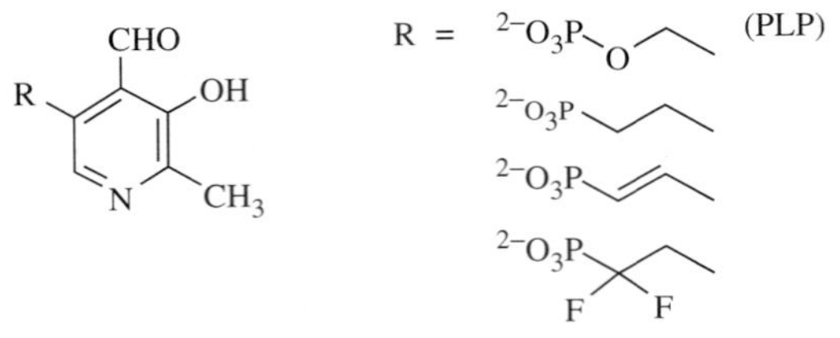

Fig. 12-10. Structures of pyridoxal-5'-phosphate (PLP) phosphonates that activate phosphorylase. The phosphonate, *trans*-vinylphosphonate, and difluorophosphonate analogs of PLP activate phosphorylase. Phosphorylase activity depends on the phosphate or phosphonate groups being dianionic, as indicated by ^{31}P NMR spectra and the absence of activity with pyridoxal-5'-sulfate.

of configuration at the anomeric carbon (C1′) to produce α-D-ribose-1-P and hypoxanthine. PNP is an inverting enzyme and functions in the steady state by a sequential kinetic mechanism involving the intermediate formation and interconversion of ternary complexes.

Isotope Effects and the Transition State

The measurement of isotope effects requires conditions under which ribosyl transfer is rate limiting. In the general case, this may require the use of a substrate analog, or conducting the reaction at a pH well removed from the optimum, or the use of a mutated form of the enzyme, or any of several other strategies for ensuring that the chemical step of interest limits the rate. In the case of PNP, arsenolysis and hydrolysis proceed with ribosyl transfer rate limiting, so the kinetic isotope effects for these reactions can be measured (Klein and Schramm, 1995).

The kinetic isotope effects measured in ribosyl transfer reactions include the primary ^{15}N and ^{14}C effects for cleavage of the C1′—N9 bond, the α-^{3}H secondary effect for the change in hybridization in the C1′—H bond, and the β-^{3}H secondary effect from the C2′—H bond. Secondary ^{3}H effects are also observed from the C5′—H bonds that are attributed to the orientation of C5′—H interacting with the oxocarbenium center in the transition state (Kline and Schramm, 1995). The ^{14}C and ^{3}H effects measured radiochemically, like ^{13}C effects measured mass spectrometrically at or near natural abundance, and are effects on k_{cat}/K_m.

Because the kinetic isotope effects report on bonding differences between the ground and transition states, their magnitudes can be used to model bonding at the transition state.

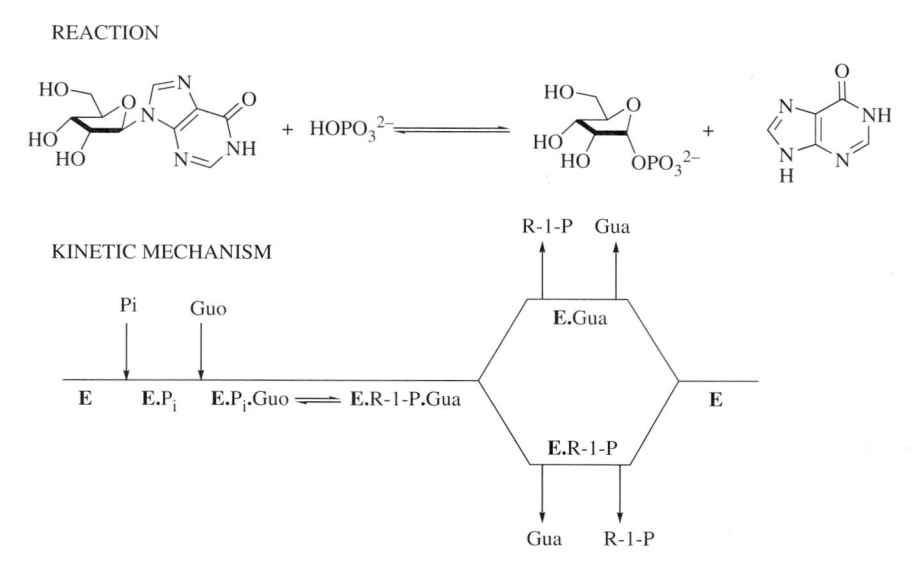

Fig. 12-11. Reaction and kinetic mechanism of purine nucleoside phosphorylase. Inosine and guanosine are the best substrates for purine nucleoside phosphorylase, and the reaction of phosphate with inosine to produce ribose-1-P and hypoxanthine is shown at the top. The kinetics is complex, in that double-reciprocal plots are often curved. A multifaceted study of the calf spleen enzyme under conditions in which double-reciprocal plots are linear indicates that the kinetic mechanism is random sequential in the direction of nucleoside synthesis and ordered in the direction of nucleoside cleavage (Porter, 1992). This mechanism is illustrated (bottom) by the reaction of guanosine (Guo) to produce ribose-1-P (R-1-P) and guanine (Gua).

The computational methods are described in reviews and references therein (Schramm, 1998). The measured kinetic isotope effects in the acid-catalyzed hydrolysis of AMP and arsenolysis of inosine by PNP are given in table 12-1. The large values of the ^{15}N kinetic isotope effects reveal that although the transition states are significantly different, the bond between the anomeric carbon (C1′) and N9 of the base is substantially cleaved in both transition states. The modeled transition states show little bonding to water or arsenate, the respective acceptors. The transition states are largely oxacarbenium ion–like. Modeling indicates that in acid-catalyzed hydrolysis the oxacarbenium ion is likely a discrete intermediate, and in the PNP-catalyzed arsenolysis there is significant residual bonding to hypoxanthine, about 40% of a bond, but little new bonding to arsenate. The oxacarbenium ion is substantially formed in the transition state, and there is little bonding to the acceptor molecule (Schramm, 1998).

The drawings in figs. 12-12A and B illustrate the structures of Michaelis and product complexes, with distances between atoms, as determined by x-ray crystallography. The drawing in fig. 12-12C illustrates the corresponding distances in the transition state as determined from the isotope effects. The results show that the main structural change occurring in the transition state is the motion of ribosyl-C1′, which moves away from hypoxanthine-N9 and toward phosphate, the acceptor, in passing through the transition state to the product (Schramm and Shi, 2001).

Immucillin-H: A Transition-State Analog

The positive electrostatic charge of the oxacarbenium ion significantly defines its chemistry. Any design of a transition-state analog must include this charge. Ideally, a transition-state analog would include an elongated bond between C1′ and N9 of hypoxanthine. Immucillin-H, shown in fig. 12-12D, incorporates a positive charge and a slightly elongated C1′—C9 bond in place of the C1′—N9 ribosidic bond. Replacement of the ring oxygen of ribose with nitrogen transforms the ring into a secondary amine, which is protonated and positively charged in neutral solutions. Immucillin-H is an extremely potent inhibitor of PMP displaying an inhibition constant of 20 pM (Kicska et al., 2001).

The structure of a complex of immucillin-H and PNP is shown in fig. 12-13 (Shi et al., 2001). This structure, when compared with those of the product complex and a model of the Michaelis complex indicates that numerous hydrogen bonds are tightened and new ones created in the transition state. Multiple hydrogen bonding interactions appear to stabilize the transition state.

Table 12-1. Kinetic Isotope Effects in Reactions of Purine Nucleosides

Reaction	Isotope Effect [a]				
	$^{15}(V/K)$	$^{14}(V/K)$	$^{\alpha T}(V/K)$	$^{\beta T}(V/K)$	$^{5T}(V/K)$
Acid hydrolysis [b]	1.030	1.044	1.216	1.077	ND
PNP-catalyzed phosphorolysis [c]	1.010	1.026	1.141	1.152	1.033

ND, no data; PNP, purine nucleoside phosphorylase; $^{15}(V/K)$, [9-^{15}N]inosine; $^{\alpha T}(V/K)$, [1′-^{3}H] inosine; $^{14}(V/K)$, [1′-^{14}C] inosine; $^{\beta T}(V/K)$, [2′-^{3}H] inosine; $^{5T}(V/K)$, [5′-^{3}H] inosine.
[a] Isotope effects on V/K refer to k_{cat}/K_m.
[b] Acid-catalyzed hydrolysis of labeled AMP (Mentch et al, 1987).
[c] Arsenolysis of labeled inosine by purine nucleoside phosphorylase (Kline et al, 1995).

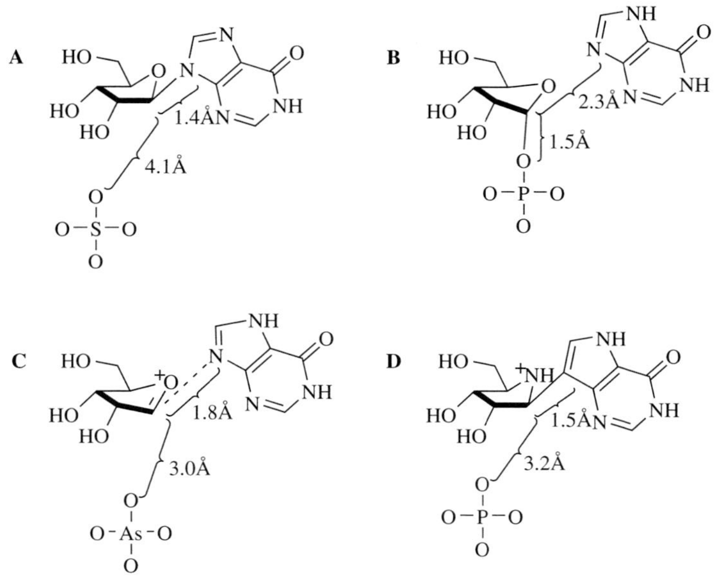

Fig. 12-12. Structures of substrate, product, and transition-state complexes at the active site of purine nucleoside phosphorylase. (A and B) The structure of the sulfate analog of the Michaelis complex at the active site of purine nucleoside phosphorylase (PNP) is compared in A with that of the product complex in B. In the product, the anomeric carbon is displaced downward relative to the Michaelis complex, and it is bonded to phosphate. The structures were determined by x-ray crystallography (Fedorov et al., 2001). (C) The structure is deduced from the bonding in the transition state, as determined by the kinetic isotope effects (Kline and Schramm, 1995). (D) The structure is that of the complex of the transition-state analog immucillin-H and phosphate with PNP, as determined by x-ray crystallography (Shi et al., 2001).

Glycosidases

Families and Structures

The hydrolysis of disaccharides, starches, cellulose, bacterial cell wall polysaccharides, chitins, xylans, dextrans, galactans, and complex polysaccharides in general is catalyzed out by glycosidases. The basic chemistry of the hydrolysis of a glycosidic linkage is similar for most glycosides, the main variation being glycosides composed of *N*-acetylaminopyranosides, many of which undergo both nonenzymatic and enzymatic hydrolysis by neighboring group participation (fig. 12-3). Most glycosidases act by the retaining and inverting mechanisms in fig. 12-4, including some acting on *N*-acetylaminoglycosides. Despite the many variations in substrate structure, the chemistry of the hydrolytic process seems to dictate the active site structures of glycosidases, which include two acidic amino acid side chains to catalyze hydrolysis, either by the retaining or inverting mechanism. An essential structural difference in the active sites of these two classes is the separation of carboxylic side chains engaged in catalysis, 5-6 Å in retaining enzymes and 9-10 Å in inverting enzymes (Davies et al., 1997). This pattern of active site structure and function is conserved in a huge variety of glycosidases, which vary greatly in global structure.

The structures and sizes of polysaccharides include small disaccharides and large polysaccharides, both linear and branched, composed of sugars and modified sugars. The glycosidases carry out hydrolysis in a variety of ways, ranging from the cleavage of a disaccharide, to the cleavage of a monosaccharide from either the reducing or nonreducing end of

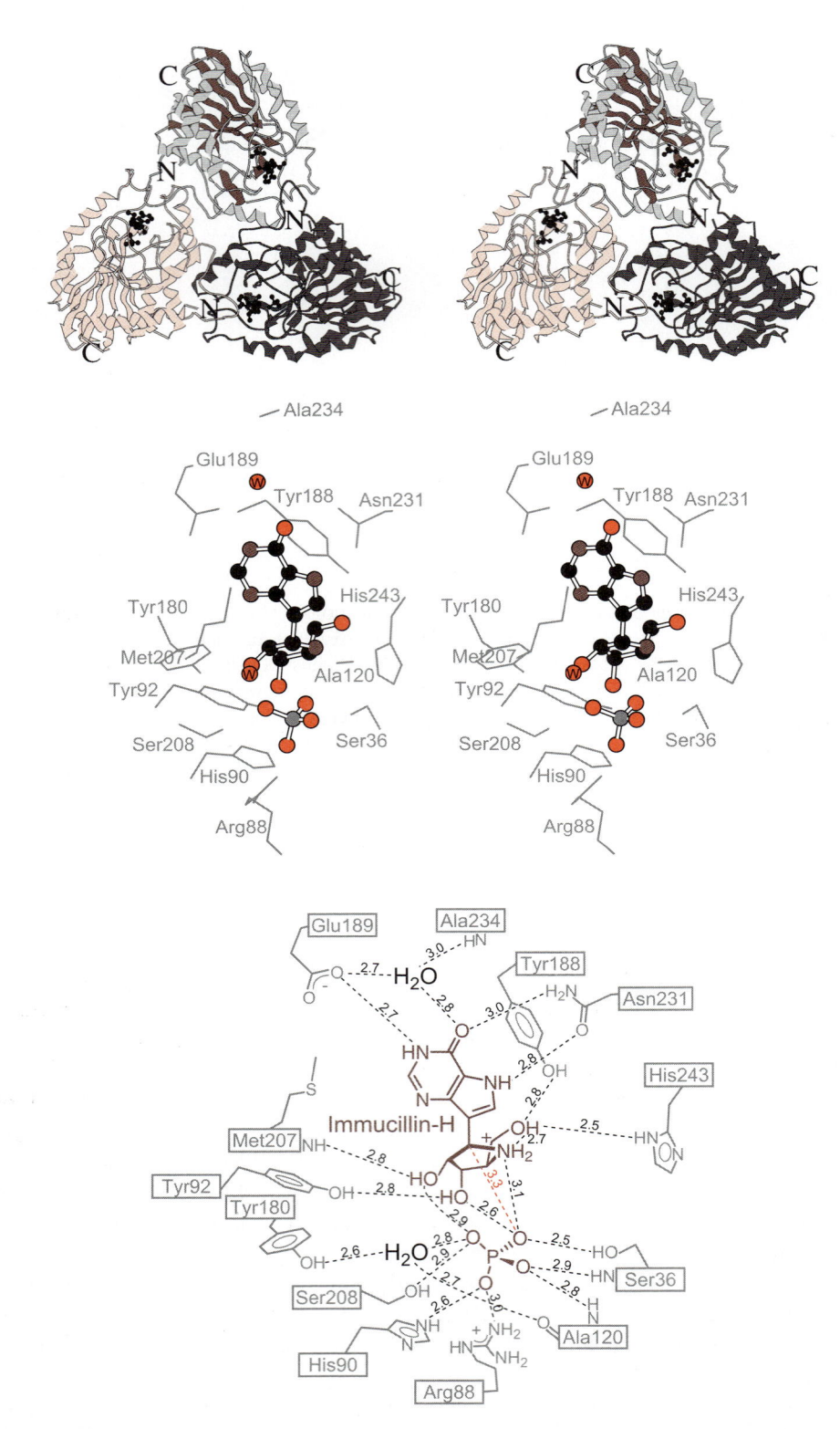

Fig. 12-13. A structure of purine nucleoside phosphorylase (PNP) with a transition-state analog. Shown is an x-ray crystallographic structure of the homotrimeric *Mycobacterium tuberculosis* PNP with immucillin-H and phosphate bound to the active site (1.75-Å resolution; PDB 1G2O; Shi et al., 2001).

a polysaccharide, to the cleavage of polysaccharides at internal sites, to the hydrolysis of glycosidic linkages at branch points. The diversity of structures and hydrolytic modes seem to be matched by the glycosidase structures, which are numerous in terms of amino acid sequences and chain folds. The glycosidases have been organized into a large number of sequence-based families, 87 to date and increasing rapidly with the characterization of genomes (Davies and Henrissat, 2003; Henrissat and Davies, 1997).

Among the 87 families, crystal structures are currently available for members of 41 families, and they range in secondary structure from essentially all helices to essentially all β-sheets (Davies et al., 1997). In view of the similarities in active sites among the retaining glycosidases and among the inverting glycosidases, the structural divergences constitute a lesson on the absence of any relationship between chain fold and active site structure and function. The families of glycosidases admirably exemplify the principle of convergent evolution.

Lysozyme

Reaction and Structure

Lysozymes (EC 3.2.1.17) catalyze the hydrolysis of bacterial cell wall polysaccharides into tetrasaccharides, as illustrated in fig. 12-14. Bacterial cell walls are constructed of cross-linked polysaccharides composed of alternating *N*-acetylglucosamine (NAG) and *N*-acetylmuramic acid (NAM) residues, which are cleaved into tetrasaccharides at *N*-acetylglucosamine sites by the action of lysozymes. Lysozymes are small enzymes, 129 amino acids in the case of hen egg white (HEW) lysozyme, the first enzyme structurally characterized by x-ray crystallography (Blake et al., 1965). Hen egg white lysozyme is in family 22 (retaining) and bacteriophage T4 lysozyme is in family 24 (inverting). The two act by different mechanisms, and lysozymes are found in other families as well. We focus on HEW lysozyme because of the intensity of research on its mechanism of action.

The first structures of HEW lysozyme documented the interactions of NAM-NAG oligomers with the active site and led to the concept of subsites A to F for binding a hexameric substrate, in which hydrolysis occurred between glycosides in the D and E sites

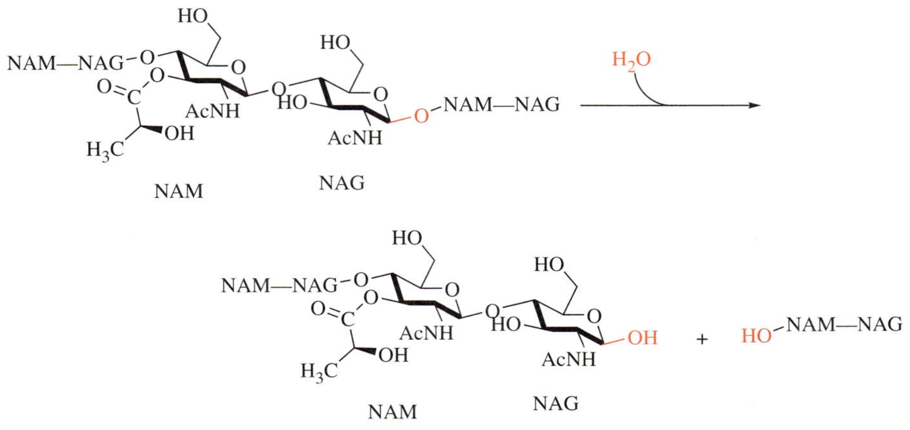

Fig. 12-14. Hydrolysis of an alternating hexaglycoside of *N*-acetylglucosamine (NAG) and *N*-acetylmuramic acid (NAM) residues by hen egg white (HEW) lysozyme. Lysozyme catalyzes the hydrolysis of the hexasaccharide at the −1 position in the D subsite, with overall retention of configuration at the anomeric carbon. The red highlights the reacting groups.

(Imoto et al., 1972). In the modern nomenclature, the saccharide units binding to subsites A to F are numbered −4 to +2, and cleavage occurs between −1 and +1:

	NAM	NAG	NAM	NAG	NAM	NAG
Subsite	A	B	C	D	E	F
Saccharide	−4	−3	−2	−1	+1	+2

Hexameric substrates proved to be cleaved as previously shown and were better as substrates than shorter oligosaccharides (Chipman et al., 1968; Rupley and Gates, 1967).

The structure of HEW lysozyme which is illustrated in fig. 12-15 for the E35Q-variant, revealed the presence of Glu35 and Asp52 in the active site. Chemical modification experiments implicated Glu35 and Asp52 in the reaction mechanism (Hoare and Koshland, 1967; Parsons and Raftery, 1969). The importance of Glu35 and Asp52 has been confirmed by site-directed mutagenesis (Malcolm et al., 1989).

Affinity Labeling by Epoxypropyl N-Acetylglycosides

The β-2,3-epoxypropyl aglycones of di- and tri-N-acetylglucosamine inactivate lysozyme in a time- and concentration-dependent reaction, in which the affinity reagent becomes covalently bonded to the enzyme (Thomas et al., 1969). The reaction is inhibited by the β-methyl glycoside of the di-N-acetylglucosamine, a competitive reversible inhibitor. The structures of lysozyme in complex with the di- and tri-N-acetylglucosamine glycosides (NAG-NAG and NAG-NAG-NAG) indicated that the epoxide group of the affinity labels could contact the active site amino acids Glu35 and Asp52. The label is attached to Asp52 (Eshdat et al., 1973; Moult et al., 1973).

Epoxides can react as alkylating agents, especially under acidic conditions, where acid is available to protonate the epoxide oxygen in the transition state, and the carboxyl groups of Asp or/and Glu can exist as mixtures of —COOH and —COO⁻. The presence of both a carboxylic acid group to serve as an acid and a carboxylate group to react as a nucleophile creates a favorable situation for alkylation at an enzymatic site. A mechanism such as in scheme 12-4 can explain the results with lysozyme.

X-ray crystallographic analysis of affinity-labeled lysozyme and lysozyme-oligosaccharide complexes, varying the structure of the glycoside, and employing site-directed mutagenesis of lysozyme, has allowed the many hydrogen bonded and van der Waals (hydrophobic) contacts in the active site to be defined (Muraki et al., 1999, 2000).

Substrate Distortion

As a retaining glucosaminidase, neighboring group participation by the 2-acetamido group might be expected according to the mechanism as in fig. 12-3, which would explain stereochemical retention. This mechanism was disproved by the observation of the lysozyme-catalyzed hydrolysis of a substrate with 2-deoxyglucose in place of NAG in the scissile position (Raftery and Rand-Meir, 1968; Rand-Meir et al., 1969). Lacking the 2-acetamido group, the 2-deoxyglucosyl residue could not have reacted with neighboring group assistance.

Analysis of the structure with substrate analogs and inhibitors bound at the active site led to the hypothesis that substrate binding was accompanied by steric distortion of the scissile pyranoside ring D (−1) toward a half-chair conformation, that is, toward the conformation of the corresponding oxacarbenium ion (Phillips, 1976). This has been a durable concept, repeatedly confirmed and extended (Strynadka and James, 1991; Radford et al., 1994).

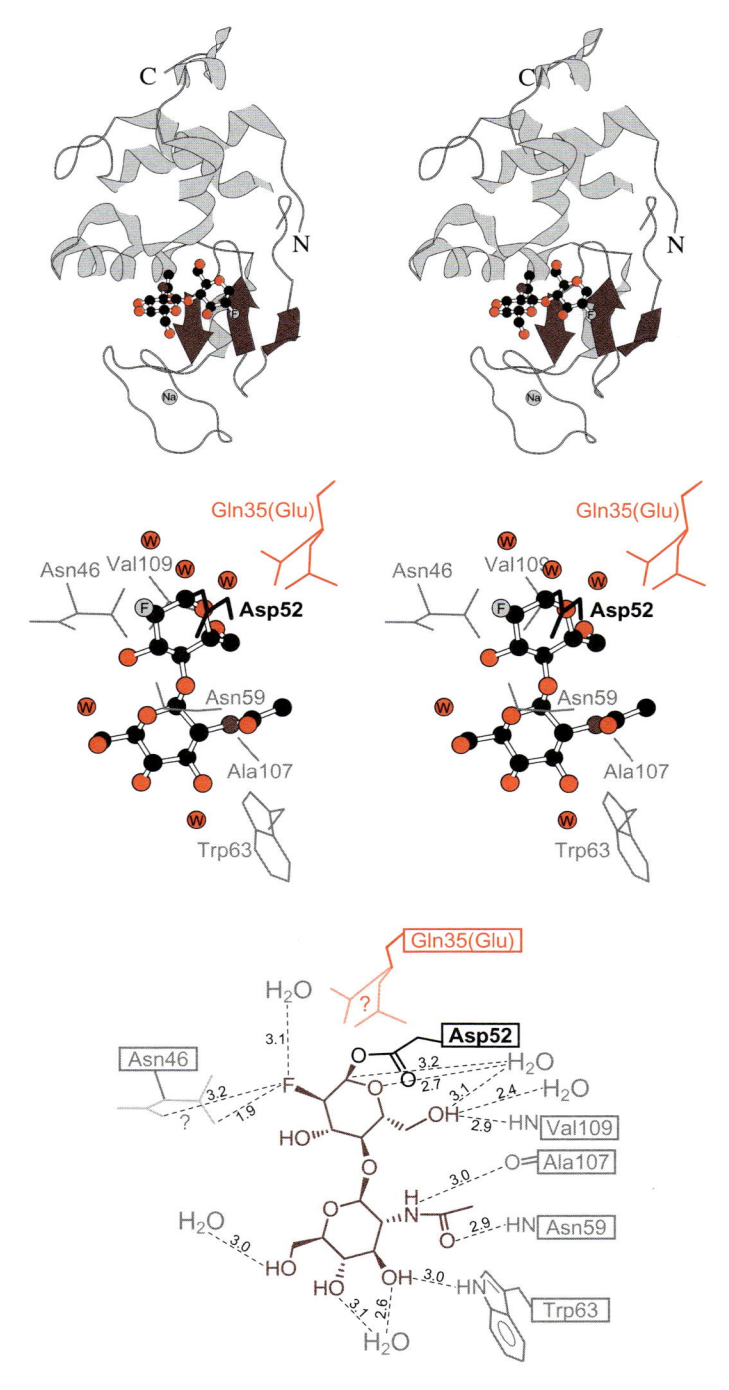

Fig. 12-15. The structure of hen egg white (HEW) lysozyme E35N variant is shown in stereo (1.65-Å resolution; PDB 1H6M; Vocadlo et al., 2001). Lysozyme reacts with glycosyl fluorides, and the enzyme is shown with *N*-acetyl-2-deoxy-2-fluoroglucosaminyl-glucosyl fluoride covalently bonded to Asp52 (bold black). The variant glutamine and Asn46 are each shown in two conformers consistent with the observed electron density.

The mechanism put forward based on this property and the involvement of Glu35 and Asp52 in the action of HEW lysozyme has fared less well.

The mechanism that became widely accepted and appeared in textbooks was inspired by the property of the D-subsite in binding the half-chair conformation of pyranosides and by the presence of Glu35 and Asp52 at the active site. This mechanism is illustrated in fig. 12-16A. In the first step, Glu35 acts as an acid catalyst to protonate the leaving group at its 4-oxygen, facilitating its departure and leading to the formation of a discrete oxacarbenium ion intermediate. Asp52 in its ionized form is postulated to stabilize the intermediate electrostatically by being held in the plane of the oxacarbenium ion, where it cannot from a covalent bond to carbon-1. The leaving group departs the active site, and a molecule of water enters in its place and reacts with the oxacarbenium ion intermediate on its β-face to form the product of retention at the anomeric carbon, which then dissociates from the site.

Isotope and pH Effects

The secondary kinetic isotope effect in the lysozyme-catalyzed hydrolysis of NAG–(β-1,4)-β-phenyl[1-^{2}H]glucopyranoside proved to be $k_H/k_D = 1.11$, a substantial effect of 11% (Dahlquist et al., 1968, 1969). This proved that the sp^3 hybridization of the anomeric carbon in the substrate underwent a change toward sp^2 hybridization in the transition state. It was regarded as proof of the formation of an oxacarbenium ion intermediate in the rate-limiting step of the mechanism in fig. 12-16A.

Detailed studies of the effects of pH on the action of lysozyme yielded bell-shaped pH-k_{cat} profiles, with p$K_1 = 3.9$ and p$K_2 = 6.8$ (Banerjee et al., 1975). Enthalpies of ionization were 0 to 2 kcal mol^{-1}, consistent with carboxylic acids. The ionizations were regarded as those of Asp52 and Glu35. They correlate well with the values of 6.7 and 3.5 for Glu35 and Asp52, respectively, measured by two-dimensional ^{1}H NMR analysis of free lysozyme (Bartik et al., 1994). The differences between the two sets of pK_a values can be attributed to the fact that the NMR values are for free lysozyme, whereas the kinetic values refer to an enzyme-substrate complex.

Glycosyl-Enzyme Intermediate

The classic mechanism in fig. 12-16A came into doubt because of the fact that the lifetime of an oxacarbenium ion is so short (10^{-12} sec) in water, and probably even shorter with an adjacent carboxylate ion. It seemed unlikely that it would persist during the time required for the dissociation of the leaving group and the binding of water in its place. Given the mobility of protein structures, it seemed questionable whether the side chain of Asp52 could be sufficiently immobile to remain in-plane with the oxacarbenium ion and unbonded to the in-flight glycosyl group. Detailed studies of many retaining glycosidases proved that the mechanisms involved glycosyl-enzyme intermediates, with the glycosyl group in flight bonded to carboxylic side chains of acidic amino acids (Davies et al., 1997). The two acidic amino acid side chains at the active sites of these glycosidases were separated by 5 to 6 Å, similar to Glu35 and Asp52 of HEW lysozyme. These facts raised questions about HEW lysozyme and whether it constituted an exception.

Studies of covalent glycosyl-enzyme intermediates in glycosidases has been greatly facilitated by the use fluorinated glycosides that serve as inactivators of retaining glycosidases. The compounds in fig. 12-17 have been employed in these studies (Davies et al., 1997). The flouro-substituents are placed in positions that allow them to serve either as the leaving group from the anomeric carbon, or as electron withdrawing substituents at C2 or C5, in position to stabilize a glycosyl-enzyme against hydrolysis. Another method for

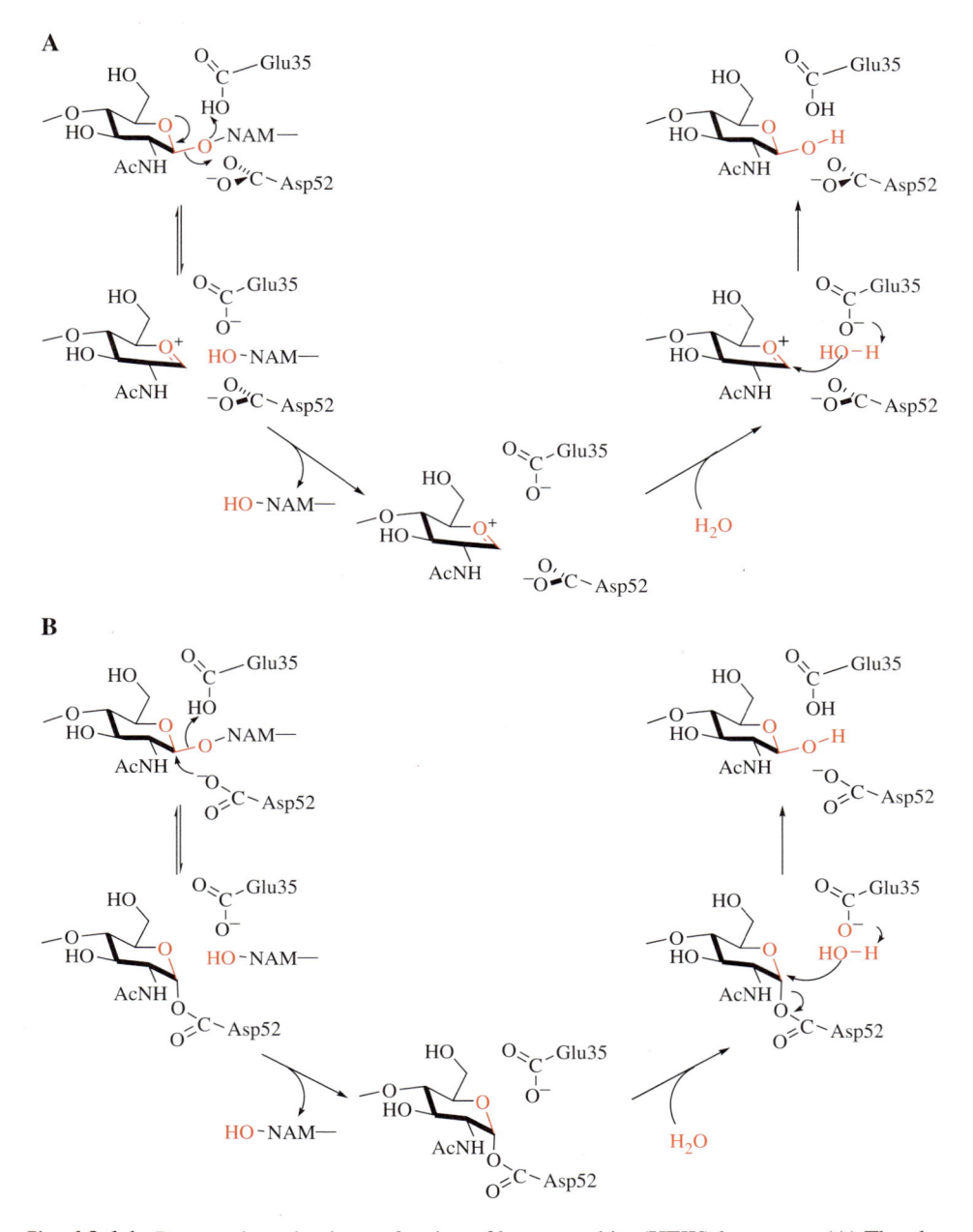

Fig. 12-16. Proposed mechanisms of action of hen egg white (HEW) lysozyme. (A) The classical mechanism involves an oxacarbenium ion intermediate that results from the Glu35-catalyzed departure of the leaving group. Asp52 stabilizes the intermediate during the dissociation of the leaving group and the binding of water in its place. The intermediate then reacts with water to form the product with retention of configuration at the anomeric carbon. (B) In the double-displacement mechanism, Asp52 displaces the leaving group and forms a covalent glycosyl-ester bond with the anomeric carbon. The leaving group departs and is replaced with water, which reacts with the glycosyl ester to displace Asp52 and form the product. In the double-displacement mechanism, overall retention results from double inversion at the anomeric carbon.

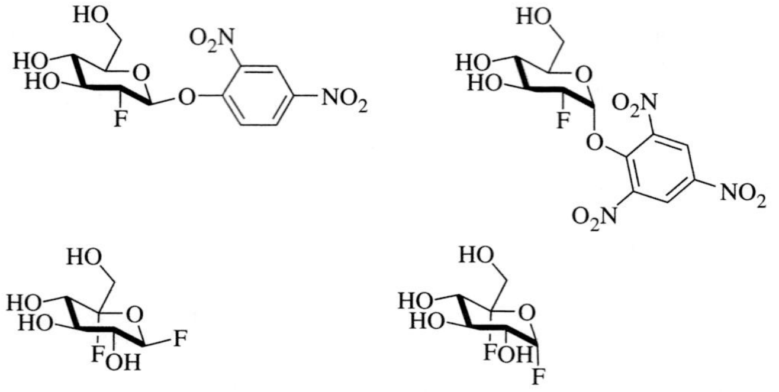

594 Enzymatic Reaction Mechanisms

Fig. 12-17. Structures of fluorinated inactivators of retaining glycosidases.

stabilizing a covalent glycosyl-enzyme is to mutate the acid-base catalyst required to hydrolyze it while employing a substrate with an excellent leaving group that does not require acid catalysis to depart. Both methods have been applied to analyze the action of lysozyme.

Reaction of β-fluoro-NAG$_2$ as a substrate for the mutated E35Q-lysozyme allowed a covalent glycosyl-enzyme to be observed in the steady state by electrospray mass spectrometry (Vocadlo et al., 2001). This intermediate proved to be kinetically competent and reacted too fast to be studied by x-ray crystallography. The intermediate from the reaction of the analogous substrate with a 2-fluoro substituent in place of the 2-acetamido group was sufficiently stable to be observed by x-ray crystallography, and it confirmed the bond linking Asp52 to the α-glycosyl group. It is most likely that the action of lysozyme takes place by a double displacement shown in fig. 12-16B, similar to the typical mechanism in the actions of other retaining glycosidases.

Transition State

The double-displacement mechanism is not inconsistent with the secondary isotope effects, which originally supported the intermediate formation of an oxacarbenium ion intermediate. The transition state for the glycosylation of Asp52 is likely to be trigonal bipyramidal, as shown in fig. 12-18, with little bonding to either the leaving group or attacking Asp52. The ring is in a half-chair conformation in the transition state, with the bonding flattened about the anomeric carbon. This would account for the secondary α-deuterium kinetic isotope effect.

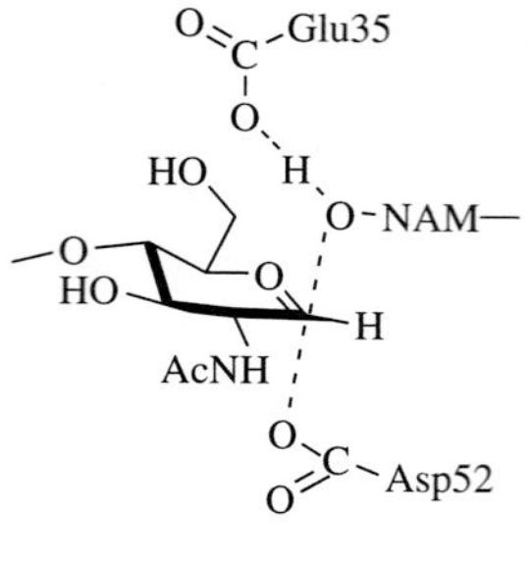

Fig. 12-18. A trigonal bipyramidal transition state for glycosyl transfer in the reaction of hen egg white lysozyme.

T4 Lysozyme

The life cycle of a bacteriophage includes self-replication through the actions of a few enzymes encoded by the phage genome and the recruitment of many bacterial enzymes to this process. The cycle is initiated by the invasion of a bacterium, which requires the penetration of the bacterium by phage particles and concludes with the lysis of the bacterial cell to release the mature phage. Bacterial cell lysis requires the action of a lysozyme to catalyze hydrolysis of the cell wall peptidoglycan.

Like HEW lysozyme, the T4 lysozyme catalyzes the hydrolysis of the alternating NAM–NAG glycan at NAG sites. However, unlike HEW lysozyme, T4 lysozyme is an inverting glycosidase. T4 lysozyme is a family 24 glycosidase with a different structure than that of HEW lysozyme. The structure shows the two catalytic carboxylic acid residues separated by 9.5 Å, typical of inverting enzymes. The reaction mechanism therefore proceeds according to mechanism B in fig. 12-4.

References

Amyes TL and WP Jencks (1989) *J Am Chem Soc* **111**, 7888.
Banerjee SK, E Holler, GP Hess, and JA Rupley (1975) *J Biol Chem* **250**, 4355.
Bartik K, C Redfield, and CM Dobson (1994) *Biophys J* **66**, 1180.
Bennett AJ and ML Sinnott (1986) *J Am Chem Soc* **108**, 7287.
Berti PJ and VL Schramm (1997) *J Am Chem Soc* **119**, 12069.
Blake CCF, DF Koenig, GA Mair, ACT North, DC Phillips, and VR Sarma (1965) *Nature* **206**, 757.
Chipman DM, JJ Pollock, and N Sharon (1968) *J Biol Chem* **243**, 487.
Coutinho PM, E Deleury, GJ Davies, and B Henrissat (2003) *J Mol Biol* **328**, 307.
Craze A, AJ Kirby, and R Osborne (1978) *J Chem Soc Perkin Trans* **2**, 357.
Dahlquist FW, T Rand-Meir, and MA Raftery (1968) *Proc Natl Acad Sci U S A* **61**, 1194.
Dahlquist FW, T Rand-Meir, and MA Raftery (1969) *Biochemistry* **8**, 4214.
Davies G, ML Sinnott, and SG Withers (1997) In Sinnott ML (ed): *Comprehensive Biological Catalysis*, vol I. Academic Press: London, p 119.
Davies GJ and B Henrissat (2002) *Biochem Soc Trans* **30** (Pt 2), 291.
Engers HD, WA Bridger, and NB Madsen (1969) *J Biol Chem* **244**, 5936.
Engers HD, WA Bridger, and NB Madsen (1970a) *Can J Biochem* **48**, 755.
Engers HD, WA Bridger, and NB Madsen (1970b) *Biochemistry* **9**, 3281.
Eshdat Y, JF McKelvy, and NJ Sharon (1973) *J Biol Chem* **248**, 5892.
Fedorov A, W Shi, G Kicska, E Fedorov, PC Tyler, RH Fumeaux, JC Hanson, GJ Gainsford, JZ Larese, VL Schramm, and SC Almo (2001) *Biochemistry* **40**, 853.
Feldman K and WE Hull (1977) *Proc Natl Acad Sci U S A* **74**, 856.
Firsov LM, TL Bogacheva, and SE Bresler (1974) *Eur J Biochem* **42**, 605.
Fischer EH, AB Kent, ER Snyder, and EG Krebs (1958) *J Am Chem Soc* **80**, 2906.
Fitting C and M Doudoroff (1952) *J Biol Chem* **199**, 153.
Gold AM, E Legrand, and GR Sanchez (1971) *J Biol Chem* **246**, 5700.
Graves DJ and JH Wang (1972) In Boyer PD (ed): *The Enzymes*, vol 7, 3rd ed. Academic Press: New York, p 435.
Hadfield AT, DJ Harvey, DB Archer, DA MacKenzie, DJ Jeenes, SE Radford, G Lowe, CM Dobson, and LN Johnson (1994) *J Mol Biol* **243**, 856.
Henrissat B and GJ Davies (1997) *Curr Opin Struct Biol* **7**, 637.
Hoare DG and DE Koshland Jr (1967) *J Biol Chem* **242**, 2447.
Hoerl M, K Feldman, KD Schnackerz, and EJM Helmreich (1979) *Biochemistry* **18**, 2457.
Hu Y, L Chen, S Ha, B Gross, B Falcone, D Walker, M Mokhtarzadeh, and S Walker (2003) *Proc Natl Acad Sci U S A* **100**, 845.
Imoto T, L Johnson, A North, D Phillips, and J Rupley (1972) In Boyer PD (ed): *The Enzymes*, vol 7, 3rd ed. Academic Press: New York, p 665.
Johnson LN, KR Acharya, MD Jordan, and PJ McLaughlin (1990) *J Mol Biol* **211**, 645.
Kaufman (2001) *Proc Natl Acad Sci U S A* **98**, 4593.
Kicska GA, L Long, H Hörig, C Fairchild, PC Tyler, RH Furneaux, VL Schramm, and HL Kaufman (2001) *Proc Natl Acad Sci U S A* **98**, 4593.

Kline PC and VL Schramm (1995) *Biochemistry* **34**, 1153.
Knier BL and WP Jencks (1980) *J Am Chem Soc* **102**, 6789.
Kokesh FC and Y Kakuda (1977) *Biochemistry* **16**, 2467.
Lee JK, AD Bain, and PJ Berti (2004) *J Am Chem Soc* **126**, 3769.
Liras JL and EV Anslyn (1994) *J Am Chem Soc* **116**, 2645.
Malcolm BA, S Rosenberg, MJ Corey, JS Allen, A de Bactselier, and JF Kirsch (1989) *Proc Natl Acad Sci U S A* **86**, 133.
McPhail DR, JR Lee, and B Fraser-Reid (1992) *J Am Chem Soc* **114**, 1905.
Mentch F, DW Parkin, and VL Schramm (1987) *Biochemistry* **26**, 921.
Mieyal JJ and RH Abeles (1972) In Boyer PD (ed): *The Enzymes*, vol 7, 3rd ed. Academic Press: New York, p 515.
Mitchell EP, SG Withers, P Ermert, AT Vasella, EF Garman, NG Oikonomakos, and LN Johnson (1996) *Biochemistry* **35**, 7341.
Moult J, Y Eshdat, and N Sharon (1973) *J Mol Biol* **75**, 1.
Muraki M, K Harata, N Sugita, and K-I Sato (1999) *Biochemistry* **38**, 540.
Muraki M, K Harata, N Sugita, and K-I Sato (2000) *Biochemistry* **39**, 292.
Palm D, HW Klein, R Schinzel, M Buehner, and EJM Helmreich (1990) *Biochemistry* **29**, 1099.
Parrish RF, RJ Uhing, and DJ Graves (1977) *Biochemistry* **16**, 4824.
Parsons SM and MA Raftery (1969) *Biochemistry* **8**, 4199.
Phillips DC (1967) *Proc Natl Acad Sci U S A* **57**, 484.
Piszkiewicz D and TC Bruice (1968) *J Am Chem Soc* **90**, 2156.
Porter DJT (1992) *J Biol Chem* **267**, 7342.
Raftery MA and T Rand-Meir (1968) *Biochemistry* **7**, 3281.
Rand-Meir T, FW Dahlquist, and MA Raftery (1969) *Biochemistry* **8**, 4206.
Rupley JA and V Gates (1967) *Proc Natl Acad Sci U S A* **57**, 496.
Schramm VL (1998) *Annu Rev Biochem* **67**, 693.
Schramm VL and W Shi (2001) *Curr Opin Struct Biol* **11**, 657.
Shi W, LA Basso, DS Santos, PC Tyler, RH Furneaux, JS Blanchard, SC Almo, and VL Schramm (2001) *Biochemistry* **40**, 8204.
Silverstein R, J Voet, D Reed, and RH Abeles (1967) *J Biol Chem* **242**, 1338.
Sprogøe D, LAM van den Broek, O Misra, JS Kastrup, AGJ Voragen, M Gajhede, and LK Skov (2004) *Biochemistry* **43**, 1156.
Stirtan WG and SG Withers (1996) *Biochemistry* **35**, 15057.
Strynadka NCJ and MNG James (1991) *J Mol Biol* **220**, 401.
Thomas EW, JF McKelvy, and N Sharon (1969) *Nature* **222**, 485.
Vocadlo DJ, GJ Davies, R Laine, and SG Withers (2001) *Nature* **412**, 835.
Voet JG and RH Abeles (1970) *J Biol Chem* **245**, 1020.
Withers SG, NB Madsen, BD Sykes, M Takagi, S Shimomura, and T Fukui (1981) *J Biol Chem* **256**, 10759.
Yan SCB, RJ Uhing, RF Panish, DE Metzler, and DJ Graves (1979) *J Biol Chem* **254**, 8263.

13
Nitrogen and Sulfur Transferases

Nitrogen Transfer

Unlike other group transfer reactions in biochemistry, the actions of nitrogen transferring enzymes do not follow a single unifying chemical principle. Nitrogen-transferring enzymes catalyze aminotransfer, amidotransfer, and amidinotransfer. An aminotransferase catalyzes the transfer of the NH_2 group from a primary amine to a ketone or aldehyde. An amidotransferase catalyzes the transfer of the anide-NH_2 group from glutamine to another group. These reactions proceed by polar reaction mechanisms. Aminomutases catalyze 1,2-intramolecular aminotransfer, in which an amino group is inserted into an adjacent C—H bond. The action of lysine 2,3-aminomutase, described in chapter 7, is an example of an aminomutase that functions by a radical reaction mechanism. Tyrosine 2,3-aminomutase also catalyzes the 2,3-amino migration, but it does so by a polar reaction mechanism. In this chapter, we consider NH_2-transferring enzymes that function by polar reaction mechanisms.

Aspartate Aminotransferase

Reaction and Kinetics

Transaminases or aminotransferases are the most extensively studied pyridoxal-5′-phosphate (PLP)–dependent enzymes, and many aminotransferases catalyze essential steps in catabolic and anabolic metabolism. In the classic transaminase reaction, aspartate aminotransferase (AAT) catalyzes the fully reversible reaction of L-aspartate with α-ketoglutarate according to fig. 13-1 to form oxaloacetate and L-glutamate. Like all aminotransferases, AAT is

598 Enzymatic Reaction Mechanisms

PLP dependent, and PLP functions in its classic role of providing a reactive carbonyl group to function in facilitating the cleavage of the α-H of aspartate and the departure of the α-amino group of aspartate for transfer to α-ketoglutarate (Snell, 1962). PLP in the holoenzyme functions in essence to stabilize the α-carbanions of L-aspartate or L-glutamate, the major biological role of PLP discussed in chapter 3. The functional groups of the enzyme catalyze steps in the mechanism, such as the 1,3-prototropic shift of the α-proton to C4′ of pyridoxamine 5′-phosphate (PMP). The steady-state kinetics corresponds to the ping pong bi bi mechanism shown at the bottom of fig. 13-1. This mechanism allows L-aspartate to react with the internal aldimine, **E=PLP** in fig. 13-1, to produce an equivalent of oxaloacetate, with conversion of PLP to PMP at the active site (**E.PMP**), the free, covalently modified enzyme in the ping pong mechanism. The pyridoxamine-enzyme then reacts with α-ketoglutarate to form L-glutamate and regenerate the internal aldimine.

The complete chemical mechanism includes many steps in addition to the binding steps in fig. 13-1 (Kirsch et al., 1984). Other important steps shown in fig. 13-2 include transaldimination between the internal aldimine (**E=PLP**) and L-aspartate to form the external aldimine (**E.PLP=Asp**), abstraction of the α-proton by Lys258 liberated in the transaldimination step to form the quinonoid α-carbanion, transfer of the α-proton to C4′ of PLP to form the external ketimine (**E.PMP=OAA**), and hydrolysis of the external ketimine to oxaloacetate (OAA) and the PMP enzyme (**E.PMP**) to conclude the first half-reaction. In the second half-reaction, α-ketoglutarate reacts with the pyridoxamine enzyme intermediate (**E.PLP**) by the same steps in reverse order, formation of the corresponding external ketimine intermediate (**E.PMP=α-KG**), abstraction of a C4′(H) by Lys258 to form

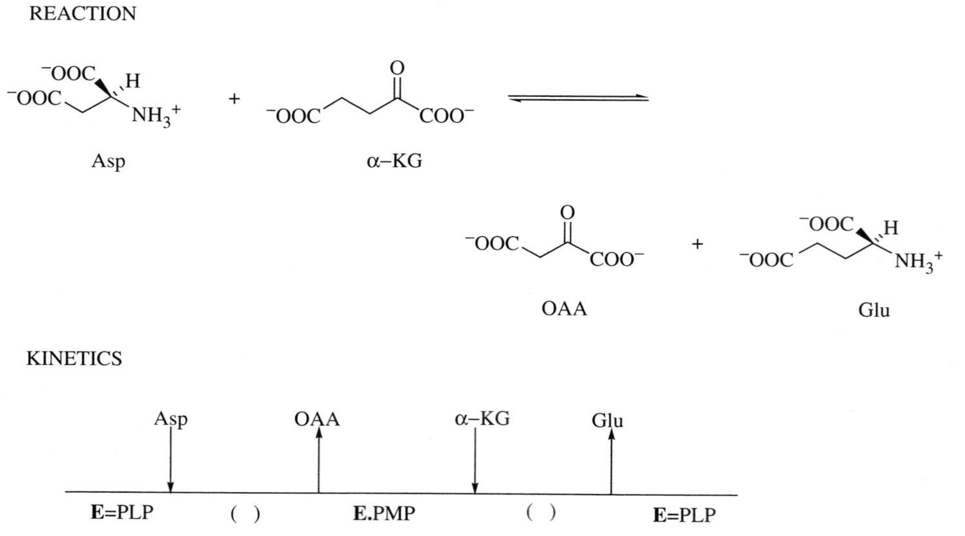

Fig. 13-1. Reaction and kinetics of aspartate aminotransferase (AAT). The reaction catalyzed by AAT is typical of pyridoxal-5′-phosphate (PLP)–dependent transaminases. The amino group of aspartate (Asp) is transferred to the α-carbon of α-ketoglutarate (α-KG), leading to the formation of oxaloacetate (OAA) from aspartate and glutamate (Glu) from α-ketoglutarate. The steady-state kinetics is classic ping pong bi bi, in which the first half-reaction consists of the transfer of the amino group from Asp to PLP in the internal aldimine complex **E=PLP** to produce OAA, which dissociates and leads **E.PMP**, the free enzyme-pyridoxamine phosphate intermediate. α-KG then binds to **E.PMP** and reacts with pyridoxamine phosphate to form Glu and regenerate the internal aldimine **E=PLP**. The parentheses represent intermediates in the conversion of **E=PLP** to **E.PMP** and back to **E=PLP**.

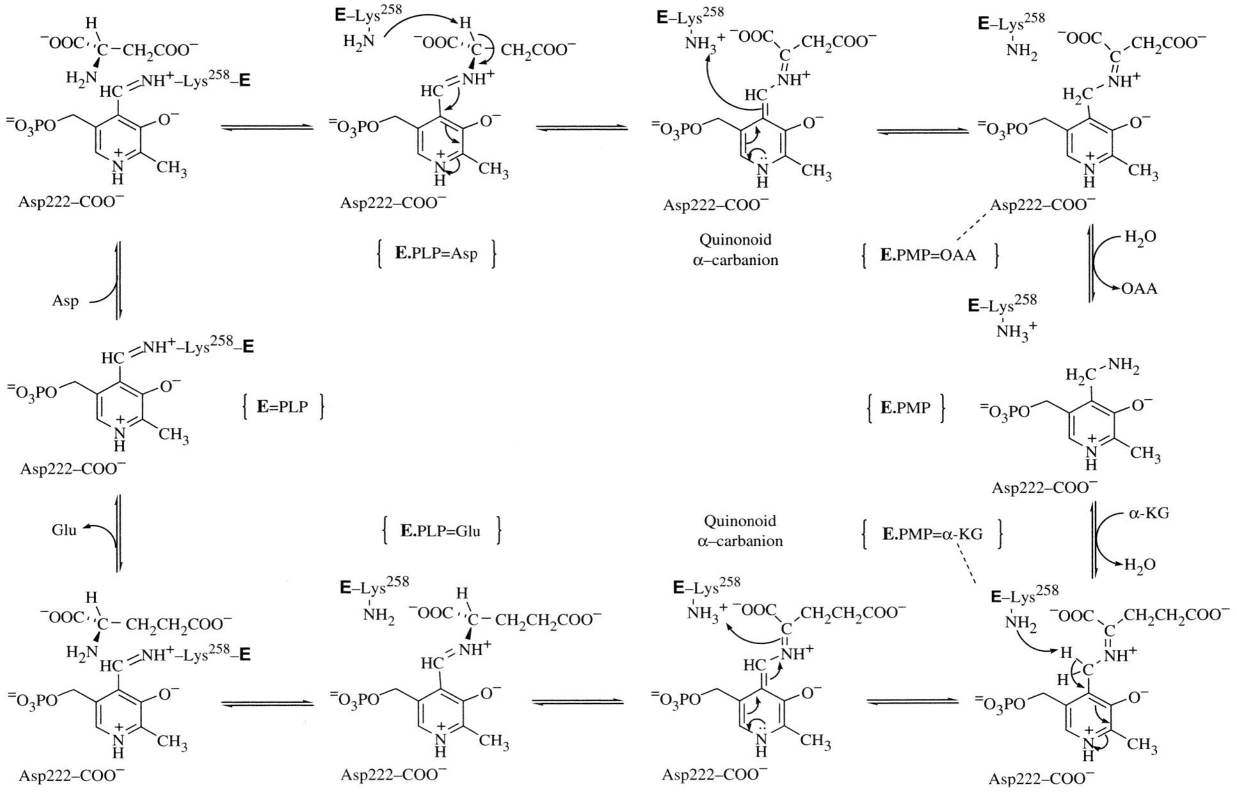

Fig. 13-2. The role of pyridoxal-5′-phosphate (PLP) in the action of aspartate aminotransferase (AAT) is demonstrated in the steps of one cycle of the mechanism of action of AAT. Structures at the center left and center right represent the internal aldimine (**E**=PLP) and the covalently modified enzyme (**E**.PMP), respectively, in the kinetic mechanism of fig. 13-1. The interactions of Lys258 and Asp222 with PLP and pyridoxamine 5′-phosphate (PMP) are also shown. The other structures represent species signified parenthetically in the kinetic scheme. The steps in the reaction of **E**=PLP with L-aspartate to produce oxaloacetate (OAA) and **E**.PMP are shown in clockwise direction at the top. The steps in the reaction of **E**.PMP with α-ketoglutarate to produce L-glutamate are essentially the same steps in reverse order and are shown in clockwise direction at the bottom.

the quinonoid carbanion, proton transfer to the a-carbon of the substrate to form the external aldimine of L-glutamate (**E**.PLP=Glu), transaldimination to the internal aldimine (**E**=PLP) and L-glutamate, and release of L-glutamate to complete the cycle.

Structure

The structures of AATs from a number of sources are available and are closely related (Arnone et al., 1982; Borisov et al., 1980; Ford et al., 1980; Jansonius and Vincent, 1987; Smith et al., 1989). Because of the ease with which the AAT from *Escherichia coli* can be overexpressed, and the ease with which specifically mutated variants can be produced, most of the detailed mechanistic work is on this species. The structure of AAT from chicken mitochondria is shown in fig. 13-3.

The classic PLP-enzymes can be grouped according to chain fold into four families. The PLP-dependent glycogen phosphorylase and lysine aminomutases fall into other families. The largest class, known as the α-family, includes AAT and many other PLP-enzymes, such as dialkylglycine decarboxylase (see chap. 8), cystathionine β-lyase (see chap. 9)

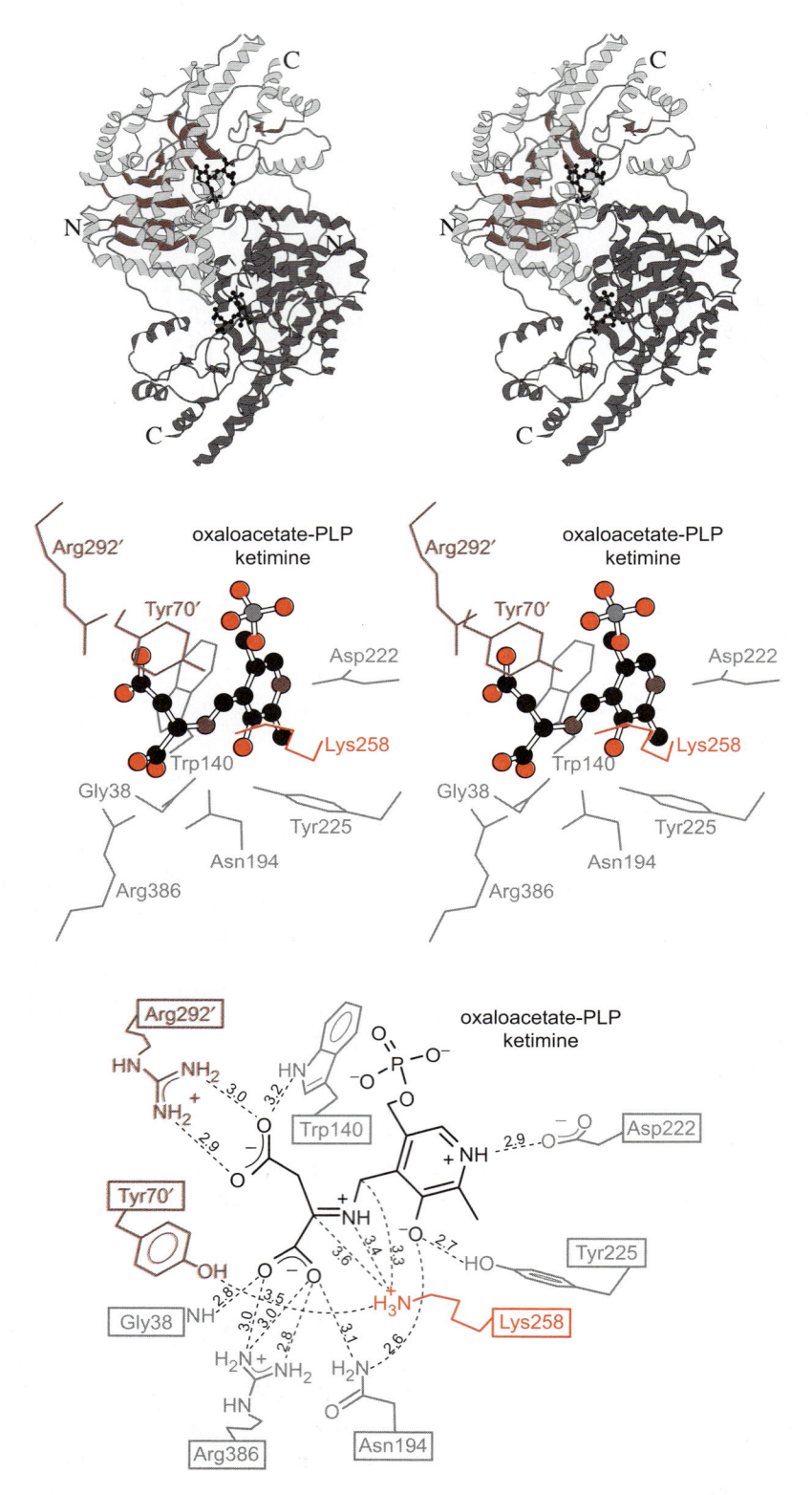

Fig. 13-3. Structure of aspartate aminotransferase. The x-ray crystallographic structure delineates chicken heart, homodimeric mitochondrial aspartate aminotransferase, with the ketimine of oxaloacetate and pyridoxamine-5′-phosphate at the active site shown in black (2.4-Å resolution; PDB 1MAP; Malashkevich et al., 1993). A more detailed stereoview of the active site is shown in the middle panel, and a two-dimensional drawing is provided at the bottom, with distances given in angstroms.

600

(Clausen et al., 1996), tyrosine phenol-lyase (Antson et al., 1993), and other aminotransferases (Hennig et al., 1997; Watanabe et al., 1989). The D-amino acid aminotransferase has a different chain fold (Sugio et al., 1995) and falls into a different family, one that includes the branched chain L-amino acid aminotranferase.

In the structure of AAT, Lys258 binds PLP through imine formation in the internal aldimine, and Asp222 forms hydrogen bonded ionic contact with N1 of the pyridine ring. Homologous residues interact similarly with PLP in all of the α-family enzymes. The β-carboxylate of L-aspartate is engaged in a hydrogen bonded ionic contact with Arg292 and the α-carboxylate with Arg386. Other close contacts include Tyr70, Arg266, and Ser107 with the 5′-phosphate and Tyr225 and Asn194 with the 2′-oxygen of PLP.

Catalytic Mechanism

The mechanism in fig. 13-2 is well supported by kinetic, biochemical, structural, and spectroscopic evidence (Goldberg and Kirsch, 1996; John, 1998; Kirsch et al., 1984). Consider first the evidence that Lys258 plays the dual role of binding PLP in the internal aldimine and serving as the base that mediates the 1,3-prototropic shift. The structure of the internal aldimine proves that Lys258 forms the internal aldimine. The evidence for the action of Lys258 as the base is necessarily less direct but nevertheless compelling. First, Lys258 is well positioned in the structure to serve as the base. Second, mutation of Lys258 to alanine decreases the activity by more than a million-fold in single-turnover experiments (Toney and Kirsch, 1993). Third, the function of K258A-AAT in the 1,3-prototropic shift is restored by exogenously added primary amines (Toney and Kirsch, 1989). This experiment introduced "chemical rescue" into the field of mechanistic analysis. The effectiveness of primary amines in rescuing the activity depends on both the basicity and the size of the amine. The basicity dependence documents the function of the primary amines as bases, and the size dependence documents the relationship between the steric requirements of the rescuing amine and the size of the cavity created in AAT by deletion of the aminopropyl group of Lys258 in K258A-AAT. The Brønsted plot of log of the rate constant corrected for the steric effect against the pK_a can be seen in fig. 1-8. d) In the reaction of L-[2-^{3}H]glutamate, the reverse direction of fig. 13-1, tritium is transferred to the C4′ of PMP in the pro-S position (Gehring, 1984). The stereospecificity implies strict enzymatic control, and a mechanism for controlling the stereochemistry is illustrated in fig. 13-4. The observation of this transfer implicates a stepwise mechanism for the 1,3-prototropic shift, as distinguished from a concerted intramolecular transfer. A stepwise mechanism requires the participation of a base, a role for which Lys258 is ideally positioned. The active site is tightly packed when a substrate is present and it is expected to limit exposure to water, so that proton exchange of the enzymatic base with water should be limited during the prototropic shift. Nevertheless, only 1.5 % of tritium could be found at C4′ of PMP. The most obvious explanation for the low incorporation is that the base is polyprotic and torsiosymmetric, so that tritium abstracted from the α-carbon would be discriminated against for transfer to PMP by a primary kinetic isotope effect and a statistical factor. Lys258 fits the prescription for such a base. On abstracting a tritium, the conjugate acid is —NH$_2$T$^+$, and transfer of a small percentage of tritium occurs in the second step.

A stepwise 1,3-prototropic shift implies the transient existence of the quinonoid α-carbanion intermediate in fig. 13-2. This is the red intermediate with the characteristic absorption band near 490 nm. The stepwise 1,3-prototropic shift implies the participation of a PLP-intermediate, and the quinonoid α-carbanion is the obvious species from a chemical standpoint. Because of its fleeting existence it is difficult to characterize spectroscopically. However, the long wavelength-chromophore of the quinonoid α-carbanion is

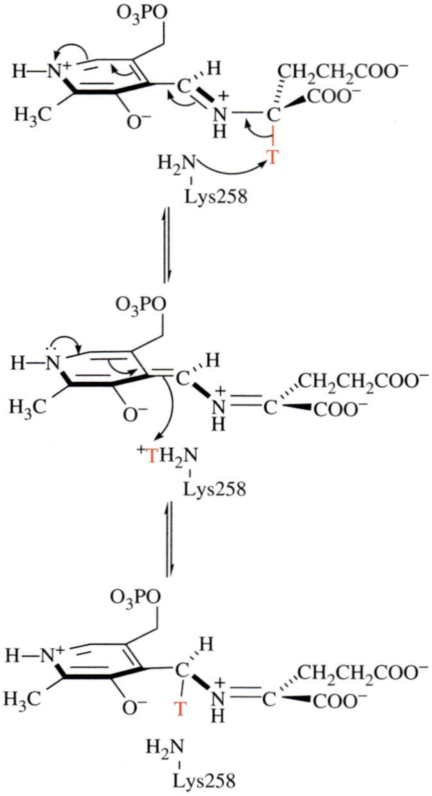

Fig. 13-4. Suprafacial 1,3-prototropic shift from Asp-Cα to pyridoxal-5′-phosphate (PLP)–C4 in the reaction of aspartate aminotransferase (AAT). Tritium at Cα of L-glutamate is transferred to the pro-S position of PLP-C4′ in the action of AAT. Because only 1.5% of tritium is transferred, and the transfer is stereospecific, this result implicates a polyprotic base of AAT and a stepwise mechanism in the 1,3-prototropic shift. If the shift were concerted, 100% of tritium would be transferred. The low yield of the transfer implicates a stepwise process mediated by a polyprotic base, which allows isotope effects and a statistical factor to discriminate against tritium transfer in the intermediate. The only such base in the active site is Lys258.

observed in the reaction of the mutated enzyme K258R-AAT in the presence of L-aspartate or L-glutamate or either of the substrate analogs L-cysteine sulfinate or L-cysteic acid (Toney and Kirsch, 1991). The chromophore appears between 485 and 520 nm, depending on the substrate and at 482 nm with aspartate. The chromophoric intermediates and absorption maxima are shown in fig. 13-5. Although the activity of K258R-AAT is very low, the long-wavelength chromophore dominates the visible absorption spectrum, presumably because the second step of the 1,3-prototropic shift is more highly compromised by mutation of Lys258 to Arg258 than is the first step. The quinonoid α-carbanion also accumulates in the reaction of the substrate analog L-*erythro*-3-hydroxyaspartate (Hayashi and Kagamiyama, 1995; Jenkins, 1961).

Tyrosine 2,3-Aminomutase

β-Amino acids appear in the structures of a number of antibiotics and drugs. Paclitaxel (Taxol), for example, incorporates α-hydroxy-β-phenylalanine, and the β-phenylalanyl group arises from phenylalanine by the action of a 2,3-aminomutase in the Pacific yew

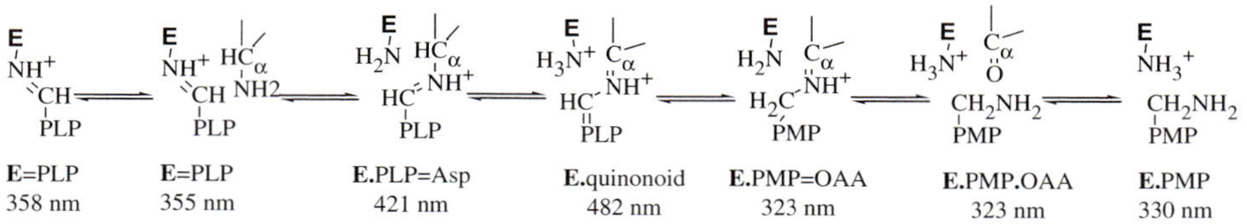

Nitrogen and Sulfur Transferases 603

$$\underset{\underset{\text{PLP}}{|}}{\overset{\text{E}}{\underset{|}{\text{NH}^+}}}\text{CH} \rightleftharpoons \underset{\underset{\text{PLP}}{|}}{\overset{\text{E}}{\underset{|}{\text{NH}^+}}}\overset{\text{HC}_\alpha}{\underset{\text{CH NH2}}{}} \rightleftharpoons \underset{\underset{\text{PLP}}{|}}{\overset{\text{E}}{\underset{|}{\text{H}_2\text{N}}}}\overset{\text{HC}_\alpha}{\underset{\text{HC NH}^+}{}} \rightleftharpoons \underset{\underset{\text{PLP}}{|}}{\overset{\text{E}}{\underset{|}{\text{H}_3\text{N}^+}}}\overset{\text{C}_\alpha}{\underset{\text{HC NH}^+}{}} \rightleftharpoons \underset{\underset{\text{PMP}}{|}}{\overset{\text{E}}{\underset{|}{\text{H}_2\text{N}}}}\overset{\text{C}_\alpha}{\underset{\text{H}_2\text{C NH}^+}{}} \rightleftharpoons \underset{\underset{\text{PMP}}{|}}{\overset{\text{E}}{\underset{|}{\text{H}_3\text{N}^+}}}\overset{\text{C}_\alpha}{\underset{\text{CH}_2\text{NH}_2}{}} \rightleftharpoons \underset{\underset{\text{PMP}}{|}}{\overset{\text{E}}{\underset{|}{\text{NH}_3^+}}}\text{CH}_2\text{NH}_2$$

| E=PLP | E=PLP | E.PLP=Asp | E.quinonoid | E.PMP=OAA | E.PMP.OAA | E.PMP |
| 358 nm | 355 nm | 421 nm | 482 nm | 323 nm | 323 nm | 330 nm |

Fig. 13-5. Structures of aspartate-intermediates in aspartate aminotransferase.

(*Taxus brevifolia*) (Walker and Floss, 1998). β-Tyrosine is the N-terminal amino acid of edeine elaborated by *Bacillus brevis* and arises through the action of tyrosine 2,3-aminomutase (Kurylo-Borowska and Abramsky, 1972). In the isomerization of tyrosine, the α-amino group is formally inserted into a β-C–H bond, and a β-H is formally transferred to the α-carbon. This does not actually occur, and a clue to the true process is the loss of the 3-pro-*S* hydrogen of tyrosine through exchange with solvent protons (Parry and Kurylo-Borowska, 1980)). The equation in fig. 13-6 shows the stereochemistry of aminotransfer and loss of tritium from C3 of tyrosine. Tritium is lost from the 3-pro-*S* position of tyrosine, and the α-amino group migrates with inversion of configuration at C3 to form (*S*)-β-tyrosine (Parry and Kurylo-Borowska, 1980). In contrast, the conversion of phenylalanine to β-phenylalanine by action of phenylalanine 2,3-aminomutase proceeds with retention of configuration at C3 (Walker and Floss, 1998).

An interesting concept for the action of tyrosine 2,3-aminomutase is that the enzyme is similar to phenylalanine ammonia lyase (see chap. 9) but functions without releasing the ammonia derived by β,α-elimination from L-tyrosine. Instead, the ammonia is held at the active site and allowed to undergo re-addition to the transiently formed *p*-hydroxycinnamate, this time to the β-carbon (C3) and in the Michael sense, to form β-tyrosine, as shown in fig. 13-7. According to this concept, tyrosine 2,3-aminomutase should contain the same cofactor as phenylalanine ammonia lyase, 4-methylideneimidazole-5-one (MIO), and this is the case (Christenson et al., 2003a, 2003b). The enzyme from *Streptomyces globisporus* also catalyzes the racemization of (*S*)-β-tyrosine at a slow rate (Christenson et al., 2003a).

The low activity of tyrosine 2,3-aminomutase from *globisporus* ($k_{cat} = 0.01$ s^{-1}) suggests that L-tyrosine may not be the natural substrate. D-Tyrosine is not a substrate, and it is likely that a tyrosyl intermediate in the biosynthesis of the antibiotic C-1027 in *globisporus* is the true substrate for the 2,3-aminomutase (Christenson et al., 2003a).

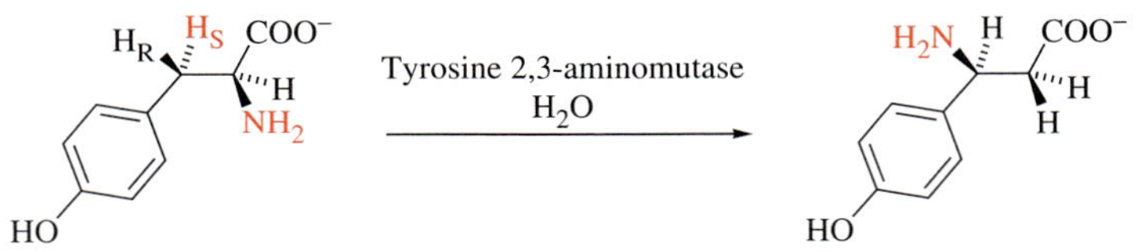

Fig. 13-6. Stereochemistry of the reaction of tyrosine 2,3-aminomutase from *Bacillus brevis*. Tritium-labeling experiments showed that the 3-pro-*S* hydrogen is lost in the reaction and the 3-pro-*R* hydrogen is retained (Parry and Kurylo-Borowska, 1980). The absolute configuration of the product is that of (*S*)-β-tyrosine. Amino group migration occurs with inversion of the configuration at C3.

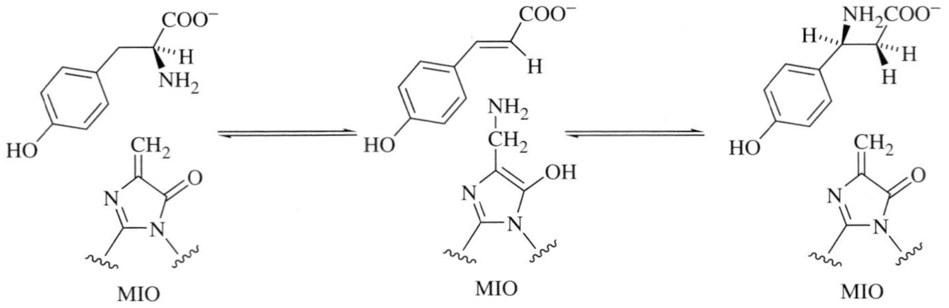

Fig. 13-7. Mechanism for the action of 4-methylideneimidazole-5-one (MIO) in the reaction of tyrosine 2,3-aminomutase. MIO is pictured functioning as an electrophilic center that accepts the amino group of tyrosine in the first step, as in the action of phenylalanine ammonia lyase. Unlike the ammonia lyase, tyrosine 2,3-aminomutase does not normally release *p*-hydroxycinnamate, nor is ammonia normally released from the intermediate amino-MIO. Instead, the amino-MIO transfers the amino group to the β-carbon of *p*-hydroxycinnamate to form β-tyrosine.

Amidotransfer

Glutamine as a Source of NH_3

The members of a distinctive group of enzymes catalyze the transfer of the amide nitrogen of glutamine to another molecule. These enzymes are sometimes known as amidotransferases but more often as synthases or synthetases. Their common feature is the utilization of glutamine as a source of ammonia, which is then channeled into reaction with a co-substrate that culminates in the incorporation of the nitrogen into a product. Ammonia arises from the hydrolysis of glutamine to glutamate and ammonia at an active site in a glutaminase domain of the structure. The glutamate is released, and the ammonia negotiates a guided tour to another site, where it reacts with the co-substrate. A selection of reactions catalyzed by enzymes in this group, among many others, is shown in fig. 13-8.

Glutamine:PRPP amidotransferase produces 5-phosphoribosylamine from 5-phosphoribosyl pyrophosphate (PRPP), after deriving ammonia from the hydrolysis of glutamine (Kim et al., 1996; Krahn et al., 1997). This is the first step in purine biosynthesis. In histidine biosynthesis, imidazole glycerol phosphate synthase catalyzes the reaction of ammonia, derived from glutamine, with N^1-(5′-phosphoribulosyl)-formimino-5-aminoimidazole-4-carboxamide ribotide to produce imidazole glycerol phosphate (IGP) and 5-aminoimidazole carboxamide ribotide (AICAR) (Chaudhuri et al., 2001; Douangamath et al., 2002; Myers et al., 3003; Omi et al., 2002). IGP is a biosynthetic precursor of histidine, and AICAR is an intermediate in purine biosynthesis. Asparagine synthetase uses ammonia from glutamine to produce asparagine from aspartate and ATP (Larsen et al., 1999). The products derived from ATP are AMP and PP_i, implying that β-aspartyl phosphate is the intermediate with which ammonia reacts to displace AMP in asparagine-formation. All of the enzymes in this group incorporate glutaminase domains with similar structures and active sites.

The enzymes catalyzing the reactions in fig. 13-8 share a property in addition to the utilization of glutamine as the source of ammonia. Their structures prevent the dissociation of the ammonia produced in the hydrolysis of glutamine and channel it into reaction with their respective co-substrates (Raushel et al., 2003). This is accomplished by the strategy of providing internal channels within their structures, through which ammonia can migrate from the active sites where it is produced to the physically separate sites at which it reacts with co-substrates. We have seen the use of tunnels (see chap. 3) in the reaction of tryptophan synthase and in the reaction of carbamoyl phosphate synthetase (see chap. 11).

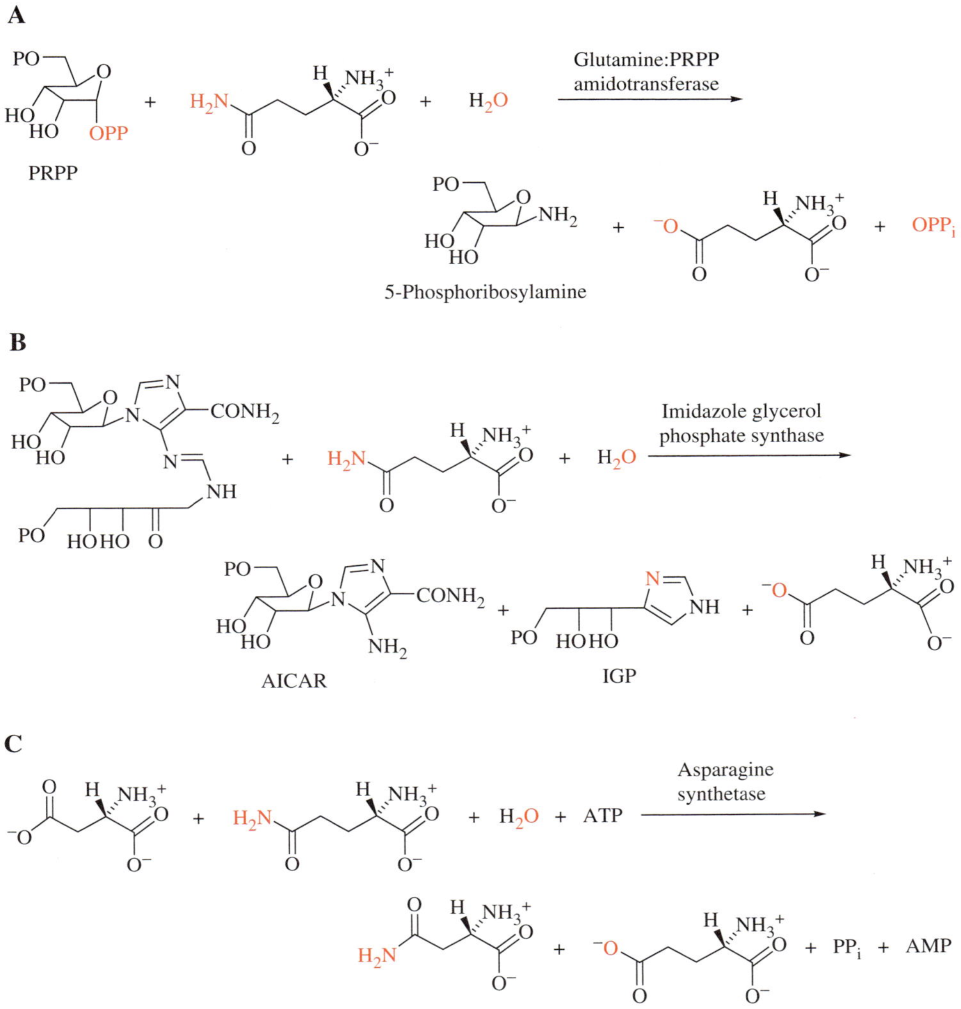

Fig. 13-8. Reactions in which the amide nitrogen of glutamine is transferred to a cosubstrate. The three reactions typify glutamine-dependent amidotransferases, in which the hydrolysis of glutamine produces ammonia as an intermediate that reacts further with incorporation into a co-substrate. (A) Glutamine:PRPP amidotransferase catalyzes the hydrolytic production of ammonia from of glutamine and uses it to displace PP_i from 5-phosphoribosyl pyrophosphate (PRPP) to produce 5-phosphribosylamine. (B) Imidazole glycerol phosphate synthase generates ammonia from glutamine and channels it into the production of aminoimidazole carboxamide ribotide (AICAR) and imidazole glycerol phosphate (IGP). (C) Asparagine synthetase uses the ammonia from glutamine to displace AMP from β-aspartyl phosphate, an intermediate produced by reaction of ATP with aspartate, leading to the formation of asparagine.

The amidotransferases all employ this strategy in preventing the diffusion of ammonia from the enzymes.

Diverse Mechanisms in Reactions of NH_3

The mechanistic and structural similarities among amidotransferases end with the generation of ammonia from glutamine. The structures of the glutaminase domains in these enzymes are similar, but the domains incorporating the active sites at which ammonia reacts with the co-substrates are unique (Raushel et al., 2003). The subsequent reactions of

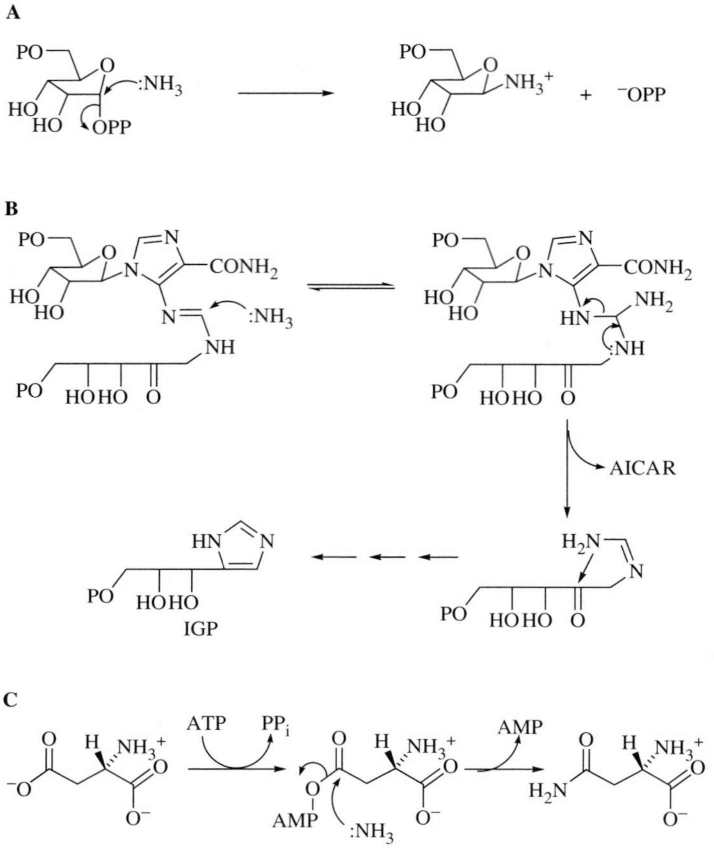

606   Enzymatic Reaction Mechanisms

Fig. 13-9. Variant mechanisms by which ammonia from glutamine may react with co-substrates. Ammonia produced from glutamine in the reactions of fig. 13-8 reacts with co-substrates, presumably by the mechanisms shown. (A) Ammonia is glycosylated by 5-phosphoribosyl pyrophosphate (PRPP) in the action of glutamine:PRPP amidotransferase. (B) Transimination with ammonia produces aminoimidazole carboxamide ribotide (AICAR) and then, in several steps, imidazole glycerol phosphate (IGP) in the action of imidazole glycerol phosphate synthase. (C) Ammonia displaces AMP from β-aspartyl phosphate in the action of asparagine synthetase.

ammonia differ in mechanism with each enzyme. In this respect the amidotransfer enzymes differ from aminotransferases, which all function by similar mechanisms. The mechanistic diversity in the use of ammonia by the three enzymes in fig. 13-8 is illustrated in fig. 13-9. The mechanisms involve glycosyltransfer in the case of glutamine:PRPP amidotransferase, transimination followed by cyclization in the reaction of IGP synthase (Chaudhuri et al., 2003), and acyl group transfer by asparagine synthetase.

Other mechanisms come into play with other members of this group. For example, in the reaction of glucosamine-6-P synthase, ammonia from glutamine reacts with fructose-6-P initially to form an imine, which undergoes isomerization to glucosamine-6-P. Glutamate synthase catalyzes a more complex reaction, in which ammonia from glutamine reacts with α-ketoglutarate to form an imine, which is then reduced by $FMNH_2$ generated by a [3Fe–4S]-mediated electron transfer from NADPH (Binda et al., 2000). The overall process is illustrated in fig. 13-10. We discuss glutamine:PRPP amidotransferase further in the next section.

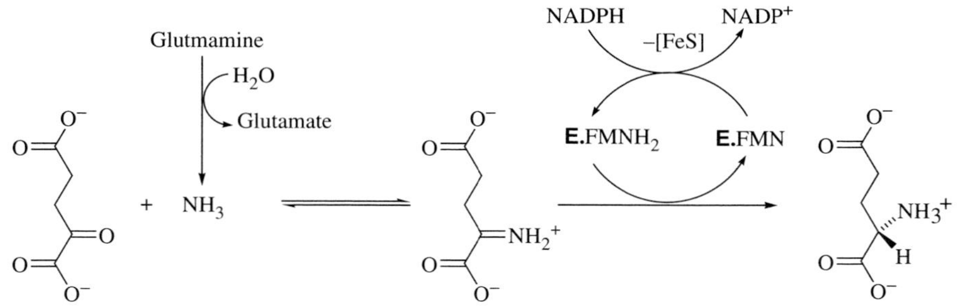

Fig. 13-10. Glutamine as the source of ammonia in the reaction of glutamate synthase. Glutamate synthase from *Azospirillum brasilense* catalyzes the reaction of glutamine with α-ketoglutarate and NADPH to produce two moles of glutamate and one mole of NADP⁺. It is a complex enzyme composed of 150- and 50-kDa subunits. A glutaminase domain in the large subunit generates ammonia, which forms the imine with α-ketoglutarate. Reduction of the imine to L-glutamate by NADPH is mediated by an iron-sulfur cluster and flavin mononucleotide.

Glutamine:PRPP Amidotransferase

The formal name of glutamine:PRPP amidotransferase (EC 2.4.2.14) is 5-phosphoribosylamine:diphosphate phospho-α-D-ribosyltransferase (glutamate amidating). The enzyme catalyzes ribosyl group transfer and not amidotransfer, despite its common name. It catalyzes the production of 5-phosphoribosylamine in the reaction of fig. 13-8A, the first committed step in purine biosynthesis. As such, the enzyme is subject to regulation through end-product inhibition by purine nucleotides. Because of its role in nucleic acid metabolism, this amidotransferase is the most extensively studied member of the group. Glutamine serves as the best source of ammonia, which is generated by a glutaminase domain. Free ammonia also serves less efficiently as a substrate in place of glutamine.

Two Classes of Amidotransferase

The amidotransferase from *Bacillus subtilis* has a [4Fe–4S]-center and an N-terminal propeptide, which is postranslationally removed, whereas the enzyme from *E. coli* does not incorporate either the propeptide or the iron-sulfide center (Chen et al., 1997). These two amidotransferases are the best known representatives of the two classes. The [4fe–4S]-center in the *B. subtilis* amidotransferase has been chemically, spectroscopically, and structurally characterized (Averill et al., 1980; Onate et al., 1989; Vollmer et al., 1983a). It is absolutely required for activity, and disruption of this center by oxidation or site-directed mutation of the cysteine ligands destabilizes the enzyme (Grandoni et al., 1989). The iron-sulfide center is physically distant from the active sites (Smith et al., 1994). In the consensus-view, the iron-sulfide center stabilizes the structure of the active enzyme and does not take part in the chemistry of amidotransferase-action. The enzymes from *B. subtilis* and *E. coli* display similar catalytic and regulatory properties.

Glutaminase Action

All amidotransferases in the group incorporate glutaminase domains that catalyze the hydrolysis of glutamine to glutamate and ammonia. The active sites in the glutaminase domains include essential cysteine residues that serve as nucleophilic catalysts in a hydrolysis mechanism that appears to be analogous to the action of cysteine proteases (see chap. 6). A mechanism for glutaminase-action similar to that in fig. 13-11A appears to pertain to

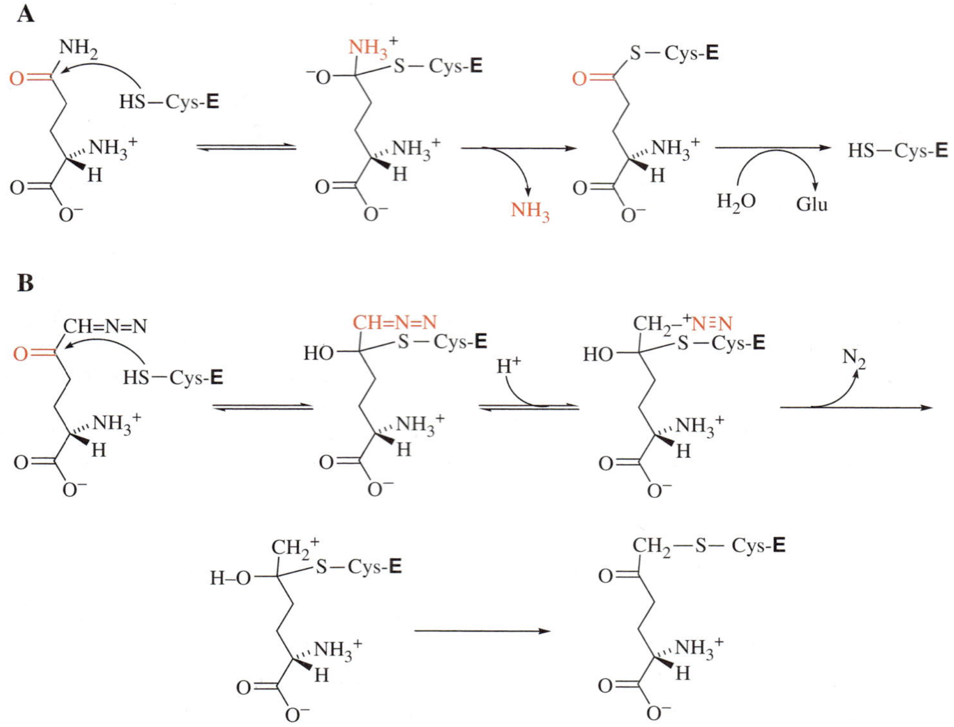

608 Enzymatic Reaction Mechanisms

Fig. 13-11. Mechanisms for glutaminase action and inactivation by diazonorleucine (DON). (A) The mechanism accounts for the role of the essential cysteine residue in glutaminase action. Cysteine functions as a nucleophilic catalyst, leading to ammonia and a covalent γ-glutamyl thioester-enzyme intermediate. Hydrolysis of the γ-glutamyl thioester leads to glutamate. (B) Inactivation of glutaminase by 6-diazo-5-oxo-norleucine (DON) is a suicide mechanism initiated by the addition of the active site cysteine to the carbonyl group of DON, analogous to the first step of the glutaminase mechanism in A. The consequent decomposition of the diazo group through protonation, elimination of N_2, and rearrangement of the resulting carbenium ion leads to the alkylation of cysteine. Reacting groups at each step are shown in red.

the actions of glutamine-dependent amidotransferases in general (Buchanan, 1973). In this mechanism, a specific cysteine residue in the active site undergoes nucleophilic addition to the amide carbonyl group of glutamine, presumably forming a tetrahedral intermediate, which eliminates ammonia to form a covalent acyl-enzyme intermediate, the thioester *S*-γ-glutamyl-enzyme. The *S*-γ-glutamyl-enzyme undergoes hydrolysis to glutamate and the resting enzyme. The ammonia generated in this process then migrates to the co-substrate to complete the transfer reaction.

The N-terminal cysteine residues of glutamine:PRPP amidotransferase from *B. subtilis* and *E. coli* serve as the active site-cysteine residues (Tso et al., 1982: Vollmer et al., 1983b). The N-termini of these enzymes are homologous and constitute the glutaminase active sites. These cysteine residues react with the affinity labeling agent, or active site-directed alkylating agent, 6-diazo-5-oxo-norleucine (DON), a structural analog of glutamine that functions as a suicide inactivator of glutaminase domains. The structure of DON and a likely mechanism by which it alkylates the active site cysteine is shown in fig. 13-11B. Alkyldiazo compounds are generally unstable, tending to eliminate N_2 and generate carbenium ions, which alkylate solvents and are thereby quenched. α-Diazoketones are stable molecules because of resonance delocalization of p-electrons into the conjugated carbonyl group (N=N=CH–C=O ↔ ⁺N–N=CH=C–O⁻). DON is a glutamine analog with a carbonyl group in place of the amido group, and the active site cysteine may undergo addition to the carbonyl group as it does in the glutaminase mechanism (fig. 13-11A). The diazo group would be destabilized in by this process, which abolishes conjugation with the carbonyl group, and the consequent acquisition of a proton on the α-carbon and elimination of N_2 would generate a transient carbenium ion, which could rearrange to alkylate the sulfur of cysteine, as illustrated in fig. 13-11B. The resulting structure is actually observed in crystal structures of glutaminases inactivated by DON.

Reaction of glutamine:PRPP amidotransferases from *B. subtilis* and *E. coli* with DON abolishes activity with glutamine as the source of ammonia. DON is covalently bonded to the inactivated enzymes. The DON-treated enzymes accept free ammonia as a substrate for reaction with PRPP to form 5-phosphoribosylamine (Tso et al., 1982; Vollmer et al., 1983b).

Structure

The structure of glutamine:PRPP amidotransferase *from E. coli* is shown in fig. 13-12 with DON bound at the glutaminase site and a carbacyclic analog of PRPP at the other site (Kim et al., 1996; Krahn et al., 1997). DON presents a convenient means of labeling the glutaminase site, and the chemically stable carbacyclic analog cPRPP allows the PRPP binding site to be identified. Moreover, the binding of cPRPP induces the formation of the tunnel connecting the glutaminase site with the PRPP site.

The N-terminal glutaminase domain (residues 1 to 230) displays the chain fold typical of glutaminase domains among this group of enzymes, two antiparallel β-sheets sandwiched between α-helices. DON is covalently bonded to the N-terminal cysteine. The C-terminal domain (residues 231 to 465) includes a five-stranded β-sheet and binding sites for PRPP and allosteric inhibitors. The tunnel allows ammonia pass from glutamine to displace PP_i from PRPP in its site in the C-terminal domain.

Regulation: End-Product Inhibition

As the first committed step in purine biosynthesis, the production of 5-phosphoribosylamine is the obvious point at which the biosynthesis of purines can be most efficiently regulated. Regulation can take place at two levels, the action of the enzyme in metabolism and the rate of enzyme expression or/and degradation. At the metabolic level, the end products of purine biosynthesis inhibit glutamine:PRPP amidotransferase by an allosteric mechanism. In particular, certain purine nucleotides serve as synergistic inhibitors, synergism being defined as inhibition by two end products that exceeds the sum of the effects exerted by the individual compounds (Chen et al., 1997). Two purine nucleotide binding sites interact to effect synergism of inhibition. ATP binds to the allosteric site A, and GMP binds in competition with PRPP to the catalytic site C. ATP increases the affinity of site C for GMP and vice versa, and this leads to synergism. Purine specificity arises from the hydrogen-bonded contacts of the purine amino group in ATP with a backbone carbonyl (Lys305) and the amino group of GMP with a serine hydroxyl group (Ser347).

Sulfur Transfer

In other chapters, we discuss a number of enzymes that catalyze sulfur transfer. Chapter 1 includes a discussion of acetoacetyl CoA:succinate CoA transferase in connection with the use of remote binding energy to enhance catalytic rates. This reaction takes place by a classic ping pong bi bi kinetic mechanism with the formation of a covalent CoA–enzyme

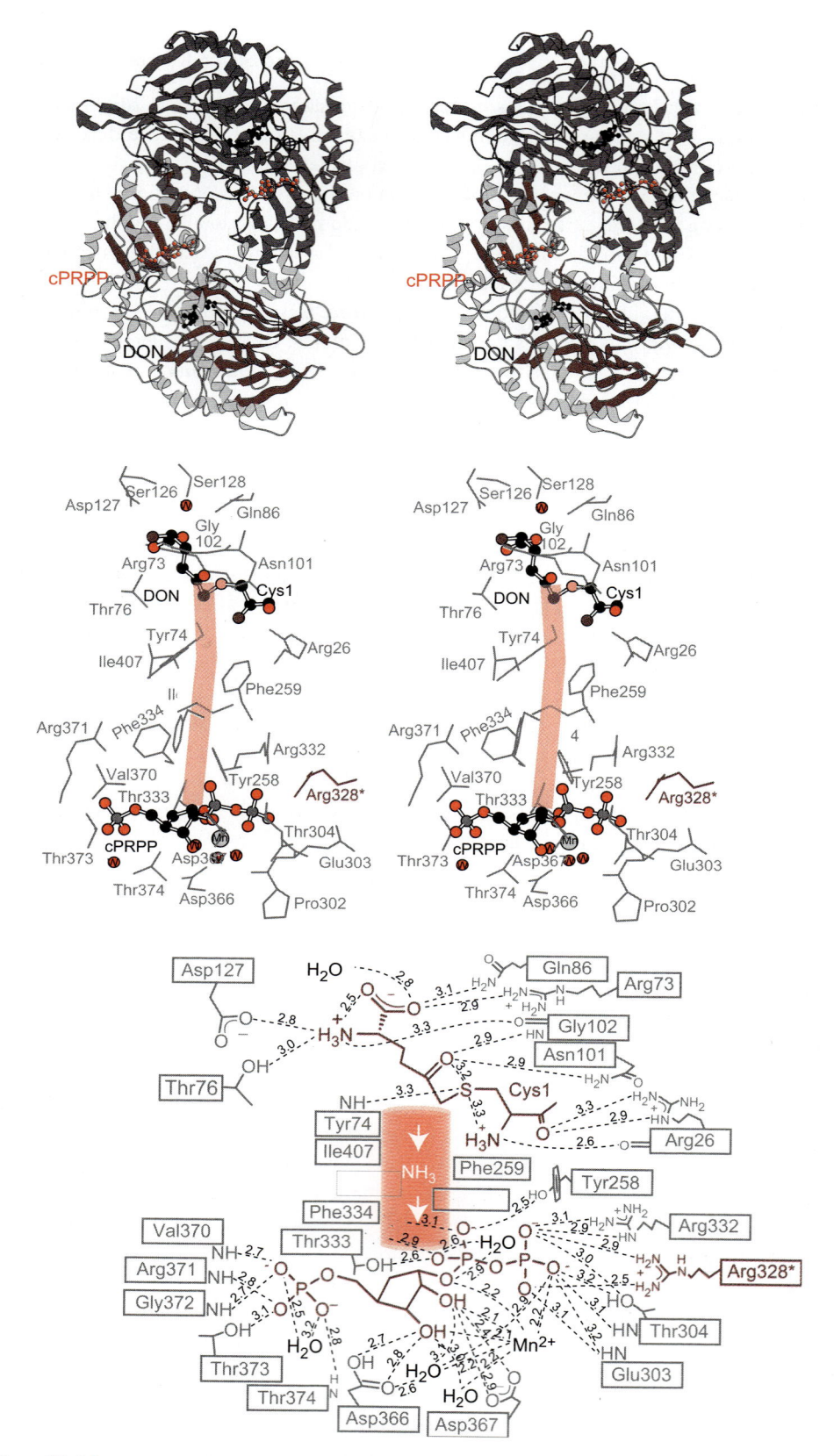

Fig. 13-12. A structure of glutamine:PRPP amidotransferase. The x-ray crystallographic structure of the enzyme from *E. coli* shows diazonorleucine (DON) covalently bonded to Cys1 (black) and cPRPP, a carbocyclic 5-phosphoribosyl pyrophosphate (PRPP) analog (red), bound at the active site (2.4-Å resolution; PDB 1ECC; Krahn et al., 1997). DON inactivates most glutamine-amidotransferases and glutaminases by the mechanism shown in fig. 13-11B. The substrate binding pockets are connected by a tunnel (pink) through which ammonia can pass without being released to solvent. Similar tunnels are present in carbamoyl-P synthetase (see chap. 11) and Trp synthetase (see chap. 3).

thioesters as an intermediate. Similar enzymes catalyze CoA transfer between acids other than acetoacetate and succinate, presumably by similar mechanisms. In chapter 8, we discuss cystathionase and cysteine sulfurylase in connection with addition and elimination reactions. All of the sulfur transfer reactions in earlier chapters proceed by polar reaction mechanisms. In this chapter, we consider enzymes that catalyze sulfur transfer by radical mechanisms.

Biotin Synthase

Reaction and Stoichiometry

In the last step of biotin biosynthesis, desthiobiotin is converted into biotin by the insertion of a sulfur atom into two unreactive C–H bonds. The reaction proceeds by abstraction of the pro-*S* hydrogen from the C6-methylene and one hydrogen atom from the C9-methyl group of desthiobiotin (Frappier et al., 1982; Parry and Kunitani, 1979; Trainer et al., 1980). The scheme in fig. 13-13 describes the current understanding of the reaction carried out by biotin synthase in soluble preparations from bacteria and by purified biotin synthase from *E. coli*, a 78-kDa, iron-sulfur protein (Birch et al., 1995; Ifuku et al., 1992; Marquet, 2001; Méjean et al., 1995; Parry, 1999; Sanyal et al., 1994). The C6(H) and C9(H) atoms in desthiobiotin are in no way activated by functional groups that could weaken their bonds to carbon. However, they are removed in the sulfur insertion process. In enzymology, such reactions proceed by radical mechanisms, and the reactions invariably involve the participation of a paramagnetic cofactor, generally one that contains a transition metal (Frey, 1990). Enzymes in this category include those depending on adenosylcobalamin, cytochrome P450, mononuclear iron/α-ketoglutarate systems, and [4Fe–4S]/*S*-adenosylmethione (SAM) systems. Biotin synthase is mechanistically related to the [4Fe–4S]/SAM systems and is a member of the Radical SAM superfamily (Sofia et al., 2000). Unlike lysine 2,3-aminomutase and a few other enzymes in this superfamily, the SAM is used as a substrate in the reaction of biotin synthase, as it is by most members of the superfamily, rather than as a coenzyme.

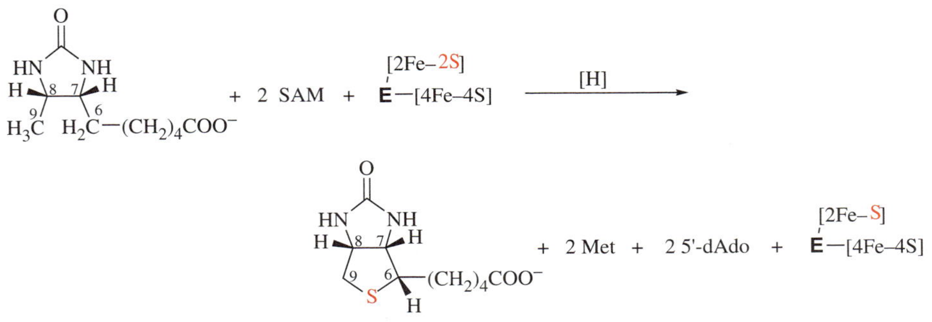

Fig. 13-13. A single turnover of biotin production is achieved by the action of biotin synthase. The enzyme contains two iron-sulfur centers: a [4Fe–4S] center characteristic of radical *S*-adenosylmethionine (or SAM) enzymes, and a [2Fe–2S] center. The four-iron center participates in the cleavage of SAM into methionine and the 5′-deoxyadenosyl radicals, which abstract hydrogen atoms from desthiobiotin. The [2Fe–2S] center serves as the immediate source of sulfide for insertion into desthiobiotin radicals. Two moles of SAM are cleaved per mole of biotin produced, one for each C—H bond cleaved by hydrogen abstraction.

The exact stoichiometry and chemistry of the reaction of biotin synthase is as yet not known. Figure 13-13 describes the present state of knowledge regarding the chemistry and stoichiometry for a single turnover of the enzyme. Two moles of SAM are required for each mole of biotin produced, the SAM is cleaved into methionine and 5′-deoxyadenosine in 1:1 stoichiometry, and the two hydrogen atoms abstracted from desthiobiotin are incorporated into 5′-deoxyadenosine (Escalettes et al., 1999; Guianvarc'h et al., 1997).

S-Adenosyl-L-Methionine and Iron-Sulfur Centers

Biotin synthase contains two iron-sulfur centers, a [4Fe–4S] center that functions in cleaving SAM into methionine and the 5′-deoxyadenosyl radical, and a [2Fe–2S] center that donates a sulfide incorporated into biotin (Brereton, 1999; Duin et al., 1997; Ollagnier-de Choudens et al., 2000; Tse Sum Bui et al., 1999; Ugulava et al., 2000). The [4Fe–4S] center is typical of radical SAM enzymes, with the cysteine-motif CxxxCxxC, in which the cysteine residues are Cys53, Cys57, and Cys60. Site-directed mutagenesis experiments implicate six cysteine residues as essential for activity and the assembly of iron-sulfur centers (Hewitson et al., 2002). In addition to those listed for the Radical SAM motif, Cys97, Cys128, and Cys188 are essential, the first two for the assembly of the [2Fe–2S] center. Oxidation of [4Fe–4S]$^+$ to [4Fe–4S]$^{2+}$ accompanies the cleavage of SAM to methionine (Ollagnier-de Choudens, 2002). Compelling evidence implicates an iron-sulfur center, specifically the [2Fe–4S] center, as the source of sulfur for insertion into biotin by the purified biotin synthase (Gibson et al., 1999; Tse Sum Bui et al., 1998; Ugulava et al., 2001). Multiple turnover catalysis has been difficult to achieve because the exact chemistry for regenerating the [2F–2S] center after the insertion of one sulfide into biotin is not known.

Structure and Reaction Mechanism

A detailed mechanism for the action of biotin synthase cannot be written at the present time. The structure in fig. 13-14 shows the overall chain fold as a β-barrel, and the binding sites for SAM and biotin place the substrates between the [4Fe–4S] and [2Fe–2S] centers, with SAM properly placed for cleavage by electron transfer from the four-iron center and biotin placed for direct interaction with the [2Fe–2S] center (Berkovitch et al., 2004). A sulfur insertion intermediate with sulfur bonded to C9 of desthiobiotin is generated with the cleavage of one molecule of SAM, and its conversion to biotin requires the cleavage of a second molecule of SAM (Shaw et al., 1998). On this basis, the cleavage of SAM by the [4Fe–4S] center, the donation of sulfide by the [2Fe–2S] center, and the structure of the active site, an emerging outline of a mechanism may be similar to that in fig. 13-15 (Ugulava et al., 2001). Much remains to be learned about the reactions of radical intermediates with sulfur and the recycling of sulfur in the [2Fe–2S] center.

Lipoyl Synthase

The coenzymatic lipoyl moieties in enzymes such as the dihydrolipoyl dehydrogenase component in the pyruvate dehydrogenase complex and the glycine reductase complex arise from sulfur insertion into C–H bonds at C6 and C8 of octanoyl derivatives. The product of the LipA gene is required for this process, and this gene encodes lipoyl synthase, a protein with a CxxxCxxC motif characteristic of Radical SAM enzymes (Hayden et al., 1992; 1993; Reed and Cronan, 1993). Purified LipA is an iron-sulfur protein with [4Fe–4S] centers (Busby et al., 1999; Ollagnier-de Choudens and Fontecave, 1999). There is evidence that the true substrates may be octanoyl enzymes that are converted by the

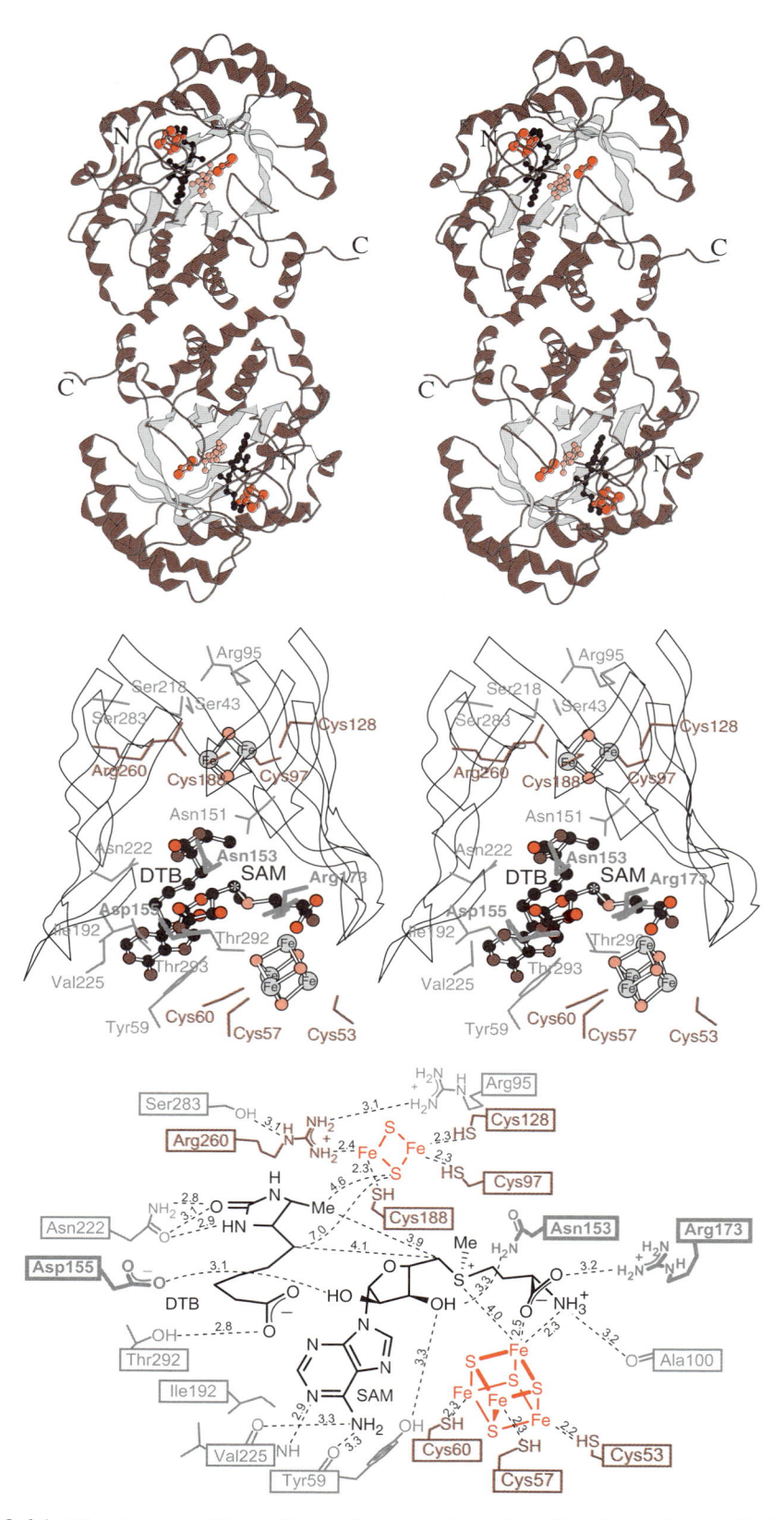

Fig. 13-14. The structure of homodimeric biotin synthase from *E. coli* contains two Fe-S clusters, one [4Fe-4S] and one [2Fe-2S] (both in red), with *S*-adenosylmethionine (SAM, black) and dethiobiotin (DTB, pink) bound at the active site (3.4-Å resolution; PDB 1R30; Berkovitch et al., 2004). Notice that SAM binds to the [4Fe–4S] center, and the [2Fe–2S] center is in position to donate a sulfur atom for the production of D-biotin. The C5′ of SAM is marked with a white asterisk.

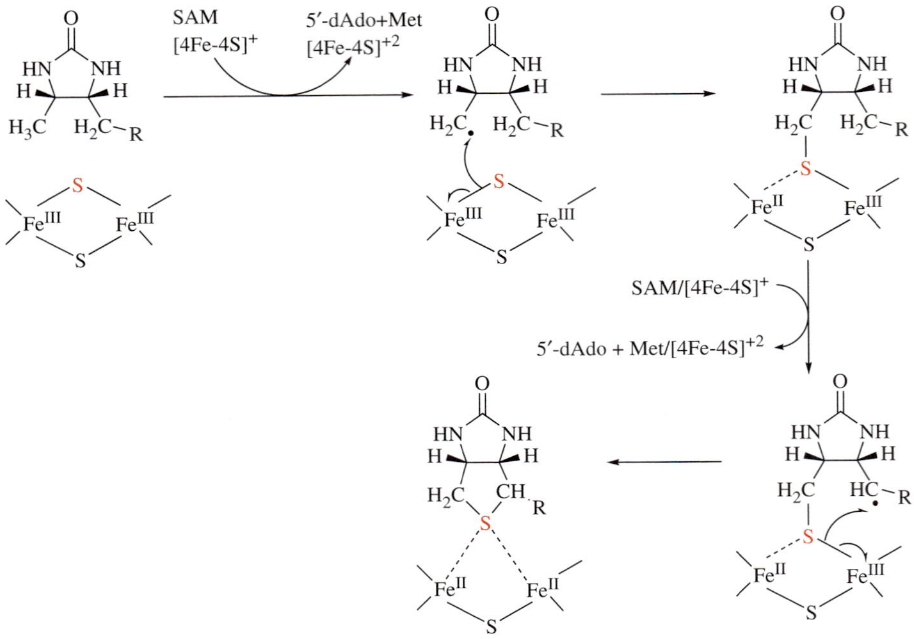

Fig. 13-15. A mechanistic outline for the insertion of sulfur into desthiobiotin in the action of biotin synthase.

action of lipoyl synthase into lipoyl enzymes. For example, lipoyl synthase activates the octanoyl pyruvate dehydrogenase component (Miller et al., 2000).

Lipoyl synthase shares properties with biotin synthase, with differences in substrate specificity, sulfur stoichiometry and iron-sulfur centers. The enzyme contains two distinct [4Fe–4S] centers (Cicchillo et al., 2004). Available evidence shows that two molecules of SAM are required for a complete reaction cycle incorporating the two sulfur atoms in a lipoyl moiety (Cicchillo and Booker, 2005). Much remains to be learned about the transformation of octanoyl into lipoyl moieties in enzymes.

References

Antson AA, TV Demedkina, P Gollnick, Z Dauter, RL Von Tersch, J Long, SN Berezhnoy, RS Phillips, EH Haratyunyan, and KS Wilson (1993) *Biochemistry* **32**, 4195.
Arnone A, PD Briley, PH Rogers, CC Hyde, CM Metzler, and DE Metzler (1982) In Griffen JF and WL Duax (eds): *Molecular Structure and Biological Activity*. Elsevier North-Holland: New York, p 57.
Averill BA, A Dwivedi, P Debrunner, SJ Vollmer, JY Wong, and RL Switzer (1980) *J Biol Chem* **255**, 6007.
Berkovitch F, Y Nicolet, JT Wan, JT Jarrett, CL Drennan (2004) *Science* **303**, 76.
Binda C, RT Bossi, S Wakatsuki, S Arzt, A Coda, B Curti, MA Vanoni, and A Mattevi (2000) *Structure* **8**, 1299.
Birch OM, M Fuhrmann, and NM Shaw (1995) *J Biol Chem* **270**, 19158.
Borisov VV, SN Borisova, NI Sosfenov, and BK Vainshtein (1980) *Nature* **284**, 189.
Brereton PS, RE Duderstadt, CR Staples, MK Johnson, and MW Adams (1999) *Biochemistry* **38**, 10594.
Buchanan JM (1973) *Adv Enzymol Relat Areas Mol Biol* **39**, 91.
Busby RW, JPM Schelvis, DS Yu, GT Babcock, and MA Marletta (1999) *J Am Chem Soc* **121**, 4706.
Chaudhuri BN, SC Lange, RS Myers, SV Chittur, VJ Davisson, and JL Smith (2001) *Structure* **9**, 987.

Chaudhuri BN, SC Lange, RS Myers, VJ Davisson, and JL Smith (2003) *Biochemistry* **42**, 7003.

Chen S, DR Tomchick, D Wolle, P Hu, JL Smith, RL Switzer, and H Zalkin (1997) *Biochemistry* **36**, 10718.

Christenson SD, W Wu, MA Spies, B Shen and MD Toney (2003a) *Biochemistry* **42**, 12708.

Christenson SD, W Liu, MD Toney, and B Shen (2003b) *J Am Chem Soc* **125**, 6062.

Cicchillo RM and SJ Booker (2005) *J Am Chem Soc* **127**, 2860.

Cicchillo RM, KH Lee, C Baleanu-Gogonea, NM Nesbitt, C Krebs, SJ Booker (2004) *Biochemistry* **43**, 11770.

Clausen T, R Huber, B Laber, H-D Pohlenz, and A Messerschmidt (1996) *J Mol Biol* **262**, 202.

Duin EC, ME Lafferty, BR Crouse, RM Allen, I Sanyal, DH Flint, and MK Johnson (1997) *Biochemistry* **36**, 11811.

Escalettes F, D Florentin, Tse Sum Bui, D Lesage, and A Marquet (1999) *J Am Chem Soc* **121**, 3571.

Ford GG, G Eichele, and JN Jansonius (1980) Proc *Natl Acad Sci U S A* **77**, 2559.

Frappier F, M Jouany, A Marquet, A Olesker, and JC Tabet (1982) *J Org Chem* **47**, 2257.

Frey PA (1990) *Chem Rev* **90**, 1343.

Gehring H (1984) *Biochemistry* **23**, 6335.

Gibson KJ, DA Pelletier, and IM Turner (1999) *Biochem Biophys Res Commun* **254**, 632.

Goldberg JM and JF Kirsch (1996) *Biochemistry* **35**, 5280.

Grandoni JA, RL Switzer, CA Makaroff, and H Zalkin (1989) *J Biol Chem* **264**, 6058.

Guianvarc'h D, D Florentin, B Tse Sum Bui, F Nunzi, and A Marquet (1997) *Biochem Biophys Res Commun* **236**, 402.

Hayashi H and H Kagamiyama (1995) *Biochemistry* **34**, 9413.

Hayden MA, IY Huang, G Iliopoulos, M Orozco, and GW Ashley (1993) *Biochemistry* **32**, 3778.

Hayden MA, J Huang, DE Bussiere, and GW Ashley (1992) *J Biol Chem* **267**, 9512.

Hennig M, B Grimm, R Contestabile, RA John, and JN Jansonius (1997) *Proc Natl Acad Sci U S A* **94**, 4866.

Hewitson KS, S Ollagnier-de Choudens, Y Sanakis, NM Shaw, JE Baldwin, E Münck, PL Roach, and M Fontecave (2002) *J Biol Inorg Chem* **7**, 83.

Ifuku O, J Kishimoto, S Haze, M Yanagi, and S Fukushima (1992) *Biosci Biotechnol Biochem* **56**, 1780.

Jansonius JN and MG Vincent (1987) In Jurnak F and A MacPherson (eds): *Biological Molecules and Assemblies*. John Wiley: New York, p 187.

Jenkins WT (1961) *J Biol Chem* **234**, 1121.

John RA (1998) In Sinnott ML (ed): *Comprehensive Biological Catalysis*, vol II. Academic Press: London, p 173.

Kim JH, JM Krahn, DR Tomchick, JL Smith, and H Zalkin (1996) *J Biol Chem* **271**, 15549.

Kirsch JF, G Eichele, GC Ford, MG Vincent, JN Jansonius, H Gehring, and P Christen (1984) *J Mol Biol* **174**, 497.

Krahn JM, JH Kim, MR Burns, RJ Parry, H Zalkin, and JL Smith (1997) *Biochemistry* **36**, 11061.

Kurylo-Borowska Z and T Abramsky (1972) *Biochim Biophys Acta* **264**, 1.

Larsen TM, SK Boehlein, SM Schuster, NGJ Richards, JB Thoden, HM Holden, and I Rayment (1999) *Biochemistry* **38**, 16146.

Malashkevich VN, MD Toney, and JN Jansonius (1993) *Biochemistry* **32**, 13451.

Marquet A, B Tse Sum Bui, and D Florentin (2001) In Begley TP (ed): *Vitamins and Hormones*, vol 61. Academic Press: San Diego, p 51.

Méjean A, B Tse Sum Bui, D Florentin, O Ploux, Y Izumi, and A Marquet (1995) *Biochem Biophys Res Commun* **217**, 1231.

Miller JR, RW Busby, SW Jordan, J Cheek, TF Henshaw, GW Ashley, JB Broderick, JE Cronan Jr, and MA Marletta (2000) *Biochemistry* **39**, 15166.

Myers RS, JR Jensen, IL Deras, JL Smith, and VJ Davisson (2003) *Biochemistry* **42**, 7013.

Ollagnier-de Choudens S and M Fontecave (1999) *FEBS Lett* **453**, 25.

Ollagnier-de Choudens S, Y Sanakis, KS Hewitson, P Roach, JE Baldwin, E Münck, and M Fontecave (2000) *Biochemistry* **39**, 4165.

Ollagnier-de Choudens S, Y Sanakis, K Hewitson, P Roach, E Münck, and M Fontecave (2002) *J Biol Chem* **277**, 13449.

Onate YA, SJ Vollmer, RL Switzer, and MK Johnson (1989) *J Biol Chem* **264**, 18386.

Parry RJ (1999) In Poulter CD (ed): *Comprehensive Natural Product Chemistry*, vol 1. Elsevier: Oxford, p 825.
Parry RJ, MG Kunitani (1979) *Methods Enzymol* **62**, 353.
Reed KE and JE Cronan Jr (1993) *J Bacteriol* **175**, 1325.
Sanyal I, G Cohen, and DH Flint (1994) *Biochemistry 33*, 3625.
Shaw NM, OM Birch, A Tinschert, V Venetz, R Dietrich, and LA Savoy (1998) *Biochem J* **330**, 1079.
Smith D, S Almo, M Toney, and D Ringe (1989) *Biochemistry* **28**, 8161.
Smith JL, EJ Zaluzec, JP Wery, L Niu, RL Switzer, H Zalkin, and Y Satow (1994) *Science* **264**, 1427.
Snell EE (1962) *Brookhaven Symp Biol* **15**, 32.
Sugio S, GA Petsko, JM Manning, K Soda, and D Ringe (1995) *Biochemistry* **34**, 9661.
Toney MD and JF Kirsch (1989) *Science* **243**, 1485.
Toney MD and JF Kirsch (1991) *J Biol Chem* **266**, 23900.
Toney MD and JF Kirsch (1993) *Biochemistry* **32**, 1471.
Trainor DA, RJ Parry, and A Gitterman (1980) *J Am Chem Soc* **102**, 1467.
Tse Sum Bui BT, D Florentin, F Fournier, O Ploux, A Mejean, and A Marquet (1998) *FEBS Lett* **440**, 226.
Tse Sum Bui B, D Florentin, A Marquet, R Benda, and AX Trautwein (1999) *FEBS Lett* **459**, 411.
Tso JY, MA Hermodson, and H Zalkin (1982) *J Biol Chem* **257**, 3532.
Ugulava NB, BR Gibney, and JT Jarrett (2000) *Biochemistry* **39**, 5206.
Ugulava NB, BR Gibney, and JT Jarrett (2001) *Biochemistry* **40**, 8343.
Vollmer SJ, RL Switzer, and PG Debrunner (1983a) *J Biol Chem* **258**, 14284.
Volmer SJ, RL Switzer, MA Hermodson, SG Bower, and H Zalkin (1983b) *J Biol Chem* **258**, 10582.
Walker KD and HG Floss (1998) *J Am Chem Soc* **120**, 5333.
Watanabe N, K Sakabe, N Sakabe, T Higashi, K Sasaki, S Aibara, Y Morita, K Yonaha, S Toyama, and H Fukutani (1989) *J Biochem* **105**, 1.

14

Carbon-Carbon Condensation and Cleavage

Chemistry

In chemistry, many methods are available to synthesize carbon-carbon bonds, and the reactions proceed by both polar and radical mechanisms. However, enzymatic ligation of two molecules through carbon-carbon bond formation invariably proceeds by a polar mechanism. Often, the reaction involves a carbanionic intermediate or a carbanion-equivalent species such as an enamine, but carbenium ion intermediates also participate in terpene biosynthesis. The only well-known enzymatic processes leading to carbon-carbon bonding by radical mechanisms are the adenosylcobalamin-dependent isomerization reactions discussed in chapter 7.

The basic mechanisms illustrated in fig. 14-1 lead to the ligation of molecules through the synthesis of carbon-carbon bonds. Fig. 14-1A depicts the addition of a stabilized carbanion to an aldehyde or ketone to form an adduct. The carbanion can itself be derived from an aldehyde or ketone, as it is in the reactions of aldolases and transketolase. In chapter 1, we discuss the mechanisms of aldolase reactions in connection with the catalytic power of metal ions and of iminium ions formed between substrate carbonyl groups and the lysyl-ε-amino groups of enzymes. In the actions of class I aldolases, X=C in fig. 14-1A is an iminium group formed between a lysyl residue of the enzyme and an aldehyde or ketone group of a substrate. In this case, the carbanion is more accurately described as an enamine, a resonance form in which the charges are not separated. In the actions of class II aldolases, X=C in fig. 14-1A is a carbonyl group (i.e., C=O) coordinated to a divalent metal ion, usually Zn^{2+}, which facilitates carbanion formation through enolization. In this case, the carbanion may be more accurately described as an enolate ion with the charge localized on metal-coordinated oxygen. Iminium ion formation and divalent metal ion ligation both lower the pK_a value of the α-C(H) by 7-10 units, thereby facilitating

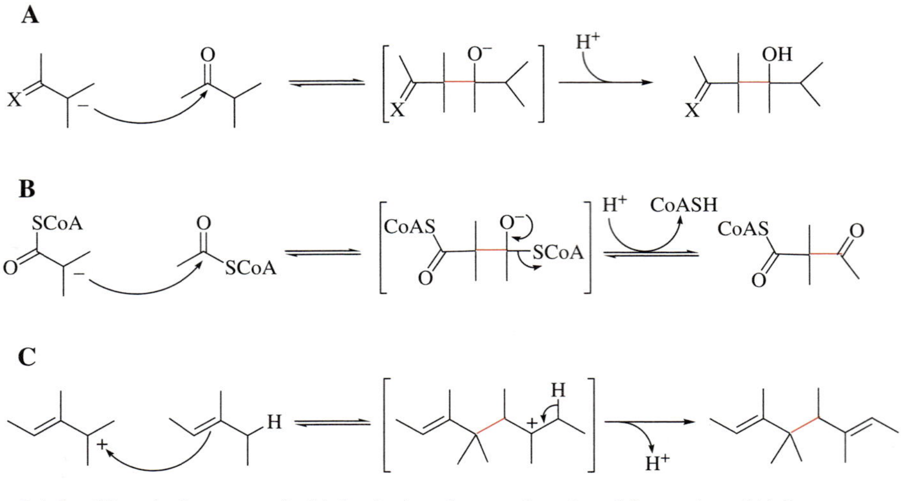

Fig. 14-1. Chemical patterns in biological carbon-carbon bond formation. (A) In many enzymatic reactions linking two molecules with new carbon-carbon bonds, a carbanion is added to the carbonyl group of an aldehyde or ketone in a two-step process through a tetrahedral intermediate, which is quenched by protonation. The requisite carbanions are stabilized by the group C=X, where X can be the carbonyl oxygen of an aldehyde or ketone in an aldolase reaction, the acyl-carbonyl group of a CoA-ester in the action of citrate synthase, the thiazole ring of thiamine pyrophosphate (TPP) in the action of transketolase, or the pyridoxamine ring in the reaction of serine hydroxymethyltransferase. (B) In reactions such as that of β-ketothiolase, a carbanionic CoA ester can add to another CoA ester and then eliminate CoASH to form a β-ketothioester. (C) In many reactions of terpene biosynthesis, allylic carbenium ions form carbon bonds by addition to carbon-carbon double bonds, and elimination of protons from the resulting carbenium ion intermediates stabilizes the new carbon-carbon bonds.

enolization and carbanion formation (see chap. 1). An enolate carbanion may also be derived from a CoA-thioester such as acetyl CoA in the reaction of citrate synthase. Once the carbanion or carbanion-equivalent is formed in an active site, its addition to an aldehyde or ketone group in an adjacent cosubstrate proceeds rapidly.

In fig. 14-1B, the carbanion is derived from a thioester such as an acyl CoA and undergoes nucleophilic addition to the acyl carbonyl group of another CoA-thioester. The resulting adduct eliminates CoASH to form the ligated product. The overall process is reversible and under cellular condition may lead to cleavage of the carbon-carbon bond, as in the action of β-ketothiolase, or to synthesis of a β-ketoacyl thioester, as in fatty acid biosynthesis.

In fig. 14-1C, an allylic carbenium ion undergoes addition to a carbon-carbon double bond, and the resulting carbenium ion loses a proton to stabilize the ligated product. This type of reaction commonly occurs in terpene biosynthesis and is particularly important in cholesterol biosynthesis. The allylic carbenium ion is derived from an allyl pyrophosphate by elimination of pyrophosphate, as in the reaction of dimethylallyl pyrophosphate with isopentenyl pyrophosphate to produce farnesyl pyrophosphate and PP_i, as subsequently discussed.

In this chapter, we discuss enzymes that catalyze the ligation of two molecules by carbon-carbon bond formation. We first consider several enzymes that function by stabilizing carbanionic intermediates, and we then consider two important enzymes that catalyze carbon-carbon ligation through carbenium ion intermediates.

Enolization of Acetyl CoA

Acetyl CoA in Ester Condensations

Acetyl CoA serves as a substrate in many carbon-carbon condensation and cleavage reactions in enzymology, ranging from the reaction of citrate synthase in the tricarboxylic acid (TCA) cycle to fatty acid biosynthesis and degradation to the biosynthesis of cholesterol and terpenes and polyketides. The equations in fig. 14-2 depict a few of them. Citrate and malate synthases typify enzymes that catalyze the addition of the α-carbon of acetyl CoA to the carbonyl group of a ketone or aldehyde to produce a hydroxyacid and CoASH. The reaction of β-ketothiolase typifies many processes in which the α-carbon of enolacetyl CoA undergoes elimination from or addition to the acyl carbonyl group of another CoA-thioester (or a phosphopantetheine ester). In the direction of condensation, the reaction produces a β-ketoacyl CoA and CoASH. In all of these reactions, acetyl CoA must undergo enolization as the first step to carbon-carbon bond formation. In this chapter, we consider aspects of the reactions in fig. 14-2 and focus on citrate synthase and β-ketothiolase as case studies. Citrate synthase functions in the direction of carbon-carbon condensation, whereas β-ketothiolase most often functions in the carbon-carbon cleavage step of the β-oxidation of fatty acids, although in certain metabolic states it may function in condensation as well (Gehring and Lynen, 1972; Spector, 1972).

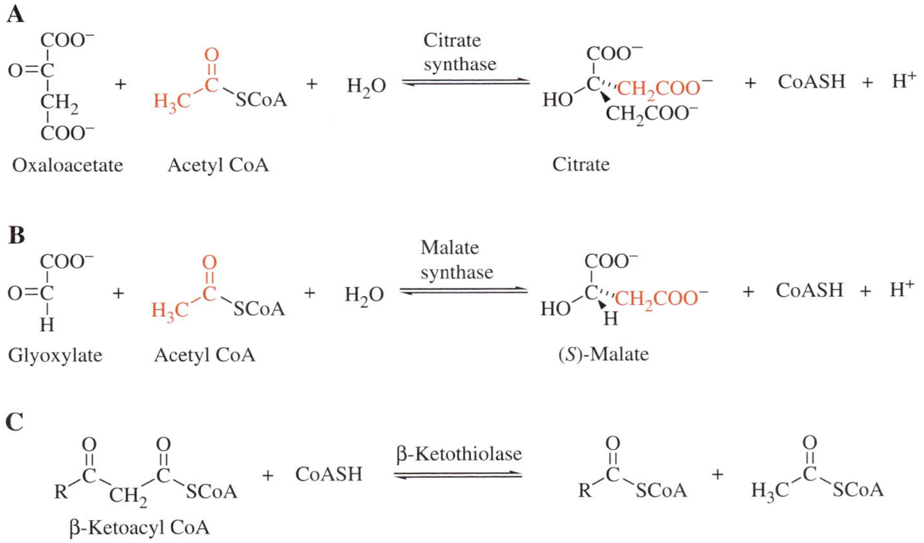

Fig. 14-2. Acetyl CoA in enzymatic carbon-carbon condensation and cleavage. (A) Citrate synthase catalyzes the condensation of acetyl CoA with oxaloacetate to form citrate and CoASH. The overall reaction includes carbon-carbon bond formation between C2 of oxaloacetate and C2 of acetyl CoA and the hydrolysis of acetyl CoA. One of the carboxymethyl groups of citrate is derived from acetyl CoA, and in the action of most species of citrate synthase, this is the pro-S carboxymethyl group (red). A few citrate synthases from anaerobic bacteria introduce the new carboxymethyl group into the pro-R position. (B) A number of other enzymes catalyze condensations of acetyl CoA to aldehydes or ketones other than oxaloacetate. In the reaction of malate synthase, acetyl CoA is condensed with glyoxylate. (C) β-Ketothiolase catalyzes a different reaction of acetyl CoA, one in which acetyl CoA is produced in the cleavage of a β-ketoacyl CoA by CoASH.

Citrate Synthase

Molecular Properties and Regulation

Two classes of citrate synthase (EC 2.3.3.1) differ in molecular size and regulatory properties (Srere, 1970). The large citrate synthases from gram-positive bacteria display molecular masses of more than 250 kDa, and the small synthases from gram-negative bacteria and higher organisms have molecular masses of about 100 kDa. The synthase from pig heart is typically small and is a dimer of identical subunits. Citrate synthases are most active in the presence of potassium or ammonium ions and are inhibited by smaller monovalent cations.

As the first enzyme to function in the TCA cycle, citrate synthase is an obvious locus metabolic regulation. It is subject to allosteric inhibition by products of the tricarboxylic acid cycle. Inasmuch as the TCA cycle is circular, end-product inhibition would be something of a misnomer. However, the TCA cycle does generate energy in the form of molecules, specifically ATP and NADH, both of which are sources of energy in metabolism. These molecules differentially inhibit citrate synthases (Srere, 1970). NADH inhibits the large species of citrate synthase, and ATP inhibits the small species. Refinements on this pattern are found among species; for example, AMP relieves NADH inhibition of some large synthases but not others.

Stereochemistry

In the first step of the TCA cycle, citrate synthase, originally known as *condensing enzyme*, catalyzes the coupling of acetyl CoA through its α-carbon to oxaloacetate to form citrate and CoASH (fig. 14-2A). Malate synthase catalyzes the same reaction with glyoxylate as the co-substrate (see fig. 14-2B). Citrate synthase has been a focus of two aspects of stereochemical analysis: the stereofacial specificity with which the carboxymethyl group of acetyl CoA is added to the carbonyl group of oxaloacetate and the stereochemical course of the replacement of an α-proton of acetyl CoA by the carbonyl carbon of oxaloacetate. Each of these studies was influential in the conceptualization of stereochemistry in the action of enzymes.

Stereospecificity

Figure 14-2A shows that citrate contains two carboxymethyl groups and is prochiral. Enzymes recognize the steric difference between the carboxymethyl groups in citrate and act on the molecule stereospecifically. Citrate synthases from most species act always on the pro-*S* carboxymethyl group of citrate. However, the synthases from a few species of anaerobic bacteria act with the opposite stereospecificity, on the pro-*R* carboxymethyl group (Gottshalk and Barker, 1967). Each citrate synthase acts stereospecifically in adding the carboxymethyl group to the *si*- or the *re*-face of the carbonyl group in oxaloacetate. The enzymes of the TCA cycle stereospecifically process only the pro-*R* carboxymethyl group of citrate.

We discussed the principle of prochirality and its recognition by enzymes in chapter 3 in connection with the stereochemistry of hydride transfer by NAD^+-dependent dehydrogenases. The stereospecific utilization of citrate in metabolism first signaled the ability of enzymes to distinguish prochiral substituents on tetrahedral carbon. For this reason, the case of citrate and its carboxymethyl groups acquired historical significance as the first clear case of prochiral recognition in enzymology (Ogston, 1948; Popják, 1970).

Inversion of (R)- and (S)-[³H,²H]acetyl CoA. Citrate and malate synthases, and all other synthases in the family, catalyze the substitution of a methyl proton in acetyl CoA with inversion of configuration. Because the methyl carbon of acetyl CoA is not a chiral center, the establishment of this fact required that samples of acetyl CoA incorporating chiral methyl groups be synthesized. Substitution of a deuteron and a triton for two protons in a methyl substituent makes it a center of chirality if the fourth substituent is not a heavy isotope of hydrogen. Chiral methyl groups can be synthesized in acetic acid and acetyl CoA and their configurations assigned by imaginative chemical and enzymatic methods (Cornforth et al., 1969; Lüthy et al., 1969).

The stereochemical course of the substitution of the α-hydrogens in (R)- and (S)-[2-³H,²H]acetyl CoA by action of malate synthase was solved by the application of the known stereochemistry of fumarase (Cornforth et al., 1969; Lüthy et al., 1969). The enzymatic transformations outlined in fig. 14-3 proved that the action of malate synthase inverted the configuration at C2 of acetyl CoA.

The analysis depended on the confluence of several prerequisites. Because tritium could not be used at high isotopic abundance and in any case was only available as a trace label, chiral methyl groups in the synthetic acetic acid had to contain almost 100% enrichment with a single deuterium. This ensured that every molecule containing tritium also contained one deuterium and one hydrogen and was therefore chiral. (Because of trace labeling with tritium, essentially no molecules contain two tritium atoms.) The synthetic methods for chiral [³H,²H]acetic acid were designed to fulfill this condition. Second, the abstraction of the α-hydrogen had to proceed with a significant primary kinetic isotope effect. This ensured that most of the chiral methyl groups would react by proton abstraction rather than deuteron abstraction. The reactions of malate and citrate synthases fulfilled this requirement and displayed values of about 1.4 for primary deuterium kinetic isotope effects. These effects were on the low side for the prospective analytical method, but it turned out that proton abstraction was not solely rate limiting and that the full isotope effects were not expressed in the kinetic parameters. The intrinsic deuterium kinetic isotope effects were between 2 and 3, which significantly facilitated the analysis. Having fulfilled the primary stereochemical conditions, a method was needed to determine the configuration at C3 of (2S)-[3-³H,²H]malate. The problem was solved by employing

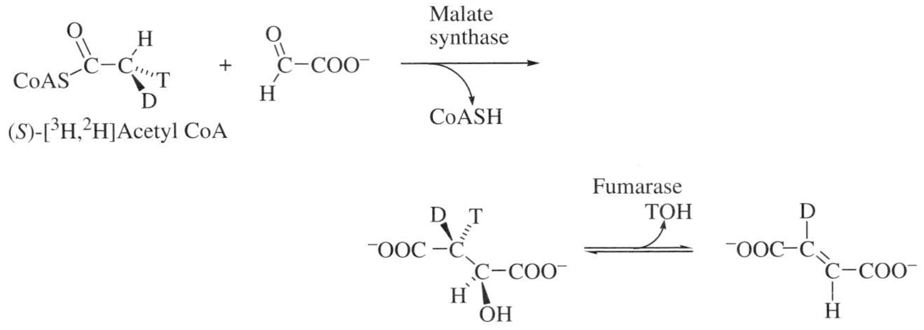

Fig. 14-3. Stereochemical inversion at the methyl group of acetyl CoA in the action of malate synthase. The malate synthase–catalyzed reaction of synthetic (S)-[³H,²H]acetyl CoA with glyoxylate led to a sample of (S)-malate containing deuterium and tritium at C3 by replacement of the α-hydrogen. Dehydration by the action of fumarase released tritium as TOH, proving that the (S)-malate must have been (2S,3R)-[3-³H,²H]malate, the product of inversion of configuration at the methyl group of acetyl CoA. The reaction of (R)-[³H,²H]acetyl CoA in the same sequence did not release TOH on formation of fumarate.

fumarase, which catalyzed the dehydration of (S)-malate with specific loss of the proton in the pro-R position to water, and this allowed the diastereotopic position of tritium at C3 to be determined. Tritium in the 3-pro-R position would be exchanged into water by action of fumarase, whereas tritium in the pro-S position would be retained in fumarate.

In the event, malate derived from (S)-[^{3}H,^{2}H]acetyl CoA lost tritium to water when treated with fumarase, so that it must have been (2S,3R)-[3-^{3}H,^{2}H]malate, the product of inversion shown in fig. 14-3. Reaction of (S)-[^{3}H,^{2}H]acetyl CoA led to (2S,3 2S)-[3-^{3}H,^{2}H]malate, which did not lose tritium to water on treatment with fumarase. An extension of this method to citrate synthase showed inversion of configuration as well (Eggerer et al., 1970). This methodology has been adapted and applied to stereochemical analysis of many enzymes acting on methyl groups (Floss and Lee, 1993).

Reaction Mechanism

The essential outline of the overall mechanism can be simply stated and related to that in fig. 14-1A (Remington, 1992; Srere, 1975). Citrate synthase catalyzes the enolization of acetyl CoA and its addition to the carbonyl group of oxaloacetate. This mechanism appears today to be essentially obvious. However, other mechanisms could be conceived. For example, it is possible to imagine acetyl CoA acetylating an enzymatic nucleophilic group to form a covalent acetyl-enzyme intermediate, which could undergo enolization and addition of the enol-methylene group to oxaloacetate. No evidence for such a mechanism could be found by searching for relevant exchange reactions (Stern, 1961). The formation of the citryl CoA intermediate in step 3 (fig. 14-4) effectively excludes a covalent acetyl-enzyme as an intermediate.

Complex steady-state kinetics in the action of citrate synthase were consistent with ordered binding of substrates accompanied by substrate inhibition (Kosicki and Srere, 1961).

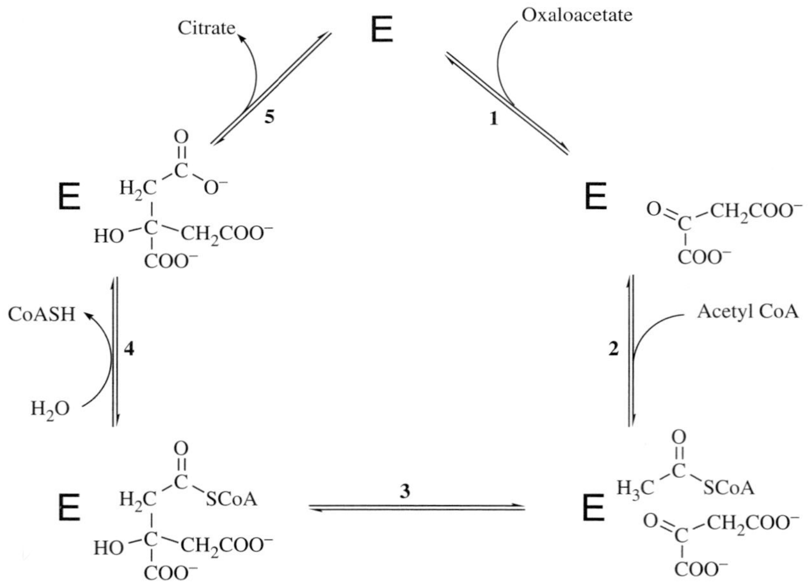

Fig. 14-4. The mechanism for the overall action of citrate synthase shows the condensation of acetyl CoA with oxaloacetate in five stages: binding of oxaloacetate, binding of acetyl CoA, reaction of the ternary complex to form citryl CoA, hydrolysis of citryl CoA to release CoASH, and dissociation of citrate.

Further kinetic analysis indicated an ordered binding mechanism, with oxaloacetate leading, synergistic binding of acetyl CoA, and substrate inhibition by acetyl CoA (Johansson and Pettersson, 1977).

The direct addition of acetyl CoA to oxaloacetate in the third step produces citryl CoA as an enzyme-bound intermediate. Evidence for this intermediate includes the utilization of synthetic citryl CoA as a substrate for citrate synthase, which catalyzes its hydrolysis into CoASH and citrate (Eggerer and Remberger, 1963). The hydrolysis occurs in the fourth step of the mechanism in fig. 14-3, and the release of citrate occurs in the final step of the catalytic cycle.

The hydrolysis of the thioester group in acetyl CoA makes the overall reaction of citrate synthase more thermodynamically favored than it would otherwise be. The equilibrium constant in the forward direction, expressed in terms of all ionic species of substrates and products and unitary activity of water, is 0.47 in neutral solutions (Stern, 1961). The reaction is readily reversible but would be disfavored if the CoA-thioester did not suffer hydrolysis.

The mechanism by which citryl CoA undergoes hydrolysis at the active site is of some interest. Hydrolysis might be catalyzed by the central carboxylate displacing CoASH to form an anhydride intermediate, which would undergo hydrolysis to citrate. This mechanism might explain the instability of citryl CoA in aqueous solutions (Eggerer and Buckel, 1969). However, citrate and citric acid resist internal anhydride formation (Glusker, 1992). Synthase-catalyzed hydrolysis in $H_2^{18}O$ proceeds with incorporation of ^{18}O in place of CoASH, as shown in fig. 14-5, and not in the central carboxylate (Suelter and Arrington, 1967; Wunderwald and Eggerer, 1969). This outcome is compatible with most hydrolytic mechanisms and does not implicate an anhydride intermediate, as incorporation into the central carboxylate would have done. Although the mechanism of hydrolysis at the active site remains an open question, available structures do not show the central carboxylate of citrate in position to interact chemically with either of the other two carboxylate groups (Remington, 1992).

Structure

A number of structures of the small forms of citrate synthases from chicken and pig heart are available with various substrate analogs bound at the active site. The structures show the proteins are largely helical, with a small and large domain in each subunit (Remington, 1992). A cleft between the domains in free citrate synthase closes around the substrates in the ternary complex. This conformational change buries the substrates and isolates them from exposure to the solvent.

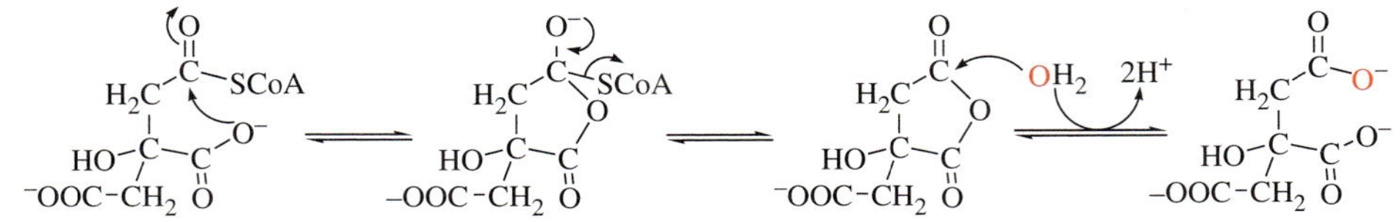

Fig. 14-5. A mechanism for the hydrolysis of citryl CoA by internal catalysis through an anhydride intermediate. In this mechanism, the C3-carboxylate displaces CoASH to form an anhydride intermediate, which is hydrolyzed by water to form citrate, a product in the action of citrate synthase. The synthase-catalyzed reaction in $H_2^{18}O$ (red) proceeds with incorporation of the label in place of the departing CoASH, not into the C3-carboxylate.

Among other features, the structures show that substrates or analogs interact with several basic amino acid side chains, and the active site is well constituted to bind citrate. The amino acid side chains in contact with the substrates at each active site are derived from both subunits of the dimer. The structure shown in fig. 14-6 is at high resolution with oxaloacetate and an analog of acetyl CoA, amidocarboxymethyldethia CoA (Usher et al., 1994). The side chains of His238, His274, His320, Arg 329, Arg401, and Arg421 are in contact with oxaloacetate or citrate. This is a remarkable arrangement for the large number of potentially positively charged groups in contact with oxaloacetate and the acetyl group of acetyl CoA.

Enolization Mechanism

In early studies, the most difficult aspect of the overall mechanism to observe experimentally was the enolization of acetyl CoA. Citrate synthase would not catalyze the exchange of deuterium from D_2O or tritium from [^{3}H]H_2O into acetyl CoA in the absence of oxaloacetate (Bové et al., 1959; Marcus and Vennesland, 1958). A small amount of exchange could be observed in the presence of oxaloacetate, but this could be explained by the overall reversibility of the reaction. The absence of exchange and the observation of inversion of configuration at the methyl group of acetyl CoA introduced the thought that enolization might not be involved, and carbon-carbon bond formation might be concerted with cleavage of the methyl C–H bond of acetyl CoA. However, a double-isotope fractionation analysis of the action of malate synthase excluded a concerted mechanism, at least for malate synthase. The reaction displayed values of 1.3 for $^Dk_{cat}/K_m$ and 1.0037 for $^{13}k_{cat}/K_m$ with the substrates [^{2}H$_3$]acetyl CoA and [aldehyde-^{13}C]glyoxylate, respectively (Clark et al., 1988). The ^{13}C isotope effect did not change in the reaction of the ^{13}C-labeled glyoxylate with [^{2}H$_3$]acetyl CoA. It would have increased to about 1.01 in a concerted mechanism. The actions of malate and citrate synthases are similar apart from their specificities for glyoxylate and oxaloacetate, respectively.

The absence of observable exchange between the methyl hydrogens of acetyl CoA and solvent protons presented a formidable hurdle to the mechanistic analysis of citrate synthase, primarily because it was a negative result. Negative findings have always frustrated mechanistic research. The situation was clarified by the observation that citrate synthase readily catalyzed this exchange in the presence of L-malate, a structural analog of oxaloacetate that could not react with enolized acetyl CoA (Eggerer, 1965). The presence of a four-carbon dicarboxylic acid that could not react as a substrate stimulated the enolization of acetyl CoA by citrate synthase. This finding implicated oxaloacetate (and L-malate) in creating the correct conformation or chemical constitution at the active site for catalysis of enolization.

Conformational effects seemed most likely for the role of L-malate; however, introducing two anionic carboxylate groups into the active site with the binding of malate might also have perturbed the chemistry of the site. For example, the proximity of dianionic L-malate might have altered the pK_a values of nearby acid or base catalytic groups. L-Malate proved to perturb the ultraviolet and fluorescence spectra and to stabilize citrate synthase against denaturation by urea, showing that its binding interactions held the enzyme in a stable conformation (Srere, 1966, 1967). These experiments proved that the free enzyme existed in less stable conformations, and L-malate induced a transition to a stable conformation. The experiments also rationalized the synergistic binding of oxaloacetate and acetyl CoA and their preferentially ordered binding in the kinetic mechanism.

A complete understanding of the enolization of acetyl CoA remains to be attained. The pK_a of a thioester such as acetyl CoA in water is very high, about 21 (Aymes and

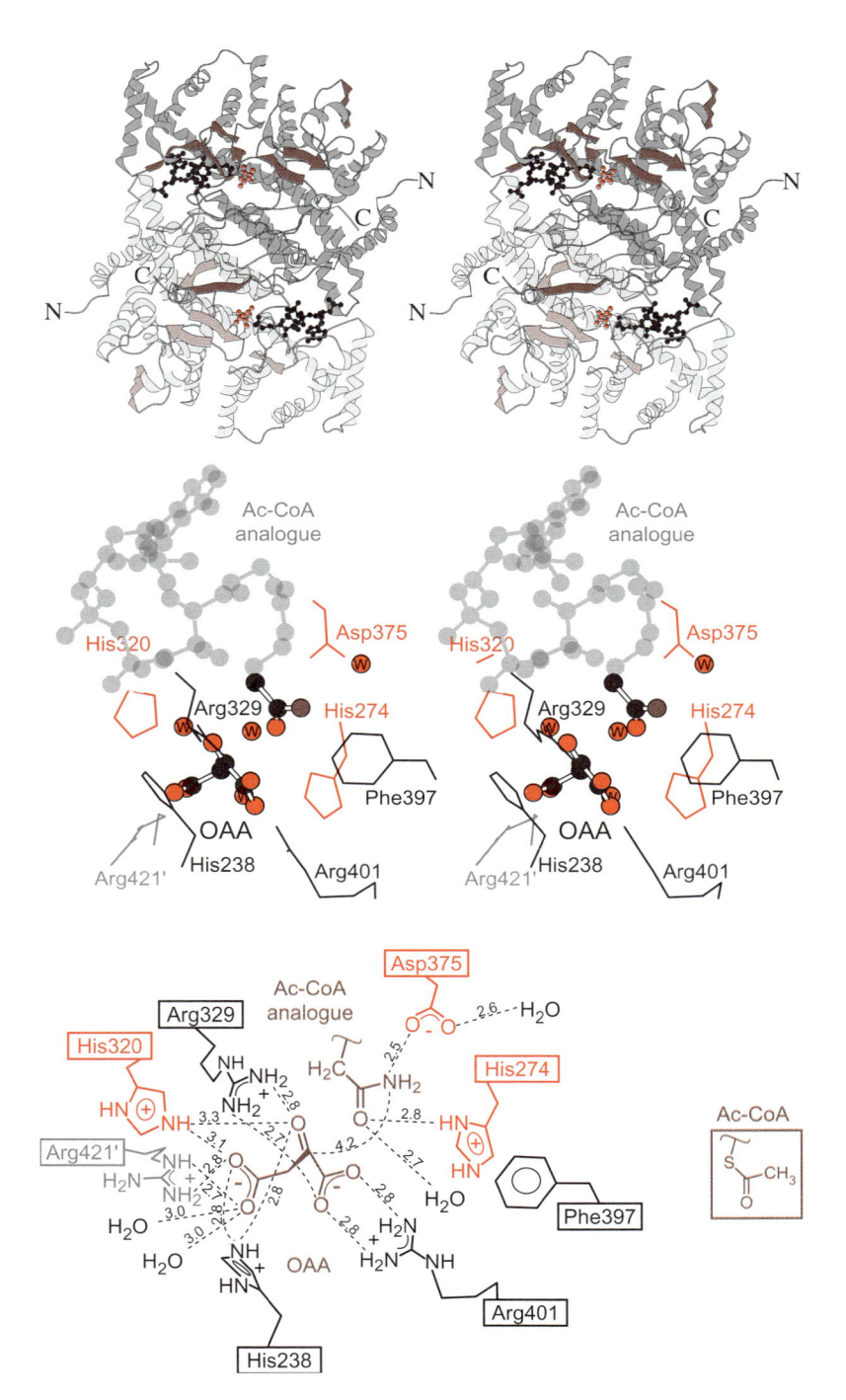

Fig. 14-6. The x-ray crystallographic structure shows homodimeric, chicken heart citrate synthase in a complex with oxaloacetate (red) and amidocarboxymethyldethia CoA (black) (1.60-Å resolution; PDB 1CSH; Usher et al., 1994). The middle stereoimage provides a closer view of the active site, with the bulk of the CoA shown in gray behind the active site. The bottom panel shows the active site in two dimensions, with the distances given in angstroms. The Ac-CoA analog is shown in brown, with the corresponding Ac-CoA atom placement shown in a box to the right of the diagram. There is a 2.5-Å distance between the Asp375 acid oxygen and the CoA analog amide nitrogen, which may result from the presence of a low-barrier hydrogen bond (Cleland et al., 1998).

626 Enzymatic Reaction Mechanisms

Richard, 1992). The acidity of the α-protons of acetyl CoA must be increased by some means to allow enolization to take place at neutral pH, and this requires a lowering of the normal pK_a from 21 to = 10.

Figure 14-7 depicts two enolization mechanisms currently under consideration. The mechanism in fig. 14-7A overcomes much of the barrier to enolization by eliminating the enolate as an intermediate (Remington, 1992). In this mechanism, enolization to enolacetyl CoA takes place in a concerted process, in which Asp375 serves as the base to abstract a methyl proton at the same time that His274 in its conjugate acid form donates

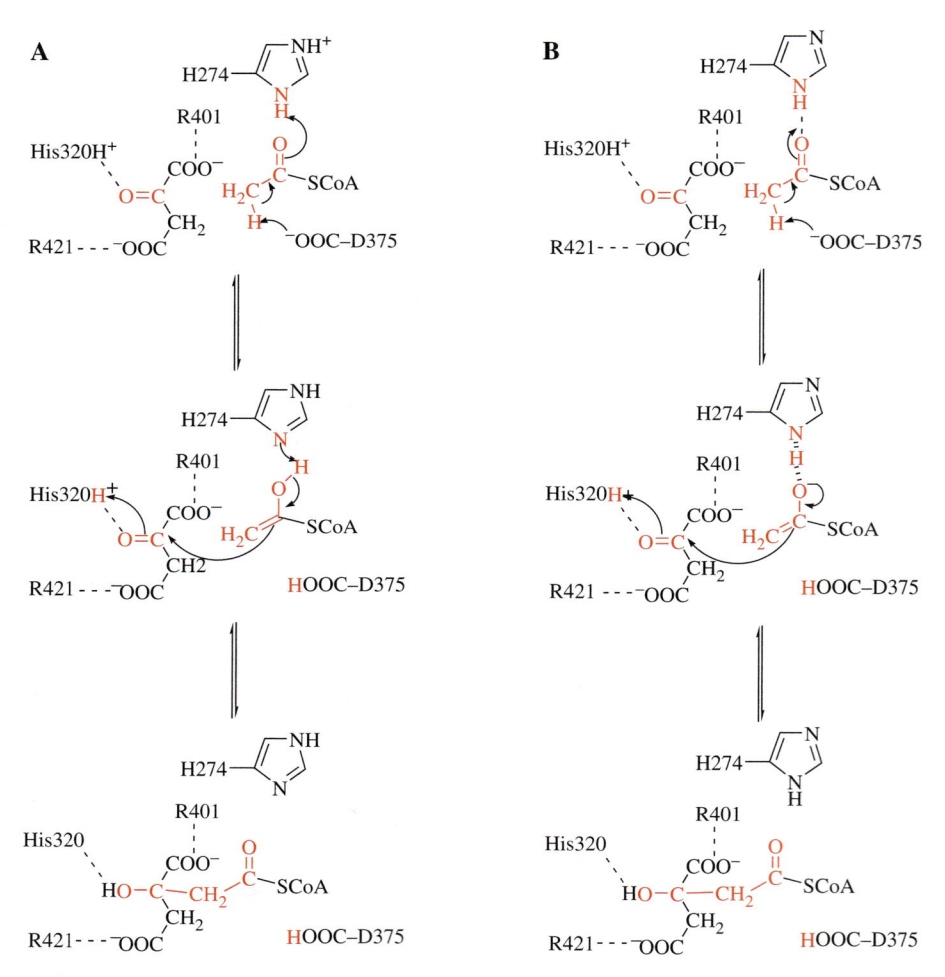

Fig. 14-7. (A and B) Two mechanisms for the enolization and reaction of acetyl CoA are being considered. Both mechanisms involve the same amino acid side chains of the enzyme. In mechanism A, enolization proceeds by two concerted proton transfers: from the α-carbon of acetyl CoA to Asp375 and from the conjugate acid of His274 to the acyl carbonyl oxygen of acetyl CoA. The condensation of enolacetyl CoA with oxaloacetate takes place with two proton transfers in concert with carbon-carbon bond formation. The proton transfers are from enolacetyl CoA to His274 and from His320 to the carbonyl oxygen of oxaloacetate. In mechanism B, enolization proceeds with proton transfer from the α-carbon of acetyl CoA to Asp375 to form the enolate of acetyl CoA, which is engaged in a low-barrier hydrogen bond with the neutral, conjugate base of His274. In the second step, the enolate of acetyl CoA undergoes nucleophilic addition to the carbonyl group of oxaloacetate, accompanied by proton transfer from His320 to the carbonyl oxygen.

a proton to the carbonyl oxygen of acetyl CoA. The resultant enolacetyl CoA then undergoes nucleophilic addition to the carbonyl group of oxaloacetate in a process catalyzed by the conjugate base of His274, which in turn abstracts the enol-proton from enolacetyl CoA, and by His320 in its acid form, which donates a proton to the carbonyl oxygen of oxaloacetate. The products then dissociate from the enzyme. This is an iso-mechanism, in which at least Asp375 and His320 are in different ionization states at the beginning and end of the catalytic cycle. The formation of enolacetyl CoA presents a barrier to carbon-carbon bond formation because it is not very reactive as a carbanion. It is postulated that this is overcome by the actions of His320, which polarizes the carbonyl group and increases its reactivity, and His274, which abstracts the enol-proton from enolacetyl CoA.

The mechanism in fig. 14-7B overcomes the barrier to enolization by stabilizing the enolate of acetyl CoA with a low-barrier hydrogen bond (LBHB) between the enolate oxygen and the conjugate base of His274 (Cleland et al., 1998). Asp375 serves as the base to abstract the α-proton. The LBHB is feasible because the pK_a values of enolacetyl CoA and His421 are likely to be similar, so that they would have similar affinities for the proton. This interaction would stabilize the enolate and allow it to undergo addition as a carbanion to the carbonyl group of oxaloacetate, with protonation of the oxygen by His320. This is also an iso-mechanism, with Asp375 and His320 in different ionization states at the beginning and end of the catalytic cycle.

The roles of Asp375 and His320 are similar in the two mechanisms. Mutation of His320 to glycine introduces interesting properties that strongly support its function in carbon-carbon bond formation (Kurz et al., 1995). The mutation modestly increases the value of K_m, but the overall catalytic activity of H320G-synthase is about 600-fold lower than that of citrate synthase. The effect is limited to carbon-carbon bond formation and not to the hydrolysis of citryl CoA, which is hydrolyzed to citrate but not cleaved to acetyl CoA and oxaloacetate by reaction in reverse. The consequences of the mutation of His320 extend to the carbonyl group of oxaloacetate in binary complexes. In its binary complex with citrate synthase, the ^{13}C-NMR signal of the carbonyl carbon is shifted downfield, indicating polarization, presumably by hydrogen bonding. Downfield perturbation of the carbonyl ^{13}C-NMR signal is absent in the binary complex of oxaloactate with H320G-synthase, directly implicating His320 in the polarization of the carbonyl group (Kurz et al., 1995).

Biochemical analysis indicates that the dithio-analog of acetyl CoA, $H_3C-(C=S)-SCoA$, reacts normally albeit slowly (Wlassics and Anderson, 1989). Spectroscopic data indicate that the enolization of $H_3C-(C=S)-SCoA$ at the active site leads to the thioenolate anion, not the neutral thioenol, and that the anion is about 5 kcal more stable than the neutral thioenol. Although this is consistent with the mechanism in fig. 14-7B, the substitution of S for O is a significant perturbation; the pK_a values for sulfur acids are about four units lower than for the corresponding oxygen acids (Frey and Sammons, 1985).

A fundamental difference between the mechanisms in fig. 14-7 is the ionization state of His274. In the LBHB mechanism, it must be neutral as in triose phosphate isomerase, which catalyzes an analogous enolization. In the concerted proton transfer mechanism of fig. 14-7A, it must be ionized as its conjugate acid. Whenever the ionization state of His274 becomes known, one of the mechanisms in fig. 14-7 can be excluded.

Thiolases

Degradative and Biosynthetic Thiolases

The reaction of thiolase in fig. 14-2C is typical of enzymatic condensations and cleavages involving β-ketoacyl CoA and CoASH. Thiolases function in metabolism and biosynthesis.

Thiolase catalyzes the carbon-carbon cleavage step in the β-oxidation of fatty acids, and the equilibrium constant of 10^5 in the direction of cleavage is compatible with this process (Gehring and Lynen, 1972). Biosynthetic thiolases such as acetoacetyl CoA thiolase (EC 2.3.1.9) catalyzes carbon-carbon condensation of two molecules of acetyl CoA to from acetoacetyl CoA in the biosynthesis of cholesterol and ketone bodies in animals and of β-hydroxybutyrate in bacteria. In biosynthesis, the unfavorable equilibrium is overcome by successive steps, beginning with reduction of acetoacetyl CoA to β-hydroxybutyryl CoA. The mechanistically related step in fatty acid biosynthesis operates in the direction of condensation because of the decarboxylation of malonyl-ACP to generate enolacetyl-ACP through decarboxylation (see chap. 18). The release of CO_2 drives the overall reaction in the direction of condensation. Most thiolases are homotetrameric enzymes with molecular masses of 140 to 170 kDa; however, some are homodimeric.

The thiolases in β-oxidation display broad specificity for β-ketoacyl CoAs and accept longer chain substrates, whereas the biosynthetic acetoacetyl CoA thiolase is much more selective for acetoacetyl CoA. Available evidence indicates that all thiolases act by the same basic mechanism.

Kinetics and Mechanism

All thiolases are extremely sensitive to cysteine-modifying reagents such as iodoacetamide and hydroxymercuribenzoate because of the presence of an essential cysteine residue at the active site (Gehring and Lynen, 1972). This cysteine exists as a thiolate in the resting enzyme, and the structure shows this cysteine in association with a histidine residue, reminiscent of the cysteine proteases and cysteine phosphatases (see chaps. 6 and 10). The biodegradative thiolase from pig heart reacts with β-ketoacyl CoAs to form covalent acyl-enzymes, and the covalent intermediates react with CoASH to form the corresponding acyl CoAs (Gehring and Lynen, 1972). Moreover, the enzyme also catalyzes the exchange of radioactivity from [1-^{14}C]acetyl CoA into the acyl carbonyl group of acetoacetyl CoA in the absence of CoASH (Gehring et al., 1968). These facts, especially the CoASH-independent acetyl CoA/acetoacetyl CoA exchange, indicate the ping pong mechanism shown in fig. 14-8, and this is the kinetic mechanism for the action of the biosynthetic acetoacetyl CoA thiolase from *Zoogloena ramigera* (Middleton, 1974). The reaction is very sensitive to competitive substrate inhibition by CoASH, a property often associated with ping pong mechanisms (see chap. 2).

Intensive studies of the biosynthetic acetoacetyl CoA thiolase define the chemical mechanism in fig. 14-8 in considerable detail (Anderson et al., 1990; Kursula et al., 2002; Masamune et al., 1989; Thompson et al., 1989; Williams et al., 1992). Cysteine 89 serves as the nucleophilic catalyst, as shown by the fact that acetylation by [^{14}C]acetyl CoA to form the [^{14}C]acetyl-enzyme, acid denaturation, and proteolytic fragmentation with trypsin leads to a single radioactive peptide, which contains Cys89 as the sole site for the [^{14}C]acetyl group. The acetyl-enzyme cannot be characterized in an undenatured from because of its hydrolytic reactivity. It is much more reactive with water than a typical thioesters, and this presumably represents enhanced chemical reactivity at the active site, a property that would be advantageous for the acetyl transfer step of the chemical mechanism in fig. 14-8. Mutation of Cys89 to serine dramatically decreases the activity, 90-fold for acetyl transfer and 2000-fold for condensation. [^{14}C]Acetyl CoA labels Ser89 in C89S-thiolase, as it does Cys89 in the wild-type enzyme. The residual activity of the mutated enzyme evidently represents the decreased activity of C87S-thiolase when Ser89 functions as the nucleophilic catalyst.

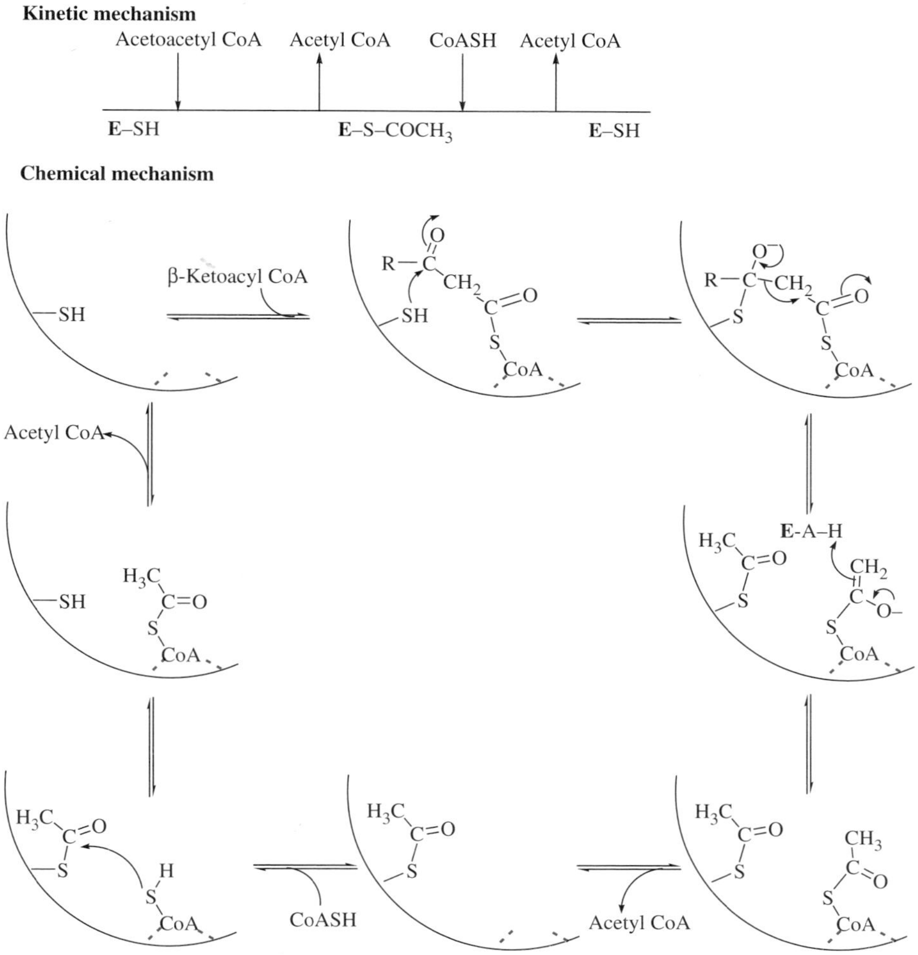

Fig. 14-8. Kinetic and chemical mechanisms for the action of thiolases. The basic kinetic mechanism for the action of biosynthetic acetoacetyl CoA thiolase is ping pong bi bi (top). One acetyl CoA is released before CoASH binds, with the formation of an intervening covalent acetyl-enzyme intermediate. The general chemical mechanism for the action of thiolase shows the dual role of the active-site cysteine (Cys89 in acetoacetyl CoA thiolase).

Dithio-acetyl CoA (CH$_3$CS–SCoA) reacts as a partial substrate for acetoacetyl CoA thiolase (Anderson et al., 1990). Dithio-acetyl CoA alone does not react as a substrate; that is, it will not react to trithio-acetoacetyl CoA (CH$_3$CSCH$_2$CS–SCoA). However, it readily reacts in the presence of acetyl CoA to form aceto-dithioacetyl CoA (CH$_3$COCH$_2$CS–SCoA). Dithio-acetyl CoA does not acetylate the enzyme, but it does react in the condensation step with the acetyl-enzyme generated from acetoacetyl CoA. This means that although dithio-acetyl CoA does not react as an electrophilic acetylating substrate, it does react as the nucleophilic thioenolate to form the carbon-carbon bond. As in the reaction of citrate synthase, dithio-acetyl CoA is readily enolized to the thioenolate anion, and this reacts with the acetyl-enzyme to form aceto-dithioacetyl CoA.

Acetoacetyl CoA thiolase catalyzes the exchange of the α-protons of dithio-acetyl CoA with deuterium in D$_2$O. In contrast, thiolases do not catalyze this exchange with acetyl CoA. The difference presumably arises from the greater acidity of the α-protons in dithioacetyl CoA relative to acetyl CoA ($\Delta pK_a \approx 7$ to 8). The reactions of dithio-acetyl

CoA catalyzed by acetoacetyl CoA thiolase suggest that the dithioenolate is the nucleophilic intermediate and not the neutral dithioenol. Ditho-acetyl CoA reacts at about 1/35 the rate of acetyl CoA, and this may be regarded as a modest difference, given the steric consequences of the replacement of C=O with C=S.

Structure of Acetoacetyl Coenzyme A Thiolase

A number of structures of the enzyme from *Zoogloea ramigera* are available, including the free enzyme and complexes with CoASH, the acetylated enzyme complexed with CoASH, the acetylated enzyme, and the acetylated enzyme complexed with acetyl CoA. Structures of the mutated enzyme C89A-thiolase and its complexes with CoASH and with acetyl CoA are available. These structures dramatically illuminate a number of mechanistic aspects in the action of thiolases. The basic structure is a peripherally associated dimer of very intimately associated dimers, described as a "dimer of tight dimers" (Kursula et al., 2002). A structure is shown in fig. 14-9, in which each subunit is folded into three domains. The N- and C-terminal domains are associated into five layers of α–β–α–β–α strands packed around two central N- and C-terminal helices. The third domain is a loop (119-249) inserted into the N-terminal domain. The active site is located at the end of a short tunnel that binds CoASH and the CoA-moiety of acetyl CoA and acetoacetyl CoA.

The active site contains the catalytic residues Cys89, Cys378, His348 and a water molecule, Wat82, which is found in all structures. Consideration of all the available structures allows two oxyanion binding sites to be identified, one for each of the carbonyl groups of acetoacetyl CoA. The site for the aceto group is formed by the amide-NH groups of Cys89 and Gly380, and the site for the acetyl carbonyl group is formed by His348-N$^\varepsilon$H and Wat82. Structures of acetyl-thiolase and the acetyl CoA complex of acetylthiolase show that the binding of acetyl CoA changes the conformation of the acetyl group of acetylthiolase. Consideration of the available structures allows the catalytic cycle to be rationalized structurally (Kursula et al., 2002). The role of Cys89 as the nucleophilic catalyst is confirmed by the structures. Cys378 is postulated to mediate the shuttling of protons. His343 is involved in oxyanions binding.

The exact chemical mechanism for the enolization of acetyl CoA remains unknown. The reaction of dithioacetyl CoA suggests that the enolized form is the anionic dithioenolate and not the neutral dithioenol. The reactivity of this species as the nucleophile in carbon condensation suggests is compatible with the participation of an anionic acetyl CoA enolate in the normal reaction. As in the case of citrate synthase, mechanisms can be written with the conjugate acid of His343 in a hydrogen bonded ion pair with the enolate and with the conjugate base of acetyl CoA in an LBHB with the enolate. The elucidation of the protonation state of His343 in the enolate complex would clarify the enolization mechanism.

Carbanionic Mechanisms

Certain carbon condensation reactions clearly proceed through carbanionic intermediates. These reactions often involve iminium intermediates formed between ketone groups of substrates and lysine-ε-amino groups of active sites. Alternatively, coenzymes such as thiamine pyrophosphate (TPP) or pyridoxal-5'-phosphate (PLP) can facilitate carbanion formation for condensation reactions (see chap. 3). We consider examples of each type in this chapter.

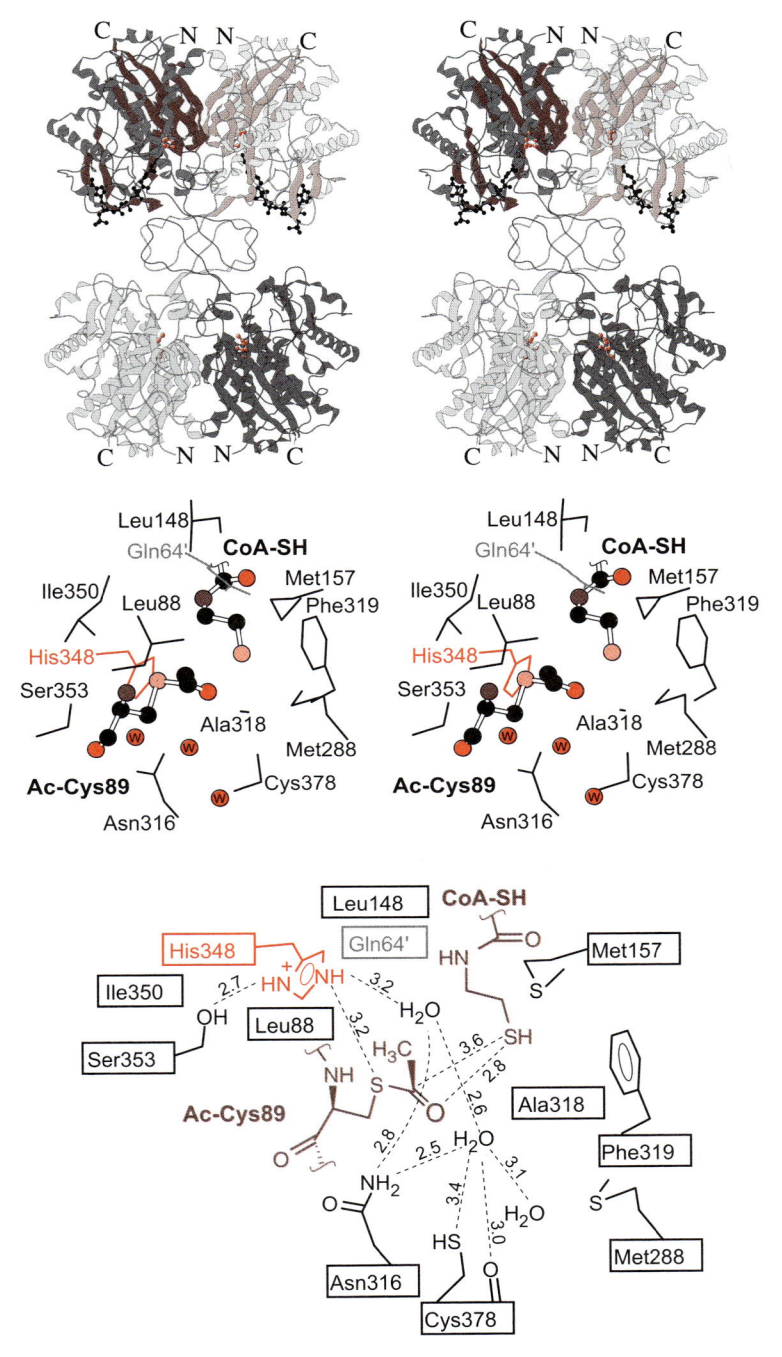

Fig. 14-9. Structure of acetoacetyl CoA thiolase. The x-ray crystallographic structure shows the homotetrameric enzyme from *Zoogloea ramigera* in a complex with coenzyme A (black) and *S*-acetylcysteine (red) (1.90-Å resolution; PDB 1NL7; Kursula et al., 2002).

Transaldolase

Transaldolase (EC 2.2.1.2) and transketolase play key roles in the pentose phosphate pathway of carbohydrate metabolism, also known as the hexose monophosphate shunt (Tsolas and Horecker, 1972). Both enzymes catalyze carbon-carbon bond formation through the intermediate formation of carbanionic intermediates. However, in both processes, carbon-carbon

632   Enzymatic Reaction Mechanisms

bonds are also broken, and the reactions essentially involve transfers of carbon-carbon bonds. Because of the similarities of the broken and newly formed bonds, the reactions are approximately energetically balanced. We first consider transaldolase, which acts by a mechanism related to that of a class I aldolase.

Reaction and Kinetics

Transaldolase catalyzes the overall reaction of sedoheptulose-7 P with glyceraldehyde-3-P to form fructose-6-P and erythrose-4-P, as shown in fig. 14-10A. The overall reaction entails the transfer of the dihydroxyacetone portion of sedoheptulose-7-P to the aldehyde group of glyceraldehyde-3-P. Cleavage of sedoheptulose-7-P by a retroaldol process would produce erythrose-4-P and a dihydroxyacetone-carbanion-equivalent, and addition of the carbanion-equivalent to glyceraldehyde-3-P would produce fructose-6-P. Transaldolase accepts a broad range of hydroxyacetone-donor substrates and an even broader range of acceptor aldehydes (Tsolas and Horecker, 1972). This property allows tranaldolase to be employed in the synthesis of a large variety of sugar phosphates (Takayama et al., 1997).

How is the dihydroxyacetone carbanion formed and transferred? Part of the key to the answer is that transaldolase is subject to substrate-dependent reductive inactivation by NaBH$_4$, a classic property of class I aldolases, and this is accompanied by the covalent attachment of dihydroxyacetone as a glyceryl substituent to a lysyl residue of the protein (Tsolas and Horecker, 1972). Transaldolase cleaves sedoheptulose-7-P to erythrose-4-P and a dihydroxyacetone enamine-iminium hybrid with the active site lysine. In addition to reduction by NaBH$_4$, imine linkage between dihydroxyacetone and the lysine at the active site also undergoes addition by cyanide ion, so that the enzyme is subject to substrate-dependent inactivation by cyanide (Tsolas and Horecker, 1972). The ping pong mechanism in fig. 14-10B likely pertains to the action of transaldolase.

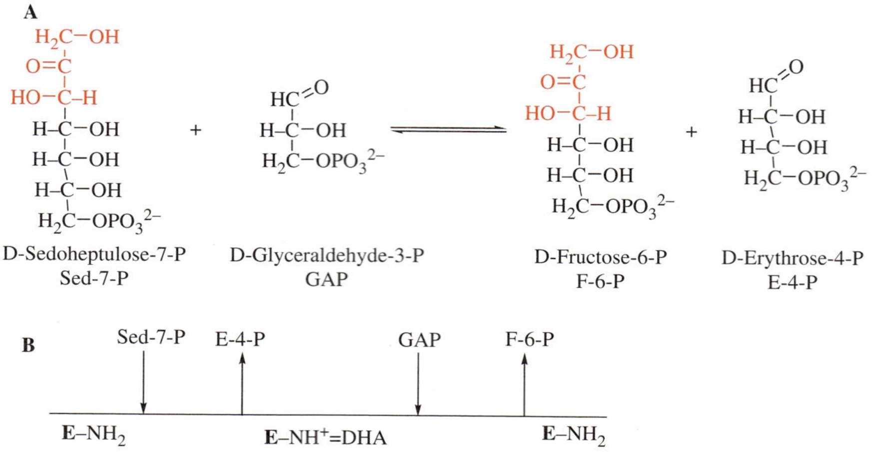

Fig. 14-10. (A) Transaldolase catalyzes the transfer of the dihydroxyacetone (DHA) moiety of sedoheptulase-7-P to the aldehyde carbon of glyceraldehyde-3-P, producing fructose-6-P and erythrose-4-P. (B) The dihydroxyacetone moiety of Sed-7-P becomes bonded in an iminium linkage to the ε-amino group of Lys132 at the active site, and this allows erythrose-4-P to be released. Glyceraldehyde-3-P then reacts with the covalent enzyme intermediate to form fructose-6-P.

Structure

Transaldolases are homodimeric enzymes with molecular masses in the range of 70 kDa (Tsolas and Horecker, 1972). The transaldolase from *Escherichia coli* can be induced, through directed mutagenesis in the subunit interface, to exist in a predominantly monomeric form that is fully active (Schörken et al., 1998). A crystal structure of transaldolase from *E. coli* reductively inactivated by NaBH$_4$ in the presence of substrate, is shown in fig. 14-11 (Jia et al., 1997). The structure is an α/β barrel, typical of aldolases, with dihydroxyacetone attached as a glyceryl-moiety to Lys132. Several conserved amino acid side chains are at the active site and in contact with the N$^\varepsilon$-glyceryl-Lys132. Asp17 and Ser176 are within hydrogen bonding range of the glyceryl-OH groups. Asn35 and Thr156 are in contact with a fixed water molecule, which in turn is near the glyceryl moiety and forms a bridge to Glu96. Mutation of any one of these residues to alanine severely compromises activity, with the main effect on k_{cat}, and in the case of Ser176 a lesser effect on K_m (Schörken et al., 2001). The single mutations to alanine have little effect on the structure, apart from depletion of electron density for the absent side chain, except for the mutations of Thr156 and Glu96, which affected the occupancy of the fixed water molecule (Schörken et al., 2001).

Reaction Mechanism

The mechanism in steps 1 through 6 of fig. 14-12 for the enzyme from *E. coli* explains the action of transaldolase in transferring the dihydroxyacetone group from a ketose phosphate to an aldose phosphate. Imine formation between Lys132 and the donor substrate facilitates cleavage of the C3–C4 bond, through the action of Asp17 in abstracting the proton from C3(OH). Lys132 stabilizes the resulting carbanion as an enamine. The carbon pK_a of an iminium group, a protonated imine, is about 8 units lower than that of an analogous ketone (see table 1-4 in chap. 1) (Bender and Williams, 1962), so that an imine of dihydroxyacetone should display a pK_a of about 10, a reasonable value for an enzymatic intermediate. There is no information regarding whether the microenvironment at the active site might lower the pK_a into the neutral range. In step 5, glyceraldehyde-3-P replaces the erythrose-4-P released in step 3. In the current working mechanism, the dihydroxyacetone-enamine reacts as a carbanion with glyceraldehyde-3-P to form the Lys132—imine of fructose-6-P, which undergoes hydrolysis to release the product and complete the catalytic cycle.

The exact role of the fixed water molecule, found in all structures of wild-type transaldolase and shown in fig. 10-12, is not known. It may mediate a proton shuttle mechanism for reversible protonation of the carbanion in step 4a of fig. 10-12, in which Glu96 acts as the acid-base catalyst. This is not an essential step in the mechanism, and mutation of Glu96 decreases the activity by only about 20-fold (Wikner et al., 1997). Alternatively, it may participate in imine formation and hydrolysis by mediating proton transfer with Glu96, a process that is unlikely to be rate limiting.

The action of transaldolase differs from other aldolases in that the enzyme does not catalyze the hydrolysis of the iminium bond linking Lys1132 to dihydroxyacetone. This group is able to react with any of a number of aldoses to form the corresponding ketoses without the complication of an intervening hydrolytic step. The imine complex is so stable toward hydrolysis that it can be isolated by column chromatography, and it will then react with glyceraldehyde-3-P or erythrose-4-P to form fructose-6-P or sedoheptulose-7-P respectively (Horecker et al., 1962). The corresponding intermediate in the action of a typical aldolase would undergo hydrolysis, as in hydrolysis to dihydroxyacetone phosphate by fructose-1,6-bisphosphate aldolase (see chap. 1).

634 Enzymatic Reaction Mechanisms

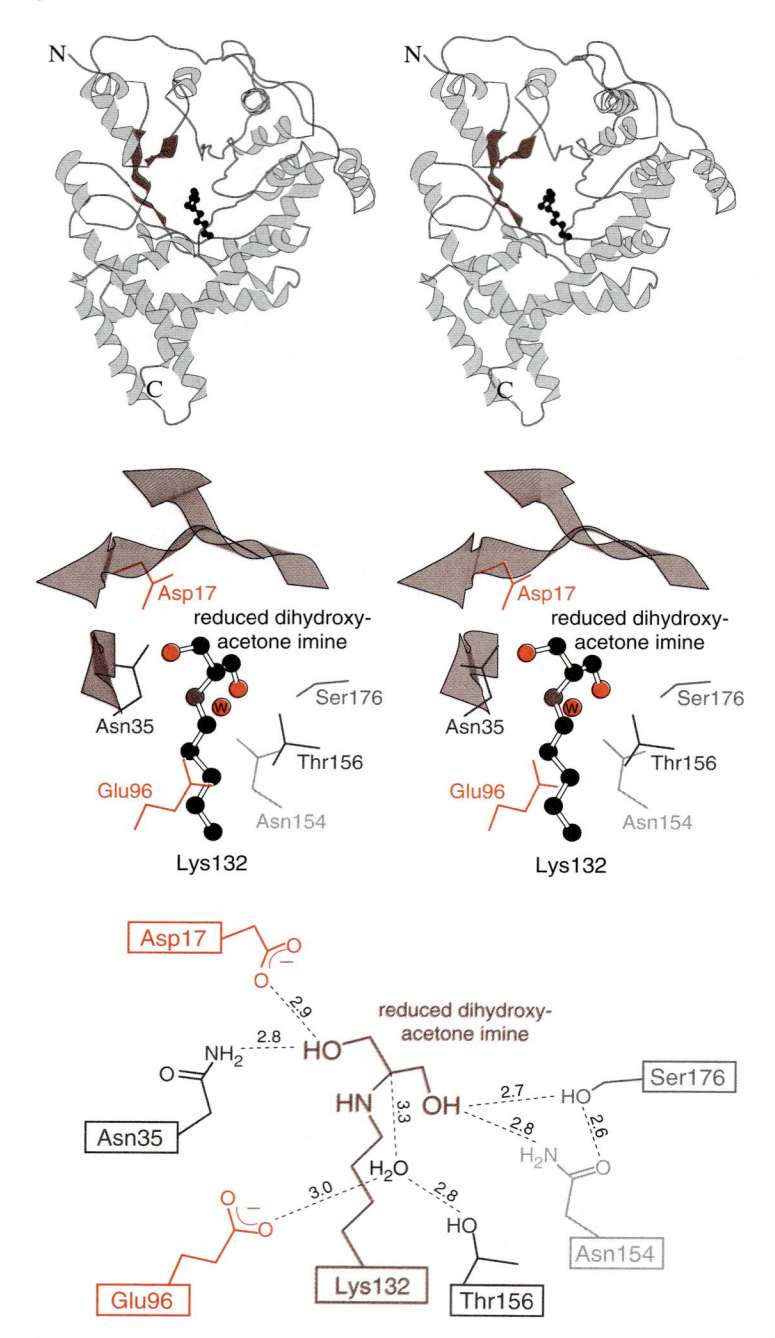

Fig. 14-11. The x-ray crystallographic structure shows transaldolase from *E. coli* with the intermediate ketimine reduced by KBH_4 (black) (2.20-Å resolution; PDB 1UCW; Jia et al., 1997).

Transketolase

Reaction and Kinetics

Sedoheptulose-7-P arises in the pentose phosphate pathway through the action of transketolase (EC 2.2.1.1), which catalyzes the transfer of a hydroxyketone group from C1 and C2 of xylulose-5-P to the aldehyde carbon of ribose-5-P, as illustrated in fig. 14-13A

Carbon-Carbon Condensation and Cleavage 635

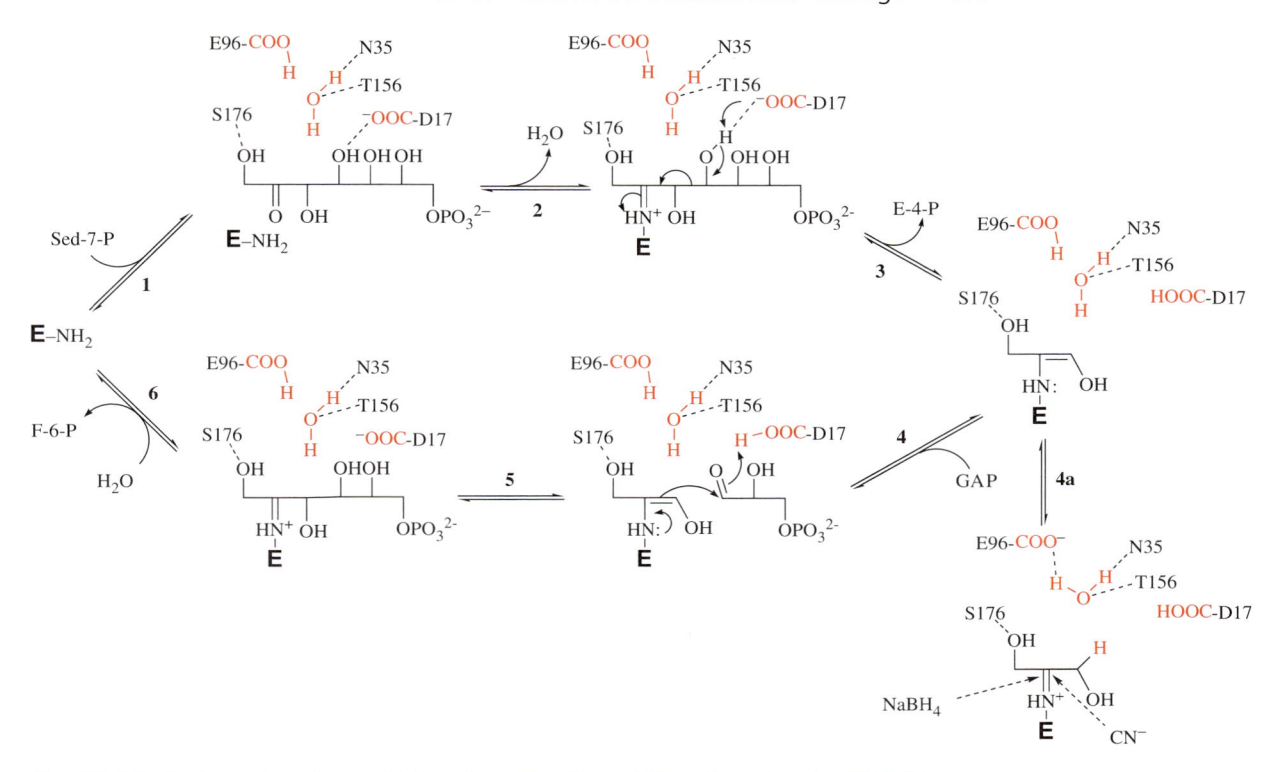

Fig. 14-12. In the action of transaldolase from *E. coli*, Lys132 at the active site (**E-NH₂**) reacts with the 2-keto group of the donor substrate, sedoheptulose-7-P, to form the protonated imine, or iminium, intermediate in steps 1 and 2. Asp17 is in position to abstract the proton from the C3(OH) and drive the cleavage in step 3 to release erythrose-4P and form the enamine-carbanion intermediate. The intermediate binds glyceraldehyde-3-P in step 4 and condenses with it in step 5 to form the imine of fructose-6-P. Hydrolysis of the imine and release of fructose-6-P in step 6 completes the cycle. Functional groups of the enzyme and the fixed water molecule are shown in red.

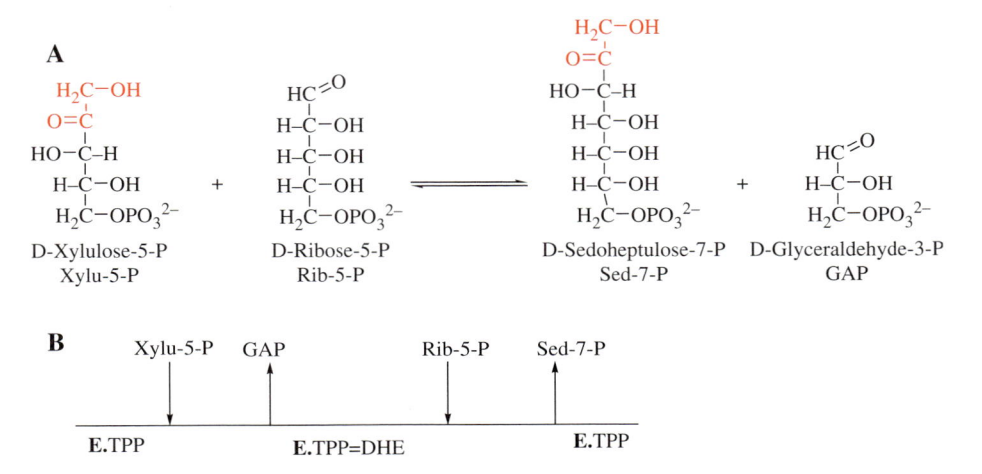

Fig. 14-13. (A) Transketolase catalyzes the transfer of the hydroxyketone moiety of xylulase-5-P to the aldehyde carbon of ribose-5-P, producing sedoheptulose-7-P and glyceraldehyde-3-P. (B) Transketolase is a thiamine pyrophosphate (TPP)–dependent enzyme, and TPP provides the chemical power to accept and transfer the hydroxyketone group. The reaction follows ping pong kinetics, in which the first half-reaction proceeds with transfer of the hydroxyketone group from xylulose-5-P to TPP to form dihydroxyethylidene-TPP at the active site (**E.TPP=DHE**), with release of glyceraldehyde-3-P. In the second half-reaction, ribose-5-P accepts the hydroxyketone group from dihydroxyethylidene-TPP to form sedoheptulose-7-P.

(Kochetov, 2001). The other product arising from xylulose-5-P is glyceraldehyde-3-P. The basic chemistry required for this process is provided by TPP, the coenzyme of transketolase, which is bound in complex with a divalent metal ion, usually calcium, although other divalent ions support activity. TPP functions in its traditional role of facilitating the cleavage of a hydroxyketone (see chap. 3). The reaction proceeds with the intermediate formation of dihydroxyethylidene-TPP at the active site (Fiedler et al., 2002; Usmanov et al., 1996). The reaction proceeds by the ping pong bi bi kinetic mechanism in fig. 14-13B (Gorbach and Kubyshin, 1989).

Transaldolase and transketolase catalyze transfers of different groups, dihydroxyacetone and glycolaldehyde respectively, to aldose acceptors. The enzymes act with varying efficiency on a range of substrates in addition to the natural ones shown in fig. 14-10 and 14-13 (Kotchetov, 2001). This property, which extends to aldolases in general, allows transketolase and transaldolase to be employed together with other aldolases in the enzymatic synthesis of novel carbohydrates (Takayama et al., 1997). Transketolase acts on simpler substrates such as hydroxypyruvate and dihydroxyacetone, and this extends its utility in chemical synthesis.

Structure

Transketolase is a homodimeric enzyme, with a molecular mass of about 150 kDa and two active sites (Cavalieri et al., 1975; Kochetov, 2001). The molecular structure is much like those of other TPP-dependent enzymes, including pyruvate decarboxylase, the E1-component of the pyruvate dehydrogenase complex, and pyruvate oxidase (Muller et al., 1993). The three enzymes have similar chain folds and binding interactions with TPP. The structures of pyruvate decarboxylase and the E1-component of the pyruvate dehydrogenase complex are shown in chapters 8 and 18, respectively.

The similarity among the structures of TPP-dependent enzymes is unlike the diversity of structures among PLP-dependent enzymes. If there is a reason for this state of affairs, it may reside in the comparative structures and mechanisms of action of TPP and PLP. TPP includes considerable structural flexibility in solution because of torsion of the pyrimidine and thiazolium rings about the central methylene group. Only one rotamer of the V-shaped molecule is found in enzymatic active sites, the one that appears to permit the pyrimidine-6-NH_2 group to participate in catalysis by the thiazolium ring (see chap. 8). The correct binding of this rotamer at all TPP sites may be the impetus for the structural similarities among TPP-enzymes. The PLP-enzymes seem different. The pyridinium ring is differentially involved in catalysis among this group of enzymes, and in any case it is an essentially rigid, planar molecule, apart from the torsional freedom of the phosphate group. It may be that the differential participation of the pyridine ring in catalysis may allow variations in binding sites and chain folds among PLP-enzymes.

The active site interactions of TPP and dihydroxyethylidene-TPP with transketolase from *Saccharomyces cerevisiae* are illustrated in fig. 14-14 and show some of the amino acid contacts and the chelation of Ca^{2+}, which binds the pyrophosphate moiety to the enzyme. Fig. 14-14A shows His30, His103, His263, and His481 in position to serve as acid-base catalysts or hydrogen bonding sites, or both, for substrates in the resting complex of enzyme and TPP. Fig. 14-14A also illustrates how Glu418 and the 6-amino group of the pyrimidine ring are thought to mediate the ionization of C2(H) in the thiazolium ring (Kochetov, 2001). The structure of the complex of transketolase with dihydroxyethylidene-TPP in fig. 14-14B shows His481 hydrogen bonded to the α-hydroxyl group and His103 hydrogen bonded to the β-hydroxyl group of dihydroxyethylidene-TPP. The 6-amino group of the pyrimidine moiety is also less than 3 Å from the α-hydroxy group. Most of

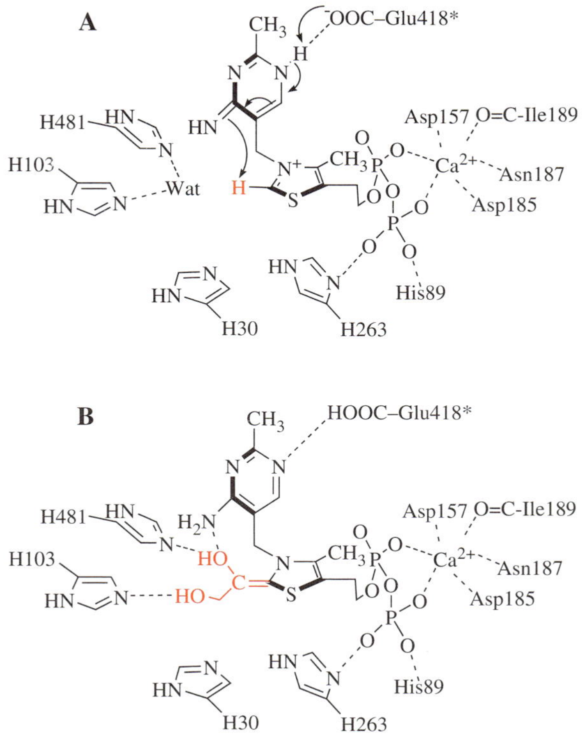

Fig. 14-14. Thiamine pyrophosphate (TPP) and dihydroxyethylidene-TPP at the active site of transketolase. (A) TPP is held in the active site through hydrogen-bonded interactions with main-chain C=O and N–H groups of the enzyme and the amino acid side chains (Schneider and Lindqvist, 1998). Ca^{2+} contributes to binding through chelation with the pyrophosphate group and the enzyme ligands. On the basis of spectroscopic evidence and the structure, it is postulated that the ionization of C2 of TPP is brought about through proton transfer to the 6-amino group of the pyrimidine ring and that Glu418 facilitates this process (Kochetov, 2002). (B) The structure of transketolase with dihydroxyethylidene-TPP bound at the active site shows His481 in hydrogen-bonded contact with the α-hydroxyl group of dihydroxyethylidene-TPP and His103 hydrogen bonded to the β-hydroxyl group (Fiedler et al., 2002).

the amino acid residues in the active site are in one subunit, but His481 and Glu418 are from the neighboring subunit.

Reaction Mechanism

An overall mechanism is based on the involvement of dihydroxyethylidene-TPP as the key intermediate, as shown in fig. 14-15. However, the detailed mechanism of proton transfer is elusive because of the many potential acid-base catalysts, especially histidine residues, in the active site. The hydrogen-bonded contact of His481 with the α-hydroxyl group of dihydroxyethylidene-TPP implies that it may function in acid-base catalysis in the addition/elimination of C2 of TPP to the 2-keto group of the substrate and product. However, mutation of this residue to alanine only modestly affects the kinetic parameters (Wikner et al., 1997), and this residue is glutamine in mammalian transketolase. It may be that the nearby 6-amino group of the pyrimidine ring catalyzes proton transfer in the addition/elimination process. The 3-hydroxyl group of the substrate interacts with both His263 and His30, and mutation of these residues leads to low activity (<1.5%) (Wikner et al., 1997).

638 Enzymatic Reaction Mechanisms

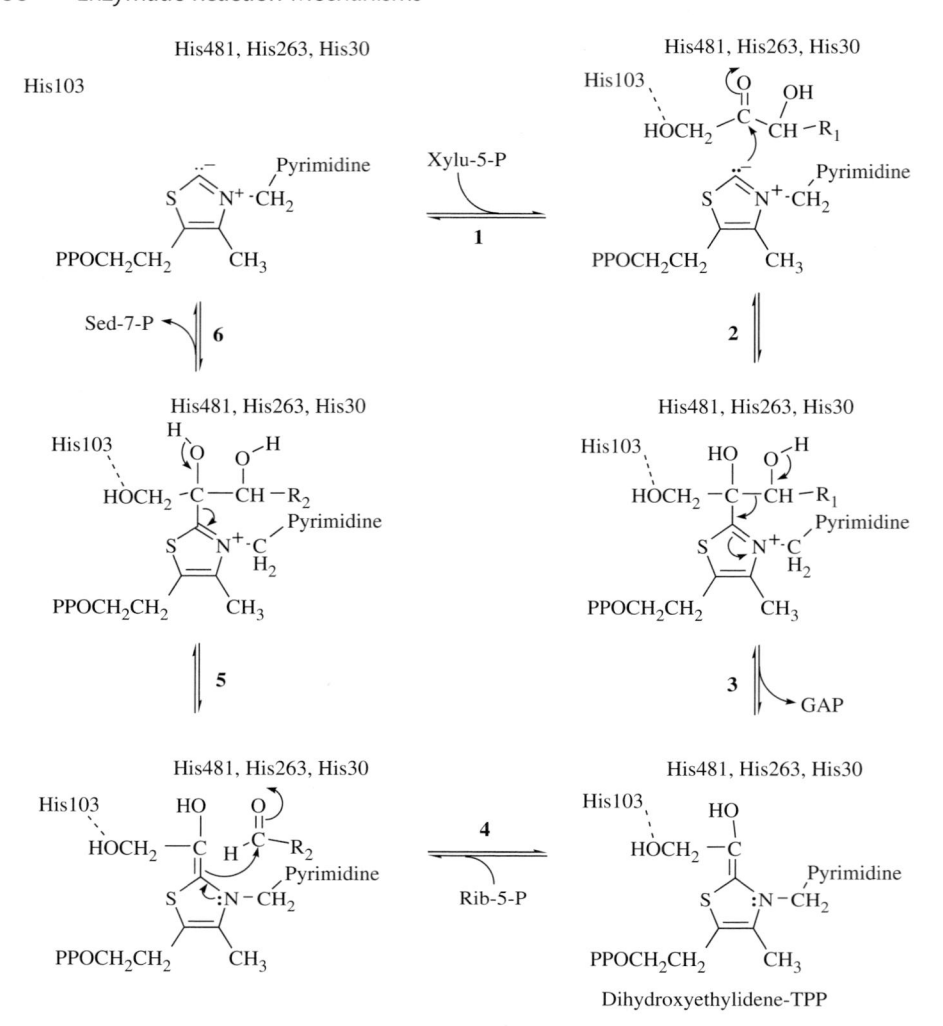

Fig. 14-15. A mechanism for the action of transketolase. This mechanism makes use of the chemical properties of thiamine pyrophosphate (TPP) to generate a carbanion-enamine resonance hybrid intermediate by cleavage of the C3-C3 bond of xylulose-5-P. The resulting dihydroxyethylidene-TPP intermediate reacts as a carbanion with the aldehyde group of ribose-5-P to form an adduct, which decomposes by elimination of TPP to generate sedoheptulose-7-P. The dihydroxyethylidene-TPP intermediate is the species **E.TPP=DHE** of the ping pong bi bi mechanism in fig. 14-13B. Structures show that His481 is in position to react with the 2-keto group of substrates and the α-hydroxyl group of dihydroxyethylidene-TPP, and His103 engages in hydrogen-bonded contact with the β-hydroxyl group of dihydroxyethylidene-TPP (Fiedler et al., 2002). His30 and His263 interact in hydrogen-bonding distance with the 3-hydroxyl group of the substrate (Wikner et al., 1997).

One or both may function in acid-base catalysis of carbon-carbon bond cleavage and formation. His69 lies near the binding site for the 1-hydroxyl group and is required for activity (Wikner et al., 1997). It may act in cooperation with His103 in substrate recognition. Transient kinetics and kinetic isotope effects may clarify the mechanism of action of transketolase in the future.

The phosphate moiety of erythrose-4-P binds through interactions with Arg359, Ser386, His469, and Arg528. The aldehyde group is hydrogen bonded to His30 and His263. Mutation of His469 to alanine has little effect on the kinetic parameters. The variants

R359A and R528A display modestly less activity (threefold to sixfold) and threefold to sixfold higher values of K_m relative to the wild-type enzyme (Nilsson et al., 1997).

Serine Hydroxymethyltransferase

Reaction and Metabolic Role

In all organisms, the biosynthesis of nucleic acids requires one-carbon units in the form of methyl groups for the biosynthesis of thymine and formyl groups for the biosynthesis of purines. Molecules such as 10-formyltetrahydrofolate, 5,10-methylenetetrahydrofolate, and *S*-adenosylmethionine embody the one-carbon units. They are derived directly or indirectly from the action of serine hydroxymethyltransferase (SHMT; EC 2.1.2.1), which catalyzes the reaction of serine with tetrahydrofolate to produce glycine and 5,10-methylenetetrahydrofolate (fig. 14-16A). SHMT also catalyzes the hydration of 5,10-methenyltetrahydrofolate to 5-formyltetrahydrofolate, which appears to play a regulatory role in one-carbon metabolism (Girgis et al., 1997; Stover and Schirch, 1991). For these reasons, SHMT is a potential target for cancer chemotherapy.

The traditional reaction of SHMT, the cleavage of serine to glycine and 5, 10-methylenetetrahydrofolate, is reversible and in some biological settings functions to produce serine. For example, in methanotrophs methane is initially oxidized to formaldehyde.

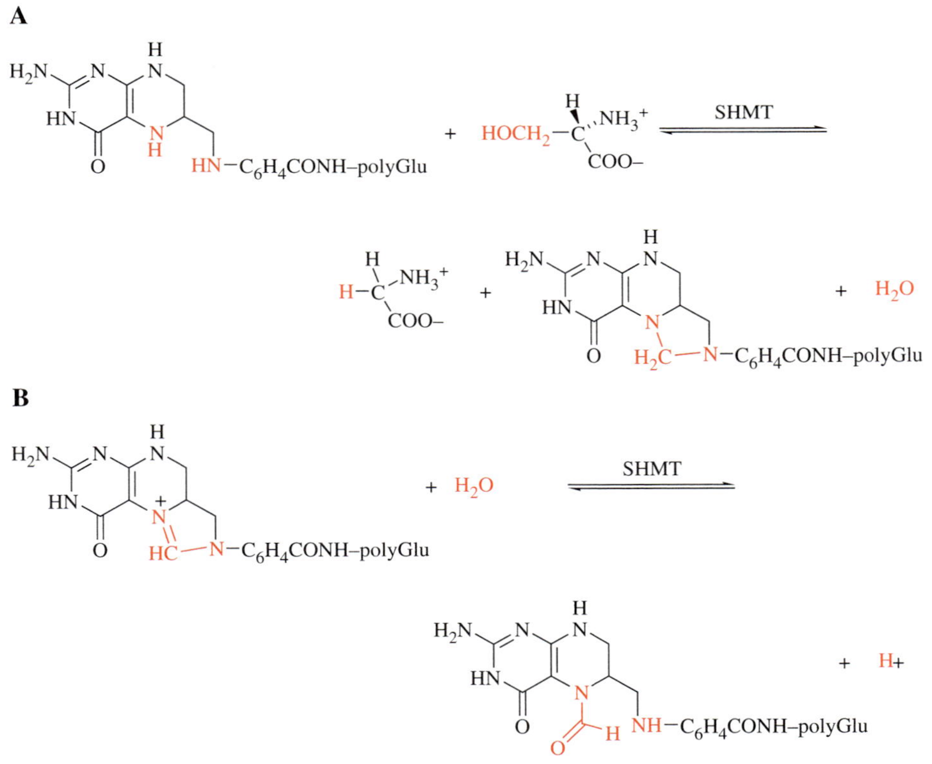

Fig. 14-16. Reactions catalyzed by serine hydroxymethyltransferase. (A) In its primary metabolic role, serine hydroxymethyltransferase (SHMT) catalyzes the reversible transfer of the hydroxymethyl group between L-serine and polyglutamylated methylenetetrahydrofolate. (B) SHMT also catalyzes the hydration of 5,10-methenyltetrahydrofolate to 5-formyltetrahdrofolate. The reacting groups are shown in red.

640   Enzymatic Reaction Mechanisms

Serine generated from formaldehyde and glycine through a methylene tetrahydrofolate intermediate serves as the carbon building block for these organisms.

Molecular Properties

SHMT is a PLP-enzyme, the action of which has attracted considerable interest (Schirch, 1982). As purified from *E. coli*, SHMT is a homodimer with an overall molecular weight of 94,000 and a subunit molecular weight of 47,000 (Schirch et al., 1985). Mammalian hydroxymethyltransferases have subunit molecular weights of about 60,000 (Martini et al., 1987, 1989).

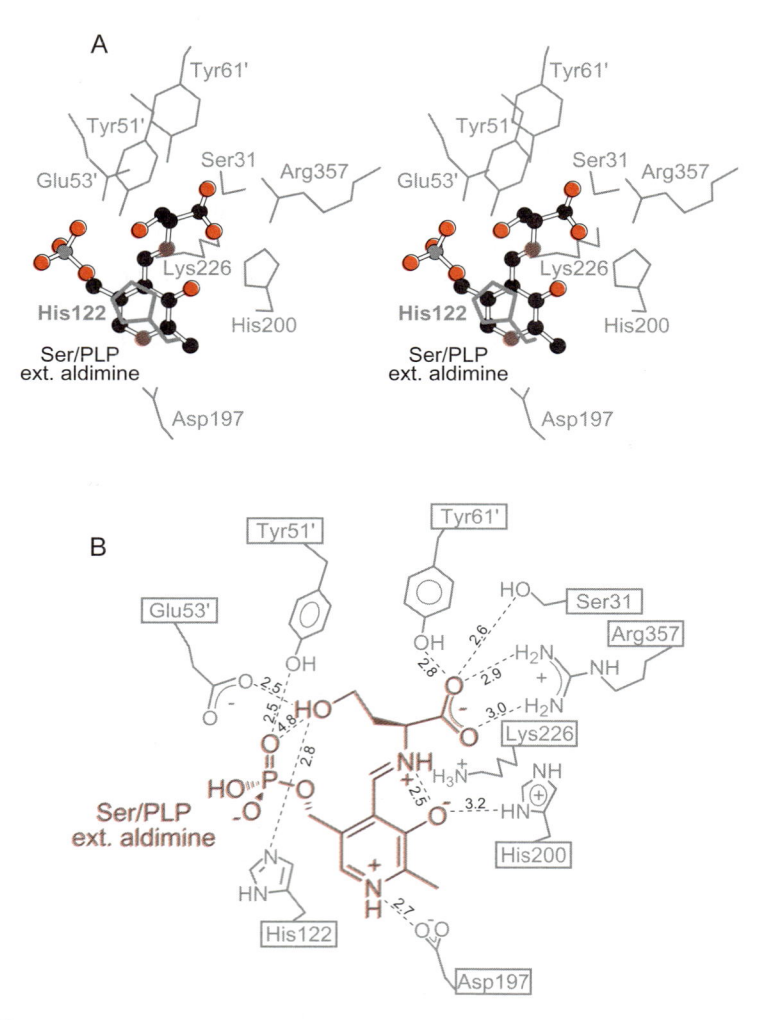

Fig. 14-17. The x-ray crystallographic structure of serine hydroxymethyltransferase (SHMT) from *Bacillus stearothermophilus* is in the form of its serine external aldimine (1.93-Å resolution; PDB 1KKP; Trivedi et al., 2002). SHMT is a member of the α-family of pyridoxal-5′-phosphate (PLP) enzymes, and its global structure and chain fold (not shown) are similar to those of aspartate aminotransferase (Fig. 13-3). (A) The active site of SHMT is shown with the external aldimine of serine with PLP. (B) The amino acid contacts with the external aldimine are shown. In the internal aldimine, PLP is bonded to Lys226. Despite the close contacts of Glu53′ and His122 with the β-OH group of serine, evidence indicates that neither of these residues serves as a base catalyst for abstracting a proton in the retroaldol mechanism of carbon-carbon cleavage.

Carbon-Carbon Condensation and Cleavage

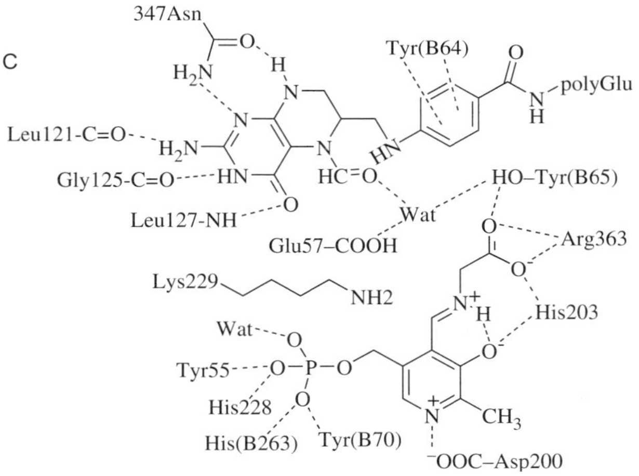

Fig. 14-17. cont'd (C) The amino acid contacts of 5-formyl-tetrahydrofolate are shown for the complex of the external aldimine of *E. coli* SHMT with glycine (Constabile et al., 2000). The numbering is different from that in B because of the species difference.

Active-Site Structure

Human, rabbit, and mouse SHMTs and *E. coli*, and *Bacillus stearothermophilus* SHMTs are similar members of the α-class of PLP-enzymes for which structures are available (Renwich et al., 1998; Scarsdale et al., 1999, 2000; Szebnyi et al., 2000; Trivedi et al., 2002). The bacterial enzymes are homodimers, and the eukaryotic enzymes are dimers of tight homodimers. The overall chain fold of a subunit is similar to that of aspartate aminotransferase (AAT) in fig. 13-3, and the differences from AAT allow for the creation of a binding site for tetrahydrofolate. As in AAT, the active site is formed by amino acid side chains from both subunits.

Crystalline SHMT from *B. stearothermophilus* forms binary external aldimine complexes with glycine or serine and ternary complexes of these with 5-fromyltetrahydrofolate (Trivedi et al., 2002). The structure at the active site of the binary external aldimine complex with serine is shown in fig. 14-17. In free SHMT, PLP is bound as an internal aldimine by a lysyl residue, and N1 of the pyridine ring in PLP is associated in an ion pair with an aspartate residue. In *B. stearothermophilus* SHMT, these residues are Lys226 and Asp197, respectively (Trivedi et al., 2002), whereas in *E. coli*, they are Lys229 and Asp200 (Shirch et al., 1993; Stover et al., 1992). K229A-SHMT from *E. coli* binds the external aldimine of serine with PLP and undergoes a single turnover with tetrahydrofolate (Schirch et al., 1993). The consequences of mutating other amino acid residues with side chains contacting PLP, glycine, serine, or 5-formyltetrahydrofolate, and of residues at the intersubunit interface, are compiled in a review (Rao et al., 2003). Site-specific mutations of residues corresponding to Lys226, Arg357, and His200 in SHMTs from various species led to essentially inactive variant proteins (Jagath et al., 1997; Talwar et al., 1997, 2000).

Reaction Mechanism

The kinetic mechanism is random sequential for substrate binding and product release (Schirch et al., 1977). PLP plays an essential and classical role in the action of SHMT by stabilizing the α-carbanionic amino acid intermediates generated in the mechanism. Three chromophores are observed in transient and steady states at saturating substrates, the

425-nm band characteristic of external aldimines, the 498-nm band characteristic of the quinonoid-α-carbanion, and a 343-nm band assigned to the *gem*-diamine intermediate in the transaldimination reaction (Schirch, 1975). These structures are shown in fig. 14-18. The quinoid chromophore at 498 nm, a characteristic of PLP-amino acid α-carbanions, is a prominent transient in the action of SHMT and is a compulsory intermediate in any mechanism for the action of SHMT.

The detailed reaction mechanism for the tetrahydrofolate-dependent cleavage of serine and formation of 5,10-methylenetrahydrofolate in steps 4a, 4b, and others of fig. 14-18 remains uncertain. Three mechanisms shown in fig. 14-19 are under consideration, retroaldol, nucleophilic C—C cleavage, and nucleophilic C—O cleavage. The retroaldol mechanism generates formaldehyde as a discrete, enzyme-bound intermediate that is trapped by tetrahydrofolate to form 5,10-methylenetetrahydrofolate. This mechanism seems likely in

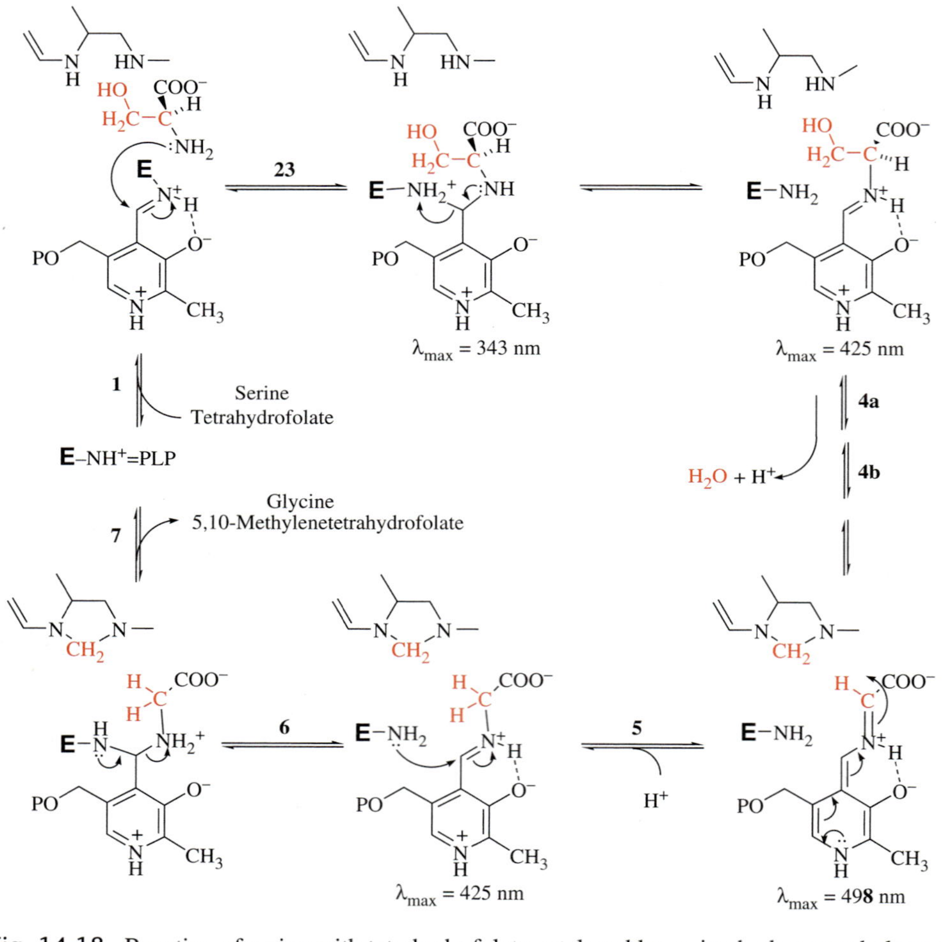

Fig. 14-18. Reaction of serine with tetrahydrofolate catalyzed by serine hydroxymethyltransferase. One cycle in the reaction of serine with tetrahydrofolate begins with binding of the substrates in step 1 and transaldimination in steps 2 and 3. Then, cleavage of the carbon-carbon bond in several steps (4a, 4b, ...) leads to methylenetetrahydrofolate and the α-carbanion of the external aldimine between glycine and pyridoxal-5′-phosphate, which displays an absorption maximum at 498 nm. Protonation of the carbanion in step 5, transaldimination in step 6, and release of glycine and methylenetetrahydrofolate in step 7 complete the cycle. The functional groups undergoing chemical change at each step are shown in red.

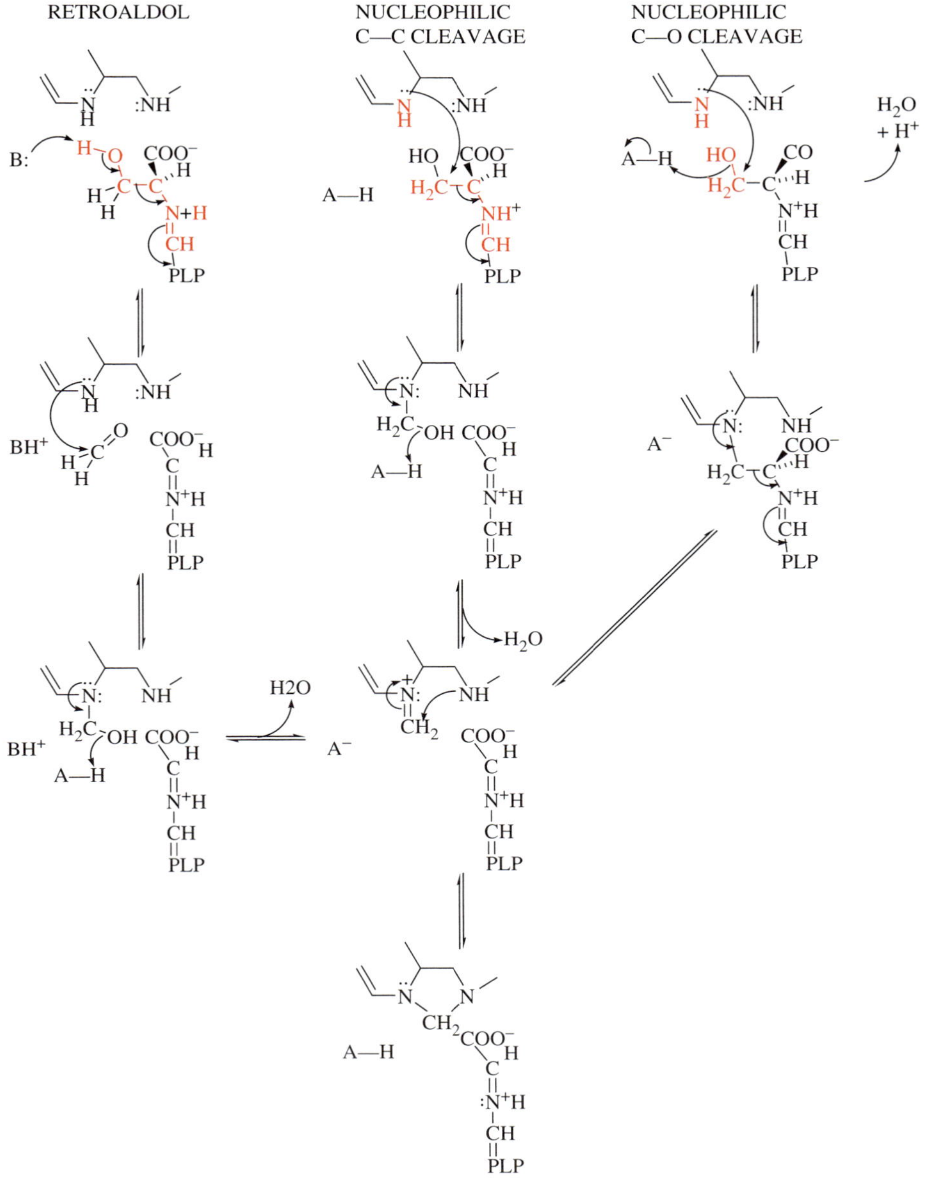

Carbon-Carbon Condensation and Cleavage   643

Fig. 14-19. Mechanisms under consideration for cleavage of serine by action of SHMT. The mechanisms refer to steps 4a, 4b, and so on in fig. 14-18. Different processes initiate the three mechanisms, and the initial reacting groups are indicated in red. All mechanisms lead to a common penultimate intermediate and then to 5-formyltetrahydrofolate and the quinoid-α-carbanion of the glycine–pyridoxal-5′-phosphate external aldimine.

view of the reactivities normally associated with PLP and the exceedingly high reactivity of formaldehyde with vicinal amines and tetrahydrofolate. However, the reaction of serine with SHMT is about 200,000 times faster in the presence of tetrahydrofolate. Moreover, the retroaldol mechanism is expected to require base catalysis to abstract the proton from the β-OH group of serine, and the hydrogen bonded contacts of Glu57 in *E. coli* SHMT (Glu53 in fig. 14-17) are interpreted to indicate that it is the conjugate acid (-γ-COOH). For these reasons, the two nucleophilic displacement mechanisms in fig. 14-19 are under consideration (Rao et al., 2003; Szebenyi et al., 2004; Trivedi et al., 2002). The acid-base catalytic groups have not been assigned, although kinetic and spectral analyses of site-directed mutated forms of the enzyme have ruled out His 228 as an acid-base catalyst, and they have ruled out the protonation of glycyl-C2 by Lys 229 (Fitzpatrick and Malthouse, 1998a; Hopkins and Schirch, 1986; Schirch et al., 1993; Stover et al., 1992). Protonation at glycyl-C2 proceeds at the pro-S position, as observed by exchange of this proton of glycine with deuterium (Fitzpatrick and Malthouse, 1998b).

The most compelling evidence for the retroaldol mechanism is the long known, tetrahydrofolate-independent reactions of threonine and *allo*-threonine and of *erythro*- and *threo*-β-phenylserine as substrates. SHMT catalyzes cleavages of these molecules to glycine and the corresponding aldehyde, either acetaldehyde or benzaldehyde, respectively (Schirch, 1984; Schirch and Diller, 1971; Schirch and Gross, 1968; Ulevitch and Kallen, 1977; Webb and Matthews, 1995). Tetrahydrofolate is not required for these reactions, and the aldehydes produced do not react with tetrahydrofolate. SHMT also catalyzes the tetrahydrofolate-independent condensation of glycine with formaldehyde to form serine. These reactions are generally conceded to proceed by the retroaldol mechanism in fig. 14-19. The steady-state kinetics of the condensation of glycine and formaldehyde, an aldol reaction, indicates an ordered sequential mechanism, with glycine binding first (Chen and Schirch, 1973).

In detailed structure-function studies of the kinetics for reactions of series of β-hydroxyamino acids, electronic effects in linear free energy relationships show that the rates of cleavage are enhanced by electron donation to the β-carbon (Ulevitch and Kallen, 1977; Webb and Matthews, 1995). These relationships indicate that a single step limits the rates for all the substrates tested, including serine, and that the rate is enhanced by an electron-donating inductive effect on the β-carbon; the transition state is more positive or less negative than the ground state. Transient kinetic experiments show that external aldimine formation is much faster than the mixing time in a stopped-flow spectrophotometer and cannot limit the rate (Chen and Schirch, 1973; Ulevitch and Kallen, 1977). The quinoid-α-carbanion of glycine also appears within the mixing time. These observations rule out steps 1, 2, 3, 5, 6, and 7 in fig. 14-18 in limiting the rate. Carbon-carbon bond cleavage limits the rate, and this corresponds to steps 4a, 4b, and others in fig. 14-18. In the retroaldol mechanism of fig. 14-19, only the first step can be sensitive to electronic effects and so must limit the rate in the tetrahydrofolate-independent cleavages.

In the tetrahydrofolate-dependent reaction of serine, the arguments supporting the nucleophilic C—C and C—O cleavage mechanisms in fig. 14-19 are based on the apparent absence of a base catalyst in the active site of SHMT. The structures are interpreted to imply a γ-COOH for Glu53 in fig. 14-17. Such a group could not abstract the proton from the β–OH of serine in the first step of the retroaldol mechanism. In the nucleophilic C—C cleavage mechanism, this group would function as an acid in the dehydration of 5-hydroxymethyltetrahydrofolate, and in the nucleophilic C—O cleavage it would function as the acid catalyst in the displacement of the β-OH group by N5 of tetrahydrofolate.

Mutation of the active site glutamate, Glu75 in rabbit cytosolic SHMT, essentially abolishes activity in the reaction of serine with tetrahydrofolate (Szebenyi et al., 2004). Other catalytic properties of rabbit E75Q- and E75L-SHMT allow the function of Glu75 to be assigned to the processing of tetrahydrofolate intermediates, and they exclude Glu75 from participation in the retroaldol mechanism. The mutated variants actually display enhanced tetrahydrofolate-independent rates. Both E75Q- and E75L-SHMT catalyze the cleavage of *allo*-threonine faster than SHMT by 4.3- and 1.1-fold, respectively.

Moreover, they catalyze the cleavage of serine faster than SHMT by 46- and 120-fold, respectively. In contrast, neither variant catalyzes the hydration of 5,10-methenyltetrahydrofolate, an important activity of SHMT, just as they are inactive in the reaction of serine with tetrahydrofolate. Nor do the variants catalyze the conversion of 5-hydroxymethyltetrahydrofolate to 5,10-methylenetetrahydrofolate. Glu75 in rabbit SHMT functions in processing tetrahydrofolate intermediates.

The retroaldol mechanism is generally accepted for tetrahydrofolate-independent cleavages of threonine and β-phenylserine, although no base has been identified for catalyzing the cleavage step in fig. 14-19. The active site glutamate is ruled out for this role by the mutagenesis experiments with Glu75 in the rabbit SHMT. Another candidate shown in fig. 14-17 is His122 of *B. stearothermophilus* SHMT, which lies near the β-hydroxy group of serine. Mutagenesis experiments raise doubt about this residue participating in catalysis. The corresponding residue in sheep SHMT is His147, and the activity of H147N-SHMT in the cleavage of *allo*-threonine is about 1/6 that of the wild-type enzyme, and the value of k_{cat}/K_m is about 1/20 that of the wild type (Jagath et al., 1997). The differences are not sufficient to implicate this histidine residue in catalysis. H147N-SHMT displays weakened binding of PLP, and in AAT, the tyrosine residue at this position is believed to participate in binding PLP.

In this connection, the absence of stereochemical specificity for threonine and *allo*-threonine and for *threo-* and *erythro*-β-phenylserine may be significant. Fig. 14-17 shows the 5′-phosphate of PLP within 4.8 Å of the β-hydroxyl group, so that it is conceivable that a modest conformational change may bring it into play as the base catalyst.

The retroaldol mechanism is not ruled out in the reaction of serine with tetrahydrofolate (Rao et al., 2003; Szrebenyi et al., 2004). The nucleophilic mechanisms are under consideration for 5,10-methylenetetrahydrofolate production because of the apparent ionization state of the active-site glutamate. Future research may clarify the mechanisms of the interesting reactions of SHMT.

Carbocationic Mechanisms

Carbocationic intermediates seem to be rare in enzymatic reactions, but this may be illusory. The best known examples are the glycosyl cations in the actions of glycosyltransferases. However, most of the many enzymes engaged in the biosynthesis of more than 23,000 terpenes remain to be studied, and many of their reactions are likely to proceed through carbocationic intermediates. The substrates generally include allylic pyrophosphates that can form allylic carbocations on elimination of pyrophosphate (Sacchettini and Poulter, 1997), The cations can react with alkenes to form new carbon-carbon bonds. The basic chemistry of this process is outlined in fig. 14-1C, and allylic pyrophosphates such as dimethylallyl pyrophosphate (DMAPP) as sources of allylic carbocations are subjects of the present discussion. The prenyl pyrophosphates exist as complexes with Mg^{2+} in cells.

Farnesyl Pyrophosphate Synthase

Reactions and Stereochemistry

Farnesyl pyrophosphate (FPP) synthase (EC 2.5.1.1) catalyzes two consecutive carbon-carbon bond formations, the conversion of DMAPP and isopentenyl pyrophosphate (IPP) into geranyl pyrophosphate (GPP), and the condensation of IPP with GPP to form FPP,

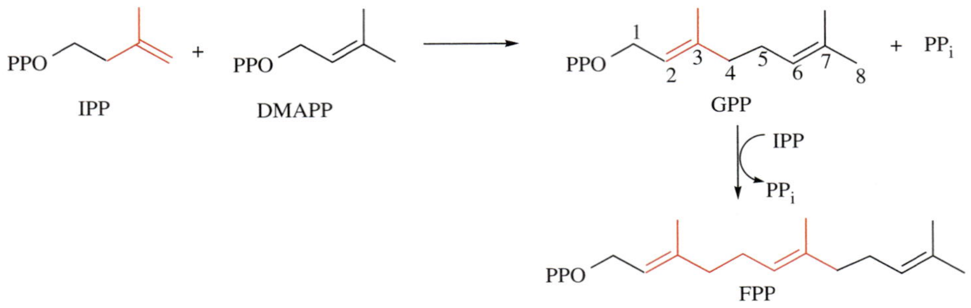

Fig. 14-20. The reactions are catalyzed by farnesyl pyrophosphate (FPP) synthase. FPP formation takes place with the addition to dimethylallyl pyrophosphate (DMAPP) of two isoprene units (red) from isopentenyl pyrophosphate (IPP). The allylic pyrophosphate moiety of the intermediate geranyl pyrophosphate (GPP) is derived from IPP, and this serves as the alkylating agent for the succeeding IPP added to form FPP.

as illustrated in fig. 14-20. The allylic reactant undergoes addition to the *si*-face of the alkene group in IPP, inverting configuration at C1, and the pro-R proton is lost from C2 of IPP to generate the new *cis*-double bond in FPP (Poulter and Rilling, 1978).

FPP is an intermediate in the biosynthesis of cholesterol and steroids, as well as in the biosynthesis of many terpenes. FPP synthases from a number of sources are homodimeric with molecular masses in the range of 70 to 90 kDa (Dolence and Poulter, 1999). Recombinant versions of these enzymes display properties identical with those of the purified enzymes.

Unlike other enzymatic carbon-carbon condensations, the reactions of FPP synthase are, for practical purposes, irreversible. A key and essential step in each step of the reaction is the essentially irreversible release of pyrophosphate.

Condensation Mechanism

Two mechanisms for the coupling of IPP with DMAPP and GPP have been discussed, and are shown in fig. 14-21. The original conceptualization of a likely mechanism, invoked an enzymatic nucleophile —X to form either a covalent bond or intimate ion pair with the carbocation resulting from the departure of PP_i (Poulter and Rilling, 1978). The covalent species are shown in Mechanism A of fig. 14-21, and the formation of such intermediates would obviate the necessity to form discrete carbocationic intermediates. A nucleophilic group has not been identified.

In Mechanism B of fig. 14-21, discrete carbocationic intermediates participate as reactive species in the mechanism. A linear free energy correlation powerfully supports the concept of carbocationic intermediates (Poulter at al, 1981). As in the mechanistic analysis of glycosidase action (see chap. 12), fluorine plays a key role in this analysis. Electron withdrawal by fluorine bonded to an adjacent carbon strongly opposes the accumulation of positive charge, and fluorine is a minimal steric perturbant, as required in all enzymatic studies. Analogs of GPP with fluorinated substituents in place of the C3-methyl group (see fig. 14-20 for numbering) react as substrates. The elimination of PP_i would generate positive charge at C3, which constitutes one end of an allylic cation. The rate constants for the chemical step are sensitive to electronic effects at C3 when the substituents are H, CH_2F, CHF_2, and CF_3. A Hammett plot, shown in fig. 14-22, correlates the enzymatic rates with those of a model, nonenzymatic solvolysis reaction. The slope of the plot (>1.0)

Fig. 14-21. Mechanisms of carbon condensation in the reaction of farnesyl pyrophosphate synthase. In mechanism A, the formation of carbocationic intermediates is avoided by postulating the participation of an enzymatic nucleophile —X to form covalent bonds with incipient cations. In mechanism B, the carbocationic intermediates are discrete species, and an essential function of the enzyme is to bind the pyrophosphate moiety and facilitate its departure to generate allylic carbocations.

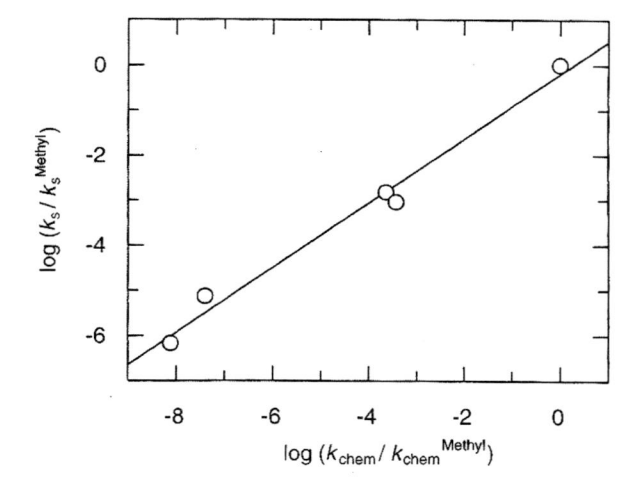

Fig. 14-22. A Hammett plot correlates catalysis by farnesyl pyrophosphate (FPP) synthase with a chemical solvolysis. Geranyl pyrophosphate (GPP) and analogs described in the text, with H, CH_2F, CHF_2, and CF_3 in place of the C3-methyl group, were the substrates for the enzymatic reaction. The corresponding methylsulfonates were the substrates for nonenzymatic solvolysis. The enzymatic rates are plotted as the ratio of rate constants for the analog and GPP (k_S/k_S^{Methyl}), and the solvolysis rate constants are plotted as the ratio of the analog and parent esters ($k_{Chem}/k_{Chem}^{Methyl}$). (From Poulter et al., 1981, with permission.)

shows that the enzymatic condensation is very sensitive to electron withdrawal by fluorine, more sensitive than the chemical solvolysis, which is known to involve allylic carbocationic intermediates. The results clearly implicate carbocation formation in the transition state.

The Hammett plot does not contain information about whether carbocations are discrete intermediates. A kinetic possibility is a concerted reaction with the departure of PP_i and carbon-carbon bond formation in the same step by way of a cationic transition state. However, consideration of the known lifetimes of the putative allylic carbocations in water strongly suggests that their lifetimes are long enough for them to serve as discrete intermediates (Dolence and Poulter, 1999).

Structure and Mutagenesis

The structure of avian FPP synthase shows a putative binding site defined at its base by the side chains of Phe113 and Phe112 (Tarshis et al., 1994). Mutations of these residues to smaller side chains lead to variant enzymes with activities comparable to wild-type FPP synthase that gave longer chain products (Tarshis et al., 1996). Mutations that constrict the binding site alter the specificity to favor GPP synthesis (Fernandez et al., 2000).

Amino acid sequence alignments reveal two conserved aspartate-rich motifs that participate in binding the divalent metal ions associated with the pyrophosphate moieties of substrates (Ashby and Edwards, 1990). Mutations of these aspartate residues to alanine severely compromises enzymatic activity (Song and Poulter, 1994).

The structure of a doubly mutated variant, F113A/F112S-FPP synthase shown in fig. 14-23 with GPP bound at the active site, reveals two aspartate residues in the first aspartate-rich motif coordinated to the two Mg^{2+} ions that form a bridge between the pyrophosphate group and the binding site. Arg126 and Lys280 are engaged in hydrogen bonded ionic contacts with the pyrophosphoryl group, together with the two Mg^{2+} ions. An assessment of electrostatic charges directly associated with the pyrophosphoryl group indicates the presence of a positive electrostatic field. This would facilitate the departure of PP_i from GPP in generating the allylic carbocation for reaction with IPP.

Squalene Synthase

In cholesterol biosynthesis, squalene synthase (EC 2.5.1.21) catalyzes the reductive 1'-1 (head-to-head) coupling of two molecules of FPP to squalene, as illustrated in fig. 14-24. The enzyme is membrane-bound and difficult to purify. A truncated version of the yeast synthase, lacking the membrane-spanning domain, is soluble and stable in water and catalyzes the reaction (Zhang et al., 1993). A structure of the intact human synthase is available (Pandit et al., 2000).

The reaction of squalene synthase proceeds in the two steps shown in fig. 14-24. First, one molecule undergoes loss of PPi and addition of the allylic carbocation to the alkene end of the other molecule of FPP, accompanied by loss of a proton, to form presqualene pyrophosphate (PSPP) (Rilling, 1966; Rilling and Epstein, 1969). In the second step, PSPP loses PP_i, and the presumptive cyclopropylcarbinyl carbocation undergoes ring opening and reduction by NADPH to squalene. The mechanisms by which the two steps produce first PSPP and then squalene are not known in detail and are of considerable interest.

The mechanisms under consideration involve carbocationic intermediates. Evidence in support of such intermediates includes the consequences of disrupting catalysis in

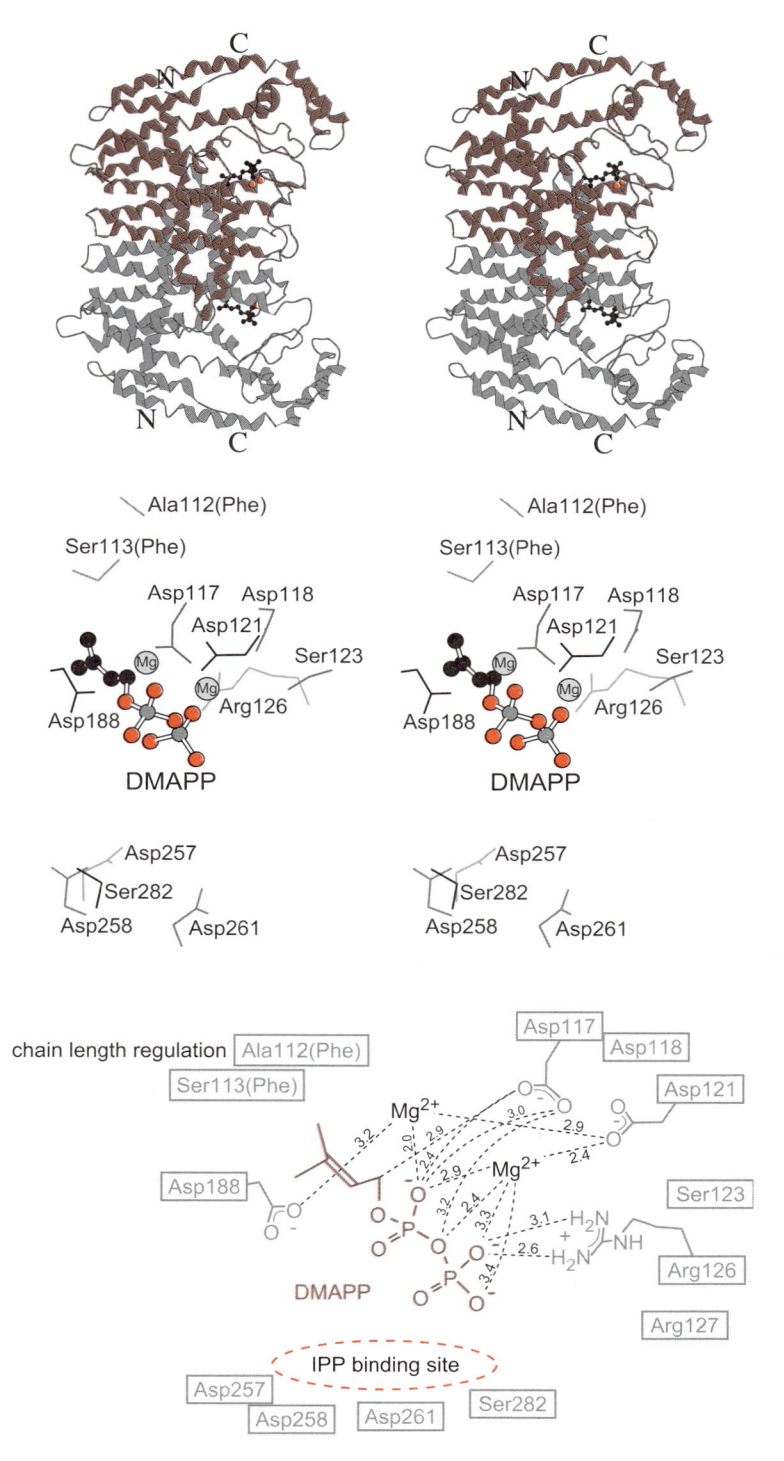

Fig. 14-23. The structure of avian farnesyl pyrophosphate (FPP) synthase incorporates a unique chain fold that is largely α-helical with turns and a few loops and that is devoid of β-structure. The structure shown is of a variant (F112A/F113S) with dimethylallyl pyrophosphate (DMAPP) bound at the active site (2.40-Å resolution; PDB 1UBY). Nascent terpene pyrophosphate chain length is limited by the two altered phenalanine residues. The putative isopentenyl pyrophosphate binding site is shown (bottom) below the site of DMAPP binding (Tarshis et al., 1996).

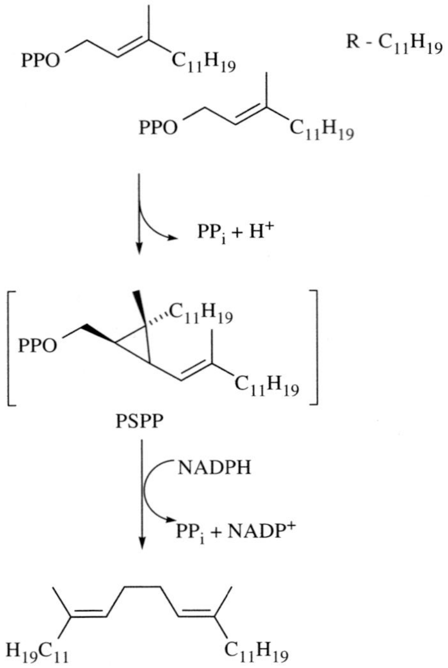

Fig. 14-24. The reaction of squalene synthase.

some way. Exclusion of NADPH leads initially to PSPP, but prolonged incubation leads to rearranged products of solvation or loss of a proton that are indicative of carbocationic species (Jarstfer et al., 2002). Substitution of the unreactive 5,6-dihydroNADPH for NADPH leads to the particularly interesting products shown in fig. 14-25 (Blagg et al., 2002). The products are dehydrosqualene, hydroxysqualene, and rillingol, a cyclopropanol.

The diversion products in fig. 14-25 are compatible with the mechanism in fig. 14-26 (Blagg et al., 2002). The steps in the elimination PP_i in the initial formation of PSPP its further reaction are likely to be irreversible, but rearrangement of carbocationic intermediates are reversible. If the reduction step is blocked by the unreactive 1,5-dihydroNADPH,

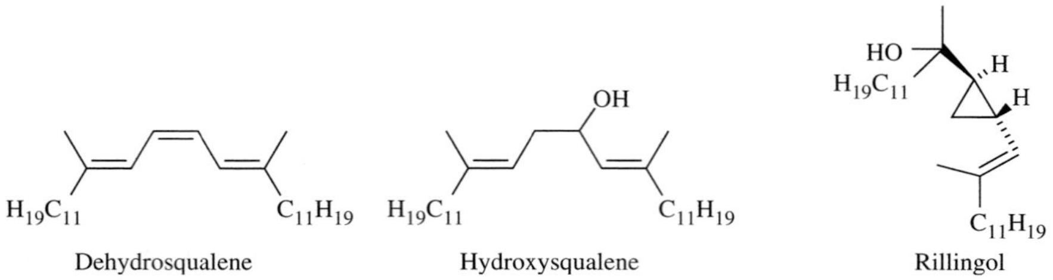

Fig. 14-25. Products of the action of squalene cyclase on farnesyl pyrophosphate (FPP) in the presence of an unreactive analog of NADPH. Substitution of NADPH by an unreactive analog, 1,5-dihydro-NADPH, in the reaction of squalene synthase prevents squalene formation and allows side reactions to occur. The side reactions lead to the products shown. All products result from side reactions of carbocationic intermediates.

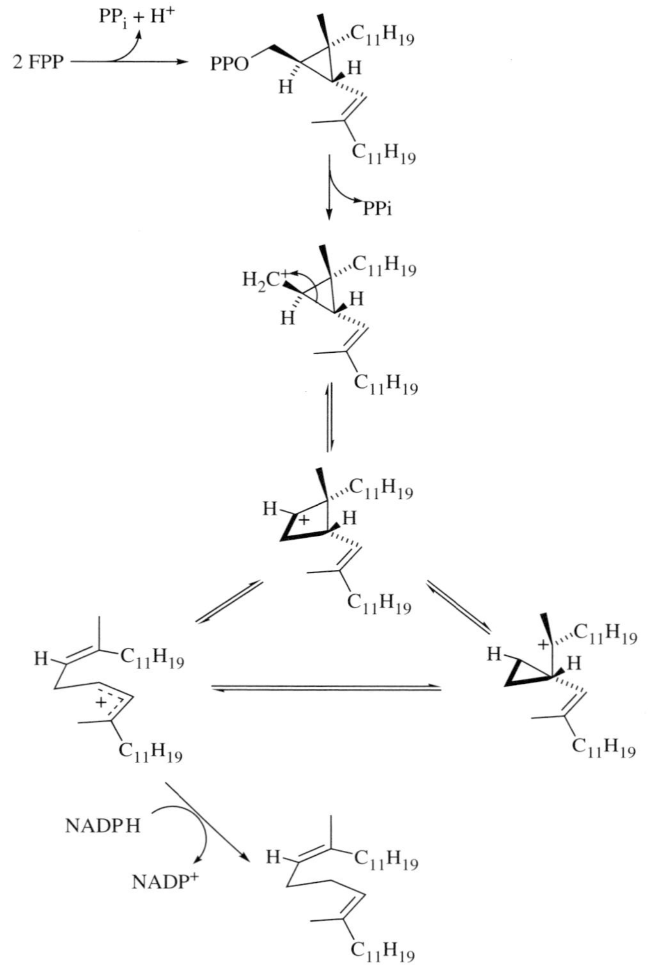

Fig. 14-26. Mechanisms for the conversion of farnesyl pyrophosphate into squalene by the action of squalene synthase.

the reversible rearrangements are likely to attain equilibrium, and prolonged lifetimes for the carbocations will expose them to side reactions. Rillingol in fig. 14-25 arises from hydrolytic quenching of the initially formed cyclopropylcarbinyl carbocation. Hydroxysqualene arises from hydrolytic quenching of the allylic carbocation precursor of squalene. Dehydrosqualene arises from the loss of a proton from the allylic carbocation precursor. The mechanism in fig. 14-26 is compatible with the side-reaction products when the reductive step is blocked by 1,5-dihydroNADPH.

References

Anderson VA, BJ Bahnson, ID Wlassics, and CT Walsh (1990) *J Biol Chem* **265**, 6255.
Ashby MN and PA Edwards (1990) *J Biol Chem* **265**, 13157
Aymes TL and JP Richard (1992) *J Am Chem Soc* **114**, 10297.
Bender ML and A Williams (1966) *J Am Chem Soc* **88**, 2502.
Blagg BSJ, MB Jarstfer, DH Rogers, and CD Poulter (2002) *J Am Chem Soc* **124**, 8846.
Bové J, RO Martin, LL Ingraham, and PK Stumpf (1959) *J Biol Chem* **234**, 999.

Cavalieri SW, KE Neet, and HZ Sable (1975) *Arch Biochem Biophys* **171**, 527.
Chen MS and L Schirch (1973) *J Biol Chem* **248**, 3631.
Clark JD, SJ O'Keefe, and JR Knowles (1988) *Biochemistry* **27**, 5961.
Cleland WW, PA Frey, and JA Gerlt (1998) *J Biol Chem* **273**, 25529.
Contestabile R, S Angelaccio, F Bossa, HT Wright, N Scarsdale, G Kazanina, and V Schirch (2000) *Biochemistry* **39**, 7492.
Cornforth JW, JW Redmond, H Eggerer, W Buckel, and C Gutschow (1969) *Nature* **221**, 1212.
Dolence JM and CD Poulter (1999) In Poulter CD (ed): *Comprehensive Natural Products Chemistry*, vol 5. Elsevier: Amsterdam, p 315.
Eggerer H (1965) *Biochem Z* **343**, 111.
Eggerer H, W Buckel, H Lenz, P Wunderwald, G Gottschalk, JW Cornforth, C Donninger, R Mallaby, and JW Redmond, (1970) *Nature* **226**, 517.
Eggerer H and U Remberger (1963) *Biochem Z* **337**, 202.
Fernandez SM Stanley, BA Kellogg, and CD Poulter (2000) *Biochemistry* **39**, 15316.
Fiedler E, S Thorell, T Sandalova, R Golbik, S König, and G Schneider (2002) *Proc Natl Acad Sci U S A* **99**, 591.
Fitzpatrick TB and JP Malthouse (1998a) *Biochim Biophys Acta* **1386**, 220.
Fitzpatrick TB and JP Malthouse (1998b) *Eur J Biochem* **252**, 113.
Floss HG, and S Lee (1993) *Acc Chem Res* **26**, 116.
Frey PA and RD Sammons (1985) *Science* **228**, 541.
Gehring U and F Lynen (1972) In Boyer PD (ed): *The Enzymes*, vol 7, 3rd ed. Academic Press: New York, p 391.
Gehring U, C Riepertinger, and F Lynen (1968) *Eur J Biochem* **6**, 264.
Girgis S, JR Suh, J Jolivet, and PJ Stover (1997) *J Biol Chem* **272**, 4729.
Glusker JP (1992) *Curr Top Cell Regul* **33**, 169.
Gorbach ZV and VL Kubyshin (1989) *Biokhimiia* **54**,1980.
Gottschalk G and HA Barker (1967) *Biochemistry* **6**, 1027.
Hopkins S and V Schirch (1986) *J Biol Chem* **261**, 3363.
Horecker BL, T Cheng, and S Pontremoli (1963) *J Biol Chem* **238**, 3428.
Jagath JR, NA Rao, and HS Savithri (1997) *Biochem J* **327**, 877.
Jarstfer MB, DL Zhang, and CD Poulter (2002) *J Am Chem Soc* **124**, 8834.
Jia J, U Schörken, Y Lindqvist, GA Sprenger, and G Schneider (1997) *Protein Sci* **6**, 119.
Johansson C-J and G Pettersson (1977) *Biochim Biophys Acta* **484**, 208.
Johansson C-J and G Pettersson (1979) *Eur J Biochem* **93**, 505.
Kochetov GA (2001) *Biochemistry (Moscow)* **66**,1077.
Kursula P, J Ojala, A-M Lambeir, and RK Wierenga (2002) *Biochemistry* **41**, 15543.
Kursula P, H Sikkila, T Fukao, N Kondo, and RK Wierenga (2005) *J Mol Biol* **347**,189.
Kurz LC, S Shah, C Frieden, T Nakra, RE Stein, GR Drysdale, CT Evans, and PA Srere (1995) *Biochemistry* **34**, 13278.
Lüthy J, J Rétey, and D Arigoni (1969) *Nature* **221**, 1313.
Marcus A and B Vennesland (1958) *J Biol Chem* **233**, 727.
Martini F, S Angelaccio, S Pascarella, D Barra, F Bossa, and V Schirch (1987) *J Biol Chem* **262**, 5499.
Martini F, B Maras, P Tanci, S Angelaccio, S Pascarella, D Barra, F Bossa, and V Schirch (1989) *J Biol Chem* **264**, 8509.
Masamune A, MAJ Palmer, R Gamboni, S Thompson, JT Davis, SF Williams, OP Peoples, AJ Sinskey, and CT Walsh (1989) *J Am Chem Soc* **111**, 1879.
Middleton B (1974) *Biochem J* **139**, 109.
Muller Y, Y Lindqvist, W Furey, GE Schulz, F Jordan, and G Schneider (1993) *Structure (Lond)* **1**, 95.
Nilsson U, L Meshalkina, Y Lindqvist, and G Schneider (1997) *J Biol Chem* **272**, 1864.
Ogston AG (1948) *Nature* **162**, 963.
Pandit J, DE Danley, GK Schulte, S Mazzalupo, TA Pauly, CM Hayward, ES Hamanaka, JF Tompson, and HJ Harwood (2000) *J Biol Chem* **275**, 30610.
Popják G (1970) In Boyer PD (ed): *The Enzymes*, vol 2, 3rd ed. Academic Press: New York, p 115.
Poulter CD and HC Rilling (1978) *Acc Chem Res* **11**, 307.
Poulter CD, PI Wiggins, and AT Le (1981) *J Am Chem Soc* **103**, 3926.

Rao N Appaji, M Ambili, VR Jala, HS Subramanya, and HS Savithri (2003) *Biochim Biophys Acta* **1647**, 24.

Remington SJ (1992) *Curr Top Cell Regul* **33**, 209.

Renwick SB, K Snell, and U Baumann (1998) *Structure* **6**, 1105.

Rilling HC (1966) *J Biol Chem* **241**, 3233.

Rilling HC and WW Epstein (1969) *J Am Chem Soc* **91**, 1041.

Sacchettini JC and CD Poulter (1997) *Science* **277**, 1788.

Scarsdale JN, G Kazanina, S Radaev, V Schirch, and HT Wright (1999) *Biochemistry* **38**, 8347.

Scarsdale JN, S Radaev, G Kazanina, V Schirch, and HT Wright (2000) *J Mol Biol* **296**, 155.

Schirch D, SD Fratte, S Iurescia, A Angelaccio, R Contenstabile, F Bossa, and V Schirch (1993) *J Biol Chem* **268**, 23132.

Schirch L (1975) *J Biol Chem* **250**, 1939.

Schirch L (1982) *Adv Enzymol Relat Areas Mol Biol* **53**, 83.

Schirch L (1984) In Blakesly RI and SJ cenkovic (eds): *Folates and Pterins*. John Wiley & Sons: New York, p 399.

Schirch L and A Diller (1971) *J Biol Chem* **246**, 3961.

Schirch L and T Gross (1968) *J Biol Chem* **243**, 5651.

Schirch LV, CM Tatum Jr, and SJ Benkovic (1977) *Biochemistry* **16**, 410.

Schirch V, S Hopkins, E Villar, and S Angelaccio (1985) J Bacteriol **163**, 1.

Schneider G and Y Lindqvist (1998) *Biochim Biophys Acta* **1385**, 387.

Schörken U, J Jia, H Sahm, GA Sprenger, and G Schneider (1998) *FEBS Lett* **441**, 247.

Schörken U, S Thorell, M Schürmann, J Jia, CA Sprenger, and G Schneider (2001) *Eur J Biochem* **268**, 2408.

Song L and CD Poulter (1994) *Proc Natl Acad Sci U S A* **91**, 3044.

Spector LB (1972) In Boyer PD (ed): *The Enzymes*, vol 7, 3rd ed. Academic Press: New York, p 357.

Srere PA (1966) J Biol Chem *J Biol Chem* **241**, 2157.

Srere PA (1967) *Biochem Biophys Res Commun* **26**, 609.

Srere PA (1970) *Adv Enzyme Regul* **9**, 221.

Srere PA (1975) *Adv Enzymol Relat Areas Mol Biol* **43**, 57.

Stern JB (1961) In Boyer PD (ed): *The Enzymes*, vol 5, 2rd ed. Academic Press: New York, p 367.

Stover P and V Schirch (1991) *J Biol Chem* **266**, 1543.

Stover P, M Zamora, K Shostak, M Gautam-Basak, and V Schirch (1992) *J Biol Chem* **267**, 17679.

Suelter CH and S Arrington (1967) *Biochim Biophys Acta* **141**, 423.

Szebenyi DME, X Liu, IA Kriksunov, PJ Stover, and DJ Thiel (2000) *Biochemistry* **39**, 13313.

Szebenyi DME, FN Musayev, ML di Salvo, MK Safo, and V Schirch (2004) *Biochemistry* **43**, 6865.

Takayama S, GJ McGarvey, and CH Wong (1997) *Annu Rev Microbiol* **51**, 285.

Talwar R, JR Jagath, A Datta, V Prakash, HS Savithri, and NA Rao (1997) *Acta Biochim Pol* **44**, 679.

Talwar R, JR Jagath, NA Rao, and HS Savithri (2000) *Eur J Biochem* **267**,1441.

Tarshis LC, PJ Proteau, BA Kellogg, JC Sacchettini, and CD Poulter (1996) *Proc Natl Acad Sci U S A* **93**, 15018.

Tarshis LC, M Yan, CD Poulter, and JC Sacchettini (1994) *Biochemistry* **33**, 10871.

Thompson S, F Mayerl, OP Peoples, S Masamune, AJ Sinskey, and CT Walsh (1989) *Biochemistry* **28**, 5735.

Trivedi V, A Gupta, VR Jala, P Saravanan, GS Rao, NA Rao, HS Savithri, and HS Subramanya (2002) *J Biol Chem* **277**,17161.

Tsolas O and BL Horecker (1972) In Boyer PD (ed): *The Enzymes*, vol 7, 3rd ed. Academic Press: New York, p 259.

Ulevitch RJ and RG Kallen (1977) *Biochemistry* **16**, 5355.

Usher KC, SJ Remington, DP Martin, and DG Drueckhammer (1994) *Biochemistry* **33**, 7753.

Usmanov RA, NN Sidorova, and GA Kochetov (1996) *Biochem Mol Biol Int* **38**, 307.

Webb HK and RG Matthews (1995) *J Biol Chem* **270**, 17204.

Wikner C, U Nilsson, L Meshalkina, C Udekwu, Y Lindqvist, and G Schneider (1997) *Biochemistry* **36**, 15643.

Williams SF, MA Palmer, OP Peoples, CT Walsh, AJ Sinskey, and S Masamune (1992) *J Biol Chem* **267**, 16041.

Wlassics ID and VE Anderson (1989) *Biochemistry* **28**, 1627.

Wunderwald P and H Eggerer (1969) *Eur J Biochem* **11**, 97.

Zhang D, SM Jennings, GW Robinson, and CD Poulter (1993) *Arch Biochem Biophys* **304**, 133.

15
Alkyltransferases

Chemistry of Alkylation

Biological Alkylations

A number of enzymes catalyze alkylation reactions, most of which are reactions of S-adenosyl-L-methionine (SAM) as a methylating agent in the biosynthesis of hormones, modification of DNA, and methyl esterification of proteins involved in signal transduction. Other examples of enzymatic alkylation include prenyl transfer reactions, adenosyltransfer from ATP to methionine in the biosynthesis of SAM, and adenosyltransfer from ATP to cob(I)alamin in the biosynthesis of adenosylcobalamin. Methyl group transfer is also the essential step in the reaction of methionine synthase, which uses 5-methyltetrahydrofolate as an alkylating agent. In an analogous reaction, an analog of 5-methyltetrahydrofolate is the methyl group donor in the methylation of coenzyme M to form methyl coenzyme M, the proximate precursor of methane in methanogenesis (see chap. 4).

Glysosyl transfer is an alkylation reaction catalyzed by a large class of enzymes, the glycosyltransferases and glycosidases. The special nature of the glycosyl compounds and their potential for undergoing glycosyltransfer places them in their own class in biochemistry (see chap. 12). The reactivity of glycosyl compounds can be attributed to the contribution of the oxygen atom directly bonded to the glycosyl carbon, the locus of alkylation. In this chapter, we consider other enzymatic alkylations.

Alkylation consists of the transfer of a carbon from a leaving group to a nucleophilic acceptor, as in eq.15-1, where R is H or an organic group.

$$\text{X:} + \text{H}-\underset{\underset{R}{|}}{\overset{\overset{H}{|}}{C}}-\text{Y} \longrightarrow \text{X}-\underset{\underset{R}{|}}{\overset{\overset{H}{|}}{C}}-\text{H} + \text{Y:} \qquad (15\text{-}1)$$

656   Enzymatic Reaction Mechanisms

The rate is controlled by the reactivity of the nucleophile X:, the stability of the leaving group Y:, and the electrophilic reactivity of the central carbon atom.

Alkylation Mechanisms

Alkylation may be regarded as one of the simplest organic chemical reactions because there are few complications in the mechanism. It is the reaction of a nucleophilic molecule with an electrophilic molecule to displace a leaving group. Enzymatic alkylations proceed by polar and not radical mechanisms. In organic chemistry, polar alkylation can occur either by an associative or one-step mechanism, as in fig. 15-1A, or by a dissociative or two-step mechanism through a carbocationic intermediate, as in fig. 15-1B. The chemical nature of the alkylating agent, the propensity of the leaving group to leave, and the polarity of the solvent determine the mechanism. Nonpolar solvents and poor leaving groups favor the associative mechanism. Polar solvents, such as media of high ionic strength, and excellent leaving groups favor the two-step mechanism. An alkylating agent in which a carbocationic intermediate is stabilized by delocalization of the positive charge also tends to react by the dissociative mechanism.

Among the biological alkylating agents discussed in this chapter, farnesyl pyrophosphate allows for delocalization of a positive charge in the allylic carbocation. However, alklyation by farnesyl pyrophosphate (FPP) need not proceed by the two-step mechanism, and delocalization of positive charge in the transition state of the associative mechanism can also promote the reaction. Well-known biological alkylating agents in which delocalization of charge is significant are glycosyl compounds that react in glycosyl transfer reactions (see chap. 12). Many biological alkylating agents cannot delocalize a positive charge, either in a dissociative mechanism or in the transition state of an associative mechanism.

In principle, an enzyme could catalyze an alkylation reaction by several other means. Binding the two molecules close together in the right orientation, increasing the reactivity of the nucleophile, and improving the leaving group would all facilitate alkylation. Enzymes certainly bind substrates in reactive orientations. In favorable cases of leaving groups like pyrophosphate or triphosphate, the microenvironment can improve the ability of a group to leave. Nucleophilic reactivity of the acceptor can be improved by general base catalysis or a microenvironment that lowers its pK_a to neutrality. We shall see that all of these factors are brought into play in the case of protein farnesyltransferases. Other alkyl transfers are subject to particular barriers.

An enzyme can often increase the nucleophilic reactivity of a heteroatom by facilitating the removal of a proton to increase the electron density on the atom. This no doubt happens in many cases of enzymatic methylation and prenylation. Much less can be done to increase leaving group ability in enzymatic alkylations. The leaving group in alkylating substrates is often intrinsically good, as in the sulfonium group of S-adenosylmethionine,

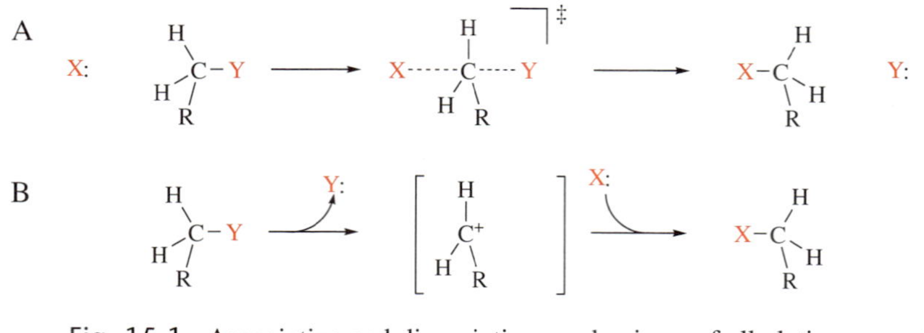

Fig. 15-1. Associative and dissociative mechanisms of alkylation.

MgPPP in alkylations by MgATP, and MgPP$_i$ in prenylation reactions by prenyl pyrophosphates. In farnesyl transfer by FPP synthase, the enzyme can provide an electrostatically positive microenvironment to facilitate the departure of MgPP. An enzyme can do little else to increase the electrophilic reactivity of the central carbon atom. All things considered, it seems that in most enzymatic alkylation reactions, binding in close proximity and in correct orientation is likely to be an important aspect in catalysis, and increasing the reactivity of the nucleophile is also likely to be important.

Enzymatic Alkylation

Protein Farnesyltransferase

Reaction

A number of eukaryotic proteins that function in molecular complexes, often at cell membranes, are alkylated with farnesyl, or geranylgeranyl groups at cysteine residues near their C-termini (Clarke, 1992). Protein farnesyl transferase (PFT, EC 2.5.1.59) and geranylgeranyl transferase catalyze the alkylation of specific cysteine residues by FPP or GGPP in these proteins. The protein substrates include, the Rho/Rac GTPases, G protein–coupled receptor kinases, γ subunits of G proteins, retinal cGMP phosphodiesterases, and the oncogenic forms of Ras GTPases among others (Schafer et al., 1989). The cysteine residues alkylated by protein farnesyltransferase (PFT) reside in C-terminal motifs that may be generalized as CaaX, where C is cysteine, a is aliphatic, and X is methionine, alanine, serine, or glutamine. The chemical equation for farnesylation of a protein is shown in fig. 15-2A. The C-terminal tetrapeptides CaaX are also good, small substrates for kinetic studies. The related geranylgeranyl transferase recognizes leucine as X in the CaaX motif. Another geranylgeranyl transferase recognizes CCXX, XXCC, and XCXC motifs (Moores et al., 1991; Omer et al., 1993).

The protein prenyltransferases are heterodimeric (αβ) and closely related. The farnesyl and geranylgeranyl enzymes from the same species share the same α subunit and have variant but related β subunits (Zhang and Casey, 1996). Genes for protein farnesyltransferases (PFTs) cloned from rat and human tissues are available; and the results of kinetic, spectroscopic, and structural studies of the recombinant PFTs from these species reveal a consistent story of the mechanism of farnesyltransfer.

Role of Zinc

PFTs are zinc metalloproteins and also require Mg^{2+} for activity. The Mg^{2+} forms part of the substrate in coordination with the pyrophosphoryl moiety of FPP and is displaced in the reaction as MgPP$_i$. As for MgATP and MgADP, we treat Mg^{2+} as implicit in the structures of prenyl pyrophosphates. Replacement of Zn^{2+} with Co^{2+} transforms the zinc binding site into a chromophore while preserving catalytic activity (Hightower et al., 1998; Huang et al., 1997). The absorption spectrum of Co^{2+} reports on ligands in the coordination sphere, and the addition of a peptide substrate introduces changes corresponding to the introduction of a thiolate ligand to Co^{2+}.

The Zn^{2+}-thiolate itself can be detected spectrophotometrically. Thiolates absorb UV light at 230-240 nm, and the extinction coefficient is increased about fourfold to 15,000 cm^{-1} M^{-1} by coordination with Zn^{2+} (Vasak et al., 1981). The thiol group is transparent in this spectral region, so that the pK_a value for a thiol group can be measured spectrophotometrically.

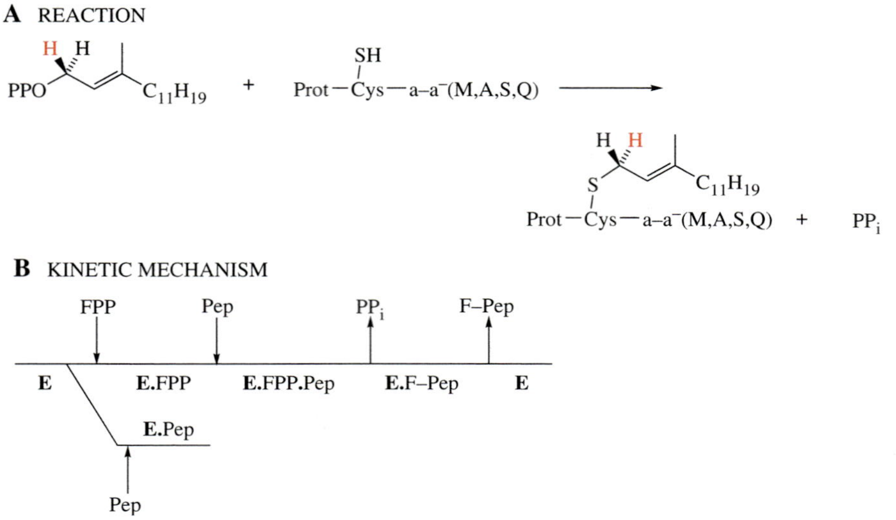

658 Enzymatic Reaction Mechanisms

A REACTION

PPO–CH(H,H)–C=C₁₁H₁₉ + Prot–Cys(SH)–a–a–(M,A,S,Q) →

Prot–Cys(S–CH(H,H)–C=C₁₁H₁₉)–a–a–(M,A,S,Q) + PP_i

B KINETIC MECHANISM

```
      FPP      Pep       PP_i     F–Pep
       ↓        ↓         ↑        ↑
   E ——— E.FPP ——— E.FPP.Pep ——— E.F–Pep ——— E
              \
               E.Pep
               ↑
              Pep
```

Fig. 15-2. The reaction and kinetic mechanism of action of protein farnesyltransferase. (A) The equation illustrates the reaction of a protein-substrate for protein farnesyl transferase (PFT) with farnesyl pyrophosphate (FPP). The C-terminal motif characteristic of substrates for PFT (CaaX) is shown here as Cys–a–a–(M,A,S,Q), where a–a is a sequence of aliphatic amino acids, and the C-terminus is methionine, alanine, serine, or glutamine. (B) The kinetic mechanism for PFT from yeast illustrates ordered binding of FPP followed by the peptide-substrate Pep. Substrate inhibition by the peptide is observed at high concentrations and is shown as binary complexation with the free enzyme to form a dead-end complex designated as **E.Pep**.

The peptide substrate AcCVIA displays this absorption band and a pK_a value of 8.88. When bound to yeast PFT the Zn^{2+}-thiol displays a value of 5.85 for the pK_a, showing that it is a Zn^{2+}-thiolate in neutral solutions (Rozema and Poulter, 1999).

Kinetics

PFT catalyzes farnesylation with inversion of configuration at C1 of the farnesyl group, as illustrated in fig. 15-2A (Mu et al., 1996). Mammalian PFTs catalyze farnesylation by an ordered kinetic mechanism with a small substrate, as shown in fig. 15-2B, with FPP binding first and then the acceptor (Furfine et al., 1995; Pompliano et al., 1993). Presteady-state transient-phase and pulse-chase kinetics of the yeast PFT demonstrate an ordered mechanism with substrate inhibition by the peptide (Mathis and Poulter, 1997). The rate is controlled by release of the farnesyl-peptide (3.5 s^{-1}), but the rate constant for farnesyl-transfer is 10 s^{-1} at 30°C. The mammalian enzymes also follow this pattern of relative rates, albeit with somewhat lower values of rate constants.

Pulse-chase experiments, also known as isotope trapping, dramatically document the ordered kinetic mechanism (1980) (see chap. 2). In this case, premixing PFT with highly radioactive [^{14}C]FPP in the pulse, followed by a chase consisting of peptide substrate mixed with a pool of unlabeled FPP at high concentration, leads largely to radioactive farnesyl-peptide. In the complementary experiment, PFT mixed with [^{14}C]peptide in the pulse chased by FPP in a concentrated pool of unlabeled peptide leads largely to unlabeled farnesyl-peptide (Mathis and Poulter, 1987). The enzyme-FPP complex is essentially captured by the peptide in the forward reaction, and the binary complex reacts much more rapidly with the peptide than it dissociates to the free enzyme and FPP. The enzyme-peptide complex dissociates more rapidly than it reacts with FPP. The dissociation rate constant is about 30 s⁻¹ for the yeast enzyme at 30°C, about an order of magnitude faster than the turnover number (k_{cat} = 3.5 s⁻¹).

Farnesyl transfer appears irreversible, as it does in carbon-carbon bond formation. A C—S bond is generally about 22 kcal mol⁻¹ weaker than a C—O bond. Microcalorimetric measurements of the reaction of PFT show that ΔH = −17 kcal mole⁻¹ for the farnesylation of cysteine in a peptide substrate (Mathis and Poulter, 1997). The somewhat lower experimental value is presumably buffered by compensatory differences in solvation energies. Overall, however, the reaction is driven by enthalpy.

Structure

The crystal structures of ternary complexes of peptide substrates with PFT and FPP or an unreactive analog reveal the overall structural fold and features of the interactions between the cysteinyl residue and Zn^{2+}, as well as the extended conformations of the peptide and FPP and contacts with the enzyme (Long et al., 1998; Strickland et al., 1998). The structure of PFT in complex with α-hydroxyfarnesylphosphonate and the peptide CVIseM, where seM is selenomethionine, is shown in fig. 15-3. The cysteinyl-Zn^{2+} interaction is clearly shown, as well as the other zinc-ligands. The structure shows interactions of amino acid side chains from both subunits with both substrates, although most contacts of α-hydroxyfarnesylphosphonate are with the α subunit and most contacts of the peptide are with the β subunit.

Single mutations of most amino acid side chains in contact with the substrates result in modest perturbations of kinetic constants, mainly weakening binding (Wu et al., 1999). Exceptions are Tyr166' and Tyr200'. The activities of variants Y166A-PFT and Y200F'-PFT are about 1% that of wild-type PFT. However, Y166F'-PFT is more active than wild-type PFT.

Complex interactions of Tyr166' in the structure of PFT appear to be important in function. This residue undergoes significant conformational reorientation on substrate binding. In the active conformation, Tyr166' engages in edge to face interactions with Arg202' and His201', which are in the active site shown in fig. 15-3. One or both of these may be π-cation interactions. π-Cation interactions are, as illustrated in fig. 15-4, electrostatic attractions between the π-electrons of aromatic rings and cations or cationic groups (Dougherty, 1996). Many such interactions of lysine or arginine with tryptophan or phenylalanine are found in protein structures. Mutation of Tyr166' to Phe166' may strengthen the π-cation interaction. Tyr166' also engages in hydrogen bonding with Tyr251 in an apparently unimportant interaction, in view of the fact that Tyr251 is not conserved in all species, and mutation of Tyr166' to Phe166' actually increases activity.

Reaction Mechanism

A linear free energy correlation analogous to that described in chapter 14 for FPP synthase indicates that positive charge accumulates on the allyl group of FPP in the transition state for the action of PFT. Substitution of fluorine for hydrogen in the methyl substituent of the allyl-PP moiety in FPP significantly decreases reactivity. The substrates for the study, which are illustrated in fig. 15-5, display significantly less reactivity as the number of fluorine substitutions for hydrogen is increased. A Hammett plot analogous to that in fig. 14-22 is linear with a slope of 0.3 (Doulence and Poulter, 1995). The slope shows that the allylic carbons carry significantly more positive charge in the transition state than in the ground state. However, the increase in charge is much less than in the action of FPP synthase, for which the slope of the Hammett plot is greater than unity. The cysteinyl-thiolate acceptor

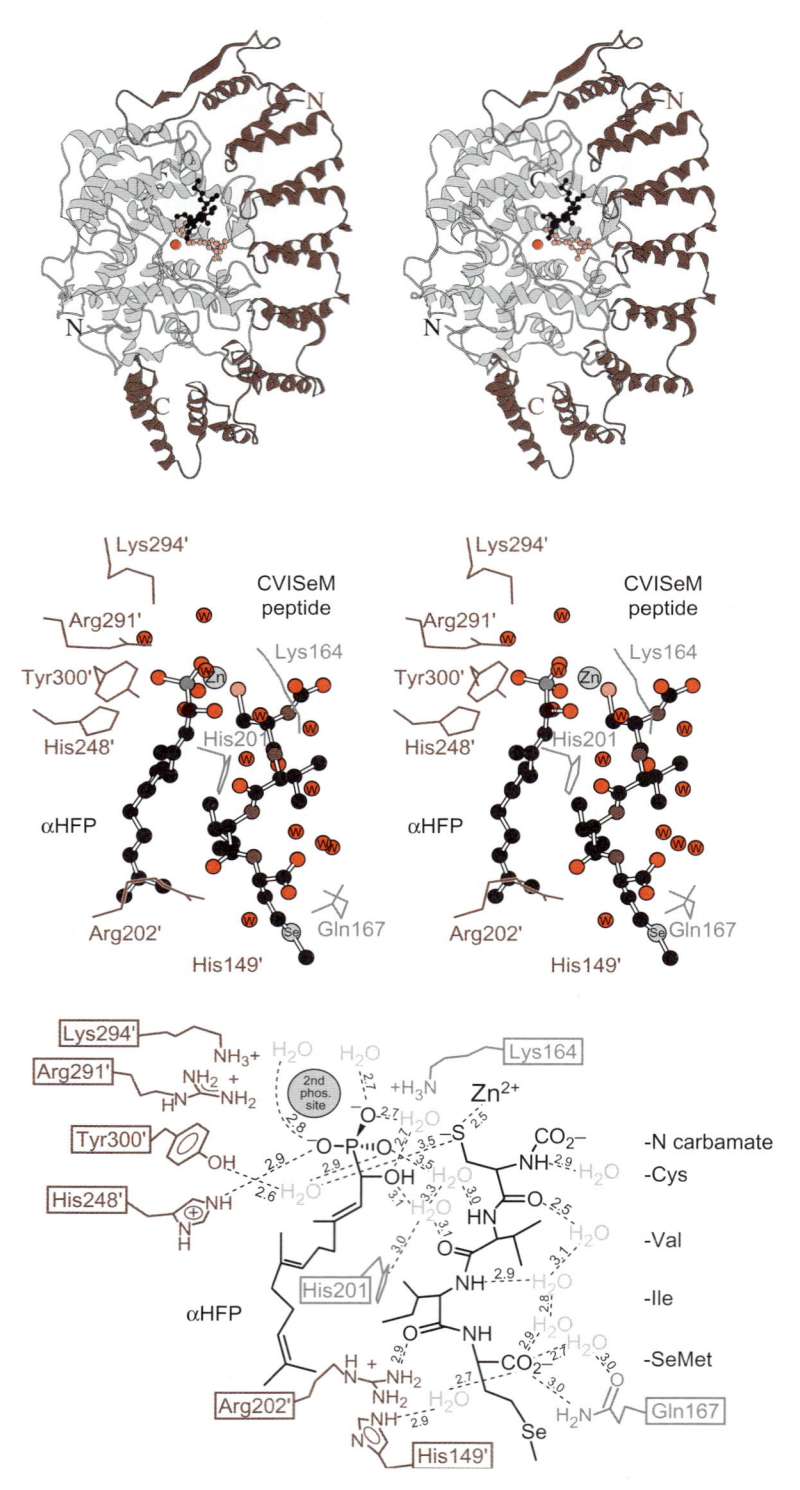

Fig. 15-3. Structure of protein farnesyl transferase (PFT) in complex with an analog of farnesyl pyrophosphate (FPP) and a peptide. The x-ray crystallographic structure reveals a complex of heterodimeric rat PFT with α-hydroxyfarnesylphosphonate and the peptide Cys-Val-Ile-SeMet bound at the active site (2.4-Å resolution; PDB 1QBQ; Strickland et al., 1998).

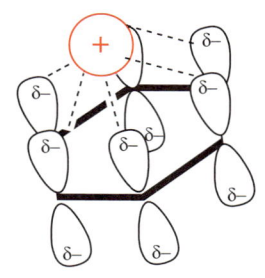

Fig. 15-4. The π-cation interaction is essentially an electrostatic attraction between the slightly negative face of an aromatic ring and a cation. The attraction is significant, and phenylalanine and tryptophan are sometimes engaged in π-cation interactions with basic amino acids in proteins.

likely prevents the formation of a discrete allylic carbocationic intermediate by forming a partial covalent bond to the farnesyl group in the transition state. The mechanism is likely to be associative, as in fig. 15-1A, with —S⁻ as X: and FPP as the alkylating agent.

FPP, like all prenylpyrophosphates, is well suited to alkylation because of the stabilization of positive charge at the alkylation center by allylic delocalization and the leaving group ability of pyrophosphate. The positive electrostatic field around the Mg-pyrophosphate moiety in the active site may enhance the reactivity of the pyrophosphoryl group as the leaving group in the transition state.

Catechol O-Methyltransferase

Chemistry of Methyl Transfer

The methyl group is the smallest and least reactive of all alkylation centers, yet it is commonly an alkylating group in biology. Methylation is essential in the biosynthesis of hormones and neurotransmitters, the maturation and species identification of DNA, the esterification of aspartate-β-carboxylate groups in signaling proteins, and the clearing of xenobiotics.

Theoretical studies show that methyl transfer in a gas phase proceeds in two steps, encounter complexation between the donor and acceptor followed by methyl transfer (Hu and Truhlar, 1996; Wang and Hase, 1997). There is no barrier to formation of the encounter complex, apart from diffusion, and methyl transfer takes place within the encounter complex more rapidly than the dissociation of the complex. The mechanism is associative and passes through a single transition state as in fig. 15-1A. In solutions the situation is different in that the methyl donor and acceptor are extensively stabilized by solvation and lie at energies well below the energy of the encounter complex (Takusagawa et al., 1998). The encounter complex is also stabilized by solvation and its energy lies well

$R = H, CH_3, CH_2F, CHF_2, CF_3$

Fig. 15-5. Substrates employed in linear free energy analysis of the reaction of protein farnesyl transferase.

below that of the transition state. The main barrier to methyl transfer in condensed media is the methyl transfer itself, and this state of affairs appears to be created by solvation effects. This conclusion is compatible with the results of computational experiments on the structure and dynamics of catechol O-methyltransferase (COMT; EC 2.1.1.6) (Zheng and Bruice, 1997).

S-Adenosylmethionine (SAM or AdoMet) is most frequently the methyl donor in biological methylation reactions. Given the low reactivity of the methyl group, its structure shown in fig. 15-6 appears optimal for a biological methyl donor. The methyl group in SAM is one of the three substituents of the sulfonium center, an intrinsically reactive alkylating species.

The structural factors that contribute to the reactivity of SAM as a methylating agent also make it a labile compound. The decomposition of SAM includes epimerization at sulfur and hydrolysis to S-methyladenosine and homocysteine. The configuration at sulfur in biologically active SAM is S by the Cahn-Ingold-Prelog convention, as shown in fig. 15-6. Epimerization at sulfur in SAM takes place with a half-time of about 12 hours in neutral solutions (Booker, 2004; Wu et al., 1984). Hydrolysis takes place with anchimeric assistance by the carboxylate group with a half-time of about 60 hours (Iwig and Booker, 2004). These reactions are not fast enough to pose a biological problem, but allowances must be made in conducting laboratory work with SAM.

Properties of Catechol O-Methyltransferase

Most tissues contain COMT, which exists in both soluble and membrane-associated forms. In brain, most of the activity is membrane-associated, whereas liver and kidney have significant soluble activity. The two forms are encoded by the same gene and arise through differential translation of the mRNA, which contains two distinct promoters and start sites and for the soluble and membrane forms (Tenhunen and Ulmanen, 1993). The rat and human soluble COMTs have 221 amino acids and molecular masses of about 24 kDa. Structures of soluble COMT are similar to the structures of other SAM-dependent methyltransferases, including DNA methyltransferases, glycine-N-methyltransferase. The SAM-dependent methyltransferases constitute a family of similar structures about the SAM binding site but differ in the binding of co-substrates (Cheng, 1995; Schluckebier et al., 1995).

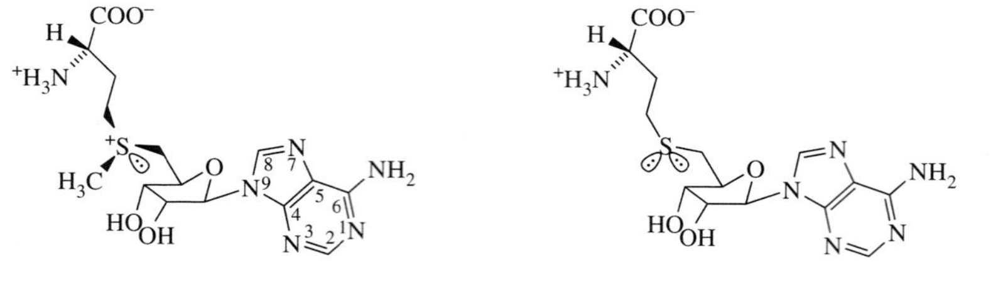

S-Adenosylmethionine (SAM or AdoMet) S-Adenosylhomocysteine (SAH or AdoHcys)

Fig. 15-6. Structure of the biologically active S-epimer of S-adenosylmethionine. The structure shows the transferable methyl group and, for later reference, the atom numbers in the adenine ring. S-Adenosylmethionine has six asymmetric centers in its atomic skeleton, four optically active carbons in the ribose ring, C2 of the methionine moiety, and the sulfonium sulfur. The optical center at sulfur is created by the orientations of the three substituents shown and the orbital occupied by the nonbonding electron pair.

A REACTION

$$\text{L-DOPA} + \text{SAM} \xrightarrow{\text{COMT/Mg}^+} \text{3-methoxy-L-DOPA} + \text{SAH} + \text{H}^+$$

B KINETIC MECHANISM

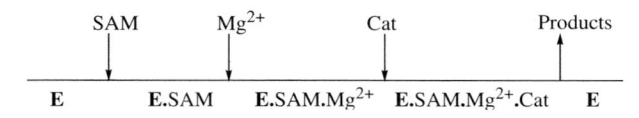

```
         SAM      Mg²⁺       Cat                Products
          ↓        ↓          ↓                    ↑
  E      E·SAM  E·SAM·Mg²⁺  E·SAM·Mg²⁺·Cat         E
```

Fig. 15-7. Reaction of catechol O-methyltransferase (COMT). (A) COMT accepts a broad range of catechols as methyl-acceptor substrates, including epinephrine and L-dihydroxyphenylalanine (L-DOPA). In the presence of Mg^{2+}, COMT catalyzes the methylation of the 3-OH or 4-OH group of L-DOPA by S-adenosylmethionine (SAM); the most reactive is the 3-OH group. (B) The reaction takes place by an ordered kinetic mechanism, in which SAM binds first, followed by the divalent metal ion and the catechol (Cat).

Reaction of Catechol O-Methyltransferase

Many catechols are methyl acceptor substrates for COMT, including epinephrine and L-DOPA, and SAM is the methyl donor. The action of COMT on L-DOPA is illustrated in fig. 15-7. The action of COMT requires Mg^{2+}, a property not universally shared by SAM-dependent methyltransferases. In keeping with the acceptance of a variety of substrates by COMT, methyltransfer takes place to either hydroxyl group of L-DOPA, but preferentially to the *meta*-hydroxyl group.

The steady-state kinetic studies show that the action of soluble COMT follows an ordered binding mechanism (fig. 15-7B). SAM binds first, followed by Mg^{2+} and then the methyl acceptor (Lotta et al., 1995). COMT is a drug target, primarily to depress its activity against L-DOPA, which is the leading therapeutic agent prescribed for Parkinson's Disease. Very potent, slow-binding inhibitors bind with Mg^{2+}-dependence (Bonifacio et al., 2002). The most efficacious inhibitors can suppress the tissue activity of COMT by more than 75% when administered to mice.

Structure

The structure of soluble COMT in complex with SAM, Mg^{2+}, and an O-nitrophenolate inhibitor is shown in fig. 15-8 (Vidgren et al., 1996). The global chain fold is similar to other SAM-dependent methyltransferases. However, unlike other members of the family, COMT incorporates a Mg^{2+} binding site anchored by Asp141, Asp169, and Asp170 on the protein side. The vicinal hydroxy groups of the catechol-inhibitor or substrate contribute apical and equatorial ligands to Mg^{2+}, and a water molecule occupies the second apical position. SAM is bound through numerous hydrogen bonded and hydrophobic contacts between the adenosyl and methionyl moieties and COMT.

The kinetic consequences of structural variations in SAM have been evaluated (Borchardt et al., 1976). The changes in k_{cat}/K_m have been mapped onto the COMT-SAM contacts as values of $\Delta\Delta G^{\ddagger}$ (Takusagawa et al., 1998). The structural changes to SAM included single atom alterations in the structure of the adenosyl moiety, and each of them modestly decreased the rate. The values of $\Delta\Delta G^{\ddagger}$ were +1.5 to +1.7 kcal mol^{-1} for the 8-aza, 7-deaza, 3-deaza, 6-NHCH$_3$, and 2'-deoxy analogs of SAM. If these values of $\Delta\Delta G^{\ddagger}$

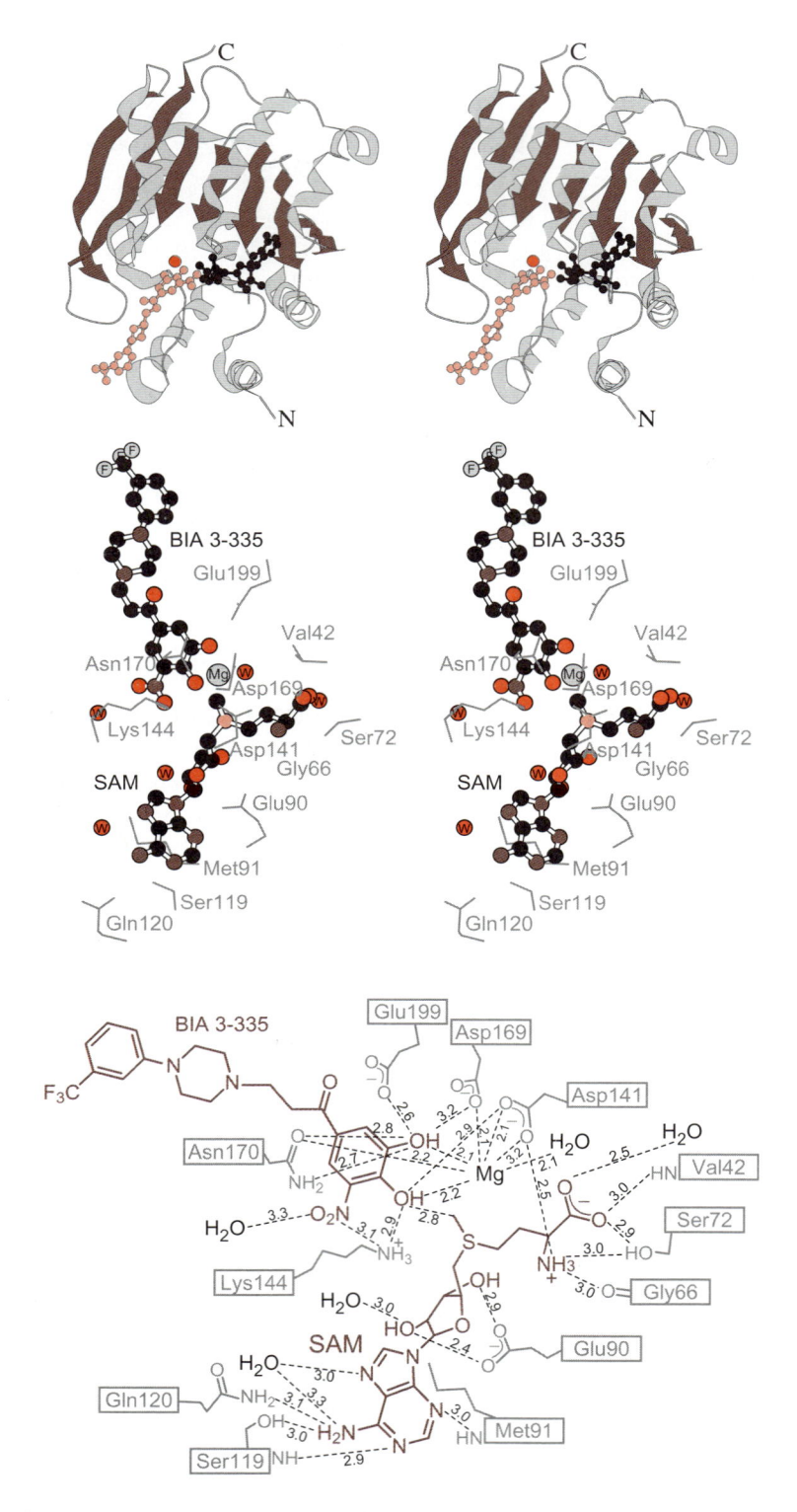

Fig. 15-8. Structure of soluble catechol O-methyltransferase (COMT) in complex with S-adenosylmethionine (SAM), Mg^{1+}, and an inhibitor. An x-ray crystallographic structure of rat liver COMT is shown with SAM (black), magnesium (red), and the inhibitor BIA3-335 (pink) bound at the active site (2.00-Å resolution; PDB 1H1D; Bonifacio et al., 2002).

are additive, one can conclude that the summation of the contributions of these contacts decreases the activation energy for catalysis by about 8 kcal mol^{-1}. The analogs with D-methionyl or carbocyclic adenosyl moieties each displayed much less activity, with $\Delta\Delta G^\ddagger = 3$ kcal mol^{-1}.

Reaction Mechanism

One aspect of catalysis by COMT that is fairly well understood is the role of Mg^{2+}. Catechols provide ligands for divalent metal ions, and the structure of COMT shows this to be the case in the ternary complex. The ligation to Mg^{2+} can lower the pK_a value of a ligand such as water by 4 units (table 4-3), so that the catechol is likely to be ionized to its mono-anion in the magnesium complex. This will facilitate the alkylation of a catechol by increasing the nucleophilic reactivity of both oxygens.

Apart from facilitating the loss of a proton from an attacking nucleophile, which is not always necessary in methyl transfer, it is not obvious how an enzyme can catalyze the process. It is difficult to imagine how the leaving group in SAM could be improved, apart from general binding interactions. The methyl group cannot be improved as an alkylating entity. In principle, an enzyme can bind SAM and the methyl acceptor together in the correct steric relationship to allow methyl transfer, and the structure in fig. 15-8 demonstrates this in the case of COMT.

It has been suggested that the binding of SAM and a nucleophilic acceptor may be accompanied by compression, and that this might facilitate the reaction (Jencks, 1969). The kinetic isotope effects appeared to provide support for this hypothesis (Gray et al., 1979; Mihel et al., 1979; Rodgers et al., 1982; Takusagawa et al., 1998; Wu et al., 1984). The value of $^{13}k_t$ in the reaction of S-adenosyl-L-[methyl-^{13}C]methionine was found to be 1.09, proving that methyl transfer was rate limiting. The value of the secondary kinetic isotope effect $^D(k_{cat}/K_m)$ was 0.84, consistent with compression in the transition state. Although the secondary isotope effect was compatible with the induction of compression by binding interactions, it did not prove it to be so. In a subsequent theoretical study of nonenzymatic and COMT-catalyzed methyl transfer, employing QM/MM methodology for the enzymatic process, both enzymatic and nonenzymatic reactions were predicted to proceed with inverse secondary deuterium kinetic isotope effects (Ruggiero et al., 2004). The study showed that the transition states of both the enzymatic and nonenzymatic reactions were compressed. Inverse deuterium kinetic isotope effects would be expected, regardless of whether binding-induced compression played a role in catalysis.

A mechanism for methyl transfer by COMT, illustrated in fig. 15-9, invokes Mg^{2+}-coordination of the catechol to lower the pK_a and increase nucleophilic reactivity, and base catalysis by Glu199 to further increase nucleophilic reactivity. All evidence point to the associative mechanism with a single transition state. The single displacement mechanism is in accord with the stereochemistry of methyl transfer, which proceeds with inversion of configuration at the methyl carbon (Woodard et al., 1980). In fig. 15-9, notice that the net electrostatic charge on the reacting atoms is zero and does not change from the ground state through the transition state and to the product.

S-Adenosylmethionine Synthetase

Reaction and Kinetics

S-Adenosylmethionine synthetase (SAM; EC 2.5.1.6) synthetase catalyzes the adenosylation of methionine to produce SAM (fig. 15-10A). In this case the 5′-CH$_2$-group of ATP is

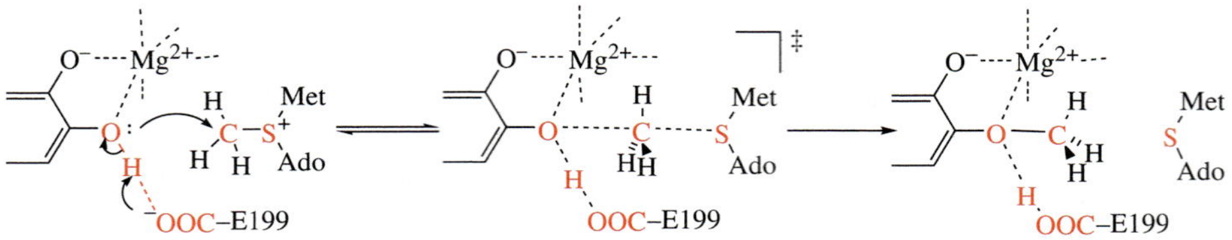

Fig. 15-9. Mechanism for methyl transfer in the action of catechol *O*-methyltransferase (COMT). The reacting atoms are shown in red. In this mechanism, Mg^{2+} is postulated to lower the pK_a value for the catechol through direct ligation to both hydroxyl groups so that one is ionized. Nucleophilic attack by one of the catechol oxygens on the methyl group of *S*-adenosylmethionine (SAM) is facilitated by Glu199, which accepts the proton from the un-ionized hydroxyl group, shown as the attacking nucleophile.

the alkylating agent, and the sulfur of methionine undergoes alkylation by the adenosyl moiety. Kinetic isotope effects show that adenosyl transfer limits the rate (Markham et al., 1987), and the reaction proceeds with inversion of configuration at the C5′-methylene carbon (Parry and Minta, 1982). The reaction also produces PP_i and P_i from the triphosphate moiety of ATP. The P_i arises exclusively from the γ-phosphate of ATP, with incorporation of ^{18}O from $H_2^{18}O$ into P_i and not into PP_i (Markham et al., 1980: McQueny et al., 2000; Mudd, 1963). The hydrolysis of PPP_i initially formed in the alkylation step presumably draws the overall equilibrium to the right, toward SAM-synthesis. In addition to the primary substrates, the action of SAM synthetase requires the presence of Mg^{2+}, and the enzyme is most active in the presence of K^+.

In the steady-state kinetic mechanism, ATP and methionine bind randomly to the free enzyme, and the adenosyl transfer takes place within the ternary complex of enzyme with ATP and methionine (Markham et al., 1980). A detailed kinetic study to evaluate rate constants by employing stopped-flow kinetics, pulse-chase kinetics, evaluation of equilibrium dissociation constants, and measurements of kinetic isotope effects, a detailed kinetic mechanism led to the mechanism in fig. 15-10B. This is a random binding mechanism for ATP and methionine, in which the further reaction of the ternary complex can follow more than one course, as illustrated in the branches. However, the values of rate constants indicate that the main course of the reaction follows the upper portion of the initial binding phase, that is, with ATP binding first and then methionine, and the main pathway for the ternary complex proceeds linearly across to the products, with mainly ordered release of P_i PP_i, and finally SAM (McQueny et al., 2002). Dissociation of PPP_i rarely occurs.

With the rate constants in hand, and knowledge of the intracellular concentrations of methionine and MgATP, a profile could be constructed giving the free energies of each intermediate in the mechanism in vivo. The profile showed that the intermediates decreased in free energy throughout the reaction, and it confirmed the rate-limiting step as adenosyl transfer from ATP to methionine (McQueny et al., 2000).

Structure and Reaction Mechanism

Structures of SAM synthetase from *Escherichia coli* and rat are available (Gonzalez et al., 2000; Komoto et al., 2004; Takusagawa et al., 1996). The crystalline protein is a dimer of tightly associated dimers, AB and CD. The structure shown in fig. 15-11 is that of the enzyme from *E. coli* co-crystallized with methionine and AMPPNP, the analog of ATP with the oxygen bridging $P_β$ and $P_γ$ replaced by NH (Komoto et al., 2004). This analog reacts normally to form the complex of enzyme with SAM and diimidotriphosphate

A REACTION

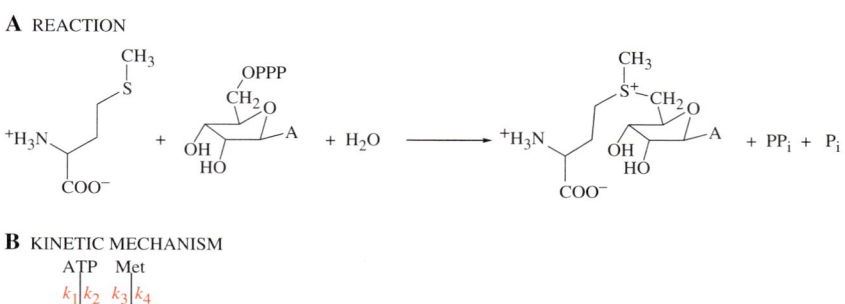

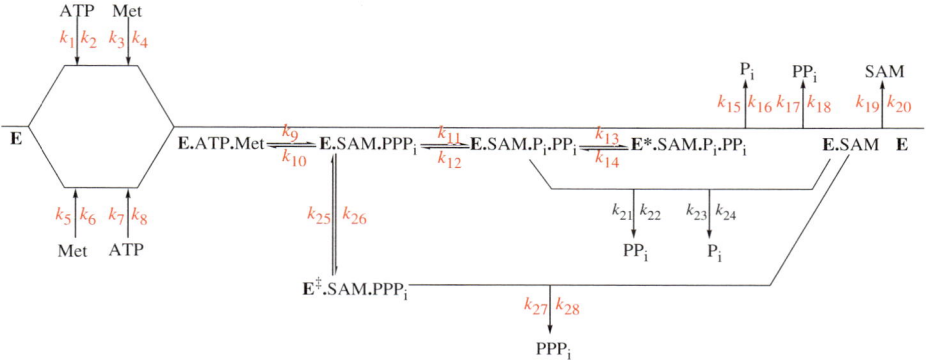

B KINETIC MECHANISM

Fig. 15-10. Reaction of S-adenosylmethionine (SAM) synthetase and a kinetic mechanism for its action. (A) The overall reaction includes the hydrolysis of ATP to PP_i and P_i, in addition to the alkylation of adenosine. (B) This mechanism was deduced from steady-state rates, equilibrium binding constants for binary complexes, stopped-flow rates for individual steps, pulse-chase experiments, and the kinetic isotope effect for adenosyl transfer (McQueny et al., 2000). Estimates were obtained for the rate constants shown in red. The enzyme species **E*** and **E‡** are isomeric forms, presumably arising from conformational changes.

(PNPNP), but it does not undergo further reactions (Markham et al., 1980). In the kinetic mechanism the next step would be hydrolysis of PPP_i (fig. 15-10A), but this does not occur with PNPNP. The structure shows the active sites at the interfaces of subunits in the tight dimer, with amino acid side chains from both subunits in contact with AMPNPNP and methionine.

In the crystal structure, the tight dimers differentially bind substrates and products. In the structural model, dimer AB binds AMPNPNP and methionine, and dimer CD binds SAM and PNPNP (Komoto et al., 2004). Figure 15-11 shows SAM and PNPNP bound to dimer CD. The chain fold in each subunit includes three domains consisting of β-sheets overlaid with α-helices (Takusagawa et al., 1996). Two domains have four β-strands in the sheet and two α-helices on the face, and one domain has three β-strands with two α-helices on the face. Two additional helices are in a connecting loop and the C-terminal loop. The domains are associated in a triangular arrangement.

The active site contacts include many hydrogen bonds between heteroatoms of the ligands and the enzyme. In addition, Phe230 forms a hydrophobic contact with the adenine ring. A striking and presumably mechanistically significant feature is the high concentration of positive charge focused on PNPNP. Two Mg^{2+} ions are coordinated to PNPNP, and the side chains of one arginyl and three lysyl residues are within hydrogen bonding range as well. This concentration of positive charges should enhance the leaving group potential of the triphosphate moiety in ATP (and AMPNPNP). The excess positive charges are neutralized by aspartate residues in contact with the basic amino acids and the Mg^{2+} ions.

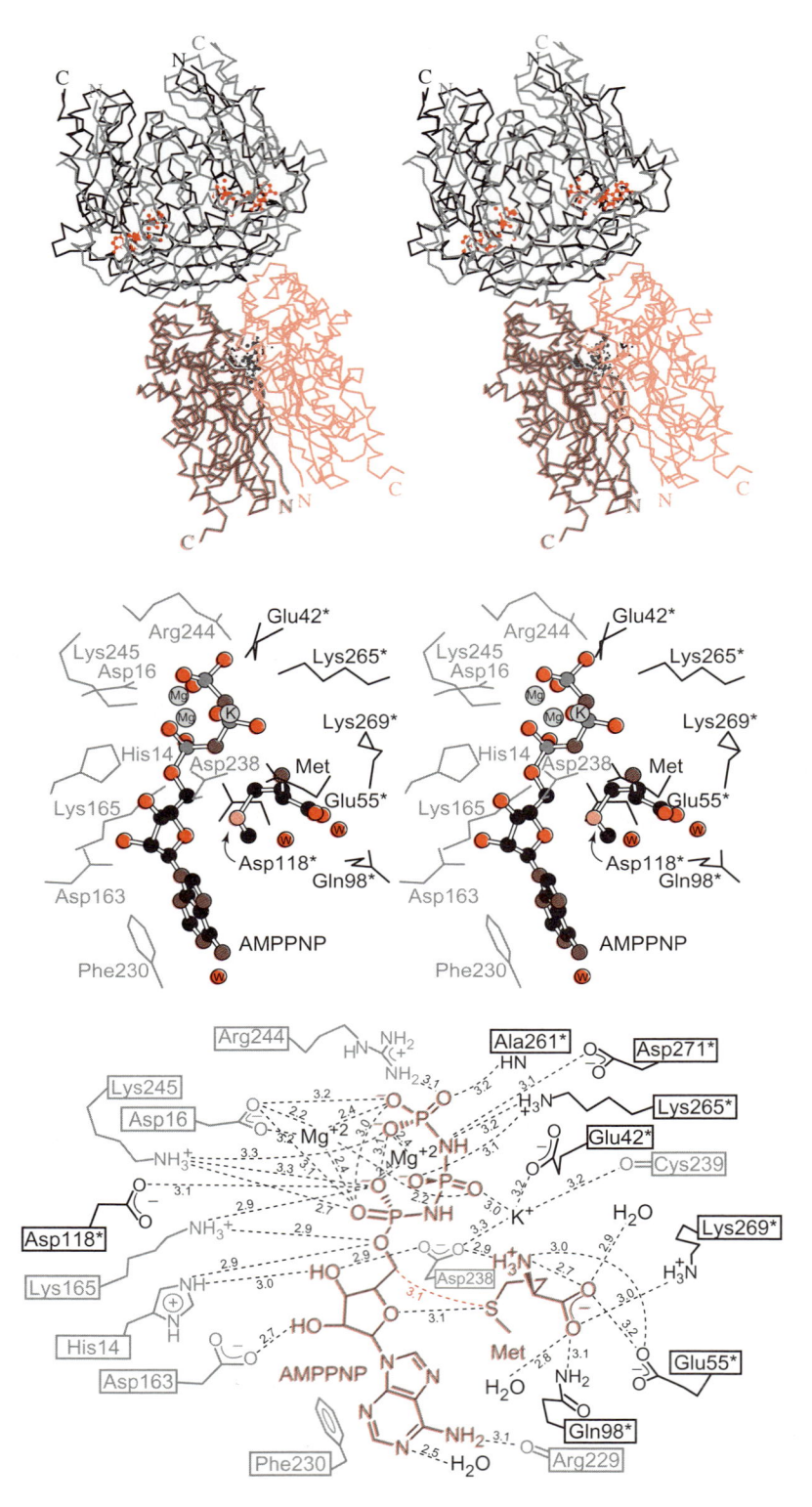

Fig. 15-11. Structure of a complex of S-adenosylmethionine (SAM) and diimidotriphosphate (PNPNP) with SAM synthetase. A Cα trace of the dimer of dimers of homotetrameric E. coli SAM synthetase is shown with SAM and PNPNP (both red) bound in both active sites of one dimer and AMPNPNP and Met (both black) bound in the sites of the other dimer (2.50-Å resolution; PDB 1P7L; Komoto et al., 2004). The Met AMPNPNP complex is shown (middle) in stereo and as a two-dimensional image (bottom). The distance from the sulfur atom of Met to the C5′ of AMPNPNP is 3.1Å.

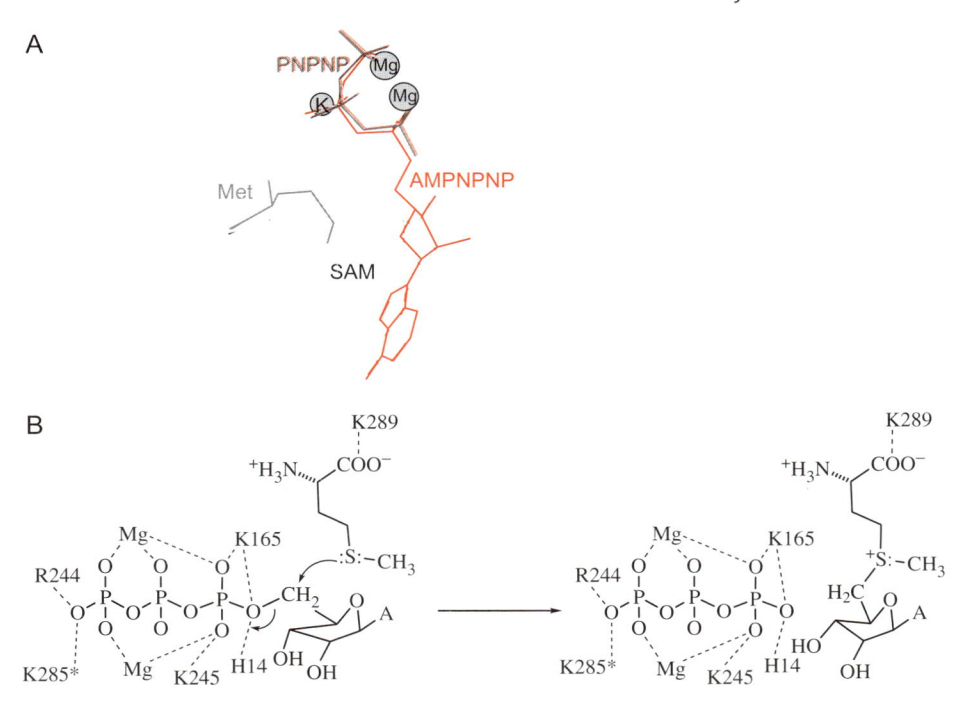

Fig. 15-12. Mechanism for adenosyl transfer by S-adenosylmethionine (SAM) synthetase. Structures of APPNP and Met and of SAM and diimidotriphosphate (PNPNP) in the active site of SAM synthetase (2.50-Å resolution; PDB 1P7L; Komoto et al., 2004) indicate that the main structural change in ATP is the conformation of the ribose ring, which changes from 4′-exo to 3′-endo, with the C5′-carbon moving toward methionine and the thiomethyl group of methionine moving toward and bonding to the C5′-carbon. (A) These structures have been aligned for comparison. (B) The mechanism is shown as an associative, nucleophilic attack on C5′ by the sulfur of methionine, with displacement of PPP$_i$ in complex with two Mg^{2+} ions, three lysine residues, and one arginine.

The structure of the complex containing APNPNP and methionine compared with that containing SAM and PNPNP show little difference except for the conformation of the ribosyl ring and the placements of C5′ and the S-methyl group. Covalent bonding between C5′ and S in SAM brings these two groups together, and this is coupled with the change in pucker within the ribosyl ring. A simple mechanism is shown in fig. 15-12, which illustrates a straightforward associative nucleophilic displacement of PPP$_i$ by the sulfur of methionine. The mechanism shown is consistent with the kinetics, the isotope effect, and the structure.

Mutations of Arg244, Lys245, Lys269, and His14 lead to dramatically lower values of k_{cat} in the variants (Reczkowski et al., 1998; Taylor and Markham, 2000). These residues are in contact with either PNPNP or methionine in the structure. Mutations of Asp16*, Asp118, Asp238*, or Asp271 (asterisk denotes a neighboring subunit) also lead to dramatically low activities in the variants (Taylor and Markham, 1999). These residues are in contact with Mg^{2+} or basic amino acids associated with PPNP in the structure. Electrostatic perturbations lead to decreased activity. In the case of His14 the mutations probably do not perturb the electrostatics, because the contacts of His14 indicate it to be in its neutral form. However, it is in close, hydrogen-bonded contact with the oxygen bridging the adenosyl moiety of AMPNPNP and with the departed PPNP (Komoto et al., 2004).

PNPNP displays slow, tight binding behavior to SAM synthetase in the presence but not in the absence of SAM (Reczkowski and Markham, 1999). The value of K_d for slow, tight

670   Enzymatic Reaction Mechanisms

binding is 2 nM, whereas PNPNP in the absence of SAM is a simple reversible inhibitor, with a K_i value of 0.3 µM. Nuclear magnetic resonance (NMR) and ultraviolet (UV) studies of slow PNPNP binding indicate reorientations of amino acid side chains associated with the binding process (Markham and Reczkoski, 2004).

Methionine Synthases

Because of the diverse roles of methylation by SAM, and because of the SAM cycle that regenerates SAM, methionine biosynthesis is metabolically connected to a variety of biological processes in all cells, both prokaryotes and eukaryotes, as well as in higher mammals including man. In the SAM cycle, transmethylation by SAM produces S-adenosylhomocysteine (SAH). The regeneration of SAM does not entail the methylation of SAH to SAM. Instead, SAH undergoes hydrolysis to adenosine and homocysteine by the action of SAH hydrolase (see chap. 3), homocysteine is methylated to methionine by action of methionine synthase, and methionine is converted into SAM by action of SAM synthetase. The functions of methionine synthase, SAH hydrolase, and SAM synthetase maintain SAM at homeostasis in metabolism, and methionine synthase plays a key role.

Methionine synthases (EC 2.1.1.13) are of two kinds: the cobalamin-dependent synthase known as MetH in *E. coli* and the cobalamin-independent form known as MetE in *E. coli*. *E. coli* contain both MetH and MetE, but humans and higher animals have only MetH. MetE is found in bacteria and lower eukaryotes. MetH from *E. coli* is the most thoroughly studied methionine synthase and is the principal focus of this section. The salient features of MetE also are elaborated.

Reactions and Properties of MetH

All methionine synthases catalyze the reaction of homocysteine with 5-methyltetrahydrofolate (methyltetrahydrofolate) to form methionine and tetrahydrofolate, as illustrated in fig. 15-13. MetH from *E. coli* is representative of methionine synthases that contain tightly bound cobalamin, which in its methylcobalamin form mediates methyl transfer. MetH from *E. coli* has a molecular mass of 136 kDa, based on the amino acid sequence deduced from the nucleotide sequence of the cloned gene (Banerjee et al., 1989).

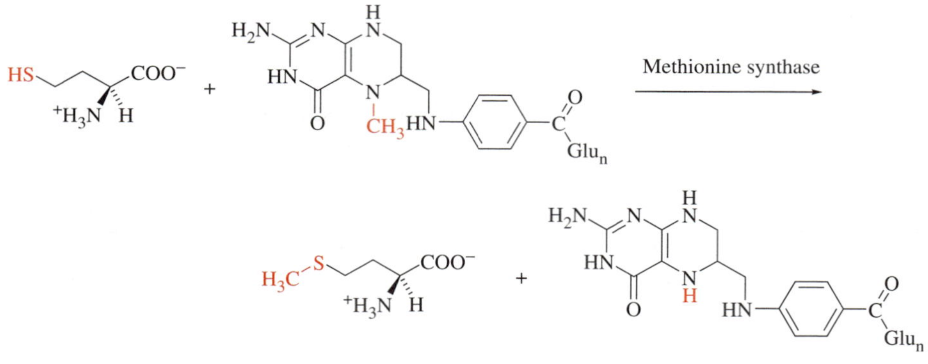

Fig. 15-13. The methyl transfer reaction of homocysteine with methyltetrahydrofolate to form methionine and tetrahydrofolate is catalyzed by methionine synthases.

Early studies of cobalamin-dependent methionine synthases revealed essential features of the reaction (Weissbach and Taylor, 1970; Banerjee et al., 1990). In its simplest form, the reaction was visualized as proceeding in two steps. First, the methyl group of enzyme-bound methylcobalamin was transferred to homocysteine to form methionine and cob(I)alamin, then a methyl group was transferred from methyltetrahydrofolate to cob(I)alamin to regenerate methylcobalamin and form tetrahydrofolate. These two steps would complete a catalytic cycle. However, several properties of the enzyme raised questions about details of the reaction mechanism. Although the enzyme-bound cobalamin appeared to mediate methyl transfer catalytically, free methylcobamin could serve stoichiometrically as the methyl donor in place of methyltetrahydrofolate. It could be shown that the endogenous cobalamin did not participate in methyl transfer when free methylcobalamin served as the methyl donor. For example, the endogenous cobalamin could be propylated with propyl iodide and a reducing agent, and this blocked methyl transfer to homocysteine from methyltetrahydrofolate. However, the propylcobalmin-enzyme could still catalyze methylation of homocysteine by free methylcobalamin. It seemed that free methylcobalamin could bypass the endogenous cobalamin to produce methionine. Moreover, SAM and a reducing agent could activate the enzyme when the endogenous cobalamin was demethylated and in the form of cob(II)alamin, but it could not activate when the endogenous cobalamin was propylated. Moreover, although SAM could activate the enzyme and restore the endogenous demethylated cobalamin to its methylated form, it could not serve as a stoichiometric methyl donor for the production of methionine. The apparent confusion raised by these facts ultimately gave way to clarity through detailed biochemical and structural experimentation.

Modular Structure

The biochemical properties of MetH are explained by its modular structure and the dynamics of the interactions of the modules (Bandarian et al., 2003; Evans et al., 2004). The catalytic functions of MetH can be explained by the reaction scheme in fig. 15-14. In the normal catalytic cycle, illustrated in the upper part of fig. 15-14, MetH catalyzes the reaction in fig. 15-13 in two steps, nucleophilic displacement of cob(I)alamin from methylcobalamin by homocysteine to form methionine, followed by methylation of cob(I)alamin by methyltetrahydrofolate to form tetrahydrofolate and regenerate methylcobalamin to complete the cycle. This process continues for 100 to 2000 turnovers. Adventitious oxidation of the intermediate cob(I)alamin by occasional encounters with molecules such as molecular oxygen leads to cob(II)alamin and inactivates the enzyme. However, reduced flavodoxin reduces endogenous cob(II)alamin to cob(I)alamin, and MetH catalyzes its methylation by SAM to regenerate methylcobalamin and reactivate the enzyme. Each of the steps in fig. 15-14 is catalyzed by a module within the structure of MetH, and to accomplish these feats the modules dynamically interact with one another.

The modules in MetH and their binding properties are illustrated in fig. 15-15 in relation to the linear amino acid sequence. Controlled proteolytic degradation releases the SAM binding and cobalamin (Cob) binding domains, indicating that the two are linked to the central core by mobile linkers that are susceptible to proteolysis (Banerjee et al., 1989; Drummond et al., 1993). These two are the 38-kDa, N-terminal, SAM binding domain (MetH$^{897-1227}$) and the neighboring 28-kDa, cobalamin binding domain (MetH$^{650-896}$). Limited trypsinolysis with methylcobalamin bound to MetH produces two fragments, 38-kDa, SAM binding domain and a fragment consisting of the other domains that retains the catalytic properties of MetH, including the gradual inactivation because of adventitious oxidation of cob(I)alamin to cob(II)alamin. However, unlike MetH, the cob(II)alamin-form

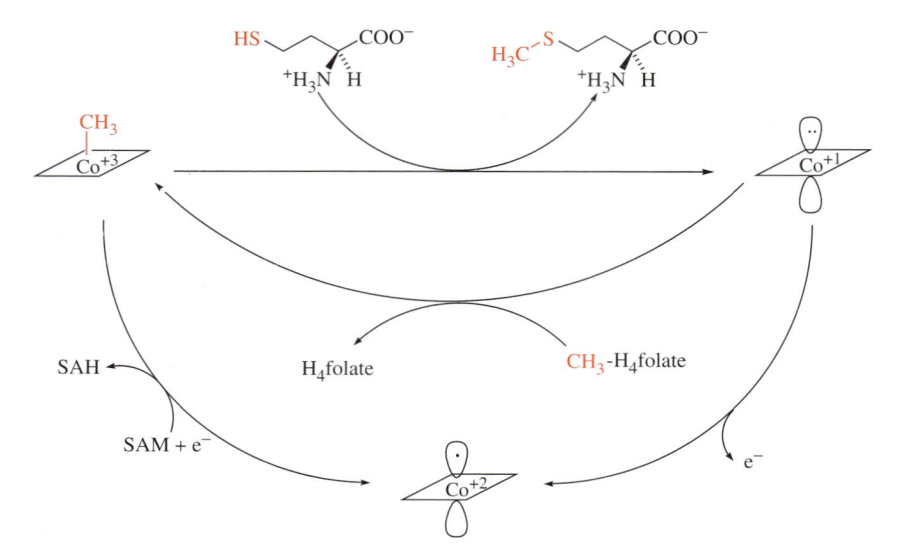

Fig. 15-14. Reactions of MetH in methionine biosynthesis. The normal catalytic cycle is represented by methyl transfer from endogenous methylcobalamin to homocysteine to form endogenous cob(I)alamin and methionine in the upper part of this scheme. Methylation of cob(I)alamin by methyltetrahydrofolate (CH$_3$-H$_4$folate) regenerates endogenous methylcobalamin and forms tetrahydrofolate (H$_4$folate). Adventitious oxidation of endogenous cob(I)alamin in the lower portion of the scheme, as by molecular oxygen, inactivates the enzyme by forming cob(II)alamin. MetH can heal itself by catalyzing the reductive methylation of endogenous cob(II)alamin by SAM and reduced flavodoxin.

of this fragment cannot be reactivated by a reducing agent and SAM, suggesting that the SAM domain is required in the reactivation.

The homocysteine (HCys) binding domain (MetH$^{2-353}$) and the 70-kDa HCys and tetrahydrofolate (Fol) binding fragment (MetH$^{2-649}$) are produced from truncated genes (Goulding et al., 1997). They catalyze the reactions shown in fig. 15-15. The 70-kDa MetH$^{2-649}$ catalyzes methyl transfer from methyltetrahydrofolate to cob(I)alamin and methyl transfer from free methylcobalamin to HCys, but the N-terminal HCys domain catalyzes only the methyl transfer from methylcobalamin to HCys. The catalytic properties of the fragments verify that the N-terminal module catalyzes the binding and methylation of homocysteine, and the intervening residues 354 to 649 are the methyltetrahydrofolate binding module. The catalytic activity of the 70-kDa HCys-Fol fragment toward free cob(I)alamin and methylcobalamin is similar to those of MetH, which also catalyzes the partial reactions shown in fig. 15-15. The truncated gene product MetH$^{650-1227}$ includes the Cob and SAM modules and catalyzes the reductive methylation of cob(II)alamin.

The structures of the SAM and Cob domains are shown in fig. 15-16, together with the structure of a 65-kDa MetH fragment encompassing the Cob and SAM domains (Drennan et al., 1994; Dixon et al., 1996; Bandarian et al., 2002). The structure of the Cob domain reveals the unanticipated mode of cobalamin binding, in which the lower axial dimethylbenzimidazole is displaced by His759, which is hydrogen bonded to Asp757 and Ser810. The resulting "dimethylbenzimidazole-tail" of cobalamin is deeply buried in the structure in what has become known as the *base-off* binding mode for cobalamin-dependent enzymes.

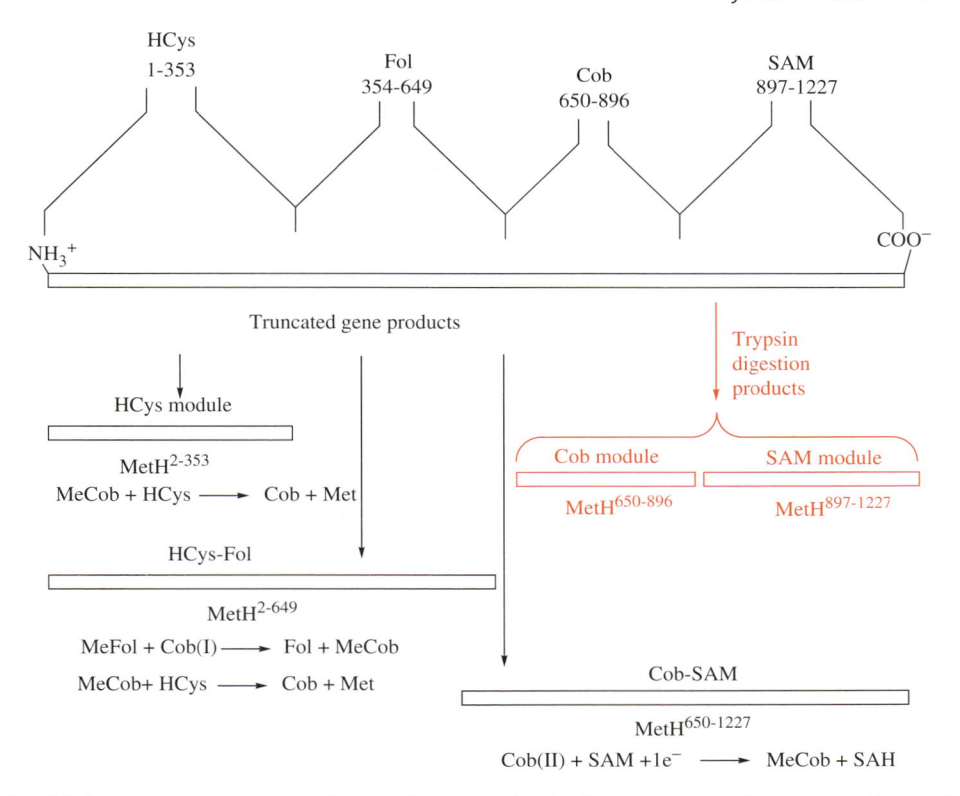

Fig. 15-15. Modular structure of MetH from *E. coli*. The linear amino acid sequence of MetH is represented by the upper bar. Trypsin digestion releases two fragments, an *S*-adenosylmethionine binding module MetH$^{897-1227}$ and a methylcobalamin (Cob) binding module MetH$^{650-896}$. The truncated gene product MetH$^{2-354}$ binds homocysteine (HCys), and the truncated gene fragment MetH$^{2-649}$ binds both HCys and methyltetrahydrofolate. The fragment MetH$^{650-1227}$ encompasses the Cob and SAM modules. Each fragment catalyzes a part of the reactions of wild-type MetH.

The structure of the 65-kDa Cob-SAM fragment in fig. 15-16 unmasks an essential element of the modular dynamics in the action of MetH. The fragment is in the activation conformation, in which cob(II)alamin is reductively methylated by SAM. A helical "cap" over cobalamin in the Cob domain is displaced 26 Å by a 63-degree rotation, creating the interface between the Cob and SAM domains and allowing access of the methyl group of SAM to cob(I)alamin. The importance of the cap displacement cannot be overemphasized. The cap shields cobalamin from interactions that could lead to futile cycling of methionine and SAM. In the activation structure, the lower axial cobalamin ligand, His759, is dissociated, making cobalt four coordinate, the coordination state for cob(I)alamin.

Figure 15-17 shows the structure of a HCys-Fol fragment of MetH from *Thermotoga maritime* (Evans et al., 2004). This structure shows that the respective active sites are not near each other, and there is an extensive interface between the HCys and Fol modules. There is no obvious possibility for dramatic, independent movements of these two domains. This arrangement is entirely compatible with the main reaction cycle of MetH in fig. 15-14, and it clarifies several properties of the fragment and of MetH itself. The active sites of the HCys and Fol modules would not interact with each other in the reaction scheme of fig. 15-14, but each would interact with cobalamin, the Fol domain in methyl transfer from methyltetrahydrofolate to cob(I)alamin and the HCys domain in methyl transfer from methylcobalamin to homocysteine. The structure allows the active sites of

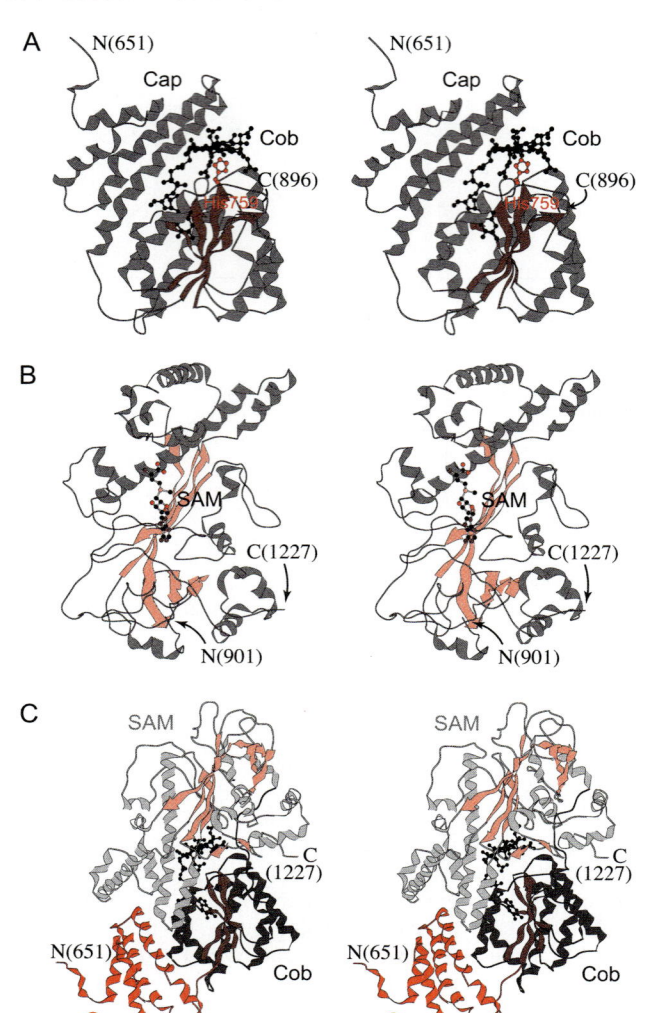

Fig. 15-16. Structures of the cobalamin (Cob) and S-adenosylmethionine (SAM) modules and the Cob-SAM fragment of MetH from *E. coli*. X-ray crystallographic structures of domains in MetH. (A) The Cob domain (3.00-Å resolution; PDB 1BMT; Drennan et al., 1994). (B) The SAM domain (1.80-Å resolution; PDB 1MSK; Dixon et al., 1994). (C) The complex of cob and SAM domains (3.75-Å resolution; PDB 1K98; Bandarian et al., 2001).

the HCys and Fol to interact with free cobalamins and presumably with cobalamins bound to the Cob domain as well. The HCys-Fol fragment and MetH itself catalyze methyl transfer with free cobalamins, either methylcobalamin as a donor or cob(I)alamin as an acceptor.

Modular Function

In the current model for modular functions, MetH exists in the four structural states centered on the Cob module, as illustrated in fig. 15-18. In state 1, the helical cap shields cobalamin in Cob, and the other modules are out of interaction range. In state 2, the helical cap is displaced by the interaction of the Cob and Fol modules. In state 3, the Cob module binds to the HCys module, the helical cap remains displaced, and the Fol and SAM modules do

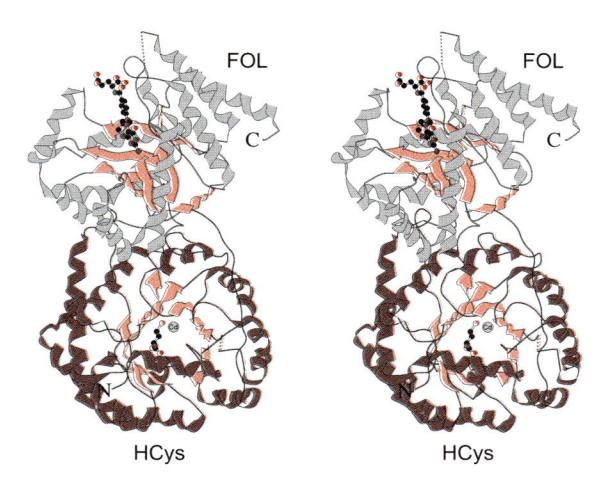

Fig. 15-17. The x-ray crystallographic structure of the homocysteine- tetrahydrofolate (HCys-Fol) module of MetH from *Thermotoga maritime* is shown with homocysteine, cadmium ion, and CH_3-H_4folate bound (1.90-Å resolution; PDB 1Q8J; Evans et al., 2004).

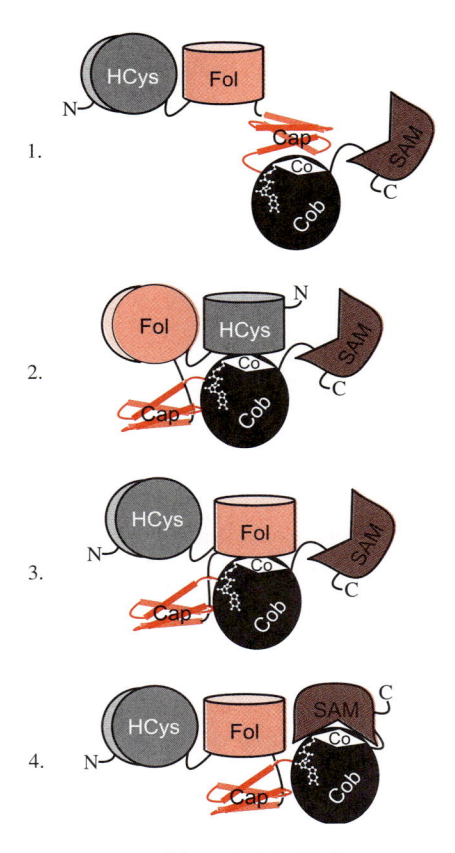

Fig. 15-18. Dynamics of modular transitions in MetH. Four states of MetH are postulated to account for the reactions catalyzed by this modular enzyme. The states differ with respect to the interactions of the modules and helical cap. In state 1, the helical cap covers the cobalamin binding site (Cob). In state 2, the cap is displaced by the homocysteine (HCys) module, which binds to methylcobalamin in the Cob module. This state brings homocysteine in contact with methylcobalamin and allows methyl transfer to form methionine. In state 3, the Cob module with cob(I)alamin binds to the active site of the tetrahydrofolate (Fol) module. Methyl transfer from methyltetrahydrofolate to cob(I)alamin takes place in this state. When cob(I)alamin is adventitiously oxidized to cob(II)alamin, the enzyme enters state 4, in which the Cob module binds to the *S*-adenosylmethionine (SAM) module, and the helical cap, HCys, and Fol modules do not have access to Cob. In this state, the SAM module catalyzes the flavodoxin and SAM-dependent reductive methylation of cob(I)alamin to methylcobalamin, and the enzyme returns to state 1.

not have access to Cob. In state 4, the Cob and SAM modules interact through their active sites, and the Fol and HCys modules and the helical cap are displaced.

In the catalytic cycle, MetH alternates from states 1 through 3. Binding of methyltetrahydrofolate to Fol signals a transition from state 1 to state 2, wherein methyl transfer to cob(I)alamin occurs. The cap over cob(I)alamin in state 1 prevents cob(I)alamin from being methylated by SAM on the SAM module, and the binding of methyltetrahydrofolate signals the transition to state 2 and prevents the formation of state 4, in which SAM could methylate cob(I)alamin. When Cob is in its methylcobalamin state, a transition to state 3 takes place, wherein methyl transfer to homocysteine occurs on the HCys module, with regeneration of cob(I)alamin on the Cob module. The system reverts to state 1 with the dissociation of methionine. Imagining starting with methylcobalamin on Cob, the system starts with state 3 and methylation of homocysteine, moves to state 1, and on binding of methyltetrahydrofolate to the Fol module, to state 2. Methylation of cob(I)alamin by methyltetrahydrofolate in state 2 completes the cycle.

An adventitious oxidation of cob(I)alamin to cob(II)alamin signals the transition to state 4, wherein cob(II)alamin is reductively methylated by SAM. In *E. coli*, the reduction of cob(II)alamin to cob(I)alamin in state 4 specifically requires *E. coli* flavodoxin (Hoover et al., 1997). The structure of state 4 is such that the dissociation of His759 from the lower axial position of methylcobalamin is facilitated, and this can be observed spectrophotometrically by the shift in absorption maximum from 325 nm to 450 nm (Bandarian et al., 2003). This spectral change allows the conformational change and factors influencing it to be monitored, and the results confirm that the interaction between SAM and Cob modules facilitates dissociation of His759 as the lower axial ligand of cob(II)alamin in preparation for reduction to cob(I)alamin. The dissociation of His759 is induced by the binding of flavodoxin to the SAM module (Hoover et al., 1997).

Reaction Mechanisms in Catalysis and Repair

The chemical mechanism of SAM-dependent repair of cob(II)alamin is reasonably well understood. Reduction of cob(II)alamin bound to the Cob module in state 4 requires reduced flavodoxin, which interacts specifically with the SAM module, and methyl transfer from the sulfonium center of SAM to cob(I)alamin is chemically well precedented in cobalamin chemistry (see chap. 4).

Methyl transfer from methylcobalamin to homocysteine in state 2 has a reasonably good precedence. Homocysteine is bound to the HCys module as the highly nucleophilic thiolate anion under physiological conditions. HCys is a zinc metallomodule, in which the Zn^{2+} ligands are Cys247, Cys310, Cys311, and homocysteine (Goulding and Matthews, 1997; Peariso et al., 1998, 2001). X-ray absorption spectrometry of MetH in the presence of homocysteine reveals four sulfur ligands to zinc, and with selenohomocysteine, three sulfur, and one selenium ligand. Site-directed mutagenesis of the cysteine residues abolishes or weakens zinc binding and inactivates the enzyme. The structure of the HCys-Fol fragment shows the coordination of three cysteine residues and homocysteine to zinc. Careful monitoring of proton release shows that the binding of homocysteine is accompanied by the release of a single proton (Jarrett et al., 1997). Coordination of homocysteine to zinc appears to lead to the zinc thiolate of homocysteine through the acid strengthening effect of Zn^{2+} (see chap. 4). Methylation of thiolate ions such as the homocysteine thiolate by methylcobalamin has a good precedence in chemistry (Matthews, 2001).

In contrast to repair and methyl transfer to homocysteine, methylation of cob(I)alamin by methyltetrahydrofolate, a tertiary amine, has no chemical precedent (Matthews, 2001). Alkylation of nucleophiles by quaternary ammonium compounds is possible, suggesting that an enzymatic process to make N5 of methyltetrahydrofolate a quaternary nitrogen might take place. This could be accomplished by a preliminary protonation of N5 to a tertiary ammonium state, which could then donate a methyl group to cob(I)alamin. A spectrophotometric analysis shows that methyltetrahydrofolate bound to MetH is not protonated on N5 between pH values 5.5 and 8.5 (Smith and Matthews, 2000). Moreover, N5-protonated methyltetrahydrofolate does not bind to MetH; and the structure does not implicate an acidic group near N5 that could donate a proton.

Whatever the mechanism, the alkylation of cob(I)alamin by methyltetrahydrofolate in state 3 takes place with protonation of N5 in tetrahydrofolate; however, there is no evidence that protonation occurs before methyl transfer. It has been suggested that proton transfer may follow methyl transfer (Matthews, 2001; Smith and Matthews, 2000). Various mechanisms by which this might occur have been discussed. For the present, the question of the mechanism of the methylation of cob(I)alamin by methyltetrahydrofolate is not known (Matthews, 2001).

MetE catalyzes the same reaction as MetH and is also a Zn-metalloenzyme. The Zn^{2+} in MetE facilitates thiolate formation in homocysteine as in MetH. However, MetE does not require cobalamin and does not have the modular structure required in the repair of Co(II)-MetH. Limited mechanistic information is available about MetE.

Outstanding questions about the actions of MetH and MetE include the mechanism by which 5-methyltetrahydrofolate becomes reactive as an alkylating agent. To serve as a methyl donor, N^5 in 5-methyltetrahydrofolate must somehow become quaternary nitrogen, presumably by transiently accepting a proton from an enzymatic group at the methyltransfer step. Recent evidence indicates that in the reaction of MetE, N^5 of 5-methyltetrahydrofolate becomes protonated in the ternary complex, that is upon binding of homocysteine to the binary complex of MetE and 5-methyltetrahydrofolate (Taurog and Matthews, 2006).

References

Bandarian V, ML Ludwig, and RG Matthews (2003) *Proc Natl Acad Sci U S A* **100**, 8156.
Bandarian V, KA Pattridge, BW Lennon, DP Huddler, RG Matthews, and ML Ludwig (2002) *Nat Struct Biol* **9**, 53.
Banerjee RV, V Frasca, DP Ballou, and RG Matthews (1990) *Biochemistry* **29**, 11101.
Banerjee RV, NL Johnson, JK Sobeski, P Datta, and RG Matthews (1989) *J Biol Chem* **264**, 13888.
Bonifacio MJ, M Archer, ML Rodrigues, PM Matias, DA Learmonth, MA Carrondo, and P Soares-Da-Silva (2002) *Mol Pharmacol* **62**, 795.
Borchardt RT, Y Shiong, JA Huber, and AF Wycpalek (1976) *J Med Chem* **19**, 1104.
Chen WJ, DA Andres, JL Goldstein, and MS Brown (1991b) *Proc Natl Acad Sci U S A* **88**, 11368.
Chen WJ, DA Andres, JL Goldstein, DW Russell, and MS Brown (1991a) *Cell* **66**, 327.
Cheng X (1995) *Annu Rev Biophys Biomol Struct* **24**, 293.
Clarke S (1992) *Annu Rev Biochem* **61**, 355.
Dixon M, S Huang, RG Matthews, and M Ludwig (1996) *Structure* **4**, 1263.
Dougherty DA (1996) *Science* **271**, 163.
Doulence JM and CD Poulter (1995) *Proc Natl Acad Sci U S A* **92**, 5008.
Drennan CL, S Huang, JT Drummond, RG Matthews, and ML Ludwig (1994) *Science* **266**, 1669.
Drummond JT, S Huang, RM Blumenthal, and RG Matthews (1993) *Biochemistry* **32**, 9290.
Evans JC, DP Huddler, MT Hilgers, G Romanchuk, RG Matthews, and ML Ludwig (2004) *Proc Natl Acad Sci U S A* **101**, 3729.
Furfine ES, JJ Leban, A Landavazo, JF Moomaw, and PJ Casey (1995) *Biochemistry* **34**, 6857.
Gonzalez B, MA Pajares, JA Hermoso, L Alvarez, F Garrido, JR Sufrin, and J Sanz-Aparicio (2000) *J Mol Biol* **300**, 363.
Goulding CW and RG Matthews (1997) *Biochemistry* **36**, 15749.
Goulding CW, D Postigo and RG Matthews (1997) *Biochemistry* **36**, 8082.
Gray CH, JK Coward, KB Schowen and RL Schowen (1979) *J Am Chem Soc* **101**, 4349.
He B, P Chen, SY Chen, KL Vancura, S Michaelis, and S Powers (1991) *Proc Natl Acad Sci U S A* **88**, 11373.

Hightower KE, CC Huang, PJ Casey, and CA Fierke (1998) *Biochemistry* **37**, 15555.

Hoover DM, JT Jarrett, RH Sands, WR Dunham, ML Ludwig, and RG Matthews (1997) *Biochemistry* **36**, 127.

Hu WP and DG Truhlar (1996) *J Am Chem Soc* **118**, 860.

Huang CC, PJ Casey, and CA Fierke (1997) *J Biol Chem* **272**, 20.

Iwig DF and SJ Booker (2004) *Biochemistry* **43**, 13496.

Jarrett JT, CY Choi, and RG Matthews (1997) *Biochemistry* **36**, 15739.

Jencks WP (1969) *Catalysis in Chemistry and Enzymology*. McGraw-Hill: New York, pp 284-285.

Komoto J, T Yamada, Y Takata, GD Markham, and F Takusagawa (2004) *Biochemistry* **43**, 1821.

Long SB, PJ Casey, and LS Beese (1998) *Biochemistry* **37**, 9612.

Lotta T, J Vidgren, C Tilgmann, I Ulmanen, K Melén, I Julkunen, and J Taskinen (1995) *Biochemistry* **34**, 4202.

Markham GD, EW Hafner, CW Tabor, and H Tabor (1980) *J Biol Chem* **255**, 9082.

Markham GD, DW Parkin, F Mentch, and VL Schramm (1987) *J Biol Chem* **262**, 5609.

Markham GD and RS Reczkowski (2004) *Biochemistry* **43**, 3415.

Matthews RG (2001) *Acc Chem Res* **34**, 681.

McQueney MS, KS Anderson, and GD Markham (2000) *Biochemistry* **39**, 4443.

Mihel I, JO Knipe, JK Coward, and RL Schowen (1979) *J Am Chem Soc* **101**, 4349.

Moores SL, MD Schaber, SD Mosser, E Rands, MB O'Hara, VM Garsky, MS Marshall, DL Pompliano, and JB Gibbs (1991) *J Biol Chem* **266**, 14603.

Mudd SH (1963) *J Biol Chem* **238**, 2156.

Mu Y, CA Omer, and RA Gibbs (1996) *J Am Chem Soc* **118**, 1817.

Omer CA, AM Kral, RE Diehl, GC Prendergast, S Powers, CM Allen JB Gibbs, and NE Kohl (1993) *Biochemistry* **32**, 5167.

Parry RJ and A Minta (1982) *J Am Chem Soc* **104**, 871.

Peariso K, ZS CW Goulding, S Huang, RG Matthews, and JE Penner-Hahn (1998) *J Am Chem Soc* **120**, 8410.

Peariso K, ZS Zhou, AE Smith, RG Matthews, and JE Penner-Hahn (2001) *Biochemistry* **40**, 987.

Pompliano DL, MD Schaber, SD Mosser, CA Omer, JA Shafer, and JB Gibbs (1993) *Biochemistry* **32**, 8341.

Reczkowski RS and GD Markham (1999) *Biochemistry* **38**, 9063.

Reczkowski RS, JC Taylor, and GD Markham (1998) *Biochemistry* **37**, 13499.

Rodgers J, DA Femec, and RL Schowen (1982) *J Am Chem Soc* **104**, 3263.

Rose IA (1980) *Methods Enzymol* **64**, 47.

Rozema DB and CD Poulter (1999) *Biochemistry* **38**, 13138.

Ruggieri GD, IH Williams, M Roca, V Moliner, and I Tunon (2004) *J Am Chem Soc* **126**, 8634.

Schafer WR, R Kim, R Sterne, J Thorner, SH Kim, and J Rine (1989) *Science* **245**, 379.

Schluckebier G, M O'Gara, W Saenger, and X Cheng (1995) *J Mol Biol* **247**, 16.

Strickland CL, WT Windsor, R Syto, L Wang, R Bond, Z Wu, J Schwartz, HV Le, LS Beese, and PC Weber (1998) *Biochemistry* **37**, 16601.

Takusagawa F, M Fujioka, A Spies, and RL Schowen (1998) In Sinnott M (ed): *Comprehensive Biological Catalysis*, vol 1. Academic Press: San Diego, p 1.

Takusagawa F, S Kamitori, S Misaki, and GD Markham (1996) *J Biol Chem* **271**, 136.

Taurog RE and RG Matthews (2006) *Biochemistry* **45**, 5092.

Taylor JC and GD Markham (1999) *J Biol Chem* **274**, 32909.

Taylor JC and GD Markham (2000) *J Biol Chem* **275**, 4060.

Tenhunen J and I Ulmanen (1993) *Biochem J* **296**, 595.

Vasak M, JHR Kagi, and HAO Hill (1981) *Biochemistry* **20**, 2852.

Vidgren J, LA Svensson, and A Liljas (1994) *Nature* **368**, 354.

Wang H and WL Hase (1997) *J Am Chem Soc* **119**, 3093.

Weissbach H and RT Taylor (1970) *Vitam Horm* **28**, 415.

Woodard RW, MD Tsai, HG Floss, PA Crooks, and JK Coward (1980) *J Biol Chem* **255**, 9124.

Wu SE, WP Huskey, RT Borchardt, and RL Schowen (1984) *J Am Chem Soc* **106**, 5762.

Wu Z, M Demma, CL Strickland, ES Radisky, CD Poulter, HV Le, and WT Windsor (1999) *Biochemistry* **38**, 11239.

Zhang FL and PJ Casey (1996) *Annu Rev Biochem* **65**, 241.

16
Oxidoreductases

Oxidoreductases constitute a very large class of enzymes. They are dehydrogenases and reductases that catalyze the removal or addition of the elements of molecular hydrogen to or from substrates. Enzymatic dehydrogenation is sometimes linked to auxiliary functions such as decarboxylation, deamination, or dehydration of the substrate, as in the actions of isocitrate dehydrogenase (decarboxylation), glutamate dehydrogenase (deamination), and ribonucleotide reductase (deoxygenation). The best known oxidoreductases are the NAD-dependent dehydrogenases, and a thorough discussion of the actions of these enzymes could easily fill a volume the size of this book. For this reason, this discussion must focus on the salient aspects of reaction mechanisms that represent the major classes of oxidoreductases. Authoritative reviews on the kinetics and structures of the main dehydrogenases are available (Banaszak et al., 1975; Brändén et al., 1975; Dalziel, 1975; Harris and Waters, 1976; Holbrook et al., 1975; Rossman et al., 1975; Smith et al., 1975; Williams, 1976). In this chapter, we emphasize the diverse oxidoreduction mechanisms and place less emphasis on auxiliary functions such as decarboxylation, the mechanisms of which are similar to the actions of enzymes discussed in earlier chapters of this book.

Discussions of several dehydrogenases not included in this chapter can be found in other chapters. These include methanol, glucose, and methylamine dehydrogenases in chapter 3, dimethylsulfoxide reductase in chapter 4, and dihydrofolate reductase and β-hydroxymethylglutaryl CoA reductase in chapter 5. Pyruvate and α-ketoglutarate dehydrogenases are discussed in chapter 18.

Enzymatic addition or removal of the elements of hydrogen to or from an organic molecule generally requires the action of a coenzyme. In principle, the process may proceed by any of several mechanisms, including the formal transfer of a hydride and a proton; or the transfer of two electrons and two protons; or the transfer of a hydrogen atom, an electron,

680 Enzymatic Reaction Mechanisms

and a proton; or any of several other sequences. Proteins alone do not efficiently catalyze these processes; coenzymes and cofactors generally provide the essential chemistry for catalysis by oxidoreductases.

Pyridine Nucleotide–Dependent Dehydrogenases

Many enzymes catalyze the dehydrogenation of an alcoholic group to a ketone or aldehyde coupled with the reduction of NAD^+ to NADH. Enzymes of this type in glycolysis and the tricarboxylic acid cycle include alcohol dehydrogenase, lactate dehydrogenase, malate dehydrogenase, glyceraldehydes-3-P dehydrogenase, isocitrate dehydrogenase, and pyruvate and α-ketoglutarate dehydrogenases. In this section, we emphasize one example of each class or type of NAD^+-dependent dehydrogenase.

A striking feature of nicotinamide nucleotide-dependent dehydrogenases is the similarity of the binding sites for $NAD^+/NADP^+$. Most members of this family incorporate the same folding motif for binding the nicotinamide nucleotide, the dinucleotide fold, often referred to as the Rossman fold (Rossman et al., 1975). An exception is isocitrate dehydrogenase, which has a completely different structure.

Alcohol Dehydrogenase

The dehydrogenation of ethanol is catalyzed by yeast or liver alcohol dehydrogenase (ADH; EC 1.1.1.1) according to the stoichiometry in fig. 16-1A. Liver ADH is dimeric, with one active site in each subunit, and yeast ADH is tetrameric. In humans and higher animals ADH exists as isoenzymes, with similar but variant subunits that can exist as homodimers or heterodimers (Briganti et al., 1989; Eklund et al., 1990). The isoenzymes act by similar catalytic mechanisms, although their selectivities for substrates other than ethanol vary. We limit the continuing discussion to mechanistic aspects of catalysis.

Reaction, Stereochemistry, and Kinetics

The reduction of NAD^+ proceeds by a formal hydride transfer to NADH, a proton is released to the solvent, and ethanol is oxidized to acetaldehyde. Unlike other NAD-dependent

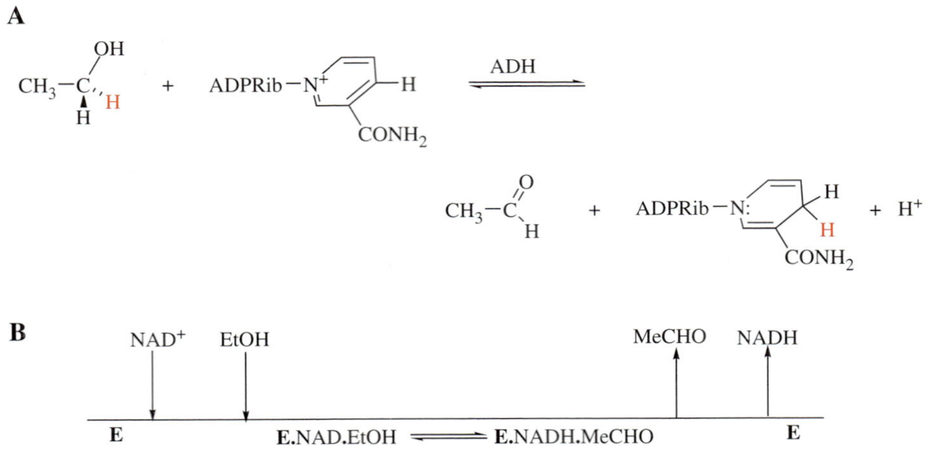

Fig. 16-1. Reaction, stereochemistry, and steady-state kinetic mechanism in the action of alcohol dehydrogenase (ADH).

dehydrogenases, ADHs are zinc metalloenzymes. Zinc plays an essential role in the mechanism of action of ADH. Other classes of NAD⁺-dependent dehydrogenases do not contain zinc, and catalysis proceeds by alternative mechanisms.

Yeast ADH is highly selective for ethanol as the substrate, although higher alcohols will react slowly. In contrast, liver ADH will accept a broad range of substrates ranging in size from ethanol to octanol and benzyl alcohol. Yeast ADH is also much more active (>200 IU mg⁻¹) than liver ADH (≈4 IU mg⁻¹). Both yeast and liver ADHs act with the stereospecificity of hydride transfer shown in fig. 16-1A. ADH specifically transfers hydride to nicotinamide-C4 on the *re*-face (A side) into the pro-*R* position of NADH. The stereochemistry of hydride transfer by nicotinamide coenzyme-dependent dehydrogenases is discussed in chapter 3.

ADH-action follows an ordered bi bi pattern, with NAD⁺ binding first and NADH dissociating last, as illustrated in fig. 16-1B (Dalziel, 1975). Under many conditions, the steady-state kinetics appears to be of the Theorell-Chance type, in which the binary complex of ADH.NAD⁺ reacts so rapidly with ethanol to release acetaldehyde that a ternary complex does not seem to be required. However, with higher alcohols as substrates, dissociation of NAD⁺ from a ternary complex can be detected in the kinetics of isotope exchange at equilibrium (Silverstein and Boyer, 1964b), and transient kinetics implicates ternary complexes.

Formal hydride transfer occurs in the interconversion of ternary complexes. However, hydride ion (H⁻) is never a free species in the mechanism. Instead, hydrogen is transferred with two electrons in a single transition state, as illustrated in fig. 16-2. In general, the rate-limiting step in the actions of yeast and liver ADHs is not the chemistry of hydride transfer. The most obvious manifestation of this fact is the absence of or very small value, depending on the substrate, of the primary kinetic isotope effect for deuterium transfer ($^{D}k_{cat}$). For example, in the action of horse liver ADH on [1-²H₂]benzyl alcohol, the value of $^{D}k_{cat}$ is 1.4 (Dworschack and Plapp, 1977). The dissociation of NADH (5.5 s⁻¹) limits the rate in the reaction of ethanol (Sekhar and Plapp, 1990). The ADH-catalyzed reaction of [1-²H₂]ethanol proceeds with no deuterium kinetic isotope effect.

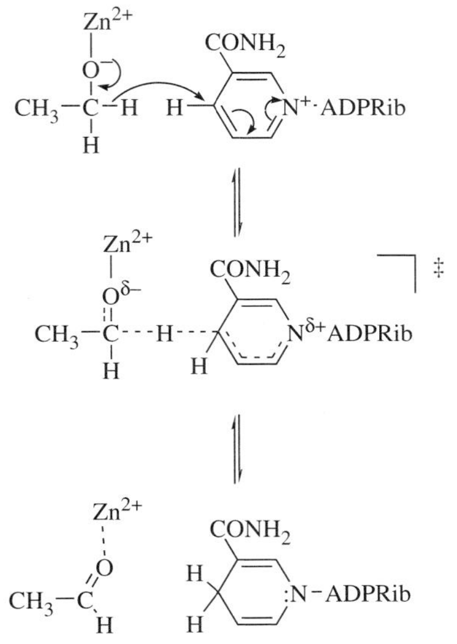

Fig. 16-2. Hydride transfer in the transition state in the reaction of ethanol with NAD⁺.

Structure

The structure of ADH reveals much about the active site, including the structural relationship of Zn^{2+} to the substrates. The stereochemistry of hydride transfer from ethanol can be accounted for on the basis of structural modeling, which indicates that Phe93 introduces a constraint into the binding site that imparts stereochemical specificity for hydride transfer (Eklund et al., 1982). Structures of the free enzyme and ternary complexes show two domains, a dinucleotide binding domain (Rossman fold) and a substrate binding domain, with an intervening cleft in the free enzyme. The cleft closes on ternary complex formation, and modeling studies suggest that this results from the substrate binding domain closing on the dinucleotide fold by a sliding motion (Colonna-Cesari et al., 1986). ADH contains two Zn^{2+} binding sites, one of which is the active site, which binds the catalytic zinc. The second Zn^{2+} binding site is remote from the active site and is regarded as a structural zinc site (Sytkowski and Vallee, 1976, 1979).

Many structures of ADHs are available. The one shown in fig. 16-3 is of the complex of liver ADH with NAD^+ and difluorobenzyl alcohol bound to the active site (Rubach and Plapp, 2002). As in other structures of ternary complexes (Eklund and Brändén, 1987), the enzyme is in its closed conformation. The alcoholic group of difluorobenzyl alcohol is directly ligated to Zn^{2+}, and the other ligands to the tetrahedrally coordinated Zn^{2+} are Cys46, Cys176, and His67. Like all zinc metalloenzymes in which the metal participates in catalysis, Zn^{2+} in the free enzyme has a molecule of water in its coordination sphere (Vallee and Auld, 1990). Zinc-coordinated water is displaced by the alcoholic group of the substrate on formation of the ternary complex.

In the NAD^+ binding site, the nicotinamide ring lies in a pocket lined with hydrophobic amino acid side chains along the face opposite the A-side, to which hydride transfer occurs. Molecular dynamics simulations indicate that contacts between the nicotinamide ring and hydrophobic residues along its B-face force the ring toward a boat conformation that facilitates hydride transfer to the A-side (Almarsson and Bruice, 1993). Mutations of Val292 in potential contact with the B side of the nicotinamide ring to decrease the steric bulk and hydrophobicity lead to significant decreases in catalytic efficiency (Rubach and Plapp, 2003). Individual mutations of Val292 to alanine, serine, or threonine increase K_m by 30- to 50-fold and decrease k_{cat} by 3- to 60-fold in the reaction of benzyl alcohol. Under conditions of subsaturating NAD^+, the effects would decrease catalytic efficiency (k_{cat}/K_m) by 90- to 800-fold. The structure of V292T-ADH is very similar to that of wild-type ADH, showing that minor differences lead to kinetic consequences. The variant V292S-ADH crystallizes as a ternary complex with pentafluorobenzyl alcohol and NAD^+ in the open conformation, like the wild-type apo-ADH, in contrast to the closed conformation of wild-type ternary complexes (Rubach et al., 2001).

Role of Zinc

In the dehydrogenation of an alcohol, two bonds to hydrogen must be broken, the C–H and O–H bonds of the alcohol. (They must also be formed in the reverse direction.) The C–H bond is cleaved on transfer of the hydride to NAD^+ (fig. 16-2). Some provision must be made for cleaving the O–H bond, and this is a fundamental mechanistic difference among the classes of dehydrogenases that operate on alcoholic groups. In the case of ADH, the problem of breaking the O–H bond is solved by the function of the catalytic Zn^{2+}.

The structure in fig. 16-3 shows the alcohol oxygen coordinated to Zn^{2+} at the active site. In the resting enzyme, a water molecule occupies this coordination site. Table 4-3 in chapter 4 showed that the pK_a of water coordinated to Zn^{2+} is 9.6, or 6 units below the

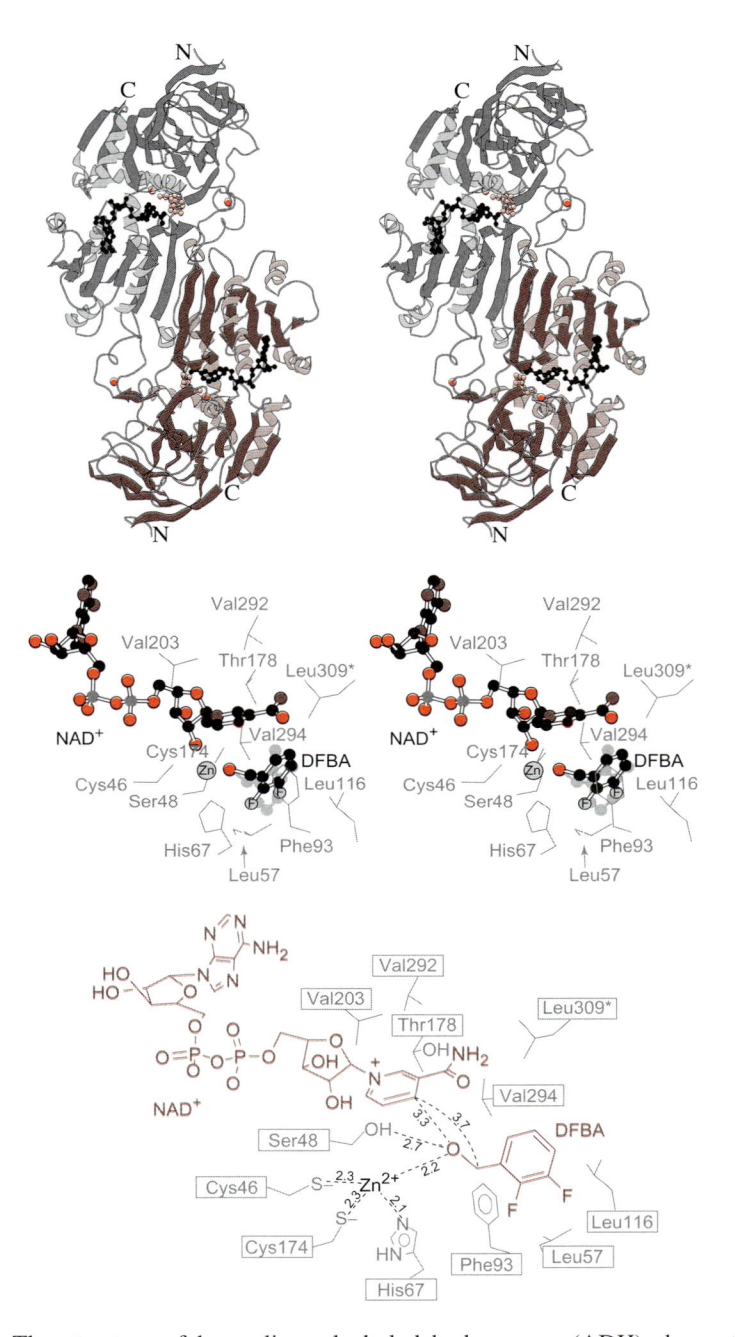

Fig. 16-3. The structure of horse liver alcohol dehydrogenase (ADH) shows the ternary complex of ADH with NAD$^+$ and difluorobenzyl alcohol (DFBA) (1.80-Å resolution; PDB 1MG0; Rubach and Plapp, 2002). At the top is the dimeric structure showing the overall chain fold for the subunits. NAD$^+$ (black), DFBA (pink), and two Zn^{2+} ions (red) are shown per monomer as ball-and-stick diagrams. In the center is a stereographic representation of the active site, showing NAD$^+$ and DFBA as ball-and-stick models. The observed electron density for the DFBA molecule was consistent with two binding modes; one is shown with typical ball-and-stick coloration, and the other is light gray. The catalytically important His51 residue was omitted for clarity because it resides directly behind the ribose moiety in this view; His51 is hydrogen bonded to the ribose-2'OH, which interacts with Ser48 adjacent to the alcohol substrate. At the bottom is a two-dimensional map of the active site, showing the contacts (distances in angstroms) of the ligands and Zn^{2+} with amino acid side chains.

value for free water. By the same acid-strengthening effect, the ethanolic OH group would display a pK_a of 9.9 when coordinated to Zn^{2+}. Zn^{2+} coordination substantially increases the acidity of the alcohol and facilitates dissociation of the proton. The acid strengthening effect of Zn^{2+} is amplified at the active site by the presence of NAD^+ in complex with ADH. The positive charge in the pyridinium ring augments the positive charge of Zn^{2+} and increases the positive electrostatic field. The pK_a of ethanol bound to the active site is even lower. The reported values of pK_a for water, ethanol, chloroethanol, and trifluoroethanol bound to the complex of ADH with NAD^+ are 7.6, 6.4, 5.5, and 4.3, respectively (Kvassman et al., 1981). These values are 8 to 9 units below the pK_a values of 15., 15.9, 14.3, and 12.4 in aqueous solution. All of the alcohols, and even water, exist as the oxyanions coordinated to Zn^{2+} at the active site when NAD^+ is bound to ADH in neutral solution. In this way, Zn^{2+} makes a decisive contribution to cleaving the O–H bond.

Transient Kinetics

The overall reaction of ADH results in the release of a proton in the direction shown in fig. 16-1A. In the closed conformation of ADH, there must be a mechanism for ushering the alcoholic proton to the aqueous medium in the forward direction and back from the aqueous medium in the reverse direction. The structure in fig. 16-3 implicates the hydrogen bonded network of substrate OH, Ser48, ribosyl-2'-OH, and His51 in this process. His51 is in position at the surface to interact with the aqueous medium, and the ribosyl-2'-OH and Ser48 connect His51 to the alcoholic OH. It is reasonable to write the proton transfer mechanism as in fig. 16-4. For a variety of reasons, including the fact that NADH-release limits the overall rate, this process is difficult to study. Mutation of His51 should inhibit proton transfer in the pathway shown in fig. 16-4, but mutation to isosteric Gln51 only modestly slows the overall rate, decreasing the value of k_{cat}/K_m by 30-fold (Plapp et al., 1991). It is possible that the true kinetic effect of mutating His51 is masked by rate-limiting release of NADH in the wild-type enzyme. There is evidence that the proton transfer process is altered by this mutation (LeBrun and Plapp, 2004).

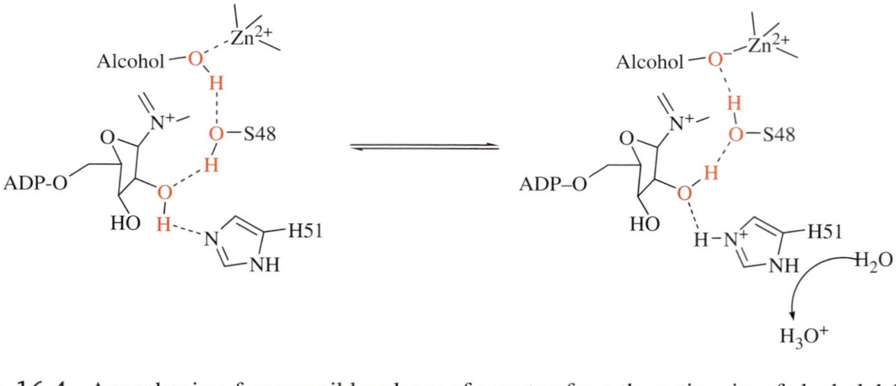

Fig. 16-4. A mechanism for reversible release of a proton from the active site of alcohol dehydrogenase (ADH). In the direction of dehydrogenation of an alcohol by the action of ADH, a proton is released to the solution, and in the reverse direction, it is taken up. The proton formally originates with the OH group of the substrate. The hydrogen-bonded network is thought to mediate the reversible proton transfer to a water molecule, with formation of a hydronium ion.

Because ternary complex formation in fig. 16-1B is faster than the release of NADH, which limits the overall rate, a stopped-flow study of ADH-action displays burst kinetics (LeBrun and Plapp, 2004; Sekhar and Plapp, 1990; Shore and Gutfreund, 1970). The reaction is monitored at 333 nm, the absorption maximum for the complex of ADH with NADH, rather than the maximum of 340 nm for free NADH. (All complexes of A side–specific dehydrogenases with NADH display a blue shift in the UV spectrum of NADH, whereas B side–specific enzymes display a smaller red shift.) Proton release takes place in the burst phase.

Transient kinetic analysis of the burst phase allowed estimates for the rate constants in the kinetic mechanism of fig. 16-5 to be obtained. The burst allowed the deuterium kinetic isotope effect ($^Dk = 6$) in the reaction of [1-^{2}H$_2$]ethanol to be observed (Shore and Gutfreund, 1970). The transient kinetics also unmasked an isomerization step of the binary ADH-NAD complex, which significantly limited the burst rate. Significant differences were found between the estimated rate constants for the wild type and H51Q-ADH, the largest being an approximately 30-fold differences in the forward and reverse rate constants for the isomerization step (k_3 and k_4) (LeBrun and Plapp, 2004).

The pH dependencies of kinetic parameters for ADH and H51Q-ADH also differ significantly. The pH-rate profiles for steady-state parameters at saturating pyridine nucleotide and varying alcohol or aldehyde are shifted toward higher pH. In the forward direction, the plot of log(V/K) for wild-type ADH is bell-shaped corresponding to two ionizations, but for H51Q-ADH it is shifted to the alkaline side and displays only one break. The pH dependence of the presteady-state burst is also significantly different for wild-type ADH and H51Q-ADH, such that an ionizing group that influences the rate in the action of ADH appears to be absent in H51Q-ADH.

The pre–steady-state rate differences between ADH and H51Q-ADH are significant, although not by the four or even two orders of magnitude often observed for the loss of acid-base catalysis on site-directed mutation. Other pathways of proton transfer may participate as well or may be recruited on mutation of His51. The step at which the proton in the overall reaction is released is postulated to follow the binding of NAD$^+$ to ADH (Eklund et al., 1974), and this is supported by kinetic data (Kovaleva, 2004). Presumably, the proton originates with the ionization of the zinc-bound water molecule on binding NAD$^+$, which lowers its pK_a. Then the binding of the alcohol substrate should involve an exchange of the alcoholic-OH group with zinc-bound OH$^-$ and release of H$_2$O, a process that itself involves proton transfer.

Isotope Effects

The deuterium kinetic isotope effect in the reaction of [1-^{2}H$_2$]benzyl alcohol is larger than with other substrates, especially $^Dk_{cat}/K_m$, which is 3.0 ($^Dk_{cat} = 1.4$). The hydride transfer is

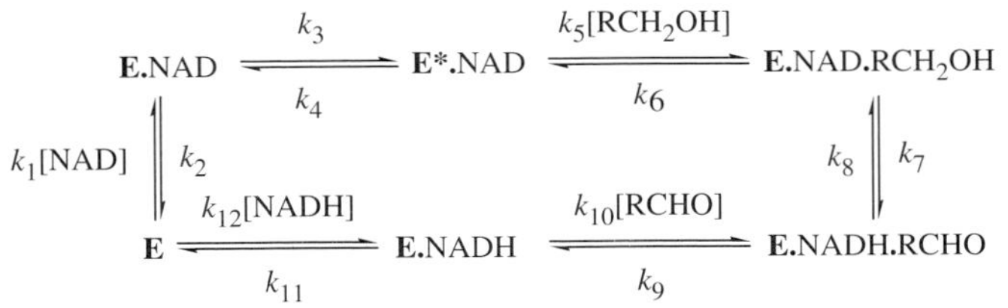

Fig. 16-5. Overall kinetic mechanism for the action of alcohol dehydrogenase (ADH). This mechanism includes an isomerization of the binary enzyme-NAD complex (k_3/k_4) uncovered in the stopped-flow analysis of the presteady-state phase. Estimates of the rate constants for wild-type and H51Q ADH have been obtained (LeBrun and Plapp, 2004; Sekhar and Plapp, 1990).

partially rate limiting in the reaction of benzyl alcohol. The ^{13}C kinetic isotope effect for the reaction of benzyl alcohol can be measured by observing the natural abundance of ^{13}C in the product. The effect is 1.5%, a small value due to partial rate limitation. The ^{13}C isotope effect is perturbed by deuterium in [1-^{2}H$_2$]benzyl alcohol or [4-^{2}H]NAD$^+$ (Scharschmidt et al., 1984). These perturbations allow the intrinsic isotope effects to be calculated by the dual isotope effect technique (Hermes et al., 1982). The intrinsic effects are 4.05 for the deuterium isotope effect on hydride transfer and 1.025 for the ^{13}C kinetic isotope effect. These effects depend on the reduction potential of the pyridine nucleotide. The range of $E^{o\prime}$ is –0.320 V to –0.258 V in the series NAD, deamino-NAD, thio-NAD, pyridinealdehyde-NAD, and acetylpyridine-NAD. The deuterium kinetic isotope effects trend upward in this series to as high as 6.25, and the ^{13}C isotope effects trend downward to 1.012. These trends are interpreted to report on a change from a late transition state in the reaction of NAD$^+$ to a more symmetrical transition state for acetylpyridine-NAD. The observation of a secondary deuterium isotope effect in the reaction of [4-^{2}H]NAD and its effect on the ^{13}C-isotope effect indicated coupled motion of protons in the transition state.

Catalysis of Hydride Transfer

All things considered, the mechanism of hydride transfer illustrated in fig. 16-2 appears to be a reasonable description of the process. Zinc coordination facilitates alkoxide formation in the substrate, and the hydrogen with two electrons is in flight to nicotinamide-C4 in the transition state. The proton arising from the formation of the substrate alkoxide is released to the solvent, most likely by the hydrogen bonded network connecting His51 to the active site. Evidence of hydrogen tunneling in the reaction of ADH is discussed in chapter 2.

Lactate Dehydrogenase

Reaction, Stereochemistry, and Properties

The reaction of lactate dehydrogenase (LDH; EC 1.1.1.27) is chemically similar to that of ADH, substituting L-lactate for ethanol and pyruvate for acetaldehyde in the equation in fig. 16-1A. The steady-state kinetic mechanism, as determined by rates of isotope exchange at equilibrium and initial rate analysis, is also analogous to that for ADH in fig. 16-1B, ordered bi bi with NAD$^+$ binding first and NADH dissociating last (Silverstein and Boyer, 1964a; Zewe and Fromm, 1965), substituting L-lactate for ethanol and pyruvate for acetaldehyde. Like ADH, LDH is A-side specific (re-side) with respect to hydride transfer to NAD$^+$. Despite these similarities, significant structural and mechanistic differences distinguish LDH from ADH (Holbrook et al., 1975). We focus on the differences.

LDH in higher animals is a tetrameric enzyme, unlike the dimeric ADH, and it is the classic case of isozyme variance. The main isozymes in humans are known as the heart (H) and muscle (M) enzymes. The enzymes are similar in many ways but differ in steady-state kinetic properties; the differences in amino acid sequences and content allow them to be separated easily by nondenaturing gel electrophoresis. As tetramers, they are known as H$_4$ and M$_4$, respectively. All possible hybrid forms exist (e.g., H$_3$M, H$_2$M$_2$) and the statistical distribution of hybrid isozymes can be observed in vitro (Markert and Moller, 1959). Isozyme patterns in the blood stream can indicate cellular damage in the heart or other organs.

Unlike ADH, LDH is not a metalloenzyme. Given the essential role of Zn^{2+} in the mechanism of action of ADH, the absence of any other divalent metal ion in LDH means that the mechanism by which the O–H bond of lactate is cleaved by action of LDH must be different from that of ADH. This difference is discussed in the next section.

Structure and Reaction Mechanism

The x-ray crystallographic structure of LDH was determined early in the development of modern structural biology (Adams et al., 1970; Holbrook et al., 1975; White et al., 1976). The structures of the apo-enzyme and ternary complexes revealed a chain fold that was at the time unique and became known as the Rossman fold, or dinucleotide binding fold, consisting of a parallel β-sheet with peripheral helices intervening between β-strands. The NAD⁺ binding site was associated with the C-terminal ends of the β-strands, as shown in fig. 16-6 for the LDH from *Plasmodium falciparum* (Cameron et al., 2004).

The dinucleotide fold has been found in nicotinamide coenzyme-binding proteins, in other nucleotide-binding proteins (Rossman et al., 1975), and in other types of proteins. For example, a class of adenosylcobalamin-dependent enzymes contains two domains, one of which is a β-barrel similar to that in triose phosphate isomerase, and the other is a dinucleotide fold similar to that in LDH. Adenosylcobalamin is bound between the two domains.

Another important feature of the LDH structure is the active site. The absence of a divalent metal ion focuses attention on the amino acid side chains that might be involved in catalyzing the removal of the proton from the O–H group of lactate in the dehydrogenation mechanism. The structures of ternary complexes of LDH with NAD⁺ (or NADH) and oxamate (or oxalate) reveal the presence of His195 in close proximity to oxamate. His195 reacts much more rapidly with diethylpyrocarbonate than does free imidazole, and the pH-rate profile for chemical modification displays a value of 6.7 for the pK_a of His195 (Holbrook et al., 1975). The same pK_a value of 6.7 controls the binding of substrates and inhibitors. Arg171 is in position to bind the carboxylate group of lactate (Wigley et al., 1992).

All evidence implicates His195 as an acid-base catalyst that abstracts the proton from lactate in concert with hydride transfer to NAD⁺, or in its conjugate acid form, that donates a proton to pyruvate in concert with hydride transfer from NADH. This mechanism is illustrated in fig. 16-7. Hydride transfer by LDH proceeds with a small but significant primary deuterium kinetic isotope effect of $^DV = 1.75$ and $^DV/K = 1.93$ in the reduction of pyruvate by [4-²H]NADH (Cook et al., 1993).

Short-Chain Alcohol Dehydrogenases

A family of NAD⁺-dependent dehydrogenases that differ from ADH and LDH in the mechanism of hydride transfer have a conserved active site sequence motif that may be formulated as YX_3K, and most members also have a serine or threonine 25 residues on the N-terminal side of tyrosine, that is $(S,T)X_{25}YX_3K$. This is known as the short chain dehydrogenase "reductase" family (Kallberg et al., 2002). The conserved motif catalyzes abstraction of the proton from the substrate grouping H—C—O—H. The family includes dihydropterin, 17β-hydroxysteroid, 3-hydroxyacyl CoA, and corticosteroid-11β dehydrogenases among others. UDP-galactose 4-epimerase and dTDP-glucose 4,6-dehydratase are members of this family and are discussed in chapters 7 and 9. In the case of UDP-galactose 4-epimerase, the active site motif consists of Ser124, Tyr149, and Lys153. Briefly stated, the ionized form of Tyr149 functions as the acid-base catalyst in hydride transfer (see chap. 7). Tyr149 displays a pK_a of 6.2, 4 units below the value for tyrosine in water. It is therefore a phenolate anion in neutral solution, and being in proper contact with the substrate,

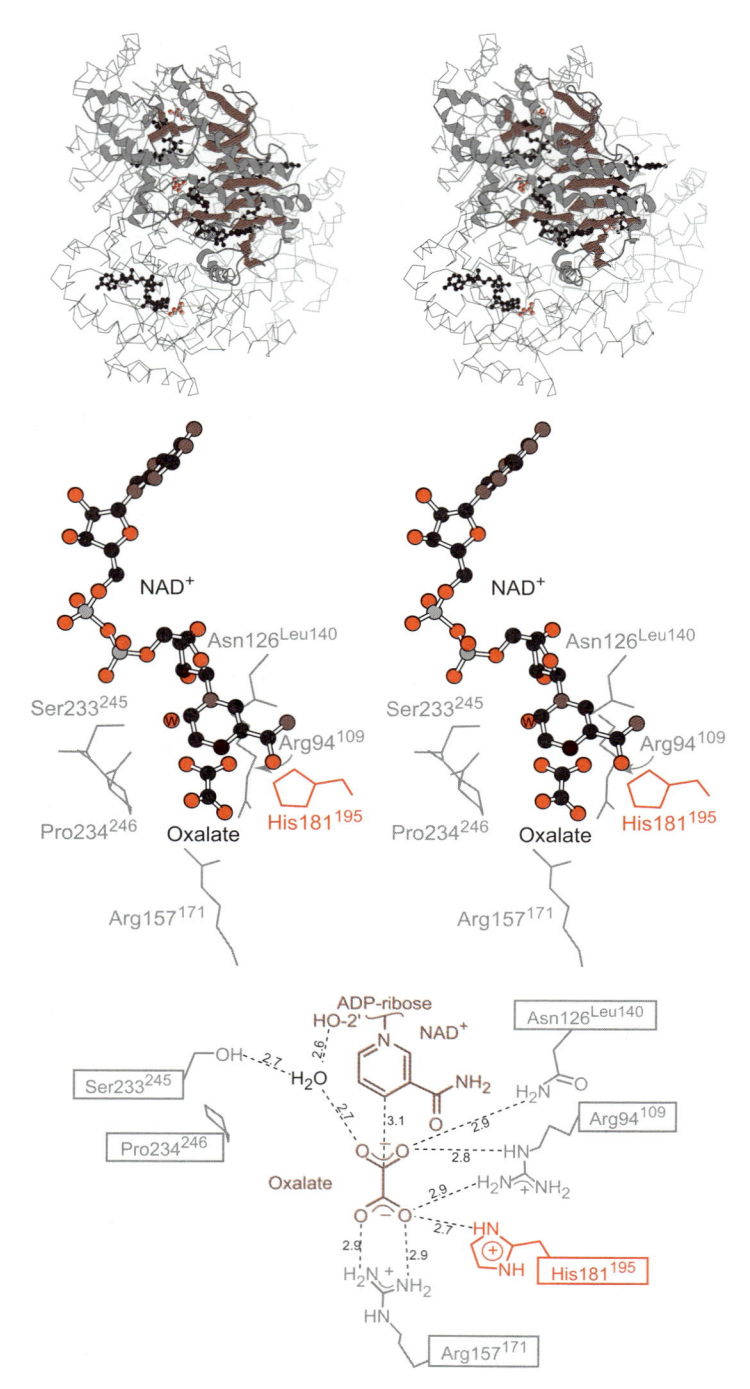

Fig. 16-6. The structure is shown for lactate dehydrogenase (LDH) from *Plasmodium falciparum* in a ternary complex with NAD$^+$ and oxalate (1.10-Å resolution; PDB 1T2D; Cameron et al., 2004). The enzyme is homotetrameric and is shown in the top panel with NAD$^+$ (black) and oxalate (red) bound in each subunit. Three of the four subunits are drawn as Cα traces and one as a ribbon diagram with brown β-strands and gray α-helices. The middle panel shows the active site, with NAD$^+$ adjacent to oxalate and the catalytically important histidine residue. The amino acid residue numbering is consistent with the Protein Data Bank (PDB) entry and *P. falciparum* LDH sequence entries, and the superscripted numbers are those used by Cameron et al. (2004) in the supporting literature. A two-dimensional scheme is shown (bottom), with the distances given in angstroms.

688

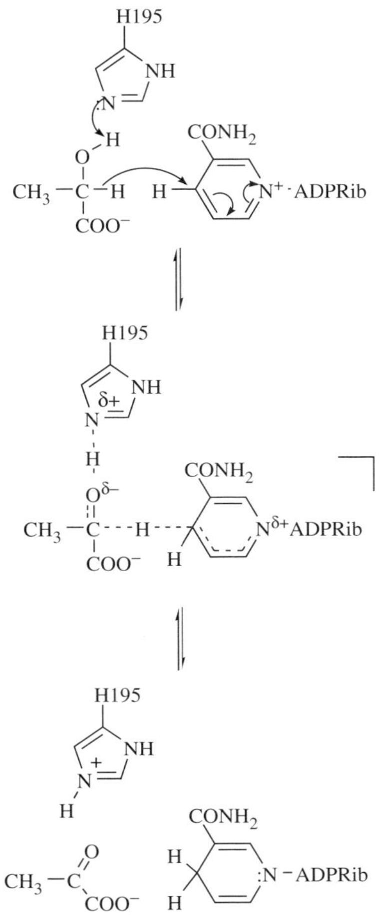

Fig. 16-7. Catalysis of hydride transfer by His195 in the action of lactate dehydrogenase.

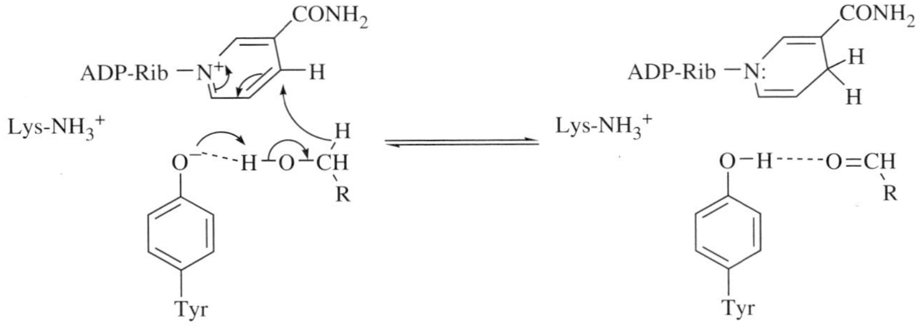

Fig. 16-8. Tyrosine is the base catalyst for hydride transfer in the action of short-chain dehydrogenases. To function as a base, tyrosine must be ionized. The normal pK_a of tyrosine is 10.2, which would make it difficult to ionize in neutral solutions. In the active sites of short-chain dehydrogenases, the pK_a of tyrosine is lowered by the positive electrostatic field created by a conserved lysine and the quaternary nitrogen in NAD^+, so that it can function as a base in the forward direction. Chapter 7 provides more detail about the action of UDP-galactose 4-epimerase.

it serves as the base for abstracting the proton from the alcoholic group in hydride transfer. A mechanism for hydride transfer in this family of dehydrogenases is illustrated in fig. 16-8. The global structure of UDP-galactose 4-epimerase includes a dinucleotide fold, to bind NAD$^+$/NADH, and a UDP-hexose binding domain (Bauer et al., 1992). All members of the family have the dinucleotide fold and the motif **YX$_3$K** in common.

Glyceraldehyde-3-P Dehydrogenase

Reaction and Stereospecificity

The reaction of glyceraldehyde-3-P with NAD$^+$ and P$_i$ catalyzed by glyceraldehyde-3-P dehydrogenase (GAPDH; EC 1.2.1.12) produces NADH and 1,3-diphosphoglycerate. The reaction mechanism follows a complex course involving carbonyl addition, thioesters formation, and acyl group transfer in addition to hydride transfer (Harris and Waters, 1976). The scheme in fig. 16-9 illustrates chemical steps in the mechanism (Segal and Boyer, 1953). Fundamental mechanistic differences from the actions of lactate dehydrogenase and related dehydrogenases, and from all families of alcohol dehydrogenases, include the initial formation of a thiohemiacetal intermediate, the formation of a covalent acyl-enzyme intermediate on hydride transfer to NAD$^+$, and the phosphorolysis of the acyl-enzyme intermediate.

In another difference from the alcohol dehydrogenases, GAPDHs catalyze hydride transfer to the B-side (*si*-face) of the nicotinamide ring in NAD$^+$, and they transfer the 4-pro-*S* hydrogen from NADH/NADPH. The conformations of NAD$^+$ at the active sites of B-side and A-side dehydrogenases differ with respect to torsional orientation about the *N*-glycosyl bond linking nicotinamide to the ribosyl moiety. In the A-side dehydrogenases, the orientation places the carboxamide group of nicotinamide *anti* with respect to the ribosyl ring, as shown in figs. 16-3 and 16-6. In B-side dehydrogenases, the orientation places the carboxamide group *syn* to the ribosyl ring. This is shown in fig. 16-10 for NAD$^+$ bound to two B-side dehydrogenases, GAPDH and phenylalanine dehydrogenase.

Kinetics

The kinetic mechanism departs from the pattern characteristic of other dehydrogenases, in that it includes ping pong components because of the properties of the covalent acyl-enzyme intermediate. NAD$^+$ stimulates the phosphorolysis of this intermediate, and NADH

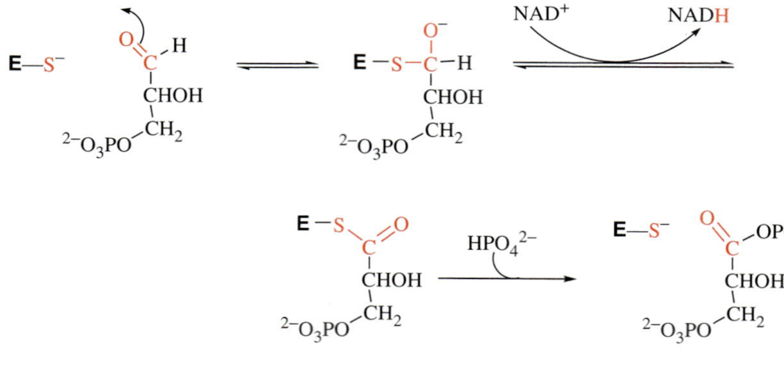

Fig. 16-9. Steps in the action of glyceraldehyde-3-P dehydrogenase.

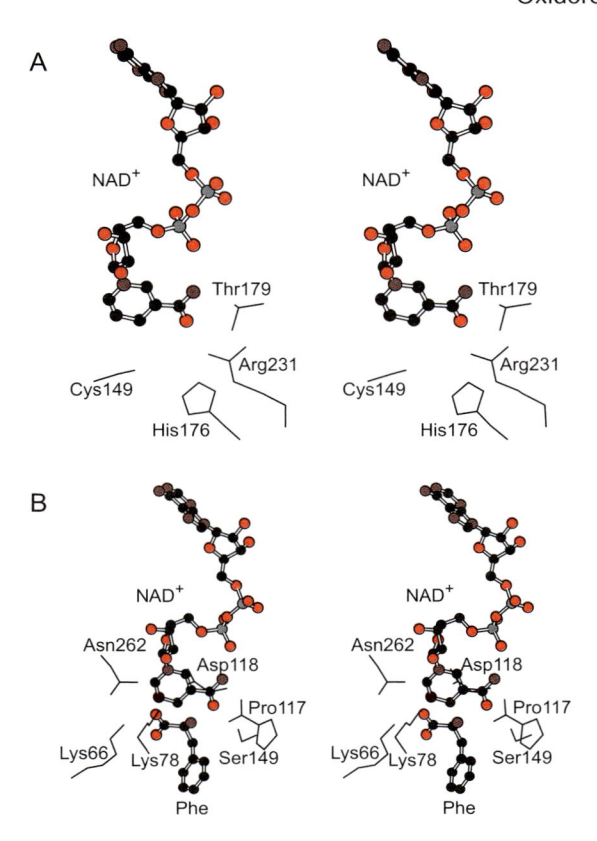

Fig. 16-10. Structures of active sites and NAD⁺ bound to two B-side–specific dehydrogenases. (A) A structure of NAD⁺ at the active site of glyceraldehyde-3-P dehydrogenase (GAPDH) from *Escherichia coli*, an enzyme that transfers hydride specifically to the B side (*si* face) of NAD⁺ (2.00-Å resolution; PDB 1DC6; Yun et al., 2000). (B) A structure of NAD⁺ at the active site of phenylalanine dehydrogenase from *Rhodococcus* sp. m4, also a B-side–specific enzyme (1.25-Å resolution; PDB 1C1D; Vanhooke and Thoden, 2000).

dissociates prior to its phosphorolysis. The kinetic scheme in fig. 16-11 describes the catalytic cycle (Duggleby and Dennis, 1976; Segal and Boyer, 1953). A key to the mechanism is the binding of NAD⁺. The resting enzyme normally exists as the binary complex with NAD⁺ at the start of the catalytic cycle. The substrate binds as the thiohemiacetal with Cys149, and hydride transfer to NAD⁺ produces NADH and the covalent 3-phosphoglyceryl-enzyme. NADH dissociates and is replaced by NAD⁺, which stimulates the reaction of P_i with the covalent intermediate to form 1,3-diphosphoglycerate and regenerate the resting enzyme.

A substantial body of biochemical evidence supports the mechanism in figs. 16-9 and 16-11 (Harris and Waters, 1976). First, resting GAPDH (**E-S⁻.NAD⁺** in fig. 16-11) displays an UV/Vis absorption band at 365 nm characteristic of charge complexation (Racker and Krimsky, 1952). This results from the proximity of the Cys149-thiolate to the nicotinamide ring. Cys149 serves as the nucleophilic catalyst in the mechanism (Harris et al., 1963). Second, acetyl phosphate acetylates the enzyme to form an *S*-acetyl-thioester enzyme that can be reduced by NADH to acetaldehyde (Krimsky and Racker, 1955). This reaction corresponds to the overall reversal of the mechanism, which acetyl phosphate reacting in place of 1,3-diphosphoglycerate. Third, GAPDH catalyzes the exchange of ³²P_i into acetyl phosphate or 1,3-diphosphglycerate and the arsenolysis (by arsenate) of acyl phosphates

692 Enzymatic Reaction Mechanisms

$$R = CH(OH)CH_2OPO_3^{2-}$$

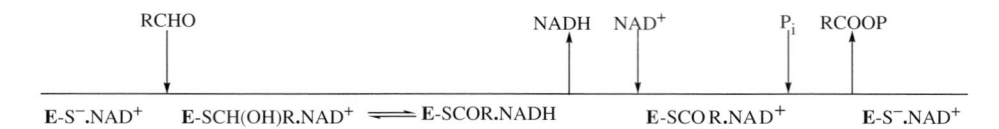

```
                RCHO                              NADH  NAD+        P_i   RCOOP
                 ↓                                  ↑    ↓           ↓     ↑
─────────────────────────────────────────────────────────────────────────────────
E-S⁻.NAD⁺    E-SCH(OH)R.NAD⁺  ⇌  E-SCOR.NADH           E-SCOR.NAD⁺       E-S⁻.NAD⁺
```

Fig. 16-11. Kinetic mechanism for the action of glyceraldehyde-3-P dehydrogenase (GAPDH). In its resting state, GAPDH binds NAD⁺ in the active site, and the reaction begins with the binding of glyceraldehyde-3-P (RCHO), which forms a thiohemiacetal adduct with Cys149. Hydride transfer proceeds from this adduct to NAD⁺ to form the covalent thioester enzyme and NADH, which dissociates. Binding of NAD⁺ in place of NADH stimulates acyl transfer to phosphate leading to 1,3-diphosphoglycerate (RCOOP), which dissociates to regenerate the resting enzyme.

(Harting and Velick, 1954; Oesper, 1954). The exchange of ^{32}P arises from the reversal of acylation, and arsenolysis arises from the hydrolytic lability of acyl arsenates formed when arsenate reacts in place of P_i. Fourth, GAPDH catalyzes the hydrolysis of *p*-nitrophenylacetate with burst kinetics in a kinetic mechanism similar to the hydrolysis of this compound by chymotrypsin. The mechanism involves the intermediate formation of the *S*-acetyl-enzyme intermediate. These reactions represent portions of the chemical and kinetic mechanisms in figs. 16-9 and 16-11.

Binding of NAD⁺

Both yeast and muscle GAPDHs are tetrameric and display cooperative binding of NAD⁺ (Harris and Waters, 1976). Yeast GAPDH binds NAD⁺ cooperatively, and binding studies implicate the concerted transition model for cooperative binding. In contrast, muscle GAPDH binds NAD⁺ with negative cooperativity (Conway and Koshland, 1968). We discussed positive and negative cooperativity in chapter 2. The detailed analysis of negatively cooperative NAD⁺ binding by muscle GAPDH served as the original proof of negative cooperativity in enzymology. Negative cooperativity could not be accommodated in the concerted transition binding model and forced the adoption of the sequential transition model in the binding of NAD⁺ by GAPDH. In any case, the binding of NAD⁺ to GAPDH is very tight, especially to the muscle enzyme, which is purified with substantial NAD⁺. Although this can be removed by various methods, the presence of NAD⁺ in the purified enzyme supports the kinetic mechanism in fig. 16-11.

Structure

The structure of the NAD⁺ binding domain in GAPDH is a dinucleotide fold, similar to those in other dehydrogenases (Rossman et al., 1975). The structure of the active site shown in fig. 16-10A shows the orientation and proximity of Cys149 with respect to the nicotinamide ring of NAD⁺. His171 is also near Cys149. A study of the interaction of Cys149 with NAD⁺ as a function of pH, by observation of the charge transfer band at 365 nm, indicates that the interaction lowers the pK_a of Cys149 from about 8 to 5.5 and verifies that Cys149 is in the thiolate form when NAD⁺ is bound to the active site (Behme and Cordes, 1967). The existence of Cys149 in its thiolate form accounts for its high reactivity in numerous chemical modification studies (Harris and Waters, 1976).

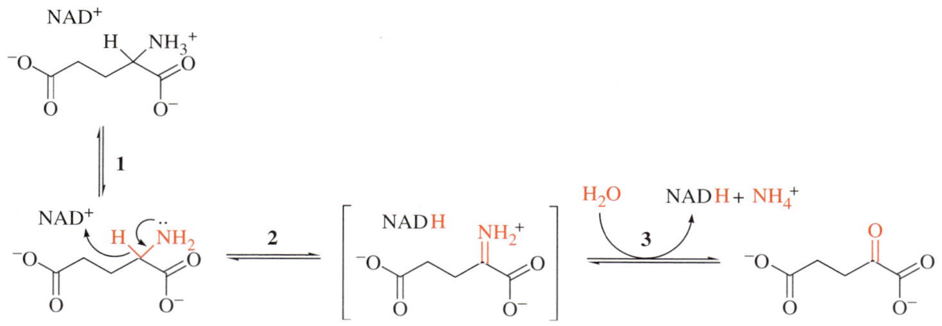

Fig. 16-12. Steps in the dehydrogenation of glutamate to α-ketoglutarate by action of glutamate dehydrogenase. The processes take place at the active site of glutamate dehydrogenase. Ionization of the 2-aminium ion of glutamate in step 1 involves proton transfer to a base at the active site. An acid-base catalyst also participates in the deamination mechanism in step 3. The hydride transfer occurs in step 2 and does not involve proton transfer in the transition state.

Glutamate Dehydrogenase

In a departure from alcoholic substrates, an amino acid dehydrogenase catalyzes hydride transfer to NAD$^+$(or NADP$^+$) from the α-carbamino group of an α-amino acid. The resulting α-imino acid undergoes hydrolysis to the corresponding α-ketoacid and ammonia. These are B-side (*si*-face) specific dehydrogenases and bind NAD$^+$ in the *syn*-conformation, as shown for phenylalanine dehydrogenase in fig. 16-10B.

Glutamate dehydrogenases (EC 1.4.1.3) catalyze the reaction of glutamate with NAD$^+$ (or NADP$^+$) and water to form α-ketoglutarate, ammonium ion, and NADH (or NADPH). Some of these dehydrogenases are specific for NAD$^+$ and others for NADP$^+$. The chemical mechanism is similar or identical for both classes and proceeds in two phases, the dehydrogenation by NAD$^+$/NADP$^+$ at C2 of glutamate to form the corresponding α-iminoglutarate intermediate, followed by hydrolysis of the imine to α-ketoglutarate and ammonia, as outlined in fig. 16-12. Hydride transfer from an amine differs mechanistically from the reaction of an alcohol, in that a proton is not abstracted from an amine in the transition state. As shown in fig. 16-12, the α-aminium group of glutamate sheds a proton in a pre-equilibrium process, and hydride transfer from the α-amine to NAD$^+$ forms the α-iminium group with no proton transfer in the transition state.

The kinetic mechanism for the NADP$^+$-dependent glutamate dehydrogenase is ordered. Both substrates and all three products bind and dissociate in strict order. In the direction of NADPH and α-ketoglutarate formation, the binding order is NADP$^+$ followed by glutamate, and product dissociation is in the order NH$_4^+$ followed by α-ketoglutarate and then NADPH (see scheme 2-11 in chap. 2).

Glutamate dehydrogenases function in a key metabolic role that links amino acid metabolism with the tricarboxylic acid cycle and with nitrogen metabolism. The production or consumption of ammonium ion links the reaction to nitrogen metabolism, and the production or consumption of α-ketoglutarate links it to the tricarboxylic acid cycle and carbohydrate metabolism. Presumably because of its central role in metabolism, glutamate dehydrogenases are regulated proteins and subject to activation and inhibition by the metabolites.

Disulfide Oxidoreductases

A family of nicotinamide coenzyme-dependent flavoproteins catalyzes the reduction of disulfides by NAD(P)H or the dehydrogenation of vicinal dithiols by NAD(P)$^+$. The enzymes are named for the direction in which the biological reaction normally proceeds. Reactions catalyzed by leading members of this family are shown in fig. 16-16. The reduction of oxidized glutathione to reduced glutathione by NADH is catalyzed by glutathione reductase, the reduction of oxidized thioredoxin is catalyzed by thioredoxin reductase, and the dehydrogenation of dihydrolipoamide by NAD$^+$ is catalyzed by dihydrolipoyl dehydrogenase. All three enzymes are flavoproteins with essential cysteinyl-disulfides at their active sites, and dihydrolipoyl dehydrogenase and glutathione reductase are homologous (Carothers et al., 1989; Williams et al., 1982). They all function by similar chemical mechanisms, but the *E. coli* thioredoxin reductase must undergo unique conformational changes to accommodate the chemistry (Williams, 1995; Williams et al., 2000). The action of dihydrolipoyl dehydrogenase serves as the mechanistic paradigm for this group.

Dihydrolipoyl Dehydrogenase

Reaction and Kinetics

Early steps in the action of the α-ketoacid dehydrogenation complexes lead to the reduction of the lipoyl moieties covalently bonded to the E2 protein. Dihydrolipoyl dehydrogenase (EC 1.8.1.4) is the E3 protein, and it catalyzes the dehydrogenation of the dihydrolipoyl moieties covalently bonded to E2. Dihydrolipoyl dehydrogenase also accepts free

A Disulfide oxidoreductases

$$\text{NADH} + \text{G-S-S-G} \xrightleftharpoons{\text{Glutathione reductase}} \text{NAD}^+ + 2\,\text{G-SH}$$

$$\text{NADPH} + \text{Thioredoxin-S}_2 \xrightleftharpoons{\text{Thioredoxin reductase}} \text{NADP}^+ + \text{Thioredoxin-(SH)}_2$$

$$\text{NAD}^+ + \text{Lipoyl-(SH)}_2 \xrightleftharpoons{\text{Dihydrolipoyl dehydrogenase}} \text{NADH} + \text{Lipoyl-S}_2$$

B Dihydrolipoyl dehydrogenase kinetics

Lip(SH)$_2$ ↓   LipS$_2$ ↑   NAD$^+$ ↓   NADH ↑

E.FAD → dihydroE.FAD → E.FAD

dihydroE.FADH$_2$

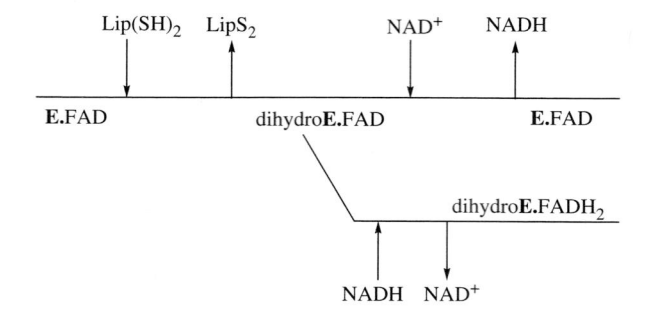

NADH ↓   NAD$^+$ ↑

Fig. 16-13. Reactions of disulfide dehydrogenases and kinetic mechanism in the action of dihydrolipoyl dehydrogenase.

dihydrolipoate or dihydrolipoamide as substrates, as well as dihydrolipoyl-peptides (see chap. 18). The dehydrogenation of a dihydrolipoyl substrate follows ping pong bi bi kinetics, as shown in fig. 16-13B (Williams, 1976). The kinetics shows that a dihydrolipoyl substrate reacts with the resting enzyme to form oxidized lipoyl substrate and an enzymatic intermediate, which must be reduced by two electrons. The reduced enzyme is designated as a dihydro form in fig. 16-13B, and this form reacts with NAD$^+$ to produce NADH and the resting enzyme.

The dihydro intermediate does not consist simply of a complex of enzyme and FADH$_2$, as shown by its electronic spectrum. The spectrum of FAD displays a prominent band at about 455 nm, whereas that of FADH$_2$ is featureless and bleached in the visible spectral region. The dihydro-intermediate displays alterations in the 450 nm region and a new, weaker band at about 530 nm. The 530 nm band is due to a charge transfer interaction between a thiolate and FAD at the active site (Wilkinson and Williams, 1979; Williams, 1976). The dihydro form of the *E. coli* enzyme is a mixture of three species differing at the active site as follows: a fluorescent species with two thiol groups and FAD, a species with a disulfide and FADH$_2$, and a species with one thiolate and FAD in charge transfer complexation. The pH independence of the absorption spectrum between pH 5 and 8 implicates a base in the active site that accepts a proton to form the thiolate in the charge transfer complex. The spectra of the three species are shown in fig. 16-14 (Wilkinson and Williams, 1979). In an equilibrium titration at pH 7.1, the dihydro-form consists of about 31% fluorescent, 52% charge transfer, and 17% FADH$_2$. In transient kinetic experiments the main species in the dihydro-form appears to be the charge transfer complex, and we refer to this as dihydro-E.FAD. In keeping with the role of the dihydro-enzyme as an intermediate, the charge transfer band is a transient that can be observed in stopped-flow experiments.

Structure and Reaction Mechanism

Much biochemical evidence implicates a catalytically functional disulfide and a base in the active site (Matthews et al., 1977; Thorpe and Williams, 1976a; Wilkinson and Williams,

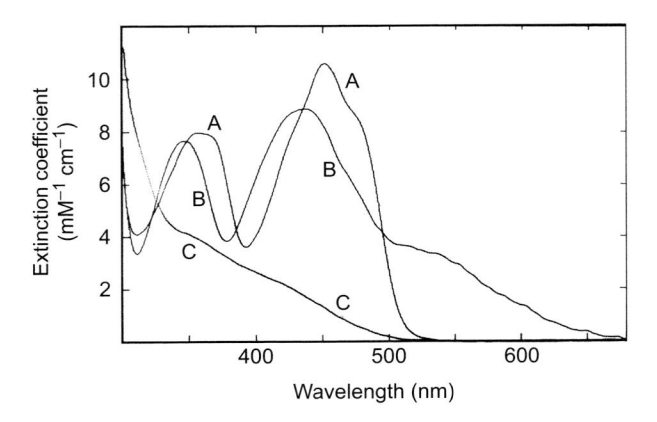

Fig. 16-14. Spectral analysis of the two-electron reduced form of dihydrolipoyl dehydrogenase from *E. coli* indicated three species, the absorption spectra of which are shown here. The bleached spectrum (C) is that of the species with a disulfide and FADH$_2$; the spectrum with maxima at 440 and 530 nm (B) is that of the thiolate-FAD charge transfer species; and the spectrum with a maximum at 450 nm (A) is that of the fluorescent species with two SH groups and FAD at the active site. The latter spectrum is similar to that of FAD. (From Wilkinson and Williams, 1979, with permission.)

1979; Williams, 1976). Important evidence includes charge transfer complexation on two-electron reduction. Other evidence includes the appearance of free sulfhydryl (SH) groups on reduction by two electrons, which display differential rates of alkylation by iodoacetamide. In the *E. coli*, dehydrogenase the disulfide is formed between Cys44 and Cys49. In the reduction by to the dihydro-enzyme, Cys44 reacts with dihydrolipoamide and Cys49 engages in charge transfer complexation with FAD (Hopkins and Williams, 1995). Mutation of either of these residues to serine practically inactivates the enzyme (0.003% and 0.012% residual activity). However, the mutated variants catalyze reactions of NADH with alternative electron acceptors such as ferricyanide and dichloroindophenol, indicating that electron transfer between NADH and FAD is not seriously impaired by mutation of these residues (Hopkins and Williams, 1995). This is consistent with the ping pong kinetics, in which electron transfer between NADH and FAD is clearly distinct from electron transfer between dihydrolipoamide and FAD.

The structure confirms the proximity of the disulfide to FAD, as shown in fig. 16-15. The structure further elucidates the interactions of FAD and NAD$^+$. The two redox active coenzymes are in face-to-face contact, with the NAD$^+$ and disulfide at opposite faces of FAD. The arrangement perfectly places FAD in position to serve as the mediator of electron transfer between the disulfide and NAD$^+$, and allows the disulfide to react essentially independently in disulfide exchange with the dihydrolipoyl substrate.

A catalytic cycle that accounts for known facts is shown in fig. 16-16. In the forward direction, dihydrolipoamide reduces the disulfide at the active site of the resting enzyme by disulfide exchange through a mechanism such as that in scheme 16-1.

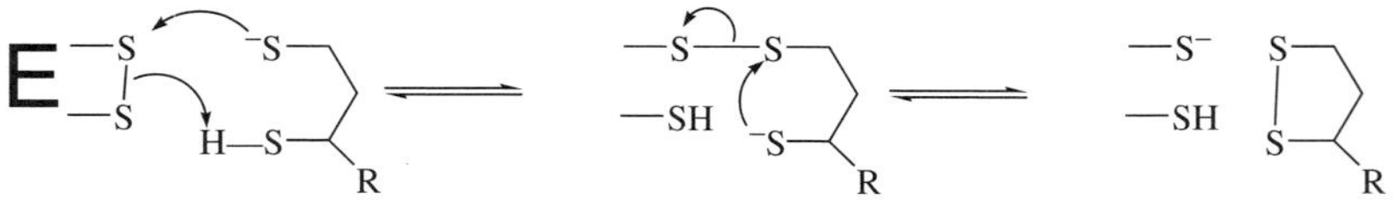

Scheme 16-1

A thiolate in the reduced disulfide engages in charge transfer complexation with FAD, which is nearby, as illustrated in fig. 16-16. Then FAD mediates electron transfer from the reduced enzyme-dithiolate to NAD$^+$ at its opposite face, and NADH dissociates to regenerate the resting enzyme. FADH$_2$ is not observed as an intermediate in the electron transfer process. Electron transfer must be accompanied by proton transfer in order to produce NADH. The mechanism of this process is not known.

A property of dihydrolipoyl dehydrogenase from *E. coli* hampers detailed studies of the reaction mechanism. The two-electron reduced enzyme, dihydro-E.FAD, readily binds NADH, which reduces FAD to FADH$_2$, as illustrated in the side reaction of fig. 16-16. The resulting complex of dihydro-E.FADH$_2$ is over-reduced, having accepted four electrons instead of two (Wilkinson and Williams, 1981). It is an inactive complex that leads to confusion in kinetic experiments. Because NAD$^+$ binds in competition with NADH, the undesired over-reduction can be prevented by the presence of NAD$^+$ at high concentrations. However, the need for high concentrations of NAD$^+$ hampers mechanistic work. The pig heart dihydrolipoyl dehydrogenase is less reactive in this respect toward NADH.

Overreduction is thought to be a biological mechanism for regulating the activity of dihydrolipoyl dehydrogenase. The enzyme normally produces NADH in the actions of α-ketoacid dehydrogenase complexes. Under metabolic conditions in which the ratio of concentrations [NADH]/[NAD$^+$] is elevated, the enzyme would be over-reduced and inactive, suppressing the formation of more NADH.

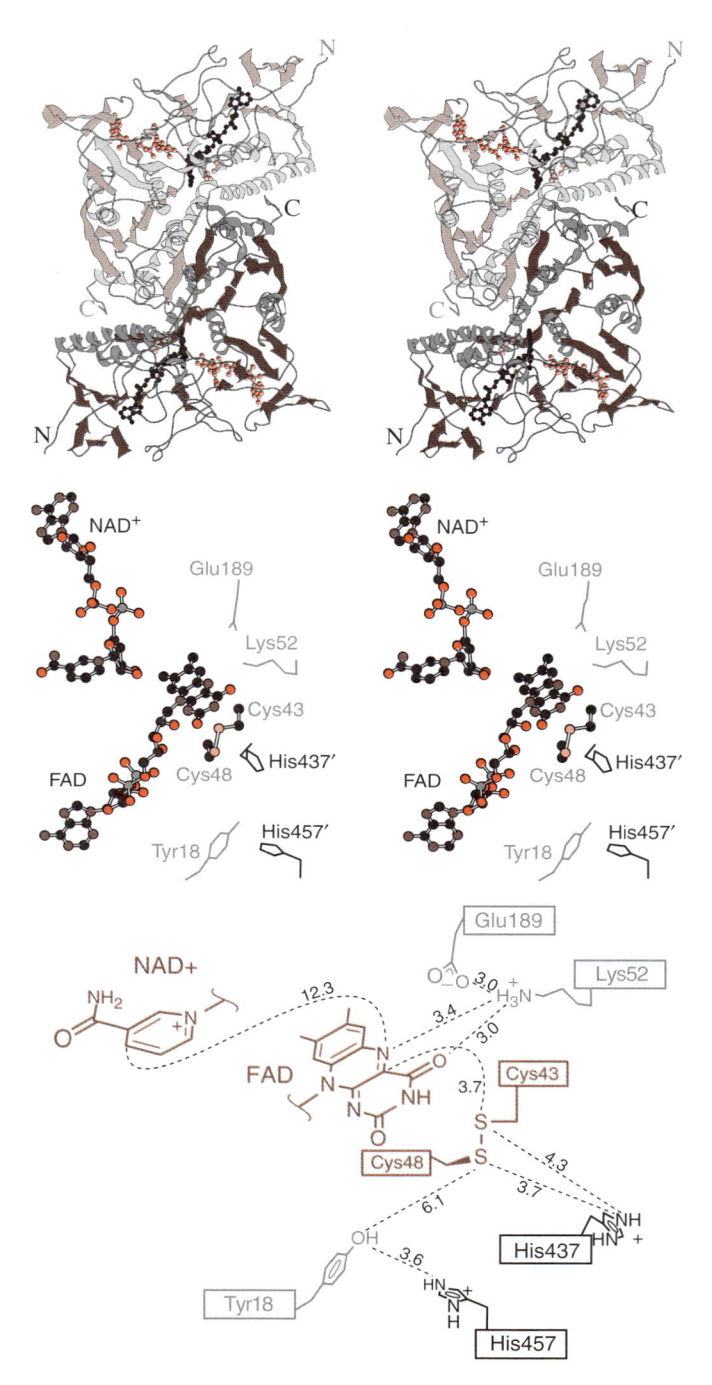

Fig. 16-15. The structure shows the homodimeric dihydrolipoyl dehydrogenase from *E. coli* in a complex with NAD⁺ (red) and FAD (black) adjacent to the two active-site cysteine residues (43 and 48 in pink) (2.45-Å resolution; PDB 1LVL; Mattevi et al., 1992).

698 Enzymatic Reaction Mechanisms

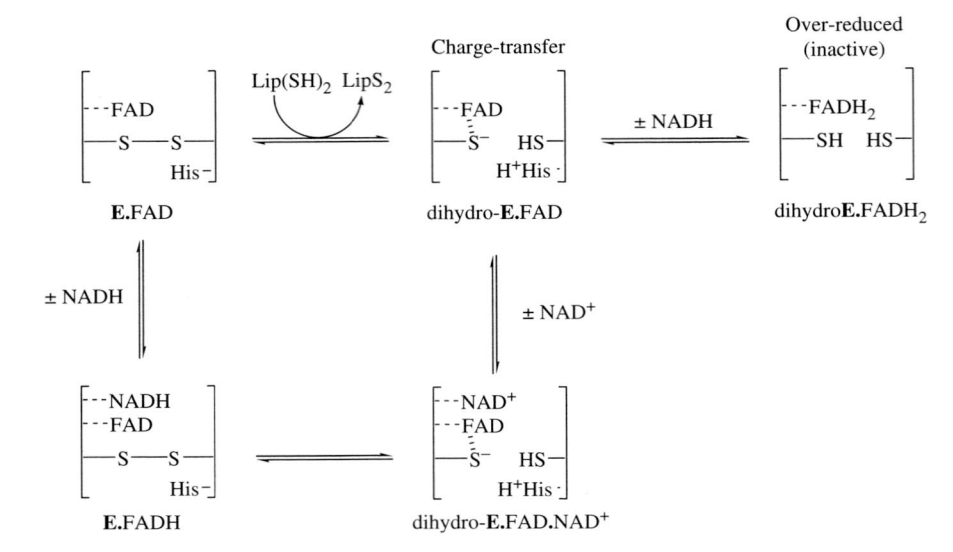

Fig. 16-16. Catalytic cycle in the action of dihydrolipoyl dehydrogenase. The resting enzyme (E.FAD) has FAD and a protein disulfide in the active site. Reduction by a dihydrolipoyl group leads to the dihydro form of the enzyme, in which the disulfide is reduced and forms a charge transfer complex with FAD that displays an absorption maximum at 500 nm. Binding of NAD^+ leads to a transitory Michaelis complex that undergoes internal oxidoreduction to form NADH, with regeneration of the disulfide. Dissociation of NADH returns the enzyme to its resting state. Overreduction of the dihydro form by NADH in competition with binding NAD^+ produces the inactive complex of the dihydroE.FADH$_2$.

Covalent Addition to FAD

One of the two cysteine-SH groups of the pig heart dihydrolipoyl dehydrogenase can be selectively alkylated by iodoacetamide after two-electron reduction of the enzyme (Thorpe and Williams, 1976a). Monoalkylation causes a reversion of the absorption spectrum from that of the charge transfer species to that of FAD. Because the disulfide cannot be restored in the monoalkylated enzyme, it does not reduce NAD^+ to NADH. However, addition of NAD^+ to the monoalkylated enzyme induces a prominent change in the absorption spectrum that signals the formation of a 4a-adduct of FAD (Thorpe and Williams, 1976b). The NAD^+-induced spectral change supports postulated mechanisms for reduction of the isoalloxazine ring of flavin coenzymes by dithiols, which involve the intermediate formation of 4a-adducts (Fisher et al., 1976; Loechler and Hollocher, 1975; Walker et al., 1970; Yokoe and Bruice, 1975). Such a mechanism is shown in fig. 16-17. In the enzymatic experiment, the binding of NAD^+ may facilitate the formation of a 4a-adduct between FAD and the unalkylated cysteinyl-SH group normally engaged in electron transfer. This would correspond to the first step of the mechanism in fig. 16-17. Subsequent steps in fig. 16-17 would be blocked by the alkyl group bonded to the other cysteine residue in the monoalkylated enzyme.

Ribonucleotide Reductases

Deoxyribonucleotides for DNA biosynthesis arise from ribonucleotides by the action of ribonucleotide reductases (RNRs), which catalyze the formal displacement of the 2′-OH in a ribonucleotide by hydrogen. From an organic chemical standpoint, the displacement

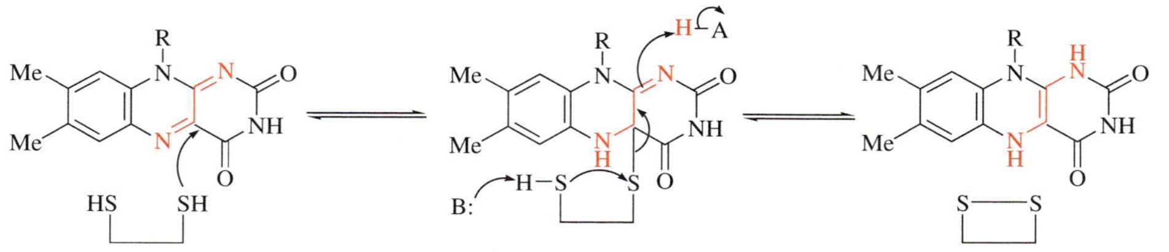

Fig. 16-17. The mechanism for reduction of a flavin by a vicinal dithiol entails the intermediate formation of a 4a-flavin adduct.

of a hydroxide group in a hydride reduction is a difficult or impossible reaction, considering the reducing agents available in either the organic laboratory or the living cell. The process can be viewed as a hydrogenation at C2, in which one atom of hydrogen replaces the 2′-OH group and the other is bonded with the OH group to form water. Molecular hydrogen itself does not come into play in the action of RNRs. Instead, the elements of molecular hydrogen are introduced in fragments through a reaction mechanism featuring radical intermediates. The initiation of the radical mechanism varies in cells of different species and in aerobic and anaerobic metabolism. In all cells, a common underlying chemistry follows the initiation process.

The different classes of RNRs specifically accept either ribonucleoside diphosphates or ribonucleoside triphosphates as substrates. However, they display broadly based activity with respect to the heterocyclic base in diphosphate or triphosphate substrates. Class I RNR accepts ADP, GDP, CDP, and UDP as substrates, whereas class II RNR accepts ATP, GTP, CTP, and UTP as substrates for reduction. The RNRs catalyze the de novo biosynthesis of deoxyribonucleotides for DNA biosynthesis. However, RNRs do not directly produce dTDP or dTTP. The dUDP or dUTP from RNR-action are dephosphorylated to dUMP, which is then transformed into dTMP by the action of thymidylate synthase (see chap. 5). Phosphorylation of dTMP by thymidylate kinase and nucleoside diphosphates kinase produces dTTP for DNA biosynthesis.

All RNRs are regulated enzymes. Each class of RNR displays a unique and complex allosteric regulation pattern. However, a common feature among the known classes is that purine deoxyribonucleotides stimulate the reduction of pyrimidine ribonucleotides, and pyrimidine deoxyribonucleotides stimulate the reduction of purine ribonucleotides. The base-pairing requirements in the structure of DNA are served by the regulatory properties of the reductases. Much confusion attended the discovery and characterization of RNRs. In retrospect, the difficulties can be traced to the diversities among RNRs in bacteria, which first served as sources of these enzymes. The individual classes of RNRs in bacteria accept either ribonucleoside diphosphates or ribonucleoside triphosphates as substrates, they require entirely different coenzymes and cofactors, and they use different reducing substrates. Much painstaking research and elegant experimentation went into the discovery of novel coenzymes and reducing substrates in the characterization of the RNRs (Fontecave et al., 1992; Jordan and Reichard, 1998; Reichard, 1993). A detailed delineation of all these experiments could not be included in this space. However, a few brief vignettes can convey an essence of the mystery that surrounded this subject in early days and of how biochemical experimentation illuminated the field. Early experiments in bacteria of various species suggested that vitamin B_{12} might be involved in ribonucleotide reduction. To determine whether this was the case in *E. coli*, the cells were grown in a defined medium containing [Co^{60}] vitamin B_{12}. RNR (now known as class I) was then purified, and the radioactivity of the protein was measured at each step. The more highly purified the protein, the less radioactivity it contained, until the pure protein did not contain a significant amount of Co⁶⁰, proving that it could not contain vitamin B_{12} (Holmgren et al., 1965). The purified RNR from *E. coli* was found by EPR analysis to contain a tyrosyl radical that was essential for activity (Ehrenberg and Reichard, 1972; Sjoberg et al., 1978). Similarly incisive experiments *in E. coli* grown anaerobically led to the discovery of a new RNR (class III) that required several factors for activity. One factor proved to be *S*-adenosylmethionine, which was required for radical formation, and another proved to be formate, which turned out to be the reducing substrate (Mulliez et al., 1995). Studies of RNR in certain anaerobic bacteria led to the discovery of the class II vitamin B_{12}-dependent RNR (Blakely, 1966). A less studied RNR, class IV, is known to contain a manganese cofactor.

In the following sections, we describe three classes of RNRs in an artificially organized manner as if the knowledge had been given in the natural course of events. Current knowledge of RNRs is hard won and arises from careful and patiently conducted biochemical research.

Classes of Ribonucleotide Reductases

We introduced a radical mechanism for the reduction of ribonucleotides in chapter 4, fig. 4-6 in connection with the adenosylcobalamin-dependent RNR. In that case, adenosylcobalamin initiates the formation of a thiyl radical in the active site (Licht et al., 1996). The original observation of thiyl radical formation in the adenosylcobalamin-dependent RNR is extended to all other RNRs, but this takes place by different mechanisms at the active sites of the other classes of RNRs (Stubbe and van der Donk, 1998; Stubbe et al., 2003). The thiyl radical-initiated mechanism by which ribonucleotides are reduced is a common feature of the actions of RNRs. A generic thiyl radical based mechanism is illustrated in fig. 16-18. Evidence for this mechanism, in addition to evidence for the thiyl radical, includes the observation of primary tritium kinetic isotope effects in reductions of [3′-³H]ribonucleotides (Stubbe and Ackles, 1980; Stuvve et al., 1981) and the observation of EPR signals for radical species in the reactions of substrate analogs in which the unpaired electron residing on substrate-based radicals in fig. 16-18 would be stabilized by delocalization (van der Donk et al., 1996, 1998).

Class I Ribonucleotide Reductase

The first RNR to be purified and characterized is that from aerobically grown *E. coli* (Jordan and Reichard, 1998; Thelander and Reichard, 1979). It has much in common with

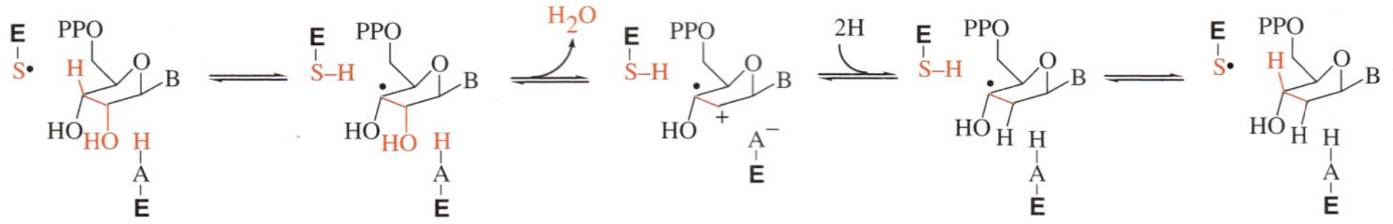

Fig. 16-18. Radical mechanism for the reduction of ribonucleotides by ribonucleotide reductases (RNRs). All RNRs generate a thiyl radical at their active sites that initiates the reduction process. In the mechanism shown, the thiyl radical abstracts the 3-H from the substrate to form a substrate radical. Elimination of water requires an acid catalyst and generates a cation radical intermediate, which undergoes reduction to the 3-radical of the deoxyribonucleotide. Abstraction of a hydrogen atom by this radical regenerates the enzyme-thiyl radical and forms the product. The RNRs differ in the mechanisms by which the thiyl radical is generated and in the reducing agent employed. Some RNRs use ribonucleoside diphosphates, and others use the triphosphates as substrates.

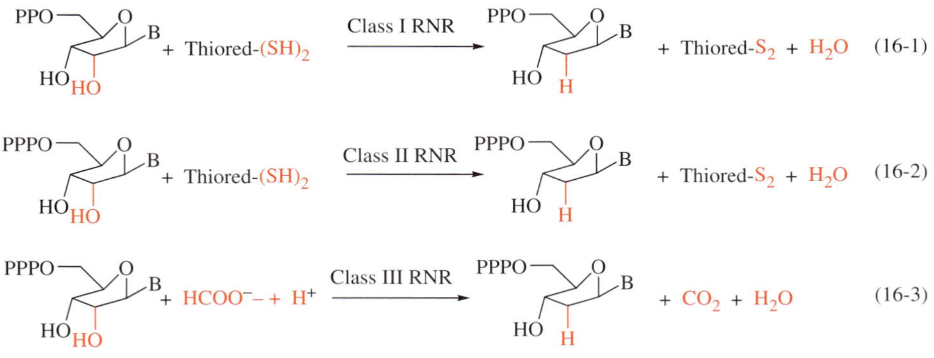

Oxidoreductases 701

$$\text{Substrate(NDP)} + \text{Thiored-(SH)}_2 \xrightarrow{\text{Class I RNR}} \text{Product(dNDP)} + \text{Thiored-S}_2 + \text{H}_2\text{O} \quad (16\text{-}1)$$

$$\text{Substrate(NTP)} + \text{Thiored-(SH)}_2 \xrightarrow{\text{Class II RNR}} \text{Product(dNTP)} + \text{Thiored-S}_2 + \text{H}_2\text{O} \quad (16\text{-}2)$$

$$\text{Substrate(NTP)} + \text{HCOO}^- + \text{H}^+ \xrightarrow{\text{Class III RNR}} \text{Product(dNTP)} + \text{CO}_2 + \text{H}_2\text{O} \quad (16\text{-}3)$$

Fig. 16-19. Ribonucleotide reduction reactions are catalyzed by three classes of ribonucleotide reductases (RNRs), which have different requirements for cofactors and substrates. Class I uses nucleoside diphosphates (NDPs) as substrates and reduced thioredoxin as the reducing agent. Class II uses nucleoside triphosphates (NTPs) as substrates and reduced thioredoxin as the reducing agent. Class III uses NTPs as substrates and formate as the reducing agent.

eukaryotic RNRs. Class I RNRs (EC 1.17.4.1) catalyze the reduction of ribonucleoside diphosphates by reduced thioredoxin according to eq. 16-1 in fig. 16-19. The class I RNR from *E. coli* is a complex composed of two homodimeric proteins R1 and R2. R1 contains the active site for binding and reducing substrates. R2 contains a di-iron complex and a tyrosyl radical, Tyr122·, which is essential for catalysis (Ehrenberg and Reichard, 1972; Sjoberg et al., 1978). R2 also contains the allosteric site that binds deoxyribonucleotides and activates the enzyme. The assembly of the di-iron complex and the creation of the radical Tyr122· in the R2-apoprotein take place in concert in a reaction that requires molecular oxygen and a reducing agent (Stubbe, 2003). After the di-iron-Tyr122· cofactor is assembled in R2, the role of the di-iron center is fulfilled, and Tyr122· initiates the radical mechanism on binding of the substrate at the active site in R1.

The structures of the di-iron center in R2 in two oxidation states of the protein as determined by x-ray crystallography are illustrated in fig. 16-20 (Logan et al., 1996; Nordlund and Eklund, 1993). Spectroscopic analysis indicates that in solution the reduced center is five-coordinate about Fe_2, with Glu204 ligated in a bidentate fashion (Yang et al., 2000).

Although the structures of R1 and R2 are known, that of the functioning complex R1R2 is not known experimentally. A docked model is available (Uhlin and Eklund, 1994), and it shows that the active site Cys439 in R1 lies some 42 Å from Tyr122· in R2. All mechanisms under consideration are based on the assumption that Tyr122· initiates thiyl radical

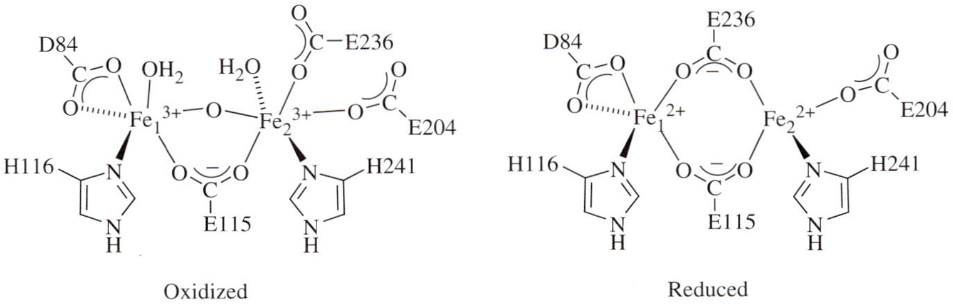

Fig. 16-20. Structures of the oxidized and reduced di-iron complex in the R2 component of ribonucleotide reductase from *E. coli*.

formation at Cys439 in R1. A mechanism must be found that allows a long-range interaction between these residues such that an electron is transferred from Cys439 in R1 to Tyr122· in R2 to generate the thiyl radical Cys439-S· in R1. The distance of 42 Å is too great to allow long-range electron transfer to occur directly between these residues on the catalytic time-scale (see chap. 4). Any electron transfer between R1-Cys439 and R2-Tyr122· must be mediated by intervening amino acid residues. Tyrosine and tryptophan in proteins are competent to mediate electron-transfer. The docking model for R1 and R2 reveals the presence of several tyrosine residues and a tryptophan residue that could mediate electron transfer. The distances are mapped in fig. 16-21. In R1 the residues Tyr730 and Tyr731 are in position to mediate electron transfer from Cys439, and mutagenesis experiments show that they are essential for activity. In R2, Trp48 lies within electron transfer distance of Tyr122· and could mediate electron transfer, and it too is essential for activity (Ekberg et al., 1996, 1998). R2-Trp48 and R1-Tyr731 are separated by 25 Å through the interface of R1 and R2, still a much longer distance than the 14 Å limit for catalytically competent electron transfer (see chap. 4). R2-Tyr356 is in a disordered region of R2 that

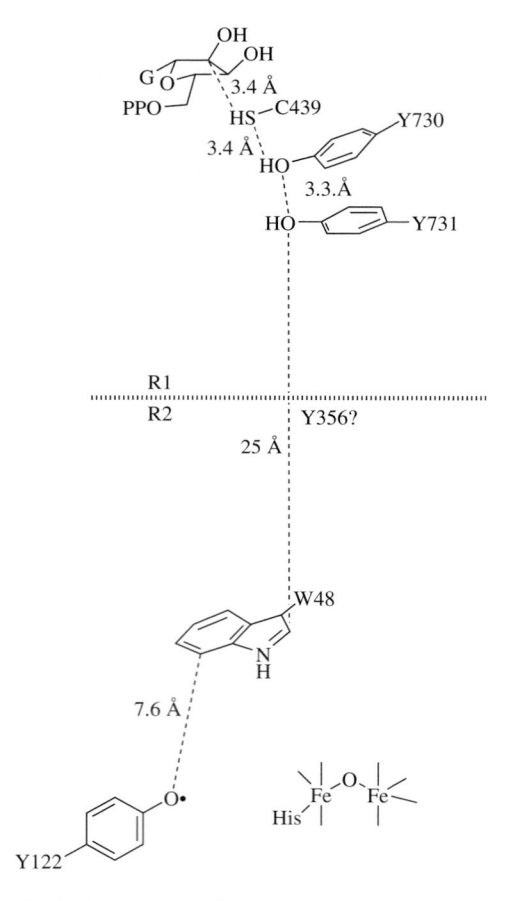

Fig. 16-21. A hypothetical electron transfer pathway for thiyl radical formation in *E. coli* ribonucleotide reductase. The hashed line at the center represents the interface between R1 and R2 in the docking model (Uhlin and Ecklund, 1994). Electron transfer between R1-Cys439 and R2-Tyr122 must be mediated by intervening amino acid side chains. The amino acids include Tyr730 and Tyr731 in R1 and Trp48 in R2. The distance spanning R2-Trp48 and R1-Tyr731 through the interface is too great for catalytically competent electron transfer. The process appears to be mediated by R2-Tyr356, which resides in a disordered peptide of R2 that is required for the interaction of R1 and R2.

could be near the interface of R1 and R2. It might bridge the electron transfer gap between R1-Tyr731 and R2-Trp48. Available evidence from the introduction of unnatural amino acids at position R2-358 indicates that Tyr358 mediates electron transfer (Stubbe, 2003).

Class II Ribonucleotide Reductases

In *Lactobacillus leichmannii*, the reduction of ribonucleoside triphosphates is catalyzed by a class II RNR (EC 1.17.4.2) according to eq. 16-2 in fig. 16-19. Unlike other RNRs, *the L. leichmannii reductase is a monomeric enzyme that contains both the active and allosteric sites*. Class II RNRs require adenosylcobalamin as the coenzyme that initiates thiyl radical formation. In fact, a thiyl radical is observed by EPR in the steady state of the reaction, and it is kinetically competent (Licht et al., 1996). The narrowing effect of deuterium on the EPR spectrum of the radical when the enzyme contains [3-^{2}H]cysteine allows the EPR signal to be assigned to a thiyl radical. In RNR from *L. leichmannii* the unpaired electron of the thiyl radical resides on Cys408 in the active site.

Homolytic cleavage of the Co—C5′ bond in adenosylcobalamin is required for thiyl radical formation and leads to cob(II)alamin (Tamao and Blakely, 1973). Paramagnetic cob(II)alamin and the thiyl radical coexist in the steady state, and inspection of the EPR signals shows that they interact magnetically. Based on a detailed analysis, the distance between them must be 5.5 to 7.5 Å in the structure of the thiyl radical intermediate (Gerfen et al., 1996).

In addition to catalyzing the reduction of ribonucleoside triphosphates, RNR from *L. leichmannii* catalyzes the two reactions in fig. 16-22A. Equation 16-4 in fig. 16-22A describes the exchange of solvent protons with the hydrogens bonded to C5′ of the adenosyl moiety of adenosylcobalamin (Beck et al., 1966; Hogenkamp et al., 1968). Equation 16-5 in fig. 16-22A describes the epimerization at C5′ of stereoselectively deuterated (5′R)-[5′-^{2}H]adenosylcobalamin (Chen et al., 2003). Both reactions are allosterically activated by dGTP. The two reactions allow the mechanism of thiyl radical formation to be dissected into steps. The C5′-hydrogen exchange requires Cys408, whereas the C5′-epimerization does not require Cys408 and is efficiently catalyzed by C408A-RNR-II (Chen et al., 2003).

The mechanism for Cys408-thiyl radical formation in fig. 16-22B explains both solvent hydrogen exchange and epimerization at C5′. Solvent exchange requires all five steps and, therefore, requires thiyl radical formation at Cys408. Mutation of Cys408 to alanine or serine blocks the exchange reaction, as well as the overall reduction of ribonucleotides. In contrast, epimerization requires only steps 1 and 2, binding of adenosylcobalamin and dGTP to RNR and reversible cleavage of the Co—C5′ bond. These steps do not involve Cys408 in a direct or essential way and so are efficiently catalyzed by C408A- and C408S-RNR-II. The requirement for dGTP in C5′-epimerization shows that the allosteric effector stimulates cleavage of the Co—C5′ bond of adenosylcobalamin, thereby initiating the radical mechanism of ribonucleotide reduction.

Class III Ribonucleotide Reductases

E. coli grown anaerobically cannot use the class I RNRs because molecular oxygen is required to generate the Tyr122· radical. When grown under anaerobic conditions, *E. coli* produce a different RNR, an anaerobic RNR known as ARR, for the de novo biosynthesis of deoxyribonucleotides (Fontecave et al., 1989). ARR catalyzes the reduction of ribonucleoside triphosphates according to eq. 16-3 in fig. 16-19. An interesting aspect of ARR action is the use of formate as the reducing agent (Mulliez et al., 1995). In chapter 8, we

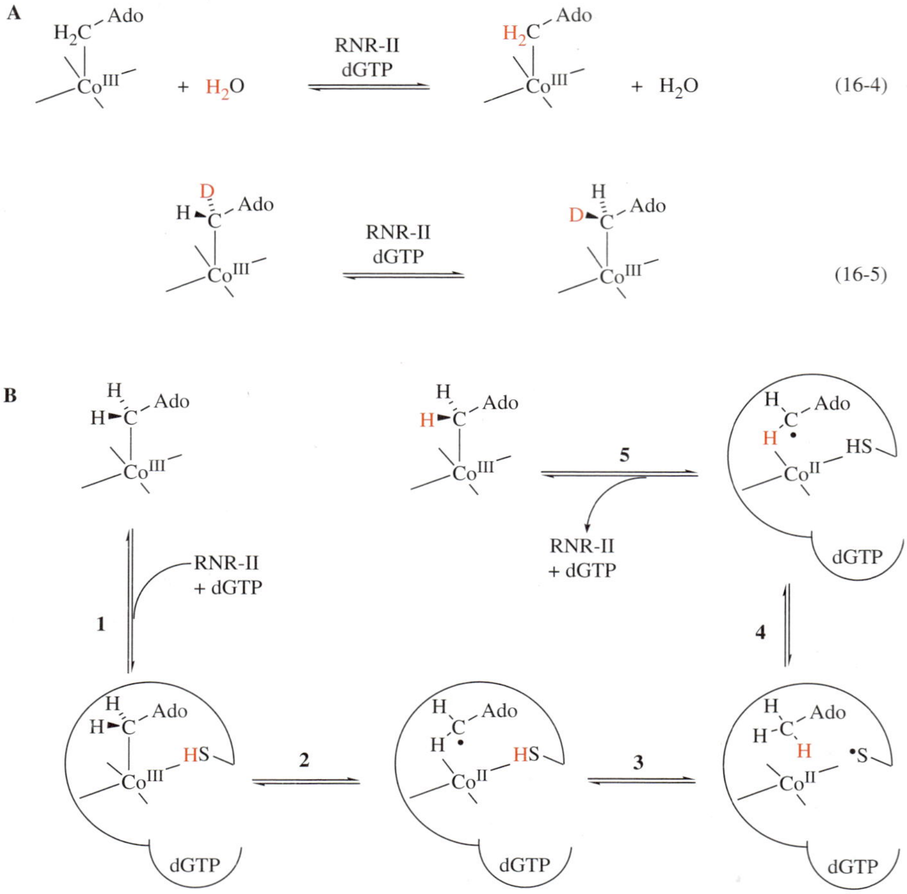

704   Enzymatic Reaction Mechanisms

Fig. 16-22. Mechanism for thiyl radical formation in the action of the class II ribonucleotide reductases (RNR) from *Lactobacillus leichmannii*. (A) RNR from *L. leichmannii* catalyzes the exchange of solvent protons with the C5-methylene hydrogens of adenosylcobalamin in a dGTP-dependent reaction shown in eq. 16-4. This reaction requires Cys408 in the active site, and mutation of this residue to alanine or serine blocks the exchange. RNR also catalyzes epimerization at C5, as shown in eq. 16-5. Epimerization is also catalyzed by C408A- and C408S-RNR at nearly the same rate as by wild-type RNR. (B) This mechanism accounts for the proton exchange and epimerization in A. The Co—C5 bond cleavage in step 2 does not involve the participation of Cys408. The cleavage occurs rapidly at the active site of RNR in the presence (but not in the absence) of dGTP at the allosteric site. The resulting 5-deoxyadenosyl radical abstracts a hydrogen atom from Cys408 to form the thiyl radical and 5-deoxyadenosine.

describe the production of formate from pyruvate by the action of pyruvate formate-lysase (PFL) in anaerobically grown *E. coli*. The metabolic functions of the two enzymes are complementary, the one producing and the other consuming formate. The metabolic linkage between ARR and PFL extends into the mechanistic sphere. Class III RNRs similar to ARR in *E. coli* are found in other organisms, including bacteriophage T4 and *Lactococcus lactis* (Logan et al., 1999; Olcott et al., 1998; Torrents et al 2000).

The ARR from *E. coli* is a homodimeric enzyme composed of 80-kDa subunits (Sun et al., 1993). It contains an amino acid sequence motif surrounding Gly681 that is reminiscent of the glycyl radical motif in PFL, and the active ARR also contains a glycyl radical, Gly681· (Mulliez et al., 1993; Sun et al., 1996). As in the case of Gly734· in PFL, exposure of the active, radical form of ARR to air leads to the cleavage of the polypeptide chain

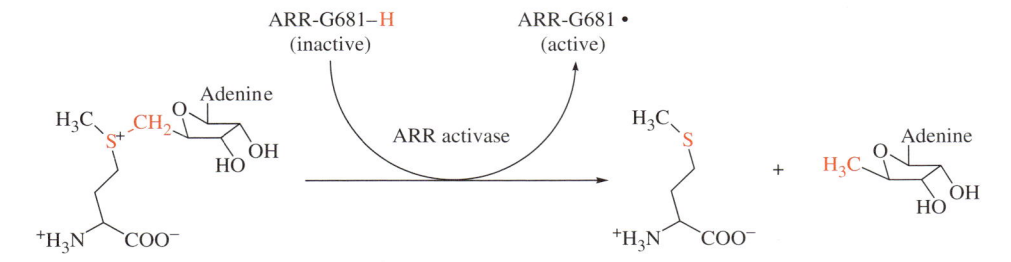

Fig. 16-23. Activation of the class III ribonucleotide reductase from *E. coli* by anaerobic ribonucleotide reductase activase.

at Gly681, transforming Gly681 into an N-terminal glyoxylate residue. Mass spectral analysis of the fragments is consistent with Gly681 at the cleavage site, and this is confirmed by site-directed mutagenesis. The radical site Gly681· is regarded as the ARR counterpart to Tyr122· in class I RNR; that is, it serves as a stable radical, a haven for the unpaired electron, and is brought into play when it initiates thiyl radical formation at the active site, presumably through long range electron transfer.

Gly681· arises through the action of ARR activase, a Radical SAM enzyme that functions similarly to the PFL activase to abstract an α-hydrogen atom from Gly681 in post-translational activation of ARR. Purified ARR activase is a homodimeric enzyme with a subunit molecular mass of 17.5 kDa (Ollagnier et al., 1996; Tamarit et al., 1999). ARR activase contains the [4Fe–4S] cluster characteristic of Radical SAM enzymes (Mulliez et al., 1993; Ollagnier et al., 1999). In activating ARR, it catalyzes the cleavage of *S*-adenosylmethionine into methionine and 5′-deoxyadenosine in concert with the abstraction of hydrogen from Gly681, transforming it into the radical center, as illustrated in fig. 16-23 (Harder et al., 1992).

Structural Relationships of Ribonucleotide Reductases

The structures of three classes of RNRs shown in fig. 16-24 reveal similarities in the core domains encompassing the active sites and in the spatial relationships between the active sites and the allosteric specificity sites (Logan et al., 1999; Nordlund et al., 1990; Sintchak et al., 2002; Uhlin and Eklund, 1994). These relationships are striking because of the great structural and chemical differences among the coenzymes required for thiyl radical formation in the three enzymes. Another difference is that the allosteric specificity sites reside at intersubunit interfaces of the dimeric class I and class III RNRs, whereas the class II RNR is monomeric. However, the spatial relationships between the active site cysteine residues and the allosteric specificity sites are very similar in the three enzymes. The essential structural feature of the subunit interface of the class I and III RNRs is effectively retained in the monomeric class II enzyme through the agency of an insert (Sintchak et al., 2002).

The core structures of RNRs are α/β_{10} barrels, with the active site cysteine residues in finger loops. In the class II RNR, the binding site for adenosylcobalamin is not near the active site Cys408. However, in the complex with adeninepentylcobalamin the structure is partially closed by a hinged motion, so that Cys408 is brought within 10 Å of cobalt. This distance is greater than the 5.5-7.5 Å in the thiyl radical enzyme. The allosteric specificity site is not occupied in these structures. Given that dGTP is required for Co—C5′ bond cleavage in the epimerization reaction (Chen et al., 2003), it is reasonable to consider that the allosteric effector might induce a further conformational closure of the active site.

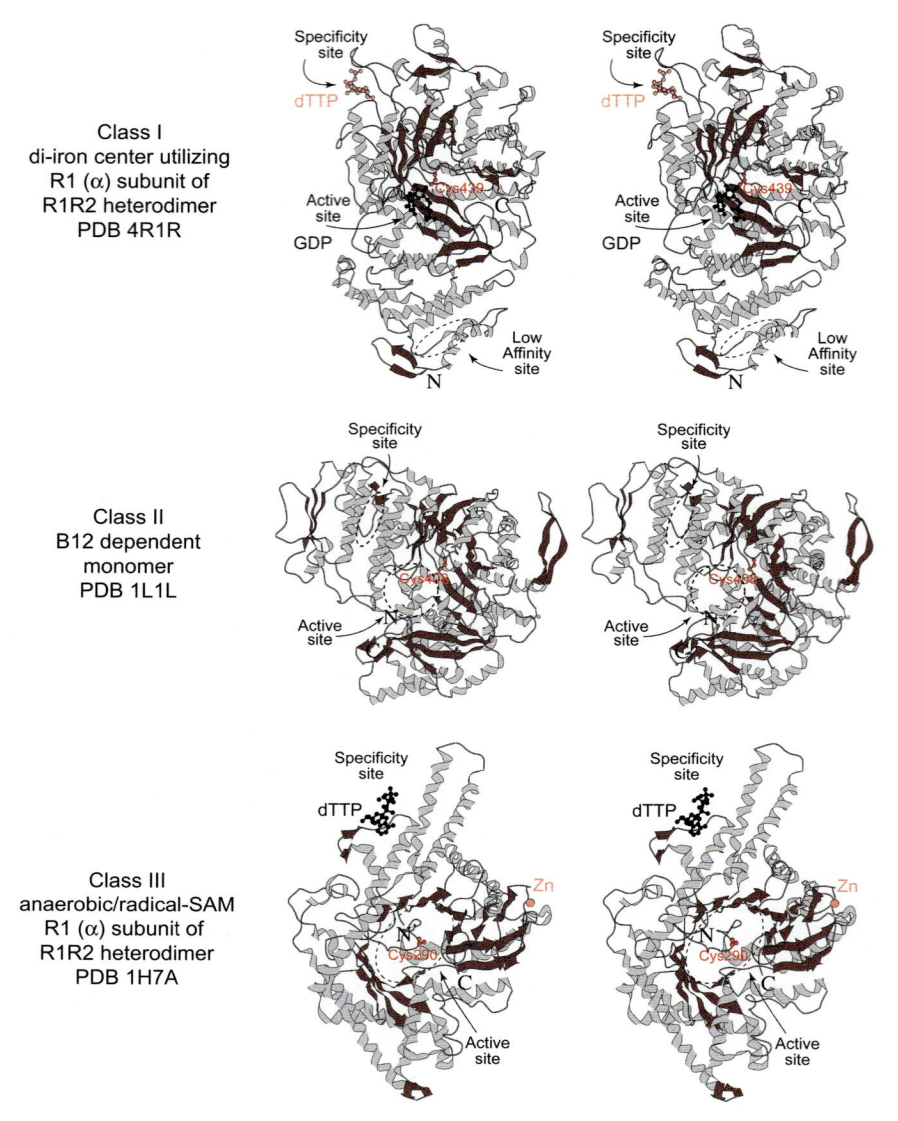

Fig. 16-24. Comparative structures of three classes of ribonucleotide reductases. A structure of the class I enzyme from *E. coli* is shown at the top (3.20-Å resolution; PDB 4R1R; Eriksson et al., 1997), of the class II enzyme from *L. leichmannii* in the center (1.75-Å resolution; PDB 1L1L; Sintchak et al., 2002), and of the class III anaerobic enzyme from bacteriophage T4 at the bottom (2.75-Å resolution; PDB 1H7A; Larsson et al., 2001).

Another striking feature among the structures is the spatial similarity among thiyl radical generation loci. In the class II and class III RNRs, the adenosylcobalamin and glycyl radical sites reside in similar loci corresponding to Tyr730 and Tyr731 of class I RNR. The two tyrosine residues in class I RNR are also in the electron transfer chain for thiyl radical formation (see fig. 16-21). These spatial relationships appear to be guided or mandated by the similar placements of the active site cysteine residues. An apparent outcome of these structural relationships is that the binding site for adenosylcobalamin in the class II RNR is not similar to other coenzyme B_{12}-dependent enzymes (Sintchak et al., 2002). Rather, the basic chain fold for the RNRs is adapted to binding adenosylcobalamin in the class II RNR, and further to cleaving the Co—C5' bond under the influence of the allosteric effector dGTP. These relationships attest to the diversities that can be imposed on a chain fold motif through evolutionary adaptation. It seems likely that the class II and III RNR predate the class I RNR in evolution, given the requirement for molecular oxygen to generate the di-iron-Tyr122• radical.

References

Adams MJ, GC Ford, R Koeckoeck, PJ Lenz Jr, A McPherson Jr, MG Rossman, IE Smiley, RW Schevitz, and AJ Wonacott (1970) *Nature* **227**, 1098.
Almarsson Ö and TC Bruice (1993) *J Am Chem Soc* **115**, 2125.
Bauer AJ, I Rayment, PA Frey, and HM Holden (1992) *Proteins* **12**, 372.
Beck WS, RH Abeles, and WG Robinson (1966) *Biochem Biophys Res Commun* **25**, 421.
Behme MTA and EH Cordes (1967) *J Biol Chem* **242**, 5500.
Blakely RL (1966) *Fed Proc* **25**, 1633.
Brändén CI, H Jörnvall, H Eklund, and B Furugen (1975) In Boyer PD (ed): *The Enzymes*, vol 11, 3rd ed. Academic Press: New York, p 104.
Briganti F, WP Fong, DS Auld, and BL Vallee (1989) *Biochemistry* **28**, 5374.
Brunhuber NM, JB Thoden, JS Blanchard, and JL Vanhooke (2000) *Biochemistry* **39**, 9174.
Cameron A, J Read, R Tranter, VJ Winter, RB Sessions, RL Brady, L Vivas, A Easton, H Kendrick, SL Croft, D Barros, JL Lavandera, JJ Martin, F Risco, S Garcia-Ochoa, FJ Gamo, L Sanz, L Leon, JR Ruiz, R Gabarro, A Mallo, and FG De Las Heras (2004) *J Biol Chem* **279**, 31429.
Carothers DJ, G Pons, and MS Patel (1989) *Arch Biochem Biophys* **268**, 409.
Chen D, A Abend, J Stubbe, and PA Frey (2003) *Biochemistry* **42**, 4578.
Colonna-Cesari F, D Perahia, M Karplus, H Eklund, CI Brändén, and O Tapia (1986) *J Biol Chem* **261**, 15273.
Cook PF, MY Yoon, S Hara, and GD McClure Jr (1993) *Biochemistry* **32**, 1795.
Dalziel K (1975) In Boyer PD (ed): *The Enzymes*, vol 11, 3rd ed. Academic Press: New York, p 1.
Duggleby RC and DT Dennis (1974) *J Biol Chem* **249**, 167.
Dworschack RT and BV Plapp (1977) *Biochemistry* **16**, 2716.
Ehrenberg A and P Reichard (1972) *J Biol Chem* **247**, 3485.
Ekberg M, S Potsch, E Sandin, M Thunnissen, P Nordlund, M Sahlin, and BM Sjöberg (1998) *J Biol Chem* **273**, 21003.
Ekberg M, M Sahlin, M Eriksson, and BM Sjöberg (1996) *J Biol Chem* **271**, 20655.
Eklund H and CI Brändén (1987) In Jurnak FA and A McPherson (eds): *Biological Macromolecules and Assemblies*, vol 3. *Active Sites of Enzymes*. John Wiley & Sons: New York, p 74.
Eklund H, P Muller-Wille, E Horjales, O Futer, B Holmquist, BL Vallee, JO Hoog, R Kaiser, and H Jornvall (1990) *Eur J Biochem* **193**, 303.
Eklund H, B Nordström, E Zeppezauer, B Söderlund, I Ohlsson, T Boiwe, and CI Brändén (1974) *FEBS Lett* **44**, 200.
Eklund H, BV Plapp, JP Samama, and CI Brändén (1982) *J Biol Chem* **257**, 14349.
Eriksson M, U Uhlin, S Ramaswamy, M Ekberg, K Regnstrom, BM Sjoberg, and H Eklund (1997) *Structure* **5**, 1077.
Fisher J, R Spencer, and C Walsh (1976) *Biochemistry* **15**, 1054.
Fontecave M, R Eliasson, and P Reichard (1989) *Proc Natl Acad Sci U S A* **96**, 2147.
Fontecave M, P Nordlund, H Eklund, and P Reichard (1992) *Adv Enzymol Relat Areas Mol Biol* **65**, 147.
Gerfen GJ, S Licht, JP Willems, BM Hoffman, and J Stubbe (1996) *J Am Chem Soc* **118**, 8192.
Harder J, R Eliasson, E Pontis, MD Ballinger, and P Reichard (1992) *J Biol Chem* **267**, 25548.
Harris JI, BP Meriwether, and JH Park (1963) *Nature* **198**, 154.
Harris JI and M Waters (1976) In Boyer PD (ed): *The Enzymes*, vol 13, 3rd ed. Academic Press: New York, p 1.
Harting J and SF Velick (1954) *J Biol Chem* **207**, 867.
Hermes JD, CA Roeske, MH O'Leary, and WW Cleland (1982) *Biochemistry* **21**, 5106.
Hogenkamp HPC, RK Ghambeer, C Brownson, RL Blakley, and E Vitols (1968) *J Biol Chem* **243**, 799.
Holbrook JJ, A Liljas, SJ Steindel, and MG Rossman (1975) In Boyer PD (ed): *The Enzymes*, vol 11, 3rd ed. Academic Press: New York, p 191.

Holmgren A, P Reichard, and L Thelander (1965) *Proc Natl Acad Sci U S A* **54**, 830.
Hopkins N and CH Williams Jr (1995) *Biochemistry* **34**, 11757.
Jordan A and P Reichard (1998) *Annu Rev Biochem* **67**, 71.
Kallberg Y, U Oppermann, H Jörnvall, and B Persson (2002) *Protein Sci* **11**, 636.
Kovaleva E (2004) PhD dissertation, University of Iowa.
Kvassman J, A Larsson, and G Pettersson (1981) *Eur J Biochem* **114**, 555.
Larsson K-M, J Andersson, B-M Sjöberg, P Nordlund, and DT Logan (2001) *Structure (Lond)* **9**, 739.
LeBrun LA, DH Park, S Ramaswamy, and BV Plapp (2004) *Biochemistry* **43**, 3014.
Licht S, GJ Gerfen, and J Stubbe (1996) *Science* **271**, 477.
Loechler EL and TC Hollocher (1975) *J Am Chem Soc* **97**, 3235.
Logan DT, J Andersson, BM Sjöberg, and P Nordlund (1999) *Science* **283**, 1499.
Logan DT, XD Su, A Aberg, K Regnstrom, J Hajdu, H Eklund, and P Nordlund (1996) *Structure* **4**, 1053.
Markert CL and F Moller (1959) *Proc Natl Acad Sci U S A* **45**, 753.
Mattevi A, G Obmolova, JR Sokatch, C Betzel, and WG Hol (1992) *Proteins* **13**, 336.
Matthews RG, DP Ballou, C Thorpe, and CH Williams Jr (1977) *J Biol Chem* **252**, 3199.
Mulliez E, M Fontecave, J Gaillard, and P Reichard (1993) *J Biol Chem* **268**, 2296.
Mulliez E, S Ollagnier, M Fontecave, R Eliasson, and P Reichard (1995) *Proc Natl Acad Sci U S A* **92**, 8759.
Nordlund P and H Eklund (1993) *J Mol Biol* **232**, 123.
Nordlund P, BM Sjöberg, and H Eklund (1990) *Nature* **345**, 593.
Oesper P (1954) *J Biol Chem* **207**, 421.
Olcott MC, J Andersson, and BM Sjöberg (1998) *J Biol Chem* **273**, 24853.
Ollagnier S, C Meier, E Mulliez, J Gaillard, V Schuenemann, A Trautwein, T Mattioli, M Lutz, and M Fontecave (1999) *J Am Chem Soc* **121**, 6344.
Ollagnier S, E Mulliez, J Gaillaard, R Eliasson, M Fontecave, and P Reichard (1996) *J Biol Chem* **271**, 9410.
Plapp BV, AJ Ganzhorn, RM Gould, DW Green, T Jacobi, E Warth, and DA Kratzer (1991). *Adv Exp Med Biol* **284**, 241.
Racker E and I Krimsky (1952) *Nature* **169**, 1043.
Reichard P (1993) *J Biol Chem* **268**, 8383.
Rossman MG, A Liljas, CI Brändén, and LJ Banaszak (1975) In Boyer PD (ed): *The Enzymes*, vol 11, 3rd ed. Academic Press: New York, p 62.
Rubach JK and BV Plapp (2002) *Biochemistry* **41**, 15770.
Rubach JK and BV Plapp (2003) *Biochemistry* **42**, 2907.
Rubach JK, S Ramaswamy, and BV Plapp (2001) *Biochemistry* **40**, 12686.
Scharschmidt M, MA Fisher, and WW Cleland (1984) *Biochemistry* **23**, 5471.
Segal HL and PD Boyer (1953) *J Biol Chem* **204**, 265.
Sekhar VC and BV Plapp (1990) *Biochemistry* **29**, 4289.
Shore JD and H Gutfreund (1970) *Biochemistry* **9**, 4655.
Silverstein E and PD Boyer (1964a) *J Biol Chem* **239**, 3901.
Silverstein E and PD Boyer (1964b) *J Biol Chem* **239**, 3908.
Sintchak MD, G Arjara, BA Kellogg, J Stubbe, and CL Drennan (2002) *Nat Struct Biol* **9**, 293.
Sjöberg BM, P Reichard, A Graslund, and A Ehrenberg (1978) *J Biol Chem* **253**, 6863.
Smith EI, BM Austen, KM Blumenthal, and JF Nyc (1975) In Boyer PD (ed): *The Enzymes*, vol 11, 3rd ed. Academic Press: New York, p 294.
Stubbe J (2003) *Curr Opin Chem Biol* **7**, 183.
Stubbe J and D Ackles (1980) *J Biol Chem* **255**, 8027.
Stubbe J, D Ackles, R Segal, and RL Blakley (1981) *J Biol Chem* **256**, 4843.
Stubbe J, DG Nocera, CS Yee, and MC Chang (2003) *Chem Rev* **103**, 2167.
Stubbe J and WA van der Donk (1998) *Chem Rev* **98**, 705.
Sun X, J Harder, M Krook, H Jörnvall, BM Sjöberg, and P Reichard (1993) *Proc Natl Acad Sci U S A* **90**, 577.
Sun X, S Ollagnier, P Schmidt, M Atta, E Mulliez, L Lepape, R Eliasson, A Grässlund, M Fontecave, P Reichard, and BM Sjöberg (1996) *J Biol Chem* **271**, 6827.
Sytkowski AJ and BL Vallee (1976) *Proc Natl Acad Sci U S A* **73**, 344.
Sytkowski AJ and BL Vallee (1979) *Biochemistry* **18**, 4095.
Tamarit J, E Mulliez, C Meier, A Trautwein, and M Fontecave (1999) *J Biol Chem* **274**, 31291.

Thelander L and P Reichard (1979) *Annu Rev Biochem* **48**, 133.
Thorpe C and CH Williams Jr (1976a) *J Biol Chem* **251**, 3553.
Thorpe C and CH Williams Jr (1976b) *J Biol Chem* **251**, 7726.
Torrents E, C Buist, A Liu, R Eliasson, J Kok, I Gilbert, A Gräslund, and P Reichard (2000) *J Biol Chem* **275**, 2463.
Uhlin U and H Eklund (1994) *Nature* **370**, 533.
Vallee BL and DS Auld (1990) *Proc Natl Acad Sci U S A* **87**, 220.
Van der Donk WA, G Yu, L Perez, RJ Sanchez, J Stubbe, V Samano, and MJ Robins (1998) *Biochemistry* **37**, 6419.
Van der Donk WA, G Yu, DJ Silva, J Stubbe, JR McCarthy, ET Jarvi, DP Matthews, RJ Resvick, and E Wagner (1996) *Biochemistry* **35**, 8381.
Walker WH, P Hemmerich, and V Massey (1970) *Eur J Biochem* **13**, 258.
White JL, ML Hackert, M Buehner, MJ Adams, GC Ford, PJ Lentz Jr, IE Smiley, SJ Steindel, and MG Rossmann (1976) *J Mol Biol* **102**, 759.
Wigley DB, SJ Gamblin, JP Turkenburg, EJ Dodson, K Piontek, H Muirhead, and JJ Holbrook (1992) *J Mol Biol* **223**, 317.
Wilkinson KD and CH Williams Jr (1979) *J Biol Chem* **254**, 852.
Wilkinson KD and CH Williams Jr (1981) *J Biol Chem* **256**, 2307.
Williams CH, LD Arscott, S Muller, BW Lennon, ML Ludwig, PF Wang, DM Veine, K Becker, and RH Schirmer (2000) *Eur J Biochem* **267**, 6110.
Williams CH Jr (1995) *FASEB J* **9**,1267.
Williams CH Jr (1976) In Boyer PD (ed): *The Enzymes*, vol 13, 3rd ed. Academic Press: New York, p 89.
Williams CH Jr, LD Arscott, and GE Schulz (1982) *Proc Natl Acad Sci U S A* **79**, 2199.
Yang YS, J Baldwin, BA Ley, JM Bollinger, and E Solomon (2000) *J Am Chem Soc* **122**, 8495.
Yokoe I and TC Bruice (1975) *J Am Chem Soc* **97**, 450.
Yun M, CG Park, JY Kim, and HW Park (2000) *Biochemistry* **39**,10702.
Zewe V and HJ Fromm (1965) *Biochemistry* **3**, 782.

17
Oxidases and Oxygenases

An oxidase catalyzes the oxidation of a substrate by O_2 without incorporating an oxygen atom into the product. A monooxygenase catalyzes oxidation by O_2 with incorporation of one oxygen atom into the product, and oxidation by a dioxygenase proceeds with incorporation of both atoms of O_2 into the product. These reactions generally require an organic or metallic coenzyme, with few exceptions, notably urate oxidase. Mechanisms of action of phenylalanine hydroxylase, galactose oxidase, and ascorbate oxidase are provided in chapter 4 in connection with the introduction of metallic coenzymes. In this chapter, we present cases of well-studied coenzyme and metal-dependent oxidases and oxygenases, and we consider one example of an oxidase that does not require a cofactor.

Oxidases

Biochemical diversity may be a characteristic of oxidases, which include flavoproteins, heme proteins, copper proteins, and quinoproteins. The actions of copper and topaquinone-dependent amine oxidases are presented in chapter 3, and in chapter 4, two copper-dependent oxidases are discussed. In this chapter, we discuss flavin-dependent oxidases, a mononuclear iron oxidase, and a cofactor-independent oxidase.

D-Amino Acid Oxidase

Reaction and Properties

Flavin-dependent oxidases catalyze the reaction of O_2 with an alcohol or amine to produce the corresponding carbonyl compound and H_2O_2. Examples include glucose oxidase, which produces gluconolactone and H_2O_2 from glucose and O_2 according to eq. 17-1.

$$\text{glucose} + O_2 \longrightarrow \text{gluconolactone} + H_2O_2 \quad (17\text{-}1)$$

A D-Amino acid oxidase (EC 1.4.3.3) catalyzes a formally similar reaction to produce an α-keto acid from the corresponding α-D-amino acid. The oxidation of an amino acid by an oxidase produces ammonium ion in addition to hydrogen peroxide and the ketoacid, and so it is formally more complex. It proceeds in the three phases described in eqs. 17-2a to 17-2c, the reduction of FAD to $FADH_2$ by the amino acid, hydrolysis of the resultant α-iminoacid to the corresponding α-ketoacid and NH_4, and oxidation of $FADH_2$ by O_2 to form H_2O_2.

$$R\text{-CH}(NH_3^+)\text{-COO}^- + \text{E.FAD} \rightleftharpoons R\text{-C}(=NH_2^+)\text{-COO}^- + \text{E.FADH}_2 \quad (17\text{-}2a)$$

$$R\text{-C}(=NH_2^+)\text{-COO}^- + H_2O \longrightarrow R\text{-C}(=O)\text{-COO}^- + NH_4^+ \quad (17\text{-}2b)$$

$$\text{E.FADH}_2 + O_2 \longrightarrow \text{E.FAD} + H_2O_2 \quad (17\text{-}2c)$$

D-Amino acid oxidase is a thoroughly studied example of a flavoprotein oxidase. The enzyme is a 84-kDa homodimer containing one molecule of FAD per subunit.

Hypothetical Mechanisms of Flavin Reduction

The mechanisms of the hydrolysis of imines and of the oxidation of dihydroflavins are discussed in chapters 1 and 3. In this chapter, we focus on the mechanism by which the amino acid reduces FAD in the first phase (see eq. 17-2a). Two mechanisms have been under consideration for more than 30 years, and one of them has been excluded. Another mechanism has not been excluded by experimental evidence, but no direct evidence implicating it has been obtained. The carbanion, hydride transfer, and free radical mechanisms in fig. 17-1 have been put forward. In the following text, we discuss them and the evidence supporting or excluding each mechanism.

In the carbanion mechanism, a base at the active site abstracts the C2(H) from the substrate as a proton to form a carbanionic intermediate, which then undergoes nucleophilic addition to FAD to form a covalent bond. In fig. 17-1 the carbanion is shown adding to C4a of the isoalloxazine ring; however, a similar mechanism can be written with carbanion-addition to N5. In the next step, the electron pair on the amine nitrogen shifts into the α-carbon, driving the electron pair from the bond linking the substrate to FAD into the flavin, to form $FADH_2$ and the α-imino-acid. The evidence supporting a carbanion mechanism consists of the behavior of D-3-chloroalanine as a substrate. fig. 17-2A shows the fate of D-3-chloroalanine in the action(s) of D-amino acid oxidase (Walsh and Abeles, 1972). The reaction takes two courses: the O_2-dependent formation of chloropyruvate, NH_3, and H_2O_2 in the normal reaction, and the O_2-independent formation of pyruvate, NH_3, and chloride ion in an alternative pathway. The alternative pathway dominates in an atmosphere of pure O_2, whereas the normal pathway is followed in the absence of O_2. The observation of two reaction paths leading to different products, in which the products are determined by the presence or absence of the second substrate (O_2), constitutes powerful

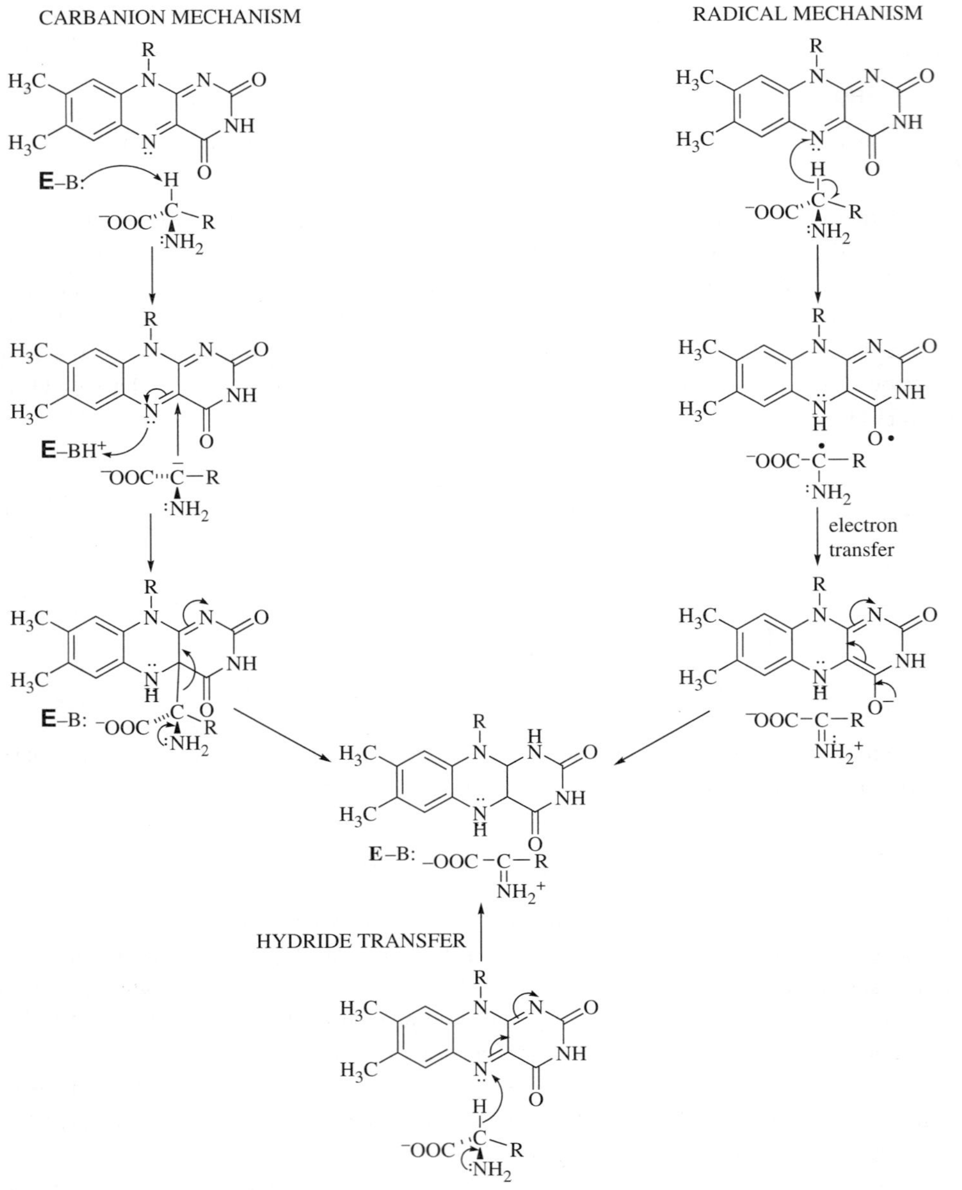

712   Enzymatic Reaction Mechanisms

CARBANION MECHANISM

RADICAL MECHANISM

Fig. 17-1. Three mechanisms for the reduction of FAD in the reaction of D-amino acid oxidase.

evidence for an intermediate that is common to the two pathways. In this case, a carbanionic intermediate can explain the results in terms of its potential ability to eliminate chloride ion, as illustrated in fig. 17-2B.

However, the carbanion mechanism must be excluded (Fitzpatrick, 2004). Direct transfer of C2(^{3}H) from a substrate to 5-deaza-FAD, bound to D-amino acid oxidase in place of FAD, is incompatible with a C2-carbanionic intermediate (Hersh and Schuman-Jorns, 1975). In reactions of 5-deaza-FAD, N5 of the isoalloxazine ring is replaced by carbon, so that deuterium or tritium transferred to C5 is stable to exchange with solvent, unlike the solvent exchange that accompanies hydrogen transfer to N5 of FAD. A linear free energy correlation does not indicate charge development at the α-carbon of substrates in the

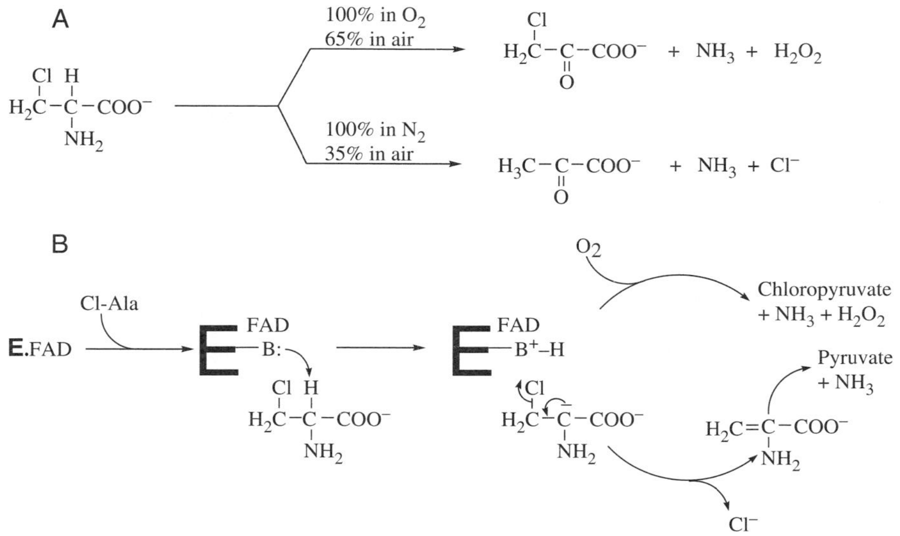

Fig. 17-2. Alternative fates of 3-chloroalanine as a substrate of D-amino acid oxidase. (A) In air, 3-chloroalanine reacts to form chloropyruvate (65%) and pyruvate (35%). In the normal oxidase reaction, chloropyruvate and H_2O_2 are formed, but the alternative elimination of chloride to form pyruvate does not lead to H_2O_2. The formation of chloropyruvate is O_2 dependent, whereas the formation of pyruvate occurs under anaerobic conditions. (B) The formation of pyruvate can be rationalized on the basis of the elimination of chloride ion from a carbanion intermediate. Reaction with O_2 to form chloropyruvate competes with elimination of chloride. This rationale can be excluded because of the absence of a base at the active site; the elimination of chloride must be explained in another way.

transition state (Pelligioni et al., 1997). The structures of D-amino acid oxidases do not reveal the presence of a base at the active site that could abstract C2(H) as a proton (Mattevi et al., 1996; Umhau et al., 2000). Moreover, the structure is compatible with a hydride transfer mechanism. The observation of chloride elimination in competition with H_2O_2 formation must be explained by another mechanism.

In the hydride transfer mechanism in fig. 17-1, the reduction of FAD proceeds in a single step by transfer of C2(H) with two electrons from the amino acid to N5 of FAD to form the α-iminoacid and $FADH_2$. This reasonable mechanism has the advantage of simplicity and is compatible with direct transfer of C2(H) to 5-deaza-FAD and with the results of the substrate structure-reactivity correlation. Moreover, structures of D-amino acid oxidases with substrates bound at the active site show that C2(H) is brought into close proximity to N5 of FAD and in the correct orientation for direct transfer. The structure in fig. 17-3 shows how D-alanine binds very near N5 of FAD. The active site is devoid of catalytic groups. The α-carboxylate of D-alanine is anchored to the active site by an electrostatic interaction with Arg285 and hydrogen bonding with Tyr238 and Tyr223.

The hydride transfer mechanism is also compatible with the pH-dependence and kinetic isotope effects. The enzyme is maximally active above pH 8, and the pH dependence of k_{cat}/K_m in the reaction of D-phenylglycine corresponds with a pK_a of 8.1 (Pollegioni et al., 1997). The absence of ionizing groups at the active site indicates that the pH dependence results from ionization of the α-amino group of D-phenylglycine, so that the reactive substrate has the α-NH_2 group. At high pH the deuterium kinetic isotope effect is about 2.4 (Pollegioni et al., 1997). At low pH values, the deuterium kinetic isotope effect is larger and accompanied by a solvent kinetic isotope effect, indicating that reaction of the zwitterionic

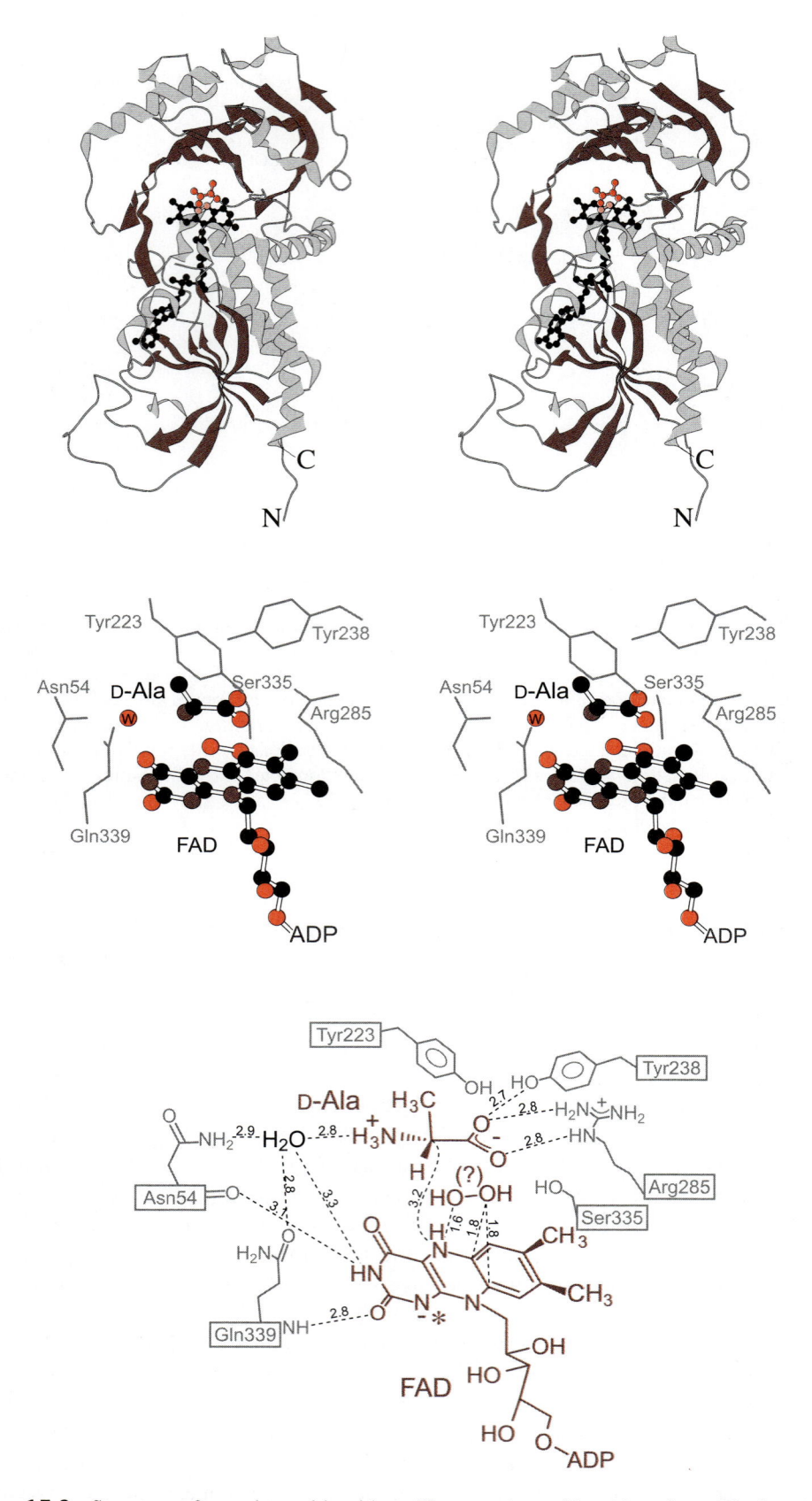

Fig. 17-3. Structure of D-amino acid oxidase. The D-amino acid oxidase from *Rhodotorula gracilis* in a complex with FAD (black), D-alanine (red), and a dioxygen species (O₂ or H₂O₂, pink) is illustrated in stereo in the top panel (1.2-Å resolution; PDB 1C0P; Umhau et al., 2000). The middle panel shows a closer view of the active site, revealing the dioxygen species sandwiched between the FAD isoalloxazine and the alanine. The active site is shown schematically in two dimensions in the bottom panel. Assignment of a discrete negative charge at the FAD nitrogen (asterisk) is based on the absence of a proton-associated bulge in the electron density, which is visible for the other protons around the aromatic system.

substrate (α-NH$_3^+$) proceeds with rate-limiting hydrogen transfer kinetically coupled with deprotonation of the aminium ion to α-NH$_2$. In a hydride transfer mechanism, the nonbonding electron pair on the α-amino group provides the driving force, as illustrated in scheme 17-1. The ^{15}N-kinetic isotope effects in the reaction of D-[^{15}N]serine, and their pH dependence, link imine formation with hydrogen transfer in the same step and are consistent with the hydride transfer mechanism (Kurtz et al., 2000).

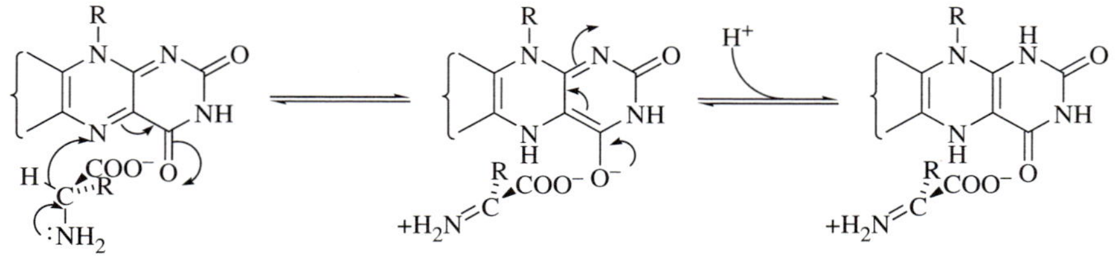

Scheme 17-1

A hydride transfer mechanism must be reconciled with the elimination of chloride in the reaction of D-3-chloroalanine. Advocates of the hydride transfer mechanism visualize the displacement of chloride by hydride transfer from FADH$_2$ in the absence of O$_2$, as illustrated in fig. 17-4A.

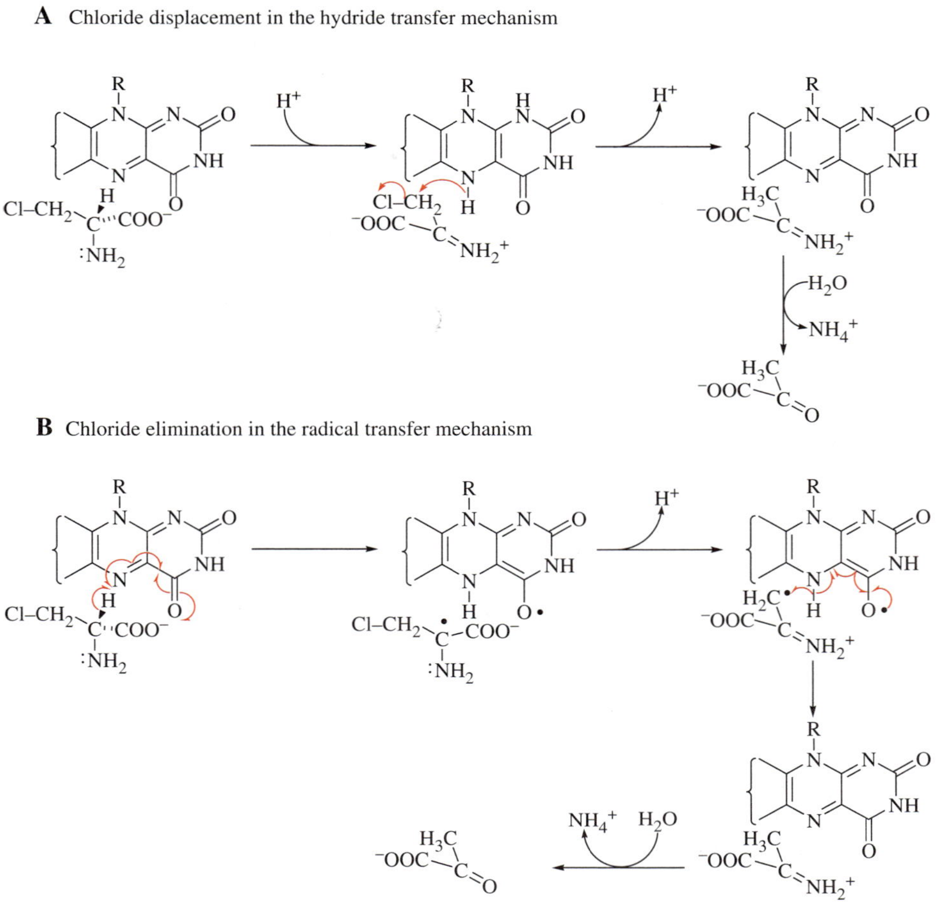

A Chloride displacement in the hydride transfer mechanism

B Chloride elimination in the radical transfer mechanism

Fig. 17-4. Mechanisms for chloride elimination in the reaction of D-chloroalanine with D-amino acid oxidase.

Available evidence does not exclude radical mechanisms such as that in fig. 17-1. Transfer of C2(H) with one electron, as a hydrogen atom to N5 of FAD would lead to a flavin semiquinone radical and the amino acid radical. Both radicals are stabilized by extensive delocalization of the unpaired electron. The flavin semiquinone radical is described as a stable species in chapter 3, and the amino acid radical is stabilized through the captodative effect, with the α-NH$_2$ as the donor and α-COO$^-$ as the acceptor in delocalization of the unpaired electron. Short-range electron transfer from the substrate radical to the flavin semiquinone would be very fast to form FADH$_2$ and the α-iminoacid. This mechanism is also consistent with direct hydrogen transfer, the linear free energy relationship, and the absence of acid-base catalytic groups at the active site. Radical mechanisms can be compatible with the isotope effects if the electron and hydrogen transfer steps are kinetically coupled.

The radical mechanism can explain chloride elimination in the reaction of D-3-chloroalanine by the mechanism in fig. 17-4B. Radicals eliminate adjacent halides very rapidly, and this process could compete with the electron transfer step in the radical mechanism and prevent the build-up of FADH$_2$. This would prevent the formation of H$_2$O$_2$. Although the radical mechanism is consistent with available information, UV/Vis and electron paramagnetic resonance (EPR) spectroscopy do not detect the putative radicals. Further experiments may exclude the radical mechanism or generate supporting evidence. As matters stand, the preponderance of evidence supports the hydride transfer mechanism.

Monoamine Oxidases

In mammals, monoamine oxidases A and B (EC 1.4.3.4) catalyze the oxidation of neurotransmitters such as serotonin and dopamine by O$_2$ to form aldehydes and H$_2$O$_2$ (fig. 17-5A). These membrane-bound flavoproteins, known as MAO A and MAO B, are drug targets for the treatment of depression and Parkinson's disease, respectively (Silverman, 2004). MAO A in the brain selectively catalyzes oxidation of serotonin and norepinephrine, and inhibitors of brain MAO A are antidepressants. MAO B selectively catalyzes the oxidation of dopamine, and inhibitors can be employed in the treatment of Parkinson's disease. MAO A and B display about 70% amino acid sequence identities in their catalytic domains.

Reaction Mechanism

As a flavoprotein oxidase, questions of the reaction mechanism and role of the flavin arise. As in amino acid oxidases, the reaction proceeds in the two phases outlined in fig. 17-5B, oxidation of the amine by the flavin to form the imine and dihydroflavin accompanied by hydrolysis of the imine, and in the second phase oxidation of the dihydroflavin by O$_2$ to form H$_2$O$_2$. The mechanistic issue, as in D-amino acid oxidase, is the initial oxidation of the substrate and reduction of the flavin. The FAD in monoamine oxidases is covalently linked to a cysteine residue through the C8α-methylene group of the isoalloxazine ring.

The mechanisms in fig. 17-6 are currently under consideration. The one electron oxidation mechanism begins with electron transfer from the amino group of the substrate to FAD to form the semiquinone and substrate cation radical transiently, and this is quickly followed by either a protonation coupled second electron transfer or by radical combination. Further steps lead directly to the FADH$_2$ anion and the substrate iminium ion. Evidence for this mechanism is the formation of cyclopropyl ring-opened products and suicide inactivation of MAO by cyclopropylamine analogs of substrates (Silverman, 1995).

A

$$RCH_2NH_2 + O_2 + H_2O + H^+ \xrightarrow{MAO} RCHO + NH_4^+ + H_2O_2$$

B

$$E\text{–}FAD + RCH_2NH_2 + H_2O + H^+ \longrightarrow E\text{–}FADH_2 + RCHO + NH_4^+$$

$$E\text{–}FADH_2 + O_2 \longrightarrow E\text{–}FAD + H_2O_2$$

C

Serotonin (MAO A)   Dopamine (MAO B)

Fig. 17-5. Reactions of monoamine oxidases A and B. (A) Oxidation of amines by MAO A and MAO B. (B) The two-phase mechanism in the oxidation of amines by monoamine oxidases. In the first phase, FAD is reduced to FADH$_2$ by the substrate, which is oxidized to the corresponding imine and then hydrolyzed to the aldehyde and ammonium ion. FADH$_2$ then reacts with O$_2$ to form H$_2$O$_2$. (C) Typical substrates for brain MAO A and MAO B.

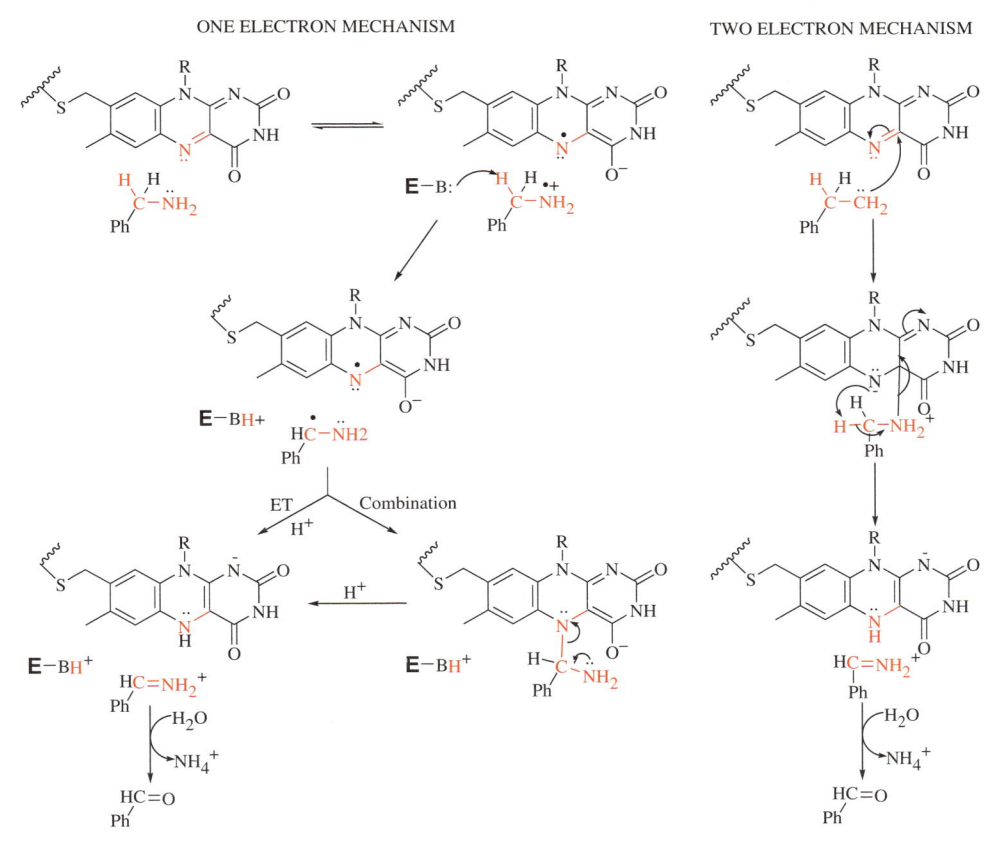

Fig. 17-6. Mechanisms under consideration for flavin reduction by substrates in the action of monoamine oxidases.

718 Enzymatic Reaction Mechanisms

The radical cationic forms of cyclopropylamines are expected to undergo ring openings to electrophilic species that would alkylate nucleophilic groups of MAO, and these compounds do inactivate MAOs by alkylation. However, no direct evidence for radical intermediates is available, and the reduction potentials for FAD in MAO B and an amine are separated by more than one volt, which may constitute a barrier for electron transfer.

In the two-electron mechanism, the amino group undergoes nucleophilic addition to FAD to form a C4a-adduct. Proton and electron transfers then lead directly to the substrate iminium ion and the $FADH_2$ anion. Substrate structure and function correlations support this mechanism by indicating charge development in the transition state that can be explained by abstraction of C1(H) as a proton from the substrate (Miller and Edmondson, 1999). In the currently postulated mechanism, flavin-N5 functions as the base to abstract the substrate-C1(H). There is no direct evidence for adduct formation.

Structure of MAO B

The outer mitochondrial membrane anchorage of MAO B arises from the C-terminal 50 residues, which form a hydrophobic, helical tail. The structure of the dimeric MAO B, consisting of 59-kDa subunits, is shown in fig. 17-7 (Binda et al., 2003, 2004). The globular catalytic portions protrude from the membrane surface and consist of FAD-binding and substrate binding domains. FAD is covalently linked to Cys397. The structure reveals a sterically distorted FAD, in which the molecule is folded slightly along the N5/N10 axis. The substrate is bound near the *re*-face of FAD, in position to react either by hydride transfer or by the one electron mechanism.

The substrate-binding site does not include an acid-base catalytic group that could abstract a proton from the substrate, a process that is required in both mechanisms. The two-electron mechanism is constructed to allow N5 of FAD in the initial adduct to serve as the base to abstract the proton. This role for N5 can also be envisioned in the one electron mechanism, as illustrated in scheme 17-2, where the semiquinone form of FAD is represented in two of its several resonance forms. The basic properties of N5 are evident, and the enhanced acidity of the substrate-C1(H) in the cation radical should allow proton abstraction.

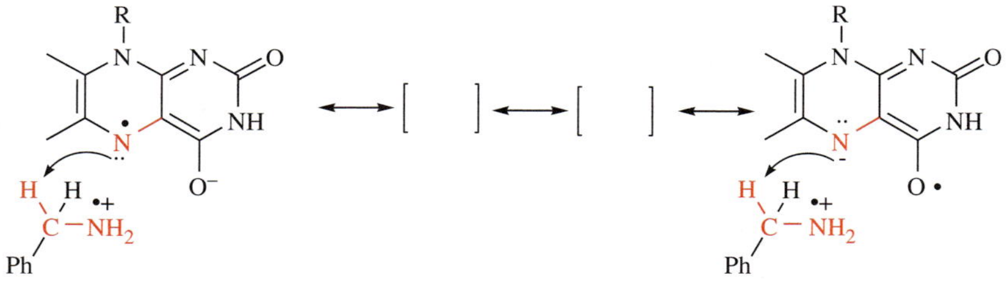

Scheme 17-2

Isopenicillin-N Synthase

The biosynthesis of penicillin begins with the assembly of a tripeptide, L-α-aminoadipoyl-L-cysteinyl-D-valine (ACV). Isopenicillin N synthase (IPNS; EC1.21.3.1), catalyzes the O_2-dependent cyclization of this molecule (Huffman et al., 1992). The equation in fig. 17-8 describing the overall reaction shows that molecular oxygen is reduced to two molecules of water. The formation of the two rings in the product result from the removal of four hydrogen atoms, the equivalent of two molecules of H_2, which are used in the reduction of O_2. IPNS may be regarded as an oxidase.

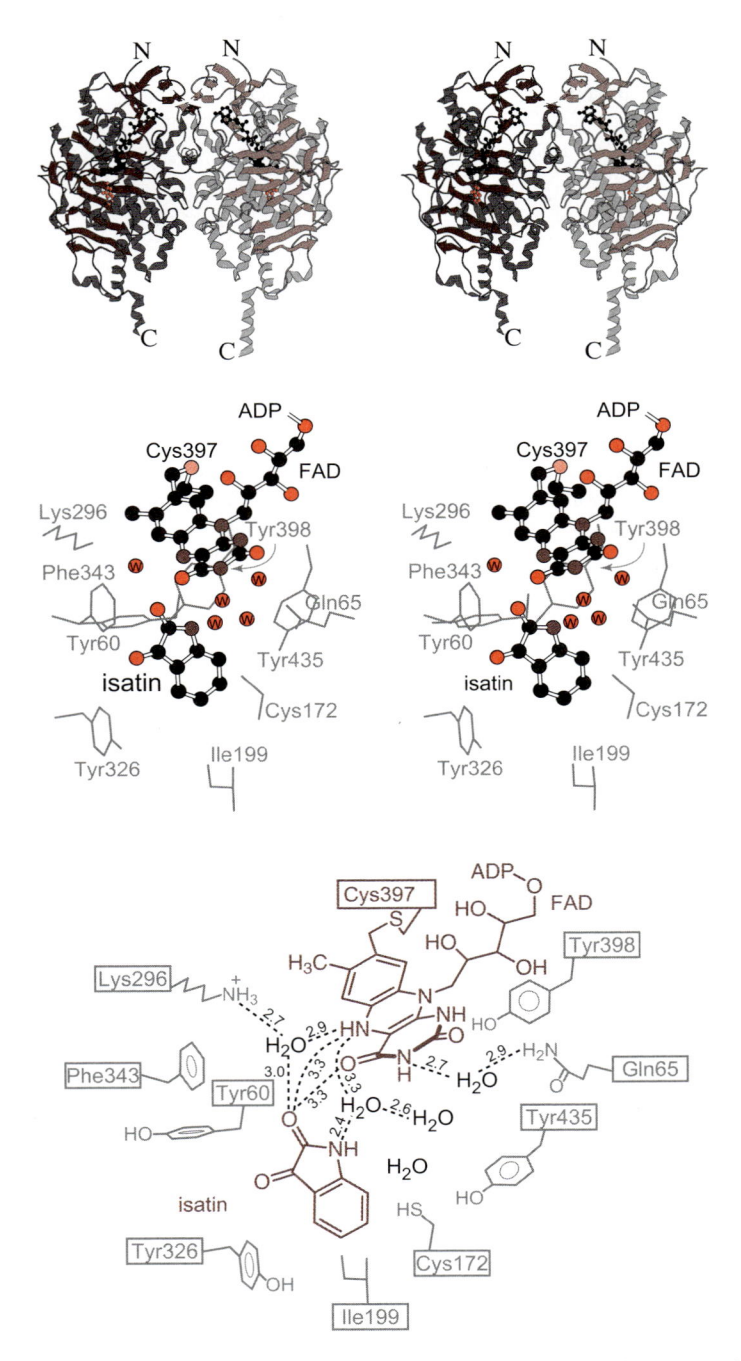

Fig. 17-7. The x-ray crystallographic structure reveals a complex of human mitochondrial monoamine oxidase B (MAO B) with FAD and the inhibitor isatin (1.7-Å resolution; PDB 1OJA; Binda et al., 2003). The homodimeric enzyme is shown in stereo in the top panel with the covalently linked FAD (black) and isatin (red) bound. The middle panel shows the active site, with the FAD linked to the enzyme by a sulfoether linkage through residue Cys397. In this structural model, the isoalloxazine ring displays considerable distortion from planarity.

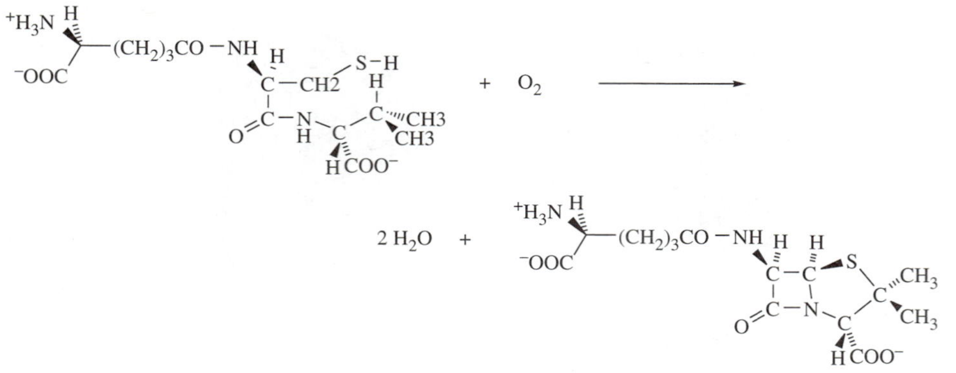

720 Enzymatic Reaction Mechanisms

Fig. 17-8. The overall reaction is catalyzed by isopenicillin N synthase (IPNS). The cyclization of L-α-aminoadipoyl- L-cysteinyl- D-valine catalyzed by IPNS proceeds in several steps that are facilitated by a mononuclear iron center. A stepwise mechanism is shown in fig. 17-9.

IPNS is a mononuclear iron protein with a 2-His, 1-carboxylate facial triad ligating the iron (Roach et al., 1995). Unlike many iron-dependent enzymes that react with O_2, an external reducing agent is not involved, and all the reducing equivalents are derived from the substrate itself. The results of careful and imaginative structure-function and crystallographic studies are the inspiration for the mechanism shown in fig. 17-9 (Elkins et al., 2003; Roach et al., 1997). The resting enzyme contains ferrous iron that binds O_2 by ligand exchange with a water molecule, as well as the cysteinyl-thiolate group. The ferrous-O_2 complex, like oxyhemoglobin (see chap. 4), may be regarded as a ferric-superoxide complex and can abstract a hydrogen atom from the β-carbon of the cysteinyl residue through a six-member cyclic transition state. The electron flow leads transiently to the thioaldehyde of cysteine ligated to ferrous-peroxide. Internal cyclization of the peptide nitrogen with the thioaldehyde, concomitant with dehydration of the ferrous-peroxide leads to the ferryl oxy species and formation of the lactam ring. The ferryl oxy species abstracts the β-hydrogen atom of D-valine, and the resulting tertiary radical attacks the thiol ligand to close the second ring and regenerate the ferrous form of IPNS.

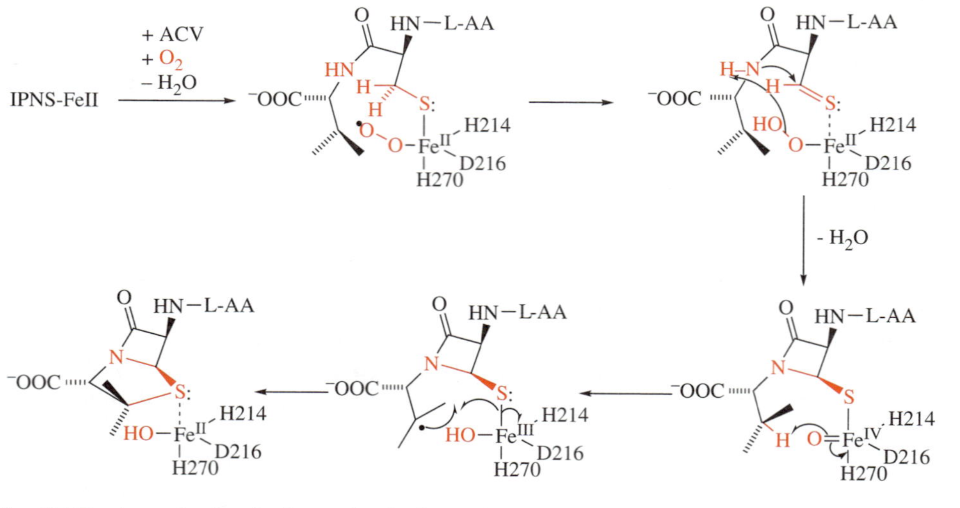

Fig. 17-9. A mechanism is shown for the isopenicillin N synthase (IPNS)–catalyzed cyclization of L-α-aminoadipoyl- L-cysteinyl- D-valine. ACV, L-α-aminoadipoyl- L-cysteinyl- D-valine; L-AA, L-α-aminoadipate.

Evidence for the ferryl oxy species has been difficult to obtain. Because of the relatively low value of k_{cat} (4.6 s^{-1}), and the still lower value for substrate analogs, evidence for the intermediacy of the ferryl-oxy species could be obtained in crystals formed anaerobically with a substrate analog containing S-methyl-D-cysteine in place of D-valine (Burzlaff et al., 1999). The crystal structure of the complex showed the tripeptide bound to Fe^{2+} through its L-cysteinyl-thiolate group. Exposure of the crystals to O$_2$ at high pressure led to the transient appearance of a yellow chromophore, possibly attributable to Fe^{IV}=O. The crystal structure obtained after O$_2$-exposure showed the lactam ring and the S-methyl-D-cysteinyl sulfoxide side chain. The S-methyl group evidently served as a chemical trap for the ferryl oxy intermediate, which followed the chemically simpler path of sulfoxidation in place of ring closure.

Urate Oxidase

Not every oxidase requires or contains a coenzyme. Urate oxidase (EC 1.7.3.3) is an exception that can be explained on the basis of the chemistry of the substrate. Urate oxidase catalyzes the oxidation of uric acid into 5-hydroxyurate according to the equation in fig. 17-10A (Kahn et al., 1997). The hydroxylation of uric acid might appear to be an oxygenation, but the 5-hydroxyl group is derived from water, and O$_2$ is converted into H$_2$O$_2$, so that it is a true oxidase. Uric acid oxidation has diverse metabolic roles in plant and animals.

Urate oxidase had been known as a copper protein until it was discovered that the removal of copper did not cause the enzyme to lose activity, and the molecular mass corresponded exactly to the amino acid composition as deduced from the translated amino acid sequence (Kahn and Tipton, 1997). The mechanism of O$_2$ reduction to H$_2$O$_2$ was postulated to be analogous to that for the oxidation of dihydroflavins (Bruice, 1984). A reasonable mechanism based on this precedent is shown in fig. 17-10B. Initial ionization of urate, followed by electron transfer to O$_2$ generates superoxide anion and a highly delocalized radical from urate. Radical coupling and protonation generates a hydroperoxide adduct.

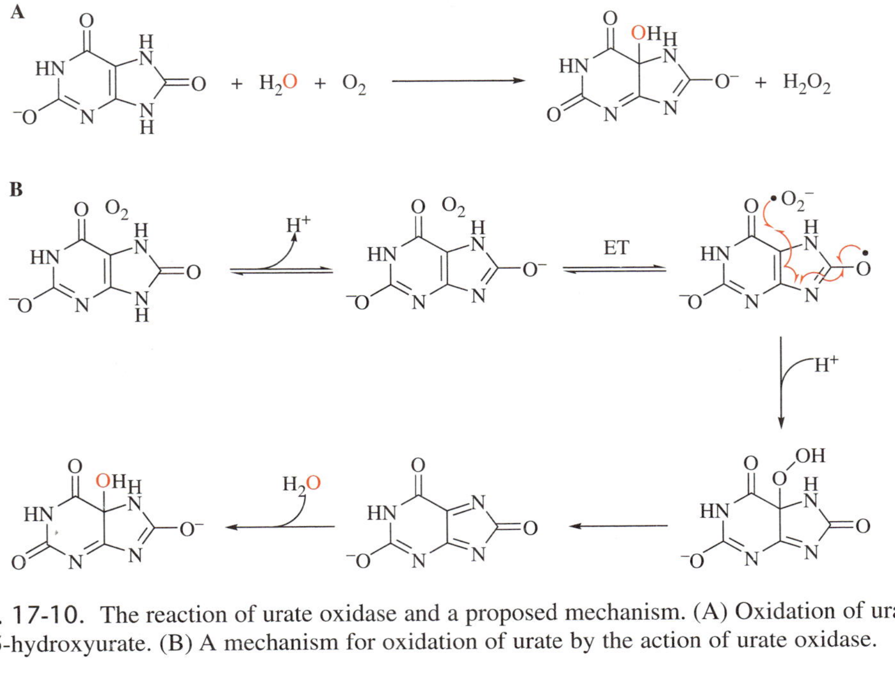

Fig. 17-10. The reaction of urate oxidase and a proposed mechanism. (A) Oxidation of urate to 5-hydroxyurate. (B) A mechanism for oxidation of urate by the action of urate oxidase.

722 Enzymatic Reaction Mechanisms

Elimination of H_2O_2 followed by addition of water produces 5-hydroxyurate. Spectroscopic results support the mechanism (Kahn and Tipton, 1998).

Monooxygenases

Lactate Monooxygenase

Lactate monooxygenase (EC 1.13.12.4) is a flavoprotein, and the mechanism of its action is in many respects similar to that of D-amino acid oxidase. This enzyme has been known as lactate oxidase, but it is a true monooxygenase because of the incorporation of on atom of oxygen from O_2 into the product. eq. 17-3 describes the overall reaction.

$$\text{lactate} + O_2 \longrightarrow CH_3COO^- + CO_2 + H_2O \quad (17\text{-}3)$$

Notice the decarboxylation and production of water in place of H_2O_2. The reaction presumably takes place by a mechanism analogous to that of D-amino acid oxidase except for the disposition of H_2O_2 and the putative oxidation product pyruvate. α-Ketoacids react readily with H_2O_2 and undergo decarboxylation and oxidation to the lower acid. The simplest rationale for the action of lactate monooxygenase is the initial oxidation of lactate to pyruvate and H_2O_2 as in a flavoprotein oxidase. However, neither pyruvate nor H_2O_2 dissociate as products. Instead, they react together at the active site according to scheme 17-3 to give the observed products.

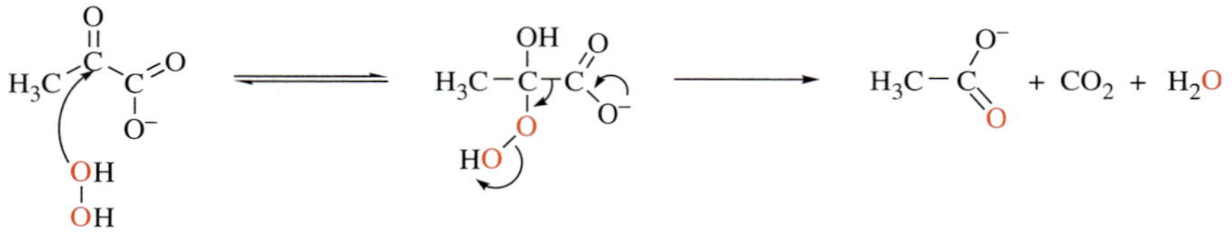

Scheme 17-3

Cytochrome P450 Monooxygenases

Hydroxylation of Alkanes

The cytochrome P450 monooxygenases are the most thoroughly studied heme enzymes (Ortiz de Montellano, 1986; Sono et al., 1996). They derive their name from the 450 nm maximum in their absorption spectra when CO is added to the reduced protein and occupies the upper axial ligand of iron. fig. 4-13B (see chap. 4) shows the spectrum of P450cam. Cytochrome P450 reductases provide the reducing equivalents for the monooxygenation of substrates by O_2. Equation 17-4 describes the overall stoichiometry in the oxygenation of a substrate by a cytochrome P450 and its associated reductase, where R– is an organic group.

$$R\text{—}H + NAD(P)H + O_2 + H^+ \rightarrow R\text{—}OH + NAD(P)^+ + H_2O \quad (17\text{-}4)$$

Cytochrome P450s somehow cleave O_2, insert one of the oxygen atoms into a C—H (or N—H) bond, and reduce the other oxygen atom to water. Labeling experiments with $^{18}O_2$ prove that one oxygen atom is incorporated into the hydroxylation product and the other into water. In many cases, the C—H bond is part of an ordinary alkane group that is in no way activated or reactive toward C—H cleavage. As in adenosylcobalamin-dependent reactions, this feature of cytochrome P450s makes them particularly fascinating subjects for mechanistic analysis.

Cytochrome P450s play important roles in the biosynthesis of steroid hormones, in the detoxification and metabolism of drugs and xenobiotics, and in the metabolism of alkanes by bacteria. Of the more than 20 distinct primary amino acid sequences per eukaryotic species, the literature describes more than 450 cytochrome P450s. Most are membrane-associated proteins, which complicates detailed mechanistic and structural analysis. The soluble P450cam from *Pseudomonas putida* is an exception, and for this reason, many of the detailed and quantitative studies have been reported for this enzyme. P450cam catalyzes the hydroxylation of camphor according to eq. 17-5.

$$\text{camphor} + \text{NADH} + O_2 + 2\,H^+ \longrightarrow \text{5-hydroxycamphor} + \text{NAD}^+ + H_2O \qquad (17\text{-}5)$$

Hydroxylation Cycle

Spectroscopic and kinetic analyses made it possible to dissect the reaction of P450cam into steps and to order the steps as shown in fig. 17-11 (Sono et al., 1996). The initial resting state of P450 contains heme in its ferric (Fe^{3+}) state with the thiolate sulfur of Cys 357 as the lower axial ligand.. The reaction begins with the binding of the substrate RH in step 1 to P450. Only after this has occurred is an electron transferred from cytochrome P450 reductase to heme iron in step 2 to produce the ferrous (Fe^{2+}) state, and only this form of heme binds O_2 in step 3. Then a second electron is transferred to the complex in step 4 to form the ferric peroxide species, which undergoes acid catalyzed dehydration in step 5 to an analog of compound I in peroxidases, with cysteine as the lower axial ligand of iron. This species is so reactive that it inserts an oxygen atom into the C—H bond of the substrate in step 6 to form the product and restore the coenzyme to its ferric heme state. Dissociation of the product returns the complex to its starting point in preparation for another catalytic cycle. The process involves low spin Fe^{3+} at the start, but on binding the substrate, the iron in heme reverts to high spin Fe^{3+} and, on accepting an electron, to high-spin ferrous heme. In the high-spin states, the iron is displaced out of the heme plane toward the lower axial Cys 357. The catalytic cycle can be short-circuited by substituting H_2O_2 for O_2 and reducing electrons from cytochrome P450, and the reaction then follows the hydrogen peroxide shunt shown in fig. 17-11.

The reducing electrons for cytochrome P450s are supplied by reductases that usher single reducing equivalents from NAD(P)H in steps 2 and 4 of the cycle in fig. 17-11. These are flavoproteins that divide the two-electron reducing equivalents from NAD(P)H into one-electron units and relay them to cytochrome P450, often through intervening iron-sulfide clusters. The diverse reductases in bacteria and eukaryotic cells interact with cognate cytochrome P450s.

Oxygenation Mechanism

The mechanism of oxygen insertion into carbon-hydrogen bonds by cytochrome P450s is under debate. The two mechanisms in scheme 17-4 have been put forward.

724 Enzymatic Reaction Mechanisms

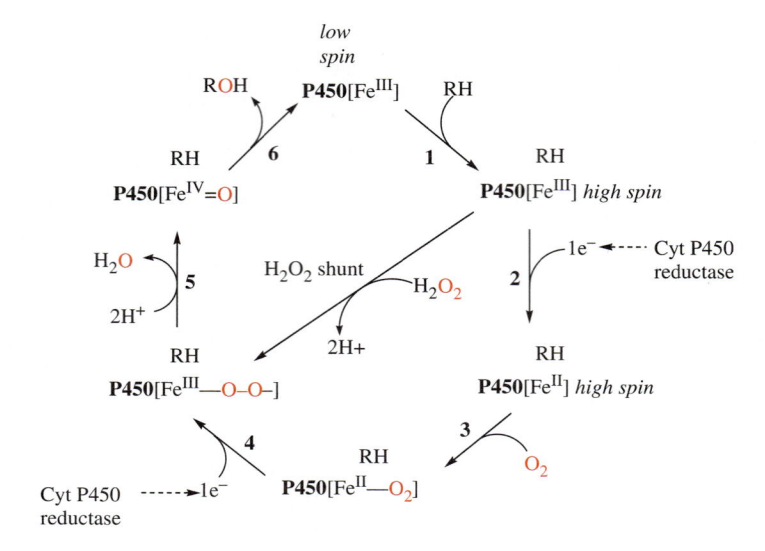

Fig. 17-11. The essential steps in cytochrome P450–catalyzed oxygenation cycle have been documented in spectroscopic and kinetic studies (Ortiz de Montellano, 1986; Sono et al., 1996). The reaction proceeds in the clockwise direction from the top. Hydrogen peroxide may be substituted for O$_2$ and the reducing electrons supplied by P450 reductase, and the reaction then follows the H$_2$O$_2$ shunt. All of the intermediates have been characterized spectroscopically and kinetically, except for the oxygenating species in the last step.

Scheme 17-4

In the stepwise mechanism, the ferryl oxy heme abstracts a hydrogen atom from the substrate to form an organic radical, which is immediately quenched by reaction with the ferryl hydroxide to form the product in what is known as the rebound step. In the concerted mechanism the oxygen of ferryl oxy heme is inserted into the C—H bond in a single step, with no discrete intermediate.

The preponderance of experimental evidence seems to favor the stepwise mechanism. Stereochemical experiments, the reactions of substrates incorporating radical clock structures, and the deuterium kinetic isotope effect for oxygenation all support the existence of radicals as oxygenation intermediates.

In the stereochemical work, stereospecifically deuterium-labeled bicyclic molecules served as the substrate for microsomal P450 (Groves and McClusky, 1978). Hydroxylation led to stereochemical randomization of the deuterium, implicating a stepwise mechanism with a trigonal carbon intermediate. Such an intermediate would be a radical, carbanion, or carbenium ion. The radical is regarded as most compatible with the chemistry of ferryl oxy heme and the relative stabilities of trigonal carbon species.

Radical clock experiments depend on the isomerization of cyclopropyl carbinyl radicals. These radicals undergo isomerization to the homoallylic radicals with a rate constant of 10^8 s^{-1}, as illustrated in scheme 17-5 for the parent reaction (Griller and Ingold, 1980).

Scheme 17-5

The reaction is reversible and favors the homoallylic species. The rate of the opening makes it a useful measure of the lifetime of radicals that incorporate the cyclopropyl ring attached to the radical in homogeneous media. It will be clear that, given the rate of this isomerization, an enzymatic intermediate incorporating this group will have a natural propensity to undergo the rearrangement in competition with subsequent steps in the mechanism. The observation of this rearrangement in the course of an enzymatic reaction constitutes evidence for the participation of a radical in the mechanism. Such an isomerization takes place in the hydroxylation of bicyclopentane by cytochrome P450, which proceeds according to scheme 17-6 (Ortiz de Montellano, 1986).

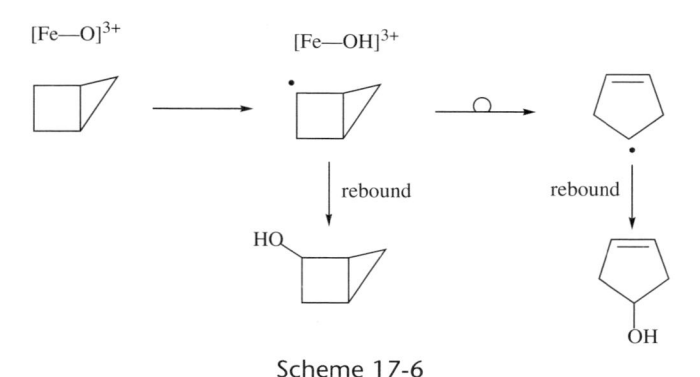

Scheme 17-6

The mechanisms in scheme 17-4 should lead to different deuterium kinetic isotope effects in the hydroxylation of substrates. The most difficult process in the stepwise mechanism is the initial abstraction of a hydrogen atom to form the radical. This should proceed through a linear transition state characterized by a large, positive deuterium kinetic isotope effect when a C—D bond is broken. The concerted reaction differs fundamentally in that it passes through a distinctly nonlinear transition state. Hydrogen transfer should proceed largely with loss of low-energy bending vibrational frequencies. Reactions of this type proceed with small deuterium kinetic isotope effects (see chap. 2). The observed effect of $k_H/k_D > 11$ strongly implies a linear transition state, unlike the nonlinear transfer in the concerted mechanism (Groves et al., 1978; Hjelmeland et al., 1976; Kadkhodayan et al., 1995).

Further studies of bicyclic substrates consolidate evidence for radical intermediates and the rebound mechanism for cytochrome P450. Oxygenation of norcarane (bicyclo[4.1.0]heptane) produces radical rearrangement and carbocation rearrangement products, in addition to the conventional oxygenation products *endo-* and *exo-*norcaranols (Auclair et al., 2002).

Structure of P450 Cam

The structure of P450cam (EC 1.14.15.1) in fig. 17-12A shows heme buried in the protein with the thiolate group of Cys357 in the lower axial position and no nearby acid-base

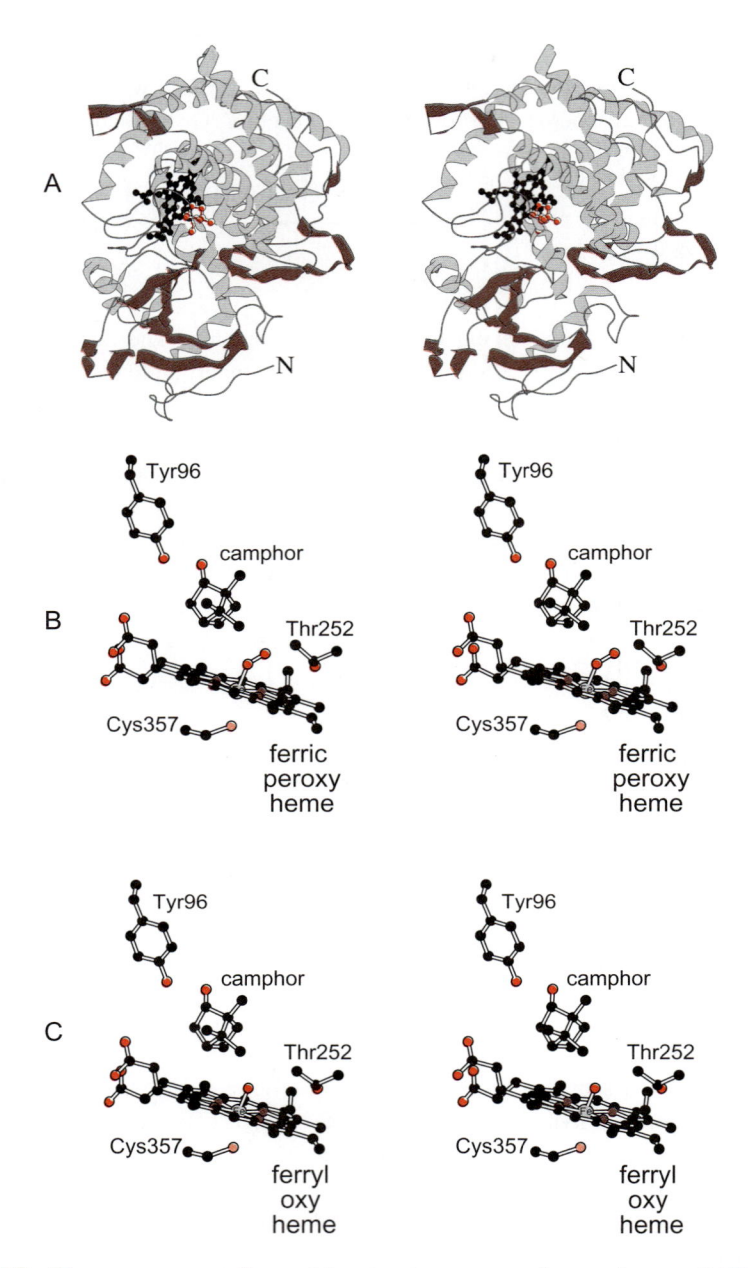

Fig. 17-12. The x-ray crystallographic structures are of cytochrome P450cam from *Pseudomonas putida*. (A) The ribbon diagram of the peptide chain of ferric P450cam in complex with camphor shows the relative locations of heme (black) and camphor (red) in the structure (1.63-Å resolution; PDB 2CPP; Poulos et al., 1987). (B) The ball-and-stick model of the active site of the ferric peroxy intermediate (1.90-Å resolution; PDB 1DZ8; Schlichting et al., 2000) shows Cys357 as the lower axial ligand to Fe, Thr252 near the peroxy group, and the camphor carbonyl group hydrogen bonded to Tyr96. (C) This is a ball-and-stick model of the active site shows the putative ferryl oxy P450 in complex with camphor (1.90-Å resolution; PDB 1DZ9; Schlichting et al., 2000).

catalytic groups (Li et al., 1995; Poulos et al., 1985, 1986). In the resting ferric-P450 with camphor bound, iron is five coordinate and about 0.3 Å above the heme plane. Iron recedes into the heme plane as the ferric peroxy and ferryl oxy species develop. Camphor is hydrogen bonded through its ketone group to Tyr96.

An atomic resolution structure (0.91 Å) of the ferric peroxy intermediate clearly reveals the structure of the peroxy group, and a structure assigned to the putative ferryl oxy intermediate is consistent with the loss of one oxygen (Schlichting et al., 2000). The side chain of Thr252 is in position to mediate proton transfer in the oxygenation of heme. The structures of the ferric peroxy and putative ferryl oxy intermediates are shown in fig. 17-12B and C. Because of the compulsory order of steps in the catalytic cycle, the ferric peroxy and ferryl oxy species could be generated in crystals by thoughtful introduction of reducing equivalents and O_2. Using dithionite as the electron source for the ferric complex of P450 with camphor, the addition of O_2 led to the ferric peroxy species by steps 1 to 3, and this could be trapped within the crystals by freezing to stop the chemistry. The structure of the ferric peroxy intermediate is shown in fig. 17-12B. Then, photoradiolytic reduction of these crystals led to the structure shown in fig. 17-12C, putatively that of the ferryl oxy heme by steps 4 and 5 of fig. 17-11. The structure in Fig 17-12C of the putative ferryl oxy species is similar to those of Compound 1 of catalase and cytochrome c peroxidase (Edwards et al., 1987; Fülöp et al., 1994; Gouet et al., 1996). The structure in fig. 17-12B is compatible with the oxygen rebound mechanism of oxygen insertion (Groves and McClusky, 1978; Groves and Subramanian, 1984) and the concerted mechanism (see scheme 17-4).

The kinetic, spectroscopic, and structural evidence support the catalytic cycle in fig. 17-11 and the oxygen insertion mechanisms in scheme 17-4. The oxygen rebound mechanism appears reasonable; however, spectroscopic evidence of the participation of a radical intermediate would consolidate the case.

Iron-Methane Monooxygenase

Hydroxylation of Substrates

Methanotrophic bacteria can live on methane as their sole source of carbon and energy. This is made possible by the action of methane monooxygenase (MMO; EC 1.14.13.25), which catalyzes the oxygenation of methane according to eq. 17-6.

$$CH_4 + O_2 + NADH + H^+ \rightarrow CH_3OH + NAD^+ + H_2O \qquad (17\text{-}6)$$

Soluble MMO (sMMO) from *Methylosinus trichosporium* or *Methylococcus capsulatus* (Bath) are similar in many respects (Waller and Lipscomb, 1996). Both consist of three main subunits, the 245-kDa hydroxylase enzyme designated MMOH, a 15-kDa protein designated MMOB, and a 40-kDa reductase containing FAD and a [2Fe–2S] center designated MMOR. MMOH contains the di-iron complex that carries out the hydroxylation of methane, MMOR is an iron-sulfur containing flavoprotein that transfers reducing electrons from NADH to MMOH, and MMOB is a coupling factor that facilitates electron transfer between MMOR and MMOH. This role of MMOB may somewhat understate its function, but it is certainly important and perhaps its main significance to the overall mechanism.

MMOH is not specific for methane and will oxygenate nearly any hydrocarbon of up to seven or eight carbons, including aromatic rings. It also oxygenates alkenes to epoxides and sulfides to sulfoxides. Acceptance of a broad range of substrates makes it possible to carry out extensive structure-function analyses in studying the mechanism. The ability of MMOH to insert an oxygen atom into unreactive C—H bonds appears to relate it mechanistically to cytochrome P450, and parallels are often drawn. Similarities and analogies in

728   Enzymatic Reaction Mechanisms

the mechanisms have become clear, as have very significant differences in the kinetic mechanisms for the generation of a ferryl oxygenating species. The highly electrophilic ferryl oxy intermediates in both cytochrome P450 and MMOH not only insert oxygen atoms into unactivated C—H bonds but into highly reactive π bonds to form epoxides and onto nonbonding electron pairs of sulfur or nitrogen to form sulfones or N-oxides. Insertion of an oxygen atom into an aromatic ring can lead to hydroxylation of the ring by a mechanism that may be similar to that of aromatic amino acid hydroxylases. For this reason, many molecules that can bind at the active sites of these enzymes will react as alternative substrates to form alternative products.

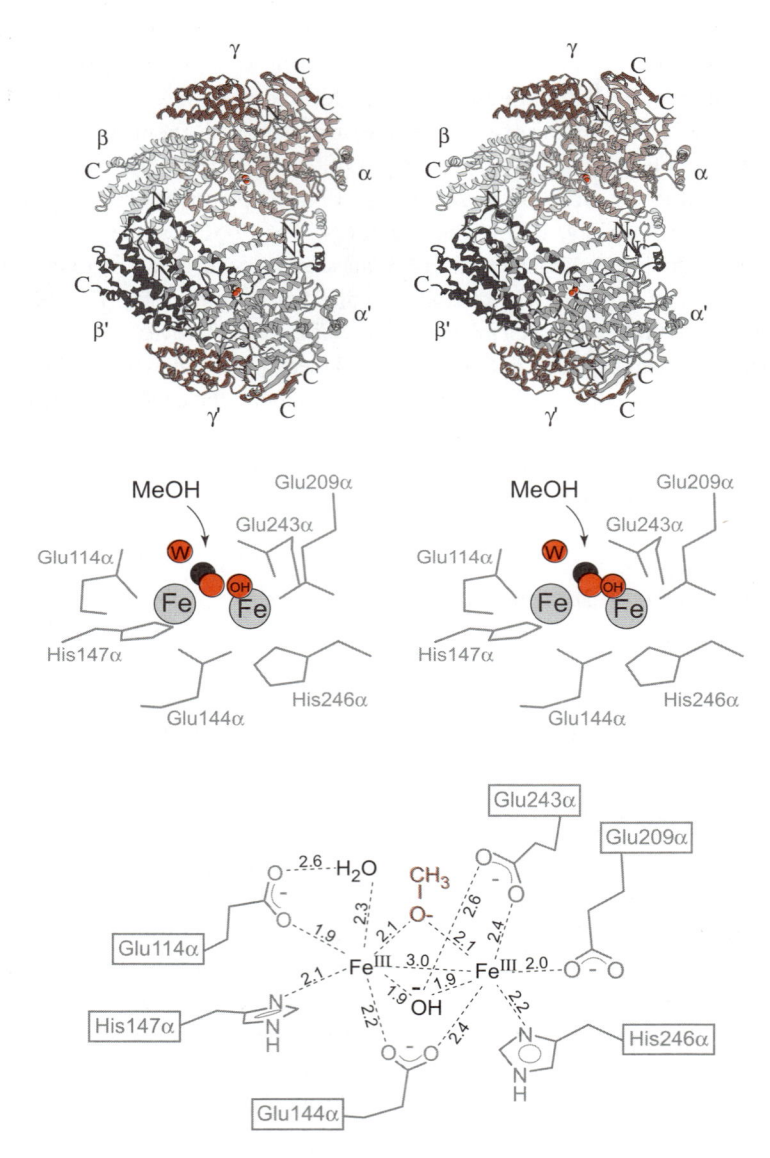

Fig. 17-13. The x-ray crystallographic structure of methane monooxygenase from *Methylococcus capsulatus* (bath) shows methanol bound at the active site near the di-iron center (2.05-Å resolution; PDB 1FZ6; Whittington et al., 2001). The enzyme is a dimer of heterotrimers $(\alpha\beta\gamma)_2$, as shown in stereo (top), with the iron atoms shown as red spheres. The diiron center is largely encompassed by the α subunit (middle). The methoxide (from methanol) and hydroxide anions are shown interacting equally with both ferric iron atoms. Contacts and distances (in angstroms) to the two ferric ions are displayed in two dimensions (bottom).

Structure

Figure 17-13 depicts the chain fold of MMOH, which is similar in *Methylococcus capsulatus* and *Methylosinus trichosporium*. MMOH is composed of two copies of each of three types of subunits, α, β, and γ, for an overall composition of $\alpha_2\beta_2\gamma_2$. The enzyme appears to be assembled as a dimer of two αβγ units. The α-chain of MMOH binds the di-iron cluster, with glutamates 114, 144, 209, and 243 and histidines 147 and 246 serving as ligands. The diferric and diferrous forms of the di-iron coenzyme are shown in fig. 17-13B and C.

Hydroxylation Cycle

A focus of research is the mechanism by which MMOH inserts an oxygen atom into a C—H bond of methane to form methanol. Unlike cytochrome P450, electron transfer to MMOH to reduce the diferric coenzyme to the diferrous state does not require the presence of the substrate. fig. 17-14 illustrates the nature of the di-iron species and the steps in the overall oxygenation of a substrate (Wallar and Lipscomb, 1996). The reaction cycle begins with the NADH-, MMOR-, and MMOB-dependent reduction of the diferric complex $\mathbf{H}^{ox}$ in MMOH to its diferrous state $\mathbf{H}^{red}$. Molecular oxygen then binds to form an association complex **O**, the transient existence of which is indicated by kinetic data. A symmetric diferric peroxy species, assigned on the basis of Mössbauer and resonance Raman data and named compound **P**, is produced from **O**. Further reaction of **P** produces compound **Q**, a yellow complex (λ_{max} 430 nm) that carries out the oxygenation of methane. Mössbauer data indicate that the irons in compound **Q** are essentially electronically equivalent and in the oxidation state Fe(IV). The complex is diamagnetic, owing presumably to antiferromagnetic coupling between the two high spin Fe(IV) atoms. Compounds **P** and **Q** are kinetically and spectroscopically distinct, and the structures of the di-iron complexes shown in fig. 17-14 are formulations based on spectrophotometric, Mössbauer and resonance Raman data (Wallar and Lipscomb, 1996).

Oxygenation Mechanism

The mechanism by which compound **Q** inserts oxygen into a C—H bond is controversial,; most experimental evidence implicates a radical mechanism. In an early study of mechanism, the reaction of 1,1-dimethylcyclopropane led to the three products in scheme 17-7 (Ruzicka et al., 1990).

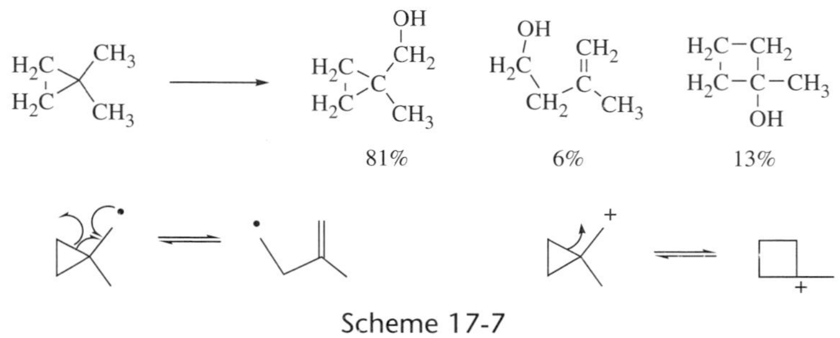

Scheme 17-7

The conventional oxygenation product dominated the mixture, but the formation of 6% homoallylic alcohol implicated the 1-methylcyclopropylcarbinyl radical as a transient intermediate that could undergo oxygenation. As discussed in the section on cytochrome P450, cyclopropylcarbinyl radicals undergo very fast ring opening to homoallylic radicals by the mechanism shown at the bottom left of scheme 17-7. When the ring opening is fast enough to compete for oxygenation of the initial cyclopropylcarbinyl radical, a ring-opened homoallylic radical is formed and undergoes oxygenation.

730 Enzymatic Reaction Mechanisms

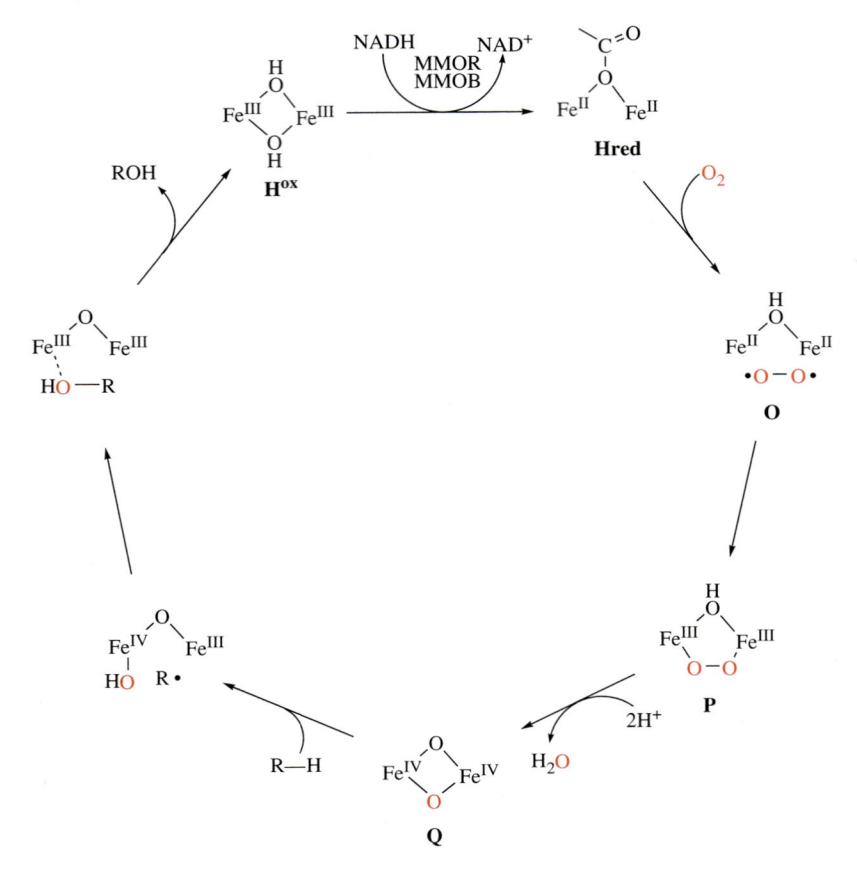

Fig. 17-14. In the hydroxylation of a substrate by methane monooxygenase, the essential steps in the oxygenation cycle have been documented in spectroscopic and kinetic studies (Wallar and Lipscomb, 1996). The reaction proceeds in the clockwise direction from the upper left. The di-iron center is reduced from the diferric to the diferrous state by NADH through the actions of MMOR and MMOB subunits. O_2 then reacts in a multistep process to form intermediate Q, the oxygenating species. The intermediate P has been detected spectroscopically, and the intermediate formation of O has been inferred from kinetic data. Intermediate Q oxygenates substrates, apparently by hydrogen atom abstraction and a radical rebound mechanism.

The formation of 1-methylcyclobutanol in a 13% yield indicated the transient formation of a carbocationic intermediate. The rearrangement characteristic of a cyclopropylcarbinyl cation is ring expansion to the cyclobutyl cation shown at the bottom right of scheme 17-7. The formation of both the radical and carbocation products suggested the possibility that both radicals and carbocations could be intermediates (Ruzicka et al., 1990). The involvement of a di-iron complex suggested that one iron-oxy species might serve in hydrogen abstraction to produce the initial radical. The other high-valent iron might accept an electron from the radical to form the carbocation, which would then capture oxygen from the di-iron complex in the production of 1-methylcyclobutane. The subsequent discovery of compound **Q** as the oxygenating species gave structural and further chemical support to this concept. In **Q** the two high-valent iron-oxy centers could well support such a mechanism.

Other oxygenation substrates also react to form products characteristic of radical and carbocationic intermediates (Brazeau et al., 2001). Soluble MMO catalyzes the oxygenation of the bicyclic norcarane (bicyclo[4.1.0]heptane) to produce the products in scheme 17-8.

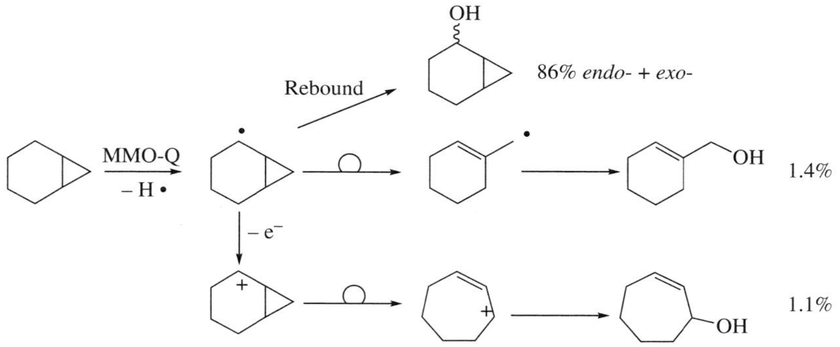

Scheme 17-8

Production of 1.4% of the radical isomerization product and 1.1% of the cation ring expansion product indicates the participation of a structurally labile intermediate that can undergo isomerization to the homoallylic radical and form the cyclohexene carbinol. Electron transfer from the same intermediate to the di-iron center leads to the carbocationic species. In this case the carbocationic rearrangement leads to the cycloheptenol. Traces of other products are formed as well. The products in scheme 17-8 are rationalized on the basis of the radical intermediate resulting from abstraction of a hydrogen atom by the **Q** species of the di-iron center (Brazeau et al., 2001).

Oxygenation of chiral ethane, (R)- or (S)-CH_3—CHDT by both species of soluble MMO gave partial racemization at the chiral methyl center (Priestly et al., 1992; Valentine et al., 1997). Partial racemization signals the participation of a trigonal C1-intermediate that has time to undergo rotational reorientation in the active site in the course of the hydroxylation process. The ethyl radical would be such an intermediate.

The observation of products derived from carbocationic intermediates in the reactions of certain substrates does not mandate the formation of CH_3^+ in the oxygenation of methane. The substrates used to trace the likely intermediate formation of radicals would more easily undergo electron transfer from the radical center than the methane radical. The methane radical should be more likely to form methanol through the rebound mechanism.

In an alternative mechanism, oxygen insertion proceeds by an asynchronous concerted mechanism analogous to that illustrated for cytochrome P450 at the bottom of scheme 17-4 (Liu et al., 1992; Valentine et al., 1999). This mechanistic proposal was inspired by the fact that product ratios in the reactions of substrates that can undergo radical isomerization through cyclopropylcarbinyl radicals are not in accord with expectations for the corresponding radicals in solution. The question of whether the environment at an enzymatic site would allow isomerization behavior characteristic of a homogeneous environment has been raised (Frey, 1997). A concerted mechanism would not lead to radical rearrangement products. Moreover, this mechanism would proceed with adjacent attack of oxygen on the C—H bond. A small deuterium kinetic isotope effect in the reaction of a deuterated substrate would be expected in a concerted mechanism because of the nonlinear transition state. The experimental deuterium isotope effects in the reactions of deuterated species of methane are very large, consistent with a linear transition state for hydrogen transfer (Nesheim and Lipscomb, 1996). The rebound mechanism allows a linear transition state. A computational study shows that the active site does not allow completely free bond rotation in the ethyl radical and that most of the reaction passes through a bound radical intermediate, although a minor fraction of the product could be formed in a concerted process (Guallar et al., 2002).

All possible mechanisms have been delineated in reference (Wallar and Lipscomb, 1996). The weight of evidence currently favors the oxygen rebound mechanism in the upper route

732 Enzymatic Reaction Mechanisms

of scheme 17-9 in the reaction of methane. This is accompanied by electron transfer to Fe(IV) in the intermediate radical complex.

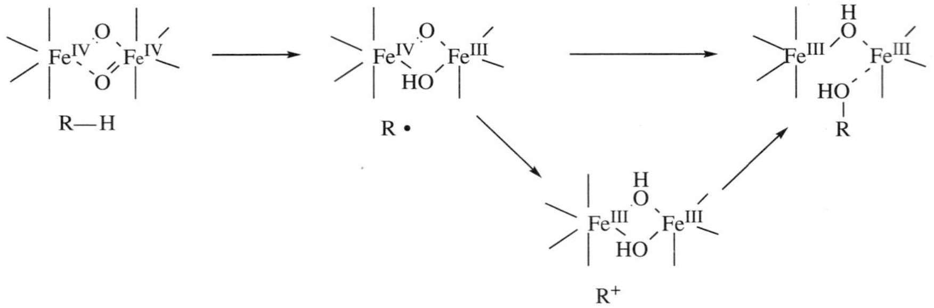

$$\begin{array}{c}\text{Fe}^{IV}\text{—O—Fe}^{IV} \\ \text{R—H}\end{array} \longrightarrow \begin{array}{c}\text{Fe}^{IV}\text{—O—Fe}^{III} \\ \text{HO} \\ \text{R}\bullet\end{array} \longrightarrow \begin{array}{c}\text{Fe}^{III}\text{—OH—Fe}^{III} \\ \text{HO} \\ \text{R}\end{array}$$

$$\begin{array}{c}\text{Fe}^{III}\text{—OH—Fe}^{III} \\ \text{HO} \\ \text{R}^+\end{array}$$

Scheme 17-9

The alternative mechanism in the lower pathway of scheme 17-9 allows electron transfer from the radical to Fe (IV) to intervene when the intermediate radicals allow sufficiently fast electron transfer. The resulting carbocation is quenched by Fe^{3+}—OH to form the product. The two mechanisms differ in the timing and pathway of electron transfer to Fe^{4+} in the intermediate complex. The lower pathway through carbocationic intermediates evidently competes with the oxygen rebound in the upper pathway when a reasonably stable carbocation can be formed from electron transfer. The oxygenation of methane is likely to proceed through the oxygen rebound mechanism.

α-Ketoglutarate–Dependent Oxygenases

α-Ketoglutarate as a Reducing System

Monooxygenation generally requires a reducing agent as a cosubstrate with oxygen; examples are cytochrome P450s and their cognate NADH-dependent reductases. An important group of mononuclear iron oxygenases use α-ketoglutarate as the reducing system (Ryle and Hausiger, 2002). These enzymes use α-ketoglutarate and O_2 to transform the mononuclear iron to a ferryl oxy species, perhaps by the mechanism in fig. 17-15. A 2-His 1-carboxylate facial triad ligates the ferrous ion and α-ketoglutarate. Ferrous iron binds oxygen and reduces it to superoxide anion in step 1 of fig. 17-15, and in step 2 the ferric superoxide undergoes radical addition to the carbonyl group of α-ketoglutarate. The resulting radical adduct is subject to fragmentation with decarboxylation coupled to reduction of ferric to ferrous iron in step 3. Heterolytic cleavage of the peroxy group to release succinate requires two electrons from ferrous iron in step 4 and leads directly to the ferryl oxy species, which oxygenates the co-substrate.

In α-ketoglutarate-dependent oxidases and oxygenases the initial generation of the ferryl oxy species occupies the face of the 2-His 1-carboxylate facial triad, and once generated it carries out the oxidation or oxygenation of the cosubstrate. As such, the process lends itself to ping pong kinetics, where the ferryl oxy species is the chemically modified intermediate.

Typical α-ketoglutarate-dependent enzymes in this group catalyze monooxygenation of substrates. Examples include prolyl-4-hydroxylase and lysine-5-hydroxylase, γ-butyrobetaine hydroxylase, and thymine hydroxylase; the latter hydroxylates the methyl group of thymine in DNA transformation. Prolyl and lysine hydroxylases are essential in collagen biosynthesis. Hydroxylation of these substrates should follow a course of hydrogen abstraction, as in the action of catalase, as in fig. 4-12, followed by oxygenation. A variation on this theme is *deacetoxycephalosporin C synthase*, which uses α-ketoglutarate in a similar way but carries out a ring expansion instead of an oxygenation.

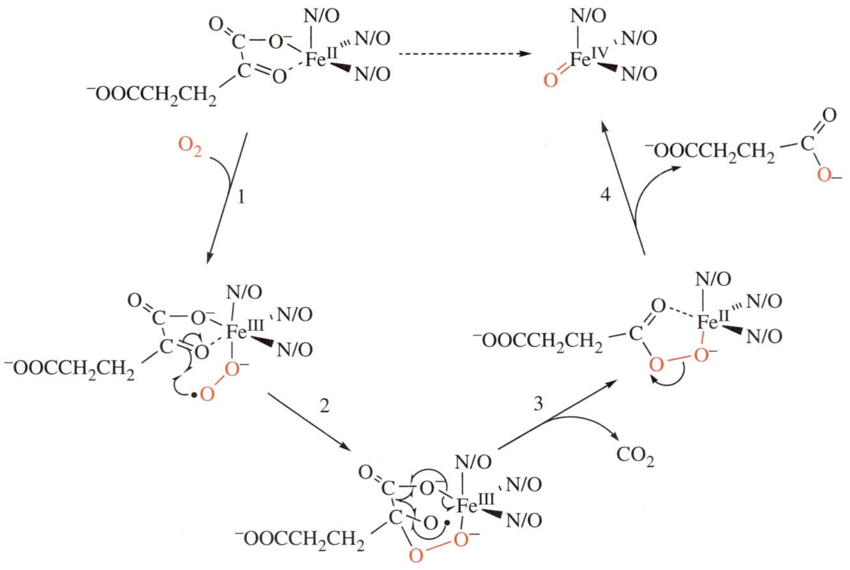

Fig. 17-15. α-Ketoglutarate is the reductant with O_2 in the generation of a mononuclear ferryl oxy species. The hypothetical mechanism shown accounts for the role of α-ketoglutarate as the source of reducing electrons and for its transformation into succinate and CO_2. It also accounts for the incorporation of an oxygen atom from O_2 into succinate and into the ferryl oxy species.

Deacetoxycephalosporin C Synthase

Deacetoxycephalosporin C synthase catalyzes two steps in antibiotic biosynthesis, the transformation in eq. 17-7 of isopenicillin N to deacetoxycephalosporin and the further hydroxylation of the methyl group in this product to deacetoxycephalosporin C (EC 1.14.20.1).

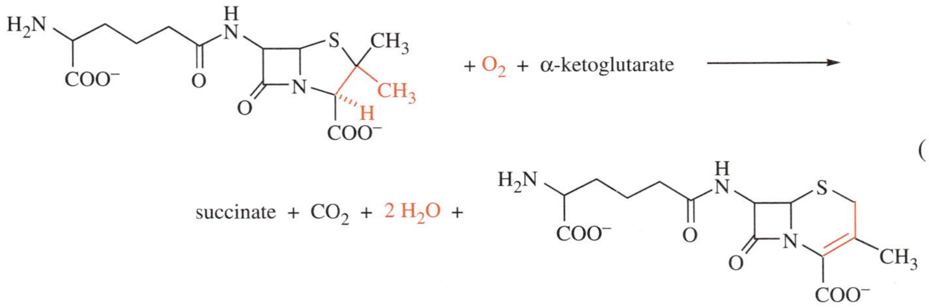

(17-7)

The structure of this enzyme in fig. 17-16 shows α-ketoglutarate bound to mononuclear iron in the 2-His 1-carboxylate facial triad.

The mechanism of ring expansion is not obvious. It can be rationalized on the basis that the ferryl oxy species of iron is generated as in fig. 17-15 and serves as a hydrogen atom abstracting species. Then a mechanism such as in fig. 17-17 can account for ring expansion. The process begins in step 1 with hydrogen abstraction by the ferryl oxy species to generate a substrate radical adjacent to the carboxyl group and ferric hydroxide. Internal fragmentation of the radical generates the double bond and a thiyl radical in step 2. The thiyl radical can abstract a hydrogen atom from one of the methyl groups to generate the allylic radical in step 3. Electron transfer from the allylic radical to ferric iron in step 4 generates the allylic carbenium ion and returns iron to its starting oxidation state. The allylic

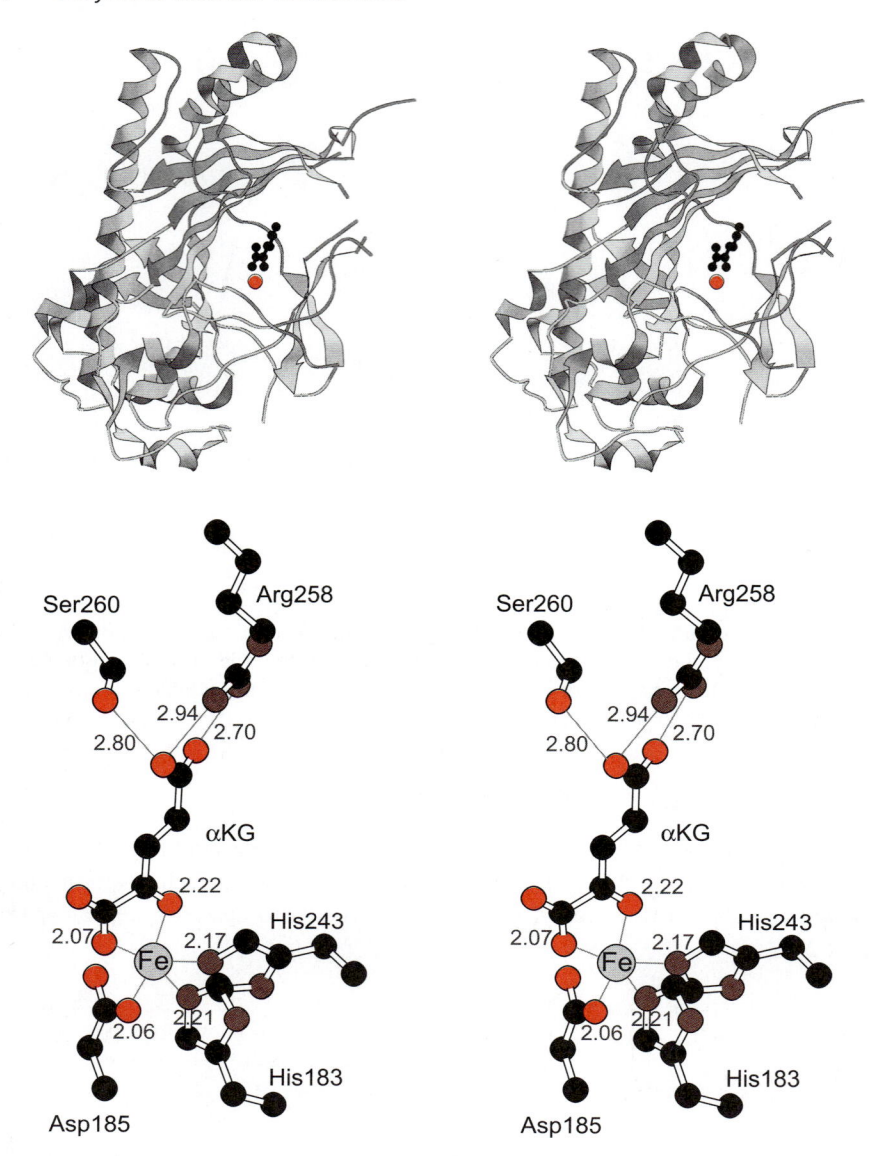

Fig. 17-16. The x-ray crystallographic structure of deacetoxycephalosporin C synthase from *Streptomyces clavuligerus* is shown with iron and α-ketoglutarate bound to the active site (1.50-Å resolution; PDB 1RXG; Valegard et al., 1998). The structure shows iron bound to the 2-histidine-1-carboxylate facial triad, composed of His183, His243, and Asp185, and to α-ketoglutarate.

carbenium ion alkylates the thiol group with release of a proton in step 5 to close the ring and complete its expansion.

Deacetoxycephalosporin C synthase also catalyzes the subsequent hydroxylation of the methyl group in deacetoxycephalosporin to form deacetoxylcephalosporin C. This second reaction is also α-ketoglutarate-dependent. It presumably proceeds through the initial α-ketogluratare/oxygen-dependent transformation of the ferrous iron center to the ferryl oxy species (Fe^{IV}=O). Oxygen insertion into a C—H bond of the methyl group presumably follows the course outlined in fig. 17-18, hydrogen abstraction to form the allylic radical and Fe^{3+}—OH, followed quickly by oxygenation to capture the radical as its alcohol.

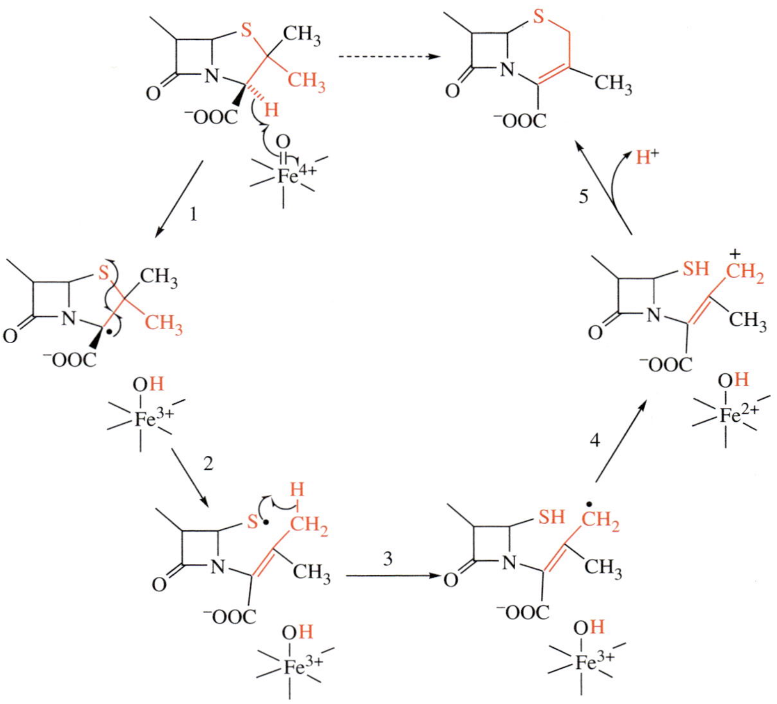

Fig. 17-17. A mechanism for ring expansion in the action of deacetoxycephalosporin C synthase.

Dopamine β-Monooxygenase

Reaction and Molecular Properties

The transformation of dopamine into norepinephrine by O_2-dependent hydroxylation is catalyzed by dopamine β-monooxygenase (DβM; EC 1.14.17.1) according to eq. 17-8. The reducing equivalents can be supplied by dihydroascorbate, which appears to be the electron donor in vivo (Freeman et al., 1993).

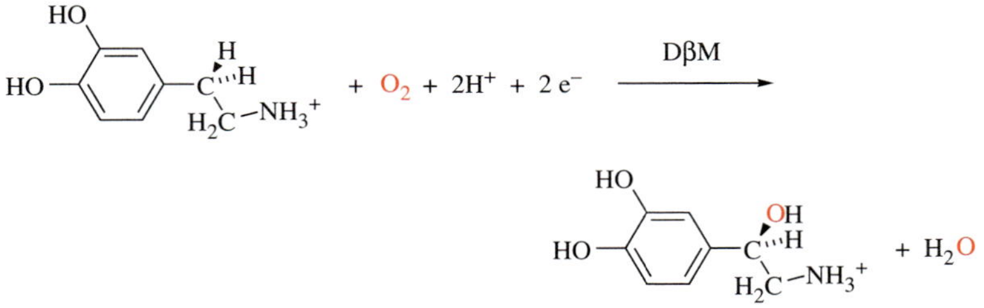

(17-8)

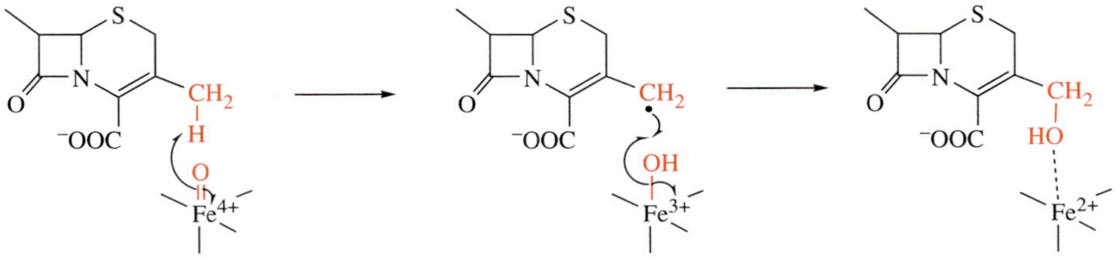

Fig. 17-18. A mechanism for oxygenation in the action of deacetoxycephalosporin C synthase.

DβM is a copper monooxygenase with two copper ions per subunit that does not appear to require any other cofactor for activity. Dopamine and norepinephrine are neurotransmitters. DβM catalyzes the β-oxygenation of a variety of phenethylamines in addition to dopamine, including phenethylamine itself, tyramine, and phenyl-substituted phenethylamines.

DβM is a tetrameric glycoprotein composed of 75-kDa subunits associated with chromaffin granules. It has not been successfully crystallized for x-ray crystallographic analysis. X-ray absorption spectroscopy and spin-echo EPR spectroscopy indicates the copper is ligated to nitrogen ligands (Blumberg et al., 1989; McCracken et al., 1988). The absence of any evidence of spin coupling between the two copper sites is taken as evidence for two mononuclear copper sites rather than a binuclear copper site (Blackburn et al., 1988; Scott et al., 1988; Brenner et al., 1989). Amino acid sequence analysis reveals the presence of histidine motifs in the primary structure. Current interpretations based on spectroscopic evidence and in conjunction with the structure of the homologous copper enzyme peptidyl-glycine-α-hydroxylating monooxygenase implicate His255, His256, His 326, and water as ligands to Cu_A and HisHis405, His407, Met480, and water to Cu_B. Cu_B is thought to be the oxygenation site for substrates, and Cu_A is thought to mediate electron transfer (Evans et al., 2003).

Kinetics and Isotope Effects

All kinetic information indicates that the reaction cycle begins with both copper ions reduced as Cu(I). The substrates bind in preferential order, dopamine followed by O_2, and oxygenation leads to water and a complex of Cu_B(II) with norepinephrine (Ahn and Klinman, 1983). Reduction of Cu_A(II) and Cu_B(II) in this complex by ascorbate is coupled with the release of norepinephrine. Cu(II) is readily observed by its characteristic EPR spectrum in the oxidized enzyme, and its reduction to Cu(I) can be observed by EPR, but putative oxygenated intermediates elude detection as transient species.

Reactions of β-deuterated substrates proceed with substantial deuterium kinetic isotope effects, typically 11 for $^Dk_{cat}/K_m$ but ranging up to 18 in the reaction of the very poor substrate p-trifluoromethylphenyl[2,2-$^2H_2$]ethylamine (Ahn and Klinman, 1983; Miller and Klinman, 1983). Hydrogen transfer proceeds with quantum mechanical tunneling (Francisco et al., 2002). Oxygenation by $^{18}O_2$ proceeds with a kinetic isotope effect ranging from 1.0028 on $^{18}k_{cat}/K_m$ for a good substrate to 1.0022 for a poor substrate (Tan et al., 1994). The observation of substantial deuterium and oxygen-18 kinetic isotope effects implicates both oxygenation and C—H bond cleavage in rate limitation. This can arise either from a concerted mechanism of C—H cleavage and oxygenation or from stepwise mechanism in which the steps are tightly kinetically coupled.

Reaction Mechanism

The chemical course of reactions of substrate analogs and suicide inactivators of DβM are consistent with the abstraction of a β-hydrogen from a substrate or analog by an oxo species to generate a benzylic radical (Bossard and Klinman, 1986; Fitzpatric et al., 1985, 1986; Fitzpatrick and Villafranca, 1987). Benzylic radicals are implicated as intermediates by the results of product analyses. All hypothetical mechanisms under consideration include abstraction of a hydrogen atom by an oxy species as a compulsory step in the oxygenation mechanism.

A search for transient copper-oxy intermediates in rapid mix-freeze-quench experiments failed to give any evidence for such a species (Evans et al., 2003). It was concluded that oxygenation and hydrogen abstraction were tightly kinetically coupled, so that such an intermediate could not accumulate to detectable levels or that such an intermediate was present but was undetectable by EPR because of internal spin coupling. All evidence indicated that the oxygenation of copper and subsequently of dopamine did not follow the same course as in cytochrome P450, and it could not follow the course of methane monooxygenase because the two copper ions are not in a binuclear complex.

A working hypothesis for the mechanism is shown in fig. 17-19 (Evans et al., 2003). The reaction cycle begins with both copper ions reduced. Dopamine binds near Cu_B and O_2 binds to Cu_B to form a $Cu_B(II)$-superoxide complex, which is postulated to abstract a benzylic hydrogen atom from the substrate to form a hydroperoxy complex and a benzylic radical. Electron transfer from Cu_A coupled with elimination of water leads to a Cu_B^{II}—O · complex that undergoes coupling with the radical to form the product ligated to Cu_B^{II}. Reduction and proton transfer release the product and regenerate the reduced enzyme for the next cycle. This mechanism can serve as a working hypothesis. Until structures of DβM become available, the structure of the homologous and mechanistically related peptidyl-glycine-α-hydroxylating monooxygenase may guide future research.

Copper-Methane Monooxygenase

Methanotrophs grown with an abundance of copper produce the membrane-bound particulate methane monooxygenase (pMMO) (reviewed in Chan et al., 2004). This enzyme has been difficult to obtain in pure form until recently with purification in the presence of

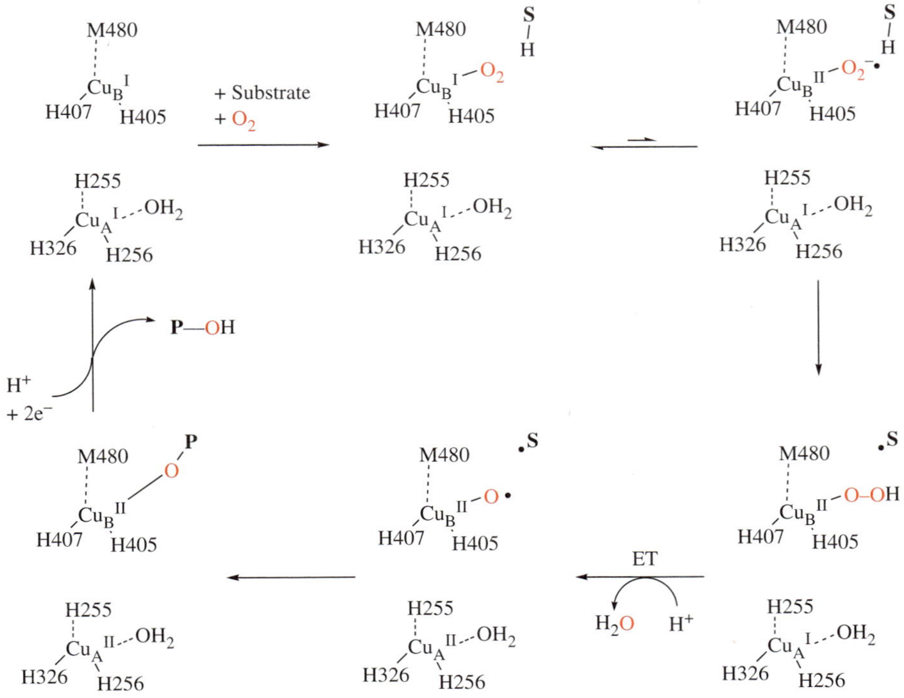

Fig. 17-19. A mechanism for oxygenation by dopamine β-monooxygenase. In this scheme, **S**–H is a phenethylamine substrate such as dopamine, and **P**—OH is the hydroxylation product, norepinephrine from dopamine. The putative CuBII—O· and **S**· intermediates have not been observed, but they are inferred from the chemistry of the action of this enzyme on substrates and substrate analogs (Evans et al., 2003).

controlled concentrations of the detergent dodecyl-β-D-maltoside. pMMO is a copper monooxygenase, and its only relation to the soluble iron-methane monooxygenase (sMMO) is that both catalyze the NADH-dependent oxygenation of methane to methanol. Like sMMO, pMMO can be purified from either *Methylosinus trichosporium* or *Methylococcus capsulatus* (Bath) when the cells are grown in a copper-rich medium. These organisms are also excellent sources of sMMO when grown in copper-poor media.

pMMO is composed of a 45-kDa α subunit, a 27-kDa β subunit, and a 23-kDa γ subunit. The subunits are separately encoded in a gene cluster (Semura et al., 1995). The amino acid sequences include membrane-spanning domains and water-exposed domains. The domain localizations inferred from sequence information are consistent with the results of partial proteolysis experiments to identify water-exposed domains. Each αβγ heterotrimer contains up to 15 copper ions, only a few of which are in Cu(II) in the protein purified in the atmosphere, the rest being Cu(I). Spectroscopic and biochemical results indicate that the copper exists in clusters of three and are of two types, two catalytic clusters (C-clusters), and three electron transfer clusters (E-clusters). The E-clusters reside in the water-soluble domains and the C-clusters in the membrane-spanning domains.

pMMO is far less promiscuous than sMMO in oxygenating substrates. It will accept small linear alkanes no larger than five carbons. The oxygenation site seems to be small and excludes larger and branched molecules. There is no evidence for the participation of radicals or any other intermediates species in the mechanism. Oxygenation of chiral centers proceeds with overall retention of configuration, consistent with a concerted oxene-insertion mechanism, unlike the racemization in reactions of sMMO indicative of discrete radical intermediates.

The reaction scheme in fig. 17-20A is under consideration (Chan et al., 2004). Electron transfer from NADH to the C-clusters in the membrane-spanning domains is mediated by the E-clusters, placing the three coppers in the C-clusters (and E-clusters) in the Cu(I)-state. Reaction of O_2 with the C-clusters transforms them into identical mixed valence μ-dioxo-$Cu^{II/III}Cu^{II}$ states. Reaction with methane at one C-cluster leads directly to methanol and a reduced C-cluster. Electron transfer to the other C-cluster transforms both C-clusters to the all-Cu(II) state with release of water. Finally, electron transfer mediated by the E-clusters returns the C-clusters to the all-Cu(I) state for the next catalytic cycle. The oxygenation step is postulated to proceed by an oxene-insertion mechanism as illustrated in fig. 17-20B. Note that this process must take place with retention of stereochemical configuration at carbon, and this is in accord with experimental observations in the case of pMMO (Yu et al., 2003).

The insertion of oxygen into C—H without racemization implies a concerted mechanism. The oxene insertion mechanism in fig. 17-20B has been proposed. In general, the absence of information implicating intermediates is taken as evidence of concertedness. In this case, stereochemical retention provides a particularly powerful case for a concerted reaction.

Nitric Oxide Synthase

Reaction and Molecular Properties

Nitric oxide (N=Ö·) is a radical (hereafter abbreviated as NO) that serves as a signaling molecule in a variety of physiological processes, including the regulation of vascular tone, vasodilation, neuronal transmission, and inhibition of platelet aggregation (Marletta, 1993). Nitric oxide synthases (NOS) catalyze the NADPH-dependent reaction of arginine with O_2 to form citrulline and NO, as illustrated in scheme 17-10.

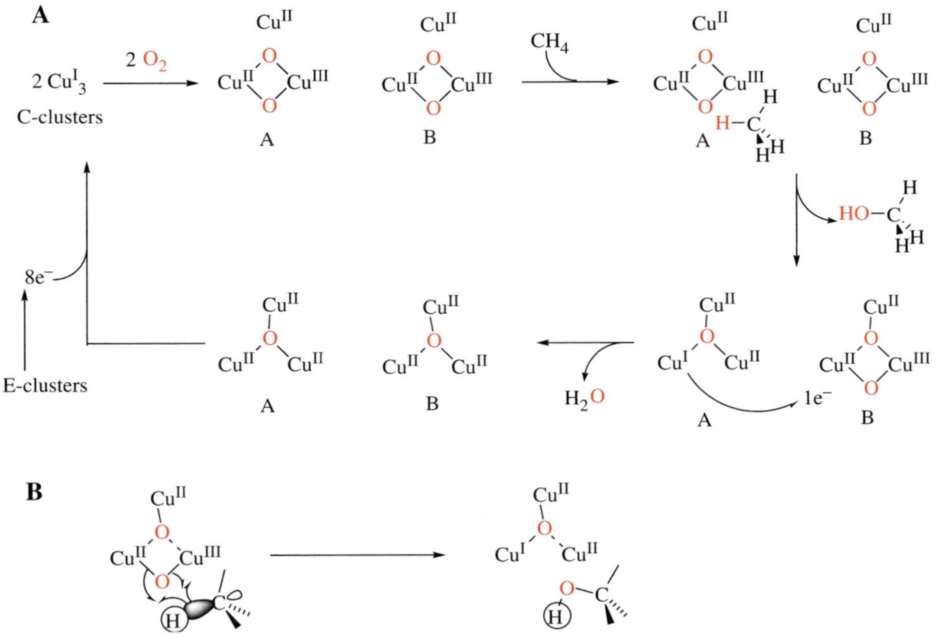

Fig. 17-20. A mechanism for a catalytic cycle of methane oxygenation by particulate methane monooxygenase (pMMO).

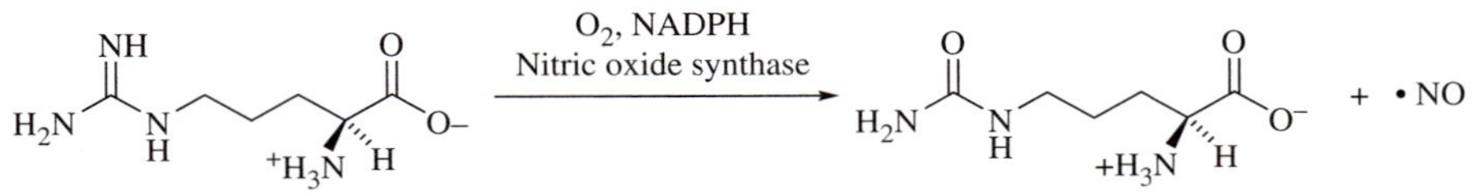

Scheme 17-10

Nitric oxide synthases (EC 1.14.13.39) are large, dimeric, heme-flavoproteins ranging in size from 130 to 150 kDa. Each subunit contains one heme, one FMN, and one FAD, and the activity is stimulated by tetrahydrobiopterin (BH$_4$). NOS exists as isoforms in various tissues, all of which appear to function by the same chemical mechanism but are subject to tissue-specific regulation (Roman et al., 2002).

Amino acid sequence information from cDNA sequences revealed the presence of domains typical of heme proteins, flavoproteins, and calmodulin (CaM) binding proteins. The heme, FMN, and FAD participate in the chemistry of NO production (Marletta, 1993; McMillan et al., 1992; Stuehr and Ikeda-Saito, 1992; White and Marletta, 1992). The calcium binding protein CaM regulates activity (Roman et al., 2002). Other regulatory mechanisms, including phosphorylation/dephosphorylation and protein/protein interactions are reported for the isoforms of NOS (Roman et al., 2002). Most or all regulation appears directed toward controlling the rate of electron transfer from NADPH through the flavin cofactors to heme (Roman et al., 2002).

Structure

Many structures of NOS isoforms and of derived domains are available (Crane et al., 1998; Raman et al., 1998; Fischmann et al., 1999; Li et al., 1999). Figure 17-21 shows the structure of one isoform. The heme and reductase domains are distinct, as is the CaM domain. Although the heme functions analogously to cytochrome P450, and heme is ligated to a cysteine residue, there is little sequence identity between the heme domain and P450s.

740 Enzymatic Reaction Mechanisms

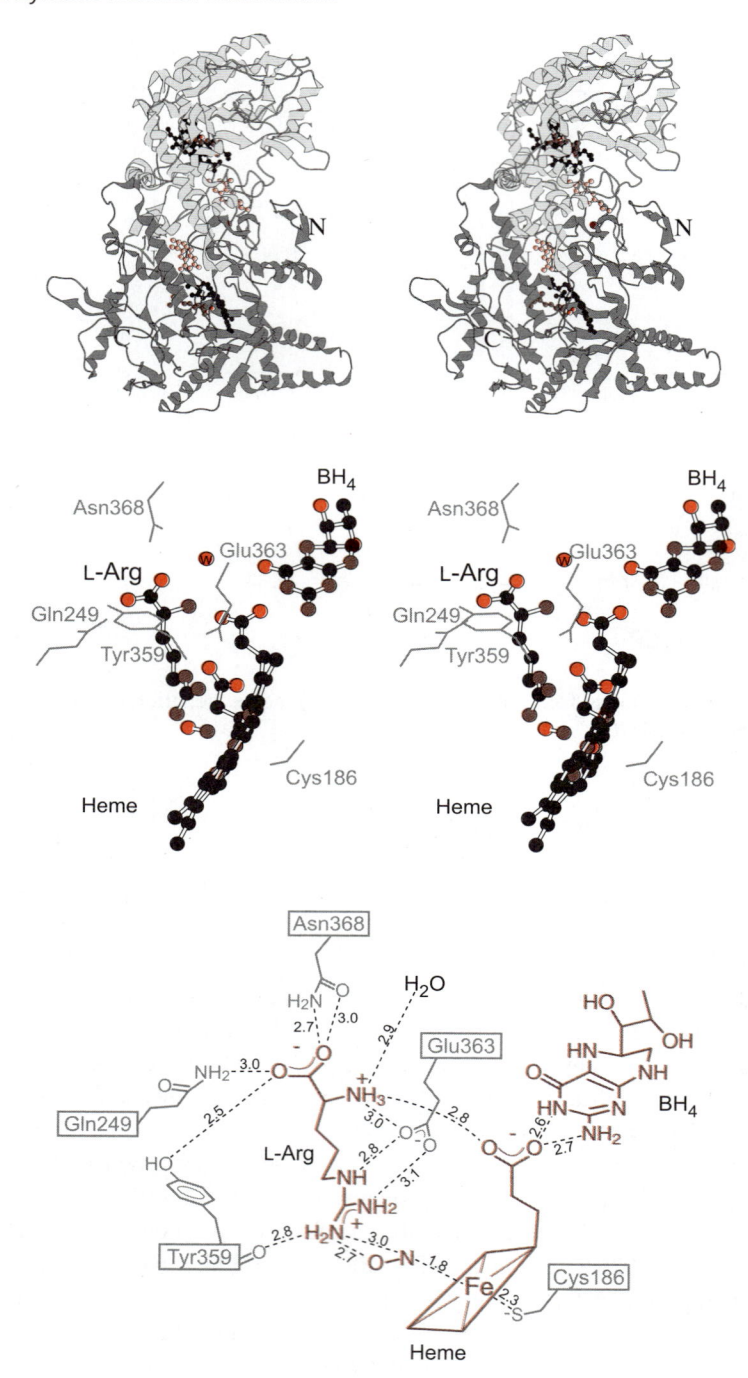

Fig. 17-21. The x-ray crystallographic structure reveals the bovine endothelial nitric oxide synthase heme domain in a complex with heme (black), nitric oxide (red), zinc (brown), arginine (brown), and tetrahydrobiopterin (pink) (2.3-Å resolution; PDB 1FOP; Li et al., 2001).

Aspects of Reaction Mechanism

The reaction of NOS requires the presence of BH_4 (tetrahydrobiopterin). This cofactor does not seem to participate in the chemistry of NO formation but seems to stabilize the enzyme in its active conformation (Marletta, 1993). NOS binds BH_4 very tightly, and purified preparations contain significant amounts of this cofactor. Activation by added BH_4

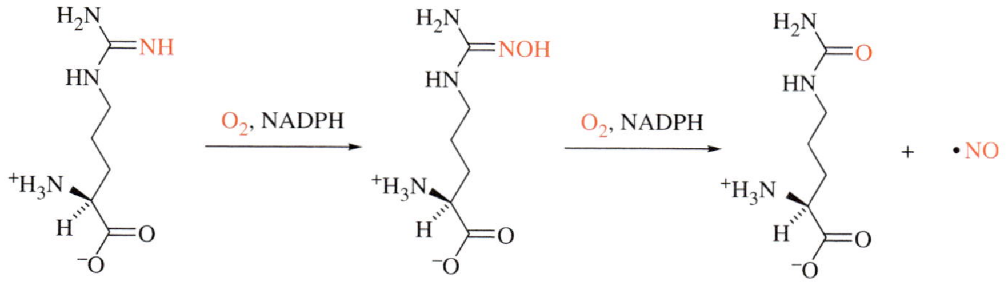

Fig. 17-22. Steps in the production of nitric oxide by action of nitric oxide synthase.

is variable. BH$_4$ bound to NOS in 1:1: stoichiometry supports maximum activity in NO production.

The chemical mechanism of NO formation by the action of NOS cannot be described in detail. A few facts are available regarding the process. An amidino nitrogen of arginine is the precursor of NO, as determined by ^{15}N-labeling, and arginine is transformed into citrulline (Iyengar et al., 1987). The origin of oxygen in NO and the ureido group of citrulline is O$_2$ itself, as determined by ^{18}O-labeling (Kwan et al., 1990; Leone et al., 1991). N^ω-Hydroxyarginine is efficiently converted into citrulline and NO by NOS, and the NO is specifically derived from the NHOH group, as determined by ^{15}N labeling (Pufahl et al., 1992; Stuehr et al., 1991). The overall reaction proceeds through the initial oxygenation of arginine followed by further oxidation and cleavage by an unknown mechanism into citrulline and NO.

The known transformations are depicted structurally in fig. 17-22. The reaction begins with the oxygenation of arginine to N^ω-hydroxyarginine in a P450-type process. Oxygenation requires reducing electrons, which are supplied by NADPH. The cytochrome P450 reductase-like domain with FMN and FAD as cofactors carries out the oxidation of NADPH, and the flavin coenzymes ferry the electrons to heme in the oxygenation domain. Further oxygenation of N^ω-hydroxyarginine produces NO and citrulline. Much remains to be learned about this interesting and important reaction.

Dioxygenases

Intradiol Dioxygenases

Protocatechuate 3,4-Dioxygenase

Protocatechuate 3,4-dioxygenase (EC 1.13.11.3) is an enzyme that carries out intradiol cleavage according to eq. 17-9.

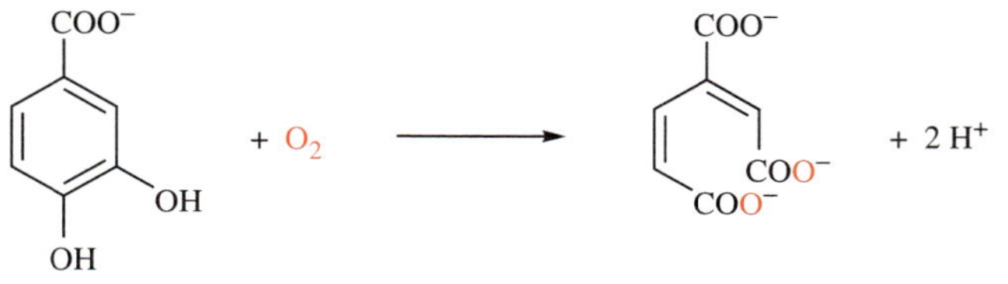

$$\text{protocatechuate} + O_2 \longrightarrow \text{product} + 2 H^+ \tag{17-9}$$

The structure of protocatechuate dioxygenase in fig. 17-23 (4-17) shows the overall chain fold, the location of iron, and its ligands. Ferric iron binds the side chains of Tyr 108, Tyr 147, His 160, and His 162 in trigonal bipyramidal geometry, with HO$^-$ as the fifth

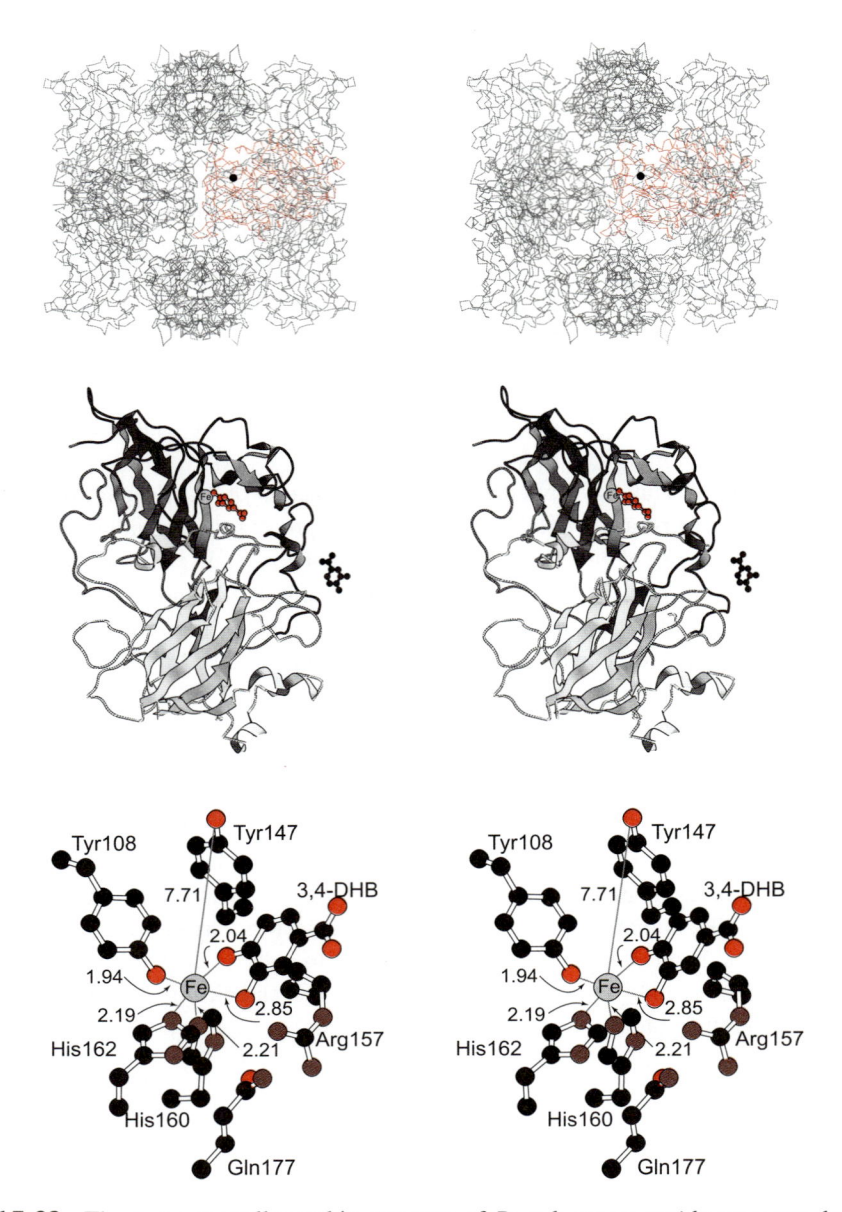

Fig. 17-23. The x-ray crystallographic structure of *Pseudomonas putida* protocatechuate 3, 4-dioxygenase shows iron coordination with 3,4-dihydroxybenzoate at the active site (2.20-Å resolution; PDB 3PCA; Orville et al., 1997). One αβ-heterodimeric unit of the large aggregate (αβ)$_{12}$ complex shown in the stereoview (top) is highlighted in red. The black dot is the site of mononuclear iron bound at the αβ-subunit interface. A stereoimage of one heterodimeric unit is shown in the center. The substrate, 3,4-dihydroxybenzoate (red, DHB), is coordinated to iron (gray Fe), as shown in the ball-and-stick model. A second molecule (black) of DHB was observed bound between the heterodimers under the high DHB concentrations used in this experiment. At the bottom is a stereoview of the active site with DHB bound to iron. Notice that Tyr147 has been displaced as an iron ligand by DHB. In the free enzyme, the iron ligands are His160, His 162, Tyr 108, and Tyr 147.

ligand in an equatorial position. EPR line broadening of high spin Fe^{3+} in $H_2^{17}O$ indicates a solvent ligand (Whittaker and Lipscomb, 1984). The extended x-ray absorption fine structure (EXAFS)–derived distances of 1.9 Å for three Fe—O bonds indicates HO⁻ rather than H_2O as the fifth ligand (True et al., 1990). The EXAFS data also require phenolate ligands from the two tyrosines rather than phenolic ligands, and this is also indicated by the ligand metal charge transfer complexation lending the enzyme its color. Tyr 147 and His 162 occupy the apical positions in the resting enzyme.

The oxygenation mechanism may follow the course outlined in fig. 17-24 (4-18). When the substrate binds, the apical Tyr 147 and HO⁻ ligands dissociate and are replaced with the OH groups of the substrate in step 1. The broadening of the iron EPR signal with ^{17}O-labeling in the substrate OH groups proves ligation of the substrate to iron.

As a rule, dioxygenases with ferric iron in their resting states activate the substrate for reaction with O_2, and those with ferrous iron in their resting states activate O_2 for reaction with the substrate. The nature of activation by iron in both cases is thought to be electron transfer, from the substrate to Fe^{3+} in ferric dioxygenases and from Fe^{2+} to O_2 in ferrous dioxygenases. Protocatechuate dioxygenase is a ferric dioxygenase that can transiently oxidize the substrate by one electron to the ortho semiquinone in step 2 (fig. 17-24).

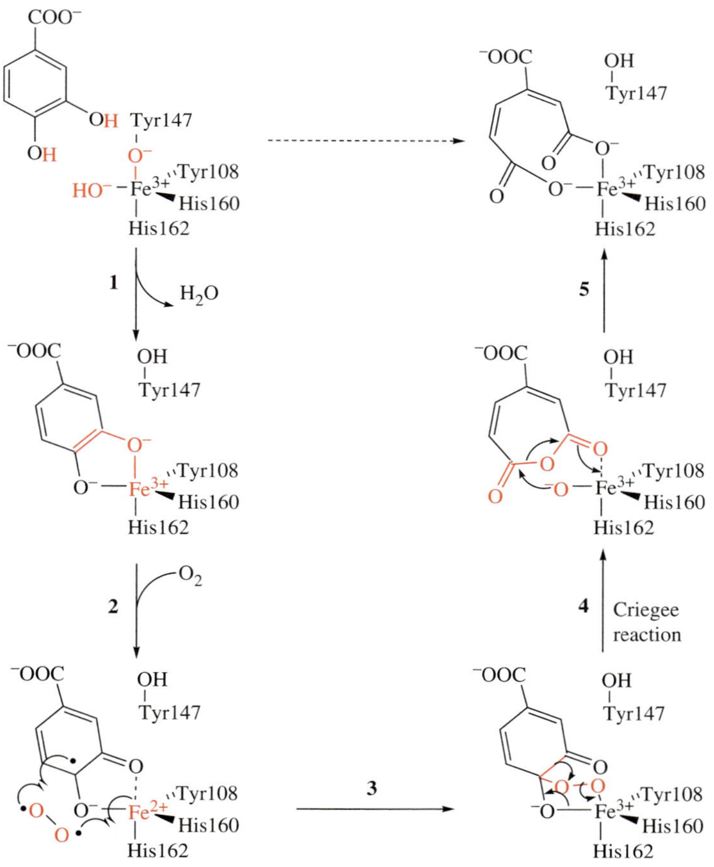

Fig. 17-24. The hypothetical mechanism for intradiol oxygenation by protocatechuate 3, 4-dioxyenase proceeds counterclockwise, beginning with step 1, in which the substrate displaces Tyr147 as a ligand to mononuclear ferric iron. Biochemical and spectroscopic evidence supports this step, as does the structure in fig. 17-23. The hypothetical steps 2 through 5 are reasonable and consistent with the oxygen labeling pattern.

744 Enzymatic Reaction Mechanisms

Being paramagnetic, the semiquinone reacts by radical coupling (electron pairing) with an unpaired electron of oxygen in step 3. At the same time, Fe^{2+} can donate an electron to pair with the other unpaired electron of oxygen, leading to the peroxy bridge between the substrate and Fe^{3+}. A Criegee-type rearrangement in step 4 leads to ring cleavage and anhydride formation, with cleavage of the peroxy group and creation of a new HO^- ligand to Fe^{3+}. Nucleophilic attack by this hydroxide group on the anhydride in step 5 cleaves it to the product.

The mechanism in fig. 17-24 (4-18) accounts for the oxygen labeling pattern and is compatible with all available facts, including that iron is in its high spin ferric state at the start and that substrate binding precedes O_2 binding (Que and Ho, 1996). Nonenzymatic chemical modeling of intradiol cleavage provides essential confirmation and further support for the mechanism (Que and Ho, 1996). In particular, $^{18}O_2$-labeling studies in the chemical models showed that a significant loss of ^{18}O-incorporation through solvent exchange supported the intermediate formation of the Fe^{3+}—OH/anhydride intermediate in step 4 of the mechanism in fig. 17-24 and was inconsistent with the intermediate formation of an oxetane adduct by O_2.

Extradiol Dioxygenases

These enzymes cleave aromatic rings adjacent to two phenolic groups. An example of such a cleavage and a reasonable mechanism for this process is depicted in fig. 17-25.

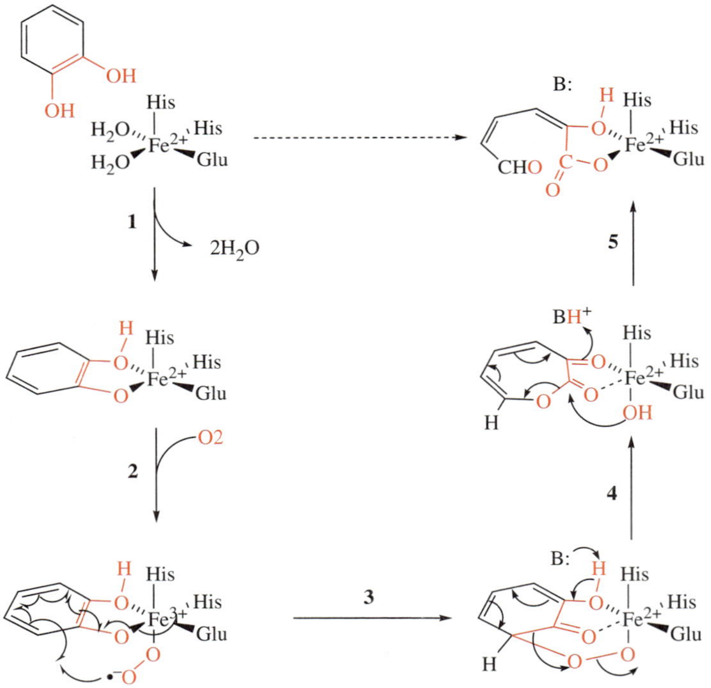

Fig. 17-25. Extradiol oxygenation starts with ferrous iron in a 2-His 1-carboxylate facial triad. The substrate binds with coordination to iron in the first step. Unlike intradiol cleavage in fig. 17-24, the O_2 reacts first with ferrous iron to from a ferric superoxide intermediate, which undergoes radical addition to the aromatic ring of the substrate in step 3. Fragmentation of the peroxy adduct inserts an oxygen atom into the aromatic ring in step 4, and hydrolysis by the resulting axial OH group produces the product. The red-coded atoms represent reacting species, not a labeling pattern. Oxygen labeling experiments show that O_2 donates one atom to the aldehydic and one to the carboxylate group of the product, in accord with the mechanism shown.

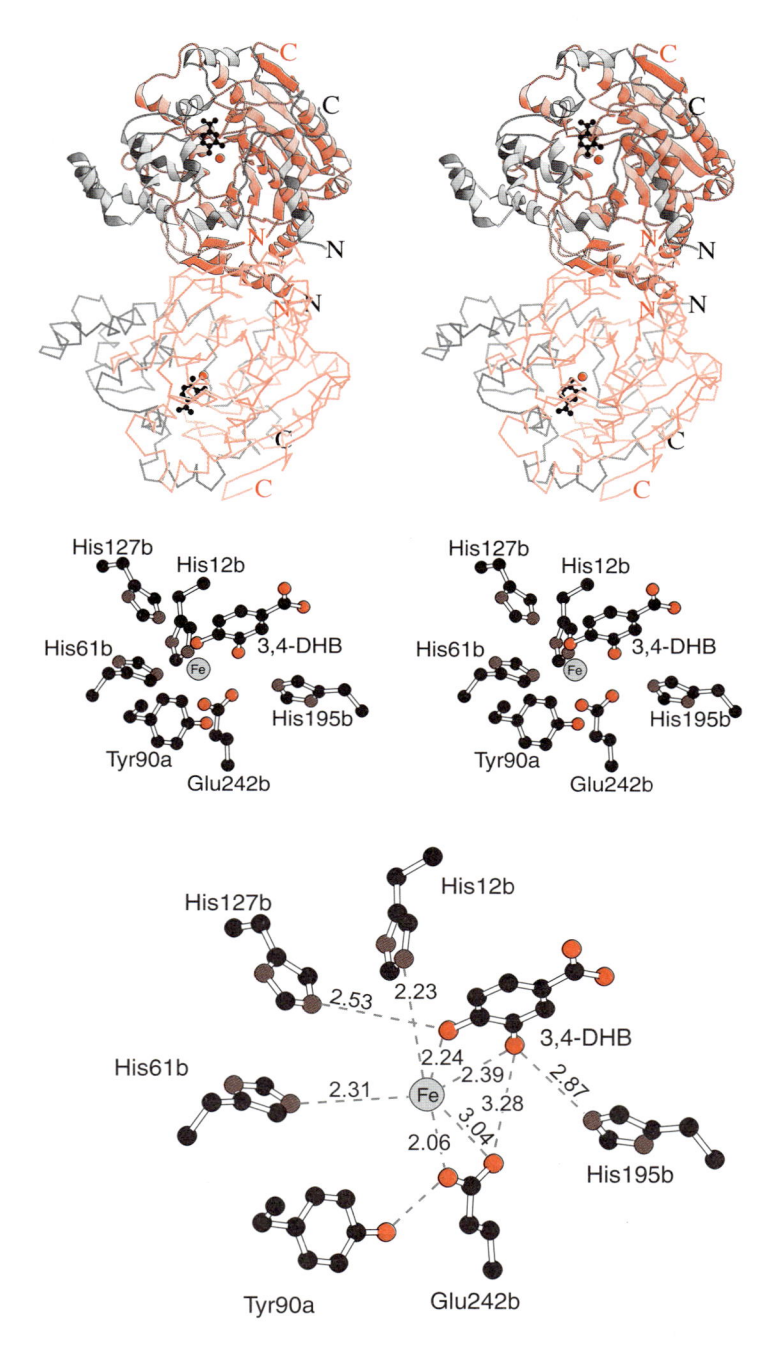

Fig. 17-26. The structure shows protocatechuate 4,5-dioxygenase, an extradiol-cleaving enzyme from *Sphingomonas paucimobilis*, with protocatechuate bound at the active site (2.20-Å resolution; PDB 1B4U; Sugimoto et al., 1999). The 2-His 1-carboxylate facial triad consists of His12b, His61b, and Glu242b. The iron is also ligated by 3,4-dihdroxybenzoate (3,4-DHB). The stereoview (top) shows one heterodimeric unit (αβ) in a ribbon diagram and the other heterodimer of the dimer of dimers (αβ)$_2$ as Cα traces, with iron as a red ball and 3,4-DHB as a black ball-and-stick model. In the center is a stereoview of the active site as a ball-and-stick model. At the bottom is a two-dimensional representation of the active site, showing the contact distances (in angstroms) between key atoms.

The ferrous iron ligands consist of two histidine residues and a glutamate, the other coordination positions being occupied by water. This coordination pattern emerges as a widespread motif for mononuclear iron oxygenases (Bugg, 2001), as we have seen in earlier sections of this chapter and in chapter 4; the pattern is known as the 2-His 1-carboxylate facial triad (Hegg and Que, 1997). Displacement of two water ligands by the substrate phenolic groups initiates the catalytic process in step 1. As a ferrous dioxygenase, the enzyme activates O_2 by binding it with electron transfer to form the ferric superoxide complex in step 2. Superoxide radical attack on the aromatic ring of the substrate adjacent to the diol is accompanied by electron transfer around the ring and re-reduction of Fe^{3+} to Fe^{2+} in step 3. Base abstraction of the phenolic proton leads to carbonyl formation at the distal phenolic group and concomitant ring expansion with oxygen incorporation into the ring, as well as peroxide cleavage leaving a hydroxide ligand on iron in step 4. Nucleophilic attack of the iron hydroxide ligand on the carbonyl intermediate in step 5 cleaves it to the aldehyde/carboxylic acid product.

Extensive spectroscopic studies of 2,3-dihydroxybiphenyl 1,2-dioxygenase and catechol 2,3-dioxygenase support the coordination scheme in fig. 17-25, which serves as the starting point for the mechanism (Que and Ho, 1996). The ^{18}O labeling pattern also supports the mechanism. Individual steps of this mechanism have not been observed and characterized. However, evidence supporting the lactone intermediate has been put forward in the case of 3-(2′,3′-dihydroxyphenyl)propionate 1′,2′-dioxyenase (Sanvoisin et al., 1995). This reaction proceeds with partial incorporation of label from $H_2^{18}O$ and partial loss of label from $^{18}O_2$, which supports the intermediate formation of the potentially exchangeable Fe^{2+}—OH species. The enzyme catalyzes the hydrolysis of an analog of the putative lactone intermediate formed in step 4 of the mechanism in fig. 17-25.

Figure 17-26 depicts the structure of an extradiol cleaving dioxygenase, protocatechuate 4,5-dioxygenase (EC 1.13.11.8) with the catechol substrate bound at the active site (Sugimoto et al., 1999). The structure clearly shows the coordination of ferrous iron by the 2-His 1-carboxylate facial triad and the two phenolic groups of the substrate.

References

Ahn N and JP Klinman (1983) *Biochemistry* **22**, 3096.
Binda C, F Hubálek, M Li, DE Edmondson, and A Mattevi (2004) *FEBS Lett* **564**, 225.
Binda C, M Li, F Hubálek, N Restelli, DE Edmondson, and A Mattevi (2003) *Proc Natl Acad Sci U S A* **100**, 9750.
Blackburn NJ, M Concannon, SK Shahiyan, FE Mabbs, and D Collison (1988) *Biochemistry* **27**, 5411.
Blumberg WE, PR Desai, L Powers, JH Freedman, and JJ Villafranca (1989) *J Biol Chem* **264**, 6029.
Bossard MJ and JP Klinman (1986) *J Biol Chem* **261**, 16421.
Brazeau BJ, RN Austin, C Tarr, JT Groves, and JD Lipscomb (2001) *J Am Chem Soc* **123**, 11831.
Brenner MC, CJ Murray, and JP Klinman (1989) *Biochemistry* **28**, 4656.
Bruice TC (1984) *Israel J Chem* **24**, 54.
Bugg TDH (2001) *Curr Opin Chem Biol* **5**, 550.
Burzlaff NI, PJ Rutledge, IJ Clifton, CMH Hensgens, M Pickford, RM Adlington, PL Roach, and JE Baldwin (1999) *Nature* **401**, 721.
Chan SI, KHC Chen, SSF Yu, CL Chen, and SSJ Kuo (2004) *Biochemistry* **43**, 4421.
Crane BR, AS Arvai, DK Ghosh, C Wu, ED Getzoff, DJ Stuehr, and JA Tainer (1998) *Science* **279**, 2121.
Edwards SL, HX Nguyen, RC Hamlin, and J Kraut (1987) *Biochemistry* **26**, 1503.
Elkins JM, PJ Rutledge, NI Burzlaff, IJ Clifton, RM Adlington, PL Roach, and JE Baldwin (2003) *Org Biomol Chem* **1**, 1455.
Evans JP, K Ahn, and JP Klinman (2003) *J Biol Chem* **278**, 49691.

Fischman TO, A Hruza, XD Niu, JD Fossetta, CA Lunn, E Dolphin, AJ Prongay, P Reichert, DJ Lundell, SK Narula, and PC Weber (1999) *Nat Struct Biol* **6**, 233.

Fitzpatrick PF (2004) *Bioorganic Chem* **32**, 125.

Fitzpatrick PF, DR Flory Jr, and JJ Villafranca (1985) *Biochemistry* **24**, 2108.

Fitzpatrick PF, MR Harpel, and JJ Villafranca (1986) *Arch Biochem Biophys* **249**, 70.

Fitzpatrick PF and JJ Villafranca (1987) *Arch Biochem Biophys* **257**, 231.

Francisco WA, MJ Knapp, NJ Blackburn, and JP Klinman (2002) *J Am Chem Soc* **124**, 8194.

Freeman JC, JJ Villafranca, and DJ Merkler (1993) *J Am Chem Soc* **115**, 4923.

Frey PA (1997) *Curr Opin Chem Biol* **1**, 347.

Fülöp V, RP Phizackerley, SM Soltis, IJ Clifton, S Wakatsuki, J Erman, J Hajdu, and SL Edwards (1994) *Structure* **2**, 201.

Gouet P, HM Jouve, PA Williams, I Andersson, P Andreoletti, L Nussaume, and J Hajdu (1996) *Nat Struct Biol* **3**, 951.

Griller D and KU Ingold (1980) *Acc Chem Res* **13**, 317.

Groves JT, GA McClusky, RE White, and MJ Coon (1978) *Biochem Biophys Res Commun* **81**, 154.

Groves JT and DV Subramanian (1984) *J Am Chem Soc* **106**, 2177.

Guallar V, BF Gherman, WH Miller, SJ Lippard, and RA Friesner (2002) *J Am Chem Soc* **124**, 3377.

Hegg EL and L Que Jr (1997) *Eur J Biochem* **250**, 625.

Hersh LB and M Schuman-Jorns (1975) *J Biol Chem* **250**, 8728.

Hjelmeland LM, L Aronow, and TR Trudell (1976) *Biochem Biophys Res Commun* **76**, 541.

Huffman GW, PD Gesellchen, JR Turner, RB Rothenberger, HE Osborne, FD Miller, JL Chapman, and SW Queener (1992) *J Med Chem* **35**, 1897.

Iyengar R, DJ Stuehr, and MA Marletta (1987) *Proc Natl Acad Sci U S A* **84**, 6369.

Kadkhodayan S, ED Coulter, DM Marynyak, TA Bryson, and JH Dawson (1995) *J Biol Chem* **270**, 28042.

Kahn K, P Serfozo, and PA Tipton (1997) *J Am Chem Soc* **119**, 5435.

Kahn K and PA Tipton (1997) *Biochemistry* **36**, 4731.

Kahn K and PA Tipton (1998) *Biochemistry* **37**, 11651.

Kurtz KA, MA Rishavy, WW Cleland, and PF Fitzpatrick (2000) *J Am Chem Soc* **122**, 12896.

Kwon NS, CF Nathan, C Gilker, OW Griffith, DE Matthews, and DJ Stuehr (1990) *J Biol Chem* **265**, 13442.

Leone AM, RM Palmer, RG Knowles, PL Francis, DS Ashton, and S Moncada (1991) *J Biol Chem* **266**, 23790.

Li H, S Narasimhulu, LM Havran, JD Winkler, and TL Poulos (1995) *J Am Chem Soc* **117**, 6297.

Li H, CS Raman, CB Glaser, E Blasko, TA Young, JF Parkinson, M Whitlow, and TL Poulos (1999) *J Biol Chem* **274**, 21276.

Li H, CS Raman, P Martasek, BS Masters, and TL Poulos (2001) *Biochemistry* **40**, 5399.

Liu KE, CC Johnson, M Newcomb, and SJ Lippard (1993) *J Am Chem Soc* **115**, 939.

Marletta MA (1993) *J Biol Chem* **268**, 12231.

Mattevi A, MA Vanoni, F Todone, M Rizzi, A Teplyakov, A Coda, M Bolognesi, and B Curti (1996) *Proc Natl Acad Sci U S A* **93**, 7496.

McCracken J, PR Desai, NJ Papadopoulos, JJ Villafranca, and J Peisach (1988) *Biochemistry* **27**, 4133.

McMillen K, DS Bredt, DJ Hirsch, SH Snyder, JE Clark, and BSS Masters (1992) *Proc Natl Acad Sci U S A* **89**, 11141.

Miller JR and DE Edmondson (1999) *Biochemistry* **38**, 13670.

Miller SM and JP Klinman (1983) *Biochemistry* **22**, 3091.

Nesheim JC and JD Lipscomb (1996) *Biochemistry* **35**, 10240.

Ortiz de Montellano PR (1986) *Cytochrome P450. Structure, Function, and Biochemistry* Plenum: New York, p 217.

Orville AM, JD Lipscomb, and DH Ohlendorf (1997) *Biochemistry* **36**, 10052.

Pollegioni L, W Blodig, and S Ghisla (1997) *J Biol Chem* **272**, 4924.

Poulos TL, BC Finzel, IC Gunsalus, GC Wagner, and J Kraut (1985) *J Biol Chem* **260**, 16122.

Poulos TL, BC Finzel, and AJ Howard (1986) *Biochemistry* **25**, 5314.

Poulos TL, BC Finzel, and AJ Howard (1987) *J Mol Biol* **195**, 687.

Priestly ND, HG Floss, WA Froland, JD Lipscomb, PG Williams, and H Morimoto (1992) *J Am Chem Soc* **114**, 7561.

Pufahl RA, PG Nanjappan, RW Woodard, and MA Marletta (1992) *Biochemistry* **31**, 6822.
Que L Jr, and RYN Ho (1996) *Chem Rev* **96**, 2607.
Raman CS, H Li, P Martásek, V Kral, BSS Masters, and TL Poulos (1998) *Cell* **95**, 939.
Roach PL, IJ Clifton, V Fulop, K Harlos, GJ Barton, J Hajdu, I Andersson, CJ Schofield, and JE Baldwin (1995) *Nature* **375**, 700.
Roach PL, IJ Clifton, CMS Hensgens, N Shibata, CJ Schofield, J Hajdu, and JE Baldwin (1997) *Nature* **387**, 827.
Roman LJ, P Martásek, and BSS Masters (2002) *Chem Rev* **102**, 1179.
Ruzicka F, D-S Huang, MI Donnelly, and PA Frey (1990) *Biochemistry* **29**, 1696.
Ryle MJ and RP Hausinger (2002) *Curr Opin Chem Biol* **6**, 193.
Sanvoisin J, GJ Lancley, and TDH Bugg (1995) *J Am Chem Soc* **117**, 7836.
Schlichting I, J Berendzen, K Chu, AM Stock, SA Maves, DE Benson, RM Sweet, D Ringe, GA Petsko, SG Sligar (2000) *Science* **287**,1615.
Scott RA, RJ Sullivan, WE Dewolf Jr, RE Dolle, and LI Kruse (1988) *Biochemistry* **27**, 5411.
Semrau JD, A Chistoserdov, J Lebron, A Costello, J Davaginino, E Kenna, AJ Holmes, R Finch, JC Murell, and ME Lidstrom (1995) *J Bacteriol* **177**, 3071.
Silverman RB (1995) *Acc Chem Res* **28**, 335.
Silverman RB (2004) *The Organic Chemistry of Drug Design and Drug Action*, 2nd ed. Academic Press: London, p 295.
Sono M, MP Roach, ED Coulter, and JH Dawson (1996) *Chem Rev* **96**, 2841.
Stuehr DJ and M Ikeda-Saito (1992) *J Biol Chem* **267**, 20547.
Stuehr DJ, NS Kwon, CF Nathan, OW Griffith, PL Feldman, and J Wiseman (1991) *J Biol Chem* **266**, 6259.
Sugimoto K, T Senda, H Aoshima, E Masai, M Fukuda, and Y Mitsui (1999) *Structure* **7**, 953.
Tian G, JA Berry, and JP Klinman (1994) *Biochemistry* **33**, 226.
True AE, AM Orville, LL Pearce, JD Lipscomb, and L Que Jr (1990) *Biochemistry* **29**, 10847.
Umhau S, L Pollegioni, G Molla, K Diederichs, W Welte, MS Pilone, and S Ghisla (2000) *Proc Natl Acad Sci U S A* **97**, 12463.
Valegard K, AC van Scheltinga, MD Lloyd, T Hara, S Ramaswamy, A Perrakis, A Thompson HJ Lee, JE Baldwin, CJ Schofield, J Hajdu, and I Andersson (1998) *Nature* **394**, 805.
Valentine AM, M-H Letadic-biadatti, PH Toy, M Newcomb, and SJ Lippard (1999) *J Biol Chem* **274**, 10771.
Valentine AM, B Wilkinson, KE Liu, S Komar-Paniucci, ND Priestly, PG Williams, H Morimoto HG Floss, and SJ Lippard (1997) *J Am Chem Soc* **119**, 1818.
Waller BJ and JD Lipscomb (1996) *Chem Rev* **96**, 2625.
Walsh CT, A Schonbrunn, and RH Abeles (1971) *J Biol Chem* **246**, 6855.
White KA and MA Marletta (1992) *Biochemistry* **31**, 6627.
Whittaker JW and JD Lipscomb (1984) *J Biol Chem* **259**, 4487.
Whittington DA, MH Sazinsky, and SJ Lippard (2001) *J Am Chem Soc* **123**, 1794.
Yu SSF, LY Wu, KHC Chen, WL Luo, DS Huang, and SI Chan (2003) *J Biol Chem* **278**, 40658.

18
Complex Enzymes

Most enzymes discussed in the preceding chapters consist of single proteins that catalyze single biochemical reactions. Many of them contain one type of polypeptide chain, although most exist as oligomers of a polypeptide, and some consist of different polypeptides that cooperate to catalyze one reaction. Increasing attention is being focused on enzymes that catalyze more complex processes and are composed of more than one enzyme or enzymatic domain, each of which catalyzes or facilitates a specific biochemical process. These complex enzymes are the subjects of this chapter. Complex enzymes are so numerous and the processes they catalyze so complex that a complete discussion would fill a book. We therefore limit this discussion to a few examples.

The first complex enzymes to be discovered were the multienzyme complexes. They included the four terminal electron transport complexes of the respiratory chain: complex I, known as NADH dehydrogenase (formerly DPNH dehydrogenase); complex II, known as succinate dehydrogenase; complex III, known as cytochrome *c* reductase; and complex IV, known as cytochrome *c* oxidase. Other multienzyme complexes discovered at about the same time were the pyruvate dehydrogenase and α-ketoglutarate dehydrogenase complexes, the fatty acid synthase complexes, and the glycine reductase complex and the anthranilate synthase complex. Later, the multimodular polyketide synthases and nonribosomal polypeptide synthetases were characterized. The ATP synthases are multiprotein complexes that function as molecular motors in catalyzing a complex reaction, the condensation of ADP with P_i driven by proton translocation to form ATP. The ribosome catalyzes the polymerization of amino acids in defined sequences specified by the nucleotide sequences in species of mRNA, and nitrogenase catalyzes the ATP-dependent reduction of molecular nitrogen to ammonia.

Some of the actions of complex enzymes link together common biochemical reactions of the types discussed in preceding chapters. Others catalyze difficult reactions through mechanistic coupling to energy-producing processes that provide driving force for otherwise unfavorable transformations. We present examples of each type.

Multienzyme Complexes

α-Ketoacid Dehydrogenase Complexes

Catalysis by an α-ketoacid dehydrogenase complex is carried out by three physically associated enzymes, a TPP-dependent α-ketoacid dehydrogenase (E1), a dihydrolipoyl transacetylase (E2), and dihydrolipoyl dehydrogenase (E3). The complexes catalyze reactions described by eq. 18-1, where R is methyl, in the reaction of the pyruvate dehydrogenase (PDH) complex, carboxyethyl in the action of the α-ketoglutarate dehydrogenase (KGDH) complex, or a branched alkyl group in the action of the branched chain α-ketoacid dehydrogenase complex.

$$RCOCOO^- + NAD^+ + CoASH \rightleftharpoons RCOSCoA + NADH + CO_2 \quad (18\text{-}1)$$

The transformations required for the overall reaction of eq. 18-1 include decarboxylation, oxidoreduction (or dehydrogenation), and acyl group transfer. We have discussed enzymatic mechanisms for these reaction types in chapters 6, 8, and 16. Each of the component enzymes catalyzes one or two of the steps in the overall reaction.

Pyruvate Dehydrogenase Complex

Stepwise Reaction Mechanism

The reaction of the PDH complex takes place in the five distinct steps described by eqs. 18-2a to 18-2e in fig. 18-1A. E1 is pyruvate dehydrogenase, E2 is dihydrolipoyl transacetylase, and E3 is dihydrolipoyl dehydrogenase. E1 catalyzes the TPP-dependent decarboxylation of pyruvate to CO_2 and hydroxyethylidene-TPP, which is a tightly bound intermediate. E1 also catalyzes the next step, eq. 18-2b, the mechanistically coupled dehydrogenation of hydroxyethylidene-TPP and transfer of the electrons and acetyl group to the lipoyl moiety on E2. The central role of the lipoyl group covalently bonded to E2 through a lysyl-ε-amino group is made clear by its participation in the reactions of eqs. 18-2b to 18-2d. It is graphically illustrated in fig. 18-1B, which shows that the lipoyl moiety physically links the active sites of the three enzymes.

The disulfide group in the dithiolane ring chemically couples the electron transfer and acyl group transfer processes and physically links them among three active sites as illustrated in fig. 18-1. It accepts electrons and the acetyl group from E1, it transfers the acetyl group to CoASH at the active site in E2, and it relays the electrons to E3.FAD. As seen in chapter 16, the reduced E3.FAD (dihydrolipoyl dehydrogenase) does not contain $FADH_2$ but is a charge transfer complex of FAD with the thiolate group of a reduced cystine-disulfide, and we describe it in fig. 18-1 as dihydro-E3.FAD. In the last step, dihydro-E3.FAD reduces NAD^+.

Composition and Structure: E1, E2, and E3

The bacterial PDH complexes are the most thoroughly studied α-ketoacid dehydrogenase complexes. They are composed of 24 subunits of E1, 24 subunits of E2, and 12 subunits

A

$$H_3C-\underset{O}{\underset{\|}{C}}-COO^- + E1\cdot TPP + H^+ \rightleftharpoons H_3C-\underset{OH}{\underset{|}{C}}=TPP\cdot E1 + CO_2 \quad (18\text{-}2a)$$

Hydroxyethylidene-TPP

$$H_3C-\underset{OH}{\underset{|}{C}}=TPP\cdot E1 + E2\text{–Lipoyl–}S_2 \rightleftharpoons E1\cdot TPP + E2\text{–Lipoyl-(SH)–S–COCH}_3 \quad (18\text{-}2b)$$

$$E2\text{–Lipoyl-(SH)–S–COCH}_3 + CoASH \rightleftharpoons E2\text{–Lipoyl–(SH)}_2 + CH_3COSCoA \quad (18\text{-}2c)$$

$$E2\text{–Lipoyl–(SH)}_2 + E3\cdot FAD \rightleftharpoons E2\text{–Lipoyl–}S_2 + E3\cdot\text{dihydro-FAD} \quad (18\text{-}2d)$$

$$E3\cdot\text{dihydro-FAD} + NAD^+ \rightleftharpoons E3\cdot FAD + NADH + H^+ \quad (18\text{-}2e)$$

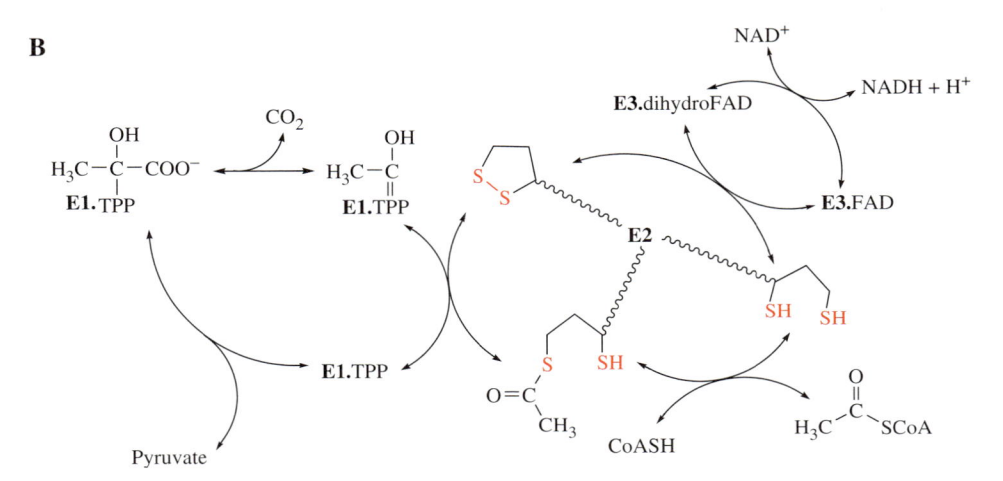

B

Fig. 18-1. Steps in the action of the pyruvate dehydrogenase (PDH) complex and the role of lipoic acid. (A) Biochemical steps in the reaction of the PDH complex. (B) The physicochemical role of lipoyl groups in coupling the actions of three enzymes in the PDH complex. Torsion about single bonds in the lipoyl-lysine conjugate (see fig. 3-15) facilitates interactions of the chemically functional portion of the lipoyl-intermediates with active sites on E1, E2, and E3, which are separated by more than 50 Å (Shepherd and Hammes, 1977). Lipoyl moieties are bonded to mobile structural domains of E2, which increase the intrinsic mobility of the lipoyl groups and facilitate their interactions with distinct active sites on E1, E2, and E3.

of E3 (Reed, 1974; Reed and Cox, 1970). E1 is a dimer composed of 102-kDa chains, E2 is composed of 24 100-kDa subunits in cubic symmetry, and E3 is a dimer of 50-kDa subunits. The subunits are organized in cubic symmetry about a central core of the 24 E2 subunits, which are assembled as trimers at the corners of the cube, as illustrated in fig. 18-2. Twelve dimers of E1 bind to the vertices of the E2-cube, and six dimers of E3 bind to the six faces (Reed, 1974; Reed and Cox, 1970; Yang et al., 1985, 1986). The overall molecular weight of 5.3×10^6 is accounted for by the sum of the molecular weights of the 60 subunits (CaJacob et al., 1985a). The mammalian PDH complex is composed of similar enzymatic components assembled around an icosahedral core.

The activities of PDH complexes are subject to metabolic regulation through modulation of the activity of E1. In the bacterial complexes, GTP is an allosteric inhibitor of E1. GTP is plentiful when ATP concentrations are high, so that the GTP concentration is a sensor for high ATP and serves as a signal of a state of sufficient energy. Flux through the

752 Enzymatic Reaction Mechanisms

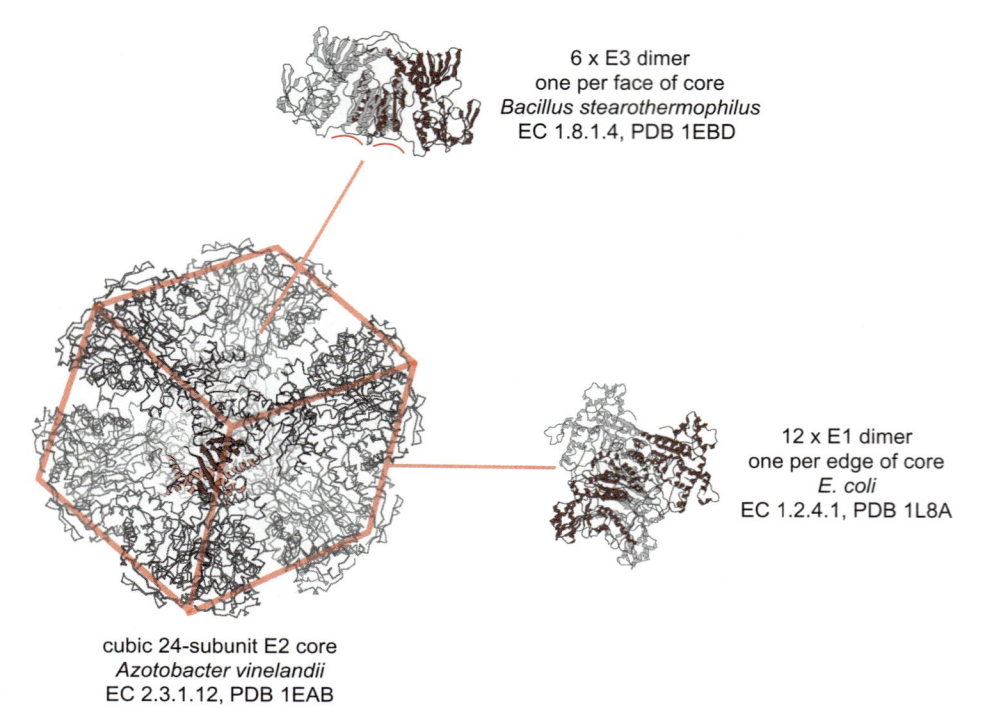

6 × E3 dimer
one per face of core
Bacillus stearothermophilus
EC 1.8.1.4, PDB 1EBD

12 × E1 dimer
one per edge of core
E. coli
EC 1.2.4.1, PDB 1L8A

cubic 24-subunit E2 core
Azotobacter vinelandii
EC 2.3.1.12, PDB 1EAB

Fig. 18-2. Structural organization of bacterial pyruvate dehydrogenase (PDH) complexes. The central core structure is E2, dihydrolipoyl transacetylase, consisting of 24 subunits arranged in cubic symmetry (Reed, 1974; Reed and Cox, 1970). The pyruvate dehydrogenase dimers (E1) bind to each of the 12 edges of the core, and the dihydrolipoyl dehydrogenase dimers (E3) bind to the six faces (along the surfaces indicated by the red marks on E3). The core E2 contains the lipoyl-bearing domains that link the actions of the three enzymes. X-ray crystallographic structures are shown for the E2-core from *Azotobacter vinelandii* (2.6-Å resolution; PDB 1EAB; Mattevi et al., 1992), E1 from *Escherichia coli* (1.85-Å resolution; PDB 1L8A; Arjunan et al., 2002), and E3 from *Bacillus stearothermophilus* (2.6-Å resolution; PDB 1EBD; Mande et al., 1996).

PDH complex generates energy through the production of NADH and acetyl coenzyme A (CoA), and inhibition by GTP is an effective control in bacteria. NADH is also an inhibitor of dihydrolipoyl dehydrogenase (see chap. 16) and may inhibit the complex when NADH is plentiful. Mammalian PDH complexes are regulated by specific protein kinase and phosphoprotein phosphatase activities associated with the E1 components. These activities regulate the activity of E1 by phosphorylation and dephosphorylation of serine residues.

The active sites of E1, E2, and E3 are separated from one another by 4 to 6 nm in the complex from *E. coli*, as determined by fluorescence energy transfer measurements (Shepherd and Hammes, 1977). The structural mobility of the lipoyl groups must be in play to allow chemical communications among the sites. The S^8 of lipoamide is separated from the E2 backbone by a maximum of 1.4 nm (see chap. 3), and torsional freedom about bonds in the lysyl and lipoyl side chains in principle allow the dithiolane ring to sweep out a sphere of about 2.8 nm in diameter. By itself, this property of the lipoyl moieties does not allow communications among active sites 4 to 6 Å apart, a problem that is overcome in two ways. Each subunit of E2 contains three similar though not identical lipoyl bearing domains, each with a single lysine residue for binding a lipoyl group (Perham, 1991). At least two lipoyl groups per E2 are potentially functional in the reaction, as shown by the fact that up to 48 [^{14}C]acetyl groups can be covalently bound to the complex from [2-^{14}C]pyruvate in the absence of CoA (Collins and Reed, 1977; Speckhard et al., 1977), or from [^{14}C]acetyl CoA in the presence of NADH (Frey et al., 1978). Two lipoyl groups can cooperate in the electron and acetyl group transfer by a reductive transacetylation mechanism such as that in scheme 18-1.

Scheme 18-1

Reductive transacetylation proceeds through the two processes in scheme 18-1, disulfide exchange and transacetylation. Although both are chemically facile, they are not fast enough to proceed at enzymatic rates without catalysis by the enzyme. The role of the E2 component in facilitating reductive transacetylation is not known.

Secondly, the lipoyl bearing domains of E2 are extended structures and highly mobile, as determined by NMR spectroscopy. Through their mobility and extended structures, they magnify the distances through which the lipoyl domains interact with active sites and one another (Perham, 1991). The structures of lipoyl-bearing domains from *E. coli* and *Bacillus stearothermophilus* are similar, and the lipoyl moieties are attached as shown in fig. 18-3 (Dardel et al., 1993; Green et al., 1995).

TPP in Decarboxylation

E1 catalyzes the TPP-dependent decarboxylation of pyruvate by a mechanism related to the action pyruvate decarboxylase discussed in chapter 8 and illustrated in fig. 18-4.

TPP-dependent decarboxylation begins with ionization of the thiazolium-C2(H) to form the reactive ylid (see chap. 3). This is brought about by a glutamate residue, Glu571 in the *E. coli* enzyme, in a process mediated by N^1 and $N^{4'}$ of the pyrimidine ring, as shown in fig. 18-4A. The ylid reacts as a nucleophile with the carbonyl group of pyruvate to form lactyl-TPP, as shown in fig. 18-4B. Lactyl-TPP readily undergoes decarboxylation by a mechanism analogous to that of a β-ketoacid (see chap. 8), where the iminium group in the thiazolium ring serves as the electron sink, to form CO_2 and hydroxyethylidene-TPP.

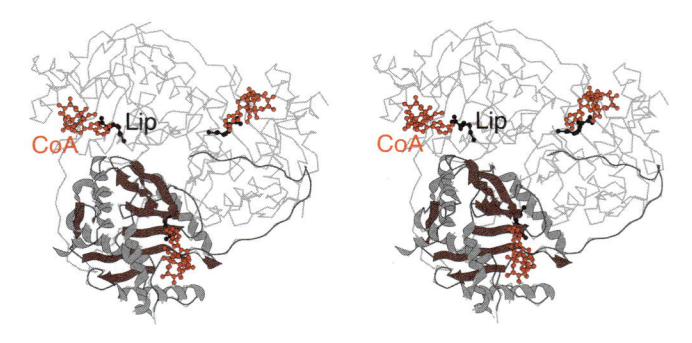

Fig. 18-3. Structure of lipoyl-bearing domain of E2 monomer in *Azotobacter vinelandii* pyruvate dehydrogenase (PDH) complex. One vertex of the cubic 24 subunit E2 core is composed of a homotrimeric unit shown here with one subunit as a ribbon diagram and the remaining two as Cα traces (2.6-Å resolution; PDB 1EAB). One CoA (pink) and one lipoate (LIP; black) bind at each subunit interface within the homotrimeric unit (Mattevi et al., 1992).

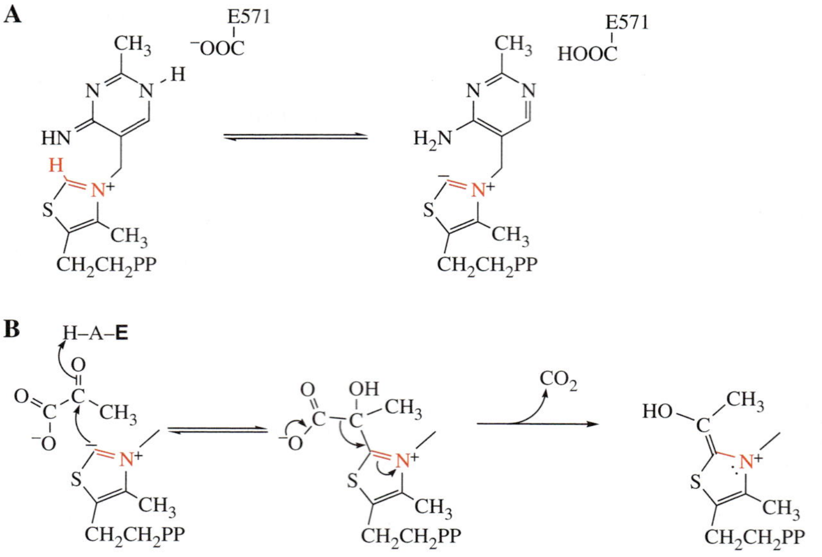

Fig. 18-4. Mechanisms for decarboxylation of pyruvate by E1 of the pyruvate dehydrogenase (PDH) complex. (A) A mechanism is shown for ionization of the thiazolium-C2 of thiamine pyrophosphate (TPP) to the catalytic ylid in an active site of the E1 component of the PDH complex from *E. coli*. The roles of Glu571 and the pyrimidine-N4′ are well supported by structural and spectroscopic evidence. (B) The mechanism by which the thiazolium ring of TPP facilitates the decarboxylation of pyruvate to form hydroxyethylidene-TPP.

Site-Site Interactions in the E1 Dimer

The E1 component displays biochemical properties that mandate communication between the two active sites in the dimeric structure. The binding of TPP to one site of an E1 dimer is much faster than to the second site (Horn and Bisswanger, 1983). A detailed structural and biochemical study of the E1 component from *Bacillus stearothermophilus* has unmasked the molecular basis for site-site communication (Frank et al., 2004). The structure reveals a polar network of acidic amino acids engaged in hydrogen bonded contacts constituting a proton wire between the pyimidine-N4′ atoms of the TPP molecules within the dimer. The structures of other TPP-dependent enzymes show analogous proton wires connecting TPP sites in the dimeric structures. The wire in the E1 component from *E. coli* is illustrated in fig. 18-5, which shows Glu571, Glu235, Glu237, and Arg606, as well as two fixed water molecules in the wire (Jordan, 2004). In the enzyme from *B. stearothermophilus*, the wire is composed of Glu28, Glu59, Glu88, Asp91, Asp180, and Glu183 from the two subunits, as well as a magnesium ion (Frank et al., 2004). The contacts form a continuous hydrogen bonded network between the pyrimidine-N1 atoms of TPP in the two subunits.

Figure 18-5 shows the two TPP molecules in different ionization states, one with the catalytically functional ylid and the other with un-ionized thiazolium-C2(H). The proton wire allows the ionization of either TPP molecule to the ylid through proton transfer between sites, which is mediated by Glu571 and the pyrimidine-N4′ at the ends of the wire. Only one of the two TPP molecules can function catalytically at a given time. Mutations of glutamate residues to disrupt the wire severely decrease the activities of the resulting variants relative to the wild-type enzyme (Frank et al., 2004). Biochemical evidence links this network with the phenomenon of "half-of-the-sites" reactivity of TPP in the two subunits. We have encountered half-of-the-sites behavior in the action of UDP-glucose

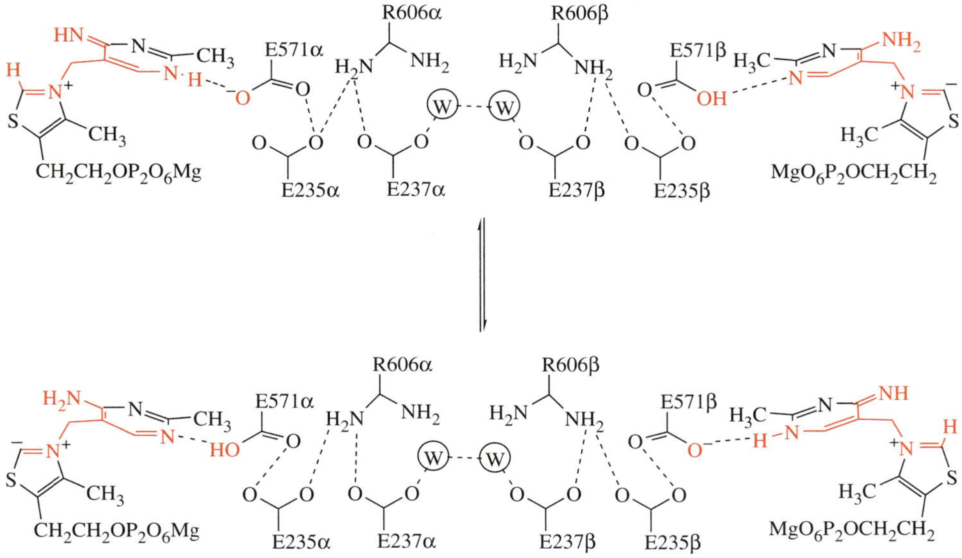

Fig. 18-5. A mechanism for proton transfer between thiamine pyrophosphate (TPP) sites of E1 in the pyruvate dehydrogenase (PDH) complex. These drawings illustrate the amino acid side chain contacts between the TPP molecules in the A and B subunits of E1 in the PDH complex from *E. coli* (Jordan, 2004). The corresponding network in the enzyme from *Bacillus stearothermophilus* consists of ten glutamates, two aspartates, and a magnesium ion. The networks constitute proton wires that relay a proton from one site to the other such that only one of the thiazolium rings can be ionized at C2 to the catalytically functional ylid form at a given time (Frank et al., 2004).

pyrophosphorylase in chapter 10, and it is widely observed in enzymology. Rarely is this phenomenon as well understood as in the action of pyruvate dehydrogenase, and by extension other TPP-dependent enzymes.

Reductive Acylation of Lipoamide

Unlike pyruvate decarboxylase, E1 does not catalyze the formation of acetaldehyde. Instead, the chemical properties of hydroxyethylidene-TPP are channeled by E1 to electron transfer and acetyl group transfer. Hydroxyethylidene-TPP is well suited to electron transfer. All TPP-dependent enzymes can be assayed by observing the reduction of ferricyanide or dichlorophenol indophenol by the hydroxyethylidene-TPP intermediates they generate. E1 prevents the protonation of hydroxyethylidene-TPP generated in the decarboxylation of pyruvate, which would lead to acetaldehyde. It channels electron transfer to the lipoamide moieties bound to E2, and it couples this process with acetyl group transfer.

The chemical mechanism by which hydroxyethylidene-TPP reductively acetylates lipoamide is not known. The process shown as eq. 18-2b in fig. 18-1 must take place in steps, but more than one reasonable mechanism can be written, and two distinct mechanisms are delineated in fig. 18-6.

In mechanism A, shown with black arrows, hydroxyethylidene-TPP reacts as a carbanionic species that cleaves the disulfide bond and forms a tetrahedral adduct of the incipient acetyl group with S^8 of lipoamide. The tetrahedral adduct eliminates the TPP-thiazolium ylid anion to form S^8-acetyldihydrolipoamide. In this mechanism electron and group transfer take place in the same step, the initial addition of the hydroxyethylidene-TPP to lipoamide.

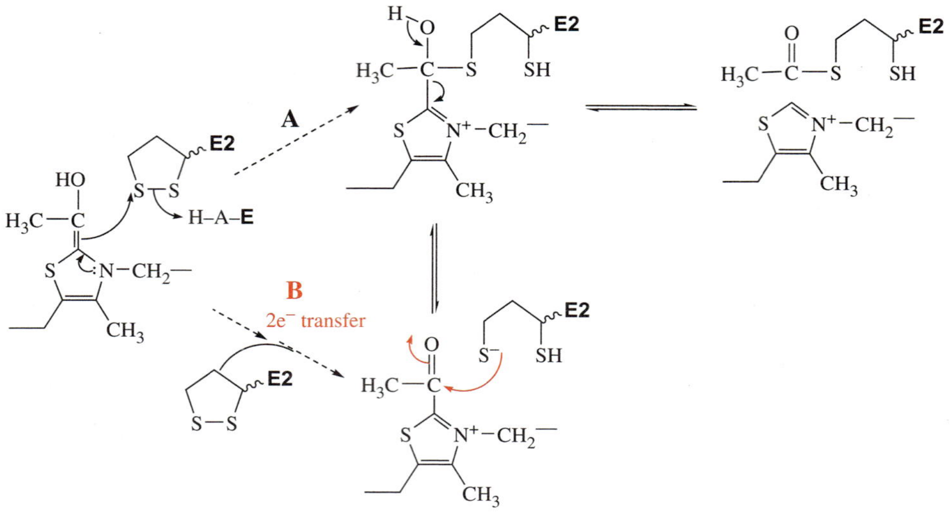

Fig. 18-6. Mechanisms for coupled electron and acetyl group transfer between E1 and E2 of the PDH complex.

In an alternative mechanism, shown as line B with red arrows in fig. 18-6, electron transfer and group transfer take place in separate steps, and acetyl-TPP is a compulsory intermediate. Electron transfer in two steps from hydroxyethylidene-TPP to lipoamide produces acetyl-TPP and the monothiolate of dihydrolipoamide in the first step. The acetyl group is then captured by the thiolate of dihydrolipoamide to form the same tetrahedral adduct as in the other mechanism, and this breaks down as before to TPP and S^8-acetyldihydrolipoamide.

The two mechanisms seem distinguishable by the compulsory involvement of acetyl-TPP in mechanism B. Evidence for the participation of acyl-TPPs in these reactions is of two types. The PDH complex catalyzes the NADH and TPP-dependent hydrolysis of acetyl CoA (CaJacob et al., 1985b), and the KGDH complex catalyzes the NADH- and TPP-dependent hydrolysis of succinyl CoA (Steginsky and Frey, 1984). The hydrolysis of acyl CoAs takes place by the overall reversal of the E3, E2, and E1 reactions (eqs. 18-2e to 18-2b in fig. 18-1) to form acyl-TPPs. In the absence of sufficient CO_2 to complete the reversal to α-ketoacids, the acyl-TPPs undergo hydrolysis. Second, [^{14}C]acetyl-TPP can be isolated from acid quenched solutions of the PDH complex in the steady-state with [2-^{14}C]pyruvate. The isolated [^{14}C]acetyl-TPP is chemically identical to authentic, synthetic acetyl-TPP (Gruys et al., 1987, 1989). The experiments prove that acetyl-TPP is formed in the course of the reaction of the PDH complex; however, they fall short of proving that it is a compulsory intermediate. In fig. 18-6, acetyl-TPP is shown in equilibrium with the tetrahedral adduct, which in principle can be reached by either route A or B, so that acetyl-TPP could be in an intermediate by mechanism B or in equilibrium with the tetrahedral adduct formed by way of route A.

Acetylation of lipoamide is regiospecific for S^8, as established in NMR spectroscopic and kinetic studies (Yang and Frey, 1986). The E2 core from *Azotobacter vinelandii* lacking the lipoyl-bearing domains has been crystallized and the structure determined (Mattevi et al., 1992). The crystal structure of the acetyltransferase domain in complex with CoA and lipoamide elegantly confirmed the location of S^8 in position to accept the acetyl group (Mattevi et al., 1993).

Fatty Acid Synthesis

Fatty acids are assembled in two-carbon units, ultimately from acetyl CoA. The chemistry of this process entails cycles of Claisen condensation, reduction, dehydration, and further reduction of the growing the fatty acid chains. The biological process begins with the acetyl CoA carboxylase-catalyzed carboxylation of acetyl CoA to malonyl CoA. In fatty acid synthesis, seven malonyl groups serve as the two-carbon units serially added, initially to the acetyl group of acetyl CoA and then to the growing chain, to assemble a molecule of palmitate. Decarboxylation of the malonyl group by one component of fatty acid synthase generates the acetyl-enolate group for each round of condensation in fatty acid synthesis. Scheme 18-2 illustrates the chemical mechanism of condensation, with the malonyl acyl carrier protein (ACP) as the source of the two-carbon extension of a fatty acid chain.

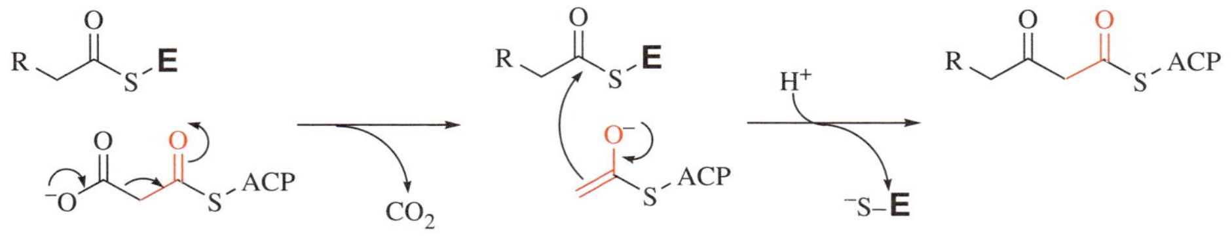

Scheme 18-2

Decarboxylation of the malonyl group generates the enolate of acetyl-ACP, which undergoes nucleophilic addition to the carbonyl group of the growing acyl thioester. The growing fatty acyl group is protein-bound as a thioester throughout the process. In this way, malonyl CoA serves as the source of most of the two-carbon units in fatty acids. We begin by considering acetyl CoA carboxylase, the enzyme that produces malonyl CoA.

Acetyl CoA Carboxylase

Reaction and Molecular Properties

Malonyl CoA arises from the carboxylation of acetyl CoA. Acetyl CoA carboxylase catalyzes this reaction and uses bicarbonate as the source of the carboxyl group (Wakil et al., 1983). In carboxylation, an enol or enolate reacts as a nucleophile with CO_2, as discussed in chapter 8; therefore, carboxylation of acetyl CoA requires the enolization of acetyl CoA. Carboxylation by bicarbonate is, however, kinetically slow and energetically uphill because bicarbonate is not electrophilic and must be converted into CO_2. ATP serves as the energy source for the dehydration of bicarbonate. In the carboxylation mechanism, the resulting CO_2 is captured by biotin to form carboxybiotin, which maintains the carboxyl group in a chemically poised state and subsequently carboxylates the enolate of acetyl CoA. The overall reaction catalyzed by acetyl CoA carboxylase is described by eq. 18-3.

$$\text{Acetyl CoA} + \text{MgATP} + \text{HCO}_3^- \longrightarrow \text{Malonyl CoA} + \text{MgADP} + \text{P}_i \qquad (18\text{-}3)$$

Acetyl CoA carboxylases are complex enzymes that exist in various molecular forms in different species. In *E. coli*, it is a multienzyme complex consisting of a homodimeric biotin carboxylase (BC) composed of 49-kDa subunits, a homodimeric biotinyl carboxyl carrier protein (BCC) composed of 19-kDa subunits, and a carboxyltransferase composed of two copies each of 35- and 30-kDa subunits ($\alpha_2\beta_2$) (Choi-Rhee and Cronan, 2003; Fall and Vagelos, 1972; Guchhait et al., 1974a; 1972; Li and Cronan, 1992a, 1992b). In mammals and birds, these three proteins are fused into a single, multienzyme or multimodular protein more than 2300 amino acids long.

Carboxylation of acetyl CoA is the first committed step in fatty acid biosynthesis, and it is the logical point for metabolic regulation of fatty acid synthesis. Mammalian and avian carboxylases exist as cytosolic and mitochondrial isoforms and are subject to regulation by complex, phosphorylation/dephosphorylation processes that are beyond the scope of this volume (Munday, 2002; Wakil et al., 1983). The bacterial enzymes are not regulated by any of the mechanisms identified for the mammalian enzymes, presumably because fatty acid metabolism is much simpler in prokaryotes (Cronan and Waldrop, 2002). Bacteria do not store fats or use fatty acids for energy but use them primarily as precursors of membrane phospholipids. In bacteria, guanine nucleotides regulate the activity of acetyl CoA carboxylase and serve as signals for cell growth.

Because of its key role in regulating fatty acid biosynthesis, acetyl CoA carboxylase is a target for pharmaceutical agents against diabetes, morbid obesity, and other diseases and disorders of fat metabolism. It is also a target for herbicides (Zhang et al., 2004). The carboxyltransferase domain is a particular target for the development of anti-obesity medications (Harwood et al., 2003; Zhang et al., 2004).

Carboxylation Mechanism

We describe the sequence of steps in terms of the roles of the three proteins in the acetyl CoA carboxylase from *E. coli*. The carboxylation of acetyl CoA follows the course outlined in fig. 18-7B. Carboxylation of biotin by either HCO_3^- in the forward direction catalyzed by biotin carboxylase or by malonyl CoA in the reverse direction catalyzed by carboxyltransferase produces $N^{1'}$-carboxybiotinyl-BCC, shown in fig. 18-7A (Guchhait et al., 1974b; Lane and Lynen, 1963).

This is known by the fact that biotin carboxylase and carboxyltransferase can accept biotin as a substrate, and they catalyze its carboxylation to a compound that is chemically indistinguishable from $N^{1'}$-carboxybiotin prepared by unambiguous chemical synthesis. The enzymatic and synthetic products undergo nonenzymatic decarboxylation at identical rates and with identical pH-rate profiles. Moreover, authentic $N^{1'}$-carboxybiotin serves as a substrate for biotinyl carboxylase in producing MgATP from MgADP and P_i, verifying in a model reaction the reverse of eq. 18-4a (fig. 18-7C), the step catalyzed by biotin carboxylase. N^1-carboxybiotinol serves as a substrate for the carboxyltransferase in the production of malonyl CoA. This is a model for malonyl CoA formation in eq. 18-4b (fig. 18-7C).

The carboxylations of biotinyl-BCC and of acetyl CoA catalyzed by biotin carboxylase and carboxyltransferase, respectively, take place independently, as illustrated in fig. 18-7B (Polakis et al., 1974). The two reactions are shown as eqs. 18-4a and 18-4b in fig. 18-7C. Purified biotinyl carboxylase catalyzes the carboxylation of biotinyl-BCC according to eq. 18-4a, as well as the isotope exchanges ATP/[^{14}C]ADP and ATP/$^{32}P_i$, in the absence of carboxyltransferase. Purified carboxyltransferase catalyzes the carboxylation of acetyl CoA by $N^{1'}$-carboxybiotinyl-BCC and the isotope exchange malonyl CoA/[^{14}C]acetyl CoA in the absence of biotin carboxylase.

The simplest mechanism for the dehydration of bicarbonate and carboxylation of biotin is that illustrated in scheme 18-3, in which bicarbonate first reacts with MgATP to form MgADP and carboxyphosphate, the anhydride of carbonic and phosphoric acids. Decarboxylation of carboxyphosphate leads to CO_2, which would readily carboxylate the "enolized" tautomer of biotin. Alternatively, the biotinyl tautomer could react as a nucleophile with carboxyphosphate to capture the carboxyl group. Evidence for the intermediacy of

A

BCC — $N^{1'}$-Carboxybiotinyl–BCC

B

Biotin Carboxylase / Carboxyltransferase

N^1-Carboxybiotin
Acetyl CoA
ADP/P$_i$
HCO$_3^-$/ATP
Biotin
Malonyl CoA
Biotin
BCC

C

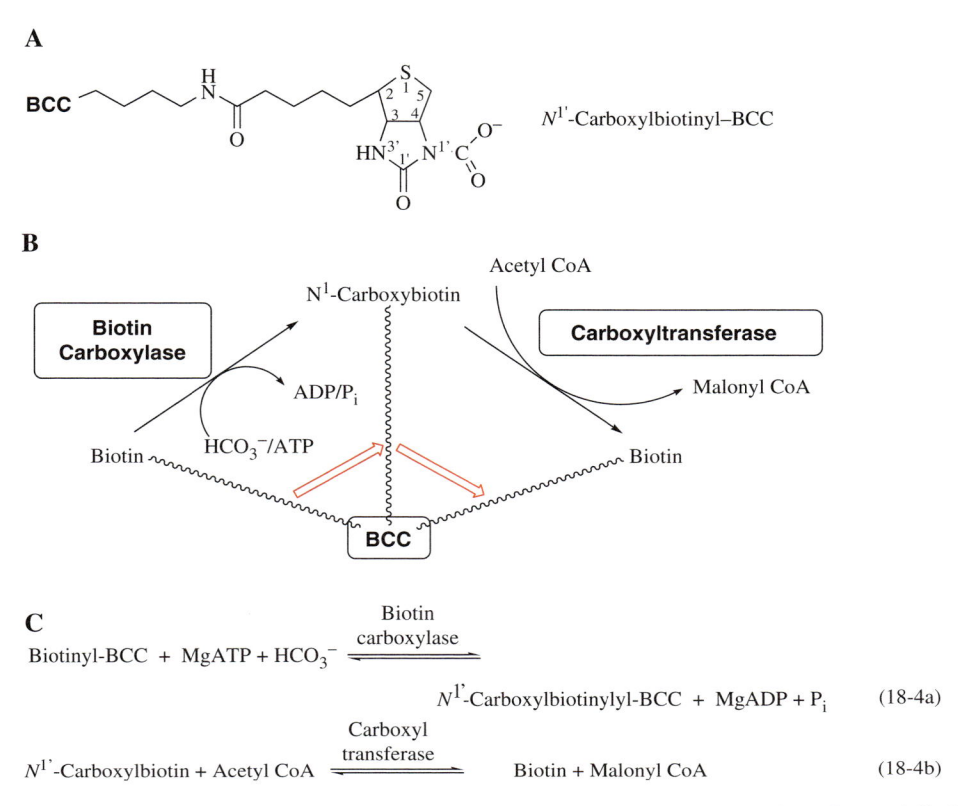

$$\text{Biotinyl-BCC} + \text{MgATP} + \text{HCO}_3^- \xrightleftharpoons{\text{Biotin carboxylase}} N^{1'}\text{-Carboxybiotinylyl-BCC} + \text{MgADP} + \text{P}_i \quad (18\text{-}4a)$$

$$N^{1'}\text{-Carboxybiotin} + \text{Acetyl CoA} \xrightleftharpoons{\text{Carboxyl transferase}} \text{Biotin} + \text{Malonyl CoA} \quad (18\text{-}4b)$$

Fig. 18-7. Role of biotin and the biotin carboxyl carrier protein in the action of acetyl CoA carboxylase. (A) Structure of the $N^{1'}$-carboxybiotinyl moiety of carboxylated biotin carboxyl carrier protein (BCC), which mediates carboxyl transfer in the action of acetyl CoA carboxylase. (B) Action of biotinyl-BCC in the overall reaction of acetyl CoA carboxylase. The drawing illustrates the independent interaction of the biotinyl moiety with biotin carboxylase and carboxyltransferase through the movement of the biotinyl moiety from one enzyme to the other. (C) Stoichiometric equations for the reactions catalyzed by biotin carboxylase and carboxyltransferase.

carboxyphosphate includes the fact that the carboxylation of acetyl CoA proceeds with the transfer of one equivalent of ^{18}O from [^{18}O$_3$]HCO$_3^-$ to form [^{18}O$_1$]HPO$_4^{2-}$ and [^{18}O$_2$] malonyl CoA (Kaziro et al., 1962). The fact that carbamyl phosphate, a chemically stable analog of carboxyphosphate, reacts in place of carboxyphosphate in the reverse of step 1 in scheme 18-3 to produce carbamate and MgATP elegantly supports the involvement of carboxyphosphate in the mechanism (Polakis et al., 1972).

Scheme 18-3

Structure

The structures of the *E. coli* acetyl CoA carboxylase complex or the mammalian multimodular enzymes are not available. Structures of *E. coli* biotin carboxylase and the biotinyl-bearing domain of BCC and of the carboxyltransferase domain of the multimodular mammalian enzyme are available (Athappilly and Hendrickson, 1995; Roberts et al., 1999; Waldrop et al., 1994; Zhang et al., 2003). Representative structures are illustrated as ribbon diagrams in fig. 18-8.

The structures of the biotinyl-bearing domain of *E. coli* BCC and the lipoyl-bearing domains of E2 in the PDH complex are remarkably similar, although there are few identities in the amino acid sequences (Roberts et al., 1999). In one mechanistic difference, the mobility of the biotinyl-lysyl (biocytin) chain seems to be restricted in biotinyl-BCC, so that the mobility of the domain itself must account for translocation of the $N^{1'}$-carboxybiotinyl moiety between biotin carboxylase and carboxyltransferase (Roberts et al., 1999).

Regardless of the minimal amino acid sequence identities between lipoyl- and biotinyl-bearing domains, respectively, the two similar structures appear to function in the same way. The lipoyl and biotinyl moieties are linked to lysyl residues in the C-terminal halves of the domains, and upstream sequences rich in proline and alanine are thought to be linkers that confer mobility. Deletion of this linker in the biotinyl-bearing domain of BCC abolished function, although the biotin moiety was correctly attached through the action of the biotinyl ligase (Cronan, 2002). Presumably, the loss of function was due to the absence of mobility. Insertion of the corresponding linker from a lipoyl-bearing domain of E2 in the PDH complex restored the function of the biotinyl-bearing BCC. The mobility of the linker rather than its amino acid sequence seems to be the essential property required for the function of BCC.

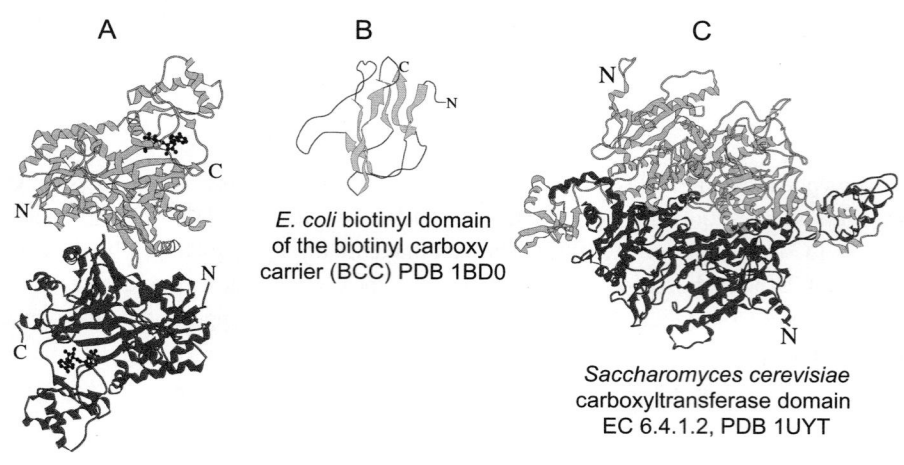

Fig. 18-8. Acetyl CoA carboxylase. (A) Ribbon diagrams are shown for the structures of the biotin carboxylase (BCC) from *E. coli* (2.5-Å resolution; PDB 1DV2; Thoden et al., 2000). (B) Biotinyl-bearing domain of BCC from *E. coli* (1.8-Å resolution; PDB 1BDO; Athappilly and Hendrickson, 1995). (C) Carboxyltransferase domain from yeast (2.5-Å resolution; PDB 1UYT; Zhang et al., 2004).

Fatty Acid Synthases

Processing Acetyl and Malonyl CoA into Palmitate

A complement of six enzymes plus the acyl carrier protein (ACP) produce palmitic acid from one molecule of acetyl CoA and seven molecules of malonyl CoA by means of the reaction sequence catalyzed by fatty acid synthases (FAS) shown in fig. 18-9. ACP contains the phosphopantetheinyl group covalently bonded to the β-hydroxyl group of a cysteinyl residue.

The phosphopantetheinyl sulfhydryl group accepts acyl groups slated for fatty acyl elongation steps and delivers the intermediates to the active sites of the participating enzymes within the complex through its structural mobility (fig. 3-32), as illustrated in fig. 18-10.

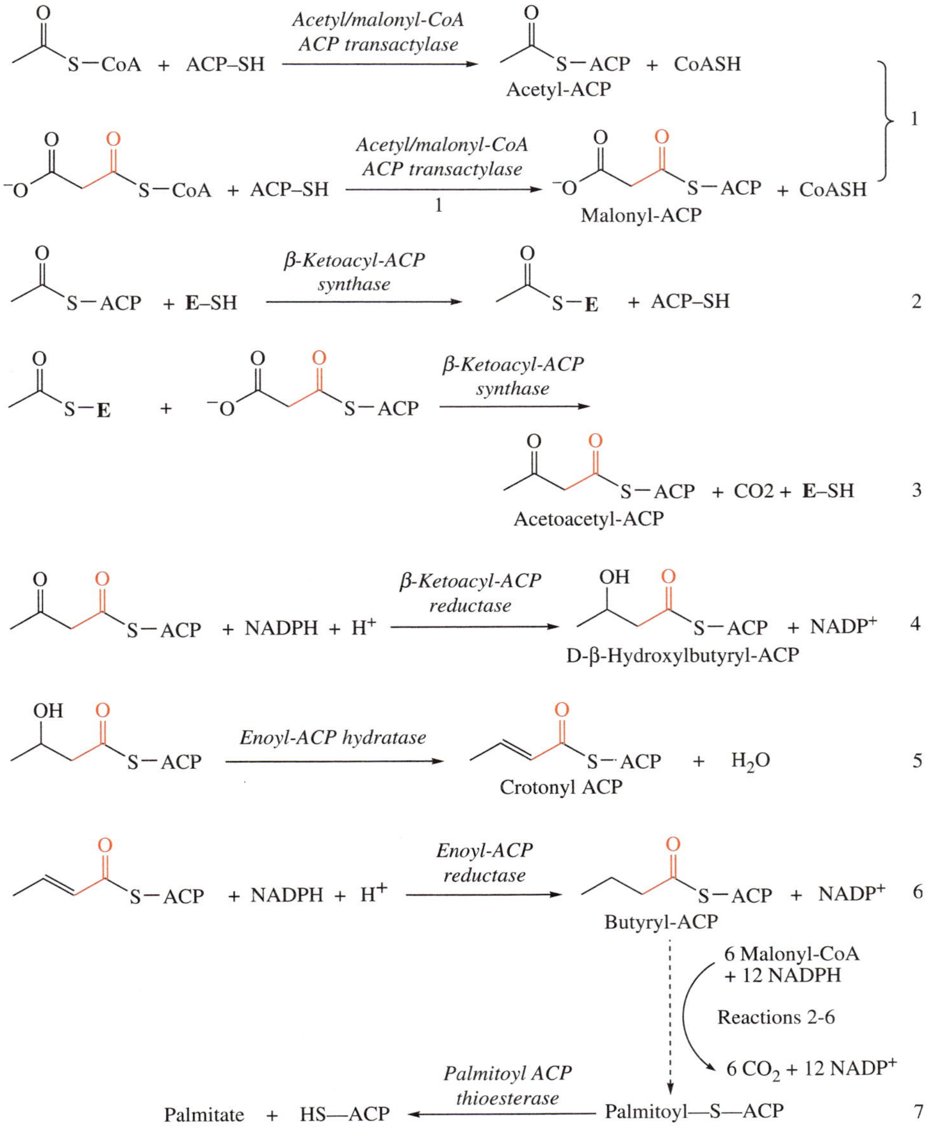

Fig. 18-9. Reactions catalyzed by the fatty acid synthase complex in the biosynthesis of palmitic acid.

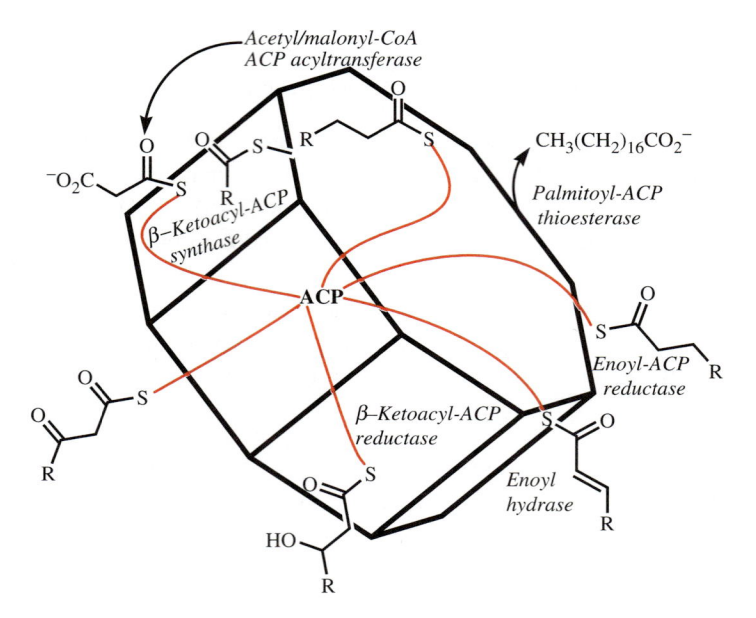

Fig. 18-10. The role of acyl carrier protein (ACP) in ferrying acyl groups among enzymes in the action of fatty acid synthases.

Many of the steps in fatty acid synthesis are analogous to steps of fatty acid degradation by the β-oxidation pathway proceeding in the reverse direction. In metabolism, intermediates of the two pathways are distinguished in the following ways. First, the enzymes of fatty acid degradation are different proteins with different substrate specificities. Second, all intermediates in fatty acid degradation are CoA-esters. Third, the enzymes of β-oxidation that catalyze the reverse of steps 4 and 5 in fig. 18-9 are specific for the L-β-hydroxyacyl group, in contrast to the D-β-hydroxyacyl group in fig. 18-9. The reductase/dehydrogenases of β-oxidation are specific for $NAD^+/NADH$, in contrast to the $NADP^+/NADPH$ required by FAS.

A FAS produces palmitate as the end product, and palmitate serves as the intermediate in the biosynthesis of higher fatty acids in reactions of other enzymes. The assembly of palmitate begins with the action of acetyl/malonyl CoA:ACP acyltransferase (AT), which produces acetyl- and malonyl-ACP from the CoA-esters. A cysteinyl-SH group of β-ketoacyl-ACP synthase (KS) accepts acyl groups, initially the acetyl group, from acyl-ACPs 2 to 14 carbons long in step 2. This frees the ACP to accept a malonyl group from malonyl CoA, and KS then catalyzes the decarboxylation and condensation of malonyl-ACP with the acyl-thioester (acetyl-S-**E** in fig. 18-9) at its active site. This releases the cysteinyl-SH group in the active site of KS, and the β-ketoacyl-ACP undergoes reduction to the corresponding acyl-ACP in a round of three reactions at the active sites of three different enzymes. β-Ketoacyl-ACP reductase (KR) catalyzes an NADPH-dependent reduction to the D-β-hydroxyacyl-ACP, the hydroxyacyl-ACP dehydratase (DH) catalyzes dehydration to the *trans*-α,β-enoyl-ACP, and the flavoprotein enoyl-ACP reductase (ER) catalyzes the NADPH-dependent reduction to the corresponding acyl-ACP. Further elongation of the acyl group by two carbons proceeds by acyltransfer to the active site cysteinyl-SH of KS, followed by another round of condensation, reduction, dehydration, and reduction in steps 2 through 6 of fig. 18-9. Palmitoyl-ACP cannot undergo further elongation but suffers hydrolysis to palmitate by the action of palmitoyl-ACP thioesterase (TE).

Multienzyme Fatty Acid Synthase Systems

Fatty acid synthases are organized differently in bacteria, yeast, and animals (Chang and Hammes, 1990; Wakil, 1989) The enzymes and ACP for fatty acid synthesis can be purified as individual proteins from bacteria. They may function individually or be associated as a complex in the cell. In eukaryotes, the enzymes are found in multifunctional proteins. In yeast, the FAS consists of a complex of six 213-kDa α subunits and six 203-kDa β subunits associated as $\alpha_6\beta_6$ that include all of the enzymatic activities. The complex contains sufficient active sites to carry out the biosynthesis of six different molecules of palmitic acid simultaneously. The α subunit includes domains for β-ketoacyl CoA synthase, β-ketoacyl CoA reductase and ACP. The β subunit contains domains for an acetyltransferase, a malonyltransferase, D-β-hydroxyacyl-ACP dehydratase, and enoyl ACP reductase.

In animals, the FASs are homodimeric enzymes, and each subunit in a dimer contains all six enzymes plus ACP as domains within the polypeptide chain. The domains appear in the order shown in fig. 18-11, with the β-ketoacyl-ACP synthase at the N-terminus and the palmitoyl-ACP thioesterase at the C-terminus. Electron photomicrographic images of fatty acid synthase complexes have been published, but no x-ray crystal structure is as yet available. Structures of several of the bacterial enzymes and catalytic domains of the eukaryotic FASs are available.

Modular Enzymes

Polyketide Synthases

In the biosynthesis of polyketide antibiotics, complex proteins employ much of the chemistry and many of the functional principles that we have seen in the action of FAS (Shen, 2000; Walsh, 2003a). The higher oxygenation states of polyketides relative to fatty acids arise from the omission of reductive and dehydration steps in fatty acid synthesis, but the carbon-carbon ligation and a few of the reductive and dehydration steps are retained. Polyketide synthases (PKSs) catalyze the assembly of the core structures of polyketide antibiotics, including tetracyclines, erythromycins, daunorubicin, doxorubicin, and many others. The PKSs include ACP and enzymes analogous to those in FAS that display variant substrate specificities. The starting acyl group in FAS is acetyl CoA, but a PKS may accept propionyl-, butyryl-, cyclohexanoyl, or benzoyl CoA or another CoA ester as the starter. In place of malonyl CoA, PKSs often accept methylmalonyl CoA, and this leads to methyl substituents in the polyketides. Because of these differences and the selective omission or inclusion of reductive and dehydration steps, the PKSs produce a highly diverse collection of products. Other enzymes further modify the core polyketides by hydroxylation,

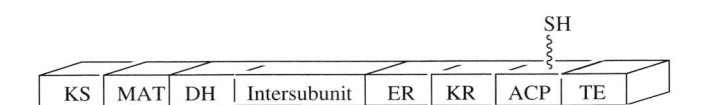

Fig. 18-11. Organization of the domains in an animal fatty acid synthase (FAS). The sequence of domains in a single subunit of the dimeric animal FAS, beginning with the N terminus, is β-ketoacyl-ACP-synthase (KS), malonyl/acetyl CoA:ACP acyltransferase (MAT), D-β-hydroxyacyl-ADP dehydratase (DH), intersubunit contact domain, enoyl-ACP reductase (ER), β-ketoacyl-ACP reductase (KR), acyl carrier protein (ACP), and terminating acylthioesterase (TE).

glycosylation, or methylation to further amplify the diversity in structure and biological activities of polyketides.

Type I Polyketide Synthase

Three classes of PKSs are known. Types I, II, and III differ with respect to their molecular composition and type of antibiotic produced. We begin with type I PKSs, which produce macrocyclic polyketides. The translated nucleotide sequences of type I PKSs reveal primary structures of large proteins that include segments with domains typical of those in FAS (Cortes et al., 1990; Donadio et al., 1991). ACP domains punctuate sequences of enzymatic domains, all of which include at least one β-ketoacyl-ACP synthase, so that each such segment can potentially elongate a polyketide. Each multidomain segment with its associated ACP domain is known as a module within a PKS protein.

The domain and modular structures of the three proteins in the 6-deoxyerythronolide synthase are illustrated in fig. 18-12. Each protein contains at least two modules with associated ACPs. Among the domains within modules, the ketosynthases (KSs), acyltransferases (ATs), and ketoreductases (KRs) dominate. There is only one each of hydroxyacyl-ACP dehydratase (DH) and enoyl-ACP reductase (ER). This composition is required for the many synthase, acyltransferase and reductase steps required in the pathway, with only a single dehydration and enoyl reduction.

Figure 18-12 demonstrates the transformations in the biosynthesis of 6-deoxyerythronolide B. The process begins with the loading module, consisting of an acyltransferase (AT) and ACP at the N-terminus of the first protein. This module catalyzes propionyltransfer from propionyl CoA to form propionyl-ACP as the starter. Module 1 contains an acyltransferase (AT), a β-ketoacyl-ACP synthase (KS), a β-ketoacyl-ACP reductase (KR), and ACP. AT catalyzes methylmalonyl transfer from the CoA ester to ACP, KS accepts the acetyl group from acetyl-ACP from the loading module at a cysteine residue in its active site and then catalyzes the decarboxylation and condensation of methylmalonyl-ACP. The resulting α-methyl-β-ketoacyl-ACP is reduced to the α-methyl-β-hydroxyacyl-ACP by action of the β-ketoacyl-ACP reductase domain. Then the α-methyl-β-hydroxyacyl group is transferred to the KS in module 2, and another round of ketosynthase and ketoacyl reductase action by module 2 leads to the acyl-ACP shown at the upper right in fig. 18-12. Further acyltransfer to module 3 leads to elongation, but in the absence of a ketoreductase, no reduction in module 3. Module 4 accepts the acyl group on its KS and catalyzes elongation, reduction, dehydration, and final reduction to the methylene group with its full complement of FAS-like enzyme domains. Modules 5 and 6 catalyze further elongation and ketoreduction but not dehydration or enoyl reduction. Module 6 also contains a termination (TE) domain, which catalyzes the cyclization by internal transesterification of the ultimate polyketide to form 6-deoxyerythronolide B. Further processing of the macrocyclic polyketide leads to the erythromycin family of antibiotics.

The drawing in fig. 18-12 implies that more than one polyketide molecule can be assembled at one time by a molecule of polyketide synthase, that is, that assembly is processive. Processivity has been proven in the case of rifamycin polyketide synthase (Yu et al., 1999).

In principle, the type I PKSs lend themselves to engineering with the aim of producing new molecules with desired pharmaceutical properties. Domains can be inserted or deleted by molecular genetic methods to potentiate the alteration of reduction and dehydration steps, as well as ring size. Because the interdomain and intermodular interactions entail complex and presumably specific domain-domain contacts and recognition, apparently straightforward approaches based on this principle are laden with complications. The several post-assembly processes of glycosylation, hydroxylation, and methylation also

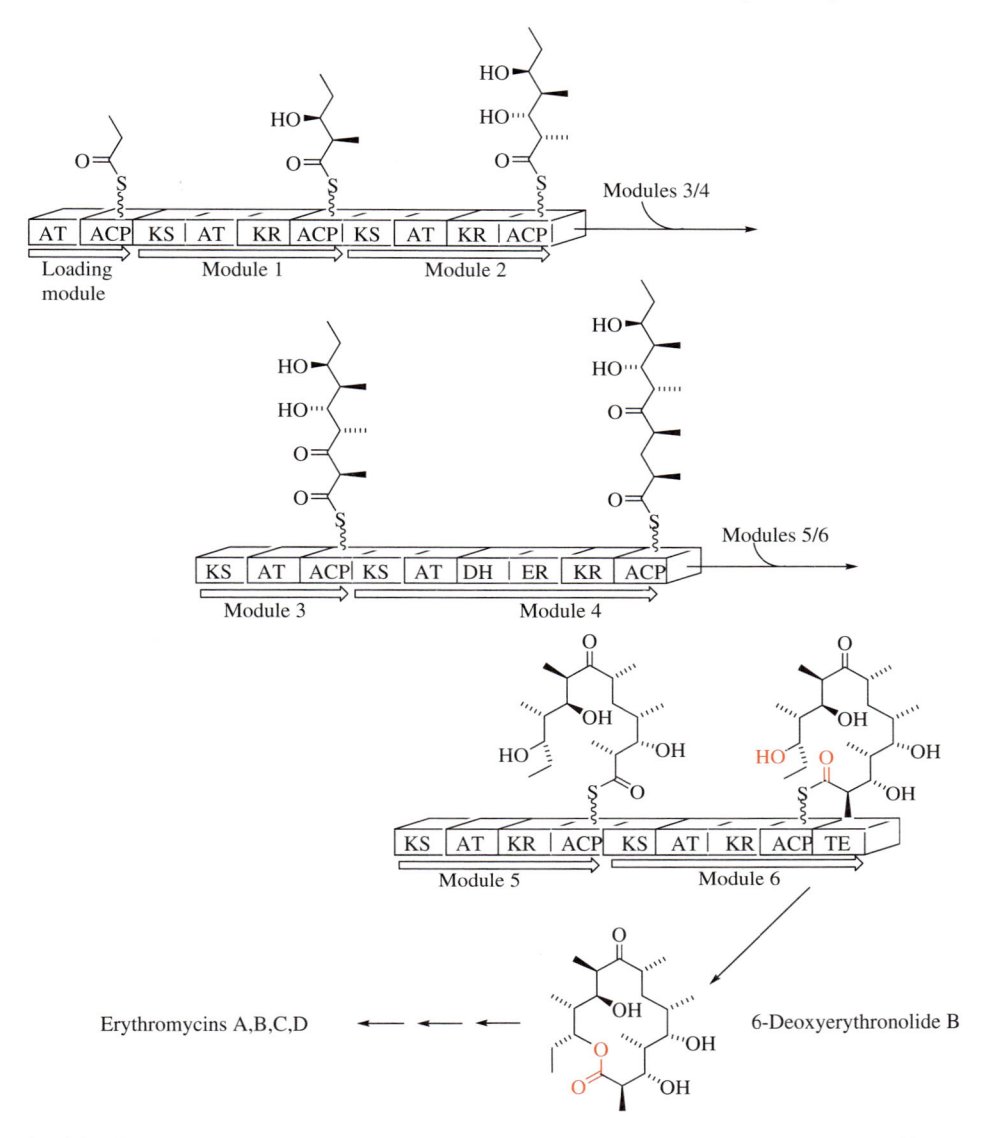

Fig. 18-12. Modular structure of a type I polyketide synthase (PKS) and its role in the biosynthesis of erythromycins. The PKS is composed of three proteins, each of which comprises two or three modules consisting of enzymatic domains and acyl carrier protein (ACP). The domains are analogous to the fatty acid synthase (FAS) domains, and the abbreviated names are the same as in fig. 18-11 with two exceptions. The acyltransferase domains (AT) shown here have different specificities from the MAT domain in fig. 18-11, and the terminating domain here (TE) catalyzes an internal transesterification to cyclize the polyketide, unlike the TE domain in FAS, which catalyzes hydrolysis of the fatty acyl-ACP.

offer opportunities to diversify antibiotic structure. The many possibilities for antibiotic engineering are a current focus of research (Hutchinson and Fujii, 1995; Khosla, 1997; Thorson et al., 2001).

Type II Polyketide Synthase

Polycyclic polyketide antibiotics such as tetracyclines, doxorubicin, and daunorubicin are assembled by type II PKSs (Walsh, 2003b). These PKSs employ the same condensation

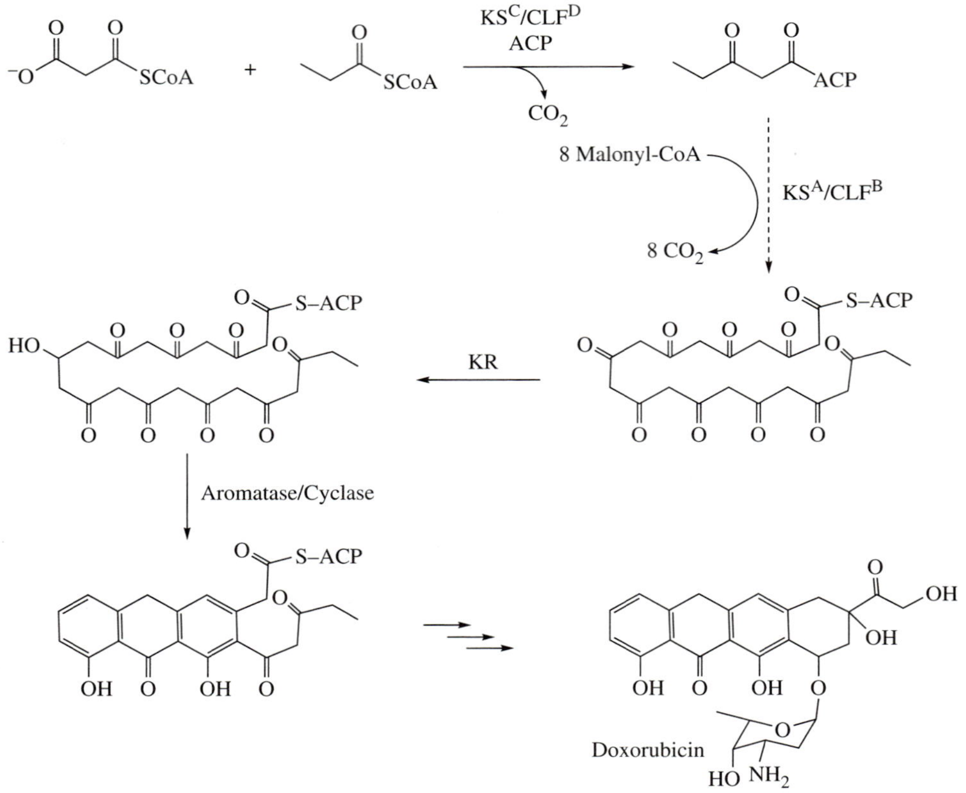

766   Enzymatic Reaction Mechanisms

Fig. 18-13. Role of a type II polyketide synthase (PKS) in the biosynthesis of Doxorubicin. The β-ketoacyl-ACP-synthase (KS)/chain-limiting factor (CLF) dimers catalyze the condensation steps and function with acyl carrier protein (ACP) as minimal PKSs. The CLF subunit is similar to the KS and is also known as KS$_\beta$, but it is enzymatically inactive. After the assembly of the 21-carbon intermediate from propionyl CoA and nine molecules of malonyl CoA, a β-ketoacylreductase (KR) catalyzes the reduction of the 9-keto group. An aromatase/cyclase, a component of the PKS gene cluster, catalyzes three cyclization steps, and auxiliary enzymes catalyze subsequent steps in the synthesis of doxorubicin.

chemistry as type I PKSs and FASs. However, the products of type II PKSs are more highly oxidized, and fewer ketoreductases are involved. In a fundamental difference from the molecular organization of types I PKSs, the growing polyketide in a type II PKS grows from a single ACP throughout the elongation process. As each ketoacyl group is added, the elongated ketoacyl group must be transferred to the β-ketoacyl synthase before the next elongation occurs, as in the action of the animal FAS or a module of a type I PKS. This is shown in scheme 18-4.

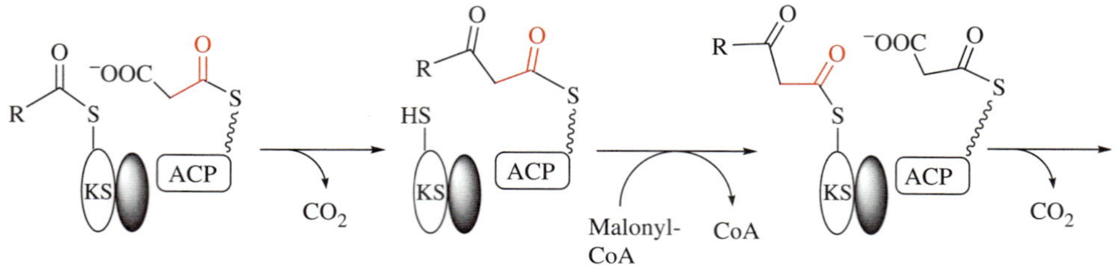

Scheme 18-4

Figure 18-13 illustrates the assembly of the tetracyclic, aromatic antibiotic doxorubicin by a type II PKS. This PKS consists of ACP, two heterodimeric minimal modules KS/CLF, a ketoreductase (KR), and an aromatase/cyclase enzyme. The two KS/CLF functional dimers consist of an active β-ketoacyl-ACP synthase and a structurally similar but inactive KS-protomer known as a *chain-limiting factor* (CLF), named for its hypothetical function in limiting the length of polyketide chain produced. KS^C/CLF^D with ACP catalyzes the initial condensation of propionyl CoA with malonyl CoA to form β-ketovaleryl-ACP. KS^A/CLF^B catalyzes eight rounds of malonyl CoA condensation to generate a 21-carbon polyketyl-ACP. After reduction of the 9-keto group by action of KR, the aromatase/cyclase catalyzes the closure of three rings, with elimination of water and aromatization in two rings. Auxiliary enzymes catalyze the final ring closure and glycosylation to doxorubicin.

Type III Polyketide Synthase and Others

A type III PKS catalyzes polyketide synthesis from acyl CoAs without the participation of ACP. Type III enzymes generate mainly mono and bicyclic polyketides that involve fewer condensations. The classifications of PKSs turn out to be oversimplified in the developing field of polyketide biosynthesis, and exceptions are turning up (Shen, 2004). In the original definition of a type I PKS, each module included domains ACP, AT, and KS. However, a recently discovered type I PKS lacks AT domains, and this function is provided by a separate AT enzyme that serves each module (Shen, 2004).

Nonribosomal Polypeptide Synthetases

Many bioactive polypeptides, such as tyrocidine, gramicidin, bacitracin, and vancomycin are synthesized by complex enzymes and not by ribosomal assembly (Konz and Marahiel, 1999; Marahiel et al., 1997; Walsh, 2003b). The basic chemistry of peptide bond formation is analogous to ribosomal assembly, but short-circuited with the avoidance of nucleic acids to specify amino acid sequences. The nonribosomal polypeptide synthetases (NRPSs) consist of enzymatic domains that catalyze each cycle of peptide bond formation. As in type I PKSs, NRPSs are organized in modules, each of which contains a phosphopantetheinyl domain, known as PCP, and an amino acid activating domain, known as AAla, Aval, and so forth. Each module activates a cognate amino acid as a PCP-thioester by the mechanism shown in fig. 18-14. Reaction of an amino acid with MgATP catalyzed by a cognate activation domain leads to the aminoacyl-adenylate, which donates the aminoacyl group to the thiol of PCP to form the aminoacyl-PCP with elimination of AMP. The activated aminoacyl-PCPs retain free amino groups that can react with neighboring aminoacyl-PCP thioesters to form peptide bonds.

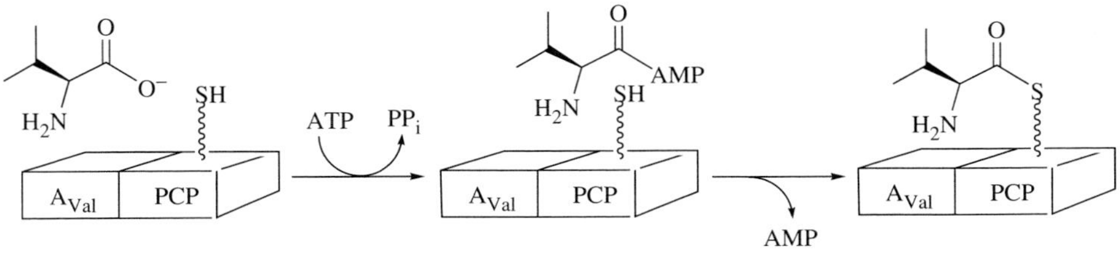

Fig. 18-14. Mechanism by which valine is activated and loaded onto a phosphopantetheinyl domain (PCP) in a nonribosomal polypeptide synthetase. A$_{val}$, amino acid activating domain.

768   Enzymatic Reaction Mechanisms

The sequence of modules defines the sequence in which the aminoacyl residues appear in the polypeptide. This arrangement would be too cumbersome for the assembly of very large polypeptides and proteins, which are more efficiently produced by ribosomal biosynthesis. However, the NRPSs lend themselves to the incorporation of many more amino acids than the standard 20 residues found in proteins, a process that is accomplished by two means. An activating domain in a module may recognize an amino acid other than one of the 20 common L-amino acids. Alternatively, a module may contain a domain that catalyzes a chemical modification of an aminoacyl group being incorporated into the polypeptide. Many polypeptides produced by NRPSs contain nonstandard amino acids such as D-amino acids, and this feature enhances the diversity of the polypeptides that can be produced by these complexes.

ACV Synthetase

Among the many bioactive polypeptides is L-α-aminoadipoyl-L-cysteinyl-D-valine (ACV), the penicillin precursor and substrate for isopenicillin N synthase, which we encountered in chapter 17. ACV contains two unusual amino acyl residues, L-α-aminoadipate and D-valine, and its synthesis illustrated in fig. 18-15 exemplifies the assembly-line function of NRPSs.

ACV synthetase has three modules, one for each aminoacyl residue in ACV. Each module contains an activating domain that specifically catalyzes the thioesterification of its cognate amino acid. The fully charged synthetase is symbolized by the drawing at the top of fig. 18-15. Assembly begins with peptide bond formation between L-α-aminoadipoyl-PCP in the N-terminal module and the α-amino group of L-cysteinyl-PCP in the neighboring downstream module to produce L-α-adipoyl-L-cysteinyl-PCP in the central module. Further peptide formation with the α-amino group of valyl-PCP in the third module produces ACV-PCP in the C-terminal module. The TE domain catalyzes the release of ACV by hydrolysis. Epimerization of the valyl group by the epimerase domain in the C-terminal module takes place either before or subsequent to formation of the last peptide bond, but in either case, the ACV contains only the D-valyl group. Therefore, the third peptidyl transfer is specific for D-valyl-PCP, or the hydrolytic release by action of the TE domain is specific for the presence of the D-valyl residue in the tripeptidyl-PCP.

Ribosomal Protein Synthesis

RNA Polymerase

Biosynthesis of a protein begins with the capture of genetic information for the amino acid sequence in DNA. Each gene specifying the primary structure of a protein stores the information in the nucleotide sequences of the DNA strands, one of which, the template strand, contains the coding sequence. RNA polymerase (RNAP) reads the coding sequence by catalyzing its transcription into a molecule of mRNA carrying the complementary nucleotide sequence, and this is translated into an amino acid sequence by ribosomal protein synthesis. The transcription of DNA coding strands by RNAP is a complex process, and because of this, RNAPs are complex enzymes.

A cell contains several kinds of RNA in addition to mRNA. There are at least 20 species of tRNA and at least three species of rRNA. DNA primase is an RNAP that initiates DNA replication by generating an RNA primer. Every cell contains at least two kinds of RNAP, and eukaryotic organisms contain more. The fully constituted bacterial RNAPs (≈400 kDa)

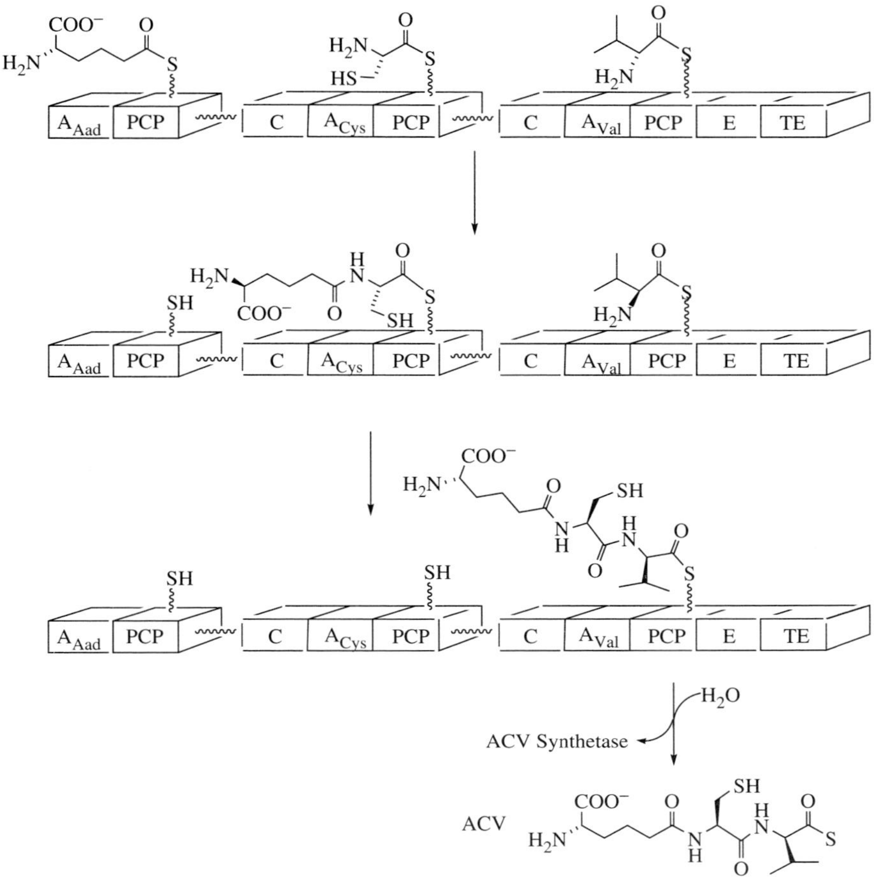

Fig. 18-15. Assembly of the isopenicillin H precursor by action of the L-α-aminoadipoyl-L-cysteinyl-D-valine (ACV) synthetase. The tripeptide precursor of isopenicillin H is assembled by the ACV synthetase, which comprises three nonribosomal polypeptide synthetase (NRPS) modules. Shown at the top are aminoacyl-loaded modules, which assemble the tripeptide by aminoacyl transfer between modules. Epimerization at the α-carbon of the valyl residue is shown in the first step, but the exact point at which the epimerase domain acts is unknown. The termination domain catalyzes the hydrolytic release of the assembled tripeptide. A$_{cys}$, A$_{val}$, amino acid activating domains; PCP, phosphopantetheinyl domain; TE, terminating acylthioesterase (TE).

are the most thoroughly studied and are heterohexameric when fully constituted. The core bacterial RNAPs consist of five subunits ($\alpha_2\beta\beta'\omega$) as purified, and a σ subunit binds to this core to form the holopolymerase that transcribes genes. Sigma is thought to be required for transcription initiation, and it dissociates from the complex of RNAP and DNA after several rounds of transcription to form a growing strand of RNA.

Complexity in the action of RNAP arises from the fact that it must carry out all the auxiliary physical functions of recognizing genes in DNA, separating the coding and complementary DNA strands at transcription sites, and recognizing transcription termination sites downstream from coding sequences. In DNA replication, other proteins carry out functions such as recognition of origination sites, separation of strands, and initiation of replication, so that a DNA polymerase can be a simple enzyme that catalyzes phosphodiester formation between dNTPs and the 3′-OH end of a growing strand. To initiate transcription, an RNAP must recognize promoter-sequences six base pairs long at −10 or −35 base pairs from a start site, it must bind and induce melting of the double strand to separate the template and complementary strands into a "bubble" about 25 base pairs long, it must recognize and transcribe only the template strand, and it must start transcription at the start site (deHaseth et al., 1998). Structural aspects of the initiation process at the level of domain function in RNAP are discussed in a review by Murakami and Darst (2003). The RNAP must then transcribe by polymerizing NTPs while sweeping the bubble through the double helix, reading the template strand, and accommodating the complementary strand from the point of melting to the point of re-annealing. It also must recognize transcription termination sequences and cease polymerization. These functions require a complex and dynamic structure.

Several crystal structures of RNAPs are available (Cramer et al., 2001; Gnatt et al., 2001; Murakami et al., 2002a, 2002b; Vassylyev, 2002; Zhang et al., 1999). The structure in fig. 18-16 is an RNAP in an initiation complex with double-stranded DNA containing a promoter. The molecule of RNAP is in the form of a claw, with the active site and metal ion at the base of the claw. The template DNA strand is internal to the structure passing near the Mg^{2+} at the active site, with the complementary strand wrapped around the molecule.

The basic biochemical mechanism of 3′,5′-phosphodiester formation is likely to be similar for DNA and RNA polymerases. The two-metal ion mechanism for the action of DNA polymerase is illustrated in fig. 10-23 and adapted to RNAP in fig. 18-17.

Both enzymes from many sources contain a divalent metal ion (Scrutton et al., 1971; Slater et al., 1971; Valenzuela et al., 1973). In the two-metal mechanism, one divalent metal ion is imported to the active site in an elongation turnover with the MgNTP, and this metal ion facilitates the departure of PP_i as $MgPP_i$. The other divalent metal ion is endogenous to RNAP, and as in the action of DNA polymerase, it is thought to facilitate phosphodiester formation by coordinating the 3′-OH group at the end of the growing chain. Metal complexation will lower the pK_a of the 3′-OH group, allowing its ionization in the formation of the 3′,5′-phosphodiester linkage.

The Ribosome

Peptide bond formation in the standard protein biosynthetic pathway requires specific interactions of 20 species of aminoacyl-tRNA with mRNA at a ribosome, which translates the coding nucleotide sequence and guides the ordered sequence of aminoacyl transfer steps in the synthesis of a protein. The ribosome binds mRNA like a thread passing through its structure and guides the binding of aminoacyl-tRNAs in the appropriate sequence corresponding to the amino acid sequence of the protein, and it facilitates peptide bond formation. The process is detailed in textbooks of biochemistry, which explain how mRNA is produced by RNAP-catalyzed transcription of the coding nucleotide sequence in DNA, how a ribosome recognizes an mRNA code for starting a polypeptide, how aminoacyl-tRNAs employ their anticodons to recognize codons at the start and succeeding sites of mRNA bound to the ribosome, and how the aminoacyl-tRNAs are aligned by the ribosome in sequence corresponding to the amino acid sequences of the protein. Here we briefly describe the enzymology of recognition and control in ribosomal action and consider the chemistry by which a ribosome catalyzes peptide bond formation between a growing polypeptide chain and the aminoacyl moieties of charged tRNAs as they are presented in order at the peptidyl transfer center (PTC) within the ribosome.

Structure

Ribosomes are ribonucleoproteins comprising three species of rRNA and a large number of proteins in two subunits that form large complexes (2.6 mDa), about one half of the size

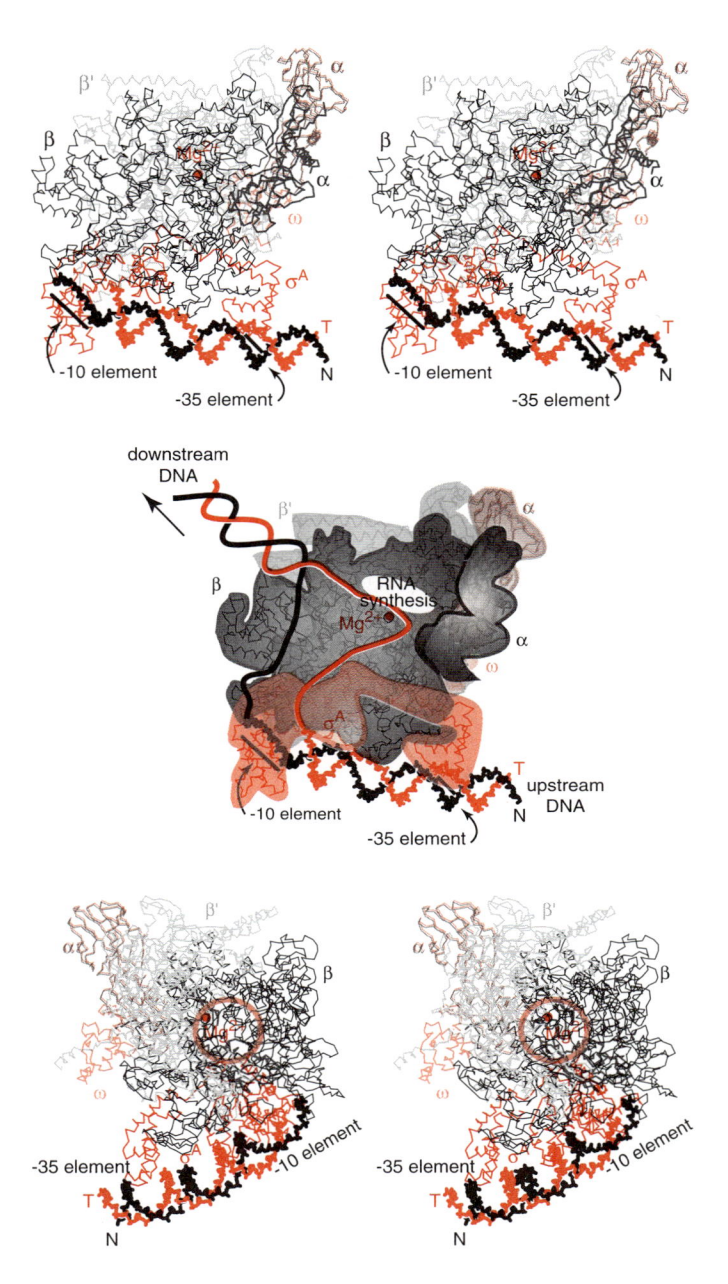

Fig. 18-16. The structural model of RNA polymerase (RNAP) shows the overall chain folds in an initiation complex of RNAP from *Thermus aquaticus* and the Mg^{2+} ion in the active site (6.5-Å resolution; PDB 1L9Z; Murakami et al., 2002b). A segment of double-stranded DNA upstream from the melted bubble is visible, with the template strand in red (T) and the complement in black (N). Although not part of this model, the strands split apart after the −10 element, and the template strand passes around a portion of the s domain and into the active site near the Mg^{2+} (red) in the center of the complex. The two strands recoil after the template strand emerges from the active site at the top of the complex (middle). The bottom stereoimage shows the active site Mg^{2+} through the passage (pink circle) from which the nascent RNA strand is extruded as it is synthesized.

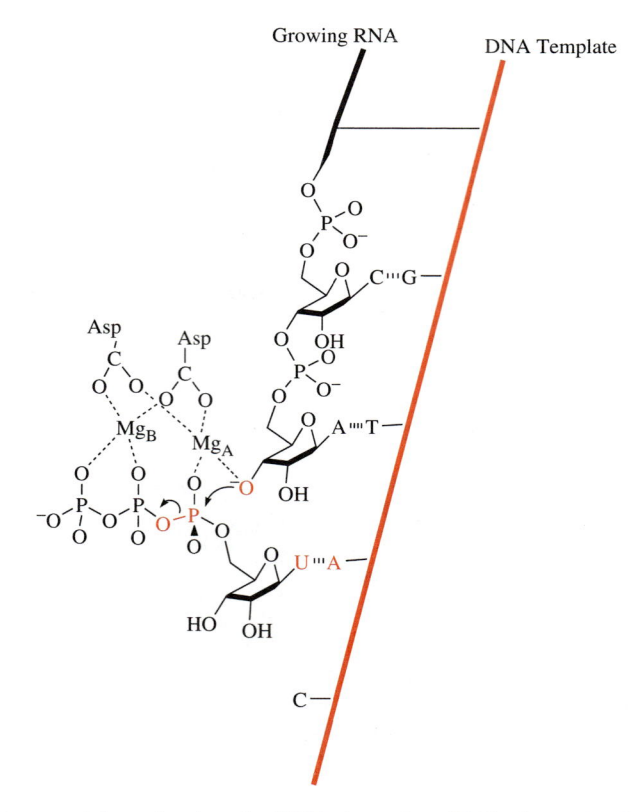

Fig. 18-17. A two-metal mechanism for DNA-dependent RNA elongation by action of RNA polymerase.

of an α-ketoacid dehydrogenase complex. A number of high-resolution x-ray diffraction crystal structures of ribosomes have been published (Ban et al., 2000; Harms et al., 2001; Hansen et al., 2002; Nissen et al 2000; Schlunzen et al., 2001). Ribosomes are the ultimate ribozymes. Ribosomes largely deprived of protein components in the large subunit catalyze peptide formation (Noller et al., 1992). A 2.4-Å crystal structure reveals the PTC in the 23s rRNA, with no protein group within 18 Å of the active site (Ban et al., 2000).

Prokaryotic and eukaryotic ribosomes are diverse in terms of detailed composition and function. We focus on the most extensively studied prokaryotic ribosomes, for which high-resolution crystal structures are available. The *E. coli* ribosome is the most thoroughly described and serves as the exemplar of ribosome function. Each ribosome has three sites that participate in translation, an A site that matches the anticodon of an aminoacyl-tRNA with its cognate triplet code in mRNA, a peptidyl transfer or P site that facilitates peptide bond formation between the amino group of an incoming aminoacyl-tRNA and the peptide carbonyl of the growing peptidyl-tRNA, and an exit or E site that accommodates the discharged tRNA after peptide formation. The mRNA passes through the ribosome in the course of peptide elongation, and the polypeptide emerges through a channel from the P site during this process. Shown in fig. 18-18 is the overall shape of an assembled ribosome; the locations of the P site, the A site, the PTC; and the channel for the growing peptide chain.

Initiation

Auxiliary enzymological functions of proteins known as initiation and elongation factors accomplish the essential ordering of aminoacyl-tRNA binding to mRNA, as well as the

Complex Enzymes 773

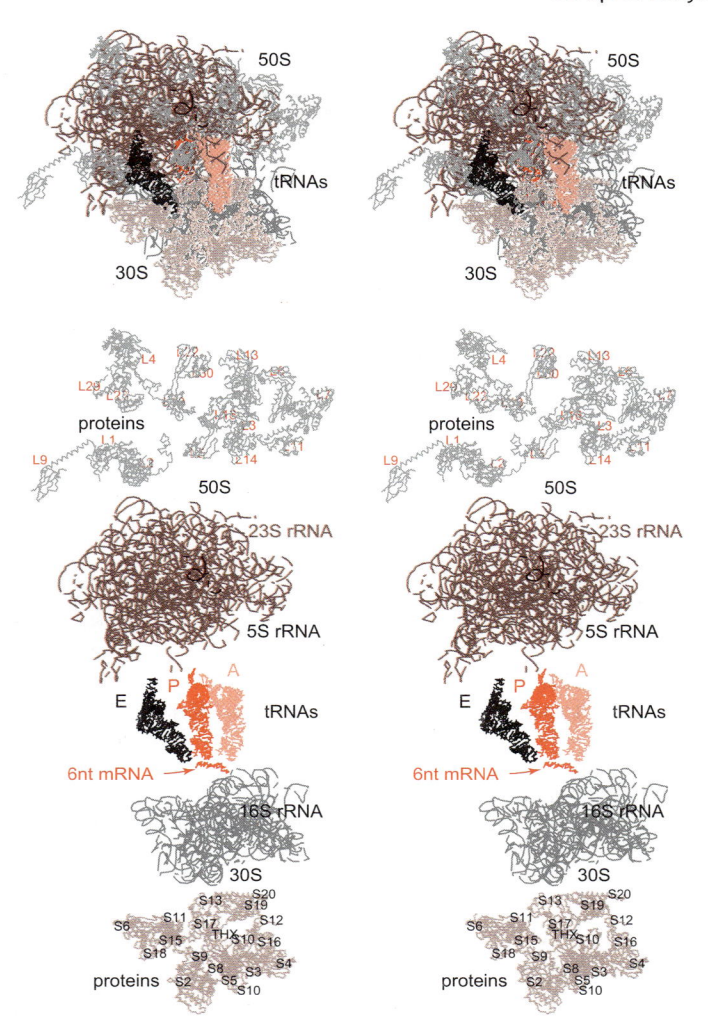

Fig. 18-18. Structure of the 70s ribosome from *Thermus thermophilus*. The top stereoimage shows the intact 70s ribosome with three tRNAs (black, red, and pink in the E, P, and A sites, respectively) bound between the large (50s) and small (30s) ribosomal subunits (5.5-Å resolution; PDB 1GIX and 1GIY; Yusupov et al., 2001). Below that image is an exploded view showing the proteins, the 23s and 5s rRNAs of the 50s subunit, the tRNAs and a 6-nucleotide mRNA, and the 18s rRNA and proteins of the 30s subunit.

translocation of ribosomes along the mRNA-chain. Translation by ribosomes in *E. coli* is the most thoroughly studied. The *E. coli* ribosomes comprise a 30s subunit and a 50s subunit that associate in the translation process into the 70s ribosome. The 30s subunit consists of a molecule of 16s RNA and a complement of ribosomal proteins. The 50s ribosome contains molecules of 23s and 5s RNA and associated ribosomal proteins. The drawing in fig. 18-19 illustrates the overall process of ribosome assembly on mRNA followed by initiation and elongation in the synthesis of a polypeptide. The overall scheme at the top of fig. 18-19 describes the steps in which the nucleotide sequence of mRNA is deciphered by recognition and regulatory interactions in the translation process.

To initiate translation, a ribosome binding site within the nucleotide sequence of mRNA and upstream of the start site for peptide initiation must be recognized by the ribosome. All start sites are encoded as the triplet AUG reading in the 5' to 3' direction. Only the AUGs in-frame within a coding sequence serve as start sites, and they must be recognized

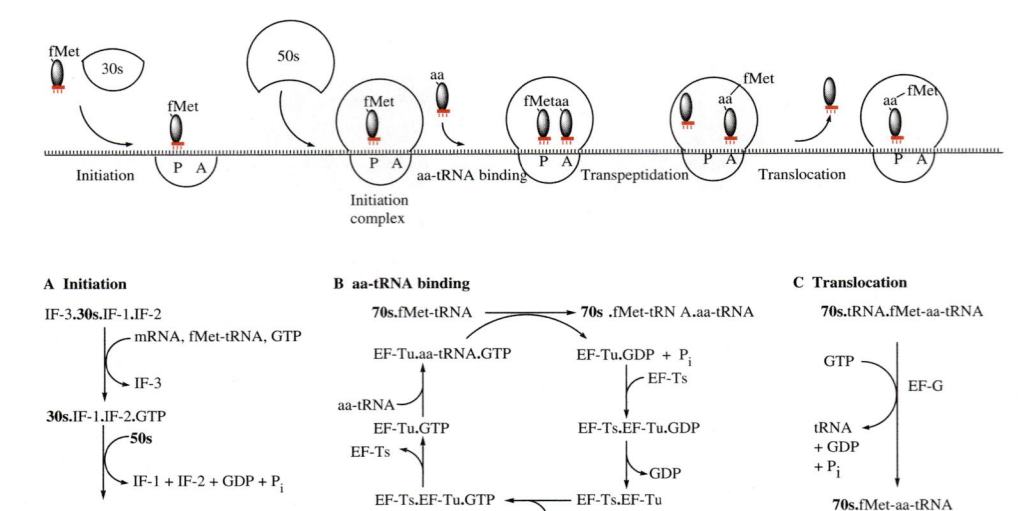

Fig. 18-19. Steps in ribosomal peptide biosynthesis. (A) Ribosomal synthesis of the first peptide bond in a polypeptide. Initiation leads to the 70s initiation complex, consisting of both ribosomal subunits and the initiating fMet-tRNA bound at the P site through its anticodon to the initiation codon AUG on mRNA. Elongation consists of binding the next species of aminoacyl-tRNA (aa-tRNA) to the A site, transpeptidation to form the fMet-aa-tRNA in the A site, and its translocation to the P site, coupled to release of the discharged tRNA. Initiation and elongation factors are proteins that facilitate these processes. The functions of initiating factors IF-1, IF-2, and IF-3 are shown. (B) Steps in binding the next aa-tRNA and the roles of Ef-Tu and Ef-Ts (i.e., elongation factors). (C) Steps in translocation and the role of elongation factor-G (EF-G).

by the ribosome. Recognition of these sites seems to involve surrounding sequence motifs that vary among species. In *E. coli*, a nucleotide sequence in mRNA, known as a Shine-Delgarno sequence, located 10 base-pairs upstream from a starting AUG is recognized by a complementary sequence in the 16s rRNA of the small ribosomal subunit (30s), which binds to that site. This places the mRNA in the correct position relative to the ribosomal P site to allow the initiation species of aminoacyl-tRNA, *N*-formylmethionyl-tRNA to bind through its anticodon to the AUG start site of mRNA, and the large ribosomal subunit (50s) binds to form the 70s initiation complex.

Initiation factors drive and control the complex, energy requiring process by which the ribosome finds the start site and initiates translation. Initiation factors IF-1, IF-2, and IF-3 exist in complex with 30s ribosomal subunits before initiation, and they are released in the initiation process as described in fig. 18-19A. IF-3 seems to prevent association of 30s and 50s ribosomal subunits. Its release takes place only on formation of a productive complex of mRNA, fMet-tRNA, and GTP with a 30s ribosomal subunit. Release of IF-3 allows a 50s ribosomal subunit to bind in a G protein–like sequence of steps culminating in the hydrolysis of GTP, the release of IF-1 and IF-2, and the formation of the initiation complex.

Elongation

Peptide bond formation is the central chemical event in elongation. However, it is a preactivated process that constitutes just one step of elongation. The three stages of elongation are multistep processes, beginning with binding the aminoacyl-tRNA specified by the triplet code adjacent to the starting AUG, followed by transpeptidation and translocation of the peptidyl-tRNA. These processes are depicted diagrammatically in the upper panel of fig. 18-19.

Binding of a specific aa-tRNA to the A site is a complex, GTP-dependent process detailed in fig. 18-19B. An aa-tRNA is brought into the ribosome in a complex with EF-Tu (an *E. coli* translational elongation factor) and GTP, and the binding of aa-tRNA to the A site is accompanied by GTP hydrolysis. The complex of EF-Tu.GDP is recharged with another aa-tRNA in a further complex process in which EF-Ts must bind to release GDP, and GTP must bind to release EF-Ts, before an aa-tRNA can bind for the next cycle of aa-tRNA binding. This is a G protein–like process, in which the aminoacyl-tRNA-binding properties of ER-Tu depend on which combination of GTP, EF-Ts, and GDP are bound to the protein, and the cycle is driven by the hydrolysis of GTP.

Transpeptidation is the chemical formation of a peptide bond in the PTC, associated with 23s rRNA in the 50s subunit between a growing peptidyl-tRNA and an aa-tRNA, which are locked on mRNA through their respective anticodons at the P and A sites. Because the peptidyl-tRNA is preactivated, transpeptidation is the one process in ribosomal action that does not require energy from GTP hydrolysis. We consider the mechanism of peptide formation in the next section.

The movement of the ribosome along the mRNA requires translocation of the growing peptidyl-tRNA from the A to the P site, and this requires an input of energy. As illustrated in part C of fig. 18-19, GTP serves as the source of free energy. Elongation factor EF-G facilitates the movement of peptidyl-tRNA, locked on mRNA through codon/anticodon pairing, to the P site as the discharged tRNA departs from a site known as the exit or E site. As the growing peptide chain emerges, it undergoes a process of folding into its biologically active tertiary structure. Protein folding in a cell is not left to chance but is guided by the actions of chaperonins, which bind to the nascent polypeptide chain and prevent interactions that would derail folding into the correct tertiary structure. The actions of chaperonins require energy that is derived from ATP hydrolysis. The exact and detailed functions of initiation and elongation factors and of chaperonins are subjects of intense interest and research.

Termination

Termination of peptide polymerization occurs when a termination codon of mRNA is reached in iteration of the steps illustrated at the top of fig. 18-19. As a rule, the most common termination codon is UGA, and the absence of species of aminoacyl-tRNA complementary to this codon disrupts the iterative cycle, during which the ribosome dissociates from mRNA, and polymerization ceases. An exception to the termination rule occurs in the biosynthesis of enzymes that include selenomethionine in their sequences. Glutathione peroxidase, among others, has selenocysteine at its active site (Bock et al., 1991; Stadtman, 1990). In the synthesis of these enzymes, a species of tRNA incorporates an anticodon that complements the stop codon UGA, and a selenocysteinyl-tRNA is produced by a specific aminoacyl-tRNA synthetase. The recognition of which stop codons are employed for incorporation of selenocysteine appears to be governed by surrounding nucleotide sequences in the mRNA (Gladyshev et al., 2004).

Peptide Formation

The chemistry underlying peptide formation between the acyl-carbonyl group of a growing peptidyl-tRNA and the amino group of an aminoacyl-tRNA in the PTC appears

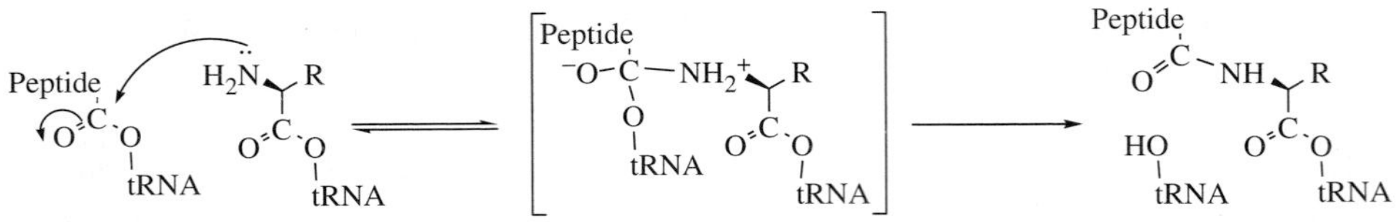

Fig. 18-20. Steps in peptide bond formation at the peptidyl transfer center site of the ribosome.

straightforward in principle; however, controversy remains regarding the mechanistic details. The likely chemical mechanism in fig. 18-20 illustrates the current concept of the steps in peptidyltransfer.

An aminoacyl-tRNA is highly activated for acyl group transfer. All evidence indicates that peptide formation is rate limiting. The rate does not vary with viscosity, proving that diffusion does not limit the rate (Sievers et al., 2004). Under conditions of equilibrium binding, the rate of peptide bond formation displays a ^{15}N kinetic isotope effect, proving that the chemical step is rate limiting (Seila et al., 2005). The amino group of an aminoacyl-tRNA displays a low enough pK_a value to be reactive as a nucleophile within the elongation complex. Values of pK_a at 25°C for most species of aminoacyl-tRNAs are not available. That of leucyl-tRNA is estimated to be 8.1 (Sievers et al., 2004). To the degree that methyl esters of amino acids can serve as models for tRNA esters, it is worth noting that the pK_a values range from 7.0 for phenylalanine methyl ester to 7.8 for glycine ethyl ester (Jencks and Regenstein, 1970). Available data indicate that amino groups of aminoacyl-tRNAs ionize in the physiological pH range and can be reactive as nucleophiles in the elongation of peptidyl-tRNAs. Proton transfer to a base at the PTC does not appear necessary to generate the amino group.

The mechanism of peptide formation in fig. 18-20 shows that proton transfer from nitrogen to the leaving tRNA must be part of the transformation of the initial tetrahedral adduct into the peptide. What is not known is whether this process is spontaneous or catalyzed by an acid/base group in the PTC. Peptide formation depends on an ionizing group with a pK_a of 7.5; however, nothing is known of the identity or function of this group (Katunin et al., 2002). The structure of the ribosome does not reveal the presence of an obvious catalytic group in the PTC, and specific mutations of nucleotide bases in the PTC do not block function (Beringer et al., 2003; Katunin et al., 2002; Polacek et al., 2001; Thompson et al., 2001). It has long been speculated that the catalysis by the ribosome essentially consists of binding the peptidyl-tRNA and aminoacyl-tRNA in the right positions to allow the reaction to occur spontaneously; that is, by lowering the entropy of activation for the reaction (Nierhaus and Cooperman, 1980). Measurements of the enthalpy and entropies of activation ($\Delta H^\ddagger$ and $\Delta S^\ddagger$) for the nonenzymatic and ribosomal peptide formation supported this hypothesis (Siemens et al., 2004). The value of $T\Delta S^\ddagger$ for the ribosomal process proved to be about +2 kcal mol^{-1}, whereas that for the nonenzymatic process was −11 kcal mol^{-1}. This difference accounted for the 2×10^7-fold rate enhancement at the ribosome. The values of $\Delta H^\ddagger$ were +12.7 and +16.0 kcal mol^{-1} for the nonenzymatic and ribosomal processes, respectively. Ribosomal catalysis involves desolvation of aminoacyl-tRNAs, protection of the growing peptide from external solvation, and alignment of the aminoacyl-tRNAs with the N-terminal amino group in the correct orientation for nucleophilic addition and the elimination of tRNA.

Although a catalytic group in rRNA has not been found, the 2′-OH group at position A76 of aminoacyl-tRNA has been found to be essential for peptide formation (Weinger et al., 2004). The exact function of this group is not known. A 2′-OH group does not display

Complex Enzymes 777

acid-base catalytic properties in neutral solution, but it could stabilize a transition state through hydrogen bonding or contribute to the structural integrity of the active site, or it could mediate proton transfer through a hydrogen bonding network. Much remains to be learned about the mechanistic details of peptide formation at the ribosome.

Energy-Coupling Enzymes

We have seen examples of energy coupling in the actions of G proteins in chapter 10 and ribosomal action in the preceding section. These are examples of what has become known as *"binding change"* on exchange of nucleoside diphosphates and triphosphates. Energy-coupling processes embrace a larger class of enzyme action, which includes any coupled action in which one step releases free energy and another requires free energy. In this section, we consider nitrogenase, cytochrome oxidase, and ATP synthase as examples of complex enzymes that function by energy coupling.

Nitrogenase

Reactions of Nitrogenase

Nitrogenases catalyze the fixation of atmospheric nitrogen and are found in anaerobes such as those that thrive in root nodules of legumes. They catalyze the reduction of molecular nitrogen to ammonia in an ATP-driven reaction. The stoichiometry in eq. 18-5 is generally quoted and is regarded as the limiting stoichiometry, with 1 mole of H_2 produced and 16 moles of MgATP hydrolyzed for each mole of nitrogen used.

$$N_2 + 8e^- + 16\,MgATP + 8H^+ \rightarrow 2NH_3 + H_2 + 16\,MgADP + 16P_i \quad (18\text{-}5)$$

The reduction of nitrogen to ammonia is not understood in mechanistic terms, although a great deal is known about the enzyme and the reaction; and available facts limit the mechanisms that can be considered (Burgess and Lowe, 1996; Howard and Rees, 1994; Kim and Rees, 1994; Seefeldt and Dean, 1997; Seefeldt et al., 2004).

Reduction of N_2 is coupled with the reduction of two protons to H_2 (Guth and Burris, 1983; Li and Burris, 1983; Simpson and Burris, 1984). The remaining six protons on the left side of eq. 18-5 are incorporated into the two moles of ammonia produced. When N_2 is absent, and no other substrate is present, nitrogenase catalyzes only H_2 formation. Nitrogenase also catalyzes the formation of HD from H_2 in an N_2-dependent reaction, which indicates the partial reversal of hydrogen formation. Any mechanism must account for these facts.

Hydrogen cyanide (HCN) is a substrate for nitrogenase and is reduced to methane, ammonia, and methylamine (Fisher et al., 2000; Hardy and Knight, 1967; Li et al., 1982; Rivera-Ortiz and Burris, 1975). The formation of methylamine as a final product and the production of ammonia in excess of methane are explained by scheme 18-5.

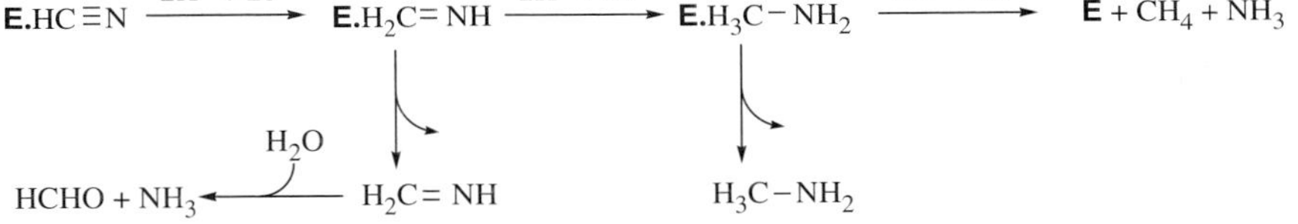

$$\mathbf{E.}HC\equiv N \xrightarrow{2H^+ + 2e^-} \mathbf{E.}H_2C=NH \xrightarrow{2H^+ + 2e^-} \mathbf{E.}H_3C-NH_2 \xrightarrow{2H^+ + 2e^-} \mathbf{E} + CH_4 + NH_3$$

$$HCHO + NH_3 \xleftarrow{H_2O} H_2C=NH \qquad H_3C-NH_2$$

Scheme 18-5

The reduction takes place in three two-electron steps, and methylamine and excess ammonia result from adventitious dissociation of methylene imine and methylamine formed as intermediates. In the reduction of N_2, the intermediates are diimide (HN=NH) and hydrazine (H_2N—NH_2); these intermediates do not dissociate from nitrogenase and are reduced to ammonia. Cyanide ion is not reduced, although it is isoelectronic with N_2, but it inhibits electron transfer and the fixation of nitrogen. Carbon monoxide is also isoelectronic with N_2, and it is an inhibitor of nitrogenase but not a substrate. Nitrogenase catalyzes the reduction of acetylene to ethylene. Because both N_2 and acetylene have triple bonds, one might think that the reactions of these molecules are similar, and they are likely to be related. However, their electronic structures are significantly different. In addition to being triply bonded, the two nitrogens in N_2 have nonbonding electron pairs, whereas the two carbons in acetylene are covalently bonded to hydrogens. This limits the interactions of acetylene with metal cofactors to π-binding, whereas N_2 may also bind end-on to a metal cofactor. In the reduction of N_2, acetylene inhibits noncompetitively, whereas N_2 competitively inhibits the reduction of acetylene (Rivera-Ortiz and Burris, 1975). These inhibition patterns indicate that these molecules bind to different enzyme forms. It has been suggested that the two substrates may bind to different oxidation states of nitrogenase (Davis et al., 1975).

Structure

Nitrogenases comprise two multi-subunit proteins, a homodimer known as the Fe-protein and a heterotetramer called the FeMo-protein. The homodimeric Fe-protein (65 kDa) has a single [4Fe–4S] cluster bridging the subunits through cysteine ligands from each subunit, and it has two MgATP binding sites, one in each subunit. The heterotetrameric FeMo-protein (230 kDa) is composed of two heterodimers [$(\alpha\beta)_2$] with two metallocofactors in each $\alpha\beta$ unit. One, known as the P cluster, has the composition [8Fe–7S] and consists of two four-iron clusters fused at one apex through a single sulfide. In the most thoroughly studied nitrogenases from *A. vinelandii* and *Clostridium pasteurianum*, the second cofactor in the FeMo-protein is a molybdenum-containing cluster known as the FeMo-cofactor and is composed as [7Fe–9S–Mo–X–homocitrate]. The ligand X is enclosed within the FeMo cluster and is postulated to be nitrogen, the nitride (N^{3-}); however, the possibility that X is carbide or oxide has not been excluded. The P cluster mediates electron transfer from the Fe-protein to the FeMo-cofactor, which carries out the reduction of N_2. In certain species, vanadium or iron occupies the site of molybdenum in the FeMo-cofactor (Eady, 1996).

Crystal structures of the FeMo-protein and the nitrogenase complex are available (Chan et al., 1993; Kim and Rees, 1992; Kim et al., 1993), the Fe-protein (Chiu et al., 2001; Georgiadis et al., 1992; Jang et al., 2000; Schindelin et al., 1997). Shown in fig. 18-21 is one structure of a nitrogenase complex, together with structures of the three cofactors.

The subunits incorporate β-sheets flanked by α-helices in the subunits. The dimeric Fe-protein is docked to the base of the heterotetrameric FeMo-protein, which is constructed as a dimer of heterodimers. The single [4Fe–4S] cluster at the interface of the Fe-protein is located near the docking surface, and the MgATP binding sites to each subunit are at the opposite end of the molecule. The P clusters in the FeMo-protein are placed between the FeMo-cofactor and the [4Fe–4S] clusters of the Fe-protein, and the distances separating the cofactors are compatible with electron tunneling rates, with the 14-Å separation between the [4Fe–4S] and P clusters being near the limit of effective enzymatic electron transfer (see chap. 4).

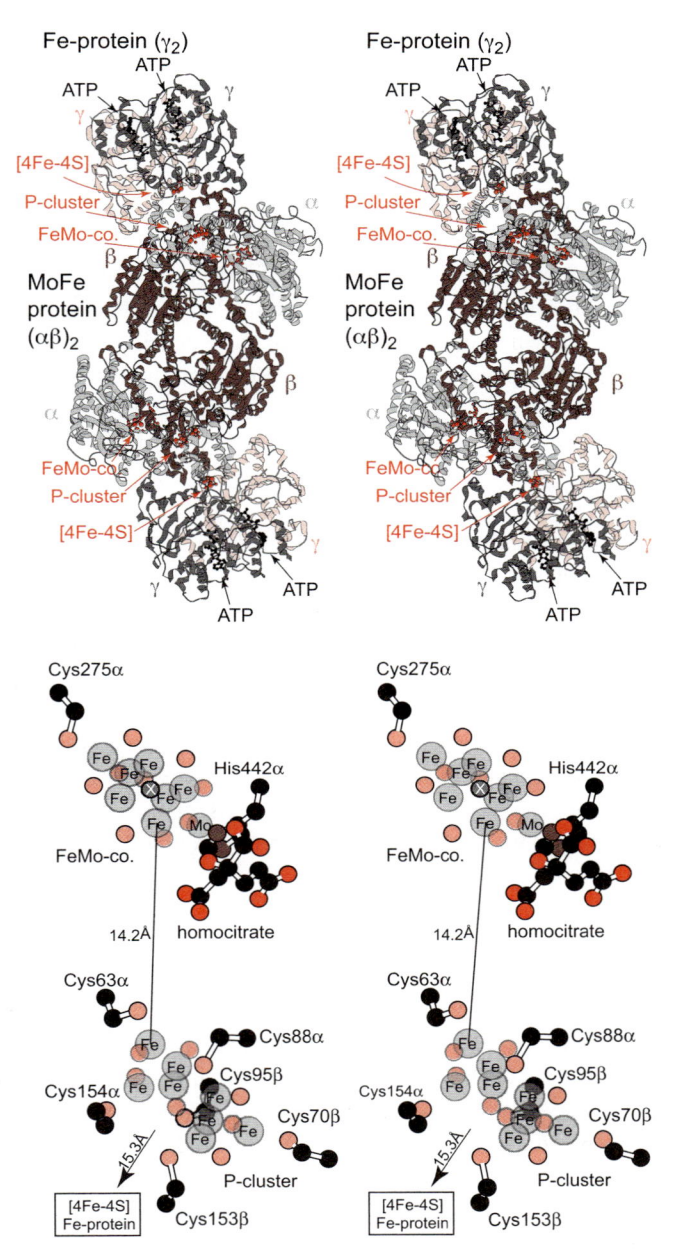

Fig. 18-21. The structures of nitrogenase and its metallocoenzymes. In the upper panel is a structure of the nitrogenase complex from *Azotobacter vinelandii*, showing the homodimeric Fe-protein docked on the heterotetrameric FeMo-protein (3.0-Å resolution; PDB 1G21; Chui et al., 2001). The ATP binding sites are located in each subunit of the Fe-protein on the opposite side from the docking interface, whereas the [4Fe–4S] cluster at the interface of subunits is located near the docking surface. The P cluster and FeMo-cofactor are in the FeMo-protein. In the lower panel are structures of the cofactors (1.16-Å resolution; PDB 1M1N; Einsle et al., 2002) shown in the order of the direction of electron flow from left to right. The atom (X) in the center of the FeMo-cofactor was observable in this high-resolution structure; its density was obscured by the adjacent iron atoms at lower resolution. The density observed for X is consistent with O, N, or C; although it seems likely to be N^{3-}, this has yet to be determined. The [4Fe–4S] and P clusters are separated by 14 Å, and the P cluster is separated from the FeMo-cofactor by 9 Å.

Much was learned about the structures of the [4Fe–4S] cluster in the Fe-protein and the P cluster in the FeMo-protein by biochemical and spectroscopic methods. The FeMo-cofactor could be extracted from the FeMo-protein by organic solvents and shown to contain molybdenum and homocitrate (Hoover et al., 1989; Imperial et al., 1989; Shah and Brill, 1977). However, elucidation of its detailed structure had to await the x-ray crystal structure of the FeMo-protein at a resolution of 1.16 Å (Einsle et al., 2002). This structure confirmed the overall array of iron, sulfide, molybdenum, and homocitrate that had been seen in the original crystal structures at lower resolutions, but it also revealed the internal mononuclear ligand shown as X in fig. 18-21 with the FeMo-cofactor. The identity of the internal ligand is not known at this writing. The electron density is compatible with C, N, or O, and the ligand is thought to be nitride (N^{3-}). However, the identity of X has not been proven.

ATP-Dependent Electron Transfer

Electron transfer in nitrogenase is ATP-dependent and results in the formation of ADP and P_i. The Fe-protein couples electron transfer with ATP hydrolysis, and the kinetic outline of this process is known (Duyvis et al., 1998; Thornley and Lowe, 1985). Each cycle of N_2-reduction requires 16 cycles of ATP-hydrolysis and electron transfer according to eq. 18-5. The Fe-protein functions by accepting electrons from the external reducing agent—generally dithionite in vitro—and transferring them to the FeMo-protein for use in reducing N_2 and H^+. The process proceeds in the four stages illustrated in fig. 18-22, association of the FeMo-protein with the reduced Fe-protein carrying two ATPs, electron transfer to the FeMo-protein coupled with ATP-hydrolysis, dissociation of the reduced FeMo-protein, and addition of one electron to the Fe-protein coupled with exchange of 2 ATP for 2 ADP.

The model in fig. 18-22 accounts for the stoichiometry in eq. 18-5 of two molecules of ATP hydrolyzed for each electron transferred. Several cycles are required to reduce each molecule of N_2, eight cycles according to eq. 18-5.

The details of the mechanistic linkage between electron transfer and ATPase action by the Fe-protein are not known in atomic detail (Rees and Howard, 2000). The stoichiometries of H_2/NH_3 production, electron transfer, and ATP hydrolysis are not even known with certainty. While eq. 18-5 is generally accepted, there are reasons to consider whether it is the last word (Rees and Howard, 2000). The stoichiometry depends on reaction conditions, and the limiting value of one half for H_2/NH_3 is attained at high pressures of N_2 (Simpson and Burris, 1984). Moreover, the stoichiometry of ATP-hydrolysis appears higher than necessary to reduce N_2 (Alberty, 1994). The oxidation state of the Fe-protein seems to

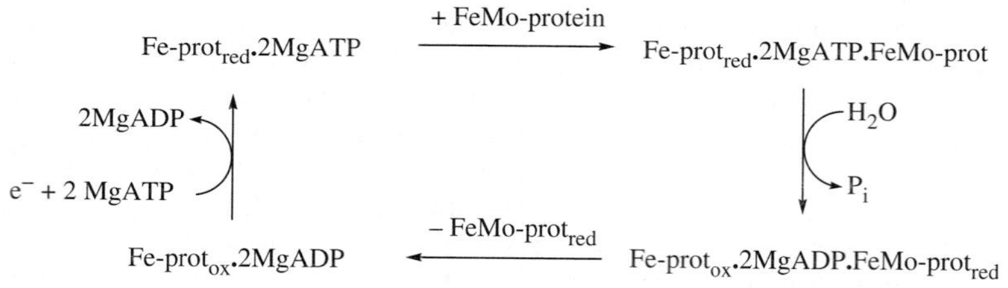

Fig. 18-22. One cycle of ATP-driven electron transfer by the Fe-protein in nitrogenase. The Fe-protein of nitrogenase catalyzes cycles of electron transfer to the FeMo-protein. One cycle proceeds in the four stages shown. Eight cycles of electron transfer and ATP hydrolysis are required for one cycle of N_2 reduction according to eq. 18-5.

affect the overall stoichiometry. In most experiments performed in vitro, employing dithionite as the reducing agent, the iron-sulfur cluster seems to cycle between the [4Fe–4S]$^{1+}$ and [4Fe–4S]$^{2+}$ states. However, the all ferrous [4Fe–4S]^0 state is also known (Watt and Reddy, 1994; Yoo et al., 1999). Reductants for the Fe-protein that can produce the all ferrous form [4Fe–4S]^0 are reported to support nitrogen reduction with lower ratios of ATP hydrolyzed to electrons transferred, such that the ratio approaches one ATP per electron (Erickson et al., 1999). In this case, two electrons would be transferred in each cycle of the action of Fe-protein in fig. 18-22.

The complexes of Fe-protein with MgATP and MgADP have different structures and presumably different docking interactions with the FeMo-protein, and the differences presumably control the rate of electron flow from the [4Fe–4S] to the P cluster. The G protein–like binding change contributes to the process as illustrated in fig. 18-22, in which the complex of Fe-protein with ATP binds to the FeMo-protein and the complex with ADP dissociates from the FeMo-protein.

The mechanism by which ATP-hydrolysis facilitates electron transfer is not known. In one hypothesis, the structural consequences of ATP binding to the reduced Fe-protein involve a reorientation of the subunits with respect to each other, leading to a structural distortion of the [4Fe–4S] cluster, such that it undergoes cleavage into two [2Fe–2S] clusters on oxidation (Sen et al., 2004). The model is inspired by the structure of a mutational variant that displays binding properties toward the FeMo-protein similar to those of the complex of Fe-protein with ATP. Deletion of Leu127, a residue in the hinge region involved in the structural change brought about by ATP, leads to increased affinity for the FeMo-protein. The structure of this mutated Fe-protein is consistent with a transition from a reduced [4Fe–4S] cluster and two [2Fe–2S] clusters in the oxidized state.

Mechanism of N$_2$-Reduction

Little is known of the detailed mechanism by which the reduced FeMo-protein carries out the reduction of N$_2$ and the production of H$_2$. The true structure of the FeMo-cofactor has only recently been revealed with its internal ligand X, which has not been identified (Einsle et al., 2002). The ligand is thought to be nitride (N^{3-}), but whatever its chemical identity, it does not exchange with external nitrogen (Lee et al., 2003). The current state of knowledge has been reviewed (Seefeldt et al., 2004), and several facts have been established. The FeMo-cofactor is in an electron paramagnetic resonance (EPR)–sensitive oxidation state designated M^N (s = 3/2; g = 4.3, 3.7, and 2.0) in the purified FeMo-protein and can be reduced to an EPR-silent state designated M^R. Molecular nitrogen binds only to the reduced state and is reduced in sequence to diimide (HN=NH), hydrazine (H$_2$N—NH$_2$), and 2 NH$_3$ through several cycles of electron transfer from the Fe-protein. The reduction of N$_2$ is accompanied by reduction of H$^+$ to H$_2$. Molecular hydrogen undergoes a N$_2$-dependent exchange with deuterium in D$_2$O to form HD.

The site at which N$_2$ binds to the FeMo-cofactor is unknown. In one school of thought, N$_2$ binds at the Mo-homocitrate; another school suggests that N$_2$ binds to a [4Fe–4S] face of the FeMo-cofactor (Seefeldt et al., 2004). Experimental support for the involvement of Mo-homocitrate in several studies show that any perturbation in the homocitrate ligand or the protein contacts in this region inhibit or inactivate reduction of N$_2$ (Imperial et al., 1989; Scott et al., 1992). Nonenzymatic reduction of N$_2$ by Mo-containing inorganic clusters to hydrazine or ammonia also supports the involvement of molybdenum (Demadis et al., 1996; Yandulov and Schrock, 2003). Experimental support for the involvement of iron in binding N$_2$ includes observations of the binding of intermediates in the reduction of acetylene or CS$_2$ to iron (Lee et al., 2000; Ryle et al., 2000).

The many difficulties attending studies of the action of nitrogenase include the necessity to work under anaerobic conditions because of the sensitivity of the iron cofactors to O_2; the fact that several cycles of ATPase action and electron transfer are required for each reduction cycle of N_2; the absence of spectroscopic methods to detect and differentiate diimide and hydrazine as intermediates; the uncertain nature of the interstitial ligand X in the FeMo-coenzyme; and uncertainties about the oxidation states of the iron-sulfur clusters. Modern spectroscopic and x-ray crystallographic methods in hand or under development may lead to an improved understanding of this important process.

Cytochrome c Oxidase

Vectorial translocation of ions arising from a biochemical reaction across a membrane creates an electrochemical potential that can serve as an energy source for another biochemical reaction when there is a coupling mechanism. The translocated ion can in principle be any ion, but most often it is K^+, Na^+, or H^+. Many examples of ion translocation and energy coupling are known, too many for all to be included here.

The most widely known and thoroughly studied cases of the creation of transmembrane electrochemical potentials are the actions of the terminal electron transport complexes such as cytochrome c oxidase (CcO), which translocate protons across membranes. The four complexes in the mitochondria of eukaryotic cells are NADH dehydrogenase, succinate dehydrogenase, cytochrome c reductase, and CcO. All of these complexes are multiprotein, integral membrane systems that relay reducing equivalents from one set of electron transfer cofactors to another, while at the same time translocating protons from the interior to the exterior of the mitochondrial membrane. Proton translocation sets up an electrochemical potential across the membrane. Relief form this potential can be brought about by any system that translocates protons from outside the membrane to the inside. Such a system is ATP synthase, which harvests the energy from the proton-electrochemical potential, the proton-motive force, by coupling the formation of ATP from ADP and P_i to the translocation of protons to the inside of the membrane.

A simplified picture of the actions of CcO and ATP synthase in a mitochondrial membrane is shown in fig. 18-23. The coupling of proton translocation with oxidation and ATP synthesis by the individual complexes are both slightly imperfect, and to the degree that coupling is incomplete the free energy lost is dissipated as heat. However, the overall process is remarkably efficient (about 50%) in producing ATP. In the following paragraphs, we consider the electron transport complex CcO, and in the next section, we discuss ATP synthase.

Function and Molecular Composition of Cytochrome c Oxidase

The last complex in most electron transport pathways, CcO, catalyzes the reduction of O_2 by four electrons relayed in the form of reduced cytochrome c (Fe^{2+}) from cytochrome c reductase, the preceding complex in terminal electron transport. The reduction of O_2 to $2H_2O$ requires four electrons from ferrocytochrome c and four protons from inside the mitochondrial membrane—or the cytosol of bacteria—and is accompanied by the vectorial translocation of up to four additional protons across the mitochondrial membrane to the intermembrane space—or the periplasm of bacteria. The overall process can be described by eq. 18-6 (Ferguson-Miller and Babcock, 1996; Wikström, 2004).

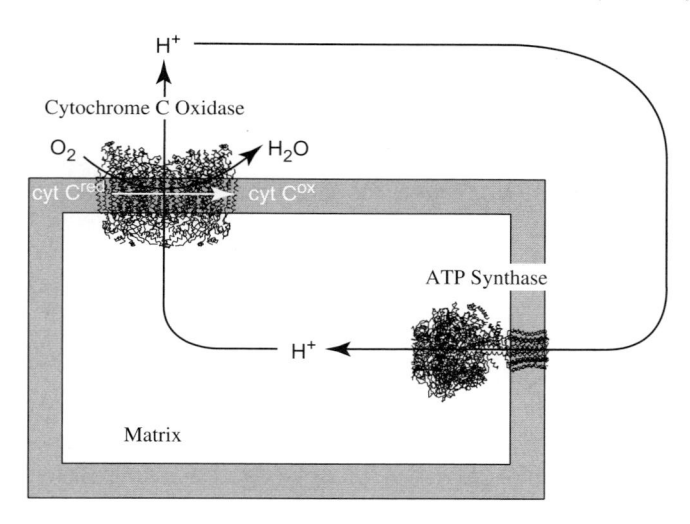

Fig. 18-23. Role of proton translocation in terminal electron transfer and oxidative phosphorylation.

$$4e^- + O_2 + 8H^+_{in} \rightarrow 2H_2O + 4H^+_{out} \qquad (18\text{-}6)$$

The latter, additional four protons, known as "pumped" or "vectorial" protons, amount to up to one proton pumped per electron transferred.

Beef heart mitochondrial CcO is a 400-kDa complex embedded in the mitochondrial membrane and consisting of 13 proteins, three large subunits I, II, and III encoded by mitochondrial DNA (57 kDa, 26 kDa, and 30 kDa, respectively), and 10 smaller subunits ranging from 5 to 17 kDa that are encoded by nuclear DNA (Azzi and Müller, 1990; Musser et al., 1995; Rizzuto et al., 1991). Bacterial CcOs consist of two or four subunits corresponding to I and II of the mitochondrial enzyme. Subunit I catalyzes the chemistry of O_2-reduction and proton translocation, subunit II mediates electron transfer from cytochrome c to subunit I, and the other subunits of mitochondrial CcOs play structural and regulatory roles.

The metallic centers carry out the chemistry of electron transfer and O_2-reduction in CcO: the centers are heme a, heme a_3, Cu_A, and Cu_B (Ferguson-Miller and Babcock, 1996; Musser et al., 1995). Subunit I contains heme a, heme a_3, and Cu_B, and subunit II contains Cu_A. The Cu_A center in subunit II mediates electron transfer from cytochrome c to heme a in subunit I, which mediates the transfer to heme a_3 and Cu_B.

Structure of Cytochrome c Oxidase

Crystal structures of the mitochondrial and bacterial CcOs are available (Iwata et al., 1995; Taukihara et al., 1995; 1996), and the structure of the mitochondrial CcO from bovine heart with its complement of metallocoenzymes is shown in fig. 18-24.

Subunit I contains heme a and the bimetallic center composed of heme a_3 and Cu_B, in which the copper and iron are separated by 3 to 4 Å (Chance and Powers, 1985; Musser et al., 1995; Scott, 1989). Heme a is bound through two histidine ligands in the axial positions, and heme a_3 is bound by a single axial histidine, leaving an axial position available for binding O_2. Two histidine residues and a third histidine chemically linked to a tyrosyl

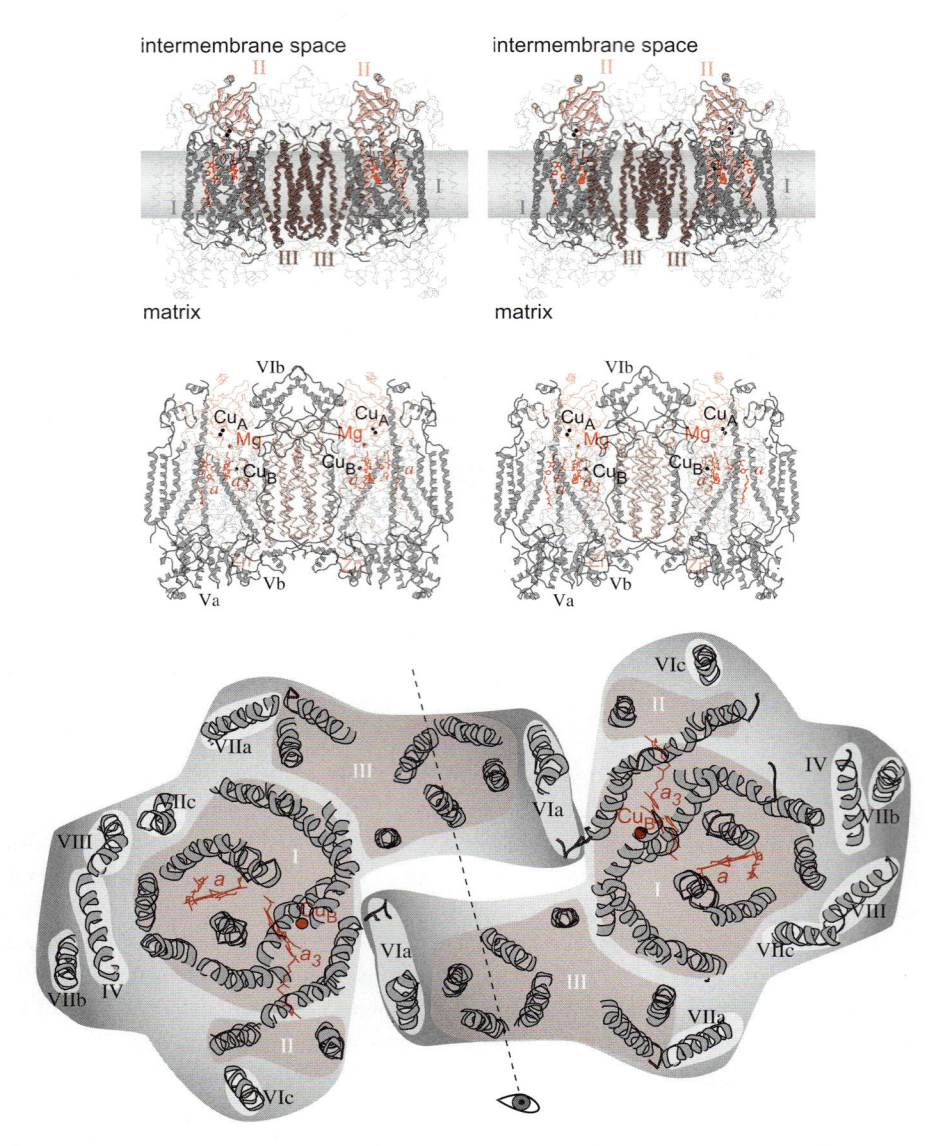

Fig. 18-24. Structure of the dimeric 13-subunit cytochrome *c* oxidase from bovine heart. This structure of cytochrome *c* oxidase shows a dimeric integral membrane protein containing two hemes (*a* and *a*$_3$), three Cu^{2+}, one Mg^{2+}, and one Zn^{2+} per 13-subunit monomer (2.8-Å resolution; PDB 1OCC; Tsukihara et al., 1996). Three core catalytic domains (I, II, and III) are encoded in the mitochondrial genome and make all of the contacts to the metal ions and heme cofactors. The core subunits (I is gray, II is pink, and III is brown, hemes are red) are shown as ribbon drawings in the top stereoview looking through a cross section of the mitochondrial membrane. In the middle panel, the other subunits (all of which are encoded in the nucleus, some of which vary by tissue type), which appeared as faint gray Cα traces in the first image, are shown as gray ribbon drawings. Only three subunits (Va, Vb, and VIb) do not contribute membrane-spanning helices, and these are labeled in the center panel along with the hemes and metal ions. The bottom panel shows a cross section of the enzyme dimer within the membrane span with all of the transmembrane helices grouped and labeled by subunit. The vantage of the two previous views is shown with the dashed line, and the positions of the hemes *a*, *a*$_3$ and Cu$_B$, which lie within the membrane span, are indicated.

residue coordinate the dinuclear copper center in the Cu_B site. The histidine-linked tyrosyl residue is thought to participate in electron transfer through radical formation (Proshlyakov, 2004). The 26-kDa subunit II contains Cu_A, a dinuclear copper complex ligated by a methionine, a glutamate, and two histidines.

The arrangement of metallocoenzymes in CcO raises questions about mechanisms and pathways of electron transfer. The Cu_A site is 19 Å from heme a and 22 Å from heme a_3, metal-to-metal, and the distance between hemes a and a_3 it is 14 Å. These distances are at or beyond the limit for long range electron transfer at enzymatic rates (see chap. 4). However, the distances between ligands in the metallic centers are much shorter, only 4.5 Å between the macrocyclic edges of hemes a and a_3. Moreover, a histidine ligand to Cu_A and a propionate side chain of heme a are bridged by the peptide linkage between a conserved Arg-Arg motif and might allow for fast electron transfer. Another ligand to Cu_A, glutamate, is connected through coordination to Mg^{2+} to a conserved histidine, which is in contact with a propionate side chain of heme a_3. Pathways for electron transfer are available, and it is unclear whether electrons pass along unique routes.

Reduction of O_2

Results of steady-state and transient spectrophotometry indicate that the Cu_A center initially accepts electrons from reduced cytochrome c and relays them to heme a in the large subunit (Geren et al., 1995; Hill, 1991). Heme a relays electrons to the heme a_3/Cu_B center, which reduces O_2 to 2 H_2O and pumps protons. Transient kinetic analyses by visible absorption, EPR, isotope-edited FTIR, and resonance Raman spectroscopies have led to the identification of several intermediates in the reduction of O_2 by the dinuclear heme a_3/Cu_B center. Their chemical natures have been inferred, partly by spectroscopic correlation with model compounds and partly by chemical logic in the transformation of O_2 into two molecules of H_2O.

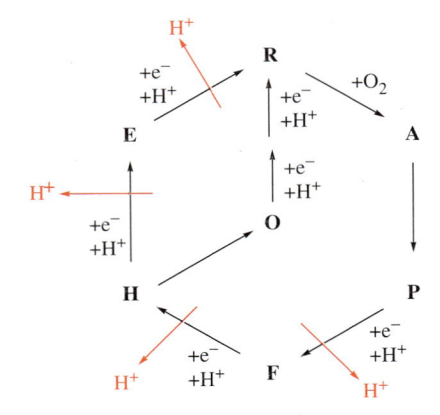

Fig. 18-25. A pathway for reduction of O_2 and pumping of protons by the heme a_3/CuB center of cytochrome oxidase. This pathway is adapted from Wikström (2004). The letters refer to species of the dinuclear center and oxygen intermediates: R is fully reduced, A is oxyheme, P is a peroxy heme, F is a ferryl, H and O are Fe^{3+}/Cu^{2+} states, and E is a Fe^{3+}/Cu^+ state. Two water molecules are released on reduction of Fe^{3+} and Cu^+ to form the E or R states. At each step of electron uptake, there is also a substrate proton taken up that is subsequently released as H_2O. The intermediate H normally proceeds to E in the presence of a reducing agent; but in an experiment starting with R and no reducing agent, H relaxes to O, which may subsequently be reduced. A proton (red) is pumped at each of four steps.

The generally recognized species are R, A, P, F, and O in fig. 18-25. The symbols represent the chemistry of the species. R stands for fully reduced (Fe_{a3}^{2+}/Cu_B^+), A for oxyheme ($Fe_{a3}^{2+}-O_2 \leftrightarrow Fe^{3+}-O_2^{-\circ}$), P for peroxy, F for ferryl ($Fe_{a3}^{4+}=O$), and O for the ferric hydroxide. Alternative pathways from F to R appear possible, as indicated in fig. 18-25. However, consensus is lacking regarding the exact structures of the intermediates and the significance of isomeric species such as tautomeric forms and forms related by short-range electron transfer between metals or between metallic centers and protein groups such as the His–Tyr ligand of Cu_B (Hill, 2004; Namslauer and Bzezinski, 2004; Ogura and Kitagawa, 2004; Proshlyakov, 2004; Schmidt et al., 2004; Wikström, 2004). In a further complication, observations of absorbance changes with time suggest the possibility that interconversions of the intermediates may follow a more complex course than a simple sequential kinetic model such as in fig. 18-25 (Szundi et al., 2004).

Proton Pumping

The protons in eq. 18-6 and fig. 18-25 are of two classes. Four substrate protons go into water formation, and four pumped protons are translocated from the inside to the outside of the membrane. There are unsettled questions and debates about the exact mechanisms, reaction sequences, and structures of intermediates in the use of substrate protons. In contrast, there is general agreement that the mechanism of proton pumping is not understood. In another point of agreement, the proton pump must include a conformational change as an essential step (Musser et al., 1995). The nature of the conformational change and the mechanism by which it directs electrons across the membrane remains to be proven.

An attractive hypothesis holds that a proton-bearing metal ligand could undergo a conformational reorientation in response to a change in the oxidation state of the metal, and such a conformational change could govern the vectorial movement of the proton (Gelles et al., 1987; Iwata et al., 1995; Wikström et al., 1994). A simple model for such a pump is shown in fig. 18-26 (Wikström, 2004). The pump model is based on an ordered sequence of electron and proton transfer steps that include a conformational change based on a ligand exchange or isomerization step. In the sequence, an electron from the low-potential reducing agent ferrocytochrome c reduces the heme a_2/Cu_B center at the inside of the membrane, and a ligand acquires an inside proton. Then, a ligand reorientation occurs linking the proton to a proton transfer pathway to the outside. On oxidation of the metal by a high-potential oxidizing agent, O_2 or a peroxide, the proton is released outside the membrane. Reorientation of the ligand returns the system to its original configuration for another round. The key ligand for the model in fig. 18-26 is not known and may be a histidine ligand to Cu_B.

ATP Synthase

The protonmotive force generated by the action of CcO and the other electron transfer complexes provides more than sufficient free energy to drive the condensation of phosphate with ADP to produce ATP, based on the principle of the chemiosmotic effect (Mitchell and Moyle, 1965). When embedded in the membrane, ATP synthase couples proton translocation into the mitochondrial matrix—or the cytosol of bacteria—to the synthesis of ATP from ADP and P_i. The mechanism of this process has been under intense investigation for more than 60 years and has been the subject of many reviews (Boyer, 1993, 1997; Capaldi et al., 1996; Fillingame, 1996; Fillingame and Dmitriev, 2002; Frasch, 2000; Penefsky and Cross, 1991; Stock et al., 2000). For each molecule of ATP

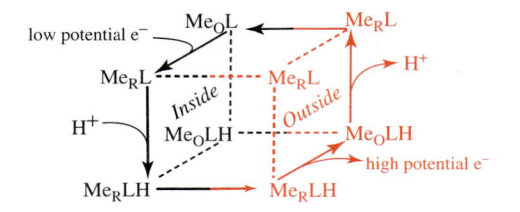

Fig. 18-26. Model for an electron transfer-coupled proton pump. This scheme is modeled after one presented by Wikstr^m (2004). It consists of chemical processes connecting the protonation states of a ligand for a metal, which can be in oxidized (Me$_O$) or reduced (Me$_R$) states, with proton translocation across a membrane. The ligand is connected to the two sides of the membrane by a conformational shift of the ligand. The ligand accepts protons from the inside while in the inside conformation (black), and it releases protons while in the outside conformation (red). The metal accepts reducing, low-potential electrons when the ligand is in the inside conformation (black) and transfers high-potential electrons only when the ligand is in the outside conformation (red). As long as the system follows these two rules, electron and proton transfers follow the solid lines, and protons are pumped. When transitions adventitiously follow the dashed lines, electrons are transferred, but protons are not pumped, and the energy from electron transfer is dissipated as heat

synthesized, three protons are translocated into the mitochondrial matrix, as described in eq. 18-7.

$$MgADP + P_i + 3H^+_{outside} \rightleftharpoons MgATP + H_2O + 3H^+_{inside} \qquad (18\text{-}7)$$

The reaction is driven in the forward direction by the protonmotive force generated by the vectorial transport of protons from the matrix into the intermembrane space by action of electron transfer complexes such as CcO. In the absence of a protonmotive force (i.e., in the absence of a reducing substrate for an electron transfer complex), the ATP synthase tends to function as an ATPase, and catalyzes eq. 18-7 in the reverse direction. This is not an important phenomenon in living cells, in which regulatory phenomena tend to inhibit and there is always a protonmotive force, but it happens in isolated mitochondrial particles deprived of substrates for the electron transfer complexes. Such substrates include NADH for NADH dehydrogenase or succinate for succinate dehydrogenase or reduced cyctochrome *c* for CcO. Succinate is the simplest substrate to supply for generating a protonmotive force in mitochondrial particles, and in the presence of succinate reaction 18-7 proceeds in the forward direction.

Molecular Properties

ATP synthases are integral membrane proteins composed of two main segments, F_1 and F_0. F_1 in mitochondrial particles can be separated and solubilized from F_0 by subjecting the mitochondria to low ionic strength (Boyer, 1997). The enzyme is often known as F_1F_0-ATPase, indicating the two major fragments F_1 and F_0 and the reaction observed in the absence of a protonmotive force. At low ionic strength F_1, also known as F_1-ATPase, becomes soluble and leaves F_0 buried in the membrane. F_0 functions in proton translocation and F_1 in catalyzing phosphoanhydride formation in the synthesis of ATP. Only the intact, membrane-bound F_1F_0-ATPase produces ATP from ADP and P_1, and then only when there is a protonmotive force. Solubilized F_1 is an active enzyme that catalyzes the hydrolysis of ATP to ADP and P_i (ATPase) by reversal of eq. 18-7, the energy-downhill process.

788   Enzymatic Reaction Mechanisms

The membrane-bound F_1F_0-ATPase harvests the free energy of the protonmotive force to reverse the chemistry and produce ATP from ADP and P_i.

All cells contain ATP synthase. The production of ATP is fundamental to life, and an organism turns over a stupendous amount of ADP and P_i into ATP, which is used for cellular, organismal, and muscular work and is recycled as ADP and P_i. In the cycling of ADP and ATP, a human produces—and consumes—more than his/her body weight of ATP in a day. The ATP synthases from various sources have been subjected to detailed studies, and the soluble forms have been referred to as MF_1 for mitochondrial, CF_1 for chloroplast, TF_1 for thermophilic, and EcF_1 for *E. coli* ATPase. We refer to the F_1-ATPases as MF_1, EcF_1, CF_1, and so forth, or generically as F_1 or F_1-ATPase; to the F_0s as MF_0, EcF_0, and so forth; and to F_1F_0-ATPases as ATP synthases.

Mitochondrial ATP synthase appears as knob-like structures 90-100 Å in diameter on the inner surface of the mitochondrial membrane. The solubilized F_1 includes the knob as well as a vestigial stalk, which is normally associated with F_0 embedded in the membrane. F_1-ATPases are complex enzymes composed of five subunits named α, β, γ, δ, and ε. The mitochondrial F_1 has a molecular mass of about 360 kDa and subunit composition $\alpha_3\beta_3\gamma\delta\varepsilon$, with subunit molecular masses of 55, 50, 31, 19, and 14 kDa, respectively. The hydrolysis or formation of ATP is carried out by the β subunits, which are closely associated with the α subunits as three heterodimers $(\alpha\beta)_3$. There are also nucleotide binding sites on the α subunits, but they are not catalytic and are thought to be regulatory. EcF_0 is also complex, with subunits a, b_2, and c_{10} and subunit molecular masses of 30, 17, and 8 kDa, respectively. F_0 couples proton translocation to ATP synthesis by F_1. The subunit assemblies of MF_1 and CF_1 are similar to EcF_1, with $(\alpha\beta)_3$-compositions, but the subunit compositions of MF_0 and CF_0 are more complex than of EcF_0.

Chemistry of ATP Synthase and the Binding Change Mechanism

As shown in biochemical studies, ATP synthase has preferential binding sites for MgATP and MgADP. Moreover, the kinetic studies of exchange reactions during hydrolysis by F_1-ATPase and ATP synthesis by ATP synthase and the effects of varying the free nucleotide concentration on the exchange kinetics, provide valuable clues to site-site interactions within the multimeric enzyme. As shown in early radiochemical experiments, F_1-ATPase catalyzes the rapid exchange of $^{32}P_i$ into "unreacted" ATP, the ATP/P_i exchange. That is, in an ATPase experiment terminated before the complete consumption of ATP, ^{32}P from $HO^{32}PO_3^{2-}$ appears in the pool of residual ATP. Such behavior is reminiscent of ping pong kinetics and the reversible formation of a covalent enzyme-substrate intermediate. However, no such intermediate exists despite the extensive studies carried out over many years in numerous laboratories in search of one. All evidence points to a direct reaction between $HOPO_3^{2-}$ and the β-phosphate of ADP in both ATPase and ATP synthase reactions. Careful measurements show the equilibrium constant for the interconversion of MgADP and MgATP (MgATP + H_2O $\rightleftharpoons$ MgADP + P_i) to be near 1.0 (O'Neal and Boyer, 1984). Moreover, under conditions of low nucleotide concentration, the interconversion of MgATP and MgADP at the active site occurs up to 400 times for each time ATP dissociates from the enzyme. Binding interactions alone, not proton translocation, potentiate the chemical reaction, and proton translocation brings about the release of ATP from ATP synthase in the overall reaction.

Reaction 18-7 produces water, which could in principle be derived from P_i or the β-phosphate of ADP. Experiments with ^{18}O showed H_2O to be derived from $HOPO_3^{2-}$ in the synthesis and hydrolysis of ATP by this system. In the ATPase reaction of mitochondrial particles, the exchange reactions P_i/ATP, total ATP/HOH, and free P_i/HOH were severely inhibited or abolished when the MgADP was recycled back to MgATP by the inclusion of pyruvate kinase and P-enolpyruvate (Kayalar et al., 1977). However, the ATPase rate was accelerated, and the exchange reactions total P_i/HOH and intermediate P_i/HOH were only slightly inhibited. Intermediate P_i/HOH refers to the exchange of water oxygen into free P_i accompanying transformation of free ATP into free P_i. ADP recycling dramatically lowered the concentration of free MgADP, and the consequences included the abolition of three exchange reactions. Those exchanges therefore required free MgADP. Inasmuch as the ATPase reaction took place rapidly in the absence of free MgADP, the role of free MgATP in facilitating the inhibited exchanges could not have been to stimulate a chemical step in catalysis and must have been to promote the release of exchanged species from the active site. These observations stimulated the invention of the alternating site model and binding change mechanism for the action of F_1-ATPase and ATP synthase. It was postulated that MgADP and P_i must bind to one β subunit before ATP could be released from another β subunit.

In the direction of ATP synthesis, exchange reactions catalyzed by ATP synthase in mitochondrial particles also respond to alterations in the concentrations of free MgATP (Boyer, 1997; Hackney and Boyer, 1978; Kohlbrenner and Boyer, 1983). The concentration of free MgATP in such experiments can be decreased to essentially zero by including glucose and hexokinase in the reaction medium, enabling all MgATP that is synthesized to be immediately transformed into glucose-1-P and MgADP. In the normal steady-state production of MgATP from [^{18}O]HPO$_4^{2-}$ and MgADP, the ^{18}O-content of residual free P_i declines with time. However, when the ATP-trapping system is included, the ^{18}O-content of free P_i is maintained. Free MgATP therefore must facilitate the [^{18}O]P_i/HOH exchange. In conventional enzyme catalysis, the reverse would occur because of product inhibition by accumulating MgATP. Moreover, during MgATP synthesis with [^{18}O]HPO$_4^{2-}$ and MgADP in the presence of the glucose/hexokinase trap, the amount of ^{18}O retained in glucose-6-P depends on the concentration of MgADP. When the MgADP concentration is 2.5 mM, most ^{18}O is retained, but when the MgADP concentration is 5 μM, the glucose-6-P retains only 14% of the ^{18}O. Therefore, an *intermediate* in the action of ATP synthase must undergo an MgADP-dependent exchange, because free MgATP is efficiently trapped as glucose-6-P. This is the intermediate [^{18}O]P_i/HOH exchange. The extent of intermediate [^{18}O]P_i/HOH exchange also increases with decreasing concentration of free P_i, which is contrary to expectations of conventional enzyme kinetics, in which the extent of exchange would decrease at lower substrate concentrations. These results support the concept of site-site interactions in the multimeric enzyme and an alternating site mechanism of catalysis.

The binding change mechanism for ATP synthase was put forward to explain the foregoing isotope exchange experiments (Hackney and Boyer, 1978; Kayalar et al., 1977; Kohlbrenner and Boyer, 1983; O'Neal and Boyer, 1986). In the original formulation, a two-site model of alternating sites sufficed to explain the results (Boyer, 1997). However, the discovery that ATP synthase contains three αβ-heterodimers rather than two or four led to revision to the three-site model. Early evidence suggested that the binding change would entail an ordered rotational alteration of conformations among the three αβ-heterodimers, as illustrated in fig. 18-27.

In the binding change model, each β subunit has a unique conformation and binding affinity for ATP or ADP and P_i (fig. 18-27). The conformational transitions are driven by proton translocations, in the forward direction by a protonmotive force and in reverse during ATPase action. Each conformational transition alters the conformations of the three subunits. The reaction sequence is shown in the clockwise direction for ATP synthesis

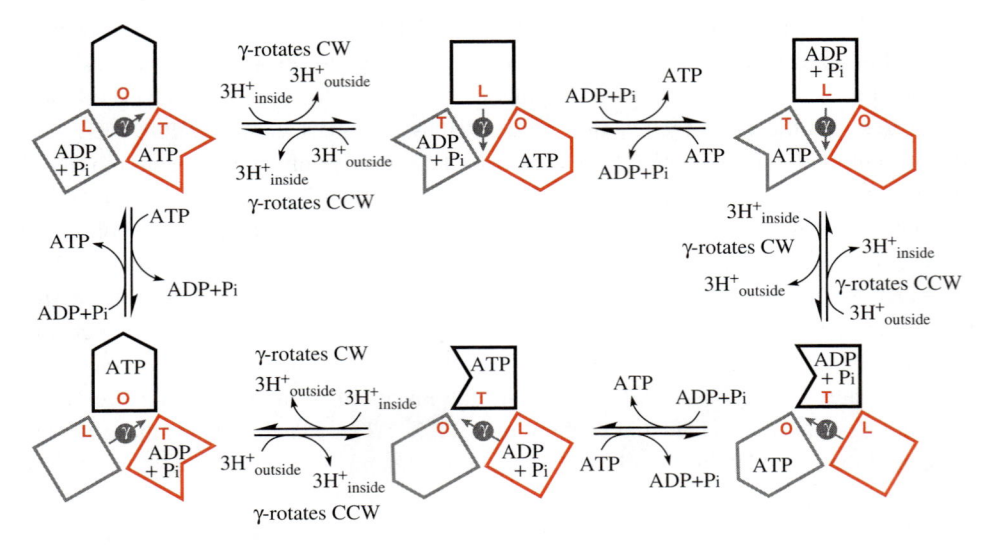

Fig. 18-27. A complete cycle of the binding change mechanism for the three α subunits of ATP synthase. The identical α subunits are shown in colors representing different binding properties. As their nucleotide ligation states change, they undergo complementary physicochemical changes in the binding change mechanism. Each color-coded form displays a unique binding affinity, which is designated O for open, T for tight, and L for loose. The T state binds ATP tightly and catalyzes its interconversion with ADP and P_i. The L state preferentially binds ADP and P_i and does not bind ATP. The O state displays little affinity for ADP or ATP and releases ATP in the synthesis direction. The transitions are driven by proton translocation through the γ subunit, which rotates clockwise (CW) in ATP synthesis (see fig. 18-28). The system synthesizes ATP because ATP does not bind to the L state, ADP and P_i bind only to the L state, and proton translocation drives the transition of the L state into the T state. The gray circle and clockwise arrow at the center represents the γ subunit, which projects through the $(\alpha\beta)_3$ domain; it rotates counterclockwise (CCW) in the hydrolysis of ATP by F_1.

in fig. 18-27. The L conformation has low affinity for MgATP but binds MgADP and P_i. The T conformation binds MgATP very tightly and catalyzes the formation of MgATP from MgADP and P_i. The O conformation releases MgATP with each conformational transition. This model explains all of the isotope exchanges.

The three β subunits illustrated in fig. 18-27 display astonishingly different affinities for MgATP. In the case of EcF_1, the affinities for MgATP vary over a range of five orders of magnitude (Weber et al., 1996). However, in the absence of Mg^{2+}, the β subunits bind ATP with identical affinities. The divalent metal ion is implicated in the differential binding of MgATP to the three β subunits, most likely through differential coordination of the metal ion by amino acid side chains of the β subunit. EPR analysis of ligation using vanadyl ($V=O^{2+}$), which functions in place of Mg^{2+}, unmasks ligand change processes in the β subunits (Frasch, 2000).

Molecular Structure

Biochemical data did not reveal how proton translocation and the protonmotive force drove the conformational changes required in the binding change mechanism. This required structural information. A hint at the role of the γ subunit is included in fig. 18-27, but this originates from structural information.

Complex Enzymes   791

Successes in structural analysis of ATP synthases have allowed mechanistic data and the binding change mechanism of ATP synthesis to be rationalized in structural terms (Stock et al., 2000). Shown in fig. 18-28 are schematic representations of an ATP synthase and a diagram of parts of the complex from the crystal structure.

Figure 18-28 shows a Cα trace of portions of F_1 and F_0 derived from a crystal structure, showing the αβ domains as the knoblike portion of F_1 and the (partially visible) γ, δ, and

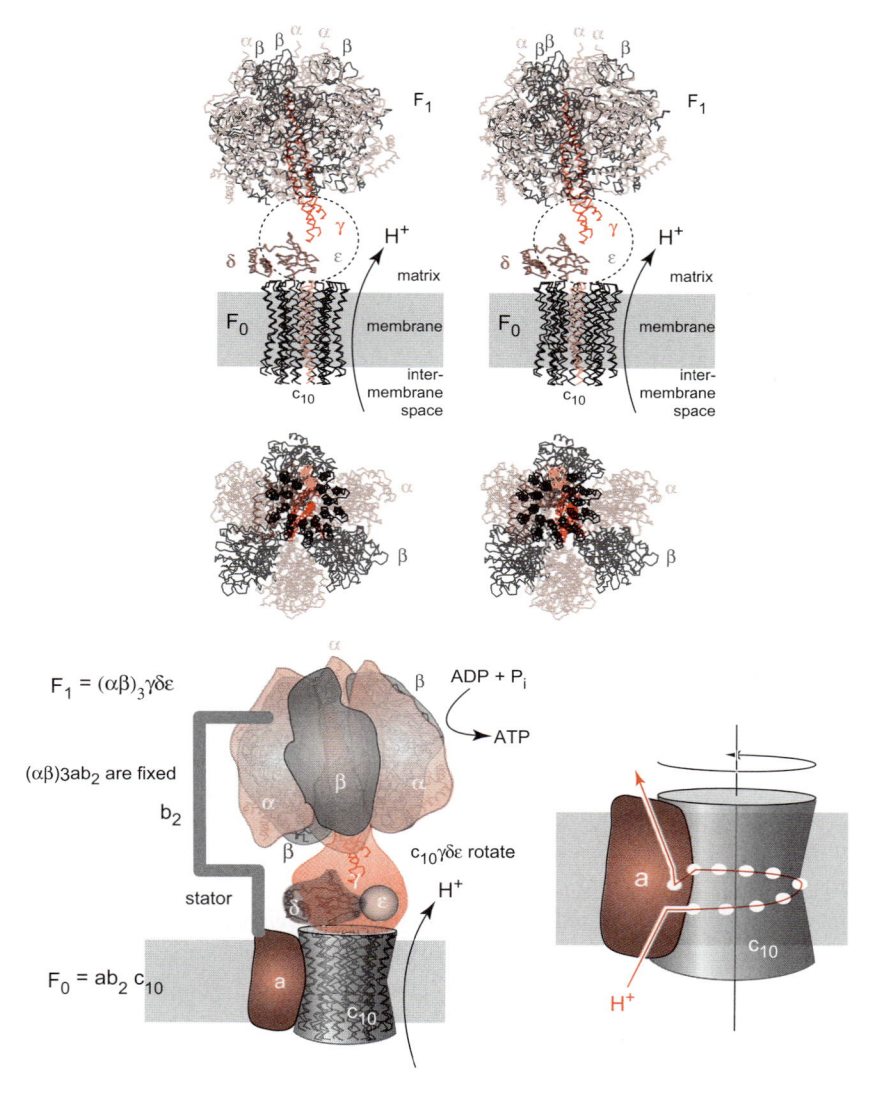

Fig. 18-28. The structure of F_1F_0 ATP synthase. The structure of yeast F_1F_0 ATP synthase shows most of the components of the functional synthase (3.9-Å resolution; PDB 1QO1; Stock et al., 1999). The F_1 subunit is composed of $(αβ)_3γδε$, and the F_0 is composed of ab_2c_{10}. Part of the δ and γ subunits and all of the ε subunit of the F_1, and both b and a subunits of the F_0 are not in the model. The top stereoimage shows the synthase from a cross section through the mitochondrial membrane. The middle image view is looking out the axis of the c_{10} "rotor" 20-helix bundle from the intermembrane space into the matrix. At the bottom left is a diagram showing the approximate position of all of the subunits, including the previously absent a, b_2 "stator," and γδε stalk proteins. To the right is a simple model for the proton gradient–driven rotation of the c_{10} rotor. In this model, the $c_{10}γδε$ subunits rotate, while the $ab_2(αβ)_3$ subunits are fixed.

ε subunits in the stalk, with the coiled α-helices of the γ subunit projecting through the center of the knob between αβ domains (Gibbons et al., 2000). The structure shows the α and β subunits arranged in alternation around the extended α-helices of the γ subunit. The β subunits exist in three conformations, which differ in their contacts with the asymmetric coiled helices of γ. As discussed in later sections, the γ subunit is a rotor, driven by proton translocation at the interface of the stalk and F_0.

The diagrammatic model in fig. 18-28 provides the components that were missing in the crystallographic model, and show how the α and β subunits are arranged around the coiled coil of α-helices from the γ subunit, with the δ, ε, and γ subunits in the stalk contacting the ring of 10 c subunits in F_0. A peripheral stalk composed of F_0 subunits a, b, d, F_6, and oligomycin sensitivity-conferring protein (OSCP) links the F_1 to F_0 and may serve as a stator that inhibits rotation of the αβ domains during rotatory action of the γ subunit. This model is derived from electron photomicrographic images and x-ray crystallographic data (Abrahams et al., 1994; Bottcher et al., 1998; Braig et al., 2000; Gibbons et al., 2000; Karrasch and Walker, 1999; Stock et al., 1999, 2000; Wilkens and Capaldi, 1998).

The central structure in F_0 is a cylinder composed of 10 c subunits associated with a- and b_2 subunits in the peripheral stalk. The structure of the c subunits is apparent from the crystal structure, and was originaly determined by NMR analysis in an organic solvent to be hairpin-like pairs of parallel transmembrane helices connected by a loop (Fillingame and Dimitriev, 2002; Girvin et al., 1998). The structure shows the proximity of Asp61 in one helix to Ala24 in the other. Asp61 is modified by the inactivator dicyclohexylcarbodi-imide (DCCD) and is required for activity. The proximity to Ala24 is significant because of the remarkable activities of the doubly mutated cA24D/cD61G and cA24D/cD61N synthases, showing that the essential aspartate could function from either helix (Miller et al., 1990; Zhang et al., 1994). Models of the cylindrical arrangement of c subunits are compatible with an essential role for Asp61 in proton translocation and the rotatory mechanism (Fillingame and Dmitriev, 2002).

Rotatory Action

The structure of F_1 showing structural differences in the β subunits and differential interactions with the coiled helices of the γ subunit implied that the conformational transitions proposed in the binding change mechanism (fig. 18-27) could arise from rotation of the γ subunit. In a direct verification, the γ subunit in an assembly of $(αβ)_3γ$ functioning as an F_1-ATPase was labeled with a fluorescent conjugate of actin and observed under a fluorescence microscope. The fluorescent labels underwent counterclockwise rotation in the presence of, but not in the absence of, MgATP (Noji et al., 1997). Moreover, the rotations proceeded in 120-degree steps corresponding to three steps for a complete rotation, consistent with the three β subunits (Yasuda et al., 1998). In ATP synthesis, the direction of rotation is clockwise.

The biochemical, structural and dynamical properties of ATP synthase characterize it as a rotary molecular motor that does chemical work in the synthesis of ATP using a proton-motive force as its source of energy.

Myosin and Muscle Contraction

The Sliding Filament Model

Classic microscopic experiments on muscle fibers led to the sliding filament theory of muscle contraction (Huxley, 1969; Huxley and Hansen, 1954; Huxley and Niedergerke, 1954).

In this theory, a muscle sarcomere undergoes shortening through differential contact interactions between filaments of myosin and actin by a process in which the catalytic domains, or heads, of myosin physically bind to actin filaments. The overall process is linked to the hydrolysis of MgATP. Biochemical experiments indicate differential interactions of myosin heads with actin depending on nucleotide ligation to myosin (Lynn and Taylor, 1971). Free myosin heads cleave MgATP, and the complex myosin. MgADP.P$_i$ binds to actin. Then, release of P$_i$ and MgADP leads to a conformational change in the myosin head that draws actin filaments together, shortening the sarcomere.

The highly simplified, two-dimensional drawing in fig. 18-29 illustrates the principle of the power stroke in shortening the sarcomere. The principal proteins in muscle filaments are polymerized actin and myosin. A protomer of myosin consists of a large catalytic domain, known as the head, and a long helical domain, known as the tail. Myosin forms a dimeric structure in which the helical tails form a coiled coil known as a rod. The myosin or thick filament arises from polymerization of the myosin rods, and the heads become projections extending outward from the filament.

The two-dimensional drawing in fig. 18-29 illustrates myosin heads as protrusions from the thick filament. In a sarcomere, the heads are projected in a helical array and contact actin filaments at many radial angles about the myosin filament. The actin molecules are smaller globular proteins that polymerize into linear helical filaments. In the contractile process, which is coupled with the action of the myosin heads in catalyzing the hydrolysis of MgATP, the myosin heads undergo a conformational change that leads to a change in their angular contact between the myosin and actin filaments. The consequent lateral displacement of actin filaments leads to sliding and to shortening of the sarcomere. Figure 18-29 illustrates this step, which is known as the power stroke, as arising from a tilting of the myosin heads accompanying the release of P$_i$ and MgADP from the myosin head. The model in fig. 18-29 also implies a highly idealized cooperative process, in which all of the myosin heads are in contact with actin. In a real muscle, a fraction of the heads are attached to actin at a given time, and others may be detached between power strokes.

Structures of Myosin Heads

A molecular or atomic model for muscle contraction is evolving from correlations of advanced experiments on the actions of muscle filaments with atomic resolution structures of actin and myosin heads (Highsmith, 1999; Holmes, 1996; Holmes et al., 2004; Huxley, 2004; Lombardi et al., 2004; Sweeney and Houdusse, 2004; Zeng et al., 2004). Myosin heads, known as subfragment 1 (S1), can be prepared by controlled proteolysis from myosin filaments. They can be crystallized and the structures determined by x-ray crystallography (Rayment, 1996). One class of structures of free myosin heads and of heads in complexes with analogs of MgATP are similar and display extended structures, and complexes with MgADP and analogs of P$_i$ display more compact structures (Coureux et al., 2003; Dominguez et al., 1998; Houdusse et al., 1999; Rayment, 1996; Rayment et al., 1993).

The structures of the two classes of myosin heads in fig. 18-30 illustrate the global differences. Free myosin heads have a globular domain and an extended domain centered by a long (≈8 nm) helix to which are associated the essential light chain and the regulatory light-chain domains. The light-chain domains are similar to the calcium binding domains of calmodulin. The globular domain contacts actin and is known as the motor domain; it has a cleft formed between the 50-kDa upper and 50-kDa lower domains. The cleft is open when nucleotides are bound and closed when unoccupied by nucleotides or analogs.

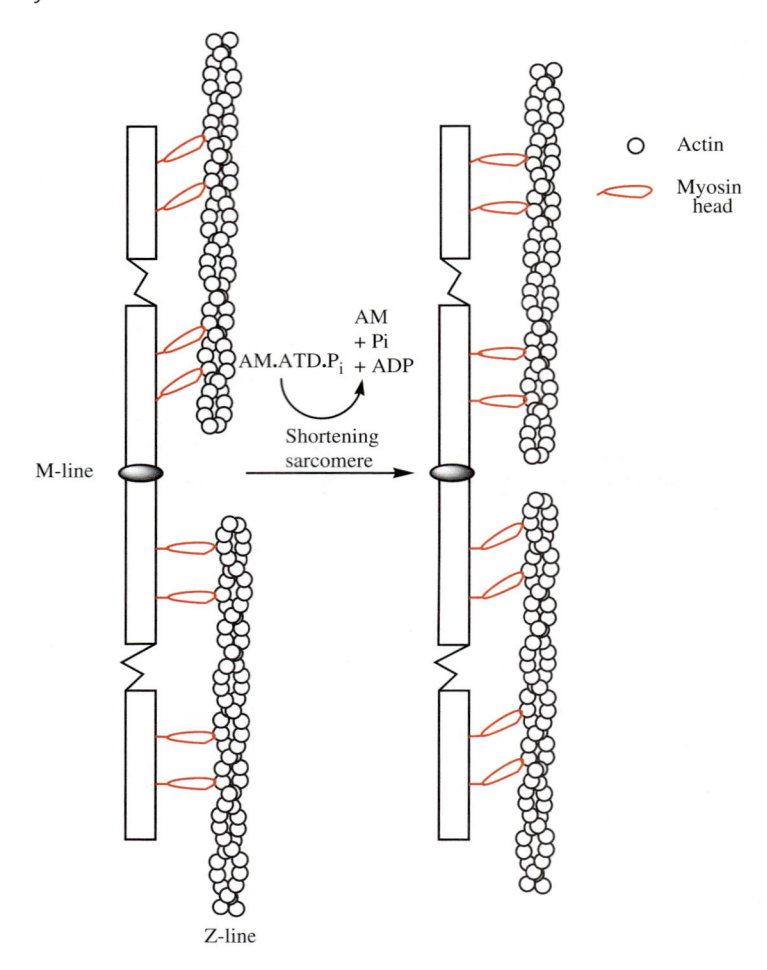

Fig. 18-29. In the sliding filament mechanism of muscle contraction, contractile muscle fibers contain aligned filaments of myosin and actin that undergo a sliding action on contraction. The thick filaments (black) consist of myosin dimers in which the coiled coil rods are polymerized to form the filament, from which the myosin heads protrude. Actin filaments (gray) consist of helically polymerized actin. The myosin heads bind to actin during contraction. The drawing illustrates the tilting bridge mechanism of filament sliding in a contracting sarcomere, in which the bridges are myosin heads that undergo a tilting motion on hydrolysis of MgATP. The nature and mechanism of this process are explained in greater detail in the text and figs. 18-30 to 18-33.

The relative orientations of the motor and extended helical domains are strikingly different in the two structures. The difference arises from a kink in the relay helix that is propagated by a converter domain to the long helix, or lever arm, and rotates it by 60 degrees. This conformational difference can be correlated with the two extreme states of the myosin heads in fig. 18-29 to explain the power stroke. The conformational difference quantitatively accounts for the 10- to 12-nm mean working stroke observed in x-ray diffraction analysis of muscle fibers subjected to mechanical lengthening (Huxley, 2004; Lombardi et al., 2004), Inasmuch as free myosin heads with open clefts are structurally similar to their complexes with MgATP analogs, the conformational change cannot be attributed to the binding of MgATP but must arise from the hydrolytic cleavage of MgATP to MgADP and P_i.

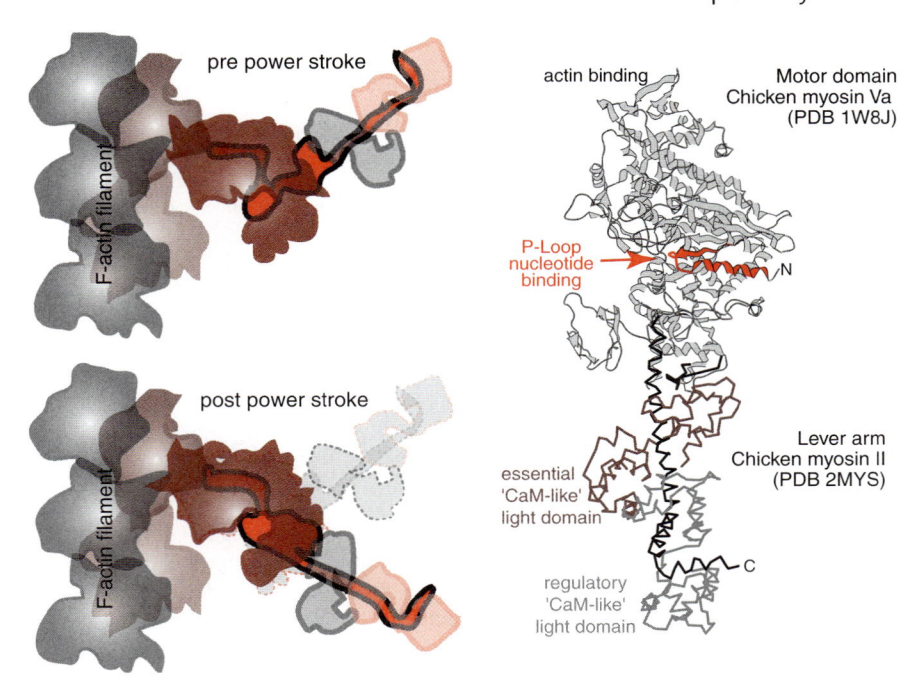

Fig. 18-30. Two structural classes of myosin heads. Crystal structures of myosin heads reveal two main classes of variant global structure, depending on the nucleotide/phosphate content in the ATPase site. All structures include a globular portion that includes the ATPase site and an elongated structure constructed around a long helix (8 nm). The globular portion, known as the motor domain, has a cleft formed between two domains, the 50-kDa upper and 50-kDa lower domains, and it binds directly to actin or actin filaments. The elongated domain is known as the lever in the lever arm mechanism of contraction. In one class of conformers found for free myosin heads and their complexes with analogs of MgATP, the long helical domain is extended. In the other class of conformers found for myosin heads in complex with analogs of MgADP and P_i, the long helical domain, or tail, is rotated by about 60 degrees. The ribbon diagram (right) is an assembly of the chicken myosin Va motor domain (Coureux et al., 2004) and the chicken myosin II lever arm (Rayment et al., 1993). (Adapted from Holmes et al., 2004.)

Kinetics of Myosin Action

Free myosin heads (M) catalyze the hydrolysis of MgATP in a process that does not involve any interaction of myosin with actin. Myosin heads bind actin in solution to form actomyosin (AM), and kinetic comparisons of AM with M illuminate the role of MgATP hydrolysis in muscle contraction (Lynn and Taylor, 1971; Takagi et al., 2004). Early observations proved that the hydrolysis of MgATP by free myosin heads proceeds by a mechanism in which the release of products limits the rate. MgATP undergoes cleavage at the active site much more rapidly than P_i and MgADP are released as products. Evidence of this included the observation of the formation of inorganic phosphate as $HOP^{18}O_3$ when the hydrolysis of MgATP was carried out in $H_2^{18}O$ (Sartorelli et al., 1966). Moreover, the hydrolysis of $Mg[\gamma\text{-}^{18}O_3]ATP$ in H_2O proceeds with loss of 79% of ^{18}O through exchange with water (Bagshaw et al., 1975). This corresponds to the loss of 2.4 of the 3 ^{18}O atoms, on average. The kinetic mechanism in fig. 18-31A was established by transient kinetic analysis, in which M, M*, and M** represent myosin heads in different conformational states detected by variations in tryptophan fluorescence.

In this mechanism, the binding of MgATP by free myosin heads is an equilibrium process with $K_{eq} = 4500$ M^{-1} at pH 8 and 21°C. The cleavage of MgATP is very fast ($k_4 \geq 160$ s^{-1}), and the conformational transition preceding P_i release, which is governed by k_6 (0.06 s^{-1}),

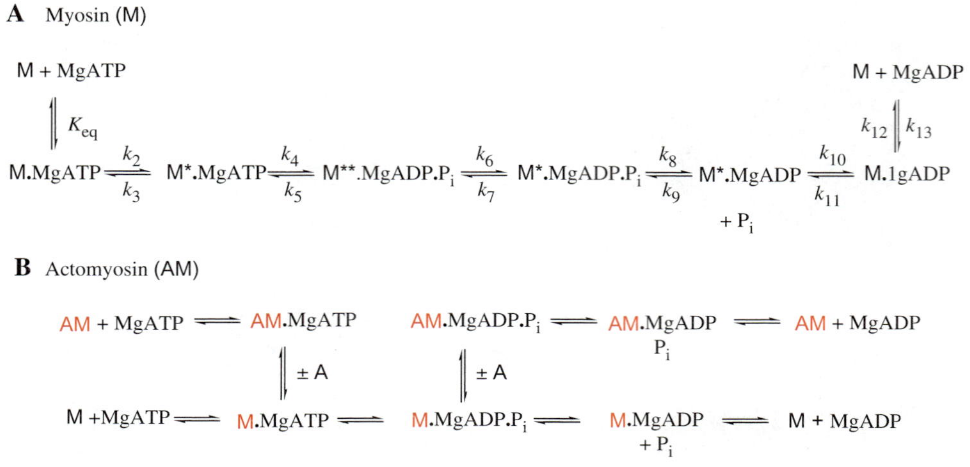

A Myosin (M)

$$M + MgATP \qquad\qquad\qquad\qquad\qquad\qquad\qquad\qquad\qquad\qquad M + MgADP$$

$$\Big\| K_{eq} \qquad\qquad\qquad\qquad\qquad\qquad\qquad\qquad\qquad\qquad\qquad k_{12} \Big\| k_{13}$$

$$M.MgATP \underset{k_3}{\overset{k_2}{\rightleftharpoons}} M^*.MgATP \underset{k_5}{\overset{k_4}{\rightleftharpoons}} M^{**}.MgADP.P_i \underset{k_7}{\overset{k_6}{\rightleftharpoons}} M^*.MgADP.P_i \underset{k_9}{\overset{k_8}{\rightleftharpoons}} M^*.MgADP \underset{k_{11}}{\overset{k_{10}}{\rightleftharpoons}} M.MgADP$$
$$+ P_i$$

B Actomyosin (AM)

$$AM + MgATP \rightleftharpoons AM.MgATP \qquad AM.MgADP.P_i \rightleftharpoons AM.MgADP \rightleftharpoons AM + MgADP$$
$$\qquad\qquad\qquad\qquad\qquad\qquad\qquad\qquad\qquad\qquad P_i$$
$$\qquad\qquad\qquad\Big\| \pm A \qquad\qquad\qquad \Big\| \pm A$$
$$M + MgATP \rightleftharpoons M.MgATP \rightleftharpoons M.MgADP.P_i \rightleftharpoons M.MgADP \rightleftharpoons M + MgADP$$
$$\qquad\qquad\qquad\qquad\qquad\qquad\qquad\qquad\qquad\qquad + P_i$$

Fig. 18-31. Mechanisms for ATPase action by myosin and actomyosin. (A) Kinetic mechanism for the ATPase action of myosin heads, as determined by transient kinetic methods. The rate-limiting step in this mechanism is the dissociation of P_i, governed by k_8. The myosin species M, M*, and M** are conformational variants detected by characteristic properties of tryptophan fluorescence. (B) Mechanism for ATPase action by actomyosin, a complex of myosin heads with g-actin. The differential burst kinetics in fig. 18-32 indicate that the principal pathway taken in the action of actomyosin ATPase is that involving complexes with actin and myosin (red). Binding of MgATP leads to dissociation of actin, the complex of M.MgATP undergoes hydrolysis to M.MgATP.P_i, actin binds again, and the products dissociate from the actomyosin complex.

limits the rate (Bagshaw and Trentham, 1974; Bagshaw et al., 1974). The reverse of the hydrolytic cleavage step is also fast relative to the conformational transition and release of P_i, with $k_7 \geq 18$ s^{-1}. The hydrolytic cleavage step undergoes many reversals for each forward step and release of P_i, and the exchange of ^{18}O with H_2O takes place in the course of reversible hydrolysis.

Myosin-Actin Interaction

Actin is a filamentous polymer (f-actin) at moderate and high ionic strengths, and this property complicates crystallography. At very low ionic strength, it exists as monomeric g-actin. DNase I adventitiously binds g-actin, and the complex can be crystallized. A structure for g-actin became available from the crystal structure of this complex, and a structure of g-actin in complex with ADP was also solved (Kabsh et al., 1990; Otterbein et al., 2001). The two structures are similar except for a short helix in the ADP complex, which is opened by contact with DNase I in the original structure.

Myosin heads bind to actin to form actomyosin. The strength of myosin-actin binding depends on the nucleotide content of myosin. Free myosin heads with closed clefts bind strongly to actin, and the binding of MgATP to actomyosin leads to the dissociation of myosin as its complex with MgATP. The complex of myosin with ADP and P_i binds strongly to actin. These differential binding interactions form the basis for a model of myosin action in muscle contraction.

Evidence for the effect of actin on the action of myosin originates with comparative kinetic properties of myosin and actomyosin as ATPases (Lynn and Taylor, 1971). Transient kinetics of P_i-production, as shown in fig. 18-32, show that the complex M.MgADP.P_i appears much more rapidly than free P_i because of the slow dissociation of P_i. The transient

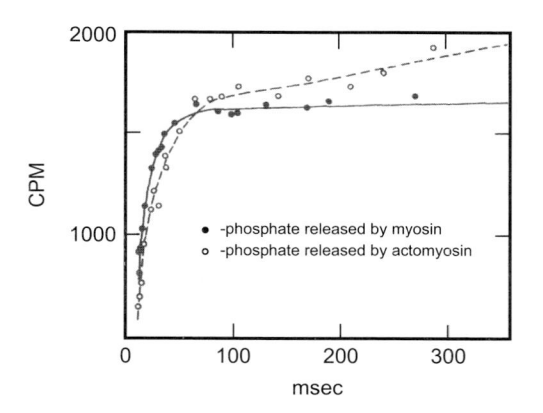

Fig. 18-32. Transient kinetics of ATPase action by myosin and actomyosin. The solid circles describe the time course of total P_i formation in the ATPase action of free myosin, and the open circles show the time course for the same concentration of actomyosin. The burst represents the hydrolytic cleavage of MgATP at the active site of free myosin, and the secondary steady-state rate represents the release of P_i from the product complexes, M.MgADP.P_i or AM.MgADP.P_i. The faster steady-state rate for actomyosin represents the 200-fold faster rate of P_i dissociation from actomyosin than from myosin. (From Lynn and Taylor, 1971, with permission.)

formation of total P_i in rapid mix-quench experiments displays burst kinetics, with the complex appearing in the burst followed by an extremely slow steady-state appearance of P_i because of its slow dissociation. The burst kinetics of actomyosin ATPase is related but different, as shown in fig. 18-32. The burst for actomyosin is of the same magnitude but slightly slower than with myosin alone, whereas the steady-state phase is much faster.

The ATPase bursts and steady-state behaviors of myosin and actomyosin are best understood on the basis of the kinetic mechanism in fig. 18-31B. This is an extension of the kinetic model for the action of myosin, modified by the consequences of actin-myosin interactions. On binding MgATP, myosin dissociates from actomyosin, and the complex of myosin and MgATP undergoes hydrolysis. The product complex is partitioned between two pathways; the dominant or faster one is shown in color in fig. 18-31B. This scheme accounts for the kinetics on the basis of the differential actin-binding properties of complexes of myosin with MgATP or MgADP.P_i, as discussed earlier. The hydrolysis of MgATP occurs on free myosin, but the release of P_i predominantly occurs from actomyosin. The myosin-MgATP complex has little affinity for actin, but the myosin-MgADP-P_i complex has high affinity for actin.

The crystal structures of g-actin and myosin heads are being incorporated into the modeling of electron micrographs of f-actin decorated with myosin heads (Holmes et al., 2004). The models show that the catalytic domains of the myosin heads are in the binding interface and that the cleft of the binding pocket for MgATP is in contact with actin. The structure of myosin in the complexes is that of the tightly binding free myosin.

Mechanochemical Coupling

Kinetic experiments unveiled a key to the mechanism by which the hydrolysis of MgATP at the active site of myosin is linked to the sliding filaments in fig. 18-29 (Lynn and Taylor, 1971; Sweeney and Houdusse, 2004; Takagi et al., 2004; Zeng et al., 2004). A model for this mechanism, in its simplest and original form, is illustrated in fig. 18-33. When the active site of the myosin head is free of magnesium, phosphate, and ADP or ATP, the actin-myosin

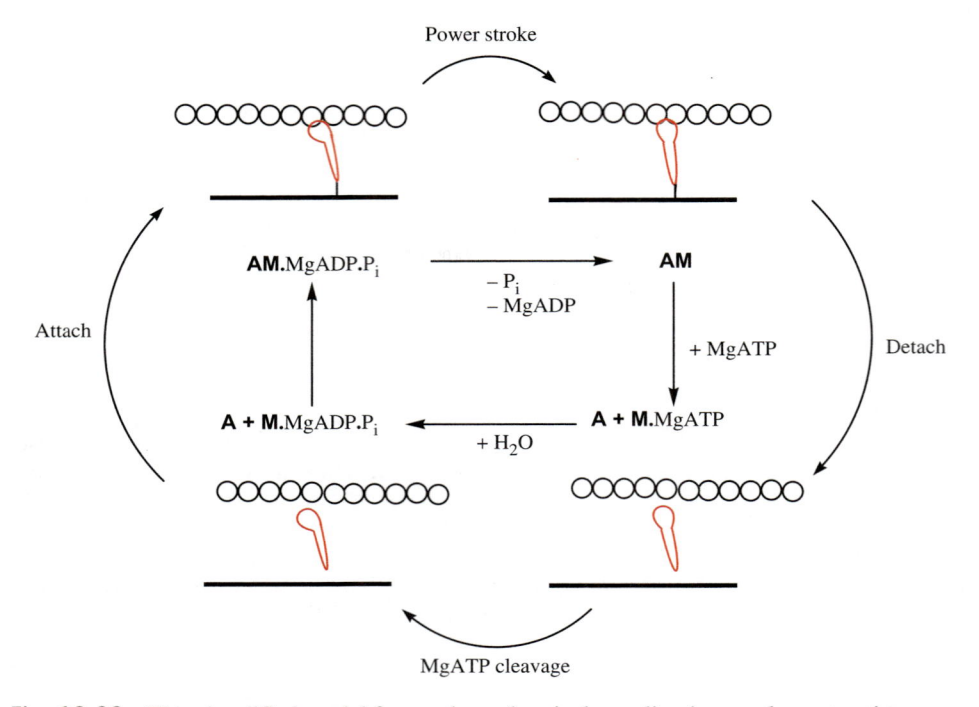

Fig. 18-33. This simplified model for mechanochemical coupling in muscle contraction correlates the hydrolytic cycle in the action of actomyosin with major steps in the sliding filament model of muscle contraction. The actomyosin cycle is shown at the center, and the action of myosin heads in the contractile process is illustrated in the peripheral diagrams. The detailed models are based on the same principles as in this simplified model but include additional steps for the conformational transitions of myosin shown in fig. 18-31A and for the stepwise release of P_i, Mg^{2+}, and ADP from actomyosin (Sweeney and Houdusse, 2004; Zeng et al., 2004).

contact is defined as the rigor state. On binding MgATP, the myosin head becomes detached because of the low affinity of M.MgATP for actin. The hydrolysis of MgATP to MgADP.P_i takes place during this detached state. The hydrolysis product M**.MgADP.P_i has a high affinity for actin because of the conformational difference between M and M**, and it again binds to the actin filament. The dissociation of P_i, Mg^{2+}, and ADP allows the conformation of myosin to relax, and this drives the power stroke that moves the actin filament along the myosin filament.

The model in fig. 18-31 illustrates the main principle in a simplified form. Current models are expanded versions with additional steps that distinguish processes collapsed into single steps in fig. 18-31 (Sweeney and Houdusse, 2004; Takagi et al., 2004; Zeng et al., 2004). For example, kinetic evidence indicates that the power stroke may proceed in a series of steps associated with the dissociation of P_i, Mg^{2+}, and ADP. The extended models allow for the known diverse behavior of myosins engaged in dynamically different contractile roles, as in skeletal muscle, smooth muscle, and free myosin. It is thought that additional structures of myosin heads will be required to fully explain the known dynamic functions of myosin (Sweeney and Houdusse, 2004).

References

Abrahams JP, AGW Leslie, R Lutter, and JE Walker (1994) *Nature* **370**, 621.
Abrahams JP, R Lutter, RJ Todd, MJ vanRaaij, AFW Leslie, and JE Walker (1993) *EMBO J* **12**, 1775.

Arjunan P, N Nemeria, A Brunskill, K Chandrasekhar, M Sax, Y Yan, F Jordan, JR Guest, and W Furey (2002) *Biochemistry* **41**, 5213.
Athappilly FK and WA Hendrickson (1995) *Structure* **3**, 1407.
Azzi A and M Müller (1990) *Arch Biochem Biophys* **280**, 242.
Bagshaw CR, JF Eccekeston, F Eckstein, RS Goody, H Gutfreund, and DR Trentham (1974) *Biochem J* **141**, 351.
Bagshaw CR and DR Trentham (1974) *Biochem J* **141**, 331.
Bagshaw CR, DR Trentham, RG Wolcott, and PD Boyer (1975) *Proc Natl Acad Sci U S A* **72**, 2592.
Ban N, P Nissen, J Hansen, PB Moore, and TA Steitz (2000) *Science* **289**, 905.
Beringer M, S Adio, W Wintermeyer, and M Rodnina (2003) *RNA* **9**, 919.
Bock A, K Forchhammer, J Heider, W Leinfelder, G Sawers, B Veprek, and F Zinoni (1991) *Mol Microbiol* **5**, 515.
Bottcher B, L Schwarz, and P Graber (1998) *J Mol Biol* **281**, 757.
Boyer PD (1993) *Biochim Biophys Acta* **1140**, 215.
Boyer PD (1997) *Annu Rev Biochem* **66**, 717.
Braig K, RI Menz, MG Montgomery, AGW Leslie, and JE Walker (2000) *Structure* **8**, 567.
Burgess BK and DJ Lowe (1996) *Chem Rev* **96**, 2983.
CaJacob CA, PA Frey, JF Hainfeld, JS Wall, and H Yang (1985a) *Biochemistry* **24**, 2425.
CaJacob CA, GR Gavino, and PA Frey (1985b) *J Biol Chem* **260**, 14610.
Capaldi RA, R Aggeler, S Wilkens, and F Grüber (1996) *J Bioenerg Biomembr* **28**, 397.
Chan MK, J Kim, and DC Rees (1993) *Science* **260**, 792.
Chang SI and GG Hammes (1990) *Acc Chem Res* **23**, 363.
Chiu H, JW Peters, WN Lanzilotta, MJ Ryle, LC Seefeldt, JB Howard, and DC Rees (2001) *Biochemistry* **40**, 641.
Choi-Rhee E and JE Cronan (2003) *J Biol Chem* **278**, 30806.
Collins JH and LJ Reed (1977) *Proc Natl Acad Sci U S A* **74**, 4223.
Cortes J, SF Haydock, GA Roberts, DJ Bevitt, and PF Leadlay (1990) *Nature* **348**, 176.
Coureux P-D, HL Sweeney, and A Hondusse (2004) *EMBO J* **23**, 4527.
Cramer P, DA Bushnell, and RD Kornberg (2001) *Science* **292**, 1863.
Cronan JE (2002) *J Biol Chem* **277**, 22520.
Cronan JE Jr and GL Waldrop (2002) *Prog Lipid Res* **41**, 407.
Dardel F, AL Davis, ED Laue, and RN Perham (1993) *J Mol Biol* **229**, 1037.
Davis LC, VK Shah, and WJ Brill (1975) *Biochim Biophys Acta* **403**, 67.
deHaseth PL, ML Zupanic, and MT Record (1998) *J Bacteriol* **180**, 3019.
Demadis KD, SM Malinak, and D Coucouvanis (1996) *Inorg Chem* **35**, 4038.
Dominguez R, Y Freyzon, KM Trybus, and C Cohen (1998) *Cell* **94**, 559.
Donadio S, MJ Staver, JB McAlpine, SJ Swanson, and L Katz (1991) *Science* **252**, 675.
Duyvis MG, H Wassink, and H Hasker (1998) *Biochemistry* **37**, 17345.
Eady RR (1996) *Chem Rev* **96**, 3013.
Einsle O, FA Tezcan, SL Andrade, B Schmid, M Yoshida, JB Howard, and DC Rees (2002) *Science* **297**, 1696.
Erickson JA, AC Nyborg, JL Johnson, SM Truscott, A Gunn, FR Nordmeyer, and GD Watt (1999) *Biochemistry* **38**, 14279.
Fall RR and PR Vagelos (1972) *J Biol Chem* **247**, 8005.
Ferguson-Miller S and GT Babcock (1996) *Chem Rev* **96**, 2889.
Fillingame RH (1996) *Curr Opin Struct Biol* **6**, 491.
Fillingame RH and OY Dmitriev (2002) *Biochim Biophys Acta* **1565**, 232.
Fisher K, MJ Dilworth, CH Kim, and WE Newton (2000) *Biochemistry* **39**, 10855.
Frank RAW, CM Titman, JV Pratap, BF Luisi, and RN Perham (2004) *Science* **306**, 872.
Frasch WD (2000) *Biochim Biophys Acta* **1458**, 310.
Georgiadis MM, H Komiya, P Chakrbarti, D Woo, JJ Kormac, and DC Rees (1992) *Science* **257**, 1653.
Geren LM, JR Beasley, BR Fine, AJ Saunders, S Hibdon, GJ Pielak, B Durham, and FJ Millett (1995) *J Biol Chem* **270**, 2466.
Gibbons C, MG Montgomery, AGW Leslie, and JE Walker (2000) *Nat Struct Biol* **7**, 1055.
Girvin ME, VK Rastogi, F Abildgaard, JL Markley, and RH Fillingame (1998) *Biochemistry* **37**, 8817.
Gladyshev VN, GV Kryukov, DE Fomenko, and DL Hatfield (2004) *Annu Rev Nutr* **24**, 579.

Gnatt AL, P Cramer, J Fu, DA Bushnell, and RD Kornberg (2001) *Science* **292**, 1876.
Green JDF, ED Laue, RN Perham, ST Ali, and JR Guest (1995) *J Mol Biol* **248**, 328.
Gruys KJ, A Datta, and PA Frey (1989) *Biochemistry* **28**, 9071.
Gruys KJ, CJ Halkides, and PA Frey (1987) *Biochemistry* **26**, 7575.
Guchhait RB, SE Polakis, P Dimroth, E Stoll, J Moss, and MD Lane (1974a) *J Biol Chem* **249**, 6633.
Guchhait RB, SE Polakis, D Hollis, C Fenselau, and MD Lane (1974b) *J Biol Chem* **249**, 6646.
Guth JH and RH Burris (1983) *Biochemistry* **22**, 5111.
Hansen JL, TM Schmeing, PB Moore, and TA Steitz (2002) *Proc Natl Acad Sci U S A* **99**, 11670.
Hardy RWF and E Knight Jr (1967) *Biochim Biophys Acta* **139**, 69.
Harms J, F Schluenzen, R Zarivach, A Bashan, S Gat, I Agmon, H Bartels, F Franceschi, and A Yonath (2001) *Cell* **107**, 679.
Harwood HJ Jr, SF Petras, LD Shelly, LM Zaccaro, DA Perry, MR Makowsi, DM Hargrove, KA Martin, WR Tracy, JG Chapman, WP Magee, DK Dalvie, VF Soliman, WH Martin, CJ Malarski and SA Eisenbeis (2003) *J Biol Chem* **278**, 37009.
Highsmith S (1999) *Biochemistry* **38**, 9791.
Hill BC (1991) *J Biol Chem* **266**, 2219.
Hill BC (2004) *Biochim Biophys Acta* **1655**, 256.
Holmes KC (1996) *Curr Opin Struct Biol* **6**, 781.
Holmes KC, RR Schroder, HL Sweeney, and A Houdusse (2004) *Phil Trans R Soc Lond B Biol Sci* **359**,1819.
Hoover TR, J Imperial, PW Ludden, and VK Shah (1989) *Biochemistry* **28**, 2768.
Horn F and H Bisswanger (1983) *J Biol Chem* **258**, 6912.
Houdusse A, VN Kalabokis, D Himmel, AG Szent-Gyorgyi, and C Cohen (1999) *Cell* **97**, 459.
Howard JB and DC Rees (1994) *Annu Rev Biochem* **63**, 235.
Hutchinson CR and I Fujii (1995) *Annu Rev Microbiol* **49**, 201.
Huxley AF and R Niedergerke (1954) *Nature* **173**, 971.
Huxley HE (1969) *Science* **164**, 1356.
Huxley HE (2004) *Phil Trans R Soc* **359**, 1879.
Huxley HE and J Hansen (1954) *Nature* **173**, 973.
Imperial J, TR Hoover, MS Madden, PW Ludden, and VK Shah (1989) *Biochemistry* **28**, 7796.
Jang SB, LC Seefeldt, and JW Peters (2000) *Biochemistry* **39**, 14745.
Jencks WP and J Regenstein (1970) In HA Sober (ed): *Handbook of Biochemistry: Selected Data for Molecular Biology*. The Chemical Rubber Co: Cleveland, pp J198-J199.
Jordan F (2004) *Science* **306**, 818.
Kabsch W, HG Mannherz, D Suck, EG Pai, and KC Holmes (1990) *Nature* **347**, 37.
Karrasch S and JE Walker (1999) *J Mol Biol* **290**, 379.
Katunin VI, GW Muth, SA Strobel, W Wintermeyer, and MV Rodnina (2002) *Mol Cell* **10**, 339.
Kaziro Y, LF Hass, PD Boyer, and S Ochoa (1962) *J Biol Chem* **237**, 1460.
Khosla C (1997) *Chem Rev* **97**, 2577.
Kim J and DC Rees (1992) *Science* **257**,1677.
Kim J and DC Rees (1994) *Biochemistry* **33**, 389.
Kim J, D Woo, and DC Rees (1993) *Biochemistry* **32**, 7104.
Knowles JR (1989) *Annu Rev Biochem* **58**, 195.
Konz D and MA Marahiel (1999) *Chem Biol* **6**, R39.
Lane MD and F Lynen (1963) *Proc Natl Acad Sci U S A* **49**, 379.
Lee HI, PM Benton, M Laryukhin, RY Igarashi, DR Dean, LC Seefeldt, and BM Hoffman (2003) *J Am Chem Soc* **125**, 5604.
Lee HI, M Sorlie, J Christiansen, R Song, DR Dean, BJ Hales, and BM Hoffman (2000) *J Am Chem Soc* **122**, 5582.
Li J, BK Burgess, and JL Corbin (1982) *Biochemistry* **21**, 4393.
Li JL and RH Burris (1983) *Biochemistry* **22**, 4472.
Li SJ and JE Cronan (1992a) *J Biol Chem* **267**, 855.
Li SJ and JE Cronan (1992b) *J Biol Chem* **267**, 16841.
Lombardi V, G Piazzesi, M Reconditi, M Linari, L Lucii, A Stewart, YB Sun, P Boesecke, T Narayanan, T Irving and M Irving (2004) *Phil Trans R Soc Lond B Biol Sci* **359**, 1883.
Lymn RW and EW Taylor (1971) *Biochemistry* **10**, 4617.
Mande SS, S Sarfaty, MD Allen, RN Perham, and WG Hol (1996) *Structure* **4**, 277.
Marahiel MA, AT Stachelhaus, and HD Mootz (1997) *Chem Rev* **97**, 2651.

Mattevi A, GE Obmolova, KH Kalk, WJH van Berkel and WGJ Hol (1993) *J Mol Biol* **230**, 1200.
Mattevi A, G Obmolova, E Schulze, KH Kalk, AH Westphal, A de Kok, and WGJ Hol (1992) *Science* **255**, 1544.
Miller MJ, M Oldenburg, and RH Fillingame (1990) *Proc Natl Acad Sci U S A* **87**, 4900.
Mitchell P and J Moyle (1965) *Nature* **208**, 147.
Munday MR (2002) *Biochem Soc Trans* **30** [Pt 6], 1059.
Murakami KS and SA Darst (2003) *Curr Opin Struct Biol* **13**, 31.
Murakami KS, S Masuda, EA Campbell, O Mussin, and SA Darst (2002b) *Science* **296**, 1285.
Murakami KS, S Masuda, and SA Darst (2002a) *Science* **296**, 1280.
Musser SM, MHB Stowell, and SI Chan (1995) *Adv Enzymol Rel Areas Mol Biol* **71**, 79.
Namslauer A and P Brzezinski (2004) *FEBS Lett* **567**, 103.
Nissen P, J Hansen, N Ban, PB Moore, and TA Steitz (2000) *Science* **289**, 920.
Noji H, R Yasuda, M Yoshida, and K Kinosita (1997) *Nature* **386**, 299.
Noller HF, V Hoffarth, and L Zimniak (1992) *Science* **256**, 1416.
Ogura T and T Kitagawa (2004) *Biochim Biophys Acta* **1655**, 290.
O'Neal CC and PD Boyer (1984) *J Biol Chem* **259**, 5761.
Otterbein LR, P Graceffa, and R Dominguez (2001) *Science* **293**, 708.
Penefsky HS and RL Cross (1991) *Adv Enzymol* **64**, 173.
Perham RN (1991) *Biochemistry* **30**, 8501.
Polacek N, M Gaynor, A Yassin, and AS Mankin (2001) *Nature* **411**, 498.
Polakis SE, RB Guchhait, and MD Lane (1972) *J Biol Chem* **247**, 1335.
Polakis SE, RB Guchhait, EE Zwergel, MD Lane, and TG Cooper (1974) *J Biol Chem* **249**, 6657.
Proshlyakov DA (2004) *Biochim Biophys Acta* **1655**, 282.
Rayment I (1996) *J Biol Chem* **271**, 15850.
Rayment I, WR Rypniewski, K Schmidt-Base, R Smith, DR Tomchick, MM Benning, DA Winkelmann, G Wesenberg, and HM Holden (1993) *Science* **261**, 50.
Reed LJ (1974) *Acc Chem Res* **7**, 40.
Reed LJ and DJ Cox (1970) In Boyer PD (ed): *The Enzymes*, vol 1, 3rd ed. Academic Press: New York, p 213.
Rivera-Ortiz JM and RH Burris (1975) *J Bacteriol* **123**, 537.
Rizzuto R, D Santona, M Brini, RA Capaldi, and R Bisson (1991) *Biochim Biophys Acta* **1129**, 100.
Roberts EL, N Shu, MJ Howard, RW Broadhurst, A Chapman-Smith, JC Wallace, T Morris, JE Cronan, and RN Perham (1999) *Biochemistry,* **38**, 5045.
Ryle MJ, HI Lee, LC Seefeldt, and BM Hoffman (2000) *Biochemistry* **39**, 1114.
Sartorelli L, HJ Fromm, RW Benson, and PD Boyer (1966) *Biochemistry* **5**, 2877.
Schindelin H, C Kisker, JL Schlessman, JB Howard, and DC Rees (1997) *Nature* **387**, 370.
Schlunzen F, R Zarivach, J Harms, A Bashan, A Tocilj, R Albrecht, A Yonath, and F Franceschi (2001) *Nature* **413**, 814.
Schmidt B, W Hillier, J McCracken, and S Ferguson-Miller (2004) *Biochim Biophys Acta* **1655**, 248.
Scott DJ, DR Dean, and WE Newton (1992) *J Biol Chem* **267**, 20002.
Scrutton MC, CW Wu, and DA Goldthwait (1971) *Proc Natl Acad Sci U S A* **68**, 2497.
Sedgwick B, JW Cornforth, SJ French, RT Gray, E Kelstrup, and P Willadsen (1977) *Eur J Biochem* **75**, 481.
Seefeldt LC, IG Dance, and DR Dean (2004) *Biochemistry* **43**, 1401.
Seefeldt LC and DR Dean (1997) *Acc Chem Res* **30**, 260.
Seila AC, K Okuda, S Nunez, AF Seila, and SA Strobel (2005) *Biochemistry* **44**, 4018.
Sen S, R Igarashi, A Smith, MK Johnson, LC Seefeldt, and JW Peters (2004) *Biochemistry* **43**, 1787.
Shah VK and WJ Brill (1977) *Proc Natl Acad Sci U S A* **74**, 3249.
Shen B (2000) *Curr Top Chem* **209**, 1.
Shen B (2004) *Curr Opin Chem Biol* **7**, 285.
Shepherd G and GG Hammes (1977) *Biochemistry* **16**, 5234.
Sievers A, M Beringer, MV Rodnina, and R Wolfenden (2004) *Proc Natl Acad Sci U S A* **101**, 7897.
Simpson FB and RH Burris (1984) *Science* **224**, 1095.
Singer TP and DE Edmondson (1980) *Methods Enzymol* **66**, 253.

Slater JP, AS Mildvan, and LA Loeb (1971) *Biochem Biophys Res Commun* **44**, 37.
Speckhard DC, BH Ikeda, SS Wong, and PA Frey (1977) *Biochem Biophys Res Commun* **77**, 708.
Stadtman TC (1990) *Annu Rev Biochem* **59**, 111.
Steginsky CA and PA Frey (1984) *J Biol Chem* **259**, 4023.
Stock D, C Gibbons, I Arechaga, AGW Leslie, and JE Walker (2000) *Curr Opin Struct Biol* **10**, 672.
Stock D, AGW Leslie, and JE Walker (1999) *Science* **286**, 1700.
Sweeney HL and A Houdusse (2004) *Phil Trans R Soc Lond B Biol Sci* **359**,1829.
Szundi I, J Cappuccio, and EinarsdÛttir (2004) *Biochemistry* **43**, 15746.
Takagi Y, H Shuman, and YE Goldman (2004) *Phil Trans R Soc Lond B Biol Sci* **359**,1913.
Thoden JB, CZ Blanchard, HM Holden, and GL Waldrop (2000) *J Biol Chem* **275**, 16183.
Thompson J, DF Kim, M O'Connor, KR Lieberman, MA Bayfield, ST Gregory, R Green, HF Noller, and AE Dahlberg (2001) *Proc Natl Acad Sci U S A* **98**, 9002.
Thorson JS, TJ Hosted, J Jiang, JB Biggins, and J Ahlert (2001) *Curr Org Chem* **5**, 139.
Tsukihara T, H Aoyama, E Yamashita, T Tomizaki, H Yamaguchi, K Shinzawa-Itoh, R Nakashima, R Yaono, and S Yoshikawa (1996) *Science* **272**, 1136.
Valenzuela P, RW Morris, A Faras, W Levinson, and WJ Rutter (1973) *Biochem Biophys Res Commun* **53**, 1036.
Vassylyev DG, S Sekine, O Laptenko, J Lee, MN Vassylyeva, S Borukhov, and S Yokoyama (2002) *Nature* **417**, 712.
Wakil SJ (1989) *Biochemistry* **28**, 4523.
Wakil SJ, JK Stoops, and VC Joshi (1983) *Annu Rev Biochem* **52**, 537.
Waldrop GL, I Rayment, and HM Holden (1994) *Biochemistry* **33**, 10249.
Walsh C (2003a) *Antibiotics*. ASM Press: Washington, DC, p 175.
Walsh C (2003b) *Antibiotics*. ASM Press: Washington, DC, p 195.
Watt GD and KRN Reddy (1994) *J Inorg Biochem* **53**, 281.
Weinger JS, KM Parnell, S Dorner, R Green, and SA Strobel (2004) *Nat Struct Mol Biol* **11**, 1101.
Wikström M (2004) *Biochim Biophys Acta* **1655**, 241.
Wilkens S and RA Capaldi (1998) *Nature* **393**, 29.
Yandulov DV and RR Schrock (2003) *Science* **301**, 76.
Yang H, PA Frey, JF Hainfeld, and JS Wall (1986) *Biophys J* **49**, 56.
Yang HC, JF Hainfeld, JS Wall, and PA Frey (1985) *J Biol Chem* **260**, 16049.
Yang YS and PA Frey (1986) *Biochemistry* **25**, 8173.
Yasuda R, H Noji, K Kinosita, and M Yoshida (1998) *Cell* **93**, 1117.
Yonath A and A Bashan (2004) *Annu Rev Microbiol* **58**, 233.
Yoo SJ, HC Angrove, BK Burgess, MP Hendrich, and E Munck (1999) *J Am Chem Soc* **121**, 2534.
Yu TW, Y Shen, Y Doi-Katayama, L Tang, C Park, BS Moore, CR Hutchinson, and HG Floss (1999) *Proc Natl Acad Sci U S A* **96**, 9051.
Yusupov MM, GZ Yusupova, A Baucom, K Lieberman, TN Earnest, JHD Cate, and HF Noller (2001) *Science* **292**, 883.
Zeng W, PB Conibear, JL Dickens, RA Cowie, S Wakelin, A Malnasi-Csizmadia, and CR Bagshaw (2004) *Phil Trans R Soc Lond B Biol Sci* **359**,1843.
Zhang G, EA Campbell, L Minakhin, C Richter, K Severinov, and SA Darst (1999) *Cell* **98**, 811.
Zhang H, B Tweel, J Li, and L Tong (2004) *Structure* **12**, 1683.
Zhang H, B Tweel, and L Tong (2004) *Proc Natl Acad Sci U S A* **101**, 5910.
Zhang H, Z Yang, Y Shen, and L Tong (2003) *Science* **299**, 2064.
Zhang Y and RH Fillingame (1994) *J Biol Chem* **269**, 5473.

Appendices

Appendix A: Haldane Relationships for Some Kinectic Mechanisms

Mechanism	Haldane Relationship
Uni uni	$K_{eq} = \dfrac{V_f K_p}{V_r K_a}$
Ordered unibi	$K_{eq} = \dfrac{V_f K_{ip} K_q}{V_r K_{ia}} = \dfrac{V_f K_p K_{iq}}{V_r K_a}$
Ordered bi bi	$K_{eq} = \dfrac{V_f K_p K_{iq}}{V_r K_{ia} K_b} = \dfrac{V_f^2 K_{ip} K_q}{V_r^2 K_{ib} K_a}$
Equilibrium random bi bi	$K_{eq} = \dfrac{V_f K_p K_{ib}}{V_r K_{ia} K_b} = \dfrac{V_f K_{ip} K_q}{V_r K_a K_{ib}} = \dfrac{V_f K_p K_{iq}}{V_r K_a K_{ib}} = \dfrac{V_f K_{ip} K_q}{V_r K_{ia} K_b}$
Ping pong bi bi	$K_{eq} = \dfrac{V_f^2 K_p K_q}{V_r^2 K_a K_b} = \dfrac{V_f K_{ip} K_q}{V_r K_{ia} K_b} = \dfrac{V_f K_p K_{iq}}{V_r K_a K_{ib}} = \dfrac{K_{ip} K_{iq}}{K_{ia} K_{ib}}$
Theorell-Chance	Same as equilibrium random bi bi

Appendix B: Inhibition Patterns for Three-Substrate Kinetic Mechanisms

Mechanism	Product	A	Sat B	Sat C	B	Sat A	Sat C	C	Sat A	Sat B
Ordered ter ter	P	NC	UC	UC	NC	NC	UC	NC	NC	NC
	Q	UC	UC	UC	UC	UC	UC	UC	UC	UC
	R	C	C	C	NC	—	NC	NC	—	UC
Bi uni uni bi ping pong	P	NC	UC	—	NC	NC	—	C	C	C
	Q	UC	UC	UC	UC	UC	UC	NC	NC	NC
	R	C	C	C	NC	—	NC	UC	—	UC
Bi bi uni uni ping pong	P	NC	UC	NC	NC	NC	NC	UC	UC	UC
	Q	UC	UC	—	UC	UC	—	C	C	C
	R	C	C	C	NC	—	NC	NC	—	NC
Hexa uni ping pong	P	NC	—	NC	C	C	C	UC	UC	—
	Q	UC	UC	—	NC	NC	—	C	C	C
	R	C	C	C	UC	—	UC	NC	—	NC

C, competitive; NC, noncompetitive; UC, uncompetitive.
From Plowman KM (1972) *Enzyme Kinetics*. McGraw-Hill: New York.

Appendix C: Equations for Number of Occupied Sites in the Binding of a Ligand to a Multisite Macromolecule

$$M + A \xrightleftharpoons{K_1} MA \qquad K_1 = \frac{[MA]}{[M][A]}$$

$$MA + A \xrightleftharpoons{K_2} MA_2 \qquad K_2 = \frac{[MA_2]}{[MA][A]}$$

$$\vdots$$

$$MA_{n-1} + A \xrightleftharpoons{K_n} MA_n \qquad K_n = \frac{[MA_n]}{[MA_{n-1}][A]}$$

Average moles of A bound per mole of enzyme
$$N_x = \frac{[MA] + 2[MA_2] + \ldots + n[MA_n]}{[M] + [MA] + [MA_2] + \ldots + [MA_n]} = \frac{\sum_{1}^{n} i[MA_i]}{[M] + \sum_{1}^{n}[MA_i]}$$

Let $L_0^* = 1, L_1^* = K_1, L_2^* = K_1 K_2, \ldots L_n^* = K_1 K_2 \ldots K_n$

Then:
$$N_x = \frac{\sum_{1}^{n} i L_i^*[M][A]^i}{[M] + \sum_{1}^{n} L_i^*[M][A]^i} = \frac{\sum_{1}^{n} i L_i^*[A]^i}{1 + \sum_{1}^{n} L_i^*[A]^i}$$

$n = 1$
$$N_x = \frac{K_1[A]}{1 + K_1[A]}$$

$n = 2$
$$N_x = \frac{K_1[A] + 2K_1 K_2[A]^2}{1 + K_1[A] + 2K_1 K_2[A]^2}$$

The symbol M designates a multiside macromolecule, usually a multisubunit protein such as an enzyme with a binding site in each subunit. Binding of a ligand such as a substrate A is governed by dissociation constants $K_1, K_2, \ldots K_n$, which differ from one another if the sites are interacting, that is, if the subunit interfaces transmit binding effects from one subunit to another. In this case, binding is cooperative or anticooperative, depending on whether $K_1 < K_2$, etc, or $K_1 > K_2$, etc. Shown is the derivation of a binding equation for the average number of ligands bound per macromolecule as a function of the dissociation constants and the concentration of the ligand A.

Appendix D: Derivation of Steady-State Kinetic Equations by the King-Altman Method

Enzymatic activities are typically measured under conditions in which the substrate concentrations are much higher than the enzyme concentrations, and this ensures the validity of the steady-state approximation. A typical kinetic mechanism includes several or many enzyme forms, such as enzyme-substrate and enzyme-product complexes or chemically modified enzyme forms, all of which are connected by ligand binding or chemical steps. The number of enzyme forms may be as few as three (E, E.S, and E.P) but typically is four or more. The kinetic mechanism describes a cyclic process, starting and ending with the free enzyme.

In general, a differential equation for the steady state approximation can be written for each enzyme form. However, the independent equations of this type are one less than the number of enzyme forms. By including the enzyme conservation equation, which equates the total enzyme concentration with the sum of the concentrations of all of the enzyme forms, the number of equations can equal the number of enzyme forms, the variables, and the simultaneous solution gives a single rate equation. The solutions can be conveniently obtained by matrix methods. Even this can be cumbersome. The King-Altman method is an algorithm that simplifies the process and leads directly to the rate equation.

The King-Altman method is applied to a simplified ping pong bi bi kinetic mechanism. It may be written as follows, where E is the free enzyme, and F is the chemically modified form:

Reaction: $A + B \rightleftharpoons P + Q$

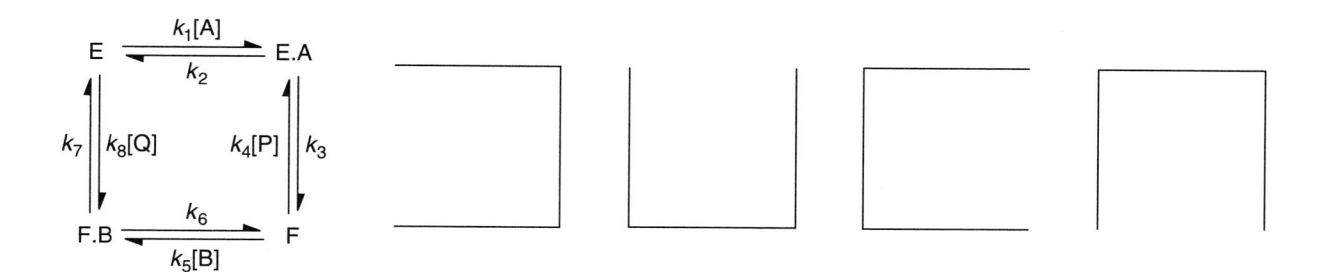

This simplified scheme does not include the E.Q or F.P complexes, but it suffices to give a valid equation; the inclusion of the product complexes gives a hexagonal instead of a square pattern, and the resulting kinetic equation has the same form with more rate constants. The writing of the basic scheme in a cyclic pattern is step 1 of the King-Altman derivation.

In step 2, all possible patterns are drawn with one side of each loop missing. In the case of the square pattern, there is a single loop, so there are four patterns with one side missing as shown above. More than one loop in a basic pattern is a complication that is dealt with later.

In step 3, the patterns identified in step 2 are used together with the rate constants and concentrations in the basic scheme to write distribution equations for the fraction of each enzyme form. The fractional enzyme forms on the left sides of the distribution equations are, where $[E_0]$ is the total enzyme concentration: $[E]/[E_0]$, $[E.A]/[E_0]$, $[F]/[E_0]$, and $[F.B]/[E_0]$. The right side of each distribution equation has a collection of rate constant and concentration terms in the numerator, and it has a denominator that is the same for all the distribution equations and is the sum of the numerators for all the equations. Each numerator term is constructed as follows. Find the location of the enzyme form in question in the first pattern. One term in the numerator of the distribution equation for that enzyme form is the product of all the rate constants and concentrations leading to that enzyme form, starting from both ends of that pattern. There is one for each pattern, and the sum of the rate constant and concentration terms is the numerator of the distribution equation for that enzyme form. For example, in the previous case, the first numerator term in the distribution equation for E is $k_6 k_4 [P] k_2$. The second term corresponding to the second pattern from the left is $k_3 k_5 [B] k_7$, and the third is $k_2 k_5 [B] k_7$, etc. The distribution equations are

$$\frac{[E]}{[E_0]} = \frac{k_6 k_4 [P] k_2 + k_3 k_5 [B] k_7 + k_2 k_5 [B] k_7 + k_7 k_4 [P] k_2}{D}$$

$$\frac{[E.A]}{[E_0]} = \frac{k_1 [A] k_6 k_4 [P] + k_8 [Q] k_6 k_4 [P] + k_5 [B] k_7 k_1 [A] + k_4 [P] k_7 k_1 [A]}{D}$$

$$\frac{[F]}{[E_0]} = \frac{k_6 k_1 [A] k_3 + k_8 [Q] k_6 k_3 + k_2 k_8 [Q] k_6 + k_7 k_1 [A] k_3}{D}$$

$$\frac{[F.B]}{[E_0]} = \frac{k_1 [A] k_3 k_5 [B] + k_8 [Q] k_3 k_5 [B] + k_5 [B] k_2 k_8 [Q] + k_4 [P] k_2 k_8 [Q]}{D}$$

$$[E] + [E.A] + [F] + [F.B] = [E_0]$$

Therefore, D = sum of numerators.

Step 4 is the elimination of zero-terms. These are terms that do not contribute to the rate being measured. For example, in an experiment to evaluate initial rates in the absence of both products, the rate equation does not include terms in [P] or [Q]; these are zero terms in the distribution equations. In an experiment to determine product inhibition, one product is absent; and another is present; then, terms in [P] may be set to zero with retention of [Q] for product inhibition. In an inital rate experiment with products absent, the distribution equations become

$$\frac{[E]}{[E_0]} = \frac{k_3 k_5 [B] k_7 + k_2 k_5 [B] k_7}{D} \qquad \frac{[E.A]}{[E_0]} = \frac{k_5 [B] k_7 k_1 [A]}{D}$$

$$\frac{[F]}{[E_0]} = \frac{k_6 k_1 [A] k_3 + k_7 k_1 [A] k_3}{D} \qquad \frac{[F.B]}{[E_0]} = \frac{k_1 [A] k_3 k_5 [B]}{D}$$

Step 5 is substitution into any one of the equations that directly give the observed rate for the formation of a product. These equations include $v = k_7 [F.B]$, the rate at which the product Q is formed, and $v = k_3 [E.A]$, the rate at which the product P is formed. Substitution for [F.B] in the former equation gives

$$v = \frac{k_{cat} [E_0]}{1 + \frac{K_m^A}{[A]} + \frac{K_m^B}{[B]}}$$

$$k_{cat} = k_3 k_7 / (k_3 + k_7)$$

$$K_m^A = \frac{k_7 / (k_2 + k_3)}{k_1 (k_3 + k_7)}$$

$$K_m^B = \frac{k_3 / (k_6 + k_7)}{k_5 (k_3 + k_7)}$$

This is the rate equation for the ping pong kinetic mechanis. Another term in product appears in the denominator when a product is present, as in a product inhibition study.

In more complex mechanisms, the number of line patterns becomes large, and there may be more than a single loop in the basic pattern. Consider the following improbable example of five enzyme forms in a pattern with two loops.

The King-Altman patterns may not contain a closed loop, so in this case, all of the King-Altman patterns must lack two sides. In finding the King-Altman patterns, it is useful to know the number in advance. There are equations to calculate the number of such patterns. Let n = the number of enzyme forms and m = the number of sides in the pattern. Then,

$$n - 1 \text{ lined patterns} = m!/(n - 1)!(m - n - 1)!$$

In the previous case, this would be 15, including those with closed loops. The number of patterns with closed loops can be calculated. Let r = the number of sides in a given closed loop. Then, the number of patterns with a closed loop of r sides is given by

$$\text{Patterns with closed loop } r = (m - r)!/(n - 1 - r)!(m - n + 1)!$$

From this equation, there are three closed-loop patterns for each of the two three-sided loops, and these must be subtracted from the total number of patterns. The number of King-Altman patterns is $15 - 3 - 3 = 9$. The nine King-Altman patterns are

Index

Acetate kinase, dual biological function, 492
  phosphorylation of enzyme, 492
  reaction, 492
  steady-state kinetics, 492
  stereochemistry, 492
Acetoacetate decarboxylase, 403–405
  active site pKa's, 13, 115–116, 404
  inhibition by acetopyruvate, 405
  type I aldolase, 25
  [^{14}C]isopropyllysine from, 403
  5-nitrosalicylaldehyde and, 404
  mechanism, 404
  reaction, 403
  reductive inactivation, 403
Acetoacetyl CoA, 49–50, 167, 609, 628
Acetoacetyl CoA thiolase, 628–631
  acetyl CoA/acetoacetyl CoA exchange, 628
  [^{14}C]acetyl-enzyme, 628
  chemical mechanism, 629
  dithio-acetyl CoA as substrate, 629
  kinetic mechanism and reaction, 629
  ping pong kinetics, 629
  structure, 631
Acetoacetyl CoA:succinate CoA transferase, 49–51, 167, 609
  anhydride intermediate, 51
  enzyme-CoA intermediate, 49
  importance of CoA-binding, 51
  two-step mechanism, 49

Acetopyruvate, as inhibitor of acetoacetate decarboxylase, 405
Acetyl CoA, 50, 163–164, 245, 619–620, 622, 624–626, 628–630
  enolization of, 619
  in ester condensations, 619
  in fatty acid synthesis, 757–759
  in PFL, 407–408
Acetyl CoA carboxylase, 757–759
  and biotin carboxylase (BC), 757
  biotinyl carboxyl carrier (BCC), 757
  and carboxyltransferase, 757
  and carboxyphosphate, 759
  carboxylation mechanism, 758–759
  reaction, 163, 757
  role of biotin, 163
  structures of components, 760
Acetyl imidazole, 18
Acetyl phosphate, by action of phosphoketolase, 145
  from phosphotransacetylase, 492
Acetyl/malonyl CoA:ACP acyltransferase (AT), component of FAS, 762
Acetylcholinesterase, 291
  and nerve function, 291
  inhibition by sarin, 292
  reaction and mechanism, 292–294
  structure with fasciculin, 293

O-Acetylserine sulfhydrylase, 451–455
  deuterium kinetic isotope effect, 453
  kinetic mechanism, 453
  PLP-dependent β–replacement, 451
  serine binding PLP, 454
  structure with methionine, 455
  transient kinetics, 453
Acetyl-TPP, 144–146
  and phosphoketolase, 144–145
Acetyltyrosine ethyl ester (ATEE), chymotrypsin substrate, 2
Acetyltyrosine-p-nitroanilide (ATNA), chymotrypsin substrate, 2
Acid and base-catalyzed reactions, 11
Acid-base catalysis, 9
Acidic amino acids, values of pKa, 13
Acids and bases, 9
Aconitase, 232–233, 442–446
  aconitate intermediate, 443
  actions converting citrate and isocitrate, 443
  activation by Fe^{2+}, 444
  ENDOR spectroscopy, 444
  [3Fe-4S]/ [4Fe-4S]$^{2+}$, 445
  inter- intramolecular tritium transfer, 443
  Lewis acid mechanism, 445
  Mössbauer spectroscopy, 444
  reaction, 232, 443
  role of [4Fe-4S] center, 233, 442–446
  structure with citrate, 447
Activated complexes, 41
Activation, by adenylylation, 559
  by phosphorylation, 548
Activation energy, binding and, 34
  energy level diagrams, 37
Active site, and catalysis, 1
  characterization of, 53
  chemical modification, 53
  competitive inhibitors, 53
  as an entropy trap, 36
  essental amino acids, 53–62
  group selective reagents, 54–55
ACV (L-α–aminoadipoyl-L-cysteinyl-D-valine), 718–720, 768–769
  structure, 720
  precursor of isopenicillin N, 718, 768
ACV assembly, 769
ACV synthetase, a NRPS, 768
N-Acylamino acid racemase, enolase superfamily, 353
Acyl carrier protein (ACP), role of in FAS, 752
S-Acyldihydrolipoamide, in action of lipoamide, 143–144
Acyl transfer, chemistry of, 297–300
Acyl-TPP, 144
Adenosylcobalamin, 192–194, 466
  Co-C bond dissociation, 192
  dependent enzymes, 193

and elimination, 466
  enzymes, 193–199
  free radical mechanisms, 193
  hydrogen transfer, 193, 196
hydrolysis and photolysis, 192
  rearrangement reactions, 194
  and ribonucleotide reducase class II, 703
  structure, 190
Adenosylcobalamin synthetase, and cob(I)alamin, 191
S-Adenosylhomocysteine (SAH or AdoHcys), 670
  structure, 662
S-Adenosylhomocysteine hydrolase, 136–137, 670
  and 3′-ketoadenosine, 137
  and 3′-keto-5′-deoxyadenosine, 137
  a mechanism, 137
  reaction, 136
  role of NAD^+, 136–137
S-Adenosylmethionine (SAM or AdoMet), 656, 675–676
  epimerization, 662
  hydrolysis, 662
  and iron-sulfur centers, 234
  structure, 662
S-Adenosylmethionine and [4Fe-4S] centers, catalytic action of, 234
  stoichiometric reactions of, 236
S-Adenosylmethionine cycle, 670
S-Adenosylmethionine decarboxylase, pyruvoyl, 172
S-Adenosylmethionine synthetase, 665–670
  active site with S-adenosylmethionine, 669
  kinetic isotope effects, 666
  reaction mechanism, 666–669
  steady-state kinetic mechanism, 667
  stereochemistry, 666
  structure with PNPNP, 668
  transient kinetics, 666
Adenylate kinase, 6, 76–77, 521
  and AMPPNP, 77
  complex with AMP and AMPPNP, 77
  conformational change, 6
  induced conformation change, 77
  initial rate pattern, 76
  stereochemistry, 77
  structure, 77
Adenylyl cyclase, 522–526
  magnisum ions, 524
  composition, 523
  G-protein mediation of hormonal activation, 523
  mechanism, 526
  reaction, 523
  structure of catalytic domains, 525

Index

Affinity labeling, 59–60
　chymotrypsin with PMSF, 59
　chymotrypsin with TPCK, 60
　His57 in chymotrypsin, 60
　kinetics, 59
Alanine racemase, 285–287
　a racemization mechanism, 286
　and bacterial cell wall biosynthesis, 285
　halide elimination, 285
　inhibition by 1-aminoethylphosphonate, 288
　reaction, 149
　structure with 1-aminoethylphosphonate, 287
　suicide inactivation by 3-haloalanines, 285
Alcohol dehydrogenase (ADH), 680–686
　hydride transfer stereospecificity, 132–134, 680
　isotope effects, 685–686
　isotope exchange at equilibrium, 681
　kinetic mechanism, 680, 685
　pK_a's of alcohols bound to zinc, 684
　reaction and stereochemistry, 680
　Rossman fold, 682
　structure with difluorobenzyl alcohol, 683
　substrates, 680
　transient kinetics, 681
　transition state for hydride transfer, 681
　and Zn^{2+}, 132, 237, 681–684
Aldehyde oxidase, and xanthine oxidase family, 226
Aldolase, and dihydroxyacetone phosphate (DHAP), 25
　and N^ε-[^{14}C]dihydroxypropyl-lysine, 25
　and glyceraldehyde-3-phosphate (GAP), 25
　and sodium borohydride inactivation, 25
Aldolase type I, acetoacetate decarboxylase, 25
Aldolase type II, 617
　divalent metals, 27
Aldose and ketose isomerases, 333–341
Aldose/ketose isomerization, chemistry, 333–334
　enolization or hydride transfer, 335
Alkaline phosphatase, 517–520
　activation by 2 Me^{2+}, 238
　biological functions, 520
　linear free energy plot, 519
　loose transition state, 519
　p-nitrophenyl phosphate as substrate, 519
　phosphate hydrogenase activity, 520
　serine phosphatase, 517
　structure, 518
Alkylation, biological, 655
　chemistry of, 655
　enzymatic, 657
　mechanisms, 656
Alkylation mechanisms, allylic carbocations, 656
　associative and dissociative, 656

Alkyltransferases, 655
Allosteric regulation, 117–123
　binding equations for cooperative systems, 120
　cooperativity, 117–123
　theory, 118
Amethopterin, methotrexate, 271
Amidocarboxymethyldethia CoA, 625
Amidotransfer, 604
Amino acid-based coenzymes, 172
Amino acid decarboxylases, 394
Amino acid decarboxylation, PLP and pyruvoyl coenzymes, 394
D-Amino acid oxidase, 710–716
　D-3-chloroalanine as substrate, 711, 713
　5-deaza-FAD, 713
　deuterium kinetic isotope effect, 713
　exclusion of carbanion mechanism, 712
　flavoprotein oxidase, 711
　hypothetical mechanisms, 711–712
　linear free energy correlation, 712
　mechanisms for chloride elimination, 715
　pH dependence, 71
　reaction in steps, 711
　structure with alanine, 714
Aminoacyl-tRNA synthetases, 561–566
　aminoacyl adenylate, 564
　burst kinetics, 564
　classes, 563
　fidelity, 564–566
　mechanism of action, 564
　and translation of genetic code, 561
p-Aminobenzoate, folate constituent, 168
L-α–Aminoadipoyl-L-cysteinyl-D-valine (ACV), see ACV
(S)-4-Amino-5-fluoropentanoic acid, fluorinated GABA, 267
Aminomutases, role of PLP, 157
Aminopterin, structure, 271
AMPNPNP, 667
AMPPNP, structure, 77
Anaerobic ribonucleotide reductase (ARR), reaction, 235
Animal FAS, module structure, 763
Anthranilate synthase complex, 749
Antiferromagnetic coupling, in galactose oxidase, 241
Aquocobalamin, B12 compound, 191
Arginase, active site structure, 239
Ascorbate oxidase, reaction, 240
Asparagine synthetase, and β–aspartyl adenylate, 549
　and glutamine, 604–605
　random kinetics, 91
Aspartate aminotransferase, 597–602
　and Brønsted catalysis law, 16
　catalytic mechanism, 601
　chemical rescue, 601
　kinetics, 598

Aspartate aminotransferase *(Continued)*
   PLP-quinonoid α–carbanion, 601
   reaction, 149, 598
   steps and role of PLP, 599
   structure, 600
   suprafacial 1, 3-prototropic shift, 602
Aspartate β–decarboxylase, ^{13}C kinetic isotope effect, 398
   mechanism, 399
   and β–carbanion, 398
   reaction, 149, 398
Aspartate transcarbamoylase (ATCase), an allosteric enzyme, 123–126, *see also* ATCase
   and PALA, 254
Aspartic proteases, mechanism of action, 320
   molecular properties, 318
   pepsin and HIV protease, 317
   two-aspartate active sites, 317
ATCase (aspartate transcarbamoylase), 123–126
   ^{13}C isotope effects, 124
   cysteine sulfinate as substrate, 124
   kinetics, 124
   reaction mechanism, 126
   structure, 125
Atorvastatin acid, structure, 283
ATP energy, cleavage to ADP/Pi, 547
   cleavage to AMP/PPi, 547
ATP synthase, binding change mechanism, 788–790
   chemistry of, 788
   F_1F_0-ATPase and F_1-ATPase, 788
   molecular properties, 787
   reaction, 787
   rotatory action, 792
   structure of F_1F_0-ATPase, 791
ATP synthases, 749, 786
ATPase, myosin and actomyosin, 796
ATP-dependent synthetases and ligases, 547
Avidin, enthalpy of biotin-binding, 196

B12 coenzyme, adenosylcobalamin, 190, *see also* Adenosylcobalamin, Methylcobalamin
   methylcobalamin, 190
Bacitracin, by action of NRPS, 767
Basic amino acids, values of pKa, 13
α, β–barrel (TIM barrel), in triosephosphate isomerase, 336, *see also* TIM barrel
Benzylsuccinate synthase, 236
Binding and activation Energy, 34
Binding and enzymatic action, 40
Binding and near attack conformation (NAC), 46
Binding of remote groups, and rate enhancement, 48

Biological alkylations, and 5-methyltetrahydrofolate, 655
   and ATP, 655
   and *S*-adenosyl-L-methionine (SAM), 655
Biopterin, 161–162, *see also* tetrahydrobiopterin
Biotin, in acetyl CoA carboxylase, 757–758
   carboxyl carrier, 163–165
   chemistry of and N_1-carboxybiotin, 164
   enthalpy of binding avidin, 196
   structure and role as a carboxyl carrier, 163
   in transcarboxylase, 83–84
Biotin carboxylase (BC), component of acetyl CoA carboxylase, 757
Biotin carboxylation, and carboxyphosphate, 165
   mechanism of, 165
Biotin sulfoxide reductases, DMSO reductase family, 227
Biotin synthase, 612–614
   reaction, 235, 611
   role of iron-sulfur centers, 612, 614
   role of *S*-adenosylmethionine, 612, 614
   structure, 613
Biotin-dependent carboxylation, mechanism of, 164
Biotinyl carboxyl carrier (BCC), component of acetyl CoA carboxylase, 757
Biotinyl carboxylation, and N_1-carboxybiotin, 163
   propionyl CoA to methylmalonyl CoA, 163
   pyruvate to oxaloacetate, 163
   β–methylcrotonyl CoA, 163
Biotinyl transcarboxylation, transcarboxylase, 163
Briggs and Haldane, kinetic mechanism, 72
   steady-state approximation, 71, 72
Brønsted acids and bases, definition, 9
Brønsted catalysis law, 15
Burst kinetics, of alcohol dehydrogenase, 685
   of aminoacyl-tRNA synthetases, 564
   of chymotrypsin, 103–104, 301–302
   of glyceraldehyde-3-P dehydrogenase, 692
   of myosin and actomyosin, 797
   of protein kinase A, 506
   of protein tyrosine phosphatase, 514
γ–Butyrobetaine hydroxylase, α–ketoglutarate-dependent, 732

cAMP-dependent protein kinase, 502–509, *see also* protein kinase A
Carbamoyl phosphate, 123
Carbamoyl phosphate synthetase (CPS I-III), 554–559
   ^{18}O-positional isotope exchange, 557
   and carboxyphosphate, 556
   and glutamine, 604–605
   chemical mechanism, 556
   kinetics, 557

Carbamoyl phosphate synthetase *(Continued)*
  partial reactions, 556
  reaction of E. coli enzyme, 155
  steps in the reaction, 556
  structure and tunnels, 558
Carbamoylaspartate, 123
Carbanionic mechanisms, 630
Carbaryl, structure of acetylcholinesterase inhibitor, 291
Carbocationic mechanisms, 645
Carbon-carbon condensation, allylic carbenium ions, 617–618
  carbanionic intermediates, 617–618
  chemistry, 617–618
Carbonic anhydrase, 462–465
  activation by Zn^{2+}, 237, 462
  free energy correlation, 463
  His64 as acid-base catalyst, 462
  kinetic parameters, 462
  Marcus parameters for proton release, 463
  a mechanism, 465
  reaction, 462
  structure, 464
Carboxybiotin, 164
Carboxylases, 418–426
Carboxylation, chemistry, 387–388
Carboxyltransferase, component of acetyl CoA carboxylase, 757
Carboxypeptidase A, 324
  activation by Zn^{2+}, 237, 324
  correlation of inhibition constants, 328
  esterase activity, 324
  kinetic constants, 325
  kinetics, 324
  a mechanism, 327
  peptide phosphonate inhibitors, 327
  structure with inhibitor, 326
  transient kinetics, 325
Caspases, calcium activated cysteine proteases, 317
Catalase, 204–206
  compounds I and II, 206
  and hydroperoxide radical, 206
  a mechanism, 206
  reaction, 204
  role of heme, 206
  structure, 205
Catalase intermediates, spectra, 207
Catalysis, and binding, 40
  binding energy in, 34
  and conformational mobility, 5
  electrophilic, 21
  enzymatic rate enhancement, 3
  hydrogen bonding in, 32
  by metal ions, 26
  of multistep reactions, 6
  nucleophillic, 21
Catalysis of enolization, 21

Catalytic antibodies, aldolase reactions, 64
  design, 64
  Diels-Alder reactions, 64
  theory, 63
Catalytic triad, in subtilisin, 312
  in chymotrypsin, 301
Catechol *O*-Methyltransferase, 661–666
  associative mechanism, 666
  L-DOPA and epinephrine as substrates, 663
  drug target for Parkinson's disease, 663
  kinetic isotope effects, 665
  kinetic mechanism, 663
  methyl transfer chemistry, 661
  reaction, 663
  reaction mechanism, 665
  role of Mg^{2+}, 665
  soluble and membrane forms, 662
  structure, 664
CDP-glucose 4, 6-dehydratase, 135
Cha, derivation of kinetic equations, 72
Charge transfer complexation, in copper amine oxidase, 181
  in dihydrolipoyl dehydrogenase, 695
  in flavin dehydrogenases, 161
  in glyceraldehyde-3-P dehydrogenase, 692
  di-iron cofactors, 218
  in protocatechuate 3, 4-dioxygenase, 743
  in purple acid phosphatase, 222
  in pyruvate dehydrogenase complex, 750
Chemical modification, group selective, 53
Chiral methyl analysis, fumarase, 621
  malate synthase, 621
Chorismate mutase, biological Claisen rearrangement, 364
  catalytic antibody-structure, 365
  kinetic isotope effects, 364
  reaction to prephenate, 364
  structure, 365
  transition state, 364. 366
Chymotrypsin, aldehyde adducts to Ser195, 308
  boronic acid adducts to Ser195, 307
  identification of Ser195, 2
  kinetic isotope effects, 305
  kinetic mechanism, 301
  kinetic parameters, 302
  the oxyanion site, 308
  partitioning of acylchymotrypsin, 302–303
  pH-dependence, 115–116
  pH-dependence and His57, 303
  postulated role of LBHB, 310
  proton inventories, 307
  rate enhancement, 2
  role of Ile16, 303–304
  specificity, 1, 2, 301
  structure with *N*-AcLeuPhe-aldehyde, 309
  transient and steady-state kinetics, 301
  trifluoromethylketone adducts to Ser195, 308

Chymotrypsin complexes, ionization
    properties, 312
Chymotrypsinogen, activation in vitro, 304
Ciprofloxacin, inhibition of topoisomerase, 534
    structure, 534
Citrate, 619, 622
Citrate synthase, citryl CoA, 622, 623
    dithio-acetyl CoA as substrate, 627
    enolization mechanisms, 626
    hydrolysis of citryl CoA, 623
    overall reaction mechanism, 622
    reaction, 619
    regulation, 620
    steady-state kinetics, 622
    stereochemistry and stereospecificity, 620
    structure, 625
Citryl CoA, 623
Cleland notation, kinetic mechanisms, 74
Cleland's rules, interpreting inhibition, 86
Cob(I)alamin, 655, 671–672, 675, 677
Cob(II)alamin, 673, 675, 676
    and B12 coenzyme reactions, 191
Cobalamin, chemical properties, 191–192
    oxidation states, 191–192
Cobalamins, spectra, 192
Coenzyme B12, and elimination, 466
Coenzyme M (CoM), in methanogenesis,
    243–244
Commitment factors, 97–99
Competitive inhibition, definition, 85
Competitive inhibitors, 53
Complex enzymes, 749
Concerted acid and base catalysis, 15
Conformational changes, substrate-induced, 5
Cooperativity, concerted transition, 118–119
    general model, 119–120
    and Hill equation, 122
    ligand-induced transition, 118–119
    positive and negative, 117
    negative cooperativity of glyceraldehyde-3-P
        dehydrogenase, 692
    theories, 118
Copper, as a cofactor, 240
Copper amine oxidase, 180–185, 241
    origin of TPQ, 181, 183
    ping pong kinetics, 183
    reaction, 180
    role of copper, 183, 185
    role of TPQ, 184, 185
    structure, 182
Copper enzymes, 240–241
Copper proteins, types I-III, 240
Copper-methane monooxygenase, 241,
    737–739
    absence of intermediates, 737–738
    a concerted mechanism, 739
    particulate, 737
    proposed oxene-insertion mechanism, 738
    retention of configuration, 738

Coproporphyrinogen oxidase, aerobic
    Mn-enzyme, 411
    a radical mechanism, 412
    radical SAM enzyme, 411
    reaction, 411
Corrin ring, in adenosylcobalamin, 190
COX inhibitors, slow- and fast-binding, 279
COX-1 and COX-2, 275, see also prostaglandin
    H synthase-cyclooxygenases
Creatine kinase, active site structure, 491
    biochemical role, 490
    kinetic mechanism, 490
    reaction, 490
    structure MgATP, 491
Crotonase, see Enoyl CoA hydratase
CTQ (cysteine tryptophylquinone), 179
Cyclic pentagonal phosphoesters, 484–486
    hydrolysis rates, 484
    preference rules, 485
    pseudorotation, 486
Cyclooxygenase, mechanism of arachidonic
    acid into prostaglandin G2, 277
    structure with ibuprofen, 276
γ–Cystathionase, cystathionine γ–lyase, 454
    hydrolysis of cystathionine, 454
    α–ketobutyrate from cystathionine, 454
    a mechanism, 456
    reaction, 149
Cystathionine β–synthase, PLP-dependent
    β–replacement, 451
    serine binding PLP, 454
Cysteine-bridged tyrosine radical, in galactose
    oxidase, 241
Cysteine phosphatases, 513–517
    phosphotyrosyl proteins, 513
    protein tyrosine phosphatases (PTPs), 513
Cysteine proteases, papain family and
    caspases, 314
Cysteine tryptophylquinone (CTQ), amine
    dehydrogenases, 179
Cysteinylflavin, monoamine oxidase, 159
    trimethylamine dehydrogenase, 159
Cytidine deaminase, and transition state
    analogs, 44–46
    structure, 47
Cytochrome c, structure, 209
Cytochrome c oxidase, 782–786
    complex IV, 749, 782
    composition, 783
    function, 782–783
    proton pump model, 786–787
    reaction, 783
    reduction of O_2, 785
    structure, 784
Cytochrome c peroxidase, and compound I,
    207
    a mechanism, 207
    reaction, 207
Cytochrome c reductase, complex III, 749, 782

Cytochrome oxidase, and copper, 241
Cytochrome P450, spectra, 207
Cytochrome P450 monooxygenases, 722–727
  compound I-like intermediate, 724
  concerted oxygen insertion, 724
  deuterium isotope effects, 724
  heme monooxygenases, 722
  hydroxylation cycle, 724
  oxygenation mechanism, 723–725
  radical clock experiments, 725
  radicals and oxygen rebound, 724
  reaction, 208, 722
  stereochemical evidence, 724
Cytochrome P450cam, reaction, 723
  structure, 726

Deacetoxycephalosporin C synthase, allylic carbenium ion intermediate, 733
  allylic radical intermediate, 733
  possible oxygenation mechanism, 735
  possible ring-expansion mechanism, 735
  reaction, 733
  structure, 734
  thiyl radical intermediate, 733
Dead-end inhibition, resolving kinetic ambiguity, 87
Decarboxylases, 388–417
Decarboxylation, chemistry, 387–388
Dehydroquinate synthase, and 3-deoxy-D-arabino-heptulosonate-7-P, 137
  a mechanism, 138
  and NAD⁺, 137–139
  and shikimic acid, 137
  and Zn^{2+}, 138
Dehydrosqualene, by-product of squalene synthase, 650
  structure, 650
5′-Deoxyadenosine, in adenosylcobalamin reactions, 198
5′-Deoxyadenosyl moiety, in adenosylcobalamin, 190
3-Deoxy-D-arabino-heptulosonate-7-P, and dehydroquinate synthase, 137
6-Deoxyerythronolide B, structure, 765
1-Deoxyribose-5-phosphate aldolase, structure, 26
Destabilization of ground states, 48
DFP (diisopropylphosphorofluoridate), 2
D-Fructose-6-P, 632
D-Glyceraldehyde-3-P, 632, 635
Dialkylglycine decarboxylase, 395–398
  aminomalonate decarboxylation, 398
  α–decarboxylation and transamination, 395
  2, 2-dialkylglycines as substrates, 395
  a mechanism, 396
  overall transamination, 396
  quinonoid intermediate, 396
  reaction, 395
  and stereoelectronic control, 397–398
  structure, 397
3, 6-Dideoxysugar biosynthesis, role of PLP, 158
Diffusion limited k_{cat}, triosephosphate isomerase, 338
Dihydrofolate, one-carbon metabolism, 168, 169
Dihydrofolate (H₂folate), 167, 168
Dihydrofolate reductase, 170–171, 271–274
  and methotrexate, 170–171, 272–273
  and tetrahydrofolate, 171
  inhibition by methotrexate, 271–273
  kinetic mechanism, 274
  methotrexate inhibition kinetics, 272
  reaction, 271
  stereochemistry, 271
  structure with methotrexate, 273
Dihydrolipoyl dehydrogenase, 694–698
  catalytic cycle, 698
  charge transfer complexation, 695–698
  dihydro intermediate, 695
  disulfide oxidoreductase family, 694
  E3 of PDH complex, 750
  kinetic mechanism, 694
  reaction, 694
  reaction mechanism, 695–698
  spectra of flavin coenzyme states, 695
  structure, 697
Dihydrolipoyl transacetylase, E2 of PDH complex, 750
3, 4-Dihydrouridine, zebularine-3, 4-hydrate, 47
Dihydroxyacetone phosphate (DHAP), and aldolase, 25
  and triosephosphate isomerase, 335
a, β–Dihydroxyethylidene-TPP, action of phosphoketolase, 145
Di-iron cofactors, 217–222
  hydrogen abstraction, 219
  oxidation states, 218
  structures, 218
  reactions of, 219
Diisopropylphosphorofluoridate (DFP), and Ser195 in chymotrypsin, 2
Dimethylallyl pyrophosphate (DMAPP), 646, 649
Dimethylbenzimidazole α–ribotide, in adenosylcobalamin, 190
Dimethylsulfoxide reductase, and molybdopterin, 222
1, 5-Dimethylthiazolium, exhange of C2(H), 141–143
  TPP model, 141–143
Dioldehyrase (adenosylcobalamin-dependent), 466–472
  adenosylcobalamin-function, 467

Dioldehydrase (adenosylcobalamin-dependent) (Continued)
  and 5′-deoxyadenosine, 466
  and 5′-deoxyadenosyl radical, 466
  deuterium and tritium kinetic isotope effects, 467
  EPR spectra of cob(II)alamin, 468
  EPR spectrum of substrate radical, 468
  mechanisms for radical isomerization, 472
  mechanistic analysis, 193–195
  reaction, 194
  stereospecificity, 46
  structure with cyanocobalamin and 1–2-propanediol, 470
  suicide inactivation, 194, 255
  translocation of 5′-deoxyadenosyl radical, 471
Dioxygenases, 217, 741–746
1, 3-Diphosphoglycerate, 690
  and glyceraldehyde-3-P dehydrogenase, 131
Disulfide oxidoreductases, 694
Dithio-acetyl CoA, 629
Divalent metal ions, activation of water, 237
  and enolization, 238
D-Lysine 5,6-aminomutase, reaction, 194
DMSO reductase, 226–227
  a mechanism, 229
  reaction, 227
  structure, 228
DNA ligase, 559–561
  chemical mechanism, 560
  exchange of NMN, 560
  kinetic mechanism, 560, 561
  rate constants for steps, 560
  structure of ATP-dependent ligase, 562
  synthesis and repair of DNA, 559
  NAD$^+$-dependent and ATP-dependent, 559
DNA methyltransferases, 662
DNA polymerase I, conformational hange, 530
  detailed kinetic mechanism, 532
  editing, 532
  isotope exchange experiments, 530
  structure, 531
  transient kinetic experiments, 530
DNA polymerases, 529–532
  a 2Mg^{2+}-mechanism, 529
L-DOPA, 663
Dopamine, structure of substrate for MAO B, 717
Dopamine β–monooxygenase, ascorbate as reductant, 735
  copper monooxygenase, 736
  isotope effects, 736
  kinetics, 736
  reaction, 735
  a reaction mechanism, 737
  structure, 735
Doxorubicin (Adriamycin), inhibition of topoisomerase II, 534
  structure, 534, 766

dTDP-4-Keto-6-deoxyglucose, product of dTDP-glucose 4, 6-dehydratase, 136
dTDP-4-Ketoglucose, and dTDP-glucose 4, 6-dehydratase, 446, 448
dTDP-4-Ketoglucose-5, 6-ene, and dTDP-glucose 4, 6-dehydratase, 136, 446, 448
dTDP-4-Ketoxylose, and dTDP-glucose 4, 6-dehydratase, 449
dTDP-Glucose 4, 6-dehydratase, and dTDP-6-deoxy-6-fluoroglucose, 449
  and dTDP-4-ketoxylose, 449
  dTDP-xylose, 449
  a mechanism, 136, 448
  role of NAD$^+$, 135–136, 446
  short-chain dehydrogenase/reductase, 446
  structure with dTDP-xylose, 450
  transient kinetic analysis, 449
  transient kinetics by MALDI TOF, 446

Edrophonium, structure of acetylcholinesterase inhibitor, 291
Electron transfer, biological, 247
  long-range, 247
  and Marcus theory, 248
Electron tunneling, and long range transfer, 249
Electrophilic catalysis, and enolization, 21
α, β–Elimination, cofactor-independent, 434–440
  PLP-dependent, 451–455
α, β–Elimination/addition reactions, 433–455
  cofactor-dependent, 440
  cofactor-independent, 434
  and methylidene imidazolone (MIO), 456
β, α–Elimination, 233
  methylidene imidazolone-dependent, 456–461
β, γ-Elimination, PLP-dependent, 454–455
ENDOR spectroscopy, of aconitase, 444
Energy-coupling enzymes, 777–797
Enolacetyl-TPP, action of phosphoketolase, 145
Enolase, dehydration of 2-phosphoglycerate, 441
  and divalent metal catalysis, 29, 441
  formation of phosphoenolpyruvate, 441
  a mechanism, 442
  2-phosphoglycerate and phosphoenolpyruvate, 28
  role of 2Mg^{2+}, 441
  structure, 29
Enolization, of acetyl CoA, 619
  catalysis of, 21
  facilitation by divalent metals, 28
  facilitation by imine formation, 27
  kinetic barrier, 23
  thermodynamic barrier, 21
5-Enolpyruvoylshikimate-3-phosphate synthase, EPSP synthase, 289, see also EPSP synthase

Enoyl CoA hydratase (Crotonase), 436–440
　kinetic isotope effects, 438
　Raman, UV-Vis and ^{13}C-NMR spectra, 440
　reaction, 436
　stepwise mechanism, 436
　structure with 4-dimethylaminocinnamoyl CoA, 439
Enoyl-ACP reductase (ER), component of FAS, 762
Enthalpy and entropy in catalysis, 36–40
Enzymatic binding domains, sizes of, 62
Enzyme inhibition, and binding, 254
　slow binding, 268
　suicide inactivation, 255
　tight binding, 269
　two-substrate analogs, 254
Enzyme-CoA intermediate, acetoacetyl CoA:succinate CoA transferase, 49
Epimerases and racemases, 346
Epinephrine, 663
EPSP synthase, importance in plants, 289
　inhibition by glyphosate, 289
　reaction, 289
　role in aromatic amino acid biosynthesis, 289
　structure with shikimate-3-phosphate and glyphosate, 289
EPR spectroscopy, dioldehydrase, 468
　glutamate mutase, 371
Equilibrium isotope effects, 91
Equilibrium ordered mechanism and rate equation, 78
Equilibrium random bi bi mechanism and rate equation, 75
D-Erythrose-4-P, 632
Ester hydrolysis, mechanism of, 298
Esterases, 328
R-[1–^{2}H$_1$] and S-[1–^{2}H$_1$]Ethanol, and alcohol dehydrogenase, 133
Ethanolamine ammonia-lyase, 472–473
　adenosylcobalamin-dependent, 473
　deuterium kinetic isotope effect, 473
　a mechanism, 473
　reaction, 194, 472
　substrate-derived C1 radical, 473
Evolution of binding sites, principle of economy, 81
Extradiol dioxygenase, in toluene metabolism, 217
Extradiol dioxygenases, 744, see also Intradiol dioxygenases

FADH$_2$/FMNH$_2$, mechanism of oxidation by O$_2$, 162
FAD/FADH$_2$, see Flavin coenzymes
Farnesyl pyrophosphate, 656
Farnesyl pyrophosphate synthase, 645–649
　allylic carbenium intermediates, 648
　carbocationic mechanism, 645
　carbon condensation mechanisms, 647
　and cholesterol biosynthesis, 646
　Hammett plot, 647
　linear free energy correlation, 646
　reactions, 646
　structure, 649
　and terpene biosynthesis, 646
Fasciculin, peptide inhibitor of acetylcholinesterase, 292
Fatty acid synthase complexes, 749, 757
Fatty acid synthases (FAS), reaction steps, 761
　six components plus ACP, 762
Fatty acid synthesis, 757
Fatty acyl desaturases, di-iron cofactor, 217
Fidelity, in aminoacyl-tRNA synthetases, 564–566
Five-member ring phosphoesters, 484
Flavin catalysis, one- to two-electron switches, 159–162
　redox reactions, 159–161
Flavin coenzymes, FMN/FMNH$_2$ and FAD/FADH$_2$, 158–162
Flavin coenzymes (FAD/FMN), mechanisms of catalysis, 159
　structures of, 158
Flavin dehydrogenases, dihydrolipoyl dehydrogenase, 161
　FAD-charge transfer complexation, 161
　glutathione reductase, 160
　transhydrogenase, 160–161
Flavin oxidases, amino acid oxidases, 161
　glucose oxidase, 161
　lactose oxidase, 161
　oxidation of dihydroflavin by O$_2$, 161
Flavin reduction, mechanism by vicinal dithiol, 699
Flavin semiquinones, 160
5-Fluoro-dUMP (FdUMP), from 5-fluorouracil, 255
　inactivation of thymidylate synthase, 255–260
Fluvastatin acid, structure, 283
FMN/FMNH$_2$, see Flavin coenzymes
Folate, biological importance, 171
　enzymes in tetrahydrofolate metabolism, 170
Folate compounds, 167–172
　one-carbon metabolism, 168
　p-aminobenzoate and, 168
　pterin and, 168
　structures and interconversions, 168–169
Formate dehydrogenases, DMSO reductase family, 227
N-Formyl-MFR dehydrogenase, in methanogenesis, 244
N-Formyl-MFR:H4MPT formyltransferase, in methanogenesis, 244

Formyl transferases, folate and purine biosynthesis, 170–171
5-Formyltetrahydrofolate, 639, 641
10-Formyltetrahydrofolate, one-carbon metabolism, 168, 169
Formyltetrahydrofolate synthetase, 168, 170
   and formyl phosphate, 170
   reaction, 170
Free energy, and reduction potentials, 202
Fumarase, carbanionic intermediate, 434
   cofactor-independent, 434
   iron sulfur variant, 232
   iso-mechanism, 434
   isotope exchange at equilibrium, 434
   kinetic isotope effects, 435
   3-nitrolactate inhibition, 435
   pH-rate profile, 435
   reaction and mechanism, 434–437
   structure, 437

GABA (γ–aminobutyric acid), 262
GABA aminotransferase, 262–268
   inactivation by fluorinated GABAs, 266–268
   inactivation by gabaculine, 263
   inactivation by γ–vinyl GABA, 264
   reaction, 262
   structure, 265
Gabaculine, suicide inactivation of GABA aminotransferase, 262–263
Galactonate dehydratase, enolase superfamily, 353
Galactose oxidase, antiferromagnetic coupling, 241
   cysteine-bridged tyrosine radical, 241
   a mechanism, 241
   reaction, 241
   structure, 241
Galactose 1-phosphate uridylyltransferase, chemical rescue, 527
   covalent uridylyl-enzyme, 526, 527
   economy in evolution of binding sites, 526
   exchange reactions, 527
   in human galactosemia, 528
   kinetics, 527
   reaction, 526
   stereochemistry, 527
   structure of uridylyl-enzyme, 528
GAP (glyceraldehyde-3-P), and aldolase, 25, see also Glyceraldehyde-3-P
General acid/base catalysis, definition, 13
Geranylgeranyl transferase, 657
Glucarate dehydratase, enolase superfamily, 353
Glucono-δ–lactone inhibition, structure, 581
Glucosamine-6-P synthase, and glutamine, 604–605
Glucose oxidase, reaction, 161, 710–711
Glutaconate CoA transferase, structure, 52

Glutamate dehydrogenase, 90, 131, 679, 693
   and α–ketoglutarate, 131
   kinetic mechanism, 693
   and $NAD^+/NADP^+$, 131, 693
   ordered kinetics, 90
   steps in the chemical mechanism, 693
Glutamate mutase, and adenosylcobalamin, 369
   and 5'-deoxyadenosyl radical, 369
   elimination/addition isomerization mechanism, 371
   glutamate-radical EPR signal, 371
   radical isomerization, 369
   reaction, 194, 369, 371
   structure with glutamate, 370
Glutamate racemase, 350–351
   analogy to proline racemase, 350
   HCl-elimination from 3-chloroglutamate, 350
   overshoot phenomenon, 350
   structure with glutamine, 351
   two base mechanism, 350
Glutamate synthase, and glutamine, 604–605
   reaction, 607
Glutamine, as source of ammonia, 604
Glutamine synthetase, biological regulation of GS I, 549
   and γ–glutamyl phosphate, 548, 554
   GS I and GS II, 548–555
   inhibition, 550
   kinetic mechanism, 549
   methionine sulfoximine, 552
   Gmethionine sulfoximine phosphate, 552
   ^{18}O-positional isotope exchange, 549–550
   phosphinothricin, 552
   and pyrrolidine-5-carboxylate, 549
   reaction and energetics, 548
   regulation in bacteria, 550
   two Me^{2+}, 549
   tyrosine adenylylation, 550
Glutamine synthetase I, active site structure, 554
   covalent regulation, 550
   a mechanism, 555
   protein structure, 553
Glutamine:PRPP amidotransferase, glutaminase action, 607
   and glutamine, 604–605
   inactivation by diazonorleucine (DON), 609
   reaction and mechanism, 608
   structure, 610
   two classes, 607
Glutathione reductase, disulfide oxidoreductase family, 694
   reaction, 694
Glyceraldehyde-3-P (GAP), and aldolase, 25
   and glyceraldehyde-3-P dehydrogenase, 130
   and triosephosphate isomerase, 335

Glyceraldehyde-3-P dehydrogenase, active site
  structure, 691
 charge transfer complexation, 691
 covalent acyl-enzyme, 690
 and 1, 3-diphosphoglycerate, 492
 hydrolysis of p-nitrophenylacetate, 692
 kinetic mechanism, 690, 692
 negative cooperativity, 692
 reaction, 130–131, 690
 steps in the mechanism, 690
 stereospecificity, 690
 thiohemiacetal intermediate, 690
Glycerol dehydrase, reaction, 194
Glycine reductase complex, 749
Glycine-N-methyltransferase, 662
Glycogen phosphorylase, 577–584
 activation cascade, 502
 glucono-δ–lactone inhibition, 581
 kinetics, 578, 580
 mechanism of action, 582–584
 nojirimycin tetrazole, 581
 ^{18}O-positional isotope exchange, 580
 reaction and stereochemistry, 577, 578
 reaction of 1-deoxy-1-fluoroglucose, 579
 reaction of 1-heptenitol, 579
 reaction of 2-deoxyglucose, 579
 reaction of D-glucal, 579
 regulation, 579
 role of PLP, 582
 structure, 583
Glycosidases, families and structures, 587
Glycoside hydrolysis, 2-acetamido
  participation, 571, 572
 acid catalysis, 570
 chemistry of, 570–573
 endocyclic cleavage, 570
 exocyclic cleavage, 570
 oxacarbenium-like transition state, 571
 primary 1–^{13}C isotope effects, 573
 secondary deuterium isotope effects, 572
 transition state, 571
 two mechanisms, 571
Glycosyl group transferases, 569
Glycosyl transfer, chemical mechanisms, 570
 chemistry, 569–573
 enzymatic, 573
 retention or inversion, 569
 stereochemistry, 569
Glycosyl transferases, generic
  mechanisms, 574
 retaining and inverting enzymes, 573–587
Glyoxylate, 620, 623
Glyphosate, herbicidal inhibitor of EPSP
  synthase, 289–291
Gramicidin, by action of NRPS, 767
Gross-Butler equation, and proton
  inventories, 307
Ground state destabilization, 48
Group selective chemical modification, 53

Haldane relationships, derivation, 84–85
Hammett plot, for farnesyl pyrophosphate
  synthase, 647
Hammond postulate, 44–45
Heme, oxygen binding and electron transfer, 209
 spin states of iron, 204
 structure, 203
Heme and oxygen, chemistry of, 201
Heme coenzymes, 201–210
Heme enzymes, 204
Hemerythrin, di-iron cofactor, 217
Hemoglobin, structure, 209
Hexokinase, glucose induced a structural
  change, 6
 structure, 2
Hill equation, and cooperativity, 122
Histidine ammonia-lyase, active site structure
  with MIO, 458
 in histidine catabolism, 456
 kinetic mechanism, 458
 methylidene imidazolone-dependent, 173,
  456–459
 2-nitrohistidine as substrate, 459
 protein structure, 457
 reaction, 456
 trans-urocanic acid, 456
 two mechanisms, 459
Histidine decarboxylase, ^{13}C- and ^{15}N-isotope
  effects, 401
 cleavage of pro-enzyme, 400
 a mechanism, 401
 PLP-independent, 399–402
 pyruvoyl, 172, 399–402
 structure with histidine methyl ester, 402
 trapping of iminium intermediate, 400
Histidine phosphatases, double-
  displacement, 510
 linear free energy correlation, 512
 a mechanism, 513
 pH dependence, 513
 δ^1-phosphohistidine, 510
 retention of configuration at P, 510
 structure with vanadate, 511
 transphosphorylation, 510
 trapping ^{32}P-enzyme, 510
Histidylflavin, succinate dehydrogenase, 159
 thiamine dehydrogenase, 159
HIV protease, hypothetical mechanisms, 322
 iso-mechanism, 320–323
 role in viral replication, 319
 structure with saquinavir, 321
HMG-CoA reductase, and cholesterol
  biosynthesis, 280
 3-hydroxy-3-methylglutaryl CoA
  reductase, 280
 inhibition by statins, 282
 a mechanism, 281, 282
 reaction, 281
 structure with fluvastatin, 284

Index

Homocysteine, 671, 672, 673
Horseradish peroxidase, compounds I and II, 208
   reaction and mechanism, 208
Hydrogen bonding, 30
Hydrogen bonding in catalysis, 32
Hydrogen bonds, strong and weak, 30
Hydrogen peroxide in oxidative decarboxylation, in lactate monooxygenase, 722
Hydrogen tunneling, experimental tests, 101
   and isotope effects, 99–101
   and rule of geometric mean, 100
Hydroxocobalamin, B12 compound, 191
Hydroxyacyl-ACP dehydratase (DH), component of FAS, 762
β–Hydroxybutyryl CoA, 628
β–Hydroxydecanoyl thioester dehydratase, 260–261
   and palmitoleyl-ACP, 260
   decynoyl cysteamine-inactivation, 260
   suicide inactivation, 261
Hydroxyethylidene-TPP, in decarboxylation of pyruvate, 142, 144, 146
   oxidation to a radical, 147
   reduction of ferricyanide, 147
   in TPP reactions, 142–144
Hydroxyethyl-TPP, in decarboxylation of pyruvate, 142
α–Hydroxyfarnesylphosphonate, 659
2-Hydroxyglutaryl CoA dehydratase, iron sulfur enzyme, 233
   reaction, 233
Hydroxysqualene, by product of squalene synthase, 650
   structure, 650

Imidazole glycerol phosphate synthase, and glutamine, 604–605
Imidazolone propionate, production by urocanase, 140
Imine formation, by lysine, 23
   mechanism, 24
Indole, from indoleglycerol phosphate, 154–155
Indoleglycerol phosphate, and tryptophan synthase, 154–155
Inhibition pattern, competitive, 86
   noncompetitive, 86
   uncompetitive, 86
   reversible, 85
Inorganic pyrophosphatase, active site structure, 239, 521
   kinetic scheme, 520
Intradiol dioxygenase, in toluene metabolism, 217
Intradiol dioxygenases, 741
Inverting and retaining glycosyl transferases, 569
Iron, nonheme mononuclear, 210

Iron, oxo-Fe_2 complexes, 217
Iron protoporphyrin IX, heme, 203
Iron sulfur centers, 227
   catalytic functions, 230
   electron transfer, 230
   net charges and oxidation states, 231
   reduction potentials, 231
   structures, 227, 229–230
Iron sulfur centers and S-adenosylmethionine, 234, *see also* Radical SAM superfamily
   catalytic action of, 234
   stoichiometric reactions of, 236
Iron-methane monooxygenase, 727–732
   deuterium isotope effects, 731
   di-iron complex, 727
   epoxidation, 727
   ferryl oxygenating species, 728
   hydroxylating cycle, 730
   hypothetical mechanism, 732
   oxygenation mechanism, 729–732
   radical clock experiments, 729, 731
   reaction, 727
   S- and N-oxygenation, 727
   stereochemical experiments, 731
   structure, 728
   substrate hydroxylation promiscuity, 727
Isocitrate dehydrogenase, 679
Iso-mechanism, citrate synthase, 627
   enolase, 442
   fumarase, 434
   glutamate racemase, 351
   HIV-protease, 320, 323
   proline racemase, 347–348
Isomerization and elimination, catalysis of, 465
   of glycols and enthanolamine, 465–473
Isopenicillin-N, structure, 720
Isopenicillin-N synthase, 718–721
   active site structure, 720
   evidence for ferryl oxy species, 721
   2-His, 1-carboxylate facial triad, 720
   mononuclear iron protein, 720
   reaction and mechanism, 720
Isopentenyl pyrophosphate (IPP), 646
Isopropylmalate synthase, iron sulfur variant, 232
Isotope effects, classes of, 91
   direct measurement method, 95
   equilibrium perturbation, 97
   internal competition, 96
   intrinsic, 97–99
   kinetic and equilibrium, 91
   magnitudes, 91, 95
   measurement of, 95
   physical basis, 93–95
   primary and secondary, 93
   remote label method, 96
   spectroscopic, 92

Isotope exchange at equilibrium, in
  fumarase, 434
 in glutamine synthetase, 549
 in lactate dehydrogenase, 686
 resolving ambiguity, 88–90

Kanamycin nucleotidyltransferase, inactivation
  by kanamycin, 522
 m-nitrobenzyl triphosphate as substrate, 522
 ^{18}O-isotope effects, 522
α–Ketoacid dehydrogenase complexes,
  142, 144, 750
β–Ketoacyl-ACP reductase (KR), component
  of FAS, 762
β–Ketoacyl-ACP synthase (KS), component of
  FAS, 762
3′-Ketoadenosine, and S-adenosylhomocysteine
  hydrolase, 137
α–Ketobutyrate, from threonine, 451
α–Ketoglutarate, a mechanism for generating
  ferryl oxy species, 733
 as a reducing system, 217, 732
α–Ketoglutarate dehydrogenase, 680
α–Ketoglutarate dehydrogenase complex,
  749, 750
α–Ketoglutarate-dependent oxygenases,
  732–735
Δ^5-3-Ketosteroid isomerase, dihydroequilenin
  and LBHB, 366–167
 dihydroequilenin inhibitor, 366
 homoenolate ion intermediate, 366
 reaction, 367
 structure with androsterone, 368
 transformations of substrate, 367
β–Ketothiolase, reaction, 619
k_{cat}/K_m, and isotope effects, 73
 physical significance, 73
Kinetic progress curves, complications, 70
Kinetics, of enzymatic reactions, 69
King and Altman, derivation of rate
  equations, 72
K_m, definition, 72

Laccase, and copper, 241
Lactate dehydrogenase, 680, 686–689
 catalysis of hydride transfer, 689
 clinical applications and isozyme
  variance, 686
 His195 as acid-base catalyst, 687–689
 isotope exchange at equilibrium, 686
 isozyme hybrids, 686
 kinetic mechanism, 686
 reaction mechanism, 687
 stereochemistry, 686
 structure, 688
Lactate monooxygenase, 722
 flavoprotein oxidase, 722
 H_2O_2 in oxidative decarboxylation, 722
 reaction, 722

Lactyl CoA dehydratase, iron sulfur
  enzyme, 233
 reaction, 233
Lactyl-TPP in decarboxylation of pyruvate,
  142–143
LBHB (low-barrier hydrogen bonds), 32
 in chymotrypsin, 34, 301, 310
 in citrate synthase, 627
 in HIV-protease, 322–323
 Δ^5-3-Ketosteroid Isomerase, 366
 in phospholipase A2, 328–329
 physicochemical properties, 32, 33
 in subtilisin and elastase, 34, 314
Leucine aminopeptidase, active site
  structure, 239
 bimetallopeptidase, 324
Lewis acids and bases, definition, 10
Ligases, ATP-dependent, 547
Ligation, and the energy of ATP, 547
Linear free energy correlation, for farnesyl
  pyrophosphate synthase, 646
α–Lipoamide, action in ketoacid
  dehydrogenation, 147–148
 conformational mobility, 148
 in electron and group transfer,
  147–148
Lipoic acid, and α–ketoacid dehydrogenase
  complexes, 142
Lipoyl synthase, reaction, 235, 612, 613
Lovastatin acid, structure, 283
Low-barrier hydrogen bonds, see LBHB
LTQ, lysyltopaquinone, 174
Lysine 2,3-aminomutase, and S-adenosylme-
  thionine/[4Fe-4S], 234
 5′-deoxyadenosyl radical, 376
 interconversion of L-α– and
  L-β–lysine, 376
 PLP, SAM, and [4Fe-4S] center, 376
 reaction, 235
 SAM-cleavage mechanism, 376
 structure with SAM, PLP, and
  lysine, 378
 1,2-amino migration mechanism, 377
Lysine 5,6-aminomutase, adenosylcobalamin
  and PLP-dependent, 194, 378
Lysine aminomutases, PLP coenzyme in,
  154–157
Lysine-5-hydroxylase, α–ketoglutarate-
  dependent, 732
Lysozyme, 589–594
 affinity labeling, 590
 fluorinated inactivators, 594
 glycosyl-enzyme, 592
 hen egg white, reaction, 589
 identification and role of Glu35 and
  Asp52, 56, 592
 isotope effects, 592
 pH-dependence, 592
 structure, 591

Lysozyme *(Continued)*
    substrate distortion, 590
    T4 lysozyme, 595
    transition state, 594
Lysyl 4-hydroxylase, reaction, 216
Lysyltopaquinone (LTQ), lysine oxidase, 174

Magnesium, *see* Divalent metal ions
L-Malate, 624
Malate dehydrogenase, 680
Malate synthase, inversion or methyl in acetyl CoA, 621
    reaction, 619
Malonyl-ACP, 628
Mandelate racemase, 352–354
    and divalent metal ion, 352
    enolase superfamily, 353
    lysine and histidine, 352
    overshoot phenomenon, 352
    structure, 354
    TIM barrel, 353
    two-base mechanism, 352
Marcus equation, 248
Marcus theory, and electron transfer, 248
Mechanism based inactivation, *see* Suicide inactivation, 255
Mechanochemical coupling, model, 798
Metallophosphatases, bis-metallo centers, 520
Metalloproteases, and mononuclear zinc proteases, 323
Metallopterin enzymes, 222
Metaphosphate anion, electrophilic [PO_3^-], 477
    in nonsolvating media, 482
    in solvation cages, 483
Methane monooxygenase (MMO), *see* Iron-methane monooxygenase, *see also* Copper-methane monooxygenase
Methanogenesis, coenzymes, 245
Methanol dehydrogenase, deuterium kinetic isotope effect, 176
    hypothetical mechanisms, 177
    ping pong kinetics, 175–176
    and PQQ, 174
    reaction, 175
    structure with PQQ, 176
5, 10-Methenyltetrahydrofolate, 639, 642
N^5, N^{10}-Methenyl-H$_4$MPT cyclohydrolase, in methanogenesis, 244
Methionine sulfoximine, inhibition of glutamine synthetase, 552
Methionine sulfoximine, structure, 552
Methionine synthase (MetH), 172, 199–200, 655, 670
    and cob(I)alamin, 191, 671
    and homocysteine, 671
    homocysteine thiolate, 676
    kinetic mechanism, 199
    MetE cobalamin-independent, 670
    MetH cobalamin-dependent, 670
    and methylcobalamin, 199, 671
    and methyltetrahydrofolate, 671
    modular structure and function, 671, 673, 675
    reaction, 172, 199–200, 672
    structures of modules, 674–675
Methotrexate, inhibition of dihydrofolate reductase, 271–273
    structure, 271
Methyl coenzyme M, 191, 244, 246–247, 655
Methyl coenzyme M reductase, hypothetical mechanism, 247
    in methanogenesis, 244
    structure, 246
Methyl coenzyme M synthase, and cob(I)alamin, 191
Methylamine dehydrogenase, a mechanism, 180
    ping pong kinetics, 179
    reaction, 178
    structure, 179
    structure in complex, 248
    TTQ coenzyme, 178
β–Methylaspartate ammonia-lyase, enolase superfamily, 353
Methylation, and *S*-adenosyl-L-methionine (SAM), 655
Methylcobalamin, 671–672, 675
    dependent enzymes, 199
Methylcoenzyme M synthase, 200
β–Methylcrotonyl CoA, biotinyl carboxylation, 163
Methyleneglucarate mutase, reaction, 194
N^5, N^{10}-Methylene-H$_4$MPT dehydrogenase, in methanogenesis, 244
5, 10-Methylenetetrahydrofolate, one-carbon metabolism, 168, 169
N^5-Methyl-H$_4$MPT:CoM-SH methyltransferase, in methanogenesis, 244
Methylidene imidazolinone (MIO), elimination and, 456
    coenzyme of histidine ammonia-lyase, 456–459
    deaminases, 173–174
    posttranslational processing, 174
Methylmalonyl CoA mutase, a radical isomerization mechanism, 374
    deuterium kinetic isotope effect, 375
    nonenzymatic model, 374
    reaction, 194, 372
    stereochemistry, 372
    structure with succinyl CoA, 373
Methyltetrahydrofolate, 671, 672, 673, 676, 677
5-Methyltetrahydrofolate, methyltetrahydrofolate, 670
    one-carbon metabolism, 168, 169

Mevaldehyde, HMG-CoA reductase-intermediate, 281
Mevaldyl CoA, HMG-CoA reductase-intermediate, 281
Mevalonate, from HMG-CoA reductase, 281
Mevalonate pyrophosphate decarboxylase, 3'-fluoro-mevalonate-PP, 406
  mechanism and reaction, 405–406
  stereochemistry, 405
Mevastatin, structure, 283
Michaelis-Menten kinetics, assumptions, 70
Modular enzymes, 763–768
Molybdopterin and tungstopterin, 222
Molybdopterin enzymes, oxidases or reductases, 223
Molybdopterin families, coordination patterns, 223
Molybdopterin structures, 223
Monoamine oxidase, structure of MAO B with isatin, 719
Monoamine oxidases (MAO A and MAO B), 716–720,
  flavoprotein oxidases, 716
  initial formation of $FADH_2$, 716–717
  mechanisms under consideration, 717
  natural substrates, 717
  reaction, 717
  structure function correlations, 718
Mononuclear nonheme iron, 210–217
Monooxygenases, 210, 722–741
Mössbauer spectroscopy, aconitase, 444
Muconate-lactonizing enzymes, enolase superfamily, 353
Multienzyme complexes, 750
Mutarotation of glucose, acid-base catalysis, 14
Mutorotase, see UDP-Galactopyranose mutase
Myoglobin, structure, 209
Myosin, structures of heads, 793–795
Myosin and actin, sliding filament model, 792–794
Myosin and actomyosin, kinetics, 795–797
  release of Pi, 796–797
Myosin and muscle contraction, 792–798
Myosin-actin interaction, 796

NAD(P)H dehydrogenases, pro-R and pro-S stereospecificities, 134
NAD/NADP, nicotinamide coenzymes, 129
  stereospecificity of hydride transfer, 132
NAD^+, activation of diptheria toxin, 130
  activation of substrates, 135
  ADP-ribosylation, 130
  as a coenzyme, 134–141
  and DNA repair, 130
  and poly-ADP-ribosylation, 130
  regulation of nitrogenase, 130
NAD^+/ $NADP^+$, as hydride acceptors, 130–131
  nicotinamide coenzymes, 129
  structures, 130

NADH, H_R and H_S in, 130, 133
NADH dehydrogenase, complex I, 749, 782
NADH/NADPH, structures, 130
Negative cooperativity, and glyceraldehyde-3-P dehydrogenase, 692
Near attack conformation (NAC), 46–47
Nickel, in methanogenesis, 243
Nickel coenzymes, 243–245
Nicotinamide adenine dinucleotides, 129
Nicotinamide coenzymes, NAD/NADP, 129
  stereospecificity of hydride transfer, 132
  structures and functions of, 129
NIH shift, in phenylalanine hydroxylase, 214
Nitrate oxidase, reaction, 226
Nitrate reductases, DMSO reductase family, 227
Nitric oxide synthase, mechanistic aspects, 740
  reaction, 739
  regulatory mechanisms, 739
  signaling NO, 738
  steps in production of NO, 741
  structure, 740
  three coenzymes, 739
Nitrogen transfer, 597
Nitrogenase, 749, 777–782
  Fe-protein and FeMo-protein, 780
  issues in N_2 reduction mechanism, 781–782
  reactions, 777–778
  reduction of acetylene, 778
  reduction of HCN, 777
  reduction of N_2, 777
  structure, 778–779
3-Nitrolactate, fumarase inhibitor, 436
p-Nitrophenyl acetate (PNPA), 102
  burst kinetics with chymotrypsin, 103
  imidazole-catalyzed hydrolysis, 17, 299–300
  reactivity with nucleophiles, 20
p-Nitrophenylphosphoester hydrolysis, rates of mono-, di-, and triesters, 484,
NMR spectroscopy, ^{31}P-NMR ^{18}O-isotope shifts, 551
Nojirimycin tetrazole, structure, 581
Noncompetitive inhibition, definition, 85
Nonheme iron, mononuclear, 210
Nonribosomal polypeptide, 749, 767
Nonribosomal polypeptide synthetases (NRPS), modular structure, 767
Norepinephrine, structure, 735
Nucleophilic catalysis, 16
  and imidazole-catalyzed hydrolysis, 17
Nucleoside diphosphate kinase, 79–82, 521
  exchange reactions, 80
  kinetics, 80
  ping pong mechanism, 80
  stereochemistry, 81
  structure, 82

Nucleotidyl group transfer, enzymatic, 521
  nucleotidyltransferases, 521–541
Nucleotidyl transfer, chemistry of, 521
Nucleotidyltransferases, 521

OMP decarboxylase, 414–418
  conserved aspartate and lysine, 415
  five mechanisms, 416
  kinetic isotope effects, 415
  OMP conformation, 417
  reaction, 414
  remote binding, 418
  structure with UMP, 417
Ordered bi bi mechanism, 78
Ornithine cyclodeaminase, hypothetical mechanism, 139
  NAD^+ coenzyme, 139
Orotidine monophosphate decarboxylase, see OMP decarboxylase
Overshoot phenomenon, racemization of L-proline in D_2O, 347
  in glutamate racemase, 350
  in mandelate racemase, 352
Oxalate decarboxylase, 412–414
  kinetic isotope effects, 414
  O_2-dependent Mn enzyme, 412
  pH-dependence, 413
  reaction and mechanism, 412–414
  structure, 413
Oxaloacetate, 619–620, 622, 625–626
Oxidases, 710–722
Oxidases and oxygenases, 710
Oxidoreductases, 679
Oxo-Fe_2 complexes, structures, 218
Oxygen rebound, radicals and cytochrome P450 monooxygenases, 724
Oxygen, structure and chemistry, 201
Oxygen and heme, chemistry of, 201
Oxygenases, α–ketoglutarate-dependent, 732

π-Cation interaction, 661
Packing densities, atoms in enzymes, 7
PALA, and aspartate transcarbamylase, 254
  ATCase inhibition constant, 254
  structure, 254
  two-substrate analog, 254
Palmitoyl ACP thioesterase (TE), component of FAS, 762
Papain, Cys25 and His159 in catalysis, 315–317
  ionizations of Cys25 and His149, 318
  kinetic isotope effects, 315
  structure with inhibitor (E64), 316
  thiolate-imidazolium ion pair, 317
Parathion, structure of acetylcholinesterase inhibitor, 291
Pepstatin, peptidic inhibitor of pepsin, 319
  structure, 320
Peroxidases, 206–208

Perturbed pKa values, in enzymes, 116
PFL activase, reaction, 235
pH profiles, frequent patterns, 112
  interpretation, 111
  measurements of, 111
  parameters, 111
  pH dependence of K_i, 111–113
  pH dependence of V and V/K, 113
  and reverse protonation, 115
N-Phenylacetyl-glycyl-D-valine, pH-dependence in hydrolysis, 12
Phenylalanine ammonia-lyase (PAL), cinnamate from phenylalanine, 461
  3-hydroxy- L-phenylalanine as substrate, 460
  mechanism, 461
  methylidene imidazolone, 173, 460–461
  ^{15}N and deuterium kinetic isotope effects, 461
  4-nitro-L-phenylalanine as substrate, 460
  pH-dependence, 461
Phenylalanine dehydrogenase, active site structure, 691
Phenylalanine hydroxylase, 161–162, 211–216
  and ferrous ion, 213
  p-methylphenylalanine substrate, 212
  NIH shift, 214
  role of tetrahydrobiopterin, 212
  structure, 212
Phenylmethane sulfonyl fluoride (PMSF), 59
Phosphate NMR, ^{18}O-isotope shifts, 551
Phosphatidylserine decarboxylase, pyruvoyl, 172
Phosphinothricin, inhibition of glutamine synthetase, 552
Phosphodiester hydrolysis, activation parameters, 483
  low reactivity, 483
  secondary ^{18}O kinetic isotope effects, 483
  solvent kinetic isotope effects, 483
  structure-function analysis, 483
  tight transition state (aka associative), 483
Phosphodiesterases, 539
Phosphodiesters, and phosphoryl group transfer, 483
Phosphoenolpyruvate, and enolase, 28
Phosphoenolpyruvate carboxylase, and carboxyphosphate, 426
  chemical mechanism, 425
  and 3-fluorophosphoenolpyruvate, 426
  kinetics, 425
  reaction, 425
  structure, 427
Phosphofructokinase, 500–501
  kinetic mechanism, 500
  reaction, 499
  regulatory properties, 500
  site-directed mutagenesis, 500
  stereochemical course, 500
  structure with fructose 1,6-bisphosphate, 501

α-Phosphoglucomutase, 341–343
  glucose-1,6-bisphosphate as intermediate and cofactor, 342
  phosphoenzyme, 342
  reaction steps, 342
  structure, 343
β-Phosphoglucomutase, 343–344
  cofactor-independent, 343
  structure, 344
Phosphoglucose isomerase, 334–335
2-Phosphoglycerate, and enolase, 28
Phosphoglycerate kinase, active site structure, 494
  conformational change, 493
  and 1,3-diphosphoglycerate, 492
  enzyme structure, 493
  and 3-phosphoglycerate, 492
  reaction, 492
  stereochemistry, 493
Phosphoglycerate mutase, 343–346
  cofactor-independent, 346
  2,3-diphosphoglycerate-dependent, 345
  stereochemical retention at P, 345
  reaction steps, 345
Phosphoketolase, and acetyl arsenate, 145–145
  and acetyl phosphate, 145–146
  and acetyl-TPP, 145
  and enolacetyl-TPP, 145
  production of glyceraldehyde-3-P, 145–145
  a reaction mechanism, 145
  reaction of xylulose-5-P, 145
  and TPP, 144
Phospholipase A2, phosphonate inhibitor and active site, 330
Phospholipase A2, reaction, 329
Phosphomonoester hydrolysis, absence of solvent isotope effect, 478
  activation parameters, 478
  2,4-dinitrophenylphosphate, 480
  electronic effects, 478–479
  ^{18}O-kinetic isotope effects, 480–481
  pH-dependence, 478
  phenylpropyldiphosphate dianion, 478
  solvent portioning, 478
Phosphomonoesterases, 509–521
  kinetic mechanism, 510
Phosphomonoesters, dissociative mechanism and discrete metaphosphate, 477
  generic phosphotransfer, 477
  loose transition state (aka dissociative), 477
  and phosphoryl group transfer, 476
  phosphotransfer to non-lyate nucleophiles, 481
  phosphotransfer transition state, 477
Phosphomutases, 341
Phosphopantetheine, in coenzyme A, 165–166

Phosphopantetheine coenzymes, 165–168
  acyl activation, 165–167
  chemical role, 165–167
  enolization, 165–167
  in fatty acid synthases, 165
  mechanism of action, 165
  physical role, 167
  in polyketide synthases, 165, 765
  in polypeptide synthetases, 165, 767
  structural mobility, 167
  structures, 166
Phosphoryl group transfer, chemistry of, 476–486
  enzymatic, 487
  single and double displacements, 487
Phosphoryl transfer to non-lyate nucleophiles, stereochemistry, 482
  structure-function, 481
  α–effect, 481
Phosphotransacetylase, in acetyl phosphate production, 492
Phosphotransferases, 489–521
Phosphotriester hydrolysis, high reactivity, 483
Phosphotriesters, and phosphoryl group transfer, 483
pH-rate profiles, 111, see also pH profiles
Physostigmine, structure of acetylcholinesterase inhibitor, 291
PIX, see positional isotope exchange
π-Cation interaction, 661
Ping pong, kinetics definition, 74
Ping pong bi bi mechanism and rate equation, 79
pKa, of amino acids, 13
  of carbon acids, 22
  of hydroxyl coordinated to metal, 132
  from pH rate profiles, 113–115
  values perturbed in enzyme active sites, 116
  of water coordinated to metal, 238
PLP, 148–158, see also Pyridoxal-5′-phosphate
  aminotransfer, 152
  bonds cleaved by, 148
  α–carbanion stabilization, 150
  β–carbanion stabilization, 150
  chromophores, 153
  α–decarboxylation, 151
  in 3, 6-dideoxysugar biosynthesis, 157–158
  enzymatic reactions of, 149
  external aldimine, 151
  internal aldimine, 151
  and lysine aminomutases, 154–157
  mechanisms of reactions, 151–158
  in radical mechanisms, 154–158
  reaction specificity, 152
  stereospecificity, 152
  structure, 148
  transient kinetics, 153
  transimination, 151

PLP-dependent α–decarboxylase, minimal mechanism, 395
PMSF (phenylmethane sulfonyl fluoride), affinity labeling of chymotrypsin, 59, 314
PNPA, see p-nitrophenyl acetate
Polyketide synthase type I, 6-deoxyerythronolide synthase, 764–765
Polyketide synthase type II, synthesis of doxorubicin, 765–767
Polyketide synthase type III, 767
Polyketide synthases, 749, 763
　modular structure, 763
Polysulfide reductases, DMSO reductase family, 227
Positional isotope exchange (PIX), in carbamoyl phosphate synthetase, 557
　and discrete metaphosphate in phosphotransfer, 482
　in glutamine synthetase, 549
　in glycogen phosphorylase, 580
　in phosphoenolpyruvate carboxylase, 426
　in UDP-N-acetylglucosamine-2-epimerase, 362
PQQ (pyrroloquinoline quinone), 174–178
　and methanol dehydrogenase, 174
　structure, 175
　tyrosine and glutamate precursors, 175
Product inhibition, resolving kinetic ambiguity, 86–87
Proline racemase, 346–350
　buffer catalysis, 349
　cysteine residues in catalysis, 348, 349
　deuterium fractionation factor, 349
　deuterium kinetic isotope effect, 347
　iso-mechanism, 347–348
　kinetic mechanism, 349
　overshoot in racemization, 347
　PLP-independent, 346
　product inhibition and oversaturation, 349
　two base-iso-mechanism, 348
Prolyl 4-hydroxylase, α–ketoglutarate-dependent, 732
　reaction, 216
Propionyl CoA to methylmalonyl CoA, biotinyl carboxylation, 163
Prostaglandin H synthase, 274–280
　cyclooxygenase and peroxidase activities, 275
　cyclooxygenase structure, 276
　and inflammation, 275
　structure of bromoaspirin-inactivated, 278
　structures with ibuprofen and flurbiprofen, 280
Protein farnesyltransferase, 657–661, 757
　directed mutagenesis, 659
　free energy correlation, 659, 661
　kinetic mechanism, 658
　Mg^{2+} requirement, 757
　reaction and mechanism, 658–661
　role of zinc, 757
　stereochemistry, 658
　structure with α–hydroxyfarnesylphosphonate, 660
Protein kinase A (cAMP-dependent protein kinase), 502–509
　action of regulatory subunit, 508
　affinity labeling, 505
　anchoring proteins, 509
　chemical modification, 505
　conformational changes, 506
　dissociation of regulatory subunits, 505
　glycogen phosphorylase activation, 502
　glycogen synthase inhibition, 503
　kinetics, 506
　peptide inhibitory sequence, 508
　and protein phosphorylation, 502
　structure of catalytic subunit, 507
　structure of regulatory subunit, 504
Protein phosphorylation, protein kinase A, 502–509
Protein prenyltransferases, 657
Protein tyrosine phosphatase, structure, 515
Protein tyrosine phosphatases (PTPs),
　^{18}O-isotope effects, 517
　burst kinetics, 514
　cysteine phosphatases, 513
　double-displacement, 514
　function as switches, 514
　a mechanism, 516
　pH dependence, 514
　p-nitrophenyl phosphate as substrate, 516
　retention of configuration at P, 514
Protocatechuate 3,4-dioxygenase, hypothetical mechanism, 743
　reaction, 741
　spectroscopic observations, 743
　structure, 742
Protocatechuate 4,5-dioxygenase, spectroscopic observations, 746
　structure, 745
Pseudorotation, in cyclic pentagonal phosphoesters, 486
Pseudouridine synthase, conserved aspartate, 380, 382
　reaction, 381
　two mechanisms, 381
Pterin, chelator of molybdenum, 222
　folate constituent, 168
Purine nucleoside phosphorylase, 584–587
　immucillin H-transition state analog, 586–587
　isotope effects, 585–588
　kinetics, 585
　reaction, 585
　stereochemistry, 585
　structure with immucillin H, 588
Purple acid phosphatase, di-iron cofactor, 218

Pyridine nucleotide-dependent dehydrogenases, 680, *see also* NAD/NADP
  Rossman fold, 680
Pyridoxal oxidase, reaction of, 149
Pyridoxal-5′-phosphate (PLP), 148–158, *see also* PLP
  mechanisms of PLP-dependent reactions, 151
  stabilization of amino acid carbanions, 149
  reactions facilitated by, 149
Pyridoxamine, structure, 148
Pyridoxine or pyridoxal, structure, 148
Pyridoxol kinase, reaction of, 149
Pyrophosphatases, bis-metallo complexes, 520
  inorganic pyrophosphatase, 520
Pyrroloquinoline quinone (PQQ), bacterial alcohol dehydrogenases, 174–178
Pyruvate decarboxylase, 142–144, 389–394
  active site structure, 392
  and hydroxybenzylidene-TPP, 394
  and hydroxybenzyl-TPP, 394
  and hydroxyethylidene-TPP, 389, 390, 393–394
  and hydroxyethyl-TPP, 389, 390, 393–394
  iminopyrimidine-TPP, 393
  and lactyl-TPP, 389, 390, 393
  a mechanism, 142, 393
  and 2-(*p*-Nitrobenzylidene)-1-hydroxyethylidene-TPP, 390
  protein structure, 391
  pyruvamide activation, 392
  role of TPP, 142, 144, 389
  substrate activation, 392
  transient kinetics, 390
Pyruvate dehydrogenase (PDH) complex, 749–756
  composition and structure, 752
  decarboxlation mechanism, 754
  E1 of PDH complex, 750
  lipoyl bearing domains, 752–753
  reductive acetylation of lipoamide, 756
  regulation, 751–752
  role of lipoic acid, 751
  site-site interaction in the dimer, 754–755
  steps in the mechanism, 751
  TPP-dependence, 750, 751, 754, 755
Pyruvate formate lyase (PFL), 407–411
  chemical model, 408
  deuterium kinetic isotope effect, 408
  and glycyl radical, 141–147, 407
  kinetic mechanism, 408
  a mechanism, 410
  proton exchange mechanism, 408
  reaction, 407
  structure with pyruvate, 409
Pyruvate formate-lyase activase (PFL activase), radical SAM enzyme, 411
  and *S*-adenosylmethionine/[4Fe-4S], 234

Pyruvate kinase, coordination of K^+, 496
  decarboxylation of oxaloacetate, 496
  enolization of pyruvate, 495
  a mechanism, 495
  oxalate as substrate, 496
  and phosphoenolpyruvate carboxykinase, 494
  phosphorylation of F^-, 496
  phosphorylation of NH_2OH, 496
  reaction, 494
  role of K^+, 495–496
  roles of $2Mg^{2+}$ and K^+, 495
  structure of MgATP, 495
  structure with oxalate and ATP, 497
  reaction, 145
Pyruvate oxidoreductases, 142, 144
  role of thiamine pyrophosphate, 146–147
Pyruvate phosphate dikinase, 498–499
  chemical mechanism, 499
  energetics, 498
  kinetic scheme, 498
  reaction, 498
  steps in the reaction, 498
Pyruvate to oxaloacetate, biotinyl carboxylation, 163
Pyruvoyl decarboxylases, 172–174
  posttranslational processing, 173
  pyruvoyl structure, 172–173
  sensitivity to carbonyl reagents, 173
Pyruvoyl enzyme, D-proline reductase, 173

Quinoproteins, 174–185

Racemases and epimerases, 346–363
Radical clock experiments, cytochrome P450 monooxygenases, 725
  iron-methane monooxygenase, 729, 731
Radical isomerization, 197, 368–378
Radical SAM superfamily, 5′-deoxyadenosyl radical, 234, 236
  cysteine motif, 236
Radical-based decarboxylases, 407
Rate enhancement, and binding of remote groups, 48
  by enzymes, 4
  factor, 3
  transition state binding, 43
Rate limitation by diffusion, triosephosphate isomerase, 338
Reaction characteristics, 101
Reduction potentials, and free energy, 202
  of iron sulfur centers, 231
Remote groups, rate enhancement by, 48
β–Replacement, a mechanism, 452
  PLP-dependent, 452
Retaining and inverting glycosyl transferases, 569
Reverse protonation, 115

Ribonuclease, identification of His12 and
    His119, 53
  reaction with iodoacetate, 53
Ribonuclease A, His12 and His119, 539–540
  ^{18}O-kinetic isotope, 540
  a mechanism, 539
  pH-dependence, 540
  structure with uridylyl-2′,5′-guanosine, 540
Ribonuclease S, and S-peptide, 539
Ribonucleotide reductase, adenosylcobalamin-
    dependent, 197
  classes I, II, and III, 197–198, 698–707
  a mechanism, 198
  reaction, 197
  thiyl radical, 197
  tritium kinetic isotope effect, 197
Ribonucleotide reductase class I,
    allosteric site, 701
  di-iron complex-structures, 701
  electron transfer pathway, 702
  reaction, 701
  thioredoxin as reductant, 701
  thiyl radical and mechanism, 700
  tyrosyl radical, 701
Ribonucleotide reductase class II, and
    cob(II)alamin, 703
  epimerization at adenosyl-C5′, 703–704
  isotope exchange at adenosyl-C5′, 703–704
  reaction, 701
  thioredoxin as reductant, 701
  thiyl radical and mechanism, 700
Ribonucleotide reductase class III, activation
    by activase, 705
  anaerobic ribonucleotide reductase, 703
  formate as reductant, 701, 703–704
  glycyl radical site, 704
  reaction, 701
  thiyl radical and mechanism, 700
Ribonucleotide reductase classes, structures
    and relationships, 706
Ribonucleotide reductases, classes of, 700
  structural relationships, 705
D-Ribose-5-P, 635
Ribosomal protein synthesis, 768–777
Ribosome, 749, 770–777
  elongation, 774
  initiation, 772
  ^{15}N kinetic isotope effect, 776
  peptide formation, 775
  steps in peptide formation, 774,
    776–777
  structure, 773
  termination, 775
Ribozymes, 534–539
  group I intron 2Me^{2+} mechanism, 539
  group I intron self-splicing, 538
  group II intron self-splicing, 538
  hairpin, 534, 536
  hammerhead, 534, 536, 537
  hepatitis delta virus (HDV), 534, 536
  RNA self-processing, 534–549
  structures, 536
Ribulose-1, 5-bisphosphate carboxylase,
    see Rubisco
Ribulose-5-P 4-epimerase, 360–362
  active site structure, 362
  1,2-hydride shift, 361
  primary kinetic isotope effects, 360
  relation to class II aldolases, 360
  retro-aldol mechanism, 360
  secondary deuterium isotope effects, 360
  Zn^{2+}-dependent, 360
Rillingol, by product of squalene
    synthase, 650
  structure, 650
RNA, self-processing, 534–549
RNA polymerase, 768–772
  structure, 771
  two Mg^{2+}-mechanism, 772
RNase P, 534
Rossman fold, alcohol dehydrogenase, 682
  pyridine nucleotide-dependent dehydroge-
    nases, 680
Rubisco (Ribulose-1, 5-bisphosphate carboxy-
    lase), 418–425
  activation by Mg^{2+}, 420
  active site structure, 423
  catalytic residues, 421
  inhibition by CABP, 421
  mechanism of action, 420–424
  oxygenase activity, 419
  partial reactions, 424–425
  protein structure, 422
  reaction, 419
  stereochemistry, 424

SAH, see S-Adenosylhomocysteine
SAM, see S-Adenosylmethionine
Saquinavir, structure of HIV protease inhibitor,
    320
Sarin, structure of acetylcholinesterase
    inhibitor, 291
Saturation kinetics, Eadie-Hofstee plot of, 71
  Lineweaver-Burk plot of, 71
Schiff's base, 23, see also Imine formation
    by lysine
D-Sedoheptulose-7-P, 632, 635
Sequential mechanisms, 75
Serine dehydratase, iron sulfur variant, 232
  PLP-dependent α,β–elimination, 451
Serine hydroxymethyltransferase, 639–645
  active site structure, 641
  carbanionic mechanism, 639
  hydration of methenyltetrahydrofolate, 639
  kinetic mechanism, 641
  linear free energy relationships, 644
  mechanisms for C-C cleavage, 643
  metabolic role, 639

Serine hydroxymethyltransferase *(Continued)*
  overall chemical mechanism, 642
  potential cancer chemotherapy target, 639
  reaction, 170–171, 639
  structure, 640
Serine phosphatases, 517–520
  metalloenzymes, 517
  transphosphorylation, 517
Serine proteases, 300, 311
  biological functions, 300
  Ser-His-Asp triads, 300
Serotonin, structure of substrate for MAO A, 717
Short-chain alcohol dehydrogenases, 687
  tyrosine as the base catalyst, 689
Short-chain dehydrogenase/reductase, family, 135, 687
Single and double phosphotransfer, 487–489
  characterization-covalent intermediates, 489
  steady-state kinetics, 487
  stereochemistry, 488
Site directed mutagenesis, complementarity with chemical modification, 57
  proof of function, 58
Slow binding inhibition, kinetics, 268–269
  methotrexate, 271–273
Slow-binding, kinetics of, 268
Specific acid catalysis, definition, 11
Specific base catalysis, definition, 12
Spore photoproduct lyase, a mechanism, 237
  reaction, 235
Squalene synthase, carbocationic intermediates, 648
  carbocationic mechanism, 648
  and cholesterol biosynthesis, 648
  cyclopropylcarbinyl carbocation, 648
  proposed reaction mechanisms, 651
  reaction, 650
  side product dehydrosqualene, 650
  side product hydroxysqualene, 650
  side product rillingol, 650
Steady-state approximation, 72
Steady-state kinetics, initial rates, 69
  Lineweaver-Burk plots, 70
  maximum velocity, 70
  one-substrate, 70
  saturation, 70
  three-substrate, 89
  two-substrate, 74
Stearoyl ACP desaturase, a chemical mechanism, 221
  fluorinated substrates, 220
  structure, 220
Strong hydrogen bonds, 30
Structural mobility, in enzymes and proteins, 6–7
Substrate distortion, in lysozyme, 590
Substrate inhibition, and ping pong mechanisms, 88

Subtilisin, and convergent evolution in serine proteases, 311
  chemical mutagenesis, 312–314
  structure with LBHB, 313
Succinate dehydrogenase, complex II, 749, 782
O-Succinylbenzoate synthase, enolase superfamily, 353
Sucrose phosphorylase, 575–577
  glucosyl-enzyme, 575–576
  reaction, kinetics, and stereochemistry, 575–576
  stereochemistry of glucosyl-enzyme, 577
Suicide inactivation, 166, 255–268
Sulfite oxidase, and molybdopterin, 222
  reaction and mechanism, 226
Sulfur transfer, 609
Swain relationship, 99
Swain-Schaad relationship, and hydrogen tunneling, 101
Synthetases, ATP-dependent, 547

T4 Lysozyme, 595
Tetrahydrobiopterin, and nitric oxide synthase, 162
  and phenylalanine hydroxylase, 162, 211–212
  structure, 161, 739
Tetrahydrofolate (H$_4$folate), 167, 168
Theorell-Chance mechanism, 78
  alcohol dehydrogenase, 681
Thermolysin, 327
  mononuclear zinc protease, 324
Thiamine, structure, 141
Thiamine pyrophosphate (TPP), bonds cleaved by, 141
  pKa of C2(H), 143
  reaction mechanism, 141
  structure, 141, 637, *see also* TPP
Thiolases, biosynthetic and degradative, 627–628
Thioredoxin reductase, disulfide oxidoreductase family, 694
  reaction, 694
Three-substrate reactions, kinetics, 8991
Threonine dehydratase (threonine deaminase), PLP-dependent $\alpha\beta$–elimination, 451
  reaction, 149
Thymidylate synthase, exchange of C5(H) in FdUMP, 257
  5-fluorodeoxyuridylate-inactivation, 170
  inactivation by 5-Fluoro-dUMP (FdUMP), 255–260
  reaction and mechanism, 170, 256–257
  structure, 256
Thymine hydroxylase, α–ketoglutarate-dependent, 732
Tight binding inhibition, kinetics, 269–270, 280

TIM barrel (αβ–barrel), in mandelate racemase, 353
  in methylmalonyl CoA mutase, 372
  in enolase, 441
  in pyruvate kinase, 497
  in lysine 5, 6-aminomutase, 379
  in orotidine-5-phosphate decarboxylase, 417
  in xylose isomerase, 340
  in triosephosphate isomerase, 336
Toluene dioxygenase, 217
Toluene monooxygenase, di-iron cofactor, 217
N-p-Toluenesulfonyl-L-phenylalanine chloromethylketone (TPCK), see TPCK
Topaquinone (TPQ), copper amine oxidases, 179–185
Topoisomerase, ciprofloxacin inhibition, 534
Topoisomerase IB, structure, 535
Topoisomerase II, doxorubicin (adriamycin) inhibition, 534
Topoisomerases, 532–534
  cleavage and religation, 533
  covalent catalysis, 533
  strand passage mechanism, 533
TPCK (N-p-toluenesulfonyl-L-phenylalanine chloromethylketone), 60–62, 116, 301
  affinity labeling of chyomotrypsin, 60–62
TPP, 141–147, see also Thiamine pyrophosphate
TPP model, 1, 5-dimethylthiazolium, 141–143
TPP-dependent decarboxylase, acetolactate synthase, 389
  benzoylformate decarboxylase, 389
  α–ketoacid dehydrogenases, 389
  oxalyl CoA decarboxylase, 412
  pyruvate decarboxylase, 389
TPQ (topaquinone), chemical characterization, 181
  structure, 181
Transaldolase, 631–635
  carbanionic mechanism, 631
  carbanionic/enamine intermediates, 631
  chemical mechanism, 635
  class I aldolase-analog, 632
  kinetic mechanism, 632
  reaction and mechanism, 632–635
  sedoheptulose-7 P as substrate, 632
  structure, 634
Transcarboxylase, tethered biotin, 83
  two-site ping pong kinetic mechanism, 83
Transient kinetics, 101–110
  burst kinetics, 102–104
  isotope partitioning, 110
  methods, 102
  nonsteady-state, 106–109
  partial reactions, 106
  rapid mix-quench, 105
  relaxation methods, 109
  stopped-flow spectrophotometry, 104

Transition state, binding theory, 42–43
  stabilization of, 41
Transition state theory, 41
Transketolase, 142, 144, 634–639
  active site structure, 637
  carbanionic mechanism, 634
  chemical mechanism, 638
  dihydroxyethylidene-TPP, 636, 637, 638
  kinetic scheme, 635
  reaction, 635
  thiamine pyrophosphate (TPP)-dependent, 634–639
Triosephosphate isomerase, 335–341
  affinity labeling, 335
  detailed kinetics, 337
  and dihydroxyacetone phosphate (DHAP), 335
  cis-enediolate intermediate, 336
  exchange kinetics, 337
  free energy profile, 338
  and glyceraldehyde-3-P (GAP), 335
  infrared spectroscopy, 339
  mechanism, 339
  possible LBHB, 339
  rate limitation by diffusion, 338
  stereochemistry, 335
  structure with DHAP, 336
  TIM barrel, 336
tRNA, contacts with aminoacyl-tRNA synthetases, 563
Trypsin, mutation of Asp102, 40
  mutation of catalytic triad, 40
  mutation of His57, 40
  mutation of Ser195, 40
Tryptophan fluorescence, quenching of by oxygen, 6
Tryptophan synthase, indole tunnel, 154–155
  a mechanism, 155
  PLP-dependent β–replacement, 154–155, 451
  reaction, 154
  serine binding PLP, 454
  structure, 155–156
Tryptophan tryptophyl quinone (TTQ), bacterial amine dehydrogenases, 178–179
  coenzyme of methylamine dehydrogenases, 178
  structure, 178
TTQ, see tryptophan tryptophyl quinone
Tungsopterin cofactor, aldehyde:ferredoxin reductase, 223
Tunneling, by electrons, 249
  by hydrogen, 99–101
Turnover number, and k_{cat}, 3
  definition, 73
Two-site ping pong mechanisms, 83
Tyrocidine, by action of NRPS, 767
Tyrosinase, and copper, 241

Tyrosine 2,3-aminomutase, 602–604
    a mechanism, 604
    and methylidene imidazolone (MIO), 603
    reaction, 603
    role of methylidene imidazolone, 604
    stereochemistry, 603
Tyrosine and tryptophan hydroxylases, 211
Tyrosine hydroxylase, and L-DOPA, 211
Tyrosine kinase, reaction, 506
    structure-function, 508
Tyrosine kinases, 503
Tyrosyl-adenylate, 106, 107
Tyrosyl-tRNA synthetase, kinetics, 106
    multiple hydrogen bonds, 33
    partial reaction kinetics, 106
    structure, 107
    and tyrosyl adenylate, 6
    and tyrosyl tRNATyr, 6
Tyrosyl-tRNATyr, 106

Ubiquitin, ATP-dependent proteolysis, 566
    cascade, 567
    ubiquityl thioester, 566
UDP-4-ketoglucopyranose, 4-epimerase
        intermediate, 134
UDP-Galactopyranose mutase, FADH$_2$ as
        cofactor, 379
    positional isotope exchange, 379
    reaction and mechanism, 380
    trapping intermediate by reduction, 379
UDP-Galactose 4-epimerase, 355–360
    conformational change, 357
    conformations of UDP-sugars, 358
    dehydrogenation at glycosyl-C4, 357
    epimerization mechanism, 355
    mechanism of hydride transfer, 359
    NAD$^+$-dependent, 355
    nonstereospecificity model, 358
    perturbation of NAD$^+$-reduction
            potential, 360
    polarization of nicotinamide, 360
    reaction of, 134–135, 355
    role of NAD$^+$, 134
    structure with UDP-glucose, 356
    tight binding of NAD$^+$, 357
    UDP-4-ketoglucose intermediate, 357
UDP-N-acetylglucosamine-2-epimerase,
        positional isotope exchange, 362
    reversible elimination, 361
    structure with UDP, 363

    two-base mechanism, 361
Uncompetitive inhibition, definition, 85
Urate oxidase, absence of cofactors, 721
    proposed mechanism, 721
    reaction, 721
Urease, active site structure, 239
Uridylyltransferases, 526–529, *see also*
        Galactose 1-phosphate uridylyltransferase
Urocanase, hypothetical mechanism, 140
    NAD$^+$ as electrophilic catalyst, 140
    reaction and stereochemistry, 140
    reaction of trans-urocanate, 140
trans-Urocanate, hydration by urocanase, 140

Vancomycin, by action of NRPS, 767
γ–Vinyl GABA, antiepileptic drug, 263
    suicide inactivation of GABA aminotransferase, 263–265
Vitamin B12 coenzymes, 190
Vitamin K, a mechanism for oxygenation, 428
Vitamin K-dependent carboxylase, carboxylation cycle, 428
    and γ–carboxyglutamyl (Gla) residues, 426
    menaquinone coenzyme, 428
    prothrombin glutamyl residues, 426
    vitamin K epoxide requirement, 428
V_m, definition, 70, 72

Water, activation by divalent metals, 237
    pKa coordinated to metals, 238
Weak hydrogen bonds, 30

Xanthine oxidase, 224–226
    and molybdopterin, 222
    reaction scheme, 225
    structure, 225
Xylose isomerase, active site structure, 340
    and 1,2-hydride shift, 335, 340
    and divalent metal ion, 335, 340
D-Xylulose-5-P, 635

Zinc
    and alcohol dehydrogenase, 132, 237,
            681–684
    and carbonic anhydrase, 237, 462
    and carboxypeptidase A, 237, 324
    and dehydroquinate synthase, 138
    and protein farnesyltransferase, 757
    and ribulose-5-P 4-epimerase, 360
    and thermolysin, 324